ADVANCED
BIOLOGY

PRINCIPLES & APPLICATIONS

SECOND EDITION

3

Acknowledgements

To all the known and unknown experimental scientists, naturalists and observers, teachers, illustrators and writers who have influenced our own understanding we gladly acknowledge our debt.

Where we have used copyright we have sought to obtain permission, and we have acknowledged our sources (pp. 679–81). However, if we have inadvertently overlooked an existing copyright and used materials that are the intellectual property of another, without acknowledgement, then please contact the publishers so that correction can be made.

We have sought opportunities to take advice and suggestions, and to discuss specific issues with numerous well-informed biologists. **For the first edition, these were:**

John Barker, King's (Chelsea College), London
Professor Harold Baum, King's College, London
Professor Peter Bell, University College, London
Dr Richard Belment, University of Manchester
Dr David Bender, University College, London
Professor Jocelyn Chamberlain, Institute of Cancer Research, London
Judith Clegg, Ranger Service, Bracknell Forest Council
Dr Jack Cohen, University of Birmingham
Lesley Cox, Wellcome Centre for Medical Science
Dr Richard Dawkins, University of Oxford
Professor John Dodge, Royal Holloway and Bedford New College
Professor Richard Doll, Radcliffe Infirmary, Oxford
Dr David Dover, King's College, London
Professor W H Dowdeswell, University of Bath
Dr Margaret Frayne, James Allen's Girls' School, Dulwich
Dr David George, Nene College, Northampton
Ann Gimson (formerly) King's College School of Dentistry, London
Professor Charles Gimingham, University of Aberdeen
Dr Glyn Hughes, Plant Breeding International, Cambridge
Professor David Hall, King's College London
Professor Kay-Tee Khaw, Addenbrooke's Hospital, Cambridge
Professor R McNeill Alexander FRS, University of Leeds
Dr Richard M Jackson, University of Surrey

Dr Richard Johnson, University of Aberdeen
Professor Neil Jones, University College of Wales, Aberystwyth
Dr Greg Peakin, University of Greenwich
Dr Ivor Simpkins, University of Hatfield
Peter Smith, Hammersmith Hospital
Dr Frederick Toates, Open University
Rhianon Walter, Faculty of Public Medicine, London
Professor Alan Wellburn, University of Lancaster
Dr Edward Wood, University of Leeds
Bertram Worrall, (formerly) Avery Hill College
Dr Paul Wymer, Wellcome Centre for Medical Science

For the second edition, these were:
Dr Persephone Borrow, Edward Jenner Institute of Vaccine Research, Newbury
Dr Paul Clegg, University of Toronto, Ontario, Canada
Dr Robin Cook, The Marine Laboratory, Aberdeen
Professor John Dodge, (formerly) Royal Holloway and Bedford New College, London
Dr Leslie Gartner, University of Maryland, Baltimore, USA
Dr Angela Ingham, Coulsdon College, Surrey
Dr Mike Jackson, IACR – Long Ashton Research Station, University of Bristol
Dr Richard P C Johnson, (formerly) Botany Department, University of Aberdeen
Dr Gary Mantle MBE, Director, Wiltshire Wildlife Trust, Devizes
Mrs Ilse Towler, Health and Safety Adviser, London Borough of Croydon

Nevertheless, any faults that remain are the sole responsibility of the authors. It is hoped that our readers will write to point out any errors and omissions they find.

At John Murray, the skill and patience of Katie Mackenzie Stuart (Science Publisher), Helen Townson (Design Manager), Gina Walker (Editor), Jenny Fleet (Designer) and Helen Reilly (Picture Researcher) have brought together text, drawings, annotations and photographs exactly as we wished, and we are most grateful to them.

Illustrations by Don Mackean
with additional illustrations by Barking Dog Art, Ethan Danielson, Richard Draper, Chris Etheridge, Mike Humphries, Linden Artists, Andrew Mackay, TechType, Lydia Umney

Layouts by Jenny Fleet
Typeset in Sabon and Helvetica Condensed by Wearset, Boldon, Tyne & Wear
Scanning and digital files prepared by Colourscript, Mildenhall, Suffolk
Printed and bound in Spain

A catalogue entry for this title is available from the British Library

ISBN 0 7195 7670 9

ADVANCED
BIOLOGY
PRINCIPLES & APPLICATIONS

SECOND EDITION

**C J CLEGG with
D G MACKEAN**

JOHN MURRAY

Contents

Introduction

ADVANCED BIOLOGY Principles and Applications, first published in 1994, quickly established itself as a leading textbook for A and AS Level Biology students whose background was in broad, balanced science of the National Curriculum. The noted features of the first edition were

▶ the text, which blended animal and plant topics, giving an integrated approach to the whole subject

▶ the beautiful illustrations, closely complementing the text, and illustrating principles and processes in context

▶ the differentiation of the text into core and extension material, and the extensive use of student questions in order to assist comprehension, with answers provided to aid self assessment

▶ the liberal inclusion of biographies of successful scientists, using the history of ideas to enliven the text

▶ the attention to the applications of science in technology and commerce, sustaining student discussion of the economic, environmental and ethical implications of scientific discovery for societies worldwide

▶ an introduction to the nature of science at the outset, a focus on the methods of science as they have developed, and finally, a discussion of the distinctive nature of scientific ideas

▶ the rigour and depth of treatment of biological ideas, presented in ways that allow something of the excitement and adventure of scientific discovery to come across.

Today, A and AS Level Biology assessment has been reorganised. The syllabuses are presented as Specifications, arranged as Modules (Units) taken in Year One, leading to Examinations for the AS Level qualification, and Modules (Units) taken in Year Two, leading to Examinations for the A Level qualification. Included with this new approach are the aims to further

▶ develop knowledge of biology, together with the skills needed to use it

▶ facilitate an understanding of scientific methods

▶ sustain an interest and enjoyment of biology

▶ recognise the value of biology in society, so that it may be used responsibly.

Another aim is to increase the breadth of students' study programmes. For example, students of languages and the humanities could include science in their studies, at least to AS Level, and vice versa.

This Second Edition of *ADVANCED BIOLOGY* meets the demands of the new Specifications (see opposite), whilst maintaining established, distinctive features, as outlined below.

1 The content has been updated to cover new material in the Specifications and essential developments in the component disciplines of biology, where appropriate. Material no longer relevant has been withdrawn. Changes have been made within the existing structure of the book, and occur in all main Sections:

▶ **Section 1 Biology Today: Environmental Biology** is the entry point to the issues of modern biology. Developments in ecology include greater emphasis on human influences in the environment, and upon pollution and conservation. Discussion of the industrial applications of biology has been extended in the areas of the traditional industries of food production, and in new biotechnological industries.

▶ **Section 2 Cell Biology** introduces the structure, function and roles of cells in the context of how this information has been discovered. Developments here include an extended examination of the uses of enzymes outside cells, in industries new and old, and in commerce.

▶ **Section 3 The Maintenance of the Organism** discusses how nutrients are obtained, metabolised, transported internally, and finally disposed of. Developments here include a new approach to photosynthesis, and extended discussion of the functioning human body under changing conditions.

▶ **Section 4 The Responding Organism** details how organisms respond to internal and external changes. Developments here include a new approach to health and disease, including more in-depth considerations of ageing and of cancer.

▶ **Section 5 The Continuity of Life** covers reproductive processes, inheritance mechanisms and evolution. Developments here include extended discussion of applications of reproductive technology in humans, as well as detailed examination of processes and products in genetic engineering, and their social consequences.

▶ **Section 6 The Nature of Science** challenges the reader to think about the process of science itself, as well as about its long-term consequences.

2 To a greater extent throughout the text essentials have been summarised in bullet points. The vocabulary, sentence length, and complexity of the text have been adjusted where necessary to make it acceptable to a wide readership. Terminology has been kept to a minimum.

3 Each chapter begins with a statement of the core issues to be developed in the text. Chapters end with a guide to further reading and/or to appropriate web sites. Examination questions are given at the end of each Section. See also the journals listed below, which are produced for, or are of particular interest to, Advanced Level Biology students.

■ Study Guide

This book is designed to be used in conjunction with a companion volume – the *Study Guide**. The chapters of the *Study Guide* (which parallel those of this book) have the following contents:

▶ **Summary** – a brief résumé of the contents (concepts and their applications) in the corresponding 'theory' chapter.

▶ **Practical activities** – a sequence of data-obtaining tasks (consideration is given to the essential laboratory practicals needed to acquire basic science skills) and computer simulations. Statistical methods are introduced in appropriate situations.

▶ **Problem-solving assignments** – opportunities to process second-hand data, interpret results and discuss scientific issues presented in realistic scenarios.

▶ **Projects** – 'starter' suggestions for longer-term investigations.

▶ A **Glossary** is also included.

■ Taking your studies further

You can keep in touch with developments in modern biology with the help of journals. The following often carry useful articles.

New Scientist is a weekly review of developments, with occasional features of special value, for example, the 'Inside Science' series. From newsagents, and in your library.

Scientific American is a series of review articles, but less of a journalistic magazine and more of a scientific review journal. The quality of the photographs and colour illustrations is high. From major newsagents, and in main libraries.

Biological Sciences Review is a relatively new journal, written directly for students of A Level Biology, and obtainable from the publishers Philip Allan, Oxford OX5 4SE (discount with bulk purchases or through the purchase of a library copy). Definitely try this one early on!

The publications of the Institute of Biology (20 Queensbury Place, London SW7 2DZ), produced for teachers and students, including the *Journal of Biological Education* and specialist booklets such as *Careers with Biology, Tackling Biology Projects* and *Living Biology in the Classroom. Biological Nomenclature* sets out an internationally agreed list of the terms, units and symbols that are used in this book and in the *Study Guide*.

C J Clegg, Salisbury, Wiltshire
D G Mackean, Welwyn Garden City, Hertfordshire
May 2000

How the contents of all GCE AS and A2 specifications are covered:

■ AQA – AEB

AS

Unit 1 (B/HB) Molecules, Cells and Systems
  Chapters 6, 7, 8, 10, 11, 17
Unit 2 (B) Making Use of Biology
  Chapters 5, 8, 9, 16, 17, 18, 27, 29
or
Unit 3 (HB) Pathogens and Disease
  Chapters 9, 17, 24, 25, 29
Unit 4 Centre-Assessed Coursework
  Chapter 1 and *Study Guide**

A2

Unit 5 (B/HB) Inheritance, Evolution and Ecosystems
  Chapters 2, 3, 4, 9, 12, 15, 16, 28, 30
Unit 6 (B) Physiology and the Environment
  Chapters 13, 15, 16, 18, 21, 23
or
Unit 7 (HB) The Human Life-Span
  Chapters 13, 17, 19, 21, 22, 23, 24, 27
Units 8 (B) or 9 (HB)
(a) Synoptic Paper
  All the above Chapters
(b) Practical/Investigation
  *Study Guide**

■ EDEXCEL

AS

Unit 1 Molecules and Cells
  Chapters 6, 7, 8, 9, 11, 29
Unit 2B Exchange, Transport and Reproduction
  Chapters 11, 13, 15, 16, 17, 18, 26, 27
or
Unit 2H Exchange, Transport and Reproduction in Humans
  Chapters 11, 15, 17, 27
Unit 3 Energy and the Environment
  Chapters 3, 4, 5, 13, 14
Practical Assessment of Coursework
  Chapter 1 and *Study Guide**

A2

Unit 4 Respiration and Coordination
Core
  Chapters 8, 15, 21, 23
+ Options:
A Microbiology and Biotechnology
  Chapters 2, 5, 14, 19, 24, 25 + *Study Guide**
B Food Science
  Chapters 5, 13, 14, 24 + *Study Guide**
C Human Health and Fitness
  Chapters 15, 17, 22, 24 + Study Guide*
Unit 5B Genetics, Evolution and Biodiversity
  Chapters 2, 3, 4, 12, 20, 28, 30
Unit 5H Genetics, Human Evolution and Biodiversity
  Chapters 2, 3, 4, 28, 29, 30
Unit 6 Synoptic and Practical Assessment
  All the above Chapters
Individual Study
  *Study Guide**

■ AQA – NEAB

AS

Unit 1 Core Principles
  Chapters 6, 7, 8, 11, 13, 15
Unit 2 Genes and Genetic Engineering
  Chapters 9, 27, 29
Unit 3a Physiology and Transport
  Chapters 15, 16, 17
Unit 3b Coursework
  Chapter 1 and *Study Guide**

A2

Unit 4 Energy, Control and Continuity
  Chapters 2, 8, 12, 15, 18, 21, 22, 23, 28, 30
Unit 5a Environment
  Chapters 3, 4, 5, 16
Unit 5b Coursework
  *Study Guide**
Unit 6 Options:
Applied Ecology
  Chapters 3, 4, 5 + *Study Guide**
Microbiology and Disease
  Chapters 5, 8, 17, 24, 25 + *Study Guide**
Behaviour and Populations
  Chapters 19, 23, 24, 27 + *Study Guide**
Essay and Data Handling
  All the above Chapters + *Study Guide**

■ OCR

AS

Module: Biology Foundation
  Chapters 3, 6, 7, 8, 9, 11
Module: Human Health and Disease
  Chapters 13, 15, 17, 24
Module: Transport
  Chapters 16, 17
Experimental Skills
  *Study Guide**

A2

Module: Central Concepts
  Chapters 2, 3, 4, 8, 9, 12, 15, 19, 20, 21, 23, 30
Module Options:
O1 Growth, Development and Reproduction
  Chapters 19, 20, 21, 25, 26, 27, 30 + *Study Guide**
O2 Applications of Genetics
  Chapters 28, 29, 30 + *Study Guide**
O3 Environmental Biology
  Chapters 3, 4, 5 + *Study Guide**
O4 Microbiology and Biotechnology
  Chapters 5, 24, 25, 29 + *Study Guide**
O5 Mammalian Physiology and Behaviour
  Chapters 13, 21, 22, 23, 24 + *Study Guide**
Module: Unifying Concepts in Biology
  All the above Chapters
Coursework with Practical Examination
  *Study Guide**

ADVANCED BIOLOGY Study Guide (1996), C J Clegg with D G Mackean, P H Openshaw and R C Reynolds. John Murray

1 Biology Today: Environmental Biology

The badger (*Meles meles*) is a nocturnal mammal with an omnivorous diet. On emergence from its underground 'set' at dusk, the badger characteristically smells the air, probably to sense possible danger. Sadly, badgers are, on occasions, illegally trapped and used for badger-baiting 'sport'

1 Biology, the science of life

- Biology is the study of life and of the activities of living things, including their respiration, nutrition, excretion, sensitivity, movement, reproduction and growth.
- Living things are made of tiny structures called cells, consisting of a nucleus and cytoplasm, surrounded by a cell membrane.
- Cells exchange materials with their environment, and grow, divide and become specialised. There is a great variety of specialised cells.
- Many organisms consist of a single cell (unicellular organisms), but in multicellular organisms cells work together. Here, groups of cells make up tissues and organs with specific functions.

- The science of biology involves observing and measuring, formulating hypotheses, predicting, designing investigations, recording data and interpreting results, drawing conclusions and communicating them.
- Biologists may be employed in laboratory work, industry, fieldwork and the environment, agriculture, horticulture, forestry, health care, work with animals, marine and fresh water biology, information science, administration, finance, management and teaching.

1.1 INTRODUCTION

The word **biology** simply means **the study of life**. Biology is concerned with the study of all living things, from bacteria to higher plants and mammals. Biology is a modern **science**, although it has ancient origins. Biology aims to provide understanding of the structure and function of organisms and how they interact with one another.

'Life' and 'science' are words that we use frequently, and without any apparent difficulty. But it is surprisingly hard to say exactly what we mean by life and by science. What do they mean? Let's start with life.

■ Life

All the activities of living things make up 'life' as biologists understand the word. It is not a fully satisfactory definition to say that life is what living things do! Each one of us, however, has difficulty defining life to any other's satisfaction.

Living things are known as **organisms**. When organisms are studied in biology in any detail it usually turns out that the underlying processes within one type of organism are very similar to those occurring within organisms of other types. All organisms have a great deal in common, even though they may look and behave very differently. Consequently, for biologists 'life' is identified as a combination of characteristics common to all living organisms and absent from non-living things. The site of life is the cell, the functioning unit structure of which living things are made.

1.2 THE CHARACTERISTICS OF LIVING THINGS

There are eight characteristic activities common to all organisms: respiration, nutrition, metabolism, excretion, sensitivity, locomotion, reproduction and growth. The possession and practice of these characteristic activities of organisms is the way biologists identify and define life.

■ Release of energy (respiration)

Respiration is the process by which sugars such as glucose and certain other substances are broken down to release chemical energy for life processes such as protein synthesis and growth. Respiration occurs in every living cell, and takes place in very many steps. Each step involves **enzymes** and at several of the stages energy is released. Some of this energy is transferred to other molecules and is available for use in other reactions.

Figure 1.1 Respiration: gaseous exchange and the release of energy

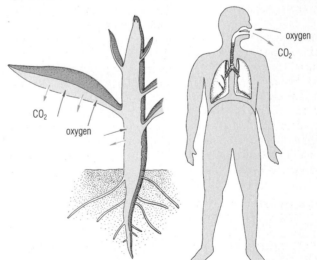

tissue respiration is usually aerobic:

oxygen + glucose → carbon dioxide + water + ENERGY

flowering plant
gaseous exchange via the air spaces of the plant and the pores in the stem and leaf surfaces

mammal
gaseous exchange via the body's transport system (the blood circulation) and the lungs

oxygen

CO_2

CO_2

oxygen

deep-growing plant roots in severely waterlogged soils respire anaerobically by alcoholic fermentation:

glucose → ethanol + carbon dioxide + ENERGY

muscle fibres during periods of prolonged and strenuous activity respire anaerobically by lactic acid fermentation:

glucose → lactic acid + ENERGY

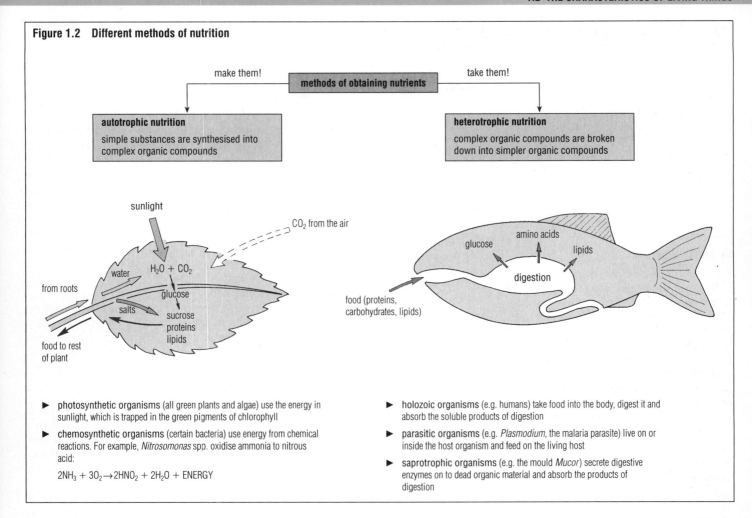

Figure 1.2 Different methods of nutrition

► **photosynthetic organisms** (all green plants and algae) use the energy in sunlight, which is trapped in the green pigments of chlorophyll

► **chemosynthetic organisms** (certain bacteria) use energy from chemical reactions. For example, *Nitrosomonas* spp. oxidise ammonia to nitrous acid:

$$2NH_3 + 3O_2 \rightarrow 2HNO_2 + 2H_2O + ENERGY$$

► **holozoic organisms** (e.g. humans) take food into the body, digest it and absorb the soluble products of digestion

► **parasitic organisms** (e.g. *Plasmodium*, the malaria parasite) live on or inside the host organism and feed on the living host

► **saprotrophic organisms** (e.g. the mould *Mucor*) secrete digestive enzymes on to dead organic material and absorb the products of digestion

---

The reactions of respiration inside cells are known as **tissue respiration** and they are dependent upon **gaseous exchange** between the organism and the environment.

Most respiration requires oxygen, and is known as **aerobic respiration**. The final breakdown products from aerobic respiration are carbon dioxide and water.

Respiration in the absence of air – **anaerobic respiration** – may take place in cells that have been deprived of air. Just a few organisms respire only anaerobically.

Respiration in the flowering plant and mammal is summarised in Figure 1.1.

■ Feeding or nutrition

Nutrition is the process by which an organism obtains the **energy** to maintain the functions of life, and **matter** to create and maintain its structures. Energy and matter are obtained from **nutrients**. Table 1.1 is a summary of the nutrients required by organisms, and Figure 1.2 shows the different methods by which nutrients are obtained.

Most plants are **autotrophs** ('self-nourishing'). This means that they make their own organic nutrients (food) from an external supply of relatively simple raw materials (inorganic nutrients), using

energy from sunlight in **photosynthesis**. Alternatively, and much more rarely, the energy source is a specific chemical reaction (**chemosynthesis**).

Autotrophs go on to make all their other requirements from the products of photosynthesis (or chemosynthesis) by combining them with each other and with additional inorganic nutrients.

Animals are **heterotrophs**. This means that they depend upon existing foods (organic nutrients), which have to be broken down before they can be used.

Questions

1 What is the ultimate source of the organic nutrients that are required by animals and all other heterotrophic organisms?
2 What is the difference between the ways in which autotrophs and heterotrophs obtain supplies of carbon?
3 Both burning (combustion) and respiration release energy from organic molecules. In what ways is the release of energy by burning similar to energy release in respiration, and in what ways is it different?
4 Where might you expect to find examples of anaerobic respiration amongst living things?

Table 1.1 The nutrients required by organisms

Inorganic nutrients	Organic nutrients
Small molecules (or ions), most not containing carbon (except CO_2). For example: ► water, makes up 70–90% of all living things ► mineral salts as ions, required for chemical activities of the organism and in the construction of chemicals and tissues ► carbon dioxide, required by chemosynthetic and photosynthetic organisms for the synthesis of organic nutrients	Relatively large molecules, all containing the element carbon, and all produced by living things. For example: ► amino acids, required to build proteins ► fats, energy-rich, and required for making lipids ► sugars, energy-rich, and required for making other sugars and polysaccharides ► vitamins, substances essential in small quantities for the correct functioning and normal growth of the organism

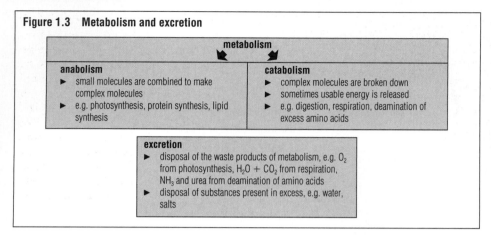

Figure 1.3   Metabolism and excretion

metabolism

anabolism
► small molecules are combined to make complex molecules
► e.g. photosynthesis, protein synthesis, lipid synthesis

catabolism
► complex molecules are broken down
► sometimes usable energy is released
► e.g. digestion, respiration, deamination of excess amino acids

excretion
► disposal of the waste products of metabolism, e.g. O_2 from photosynthesis, $H_2O + CO_2$ from respiration, NH_3 and urea from deamination of amino acids
► disposal of substances present in excess, e.g. water, salts

■ Metabolism

The **metabolism** of an organism is the total of all the chemical processes of life that take place within it. Living things have the ability to make chemical substances, including many different types of molecules and some very large molecules. Living things also break down substances. So we can divide metabolism into processes that involve building-up molecules, known as **anabolism**, and those that involve breaking molecules down, known as **catabolism**.

In anabolism simple chemicals are combined to make complex molecules, some of which are stored as energy sources for later use. Examples of anabolism include starch synthesis from glucose, and protein synthesis from amino acids. In catabolism complex compounds are broken down with release of energy. Respiration, in which glucose is broken down and energy is released, is an example of catabolism.

The reactions of metabolism are not haphazard or accidental, but are controlled by the organism. For example, each step is catalysed by a different enzyme. The cell nucleus controls which enzymes are present in a cell. Consequently the cell nucleus indirectly controls metabolism.

Metabolism also results in the production of substances of no further use to the organism. Some of them are harmful and are excreted. Metabolism and excretion are summarised in Figure 1.3.

■ Excretion

Excretion is the elimination of the unwanted products of metabolism and of substances present in excess within the organism. For example, respiration produces excretory products in every living cell. Aerobic respiration produces carbon dioxide and water.

Green plants produce oxygen in the light as an excretory product of photosynthesis.

Animals excrete nitrogenous waste products such as ammonia, urea or uric acid. These or similar substances are produced by the deamination (p. 508) of any excess protein that has been eaten. Excretion also involves eliminating any excess or toxic substance taken in with the diet, including water and salts.

■ Responsiveness or sensitivity

The ability of a plant or animal to respond to a stimulus is known as **sensitivity**. A **stimulus** might be an external event such as change in light intensity, or it might be an internal change like the arrival of food in the stomach. In animals a stimulus is detected by its sensory organs, such as its eyes; in plants there are no obvious sensory organs but certain cells are able to detect specific stimuli – gravity or light, for example.

The **response** to a stimulus might be the movement of the whole animal, perhaps in pursuit of food or away from danger (Figure 1.4), or of part of the animal, such as a hand jerking away from a hot object. In plants, the response is always made by an organ such as a leaf stalk, stem or tendril (Figure 20.3, p. 419).

A response to an internal stimulus might be the secretion of enzymes in response to food, or the production of antibodies in response to an invading microbe.

In general the responses of organisms are adaptive, in that an organism alters its environment by moving from a less favourable to a more favourable position. Alternatively, the organism may make modifications to the way it works (physiological changes), to the way it behaves (behavioural changes) or to its structures (structural changes), and as a result be better suited to the environment. Some organisms can tolerate adverse conditions or become adapted more readily than others.

In most animals there is a system of **receptors** or **sense organs** that detect changes, together with **nerves** that carry

Figure 1.4   Sensitivity and response: a European tree frog (*Hyla arborea*). A leap to safety, or in a search for food

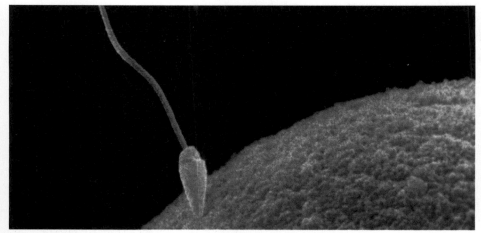

Figure 1.5 Sexual reproduction: the male gamete (a sperm, small and motile) at the surface of the female gamete (egg cell, or ovum)

electrical impulses to organs, which make a response. Responses are brought about by **effector organs** such as muscles or glands.

Plants lack such a sophisticated sensory and nervous system, yet their sensitivity is well developed. Plant responses tend to be comparatively slow growth responses, brought about by growth substances or **hormones**. Plants are extremely sensitive to changes in the environment, however, even though they lack the complex sense organs of some animals. In fact almost all organisms, including some single-celled organisms of quite simple structure, are highly sensitive.

■ Movement and locomotion

Movement is one of the most obvious characteristics of living things. The movements of many animals are speedy and efficient, but most plants and some animals lead a generally stationary life.

Animal movements are usually associated with obtaining food, finding mates or escaping from predators. In contrast, most plant movements are relatively slow growth movements, involving only part of the organism. They occur in response to an environmental stimulus.

■ Reproduction

Reproduction is the ability to produce other individuals of the same species. Reproduction may involve a sexual process, or it may be asexual.

In **sexual reproduction** special sex cells or **gametes** are formed. Two gametes fuse in a process known as **fertilisation** to form a new cell called a **zygote** (Figure 1.5). The zygote grows into a new individual. Some species have elaborate mechanisms of parental care of the developing offspring or **progeny**. In sexual reproduction the progeny are similar to their parents, but not identical to them.

Asexual reproduction (Figure 1.6) is the name for reproduction that does not involve a sexual process. In some organisms asexual reproduction is achieved by a simple division (**fission**) of a parent cell or body into two or more offspring, or fragmentation of the parent body to yield parts that are capable of developing into new individuals. Other organisms produce special spores. In yet others, individuals bud off from the parent organisms. These individuals are genetically identical to each other and to their parent.

■ Growth and development

New organisms produced by reproduction grow to roughly the same size as their parents. Growth involves an increase in size and is accompanied by an increase in complexity, described as development.

Growth is defined as an irreversible increase in size of the organism and is usually accompanied by an increase in solid material (that is, it is not just due to water uptake) and in the amount of cell materials (cytoplasm, p. 396).

Development involves the change in shape and form of an organism as it matures.

Most animals have a limited period of growth and development, which results in an adult of characteristic size and form. Most plants and a few animals grow throughout their entire life, at least during particular seasons of the year. All organisms are constantly breaking down and rebuilding their structures, but the rates of breakdown and build-up of cell components can vary over the life of the organism.

Questions

1 What living characteristics are represented in Figure 1.4?
2 What type of reproduction results in identical offspring?
3 a A fresh bean seed before soaking weighs x g. After soaking it weighs y g. Can this increase be called 'growth'?
  b Ten days later the seed has germinated and looks like a young seedling, but still weighs only y g. Can this change be called growth? Justify your answers.
4 State which of the following are examples of catabolism and which of anabolism: oxidation of carbohydrate, digestion of protein, production of starch from glucose, deamination of protein, protein formation from amino acids, photosynthesis, anaerobic respiration.

Figure 1.6 Some forms of asexual reproduction: (left) fission in an *Amoeba*, (right) spore production by a fungus, the common puffball (*Lycoperdon perlatum*)

1.3 LIVING THINGS ARE MADE OF CELLS

Another characteristic of living things is that they are made of cells. The **cell** is the basic unit of life, and it is the smallest unit of matter that we can meaningfully say is alive. Organisms consisting of a single cell are described as **unicellular**. Organisms consisting of many cells are **multicellular**. There are also a few multi-nuclear organisms, which are not divided into separate cells (p. 150).

Cells are extremely small structures. The unit of size usually used for cells is the micrometre (μm). One μm is one thousandth of a millimetre. Cells are normally between 10 and 150 μm in diameter, although a few are smaller and some are much larger.

■ The features of cells

A cell consists of the **nucleus** surrounded by **cytoplasm** contained within the **cell membrane**. The nucleus is the structure that controls and directs the activities of the cell. The cytoplasm is the site of the chemical reactions of life.

Cells exchange materials with their surroundings. The cell membrane is the barrier controlling entry to and exit from the cytoplasm. The plant cell has a cell wall outside the membrane; the animal cell does not. The cell wall is a non-living structure, made of cellulose and other carbohydrates, secreted by the cytoplasm. The main features of animal and plant cells are introduced in Figure 1.7, and their structure is described in detail in Chapter 7.

Newly formed cells can grow and enlarge. **A growing cell can normally divide** into two cells. Cell division is usually restricted to unspecialised cells, before they become modified for a particular task.

Cells also develop and specialise in their structure and in the functions they carry out. Many fully specialised cells are no longer able to divide (Figure 1.8).

As a consequence of development and specialisation, **cells show variety in shape and structure.** The great variety of cell structure reflects the evolutionary adaptations of cells to different environments, and to different specialised functions within a multicellular organism. A selection of various cells of plants and animals is illustrated in Figure 7.3.

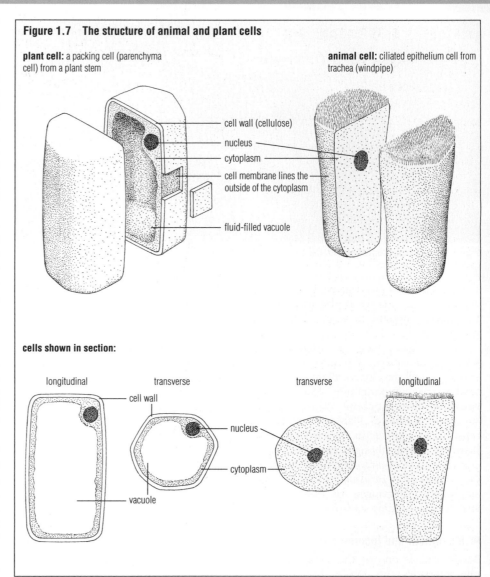

Figure 1.7 The structure of animal and plant cells

plant cell: a packing cell (parenchyma cell) from a plant stem

animal cell: ciliated epithelium cell from trachea (windpipe)

cell wall (cellulose)
nucleus
cytoplasm
cell membrane lines the outside of the cytoplasm
fluid-filled vacuole

cells shown in section:

longitudinal   transverse   transverse   longitudinal

cell wall
nucleus
cytoplasm
vacuole

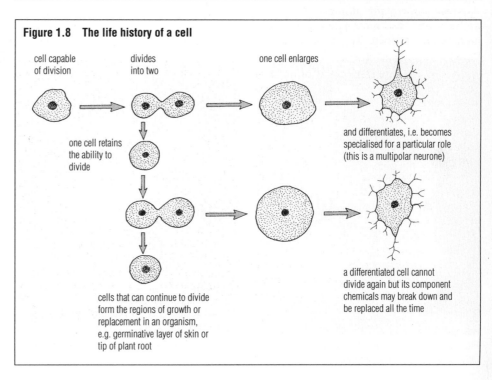

Figure 1.8 The life history of a cell

cell capable of division   divides into two   one cell enlarges

one cell retains the ability to divide

and differentiates, i.e. becomes specialised for a particular role (this is a multipolar neurone)

cells that can continue to divide form the regions of growth or replacement in an organism, e.g. germinative layer of skin or tip of plant root

a differentiated cell cannot divide again but its component chemicals may break down and be replaced all the time

1.4 CELLS AND ORGANISMS

Living things, although they are made of similar basic units (cells), exist in a vast range of different types, varying in shape, size and structural complexity.

The smallest unicellular organisms are bacteria, which are in the size range 0.8–10.0 μm. Other very small 'animals' and 'plants' are classified as part of the Protoctista (p. 28), and vary in size according to species, ranging from 10 μm to 6 m! The Protoctista that feed holozoically (that is, like animals) are called protozoa. Protozoa are unicellular. Protoctista that feed phototrophically (like plants) are algae. Some algae are unicellular, others are multicellular.

At the opposite end of the scale, the largest multicellular organisms are the eucalyptus trees of Australia, which can reach 155 metres (500 feet) above soil level, and the blue whale of Arctic and Antarctic waters, which weighs up to 150 tonnes.

Multicellular organisms show a range of structural complexity, from the simplest multicellular colonies with little or no coordination of component cells to multicellular individuals consisting of millions of cells, which are organised into groups or tissues, specialised to perform particular functions for the whole organism.

■ Unicellular organisms

Unicellular organisms such as the protozoan *Amoeba* and the unicellular alga *Chlorella* are examples of structurally simple organisms (Figure 1.9).

Unicellular organisms perform all the functions of living organisms within a single cell. After cell division the new cell grows to full size (provided the environment is favourable). The cell feeds, respires, excretes, exhibits sensitivity to internal and external conditions, possibly moves and, sooner or later, divides (or reproduces). In multicellular organisms specific functions are carried out by groups of specialised cells. In unicellular organisms these different functions are carried out either by the whole cell or by specialised regions of cytoplasm, such as the contractile vacuole in *Amoeba*.

Unicellular organisms form a very large and diverse group. They include the vast majority of bacteria, many cyanobacteria, nearly all the protozoa, and certain algae.

Figure 1.9   Two unicellular organisms common in freshwater ponds

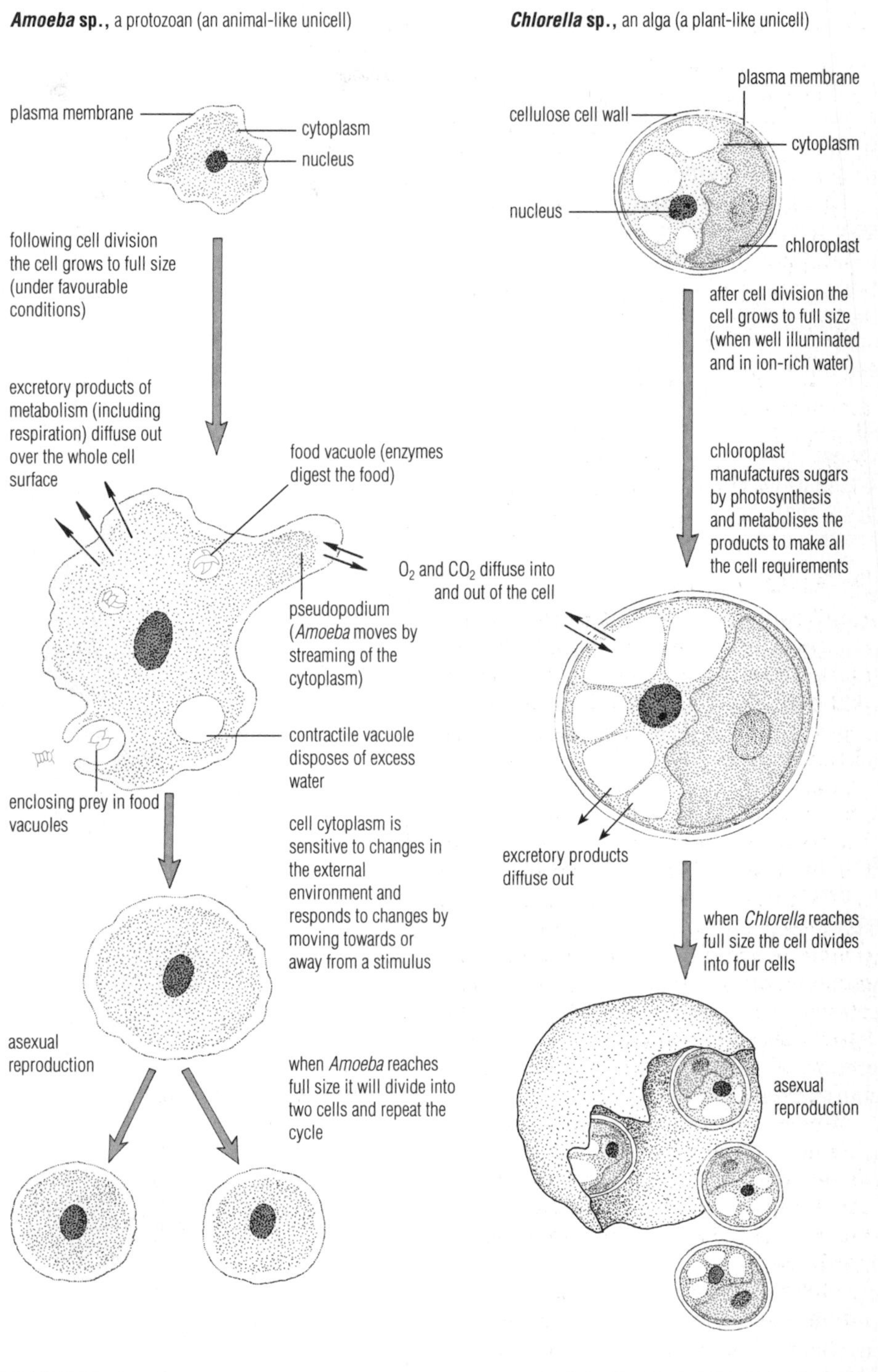

Amoeba sp., a protozoan (an animal-like unicell)

plasma membrane — cytoplasm — nucleus

following cell division the cell grows to full size (under favourable conditions)

excretory products of metabolism (including respiration) diffuse out over the whole cell surface

food vacuole (enzymes digest the food)

O_2 and CO_2 diffuse into and out of the cell

pseudopodium (*Amoeba* moves by streaming of the cytoplasm)

contractile vacuole disposes of excess water

enclosing prey in food vacuoles

cell cytoplasm is sensitive to changes in the external environment and responds to changes by moving towards or away from a stimulus

asexual reproduction

when *Amoeba* reaches full size it will divide into two cells and repeat the cycle

Chlorella sp., an alga (a plant-like unicell)

plasma membrane — cellulose cell wall — cytoplasm — nucleus — chloroplast

after cell division the cell grows to full size (when well illuminated and in ion-rich water)

chloroplast manufactures sugars by photosynthesis and metabolises the products to make all the cell requirements

excretory products diffuse out

when *Chlorella* reaches full size the cell divides into four cells

asexual reproduction

Questions

1 What differences in the behaviour or functioning of plants and animals may be linked to differences in cell structure?
2 State what particular features of cell structure you might expect to observe in cells that are specialised for:
    a absorption of useful materials from the environment
    b movement by contraction of the cells
    c support of a large flat structure such as a leaf
    d transmission of nervous impulses over some distance.
3 The living cell consists of a nucleus surrounded by cytoplasm. What is likely to happen to a cell if the nucleus is artificially removed?
4 What are the likely roles of the surface membrane in the living cell?

■ Colonial forms of multicellular organisation

When individual cells remain clumped together after cell division the resulting multicellular structure may be known as a **colony**. Cells of a colony may be stuck together by their surfaces or held in a gelatinous envelope secreted by the cells.

Clumped cells persisting together may continue to be totally functionally independent. This is seen in fast-dividing yeast cells, and in clumps or chains of bacteria. There is no coordination of activities between the cells of these groupings.

In other colonial forms, the cells of the colony coordinate activities to some extent. When cells cooperate, the organism is more efficient in performing functions than are the individual cells. The result is a truly multicellular organism.

The simplest organisms of this type consist of many cells arranged in a regular and characteristic manner, giving a recognisable shape to the whole organism. The individual cells, however, show only limited specialisation for individual tasks, and the whole organism functions with restricted coordination of activities (Figure 1.10).

■ Multicellular organisms with marked division of labour

The great majority of multicellular organisms are made of visibly different parts, in which groups of cells are specialised to perform different functions. The structural differences between groups of cells are related to the specialised functions performed.

Where cells carry out specialised functions more efficiently the result is referred to as **division of labour**. Division of labour among cells leads to increased efficiency by avoiding duplication of effort. It also causes greater dependence of specialised cells on the activities of other cells. For example, muscle cells specialise in contraction but rely on cells of the blood circulation and the liver for supplies of oxygen and glucose. Most cells of complex multicellular organisms are highly specialised to carry out particular tasks. In fact, groups of cells make up tissues and organs that more or less exclusively undertake specific functions for the whole organism.

Tissues and organs of multicellular organisms

A **tissue** is a group of cells of similar structure that perform a particular function. Examples include muscle tissue of

Figure 1.10 Simple multicellular organisation

Spirogyra sp. is an unbranched filamentous alga occurring in freshwater ponds, lakes and streams. The first cell of the filament is a holdfast cell that adheres to rocks and stones. Apart from the holdfast cell, all cells are identical and apparently all are independent.

The filament is surrounded by mucilage, which may hold the long threads as a matted blanket. Bubbles of gas (mostly oxygen given off in the light from photosynthesis) can become trapped and support the mass of *Spirogyra* at the surface.

Sponges, e.g. *Leucosolenia* sp., grow attached to rocks in freshwater and marine environments. The body is multicellular yet the cells are comparatively independent of each other. The outer layer consists of covering cells and pore cells; the inner layer consists of collar cells, each of which has a flagellum that drives the feeding current of water in through the pores, and a cytoplasmic collar that engulfs food particles. Between them is a jelly-like or calcareous layer (depending on the species) with some amoeboid cells. There are no nerve cells or sense cells; all cells are sensitive but transmit impulses only to the cells immediately around them. The sponge is extremely insensitive to changes, and reacts very slowly as an organism.

Sponges are unique multicellular animals in several ways, not least in forming a definite functioning structure of quite large dimensions that yet shows a low degree of coordination between the four types of cell of which it is made.

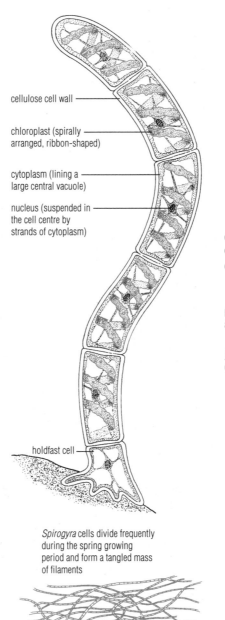

- cellulose cell wall
- chloroplast (spirally arranged, ribbon-shaped)
- cytoplasm (lining a large central vacuole)
- nucleus (suspended in the cell centre by strands of cytoplasm)
- holdfast cell

Spirogyra cells divide frequently during the spring growing period and form a tangled mass of filaments

feeding current of water

- covering cell
- collar cell
- central cavity
- pore cells
- amoeboid cell
- jelly or calcareous layer

sponge growing on the surface of a rock

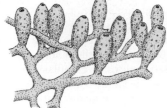

mammals, and parenchyma tissue of flowering plants.

An **organ** is part of an organism, made of a collection of tissues but having a

definite form and structure, which performs one or more specific functions. Examples include the heart of a vertebrate, and the leaf of a flowering plant.

1.5 BIOLOGY AS A SCIENCE

We noted earlier in the chapter that 'science', like 'life', is hard to define.

Science begins with curiosity about the world and a desire to understand it. Science involves testing ideas about how things work and how they are made, thereby modifying ideas and developing knowledge. Through the processes of questioning and observation of ourselves and the things about us, similarities, differences and patterns are perceived, investigated, measured and classified.

In Europe since the fifteenth century science has been on the advance. Developments have been based on accurate observations, carefully devised experiments and the collection of numerical data that is then handled mathematically. The word 'science' is itself a relatively recent one, originating in the nineteenth century. The concept of science is much older than that, however.

■ When did science start?

The beginnings of science may date back to the earliest human life! Our earliest ancestors survived at the expense of the wild plants and animals around them. From the outset, the quality of human observations played a key part in survival.

The dramatic cave paintings produced by some of the earliest human communities, 25 000 to 10 000 years ago, are described in Chapter 31. Palaeolithic cave artists observed the animals around them in scientific detail. So accurate were their representations that we can identify precisely many of the organisms they painted (Figure 1.11).

Not everyone will agree that this cave art is also 'science', but we do know that accurate observational skills such as those shown by the palaeolithic artists are also essential to practical science today.

■ The growth of formal biology

Very early in human history organisms were named, and elementary systems of **classification** of living things developed. Present-day hunter-gatherer societies, with a lifestyle that has changed little since the earliest human communities, have their own systems of classification of the plants and animals around them.

We can assume that only very much later did more formal sciences develop, under such headings as **botany** (the study of plants) and **zoology** (the study of animals). Early botany and zoology were concerned with the external form and structure of living things. Within biology, the study of the functioning of organisms, known as **physiology**, probably came very much later still. These and many other sorts of scientific activity have developed in communities and cultures all over the world.

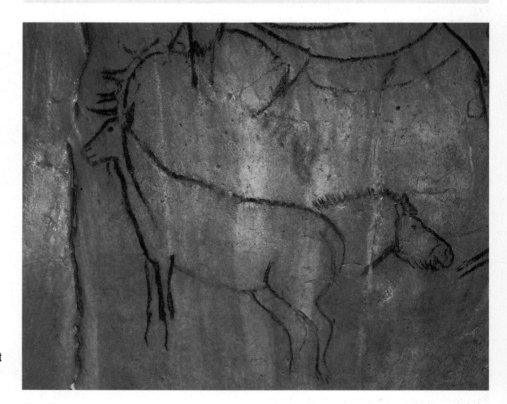

Figure 1.11 A stag engraved on a cave wall at Niaux in the Pyrenees: one of a series of cave wall paintings dating from prehistoric times

Questions

1 What features of palaeolithic cave art suggest that it can be classified as a scientific activity?
2 State in what ways the roles of the following cells differ: a unicellular organism, a cell of a multicellular colony, a cell of a multicellular organism showing marked division of labour between cells.
3 What advantages, and what disadvantages, might arise from division of labour between cells in multicellular organisms?
4 Make a list of the skills that a practical scientist uses while carrying out a major investigation or study. Which, if any, of these skills are practised by non-scientists, too?
5 What are the differences between a tissue and an organ?

1.6 PIONEERING BIOLOGISTS

In the history of biology, certain individuals started revolutions in ideas. On the following pages are introduced five people, all biologists, whose work has changed the ways we think about our world. To appreciate their achievements fully, we must bear in mind the context of their own times and the climate of ideas they absorbed.

■ Aristotle and the foundation of biological science

The foundations of biology can certainly be traced to the surviving writings and diagrams of Aristotle, who lived in Greece between 384 and 322 BC (see panel). Aristotle was not the first biologist, but in Aristotle's work there is, for the first time, a record of biological studies on a vast scale.

The story of western philosophy, too, begins with the Greeks who produced the first and some of the greatest western philosophers, including Aristotle. Philosophy is the study of the nature and limits of human knowledge. It is concerned with the principles underlying all forms of understanding. Aristotle was a pupil of Plato (427–347 BC), and Plato had been a pupil of Socrates (470–399 BC). Early science and philosophy were tightly bound together because both philosophy and science are attempts to understand the world by a disciplined use of reason. Until the eighteenth century, which saw the growth of many sciences and the development of new analytical techniques, no one distinguished science from philosophy.

■ Leeuwenhoek and the power of the single lens

Leeuwenhoek (1632–1723) was a Dutch linen-draper with no formal education in science. It is not certain how he first became involved in specialised microscopy, and thereby came to make such a distinguished contribution to science. With very simple equipment Leeuwenhoek made many important discoveries of structures too small to be seen with the naked eye. He also made observations on the functioning of living things.

With his little microscope Leeuwenhoek was able to see a vast range of structures, from some of the larger plant and

ARISTOTLE

Life and times

Aristotle was the son of the physician to Philip II, King of Macedon. When seventeen years old he became a pupil of Plato in Athens, and studied with him for several years. After Plato's death he spent the next twelve years in political exile. Here he undertook the studies in biology on which his scientific reputation is based. Later he was tutor to Alexander, Philip's son. Aristotle taught for many years at his Lyceum or 'school' in Athens. He retired to obscurity on the death of Alexander in 323 BC.

Aristotle was a tireless scholar, and a controversial public figure living a turbulent life in turbulent times. He was driven by a desire for knowledge. He wrote the equivalent of 50 substantial volumes (of which most have been lost) – a vast output that was remarkable for its scope, variety and quality.

Contributions to biology

Aristotle's vast studies in biology included:

► records of the life and breeding habits of over 500 species of animal
► embryological investigations of the development of the chick
► detailed study of the structure and behaviour of the octopus and squid, an investigation that has only recently been surpassed
► descriptions of the four-chambered stomach of ruminants, the ducts and blood vessels of the mammalian urinogenital system, and the mammalian nature of porpoises and dolphins
► an account of development in certain species of fish, such as the dogfish
► drawing attention to the importance of the heart and blood vascular system of vertebrates
► the use of diagrams to illustrate complex anatomical structure
► the realisation that different kinds of living things can be arranged in a 'ladder of nature', giving 'from plants a continuous scale of ascent towards animals' (although he cannot be claimed as an evolutionist)
► (erroneous) support for the 'spontaneous generation' of certain insects in the moisture on leaves, in mud and dung, in some wood, in hair, and in some animal flesh.

Influences

Aristotle virtually initiated biology. He worked like a modern naturalist, making first-hand observations on the whole range of living things known at that time. Certain of his books survived in the Arab world and then re-entered Europe in Latin translations in the twelfth and thirteenth centuries. They were accepted as authoritative throughout the Middle Ages, down to the eighteenth century. He made occasional crude errors, however, and he had no experimental method of 'controlled enquiry' or 'repeated observations'. He largely ignored the importance of measurement, and favoured qualitative rather than quantitative data.

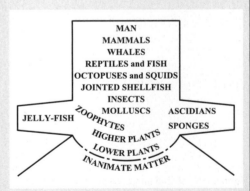

Aristotle's drawing of his dissection of the urinogenital system: the terms in brackets are modern; the dotted line represents the limits of Aristotle's diagram

MAN
MAMMALS
WHALES
REPTILES and FISH
OCTOPUSES and SQUIDS
JOINTED SHELLFISH
INSECTS
MOLLUSCS
JELLY-FISH   ZOOPHYTES   ASCIDIANS   SPONGES
HIGHER PLANTS
LOWER PLANTS
INANIMATE MATTER

Aristotle's 'ladder of nature'

animal cells to subcellular organelles. His studies of the structure of flowering plants made him a founder of plant anatomy. He also published calculations of the force required to lift water to the top of a tree, and he experimented with biological materials as well as studying structures. His experimental approach, and his use of an instrument to extend his powers of observation at this time, make Leeuwenhoek very much one of the founders of modern biological science. His contribution to biology is summarised on p. 148.

CHARLES DARWIN

Life and times

Charles Darwin, born in 1809, was to have been a doctor like his father and grandfather before him. He gave up this career after witnessing surgical operations conducted on children, during which he longed to rush out of the operating theatre. He wrote to a friend some years later:

Nor did I ever attend again, for hardly any inducement could have been strong enough to make me do so; this being long before the blessed days of chloroform. The two cases fairly haunted me for many a long year.

Darwin also abandoned his second choice of career: the Church. At 22 he became resident naturalist (unpaid) on an Admiralty-commissioned expedition to the southern hemisphere, with the reluctant approval of the captain of the ship, HMS *Beagle*. From his experiences on the five-year voyage, and from later studies, he developed his idea of organic evolution by natural selection.

Two years after his return from the expedition Darwin married his cousin, Emma Wedgwood. Within a few years they moved to Down House, near Bromley, Kent, where they lived quietly for forty years, bringing up ten children. Darwin continued his studies and experiments, and discussed his theory only with a few close friends, for many years before he published his conclusions.

Contributions to biology

Darwin's published works included studies of the effects of light on plant growth (which led, later, to the discovery of plant hormones), climbing and insectivorous plants, earthworms and soils, orchids, insects and plants of various habitats, aspects of geology, and emotion in humans and animals.

During the voyage of the *Beagle* he accumulated geological and fossil evidence that supported the idea that life changes with time. He also studied the flora and fauna of mainland South America and of some surrounding islands, including the Galapagos Islands. For example, in the birds of the Galapagos he found evidence of adaptive radiation (the development from a single ancestral group of a variety of forms adapted to various environments).

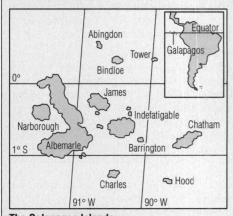

The Galapagos Islands

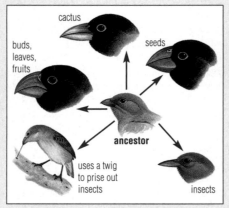

Five finches found in the Galapagos Islands: their beaks are adapted for finding different kinds of food, as indicated

Later he read Malthus's *Essay on Population*, emphasising a struggle for survival in the contest for food. Darwin concluded that a similar struggle exists among all living things. All species, he reasoned, show variation with time. Some variations confer advantages in the struggle for existence. Organisms with favourable variations are more likely to breed and to pass on their favourable characteristics. In this way new species arise from existing forms.

Darwin was aware of defects in his argument and the lack of evidence for it. Although he realised that variation was the raw material of natural selection, the inheritance process then generally accepted (blending inheritance) would actually tend to reduce variation. Moreover, there was no evidence of natural selection in action, and Darwin relied upon domestication through artificial selection to demonstrate the selection process.

Influences

Before Darwin, biologists in general believed that species had continued without change since their creation, or that the characteristics acquired in the life of an organism could be inherited by the offspring. Darwin's theory provided a mechanism for evolution that could be tested, and implied that evolution was still taking place.

Darwin's theory also implies that humans share a common ancestor with apes – this caused a controversy that has now largely subsided. A further implication is that life itself has evolved from non-living, inorganic resources of the Earth, and today this is a field of speculation and enquiry. Evolution by natural selection is an organising principle of modern biology.

■ Darwin and the theory of evolution by natural selection

The studies of Charles Darwin laid the foundations for a revolution in biology that affects us still (see panel). A careful observer and a fine naturalist, his published work showed his desire to reach general theories in biology, combined with caution. He is best remembered for his theory of evolution by natural selection.

For Darwin, there was tension between his evolutionary ideas and the values of the family social circle in which he lived. Within society at large, his theory of evolution became a focus for the developing uneasiness between science and religion. Darwin's suggestion that humans share a common ancestor with apes raised a furore that lasted for decades. Darwin himself, an intensely shy and diffident man, hated and avoided public controversy and left the arguing of his case to his great friend, the powerful propagandist T H Huxley, whose outspoken energies ensured the maximum impact of the theory.

Gradually the beliefs that had led Darwin to contemplate the life of a clergyman faded. In his later years he wrote in his autobiography:

Disbelief crept over me at a very slow rate, but was at last complete. The rate was so slow that I felt no distress, and I have never since doubted even for a single moment that my conclusion was correct.

In view of this, it is perhaps ironic that when Darwin died in 1882, Victorian Britain conferred on him the honour of burial in Westminster Abbey.

Questions

1 Aristotle is said to have worked rather like a modern naturalist does. What does this mean?

2 One difference between the ideas of Aristotle and those of Leeuwenhoek was their views on what is called 'spontaneous generation'.
  a What is meant by 'spontaneous generation'?
  b What evidence was available to Leeuwenhoek to support his doubt about spontaneous generation, which people in Aristotle's times did not have?

3 Why has Darwin's theory of natural selection influenced studies in so many aspects of biology?

■ Pasteur and applied biology

Pasteur was a gifted and hard-working scientist of whom France is justly proud. He was also a self-confident man, rather difficult to work with. Born in 1822, he worked in the second part of the nineteenth century when, for the first time, the findings of experimental science were being converted into practical applications on a large scale. Pasteur was one of the scientists most instrumental in bringing about this achievement.

Pasteur's first research was in the field of crystallography, with substances that exist as optical isomers (p. 133). Much of the rest of his work was with bacteria and fungi, including the economically important yeasts. His studies of industrial fermentation led him to establish the importance of air-borne microorganisms and their spores, and to the discovery that fermentation is a form of respiration. In his later career his studies of diseases in animals, including humans, established the germ theory of disease and the value of vaccination. All in all, Pasteur's work indirectly gave great status to applied biology. His contributions to biology are summarised in the panel on p. 527.

■ Crick and the foundation of molecular biology

Francis Crick, born in 1916, and James Watson, born in 1928, laid the foundations of a new branch of biology, and achieved this whilst they were still young men. Crick started his academic career as a physicist, but when an opportunity for a fresh start arose in 1945 he chose to work in the field we now call molecular biology. Watson, a biologist, joined him shortly after his arrival at the Cavendish Laboratory at Cambridge University. They produced a chemical model of the unit of inheritance, the gene (p. 210), which is made of the nucleic acid deoxyribonucleic acid, or DNA. They achieved this together in only about two years!

The work of Crick and Watson represents a revolution in biology because it provides a theoretical explanation of the mechanism of inheritance in genetics and a biochemical mechanism for the day-to-day management of the cell (cell biology). Their contribution to establishing molecular biology is summarised in the panel on p. 203.

1.7 THE PROCESSES OF PRACTICAL SCIENCE

People often talk about 'the scientific method' as if there is a single, unique method of science, distinctly different from other methods of knowing and investigating. In fact, there are many successful methods in science. You will appreciate this point when you know more about the methods of Darwin, Pasteur and Crick, for example, or if you compare the ways in which other, and sometimes less famous, scientists have worked. In this book you will find the practical approach of scientists introduced as part of our current understanding of many different topics.

Nevertheless, all science includes some common features, as follows:

▶ observing and measuring
▶ hypothesising and predicting
▶ designing and planning investigations
▶ carrying out explorations
▶ recording data
▶ interpreting results and drawing conclusions
▶ communicating.

Crucial to these is the concept of a **hypothesis**. A scientific hypothesis is a tentative explanation capable of being tested by observations or experiments. Sometimes a hypothesis is formulated by examining data or from observations. It is equally likely to be formulated from an inspired guess or hunch (Figure 1.12).

SCIENCE PROCESSES: SOME IMPORTANT TERMS

Observation is the process by which perceptions of objects or events are selected, interpreted and their significance judged.

A **hypothesis** is a tentative explanation for an observed event. To be scientific a hypothesis must be testable by means of predictions and investigations.

Qualitative data consist of observations made without recourse to measurements, e.g. 'moorhens lay more eggs than robins do'.

Quantitative data are observations made more precise by measurement, e.g. 'moorhens lay ten eggs on average'.

An **inference** is a tentative conclusion drawn from a series of observations. It may lead to a hypothesis being formulated.

The **independent variable** is the condition that is changed systematically in an experiment, e.g. in an experiment in which seedlings are grown at different temperatures, temperature is the independent variable.

The **dependent variable** is the effect or outcome you choose to measure, e.g. the rate of growth of seedlings at different temperatures.

Controls are the variables that must be controlled by you in the investigation and not allowed to vary at all, e.g. light and moisture in the experiment on seedling growth. A **control experiment** is often used in biology. In an investigation of the effect of light on plants the control might be a plant kept in the dark. Thus a control is one value of the independent variable.

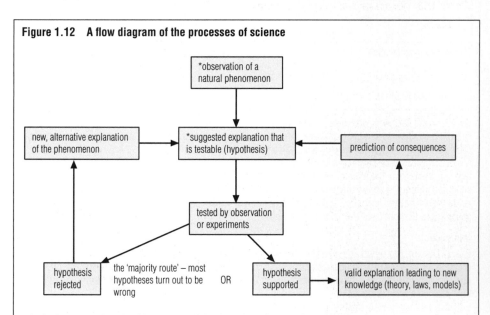

Figure 1.12   A flow diagram of the processes of science

*observation of a natural phenomenon

new, alternative explanation of the phenomenon → *suggested explanation that is testable (hypothesis) ← prediction of consequences

tested by observation or experiments

hypothesis rejected — the 'majority route' – most hypotheses turn out to be wrong — OR — hypothesis supported → valid explanation leading to new knowledge (theory, laws, models)

*A scientist may start from data (observations, results of experiments) that call for explanation, or a scientist may start with an idea (a hunch or belief, in fact) that something happens in a particular way and then make observations or conduct experiments in the hope that the idea (a hypothesis) may be supported.

FROM INITIAL OBSERVATIONS TO ACCEPTED KNOWLEDGE

Observation Birds of any one species usually lay the same number of eggs. For instance, robins lay five or six eggs, moorhens lay about ten.

Hypothesis (1) Female birds form only a limited number of eggs in their bodies. When these are laid no more are produced.

Prediction If eggs are removed from a bird's nest before the clutch is complete, the final clutch size will be less than usual for the species.

Experiment Eggs were removed from moorhens' nests with less than five eggs. The nest was visited each day until the clutch size increased no more.

Result The clutch size was still the same as usual. The same applied to other species of bird.

Outcome The results do not support the prediction and therefore the hypothesis is unsatisfactory and must be modified.

Hypothesis (2) Birds form many eggs in their bodies but stop laying when they have laid a specific number. (The implication is that they can 'count' or, rather, respond to a specific number of eggs.)

Prediction If eggs are removed from a bird's nest before the clutch is complete, the bird will continue to lay a full-sized clutch.

Experiment Four eggs were removed from a clutch of six in a moorhen's nest.

Result The final clutch size was eleven.

Outcome The prediction was correct, the hypothesis is supported. But the hypothesis is *not proved*. There could be a better, simpler or more universal explanation. For example, the number of eggs a bird lays may depend on how many young she thinks she can feed, or how big the nest is, or whether the young become independent quickly. The results might apply to robins and moorhens but not to sparrows and pheasants.

A further prediction from hypothesis (2) is that adding eggs to an incomplete clutch will not affect the final clutch size (i.e. the birds will stop laying when the clutch is up to normal size). An experiment to test this prediction yielded positive results. This strengthens the hypothesis but still does not *prove* it.

■ Hypothesis, prediction and experiments, in practice

The practical sequence of steps from initial observation to currently accepted knowledge is illustrated on the left (see panel). Note that under the Countryside and Wildlife Act of 1981, it is an offence to take eggs of many wild birds. The experiments described were conducted long before 1981.

Notice that the product of the scientific process is knowledge consisting of convincing hypotheses. It was Karl Popper, a great philosopher of science (p. 677), who established that a hypothesis is never finally proved but can be disproved at any time by a single contrary observation. He maintains that all explanations in science are of this type. That is, scientific knowledge is always only tentative. It is the best available explanation we can offer at that time. Any explanation may prove wrong or incomplete, and often does, sooner or later!

■ Variables and controls

In a biological experiment many conditions or factors (light, temperature, pH or day-length, for example) can vary and may be varied. They are referred to as **variables**. During the experiment some variables may be measured, some subtly varied and some kept constant.

In Figure 1.13 a simple and perhaps familiar experiment to investigate the production of gas by an aquatic plant is shown. (You will learn about photosynthesis, the process by which this happens, in Chapter 12.) In this experiment:

► **oxygen produced** is the variable that is accurately measured – the **dependent variable** (it depends on the effects of changing the independent variable)

► **light** is the variable that is changed – the **independent variable** (it is altered by the experimenter, that is, it is independent of other conditions)

► the **volume of water** and its **composition**, the **pondweed** and the **temperature** are other variables that might have affected the outcome and therefore were kept constant – the **controlled variables**.

Assuming that the **hypothesis** in this experiment is 'Plants need light to photosynthesise (and produce oxygen)', then the **prediction** is that if the light is not present, a plant will not produce oxygen. So the plant in the dark is the **experiment** and the plant in the light is the **control**.

Figure 1.13 An example of a control experiment

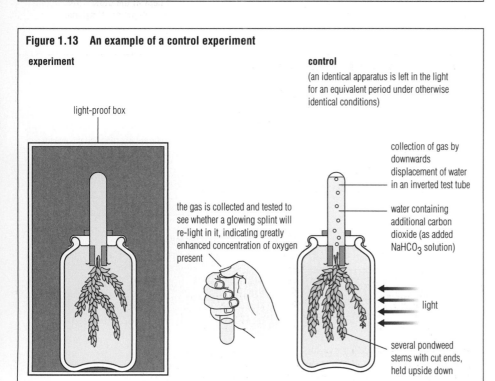

experiment

light-proof box

the gas is collected and tested to see whether a glowing splint will re-light in it, indicating greatly enhanced concentration of oxygen present

control

(an identical apparatus is left in the light for an equivalent period under otherwise identical conditions)

collection of gas by downwards displacement of water in an inverted test tube

water containing additional carbon dioxide (as added NaHCO$_3$ solution)

light

several pondweed stems with cut ends, held upside down

1.8 MODERN BIOLOGY

■ Progress in research

As you read this book you will learn more and more about the processes of biology. In the meantime, what are the factors that may lead to real progress in biological research? Here are some pointers.

The experimenter uses a novel technique or apparatus to provide data not previously obtainable. For example:

▶ the use of a radio-isotope in the work of Calvin on the path of carbon in photosynthesis (p. 262)

▶ the use of the microcomputer to develop numerical classification (p. 26).

The experimenter is lucky in selecting materials or a method that, exceptionally, yields results that can be interpreted. For example:

▶ the work of Mendel on the inheritance of particular characteristics of garden peas (p. 605)

▶ Fleming's discovery of the antibiotic action of the mould *Penicillium* against colonies of staphylococci (p. 531).

The experimenter devises proper controls to ensure a fair test; experimental batches and samples must be large enough for the effects of the individual differences to be excluded, and for the results to be statistically significant. For example:

▶ the experiments with coleoptiles and the hormone control of tropisms (p. 423)

▶ Bernard's evidence that the mammalian liver constantly exports heat (p. 509).

The experimenter reinterprets existing results; results are of a tentative nature and require interpretation. For example:

▶ Crick and Watson reinterpreted other people's results to suggest that DNA exists as a double helix (p. 203)

▶ the reinterpretation of Mendelian factors as genes on chromosomes (p. 608).

The experimenter makes a prediction about the existence of structures or mechanisms before these are known or observed. For example:

▶ Harvey's postulation of the complete circulation of the blood before the capillaries had been observed (p. 346)

▶ the prediction of the existence of viruses from the discovery of a 'filterable' agent causing disease (p. 528).

The experimenter finds an explanation that leads to the formulation of mental models; the theoretical model may represent a new way of looking at the phenomenon, and it may stimulate further fruitful investigations. For example:

▶ the theory of evolution by natural selection (p. 656)

▶ the fluid mosaic model for the structure of the plasma membrane (p. 158).

The experimenter puts forward new scientific explanations leading to change in attitudes and outlook in other people. For example:

▶ Pasteur's germ theory and the consequences for food hygiene and medicine (p. 527)

▶ Darwin's and Wallace's joint publication of the idea of evolution by natural selection (p. 656).

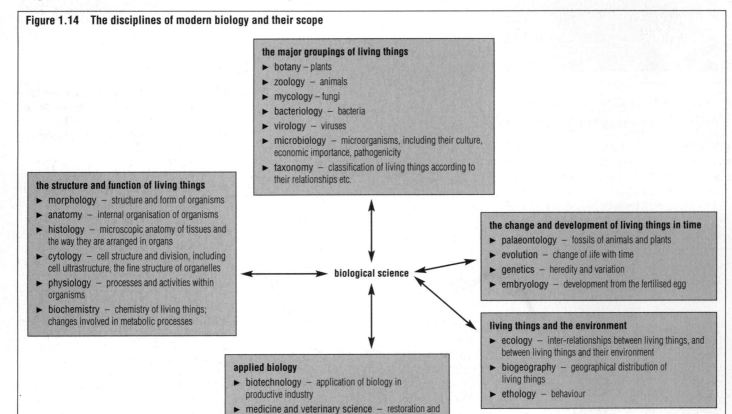

Figure 1.14 The disciplines of modern biology and their scope

the major groupings of living things
▶ botany – plants
▶ zoology – animals
▶ mycology – fungi
▶ bacteriology – bacteria
▶ virology – viruses
▶ microbiology – microorganisms, including their culture, economic importance, pathogenicity
▶ taxonomy – classification of living things according to their relationships etc.

the structure and function of living things
▶ morphology – structure and form of organisms
▶ anatomy – internal organisation of organisms
▶ histology – microscopic anatomy of tissues and the way they are arranged in organs
▶ cytology – cell structure and division, including cell ultrastructure, the fine structure of organelles
▶ physiology – processes and activities within organisms
▶ biochemistry – chemistry of living things; changes involved in metabolic processes

biological science

the change and development of living things in time
▶ palaeontology – fossils of animals and plants
▶ evolution – change of life with time
▶ genetics – heredity and variation
▶ embryology – development from the fertilised egg

living things and the environment
▶ ecology – inter-relationships between living things, and between living things and their environment
▶ biogeography – geographical distribution of living things
▶ ethology – behaviour

applied biology
▶ biotechnology – application of biology in productive industry
▶ medicine and veterinary science – restoration and maintenance of health in humans and in animals
▶ agriculture – farming of crops and livestock for food
▶ forestry – trees

Figure 1.15   Some titles of biological journals – a few of the many titles of biological journals published regularly

Scientific journals fulfil specialised roles.
For example:

1 papers on recently-conducted experimental work occur in many journals, including:

 ANNALS OF BOTANY — papers on discoveries about the structure and functioning of plants

 British Journal of Experimental Biology — papers on experimental physiology

 NATURE — papers at the forefront of developments across the sciences

2 reviews of particular topics are found in 'review' journals, including:

 Annual Reviews of Biochemistry — reviews of developments in biochemistry

 BACTERIOLOGICAL REVIEWS — reviews of developments in bacteriology

 BIOLOGICAL REVIEWS — reviews of topics in biological science generally

3 papers in specialised fields (the contents are evident from the titles of these and many other specialised journals):

Cancer Research · DEFENSE RESEARCH · Experimental Parasitology · FOOD RESEARCH · JOURNAL OF ECOLOGY · Soil Science · Transactions of the Royal Entomological Society · Methods in Enzymology

4 papers in applied biology, including:

Journal of Agricultural Science · JOURNAL OF HYGIENE · Quarterly Journal of Forestry · VETERINARY MEDICINE

■ The pace of biological enquiry

Another feature of modern biology is the pace of scientific enquiry. The subject is growing at an enormous rate.

Scientists report their experiments and explanations in concise research reports called **papers**. Papers, though brief, give exact details of the methods adopted in sufficient detail to permit repetition of the experiment, the results obtained, and a discussion of the significance of the outcomes.

The speed of growth of modern biology is reflected in the increasing numbers of papers published, and in the growing range of academic journals in which these papers appear (Figure 1.15). Copies of these are held in the libraries of universities, teaching hospitals and research institutes and foundations belonging to particular industries and companies, as well as Government departments and agencies.

In addition to biological journals papers are also read to conference meetings, often called **symposia**, of interested researchers. The papers and subsequent discussion are often published as special conference reports, generally with an international circulation.

■ The breadth of modern biology

Modern biology is an enormous subject, consisting of numerous important disciplines (or branches) of study. Certain of these disciplines are described as **pure** science because they are concerned with principles rather than with their applications. Examples include the study of cell structure, known as **cytology**, and the study of the structure of the bodies of organisms, or **anatomy**. Other aspects of biology are **applied** to particular technologies such as medicine or agriculture. Applied disciplines include **pathology**, the study of diseases of organisms, and many aspects of **immunology**, the science of organisms' resistance to disease.

The divisions within biology arise partly from the breadth of the subject matter, due to the great diversity of living things.

Divisions are also due to a very distinctive feature of scientific investigation. This is the practice of breaking down complex problems into component issues which are simple enough to investigate practically. There are numerous examples of this. Science develops by simplification and subdivision of complex and otherwise apparently intractable problems. For example, the study of the feeding of other animals and humans begins as the main issue in **nutrition**. From nutrition has grown **enzymology**, the study of enzymes, including digestive enzymes, and **dietetics**, the study of human diet. A direct consequence of progress by simplification is this great range of component disciplines within biology, many with specialised techniques of investigation (Figure 1.14).

Another important source of the new disciplines within biology arises from the links between biology and other sciences. Biology has strong links with physics, chemistry, mathematics and computing, geography and geology, for example. Developments in other sciences are also the bases of advances in biology. As a result we now have important disciplines such as **biochemistry**, **biophysics** and **biogeography**.

Questions

1 In Figure 8.17 (p. 182), an investigation of the effect of substrate concentration on the rate of enzyme action is described. Identify the independent variable, the dependent variable and the controlled variables.

2 A student observes that the grass beneath a chain-link fence fails to grow where the fence traverses an otherwise healthy lawn. She also notices that when it rains, rainwater collects on the fence and drips on to the ground beneath. The fence is made from galvanised steel wire (steel wire coated with zinc).

Suggest at least two hypotheses that might explain this observation, and suggest experiments that would help to select the better one.

3 A standard GCSE experiment is to part-cover a leaf with a stencil and show that the covered parts fail to produce starch when the leaf is exposed to light. Suggest two possible interpretations of the results.

4 In an experiment to investigate the role of the nucleus in a frog's egg, a microscopic glass hook is used to penetrate the cell membrane and remove the nucleus. The egg fails to develop normally. What control experiment should the scientist do, and what would be its purpose?

1.9 WHY STUDY BIOLOGY?

Biology plays such a valuable part in general education that there is hardly any need to justify its study further. Issues that are biologically based crop up in everyday experiences; for example:

▶ when we know why we should have thanked a green plant today!

▶ as we seek to understand the behaviour of our pets, and our fellow human beings

▶ when we read that a toddler has mum's eyes, dad's ears and the milkman's teeth!

▶ as we relax in a nature reserve, work in the garden or walk in the park

▶ when we choose food and drink from the supermarket and collect a prescription from the chemist

▶ when we support a Greenpeace campaign, or argue about defence and nuclear weapons

▶ when we read about developments in biotechnologies new and old

▶ when we discuss the ethics of transplant surgery or embryo experimentation

▶ and when we communicate concern about increasing atmospheric levels of carbon dioxide, about the destruction of the ozone layer, and the issues of world hunger and human lifestyles, east and west, north and south.

Biology may in fact be directly useful to you in finding employment. Biologists are employed in laboratory work, industry, fieldwork, agriculture, horticulture, forestry, health care, work with animals, marine and freshwater biology, information science, publishing, management, financial services and in teaching or lecturing.

All these opportunities arise through some form of further or higher education, following your Advanced or GNVQ level course. The panel below offers an intro-

duction to some careers; some are more accessible than others, but there is valuable work to be undertaken in all of them.

Even if you do not go on to further study, your biological background will still be valuable. And remember, whether or not you work in biology in adult life your work experience is likely to be very different from your school work. Study of a subject post-GCSE is a marvellous education and training, and in no way prejudices your opportunity for a varied and fascinating career subsequently – quite the reverse, in fact. Very few of us are stuck at a lab bench for ever, or rooted to one spot! So enjoy biology now . . . and later!

> **Careers guidance**
>
> *Careers with Biology. A guide for school leavers.* Institute of Biology, 20 Queensberry Place, London SW7 2DZ

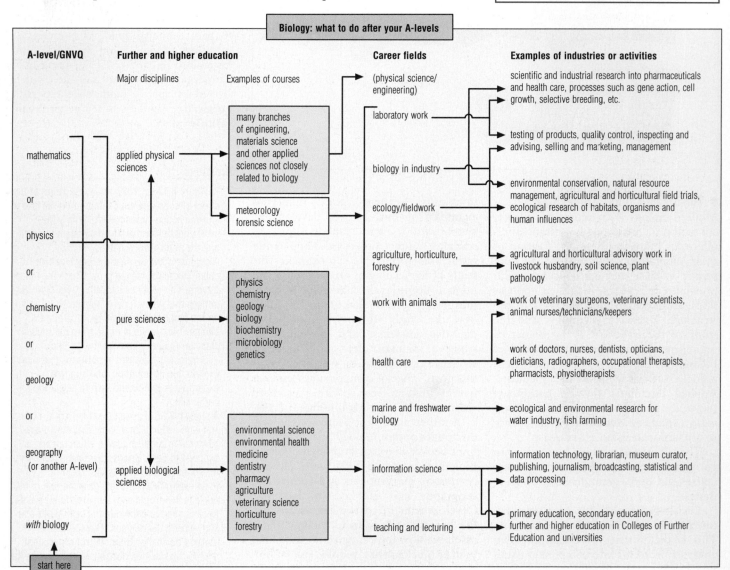

Biology: what to do after your A-levels

A VACCINE FOR AIDS?

**Dr Persephone Borrow,
Viral Immunology Group Leader,
Edward Jenner Institute for Vaccine
Research**

Career experience:

- at school, took A-levels in Biology, Chemistry and Maths with Statistics
- a 'gap year' in the immunological laboratories at ICI Pharmaceuticals
- a degree in Natural Sciences at Cambridge University
- a PhD on the immune response to a nervous system virus infection
- postdoctoral research at The Scripps Research Institute in California, studying virus–immune system interactions
- research success led to independent funding and an Assistant Professorship at the Institute
- after eight years in the USA, returned to the newly-established Edward Jenner Institute for Vaccine Research, now leading the Viral Immunology group.

My work:

My group at the Edward Jenner Institute studies the immune response to viruses that establish persistent infections in humans, and for which, as yet, there are no effective vaccines. These include human immunodeficiency virus (HIV-1), which causes AIDS, and hepatitis C virus, which produces chronic liver disease. These viruses are highly variable, and it is difficult to design a vaccine that recognises all the variants of the virus with which people become infected. Also, these viruses have evolved evasion strategies that enable them to 'hide' from the immune system. We are trying to understand how they avoid the host immune response, so that effective vaccines can be produced. Our research approaches involve:

- obtaining blood samples from infected patients and investigating what immune responses have occurred, and how successful these have been
- infecting cells from the blood of healthy individuals with virus *in vitro*, and studying how the cells respond
- carrying out experiments using animal virus infection models, allowing cause–effect relationships to be directly tested.

Our laboratory experiments involve many cellular and molecular biological techniques, both established and experimental. We collaborate with groups within and outside our Institute to share expertise.

As a group leader, I have less time for experimental work now. In a typical week I spend time helping members of the group with unfamiliar techniques, or in trying to give advice when something hasn't worked. I hope there will be time for a whole series of experiments of my own when there is less administrative work to be done.

At the moment, administration concerns the setting up of a facility to work with mice infected with viruses that are serious human pathogens. For this I deal with issues such as ethical and legal approval for the proposed experimental work, and with ensuring safety at all times.

The remainder of my time is spent much more rewardingly. I discuss experimental results with the members of the group and our external collaborators, and we decide what should be done next. It's very exciting when you finally get an answer to a particular question, especially if it's not what you expected.

A SUSTAINABLE FUTURE FOR WILDLIFE AND PEOPLE

**Dr Gary Mantle MBE,
Environmentalist,
Director of the Wiltshire Wildlife Trust**

Career experience:

- at school, my favourite subject was Rural Studies, with my spare time given to our Conservation Corps and to the local Woodland Preservation Society
- took A-levels in Biology, Chemistry and Geography
- my first career preference was to be a river-keeper
- instead, I chose to study for a degree in Ecology at the University of Lancaster
- became a Research Assistant with the Open University, studying for my PhD in system theory applied in conservation. I experimented on aspects of fish farming
- took a year out as a cabinet maker
- became Head of the Urban Ecology Unit at the University of Greenwich
- moved to Richard Branson's UK 2000 Environmental Initiative, for four years; this was a move to bring together non-governmental agencies (such as the Wildlife Trusts) with industry and government to improve the environment
- became Director of the Wiltshire Wildlife Trust.

The Trust and my work:

The Wiltshire Wildlife Trust is a charity. It is a member of The Wildlife Trusts, a partnership of all local Wildlife Trusts, working together for the benefit of wildlife nationally. The Wiltshire Trust is the largest organisation of its kind in the county, with over 10 000 members. We rely on membership subscriptions and donations to fund the majority of our work. We are aided by sponsorship from business and grants from organisations like English Nature and local councils.

We manage about two thousand acres of land on 44 nature reserves, spread throughout Wiltshire. These include ancient woodland, hay meadows, disused water meadows and heathland, as well as chalk downlands for which the county is famous. Over a thousand volunteers give their time and services free to help wildlife conservation. Our volunteers are people of all ages, working in our offices and fund raising, as well as working on biological recording and site management.

The Wiltshire Wildlife Trust spends annually over £1 million on caring for Wiltshire's countryside and encouraging others to do the same. Over 75% of our income goes to the direct benefit of the countryside; for example, on land purchase, and on the management of our nature reserves. Our education staff work with young people learning about wildlife and the environment, in their schools and through our network of Wildlife Watch groups. We work with farmers and landowners, wherever help is welcomed to manage land in sympathy with the needs of wildlife, and give advice on grants available for environmentally-friendly farming methods. Some need specialised help with the management of important wildlife sites.

My team of staff and myself are dedicated to the idea of a sustainable Wiltshire, where everyone enjoys a satisfying quality of life, based on a fair share of the world's resources, delivered without destroying the birthright of future generations. We all depend on the world of wildlife and nature as our life support system. Yet the destruction of the past fifty years has taken us close to a point of no return, we feel. The challenge is to make the protection of biodiversity an integral part of working towards a better quality of life for all.

2 The range of living things: systematics

- There are vast numbers of living things, and almost unlimited diversity. Biological classification attempts to classify living things according to how closely related we believe them to be.
- There are two types of cell organisation in the living world: the prokaryotes (e.g. bacteria) have very small cells with no true nucleus; the eukaryotes (e.g. plants, animals and fungi) have a nucleus separated from the cytoplasm by a nuclear membrane.

- In the binomial system each organism has a scientific name of two words: the first name is the genus, the second is the species. The species is the first level of the hierarchical scheme of classification.
- Living things are grouped into five kingdoms (the largest categories into which organisms are placed). The kingdoms are Monera (bacteria and cyanobacteria), Protoctista (single celled organisms and organisms closely related to them), Fungi, Plantae and Animalia.
- The great diversity of living things often makes identifying them a challenging activity, but the problems can be overcome.

2.1 THE VARIETY OF LIFE

There are vast numbers of living things – almost an unlimited diversity, in fact. No other aspect of life is more characteristic than this great variety of different organisms. The total number of different types of organism or **species** that have been named and classified is uncertain because there is, as yet, no central catalogue of named species.

A species is defined as a group of individuals of common ancestry that closely resemble each other, and are normally capable of interbreeding to produce fertile offspring. This definition is explained more fully later in this chapter.

Estimates of the total number of species on Earth today range from as low as 3 to 5 million to around 100 million, or even more. An unknown number of species remains to be discovered. These also await eventual classification! When the numbers of fossil organisms that have been discovered are added, perhaps some 500 million organisms have to be fitted into a fully comprehensive scheme of classification of all living things.

In Figure 2.1 the relative numbers of species of living things are illustrated by means of pie charts. This analysis immediately imposes some order upon the great diversity of life, because it shows that the vast majority of living things belong to only two groups of organisms: the insects and the flowering plants. This contrasts with the importance given to mammals in general, and to primates in particular, by the human animal! The biology of insects and that of flowering plants are obviously important in our understanding of all living things.

■ The total numbers of living things

There are huge numbers of individuals of most species of living things. Except for the thousand or more rare species that are now reduced to a few remaining individuals (and are therefore very close to extinction), it is impossible to make a sound estimate of the total number of all the individuals of any particular species. Consequently we have no idea at all about the total number of living things; there are just too many to study in that way.

■ Fundamental differences amongst living things

Living things have traditionally been divided into two major groupings: **animals** and **plants**. But the range of biological organisation is more diverse than this.

We know that organisms consist of cells; the simplest organisms consist of a single cell or just a very few cells, whereas the larger organisms consist of very many cells. The use of the electron microscope in cytology (p. 154) has led to the discovery of two types of cellular organisation: **prokaryotes** and **eukaryotes**.

Prokaryotes are much smaller than eukaryotes and structurally less complex, but they should not be thought of as inferior. Prokaryotes have a long evolutionary history. Some of the earliest fossil remains of simple plants appear almost identical to modern cyanobacteria. Prokaryotes have probably been intimately involved in the evolution of eukaryote cells.

Moreover, prokaryotes may exhibit complex biochemistry. Thus many cyanobacteria and certain bacteria can 'fix' atmospheric nitrogen gas (p. 333); no eukaryotes can apparently do so on their own.

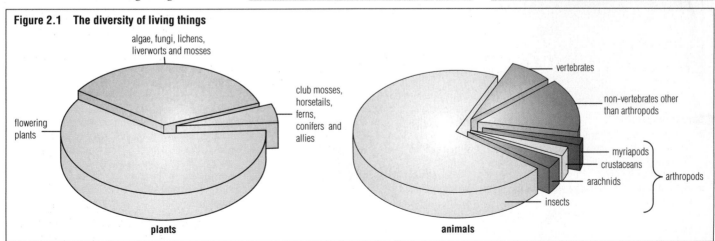

Figure 2.1   The diversity of living things

algae, fungi, lichens, liverworts and mosses

flowering plants

club mosses, horsetails, ferns, conifers and allies

plants

vertebrates

non-vertebrates other than arthropods

myriapods
crustaceans
arachnids
insects

arthropods

animals

PROKARYOTES, EUKARYOTES AND VIRUSES

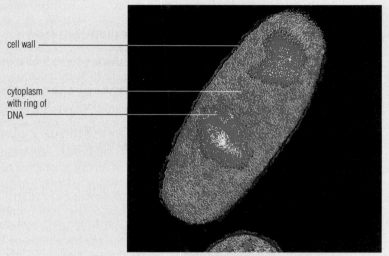

cell wall

cytoplasm
with ring of
DNA

electronmicrograph of a bacterium (*Pseudomonas* sp.) (×40 000)

Prokaryotic organisms

Bacteria and cyanobacteria: the name 'prokaryote' derives from the Greek *pro* = before, *karyon* = nucleus, signifying the absence of a true nucleus from these cells.

The electron microscope has helped show:

► the cell has no distinct nucleus
► the cell has a single circular chromosome in the form of a ring of deoxyribonucleic acid (DNA) in the cytoplasm, not contained in a nuclear membrane
► the cell lacks mitochondria, chloroplasts and complex structured flagella
► the cell has a wall of distinct composition, containing mucopeptides (of protein origin)
► the cells are extremely small (1–10 μm in diameter).

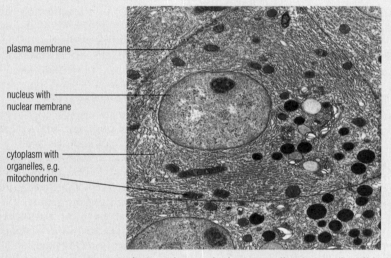

plasma membrane

nucleus with
nuclear membrane

cytoplasm with
organelles, e.g.
mitochondrion

electronmicrograph of a mammalian pancreatic cell (×6000)

Eukaryotic organisms

Plants, animals, fungi, i.e. the majority of living things: the name derives from *eu* = good, *karyon* = nucleus. The electron microscope has helped show:

► the cell has a nucleus separated from the cytoplasm by its nuclear membrane
► the nucleus contains chromosomes in which DNA occurs, supported by a special protein
► the cytoplasm is compartmentalised by membranes into regions called organelles; these include endoplasmic reticulum, mitochondria and chloroplasts
► eukaryotic flagella and cilia are made of a series of tubes arranged in a cylindrical manner
► the cell wall (when present) usually includes cellulose or chitin
► cells are 10–100 μm in diameter.

envelope of
lipoprotein

spirally coiled
nucleic acid,
surrounded by
proteins

model of the influenza virus (×150 000)

Viruses

Agents of disease that require to be transmitted from host to host. They are not 'alive' outside the host cell, i.e. not respiring, excreting, growing, etc.

► Many viruses are constructed from protein and nucleic acid alone, although some contain carbohydrates and lipids.
► Viruses are capable of replication only within specific host cells.

Questions

1 Why are the viruses hard to include in the living or non-living categories of matter, yet fungi are clearly defined as living things?
2 How can we possibly 'know' that there are 'undiscovered' species?
3 If you encountered an unfamiliar unicellular organism, how would you (in principle) decide whether it was a prokaryote or a eukaryote?

2.2 SYSTEMATICS: THE STUDY OF BIOLOGICAL DIVERSITY

Systematics is the scientific study of the diversity of living things. It is about the way organisms are classified by biologists. A classification system imposes order and a general plan upon the immense diversity of living things. Classification is essential to biology because there are far too many different types of living things to sort out and compare unless they are organised into manageable categories.

Any useful scheme of classification must be flexible, allowing new organisms to be added into the scheme as they are identified and named. Similarly, it must be possible to add extinct organisms, discovered as fossils, once it becomes clear how they relate to living forms. A functional scheme of classification must also allow biologists to search for related and similar organisms without having to refer to the entire collection of organisms.

■ The earliest schemes of classification

People have always classified living things, and they continue to do so. At an early stage in observation of their surroundings humans recognised the difference between plants and animals. Then animals and plants were divided into some obvious and useful groupings.

Early human communities classified the plants and animals useful for food. Food plants were differentiated from plants that were exploited as sources of drugs or medicine, and from others that were poisonous.

Early zoologists classified animals with backbones separately from those without, and then increasingly subdivided these major groupings in various ways.

In more recent times botanists have classified plants as herbaceous or woody, and horticulturalists have classified plants as annuals, biennials or perennials. The process continues without ceasing, according to prevailing needs and interests.

■ What classification involves

A classification system imposes an agreed name to each organism – this is essential for effective communication – and it imposes a general plan upon the diversity of living things.

Naming organisms

Observations of plants and animals are useless unless the observers can tell other people what it is they have discovered. Many organisms have local names, however. Local names differ in different parts of the country and in different countries around the world. For example, in America the name 'robin' refers to *Turdus migratorius*, a bird the size of the European blackbird and a different species from the European robin (*Erithacus rubecula*).

The results of biological enquiries cannot be compared unless they are known to refer to exactly the same types of organism. If biologists throughout the world are to communicate effectively, each different type of organism must have a unique name that is recognised worldwide. The **binomial system**, which names each organism as a member of a genus and a species, is used by international agreement.

Naming organisms by the binomial system

In the binomial system each organism has a scientific name consisting of two words in Latin. The first (a noun) designates the **genus**, the second (an adjective) the **species**. For example:

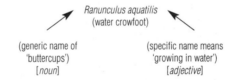

Ranunculus aquatilis (water crowfoot)

(generic name of 'buttercups') [*noun*]

(specific name means 'growing in water') [*adjective*]

The species represents the first level of classification. Members of a species are individuals of common ancestry, which closely resemble each other structurally and physiologically, and are capable of interbreeding to produce fertile offspring.

The generic name (name of a genus) is shared with other (related) species considered to be sufficiently similar to be grouped in the same genus. A genus may contain two or more species, or it may consist of only one species.

Table 2.1 Examples of the use of the binomial system

common oak	*Quercus robur*
sessile oak	*Quercus petraea*
garden snail	*Helix aspersa*
Roman snail	*Helix pomatia*
common buttercup	*Ranunculus acris*
creeping buttercup	*Ranunculus repens*
human (modern)	*Homo sapiens*
handy Man (Stone Age)	*Homo habilis*

The generic name is capitalised, the species name is not; both are italicised (underlined in handwritten material). Examples of this system of naming organisms are shown in Table 2.1.

Grouping organisms in a hierarchical scheme

In classification, species are placed in groupings, which in turn are gathered into larger groups. All systems of classification are **hierarchical**, each successive group containing more and more different kinds of organism. Classification aims to use as many characteristics as possible in placing similar organisms together and dissimilar ones apart.

Classifications are not inflexible schemes imposed from outside, however. Rather, they are the inventions of biologists, based on the best available evidence, and using principles that change as our understanding changes. At first sight the processes of classification appear simple and straightforward. In practice biological classification is a difficult discipline. Schemes of classification generate controversy, as we shall see!

Taxonomy and the system of taxa

Named organisms are placed into groups on the basis of observed, shared features. For example, individuals with almost every feature in common belong to the same species. Different but similar species with very many features in common share the same genus.

The general name for a classificatory grouping is a **taxon** (plural = taxa). The science of classification is known as **taxonomy**. The lowest or most exclusive taxon is the species; the highest or most

Table 2.2 The taxa used in taxonomy

species	a group of organisms capable of interbreeding to produce fertile offspring (occasionally a species is subdivided into subspecies or varieties, and varieties may be divided into strains)
genus	a group of similar and closely related species
family	a group of apparently related genera
order	a group of apparently related families
class	a grouping of orders within a phylum
phylum	organisms constructed on a similar plan
kingdom	the largest and most inclusive grouping, e.g. plants, animals, fungi, etc.

Some taxa are inconveniently large, so large as to be fairly meaningless. This problem is overcome, where it arises, by subdivisions such as subphyla and subclasses.

inclusive is the kingdom. Taxa have been standardised throughout the world as a result of the International Code of Botanical Nomenclature and the International Committee on Zoological Nomenclature. The categories in common use are listed in Table 2.2.

In this hierarchical scheme of classification, the higher the category the more diverse are the species included. This means that a large unit such as a **phylum** contains a wide range of organisms with one or two fundamental features in common. For example, all members of the phylum Arthropoda have a hard outer cuticle and jointed legs. No members of any other phylum have these features. In contrast, a **family** is a closely connected group of organisms with many common features (Table 2.3).

Questions

1 Place the following taxa in the correct sequence, starting with the most exclusive taxon: phylum, species, family, kingdom, genus, class.
2 Scientific names of organisms are sometimes difficult to pronounce or remember. Why are they used?
3 Suggest a method of classifying plants that might be useful to an enthusiastic gardener.

Table 2.3   The classification of three organisms using the seven principal taxa

	Eukaryota		
Kingdom	Plantae the plants	Fungi the fungi	Animalia the animals
Phylum	Angiospermophyta the flowering plants; have ovules enclosed in an ovary	Ascomycota the ascomycetes or sac fungi; sexual reproduction involves spores (ascospores) produced inside a special structure (ascus)	Chordata includes the subphylum **vertebrates**; have well-developed central nervous system including brain, an internal skeleton, two pairs of limbs or fins
Class	Dicotyledonae embryo in seed has two seed-leaves (cotyledons)	Hemiascomycetidae ascus is naked, not surrounded by cells or hyphae	**Mammalia**, the mammals feed young on milk, body covered with hair
Order	Fagales trees or shrubs, leaves alternate and simple: birches, alders, hornbeams, hazels, sweet chestnut, beech, oaks	Endomycetales zygote, derived by fusion of two cells, is directly transformed into the ascus	Rodentia the rodents; gnawing animals
Family	Fagaceae fruit is a one-seeded nut surrounded or enclosed by a scaly 'cupule': beech, sweet chestnut, oak	Saccharomyces the yeasts; body consists of cells rather than hyphae	Muroideae rats and mice
Genus	*Quercus*	*Saccharomyces*	*Rattus*
Species	*robur* (common oak) Native species to UK since last Ice Age. Occurs in woods and hedges. Dominant tree of heavy, basic soils (clays and loams)	*cerevisiae* (common yeast of commerce) Yeasts occur on surface of other organisms (e.g. fruits) and in the soil. Yeast was first microorganism to be exploited by humans; earliest record of brewing is from Egypt, 6000 BC	*rattus* (black rat) First recorded in UK in 1187. This species now under threat from the brown rat, *R. norvegicus*. Black rat was major reservoir of bubonic plague (disease transmitted by rat fleas), which reached UK in 1348

■ The history of classification

Aristotle (384–322 BC, p. 10) carried out the only systematic studies of animals that have come down to us from the classical world of the ancient Greeks. He took great trouble to observe living animals, and his writings are full of accurate observations. From these observations he produced a credible classification scheme.

Aristotle's classification involved about 500 animals, and was based upon a major division between those animals with red blood (roughly the vertebrates) and those without red blood (roughly the animals without a backbone). He used features of embryology, behaviour and ecology in classifying organisms, as well as structural and morphological features. He did not produce a completely consistent classification; for example, he occasionally used a single arbitrary characteristic, such as whether an organism was terrestrial or aquatic, in order to separate two groups.

Aristotle's classification scheme persisted in use for nearly 2000 years after his death. But by the seventeenth century so many unknown organisms had been discovered by explorers and travellers that the inadequacy of the Greek scheme had become apparent, and Aristotle's classification was at last replaced by a new scheme.

The scheme of classification in general use today is the **binomial system of nomenclature** (p. 20). This was originated by **Linnaeus** (1707–78; see below), and published in 1735. Linnaeus' interest was in devising a practical scheme of classification that was also flexible enough to absorb new species and fresh discoveries as they appeared. He also reflected the ideas of his time in believing he was uncovering a 'plan for Nature' that was in the mind of God the Creator, and was implied in the Bible in *Genesis*, Chapter 1.

In Linnaeus' time Latin was the scholarly language universally in use, and so the names given to organisms and to groups of organisms were latinised. The biologist who first described any organism and assigned its 'binomial' is today recorded in the name, at least when the full title is given. Very often that biologist was Linnaeus, and for these organisms his name, short-ened to 'L.', is placed after the specific name. Of course, many organisms have been described since Linnaeus' time.

Charles Darwin (see pp. 11 and 656) was interested in the similarities between groups of organisms as evidence of evolution. Darwin believed that natural taxonomic groups exist because members have evolved from a common ancestor. When characteristic and fundamentally similar structures are present in several different species – the pentadactyl (five-fingered) limb of vertebrates, for example – it does suggest the possibility of a common ancestor. The phenomenon of similarity of structure due to common ancestry is known as **homology** (Figure 2.2).

Georges Cuvier (1769–1832) was fascinated by natural history from childhood. From 1785 onwards, he taught at the Museum of Natural History, Paris, the largest scientific establishment at that time. Cuvier extended Linnaeus' scheme to include the phylum. He also developed comparative anatomy and the technique of reconstructing the likely form of an extinct species from only a few bones.

CARL LINNAEUS, THE GREAT CLASSIFIER

Life and times

Carl Linnaeus was born in Sweden. As he grew up he helped his father, a clergyman, tend their well-stocked garden, itself surrounded by a profusion of wild plants of marshland and water meadows. He developed a deep interest in plants at an early age, and he strove (with difficulty!) to learn their names. His formal schooling was concerned more with the study of Latin than of science. He disappointed his parents in not showing the interest and aptitude to become a clergyman. Meanwhile, a local doctor encouraged him towards botany and medicine, and introduced him to the sex-

uality of plants. This had a dramatic effect on his adolescent imagination. It led him to study plants in great detail, paying attention to the numbers of stamens (male parts). These observations were the basis of his later 'sexual system' of classifying plants on the basis of the numbers of their stamens and stigmas.

Linnaeus' parents were poor and found it difficult to support his studies. But at every crisis Linnaeus' ability, industry and enthusiasm brought influential and wealthy people to his aid. His medical and botanical training was initially at Lund, later at Uppsala in Sweden, and finally in Holland where he trained as a doctor. Subsequently he earned his living as a doctor, and finally became Professor of Medicine and Botany at the University of Uppsala.

Contributions to biology

Linnaeus was an extremely stimulating teacher. He had a great number of enthusiastic pupils, many of whom went on expeditions to remote places and discovered new species.

Linnaeus' binomial system of nomenclature (p. 20), in which every organism has a double name consisting of a latinised generic name followed by a specific adjective, is the basis of our modern system.

Linnaeus produced a concise, usable survey of all the world's known plants and animals (then 7700 species of plants and 4400 species of animals) which established and standardised his binomial nomenclature immediately. His passion for classification led him to gather related genera into classes, and classes into orders. (The French naturalist Georges Cuvier later grouped related orders into phyla.)

Linnaeus took from pharmacy the signs ♂ and ♀ and used them as symbols for male and female respectively.

Influences

Linnaeus' most lasting contribution was the introduction of binomial nomenclature. But by concentrating his studies and those of his followers on the external parts of organisms valuable in classification, he diverted interest from internal structure and the working of the living organism. The search for 'new' organisms became the chief aim for many years, to the neglect of anatomy and physiology.

The immense appeal of Linnaeus in his time, and later, stemmed from his appreciation of wildlife. This tradition is particularly strong in Britain. Natural history had long interested many country gentry and clergy, a tendency that was reinforced by the scientific interests of the rising and wealthy industrial classes during and after the industrial revolution.

The Linnaean system helped pave the way towards the notions of evolution, even though Linnaeus insisted that since the creation of the world no new species had been formed and none had become extinct. Linnaeus laid essential foundations for the longer-term development of biology.

Figure 2.2 Homology and analogy

The forelimbs of vertebrates have a similar structure but are adapted for different purposes. They are **homologous**. The wing of a fly is an organ of flight, but its structure and origin are different from those of a bird's wing. These organs are **analogous**.

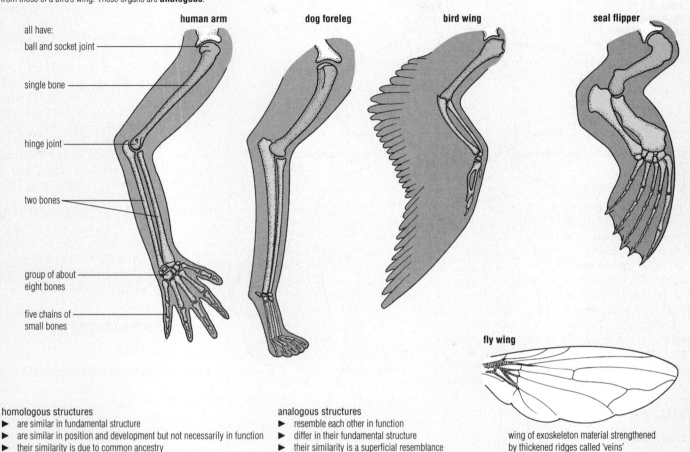

all have:
ball and socket joint
single bone
hinge joint
two bones
group of about eight bones
five chains of small bones

human arm   dog foreleg   bird wing   seal flipper

fly wing

wing of exoskeleton material strengthened by thickened ridges called 'veins'

homologous structures
► are similar in fundamental structure
► are similar in position and development but not necessarily in function
► their similarity is due to common ancestry

analogous structures
► resemble each other in function
► differ in their fundamental structure
► their similarity is a superficial resemblance

■ What is a species?

There was no problem in Linnaeus' day in defining a 'species', because it was believed that each species was derived from the original pair of animals created by God. Since species had been created in this way they were fixed and unchanging. For example, Linnaeus was certain that the species of animals he listed – well over four thousand – were absolute groupings, not to be subjected to revision by humans.

In fact, the fossil record suggests to us that changes do occur in living things over long periods of time. We now believe that species have evolved, one from another, in the course of the history of life on Earth. The concept of species has been modified. We no longer have a simple definition of 'species'.

Defining a species

Taxonomists now use as many different characteristics as possible in order to define and to identify a species. The three main types of characteristic used are external and internal structure (**morphology** and **anatomy**), **cell structure** (for example, whether cells are eukaryotic or prokaryotic) and **chemical composition** (for example, comparisons of nucleic acids or proteins and the immunological reactions of organisms, p. 366).

A species is best defined as consisting of **organisms of common ancestry that closely resemble each other structurally and biochemically, and which are members of natural populations that are actually or potentially capable of breeding with each other to produce fertile progeny, and which do not interbreed with members of other species.** The last part of this definition cannot be applied to self-fertilising populations or to organisms that reproduce only asexually. Such groups are species because they are very similar to each other morphologically, and in all other features.

Questions

1 In what way has the work of Linnaeus helped the notion of evolution, even though he was convinced all species had been created?

2 Why do you think the classification scheme of Aristotle survived in use until the seventeenth century AD?

3 Name organs that are respectively homologous and analogous to the wings of a bird.

4 a Alsations, pekinese and dachshunds are very different in appearance and yet are all classified as members of the same species. How can this be justified?

  b Lions and tigers can interbreed, though their offspring are infertile.

    Some plant species can be crossbred and produce fertile offspring (hybrids).

    How do these observations make a single definition of 'species' difficult?

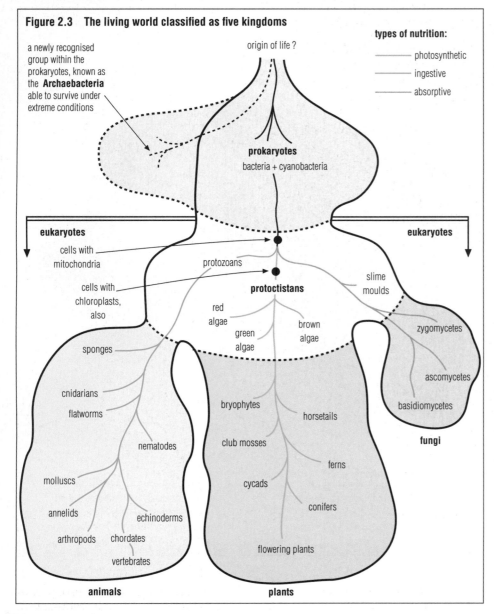

Figure 2.3 The living world classified as five kingdoms

a newly recognised group within the prokaryotes, known as the **Archaebacteria** able to survive under extreme conditions

origin of life ?

types of nutrition:
— photosynthetic
— ingestive
— absorptive

prokaryotes
bacteria + cyanobacteria

eukaryotes

cells with mitochondria

protozoans

cells with chloroplasts, also

protoctistans

red algae

green algae

brown algae

eukaryotes

slime moulds

zygomycetes

ascomycetes

basidiomycetes

fungi

sponges

cnidarians

flatworms

nematodes

molluscs

annelids

arthropods

echinoderms

chordates

vertebrates

animals

bryophytes

club mosses

horsetails

ferns

cycads

conifers

flowering plants

plants

■ How many kingdoms of living things?

Classification schemes divide living things into several large groups or kingdoms. A **kingdom** is the largest category into which organisms are placed. A major issue in classification is how best to divide the living world into kingdoms. For example, fungi were once usually classified together with plants in the Plant Kingdom. In fact the fungi have distinctive characteristics, part plant-like and part animal-like. The fungi are now recognised as a separate kingdom. Similarly, the electron microscope has established that bacteria and cyanobacteria exhibit a separate level of biological organisation, fundamentally different from those of plants, fungi and animals. The prokaryotes are a distinct kingdom.

Until quite recently the living world was generally divided up into four

kingdoms. The **four kingdom classification system** recognised:

Prokaryota:
▶ Kingdom of Bacteria and Cyanobacteria (Monera).

Eukaryota:
▶ Kingdom of Plants, with photosynthetic nutrition
▶ Kingdom of Fungi, with absorptive nutrition
▶ Kingdom of Animals, with ingestive nutrition.

The five kingdom scheme of classification

In 1988 two workers, L Margulis and K V Schwartz, published a new scheme of classification that provides the theoretical basis for the five kingdom scheme, which is currently in general use for day-to-day purposes.

A point of difference between schemes concerns the unicellular organisms. Some schemes of classification have regarded all the single-celled (unicellular) eukaryotes as a separate and distinctive Kingdom of Protista. The protista are clearly a separate level of organisation, but are they a separate kingdom?

In the five kingdom scheme the single-celled organisms, protista, are linked with certain of the multicellular organisms, in particular the algae, that appear to have a relatively simple assemblage of cells. This kingdom is known as the protoctista. The Kingdom of the Protoctista contains the eukaryotic algae and the protozoa, together with the slime moulds.

The five kingdom scheme of classification is adopted in this book, and is summarised in Figure 2.3. It is further illustrated in the check-list of organisms frequently met with at this level of study, illustrated later in this chapter.

2.3 ISSUES IN MODERN SYSTEMATICS

■ Why is taxonomy controversial?

A problem arises in taxonomy when a taxonomist has to choose which characteristics are important and which are not. An apparently simple issue becomes a matter of personal judgement, and therefore a source of potential disagreements. It is often difficult for a group of taxonomists to agree the best principles on which to take such a decision. Yet at first sight the problems do not seem difficult to resolve. Here we will look at two of them.

Number of characteristics

Should one characteristic or many characteristics be used in separating organisms? When classifying a group of organisms taxonomists may choose to consider only one characteristic or, alternatively, many characteristics in dividing the group into different taxa.

When one characteristic alone is chosen the resulting system is known as an **artificial classification**. This is because the classification is based on a single characteristic, usually chosen without any special theoretical considerations other than those of speed and convenience. An example is the arbitrary grouping together of all flowering plants with two stamens (stamens are the male parts of the flower, p. 571), a criterion applied by Linnaeus at one stage.

Figure 2.4   An example of an artificial classification: three plants of grassland and wasteland with one structural feature in common – they have two stamens to each flower (stamens are the male parts of the flower)

meadow clary (*Salvia pratensis*)
a plant of wasteland and grassland on calcareous soil in much of England

common fumitory (*Fumaria officinalis*)
a common weed of cultivated soils, but avoiding heavy clay, throughout the British Isles

sweet vernal grass (*Anthoxanthium odoratum*)
a plant of meadow and moorland, whether acid or alkaline, throughout the British Isles

Applying this criterion to flowering plants of grassland and wasteland would bring together into one group the following:

meadow clary   *Salvia pratensis*
common fumitory   *Fumaria officinalis*
sweet vernal   *Anthoxanthium*
  grass   *odoratum*

But these three plants (Figure 2.4) have little else in common. This becomes clear if we consider differences in the roots, stems and leaves and other structural features of the flowers, in addition to the number of stamens.

Meadow clary:
▶ square stem
▶ simple leaves oppositely arranged on the stem
▶ aromatic herb
▶ colourful flowers pollinated by long-tongued insects

These are characteristics of the dead nettle family.

Common fumitory:
▶ thin brittle or climbing stem
▶ finely divided compound leaves, alternately arranged on the stem

These are characteristics of the fumitory family.

Sweet vernal grass:
▶ fibrous root system
▶ stem branched at base
▶ leaves solitarily arranged, with cylindrical base, ligule and blade
▶ green flowers, wind pollinated

These are characteristics of the grass family.

In a **natural classification** taxonomists group organisms together after comparing many characteristics. The resulting groups are considered natural because the organisms classified together have many features in common, and consequently they are more likely to share common ancestors and to be related to each other.

Vertebrates and 'invertebrates'

The vertebrates are a natural grouping of animals within the phylum chordata. In addition to the backbone, all members have a brain enclosed in a cranium (p. 484). The term 'invertebrate' has long been used as a popular name for 'animals without backbones'. These groups of animals have nothing in common, however, apart from the absence of a vertebral column. Thus 'invertebrates' are a good example of an artificial grouping. The term is now obsolete.

Questions

1   Suggest firstly an artificial classification system, and secondly a natural classification system, for distinguishing between sheep and cats.
2   What features of the fungi require them to be classified in a kingdom separate from the kingdoms of animals and plants?

Structure and affinity

Are differences in structure more important than affinity (how closely related organisms are) in separating organisms?

1 We can study **the physical appearance of organisms** (their phenotypes, p. 608), and group organisms together according to the extent to which they look alike. Features of biochemical composition and embryological development are ignored. The degree of resemblance, known as **phenetic** resemblance, is based upon structural features alone.

2 We can study **the extent to which organisms are related** to one another, so that organisms with common ancestors are classified closely together and those that are more distant relatives are classified further apart. In the issue of degree of relatedness (**phylogenetic** resemblance), biochemical features of protein and DNA composition and the structural stages occurring in embryological development are regarded as more important than physical resemblances.

The difference between these alternative approaches is illustrated in Figure 2.5. In fact the classification of the reptiles touched on here highlights the continuing controversies in taxonomy.

Figure 2.5 Two ways of classifying birds, lizards and crocodiles: lizards and crocodiles resemble each other more than either resemble birds, but crocodiles and birds share a common ancestor

classification by phenetic resemblance: the grouping together of organisms that look most alike, to produce a **dendogram**

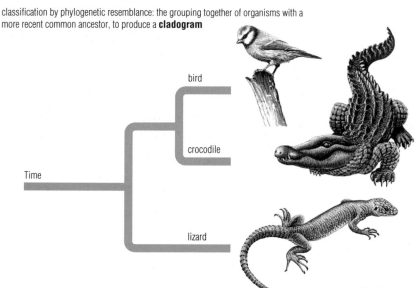

classification by phylogenetic resemblance: the grouping together of organisms with a more recent common ancestor, to produce a **cladogram**

■ Different schools of thought in modern taxonomy

The different ways of working in modern taxonomy are represented by three distinct schools of thought and practice amongst taxonomists.

Numerical taxonomy uses a very large number of phenetic characteristics to classify organisms. Possible evolutionary relationships are ignored.

The aim of numerical taxonomy is to produce a natural classification on the basis of the overall likeness of organisms by comparing as many characteristics as possible. At least one hundred characteristics are sought, and each is given the same importance as any other. The closeness between groups is determined by the total number of observable characteristics they have in common, whether morphological, physiological, biochemical or behavioural. Numerical taxonomy was made possible by the advent of computers.

Cladism uses phylogenetic relationships to classify organisms, in the belief that classification should be based on ancestry. The point where a new group branches from a common ancestor is the decisive point when a new species has arisen by evolution. Therefore a cladistic classification based on the genealogy of groups is the most natural classification.

Orthodox taxonomy uses a mixture of phenetic and phylogenetic relationships to classify organisms on the basis of their evolutionary relationships, as far as these are known. A natural classification system based upon evolutionary relationships is aimed for in biology, but it is extremely hard to establish. Evolutionary connections are mostly only guessed at. Consequently our current classification system is a compromise.

Questions

1 Why has the application of cytological techniques, such as electron microscopy, and many biochemical techniques been important in systematics?

2 What sort of information would you need to collect about a group of similar animals in order to suggest which are most closely related?

3 How might a biologist attempt to decide whether two organisms are closely related, as opposed to merely looking alike?

2.4 A CHECK-LIST OF ORGANISMS

The classification of representative organisms most frequently referred to in this book is outlined in the following illustrated check-list.

VIRUSES

Viruses cannot be fitted into a classification of living things because they are **not cellular**.

Viruses consist of a strand of **nucleic acid** and a **protein coat**. They are classified separately according to their chemical and physical properties. All viruses are:

- **ultramicroscopic** (20–300 nm) in size (too small to be seen through the light microscope)
- **able to pass through filters that** retain bacteria
- **disease-causing agents** that can reproduce themselves only inside specific living cells; that is, they are obligate parasites (p. 548). The infected cells ultimately break down and die.

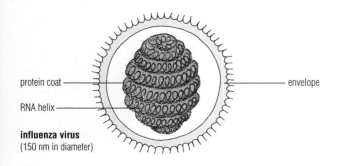

protein coat

RNA helix

envelope

influenza virus
(150 nm in diameter)

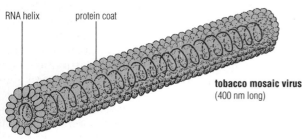

RNA helix   protein coat

tobacco mosaic virus
(400 nm long)

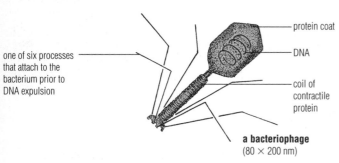

one of six processes that attach to the bacterium prior to DNA expulsion

protein coat

DNA

coil of contractile protein

a bacteriophage
(80 × 200 nm)

Virus diseases of animals: smallpox, influenza, rabies, poliomyelitis.

Virus diseases of plants: tobacco mosaic virus, turnip yellow mosaic virus.

Viruses attacking bacteria are known as bacteriophages.

Structure and replication, p. 548; viral diseases, p. 528; bacteriophage, p. 549.

PROKARYOTES

Prokaryotes are unicellular or filamentous organisms, consisting of **extremely small cells**, approximately 0.5–5.0 μm.

All prokaryotes **lack nuclei** organised within a nuclear membrane.

(The important differences between prokaryotes and eukaryotes are listed on p. 169.)

Kingdom Monera

This kingdom contains all bacteria and the cyanobacteria.

Genetic material is **circular DNA**, occurring naked in the cytoplasm, and normally attached to the cell membrane.

Cytoplasm contains **few organelles,** and none are surrounded by a two-membrane envelope. Ribosomes (p. 161) are present, but are small compared with those of eukaryotes.

The **cell wall** is a rigid structure, made of polysaccharide with amino acid residues combined in it.

Some organisms have the ability to **fix atmospheric nitrogen** into organic compounds (amino acids).

In advanced textbooks this kingdom is divided into many phyla. Here, for simplicity, it will be considered to consist of only two.

Phylum Cyanobacteria

Spherical or rod-shaped cells of unicellular structure, or arranged in simple or branched multicellular filaments.

Photosynthetic prokaryotes, with photosynthetic pigments held in intucked plasma membrane around the periphery of the cell.

Photosynthetic pigments are chlorophyll (green), phycoerythrin (red) and phycocyanin (blue).

Certain species of cyanobacteria also live mutualistically with fungi, forming lichens (p. 67).

Many species contain gas vacuoles, and the cells float at the surface of the water in which they live.

(At one time these prokaryotes were erroneously classified in the plant kingdom as **blue-green algae**.)

▶ *Anabaena* sp.
Structure and nitrogen fixation, p. 333; role in evolution, p. 676.

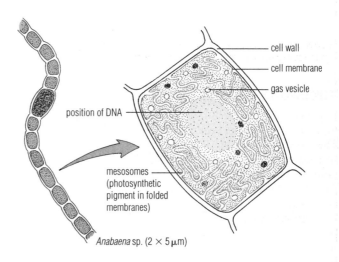

cell wall

cell membrane

gas vesicle

position of DNA

mesosomes (photosynthetic pigment in folded membranes)

Anabaena sp. (2 × 5 μm)

Phylum **Bacteria**

Unicellular prokaryotes, but the cells often remain attached after division, forming clumps, chains or simple filaments.
Bacteria occupy virtually all habitats.
Nutrition is varied; many bacteria are heterotrophs by absorption of externally digested food, others are autotrophs by chemosynthesis (p. 269) or photosynthesis.
Many bacteria cause decay and, with fungi, facilitate **recycling of nutrients.**
Certain bacteria are parasitic on other organisms, and cause specific diseases such as food poisoning, typhoid and TB in humans, and soft rots of various vegetable plants.
Bacteriologists divide the bacteria into several phyla on the basis of differences in metabolism and cell structure.

▶ *Bacillus* sp.

Structure and reproduction, p. 550; growth, p. 397; roles in cycling of nutrients, p. 332; role in disease, p. 530; roles in biotechnology, p. 112.

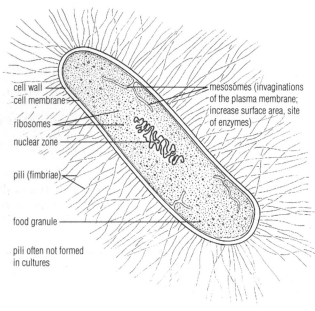

Escherichia coli (1 × 10 μm)

EUKARYOTES

Eukaryotes are unicellular, filamentous or truly multicellular organisms with large cells (10–150 μm).
The genetic material of eukaryotic cells consists of linear strands of DNA attached to protein in **chromosomes** contained within a **nucleus.**
The cytoplasm contains many membranous **organelles** and large ribosomes.
Flagella (or cilia), when present, are constructed from a system of 9 + 2 microtubules.
A cell wall, when present, contains **cellulose** (or **chitin,** in fungi).
Very few, if any, eukaryotes are believed to have the ability to fix atmospheric nitrogen into organic molecules (amino acids).
(The important differences between eukaryotic cells and the cells of prokaryotes are listed on p. 169.)

Kingdom Protoctista

This kingdom contains the single-celled eukaryotes and also all the multicellular organisms closely related to them.
The most important groupings or subkingdoms of the Protoctista are the **protozoa** and the **algae**; these very diverse organisms are classified together on the basis of a **relatively simple level of organisation.**
Here, ten phyla are selected to represent this kingdom.

Subkingdom Protozoa

The protozoa include a great diversity of minute, eukaryotic organisms. Most are single-celled animal-like organisms.
Nutrition is mostly heterotrophic, but the mode of nutrition may change according to the food available.

Phylum **Rhizopoda** (rhizopods)

Protozoans having pseudopodia used for locomotion and for feeding movements.

▶ *Amoeba*

Structure, p. 151; nutrition and movement, pp. 291 and 499; sensitivity, p. 442; reproduction, p. 556.

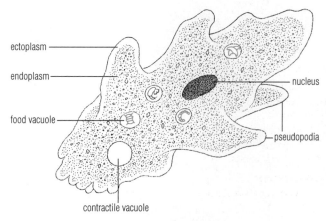

Amoeba (0.8 × 0.4 μm)

Phylum **Zoomastigina** (flagellates)

Protozoans having at least one flagellum for locomotion.
Nutrition is heterotrophic.

▶ *Trypanosoma*

Causes the human disease African sleeping sickness, which is transmitted by the tsetse fly, p. 534.

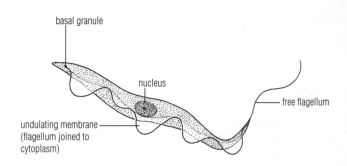

Trypanosoma (15 × 1 μm)

Phylum **Apicomplexa** (sporozoans)

Protozoans that are mostly parasites with complex life cycles involving more than one host species, and with multiple fission stages in the life cycle.

▶ *Plasmodium*

Causes malaria in humans, and involves the adult mosquito as the vector and secondary host, p. 535.

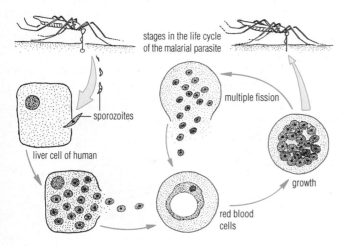

Plasmodium (2 μm long)

Phylum **Ciliophora** (ciliates)

Protozoans covered by rows of short, flexible flagella called **cilia**. Cilia are concerned with movement and with the capture of food.

Nutrition of ciliates is heterotrophic.

These organisms contain two nuclei per cell, a macronucleus and a micronucleus.

▶ *Paramecium*

Movement, p. 500; feeding, p. 291; reproduction, p. 556.

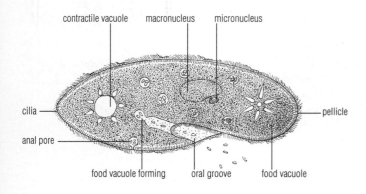

Paramecium (240 × 80 μm)

Phylum **Euglenophyta** (euglenoid flagellates)

A group of organisms seen as intermediate between the protozoa and the algae, some photosynthetic and some non-photosynthetic.

All species have one or two long, conspicuous flagella used in locomotion.

All are without a cell wall. The body is covered by a tough protein layer, the **pellicle**.

Photosynthetic species with chloroplasts containing the pigments chlorophylls *a* and *b*, carotene and xanthophyll.

Food reserves are stored as a carbohydrate called paramylon.

▶ *Euglena*

Movement, p. 499.

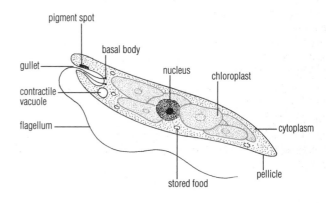

Euglena (130 × 50 μm)

Subkingdom Algae

Photosynthetic eukaryotic organisms with a cell structure similar to that of green plants.

Algae show great diversity in structure, ranging from unicellular and simple filamentous forms to huge seaweeds.

Algae occur in water or on very damp surfaces.

Phylum **Bacillariophyta** (diatoms)

Diatoms have unusual and beautiful walls made from celluloses and impregnated with silica. The walls form in two halves called **valves**, and fit together like a pill box or Petri dish.

Chloroplasts contain the pigments chlorophylls *a* and *c*, carotene and the xanthophyll fucoxanthin. Oil is stored as the food reserve.

▶ *Navicula*

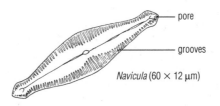

Navicula (60 × 12 μm)

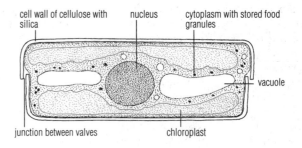

Navicula in section

Phylum **Chlorophyta** (green algae)

The green algae are very similar to green plants in cell structure and biochemistry.

Chloroplasts contain the pigments chlorophylls *a* and *b*, carotene and xanthophyll.

Starch is the stored food. The cell walls are made of cellulose.

Plant body is single-celled, or a filament, or a colonial form shaped as a flattened **thallus**.

Most green algae are freshwater inhabitants.

▶ *Chlamydomonas* (unicellular)

▶ *Spirogyra* (filamentous)
  Structure and life cycle, p. 554.

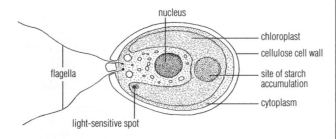

Chlamydomonas (20 × 10 μm)

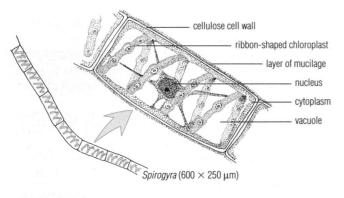

Spirogyra (600 × 250 μm)

Phylum **Phaeophyta** (brown algae)

The brown seaweeds of the intertidal zone of a rocky shore. The plant body is a multicellular thallus with a degree of tissue differentiation. No water-conducting cells have evolved.

Chloroplasts contain the pigments chlorophylls *a* and *c* with carotene pigments, masked by a brown pigment, fucoxanthin. Food reserves include a carbohydrate, laminarin.

Reproduction is sexual, and many brown seaweeds have haploid and diploid **alternation of generations** (p. 562).

▶ *Fucus*
  Structure and life cycle, p. 555; ecology, p. 53.

plant body called a thallus { holdfast / stalk (stipe) / blade (lamina) }

rock

thallus branches dichotomously

mid-rib

tip of thallus becomes swollen with conspicuous fertile conceptacles containing reproductive organs

Fucus (up to 15–20 cm)

Phylum **Rhodophyta** (red algae)

The red algae are marine plants of the lower part of the intertidal zone. Chloroplasts contain chlorophyll *a*, carotene and xanthophyll pigments, all masked by the red pigment phycobilin, which absorbs the wavelengths of light penetrating deeper sea water.

Plant body composed of branched filaments or complex aggregations of filaments.

The life cycles of red algae are complex, usually with alternation of generations.

▶ *Chondrus*
  Ecology, p. 53.

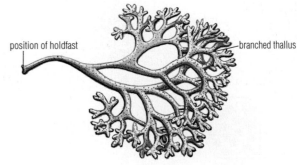

position of holdfast

branched thallus

Chondrus (8–15 cm)

Phylum **Oomycota** (oomycetes)

Plant body consists of hyphae, which have no crosswalls (non-septate).

▶ *Phytophthora*
  Structure and life cycle, p. 300.

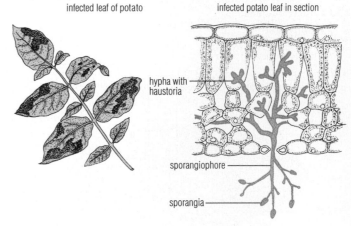

infected leaf of potato

infected potato leaf in section

hypha with haustoria

sporangiophore

sporangia

Phytophthora (hyphae 10 μm diameter)

Kingdom Fungi (fungi)

The fungi are eukaryotic organisms with protective walls composed of **chitin** (p. 136).

'Plant' body a **mycelium** of thread-like **hyphae**. The hyphae may be divided by crosswalls (**septa**) into short, multinucleate sections.

The fungi lack chlorophyll and their nutrition is non-photosynthetic. Some feed as parasites, others are mutualistic, but most are saprotrophic (absorptive, secreting enzymes on to food and absorbing the products of external digestion).

Most fungi are known to reproduce sexually and asexually. The gametes and spores produced are without flagella.

Phylum **Zygomycota** (zygomycetes)

Saprotrophic fungi widespread in soils and dung, and common in decaying food.

The mycelium is composed of hyphae without crosswalls.

Sexual reproduction results in a multinucleate resistant spore, the **zygospore**, produced by sexual conjugation of two compatible hyphae.

▶ *Mucor*
  Structure, reproduction and life cycle, p. 558; nutrition, p. 296.

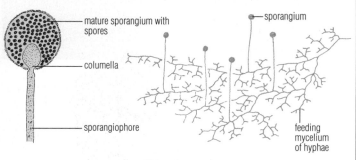

Mucor (hyphae 5 μm diameter)

Phylum **Ascomycota** (ascomycetes, the sac fungi)

The yeasts, cup-fungi, morels, truffles and powdery mildews.
Sexual reproduction involves spore (**ascospore**) production inside a container (**ascus**). In most species there is a cup-shaped or flask-shaped fruiting body in which the asci are formed. Asexual reproduction also occurs.

Mycelium of hyphae partially divided into multinucleate compartments by septa. The septum has a central pore, and cytoplasm is continuous throughout the mycelium.

▶ Yeast *(Saccharomyces)*
  Structure and economic importance, p. 298.

▶ *Neurospora*
  One gene, one enzyme experiments, p. 210.

▶ *Sordaria*

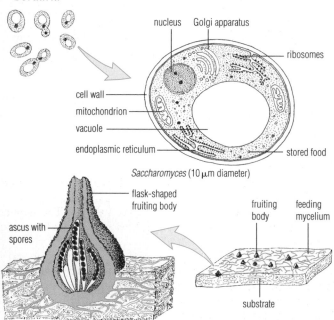

Saccharomyces (10 μm diameter)

Sordaria (1 × 0.3 mm)

Phylum **Basidiomycota** (basidiomycetes or club fungi)

The mushrooms, toadstools, puffballs, rusts and smuts, and bracket fungi. These organisms are composed of septate hyphae.

Sexual reproduction involves spore (**basidiospore**) production occurring externally on the club-shaped tip of a special hypha, the **basidium**.

▶ *Agaricus*
  Structure and life cycle, p. 560.

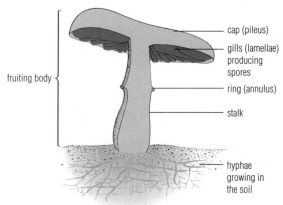

Agaricus (fruiting body 3–10 cm)

Phylum **Mycophycophyta** (lichens)

Lichens are difficult to classify within the existing scheme.
Lichens are dual organisms, formed by an association between an alga (green alga or cyanobacterium) and a fungus (usually an ascomycete).

The lichen thallus consists of compact fungal hyphae surrounding algal cells and taking the form of an encrusting layer, a leaf-like thallus or a tiny 'shrub'.

Each component is dependent upon the other (mutualism, p. 67); the algal cells manufacture sugar, and the fungi retain ions as and when they become available.

Lichens are first colonisers of bare rocks, and they have the capacity to survive extremely adverse conditions. They are susceptible to air-borne pollution, however.

▶ *Cladonia*
  Lichens as an example of mutualism, p. 67.

lichens growing on a trunk

Kingdom Plantae (green plants)

The green plants are multicellular eukaryotes with cell walls containing **cellulose** and other polysaccharides.

Nutrition is **autotrophic**, and the majority are photosynthetic. Photosynthesis occurs in **chloroplasts,** which contain the pigments chlorophylls *a* and *b*, xanthophyll and carotene.

The life histories of green plants involve **two generations,** which **alternate** with each other. First, a diploid generation occurs, known as the **sporophyte**, during which spores are formed.

This is followed by a haploid generation, known as the **gametophyte**, during which sex cells (gametes) are produced.

Phylum **Bryophyta** (bryophytes)

Green plants in which the gametophyte is the conspicuous and dominant generation, and which lack water-conducting (xylem) and elaborated food-conducting (phloem) tissue.

Most bryophytes grow on land, but they are largely restricted to damp environments.

There are no true roots; the plant is anchored by filamentous rhizoids.

Class **Hepaticae** (liverworts)

Liverworts have either a flat thallus or a stem with three ranks of leaves, anchored by unicellular rhizoids.

Spores are released by the wall of the capsule splitting into four valves.

▶ *Pellia*
Life cycle, p. 563.

spore capsules in *Pellia*

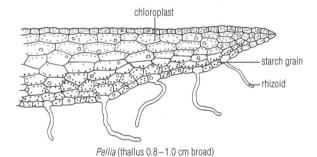

Pellia (thallus 0.8–1.0 cm broad)

Class **Musci** (mosses)

Mosses are small, leafy structures with spirally arranged leaves, anchored by multicellular rhizoids.

The spore capsules of mosses have an elaborate spore-dispersal mechanism.

▶ *Funaria*
Structure and life cycle, p. 564.

Funaria (1.5 cm tall)

The remaining phyla of green plants are collectively known as the **tracheophytes**.

Tracheophytes have a conspicuous and dominant sporophyte generation in the form of a plant differentiated into stem, leaf and root.

Vascular tissue (xylem and phloem) is present, as are lignified cells.

The leaves are elaborate structures with a waterproof cuticle.

Phylum **Lycopodophyta** (club mosses)

Plants with small, spirally arranged leaves. Sporangia usually occur in cones.

▶ *Selaginella*

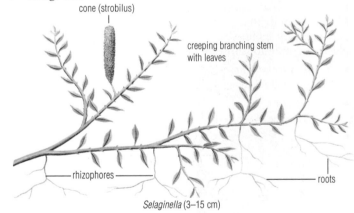

cone (strobilus)

creeping branching stem with leaves

rhizophores

roots

Selaginella (3–15 cm)

Phylum **Sphenophyta** (horsetails)

Plants with whorls of leaves along the stem. Sporangia are produced in cones.

▶ *Equisetum*

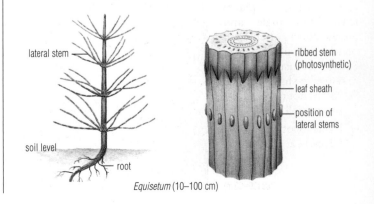

lateral stem

soil level

root

ribbed stem (photosynthetic)

leaf sheath

position of lateral stems

Equisetum (10–100 cm)

Certain other phyla of tracheophytes are of specialised interest in tracing the evolution of the flowering plants, and are referred to briefly in Chapter 30. These phyla include:

▶ Phylum **Cycadopsida** e.g *Cycas*

▶ Phylum **Ginkgopsida** e.g. *Ginkgo*

Cycas (30–40 m)

Phylum **Filicinophyta** (ferns)

Leaves are formed tightly coiled, and uncoil in early growth and development.
Sporangia occur on the lower surface of the leaves in clusters called **sori**. Each sorus contains many stalked sporangia.

▶ *Pteridium*

▶ *Dryopteris*
Structure and life cycle, p. 566.

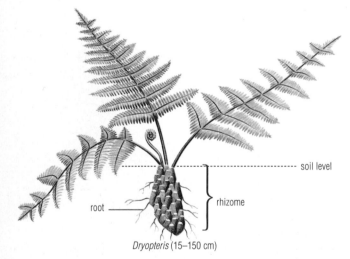

Dryopteris (15–150 cm)

Phylum **Coniferophyta** (conifers)

Cone-bearing plants with both male and female cones, often both on the same plant.
Leaves are usually waxy and needle-shaped.
Seeds are not enclosed by an ovary wall.

▶ *Pinus* (Scots pine)
Life cycle, p. 568.

Pinus (30–50 m)

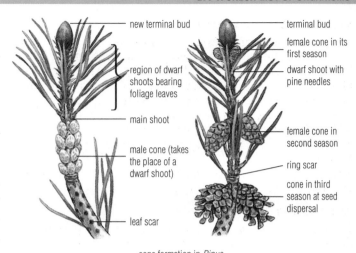

cone formation in *Pinus*

Phylum **Angiospermophyta** (flowering plants or angiosperms)

The angiosperms are the dominant group of land plants.
The angiosperms are seed-bearing plants, with the seed enclosed in a fruit formed from the ovary.
Flowers are unique to the angiosperms, and the development of flowers has been associated with the evolution of complex mechanisms for pollen transfer and seed dispersal, sometimes involving insects, birds and mammals, and the agency of wind and water.

Class **Monocotyledoneae** (monocotyledons or monocots)

The leaves of monocotyledons mostly have parallel veins.
The embryo plant in the seed has a single cotyledon (**seed-leaf**).

▶ *Poa* (annual meadow grass)
Grass flower structure and pollination, p. 571.

▶ *Triticum* (wheat)

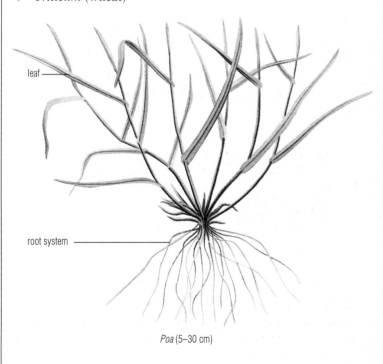

Poa (5–30 cm)

33

Class **Dicotyledoneae** (dicotyledons or dicots)

The leaves of dicotyledons have net veins.

The embryo plant in the seed has two cotyledons (**seed-leaves**).

▶ *Ranunculus* (buttercup)

Flower structure, p. 570; life cycle, p. 575; success of the angiosperms, p. 577.

▶ *Quercus* (oak tree)

Structure of woody dicots, p. 411; woodland ecology, p. 57.

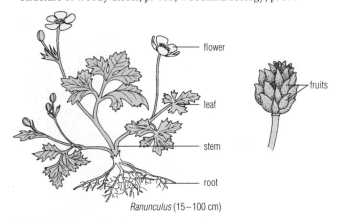

Ranunculus (15–100 cm)

Kingdom Animalia (animals)

Animals are multicellular eukaryotic organisms. Most animals show a high level of tissue differentiation and many have specialised body **organs**.

Most animals have a **nervous system** used to coordinate their body actions and responses.

Animal nutrition is **heterotrophic**. No animals are photosynthetic; they lack photosynthetic pigments. Animal cells do not have cell walls.

In **sexual reproduction** animals produce **haploid** (p. 196) male gametes (**sperms**) and female gametes (**ova**). After fertilisation the zygote divides repeatedly and the new cells produced form into a hollow ball of cells, the **blastula**.

Phylum **Porifera** (sponges)

The sponges are animals of very simple structure, little more than a colony with very little division of labour between the cells.

The sponges are the only animal phylum whose members lack any nervous system.

▶ *Leucosolenia*

Structure, p. 8.

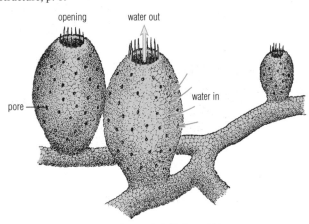

Leucosolenia (1–2 cm wide)

Phylum **Cnidaria** (cnidarians or coelenterates)

All coelenterates are aquatic animals, mostly marine. The body is radially symmetrical: longitudinal sections passing through any diameter give identical halves.

The name coelenterate means 'hollow gut'. Coelenterates have a sac-like body cavity, the gut or **enteron**, with a single opening to the exterior for ingestion and egestion.

The body wall consists of two layers (**diploblastic**): an outer **ectoderm** and an inner **endoderm**, separated by a non-cellular jelly layer, the **mesogloea**.

Many coelenterates exhibit **polymorphism** in that they exist in two different body forms, a **polyp** and a **medusa**, and these forms typically alternate in the life cycle.

The ectoderm, particularly that of the tentacles, contains stinging cells (**cnidoblasts**) which, when triggered by prey, explosively penetrate the prey and inject poison, paralysing and holding the prey prior to ingestion.

Class **Hydrozoa**

These coelenterates have a dominant polyp body form and a reduced or absent medusa.

▶ *Hydra*

Structure and nutrition, p. 294; nervous system, p. 468; reproduction, pp. 501 and 583.

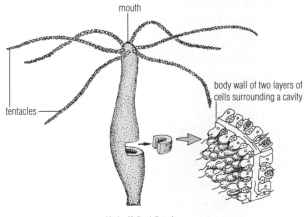

Hydra (0.5–1.5 cm)

Class **Scyphozoa**

These coelenterates have a reduced polyp body form and a dominant medusa body form.

▶ *Aurelia*

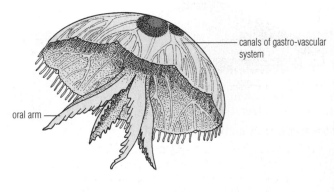

Aurelia (5–15 cm)

Class **Anthozoa**

These coelenterates exist only as polyp forms. The enteron is divided by large radial partitions called **mesenteries**.

► *Actinia*
Rock pools of the intertidal zone, p. 52.

Animals of subsequent phyla generally show the following characteristics.

1 They are **triploblastic** organisms, meaning that three body layers form from the blastula as the embryo develops:

- outer **ectoderm** from which is formed the epidermis and the nervous system
- inner **endoderm** from which is formed most of the gut
- **mesoderm** between these layers, which forms the muscles, skeleton and other internal organs.

2 The body is **bilaterally symmetrical**: it can be sectioned in only one plane to give identical halves. These animals have an **anterior end** and a **posterior end** (head and tail), and a **dorsal side** and a **ventral side** (back and belly).

Phylum **Platyhelminthes** (flatworms)

Flatworms are flat, unsegmented animals. Most have a mouth and a gut consisting of numerous cul-de-sac branches, but no anus.
Flatworms have flame cells in the mesoderm for excretion and osmoregulation.
Mostly the flatworms are hermaphrodite (p. 583) with a complex reproductive system that minimises the chances of self-fertilisation.
The phylum contains many important parasites.

Class **Turbellaria** (turbellarians)

Free-living aquatic flatworms with a ciliated outer surface.

► *Planaria*
Gaseous exchange, p. 343.

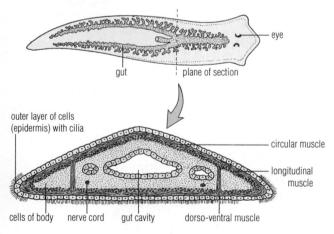

Planaria (5–15 cm)

Class **Trematoda** (trematodes or flukes)

Most flukes are parasites living inside their host.
The outer body surface has a thick, protective cuticle.
Flukes have suckers for attachment to host tissues.

► *Fasciola* (liver fluke)
Life cycle, p. 416; disease caused, p. 536.

► *Schistosoma* (blood fluke)
Life cycle, and the disease bilharzia, p. 536.

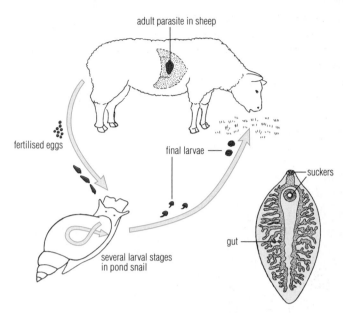

Fasciola (adult fluke – in sheep – 2.5 × 1.3 cm)

Class **Cestoda** (cestodes or tapeworms)

All tapeworms are internal parasites with a complex life history involving two vertebrate hosts.
No mouth or gut is present. Digested food is absorbed over the body surface.
The anterior end of the body (the **scolex**) has suckers and hooks, effective in anchoring to host tissues.
The long, flattened body is divided into **proglottids**, which contain the organs of sexual reproduction, and which break off from the tapeworm and are dispersed.

► *Taenia* (sheep and cattle tapeworms)
Life cycle, p. 299; disease caused, p. 536.

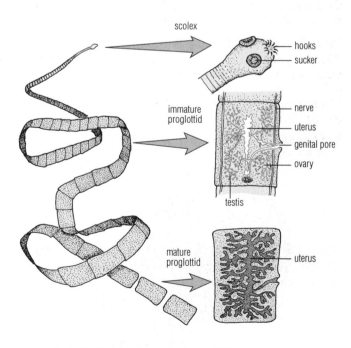

Taenia (2.5–3.5 m long, 6 mm wide, 1.5 mm thick)

Animals of subsequent phyla all have an **anus** as well as a mouth.

Phylum **Nematoda** (nematodes or roundworms)

Nematodes have narrow bodies, round in cross-section, and pointed at both ends. They are unsegmented worms.

The body is covered in a thin, elastic, protein cuticle.

The body contains a long unbranched gut with mouth and anus.

Nematodes reproduce sexually, and the sexes are separate.

Most nematodes are free-living, but others are parasites of plant or animal species.

▶ *Ascaris*

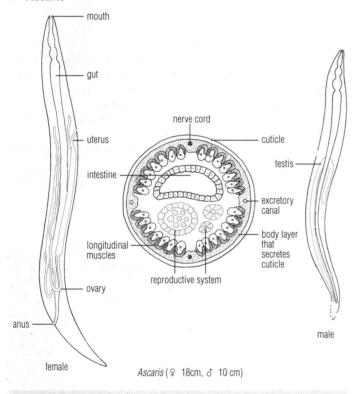

Ascaris (♀ 18cm, ♂ 10 cm)

Animals of the subsequent phyla show the following characteristics:

1 a **coelom** or fluid-filled cavity within the mesoderm (p. 345)
2 **metameric segmentation;** the animal's body consists of a series of compartments in which certain structures are serially repeated (the molluscs are an apparent exception to this)
3 a **circulatory system** containing an oxygen-carrying pigment and facilitating internal transport.

Phylum **Annelida** (annelids or segmented worms)

The annelids are worm-like animals with pronounced metameric segmentation. The segments are fully visible externally as rings. Internally, the segments are separated by septa.

The body is externally protected by a thin, elastic, protein cuticle. Excretion and osmoregulation are achieved by segmentally arranged ciliated tubules, called **nephridia.**

Central nervous system of anterior oesophageal ganglia connected to a solid, ventral nerve cord with segmental nerves.

The larval stage is a trochophore larva (p. 415).

Class **Polychaeta** (polychaete worms)

Marine worms with a distinct head and with numerous pairs of chaetae or bristles, protruding on projections from each segment.

▶ *Nereis* (ragworm)
▶ *Arenicola* (lugworm)

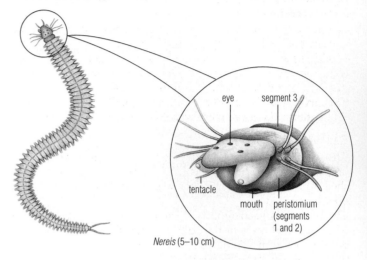

Nereis (5–10 cm)

Class **Oligochaeta** (oligochaetes or earthworms)

Freshwater worms or earthworms with no distinct head and with only a few chaetae to each segment.

▶ *Lumbricus* (earthworm)
   Movement, p. 480; nutrition, p. 295; soil formation, p. 60; gaseous exchange, p. 304; blood circulation, p. 344; nervous system, p. 469.

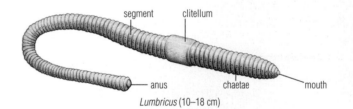

Lumbricus (10–18 cm)

Class **Hirudinea** (leeches)

Solid- and liquid-feeding predators or ectoparasites, attached to the animals on which they feed by anterior and posterior suckers.

No distinct head. No chaetae.

▶ *Hirudo* (medicinal leech)

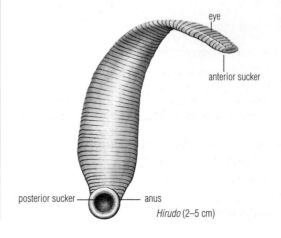

Hirudo (2–5 cm)

Phylum **Mollusca** (molluscs)

The molluscs are a very diverse and successful group of animals. They are the second largest group of animals in total numbers of species. They also have a very ancient fossil record.

Molluscs are soft-bodied animals with little or no evidence of segmentation. The body is divided into head, muscular foot and visceral mass or hump. Over the hump is formed a mantle, which secretes a shell. The space between the mantle and the body wall forms the mantle cavity.

Molluscs have gills (**ctenidia**) located in the mantle cavity, and some use these for gaseous exchange. Most molluscs have a rasping, tongue-like **radula** used for feeding.

The larval stage is a trochophore larva (p. 415).

Of the seven classes of mollusc, three are described here.

Class **Gastropoda** (gastropods)

These terrestrial and aquatic molluscs are the snails, slugs, limpets, winkles and whelks. They typically possess a spiral shell into which the animal can withdraw. In slugs the shell is greatly reduced.

The body of a gastropod is asymmetrical due to the twisting of the visceral mass in the coiling of the shell. As a result the anus is anterior. A large flat foot is present, and also a well-developed head with tentacles and eyes.

The radula is present and is used in feeding.

Land gastropods have lost their gills. The mantle cavity functions as a lung for terrestrial gaseous exchange.

▶ *Helix* (garden snail)
  Feeding, p. 294.

▶ *Limax* (slug)

▶ *Littorina* (winkle)

▶ *Buccinum* (whelk)
  Ecology of winkle and whelk, p. 55.

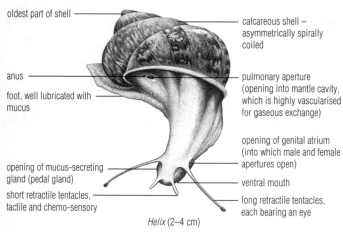

Helix (2–4 cm)

Class **Pelycopoda** (lamellibranchiata or bivalves)

These aquatic molluscs are the bivalves such as mussels, clams and oysters. They are typically burrowers in mud and sand.

The body of a bivalve is laterally compressed and is enclosed by a shell consisting of two parts. The body is bilaterally symmetrical.

The head is reduced in size and tentacles are absent.

The gills are sheet-like, and are used in filter feeding.

▶ *Mytilus* (marine mussel)
  Feeding, p. 291.

▶ *Anodonta* (freshwater mussel)

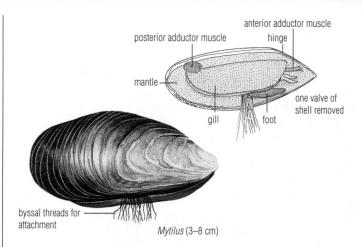

Mytilus (3–8 cm)

Class **Cephalopoda** (cephalopods)

These aquatic molluscs include the octopus and squid. They are adapted for swimming, and are active animals, with well-developed sense organs and nervous system. Some cephalopods are the largest non-vertebrate animals.

The foot is incorporated into the head, which has 8–10 conspicuous tentacles bearing suckers.

The shell is greatly reduced in size, and is internal or completely absent.

A radula and horny beak are present and are used in feeding.

▶ *Octopus* (octopus)

▶ *Loligo* (squid)

▶ *Sepia* (cuttlefish)

Octopus (20–35 cm)

Phylum **Arthropoda** (arthropods)

The arthropods are the most numerically successful group of animals.

The body of an arthropod is **segmented**, and covered by a hard **exoskeleton** made of chitin.

The exoskeleton is shed periodically (**ecdysis** or moulting) in the course of growth.

Arthropods also have **jointed legs**, typically one pair per segment.

Arthropods have an **open circulatory system**. The coelom cavity is much reduced and the body contains a second cavity, the **haemocoel**, which surrounds the body organs. A heart pumps blood into the haemocoel so that the blood supply bathes the organs before returning to the heart.

This huge grouping of organisms is divided into several important classes.

Subphylum 1: **Arthropods with two pairs of antennae**

Superclass **Crustacea** (crustaceans)

The Crustacea include a wide range of animals, mostly aquatic, showing great variation in form, but all with the dorsal body surface covered by a shield-like **carapace**. Compound eyes present.

► *Daphnia* (water flea)
► *Gammarus* (freshwater shrimp)
► *Semibalanus* (barnacle)
► *Astacus* (crayfish)
► *Armadillidium* (woodlouse)

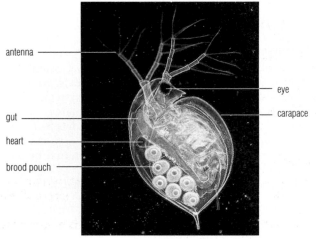

Daphnia (1–3 mm)

Gammarus (0.5–1.5 cm)

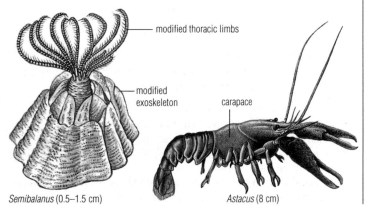

Semibalanus (0.5–1.5 cm)

Astacus (8 cm)

Subphylum 2: **Arthropods with one pair of antennae**

Class **Chilopoda** (centipedes)

Dorsi-ventrally flattened body with a distinct head bearing simple eyes and a pair of poison claws.
The body consists of numerous leg-bearing segments, one pair of legs per segment.

► *Lithobius* (garden centipede)

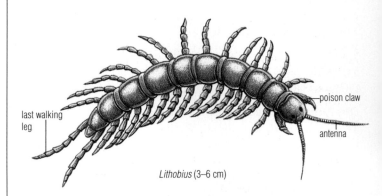

Lithobius (3–6 cm)

Class **Diplopoda** (millipedes)

Cylindrical body with simple eyes but no poison claws.
Body consists of numerous leg-bearing segments with two pairs of legs per segment.

► *Iulus* (millipede)

Iulus (2–6 cm)

Class **Insecta** (insects)

The body is divided into a **head** of six fused segments, **thorax** of three segments, each of which bears a pair of legs, and an **abdomen** of eleven segments.
Most adults have two pairs of wings on the thorax.
Insects have a pair of compound eyes.
Air reaches the tissues via a system of tubes called the **tracheal system**.

Subclass Wingless insects

► *Lepisma* (silver fish)

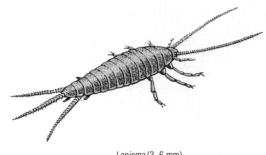

Lepisma (3–6 mm)

Subclass Winged insects

1 Insects with **gradual metamorphosis** (egg → larva → adult). Wings develop externally.

▶ *Chorthippus* (grasshopper)
Feeding, p. 295; locomotion, p. 482; nervous system, p. 470.

▶ *Locusta* (locust)
Life cycle, p. 417; copulation, p. 584; gaseous exchange, p. 305.

▶ *Periplaneta* (cockroach)

2 Insects with **complete metamorphosis** (egg → larva → pupa → adult). Wings develop internally.

▶ *Pieris* (cabbage white butterfly)
Feeding, p. 292; life cycle, p. 417.

▶ *Musca* (house fly)
Feeding, p. 293.

▶ *Apis* (honey bee)
Life cycle and behaviour, p. 524.

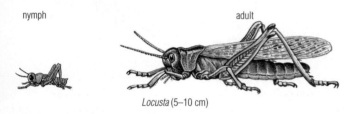

Locusta (5–10 cm)

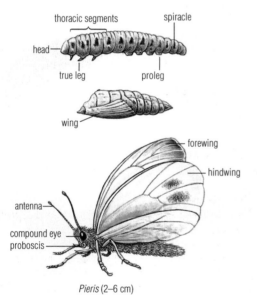

Pieris (2–6 cm)

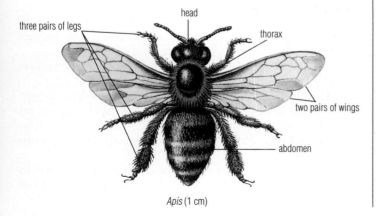

Apis (1 cm)

Subphylum 3: **Arthropods with no antennae**, but with pincers (chelicerae) present in front of the mouth

Class **Arachnida** (arachnids)
Body divided into cephalothorax (head and thorax of six segments) and abdomen (thirteen segments).
Four pairs of legs attached to the cephalothorax.

▶ *Scorpio* (scorpion)

▶ *Araneus* (garden spider)
Feeding, p. 293.

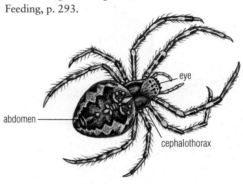

Araneus (1–3 cm)

Phylum **Echinodermata** (echinoderms)
The echinoderms are marine organisms living on the sea bottom. The phylum includes starfish, brittle stars and sea urchins.
The adult echinoderm has five-way radial symmetry, but the larva is bilaterally symmetrical.
The mouth is on the lower surface and the anus is on the upper surface.
Echinoderms have no proper circulatory system, but a water vascular system exists and tube feet are part of it. The tube feet are used for locomotion and food collection.

▶ *Asterias* (starfish)

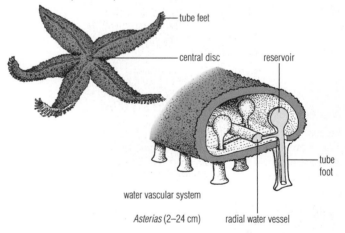

Asterias (2–24 cm)

Phylum **Chordata** (chordates)
Chordates are animals showing the following five characteristics at some stage in their development:

• a tubular, hollow dorsal **nerve cord**
• a dorsal, flexible supporting rod called a **notochord**
• a post-anal **tail**
• a set of gill slits in the throat, known as **pharyngeal** or **visceral clefts**
• a **blood circulation** system in which blood flows back down the body dorsally and forwards ventrally.

Subphylum **Vertebrata** (vertebrates or craniates)

The notochord is replaced by a vertebral column or backbone. A brain is present and is enclosed in a cranium.

Superclass **Agnatha** (jawless vertebrates)

Eel-like vertebrates without jaws, but with a round, suctorial mouth and rasping tongue.
Ectoparasites on fish.

▶ *Lampetra* (lamprey)

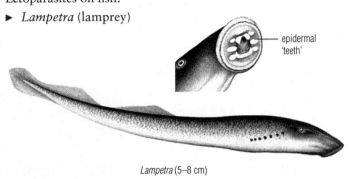

Lampetra (5–8 cm)

Superclass **Gnathostomata** (vertebrates with jaws)

Class **Chondrichthyes** (cartilaginous fish)

Fish with a skeleton of **cartilage**.
The mouth is in a ventral position on the head, and the gill openings are separate gill slits.
The fins are fleshy structures.

▶ *Scyliorhinus* (dogfish)

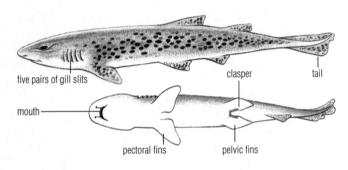

Scyliorhinus (20–30 cm)

Class **Osteichthyes** (bony fish)

Fish with a skeleton of **bone**.
The mouth is in a terminal position on the head, and the gills are covered by a bony flap, or **operculum**.
The fins are membranous structures with bony rays.

▶ *Clupea* (herring)
Respiration, p. 306; osmoregulation and excretion, p. 379; movement, p. 496.

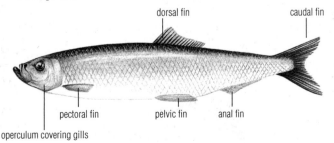

Clupea (20–30 cm)

Class **Amphibia** (amphibians)

The first land vertebrates, but amphibious – partly terrestrial and partly aquatic.
Amphibians breed in water. Fertilisation is external, and the larval stage (tadpole) is aquatic.
The tadpole undergoes metamorphosis into a terrestrial adult, which has paired limbs as part of the strong bony skeleton.
Amphibians have moist skin that is used as a supplementary respiratory surface, but gaseous exchange is by gills in the tadpole, and by lungs and skin in the adult.

▶ *Rana* (frog)
Feeding, p. 295; reproduction, p. 584.

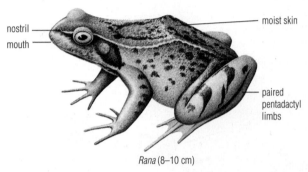

Rana (8–10 cm)

Class **Reptilia** (reptiles)

Reptiles are mainly terrestrial vertebrates, with dry, impermeable skin, covered by overlapping scales.
Reptiles typically have pentadactyl limbs, but in snakes the limbs are reduced or absent.
Gaseous exchange occurs in the lungs.
Fertilisation is internal. Typically the fertilised eggs are laid with an impermeable shell, but in certain species the eggs are retained until hatching. There is no larval stage in development.

▶ *Lacerta* (lizard)
Roles of limbs in locomotion, p. 495.

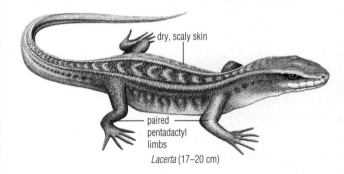

Lacerta (17–20 cm)

Class **Aves** (birds)

Birds are described as 'feathered reptiles that have evolved the power of flight'. The skin of the body is covered by feathers but the legs have scales.
Birds have a strong, light skeleton; the limb bones are hollow, with internal strutting and spaces filled by air sacs.
The fore-limbs are modified as wings. The sternum of the pectoral girdle is extended and the massive flight muscles are anchored to it.
Birds lack teeth or heavy jaws. The jaws are extended into a horny beak.
A high body temperature is maintained at a constant level by internal regulation (endothermic). Birds maintain a high metabolic rate.

Fertilisation is internal. Fertilised egg(s) are laid with a yolk (food store) and with a hard calcareous shell. There is no larval stage in development.

▶ *Columba* (pigeon)
  Respiration, p. 307; movement and support, p. 497; parental care, p. 586.

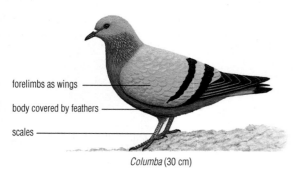

forelimbs as wings

body covered by feathers

scales

Columba (30 cm)

Class **Mammalia** (mammals)

The skin of mammals has hair that covers the body and, typically, two types of gland.
The body cavity is divided by a muscular **diaphragm** between thorax and abdomen.
The body of a mammal is maintained at a relatively high and constant temperature by means of internal regulation.
Fertilisation is internal and the young, when born, are fed on milk from **mammary glands** (modified skin glands).

Subclass **Monotremata** (monotremes or egg-laying mammals)

Primitive egg-laying mammals.

▶ *Ornithorhynchus* (duck-billed platypus)

Ornithorhynchus (45 cm)

Subclass **Marsupialia** (marsupials or pouched mammals)

Young marsupial mammals are born in an immature stage and migrate into the pouch on the mother's body and complete development there.

▶ *Megaleia* (kangaroo)

Megaleia (1–2 m)

Subclass **Eutheria** (true mammals or placental mammals)

Eggs develop in the uterus and are nurtured by the maternal blood circulation via the **placenta**.

▶ *Homo* (human)
  Human evolution, p. 668.

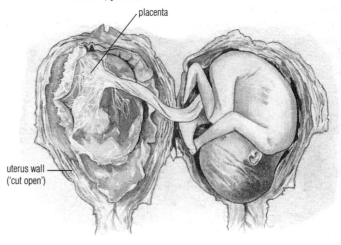

placenta

uterus wall ('cut open')

human fetus at 20 weeks

2.5 WORKING WITH THE DIVERSITY OF LIFE

When an unfamiliar organism is observed, questions are generated. Why has it appeared here? What organisms does it associate with? What are the special features in its biology?

At an early stage we need to know what an unfamiliar organism is called. Once we know its name we can begin to find out what has already been learned about it. Later, we can communicate our own discoveries to others.

Finding out the name of an unfamiliar organism is often difficult, frustrating, and extremely time-consuming – occasionally so time-consuming as to become an end in itself! Naming organisms is particularly a problem when we move into a new habitat or start laboratory work with a previously unfamiliar group of organisms.

What can we do about unknown organisms?

■ We can identify by comparison

We identify by comparison by accurate matching of the specimen with photographs, drawings or preserved specimens (a **reference collection**). Or we could match details of habitat and life cycle already recorded, with information that can be gleaned from the specimen. This method can be quick and effective, but it can also mean that we learn little biology of the specimen during the process. (A classified list of useful identification texts is given in the companion *Study Guide*, Chapter 2.)

Identification by comparison can fail in special cases when two organisms closely resemble each other although they are not related at all. For example, hoverflies mimic wasps (convergent evolution). Close matching of the 'mimic' and the 'model' may totally confuse identification by comparison.

■ We can identify using keys

We can identify unfamiliar organisms using a key. The advantage of using a key is that it requires careful observation of a specimen to determine whether a particular structure or feature is present or absent. As a result we become familiar with the anatomy of the organism to be identified.

One example of a key is shown in the panel. This is a key to the subfamilies of ants of the UK. In total, about 10 000 species of ants are known. In the UK there are only 36 species, yet these are quite difficult to identify because of their small size. UK ants can be assigned to one of four subfamilies using external features, described in this key, which is an example of a **single-access key**.

In single-access keys a contrasting or mutually exclusive characteristic (just one, if possible) is used to divide the group of organisms to be classified into two groups (a **dichotomous key**, named from Greek words meaning 'dividing into two'). Then characteristics are selected that can be used to separate each of these groups into two smaller groups, and the process is continued until the groups are sufficiently small and homogeneous. In most cases a species is reached, but keys to families or genera can also be very useful.

Most keys are of single-access construction. The alternative type of key is some form of multi-access key, such as a punched card system, in which any one of several observed features may be used to identify unknown organisms. Examples of keys and how they are constructed are given in the *Study Guide*, Chapter 2.

■ We can ask someone else

Early on in our enquiry this may be the only way to get over a hurdle. It is helpful if the expert also points out *why* an organism is classified in a particular way. Help of this sort might come from members of local or county Natural History Societies, contacted via the reference department of your local library.

■ We can perhaps avoid the problem

The soil community consists of vast numbers of organisms. Suppose you were investigating the way soil is affected by farming practices. For example, you might need to know what effects the application of a new fertiliser, pesticide or soil cultivation technique has on soil wildlife. If you were committed to taking your results as the differences in the numbers of *every* individual species of the soil community you would have launched yourself into a life-long experiment, identifying *all* the living things in your samples. Meanwhile, the broad results might be needed urgently.

A KEY TO THE IDENTIFICATION OF ANTS OF THE UK

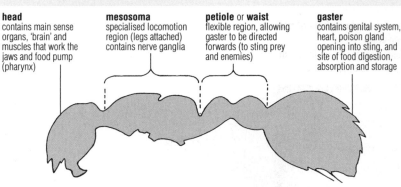

head
contains main sense organs, 'brain' and muscles that work the jaws and food pump (pharynx)

mesosoma
specialised locomotion region (legs attached) contains nerve ganglia

petiole or **waist**
flexible region, allowing gaster to be directed forwards (to sting prey and enemies)

gaster
contains genital system, heart, poison gland opening into sting, and site of food digestion, absorption and storage

Silhouette of a worker ant of *Myrmica rubra*, cut in longtitudinal section, showing the divisions of the body and the chief roles of each region

Key to subfamilies

1	Waist of two small segments	**Myrmicinae**
	Waist of one segment	**2**
2	Gaster constricted between segments 1 and 2, with a well-developed sting	**Ponerinae** (two British genera; each with one species)
	No constriction, no sting	**3**
3	Five segments of the gaster visible from above, a circular orifice for ejecting venom fringed by guide hairs	**Formicinae**
	Four segments of gaster visible from above, no circular orifice but a slit through which viscous defensive fluid is passed, no hairs	**Dolichoderinae**

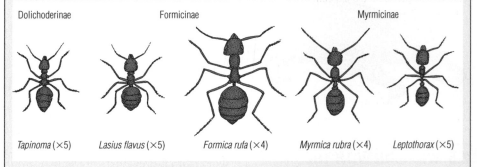

Dolichoderinae

Formicinae

Myrmicinae

Tapinoma (×5)   *Lasius flavus* (×5)   *Formica rufa* (×4)   *Myrmica rubra* (×4)   *Leptothorax* (×5)

Examples of the three main subfamilies of ants

Two examples of such enquiries are illustrated in the panel opposite. These experiments were part of a larger agricultural research project undertaken at Rothamsted Experimental Station. In one case, the effects of ploughing on the non-vertebrates of the soil community were assessed by comparing the numbers of non-vertebrates found in ploughed soil with the numbers found in unploughed (permanent pasture) land. In the other example, the effects of two insecticides on the masses of particular arthropods were assessed. In both cases, interesting and significant results were obtained without identifying *every* soil inhabitant.

Questions

1 List the potential disadvantages in attempting to name an unknown organism by:
   a asking someone you consider to be an expert
   b comparing one specimen with pictures or photographs of similar organisms.

2 Some famous biologists of earlier times set up reference collections of preserved animals or plants to aid in identification, and these collections have survived in museums and in the care of learned societies. What special uses can these collections be put to by present-day workers other than in helping with identification, which was the original aim of the collection?

SOME PRACTICAL METHODS OF INVESTIGATING THE EFFECTS OF CULTIVATION TREATMENTS ON SOIL WILDLIFE

1 Using the total numbers of non-vertebrates in cores (samples) taken from the soil

Ploughing prepares land for fresh cultivations, and may be part of the way weed growth is suppressed.

A soil sampler was used to obtain soil cores 5 cm wide, down to a depth of 30 cm. The vertical distribution of non-vertebrates in these cores was counted.

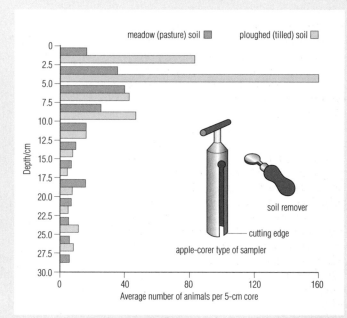

This study showed that soil non-vertebrates were concentrated near the surface, and that their numbers were also reduced by ploughing.

2 Using the total mass of non-vertebrates of selected groups

DDT and aldrin are both insecticides of a type known as chlorinated hydro-carbons. The impact of the use of insecticides on the wider environment has become an important issue. Part of the problem is that many of these substances persist in the environment for long periods, rather than being quickly degraded.

In this study, either DDT or aldrin was applied as a single dose to sample areas of soil.

The masses of four soil arthropods that survived this treatment were compared to the masses of these organisms in samples of untreated soil.

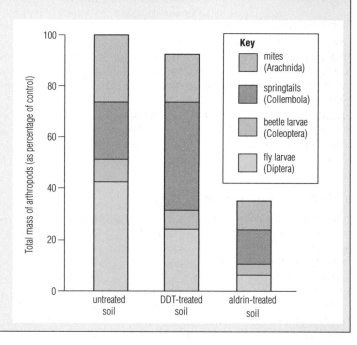

FURTHER READING/WEB SITES

G Monger and M Sangster (1988) *Systematics and Classification*. Longman
The Linnean Society of London: **www.linnean.org.uk**
London Zoo: **www.londonzoo.co.uk**
Royal Botanic Gardens, Kew: **www.rbgkew.org.uk**

3 Introduction to ecology

- Ecology is the study of relationships between living things and their environment. An ecosystem is a natural unit of living components together with non-living surroundings, such as the organisms of a woodland and the soil and atmosphere in which they occur. It is a largely self-supporting unit.
- Within any ecosystem, the green plants are 'producers' due to their autotrophic mode of nutrition. Animals and other organisms are 'consumers', dependent upon plant nutrition, directly or indirectly. Plants themselves are dependent upon the 'detritivores' and decomposers, which feed on dead organic matter, ultimately releasing inorganic nutrients for re-use.
- Thus, within an ecosystem there is a flow of energy from sunlight, through the food chain, finally becoming lost to space as heat. The nutrients, on the other hand, are endlessly recycled.

- Within a terrestrial ecosystem the physical (abiotic) factors of importance are due to climate, soil and topography of the land. These factors play an important part in determining what living things occur.
- The interactions of the organisms of the ecosystem are known as biotic factors and include competition, predation and parasitism. Humans have become the most influential organisms in many habitats.
- A population consists of the individuals of one species living in a particular habitat, and a community is the sum of all the populations there. When new land is exposed, it is relatively quickly colonised by a succession of populations, normally culminating in a stable climax community.
- The biological productivity of an ecosystem is the rate at which biomass is produced there. The net productivity of an ecosystem is the total amount of organic matter produced, after respiration and the rest of metabolism have been sustained.

3.1 INTRODUCTION

About 150 years ago the common approach to biology was to study a range of different species, each in isolation, as examples of different levels of biological organisation. This approach established a great deal of information about the structure (morphology and anatomy) and function (physiology) of individual organisms. It largely ignored the ways in which organisms interact with each other.

Environmental biology or **ecology** is the study of relationships between living things and their environment. Today, ecology is one of the most important components of biology. It is also a branch of biology in which a beginner can make original discoveries, and much field study can be carried out without expensive or elaborate instruments.

Table 3.1   An introduction to ecological terms

Biosphere	Only a part of planet Earth is inhabited, from the bottom of the oceans to the upper atmosphere. This restricted zone that living things can inhabit is called the biosphere. The majority of organisms occur in an even narrower belt, in fact, from the upper layers of soil to the lower atmosphere, or, if marine, many occur near the ocean surface.
Ecosystem	An ecosystem is a stable, settled unit of nature consisting of a community of organisms, interacting with each other and with their surrounding physical and chemical environment. Examples of ecosystems are ponds or lakes, woods or forests, seashores or saltmarshes, grassland, savanna or tundra. Ecosystems are necessarily very variable in size.
Population	A population consists of all the living things of the same species in a habitat at any one time. The members of a population have the chance of interbreeding, assuming the species concerned reproduces sexually. The boundaries of populations are often hard to define, but those of aquatic organisms occurring in a small pond are clearly limited by the boundary of the pond.
Community	A community consists of all the living things in a habitat – the total of all the populations, in fact. So, for example, the community of a well stocked pond would include the populations of rooted, floating and submerged plants, the populations of bottom-living animals, the populations of fish and non-vertebrates of the open water, and the populations of surface-living organisms – typically a very large number of organisms, in fact.
Habitat	The habitat is the locality in which an organism occurs; it is where the organism is normally found. If the area is extremely small we call it a **microhabitat**. The insects that inhabit the crevices in the bark of a tree are in their own microhabitat. Conditions in a microhabitat are likely to be very different from conditions in the surrounding habitat.
Niche	The niche of an organism is how it feeds and where it lives. For example, the sea birds known as cormorants and shags feed in the same water and nest on the same cliffs and rocks, so they share the same habitat. However, the cormorant feeds on fish living on the sea bed, such as flatfish, whereas the shag feeds on surface-swimming fish such as herring (Figure 3.37, p. 65). Since these birds feed differently they have different niches.
Competition	Resources of every sort are mostly in limited supply, and so organisms must compete for them. For example, plants may compete for space, light and mineral ions. Animals may compete for food, shelter and mates. When a resource is in short supply and preventing unlimited growth, it is known as a **limiting factor**.
Environment	Environment is a term we use for 'surroundings'. We talk about the environment of cells in an organism, or the environment of organisms in a habitat. It is our term for the external conditions affecting the existence of organisms, too, so 'environment' is a rather general, unspecific term – but useful, nonetheless.

The biosphere and biomes

The part of the Earth that is able to support life is known as the **biosphere** (Figure 3.1). The biosphere extends from the bottom of the oceans to the upper atmosphere, and amounts to a relatively narrow belt around the Earth.

Ecology sets out to explain, amongst other things, the distribution patterns of living things within the biosphere. Important patterns, often visible on a global scale, are the large, stable vegetation zones of the Earth. Examples include the tropical rain forests, temperate grasslands, deciduous woodlands and coniferous forests. These zones are called **biomes** (Figure 3.2). A biome is a major life zone characterised by the dominant plant life present.

We will discover later that in a biome the dominant plants set the way of life for many other inhabitants, largely by providing the principal source of nutrients, by determining the sorts of living spaces or habitats that exist, and by influencing the environmental conditions. In fact, the major biomes correspond broadly with major climatic zones. We will return later to the ways in which living things may determine the physical environment, as opposed to being dominated by it.

Figure 3.1   The biosphere: the occupied parts of our Earth and atmosphere

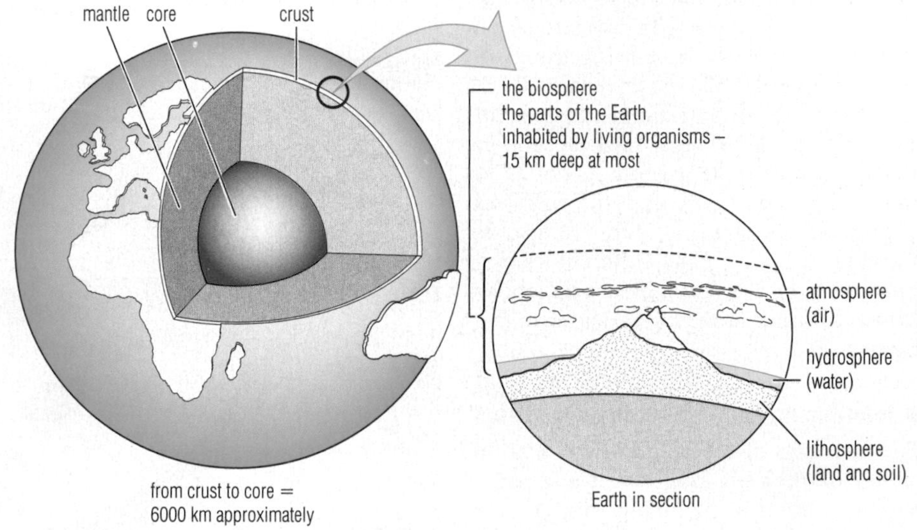

mantle core   crust

the biosphere
the parts of the Earth
inhabited by living organisms –
15 km deep at most

atmosphere (air)

hydrosphere (water)

lithosphere (land and soil)

from crust to core = 6000 km approximately

Earth in section

Ecosystems

The biomes of the world extend over huge geographical areas. They are not suitable units for our ecological investigations at this stage because they are too large. An ecosystem is a much smaller unit.

An **ecosystem** is a natural unit of living (or **biotic**) components, together with the non-living (or **abiotic**) components through which energy flows and nutrients cycle. The biotic and abiotic components influence each other in many ways, interacting to form a relatively stable system. An ecosystem such as a pond or lake, or a deciduous woodland, is a unit small enough for practical study.

An ecosystem can be divided up into various component parts. Thus individual organisms live in a small part of the ecosystem known as their **habitat** or, if the area is very small, as a **micro-habitat**. When we refer to how an organism lives as well as where it lives we are describing a **niche**. These terms are illustrated by reference to a pond in Figure 3.3 and defined in Table 3.1.

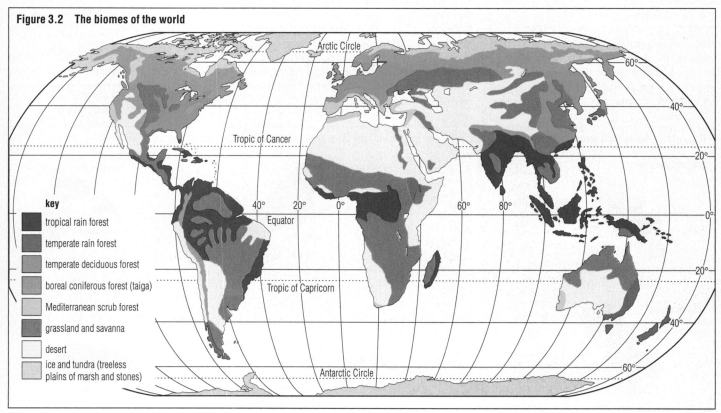

Figure 3.2   The biomes of the world

key

- tropical rain forest
- temperate rain forest
- temperate deciduous forest
- boreal coniferous forest (taiga)
- Mediterranean scrub forest
- grassland and savanna
- desert
- ice and tundra (treeless plains of marsh and stones)

Arctic Circle

Tropic of Cancer

Equator

Tropic of Capricorn

Antarctic Circle

■ Biotic factors of an ecosystem

An ecosystem contains many different plants and animals living together. In ecology we refer to these living things in terms of populations and communities of organisms. A **population** consists of living things of the same species, such as all the stickleback fish in a pond. A **community**, on the other hand, consists of populations of all the different species present, such as all the different organisms living in a pond. Their interactions with each other are discussed later in this chapter.

■ Interdependence within the ecosystem

The members of a community in a pond or woodland are engaged in the characteristic activities of living things, for which a supply of energy is essential. How do the organisms of an ecosystem obtain their energy?

Plants are **autotrophic**. They make their own organic nutrients or food from an external supply of inorganic nutrients, employing energy from sunlight in **photosynthesis** (p. 246). In contrast, animals depend upon existing nutrients, which they eat and digest before the food can be used. In fact, animal nutrition is dependent upon plant nutrition either directly or indirectly. Animal nutrition is described as **heterotrophic**.

So living things can be usefully re-classified by their feeding relationships as far as ecology is concerned. The photosynthetic green plants are **producers** and the animals are **consumers**. The consumers of the ecosystem are a diverse group. Some, like the water snail browsing on the green algae of the pond, feed directly on plants. We call these

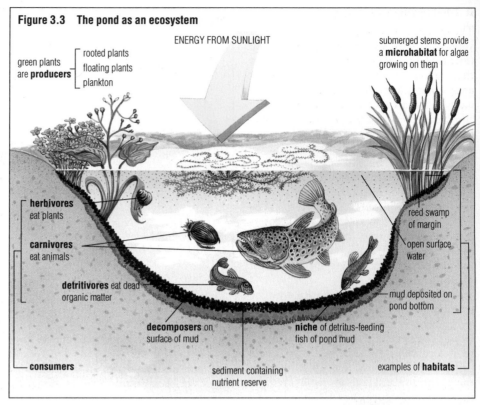

Figure 3.3 The pond as an ecosystem

ENERGY FROM SUNLIGHT

green plants are **producers** — rooted plants / floating plants / plankton

submerged stems provide a **microhabitat** for algae growing on them

herbivores eat plants

carnivores eat animals

detritivores eat dead organic matter

consumers

decomposers on surface of mud

sediment containing nutrient reserve

niche of detritus-feeding fish of pond mud

reed swamp of margin

open surface water

mud deposited on pond bottom

examples of **habitats**

herbivores. Others, the **carnivores**, feed on animals; these include the sticklebacks that feed on water fleas. Again, other animals feed on both animals and plants, and are therefore **omnivores**; caddis fly larvae, for example, feed on both plants and on smaller animals. Organisms that feed on dead plants and animals and on the waste matter of animals are another category of feeders, known as **detritivores** – earthworms feeding on dead leaves, for example – and **decomposers**, such as the many bacteria and fungi that cause the decay of dead material, releasing inorganic nutrients.

The green plants require an input of energy (sunlight). All other organisms are dependent on green plants for nutrients. Green plants depend on the activity of decomposers for much of their supply of inorganic nutrients, although part is derived from the weathering of the minerals in the soil. This interdependence of the producer, consumer and decomposer organisms means that an ecosystem has the potential to be a self-supporting ecological unit. Almost all ecosystems, however, have inputs from neighbouring systems, and outputs to them.

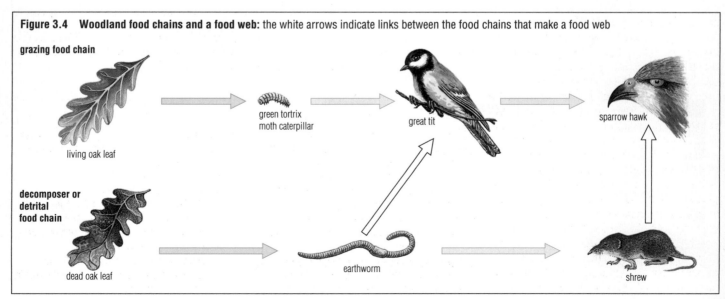

Figure 3.4 Woodland food chains and a food web: the white arrows indicate links between the food chains that make a food web

grazing food chain

living oak leaf → green tortrix moth caterpillar → great tit → sparrow hawk

decomposer or detrital food chain

dead oak leaf → earthworm → shrew

■ Food chains and food webs

The feeding relationship in which a carnivore eats a herbivore, which has been eating plants, is an example of a **food chain**. At each step in a food chain energy-containing materials are transferred. The steps of the food chain are recognised as feeding levels or trophic levels. A **trophic level** is a stage in a food chain at which organisms obtain their food in the same general manner; for example, all the herbivores are at the same trophic level of 'primary consumers'.

Within the woodland ecosystem an example of a four-step food chain is shown below.

		Trophic level
oak leaf ↓	**primary producer**	1
leaf-feeding caterpillar ↓	**primary consumer**	2
insectivorous bird ↓	**secondary consumer**	3
bird of prey	**tertiary consumer**	4

Food chains are of two types. If the chain is based on living plants it is known as a **grazing chain**; if based on dead plant material it is a **decomposer chain** or a **detrital chain**. Food chains of a deciduous woodland are shown in Figure 3.4.

Most food chains connect with other chains, since most organisms are the prey of more than one predator. Food chains are linked together to form a **food web** (Figures 3.17 and 3.32, pp. 56 and 64).

Only a small portion of the energy-containing materials obtained by a consumer as it feeds becomes built into the organism itself. This is partly because much of the food is undigested, and partly because most of the remainder is used to provide energy for processes such as movement, digestion, excretion and reproduction (Figure 3.5). Of the small proportion of material retained, some is used as an energy store and some is built into new tissue during growth and replacement. It is only these last components of the energy-containing material that are available to be passed on to the next stage in the food chain.

The mass of the organisms present at each stage of a food chain is known as **biomass**. At the start of a food chain a large mass of producer (green plant) material supports a much smaller mass of herbivores, which in turn supports a smaller mass of carnivores. Thus a pyramid of biomass (Figure 3.6) exists in a food chain (pyramids of biomass are discussed further on p. 72).

Figure 3.5   The use of energy by organisms

Figure 3.6   The pyramid of biomass: another way of representing food chains such as those shown in Figure 3.4

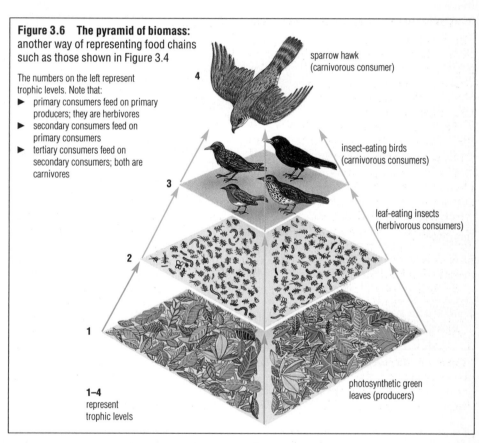

The numbers on the left represent trophic levels. Note that:
► primary consumers feed on primary producers; they are herbivores
► secondary consumers feed on primary consumers
► tertiary consumers feed on secondary consumers; both are carnivores

1–4 represent trophic levels

Questions

1 The biosphere extends to the upper atmosphere. What sorts of organism might occur so far from the Earth's surface?

2 Suggest three important ways in which a tropical rain forest differs from a temperate deciduous woodland (p. 57).

3 Why might the dominant plant species of a biome more-or-less determine the types of animal life present?

4 Name one ecosystem in your locality that you might study, and list three abiotic factors affecting living things in that ecosystem.

5 Use one or more of the terms listed below to describe each of the following features of a freshwater lake:
   ► the whole lake
   ► all the frogs in the lake
   ► the flow of water through the lake
   ► all the plants and animals present
   ► the total mass of vegetation growing in it
   ► the mud of the lake
   ► the temperature variations in the lake.
   Population, ecosystem, habitat, abiotic factor, community, biomass.

■ Abiotic factors of an ecosystem

Within an ecosystem, physical (abiotic) factors influence the living organisms.

The important abiotic factors are **climatic factors** (such as light, temperature, water availability and wind) and **edaphic factors** (the soil and its texture, nutrient status, acidity and moisture content). **Topographic factors** (angle and aspect of slope – whether north- or south-facing – and altitude) operate by their influence on local climatic and edaphic conditions. All these factors interact with each other and with the living organisms present (Figure 3.7).

Abiotic factors play an important part in determining which organisms can survive in a habitat. An organism that survives and persists in a particular habitat within the ecosystem is likely to be well adapted to the environmental conditions experienced there. Examples of special adaptations to physical features are described in the next section, where the seashore as an ecosystem is examined.

Questions

1 At the start of a food chain a large amount of green plant material supports a smaller mass of herbivores and an even smaller mass of carnivores. Why is this so?
2 What is the main difference between the ecological terms 'niche' and 'habitat'?

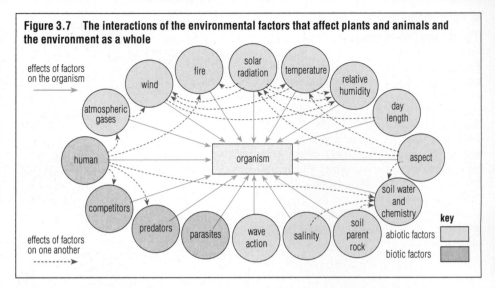

Figure 3.7 The interactions of the environmental factors that affect plants and animals and the environment as a whole

3.2 THE SEASHORE AS AN ECOSYSTEM

The seashore is chosen first because it illustrates well the influence of fairly extreme abiotic factors on living things. Biotic interactions are better illustrated by a woodland ecosystem.

The seashore or **littoral zone** is a part of the coastline of continents and marine islands affected by the tides. Tides are the periodical rise and fall of the sea level due to the attraction of Moon and Sun. The tidal rhythm is usually obvious and enables us to define the limits of the littoral zone, which extends from the lowest point of land uncovered by the sea at low tides up to the highest point of land washed or splashed by the sea at high tides (Figure 3.8).

The littoral zone is an area very rich in living things. The plants and animals of the seashore are almost all of marine origin; virtually none have come from dry land to colonise the shore. Thus the seashore is really an extension of the sea community.

The diversity of animal species of the shore is exceptionally great; all the groups, except the higher vertebrates and the insects, are well represented. Amongst plants, representatives of all phyla of algae occur but virtually no other group of plants is commonly found on the seashore, apart from certain lichens of the splash zone.

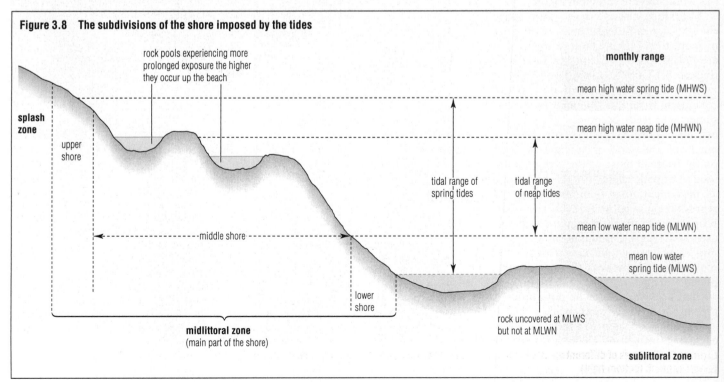

Figure 3.8 The subdivisions of the shore imposed by the tides

■ The junctions between ecosystems: ecotones

The transition between ecosystems is referred to as the **ecotone**. At this junction zone there may be an 'edge effect' between the ecosystems, as the environment of an ecosystem changes to that of the other rather abruptly.

The junction between open sea and shore is a broad and ill-defined band. This is because many of the organisms of the lower shore (sublittoral zone) extend out into the sea. For example, the anchored seaweeds extend as far as sufficient light penetrates to the sea bed to support photosynthesis. Above the intertidal zone, however, marine life ends abruptly. Shore animals become inactive when temporarily exposed to air by tides, but resume activity when reimmersed in sea water. Consequently no marine life extends to regions of more-or-less permanent exposure to air.

The types of seashore

No two parts of the coastline look exactly the same, but there are only four main types of shore, depending on the physical conditions prevailing (Figure 3.9). The physical environment of the shore influences profoundly the types of plant and animal life present.

Shingle beaches, also known as storm beaches, support many fewer organisms than do other types of beach. They consist of pebbles carried up from the sea bed by wind-blown waves and rolled up over each other, generating a steep slope. The run-off of water, now low in energy, carries only the smallest particles, mostly sand, back towards the sea. Consequently at the foot of the storm beach there is a sandy shore.

Sandy shores consist of tiny fragments of very hard rock, usually quartz. The slope of a sandy beach is quite gentle. The particles are small and water is retained by capillary action within the spaces between the grains. This water binds the sand together, reducing movement. Animals live in the sand, effectively in the sea water, after the tide has gone out. Consequently there is abundant although rather specialised marine life within the sand.

Muddy shores are made up of very fine particles of both organic and inorganic materials. Here there is virtually no slope, and the shore is best described as mudflats! Numerous animals make permanent burrows in the mud. Mudflats occur where there is complete shelter from wave action, together with a steady supply of silt and clay (such as comes down a river) and abundant organic matter. The organic matter comes partly from decaying seaweed and the remains of animal life of the shore, but mainly from the land, including sewage effluent and leaves and other debris from terrestrial woodland close to the shoreline.

Solid rock provides the most stable environment. Rocky shores are also the most variable, depending on the type of rock, on the slope and on the presence of rock pools. Many species of anchored seaweeds occur, and support a great diversity of animals. Most animals on rocky shores are permanently or temporarily anchored to the rock (sessile or semi-sessile), such as sponges, anemones, barnacles and limpets.

Figure 3.9 Features of different types of shore: sandy shore (top left), shingle or storm beach (top right), muddy beach (bottom left) and rocky beach (bottom right)

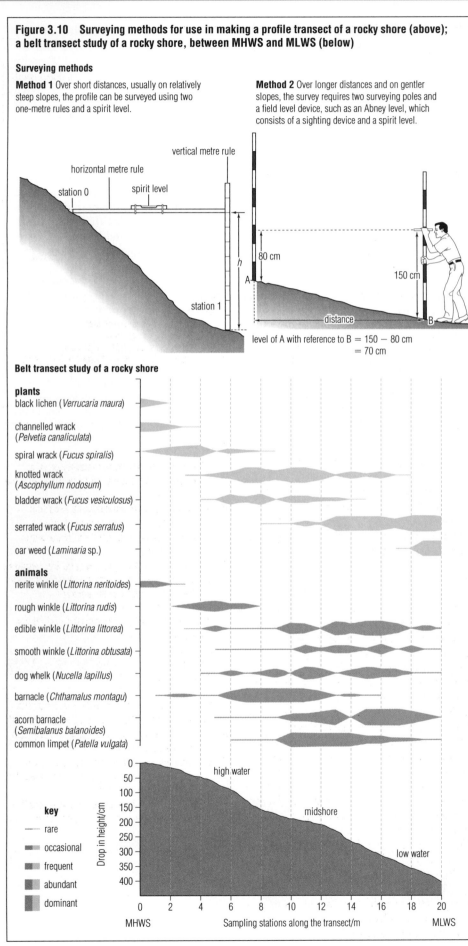

Figure 3.10 Surveying methods for use in making a profile transect of a rocky shore (above); a belt transect study of a rocky shore, between MHWS and MLWS (below)

Surveying methods

Method 1 Over short distances, usually on relatively steep slopes, the profile can be surveyed using two one-metre rules and a spirit level.

Method 2 Over longer distances and on gentler slopes, the survey requires two surveying poles and a field level device, such as an Abney level, which consists of a sighting device and a spirit level.

vertical metre rule
horizontal metre rule
station 0
spirit level
station 1
h

80 cm
150 cm
A
distance
B

level of A with reference to B = 150 − 80 cm
= 70 cm

Belt transect study of a rocky shore

plants
black lichen (*Verrucaria maura*)
channelled wrack (*Pelvetia canaliculata*)
spiral wrack (*Fucus spiralis*)
knotted wrack (*Ascophyllum nodosum*)
bladder wrack (*Fucus vesiculosus*)
serrated wrack (*Fucus serratus*)
oar weed (*Laminaria* sp.)

animals
nerite winkle (*Littorina neritoides*)
rough winkle (*Littorina rudis*)
edible winkle (*Littorina littorea*)
smooth winkle (*Littorina obtusata*)
dog whelk (*Nucella lapillus*)
barnacle (*Chthamalus montagu*)
acorn barnacle (*Semibalanus balanoides*)
common limpet (*Patella vulgata*)

high water
midshore
low water

key
rare
occasional
frequent
abundant
dominant

Drop in height/cm
0 50 100 150 200 250 300 350 400

0 2 4 6 8 10 12 14 16 18 20
MHWS   Sampling stations along the transect/m   MLWS

The community of a rocky shore

If you walk down a rocky seashore from dry land to the breaking waves, you will see a changing pattern of plant and animal life. There are changes in the physical conditions of the shore, too. To understand this ecosystem we first have to analyse the changing community of organisms and the physical environment as we pass from dry land to sea.

Analysing the shore community by transects

A straight line, such as a measuring tape, laid down the beach, provides the most immediate way of sampling the organisms present. This line we call a **line transect**, and we record the presence of the organisms touched or covered by the line. This provides a fairly approximate guide to the shore community.

A **profile transect**, on the other hand, also includes a record of the changing level of the ground, constructed from measurements taken at frequent and regular intervals along the transect line. The process of drawing up a profile transect involves surveying (Figure 3.10).

A more time-consuming but more representative study is required to produce a **belt transect**. Here the ground to be studied follows the transect line but the sample is made up of the organisms in a series of **quadrats** or squares, typically up to a metre wide, analysed in sequence to produce a record of a representative belt along the line. The plants and animals present are listed and the **relative abundance** of each is estimated and recorded on a scale that runs: D = dominant, A = abundant, F = frequent, O = occasional, R = rare.

The methods of taking line, profile and belt transects are detailed in the *Study Guide*, Chapter 3. The results of a belt transect of a rocky shore are given in Figure 3.10.

Zonation as a feature of the seashore

The profile and belt transects of one rocky shore may be quite different from those of another. For any shore, however, we find that the organisms that thrive near the top of the shoreline usually fail to occur on the lower shore, or are much less common. Figure 3.10 shows this clearly for one rocky shore. In other words, we usually observe different zones as we move vertically down the shore, as the dominant and abundant species change from the upper shore to the lowest tidemark.

In fact, on the seashore there are three basic zones each inhabited by similar life forms, though the actual species may vary. The zones, shown diagrammatically in Figure 3.8, p. 48, are the uppermost zone or **splash zone**, the **midlittoral zone** and the **sublittoral zone**. The very distinctive zonation in the physical conditions between high- and low-water marks is created by the effects of the tides and waves. The variations in the physical environment in turn determine the community in each zone.

■ The abiotic components of the shore ecosystem

Tides

The periodic rises and falls of the sea level we call **tides** are created by the attractive forces of the Sun and Moon. The strength of these gravitation forces varies with the relative positions of Earth, Moon and Sun. As a result, the pattern of tides is always changing. For example:

▶ As the Earth rotates in relation to the Moon every 24 hours, in most places along the coast high tides occur approximately every 12 hours. These are known as the **semi-diurnal tides**.

▶ When the Sun, Moon and Earth are in line ('full Moon' and 'new Moon') their gravitational forces reinforce, making the high tides extra high and the low tides extra low. These **spring tides** occur every 28½ days. Approximately 14 days later, when the Sun and Moon form a right angle with the Earth (we see a half Moon), the high tides are less high and the low tides are not as low. We call these **neap tides**.

▶ As the equinoxes approach (mid-March and mid-September) the spring tides become progressively larger. These are the **equinoctial tides**.

Waves

Anything that disturbs the surface of the water will produce a wave, but most waves are caused by the wind. The uninterrupted distance over which winds build up waves is known as the 'fetch'. Across some oceans the fetch may be more than a thousand miles, and with consistent and strong winds blowing, very large waves may form.

As waves come to shallow water they are slowed down, and as a result they come closer together and become steeper. Then, with the friction of the shore below, the top of the wave spills over and plunges on to the land. All waves tend to extend the effect of the tide by splashing the shore beyond the current tide level.

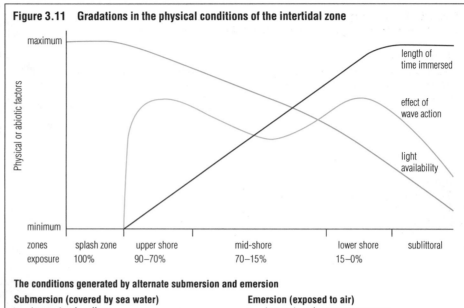

Figure 3.11 Gradations in the physical conditions of the intertidal zone

Physical or abiotic factors (vertical axis, from minimum to maximum)

length of time immersed

effect of wave action

light availability

zones	splash zone	upper shore	mid-shore	lower shore	sublittoral
exposure	100%	90–70%	70–15%	15–0%	

The conditions generated by alternate submersion and emersion

Submersion (covered by sea water)
▶ temperature is uniform
▶ there is no water loss problem
▶ water contains dissolved O_2, CO_2, inorganic ions and organic debris
▶ there is a danger of dislodgement by waves, as the tide covers the land
▶ there is danger of predation from sea organisms
▶ light penetration to lower levels is reduced

Emersion (exposed to air)
▶ temperature changes are enormous
▶ desiccation is a major danger
▶ heavy rain can cause sudden fall in salinity
▶ evaporation of water can lead to steadily rising salinity

The breaking wave, driven by high wind, is a grinding mill, pounding and tearing at the rocks, pebbles or sand, and the living things present too. The greatest damage is done by pockets of air trapped by the largest waves, when driven at high speed in storm conditions, as they plunge and strike the land.

The effects of tide and wave action

The tide imposes alternate periods during which organisms are covered by sea water (**submersion**) or exposed to air (**emersion**). The contrast between these conditions is extreme (Figure 3.11). Organisms are exposed to air for longer periods at a time the higher one goes up the shore. Consequently, physical conditions are more extreme near the top of the shore. Additionally, any part of the intertidal zone reached by the breaking waves experiences, during periodic stormy conditions at least, physical forces sufficient to dislodge and perhaps destroy all but the most firmly anchored, robust and compact organisms. Continuous pounding by waves persists longer at the high and low tide marks because the sea remains for longer at these levels than it does at the mid-tidal region.

Conditions are thus not uniform across all parts of the shore; there is a gradation in physical conditions in the intertidal zone (Figure 3.11).

Questions

1 Why might the belt transect analysis of a rocky shore community be superior to using a line transect?

2 A rocky shore is described as an eroding substrate but a mudflat is described as a depositing substrate. What do these terms suggest to you?

3 For what reasons do you imagine the seashore is colonised by certain marine organisms but not by many terrestrial organisms?

4 From the table of figures, construct a scale drawing of a profile transect of a rock shore (values in metres):

Horizontal:	0	2	4	6	8	10	12	14
Vertical:	5.6	5.1	4.4	3.6	2.4	1.6	0.9	0.0

5 What factors cause reduced light penetration to the lower regions of the submerged shore?

6 Why is there vertical zonation in the physical conditions of the seashore?

7 Explain why terrestrial organisms may be subjected to greater temperature changes than aquatic organisms (p. 58).

8 What stresses will be experienced at low tide by organisms living on the upper shore
  a on a hot day
  b during rain storms?

ROCK POOLS AS MICROHABITATS?

Rock pools are localities on the shore from which sea water does not recede as the tide flows out. They vary in size, but are common on any rocky shore made as a wave-cut platform.

On the ebb tide, a rock pool may offer a home to temporarily stranded marine animals as well as residents (anemones, crabs). But it is never a really safe haven. The higher they are up the shore, the more vulnerable are the occupants of these temporary microhabitats. Prolonged exposure in hot, sunny weather results in a steady rise in water temperature, which may reach lethal levels. At the same time evaporation brings about a rise in salinity, possibly to osmotically dangerous concentrations. In heavy rain, however, the salinity may drop to equally dangerous concentrations for marine organisms adapted to sea conditions. In very cold weather the pool may freeze up completely. And on the flood tide the salinity and temperature are suddenly and instantly returned to normal.

the production of a wave-cut platform

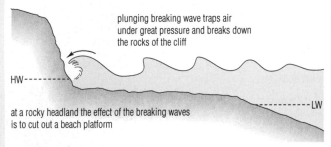

plunging breaking wave traps air under great pressure and breaks down the rocks of the cliff

HW

at a rocky headland the effect of the breaking waves is to cut out a beach platform

LW

wave-cut platform at low tide: rock pools are exposed

Rock pools exposed!

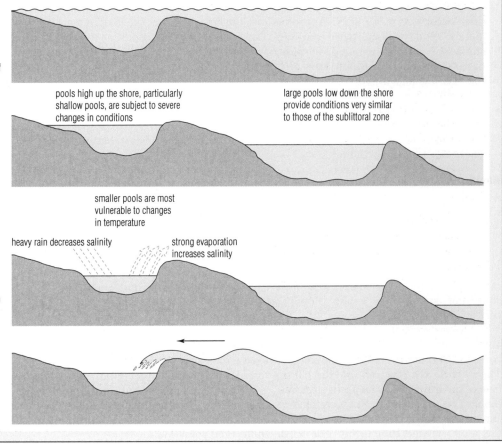

Submerged, the rock pool is merely a depression of the sea bed.

pools high up the shore, particularly shallow pools, are subject to severe changes in conditions

large pools low down the shore provide conditions very similar to those of the sublittoral zone

As the tide ebbs, the rock pool is a temporary microhabitat for the motile aquatic animals trapped there, and for all the attached animals and plants.

smaller pools are most vulnerable to changes in temperature

heavy rain decreases salinity

strong evaporation increases salinity

During emersion change is gradual. Potentially the most important changes are in temperature and in salinity. These changes may be severe enough to kill some of the animals present.

On return of the tide, the most abrupt changes of all will occur if the salinity and temperature of the pools have changed markedly during emersion. The incoming tide returns conditions to those of the sea at the crash of a wave!

■ Adaptation to the conditions of the intertidal zone

The zonation of living things on the shore largely reflects the capacity of these organisms to withstand the extremes of physical conditions experienced. Many littoral organisms show structural or physiological **adaptations** that may help them to overcome specific environmental hazards. Some examples are described below.

Seaweed zonation

In the open sea tiny, free-floating green plants, known as **phytoplankton**, are the producers in the food chain. In addition, anchored seaweeds are important producers as far down the shore as sufficient light penetrates.

The availability of light changes drastically with depth of water. Sunlight penetrates sea water as far down as 100–140 metres, but there its intensity is only 1% of that at the surface. The spectral composition of the light reaching the anchored plants changes with depth, too. This is because the sea water itself absorbs light, largely at the red end of the spectrum.

All seaweeds have the chlorophyll and carotene pigments that occur in green plants (p. 248), which trap light energy for photosynthesis. On the upper shore the green algae predominate, giving way to browns and reds further down the shore. Brown and red seaweeds, anchored at depth, survive and photosynthesise because they have additional pigments, which are capable of absorbing the predominant incident light here (Figure 3.12). For example, the additional pigments of red algae (which reflect red light) absorb the blue light that is the strongest spectral colour at depth. This light energy is passed on, via chlorophylls, and is used in photosynthesis (p. 260).

Resistance to desiccation

Water makes up the bulk of living things. Severe reductions in the water content of living cells alter the chemical balance in the cells, and lead to death. Consequently, excessive loss of water by evaporation is a hazard for both plants and animals. This is a major issue for organisms living high up the shore.

Seaweeds of the upper and midshore

All seaweeds have a covering of mucilage, which acts as a reservoir of water. This mucilage is produced in pit depressions in the surface of the seaweed thallus (Figure 25.16, p. 555). The algae

Figure 3.12   The zonation of algae in relation to the ability to photosynthesise in light of changed spectral composition: it has recently been observed that not all the deepest seaweeds are red algae, and that light is only one of the many factors influencing the depth zonation of seaweeds

green alga: *Ulva*

brown alga: *Fucus vesiculosus*

brown alga: *Laminaria digitata*

red alga: *Chondrus*

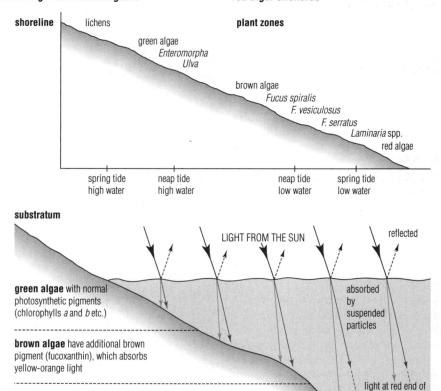

of the upper shore are better able than others to withstand desiccation because they contain more water when fully hydrated and because they lose it more slowly when exposed (Figure 3.13).

Seaweeds adapted to life on the upper shore also have the ability to recover after prolonged desiccation. This is illustrated by data from samples of three common seaweeds (Table 3.2).

Table 3.2   Data from three seaweeds

Seaweed species	Zone of shore normally colonised	% water loss from which survival is possible
channelled wrack (*Pelvetia caniculata*)	upper shore	60–80
serrated wrack (*Fucus serratus*)	midshore	40
oar weed (*Laminaria digitata*)	lower shore	20–30

Loss of water by shore animals

Intertidal animals are also in danger from the threat of desiccation, and must conserve water or prevent excessive loss of water, or both, if they are to survive prolonged exposure.

1 The limpet (*Patella vulgata*) occurs over a wide band of the shore, and is particularly common on the middle shore. This organism has a thick, largely impervious shell (Figure 3.14), and it adheres tightly to a chosen rock surface during periods of exposure. The muscular foot holds the shell to the rock surface. Individual limpets that have colonised exposed sites near the top of the shore have much thicker shells than those of limpets on less exposed sites. The thick shell protects them not only from attack by predators, but also from evaporation.

When submerged in sea water the limpet moves about to feed, rasping off the algae that grow on the rock surfaces. Prior to exposure by the receding tide the limpet returns to its chosen spot for the period of potential desiccation.

2 Other shore animals, such as **the acorn barnacle** (*Semibalanus balanoides*) **and the edible winkle** (*Littorina littorea*), also have structures that reduce water loss during periods of exposure (Figure 3.15). The common species of winkle also show adaptation to exposure involving differences in metabolism. The excretion of the waste products of protein metabolism requires varying amounts of

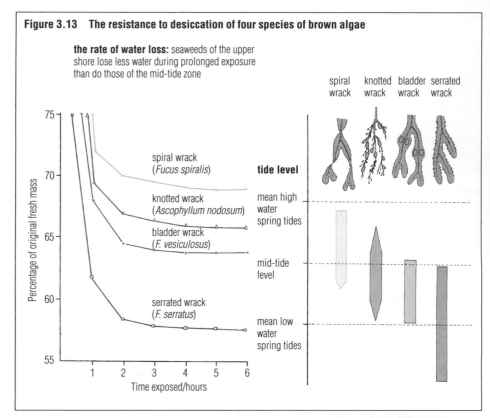

Figure 3.13   The resistance to desiccation of four species of brown algae

the rate of water loss: seaweeds of the upper shore lose less water during prolonged exposure than do those of the mid-tide zone

Figure 3.14   The limpet (*Patella vulgata*): when submerged the limpet grazes on the attached algae (above); prior to emersion the limpet returns to a 'home' site where the rock is worn by contact with the thick shell (below)

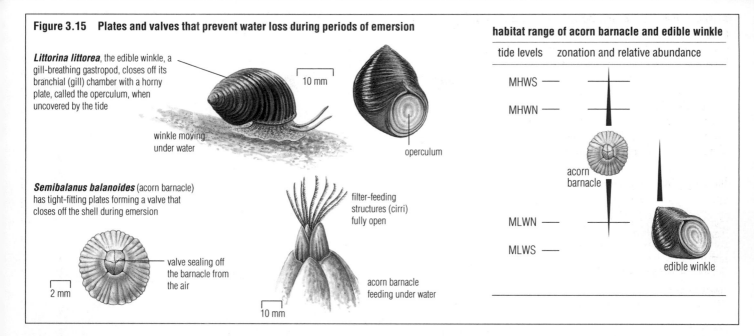

Figure 3.15   Plates and valves that prevent water loss during periods of emersion

Littorina littorea, the edible winkle, a gill-breathing gastropod, closes off its branchial (gill) chamber with a horny plate, called the operculum, when uncovered by the tide

winkle moving under water

operculum

10 mm

Semibalanus balanoides (acorn barnacle) has tight-fitting plates forming a valve that closes off the shell during emersion

valve sealing off the barnacle from the air

2 mm

filter-feeding structures (cirri) fully open

acorn barnacle feeding under water

10 mm

habitat range of acorn barnacle and edible winkle

tide levels	zonation and relative abundance
MHWS	
MHWN	
	acorn barnacle
MLWN	
MLWS	
	edible winkle

water for disposal according to whether it is largely ammonia, urea or uric acid that is excreted. For the safe disposal of ammonia (soluble and toxic) very much more water is required than for that of uric acid (less soluble and less toxic). Ammonia is the common nitrogenous excretory compound in aquatic animals. Some species of winkle (*Littorina* spp.) produce mostly uric acid and others rely on ammonia (p. 377). Those that produce most uric acid occur high on the shore, whereas those producing ammonia are common on the lower shore.

■ Secure attachment to the substratum

Many organisms of the seashore have secure attachments to the substratum in response to the mechanical effects of breaking waves. Bladder wrack (*Fucus vesiculosus*) and the dog whelk (*Nucella lapillus*), both species of the midshore zone, are examples (Figure 3.16).

Figure 3.16   Structural adaptations to exposed and sheltered positions on the shore

Dog whelks (*Nucella lapillus*) move about when submerged, feeding mainly on barnacles. As the tide goes out they become secured in sheltered cracks in rocks, using their foot to hold on with.

Bladder wrack (*Fucus vesiculosus*) is a brown seaweed on the midshore zone that grows anchored to rocks and breakwaters. When submerged the air-bladders cause the thallus to float close to the surface.

Nucella lapillus

On exposed parts of the shore:
Dog whelks that survive on exposed parts of the shore have a large shell with a wide aperture. This accommodates a large foot. With this foot extended the dog whelk has a better chance of remaining attached whilst experiencing the breaking waves.

Bladder wrack growing on exposed shores has a short, little-branched thallus with few air-bladders.

On sheltered parts of the shore:
Dog whelks that survive on the sheltered shore have a smaller foot and a narrow aperture.

Bladder wracks of sheltered places are large plants, much branched, and with numerous air-bladders.

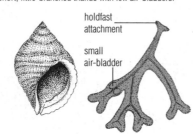

holdfast attachment

small air-bladder

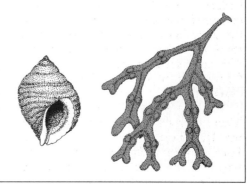

Questions

1 What differences in the seaweeds of the upper and lower shore can be related to the environmental conditions experienced by these plants?

2 What advantage may be gained by the limpet in returning to the same position for each period of exposure at low tide?

3 From the graph in Figure 3.13, calculate the difference in rate of water loss between *Fucus vesiculosus* and *F. serratus* after three hours' exposure.

4 In principle, in what ways do animals of the upper shore avoid desiccation when exposed
a by their behaviour
b by their structure?

5 What is meant by the term microhabitat? Give three examples of seashore microhabitats.

6 Construct a food web from the following organisms:
limpet; sea bird; dog whelk; seaweed; barnacle; shore crab; plankton; winkle.

7 In what ways does a rock pool differ from a pond as an ecosystem?

8 How do the features of bladder wrack on an exposed shore reduce the chances of it being damaged by wave action?

■ Interactions of biotic and abiotic factors

In many ecosystems the dominant organisms present have a direct effect on the abiotic factors of the ecosystem. Plants in particular often help to create the physical conditions that all organisms in the same ecosystem experience.

Compared with other ecosystems, the plants of the shore (seaweeds) and other shore organisms have limited influence over the stark physical conditions of the shore environment. Nevertheless, relationships between organisms of the shore are varied and complex. There are numerous examples of competition, predation and symbiosis (p. 66) to be observed. A typical seashore food chain is shown below:

young seaweed
e.g. sea lettuce (*Ulva*)
↓
herbivorous mollusc
e.g. limpet (*Patella*)
↓
carnivorous mollusc
e.g. whelk (*Nucella*)
↓
sea bird
e.g. herring gull (*Larus*)

A marine/littoral zone food web is shown in Figure 3.17. The range of biotic interactions is discussed on p. 64.

■ Human activities and the sea

No part of the Earth's surface is free of some human influence. Human activities can affect the structural features of the shore through the building and maintenance of sea walls, breakwaters and harbours. These are all means of reducing the destructive effects of wave action on a particular part of the shore. Attempts to reduce erosion and longshore drift by these means have direct effects on the habitats of the shore.

Most human effects on sea and shore are more indirect and insidious. For example, we often use the sea as a 'safe' dustbin for waste materials. Short-term waste disposal includes the dumping of raw sewage straight into the sea, and the depositing of sewage sludge at a greater distance. The long-term disposal of dangerous or poisonous industrial or chemical waste is made in supposedly water-tight containers at points of great depth. The steady accumulation of toxic substances in the seas, even if in very deep water, poses a long-term threat to

Figure 3.17 **A marine/littoral zone food web:** 'top' carnivores (common seal, herring gull) are at trophic levels 3–5 according to the prey chosen

common seal · herring gull · carnivorous fish (pollack) · lobster · common prawn · edible crab · dog whelk · common mussel · herbivorous fish (grey mullet) · flat winkle · limpet · sea urchin · barnacle · large seaweeds · tiny seaweeds and surface layers of microscopic plants · plankton

primary producers – trophic level 1

Figure 3.18 **Human activities and the sea:** oil tanker wreck, leaking oil on to coastal environments (above); seashore with rubbish from human terrestrial activities, piled up along the tide line (below)

all living things. Meanwhile, the accidental deposition of crude oil through spillage from giant tankers is currently a major hazard for the wildlife of seashores (Figure 3.18).

We also see the sea as a storehouse. Fishing of various types is carried on extensively, sometimes to the point of dangerous depletion of animal stocks. About 80% of our marine harvest comes from quite restricted areas of the oceans. Similarly the sea and sea bed may be a valuable source of minerals. These issues are discussed on pp. 93 and 103.

3.3 WOODLAND AND FOREST ECOSYSTEMS

Three-quarters of the Earth's surface is covered by the waters of the major ocean systems; of the remaining surface, not all is habitable. Of the total land area, approximately two-thirds is covered by forests or woodlands. The locations of the three major types of forest ecosystem – **boreal forest**, **temperate deciduous forest**, and **tropical rain forest** – are shown in Figure 3.2 (p. 45). The size and longevity of trees make forest ecosystems very impressive. Many trees are 20–30 metres high when mature and others very much higher; many survive for several hundred years.

Forests and woodlands are some of the most important ecosystems of the terrestrial scene. Forest ecosystems are important not only because of the huge land area they clothe, but also because of their effects on climate and the biosphere as a whole. In woodland ecosystems the distribution of trees largely determines the presence of other organisms, and influences their behaviour.

Forests are ecosystems showing layering, or stratification, into four major, more-or-less distinct layers (Figure 3.19). Each layer is defined by the dominant plant life, but carries a distinctive fauna too.

■ The interactions of environmental factors

No organism exists in isolation; each exists in the context of a range of environmental factors. At any one time or place one set of environmental factors may be more influential than another. In practice, however, all environmental factors are interrelated (Figure 3.7, p. 48). It is usually difficult and often impossible to isolate the influence of a particular factor.

■ Abiotic factors and terrestrial ecosystems

Light

Light is the ultimate source of energy, and consequently the amount of incoming solar radiation determines the structure and functioning of an ecosystem (Figure 3.20). The light energy available to ecosystems may vary greatly. For example, Figure 4.6 (p. 80) shows the changing solar energy flux at the Earth's surface between the tropics and the northern boreal region.

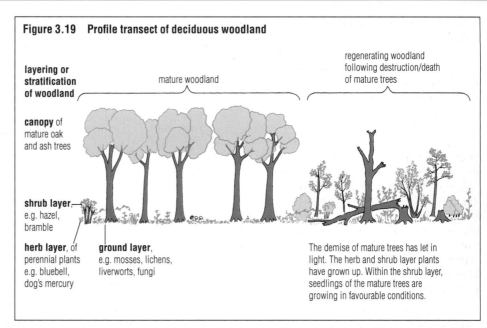

Figure 3.19   Profile transect of deciduous woodland

layering or stratification of woodland

mature woodland

regenerating woodland following destruction/death of mature trees

canopy of mature oak and ash trees

shrub layer, e.g. hazel, bramble

herb layer, of perennial plants e.g. bluebell, dog's mercury

ground layer, e.g. mosses, lichens, liverworts, fungi

The demise of mature trees has let in light. The herb and shrub layer plants have grown up. Within the shrub layer, seedlings of the mature trees are growing in favourable conditions.

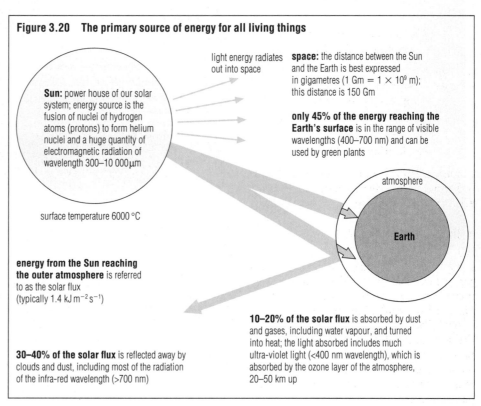

Figure 3.20   The primary source of energy for all living things

light energy radiates out into space

Sun: power house of our solar system; energy source is the fusion of nuclei of hydrogen atoms (protons) to form helium nuclei and a huge quantity of electromagnetic radiation of wavelength 300–10 000 μm

surface temperature 6000 °C

space: the distance between the Sun and the Earth is best expressed in gigametres (1 Gm = 1 × 10⁹ m); this distance is 150 Gm

only 45% of the energy reaching the Earth's surface is in the range of visible wavelengths (400–700 nm) and can be used by green plants

atmosphere

Earth

energy from the Sun reaching the outer atmosphere is referred to as the solar flux (typically 1.4 kJ m⁻² s⁻¹)

10–20% of the solar flux is absorbed by dust and gases, including water vapour, and turned into heat; the light absorbed includes much ultra-violet light (<400 nm wavelength), which is absorbed by the ozone layer of the atmosphere, 20–50 km up

30–40% of the solar flux is reflected away by clouds and dust, including most of the radiation of the infra-red wavelength (>700 nm)

Green plants, the primary producers in every ecosystem, support the nutrition of all living things, directly or indirectly. We demonstrate this dependence when we produce food chains and food webs.

Light is also the stimulus for the timing of daily (**circadian**) and seasonal rhythms. For example, the breeding season for some plants and animals is triggered by the organism's response to day length (p. 434).

Sunshine is also an important source of heat.

The duration of daily illumination and the intensity or energy content depends on latitude, season, time of day and the extent of cloud cover. Measurements of light intensity present many problems because the level of illumination fluctuates so much. Light measured in woodland shade is best expressed as a percentage of the light provided by full sunlight. The equipment and techniques for investigating light intensity in a habitat are outlined in the *Study Guide*, Chapter 3.

We have seen that light passing through sea water may determine which seaweeds survive on the lowermost parts of the seashore (Figure 3.11, p. 51). Similarly, in a woodland, the amount of light reaching the field layer determines the plant life able to thrive there (Table 3.3).

Leaves are often arranged in a way that minimises overlap (leaf mosaics, Figure 12.17, p. 254). On large trees the leaves of the upper canopy grow in full light, whilst leaves of the lower branches are often permanently shaded. Leaves that grow in the full sunlight are structurally different from shade leaves (Figure 3.21). These structural differences can be correlated with the different light regimes the leaves experience.

Table 3.3 Light intensity and the field layer of woodlands

Tree cover	% daylight penetrating to ground level	Typical species in herb and shrub flora	
Oak (*Quercus robur*)	35–52 (summer) 48–63 (winter)	bramble (*Rubus fruticosus*) honeysuckle (*Lonicera periclymenum*) dog rose (*Rosa canina*) elder (*Sambucus niger*)	willow (*Salix* sp.) rose bay willow herb (*Chamaenerion angustifolium*) stinging nettle (*Urtica dioica*) bracken (*Pteridium aquilinum*)
Beech (*Fagus sylvatica*)	20–25 (summer) 28–35 (winter)	mosses only	
Pine (*Pinus sylvestris*)	10–14 (summer) 13–18 (winter)	fungi only	

Figure 3.21 Sun and shade leaves of beech: transverse sections of lamina

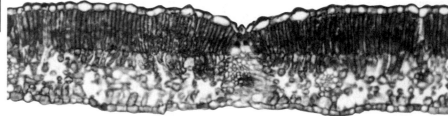

sun leaves
► thick leaves with a palisade mesophyll layer often two or three cells thick
► chloroplasts are mostly restricted to the palisade mesophyll cells
► a sun leaf can absorb much of the light available to the cells when exposed to high light intensities

shade leaves
► thin leaves with a single layer of palisade mesophyll cells
► chloroplasts occur throughout the mesophyll, with almost as many chloroplasts in spongy mesophyll as in the palisade layer
► a shade leaf can absorb the light available at lower light intensities; if exposed to high light, most would pass through

beech (*Fagus sylvatica*): the canopy seen from below, with sun and shade leaves

Figure 3.22 Annual rainfall, mean temperature and vegetation (boundaries between the biomes are approximate)

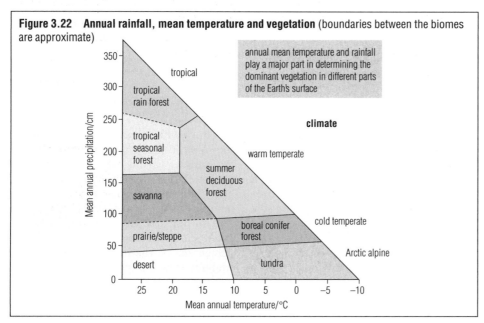

annual mean temperature and rainfall play a major part in determining the dominant vegetation in different parts of the Earth's surface

Temperature

The effect of temperature on organisms is direct, for temperature influences the rate of all biochemical reactions (p. 180). Temperature also has an indirect effect on other environmental factors, such as the rate of evaporation. Temperature can vary a great deal in terrestrial situations, but it varies much less in large volumes of water because of the high specific heat capacity of water. Temperature varies on land both seasonally and diurnally, and with latitude, topography and distance from oceans (Figure 3.22).

Techniques for investigating temperature as an abiotic factor of significance in terrestrial ecosystems are outlined in the *Study Guide*, Chapter 3.

Water availability from rainfall

Water is made available to the land by rainfall. The pattern of rainfall is largely determined by global 'weather systems'. In principle, moisture-laden air is blown on to the land from oceans. Rain is deposited on to ocean-facing slopes of hills and mountains. As air moves on over the land it may be recharged with moisture by transpiration (p. 321) and evaporation from lakes and soils, and further rainfall is possible (water cycle, Figure 3.54, p. 75).

Rainfall distribution over the year is an important limiting factor for organisms (Figure 3.23). In parts of the tropics there are well-defined wet and dry seasons, and in such regions organisms must be adapted to withstand prolonged droughts on a regular basis. In temperate zones and in the wet tropics rainfall tends to be more evenly distributed. Some of the adaptations that help organisms to survive drought are listed in Table 3.4.

Humidity

Humidity, the amount of water vapour in air, is important for its effect on the rate of water loss by plants and animals: the lower the humidity, the faster water evaporates from moist surfaces. Usually, humidity is higher at night and lower during the day. Organisms that have a relatively large, moist surface area and lose water easily may be restricted to humid areas, or be active only at more humid periods. Examples are frogs, slugs, liverworts and some mosses.

Woodlice, too, regulate their activities to avoid dehydration. During daylight, woodlice are normally found in humid woodland microhabitats: under stones or in rotting tree stumps. In fact, different species of woodlouse occupy different niches in woodlands according to their ability to withstand desiccation. Studying the responses of these species to humidity (and light and other factors) is undertaken using a simple choice chamber (*Study Guide*, Chapter 23).

Topography

Topography (Figure 3.24) is concerned with the altitude and the shape of the ground. For example, the angle of slope, and the direction it faces (its 'aspect') determine the amount of solar illumination received. The gradient of land influences water movements and the amount of erosion. All these features of land surface may modify the climatic environmental factors discussed above. They may also influence the formation and erosion of soils.

Wind

Wind is an important environmental factor because, indirectly, wind affects transpiration in plants by removing moist air from around the leaves, thereby allowing further evaporation. Trees growing in persistently windy positions may be stunted in their growth as a result of the physiological stress imposed by high transpiration.

Figure 3.23   Availability of water and the response of plants and animals

Root systems of plants respond to water sources at different depths: spurge (*Euphorbia* sp.) growing (left) in a dry soil with a very deep water table, (right) in gravel soil on a mountain side with high rainfall

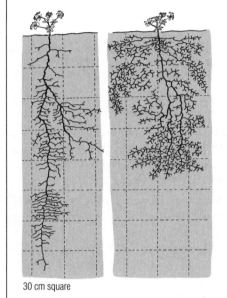

30 cm square

Slugs are sensitive to changes in the water content of their surroundings: adults live in damp, shady places and lay eggs in moist soil

common field slug (*Agriolmax reticulatus*)

Five mature slugs placed on soil samples of different percentage saturation laid eggs over a five-day period.

Percentage saturation	10	25	50	75	100	Total
Days	Number of eggs laid					
1	0	0	32	41	0	73
2	0	0	48	72	0	120
3	0	45	62	36	0	143
4	0	0	0	51	41	92
5	0	56	23	0	0	79
Total	0	101	165	200	41	507

No eggs are laid in relatively dry soils.

Figure 3.24   Topography affects other environmental factors: altitude, slope and aspect affect local climate, drainage and weathering

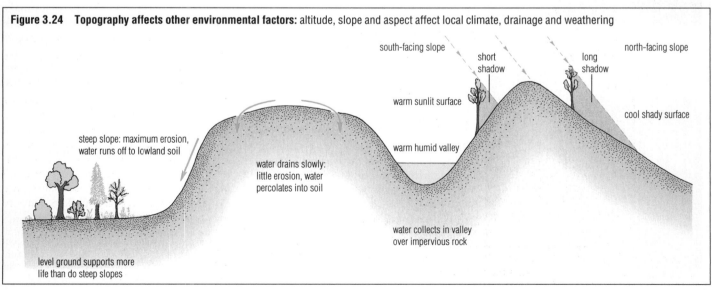

south-facing slope

short shadow

north-facing slope

long shadow

warm sunlit surface

cool shady surface

warm humid valley

steep slope: maximum erosion, water runs off to lowland soil

water drains slowly: little erosion, water percolates into soil

water collects in valley over impervious rock

level ground supports more life than do steep slopes

Table 3.4   Adaptations for survival of water shortage

Strategy	Examples of adaptations
tolerance of water loss and partial desiccation	lichens (p. 67), epiphytic algae like *Pleurococcus* (p. 553), and many mosses growing clustered together in cushion habit
	camels (p. 390) may go without drinking water for 6–8 days, if necessary, and often lose about 25% of their body weight due to water loss in the process; earthworms may lose over 50% of their body weight due to water loss in a drought, and survive
biochemical production of water	fats are oxidised to carbon dioxide and a lot of water, and this internally-produced water is essential to survival in some desert animals, such as the desert rat and the camel (p. 390)
reduction of water loss	in plants, by xeromorphic features, summarised in Table 18.2, p. 375
	excretion of nitrogenous waste as uric acid in a solid paste, in many terrestrial insects (p. 378) and birds (p. 391)
	toleration of raised body temperatures in daytime, involving limited water loss by sweating, seen in desert animals such as the camel
evasion	many plants avoid exposure by passing the unfavourable season as seeds (p. 574), or as underground structures in a dormant condition (p. 578), or by leaf fall and dormancy (p. 483)
	some animals withdraw to sheltered and protected positions; for example, by burrowing into deep soil layers (e.g. earthworms) or sealing themselves into a shell (e.g. snail aestivating behind a mucus sheath)

Figure 3.25   The windiness of Britain (right), the destruction of trees by 'windsnap' (above left) and 'windthrow' (below)

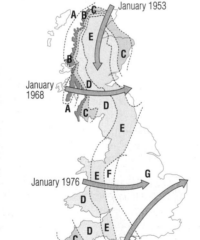

The relative windiness of Britain on a seven-point scale (A = very windy, G = calm), showing some of the catastrophic gales that occur from time to time

January 1953

January 1968

January 1976

October 1987

Wind can also cause physical damage, particularly to forest trees. High winds are a particular threat to broad-leaved trees when in full leaf. Normal winter gales, with gusts of 60–110 km h^{-1}, will cause 'windthrow' (trees blown over) and 'windsnap' (breaking off) in tall trees and those weakened by parasites, for example. Occasional catastrophic gales, with gusts of 110–160 km h^{-1}, may destroy whole woodlands of strong and healthy trees (Figure 3.25).

Edaphic factors: the soil

The soil is the upper, weathered layer of the Earth's crust. It consists of disintegrated rock, organic matter, air, water, dissolved minerals and various living organisms. The study of soil is a huge discipline in its own right, known as **pedology**.

Soil is a key component of terrestrial ecosystems because it influences the lives of all terrestrial organisms in numerous ways. Soil provides plants with anchorage for roots, supplies of water and inorganic nutrients, and the essential air for root growth. Soil is also home to a vast number of microorganisms and small non-vertebrates that live in the top few centimetres of the soil.

The structure and fertility of soil depend on the nature of the mineral matter (sand, clay and so on), and on the flora and fauna it contains. At the same time it has a major role in determining the plant and animal communities that occur. The soil's properties are also influenced by the climate and topography. Consequently it is hard to isolate the influences of edaphic factors from the other important ecological factors.

Soil components

Mineral skeleton

Soil is formed by the disintegration or **weathering** of rock. Weathering involves physical processes such as breakdown by frost and ice, and by the abrasive actions of water, wind and gravity. Chemical weathering includes the action of rainwater containing dissolved carbon dioxide; this forms carbonic acid, a weak acid, which causes hydrolysis, hydration and solution of minerals.

The resulting mineral* particles make up the largest component of most soils (Figure 3.26). Soil particles are of very diverse sizes. They are classified by the Soil Survey of England and Wales on the basis of diameter alone (although particles are of differing chemical composition) as follows:

stones	more than 2.0 mm
coarse sand	0.6–2.0 mm
medium sand	0.2–0.6 mm
fine sand	0.06–0.2 mm
silt	0.002–0.06 mm
clay	less than 0.002 mm

* Not to be confused with 'mineral salts' (containing metal ions like Ca^{2+} and K^+, and non-metal ions, NO_3^- and Cl^-), which are the inorganic constituents of plants and animals.

Table 3.5 Comparison of soil properties

Clay soil	Sandy soil
particle size less than 0.002 mm (2 µm)	particle size from 0.06 mm to 2.0 mm
small air spaces between particles: poor aeration	large air spaces between particles: good aeration
poor drainage: soil easily compacted	good drainage: soil not compacted
good water retention (waterlogging possible)	poor water retention (no waterlogging)
a wet soil: evaporation of water causes it to be cold	less water evaporation, therefore warmer
particles attract many mineral ions: nutrient content is high	minerals are easily leached so mineral content is low
particles aggregate to form clods: the soil is heavy and difficult to work	particles remain separate: the soil is light and easy to work
Loam the properties of loam soils are intermediate to those of clay and sandy soils	

The **mineral composition of soil** is investigated by mechanical or quantitative analysis (*Study Guide*, Chapter 3). The proportions of sand, silt and clay vary from soil to soil, and the resulting mix gives a particular soil its distinctive texture. For example, a coarse-textured soil has a high proportion of sand, and a fine-textured soil has a high percentage of clay particles. A medium-textured soil (a loam) contains a mixture of coarse and fine particles. In turn, texture influences the important soil properties such as aeration, water-holding capacity and drainage, soil temperature, nutrient supply and ease of root penetration (Table 3.5).

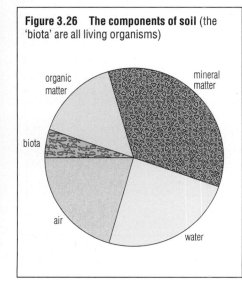

Figure 3.26 The components of soil (the 'biota' are all living organisms)

Organic matter

The organic matter in soil originates from the dead remains of organisms and from animal faeces. The breakdown of organic matter in soil is brought about partly by detritivores, such as earthworms, feeding on the dead remains of other organisms, and partly by decomposers such as bacteria and fungi. As a result, the decaying plant and animal remnants at the soil surface become an amorphous, colloidal black substance called **humus**, coating the soil mineral particles.

Humus has a complex chemical composition. It helps soil to retain nutrients in a form that slowly becomes available to plant life. Meanwhile, the mineral particles of the soil are bound together by humus into larger aggregates called soil crumbs. Soil with a crumb structure is well aerated and drained.

Humus is gradually broken down aerobically to carbon dioxide, water, ammonia and various anions (such as nitrates and phosphates) and cations (such as calcium and magnesium). All these breakdown products are essential for plant growth. In persistently waterlogged soils the dead organic matter cannot decay, but becomes compressed together as **peat** (Figure 3.27).

Figure 3.27 A mountainside environment where peat bogs have formed from *Sphagnum* moss and accumulated, undecayed plant remains: the low pH, high rainfall, low temperatures and poor drainage all contribute to the formation of peat

Questions

1 Which abiotic factors are important at field- and ground-level zones of a woodland?

2 What does the term 'solar flux' mean, and why is it important to biology?

3 From the data in Table 3.3, p. 58, it appears that light reaching a woodland floor determines the composition of the field layer. What other factors might be involved?

4 How may the composition of the animal community at the ground level of regenerating woodland differ from that of mature woodland?

5 What effect would you expect a rise in humidity to have on the rate of transpiration (loss of water vapour) by plants? What would be the effect of a rise in wind speed?

6 In what ways may the topography of a particular habitat influence the formation and conditions of the soil there?

7 What properties of a loam soil are advantageous to root growth?

8 What do we mean by the terms 'physical weathering' and 'chemical weathering' of rocks?

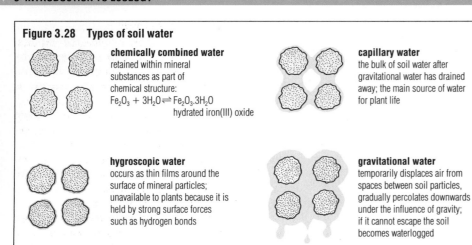

Figure 3.28 Types of soil water

chemically combined water
retained within mineral substances as part of chemical structure:
$$Fe_2O_3 + 3H_2O \rightleftharpoons Fe_2O_3.3H_2O$$
hydrated iron(III) oxide

capillary water
the bulk of soil water after gravitational water has drained away; the main source of water for plant life

hygroscopic water
occurs as thin films around the surface of mineral particles; unavailable to plants because it is held by strong surface forces such as hydrogen bonds

gravitational water
temporarily displaces air from spaces between soil particles, gradually percolates downwards under the influence of gravity; if it cannot escape the soil becomes waterlogged

Soil water

Plants take from soil greater quantities of water than of any other single substance. Consequently, the wetness of soil and its capacity for retaining moisture are very important qualities.

Soil water is added to by rainfall, and lost by evaporation from the soil surface, by uptake by plants and by drainage. When the soil is over-supplied with water some drains away under the influence of gravity (**gravitational water,** Figure 3.28). Water draining away through soils contributes to underground reserves (referred to as aquifers, p. 98). The bulk of water retained around soil particles is freely available to plants to replace water lost by transpiration, and is called **capillary water**. When the soil holds the maximum possible quantity of capillary water it is at **field capacity**.

Soil air

Soil air provides the oxygen required for aerobic respiration by plant roots, soil non-vertebrates and soil microorganisms. Consequently the air in the pores in the soil is a vital component for the soil fauna and flora. In waterlogged soil, with the soil air displaced by water, plant roots are deprived of oxygen. Plants may die if the condition persists.

The composition of soil air is similar to that of the atmosphere, with which it is in continuity by diffusion. The volume of air in soil largely depends on the shape and size of the mineral particles. For example, the air content is much less in clay soils with tiny close-fitting particles than in sandy soils with a majority of larger, ill-fitting particles in the mineral skeleton.

Soil pH and soil nutrients

Soil nutrients are the ions in soil that arise from the weathering of rock and the decay of humus. They are essential for plant growth (Figure 16.16, p. 328). Some are in solution in the soil water, and freely available to plants. These are positively charged **cations** such as those of calcium, potassium and sodium, and negatively charged **anions** such as nitrate and phosphate (p. 127). Other ions are adsorbed on to clay particles and humus. Clay and humus are negatively charged and attract cations; thus cations are less likely to be leached out of soils. Instead they are released by cation exchange; that is, they are replaced by hydrogen ions from root cells (Figure 16.17, p. 329). Other minerals are locked away in unweathered rock or undecayed humus.

The pH of soil (that is, the degree of acidity or alkalinity of the soil solution – see p. 128) affects the availability of ions that otherwise exist in unavailable forms (Figure 3.29). Most soils in Britain have a pH between 3.5 and 8.5. Climate influences soil pH because of its effects on decomposition of humus and on leaching out of ions. The rock from which soils are formed is also important: chalk and limestone form alkaline soils, sandstones form acidic soils. Plant life influences soil pH too by its uptake of ions and by the type of humus it forms.

Soil organisms

The community of soil organisms consists of a potentially huge range of nonvertebrate animals ('minibeasts') and many microorganisms, protozoa, algae, fungi and bacteria. The actual diversity of organisms depends upon specific conditions such as water content, pH and oxygen, as well as the availability of specific nutrients required by particular groups. Organisms that occur in the soil water include the microorganisms, nematode worms, flatworms and rotifers. Others, including arthropods such as mites and springtails, are dwellers in the tiny air-spaces of soil. The burrowers create their own tunnels; they include mammals such as rabbits, moles and badgers, annelids such as earthworms, and arthropods such as beetles, centipedes and millipedes. Techniques for investigating the soil community of organisms are referred to in the *Study Guide*, Chapter 3.

The activities of earthworms play a key role in improving soil fertility for plants. This is illustrated by the experiment outlined in the panel opposite, in which a winter wheat crop was grown in loam soil having a low population of earthworms. Earthworm burrows increase aeration and drainage of soil, and their surface 'casts' bring deep soil to the surface. Some worms drag organic matter (dead leaves, for example) down into the soil. Earthworms digest organic matter and release nutrient ions into soil in their alkaline secretions.

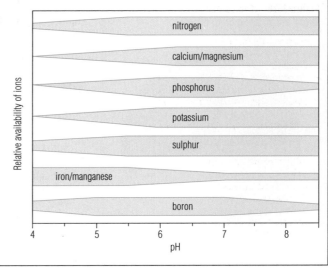

Figure 3.29 pH and the availability of soil nutrients: many important nutrients become less available for absorption by plants as the pH falls below neutral – the more acidic the soil, the less available the nutrients become

EARTHWORMS AND SOIL CONDITIONS

This study of the effects of earthworms on plant growth involved 12 plots, each of 500 m², arranged at random in a field of loam soil. Some plots were pretreated (September) as shown, before winter wheat was sown (October). Finally chopped straw was distributed evenly over all plots (November) and earthworms were added to some plots as shown.

A (3 plots)	B (3 plots)	C (3 plots)	D (3 plots)
pretreatment: earthworms killed (by fumigation)	pretreatment: earthworms killed	pretreatment: earthworms killed	no pretreatment, no fumigation
November: 300 deep-burrowing earthworms added	November: 300 shallow-burrowing earthworms added	November: no earthworms added	November: no earthworms added

Results	Plots	A	B	C	D
Mean number of earthworm casts/m²		165	73	11	21
Mean dry mass of root material/g m⁻² of row		13.3	6.4	3.6	3.4
Mean grain mass/g m⁻²		8.6	7.0	6.2	6.2
Mean dry mass of straw on surface/g m⁻²		94	217	236	219

Questions to answer

1 What effects do earthworms appear to have on the growth of wheat?
2 How do earthworm activities cause these effects?
3 What other organisms of the soil may have been affected by the pretreatment?

Soil profiles and soil types

Soils are formed from mineral particles produced by disintegration and erosion of parent rock (weathering).

Partly as a result of the way it is formed, a soil is stratified into layers or **horizons** when seen in profile (Figure 3.30). A soil profile, a vertical section through soil, is exposed by digging. The layers of the profile are recognised as (from the soil surface) the organic layer or horizon (O), the topsoil horizon (A), the subsoil (B) and parent rock (C).

Water may dissolve soluble substances from surface layers and transport them down through the soil. This is called **leaching**. Leaching occurs where rainfall exceeds surface evaporation. Where surface evaporation equals or exceeds rainfall, soluble nutrients tend to remain at higher levels in the soil.

In Britain, two main types of soil profile (known as soil types) predominate, and are illustrated in Figure 3.31. Brown earth soils are associated with deciduous forest flora. Podsols are associated with coniferous forest, or with cold, wet moorland soils where acidic plant remains accumulate. We cannot classify garden or farm soils by examining the profile because digging and ploughing constantly mix the layers and speed up the integration of components.

Questions

1 How might you set up investigations to show that it is the actions of earthworms that cause the mixing of soil components and cause surface objects to enter the soil?
2 From the data in Figure 3.29, what pH seems to give the optimum availability of ions?

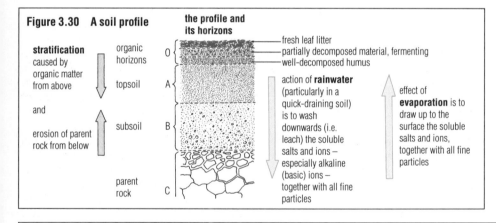

Figure 3.30   A soil profile

the profile and its horizons

stratification caused by organic matter from above

and

erosion of parent rock from below

organic horizons — O

topsoil — A

subsoil — B

parent rock — C

fresh leaf litter
partially decomposed material, fermenting
well-decomposed humus

action of **rainwater** (particularly in a quick-draining soil) is to wash downwards (i.e. leach) the soluble salts and ions – especially alkaline (basic) ions – together with all fine particles

effect of **evaporation** is to draw up to the surface the soluble salts and ions, together with all fine particles

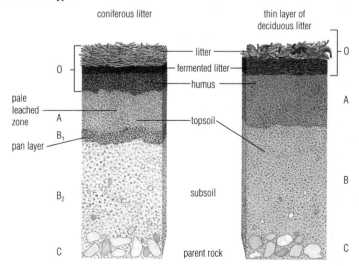

Figure 3.31   Podsol and brown earth soil types

Podsol type:
▶ occurs in wetter, colder parts of UK, the north and west
▶ formed on a freely drained substrate under conditions where rainfall exceeds evaporation
▶ the low temperature and the leaching away of cations checks the disintegration of plant debris and an acid humus (mor) accumulates
▶ this topsoil is poor in basic ions, low in nitrates, and is pale ash colour, because the humus and hydrated iron(III) oxide (a brown compound) have been washed into the subsoil; they may form a hard pan layer restricting drainage and root growth.

coniferous litter

thin layer of deciduous litter

litter
fermented litter
humus

pale leached zone

pan layer

topsoil

subsoil

parent rock

O
A
B₁
B₂
C

O
A
B
C

Brown earth type:
▶ occurs in drier, warmer parts of UK, for example in the south and east
▶ formed under conditions where rainfall is equal to evaporation
▶ moist, well-aerated soil in which bacterial decay forms an alkaline-to-neutral mild humus (mull)
▶ this topsoil is rich in cations; nitrates are present and the rich brown colour is formed by hydrated iron(III) oxide, derived from iron-containing aluminosilicates in the parent rock, liberated by weathering.

3.4 BIOTIC FACTORS

The effects of organisms on each other are known as **biotic factors**, and include competition, predation and parasitism (one form of symbiosis). Competition occurs when two or more individuals strive to obtain the same resources when these are in short supply.

Biotic factors are often more influential than physical (abiotic) factors in determining the distribution of organisms. An extreme example is the effect of the human animal.

■ Competition between organisms

Resources of every sort are in limited supply, and so organisms must compete for them. For example, plants compete for space, light and mineral ions; animals compete for food, shelter and a mate.

Competition between individuals of the same species (known as **intraspecific competition**) is the basis of the origin of species by natural selection (p. 657). Competition between individuals of different species (**interspecific competition**) is illustrated by competition between dormice and squirrels for hazelnuts, for example.

Predation and grazing

A **predator** is an organism that feeds on living species. Predators are normally larger than their prey, and tend to kill before they eat. The term 'predator' usually implies that the organism preyed upon is an animal. The eating of plants by herbivorous animals is really also a case of predation, although it is more commonly referred to as **grazing** or **browsing**. Examples of predation and grazing by woodland organisms are shown in the woodland food web (Figure 3.32).

The abundance of prey is a factor limiting the numbers of the predator. In fact, within a food chain, a **prey–predator relationship** causes both populations to oscillate. A laboratory experiment with the protozoan predator *Paramecium aurelia* and the yeast *Saccharomyces exiguus* illustrates this (Figure 3.33). In the presence of excess yeast *Paramecium* feeds well and the predator population increases. Continued and increasing feeding on yeast causes the prey population to fall. A decrease in the yeast population leads to a decrease in the *Paramecium* population. With fewer predators the prey population increases again.

It is believed that prey–predator oscillations occur in nature too. Evidence

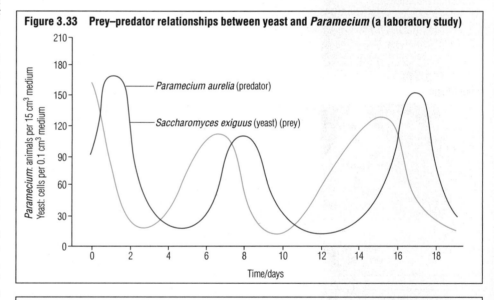

Figure 3.32   A woodland food web: the 'top consumers' are at trophic levels 2–4 according to their diet at any particular moment

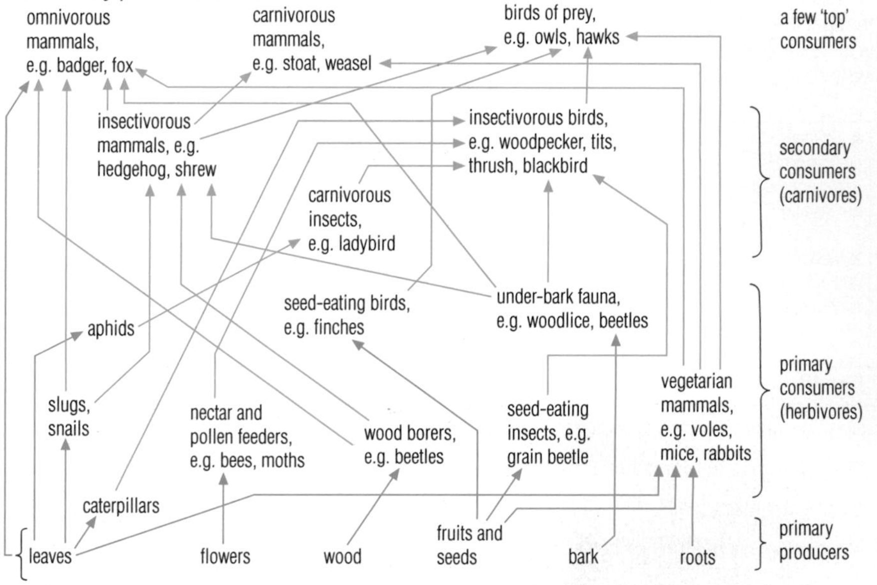

Figure 3.33   Prey–predator relationships between yeast and *Paramecium* (a laboratory study)

Figure 3.34   Prey–predator relationships: fluctuations in the numbers of pelts received by the Hudson's Bay Company for lynx (predator) and snowshoe hare (prey) over a hundred-year period

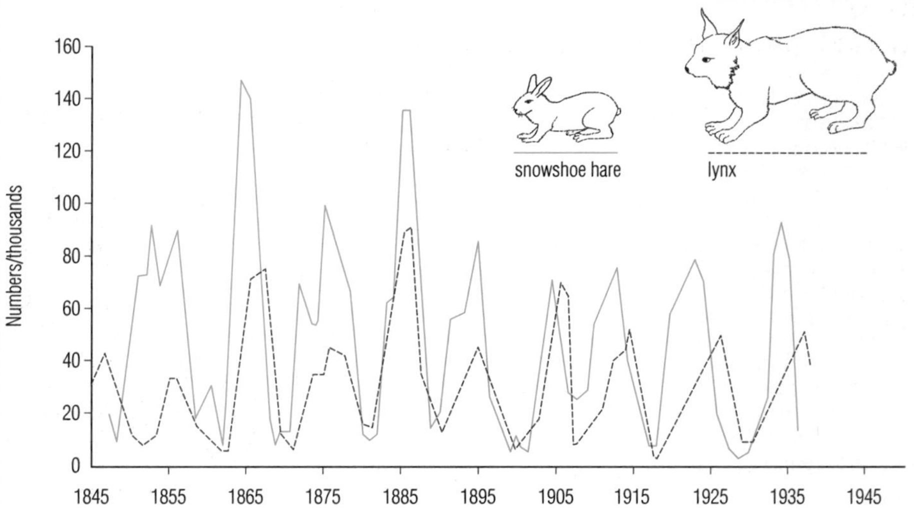

Figure 3.35 Negative feedback control of population

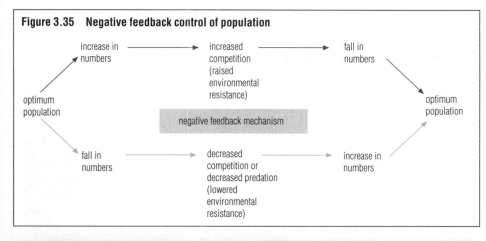

optimum population → increase in numbers → increased competition (raised environmental resistance) → fall in numbers → optimum population

negative feedback mechanism

optimum population → fall in numbers → decreased competition or decreased predation (lowered environmental resistance) → increase in numbers → optimum population

camouflage

defence

mimicry

group defence

Figure 3.36   Defence mechanisms in prey species

Figure 3.37   Competition and niche: the shag (*P. aristotelis*) and the cormorant (*Phalacrocorax canbo*) appear to feed in the same waters and nest on the same cliffs

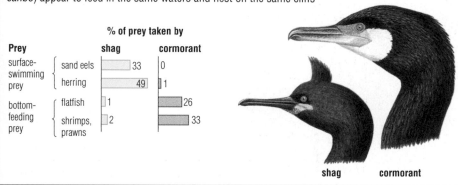

Prey		% of prey taken by	
		shag	cormorant
surface-swimming prey	sand eels	33	0
	herring	49	1
bottom-feeding prey	flatfish	1	26
	shrimps, prawns	2	33

shag   cormorant

comes from the records of pelts (skins) purchased by the Hudson's Bay Company over a hundred-year period for lynx (predator) and snowshoe hare (prey) (Figure 3.34). Thus populations of predator and prey are regulated by a process known as **negative feedback** (Figure 3.35). This is an automatic regulatory process that maintains the numbers of organisms in the food chain balanced at levels that the environment can support.

An evolutionary (long-term) effect of predation is that prey organisms develop structures, chemicals or behaviour that may minimise predation. Some examples of these measures, which include camouflage (disruptive coloration), defence, mimicry and group defence, are illustrated in Figure 3.36. Other examples are speed (flight), behaviour ('freezing'), shields (shells), weapons (spines) and chemicals (nasty taste). The chemical defences of a plant leaf tissue are illustrated in the *Study Guide*, Chapter 3.

Competition and niches

Competition is most intense between species occupying identical or similar niches (p. 44) in the same ecosystem. The results may be

▶ an equilibrium situation in which neither competitor succeeds as well as it might
▶ one competitor declines to the point of extinction (referred to as **competitive exclusion**, and first reported by the Russian biologist Gause in 1934)
▶ separation of niches occurs to avoid direct competition; for example, the fish-eating sea birds cormorant and shag appear to feed in the same water and nest on the same cliffs – the shag feeds on fish of the surface waters, however, whereas the cormorant preys on bottom-dwelling fish (Figure 3.37).

Questions

1 In what ways may competition influence the life and survival of an annual weed seed such as shepherd's purse and a bird such as the robin?

2 By how much time does the population of *Paramecium* lag behind that of *Saccharomyces* in Figure 3.33?

3 In what ways does the behaviour of cormorants and shags lead to avoidance of direct competition? They live in the same habitat, but do they occupy the same niche?

4 What is meant by the term 'negative feedback' in the context of population regulation?

■ Symbiosis

Symbiosis occurs when two or more organisms of different species live in intimate association with each other. Depending on the nature of the association the symbiotic relationship is further defined as parasitism, commensalism or mutualism.

Parasitism

Parasitism is an association in which one organism, the **parasite**, lives on or in another organism, the **host**, for all or much of its life cycle. The parasite depends on the host for food, and the host receives no benefit at all. Parasites usually attack a single host species or a very narrow range of host species. The harm done to the host comes from damage to body tissues, by taking food needed by the host, and by secretion of toxins, which poison the host.

Parasites that live within the host body are called **endoparasites**; examples of endoparasites are the tapeworm *Taenia* sp., parasitic in various mammals including humans (Figure 14.22, p. 299), and the fungal parasite of potato plants, *Phytophthora* sp. (Figure 14.23, p. 300). Other parasites, known as **ectoparasites**, occur on the surface of the host. Examples are the colonies of aphids that live on the surface of green plant stems and feed by tapping the phloem sieve tubes for sap (Figure 16.34, p. 338), and the head lice of humans (*Pediculus humanus*) (Figure 3.38), which live on the body and feed by sucking blood.

Very many organisms are parasitised for at least part of their life cycles. Bacteria are parasitised by bacteriophages (p. 549). Plants are parasitised by bacteria, fungi, viruses, nematodes and insects, and by other plants. Animals are parasitised by bacteria, fungi, viruses, protozoa, platyhelminthes (tapeworms and flukes), nematodes, insects, crustacea and mites (arachnids).

Figure 3.39   A rabbit suffering from myxomatosis

Myxomatosis is caused by the myxoma virus, which is transmitted from diseased to healthy rabbits by the rabbit flea. Initially the virus killed around 99% of the rabbit population. Over several decades the rabbit and virus have co-evolved; rabbits are resistant to the virus, and the virus is less virulent. The behaviour of the rabbit has also changed; rabbits survive above ground rather than in burrows, making them less likely to pick up fleas.

Figure 3.40   Dutch elm disease: many insect species can be contaminated with fungal spores but only bark beetles are common vectors of Dutch elm disease because they move from diseased trees to healthy ones

The bark beetle (*Scolytus*) lays eggs under the bark of a diseased elm tree.

main gallery made by adult beetle

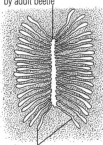

feeding galleries made by hatched larvae

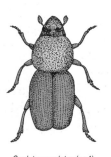

Scolytus scolytus (×4)

When the young adults emerge they carry spores of the pathogenic fungus *Ceratocystis* with them to new host trees.

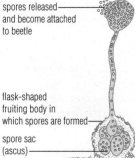

spores released and become attached to beetle

flask-shaped fruiting body in which spores are formed

spore sac (ascus)

hypae

elm wood

Elm trees killed by Dutch elm disease. The fungal hyphae grow in the xylem vessels, which become completely blocked. Dutch elm disease begins with the wilting of leaves in May–July. A virulent strain of the fungus has killed virtually all mature elm trees in the UK

Figure 3.38   *Pediculus humanus*: the body louse of humans

Pediculus humanus exists in two varieties, one occurring on the scalp and the other on the body.
The eggs of the head louse (nits) are laid attached to hair; those of the body louse are laid in clothing.
The eggs hatch into larvae (nymphs), which feed, grow and moult three times before they become adult lice. The louse has mouth parts adapted for piercing the skin and sucking blood.

Figure 3.41 Lichens as an example of mutualism

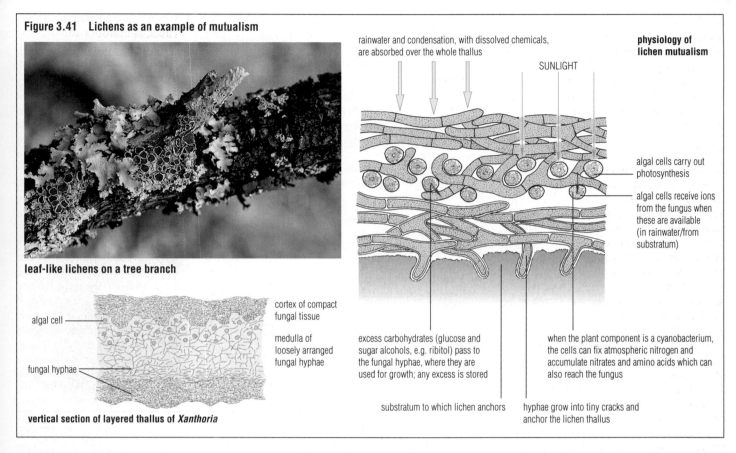

leaf-like lichens on a tree branch

rainwater and condensation, with dissolved chemicals, are absorbed over the whole thallus

SUNLIGHT

physiology of lichen mutualism

algal cells carry out photosynthesis

algal cells receive ions from the fungus when these are available (in rainwater/from substratum)

algal cell

fungal hyphae

cortex of compact fungal tissue

medulla of loosely arranged fungal hyphae

excess carbohydrates (glucose and sugar alcohols, e.g. ribitol) pass to the fungal hyphae, where they are used for growth; any excess is stored

when the plant component is a cyanobacterium, the cells can fix atmospheric nitrogen and accumulate nitrates and amino acids which can also reach the fungus

substratum to which lichen anchors

hyphae grow into tiny cracks and anchor the lichen thallus

vertical section of layered thallus of *Xanthoria*

Adaptations for parasitism

The common adaptations to a parasitic mode of life include

▶ devices to gain entry to the host tissue
▶ devices to keep the parasite attached to the host
▶ mechanisms to resist counter-attacks by the host's immune system
▶ degeneration of unnecessary organ systems, leading to a simplified digestive tract, locomotory organs or nervous system
▶ mechanisms to overcome the problem of transfer to a new host, such as the involvement of a secondary host and the production of vast numbers of potential progeny.

Harm of a parasite

Parasitism can be a limiting factor for the host species, particularly where parasitic attack harms the tissues, or causes early death of the host. Examples of this type of host–parasite relationship include the myxoma virus and the European rabbit (Figure 3.39), and Dutch elm disease (a fungal infection) and the elm (Figure 3.40). In fact, parasitic associations show a complete gradation from these, which normally kill the host, to associations so bland it is hard to detect the harm experienced by the host; an example of a virtually harmless

parasite is the protozoan *Monocystis*, an inhabitant of the seminal vesicles of the earthworm (p. 301).

Commensalism

Commensalism is the relationship where two species live in close association and in which one organism, the commensal (usually the smaller), benefits. The other organism, the host, is neither harmed nor helped. Some species of bacteria in the huge gut flora of the mammalian large intestine (p. 288) are commensals.

Mutualism

Mutualism is a form of symbiosis in which two organisms of different species live in an intimate association that offers some advantage to both. For example, the mycorrhizal association between fungal hyphae and the roots of plants, particularly the roots of trees, is mutualistic. Mycorrhizal relationships are illustrated in Figure 16.30 (p. 336).

The association between the two organisms can be a permanent one leading to the formation of a new species, as in the case of the lichens. Lichens are mutualistic partnerships between fungi and algae (Figure 3.41). Mutualism has enabled two species to make a success of habitats that would otherwise be hard to colonise.

Questions

1 What is meant by the word 'symbiosis'?
2 List four human diseases caused by parasites.
3 What might be the long-term effect of lichens as early colonisers of a rocky surface for other organisms that may follow?
4 What are the likely long-term effects of Dutch elm disease on the ecology of the elm tree?
5 Below are organisms of a food web.

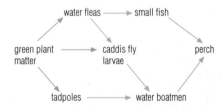

a What is the difference between a food chain and a food web?
b What is the ultimate source of energy of a food web?
c From the food web above give one organism fitting each of the following categories: primary producer; omnivore; primary consumer; carnivore; secondary consumer; tertiary consumer.

3.5 POPULATIONS, COMMUNITIES AND CHANGE

A **population** is defined as all the individuals belonging to a single species occupying a particular local area or space. If we are studying a woodland ecosystem we could consider the population of one of the field layer plants such as wood sorrel, or a population of animals such as blue tits.

A **community** consists of all the populations of that area or space. Thus populations and communities are functional groupings of organisms that interact together. At the fringes of the area or space, communities interact with other communities.

■ Growth of populations

The S-shaped curve of the growth in numbers of a population of microorganisms when inoculated into fresh medium is shown in Figure 19.7 (p. 401). The same steps or phases of growth may occur in the growth of many individual multicellular organisms and, incidentally, the growth of a population when a species, perhaps an 'escape' species, moves into a new geographical area.

For example, the collared dove (*Streptopelia decaocto*) is a recent addition to the British list of breeding birds. At the start of this century this bird was a rare visitor. It spread across northern Europe, and breeding pairs were first seen in Britain in the early 1950s. Nesting pairs were seen in Scotland by the end of that decade. The collared dove is now widespread throughout Britain. Figure 3.42 is a record of the growth of its population here.

■ Factors in population change

In practice, most populations in an ecosystem are already established, but may show fluctuations in numbers over a period of time. The size of a population of a particular species is usually expressed as the number of individuals per unit area (**population density**), or as the number of samples (quadrats of a given size, for instance) in which the species occurs (**frequency**). Density or frequency may vary for any of the following reasons:

▶ The birth rate (natality) rarely remains constant.

▶ Death rates (mortality) are variable too.

▶ Mobile organisms may move away from a population to new habitats (emigration).

▶ New individuals of the species may arrive from elsewhere (immigration).

▶ The effects of other species (biotic factors) in the form of competition, predation and grazing, and parasitism may cause the population to be depleted. These factors will increase with increasing population density, and are therefore referred to as **density-dependent factors**.

▶ Sudden and possibly violent change in one or more of the abiotic factors of the environment, such as light intensity, temperature, water availability, humidity or wind, may bring indiscriminate death and destruction to a community. These factors are unrelated to the density of the population and are therefore known as **density-independent factors**.

■ Environmental resistance and carrying capacity

We have seen from studies of the growth of populations, starting with small numbers of individuals growing in very favourable conditions, that there is a period of **exponential growth**. This applies to all populations whether of a species of microorganism, mammal or bird. In exponential growth a population doubles in each generation; one individual becomes 2, then 4, 8, 16, 32, 64, 128, 256, and so on. This type of increase in a population is quick, but is not maintained, simply because there are insufficient resources for it. As a population increases it begins to **experience environmental resistance**, since space and resources per individual are reduced and competition for them increases. Instead, the population tends to stabilise at a level that ecologists call the **carrying capacity** of the habitat in which the organisms exist. If the numbers start to increase above carrying capacity, then shortage of resources reduces the numbers of offspring produced, and the population regulates itself at the carrying capacity. If the population is reduced, say by heavy predation, then the additional resources available lead to an increase in reproductive rate, and the carrying capacity is again reached. In other words, populations tend to be naturally self-regulating. Note that it is density-dependent factors that play the major part in regulating the population size. This form of negative-feedback regulation is shown in Figure 3.35, p. 65.

■ Estimating population density or frequency

After managing to name a newly discovered species (p. 20), the next most difficult task is to estimate the population size. A total count of all members of a population is called a **census**. It is only very rarely practised in ecology. It is almost always impractical with animals because of movement. With plants it is generally appropriate only where it is easy to distinguish between separate individuals and where population numbers are not too great.

Capture, mark, release and recapture technique

This is a method of estimating population size used with mobile animals of which a random sample can be captured, and marked with a ring, tag or dab of paint. It is satisfactory provided no significant immigration or mortality occurs during the investigation, including

---

Figure 3.42 The growth of the collared dove (*Streptopelia decaocto*) population in the United Kingdom since 1950

This graph is very similar to the S-shaped growth curve shown in Figure 19.7 with
1 lag phase (adaptation to new habitat)
2 log phase (period of exponential growth)
3 linear phase (limits to growth are operating)
4 stationary phase (the optimum population the environment will support)

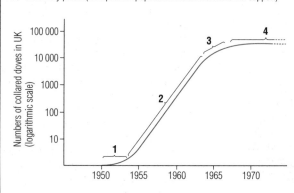

selective mortality of the marked individuals. It is also necessary to be able to trap relatively large samples if the results are to be significant (Figure 3.43).

For example, the technique might be applied to estimate the population of woodlice in a woodland microhabitat. A description of how this could be done is shown in Figure 3.43 (this description is given to illustrate the method, but this sample would be too small to be significant). The argument is summarised in the equation

$$N = \frac{n_1 \times n_2}{n_3}$$

where

N = population

n_1 = numbers captured, marked and released

n_2 = total numbers captured on the second occasion

n_3 = number of marked individuals recaptured.

This estimate of population is known as the **Lincoln Index**.

Estimating a plant population

Quadrats are used when estimating the population of terrestrial plants of the field or herb layer. The **quadrat** consists of a square frame made of metal, plastic or wood (Figure 3.44). Quadrats vary in size; $0.01\,m^2$ may be suitable for moss communities, whereas quadrats of 0.1, 0.25, 0.5 and 1 or $4\,m^2$ are commonly used with larger plants. Quadrats are placed at random (*Study Guide*, Chapter 3), and the different species present each time are identified and their presence recorded on a grid.

Estimating density

Density is the mean number of individuals per unit area. In theory this should be easy to count, but the fact that so many plants reproduce vegetatively means it is often difficult to determine just what an 'individual' is. Where plants occur as separate units (trees, annual plants) density is easily measured. Density is valuable for comparing given species in different habitats.

Estimating percentage frequency

Frequency is the number of quadrats in which a species occurs, as a percentage. Frequency is rapidly assessed, and is a useful means of comparing two similar plant communities (provided the same quadrat size is used in each).

Estimating percentage cover

Cover is the percentage of the ground covered by a species within the area sampled. Cover is a valuable way of considering the relative contribution of the different species in a given community or in different communities.

■ Autecology

Autecology is the study of an individual species. It involves measurement and observations on a sizeable sample of members of a population, and possibly of different populations, if the results are to be significant. Autecological studies might concern the relationship between the population and its environment. The distribution of the species, its growth rates, the population size and its stability may be studied. The size of a population and its stability might underpin conservation work (p. 94). The fluctuations of a population and the effects of parasites and predators might underpin economic control of pests or weeds, when contrasting biological control measures (p. 92) with the use of conventional pesticides.

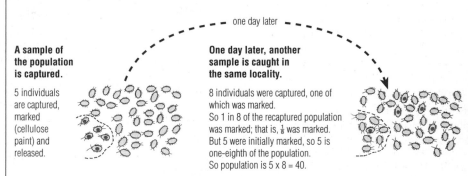

Figure 3.43 Estimating the size of a population of woodlice by capture and recapture: the method assumes that marked individuals are not harmed, and that they distribute themselves randomly in the population

one day later

A sample of the population is captured.

5 individuals are captured, marked (cellulose paint) and released.

One day later, another sample is caught in the same locality.

8 individuals were captured, one of which was marked.
So 1 in 8 of the recaptured population was marked; that is, $\frac{1}{8}$ was marked.
But 5 were initially marked, so 5 is one-eighth of the population.
So population is 5 x 8 = 40.

But note that if 2 marked individuals had been recaptured the population estimate would be only 20. So relatively large samples must be used; a recapture of one or two individuals is no good.

Questions

1 **a** What combination of factors might lead to an increase in a population of blue tits in a wood?
   b What factors might eventually limit the growth of the population?
2 In Figure 3.42, what was the approximate rate of growth of the collared dove population in Britain between 1955 and 1960?
3 In what ways may the Lincoln Index method cause disturbance to a population and thereby give misleading results?
4 Give an example of a density-dependent factor and a density-independent factor that may influence population density. How would you define these terms?

Figure 3.44 Estimating plant populations using a quadrat

■ Community ecology

Primary succession

New land is formed on the Earth's surface from time to time. This may happen in a river delta, in wind-formed sand dunes, and from cooled volcanic lava, for example. There is no soil initially, although much of the mineral skeleton of soil may be present.

The first plants to appear are known as the pioneer populations. Pioneer plants on new substrata may include algae, lichens, mosses and some specialised higher plants. Since lichens can withstand desiccation, extremes of temperature and very low levels of nutrient ions, they are ideal early colonisers.

Largely due to the presence of the first plant life, organic matter is added to the mineral particles. Simple soil forms. Cushions of moss appear, and after them tiny herbaceous plants. The shade these plants cast starts to disadvantage the lichens, and their growth is checked. The level of organic matter and mineral particles builds further, and larger plants are supported as more moisture is retained. Herbaceous plants and shrubs are established. The growth of larger plants and the activities of the fauna they support all help in the build-up of soil. Amongst the larger plants tree seedlings take root and grow. Eventually the land supports mature woodland! This sequence is called a **primary succession**. Primary succession is an example of living things (largely the plants) altering the abiotic environment.

Types of primary succession

Primary successions are named according to the conditions in which they start. Normally, they form from extremely dry conditions or from open, shallow, fresh or brackish waters (Figure 3.45).

A succession beginning on a very dry substratum, such as bare rock, sand dune or cooled volcanic lava, is a **xerosere**. This succession is associated with the build-up of soil and of soil moisture. The pioneer plants are adapted and specialised to withstand water shortage. These types of plant are described as **xerophytes** (p. 375).

A succession that begins with water is a **hydrosere**. The plants at the pioneer stage live in or on open water. They are known as **hydrophytes**. Around the margin of the pond the reed swamp plants flourish. The effect of their growth is to consolidate and dry out the debris at the pond margin, leading to soil formation there. Floating plants colonise more

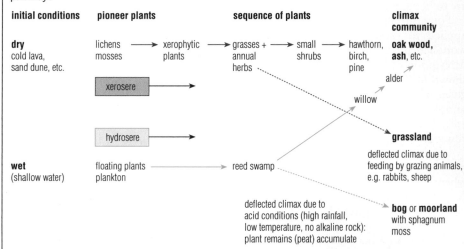

Figure 3.45 Types of primary succession: the idea of a fixed climax community has been challenged; the sequences shown below are not rigid ecological progressions but rather possible pathways

open water, and reed swamp plants extend further out into formerly open water. Progressively the pond is filled by plant life and organic debris. Eventually the pond is reduced to a marshy area with a stream flowing through it, with trees of dry land replacing the alders and willows. The succession results, finally, in the build-up of aerated and drained soil (Figure 3.47).

Seral stages and climax community

The woodland in these cases is the final, stable stage in a series of communities that make up a primary succession, and is therefore referred to as the **climax community**. The stages in a succession are referred to as **seral stages**; the whole succession is a **sere**.

The climax community is normally relatively unchanged for a long period. Climax communities survive as long as climate and other abiotic factors allow, but are vulnerable to many abiotic and biotic factors. In the case of woodland a major threat comes from high wind (Figure 3.25, p. 60), and from the danger of destruction by humans (Figure 3.46) or of attack by a virulent parasite such as the fungus that attacked European elm trees some decades ago (*Ceratocystis*, Figure 3.40, p. 66).

Secondary succession

Successions that occur on soils that have already been formed but which have suddenly lost their community are referred to as **secondary successions**. One example is the series of changes that follows upon fire in woodlands and forests.

Lightning may start a fire. Humans deliberately do so for various reasons. Immediately after fire there is bare ground and scorched earth. The ground layer plants are killed, and the heat destroys shallow roots and seeds buried in the surface soil. Fire has abruptly changed the ecosystem, but only in the short term.

Figure 3.46 Human activities have shaped a landscape once dominated by broad-leaved woodland

Figure 3.47 From pond to woodland: a hydrosere in action

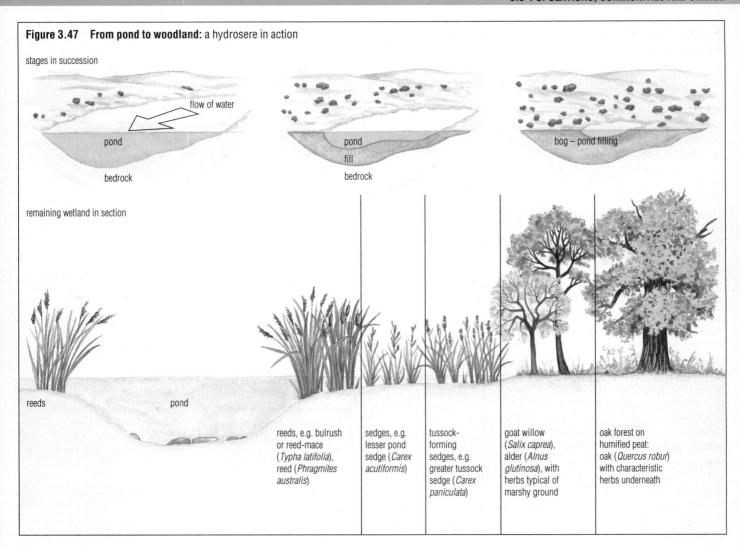

stages in succession

pond
bedrock

pond
fill
bedrock

bog – pond filling

remaining wetland in section

reeds

pond

reeds, e.g. bulrush or reed-mace (*Typha latifolia*), reed (*Phragmites australis*)

sedges, e.g. lesser pond sedge (*Carex acutiformis*)

tussock-forming sedges, e.g. greater tussock sedge (*Carex paniculata*)

goat willow (*Salix caprea*), alder (*Alnus glutinosa*), with herbs typical of marshy ground

oak forest on humified peat: oak (*Quercus robur*) with characteristic herbs underneath

A succession of plants appears on the fire site. Often the first plant on burnt wood ash is the moss *Funaria hygrometrica*, which forms extensive carpets. Within the moss carpet the seeds of herbaceous and woody plants lodge and germinate. A new herbaceous layer grows, followed by shrubs and trees. Each dominant plant community has associated with it various animal populations. Quite quickly the burnt patch is well on the way to mature woodland. The profile transect of mature woodland shown in Figure 3.19 (p. 57) also illustrates an area where a secondary succession is in process.

A similar secondary succession to that on burnt woodland occurs when ploughed agricultural land is left untended for a while. Generally all secondary successions take less time to reach the climax community than a primary succession does. This is largely because in the former the soil is already formed and supports growth of a wider range of plants immediately.

Species diversity and succession

In a stable climax community several different species are typically present, many in quite large numbers. Thus the diversity of species present in a habitat is a possible indicator of the stability of the community.

Species diversity of a plant community may be measured by applying the formula

$$\text{diversity} = \frac{N(N-1)}{\Sigma n(n-1)}$$

where N = the total number of plants, and n = the number of individuals per species. Data is obtained by analysing samples of the plant community, counting all the plants present and the numbers of each different species.

Note that a community early in a seral succession is likely to comprise relatively few species, although typically some are present in large numbers. If ill befalls any of these species, the composition of the whole community is vulnerable in a way that a diverse climax community is not.

■ Studying communities: synecology

Synecology is the study of a group of organisms associated together as a community. Synecology involves analysis of the abiotic and biotic aspects of a community in the environment in which they occur.

In a synecological study the area is mapped, the identities and approximate numbers of the species present are estimated, and the abiotic factors measured.

Questions

1 Why do you think that thistles, dandelions and willow herbs are early colonisers of bare soil?
2 In Figure 3.45, why do you think that sheep and rabbits cause the xerosere to be deflected to grassland?
3 In what ways do the initial conditions for a primary succession differ from those for secondary successions on land?
4 Consider a pond like the one illustrated in Figure 3.47. What abiotic factors may help to maintain it as a pond?

3.6 ENERGY FLOW AND THE CYCLING OF NUTRIENTS

Green plants, the primary producers, supply chemical energy to virtually all other living things, directly to herbivores and indirectly to carnivores (Figure 3.6, p. 47). At each stage in a food chain materials taken in as food are used for a variety of purposes (Figure 3.5, p. 47), but only a tiny proportion is retained over a long term for building up new tissues. The ecological pyramid in Figure 3.48 summarises the path of energy flow and cycling of nutrients in ecosystems in general. It is a descriptive or qualitative statement.

The quantification of feeding relationships within a specific ecosystem involves obtaining numerical data on the accumulation of organic matter by the producers, consumers, detritivores and decomposers of the ecosystem. This is attempted to allow comparisons of different ecosystems, and to analyse an ecosystem's response to abnormal conditions such as pollution. It also permits assessment of the potential of an ecosystem for food production. This field of study owes much to the work of two twentieth-century ecologists: an Englishman, Charles Elton, and an American, Raymond Lindeman.

■ Pyramids of numbers

The quantification of relationships between predator and prey animals by means of a pyramid of numbers was first proposed by Elton. A pyramid of numbers is a bar diagram indicating the relative numbers of organisms in a food chain. Elton's original observations of the abundance of animals noted that

► ecosystems are populated by a very large number of small animals and a progressively smaller number of larger animals
► predators are larger than the prey, and are able to catch or overpower the prey relatively easily
► the prey is never so small that it takes the predator a long time to collect sufficient food
► small creatures reproduce much faster than larger creatures do; there is a surplus of small organisms to 'power' a food chain.

Elton suggested that there is a quite limited range of food sizes for each predator. He analysed animal communities as a pyramid of numbers. Organisms

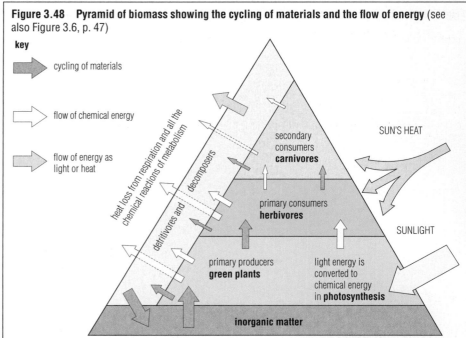

Figure 3.48   Pyramid of biomass showing the cycling of materials and the flow of energy (see also Figure 3.6, p. 47)

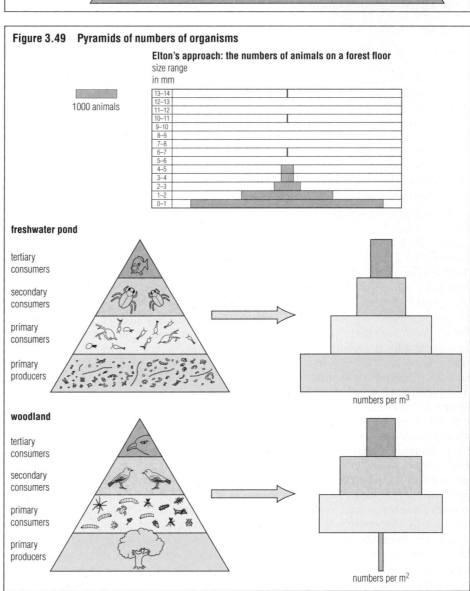

Figure 3.49   Pyramids of numbers of organisms

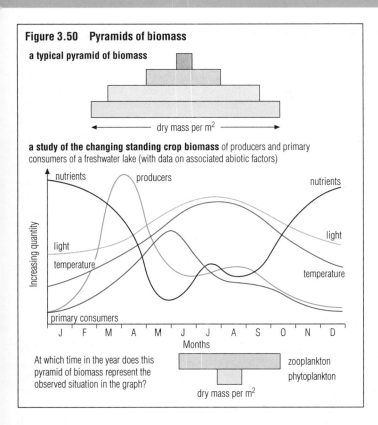

Figure 3.50 Pyramids of biomass

a typical pyramid of biomass

dry mass per m²

a study of the changing standing crop biomass of producers and primary consumers of a freshwater lake (with data on associated abiotic factors)

At which time in the year does this pyramid of biomass represent the observed situation in the graph?

zooplankton
phytoplankton

dry mass per m²

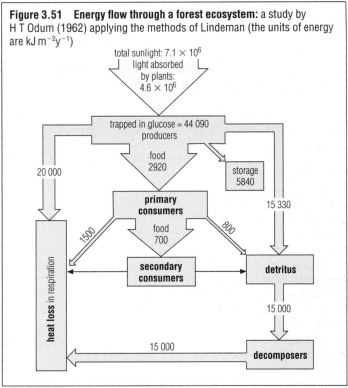

Figure 3.51   Energy flow through a forest ecosystem: a study by H T Odum (1962) applying the methods of Lindeman (the units of energy are kJ m^{-2}y^{-1})

of an ecosystem were counted or estimated, and the number in each group represented by a rectangle with an area proportional to the number of organisms. The primary producers outnumber the primary consumers, which in turn outnumber the secondary consumers. The result is a pyramid-shaped bar chart (Figure 3.49).

The data needed to construct pyramids of this type may be relatively easy to collect. The problem with a pyramid of numbers is that each organism counts as one, irrespective of its size. An oak tree counts the same as a single phytoplankton cell! This creates strange shapes, such as the woodland chain shown in Figure 3.49. The problem can be overcome by using pyramids of biomass.

■ **Pyramids of biomass**

An alternative approach is to estimate the dry mass (biomass) instead, because this is more likely to produce the pyramid-shaped chart expected, and to permit comparisons between ecosystems (Figure 3.50). The fieldwork involves estimating the population size for each member of the food chain, weighing representative samples (wet mass) and determining the dry mass by destructive analysis to find the percentage of water.

The dry matter estimated is the biomass at one particular moment only. It is called the **standing crop**. A problem arises with

organisms with short life cycles; marine phytoplankton, for example, may be outlived by their predators in the zooplankton. Then, for a time, the standing crop of phytoplankton may be smaller than that of its predators. This anomaly arises because no account is being taken of the rate at which the biomass is produced, but only the quantity present at one particular moment. It can be avoided by producing a pyramid of energy.

■ **Pyramids of energy**

Lindeman proposed that organisms of the ecosystem should each be identified as belonging to a particular trophic level. Then the energy content of the organisms at each trophic level should be estimated. The resulting energy diagram (Figure 3.51) represents the energy available at each trophic level. This is a more laborious approach, since all the samples previously dried and weighed (to find dry mass) have also to be combusted to estimate their energy content.

There is also an important difficulty with this approach in that each kind of organism has to be assigned to a single trophic level. But, for many organisms, the trophic level varies with each item of food consumed. For example, in the woodland food web (Figure 3.32, p. 64), and in the marine and littoral zone food web (Figure 3.17, p. 56) the organisms at the top of the food chains (such as badgers and birds of prey in the woodland, and sea birds of the shore) are seen to feed at a range of trophic levels (from level 2 to level 5, in fact). It is also hard to fit the consumers of dead organic matter (detritivores and decomposers) into one trophic level. Yet 80% of all plant matter is consumed as dead material by the detritivores and decomposers.

As a result, for ecologists working experimentally, the quantification of feeding relationships has many unresolved difficulties.

Questions

1 a Suggest some examples of predators that are not larger than their prey.

   b Suggest some examples of predators that feed on exceedingly small prey.

2 According to Figure 3.51, what percentage of the light energy absorbed by plants is converted to chemical energy in glucose?

3 What trophic levels do humans occupy? Give examples.

4 What nutrients were likely to have been measured in the freshwater lake of Figure 3.50? What may be causing the level of these nutrients to vary in the way shown?

73

3.7 ECOLOGICAL ENERGETICS

Biological productivity is the rate at which biomass is produced by an ecosystem. It has two components: **primary productivity**, the production of new organic matter by green plants (autotrophs), and **secondary productivity**, the production of new organic matter by consumers (heterotrophs). Both of these can be divided into **gross productivity** (the total amount of organic matter produced) and **net productivity** (the organic matter of the organism after respiration and metabolism have been fuelled).

The methods of measuring productivity are summarised in the panel.

■ Primary productivity of natural ecosystems

Reliable estimates of productivity of natural ecosystems are very difficult to make. An attempt made by an ecologist is shown in Figure 3.52.

The primary productivity of a natural ecosystem is dependent upon the intensity of solar radiation, temperature, and water and mineral supplies. A large part of the Earth's surface is of very low primary productivity, mostly because of the lack of water and nutrients to support plant growth (deserts, for example). Productivity of the oceans is even lower than that of the land because water reflects or absorbs much of the light energy before it can reach the plants, and because the nutrient concentrations are low. Moreover, nutrients released by decay are frequently not available in the surface waters where plants grow.

Primary productivity in agricultural crops

Agriculture is based upon the primary productivity of various crops, mostly cultivated grasses. Crop yields are improved by irrigation, removal of pests and the application of fertilisers. Improvements to biological productivity in 'industrialised' agriculture make it about ten times more productive than subsistence agriculture. Yet this is sustained at the expense of fossil fuels used to build and drive farm machinery, produce pesticides and fertilisers, pump water and breed high-yielding varieties. There is a huge energy subsidy to industrial agriculture that may be more apparent when fossil fuel becomes scarce.

MEASURING PRODUCTIVITY

Relative growth rate (R) is the gain in mass of plant tissue per unit time.

$$R = \frac{\text{increase in dry mass in unit time}}{\text{original dry mass of the plant}}$$

Net assimilation rate (NAR) relates increase in dry mass to leaf area.

$$NAR = \frac{\text{increase in dry mass in unit time}}{\text{leaf area}}$$

Leaf area index is a measure of the total leaf area of a given plant per unit of ground area.

Biomass is the total dry mass of all organisms in an ecosystem.

The **harvestable dry mass** is the dry mass of the crop useful for commercial purposes.

■ Secondary productivity

Secondary productivity is the accumulation of organic matter in the consumers in the food chain. It is illustrated for the bullock in Figure 3.53. More than half the food eaten passes straight through. Most of the food assimilated is used in respiration; only about 11% is incorporated into new tissue.

The amount of energy passed on at each step in the food chain is about 10%, so the amount of energy available at each step decreases exponentially. The implication for us is that where food is short many more vegetarians can feed than those who must have meat.

■ The cycling of nutrients

The decomposition of organic matter by bacteria and fungi liberates nutrients into the non-living (geochemical) environment. In soil and in water, the nutrients released tend to be taken up and re-used by green plants, the primary producers, and built up into new and repaired cells and tissues. When plants are eaten by animals, nutrients pass on into secondary and tertiary consumers, but sooner or later pass back into the non-living environment, on death and decay of the consumers.

So, all the chemical elements of living things circulate between living things and the environment. This exchange or 'flux' (back and forth movement) of materials has a biological component and a geochemical component, and the exchange

---

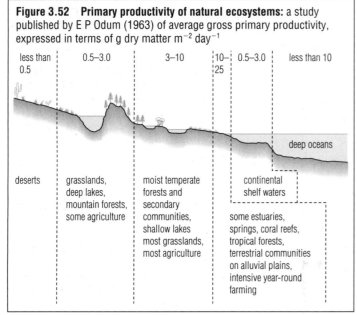

Figure 3.52 Primary productivity of natural ecosystems: a study published by E P Odum (1963) of average gross primary productivity, expressed in terms of g dry matter m^{-2} day^{-1}

| less than 0.5 | 0.5–3.0 | 3–10 | 10–25 | 0.5–3.0 | less than 10 |

deep oceans

| deserts | grasslands, deep lakes, mountain forests, some agriculture | moist temperate forests and secondary communities, shallow lakes most grasslands, most agriculture | continental shelf waters | some estuaries, springs, coral reefs, tropical forests, terrestrial communities on alluvial plains, intensive year-round farming |

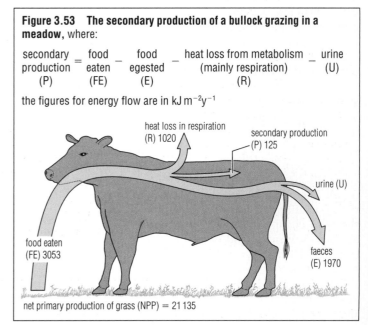

Figure 3.53 The secondary production of a bullock grazing in a meadow, where:

$$\begin{array}{ccccc} \text{secondary} & = & \text{food} & - & \text{food} & - & \text{heat loss from metabolism} & - & \text{urine} \\ \text{production} & & \text{eaten} & & \text{egested} & & \text{(mainly respiration)} & & \text{(U)} \\ \text{(P)} & & \text{(FE)} & & \text{(E)} & & \text{(R)} & & \end{array}$$

the figures for energy flow are in kJ m^{-2}y^{-1}

heat loss in respiration (R) 1020

secondary production (P) 125

urine (U)

food eaten (FE) 3053

faeces (E) 1970

net primary production of grass (NPP) = 21 135

Figure 3.54 The water cycle

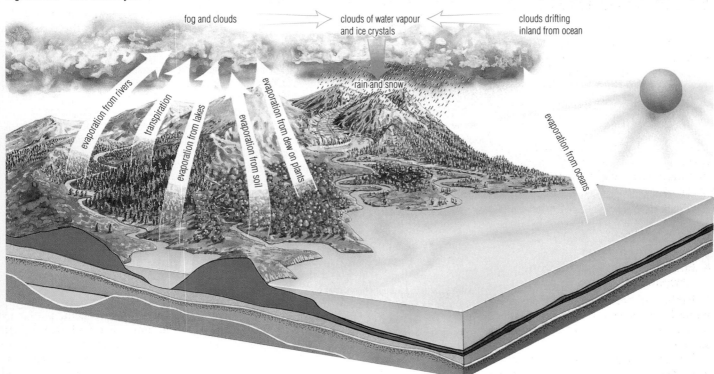

takes a more-or-less circular path. The cycling of nutrients is thus referred to as **biogeochemical cycling**.

The nutrients involved are the macronutrients (p. 247) of carbohydrates, fats, proteins, nucleic acids and energy-transfer molecules such as ATP (p. 147). These include the elements carbon, hydrogen, oxygen and nitrogen, which together make up about 95% of the dry mass of living things. Other important elements are calcium, potassium, phosphorus and sulphur. Many other nutrients are required, but only in tiny quantities; these are the micronutrients (Table 16.1, p. 328).

Types of biogeochemical cycle

Each biogeochemical cycle has two 'pools' of nutrients. The **reservoir pool** is large, non-biological and slow-moving (an example is the gaseous nitrogen of the atmosphere in the nitrogen cycle). The **exchange pool** is much smaller, more active, and sited at the point of exchange between the living and non-living parts of the cycle (in the nitrogen cycle, the nitrates and ammonium salts are dissolved in soil moisture around plant roots).

In **global cycles** some of the nutrients pass through a major gaseous phase, and the atmosphere is a major reservoir. Other nutrients pass through a **sedentary cycle**, which does not involve the atmosphere; the nutrient passes between soil and plants and other organisms only.

An example of a global cycle is the **water cycle**, illustrated in Figure 3.54. Other examples are the carbon cycle (Figure 12.2, p. 247), and the nitrogen cycle (Figure 16.28, p. 334). Examples of sedentary cycles are the phosphorus cycle (Figure 16.24, p. 332) and the sulphur cycle (Figure 16.23, p. 332).

The water cycle

In the water cycle (Figure 3.54), the water reservoir is the liquid water deep in the oceans and the ice of the polar ice caps. The exchange water is the water and water vapour in and around all living things.

The water cycle is driven by solar energy. The Sun causes evaporation of water from organisms and from the surfaces of soil, rivers and oceans. The movements of moisture-laden air masses are caused by atmospheric temperature gradients due to differential heating of the Earth's surface. As warm, moist air masses cool, the water vapour turns to water and falls as rain or snow.

Human influences on biogeochemical cycles

In many cases human activities have speeded up or actually disrupted the balance of biogeochemical cycles. For example, the winning of fossil fuels from mines and wells and their burning to carbon dioxide and water vapour has started to change the composition of the atmosphere, which may result in significant global warming from the greenhouse effect (p. 84).

The demand for phosphate for fertilisers has caused phosphate-rich rock to be mined, leading to localised pollution in the spoil heaps of the mines. Subsequent excessive use of phosphates has led to pollution of natural waters. The excess of phosphate (and other nutrients such as nitrates, as well) has led in turn to eutrophication of fresh waters, and the reduced quality of drinking waters generally (p. 100).

Questions

1 Why is secondary productivity always lower than primary productivity?
2 What is the source and ultimate fate of most energy on the Earth?
3 Which part of the water cycle involves organisms? Which is largely independent of organisms?
4 What is the fundamental difference between the flow of energy and the flow of nutrients in the biosphere?

FURTHER READING/WEB SITES

P Chenn (1999) *Advanced Biology Readers: Ecology*. John Murray

4 Humans and the environment

- Humans are relatively recent arrivals in Earth's history, but today their influence on the environment is very great. An explosive growth in the human population, and in the scale of humans' activities, has had dramatic consequences for the environment and for living things.
- Earth's climate has changed profoundly in past times, with warmer-than-average times interspersed between ice ages. Currently we are towards the end of a warm spell, in an interglacial period. Climate change may also be influenced by human activities.
- Human achievements that can help maintain and enhance the environment include our ability to improve food production capacity, our communications and transport facilities, and our ability to overcome infectious and degenerative diseases.

- Much human environmental impact is harmful, however, including the degradation of soil and atmosphere that accompanies deforestation, and the effects on the environment of excessive use of pesticides in maintaining agricultural production.
- The quality of Earth's atmosphere and environment is threatened by pollution. Harmful effects include an enhanced greenhouse effect due to increasing levels of CO_2 and other atmospheric pollutants, acid rain, and the destruction of the high-level ozone layer.
- Attempts to engineer a sustainable level in the use of resources form an important component of conservation. Conservation involves application of the principles of ecology to manage present and future environmental change.

4.1 INTRODUCTION

In the long history of the Earth, humans are recent arrivals. Consequently their effects on the Earth's environment are a relatively new feature of the planet's history. Nevertheless, human influences are very great indeed.

The Earth originated as part of the Solar System some 4600 million years ago (mya), but evidence of life in the form of the oldest fossils dates from only about 3200 million years ago. These vast periods of time are hard to visualise. One way to get an impression of them is to represent the history of life on the face of a twenty-four hour clock (Figure 4.1). On this scale the first human appeared at 23.59, and recorded human history began at a quarter of a second before midnight!

■ Earth's climate has changed during the past

There is evidence that the climate and environment of the Earth were not always constant and unchanging.

Warmer than average times

The most recent period when the climate was much warmer than it is now was the middle of the Pliocene epoch (p. 649), about 3 million years ago. Then forests grew as far north as the arctic coast of Greenland. The ice caps covering the poles were thinner, and sea levels were about 35 m higher than they are today. Study of the conditions on Earth during warmer periods provides clues to our future environment if, for example, global warming is an outcome of the 'greenhouse effect' (p. 84).

Ice ages

There is good evidence that through much of geological time ice caps were absent from the poles of the Earth. The relatively recent continental glaciations ('ice ages') represent exceptional conditions in which world-wide temperatures were lower than normal. The first ice age commenced around 2300 million years ago, and a regular rhythm of changes has developed. Each ice age, roughly 100 000 years long, is succeeded by a slightly warmer interval, called an interglacial, lasting 10 000–20 000 years. Today we are living near the end of a natural warm spell.

During ice ages enormous quantities of water are locked up in ice that extends outwards from the poles, and sea levels fall by up to 60 m.

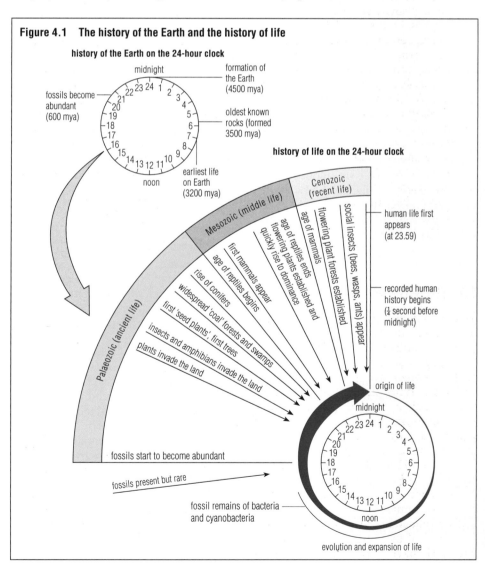

Figure 4.1   The history of the Earth and the history of life

history of the Earth on the 24-hour clock

midnight

formation of the Earth (4500 mya)

fossils become abundant (600 mya)

oldest known rocks (formed 3500 mya)

noon

earliest life on Earth (3200 mya)

history of life on the 24-hour clock

Cenozoic (recent life)

Mesozoic (middle life)

Palaeozoic (ancient life)

social insects (bees, wasps, ants) appear

flowering plant forests established

age of mammals

age of reptiles ends flowering plants established and quickly rise to dominance

first mammals appear

age of reptiles begins

rise of conifers

widespread 'coal' forests and swamps

first 'seed plants', first trees

insects and amphibians invade the land

plants invade the land

human life first appears (at 23.59)

recorded human history begins (¼ second before midnight)

origin of life

midnight

noon

fossils start to become abundant

fossils present but rare

fossil remains of bacteria and cyanobacteria

evolution and expansion of life

How do we know about climate so long ago?

Climates in the past have been reconstructed from clues in both the fossil and rock records. For example

▶ prolonged periods of warm climate can support a more varied animal and plant life than colder periods do; consequently the fossil record in rocks formed at the time reflects prevailing conditions

▶ measurements of the proportions of the stable isotopes of oxygen (^{16}O and ^{18}O, p. 261) in marine sediments allow estimates of sea temperature to be made to within 2 °C (the ratio of ^{18}O to ^{16}O in water molecules – and in the shelled animals growing in the sediments – changes with temperature, because $H_2^{18}O$ accumulates in seas as they warm)

▶ sediments in the rocks show whether they were deposited in desert conditions, under ice or in tropical environments

▶ changes in the sea level for prolonged periods leave their marks in the form of raised shorelines, which can be observed by geomorphologists in the field and recorded by satellite photography of ocean coastlines

▶ because of continental drift (plate tectonics, p. 652), continents converge and separate at rates of centimetres a year; the positions of continents have a strong impact on climate because land masses affect the circulation of air and ocean currents.

4.2 THE HUMAN POPULATION

Modern humans, *Homo sapiens*, arose at some time during the last ice age, between 1 million and 40 000 years ago. Early populations of *Homo sapiens* and related human species were probably scavengers at first, later living by hunting wild animals and collecting wild plants. This food-gathering economy cannot support a large population. Humans were not an abundant species during the Stone Age. Throughout the Pleistocene period (p. 649) it is likely that the total world population did not exceed a quarter of a million people.

Environmental conditions improved at the end of the Pleistocene period, and this may have favoured the gradual

METHODS OF FOOD PRODUCTION, AND THE LAND THEY SUPPORT

Method of food production	Area of land required per person
Hunting of wild animals, and collecting plant products; almost invariably this form of feeding involves a nomadic life style.	30 km²
Domestication of a wild herd of large herbivorous mammals (e.g. reindeer, deer) involves a nomadic life, following the herd between winter and summer feeding grounds, for example.	0.5 km²
Slash-and-burn cultivation involves the cutting down and burning of the natural vegetation to support the cultivation of crops until the soil fertility is exhausted, followed by eventual movement to new territory where slash-and-burn is repeated.	0.125 km²
Settled agriculture with some form of crop rotation; settled agriculture stimulates population growth.	0.008 km²
Modern agriculture using high-yield varieties, bred to be resistant to local parasites, together with use of additional fertilisers, pesticides and irrigation.	(Continues to reduce the area of land needed to support one person. Most of these developments are directly or indirectly dependent on industrialisation, the development of transport systems and the lavish use of fossil fuels.)

change in human life style from food-gathering to food-producing. This transition, known as the **Neolithic Revolution**, took place between six and ten thousand years ago. With the change to settled life style came the need for more people to work the land. At first the agriculture would have been at subsistence levels. Steadily the techniques would have improved and resulted in increased productivity. The various forms of human life style, from nomadic hunter to modern agriculture, are shown in the panel above.

■ Human population growth to modern times

The slow and fairly steady increase in the human population that began with the Neolithic Revolution continued over the next tens of thousands of years. World population had risen to some 100 million people by the time of the Old Kingdom of Egypt, around seven thousand years ago, and on to some 500 million by about 1650. In the two centuries between 1650 and 1850 the population probably doubled to 1 billion. Eighty years later the population had again doubled and then stood at 2 billion. Within forty-five years more (1975) the world population had doubled again, to 4 billion. It is projected to be at 8 billion by 2010 (Figure 4.2).

Figure 4.2 Estimates of the human population of the world

Year (AD)	Population (billions)	Number of years to double
1	0.25(?)	1650(?)
1650	0.50	200
1850	1.1	80
1930	2.0	45
1975	4.0	35
2010	8.0	?
2025	8.5	?

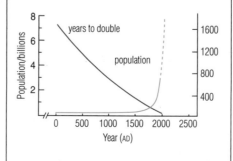

Questions

1 Look at the circular arrow on the twenty-four hour clock in Figure 4.1. Why do you think it gets wider and wider?

2 What do you think were the first forms of life on Earth? How could they have obtained their food?

3 Suggest reasons why a hunting–gathering community needs a relatively large area of land to support it.

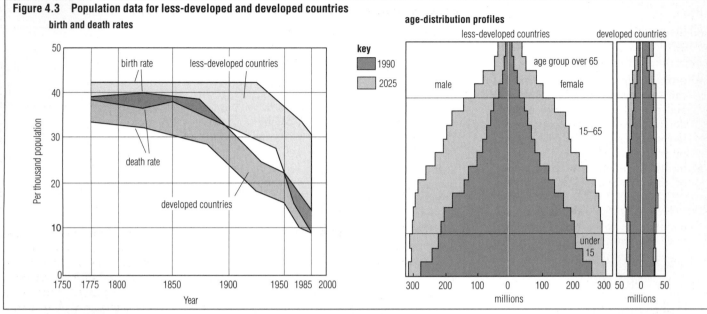

Figure 4.3   Population data for less-developed and developed countries

How long can this population explosion be sustained?

Virtually all species of plants and animals have a reproductive potential which, if unchecked, would overpopulate the Earth within a few generations. Sooner or later, however, this growth is checked by one or more of various factors: disease, the limitation of food supply, competition *within* species (intraspecific competition) and *between* them (interspecific competition) and, in the case of humans, the deliberate decisions of individuals.

Industrialisation in the developed regions of the world during the nineteenth century was eventually accompanied by a fall in the death rates, and later by a fall in the birth rates. In the less-developed countries a dramatic fall in death rates has begun recently, and is yet to be accompanied by a similarly dramatic decline in the birth rate (Figure 4.3).

A result of the regional differences in birth rates is that the age distributions of the populations of developed and less-developed regions of the world are completely different. In 1990 in the less-developed countries, 37% of the population was younger than 15 years. This means that even if the growth rate of these populations continues to decline, the numbers of young people reaching childbearing age in the population of the developing countries will continue to rise steadily, so that the absolute population numbers of young people will go on soaring for many years yet.

The pattern is quite different for the developed countries now. But, of course, these countries were themselves 'less-developed' about 150 years ago. The age-distribution pattern of their populations then was similar to that of the less-developed countries now.

Discussion point

Why did industrialisation lead to a reduction in population growth? Perhaps increased material security, shared throughout the population, encouraged people to rely less on children as a form of wealth and support? Children cease to be a 'labour force' and, instead, are seen as a responsibility.

4.3 HUMAN IMPACT ON THE ENVIRONMENT

Homo sapiens, as an animal, is influenced by the same regulatory forces that determine population size in other organisms, including those due to other living things (biotic factors) and those due to the physical environment (abiotic factors). But humans are also in a unique position with respect to other animal species, largely as a result of the dramatic expansion of their intellectual powers during evolution.

Humans' unique intelligence has led, amongst other things, to the ability to enquire and investigate, to remember and reason, to design and invent, and to produce plans in the abstract. As a result, humans have fashioned increasingly sophisticated tools, and have passed on their knowledge, skills and attitudes from generation to generation. Taken together, these developments have given humans immense powers over the world around them.

Through these achievements and because of recent population growth, a degree of control of the environment has been developed and has led to the modification of natural ecosystems. We live in a world that is more or less dominated by our technology. As a result we are now presented with the consequences of accelerating environmental changes. Not all changes are harmful and disadvantageous to the environment and to our future, however. Here we shall look at the potential of the scientific and technological revolution for the eventual mastery over environmental problems.

Transport, communication and remote sensing

Air travel permits almost any part of the world to be reached in a matter of hours. Electronic telecommunications systems allow countries to be in touch instantaneously. We live in an interdependent world, a **global village**. We can know of the ideas that are talked about on the other side of the Earth, and of events that occur anywhere on its surface. Remote sensing by satellite supports daily weather forecasting, and provides advance information of conditions that might lead to famine. Satellite images revealed the massive algal blooms of the seas of the Northern Hemisphere (Figure 4.14, p. 85), and may be used to obtain evidence of any shrinking of the polar ice caps as a result of the greenhouse effect (p. 84). Satellite images are used to pinpoint possible sites for new deep wells to provide clean and dependable supplies of water. In fact, satellite images speed up many aspects of exploration.

It is a feature of modern scholarship that the results of enquiries are published internationally and attract the widest possible response. Data and predictions concerning environmental and health issues are circulated in this way. Opinions are shared by international pressure groups such as Greenpeace, and proposals are made for international action by agencies such as the World Health Organization (WHO), the Food and Agriculture Organization (FAO) and the United Nations Children's Fund (UNICEF) supported by the United Nations from New York. There is every opportunity for a coordinated response to global problems.

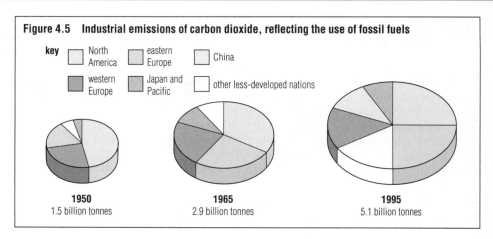

Figure 4.5   Industrial emissions of carbon dioxide, reflecting the use of fossil fuels

key: North America, western Europe, eastern Europe, Japan and Pacific, China, other less-developed nations

1950 1.5 billion tonnes   **1965** 2.9 billion tonnes   **1995** 5.1 billion tonnes

■ Food production capacity

We have a prodigiously productive agriculture; world food production is increasing faster than the population (Figure 4.4). Hunger is more often a problem of war or political instability than of widespread failure of production. The proportion of undernourished people in the world has fallen since 1974, but there are still many countries where there is a low income and a food deficit. (Half of the food deficit countries are in Africa, a quarter in Asia.)

On the demand side, the world population is expected to be 8.5 billion by 2025 and 10 billion by 2050. Much of this rise will occur in countries already carrying a food deficit and low incomes. In addition, many less-developed countries discriminate against their farmers, maintaining low prices for farm-produced commodities. Instead, they invest in urban infrastructure, so 'fuelling' the drift to the towns. At the moment it is often cheaper for these countries to import food than to grow it themselves. New policies must help the farmers to adapt, so as to improve production for the future.

■ Alternative sources of energy

Fuels such as coal, oil and gas are called **fossil fuels** because of their origins. They are also described as non-renewable, because the reserves are finite and are not being regenerated. Nuclear power also comes from a non-renewable fuel, uranium (although this is not a fossil fuel). Because of the harmful effects of carbon dioxide, oxides of nitrogen and other pollutant emissions produced when fossil fuels burn, these fuels have a significant and harmful environmental impact. Nuclear power raises the problematic issues of the disposal of radioactive waste and the devastation that might result from a nuclear accident, however unlikely this may seem to be.

The world is using energy at an ever-increasing rate (Figure 4.5). The countries of the developed world consume the greater share of fuel although they represent only about 25% of the total population, but energy use is growing very rapidly in less-developed countries too. As a result, sources of non-renewable energy are becoming exhausted.

So-called **renewable energy** comes from the exploitation of wave power, wind power, tidal power, solar energy, hydroelectric power and biological sources including biomass (p. 118). Many of these forms of renewable energy have a low environmental impact, and may be relatively 'environment-friendly'. Research into alternative sources of energy has often been given a low level of priority by governments and by the existing fuel industries. The development of renewable forms of energy is certain, however, to be an increasingly important feature of the power industry in future.

■ Medical advances and health

We have knowledge of our own health and can investigate and overcome diseases and defects of our bodies. Over the centuries, views of the causes of human disease have changed radically. We now know which diseases are caused by bacteria, viruses or (rarely) fungi. We have learned to view mental ill-health as a form of disease. We have the means to control most contagious diseases, to prevent their spread and to cure people who have been infected. The importance of diet, life style and a supply of clean water in the prevention of disease is appreciated. Most importantly, we understand the dangers of human population growth and have the contraception technology to overcome this problem (p. 600).

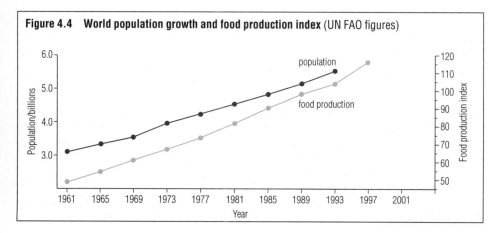

Figure 4.4   World population growth and food production index (UN FAO figures)

Questions

1 Use the graph of birth and death rates in Figure 4.3 to decide what was the period in which the population of the less-developed countries was growing most rapidly. Explain how you have reached this conclusion.

2 From the age-distribution profiles in Figure 4.3, work out approximately how many people in the world were under the age of 25 in 1990.

3 If world food production has increased faster than the world's population has grown, why do malnutrition and starvation still occur?

4.4 HUMAN ACTIVITIES THAT HARM THE ENVIRONMENT

■ Deforestation

Woodland and forest are the dominant terrestrial biomes; more inhabitable dry land supports trees as the climax community than supports other plant life. Figure 3.2 (p. 45) illustrates just how extensive are the forest zones, whether of northern coniferous forest, temperate deciduous woodland, Mediterranean scrub forest or temperate or tropical rain forests.

The problem is that the trees of forest and woodlands are being cut down faster than they can be replanted or regenerate naturally. This has happened in the past, and continues extensively today. Deforestation is an environmental catastrophe because trees have a favourable and stabilising influence on the composition of the atmosphere, on the climate, and on the soil.

The reasons for the destruction of natural woodland and forest are various. They include

▶ the release of valuable timber for sale as a realisable asset to meet the costs of imports, including oil
▶ the freeing of land for alternative uses, including stock rearing, arable cropping or the planting of fast-growing species such as conifers, eucalyptus or rubber trees
▶ clearing land for roads, housing, industrial estates or smallholdings
▶ supplying firewood as a fuel
▶ destruction of trees as a consequence of atmospheric pollution ('acid rain', p. 86).

Deforestation is a problem not only because of its effects on the composition of the atmosphere (p. 247), with possible unfavourable consequences for the climate, and on soil erosion (p. 83); there are additional dangers too. Removal of vast areas of forest is destroying the habitats of numerous species of animals and plants, which are now therefore in decline, and even in many cases in danger of extinction in a relatively short period, thereby threatening greatly to reduce genetic diversity (p. 645).

Deforestation is currently a major cause for concern but it is not a new problem nor is it a threat to the existence of tropical rain forests alone. It is a far more extensive phenomenon.

Deciduous woodland of Britain

Over most of the British Isles, on land at or below about 900 m, the natural climax community was broad-leaved deciduous woodland. These forests became established following the last great ice age, and have subsequently disappeared. We know this to be the case from examining the pollen record in peat (p. 650), for example.

The natural forests of the British Isles were almost totally destroyed long ago. This process of destruction was started around 3000 BC by the Neolithic farmers who migrated here from Europe. Later communities continued and extended forest clearing, claiming more and more space for arable crops and grazing land (grassland and heath). Humans used wood for domestic cooking, making charcoal for the smelting of iron ore and the building of homes, farms and ships. By the sixteenth century the energy crisis was so great that coal was developed as an alternative source of energy.

In recent times the pressure on existing excellent arable land for transport (trunk roads and motorways), housing (new towns and overspill estates) and green-field site industrial development has resulted in more and more upland and inaccessible land, much of which was tree-clad, being cleared for arable use. Also, in 'good times' for agriculture the financial rewards have encouraged hill farmers to extend arable farming to higher altitudes. Lowland heathlands are similarly being 'reclaimed'. In many parts of the countryside the extent of hedges between fields has been reduced to increase the available arable land, and to make the field units accessible to larger agricultural machines such as giant combine harvesters. Hedges are still being lost.

Britons have developed the use of the land in much the same way as many less-developed nations are developing theirs today. Nothing remains of our original forests except for some fragmented woodlands, largely confined to steeply sloping gorges and to boulder-strewn and rocky ground, all of little agricultural potential. Less than 5% of the total land surface of Britain is covered by deciduous woodland now.

Figure 4.6   Northern boreal forest zones and the solar energy received: the common trees in these forests are spruces (*Picea* spp.), pines (*Pinus* spp.), firs (*Abies* spp.), hemlock (*Tsuga* spp.) and larch (*Larix* spp.)

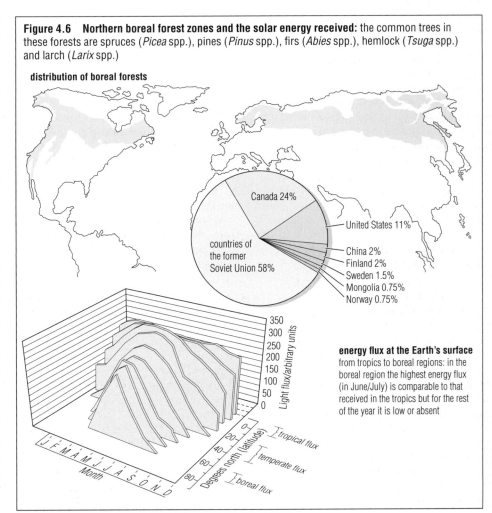

distribution of boreal forests

countries of the former Soviet Union 58%
Canada 24%
United States 11%
China 2%
Finland 2%
Sweden 1.5%
Mongolia 0.75%
Norway 0.75%

energy flux at the Earth's surface from tropics to boreal regions: in the boreal region the highest energy flux (in June/July) is comparable to that received in the tropics but for the rest of the year it is low or absent

Figure 4.7   A plantation of Norway spruce (*Picea abies*): in Scotland such plantations have replaced mixed forests of pine and deciduous trees

Boreal forests

The boreal region around the North Pole supports an immense band of forest, a mixture of coniferous trees with some broad-leaved species such as birch, aspen and rowan. These trees are well adapted to the relatively hostile environment; for example, they are able to grow and thrive even though the input of solar energy to the boreal region is very limited throughout most of the year (Figure 4.6).

Boreal forests are becoming increasingly important to the economy of the countries in which they occur. Progressively more of the boreal forest is being felled, often huge areas at a time. Much wildlife is disturbed; most does not return because the land is also drained, and replanting, where it occurs, is of dense plantations of single species of particularly fast-growing trees that can be felled again within a few decades (Figure 4.7).

Trees as a source of fuel

In the less-developed countries cooking takes up a major part of the energy consumed. The main energy source available to most urban and rural poor people in these countries is firewood (Figure 4.8); agricultural waste and cow dung are also used. In a country such as India (the tenth largest industrial power), for example, these forms of biomass energy provide about 85% of all cooking energy. This is not being used by a largely rural population, as over 55% of the domestic energy in India is consumed in cities.

The enforced reliance of poor farmers and city fringe dwellers upon firewood contributes, in the longer term, to deforestation.

Figure 4.8   Biomass as a source of energy: pie charts indicate the relative amounts of energy used

Global energy use

world

developed countries

less-developed countries

key

biomass

fossil fuels

wood as a source of fuel: in many less-developed countries women and children (often the major labour force) may spend the largest part of each day walking to collect fuel, because the land nearby has been stripped of trees

The tropical rain forests

The extent of the tropical rain forest biome can be seen in Figure 3.2 (p. 45). Tropical rain forests cover about 7% of the Earth's surface.

Today, the rain forests are being cleared at an alarming rate (Figure 4.9). These forests are seen as a valuable resource for timber, wood pulp and fuel (for which the forests are being cleared), and for fodder, fruits, game, chemicals, dyes, drugs and oils (for which forests are preserved). Government policies often encourage short-term exploitation. What is happening to the rain forests is best seen from the air; monitoring of the destruction of the forests is by satellite, aeroplane and helicopter surveys.

The accessible lowland rain forests are destroyed most quickly. Between 16 and 20 million hectares are lost each year as trees are cut for valuable hardwood

timber and to create space for agricultural developments, typically for beef ranching or plantation crops such as rubber or oil palm. The largest increases in lost forest recently have been in Brazil, India, Indonesia, Burma, Vietnam, Thailand, the Philippines and Costa Rica. It is estimated, for example, that at current rates the Indonesian forests will be effectively logged completely within 40 years.

Deforestation of tropical rain forest is leading to ecological disaster for four major reasons.

1 Rain forests are **species-diverse zones** where many still-unknown organisms exist. In fact the tropical rain forests are home to more than 50% of the world's species. When rain forest is cut down very many more species are in danger than when other habitats are threatened. Countless animal and plant species are being driven to extinction.

Figure 4.9   The destruction of a rain forest

Sustainable forestry?

In the countries concerned, forestry and associated industries provide employment for millions of poor people. Their problems of poverty, population growth and landlessness cannot be ignored if a solution is to be found. Huge world export markets demand raw materials, too.

To overcome the appalling destruction of the rain forests, alternative, sustainable forestry projects to supply world markets and provide employment are needed. Sustainable forestry allows a maximum of (say) 10 large trees to be removed per hectare. Undamaged, medium-sized trees must remain, to grow to full size in the next 30 years, before further controlled logging may occur. Can this be implemented without destruction to the remaining trees and to the forest soil, given the massive logging machinery that is used (Figure 4.9)?

2 Rain forest soils are poor in nutrients, and are unable to retain, for any length of time, those nutrients that are released into the soil. In the rain forest virtually all the nutrients are in the organic matter of the forest canopy. When released by decay of dead matter nutrients are quickly reabsorbed into the plants of the living forest and are recycled.

If the forest is felled and the vegetation dried and burnt, the soil does not maintain its fertility for long. The ash from the burned forest provides a short-lived supply of nutrients. After the first crop has been harvested the soil does not even sustain subsistence farming. Soil erosion follows (see opposite).

3 Deforestation has profound **effects upon regional climate**, mainly by reducing rainfall, and thus accelerating desertification. When the forest trees are cut down the water cycle (Figure 3.54, p. 75) is disrupted. The reduction in transpiration results in fewer clouds and less rainfall in the vicinity. Surrounding forests are threatened by desiccation. As the land becomes hotter and drier more of the soil is eroded.

4 Carbon dioxide is an essential raw material for photosynthesis by green plants (p. 246). Atmospheric carbon dioxide levels are on the increase, and this is contributing to the 'greenhouse effect' (p. 84). Deforestation is reducing the effectiveness of a most significant 'sink' for atmospheric carbon dioxide, second only in importance to the marine phytoplankton.

Figure 4.10   Deforestation of the watershed causes enhanced erosion and lowland flooding

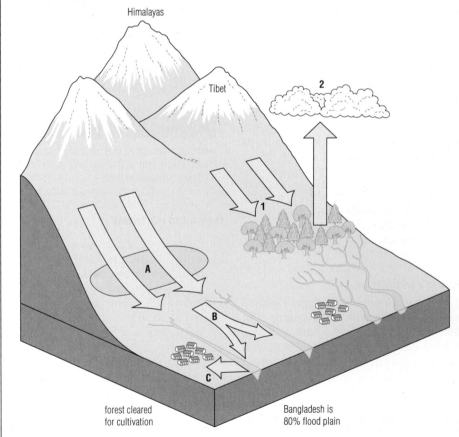

forest cleared
for cultivation

Bangladesh is
80% flood plain

deforested
A   flood water runs across deforested area without being absorbed
B   topsoil from foothills is eroded and carried into rivers
C   rivers become silted and burst their banks; sanitation facilities become crippled, disease spreads, killing crops and causing famine

forested
1   flood water from melting snow caps runs into forested area and is absorbed into soil and evaporates into the atmosphere
2   clouds carry moisture to arid regions

Flooding as a consequence of deforestation

The slopes of mountains and hillsides are water catchment zones. Water drains down these slopes and forms streams and rivers. It is usual for plant communities to cover the slopes and to stabilise the soils there. On the lower slopes the trees, shrubs, epiphytes (p. 565), plants of the field and ground layer, and the layer of leaf-litter at the soil surface all function as a sponge. In heavy rain most of the water is soaked up, and is only gradually released into the soil.

In the Himalayas, the rain forest communities on the slopes of the foothills regulate the supply of water to the major rivers of the Indian subcontinent, the Indus, Ganges and Brahmaputra. These rivers serve vast areas of northern India, together with much of Pakistan, Burma and Bangladesh.

Recent years have seen an accelerating rate of damage caused by monsoon rains sweeping exposed soil to the flood plains below. Soil blocks up the rivers and reservoirs. Flooding is more frequent and lasts for longer, over huge areas. Homes are destroyed; good agricultural land is washed away. Sometimes thousands of people are drowned. Many experts argue that the cause is the deforestation of the slopes of the Himalayas, as suggested in Figure 4.10. Other experts believe the activities of the hill farming communities, which grow crops on carefully managed terraces, have decreased the loss of soil from slopes that were once tree-clad.

■ Soil erosion and desertification

Soils erode naturally, but erosion is a selective process. The fine particles of organic matter, clays and silts, rich in nutrients, go first. This leaves the coarse sandy particles. Good soil is thus turned to desert, a process that has become known as **desertification**.

Rain washes soil down slopes and wind blows dry soils away. So soil

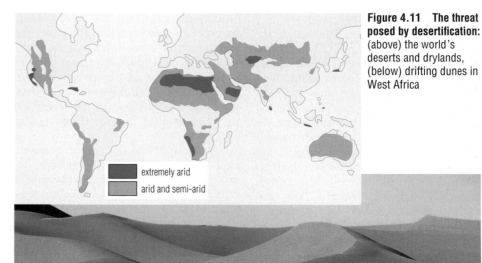

Figure 4.11 The threat posed by desertification: (above) the world's deserts and drylands, (below) drifting dunes in West Africa

extremely arid

arid and semi-arid

erosion is favoured by very heavy rains and by dry and windy conditions. Changing the way land is used also alters the speed of erosion (Table 4.1). Growing appropriate plants with strong roots is an effective method of trapping soil and retaining a firm soil surface. Abrupt removal of the plant cover exposes soil to erosion. Growing plants that are not adapted to climate and soil conditions facilitates erosion too.

The threat of desertification is greatest in the farmed margins of existing dry land and deserts. Some irrigation schemes in regions like these have proved inappropriate, and have accelerated soil erosion and desertification. In the course of irrigation the water delivers a dilute solution of salts to the soil. When the water evaporates from the soil the salts are left behind. Frequent repetition of this process leads to a build-up of salts (**salination**) in the soil to a level that kills the crop plants.

Studies funded by the United Nations Environment Programme, launched in the wake of the 1977 drought that devastated the Sahel region of west and central Africa, have found evidence that desertification threatens about 35% of the land surface of the Earth (Figure 4.11). The areas most threatened are in Africa, the Indian subcontinent and South America. Desertification is thus a global problem.

Questions

1 Why is a coniferous plantation less environmentally desirable than the same area of mixed woodland?

2 What considerations might persuade a land-owner to plant deciduous woodland?

3 Suggest ways in which large tracts of forest might influence the climate.

4 When fossil fuels become scarce which sources of energy (Figure 4.8) might then be used in
   a developed countries
   b less-developed countries?

5 Explain why the destruction of tropical forest could affect the concentration of carbon dioxide in the atmosphere.

6 Make lists of the principal causes of deforestation and of its harmful effects.

7 Explain why felling a large area of forest may lead to flooding hundreds of miles away.

8 For what possible reasons might peoples of rain forest zones be slow to accept outside advice on the best treatment of the rain forests in the immediate future?

Table 4.1 Soil erosion and the effects of ground cover: a study of erosion of loam soil from a 10% slope

Type of ground cover	Soil removed annually per 1000 m²/kg	Number of years needed to erode 18 cm topsoil at this rate
Virgin forest	5	500 000
Grass	775	3 225
Rotation	35 800	70
Cotton	79 000	32
Bare ground	166 000	15

■ The 'greenhouse effect'

Radiant energy from the Sun warms the surfaces of sea and land. Most of the incoming solar radiation is in the visible part of the spectrum. Most visible light passes through the atmosphere without being absorbed. As the earth warms up, it radiates heat (infra-red radiation; see Figure 12.9, p. 250) back out towards space. Much of this radiation does not escape into space, however; most is trapped by certain gases in the atmosphere, in particular water vapour and carbon dioxide. Water vapour absorbs radiation in the wavelength range of 4–7 µm and carbon dioxide in the range 13–19 µm. As a result, the gases of the lower atmosphere are warmed up. Some of this heat energy escapes into space, but much is radiated back to the Earth's surface, warming the planet further (Figure 4.12). This basic 'greenhouse effect' plays a key part in maintaining favourable temperatures for life on Earth.

The concentration of atmospheric carbon dioxide

Accurate measurements of the amount of carbon dioxide in the atmosphere began in 1957, as part of the International Geophysical Year initiative. The results (Figure 4.13) revealed

▶ an annual rhythm in the atmospheric carbon dioxide concentration, low in the summer months, higher in the winter months; this is correlated with the seasonal changes in the amount of photosynthetic fixation of carbon dioxide by plants of the Northern Hemisphere
▶ a rising trend in the atmospheric carbon dioxide concentration; in 1957 it was at 315 ppm (parts per million) by volume, but by 1988 it had reached 350 ppm.

Determination of the atmospheric carbon dioxide concentration in the recent past has come from analyses of air locked away in old instruments such as precision telescopes of known dates of manufacture. From these we know that, about 150 years ago, the concentration of atmospheric carbon dioxide was 270 ppm.

Ice cores drilled out of the polar ice caps have extended our knowledge. By analysis of the air extracted from the bubbles trapped in the polar ice it is possible to determine atmospheric carbon dioxide at the times the air was trapped. These analyses show that before the Industrial Revolution the carbon dioxide concentration averaged between 250 and 290 ppm.

Other 'greenhouse gases'

Other gases also absorb infra-red radiation and contribute to the greenhouse effect. These substances include ozone (p. 87), CFCs (p. 89) and dinitrogen oxide (nitrous oxide, p. 86). Methane, too, is a greenhouse gas of importance.

Increasing concentration of greenhouse gases

Carbon dioxide

The main cause for the rise in concentration of carbon dioxide in the atmosphere is the combustion of fossil fuels, largely coal and oil. For every tonne of carbon burnt, about 4 tonnes of gaseous carbon dioxide are produced. It has been estimated that in the hundred-year period beginning in 1850, when the consumption of fossil fuel in the Industrial Revolution was starting to build up, about 60 thousand million tonnes (60 gigatonnes) were burnt. Today, the same quantity of fossil fuels, now about half of it as oil, is burnt every twelve years.

The atmospheric carbon dioxide does not increase by the amount of carbon dioxide released. This is because roughly half the carbon dioxide released is absorbed by various natural 'sinks'. These include the dissolving of carbon dioxide in the water of oceans, rivers and lakes to form carbonic acid. Most of the carbon dioxide removed from the air each year, however, is taken up by green plants – both terrestrial and marine – in the process of photosynthesis (p. 246). The fixation of carbon dioxide by terrestrial plants such as those of the tropical rain forest is very important; the decay of deposited organic matter on and in the soil releases carbon dioxide into the air, however.

Much of the carbon dioxide taken up by aquatic plants is fixed by the photosynthesis of marine phytoplankton floating just below the surface of the seas. Recent satellite image studies of the oceans using special colour recording equipment have disclosed algal blooms in coastal regions of temperate and high-latitude waters in springtime (Figure 4.14). Winter storms in these regions bring nutrient-rich water to the surface. These nutrients support vigorous algal growth. Later almost all of the organic matter produced in the blooms sinks to the bottom and becomes incorporated into the ocean bed. This algal growth is a most important sink for the excess carbon dioxide of the atmosphere.

Methane

Methane is very much more efficient than carbon dioxide at trapping infra-red radiation and may come to be an important greenhouse gas for the future. The current methane concentration of the atmosphere is low, but it is rising steadily.

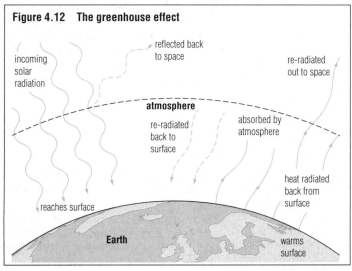

Figure 4.12 The greenhouse effect

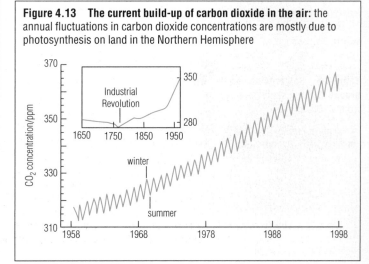

Figure 4.13 The current build-up of carbon dioxide in the air: the annual fluctuations in carbon dioxide concentrations are mostly due to photosynthesis on land in the Northern Hemisphere

About 20% of the methane in our atmosphere has come from long-term deposits such as coal seams, natural gas wells and pipelines, and from oil deposits currently being tapped. The remainder is produced by the action of certain bacteria, which excrete methane as a waste product. These live under anaerobic conditions; many occur in the mud of bogs, marshes and rice paddy fields, others in the guts of ruminants such as cattle and insects such as termites.

An enhancement of the greenhouse effect leads to global warming?

The possible imminent rise in the temperature of the Earth's atmosphere is referred to as **global warming**. Despite numerous predictions, including those produced by exceedingly elaborate computer programs, no one can yet say for certain how the Earth's average temperature and weather patterns will be affected now and in the future.

Some scientists predict that the greenhouse gases will be at such a high concentration by the years 2030–2050 that the average temperature could rise by 2–5 °C. They predict this will cause the water in the oceans to expand and some of the polar ice to melt, raising sea levels all over the world. Sea levels rose by about 15 cm during the twentieth century. Most of this rise was due to the thermal expansion of water. A global warming of about 2 °C over the next 40 years would increase sea levels by a further 30 cm or so. Changes in prevailing winds and in the distribution of rainfall may result in the effects of warming being greatest nearer the poles, and cause lower rainfall in the centre of continental land masses.

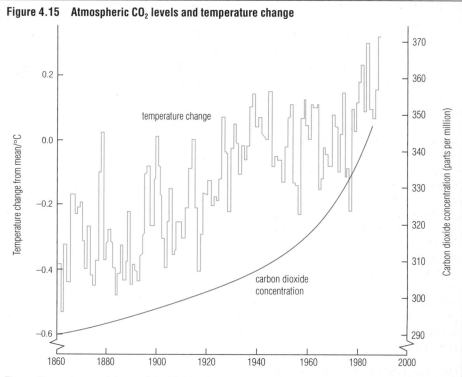

Figure 4.15 Atmospheric CO$_2$ levels and temperature change

The recent temperature record shows a slight global warming, as traced by workers at the Climatic Research Unit of the University of East Anglia. Whether the accompanying build-up of carbon dioxide in the atmosphere caused the half-degree warming is hotly debated.

At the time of writing the evidence for global warming is slight. There has been a small rise in mean temperature, and this correlates with the changes in carbon dioxide levels (Figure 4.15). However, the scientists from the University of East Anglia Climate Research Unit who produced this data hotly debate among themselves its significance. The connection is not proved by the data alone. Other scientists agree about this uncertainty. The temperature changes so far are probably indistinguishable from natural climatic fluctuations, in fact. There is no evidence of consistent polar ice melt from satellite images as yet. On the other hand, even the slight rate of rise, perhaps 0.5 °C over the past 100 years, if maintained, could produce quite radical changes in global climate.

International action on global warming

International efforts are being made to halt the rise and then reduce the levels of greenhouse gases in the atmosphere. For example, the Climate Change Convention (governments and non-governmental organisations working together) held first in 1992 required rich nations to halt their rising CO$_2$ emissions. International action is sought to decrease reliance on fossil fuels, introduce energy-saving measures, and to have more trees planted to act as 'sinks'. Real progress is slow.

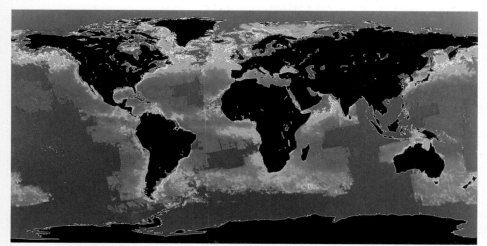

Figure 4.14 The discovery of extensive seasonal algal blooms by satellite images: the red-orange areas are those where there is most plankton, and the purple areas those with least; the extent of the spring bloom is evident

Questions

1 Make a list of the so-called 'greenhouse gases'. State, as far as possible, where each of these comes from.

2 Before the discovery of coal and oil, the main fuel was wood. Was the burning of this wood likely to raise the atmospheric carbon dioxide level? Explain your answer.

■ Acid rain

The pollutants involved in acid rain are sulphur dioxide (SO_2), oxides of nitrogen (NO_2 and NO – known as NO_X), ammonia (NH_3) and ozone (O_3), and various volatile organic vapours given off from petrol stations, oil refineries and chemical works, and by many industrial and household solvents. The effects of all these substances have become a major environmental issue. But the term 'acid rain' is not new. It originated in Manchester about a century ago, when an environmental inspector referred to the discharge from industrial chimneys as being not only black with soot but so acid as to be dissolving iron and stone work. Nothing more was heard of the term until very recently (Figure 4.16).

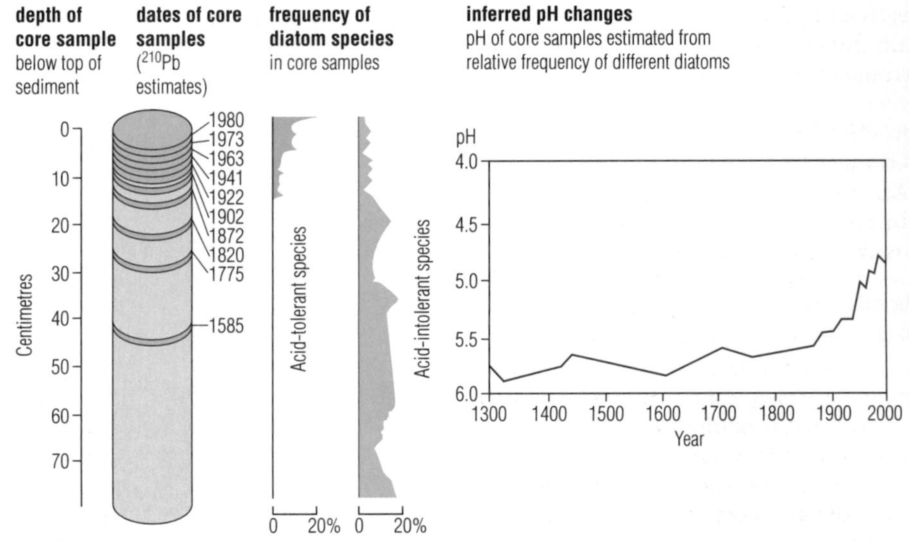

How acid is rain?

Even in unpolluted air, rain is naturally slightly acidic because it is a dilute solution of carbonic acid, formed as water passes through the air, dissolving some of the carbon dioxide present. Clean rainwater has a pH of 5.6.

Rain polluted with acidic gases typically has a pH 4.0–4.5. In the last great London smog in 1952 the highest acidity detected was pH 1.6.

Most atmospheric pollutants come from fossil fuels

Virtually all our common atmospheric pollutants come from the burning of fossil fuels for power, energy, manufacturing and transport. Sulphur dioxide is fairly typical in this: from about 50% to

Figure 4.17   Evidence of a pH change in Scottish lakes since the Industrial Revolution: the frequency of acid-tolerant and acid-intolerant species of diatoms in the sediments of Scottish lakes

85% of the sulphur dioxide discharged into the atmosphere has come from these sources (there is a close parallel between measurements of the acidity of rain and the quantities of sulphur dioxide in the emissions from power stations). The remainder comes from rotting vegetation, from natural sources such as volcanoes and from plankton – certain marine phytoplankton, part of the spring 'blooms' in the seas of the Northern Hemisphere, discharge dimethyl sulphide into the air.

Sulphur dioxide (SO_2) dissolves in water forming sulphurous acid (H_2SO_3).

Under most conditions, and in the presence of a suitable catalyst, sulphur dioxide is oxidised to sulphur trioxide (SO_3) and dissolves to form sulphuric acid (H_2SO_4). Catalysts of this oxidation occur in industrial atmospheres in tiny but effective quantities. They include hydrogen peroxide, ozone, ammonia and particles of iron and manganese.

The sulphur dioxide from the large centres of industry in western Europe is carried by the prevailing wind for varying distances before it becomes deposited. Currently, successful attempts are being made to reduce the sulphur dioxide

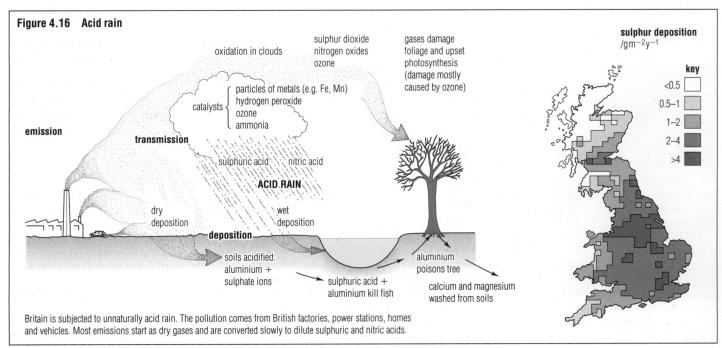

Figure 4.16   Acid rain

Britain is subjected to unnaturally acid rain. The pollution comes from British factories, power stations, homes and vehicles. Most emissions start as dry gases and are converted slowly to dilute sulphuric and nitric acids.

emissions when fossil fuels are burnt. In the period 1972–82 there was a significant reduction in the emissions of sulphur dioxide in Britain. In the same period rain became less acid, paralleling the changes in emissions of acidic gas.

How 'acid rain' may be harmful

Acidification of soil

Changes in soil pH change the relative solubility of the ions present (Figure 3.29, p. 62). Persistent acid rain causes the pH of soil to fall. In soil at pH 6.5 there is maximum availability of minerals to plants, including trace elements, but at pH 5.5–6.0 potassium, calcium, magnesium and trace elements start to be lost by leaching. Also, at low pH phosphate becomes bound to clay particles and is not available to plants. At and below pH 4.5 most valuable nutrients are rapidly leached away, and aluminium ions may appear in solution at poisonous concentrations. It also seems that low pH interferes with the trees' root growth and makes them more susceptible to drought stress.

Acidification of lakes

In recent years the acidity of many of the lakes of industrialised countries has risen sharply, to a level at which aluminium ions become stable in solution. Aluminium ions have reached poisonous concentrations, and as a result many fish have ceased to live in lakes and rivers.

Was acidification a slow, natural process as lakes steadily acquired ions from surrounding acidic soils, or has there been a precipitous change in acidity since the Industrial Revolution created acidic rain?

Ecological experiments have resolved this question, for it has been demonstrated that, as the pH of a lake changes, the populations of diatoms (a kind of phytoplankton) change too. The species that grow in neutral waters are quite different from those adapted to acidic conditions. The walls of diatoms are impregnated with silica and persist in pond mud, rather than decaying quickly as cellulose walls might. Thus it is possible to infer the pH of lakes in the past by analysing past populations of diatoms in lake sediments.

Studies of Scottish lakes that are now highly acidic show that the rise in acidity is quite recent: the pH of lakes has fallen since 1850 (Figure 4.17).

Low-level ozone and damage to plants

Certain of the adverse effects of atmospheric pollution on plant and animal life

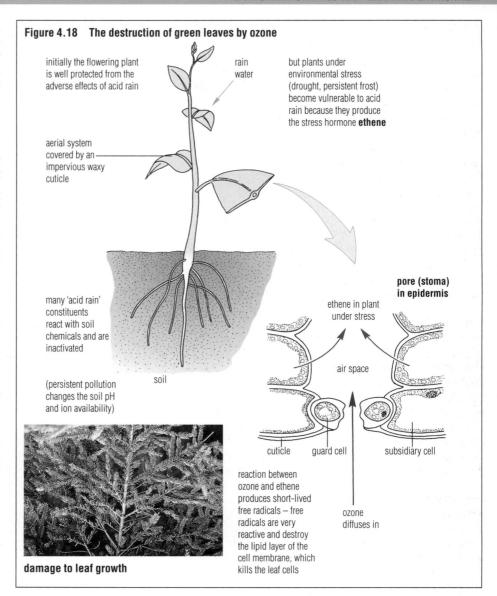

Figure 4.18 The destruction of green leaves by ozone

initially the flowering plant is well protected from the adverse effects of acid rain

rain water

but plants under environmental stress (drought, persistent frost) become vulnerable to acid rain because they produce the stress hormone **ethene**

aerial system covered by an impervious waxy cuticle

many 'acid rain' constituents react with soil chemicals and are inactivated

(persistent pollution changes the soil pH and ion availability)

soil

damage to leaf growth

pore (stoma) in epidermis

ethene in plant under stress

air space

cuticle   guard cell   subsidiary cell

reaction between ozone and ethene produces short-lived free radicals – free radicals are very reactive and destroy the lipid layer of the cell membrane, which kills the leaf cells

ozone diffuses in

attributed to acid rain are due to low levels of ozone and other non-acidic pollutants.

Ozone (trioxygen, O_3) is a highly reactive gas. It is a valuable component of the upper atmosphere, but it is also formed in the lower atmosphere. Here it is a pollutant, contributing to both acid rain and photochemical smog. Ozone readily decomposes to molecular oxygen (O_2) at 20 °C, which makes it a powerful oxidising agent. It reacts with many substances important to biological processes, often forming by-products that are more harmful than ozone itself.

Low-level ozone is formed by chemical reactions of oxides of nitrogen, sulphur dioxide, carbon monoxide and hydrocarbon vapours emitted as a result of incomplete combustion of fossil fuels; these reactions only take place in sunlight. Consequently, the concentration of low-level ozone is lower at night and

higher during the day. At higher altitudes (where conifer trees most commonly grow in the northern hemisphere) fluctuations in ozone concentrations are much smaller.

The way ozone harms trees, discovered by a team at the University of Lancaster's Institute of Environmental and Biological Sciences, is summarised in Figure 4.18.

Questions

1 What are the acid components of 'acid rain'? What are the natural sources of these substances? What are their industrial sources?

2 What effects does an increase in acidity have on the mobility and availability of ions in the soil?

3 What do car emissions contribute towards the production of low-level ozone?

4 Look at Figure 4.17. How did the diatom community of Scottish lakes change after 1900?

■ The ozone layer

In a band of the upper atmosphere, known as the stratosphere, there is a protective layer of ozone gas, which is present at a concentration of only a few parts per million (Figure 4.19). This layer, 15–40 km above the Earth's surface, absorbs ultra-violet (u.v.) radiation. Ultra-violet light is a component of the electromagnetic radiation from the Sun reaching the Earth's surface (Figure 12.9, p. 250). When sunlight passes through the ozone layer much of the u.v. light is absorbed. As a consequence, living things are well protected from radiation that otherwise would destroy them – by changing the chemistry of their nucleic acids, for example.

The environmental problem is that the ozone layer is now being damaged by atmospheric pollution.

The maintenance of the ozone layer

The ozone in the stratosphere is in dynamic equilibrium with molecular oxygen. This equilibrium is maintained by a balance of the processes that form and remove it.

Ozone is produced in the stratosphere by the action of u.v. radiation on molecular oxygen, which is split into two highly reactive atoms of oxygen. Each oxygen atom may react with a molecule of oxygen to form a molecule of ozone:

$$O_2 \xrightarrow{\text{light energy}} O + O$$
$$O + O_2 \longrightarrow O_3$$

At the top of the stratosphere there is much incoming radiation but little oxygen. At the surface of the Earth there is much oxygen but little radiation to act on it. Consequently the highest concentration of ozone occurs between these extremes, at about 23 km.

Once formed, each molecule of ozone may be broken down by u.v. radiation to an atom of oxygen and a molecule of oxygen. The reactive atom of oxygen is then most likely to react with a molecule of oxygen to re-form ozone. This ceaseless cycle of changes leaves the composition of the stratosphere unchanged. Most of the incoming u.v. radiation is absorbed and the atmosphere is warmed. Most of the heat energy generated is radiated out into space.

Pollution destroys the ozone layer

A molecule of ozone is destroyed when it reacts with an atom of oxygen:

$$O_3 + O \longrightarrow O_2 + O_2$$

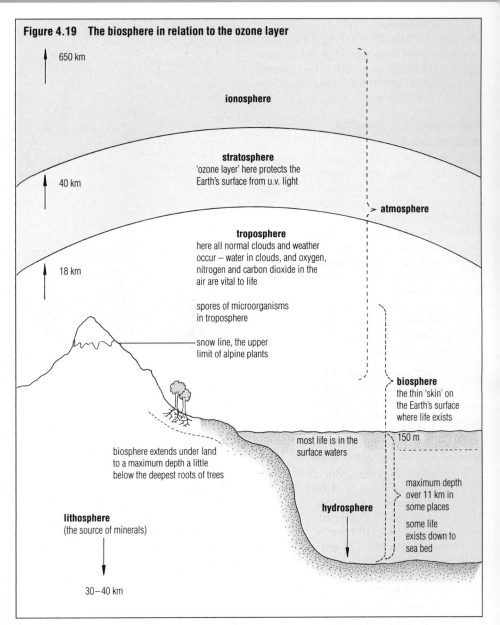

Figure 4.19   The biosphere in relation to the ozone layer

650 km

ionosphere

40 km

stratosphere
'ozone layer' here protects the Earth's surface from u.v. light

> **atmosphere**

18 km

troposphere
here all normal clouds and weather occur – water in clouds, and oxygen, nitrogen and carbon dioxide in the air are vital to life

spores of microorganisms in troposphere

snow line, the upper limit of alpine plants

biosphere
the thin 'skin' on the Earth's surface where life exists

150 m

most life is in the surface waters

biosphere extends under land to a maximum depth a little below the deepest roots of trees

maximum depth over 11 km in some places

some life exists down to sea bed

hydrosphere

lithosphere
(the source of minerals)

30–40 km

However, ozone loss is speeded up enormously when triggered by a series of highly reactive particles known as radicals.

A radical (R·) is a short-lived, intermediate product of a reaction. A radical may exist only for a split second. Radicals are formed when a covalent bond (p. 129) between two atoms breaks, with one of the pair of bonding electrons going to each of the two atoms. This type of reaction usually occurs only in gases, and is activated by light or by other radicals.

Radicals speed up the destruction of ozone because they constantly re-emerge to trigger another reaction:

$$R\cdot + O_3 \longrightarrow RO + O_2$$
$$RO + O \longrightarrow R\cdot + O_2$$

The radicals that destroy the ozone layer are NO· and Cl·. They are formed by the action of u.v. radiation in the stratosphere on specific pollutants that percolate up from the Earth's surface. Air cycling between poles and Equator carries pollutants into the troposphere, but they are taken on to the stratosphere by upward air currents generated by occasional violent tropical storms. It may take as long as five years for surface-generated pollutants to reach the ozone layer. This delay in upward movement means that, even if production of pollutants was stopped now, the full impact of their disappearance on the ozone layer would not be noticed for some decades.

One pollutant, chloromethane, is produced from rotting vegetation by decomposer microorganisms. It is also a product of the fires of tropical 'slash-and-burn' agriculture. This source of chlorine radicals is now dwarfed by those from synthetic compounds such as chlorofluorocarbons (CFCs). The other source of radicals, growing in significance, is the oxides of nitrogen.

Oxides of nitrogen

The relatively inert oxide of nitrogen, dinitrogen oxide (nitrous oxide, N_2O), is produced by denitrifying bacteria (p. 335) acting on artificial fertilisers applied to poorly aerated soils. Dinitrogen oxide is also formed during fossil fuel combustion.

In the stratosphere, in the presence of u.v. radiation, dinitrogen oxide is no longer inert. It reacts with atomic oxygen to form two nitrogen monoxide (nitric oxide, NO·) radicals. These then react with ozone molecules, converting them to molecular oxygen, and re-forming nitrogen monoxide radicals.

Chlorofluorocarbons (CFCs)

Chlorofluorocarbons are organic chemicals manufactured for use in refrigeration systems and aerosol sprays, and for making the bubbles in foamed plastic. The special quality of CFCs is that, in our atmosphere, they are inert. That is, they do not react with other substances and are therefore harmless. But they escape into the stratosphere, and here the u.v. radiation breaks them down, releasing chlorine radicals. Chlorine radicals (Cl· – free chlorine atoms) react with ozone, forming molecular oxygen and more chlorine radicals.

Evidence for the destruction of the ozone layer

In 1974 researchers first proposed that CFCs were a potential source of chlorine radicals that would deplete the ozone layer. Evidence for the depletion was first reported in 1985 from the British Antarctic Survey. Using a special spectrophotometer they measured the total amount of ozone in the column of air above the instrument. For several years NASA satellites had signalled a dramatic drop in the total ozone over Antarctica during the spring (Figure 4.20), but their computers were programmed to ignore such large reductions on the assumption that they could only be due to instrument error. When the British results were published, records were rechecked and found to agree. Regular sampling is now undertaken from special high-flying aircraft, as well. The decline in ozone is continuing.

Most ozone is formed in the stratosphere over the Equator, where the Sun's energy is greatest. Winds in the stratosphere spread the ozone around the globe. The thinning in the ozone shows first over Antarctica because winter there is specially cold. Icy particles in polar stratospheric clouds act as reaction surfaces. On these surfaces CFCs are changed to compounds that quickly release chlorine radicals when sunlight returns in spring. Then the ozone is rapidly destroyed.

The threat to life

The destruction of the ozone layer results in a significant increase in u.v. radiation reaching the Earth's surface. Ultra-violet radiation is absorbed by the purine and pyrimidine bases in DNA (p. 204) and modifies them, with consequences affecting the expression of genetic information. In humans this results in a higher incidence of skin cancer.

In higher plants grown as crops the effect of increased exposure to u.v. radiation is to reduce yields. Aquatic organisms, including fish larvae and plankton, are very sensitive to u.v. light. Disruption of the ecological balance in the ocean may affect not only marine food chains but the contribution of the ocean to the absorption of carbon dioxide (p. 82).

International action to protect the ozone layer

The Montreal Protocol came into force in 1989 when 24 countries signed an agreement to reduce emissions of ozone-depleting chemicals in a series of staged controls. As a consequence, most developed countries, particularly the USA and the countries of the EU (major sources of CFCs since 1945), ended production of CFCs in 1996. China and India have until 2006 to comply. There are slower timetables for exclusion of other pollutants, including methyl bromide. This widely-used fungicide is the principle source of atmospheric bromine.

Following these moves, there has been some progress in decreasing the concentrations of ozone-depleting chemicals in the lower atmosphere. The rate of loss of the ozone shield appears to have been decreased.

However, temporary ozone 'holes' (the popular name for the reduced ozone concentrations) continue to appear over the South and North Poles each spring. Tighter controls, enforced by all countries, are really needed. Another problem is the world's growing fleet of high-flying aircraft. These leave trails of icy particles in the stratosphere, contributing to the depletion of ozone.

If the use of all CFCs were to stop immediately, the amount of ozone in the stratosphere would continue to decrease for at least the next 20 years. This is because of the amounts of CFCs currently around us, and the time taken for these substances to reach the stratosphere.

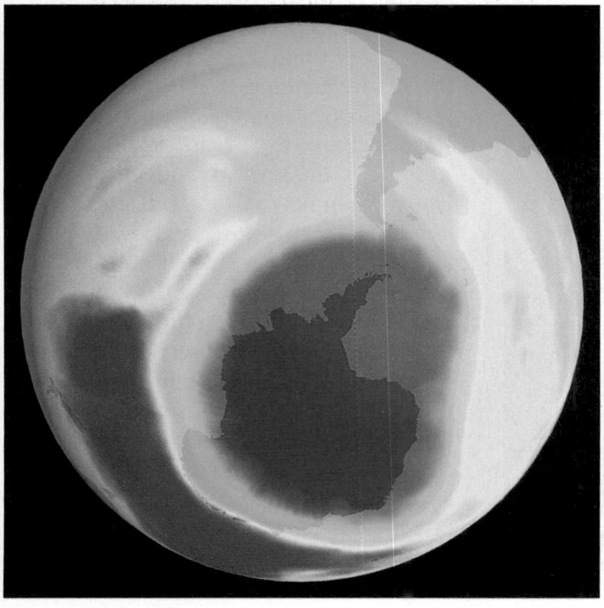

Figure 4.20   Ozone depletion over Antarctica, in 1998. The data are supplied by NASA, and have been collected by Nimbus-7 weather satellites. Note that ozone concentration is measured in Dobson units (one Dobson unit is equivalent to an ozone concentration of about one part per billion). In this illustration, different colours represent different ozone concentrations. On 30 September 1998, ozone concentration over Antarctica fell to 90 Dobson units, one of the lowest levels on record

Questions

1   Explain why ozone can be both harmful and beneficial, depending on where it occurs.
2   Outline the reactions by which both dinitrogen oxide and CFCs deplete the ozone in the atmosphere.
3   Write down the principal sources of
    a   dinitrogen oxide,
    b   CFCs.

■ Pesticides

Pesticides are substances used to control organisms considered harmful to agriculture or horticulture, or organisms involved in disease transmission. They are biocides (chemicals designed to kill organisms). Most pesticides act by interfering with fundamental biochemical and physiological processes common to a wide range of organisms, including humans. As a consequence pesticides are economic necessities, serious threats to biodiversity and potentially serious pollutants. The on-going dilemma created by the need for and use of pesticides is illustrated in Figure 4.21.

There is a great variety of pesticides on the market today, some 500 in all. They are classified according to the group of organisms they are designed to control, such as insects (insecticides), plants (herbicides), fungi (fungicides), molluscs (molluscicides), spiders (arachnicides), nematodes (nematocides) and rodents (rodenticides).

Rachel Carson's alarming book *Silent Spring*, published in 1962, alerted public opinion to the possible effects of pesticides on wildlife, on birds in particular, and on human life. Today there is a general recognition of the dangers of excessive use of pesticides to humans and to the environment. As a result, more controls on the use of pesticides are imposed. There is evidence that pesticides can in some circumstances be used selectively with no significant loss of productivity when compared with their more indiscriminate use.

Pesticides are widely used, but there is also a growing commitment on the part of some farmers and growers to avoid use of any additional chemicals in food production, a mode of agriculture known as 'organic farming'.

Concern over pesticides has centred on synthetic chemicals, but the earliest pesticides were poisons extracted from plants. Nicotine from ground-up tobacco leaves and pyrethrum from certain types of daisy were used as insecticides. The earliest synthetic pesticides were fungicides and insecticides based on organic salts of copper, mercury and lead. One example is Bordeaux mixture, a suspension of insoluble copper hydroxide, which has fungicidal properties. This was originally used on grapes, first to counteract theft, and then as a fungicide when this property was discovered. Most famously, it was later adapted for use against **potato blight**, (p. 543), but too late to prevent the potato famine in Ireland in 1845.

Insecticides

The revolution in pesticides began in 1939 with the recognition of DDT as a pesticide (this compound, 1,1-diphenyl-2,2,2-trichloroethane, was first synthesised in 1874 but was not then employed as an insecticide). DDT was the first of a family of organochlorine insecticides, so named because their chemical structure includes chlorine atoms attached to a hydrocarbon ring. The family also includes aldrin and dieldrin.

Other families of insecticides have emerged, including the organophosphates such as parathion and malathion, carbamates such as aldicarb, and synthetic pyrethroids similar to the naturally occurring pyrethrum.

Recognition of the problems

DDT acts on the sensory nerves of insects, causing rapid death. It is effective in low concentrations. At first, this poison did not seem toxic to humans. Between 1939 and 1945 DDT was used with spectacular success, particularly in the control of malaria by killing mosquitoes (p. 535). Consequently, organochlorine insecticides were widely used from 1945 onwards, throughout the world. But their very success often led to their indiscriminate use.

DDT does not break down easily; it stays lethal for years. Initially this 'stability' was seen as a great virtue but later, when its harmful effects became apparent, the word used was 'persistence'. Scientists started to discover organochlorine insecticides in other animals in the food chain. The problem arose because these insecticides are fat-soluble and accumulate in fatty tissues. Progressively they accumulate in organisms at the top of food chains, at a concentration far above that of the original application. In high doses organochlorine insecticides are extremely toxic to many organisms. Birds, fish and non-vertebrates are worst affected.

Figure 4.21   Pests and the use of pesticides: an ongoing dilemma

The **grey field slug** (*Deroceras reticulatum*) is probably the most abundant slug in lowland areas (agricultural land, hedges, gardens and grassland). This slug is a major pest of many crops – there may be more than a million to the hectare, if the soil remains damp.

However, other slugs and snails abound: most cause little or no damage to crops; many feed on dead organic matter, animal faeces etc., and have **vital roles in cycling of organic matter**. The molluscicides available to attempt to control the number of grey field slugs are also effective toxins for other species, i.e. **pesticides threaten biodiversity**.

Life cycle/feeding pattern

hermaphrodite – mating (sperm sac exchange) occurs most commonly in spring and autumn

↓

eggs are then laid in small clutches and small slugs emerge in 2+ weeks (depending on the temperature)

↓

smallest slugs browse on soil-surface algae – larger slugs feed on seedlings and leaves, foraging twice (early evening and again before dawn) in each 24 hours

↓

daylight is spent in cracks in the soil

Typical farming where *Deroceras* prospers: monocultures of oil seed rape (left) followed by winter wheat (right); the soil may be treated with molluscicides

For example, in a Californian lake, where gnats swarmed and were a nuisance to fishermen, an organochlorine insecticide was applied to the water for three years running at a concentration of 1 part of insecticide to 50 million parts of water, a dose calculated to kill the gnats at the larval stage whilst being harmless to other creatures. (Actually, the population of gnats recovered almost immediately.) Later, when the grebe colony in the lake began to die out, the pesticide residues in the whole food chain were investigated. At each step it was found that the insecticide was more concentrated. At the top of the food chain were the fish-eating grebes. Their bodies contained, on average, about 166 ppm of the insecticide.

Public concern was aroused by an increasing number of such cases of wildlife poisoning related to the use of DDT. Most organochlorine pesticides cannot now be sold in the USA or western Europe.

Health risks to humans

Workers who have to apply pesticides may be at particular risk. Additionally, the public at large is at risk from residues in food resulting from excessive use of pesticides. Studies of dietary intake show that curbing excessive use of pesticides has reduced the human intake of these chemicals (Figure 4.22). But Ministry of Agriculture scientists, who regularly screen food samples, nevertheless report numerous instances of pesticide residues in food exceeding the maximum residue levels set by the European Community and by the World Health Organization.

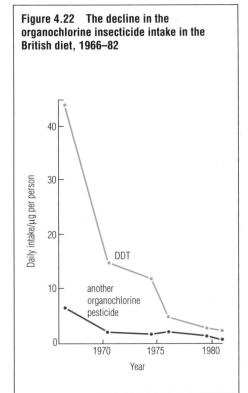

Figure 4.22 The decline in the organochlorine insecticide intake in the British diet, 1966–82

Herbicides

Herbicides were also developed in the 1930s. This followed upon the discovery that indoleacetic acid (IAA, p. 422), was a constituent of higher plants, with a role as growth hormone. Many substances related to IAA were found that had growth-regulatory properties at low concentration, but which were highly toxic to plants at higher concentration. The structures of the commonly used synthetic hormones are shown in Figure 20.36 (p. 440). The best known, 2,4-dichlorophenoxyacetic acid (2,4-D), is highly toxic to dicotyledonous plants (p. 404). 2,4-D is rapidly taken up by the leaves (it is a systemic pesticide) and transported to the growing points, causing death (Figure 4.23). Dalapon (2,2-dichloropropionic acid) is a systemic herbicide that is selectively toxic to monocotyledons (p. 406).

Many of these hormone weedkillers are now largely superseded. Quite a number of herbicides of different kinds (some selective, others not) are in common use. The herbicides paraquat and diquat destroy green plants by contact, but in the soil they are not active and do not harm soil animals. They are toxic to humans.

Herbicides are now used by British farmers in far greater quantities than all other types of pesticides combined.

EGG SHELL THICKNESS CHANGE DUE TO ORGANOCHLORINE INSECTICIDES

A feature of the toxicity of organochlorine insecticides to birds was evident in the deaths caused by shell thinning and egg breakage, and by embryo deaths. Birds of prey are particularly vulnerable.

Egg shell thickness was compared year by year, using museum samples to provide data on shell thickness before pesticides were available. This study revealed that shells had become abruptly thinner from 1947.

Several species of predatory birds suffered from this effect, with marked population declines following the widespread introduction of organochlorine pesticides. As the amounts of the pesticides used have reduced in recent years, their residues in birds have declined, egg shell thickness has begun to rise, breeding success has improved and populations have recovered.

Egg shell thickness studies of British sparrow hawks, 1870–1980: comparisons of relatively recent shells with those in museums

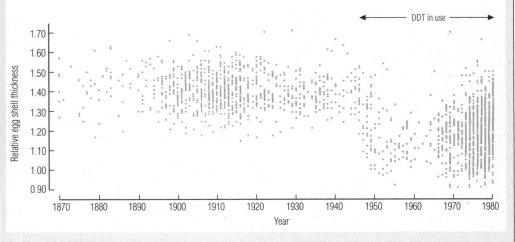

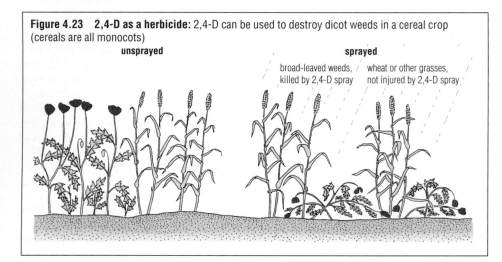

Figure 4.23   2,4-D as a herbicide: 2,4-D can be used to destroy dicot weeds in a cereal crop (cereals are all monocots)

unsprayed

sprayed

broad-leaved weeds, killed by 2,4-D spray

wheat or other grasses, not injured by 2,4-D spray

Initially it appeared that these herbicides, which were effective as weedkillers in low doses, left only short-lived residues in the soil, and were generally not harmful to animals. Some dangers due to herbicide toxicity have subsequently emerged, however, but the ecological harm done by herbicides is due not to their toxicity but to their efficiency. Disastrous ecological damage occurs when herbicide sprays drift on to hedgerows and meadows, while the deliberate suppression of 'weeds' on road verges and railways has reduced the population of wild flowers and the insects that depend on them.

Biological control

Biological pest control methods exploit natural enemies to regulate the population of pest species, rather than using chemical agents to destroy them. The oldest record of biological control comes from the Chinese who, 400 years BC, used colonies of predatory ants to control foliage-feeding pests in orchards.

A modern form of biological control exploits the chemical messengers known as **pheromones**, produced by many animal species. Pheromones (p. 477) are volatile chemicals that males or females of a species use to attract partners for mating. Synthetic pheromones are used to lure pests to traps laced with pesticides. Traps are sited in areas well away from crops, where the pest can be eliminated without harm to its natural enemies.

The impact of biological methods has largely been overrated. There have been successes, however. For example, the environmentally controlled conditions in greenhouses favour the pests spider mite, whitefly and aphids as well as the crop plants. But these pests have become resistant to pesticides. Predatory mites from South America have been imported as control agents for one of the common spider mite pests in greenhouses, and a parasitic wasp effectively controls greenhouse whitefly.

Another instance concerns aphids, which lower the productivity of wheat by tapping the phloem sieve tube contents (p. 338). Insecticides, if used, kill the natural predators of aphids more effectively than the aphids and, in addition, may leave residues in the wheat flour. There is a possibility of using biological methods to control the green aphid that feeds on the wheat crops.

The female hoverfly is a natural predator of aphids. If strips of 'weed' are left in and around wheat crops (as achieved by smaller fields with hedgerows!) the hoverfly obtains the pollen diet it needs. The fly then lays eggs on wheat stems below the aphids. When the hoverfly larvae hatch out they feed on the aphids. In most years the fly larvae alone are sufficient to suppress the aphids, without use of insecticides.

Fungal pathogens specific to persistent weeds of corn and other cash crops have been used as 'herbicides'. These are sprayed on as suspensions of viable spores under conditions favouring fungal germination and growth. The pathogen specifically attacks the weed species, causing the death of the weeds before they can flower and set seeds that would contaminate the harvest. No chemical residue remains in the soil subsequently.

A bacterium called *Bacillus thuringiensis* is available as a spray and may be applied to cabbage plants infested with caterpillars. The caterpillars are killed by this bacterium, which has no effects on other insect species present.

Pesticides today

Concern about the long-term consequences of using pesticides has arisen for several ecological reasons. For example:

1 Spraying aphids on wheat may also kill the predators that control other pests, or the bees that pollinate fruit trees.

2 Pesticides may kill natural enemies of the pest species, such as predators and parasites, more effectively than they kill the pest. Then the pest, once it is re-established, may become even more of a problem, in the absence of its natural predators and parasites.

3 Pest populations may eventually become resistant to pesticide. Repeated low doses of pesticide kill susceptible individuals but leave a few resistant organisms unharmed. It is from these resistant forms that a new population is established.

4 Residues of the pesticides may accumulate in humans and be harmful to them. Pesticides often accumulate in wildlife, causing many deaths.

Consequently, a better form of pest control, known as **integrated pest management**, involves combining methods such as

- ▶ varying cultivation methods
- ▶ rotation of crops
- ▶ minimal, well-targeted application of selective pesticide
- ▶ use of a biological control mechanism, where this is known.

Questions

1 Explain how a pesticide that is used at a harmless concentration to wildlife can nevertheless cause the death of many animals at the top of the food chain.

2 Some more recently developed pesticides break down to simpler substances when in contact with the soil. Why has this property been introduced by agricultural chemists?

3 What would be the attributes of
   a an ideal insecticide
   b an organism used for biological control of a pest?

4 What advantages might the natural production of nicotine within the leaves of the tobacco plant confer on that species?

■ The disposal of waste

Urban communities everywhere are producing huge quantities of household and commercial waste. In Britain, a city of half a million people produces more than 150 000 tonnes of rubbish per year. The traditional method of disposal is by dumping the rubbish in landfill sites such as disused quarries and gravel pits. Problems arise in three ways:

► insufficient landfill sites exist to take all the rubbish

► compacted rubbish, containing much carbon-based material, decays in the absence of air, producing methane gas, which escapes into the atmosphere; this methane contributes to the 'greenhouse gases' (p. 84) and in some places has caused explosions

► toxic substances may be leached into groundwater.

Measures to overcome the rubbish crisis

The composition of typical municipal rubbish is shown in Table 4.2. In fact, of course, the composition of each household's waste as collected by the waste disposal authorities varies greatly – for example, a home with solid fuel heating will produce more ash than the average.

Paper started to be a component of rubbish in Britain's dustbins about 100 years ago. A sustained campaign to cut down the excessive use of paper and plastic – in packaging and junk mail, for example – would reduce the rubbish burden. The remaining rubbish needs careful sorting to isolate materials for recycling (such as paper and some plastics) from the materials that have no further use. Several different metals can be reclaimed at this stage.

Table 4.2   The components of domestic waste in the UK

	Proportions of household waste /% dry mass	Calorific value of waste /kJ kg^{-1}
Fine dust	15–25	9 600
Paper	24–32	14 600
Vegetables and decomposing organic matter	15–30	6 700
Metals	6–12	–
Glass	6–11	–
Rag	2–5	6 300
Plastic	2–5	37 000
Other (timber, etc.)	2–6	17 600
		UK average = 9 000– 11 000 kJ kg^{-1}

The rubbish that cannot be used can be burnt in special incinerators where the heat energy can be reclaimed and used in heating or to generate electricity (Figure 4.24). Alternatively rubbish may be converted to pellets of refuse-derived fuel.

Toxin production by the incinerator

The chemical changes occurring during incineration are complex. If combustion does not occur at a high enough temperature the burning process may create toxins, such as dioxin, that would be discharged into the environment with the smoke. This danger arises when certain plastics are incinerated.

Industrial pollutants

Some substances released by industry do occur naturally, but are damaging at high concentrations. Among these are the heavy metal ions in mining waste tips, and crude oil.

Crude oil escapes from damaged tankers and pipelines, seeps from marine and coastal installations, and is flushed from tankers when ballast water is pumped out. The oil floats on water as a slick, and does harm to wildlife: for example, it damages marine birds by matting their plumage (Figure 4.25). Flight and feeding dives become impossible. When a slick is driven onto rocky shores, crude oil also destroys seaweeds, molluscs and crustaceans.

After initial losses, however, oil spills are rapidly dispersed by bacterial decomposition, because crude oil is quite quickly biodegradable. The use of chemical dispersants, whilst removing the crude oil scum from sight, have been found to do more harm to the whole marine food chain. Many dispersants are toxic and non-biodegradable. Sadly, very few of the sea birds that are 'rescued' and cleaned up after an oil spill usually survive for long on return to the wild.

Figure 4.25   Wildlife damaged by oil

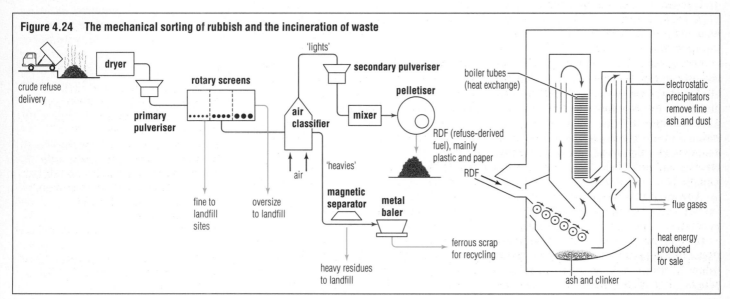

Figure 4.24   The mechanical sorting of rubbish and the incineration of waste

crude refuse delivery

dryer

primary pulveriser

rotary screens

'lights'

secondary pulveriser

air classifier

mixer

pelletiser

fine to landfill sites

oversize to landfill

air

'heavies'

magnetic separator

metal baler

RDF (refuse-derived fuel), mainly plastic and paper

boiler tubes (heat exchange)

electrostatic precipitators remove fine ash and dust

RDF

flue gases

ferrous scrap for recycling

heavy residues to landfill

ash and clinker

heat energy produced for sale

4.5 CONSERVATION

Human activities are changing the environment worldwide in dramatic and far-reaching ways. The need for conservation of the natural environment has become increasingly apparent since the second half of the nineteenth century because of the environmental degradation that is occurring on a global scale. Conservation involves **applying the principles of ecology to manage the environment** so that, despite human activities, a balance is maintained. The aims of conservation are to preserve and promote habitats and wildlife, and to ensure that natural resources are used in a way that provides a sustainable yield.

Conservation is an active process, not simply a case of preservation, and there are many different approaches to it.

■ Maintenance of representative ecosystems

Human influences on the environment have been so profound that very few completely natural ecosystems remain. Conservation measures are designed to preserve the full range of semi-natural habitats now available to us. Habitats are damaged as a result of developments in agricultural technology (increased mechanisation and application of pesticides and fertilisers) and industrialisation and urbanisation (pollution processes and the conversion of land to use in transport systems and for buildings). The deliberate or accidental introduction of species into new habitats has also been harmful; the introduction of the rabbit to Australia is an example.

The setting aside of land for restricted access and specialised uses is not new. Hunting reserves are one example. The New Forest was designated a Royal hunting forest about 900 years ago.

Biological conservation emerged via the movements to protect large areas of land from inappropriate development and from the pressures on wildlife and habitats that result from the collecting of biological and geological specimens. For example, the **Yellowstone National Park** was set up in North America in 1872. Similar National Parks were established in Australia (1886) and New Zealand (1894). The development of National Parks in Britain had a slow beginning.

By Act of Parliament the **National Parks of England and Wales** were established in 1949. Today there are eleven National Parks (Figure 4.26). These are under the control of the **Countryside Agency**, which also designates **Areas of Outstanding Natural Beauty** (AONBs) for special treatment by the community. National Parks and AONBs occupy just over 20% of the land surface of England and Wales. Ecosystems are also maintained through creation of National Nature Reserves (some 150 of them) and **Sites of Special Scientific Interest** (over 3000 of these) under the supervision of **English Nature**, the **Countryside Council for Wales** and **Scottish Natural Heritage**. Scotland has **National Scenic Areas** rather than National Parks. These give some limited protection from development. In fact, a better defined and more vigorously enforced system of rural protection is needed in Scotland.

Now the national bodies listed above come under an 'umbrella' agency, the **Joint Nature Conservancy Committee**. Government funds are also available for **Environmentally Sensitive Areas** (ESAs), of which the South Downs in Sussex are a good example. In addition there are two statutory bodies that have roles in conservation, namely the **Forestry Commission** (set up in 1919) and the National Rivers Authority (1989), now the **Environment Agency** (1996).

Today, there are also European Environmental initiatives, including a move to identify and protect **Sites of Community Interest**.

Local and voluntary initiatives are also highly significant. Most counties and large urban areas have a Trust for Nature Conservancy, or County Naturalist Trusts, which preserves sites. Local trusts are (for example, Wiltshire Wildlife Trust, p. 17) affiliated to the Royal Society for Nature Conservation (the junior club is WATCH). The British Trust for Conservation Volunteers represents the voluntary groups active in practical conservation. In addition, through the reserves of the Royal Society for the Protection of Birds and the properties of the National Trust, many thousands of hectares of countryside are protected (although this is still only a tiny proportion of available land).

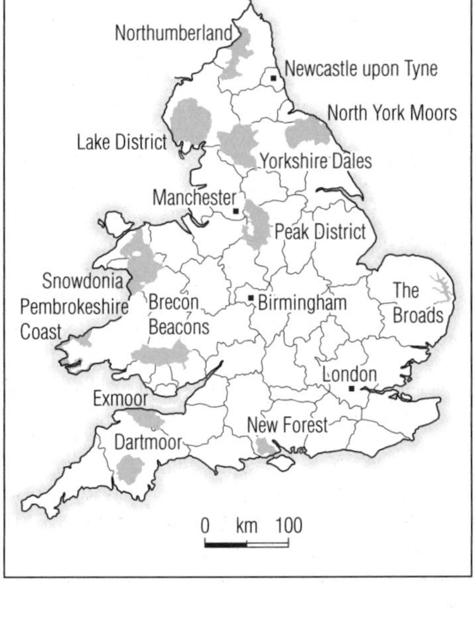

These sites are managed to balance the needs for conservation of wildlife and for research and experimentation with the needs of the public for recreational outlets, and those of land users.

On the international level, the World Conservation Strategy statement was devised and supported by the World Wide Fund for Nature working with the International Union for Conservation of Nature and with the UN Environment Programme. These organisations work for the sustainable use of Earth's resources.

An international challenge

The problem of the rain forests is especially pressing (p. 81). The systematic logging of huge areas of forest yields the timbers that the developed countries are ready to purchase. These developed countries need to review their demands for hardwoods, and seek alternatives or other sources for wood.

The rain forests occur in the less-developed countries of the tropics and subtropics. In these countries there are growing populations and a hunger for land. There is also a need for foreign currency to pay for imports of oil and manufactured goods. Forests are cut down to make way for subsistence farming. They also give way to vast monocultures of sugar cane, coffee, rubber or beef cattle, which can be exported for foreign exchange.

This deforestation is disastrous for the environment, and eventually for the farmers. It also removes an important sink for atmospheric carbon dioxide. The grassland that replaces forest does not maintain or improve soil fertility. Soil erosion follows. The bacteria in the gut of the cattle (ruminants, p. 289) that feed on the grass are extremely effective producers of the 'greenhouse gas' methane (p. 84).

Figure 4.26 The National Parks of England and Wales

At the heart of the tropical rain forest problem is the giving of nothing more than lip-service to the concept of a sustainable forest (p. 82). The developing countries that are home to the rain forests need aid arrangements (including the management of debt repayment) that facilitate economic development that is not based on clear-felling of the remaining forests.

■ Supervision of the multiple use of sites

In a countryside development such as the planting of conifers on uplands a conflict of interests may occur between the needs of wildlife, the public, a water authority and the landowner. Conservation organisations can advise on the establishment of wildlife refuges (zones supporting a diversity of trees including broad-leaved species), on the provision for the recreational needs of the public (footpaths, viewpoints, picnic facilities) and on water catchment zone management to meet the needs of water authority reservoirs. The economic return to the landowner is central to the development if the managed environment is to survive, but the interests of the landowner are often best served by being coordinated with others' needs.

■ Conservation of species

If we destroy a habitat there is a chance that it can eventually be restored. If we destroy a species it can never be replaced.

Species must be preserved; we have no right to eliminate them. They are as entitled to survive as we are. Incidentally, they are often a source of pleasure to us, and many of them are useful. For example, many wild organisms are more efficient energy-converters than existing crops and herds. Wild organisms contribute to a pool of genetic diversity (p. 645) useful to genetic engineers, and many wild plants are sources of compounds of medicinal value as new drugs. Above all, it is vital to maintain a diverse flora and fauna if we are to ensure the continuation of the process of evolution of new types and species in response to the ever-changing (and more rapidly changing) environments of the Earth.

Survival of endangered species is sought by maintaining habitat diversity, as discussed earlier, and by ensuring that all endangered species are protected by legislation. At the same time local, regional, national and international actions are necessary to ensure protection. As a very last resort, practical

THE CONSERVATION OF THE BARN OWL (*Tyro alba*): A CASE STUDY

a male barn owl (about 34 cm long): in their silent flight, barn owls show the white undersides of their wings

Barn owls hunt for small mammals (shrews, mice, voles and bats), small birds, beetles, moths and frogs. In its hunting the barn owl uses hearing as much as its sight, or even more.

Nests are not built. The young are reared on a bed of disgorged pellets, on a ledge in a barn or in a hollow tree.

Barn owls have been in decline in the UK (most northerly edge of their range, see map) since 1945. Today, barn owls hunting at dusk fly low over hedges, and may be seen in the beams of car headlamps at night. In fact, the major cause of barn owl deaths is by collision with speeding cars.

the distribution of barn owls in the world

The causes of the decline

▶ Changed farming practices have destroyed supplies of prey. The combine harvester has taken away the need to store unthreshed grain in ricks, drastically reducing the numbers of rodents around farms.

▶ The combine harvester and other farm machinery need huge fields with fewer hedges if they are to work economically. This has substantially reduced the habitats for owls and their wild prey in the countryside.

▶ The liberal use of organochlorine insecticides, such as DDT, in farming between 1945 and the 1960s has harmed a wide spectrum of wildlife. These insecticides are stable, and they dissolve in fat. They thus accumulate in animal bodies, and get passed from prey to predator, concentrating at successive steps in the food chain. At sublethal doses in birds of prey, like owls, DDT causes shell thinning, egg breakage and embryo deaths. Fewer owls were reared.

▶ The numbers of cars on the roads have increased dramatically. The roads take up land, fragment habitats and are the site of the destruction of young owls attracted to the verges as hunting grounds.

▶ The draining of farm ponds deprives the barn owl of suitable sites for bathing. This is the practice of the female after the long period of incubation. The cow drinking troughs now being visited by owls have steep or vertical sides; unable to escape, the owls drown as soon as the water is taken up by the plumage.

Conservation measures needed

It is unlikely that the species will return to the numbers common in the 1930s within the foreseeable future, because the environment has changed so much. Extinction has been avoided and the population stabilised by

▶ the reduced use of pesticides
▶ the provision of appropriate nesting sites
▶ rearing of the young of disabled or dead parents and their supervised release into suitable sites that have not been recolonised naturally.

preservation also involves the maintenance of representative organisms in zoos and botanical gardens. This may have special value when a species becomes endangered and may become extinct in the wild (Figure 4.27).

An example of action from local to national level to support an endangered species is the current campaign to protect the barn owl (see panel, p. 95).

Examples of action for endangered species at international level are the **Red Data Books** of endangered species, initiated by the late Sir Peter Scott of the Wildfowl Trust, Slimbridge, in the 1960s. Each contained a long list of threatened species. Today the updating of the lists is coordinated from Cambridge, on behalf of the International Union for the Conservation of Nature and Nature Reserves, whose headquarters are in Switzerland.

The Red Data Books are invaluable sources of biological information. Their value lies in providing ready and accurate data as the basis of scientific advice when the ecological status of threatened sites is being evaluated. Red Data Books have been produced by biologists of many countries, who have pooled observations on an international basis. Three categories of organism whose survival gives concern are recognised:

▶ **rare** = at risk
▶ **vulnerable** = endangered in the near future if no change in circumstances occurs, and
▶ **endangered** = a future is unlikely.

An outcome of this initiative is CITES (Convention on International Trade in Endangered Species) by which participating countries agree not to import or export products of endangered species.

■ Pollution control

International actions to avoid long-term damage to the environment as a consequence of 'acid rain' (p. 86), the 'greenhouse effect' (p. 84) and possible global warming (p. 85), and the threat to the ozone layer (p. 88) have begun to be agreed and controls enacted.

But much remains to be done by individuals and by organisations to avert a human-generated environmental catastrophe. Conservation bodies play an important part in collecting evidence, proposing solutions and publicising the issues. Problems really arise from the human consumption of resources such as fossil fuels and raw materials. The important underlying issue is 'human life styles', which involves us all.

■ The control of human population growth

People who make a study of populations and changes in populations (demographers) predict that by 2050 the Earth may have 10 billion human inhabitants. This is about twice as many as it has now. The projections suggest that the world population will stabilise at this level.

The critical question is: in the presence of such a huge human population, can the natural environment be maintained in a condition that will supply all living things with the required food, water and space? Will the total population of the world be sustained by the Earth's resources? What then will be the state of fragile habitats and threatened species?

An immediate task must be to reduce the birth rate. The mechanisms for birth control exist (p. 600). Do we invest sufficient resources to encourage limitation of family size?

The drastic population control programmes adopted in India in the mid-1970s concentrated on male sterilisation. As a result, strong resistance developed to the concept of family planning.

Modest but sustained reductions in the birth rate in countries like Tunisia and Jamaica have involved the provision of freely available family planning advice and materials. At the same time practical steps are taken to improve the health of mother and child by preventive medicine. This is accompanied by the introduction of education schemes and a general raising of the economic and social status of women. These are cases of national investment in resources to reduce population growth. Collectively, can the world find the resources to attempt a broad approach to family planning that may lower the birth rate quickly?

Questions

1 What influences might an upland afforestation development have on the future supply of water to a nearby reservoir?
2 What use are the Red Data Books to a conservationist advising a local authority considering the siting of a proposed housing development on heathland?
3 On average, what constitutes 50% of domestic waste? How might the volume of domestic waste be reduced?
4 What are the potential hazards respectively of landfill and incineration as methods of disposing of domestic waste?
5 What steps can a developed country such as Britain take to help conserve tropical forests?
6 In what ways have human activities in industrialised countries come to influence the ecosystems of, firstly, the open oceans and, secondly, dry deserts?

Figure 4.27 Elephants are endangered species: African elephants (left) are illegally hunted for their tusks (ivory); Asian elephants (right) are losing their habitats as humans cultivate the land that was forest

■ Conservation overview

The sustainable use of resources

Human life styles of the future will determine the environment of living things. The issues of how we choose to live and what we find necessary to consume are critical.

In the 'global village', we each influence the conditions and life styles of our neighbours. Currently the developed countries, with about 20% of the world population, consume more than 70% of the resources, and 75% of the energy, that are used each year.

Effective energy conservation measures in developed countries could substantially reduce the consumption of energy. If nothing is attempted, and if less-developed countries follow the same pattern of consumption in future, then the environmental consequences are likely to be catastrophic.

Appropriate methods of development

Many aid and development schemes have failed to touch the lives of the very poor, who are rarely helped by large-scale projects such as building dams, roads and industries. Smaller-scale projects are more likely to succeed in improving the lot of the poorest people, for whom poverty is a shortage of biomass, not a shortage of cash.

For example, the use of local crop species that grow satisfactorily under existing conditions will feed the poorest families. The farming of imported plant varieties is dependent on high inputs of chemical fertilisers, fossil fuels and expensive machinery. These can be provided only by the wealthy landowners. Rarely do western agricultural practices lead to an improvement in the diet of the poor.

When clean water, sufficient food and productive activity for the poor has been achieved the local environment is also improved. With basic security and future prospects the birth rate starts to fall. In these ways existing communities, under their own local direction and management, may be able to do most in the developing countries to reduce the environmental crisis.

Questions

1 In what ways does the human population explosion represent a threat to the Earth's wildlife?

2 Why is it that the introduction of high-yield varieties and advanced agricultural technology to a relatively poor country often mostly improves the wealth of the developed country that supplied them?

WORLD TRADE AND POVERTY

The traders: the proportions of people living in absolute poverty

key
% impoverished population

■ up to 75%
□ up to 50%
□ up to 25%

The countries whose trade is based on only a few products

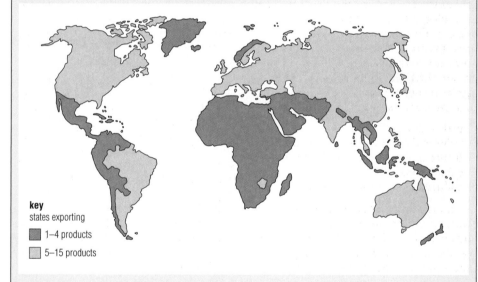

key
states exporting

■ 1–4 products
□ 5–15 products

THE ORIGINS OF HUNGER

The basic problems causing world poverty are political, social and economic, rather than technical and scientific. Despite the huge world population, we produce sufficient food for everyone to be well nourished.

In this situation poor people are hungry because they lack the means to grow enough food or to buy it. The international food and materials market trades in dollars. To obtain the foreign currency many countries grow cash crops for export, such as coffee and sugar.

These sell for prices the producers find difficult or impossible to control themselves. When prices for the developing countries' major products fluctuate and fall, more land has to be given over to growing cash crops. Less and less land of agricultural worth becomes available for use by the truly poor. The fewer products a country exports, the more vulnerable it is to fluctuations in world markets. The distribution of countries exporting four products or less corresponds closely to the distribution of people living in poverty.

FURTHER READING/WEB SITES

Global change research information: **www.ncdc.noaa.gov**
The World Wide Fund for Nature UK: **www.wwf-uk.org/home.shtml**
CITES: **www.cites.org**

5 Biology and industry

- Clean water is an essential but relatively scarce resource, endlessly recycled for domestic and industrial purposes. Sewage treatment removes the dissolved organic matter by the action of aerobic bacteria. Separated solids are digested anaerobically, releasing methane gas as a fuel. The residue is burnt or used as a soil fertiliser.
- Commercial fishing occurs mostly in coastal waters, where fish occur in greatest densities. Most fishing grounds of the world have been over-fished, and attempts to regulate this industry continue. Marine and freshwater fish farms are one response to maintaining the supply.
- Modern agriculture continues to increase its productivity, largely based on application of developments in science and technology in the areas of plant and animal breeding, and in plant cultivation and animal husbandry techniques.
- In food industries many raw materials are processed, preserved, stored and marketed in ways that may minimise preparatory work for the cook without significant loss of food value.
- A range of new biology-based industries, known as biotechnology, exploit organisms (mostly microorganisms) or their products, to produce new foods, drugs, enzymes, or chemicals for other industries.
- Biotechnology has ancient origins (e.g. cheese making, brewing) but today it is often linked with genetic engineering (Chapter 29) and enzyme technology (Chapter 8).

This chapter illustrates some of the ways in which biology underpins important areas of commercial and industrial activity. The approach is by selected case studies, rather than through an exhaustive listing of all the applications and connections between biology and these activities. Many links are made to later chapters. You may wish to read Chapter 24 ('Health and disease') and Chapter 29 ('Applications of genetics'), next.

5.1 BIOLOGY AND THE WATER INDUSTRY

■ The water cycle

The waters of the Earth are constantly circulating. The water cycle (Figure 3.54, p. 75) is rapid: on average, a molecule of water evaporating from the sea has returned to the sea within about 14 days.

Solar energy drives the water cycle, causing evaporation from the seas and fresh waters and from living things, particularly from plants (transpiration, p. 324). Solar energy also drives the winds that carry water vapour and clouds (which are condensed water vapour) over the land. The bulk of rain falling on the land has come from evaporation from the oceans.

■ The supply of clean water

An adequate supply of water is a vital requirement of a human community. Today, in Europe, the average domestic consumption of water is about 230 litres per person per day. Of this, only about 1.5 litres are used for drinking. The remainder is mostly used in washing and cleansing processes. Much water is wasted.

A great deal of water is used in industry, including agriculture and horticulture. In industrial manufacturing processes most water is used for cooling. In agriculture most water is used in irrigation.

The current sources of water for domestic and industrial uses are **surface water** taken from rivers, lakes and reservoirs, and deep **groundwater** brought up to the surface via wells and springs (Figure 5.1). Most if not all of the water taken from these sources must be treated before it enters the domestic supply.

■ Groundwater

The soil and rock of the upper part of the Earth's surface contain many pores. These are created by the spaces between the mineral particles, and are filled by groundwater. If the pores are extremely small, as in clays, or the pores do not connect up, the rock is impervious (impermeable).

Figure 5.1 Our water supply comes from surface water (rivers, lakes and reservoirs) and deep groundwater (springs and wells)

Lakes and **reservoirs** are important sources of drinking water, although they are often situated far from the consumer. Water is piped by mains; these must be maintained in order to avoid contamination from the soil.

Rivers are major sources of drinking water. Water drains from hills, fields and roads, and is added to by industrial waste water and sewage effluent. River water may contain high concentrations of dissolved salts and bacteria.

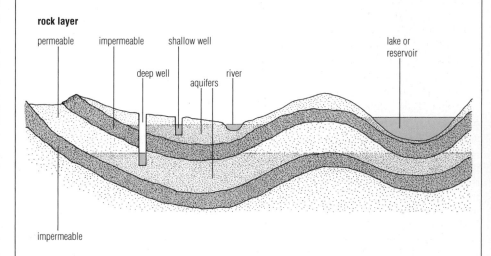

Wells may be deep or shallow. Deep wells usually provide water that has percolated through rocks. The process filters out bacteria and suspended particles (except for wells in chalk and limestone, because of the cracks and fissures in these rocks). Shallow wells are a health hazard, because water collects in them from the nearby surface. In artesian wells, water may come to the surface without pumping because of pressure of water at higher levels.

Deep springs occur where a geological fault brings deep groundwater to the surface. Shallow springs occur where impermeable strata come to the surface.

In permeable rocks the pores are large and continuous, and the water passes easily through them. An **aquifer** is a bed of permeable rock that retains water because it overlies impervious rock. Water moves freely in an aquifer, and the water level is known as the **water table**. Aquifers are recharged by rainwater percolating down from the surface. Where an aquifer reaches the soil surface and the water table is high enough, water flows out at a spring or seeps out into a stream. The water table will fall if water is regularly pumped out of wells sunk into an aquifer or – especially – as a result of several successive dry seasons.

There are approximately 50 million km^3 of groundwater in the world, of which only about 4 million km^3 are fresh water, suitable for drinking. Yet even this volume is vastly more than the water in rivers and lakes from which so much of our water supply is drawn.

■ Water supplies for the future

Undoubtedly human demands for fresh water for industrial and personal use will continue to grow. There are estimates that in the twenty-first century the amount of water needed will exceed the volume that runs off the land to the sea. The problem of water supply can be overcome, in part, by making better use of the water available. Much of this is currently untapped by humans.

An alternative source of water is sea water or brackish waters, provided that the dissolved salts can be removed. Removal of salt is known as **desalination**. This is a practicable process only where there is a source of cheap energy (see the *Study Guide*, Chapter 5).

■ Bacterial contamination of water

Most soils have a huge microflora of bacteria, largely associated with the dead organic matter present in the soil, including animal faeces.

It is potentially very dangerous to use water that is contaminated by bacteria from sewage as drinking water. This is because several serious diseases (typhoid and cholera, p. 531, among them) can be caused by certain intestinal bacteria. Tests for bacterial contamination of water look for bacteria that originate in human faeces. These bacteria are mostly commensals. The bacillus *Escherichia*

Figure 5.2   Testing a water sample for bacterial contamination

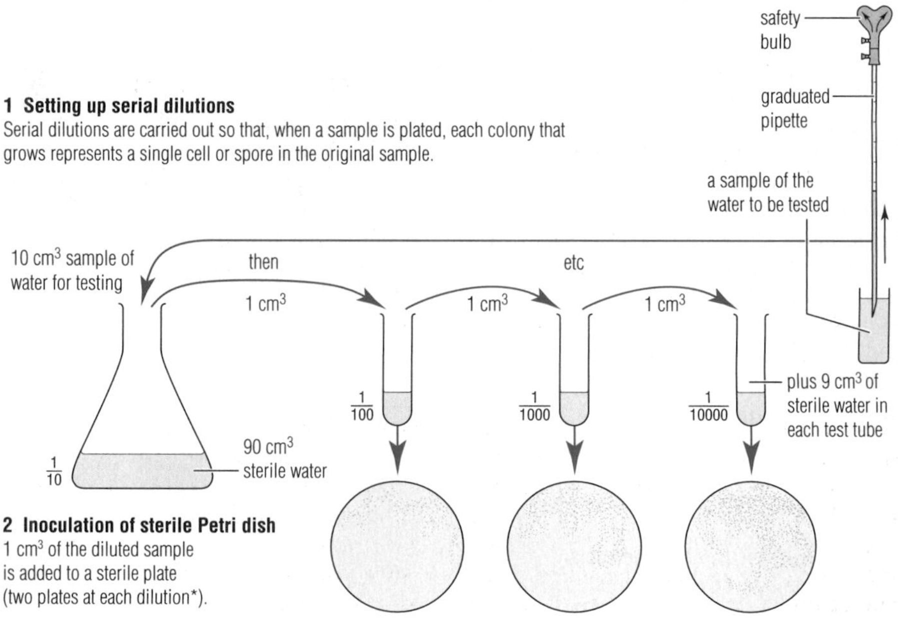

1 Setting up serial dilutions
Serial dilutions are carried out so that, when a sample is plated, each colony that grows represents a single cell or spore in the original sample.

safety bulb

graduated pipette

a sample of the water to be tested

10 cm^3 sample of water for testing

then

etc

1 cm^3

1 cm^3

1 cm^3

$\frac{1}{100}$

$\frac{1}{1000}$

$\frac{1}{10000}$

plus 9 cm^3 of sterile water in each test tube

$\frac{1}{10}$

90 cm^3 sterile water

2 Inoculation of sterile Petri dish
1 cm^3 of the diluted sample is added to a sterile plate (two plates at each dilution*).

3 Setting up nutrient agar plates
A sample of sterile nutrient agar is added to each dish, stirred to disperse the water sample, and allowed to cool and solidify. (For nutrient agar see Figure 5.17, p. 114.)

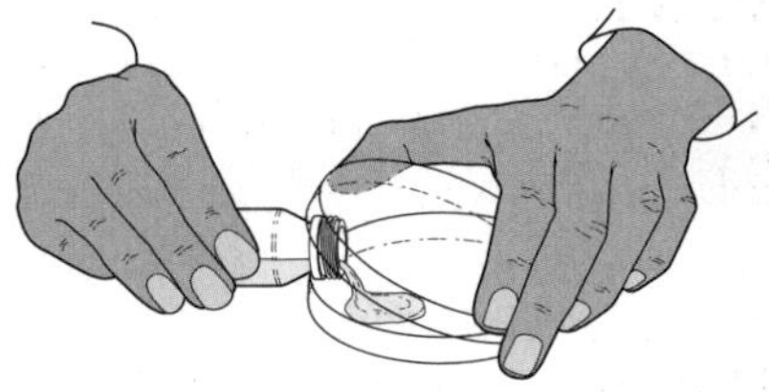

4 Culturing the colonies
Sets of inoculated dilution plates are cultured in incubators, one at 22 °C and one at 37 °C*. Any bacteria present will reproduce rapidly. Each bacterium will form a viable colony.

5 Observing the difference
When the colonies grown on the incubated plates have been counted, an *estimate* of bacterial contamination of the original water sample is made by multiplication by the dilution factor. The ratio of the count of bacterial colonies at 22 °C to those at 37 °C indicates the degree of sewage contamination. The higher the ratio, the more likely it is that the bacteria come from soil*.

6 Disposing of plates
Plates are first sterilised before disposal. Plates are only ever opened in a special inoculation chamber – *never* in a general laboratory.

*Bacteria adapted to thrive in the human body grow best at 37 °C. Bacteria that live on the dead organic matter of soil thrive best at a much lower temperature. To differentiate the numbers of bacteria that have originated from sewage from those that normally inhabit soils, water samples are cultured on agar plates, some at 22 °C and some at 37 °C.

Sterilisation
Before use, all equipment, apparatus and media are sterilised in a pressure cooker or autoclave to kill any stray microorganisms (see Figure 5.17, p. 114).

coli is an extremely common inhabitant of the human intestines. Checks of water purity test for the presence of this bacterium as evidence of contamination of drinking water by sewage. Although *E. coli* itself is harmless, its presence in water indicates that other, more harmful, intestinal bacteria might also be present.

The process of testing water samples for contamination by sewage is illustrated in Figure 5.2.

■ What is pure water?

Chemically pure water is unknown in nature. From the moment water vapour condenses to droplets in the atmosphere it takes into solution small quantities of the atmospheric gases, including carbon dioxide. As a result rainwater is, at the very least, a weak solution of **carbonic acid**. In the soil, water dissolves inorganic ions and possibly some organic matter. It may also pick up bacteria.

■ Treatment of drinking water

To make water safe to drink it is treated to remove suspended solids and bacteria. The stages in purification, physical and chemical, are illustrated in Figure 5.3.

■ Delivery and use of water

In developed countries water is piped to individual consumers. Here a large quantity of water is used on a daily basis, but only a tiny part of this purified water is used for drinking. In developing countries very much less water is consumed, partly because it is rarely piped to individual homes, and partly because the industrial demands are much less there.

Domestic water consumption varies with the method of transport (Table 5.1).

Table 5.1

Transport method	Average daily consumption per head
Carried by hand	10 litres
Carried from standpipe	25 litres
Single tap per household	47 litres
Several household taps	>160 litres

■ Contamination of water with inorganic ions

When rainwater reaches the land it dissolves soluble inorganic ions from the soil. As this water drains into the rivers it becomes contaminated with effluent from industries (including agriculture) and with the sewage effluent of towns. As a result, 'fresh water' often contains large amounts of dissolved ions, especially nitrate (NO_3^-) and phosphate (PO_4^{3-}) ions, together with smaller quantities of other ions.

The nitrate ions originate largely from intensive use of the land in agriculture. The phosphates come mainly from fertilisers and from the domestic and industrial use of detergents. These ions have accumulated in fresh water on a large scale in developed countries, particularly in the water draining from large cities and from the grain fields. Nitrate ions, above certain levels, may be harmful to people drinking the water. In infants, they interfere with normal haemoglobin function; in adults, they are a possible cause of stomach cancer.

An unfortunate consequence of the excess of phosphate and nitrate ions in fresh water is the enhancement of the natural process of eutrophication. Human-enhanced eutrophication occurs in lakes and reservoirs as well as in rivers, where the groundwater is enriched with nitrates and phosphates.

Figure 5.3   The steps to water purification

Water piped to homes is regarded as 'pure' if four criteria are met:
► it cannot transmit water-borne diseases (e.g. typhoid, cholera)
► it is bright and clear (suspended particles have been removed)
► it is free from chemical poisons
► it is reasonably 'soft' (forms a lather with soap).

Source
river →

Treatment
storage reservoir
settling out of particles, including suspended organic matter
↓

aeration by fountain or waterfall
– to remove CO_2 and H_2S (from decay of organic matter – eutrophication)
– to remove volatile organic solvents (from industrial effluents)
– flocculation of fine particles (clays, organic matter)

lake or reservoir →

↓

filtration by slow flow through fine sand beds, on which is a jelly-like film of bacteria and protozoa; this coating is largely responsible for the filtration
↓

deep underground water (may also require aeration, flocculation etc.) →

disinfection by contact with chlorine at 0.5 ppm for at least 30 minutes to kill any residual bacteria
↓

reduction of chlorine concentration by SO_2 treatment to a residual concentration of about 0.2 mg dm^{-3} to maintain disinfection during distribution
↓

covered service reservoir
↓

distribution

■ Eutrophication

In water enriched with inorganic ions beneficial to plant growth (p. 328) aquatic plants grow luxuriantly. In summer, in particular, algae may undergo a population explosion ('**algal bloom**', p. 85). Later, the organic remains of plants are decayed by saprotrophic aerobic bacteria. As a result, the water becomes deoxygenated. Anaerobic decay follows, and hydrogen sulphide is formed. The absence of oxygen (and the presence of H_2S) cause the death of other organisms, including fish. The initial enrichment may be caused by accidental pollution by raw sewage, by run-off from silage clamps, or from stock-yard effluent. (By contrast, a lake with water low in ions and organic matter is known as oligotrophic.)

Where does the NO_3^- in drinking water come from?

The process of 'top dressing' with inorganic fertilisers is often blamed for the occurrence of NO_3^- in drinking water, but if chemical fertilisers are applied at the time of most active plant growth, using the correct dosage, leaching away of nitrate ions to groundwater is minimal. Application at the wrong time, in overdose, however, does lead to the addition of nitrates to groundwater.

Organic fertilisers (such as farmyard manure) decay slowly, releasing essential nutrients for plant growth. In warm, moist weather in high summer, however, maximum decay may occur – perhaps at a time when the period of peak ion uptake by crop plants has passed. So much of the nitrate leaching to ground water may

WATER-BORNE PROBLEMS
For the developed world

For many years, effluents have been poured down the rivers of western Europe into the North Sea. Major sources of pollution are shown in Figure 5.7, p. 103. The North Sea is a fairly shallow basin, and the pollutants become concentrated in continental waters. The build-up of pollutants and increasing eutrophication are environmental hazards to life in the area.

For less-developed countries

The World Health Organization estimates that about 80% of all sickness and disease can be attributed to inadequate sanitation and impure water. Water-borne disease (p. 532) causes an estimated 25 million deaths every year. Most of this disease could be prevented.

Figure 5.4   How modern sewage treatment works

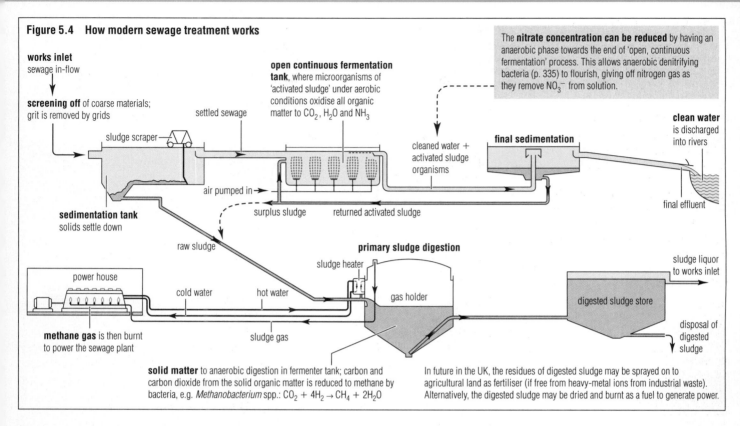

The **nitrate concentration can be reduced** by having an anaerobic phase towards the end of 'open, continuous fermentation' process. This allows anaerobic denitrifying bacteria (p. 335) to flourish, giving off nitrogen gas as they remove NO_3^- from solution.

works inlet
sewage in-flow

screening off of coarse materials; grit is removed by grids

open continuous fermentation tank, where microorganisms of 'activated sludge' under aerobic conditions oxidise all organic matter to CO_2, H_2O and NH_3

settled sewage

sludge scraper

final sedimentation

clean water is discharged into rivers

air pumped in →

cleaned water + activated sludge organisms

final effluent

sedimentation tank solids settle down

surplus sludge   returned activated sludge

raw sludge

primary sludge digestion

sludge heater

sludge liquor to works inlet

power house

cold water   hot water

gas holder

digested sludge store

methane gas is then burnt to power the sewage plant

sludge gas

disposal of digested sludge

solid matter to anaerobic digestion in fermenter tank; carbon and carbon dioxide from the solid organic matter is reduced to methane by bacteria, e.g. *Methanobacterium* spp.: $CO_2 + 4H_2 \rightarrow CH_4 + 2H_2O$

In future in the UK, the residues of digested sludge may be sprayed on to agricultural land as fertiliser (if free from heavy-metal ions from industrial waste). Alternatively, the digested sludge may be dried and burnt as a fuel to generate power.

---

come from soil humus, rather than from chemical fertilisers. Nevertheless, nitrate contamination of groundwater has been a feature of the period of maximum inorganic fertiliser application.

■ Waste water and industrial effluent treatment

The waste water from domestic use, like much of the effluent from commercial and industrial plant, is piped to a sewage treatment works for treatment and disposal. The main object of a modern sewage works is to turn the water, rich in organic matter and ammonium ions, into cleaner water.

Industrial effluent may be contaminated with poisonous substances. For example, chromium ions are released by leather tanneries, cadmium and nickel by factories producing long-life batteries, and copper and zinc by some metal-plating works. There are methods of recovering all these poisonous heavy metals from the effluent before they are discharged, but these are often regarded as being too expensive to operate. Effluent from photographic works contains silver ions, but these are always reclaimed. Poisonous substances from industry can damage the sewage treatment process because they kill the living organisms, including bacteria and other microorganisms, which are used to turn

sewage into re-usable and inoffensive substances.

Many coastal towns worldwide discharge raw or partly treated sewage into the sea. The results are eutrophication of the shallow coastal waters, and a health hazard for those who use the beaches.

Certain sea birds feed at sewage outfall pipes but often roost on inland reservoirs, and may transfer human parasites (parasitic protozoa and nematode worms, for example) from sewage to the water supply.

Sewage is purified by biological processes. Various methods have evolved, ranging from the simple cess pit to the modern activated sludge process coupled with anaerobic digestion (Figure 5.4). The sewage treatment process is another example of biotechnology. The treatment vessels are a form of fermentation tank (p. 112). As sewage is passed through, a range of microorganisms with the metabolism to degrade the organic matter present turn it into useful and sometimes re-usable products (carbon dioxide, water, ammonia and various soluble ions, for example). At the same time water-borne disease organisms, such as those of typhoid, cholera and dysentery, are eliminated. The biotechnology of sewage treatment delivers clean water to the rivers and reduces the risk of transmitting infectious diseases.

■ Biological oxygen demand (BOD)

The BOD of water is a measure of the organic pollution present. BOD is defined as the amount of oxygen taken up by water over 5 days when incubated in the dark at 20 °C. Clean water takes up less than 5 ppm; heavily polluted water, 10 ppm or more. A method of comparing the BOD of water samples, using methylene blue, is described in the *Study Guide*, Chapter 5.

Questions

1 A water sample gives more bacterial colonies when incubated at 37 °C than it does at 22 °C. How do you interpret these results?

2 In the procedure described in Figure 5.2, if five bacterial colonies grow on the 1/1000 dilution plate, what was the concentration of bacteria in the original 10 cm³ sample?

3 Nutrient agar plates, after inoculation and incubation, are not opened in the laboratory. Normally they are sterilised before the contents are disposed of. Why is this a wise precaution?

4 Explain the steps by which excess nitrate and phosphate entering rivers can lead to the death of fish.

5 What effect can industrial effluents have on
   a sewage treatment
   b sludge disposal?

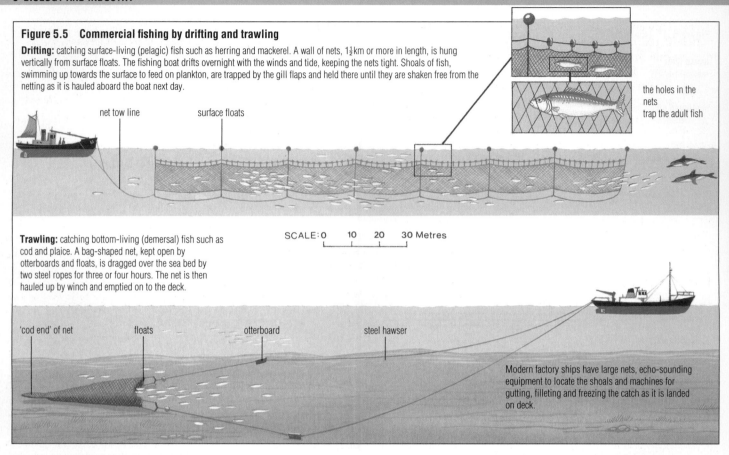

Figure 5.5 Commercial fishing by drifting and trawling

Drifting: catching surface-living (pelagic) fish such as herring and mackerel. A wall of nets, 1½ km or more in length, is hung vertically from surface floats. The fishing boat drifts overnight with the winds and tide, keeping the nets tight. Shoals of fish, swimming up towards the surface to feed on plankton, are trapped by the gill flaps and held there until they are shaken free from the netting as it is hauled aboard the boat next day.

net tow line   surface floats

the holes in the nets trap the adult fish

Trawling: catching bottom-living (demersal) fish such as cod and plaice. A bag-shaped net, kept open by otterboards and floats, is dragged over the sea bed by two steel ropes for three or four hours. The net is then hauled up by winch and emptied on to the deck.

SCALE: 0   10   20   30 Metres

'cod end' of net   floats   otterboard   steel hawser

Modern factory ships have large nets, echo-sounding equipment to locate the shoals and machines for gutting, filleting and freezing the catch as it is landed on deck.

5.2 BIOLOGY AND THE FISHING INDUSTRY

Fish feed directly or indirectly at the expense of plankton (p. 53). They are relatively efficient converters of plankton into flesh because, as ectothermic animals (p. 510), they do not use a share of the energy intake to maintain a high and constant body temperature. Fish are therefore able to use more of their food intake for growth than can endothermic animals (the mammals and birds). Much of the body of a fish is made up of muscle blocks (p. 496), a concentrated form of protein. The flesh is rich in vitamins A and D (p. 272).

The fishing industry is an ancient one. In island countries, like Britain, fish from the sea can make up a significant part of the diet. Marine fish are separated into two categories according to whether they occur near the surface of the sea (**pelagic fish**) or on the sea bed (**demersal fish**). As a result, the fishing boats are of two types: drifters fish for pelagic fish such as herring, while trawlers fish for demersal fish such as cod (Figure 5.5).

■ The world's fishing grounds

Fish stocks are not spread evenly throughout the world's oceans; indeed, very few fish are caught in the vast expanses of the open sea. Instead, fish tend to congregate in quite restricted parts of the sea, mostly where their food occurs.

Marine plankton is sustained by useful minerals dissolved in sea water.

Most minerals occur only in low concentration in sea water, however, except in two special areas. One of these is where wind and currents naturally combine to cause upwelling of water from the sea bed, which brings with it the minerals released by decay of dead organisms at

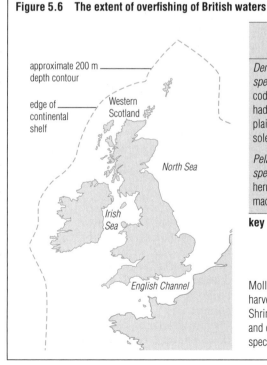

Figure 5.6 The extent of overfishing of British waters

approximate 200 m depth contour

edge of continental shelf

Western Scotland

North Sea

Irish Sea

English Channel

	North Sea	Western Scotland	Irish Sea	English Channel
Demersal species:				
cod	3	3	3	
haddock	2	3	?	
plaice	3		1	1
sole	3		3	2
Pelagic species:				
herring	3	?	3	
mackerel	**collapsed**			

key   **1** = fished, but not to maximum yield
2 = exploited at (restricted) maximum sustainable yield
3 = over-exploited

Molluscs and crustaceans are also part of the marine harvest, and many species have been over-exploited. Shrimps and prawns are caught by trawlers, scallops and oysters by dredging, and crabs and lobsters in special pots.

depth. **Upwelling zones** form approximately 0.1% of the oceans, yet 50% of the marine harvest comes from these areas. The Peru current off South America sustains one of the richest fishing areas in the world.

The other favourable areas for plankton growth are the shallow seas of continental shelf zones, where human activities and natural erosion cause minerals to flow out into the sea, and from which comes the remaining 50% of the marine harvest. Figure 5.6 shows the continental shelf around Britain, of which the North Sea is Britain's most important fishing ground.

■ Marine pollution

All too often, waste materials such as sewage are deliberately dumped in the sea; others find their way into the sea down rivers. These pollutants include harmful ions (mainly from heavy industry) and nitrates (mainly from agriculture) (Figure 5.7). Shipping itself has frequently been responsible for spillages of oil and other chemicals, and there is increasing use of toxic chemicals in the 'anti-fouling' paints applied to the hulls of ships. All these processes are introducing harmful substances into sea water, damaging the wildlife of the seas, and accumulating in food chains.

■ The need for conservation

The scale of fishing in the seas around Britain has been so intensive as to have depleted the fish stocks. Many marine species are an over-exploited resource. Fishing nations have experimented with bans on fishing of particular species; for example, in 1976 herring fishing in the north-eastern Atlantic Ocean was banned. These measures, when adhered to, are successful in enabling the breeding stocks to recover; herring fishing was allowed to recommence in certain areas of the Atlantic in 1984. Currently the European Commission is attempting to limit the size of fishing fleets, or the numbers of days spent at sea, and has regulations controlling the mesh size of nets. These have been less successful, because commercial fishermen naturally resist attempts to reduce or limit their income or (as they see it) their job prospects.

■ The need for marine biological research

The long-term commercial interests of the fishing industry demand careful attention to the conservation of fish stocks. To facilitate the protection of sea life, information is needed from marine biological research on the principal spawning grounds that need protection, the reproduction biology of fish and plankton, the feeding relationships and growth rates of fish, the effects of toxins on marine organisms, and the seasonal migratory movements of commercial species, some of which travel vast distances during their natural lives. Much of this research still remains to be done.

■ Whales and whaling

Whales are a diverse group of marine mammals that have evolved, apparently, from a single terrestrial mammal, and now divide into two groups. Those that have teeth in their jaws include the dolphins, porpoises, killer and sperm whales. Those that instead have a fibrous plate called 'baleen' that is used to filter-feed include the humpback, blue and minke whales. Some species, the 'great' whales, are huge, including all the baleen whales and the sperm whale. For example, the blue whale is the largest animal to have lived, measuring up to 30 metres and weighing up to 150 tonnes.

Conservation of whales is supervised. The 'International Convention for the Regulation of Whaling', set up in 1946, governs the conduct of whaling. It may designate whale sanctuaries, set limits on the maximum numbers taken per season, and prescribe 'open' and 'closed' seasons. It promotes studies into humane whaling. It has maintained a world-wide ban on whaling of great whales since 1985, but Norway objected and resumed commercial whaling (for minke) in 1993. Japan also objected and maintains 'scientific whaling' activities. The stocks of individual whale species vary, but none of the great whales is in imminent danger of extinction. Some species are identified as 'vulnerable', and the blue whale is one that is 'endangered'. The UK ceased whaling in 1963, and funds research into whale biology at St Andrews University, Scotland.

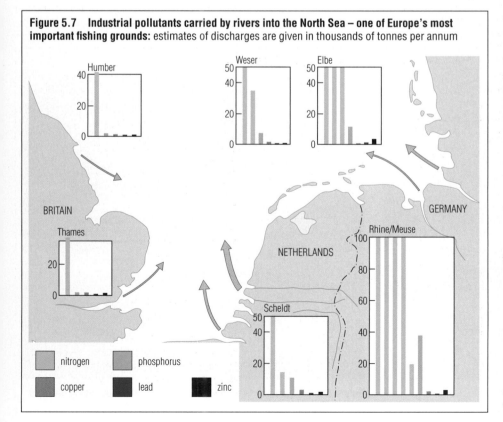

Figure 5.7 Industrial pollutants carried by rivers into the North Sea – one of Europe's most important fishing grounds: estimates of discharges are given in thousands of tonnes per annum

Questions

1 How may heavy-metal ions such as lead, released in the liquid effluent by inland industry, come to accumulate in the bodies of fish caught for human consumption?

2 **a** Suggest methods by which fish stocks can be conserved.

   b What role might a marine biologist play in this?

3 What features of the oceans determine the most favourable areas for catching fish?

■ Fish farming

Currently, over 60% of the world's fisheries are seriously over-fished, many to the point of collapse of natural fish populations. It is calculated that the ability to meet world demand for fish from natural stocks has reached a peak and is now declining. The prediction is that because of the growing human population and the dwindling stocks of naturally occurring fish, we will increasingly have to rely on fish farming (aquaculture). Only in this way can future food requirements be met without ruining global aquatic resources, it is argued.

Fish farms may occur in fresh waters, producing species such as the trout (for example, brown trout, *Salmo trutta fario*). Alternatively, fish may be net penned in sea lochs (for example, salmon, *Salmo salar*) or in sheltered bays in the open sea, in some regions (Figures 5.8 and 5.9).

The **advantages of fish farming** include

▶ avoidance of the physical perils of commercial fishing; fishing is a form of hunter–gatherer harvesting, which is often dependent on highly unpredictable weather conditions, and the efficient locating of suitable fish stocks, followed by speedy delivery of the catch to markets in as fresh a condition as possible

▶ the ability to arrange production to correspond to demand, and to deliver the required product exactly when it is needed, in a near perfect state of preservation (for example, frozen), if not as fresh fish

▶ the ability to genetically engineer species to have superior growth rates (more efficient conversion of feed into muscle tissue) and final weights, and other desired features, in relatively low temperature environments; also, the possibility of manipulating breeding seasons to extend the period in which the crop is available

▶ the ability to develop fish 'feed' from different sources, delivering the required nutrients at the lowest possible cost, often from plant based sources

▶ the ability to exploit the very high fecundity typical of fish, whereby thousands or millions of eggs are laid each season; in the wild, many eggs may fail to be fertilised and many that are fertilised fall to predators at every stage of development, whereas in the farm hatchery, survival rates are very high.

Disadvantages of fish farming include

▶ the use of pesticides to kill parasites and suppress growth of algae on netting, which damages the local environment

▶ the fact that it renders the efficient supply of fish protein a 'high-technology' industry, no longer readily available to communities that cannot maintain the sophisticated facilities required; the costs of fish protein are likely to rise

▶ the use of artificial feed at high levels in natural waters, fed to large numbers of fish, which may lead to greatly enhanced eutrophication of marine or fresh waters, due to wasted feed and the excretions of the fish, with the likely disruption of surrounding habitats; fishing becomes one more human activity that may devastate local environments, if it is not very carefully conducted

▶ the density of cultivation of fish in restricted spaces, which increases the chance of disease outbreaks spreading rapidly through the farm population and into the remaining natural stocks around; where pens are located along migratory routes, the damage to the wild stocks may be disastrous.

Figure 5.8   Salmon fish farming in a Scottish sea loch

Figure 5.9   A trout fish farm in artificial tanks and ponds, using diverted river water

5.3 BIOLOGY AND AGRICULTURE

Modern farming, as practised in many developed countries, is highly productive. Improvements in productivity are based on the selective breeding of plants and animals, the mechanisation of farming processes and the extensive use of fertilisers and pesticides. The farmer is a manager of a high-yield system, drawing upon technical developments in biology, chemistry, engineering and economics. The trend in most developed countries is towards fewer, larger farms. Typically, the modern farm is a rural factory, and farming is often referred to as **agribusiness**. In this system, sufficient profit must be made to meet the costs of fossil fuels and manufactured fertilisers, pesticides and animal feeds, and to maintain and modernise the expensive equipment required. The produce must be available to food retailers at prices they are willing to pay. Sometimes large profits are made and sometimes the profit margin is small. Profits can be wiped out by unseasonable weather or sudden outbreaks of disease.

■ The farming of crops

The biological principles of successful arable agriculture are

▶ optimum soil conditions that supply the essential minerals (p. 328), water and air
▶ the quality of plants grown, expressed by their yield and disease resistance
▶ reduction of competition by means of biological control (p. 92) and the use of pesticides.

The fifteen crop plants that make up the bulk of staple diets of humans and livestock are listed in Table 5.2. The impressive improvements in agricultural efficiency in developed countries since 1945 are illustrated by the rise in the annual wheat yield per hectare of land (Figure 5.10), and by the achievements in animal productivity through factory farming and via the mechanical milking of cows (p. 108).

Selective breeding

The selection imposed by humans that has produced the domesticated plants and animals used in agriculture is described in Chapter 29, p. 644.

The application of fertiliser

The growth of crops leads to depletion of the mineral elements in the soil. The role of inorganic ions in plant growth was discovered relatively recently (Figure 16.6, p. 328). Knowledge of the nutrient requirements of plants and the ability to measure the available soil ions has meant that the loss of plant nutrients from soil can be offset by the addition of compost, manure or artificial fertiliser (p. 331).

The use of fertiliser has increased productivity of farmland, but only where the correct amounts are applied. Too little fertiliser may result in a poor yield and insufficient profit to pay for the costs of mechanical harvesting. Too much fertiliser means that expensive chemicals are wasted, and the environment is harmed. There is a limit to how much extra growth and production a fertiliser can stimulate.

The application of pesticides

Pesticides (chemicals used to control animal and plant pests by killing many of them) are discussed in Chapter 4, p. 90. Pesticides are used to increase agricultural productivity by reducing competition from non-crop plants, by limiting damage by insect and nematode pests, and by largely eliminating diseases caused (mainly) by fungi. Sprays that kill pests must be applied at the correct stage, according to whether it is the eggs, larvae or adults that are most vulnerable or the most harmful, and also at the correct time, either to coincide with the arrival of the pests or before they have become too numerous to manage. Too early an application may wear off before the pest has started to become established. Too late an application may allow the crop to be harmed before the pest is checked. Either way the farmer faces the double loss of productivity and the costs of ineffective pesticide application.

Table 5.2   The crop plants that make up the bulk of our staple diets

Crop	Country of origin	Where now grown	Proportion of human diet*
Cultivated grasses (p. 644)			
wheat	Central Asia	North and Central America, Russia, Europe, India, China	13.5
barley	Central Asia	USSR, western Europe, North America	6.0
oats (also rye, millet)	Central Asia	Europe, North America	1.5
rice	South-east Asia	Asia	11.0
maize	tropical America	Mid-west USA, southern Europe, South America, Africa	11.0
sorghum	Africa	USA, India, China, South America, Africa	2.0
sugar cane	South-east Asia	tropical lowlands of Africa, South and Central America	22.0
Legumes† (p. 334)			
soyabean	South-east Asia	USA, China	3.0
Root crops and fruits			
potato	Peru	eastern and western Europe, North and South America	8.0
sugar beet	northern Europe	Europe, USSR, North America	11.0
cassava	South America	tropical lowlands of South America, Africa	3.0
sweet potato	Central America	tropical lowlands of China, Africa, South America	3.0
tomato	South America	Europe, North America	1.0
grape	southern/central Europe	Europe, North America	2.0
banana	South-east Asia	South and Central America, Africa, India	2.0

*Given as the percentage of worldwide human food produced directly from plants
†Includes peanuts and pulses (peas and beans)

Notes to table
1 All the above are flowering plants.
2 The largest crop grown is grass, with legumes such as clover, grown as forage for feeding animals – i.e. indirectly for humans.

Figure 5.10 Wheat yields in the United Kingdom

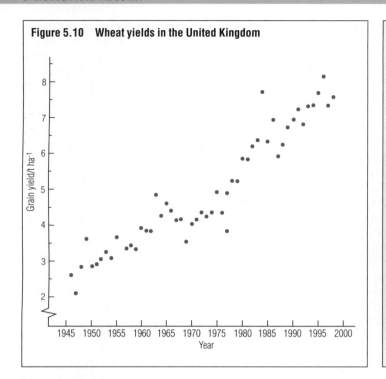

Field trials at Rothamsted Experimental Station

SOYABEAN (*Glycine max*)

The soyabean, *Glycine max*, is one of the oldest cultivated crops. Its cultivation in China is recorded about 5000 years ago, along with another species with which *Glycine* hybridises naturally. Soya was brought to Europe about 200 years ago, and is now grown as extensively as maize in the USA. It is grown for oil, as well as protein feed cake for animal feed. Soya flour is now added to wheat flour to improve protein content in human foods. Soya is one of the most genetically modified of food crops, so far.

the soyabean plant

Figure 5.11 Minimal cultivation in the planting of cereals

Mechanisation

Most of our countryside is used for the growth of arable crops and grass. The increasingly profitable production of these crops has followed the development of large, powerful and fast-working machines. All stages of the production cycle have been mechanised, from soil preparation, seed sowing and fertiliser and pesticide application, harvesting and storage of the crops. Machines have replaced many farm workers.

The search for more productive techniques often leads to the development of new machinery. An example is the development of a process of cultivation of arable crops without the plough.

For centuries the mouldboard plough has been the basic tool of agriculture, breaking and turning the soil and checking the growth of weeds, as the first stage of tilling the land. Today, the plough may be replaced by a system of cultivation in which the seeds of the next crop are planted in soil some time after the previous crop has been harvested, without ploughing (Figure 5.11). Weed control is achieved by the use of herbicides. Fewer machines and less fossil fuel are used, and cultivation is quicker. The danger of erosion is reduced.

Questions

1 Which family of flowering plants contributes most to human and livestock staple diets? What features of a common weed would enable you to identify it as a member of the same family (pp. 571 and 577)?

2 Explain the biological principles by which fertiliser, pesticide, selective breeding and irrigation can respectively increase the productivity of crops.

3 In what ways might knowledge about the physiology of a crop plant lead to increased food production?

■ Developments in animal husbandry

The biological principles of animal husbandry are

▶ selective breeding for growth, fecundity and resistance to disease
▶ optimising food and diet
▶ reducing losses as a result of disease
▶ minimising energy losses in movement and heat production.

Many of us would like to believe that farm animals are reared in traditional ways, moving freely in fields, supervised by a herdsman who has time to observe the needs of individual animals in relation to the changing seasons. In the case of hill sheep flocks this may be true in part, but today most other farm animals are reared intensively for most of their life cycle.

Factory farming of livestock

Farm animals reared indoors, under controlled environmental conditions and feeding regimes, grow rapidly. Animals are held at high densities with restricted movement. Since movement is a feature of animal behaviour, the factory farm gives rise to public concern that the conditions are stressful for the animals.

There is a risk of disease spreading in crowded conditions. As a result, broad-spectrum antibiotics (p. 530) are often added to feedstuffs in an attempt to prevent disease. This prophylactic use of antibiotics increases the cost of rearing. These substances, and certain growth hormones too, have been found to promote the animal's weight gain and final size, and also leanness of carcase. They are often given simply to guarantee a high yield. The use of drugs introduces the danger that poisonous residues, harmful to human health, may remain in the meat products.

EGG PRODUCTION

Battery house

Deep-litter house

Free-range

The case against and for battery house egg production:

'Animal rights' opinions

▶ it may be cruel to retain in cramped conditions animals that would wander and forage if free
▶ birds kept in groups in small metal compartments have to have their beaks clipped to prevent cannibalism
▶ the comfort and well-being of animals is sacrificed in the pursuit of lower costs and greater profits

▶ the health of animals can be maintained only by rearing them on feedstuffs containing antibiotics
▶ animals and animal products may contain residues of pesticides, antibiotics and growth-promoting hormones.

The producers' case

▶ conditions in battery houses (and under intensive rearing generally) are not cruel or stressful; if livestock were stressed, growth and yield might not be maintained
▶ other systems of livestock rearing are more expensive
▶ the majority of the public would much rather have cheap food than pay more in the interest of animal 'welfare'.

Mechanisation of milk production

Dairy farm production of milk has many of the features of modern factory farming. Cattle feed on grass in the fields in summer, and sometimes on crops such as kale grown for grazing in winter, provided the climate and soil type permits. Most winter feeding of the milking herd is by **silage**, however. This is a fermented grass product made in early summer from wilted grass stored where the air content, pH and temperature are regulated to produce nutritious and palatable feed. The grass is prevented from turning into compost either by baling it and sealing the bales in huge plastic bags, or by tipping the grass into a clamp which is quickly filled and rolled to exclude air. Conditions in the clamp are kept acidic (at a pH of 4), often by the addition of an acidic silage enhancer preparation. The microorganisms that produce compost are inhibited.

Dairy cows are milked in modern concrete and steel 'parlours' by a system in which suction tubes withdraw the milk from the udder, and pipe it to a cooled bulk tank. The milk is transported daily by bulk tanker, unrefrigerated, to a dairy factory. Here it is pasteurised (p. 527) and prepared for sale as liquid milk or manufactured into a range of dairy products including butter, cheese (p. 112) and yoghurt (p. 113).

Milk production in the cow is under the control of a hormone called **bovine somatotrophin** (BST) produced by the pituitary gland (p. 474). If cows are treated with extra BST on a regular basis milk production is enhanced. The health of the cow, and that of the humans drinking the milk, appear to be unaffected. There is current research and some controversy about this new development in factory farming.

MILK PRODUCTION AND THE MILKING PARLOUR

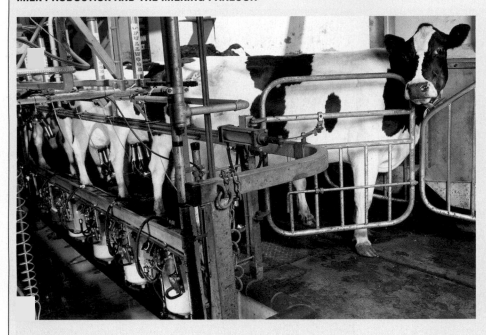

There is a milking machine (set of suction cups that fit over the cow's teats) to each two cows. Milk is withdrawn into storage bottles behind the cows' rear feet (the milk volume can be measured) and then transferred to the bulk tank in the dairy nearby. The clusters can be removed automatically when the flow of milk stops.

The life cycle of the cow, and the sequence of insemination–birth–lactation by which milk and calves are produced (male calves are used for veal and beef, female calves for milk production)

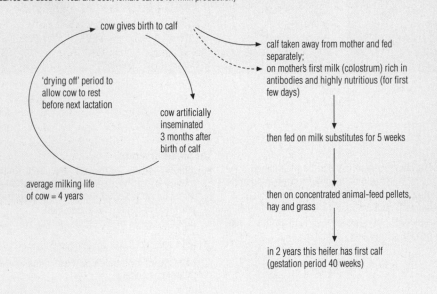

cow gives birth to calf

calf taken away from mother and fed separately;

on mother's first milk (colostrum) rich in antibodies and highly nutritious (for first few days)

'drying off' period to allow cow to rest before next lactation

cow artificially inseminated 3 months after birth of calf

then fed on milk substitutes for 5 weeks

average milking life of cow = 4 years

then on concentrated animal-feed pellets, hay and grass

in 2 years this heifer has first calf (gestation period 40 weeks)

Cows in oestrus being inseminated by artificial insemination (AI)

5.4 BIOLOGY AND THE FOOD INDUSTRY

We eat in order to provide ourselves with energy to maintain life and raw materials to build and repair the body. What constitutes a balanced intake of food is detailed on p. 271. During human history our diet has changed with the move from hunting and gathering food to settled agriculture, and again as agriculture has become more productive. Here, two modern issues of human nutrition are outlined: preservaton and food additives.

■ Preserving and storing foods

The need for food is continuous, but the production of food is seasonal. Most crops are harvested only at particular times of the year. To ensure a steady supply, some of the food produced at harvest time must be preserved and stored. Processing must prevent food loss and decay, yet must leave the food wholesome and attractive to eat. Processing also involves precooking, and the combining of various components of a whole dish.

The method of processing of food and the mechanism of storage depends upon the type of food to be stored. For example, living material may be harmed by physiological deterioration such as respiration of fruit in store and sprouting of potato tubers. Here storage conditions that retard natural processes but do not kill the cells are required. The relatively long-term storage of most foods, however, requires the cells to be killed and the tissues rendered incapable of supporting the growth of microorganisms. Many different methods are used (Figure 5.12).

Dehydration

The oldest method of preserving food is by dehydration, the natural method of preserving grain and many fleshy fruits since the earliest days of agriculture. Dried food is preserved because microorganisms require water to be active, and because the natural enzymes in the dehydrated tissues are inactivated. Drying does not destroy the microorganisms on the food, however. They are present as spores and will become active the moment the food is rehydrated, or if it becomes moist in storage. Dried food is light, and is easily transported and stored. On the other hand, drying often brings about a change in texture and sometimes a loss in palatability as well.

Drying may be carried out in the air, using the energy of the sunlight and the wind to carry away water vapour, or artificially by passing warm dry air over the product. Foods such as egg and milk are dehydrated completely under reduced pressure and lower temperatures, so that the proteins present are not denatured (p. 180). In freeze-drying (in the manufacture of instant coffee, for example) the frozen food is subjected to high vacuum. Water becomes ice, and then evaporates as water vapour (sublimation), without permitting the volatile compounds responsible for flavour and aroma to be driven off by heat.

Canning

Canning exploits the fact that if food is adequately cooked and then sealed while hot it will remain sterile and be palatable for a long time. In canning, heat-sterilised food is sealed in air-tight containers (Figure 5.13). Pathogens and food-spoiling bacteria are destroyed and the enzymes in the food are inactivated. Canning is relatively cheap and, because of the convenience, remains a popular means of preserving food. Although cans are heavy and bulky they are easily stored. Canning keeps food sterile for long periods, although eventually cans may corrode.

There is a need for careful handling during canning, however, because of the danger of contamination by spores of the bacterium *Clostridium botulinum*. These can survive in food that has been only mildly heated, which is alkaline or only slightly acidic (pH 5.0 or above) and which is stored at room temperature. Under these conditions the toxin botulin may form. Botulin is exceedingly poisonous, and can kill humans at very low concentrations. It is not formed in food that contains sodium chloride at a concentration of 8% or more, or sugar at a concentration of 50% or more. It is inactivated by heating. For safe canning, therefore, the foods generally need to be very thoroughly cooked. Consequently there is a loss of quality in canned foods.

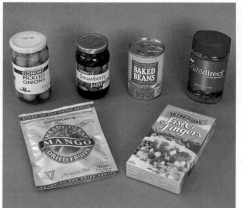

Figure 5.12 Various methods of preserving food

Food may be preserved either by killing all the microorganisms present (bacteriocidal methods; for example canning, ionising radiation), or by stopping the activities of the microorganisms (bacteriostatic methods; for example dehydration, freezing). Nutrient loss is greatest when foods are heated or dehydrated in order to preserve them.

Figure 5.13   The canning process

Freezing

Chilling has long been used to slow down the rate of food deterioration. Modern fast-freezing is achieved by exposure of food in small quantities to temperatures of −29 to −40 °C. Food is cooled to −18 °C in thirty minutes or less (Figure 5.14). Quick-freezing produces tiny ice crystals which cause little mechanical damage in the cells and tissues. Micro-organisms are still present in frozen food, but biochemical reactions are prevented and liquid water is unavailable. In the developed countries quick-freezing is a popular method of preserving certain meats, fruits and vegetables, and now supports a growing trade in ready-prepared complete meals.

Frozen foods are very convenient, provided that the special equipment required to freeze food and hold the frozen food in safe storage is available. The domestic fridge does not reach low enough temperatures for storage of frozen food (Figure 5.15), although the ice box is normally below 0 °C. In frozen foods the loss of texture and flavour is minimal, except with foods with very high water content such as lettuce or strawberries, and the nutritional value remains high.

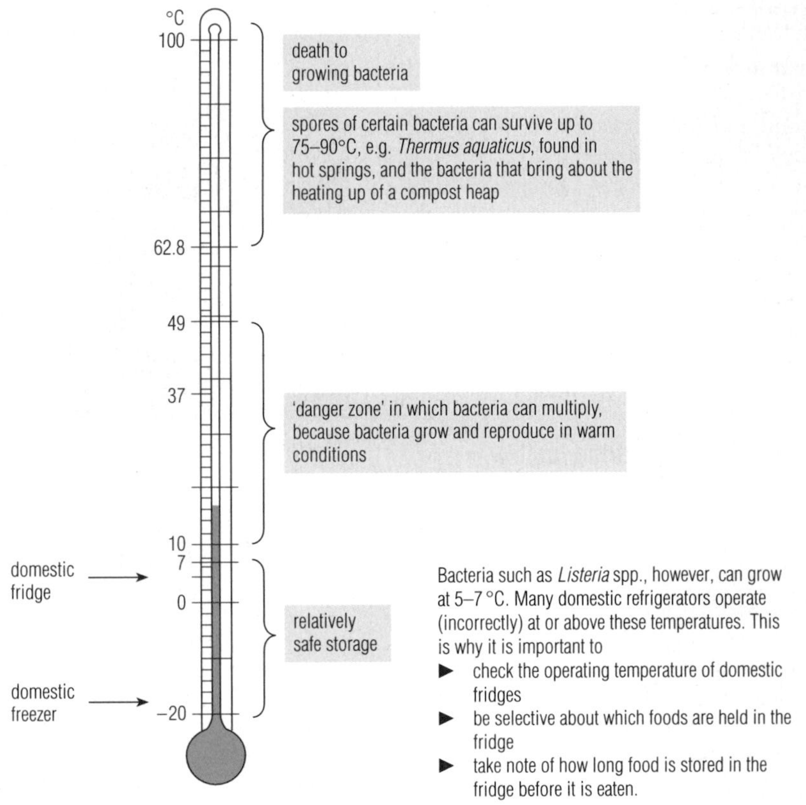

Preservation by chemicals

Many naturally occurring chemicals such as salt, sugar and vinegar will preserve foods.

Common salt, sodium chloride, is probably the oldest preservative, used especially with fish and meats. Tradition-ally, foods were soaked in a concentrated solution of rock salt. Sodium chloride (the major ingredient of the rock salt) dissociates to form sodium ions and chloride ions, and these ions strongly attract water molecules (p. 232). Water migrates from the cells and tissues to the brine solution, and the tissues are dehy-drated. Any bacteria present are also destroyed by dehydration, because the salts penetrate deep into the meat.

Rock salt contains small amounts of nitrates. Bacteria in meat turn nitrates to nitrites, and nitrite ions combine with haemoglobin in the blood (p. 358), forming nitrosohaemoglobin. This sub-stance gives the meat a strong pink colour, even when cooked. The perma-nent pink colour of ham and bacon, for example, shows that salts have reached all the tissues, and the meat is safe to eat. Nitrites and nitrates are regularly used in the preservation of meat products.

Sugars are also used as a preservative of fruits, especially in jams. The high concentration of sugar used causes de-hydration by osmosis and thus destroys any microorganisms present.

When a high concentration of sugar in foods is viewed as unhealthy an altern-ative preservative for fruits is the chemical sulphur dioxide. Sulphur dioxide is a reducing agent, which makes it an effect-ive preservative. Sulphur dioxide is volatile and is driven off in cooking.

Pickling food in vinegar (ethanoic acid solution) is another traditional method of preserving foods. In this case the very low pH (p. 128) inhibits the growth of microorganisms, particularly bacteria.

Figure 5.15 Temperature and bacterial growth

°C
- 100 — death to growing bacteria
- spores of certain bacteria can survive up to 75–90°C, e.g. *Thermus aquaticus*, found in hot springs, and the bacteria that bring about the heating up of a compost heap
- 62.8
- 49
- 37 — 'danger zone' in which bacteria can multiply, because bacteria grow and reproduce in warm conditions
- 10
- domestic fridge → 7
- 0
- relatively safe storage
- domestic freezer → −20

Bacteria such as *Listeria* spp., however, can grow at 5–7 °C. Many domestic refrigerators operate (incorrectly) at or above these temperatures. This is why it is important to
▶ check the operating temperature of domestic fridges
▶ be selective about which foods are held in the fridge
▶ take note of how long food is stored in the fridge before it is eaten.

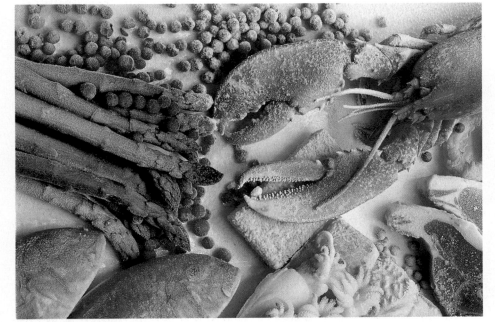

Figure 5.14 A wide range of foods can be preserved by freezing

EXAMPLES OF FOOD ADDITIVES

Some additives are obtained from natural sources

E101 riboflavin (vitamin B_2)
E140 chlorophyll (plant pigment)
E161 xanthophyll (plant pigment)
E170 anthocyanin (plant pigment)
E170 calcium carbonate (chalk)
E300 L-ascorbic acid (vitamin C)
E330 citric acid
  551 silicon dioxide (sand)

PRAWN COCKTAIL

INGREDIENTS
Maizemeal, Vegetable Oil
and Hydrogenated
Vegetable Oil, Starch, Prawn
Cocktail Flavour (Acidity
Regulator—E262), Flavour
Enhancer (621), Citric Acid,
Flavouring, Artificial
Sweetener (Saccharin),
Sugar, Salt, Colours
(E110, E160b).

additives as recorded on a common item of food

Some substances may be harmful if they accumulate

Additive		Uses	Possible adverse effects
E110	sunset yellow – a synthetic dye	used in orange squash, marzipan, apricot jam, lemon curd, etc.	allergic reactions in some people
E102	tartrazine – a synthetic dye	used in smoked haddock, cheese rind, lime and lemon squash	allergic reactions in some people
E124	Ponceau 4R – a synthetic dye	a red colour used in packet soups, seafood dressings, tinned fruits, jelly	allergic reactions in aspirin-sensitive people and asthmatics
E230	biphenyl – manufactured from benzene	used as an antifungal preservative on the skins of oranges, lemons and grapefruits	nausea in workers exposed to it
E250	sodium nitrite – naturally occurring salt	a meat-curing salt used to inhibit the growth of *Clostridium botulinum*	nausea and asthma in some people; banned from baby foods
621	sodium hydrogen glutamate (monosodium glutamate)	used as a flavour enhancer in protein-containing foods	nausea and migraine in some people

Ionising radiation

In the preservation of food by irradiation, short-wave ionising radiation from electrons, X-rays or gamma-rays passes through the pre-packaged food as it is carried along a conveyer belt. At the correct strength, the radiation kills the bacteria without making the food radioactive. As the food is already sealed, bacteria cannot re-enter.

Patients in intensive care whose immune systems have been medically suppressed receive food sterilised in this way, but otherwise ionising radiation has only a limited use in preserving foods. This is because low doses of ionising radiation do not inactivate enzymes or viruses, and irradiation does not destroy the toxins left behind by bacteria. Doses large enough to sterilise tend to produce odours and change the texture of food.

■ Food additives

Agriculture is the industry that produces our food. The food industry is involved in preserving, processing (precooking), packaging and distributing the food. The aspects with a biological component are preserving and processing.

Sugar and salt are common food additives, important in the preservation of food (see opposite), but also present (in similar quantities) to alter taste and flavour. The other food additives that are generally considered to be safe to use,

provided that they are present at low enough concentrations, have been listed and given 'E numbers' within the European Community. This regulation is part of the move to harmonise laws within the Community, and to facilitate the movement of food between countries. Some examples of both naturally occurring and artificial additives, most of which have been given E numbers, are given in the panel above. All food additives in current use are listed in M Hassen (1984), *E is for Additive: the Complete E Number Guide*, Thorsons Publishers Ltd.

Questions

1 What are the likely advantages and disadvantages of rearing bacon pigs outdoors on extended range rather than intensively in a pig house?

2 How is mown grass turned into silage and prevented from becoming compost?

3 At what stage in the growth and development of a cow can milk production commence?

4 What are the likely problems involved in storing a dried grain crop for poor people in a less-developed country?

5 Food preservation depends on destroying microorganisms in food or, at least, preventing their multiplication. Say how each of the following achieves either of these objectives: canning, refrigeration, dehydration.

6 Name two types of preserved food that you expect to contain antioxidants, two that may contain emulsifiers, and two that may contain salt used as a preservative.

The principal categories of additives are as follows:

▶ **Permitted colours,** added to make the product look attractive, by making it appear similar to the fresh product; some dyes occur naturally, others are synthetic.

▶ **Preservatives,** added to prevent or retard the growth of microorganisms in food (mechanisms of food preservation are discussed on pp. 109–11).

▶ **Antioxidants** retard the oxidation of fats in food exposed to air. Unsaturated fatty acids are susceptible to oxidation at their double bonds (p. 130). The products of oxidation are shorter-chain fatty acids, which mostly have unpleasant odours. Oxidised fat is described as being 'rancid'. Vitamin C is a naturally occurring antioxidant.

▶ **Emulsifiers and stabilisers** maintain the natural fats and oils present as stable suspensions, which are known as emulsions, rather than allowing components of a food mixture to separate out and solids to precipitate.

Other additives include sweeteners, and thickeners such as modified starches.

5.5 BIOTECHNOLOGY

Biotechnology is defined as the industrial application of biological processes. Biotechnology is often presented as one of the youngest biological sciences, but its origins go back for many centuries in the forms of cheese production, brewing and winemaking, and baking.

The birth of the modern biotechnological revolution came in the 1960s, with the development of molecular biology (Chapter 7, p. 149). The understanding of cells in terms of their molecular structures, and in terms of the chemical changes occurring during biological processes, are the bases of modern biotechnology. Today many aspects of biology contribute to biotechnology. These include microbial fermentation and microbiology generally, enzymology (p. 178), genetics and the manipulation of the hereditary material (p. 604), immunology (p. 366) and plant biology.

■ Fermenter technology

For many biotechnological processes the key steps are conducted in a fermenter vessel. The biotechnologist's purpose is to optimise the growth of microorganisms and the chemical changes they carry out. To do this an appropriate energy source and other essential nutrients must be present, and waste products efficiently removed. Within the fermenter vessel the physicochemical conditions are monitored and favourable conditions maintained. In all forms of fermenter the aim is to ensure all parts of the vessel are subject to the same conditions.

Definition of fermentation

Fermentation as a biotechnological process involves culturing microorganisms in an aqueous suspension in a culture vessel known as a fermenter. The word fermentation used in this context is, strictly speaking, a misnomer. This is because the term has long been used by biochemists with reference to the anaerobic breakdown of glucose to ethanol or lactic acid (p. 316). The word is now also used more loosely, however, and a fermenter has become any vessel used for the cultivation of microorganisms.

■ Open fermentation systems

Fermenter tanks may be simple non-sterile systems, open to the atmosphere. These are used when it is not essential to operate with a pure culture of microorganisms – for example, in brewing vats and in the activated sludge aeration tanks in the sewage works (Figure 5.4), as well as in cheese manufacture and in the bacterial extraction of metals from ore dumps.

Cheese manufacture

In milk exposed to the air, contaminating bacteria convert milk sugar (lactose, p. 134) to lactic acid. Acidic conditions cause milk proteins to solidify.

In cheese production, the protein (casein, p. 282) is precipitated from the liquid milk by the action of a starter culture of bacteria and by the addition of rennet, an enzyme preparation originally produced from the stomachs of calves. Casein coagulates to a solid curd, leaving a watery whey. The curd is separated from the whey, pressed to squeeze out liquid, then wrapped in a cloth to dry. The characteristic flavour and texture of cheese develop in the ripening processes as a result of enzymic changes caused by the microbial population retained in the curd (Figure 5.16).

Cheese was probably the first fermented food. Approximately 10 kg of milk are needed to make 1 kg of cheese, and the resulting cheese will normally keep for long periods. Cheese is therefore a way of storing milk as a food, turning a seasonal glut of milk into a preserved food supply.

Figure 5.16   The process of cheese production

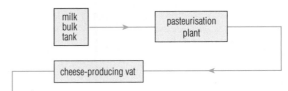

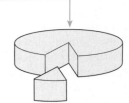

steps:
1 starter culture of lactic acid bacteria added to warm milk
   – pH lowered
2 protein-coagulating enzyme preparation added
   – milk turned to curd (solid) and whey (liquor) using (originally) 'rennet', a preparation of rennin from the stomach of calves, now with a fungal protease, or with chymosin from genetically-modified yeast (carrying the rennin gene)

3 cut to release whey, salt is added as a preservative, then curd warmed and pressed into mould lined with muslin (cheese cloth) where it is left to 'cure' for months

cheese manufacture

MICROBIAL MINING

The deliberate use of microorganisms to extract metal from extremely low-grade ores is a recent development. It has become necessary because the demand for metals for industrial uses has outstripped the supply of accessible, high-grade deposits of ores. For example, by means of the bacterial leaching of copper the metal is being economically recovered from old mine waste containing only 0.25–0.50% of copper. The process is outlined below. The bacteria concerned obtain their energy to synthesise sugar from a chemical reaction and are therefore described as chemosynthetic (see Chapter 12, p. 269).

Copper extraction by bacterial action

The tiny, rod-shaped *Thiobacillus ferro-oxidans* occurs naturally in ore dumps in the absence of light and in an extremely acid medium. The bacterium 'eats rock' – it is chemolithotrophic. It oxidises sulphides to sulphates and iron(II) to iron(III) ions:

$$CuFeS_2 + 2Fe_2(SO_4)_3 + 2H_2O + 3O_2 \xrightarrow{\text{bacterial action}} CuSO_4 + 5FeSO_4 + 2H_2SO_4 + ENERGY$$

insoluble ores → weak copper sulphate solution, sulphuric acid

The bacterium consumes the ore by transferring electrons from iron ions and sulphide ions, making the ore more soluble. The electrons are transferred to oxygen to produce water and give up energy, which is coupled to ATP synthesis.

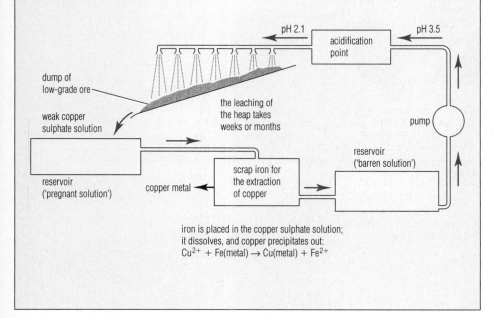

iron is placed in the copper sulphate solution; it dissolves, and copper precipitates out:
$$Cu^{2+} + Fe(metal) \rightarrow Cu(metal) + Fe^{2+}$$

YOGHURT PRODUCTION

Pasteurised milk is first heat-treated (at 90 °C for 20 minutes) to denature and coagulate the milk protein. The milk is then homogenised, cooled and inoculated with a 'starter culture' of two bacteria, *Lactobacillus bulgaricus* and *Streptococcus thermophilus*. The mixture is incubated at about 40 °C for up to 6 hours. Bacterial metabolism produces several substances that give the yoghurt its characteristic consistency, aroma and flavour, including diacetyl, methanoic (formic) acid, ethanal (acetaldehyde) and lactic acid. The product is cooled and sometimes flavoured and coloured before packaging. Some yoghurts are further heat-treated to kill bacteria.

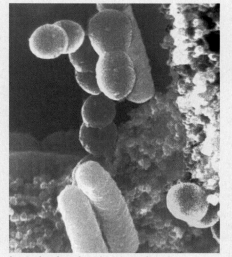

bacteria of yoghurt 'starter culture'

■ Closed fermentation systems

Certain fermentation processes exploit a particular microorganism, and therefore the possibility of contamination by other organisms must be excluded .

Handling microorganisms

Although very many species of bacteria are harmless, the bacteriologist handles all cultures as if they were dangerous. The techniques are known as **sterile** or **aseptic**, and they are designed to prevent contamination of cultures by the environment (and of the environment by the bacterium). Microorganisms used are obtained in the laboratory from sterile cultures.

All the nutrients used are sterilised, as are the culture vessels and other equipment, both before and after use for growing microorganisms. Sterilisation is normally brought about by heating at raised pressure, carried out in an autoclave. This is a laboratory instrument that functions in much the same way as a domestic pressure cooker.

Solid or liquid preparations called **media** (singular 'medium') contain the nutrients required by microorganisms for growth. A nutrient supplies a source of carbon and energy (for example, glucose, fatty acid), a source of nitrogen (combined nitrogen as amino acids or as nitrates or ammonium ions), essential growth factors and vitamins, mineral salts (macronutrients and micronutrients, p. 328) and water. A liquid medium is called a **broth**. A solid medium is produced by addition of a gelling agent called **agar**, which is a more-or-less transparent polysaccharide obtained from seaweeds (p. 30). It is not normally digested by microorganisms. A 1–2% solution of agar melts at 95 °C and sets at about 44 °C.

Liquid media are used for growing microorganisms in bulk, and for studying their growth rates. Solid media (that is, agar plates in Petri dishes) are used for holding reference cultures from which liquid cultures can be inoculated. They are also used to isolate individual species of bacteria from a mixed culture.

Preparation of sterilised nutrient media for laboratory handling of microorganisms is introduced in Figure 5.17. The technique of inoculating an agar plate with microorganisms, prior to incubation, is shown in Figure 5.18.

Figure 5.17 Preparing nutrient media for culturing microorganisms

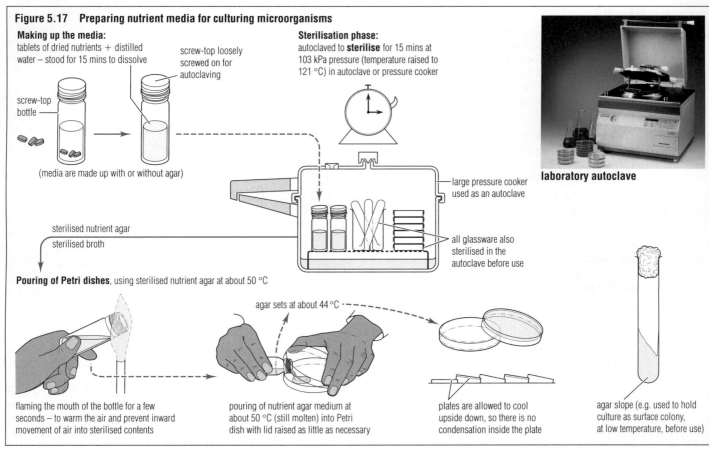

Making up the media:
tablets of dried nutrients + distilled water – stood for 15 mins to dissolve

screw-top loosely screwed on for autoclaving

screw-top bottle

(media are made up with or without agar)

Sterilisation phase:
autoclaved to **sterilise** for 15 mins at 103 kPa pressure (temperature raised to 121 °C) in autoclave or pressure cooker

laboratory autoclave

large pressure cooker used as an autoclave

all glassware also sterilised in the autoclave before use

sterilised nutrient agar

sterilised broth

Pouring of Petri dishes, using sterilised nutrient agar at about 50 °C

agar sets at about 44 °C

flaming the mouth of the bottle for a few seconds – to warm the air and prevent inward movement of air into sterilised contents

pouring of nutrient agar medium at about 50 °C (still molten) into Petri dish with lid raised as little as necessary

plates are allowed to cool upside down, so there is no condensation inside the plate

agar slope (e.g. used to hold culture as surface colony, at low temperature, before use)

Figure 5.18 The procedure for inoculation of a Petri dish

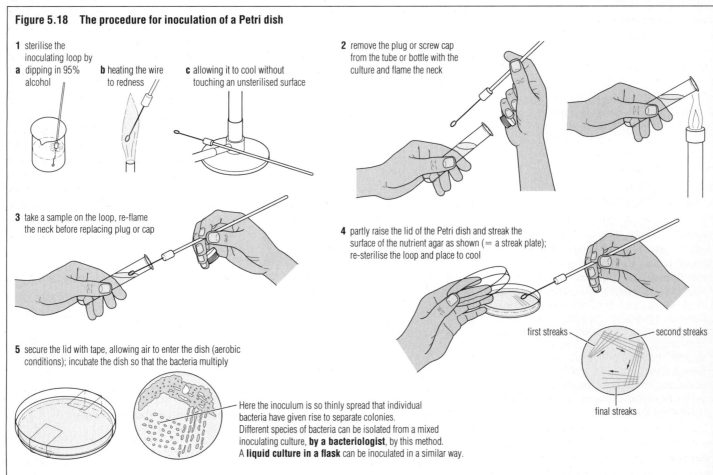

1 sterilise the inoculating loop by
a dipping in 95% alcohol
b heating the wire to redness
c allowing it to cool without touching an unsterilised surface

2 remove the plug or screw cap from the tube or bottle with the culture and flame the neck

3 take a sample on the loop, re-flame the neck before replacing plug or cap

4 partly raise the lid of the Petri dish and streak the surface of the nutrient agar as shown (= a streak plate); re-sterilise the loop and place to cool

first streaks

second streaks

final streaks

5 secure the lid with tape, allowing air to enter the dish (aerobic conditions); incubate the dish so that the bacteria multiply

Here the inoculum is so thinly spread that individual bacteria have given rise to separate colonies.
Different species of bacteria can be isolated from a mixed inoculating culture, **by a bacteriologist**, by this method.
A **liquid culture in a flask** can be inoculated in a similar way.

The principles of closed fermentation

When a closed fermentation system is adopted it is possible to culture a single species or new strain of microorganism, and exclude contamination by other organisms. As a result, the finished products are more closely controlled.

1 Finding appropriate organisms

The microbiologist may screen soil samples and other sources for new organisms or different strains that are capable of manufacturing a particular metabolite as part of their biosynthesis pathways (for example, an antibiotic).

2 Laboratory culture

The microorganism has to be cultured in laboratory conditions to find the nutrient requirements and environmental conditions that lead to the desired product being formed in significant quantities.

3 Scale up

Once the laboratory process has been identified, the production process has to be developed on increasingly larger scales to test that production proceeds as desired. Often, particular conditions require adaptation to deliver a result similar to that achieved on the laboratory bench. This phase may take time as 'teething problems' are overcome.

4 Fermenter vessel as a bioreactor

In the fermenter vessel (Figure 5.19), optimum conditions can be precisely maintained. Conditions within the bioreactor, (for example, pH, temperature, oxygen concentration, concentration of metabolites) are measured continuously, and substances can be added to maintain constancy. The contents are continuously stirred to achieve even distribution and uniform conditions.

5 Fermenter systems

In **batch culture**, an initial, fixed volume (culture medium and microorganisms) is processed in the fermenter with few additions, until the maximum product has accumulated. Then the products are harvested and the fermenter prepared for re-use. In **continuous culture**, the fermenter is run for an extended period, with fresh nutrient added and products (cells or dissolved substance) harvested at a steady rate.

6 'Downstream processing'

The product of fermentation may be a dissolved substance that accumulates in the medium; an example is the antibiotic penicillin, produced by the fungus *Penicillium*. In this case the used culture medium, after filtering, is retained and processed. On the other hand, in processes such as single-cell protein

Figure 5.19   A large-scale closed fermenter vessel

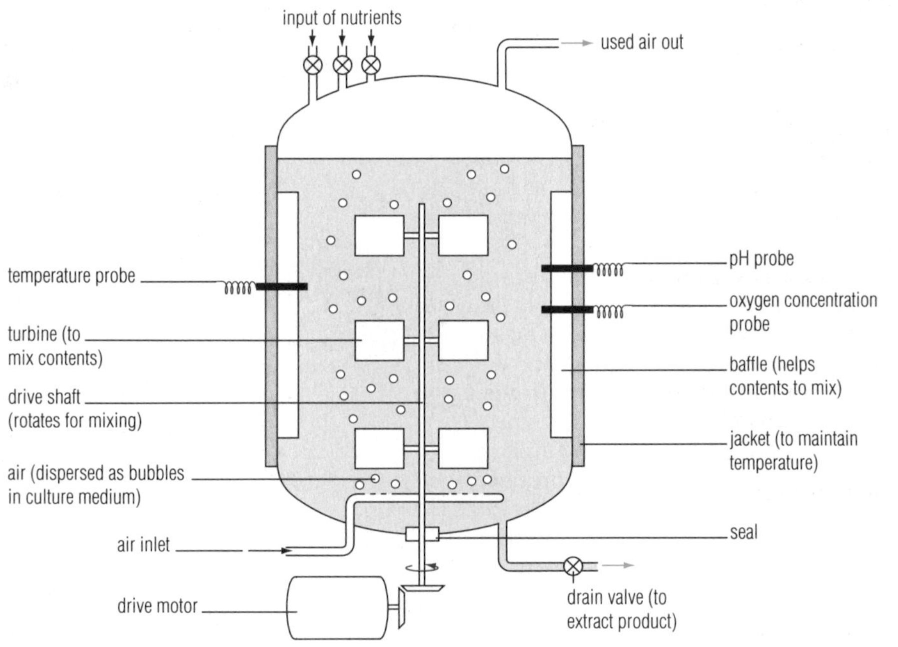

input of nutrients

used air out

temperature probe

turbine (to mix contents)

drive shaft (rotates for mixing)

air (dispersed as bubbles in culture medium)

air inlet

drive motor

pH probe

oxygen concentration probe

baffle (helps contents to mix)

jacket (to maintain temperature)

seal

drain valve (to extract product)

manufacture (p. 117) the product is the microorganism itself. In this case the medium may be recycled and the solid processed.

Harvesting the product on a bulk scale may involve more than purification of the product. Sometimes it is also necessary to modify and improve the product, perhaps using enzyme technology (Figure 5.20).

Examples of batch culture

Antibiotic production

Most antibiotics are produced by culturing certain fungi or bacteria of the actinomycete group that occur naturally in the soil. Large-scale production first started in the 1940s. The earliest antibiotics were natural products; today most are chemically modified and therefore semi-synthetic. Production starts in huge, closed fermentation vessels holding from 100 000 to 200 000 litres of nutrient solution inoculated with a starter culture of an antibiotic-secreting microorganism, under carefully controlled conditions. The antibiotics are secreted into the culture solution by the fungus after the initial growth phase is over.

After fermentation the antibiotic is extracted by filtration (Figure 5.20). The natural antibiotic is often modified to increase its effectiveness by passing it over immobilised enzymes (p. 189).

Antibiotic production is a huge and profitable industry. The cost of production of a product such as penicillin is lowered by the development of new strains of microorganism by genetic engineering (Chapter 29, p. 628), and by research into alternative culture media.

Figure 5.20   A flow-diagram of penicillin production

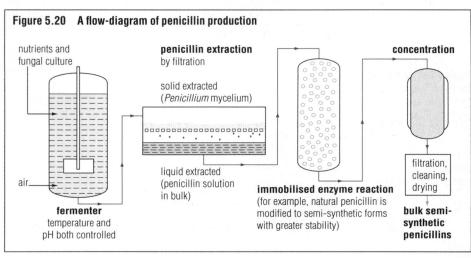

nutrients and fungal culture

air

fermenter
temperature and pH both controlled

penicillin extraction by filtration

solid extracted (*Penicillium* mycelium)

liquid extracted (penicillin solution in bulk)

immobilised enzyme reaction (for example, natural penicillin is modified to semi-synthetic forms with greater stability)

concentration

filtration, cleaning, drying

bulk semi-synthetic penicillins

Citric acid

Citric acid (citrate) is widely used in jams, drinks and sweets, and in other manufactured foods, mostly as a flavour enhancer and as an antioxidant. It is also a component of cosmetics and pharmaceuticals, where it may be added to adjust pH. In heavy industry, its ability to bind with metal ions is exploited in various ways, and it is also used in the tanning of leather.

Citrate is one of several organic acid ions manufactured microbially. Batches of waste carbohydrate liquor, typically molasses from sugar refining, are cultured in fermenters with the fungus *Aspergillus niger*. Citrate is an intermediate of the Krebs cycle (Figure 15.22, p. 315), so the fungus not only produces this acid, but naturally metabolises it too. Further conversion of citrate is slowed by the exclusion of Fe^{2+} ions from the culture medium. Citrate accumulates (Figure 5.21).

Examples of continuous culture

Biodegradable plastic

Plastic is a major component of rubbish (p. 93). Most plastics are manufactured from valuable oil or coal deposits, and after use accumulate in the biosphere, being slow to decay (Figure 5.23). A bacterium, *Alcaligenes eutrophus* (Figure 5.22), accumulates granules of a plastic called polyhydroxybutyric acid (PHB) in

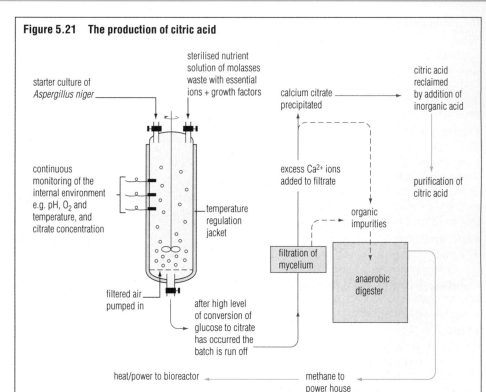

Figure 5.21   The production of citric acid

the cytoplasm, when cultured with adequate carbohydrate nutrient, but in conditions of nitrate and phosphate deficiency. PHB is held by the cells as a food reserve. When PHB is extracted and used to manufacture plastic (known as Biopol), the product is strong, but quickly biodegradable.

PHB is obtained commercially by culture of *Alcaligenes eutrophus* in a fermenter by continuous culture.

The medium is periodically withdrawn, the bacterial cells filtered off, and the nutrients recycled to support further growth. From the harvested bacteria, PHB is extracted.

Figure 5.22   Biological sources of plastics

TEM of the bacterium *Alcaligenes eutrophus*, showing naturally occurring granules of the polymer polyhydroxybutyric acid (PHB). This plastic can comprise up to 85% of the dry weight of the cell

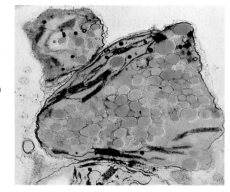

Thale cress *Arabidopsis thaliana*, has been genetically engineered to produce granules of the polymer PHB in its chloroplasts, in the light. The genes added to the thale cress come from the bacterium *Alcaligenes*

Figure 5.23   Making plastics biodegradable: plastics that are otherwise only very slowly broken down in the soil by microbial action (of saprotrophic fungi and bacteria) can be made to biodegrade more quickly by the addition of starch granules (seen here in a section through the plastic film)

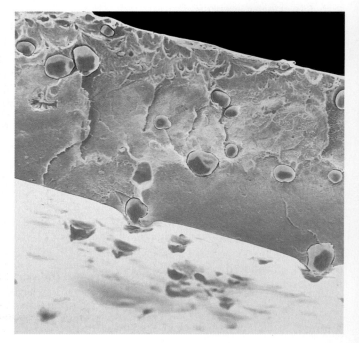

Figure 5.24   Quorn, a nutritious product: the mycoprotein has been made fibrous to resemble meat protein

Single-cell protein

Single-cell protein (SCP) is produced by the growth of microorganisms. The resulting biomass contains carbohydrates, fats, nucleic acids and vitamins as well as proteins. Since protein is usually the scarce and expensive constituent of the diet the protein component in the product is valuable. The organisms used are bacteria, cyanobacteria, yeasts and certain filamentous fungi. The product is biomass suitable for feeding to livestock, or even to humans. The pros and cons of using microorganisms for protein production are summarised in the panel below.

An example of an SCP food is 'Quorn', which is produced from the filamentous fungus *Fusarium graminearum* (Figure 5.24). The fungus is grown in a liquid medium containing glucose obtained from hydrolysed starch, or molasses. The 'mycoprotein' product has the additional attraction of being low in saturated fat and high in dietary fibre. Your local supermarket may stock 'Quorn' and foods manufactured from it.

SINGLE-CELL PROTEIN PRODUCTION

Advantages

▶ Microorganisms can grow rapidly when conditions are suitable. For example, the times taken by various organisms to double their mass are compared in the table below.

Organism	Time required to double mass
Bacteria and yeasts	20–120 minutes
Algae	2–6 hours
Grass	1–2 weeks
Chickens	2–4 weeks
Pigs	4–6 weeks
Cattle (young)	1–2 months

▶ Prokaryotes are more easily modified genetically than are plants and animals (recombinant genetics, p. 628).
▶ The protein content of microorganisms is generally high.
▶ Microorganisms can be grown in fermenter vessels by continuous processes.
▶ Microorganisms can grow on a wide range of raw materials, including waste materials.

Disadvantages

▶ The ratio of nucleus to cytoplasm in unicells may lead to toxic levels of nucleic acids in the product (which are then corrected).
▶ The initial product is colourless, odourless and tasteless.

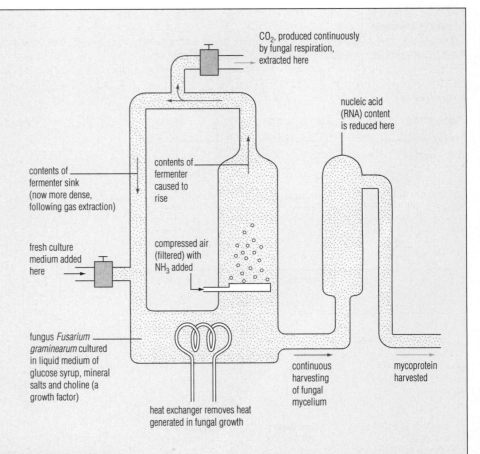

Mycoprotein manufacture

An example of a continuous 'air-lift fermenter', designed jointly by Rank Hovis McDougall and ICI, in which a filamentous fungus is cultured. From the fungus a nutritious product is produced, which is marketed under the trade name 'Quorn'.

5.6 BIOLOGICAL FUEL GENERATION

The need for renewable sources of energy has been established (p. 79). Biological sources of power directly exploit the ability of green plants to photosynthesise (Chapter 12, p. 246). The percentage of the world's energy obtained from biomass is shown in Figure 4.8 (p. 81). Biomass in the form of wood and farm waste is the most popular form of fuel worldwide; much is burnt in open fires for heating and cooking. This is very inefficient, but is the only way the very poor can obtain fuel.

One important example of the use of biomass is the energy farming of green crops for production of ethanol, a liquid fuel suitable for the motor car engine. Another is the conversion of organic waste matter to methane, a gaseous fuel suitable for cooking, heating and power generation. Additionally, biotechnological research in developed countries seeks to adapt the photosynthetic mechanism of the chloroplasts of green leaves for hydrogen gas production. Hydrogen is a gaseous fuel that poses no environmental problems.

■ Energy farming: the Brazilian 'gasohol' project

In this Brazilian scheme, devised following the steep rise in oil prices during the 1970s, the energy crop is sugar cane. Sugar cane is a C_4 plant (p. 266), and has high photosynthetic efficiency compared with many other crops. The Brazilian climate is suitable for growing sugar cane on a vast scale. In many other plants the main store of energy from green plants is in starch, and therefore they are less useful for alcohol production. (However, genetic engineering, discussed on p. 628, is being applied to develop yeasts that also produce amylase to turn starch to sugar.)

Sugar extracted from the cane is fermented by yeasts to ethanol, producing at best a 15% solution. This solution does not burn. Cars can be run on a 95% solution, or on an ethanol/petrol mixture (Figure 5.25). To obtain a practical fuel, distillation of ethanol is necessary; this is powered by burning the dried remainder of the sugar cane crop.

SUGAR CANE AS AN ENERGY CROP

The sugar cane industry, at present in decline, holds great potential as an energy crop industry, which is yet to be exploited.

Major cane-growing areas of the world

Cane is found throughout most of the tropics and subtropics, as far north as southern Spain and as far south as Australia. The highest yields are obtained in areas where there is a long warm growing season followed by a cooler and drier ripening period free of frosts.

Extraction of sugar

On harvesting, the cane must go to a local factory or mill where the sugar can be extracted immediately. Bacteriological deterioration is otherwise a major problem.

the sugar cane plant

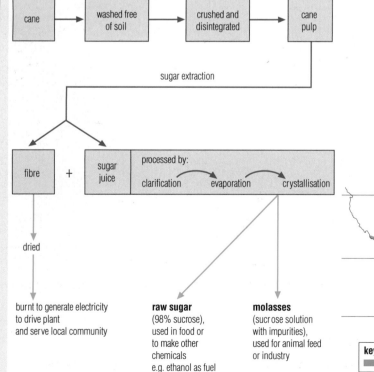

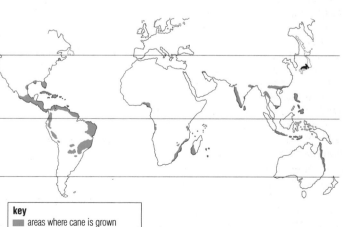

key
■ areas where cane is grown

There are drawbacks to the 'gasohol' scheme. For example, monocultures of sugar cane occupy vast areas of land, with harmful ecological effects. Land used to grow 'fuel' crops is not available for food production; the people employed to grow fuel crops may have an immediate need of land for food production. At the same time, keeping more cars on the roads creates many environmental problems (pp. 84–5).

■ Methane from biomass

Conversion of organic matter to methane occurs naturally in the absence of oxygen. Small methane fermenter plants, using animal dung and water held under anaerobic conditions, have been built throughout India, China and Pakistan, and produce, in total, a huge volume of methane (Figure 5.26). These are efficient at village levels, but as large-scale plants they are not economic. The cost of storing and transporting gaseous fuel is high. Methane is not a useful fuel in the motor vehicle.

In the developed world, much domestic waste is disposed of at landfill sites. The refuse biodegrades under anaerobic conditions, yielding a gas mixture rich in methane. This gas may be collected for use as a fuel. The approach is economically viable, and reduces both the unpleasant smells released and atmospheric pollution by methane (a 'greenhouse gas', p. 84).

Figure 5.25   An ethanol-powered car in Brazil, developed as a result of the Alcool (gasohol) project

Figure 5.26   A domestic 'biogas' digester in an Indian village, using animal manure and producing methane for cooking and heating water

Questions

1   What chemical and physical changes are involved in cheese making?
2   What is the importance of routine measurements of temperature, pH and oxygen concentration in a bulk fermenter producing penicillin?
3   In the extraction of copper from low-grade ores, how does the bacterium *Thiobacillus* benefit from the chemical changes it causes in its immediate environment?
4   What improvements to the effectiveness of an antibiotic might reasonably be sought by the pharmaceutical industry?
5   What are the advantages and the disadvantages to a less-developed country of being able to produce a substitute for petrol from a locally grown plant crop?
6   What is meant by 'recombinant genetics' (p. 628)? Give an example of the application of recombinant genetics to a biotechnological process.

FURTHER READING/WEB SITES

P Chenn (1997) Advanced Biology Readers: *Microorganisms and Biotechnology*. John Murray
Food and Farming Educational Service Fact Sheet: **www.cla.org.uk/ffes**
National Centre for Biotechnology Education: **www.ncbe.rdg.ac.uk**

EXAMINATION QUESTIONS SECTION 1

Answer the following questions using this section (chapters 1 to 5) and your further reading.

1 The diagram below shows the structure of a bacterial cell as seen using an electron microscope.

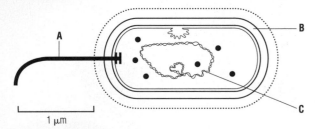

1 μm

a Name the parts labelled **A**, **B** and **C**. (3)

b Give **two** reasons why viruses are an exception to the cell theory. (2)

London A Biology and Human Biology Module Test B/HB1 January 1997 Q2

2 Hydra is a diploblastic, radially symmetrical animal. The diagram shows a longitudinal section of part of the body wall of a hydra.

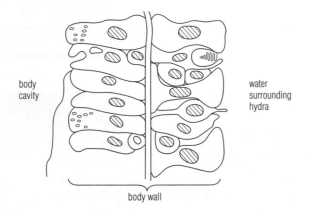

body cavity

water surrounding hydra

body wall

a Use information in the diagram to explain what is meant by the term *division of labour*. (1)

b Explain how radial symmetry may be an advantage to a sessile animal. (2)

c Larger animals, such as members of the phylum Chordata, possess blood systems while smaller animals, such as members of the phyla Platyhelminthes and Cnidaria, do not. Explain the link between body size and the possession of a blood system. (2)

AEB A Biology Module Paper 3 June 1998 Q2

3 a Define the following terms:

   i *habitat* (1)

   ii *niche* (1)

   iii *community* (1)

   iv *population* (1)

In the United Kingdom, deciduous trees lose their leaves in October–November and new leaf growth takes place in April–May of the following year.

Figure 1 shows a pyramid of biomass for a deciduous woodland in July in the United Kingdom.

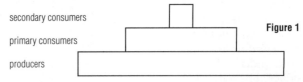

secondary consumers

primary consumers

producers

Figure 1

b Explain

   i how the biomass of the second trophic level would be determined; (4)

   ii why the biomass decreases at each trophic level. (2)

Figure 2 represents a pyramid of biomass for the same woodland in January. This pyramid is drawn to the same scale as Figure 1.

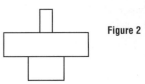

Figure 2

c With reference to Figure 1 and Figure 2, explain the changes in biomass of each of the trophic levels, between the woodland in July and in January. (4)

UoCLES AS/A Modular: Central Concepts in Biology June 1998 Section A Q1

4 The diagram below shows a number of stages in an ecological succession in a lake.

a Use information in this diagram to help explain what is meant by an ecological succession. (2)

b Give **two** general features which this succession has in common with other ecological successions. (2)

c A number of small rivers normally flow into this lake. These rivers flow through forested areas. Explain how deforestation of the area might affect the process of succession in the lake. (2)

AEB A Biology Paper 1 June 1998 Q8

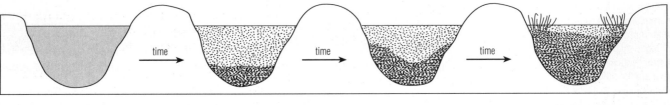

deep, clear, nutrient-poor water with very little aquatic life

nutrients and sediment begin to accumulate; increasing populations of aquatic life appear

nutrient-rich, relatively shallow water with much plant growth and many other aquatic organisms

oldest stage of a lake; very shallow, overgrown with emerging rooted plant life

5 The diagram shows a fermenter developed for the production of protein from bacteria.

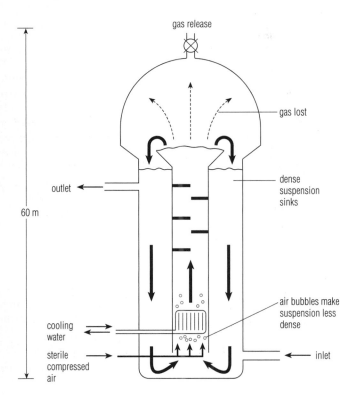

a i Suggest **two** functions of the compressed air.   (2)
   ii Explain why the contents of the fermenter are continually circulated.   (2)
b Some fermenters are designed for continuous production.
   i Give **two** differences between continuous and batch production.   (2)
   ii Give **two** advantages of continuous production.   (2)

NEAB AS/A Biology Microorganisms and Biotechnology (BY06) Module Test March 1998
Section A Q2

6 The diagram on the right shows the mean annual percentage losses of ozone in the Northern and Southern Hemispheres over a ten year period.
a With reference to the diagram,
   i describe the distribution of ozone loss in the Southern Hemisphere;   (3)
   ii compare this distribution with the losses in the Northern Hemisphere.   (3)
The ozone layer in the atmosphere reduces the amount of harmful ultra-violet light reaching the surface of the Earth.
b Outline why the maintenance of the ozone layer is biologically important.   (3)
c Explain how chlorofluorocarbons (CFCs) are thought to damage the ozone layer.   (4)
The damaging effects of CFCs were discovered during the 1970s, but it was several years before many governments finally accepted the link between CFCs and ozone depletion.
d Outline the steps that have been taken by the international community to reduce the release of CFCs into the atmosphere.   (2)
e State **two** reasons why it will be several years before any effective reduction in the CFC level in the atmosphere can be shown.   (2)

UoCLES A Modular: Ecology and Conservation June 1998 Section A Q4

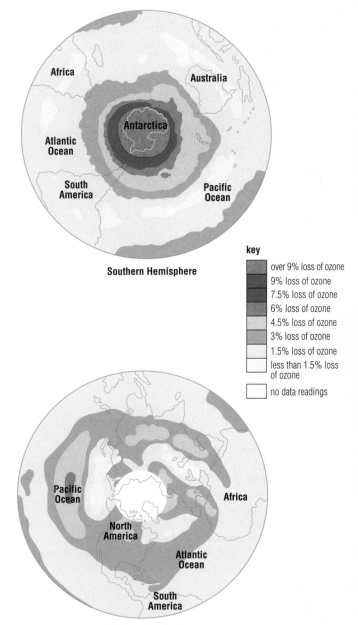

Southern Hemisphere

key

over 9% loss of ozone
9% loss of ozone
7.5% loss of ozone
6% loss of ozone
4.5% loss of ozone
3% loss of ozone
1.5% loss of ozone
less than 1.5% loss of ozone
no data readings

Northern Hemisphere

7 Lactase is an enzyme used in the dairy industry to break down the sugar, lactose, in milk into simpler sugars. Lactase is produced commercially using a species of yeast, *Kluyveromyces fragilis*. This yeast is grown in a medium containing lactose as the only energy source. After fermentation, the yeast cells are broken open and the resulting suspension is centrifuged. The liquid produced is filtered and then the lactase is precipitated.
a i What name is given to the series of methods used after fermentation to isolate the enzyme?   (1)
   ii Why is the suspension of broken yeast cells centrifuged?   (1)
   iii Explain how the information in the passage suggests that this yeast does not digest lactose outside its cells.   (1)
b Explain how an enzyme, such as lactase, breaks down its substrate.   (3)
c Lactose is the only energy source for the yeast in this fermentation. Suggest a reason for this.   (1)

AQA (NEAB) AS/A Biology: Microorganisms and Biotechnology (BY06) Module Test
June 1999 Q1

8 The diagram below shows the energy flow through a salt-marsh ecosystem. The figures represent units of energy.

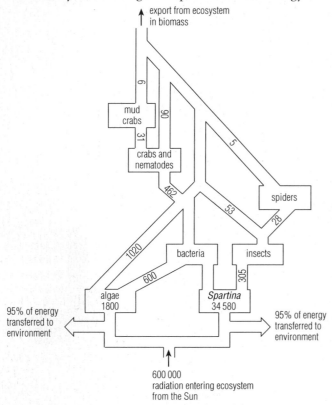

a i Name a tertiary consumer in this ecosystem. (1)
  ii Energy is exported from this ecosystem in biomass. Suggest **one** way in which this biomass is exported. (1)
b i The diagram shows the value of the solar radiation entering the ecosystem as 600 000. Suggest suitable units for measuring this value. (1)
  ii Calculate the percentage of the energy consumed by spiders which is lost in respiration. Show your working. (2)
c Suggest **two** reasons why so little of the light energy falling on the producers is used for photosynthesis. (2)

NEAB AS/A Biology Processes of Life (BYO1) Module Test March 1998 Section A Q4

9 a What is meant by the term *eutrophication* when applied to a lake or pond? (2)
b Human activities may accelerate eutrophication. State two such activities and outline, in each case the reason for the effect. (2, 2)
c Name **one** *abiotic* factor which may alter as a result of eutrophication and state how this affects the *biotic* component of the pond or lake. (2)

OCR (Oxford Modular) A Biology: Environmental Biology March 1999 Q1

10 Explain the techniques involved in the use of each of the following in **fieldwork**:
a i quadrats; (3)
  ii mark-release-recapture technique. (3)
b Explain a **named laboratory technique** which you have used in connection with fieldwork. (3)

OCR (Oxford Modular) A Biology: Environmental Biology March 1999 Q2

11 The number of earthworms in a field may be estimated by using frame quadrats. The quadrats are placed at random on the surface of the area being sampled. The ground is then watered with a very dilute solution of formalin. The earthworms which come to the surface are collected and washed.
a i Explain why the quadrats should be placed at random. (1)
  ii Throwing a quadrat does not ensure a random distribution. Describe a method by which you could ensure that the quadrats would be placed at random. (2)
b Give **one** advantage of describing the size of the population in terms of biomass rather than as the number of earthworms collected. (1)
c Similar sized populations of earthworms were kept in soils at different temperatures. The earthworms were fed on discs cut from leaves. The table shows the number of leaf discs eaten at each temperature.

Temperature/°C	Number of leaf discs eaten
0	0
5	178
10	204
15	174
20	124

Using information in the table, explain how mean soil temperature and feeding activity might affect the size of the earthworm population. (3)

AQA (NEAB) AS/A Biology: Ecology (BY05) Module Test March 1999 Q2

12 The diagram shows how organisms may be separated into five kingdoms.

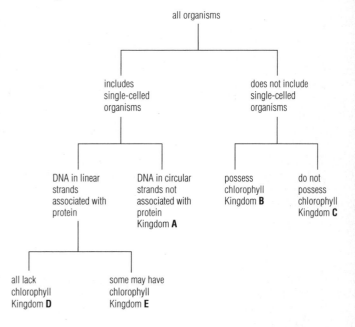

a i Name kingdom **B**. (1)
  ii Give **one** characteristic, other than the possession of chloroplasts, which could be used to distinguish cells of organisms in kingdom **B** from those of organisms in kingdom **C**. (1)
b Which of kingdoms **A, B, C, D** or **E** represents the Fungi? (1)
c *Microactinium* is a single-celled eukaryotic organism. It is an autotroph. Which of kingdoms **A, B, C, D** or **E** includes *Microactinium*? (1)

AQA (AEB) A Human Biology: Paper 1 June 1999 Q7

2 Cell Biology

Work in a cytology laboratory where DNA, extracted from human cells, is analysed by techniques including electrophoresis and autoradiography

6 The chemistry of living things

- Water is a polar molecule, electrically neutral overall, but with a negative charge on the oxygen atom and a positive charge on the hydrogen atoms. As a result, hydrogen bonds form, giving water unique properties.
- Carbon forms bonds with itself, producing a huge range of straight-chain, branched-chain and ring molecules. Carbon also forms covalent bonds with hydrogen, oxygen, nitrogen, sulphur and phosphorus. Carbon compounds with atoms of these elements form the bulk of the compounds of cells.
- Carbohydrates are molecules of carbon, hydrogen and oxygen, in which the ratio of hydrogen to oxygen is usually 2:1. Carbohydrates are simple sugars, or compounds built from sugars, condensed together.
- Lipids also contain carbon, hydrogen and oxygen, but in lipids the proportion of oxygen is much smaller than in carbohydrates. Lipids include fats, oils and steroids, and are insoluble in water.
- Proteins contain nitrogen as well as carbon, hydrogen and oxygen. Proteins are large molecules, built of a linear sequence of amino acids, but taking up complex, three-dimensional shapes.
- Nucleic acids are built from nucleotides, each consisting of an organic base (adenine, thymine, guanine or cytosine) combined with a five-carbon sugar and a phosphate. Nucleic acids are the 'information' molecules of cells.

6.1 THE CHEMICAL MAKE-UP OF LIVING THINGS

Elements are the basic units of matter. The Earth is composed of about 92 stable elements in varying abundances, including substances like hydrogen, oxygen, carbon, nitrogen and calcium. Living things are built from only 16 of these. In fact, the chemical composition of living things is quite different from that of their surroundings (Table 6.1).

The bulk of the Earth's crust consists of the elements oxygen, silicon, aluminium and iron. Of these elements, only oxygen forms a substantial part of living things. An important point about the chemistry of life is that cells are highly selective in what is taken up from the surroundings. We shall return to this feature in Chapter 11.

■ The chemical similarity of living things

About 99% of the substance of all living things is made up of the four elements **hydrogen, oxygen, carbon** and **nitrogen**. The chemical composition of plants, fungi, protoctista and prokaryotes is similar to that of the human body and of all other animals. These elements predominate in living things because

▶ living things contain large quantities of water

▶ most other substances present in cells are compounds of carbon combined with hydrogen and oxygen.

Carbon chemistry, also known as **organic chemistry**, is a special branch of chemistry, and of great importance to biology.

■ The elements of the chemistry of life

Oxygen is the heaviest of the major elements in living things but it is the lightest of the common elements of the Earth's crust. In fact, the elements that make up the bulk of living things are among the smaller and lighter of the elements that form the crust.

Generally speaking, smaller atoms form stronger and more stable bonds with other atoms than do larger atoms.

Table 6.1 The chemical composition (%) of our bodies, sea water and the Earth's crust

Human body		Sea water		Earth's crust	
H	63	H	66	O	47
O	25.5	O	33	Si	28
C	9.5	Cl	0.33	Al	7.9
N	1.4	Na	0.28	Fe	4.5
Ca	0.31	Mg	0.033	Ca	3.5
P	0.22	S	0.017	Na	2.5
Cl	0.03	Ca	0.006	K	2.5
K	0.06	K	0.006	Mg	2.2
S	0.05	C	0.0014	Ti	0.46
Na	0.03	Br	0.0005	H	0.22
Mg	0.01			C	0.19
All others < 0.01		All others < 0.1		All others < 0.1	

Only most commonly occurring elements shown. Figures are rounded, and consequently totals do not amount to 100.

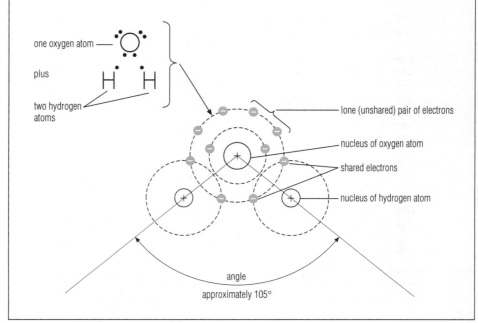

Figure 6.1 Introducing atoms and molecules: the water molecule

An **atom** is the smallest part of an element that can take part in a chemical change.
An atom consists of a **nucleus** of **protons** (positively charged particles) and, usually, **neutrons** (uncharged particles), with small, negatively charged **electrons** moving around the nucleus in orbits or shells.
An atom of oxygen combines with two atoms of hydrogen to form a molecule of water.

A **molecule** is the smallest part of an element or compound that can exist under normal conditions (e.g. H_2O, CO_2, O_2). When atoms combine to form molecules a **chemical bond** is formed (Figure 6.6).

one oxygen atom

plus

two hydrogen atoms

lone (unshared) pair of electrons

nucleus of oxygen atom

shared electrons

nucleus of hydrogen atom

angle approximately 105°

The elements hydrogen, oxygen, nitrogen and carbon form strong, stable **covalent bonds** by sharing one, two, three and four electrons respectively (Figure 6.6). Oxygen, nitrogen and carbon are among the few elements that also form double and triple bonds (p. 130). In addition, carbon atoms can link together to form straight-chain, branched-chain and ring molecules (Figure 6.8).

The combination of carbon atoms with atoms of other elements provides a huge variety of molecular structures. Molecular strength and biochemical diversity underpin the diversity of living things.

We will now examine water because of its importance to life. At the same time we can revise the underlying chemical principles we need in order to understand the chemistry of living things.

6.2 WATER

Most living things are solid, substantial objects, yet water forms the bulk of their structures. Water makes up between 65 and 95% by mass of most plants and animals. For example, about 80% of an individual human cell consists of water.

Water is formed when two hydrogen atoms combine with an oxygen atom by sharing electrons. The result is a stable molecule, which is relatively unreactive but which has some unique properties. The shape of the water molecule is triangular rather than linear; the angle between the nuclei of the atoms is approximately 105°. Overall the molecule is electrically neutral, but in both of the oxygen–hydrogen bonds the oxygen nucleus draws electrons (negatively charged) away from the hydrogen nucleus (positively charged). Thus there is a net negative charge on the oxygen atom and a net positive charge on the hydrogen atom. A molecule that carries an unequal distribution of electrical charge is called a **polar molecule**.

Because of this charge separation in an overall neutral molecule, water molecules form relatively weak bonds called **hydrogen bonds** with other water molecules nearby. Hydrogen bonds are also formed with any charged particles that dissolve in water, and with charged surfaces in contact with water. Hydrogen bonds account for the unique properties of water (Figure 6.2).

Figure 6.2   Hydrogen bonds formed by water molecules

the water molecule is polar: this is a simplified representation of the molecule shown in Figure 6.1

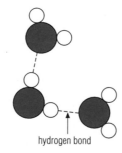

− small negative charge

\+ small positive charge

electrostatic attraction between a positively charged region of one molecule and a negatively charged region of a neighbouring one forms weak bonds called hydrogen bonds

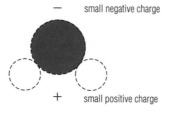

hydrogen bond

the angle between the covalent or hydrogen–oxygen bonds in water is very close to the angles of a perfect tetrahedron

in ice the water molecules form a regular tetrahedral arrangement

in liquid water the arrangement is more irregular but an enormous number of hydrogen bonds remain intact

hydrogen bonds

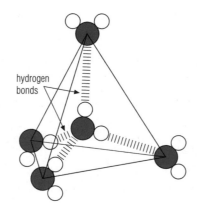

water is sometimes called the 'universal solvent' because it dissolves a greater variety of substances than any other liquid:

ionic substances (e.g. sodium chloride) dissolve in water because a shell of orientated water molecules is formed around each ion

substances such as simple sugars and alcohols dissolve due to hydrogen bonding between polar groups (e.g. —OH, the hydroxyl group) in their molecules and the polar water molecules

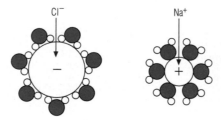

Cl^- Na^+

water molecules take up ordered positions at charged surfaces

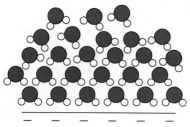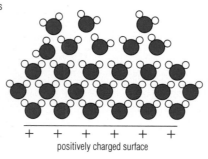

− − − − − − −
negatively charged surface

\+ + + + + + +
positively charged surface

Questions

1   Explain what is meant by the following terms: element; atom; molecule; ion (p. 126); electron; chemical bond (p. 128).

2   **a**   Name three elements that are abundant in living organisms but sparse in the Earth's crust.

   b   Name a group of substances, common in the human body but *not* common in sea water, in which hydrogen occurs.

3   The angle between the oxygen–hydrogen bonds in water is about 105°. Why does the existence of this angle cause the water molecule to be polar (although overall it is electrically neutral)?

■ The properties of water

Water (relative molecular mass $M_r = 18$) is a liquid at room temperature, in contrast with other substances with small molecules such as methane ($M_r = 16$), ammonia (17), hydrogen sulphide (34) and carbon dioxide (44), which are gases. In gases the molecules are widely spaced and free to move about independently of each other, whereas in liquids the molecules are close together. In water, the molecules are held close together by hydrogen bonding (Figure 6.2). It is the hydrogen bonds that cause water to exist as a liquid at the temperatures and pressures that normally prevail on the Earth's surface. Life as we know it depends on this property.

Water has a high heat capacity; that is, a great deal of heat energy is required to raise the temperature of water. This is because much of the energy is used to break the hydrogen bonds, which restrict the mobility of the molecules. As a result, water is relatively slow to heat up or cool down. In fact, the specific heat capacity of liquid water is the highest of any known substance.

The latent heat (enthalpy) of fusion of water (the heat energy needed to melt ice) is unusually high. By the same token, relatively large amounts of heat energy must be extracted from liquid water before it freezes. The latent heat of vaporisation of water (the heat energy required to vaporise liquid water) is also unusually high. Thus the evaporation of water requires a great deal of energy and has a remarkable cooling effect.

Most liquids contract on cooling, reaching their maximum density at their freezing point. Water is unusual in reaching its maximum density *above* its freezing point – at 4 °C. So when water freezes the ice formed is less dense than water and floats on top. Ice on the surface effectively insulates the water below, thereby making it less likely that the bulk of water (sea, pond or lake) will freeze up even if the air above is very cold.

At the surface of water the molecules are orientated so that most hydrogen bonds point inwards towards other water molecules. This gives water a very high surface tension, higher than any other liquid except mercury. Despite this, water molecules slide past each other relatively easily, and water has a remarkably low viscosity.

Compared with other liquids, water has extremely strong adhesive and cohesive properties that prevent it breaking under tension. Water adheres strongly to most surfaces, and can be drawn up through narrow tubes without the water column breaking.

Water has very powerful solvent properties and it dissolves more substances than any other common liquid (Figure 6.2).

■ Water and living things

Life is believed to have originated in an aqueous environment. In the course of evolution all living things have exploited the physical and chemical properties of water. Table 6.2 sets out the significance of the properties of water for life.

■ The ionisation of water

In liquid water a very small proportion of the molecules form **ions**. Ions are charged particles produced when electrons are completely transferred from one atom to another. The atom that has lost the electron will have a positive charge because the number of positively charged protons in its nucleus will now outnumber the remaining negatively charged electrons. Positively charged ions are called **cations**. Negatively charged ions are called **anions**. Anions are the product of atoms gaining electrons.

In pure water, very few molecules form ions (are ionised or **dissociated**); that is, there are relatively few hydrogen ions (H^+) and hydroxide ions (OH^-) present.

$$H_2O \rightleftharpoons H^+ + OH^-$$

The hydrogen and hydroxide ion concentration of pure water is 0.000 000 1 (or 10^{-7}) moles dm^{-3}. The few ions that are present move about independently amongst the water molecules.

Table 6.2 The significance of the physical properties of water

Properties of water	Significance for living things
liquid at room temperature	liquid medium for living things and for the chemistry of life
much heat energy is needed to raise the temperature of water (very high specific heat capacity)	aquatic environments are slow to change temperature
	bulky organisms have a stable temperature in the face of fluctuating external temperatures
evaporation of water requires a great deal of heat (very high latent heat of vaporisation)	evaporation of water in sweat or in transpiration causes marked cooling
	much heat is lost by the evaporation of a small quantity of water
much heat must be removed before freezing occurs (very high latent heat of fusion)	contents of cells and aquatic environments are slow to freeze in cold weather
ice is less dense than water, even very cold water (maximum density at 4 °C)	ice forms on the surface of water, insulating the water below
	when surface water does freeze aquatic life can survive below the ice
water molecules at surface with air orientate so that hydrogen bonds face inwards (very high surface tension)	water forms droplets on surfaces and runs off
	certain small animals exploit surface tension to land on and move over the surface of water
water molecules slide over each other very easily (very low viscosity)	water flows readily through narrow capillaries
water molecules adhere to surfaces (strong adhesive properties)	with low viscosity, capillarity becomes possible; water moves through extremely narrow spaces, e.g. between soil particles, and in cell walls
water column does not break or pull apart under tension (high tensile strength)	water can be lifted by forces applied at the top, e.g. movement of water up the xylem of tall trees
water dissolves more substances than any other common liquid ('universal solvent')	medium for chemical reactions of life
water is colourless (high transmission of visible light)	plants can photosynthesise at depth in water
	light may penetrate deeply into living tissues

IONS

Water

Water ionises slightly. In pure water very few molecules dissociate:

$$H—O—H \rightleftharpoons H^+ + OH^-$$

by $\quad H - e^- \rightarrow H^+ \quad$ (loss of electron) $\quad$ the hydrogen ion (a proton) is a positively charged ion (**cation**)

$\quad OH + e^- \rightarrow OH^- \quad$ (gain of electron) $\quad$ the hydroxide ion is a negatively charged ion (**anion**)

Potassium nitrate

The salt potassium nitrate ionises strongly and dissociation is complete:

$$KNO_3 \rightarrow K^+ + NO_3^-$$

by $\quad K - e^- \rightarrow K^+ \quad$ (loss of electron) $\quad$ the potassium ion is a **cation**

$\quad NO_3 + e^- \rightarrow NO_3^- \quad$ (gain of electron) the nitrate ion is an **anion**

■ Acids and bases

We can think of an acid as a compound that dissociates in water to liberate hydrogen ions. The stronger an acid, the more hydrogen ions it yields per mole. Inorganic acids, such as hydrochloric, nitric and sulphuric acids, are strong acids because they dissociate in aqueous solution.

$$HCl \rightarrow H^+ + Cl^-$$

Both inorganic and organic acids turn blue litmus red and neutralise bases. Organic acids, such as ethanoic acid, are weak acids, however, because relatively few of their molecules are dissociated in solution. The acidic group of organic acids is the carboxyl group. In carboxylic acids the following equilibrium lies far to the left.

A base is a compound that can neutralise an acid; that is, it can combine with hydrogen ions liberated by an acid. (A base that is soluble in water is known as an alkali.) Sodium and potassium hydroxides are examples of strong bases (that is, they are fully ionised in solution), and are also alkalis. Ammonia is only partially ionised in solution and is thus a weak base:

$$NH_3(aq) + H_2O \rightleftharpoons NH_4^+(aq) + OH^-$$

■ Chemical equations

In a chemical reaction, the reacting substances (reactants) are converted into new substances (products). An equation summarises the chemical changes in a reaction. Chemical formulae are used rather than chemical names. The equation indicates the numbers of moles of reactants and products. Often information is given about the physical state of each reactant and product by adding either g (gas), l (liquid), s (solid) or aq (aqueous, meaning in dilute solution in water) to the formula.

Constructing a chemical equation has four steps:

► Step 1: the word equation is written. For example:

hydrogen + oxygen → water

► Step 2: the words are replaced by the formulae of the substances involved:

$$H_2 + O_2 \rightarrow H_2O$$

► Step 3: the equation is balanced. Matter cannot be created or destroyed, so the numbers of atoms of each element on either side of the equation must be equal:

$$2H_2 + O_2 \rightarrow 2H_2O$$

► Step 4: the physical states may be added:

$$2H_2(g) + O_2(g) \rightarrow 2H_2O(l)$$

■ Moles

The **mole** (symbol, mol) is the scientific unit for the amount of substance. Atoms are extremely small, and the mass of a single atom is an incredibly tiny fraction of a gram. It is easier to express atomic mass by comparison with a standard. The **relative atomic mass** of an element is the ratio of the mass of an atom of the element to the mass of one particular atom, namely carbon, whose relative atomic mass is taken as 12. Against this standard the relative atomic mass of hydrogen is 1, that of oxygen is 16, and that of nitrogen is 14.

The amount of substance in a mole of atoms is given by the relative atomic mass expressed in grams. We can have a mole of atoms, or of molecules or of ions (but we must specify which). For example, a mole of methane molecules (CH_4) has a mass of 16 g. Sodium chloride consists of equal numbers of Na^+ and Cl^- ions (Na = 23, Cl = 35.5). So the mass of a mole of sodium chloride is 58.5 g. This mass (58.5 g) of sodium chloride dissolved in distilled water and made up to a litre (dm^3) is a **molar solution** of NaCl.

The usefulness of the mole lies in the fact that one mole of any substance contains the same number of atoms (or molecules or ions). So we will refer from time to time to an amount as (for example) 1 mole of oxygen molecules, or to a concentration as (for example) 1 mole of hydrogen ions per litre ($1 \, mol \, dm^{-3}$)

Questions

1 On a rocky shore, suggest reasons why the rocks exposed to intense sunlight at low tide may become too hot to stand on while the pools remain relatively cool.

2 Outline four of the unusual properties of water that can be ascribed to the effect of hydrogen bonds, and which are of significance to living things.

3 Write a balanced equation for the reaction between a strong acid (e.g. HCl) and a strong base (e.g. NaOH).

4 You are asked to prepare molar solutions of glucose ($C_6H_{12}O_6$) and potassium nitrate (KNO_3) (1 litre of each). How much of each substance should you weigh out? (Relative atomic masses: H = 1, C = 12, N = 14, O = 16, K = 39.)

6.3 pH

The concentration of ions in pure water is chosen as the midpoint on a scale of acidity and alkalinity known as the **pH scale** (Figure 6.3). The scale runs from 0 to 14. The lower values of pH (<7.0) refer to acid conditions, with more hydrogen ions than hydroxide ions in solution. The higher values (>7.0) refer to alkaline (basic) conditions, with fewer hydrogen ions than hydroxide ions in solution.

At pH 7 the concentration of hydrogen ions (represented symbolically by [H⁺]) is equal to the concentration of hydroxide ions ([OH⁻]), and the solution is neutral. The pH value of a liquid is the logarithm to the base of 10 of the reciprocal of the hydrogen ion concentration expressed in moles per litre. In pure water or in a neutral solution:

$$[H^+] = [OH^-] = 10^{-7} \text{ moles per litre}$$
$$(\text{mol dm}^{-3})$$

Therefore:

$$pH = \log \frac{1}{[H^+]} = \log \frac{1}{10^{-7}} = 7$$

The pH scale is a convenient way of representing concentrations of hydrogen ions, which are expressed in negative powers of 10. Because the scale is logarithmic (a change in one unit of pH corresponds to a tenfold change in

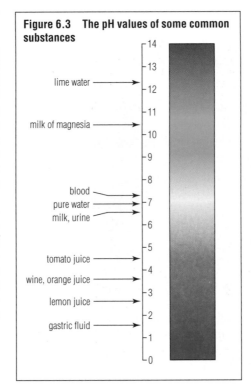

Figure 6.3   The pH values of some common substances

- lime water — 12
- milk of magnesia — 10
- blood — 7.5
- pure water — 7
- milk, urine — 6.5
- tomato juice — 4.5
- wine, orange juice — 4
- lemon juice — 3
- gastric fluid — 2

(scale 0–14)

hydrogen ion concentration), pH values cannot be added up and expressed as an average.

pH is determined experimentally using either coloured indicators (Figure 6.4) or an electronic pH meter, which is equipped with a probe sensitive to hydrogen ions (Figure 6.5). (See *Study Guide*, Chapter 3.)

■ The importance of pH in biology

The pH of the external environment is one of the most important ecological factors; the availability of ions is related to the pH of the soil (Chapter 3, p. 62), and the distribution of many species may be altered by a change in pH. Slight changes in the pH of the body's internal environment can adversely affect the internal physiology (Chapter 18, p. 386). The shapes of protein molecules and the functioning of enzymes are profoundly affected by pH change (Chapter 8, p. 180).

6.4 IONIC AND COVALENT COMPOUNDS

We know that chemical elements consist of atoms. Atoms are tiny particles consisting of a dense nucleus of neutrons and protons (the nucleus of hydrogen is a proton only, see Figure 6.1), surrounded by one or more shells of electrons. Atoms combine together ('bond') to form molecules in ways that produce a stable arrangement of electrons in the outer electron shell of each atom.

In **ionic bonding**, one or more electrons are *transferred* completely from one atom to another atom, in a way that produces stable arrangements of electrons in both atoms. This results in the formation of stable ions. For example,

---

Figure 6.4   Measuring pH colorimetrically using BDH soil indicator

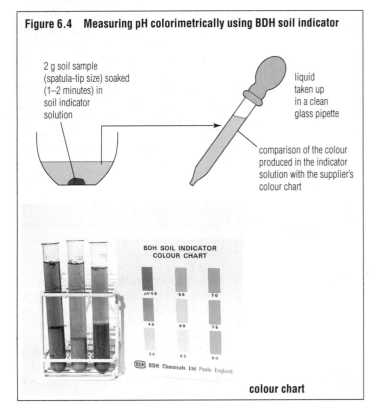

2 g soil sample (spatula-tip size) soaked (1–2 minutes) in soil indicator solution

liquid taken up in a clean glass pipette

comparison of the colour produced in the indicator solution with the supplier's colour chart

BDH SOIL INDICATOR COLOUR CHART

BDH Chemicals Ltd Poole England

colour chart

Figure 6.5   Measuring pH electrometrically

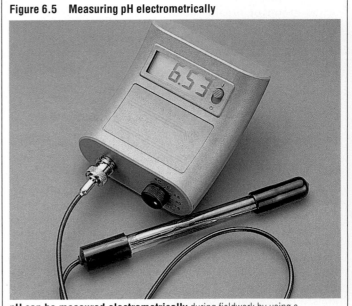

pH can be measured electrometrically during fieldwork by using a battery-operated pH meter. This carries a special electrode probe, which is placed in the sample of unknown pH (such as soil or water). The meter measures the potential difference produced in the probe unit; this potential difference is proportional to the hydrogen ion concentration. The electrode is standardised in a buffer solution of known pH just before it is used in the field

Figure 6.6   Ionic and covalent bonding

When atoms combine together to form molecules (chemical bonding), a more stable electron configuration results. Electrons in atoms are arranged in successive shells around the nucleus. The first shell (that is, the nearest to the nucleus) is the smallest, and holds up to 2 electrons. The second shell holds up to 8 electrons, the third up to 18, and the fourth up to 32. When the outermost shell of an atom is full the arrangement is particularly stable, and the atom is unreactive and at lowest energy.

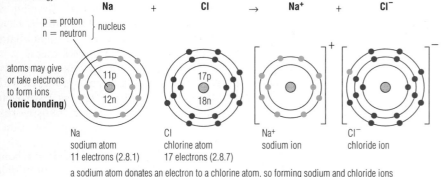

p = proton } nucleus
n = neutron

atoms may give or take electrons to form ions (**ionic bonding**)

Na   Cl   Na⁺   Cl⁻
sodium atom   chlorine atom   sodium ion   chloride ion
11 electrons (2.8.1)   17 electrons (2.8.7)

a sodium atom donates an electron to a chlorine atom, so forming sodium and chloride ions (the compound sodium chloride)

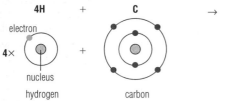

atoms may share electrons (**covalent bonding**)

4H   +   C   →   CH₄

electron

4×

nucleus

hydrogen   carbon   methane

four hydrogen atoms each share their single electron with one carbon atom, so filling its outer shell; ions are not formed

Figure 6.7   Methane, the simplest organic compound

Carbon forms bonds by the sharing of two electrons, one from the carbon atom and one from the other atom. Carbon has four electrons available for sharing, so it forms four bonds in this way.

dot and cross formula of methane:
nucleus of atom represented by symbol, electrons involved in bonding shown by dots and crosses

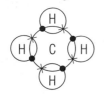

a bond can also be represented by a line:
this gives a two-dimensional representation of methane (structural formula)

```
    H
    |
H — C — H
    |
    H
```

In the methane molecule the four bonds are arranged symmetrically as if within a tetrahedron.

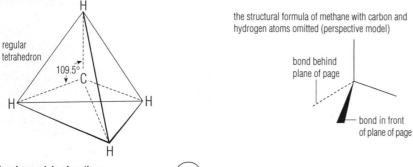

regular tetrahedron

109.5°

the structural formula of methane with carbon and hydrogen atoms omitted (perspective model)

bond behind plane of page

bond in front of plane of page

Molecular models of methane

ball and spring: different coloured balls represent different atoms; holes in the balls correspond to the numbers of bonds the atom can form, and stiff springs fit in the holes to represent the bonds

here the bonding arrangements are easily seen

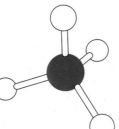

space-filling model: shows how near together the atoms of a compound are

space-filling models are the most realistic

sodium ions and chloride ions are much more stable than sodium atoms and chlorine atoms (Figure 6.6). These ions exist in sodium chloride crystals, as well as when the crystals are dissolved in water to form a solution.

In an alternative form of bonding, **covalent bonding**, electrons are *shared*. Bonding of this kind is common in molecules of non-metals such as hydrogen (H_2), nitrogen (N_2) and oxygen (O_2) and in compounds of non-metals, such as water (Figure 6.1), methane (CH_4) and carbon dioxide (CO_2).

6.5 THE CHEMISTRY OF CARBON

Organic chemistry is the chemistry of the element carbon and its compounds. (The chemistry of all other elements and their compounds is called inorganic chemistry.) At least two and a half million compounds containing carbon and hydrogen are known to organic chemists. This is more than the total of known compounds made from all the other elements. The great diversity of carbon compounds has made possible the diversity of life as we know it.

An atom of carbon has six electrons: two in the inner or first shell, and four in the outer (second) shell. The second shell can accommodate a maximum of eight electrons. The carbon atom can acquire four additional electrons to complete the outer shell by sharing electrons with other atoms; that is, it can form four covalent bonds (Figure 6.6).

The simplest organic compound is methane, CH_4. The carbon atom at the centre of the molecule forms strong, stable, covalent bonds with the four hydrogen atoms. Figure 6.7 shows various ways of representing the structure of methane.

Questions

1 Explain, in terms of hydrogen ion concentration, why solutions of high acidity have a low pH value.

2 What is the difference between an ion and an atom?

3 At four points along a woodland transect the pH of the soil was measured, and values of 6.1, 7.2, 6.5 and 6.8 were recorded. Why is it incorrect to attempt to find a mean value from these readings?

Figure 6.8   The carbon 'backbones' of some biologically important molecules: for convenience, hydrogen atoms linked directly to carbon atoms in ring structures are often omitted

1 straight carbon chains

e.g. palmitic acid (a saturated fatty acid), $C_{15}H_{31}COOH$

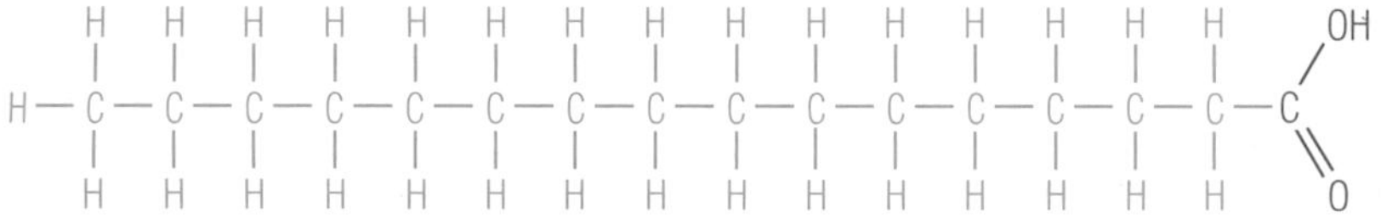

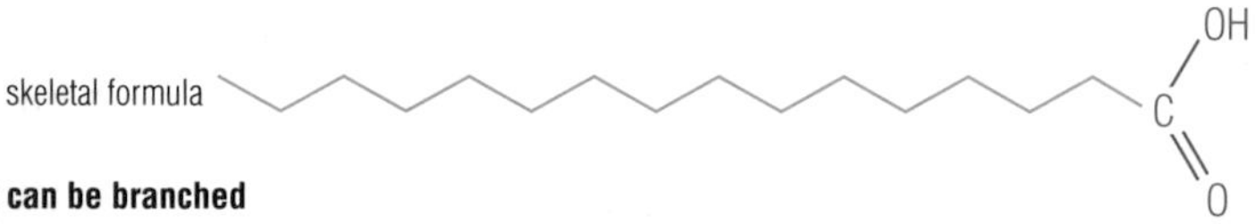

skeletal formula

carbon chains can be branched

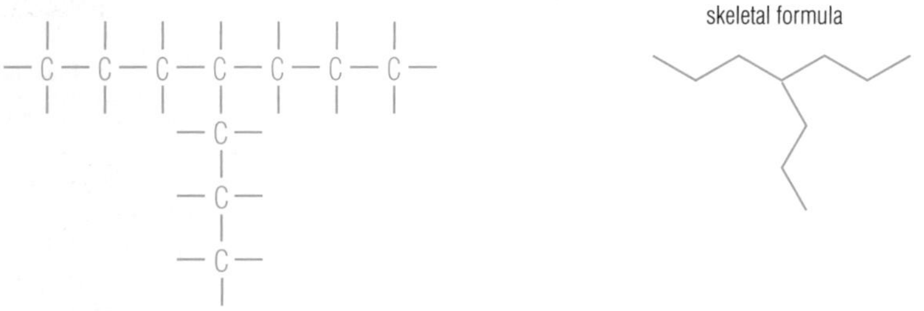

skeletal formula

2 a carbon ring

e.g. benzene, C_6H_6 (benzene is toxic)

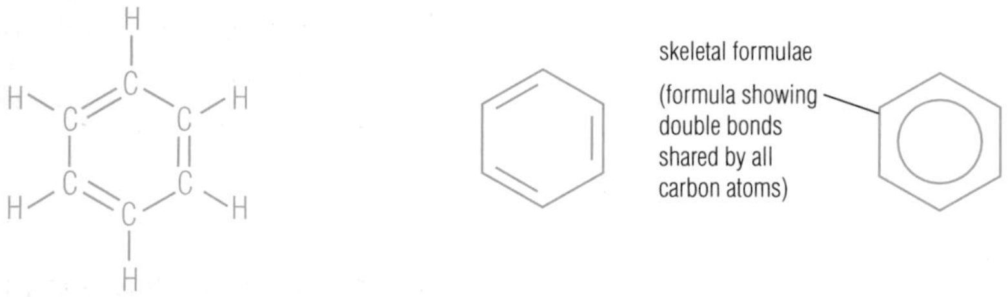

skeletal formulae

(formula showing double bonds shared by all carbon atoms)

3 rings containing atoms of other elements

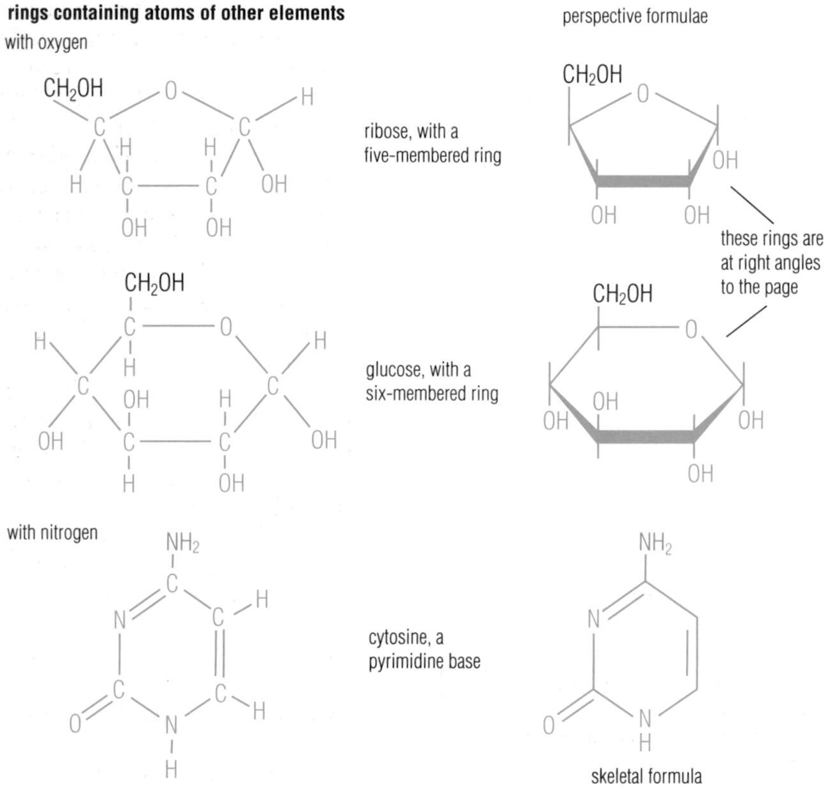

with oxygen

perspective formulae

ribose, with a five-membered ring

these rings are at right angles to the page

glucose, with a six-membered ring

with nitrogen

cytosine, a pyrimidine base

skeletal formula

The important chemical properties of carbon

Carbon is a relatively small atom, of low mass, able to form four strong, stable covalent bonds.

Carbon has the unusual ability to bond with itself. Repeated bonding of **carbon atom to carbon atom** is the basis of a huge range of straight-chain, branched-chain and ring molecules.

The electrons of carbon not shared with other carbon atoms are shared with atoms of elements such as hydrogen, oxygen, nitrogen, sulphur, phosphorus, certain metals and a very few other elements.

Covalent bonds like those in methane abound in organic molecules, but double or triple bonds are also found. In these, two or three pairs of electrons are shared, instead of a single pair. A carbon atom can form a strong double bond with another carbon atom, or with oxygen or nitrogen.

$$C{=}C, \; C{=}O, \; C{=}N$$

Carbon can also form a triple bond with carbon or with nitrogen:

$$C{\equiv}C, \; C{\equiv}N$$

although triple bonds are rare in nature.

Chains and rings of carbon atoms (together with atoms of the elements listed above) form the bulk of the compounds that make up the cell components (mainly carbohydrates, lipids, proteins and nucleic acids) (Figure 6.8).

Homologous series and functional groups

Most of the two or three million compounds of carbon known can be classified as members of a comparatively small number of families of very similar compounds, known as **homologous series**. Within such an ordered series each member differs from the preceding one by the addition of a —CH_2— group. There is a gradual variation in the physical properties of the members of a series with their increasing relative molecular mass.

All members of a homologous series have the same **functional group** (Figure 6.9); that is, a particular arrangement of a relatively few atoms, which has characteristic chemical properties. Thus all members of the series will react in similar ways.

Figure 6.9 Families of organic compounds

This is part of the homologous series of alcohols:

CH_3OH methanol (methyl alcohol)
CH_3CH_2OH ethanol ('alcohol' or ethyl alcohol)
$CH_3CH_2CH_2OH$ propan-1-ol (*n*-propyl alcohol)

ROH   This is the **general formula** for this series of alcohols;
                  OH is the functional group.

Other examples of homologous series of organic compounds include the carboxylic acids and aldehydes:

Carboxylic acids		Aldehydes

Carboxylic acids
Monocarboxylic acids / Dicarboxylic acids / **Aldehydes**

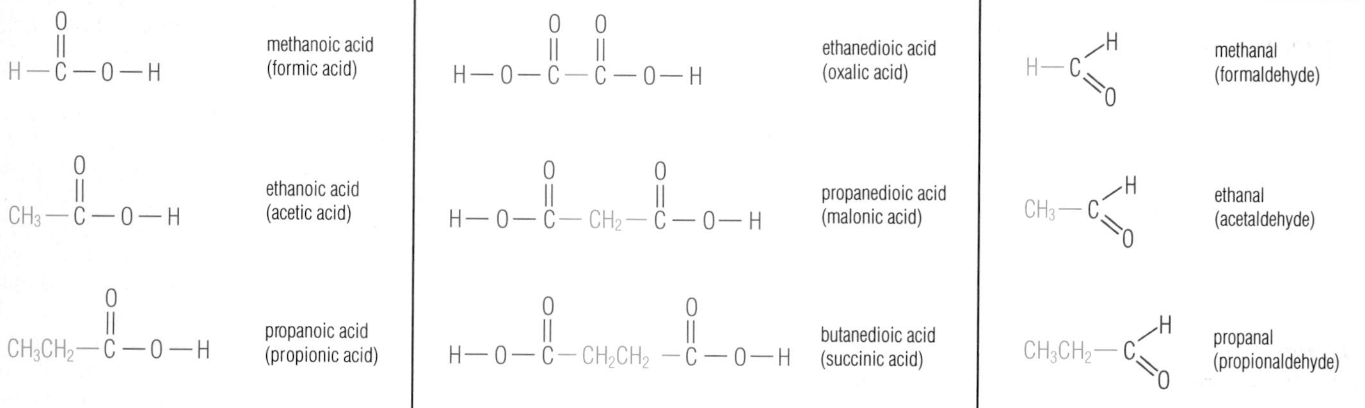

methanoic acid (formic acid) — $H-\underset{\underset{O}{\|}}{C}-O-H$

ethanoic acid (acetic acid) — $CH_3-\underset{\underset{O}{\|}}{C}-O-H$

propanoic acid (propionic acid) — $CH_3CH_2-\underset{\underset{O}{\|}}{C}-O-H$

all can be represented by
R — COOH

— COOH is the functional group

ethanedioic acid (oxalic acid) — $H-O-\underset{\underset{O}{\|}}{C}-\underset{\underset{O}{\|}}{C}-O-H$

propanedioic acid (malonic acid) — $H-O-\underset{\underset{O}{\|}}{C}-CH_2-\underset{\underset{O}{\|}}{C}-O-H$

butanedioic acid (succinic acid) — $H-O-\underset{\underset{O}{\|}}{C}-CH_2CH_2-\underset{\underset{O}{\|}}{C}-O-H$

all can be represented by
HOOC — R — COOH

methanal (formaldehyde) — $H-C\overset{H}{\underset{O}{\diagdown}}$

ethanal (acetaldehyde) — $CH_3-C\overset{H}{\underset{O}{\diagdown}}$

propanal (propionaldehyde) — $CH_3CH_2-C\overset{H}{\underset{O}{\diagdown}}$

all can be represented by
R — CHO

— CHO is the functional group

6.6 CARBOHYDRATES

The carbohydrates are a large group of organic compounds, made up of the elements carbon, hydrogen and oxygen. In carbohydrates the ratio of hydrogen to oxygen atoms is usually 2:1, as in water. In fact, many carbohydrates have the general formula $C_x(H_2O)_y$, where x is approximately equal to y; they were named by early sugar chemists who thought they were, literally, hydrates of carbon. **Simple sugars** are known as **monosaccharides**, and glucose and fructose are examples (Figure 6.10). Simple sugars may link together to form **compound sugars** such as **disaccharides, trisaccharides** and **polysaccharides**. These yield monosaccharides on hydrolysis.

Carbohydrates are the major constituents of most plants, making up 60–90% of the dry mass. Plants use carbohydrates (manufactured via photosynthesis, Chapter 12) as an energy source, as a means of storing energy (as starch) and as supporting material in the form of their cell walls. Animals consume carbohydrates as an important component of a balanced diet (p. 271). Humans not only eat carbohydrates (the cheapest foods, and making up 70% by mass of the average diet) but also use them in clothing (cotton, linen, rayon), housing (wood), fuel (wood), and in paper, books and packaging (wood). (Wood consists of carbohydrates of the cell walls impregnated with lignin; lignin, p. 147, is not itself a carbohydrate.)

Questions

1 How may the properties of the carbon atom have contributed to the diversity of living things?
2 a What does the symbol 'R' represent in a generalised formula of an organic compound?
  b What are the functional groups of, respectively, alcohols, organic acids, aldose sugars and ketose sugars?

Condensation and hydrolysis reactions

When monosaccharides combine together to form compound sugars, a molecule of water is released each time a bond is formed. This is known as a **condensation reaction**, and the bond formed is called a **glycosidic bond** (Figure 6.13). Synthesis of a compound sugar involves a nucleotide sugar (such as ADP-glucose or UDP-glucose, p. 177), which donates a sugar residue. In this way the necessary energy for the condensation reaction is made available.

Compound sugars can be broken down to their constituent monosaccharides in a reaction requiring water. It is known as a **hydrolysis reaction**. In cells specific enzymes catalyse the reaction. Compound sugars may be hydrolysed in a test tube by heating them with concentrated acid.

monosaccharide + monosaccharide

hydrolysis ⇅ condensation

disaccharide + water

■ Monosaccharides

The commonly occurring monosaccharide sugars contain between three and seven carbon atoms in their carbon chains. All the carbon atoms except one have hydroxyl groups attached. The remaining carbon atom is either part of an aldehyde group (an **aldose sugar**) or part of a ketone group (a **ketose sugar**).

The monosaccharides most frequently found are the six-carbon sugars, known as **hexoses**, such as glucose and fructose.

Hexose sugars

Hexose sugars have a **molecular formula** of $C_6H_{12}O_6$. (The molecular formula expresses the number of atoms in a molecule but does not convey the structure of the molecule.) The **structural formulae** of glucose and fructose are shown in Figure 6.10.

The relative dispositions of the —H and —OH groups along the carbon chain in hexoses is significant. Theoretically, eight possible structural isomers (see below) of the aldose form of $C_6H_{12}O_6$ exist. Only three occur commonly in nature: glucose, mannose and galactose. Fructose is the only ketohexose to occur in nature.

Ring structure of glucose and fructose

Glucose can be written as a straight-chain structure, but in fact it normally exists in a ring or cyclic form. The ring contains five carbon atoms and one oxygen atom – a **pyranose** ring. The ring structure commonly existing in fructose contains four carbon atoms and one oxygen atom (a **furanose** ring, Figure 6.10), but fructose may also occur in pyranose form.

Isomers

Compounds that have the same chemical formula (the same component atoms in their molecules) but which differ in the arrangement of the atoms are known as **isomers**. Many organic compounds exist in isomeric forms. Because of this it is important to know the structure of an organic compound as well as its composition. The names of some organic compounds are complex because they give this information.

Structural isomers have the same molecular formula, but their atoms are linked together in different sequences. The monosaccharide sugars glucose and fructose are structural isomers (Figure 6.10).

A further kind of isomerism is possible in the ring form of glucose, since interchanging the positions of the —H and —OH on carbon atom 1 gives rise to two different structures, known as α- and β-glucose (Figure 6.10). The isomers have very similar properties, but the differences between them give rise to the markedly different properties of the glucose polymers cellulose (p. 135) and starch (p. 136).

Optical isomers

An organic compound that has four different chemical groupings attached to a single carbon atom (called an **asymmetric carbon atom**) shows yet another type of isomerism. This is because a three-dimensional model of such a molecule can be built in two ways, which are identical in every way except that they are mirror images of each other. The three-carbon

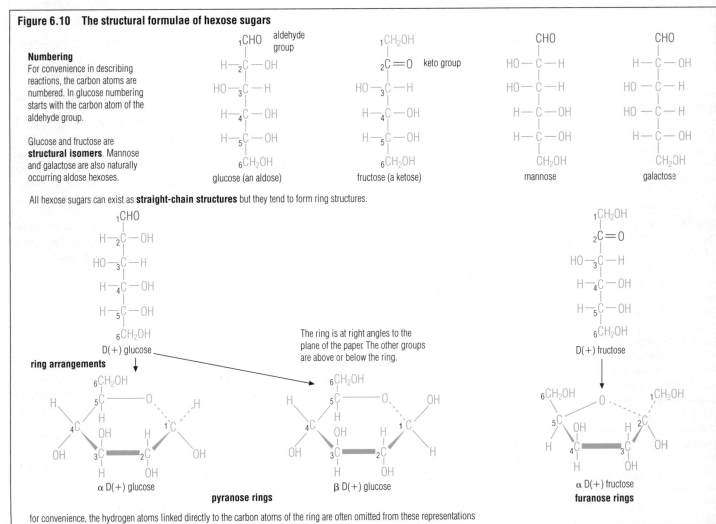

Figure 6.10   The structural formulae of hexose sugars

Numbering
For convenience in describing reactions, the carbon atoms are numbered. In glucose numbering starts with the carbon atom of the aldehyde group.

Glucose and fructose are **structural isomers**. Mannose and galactose are also naturally occurring aldose hexoses.

glucose (an aldose)

fructose (a ketose)

mannose

galactose

All hexose sugars can exist as **straight-chain structures** but they tend to form ring structures.

D(+) glucose

ring arrangements

The ring is at right angles to the plane of the paper. The other groups are above or below the ring.

α D(+) glucose

β D(+) glucose

pyranose rings

D(+) fructose

α D(+) fructose

furanose rings

for convenience, the hydrogen atoms linked directly to the carbon atoms of the ring are often omitted from these representations

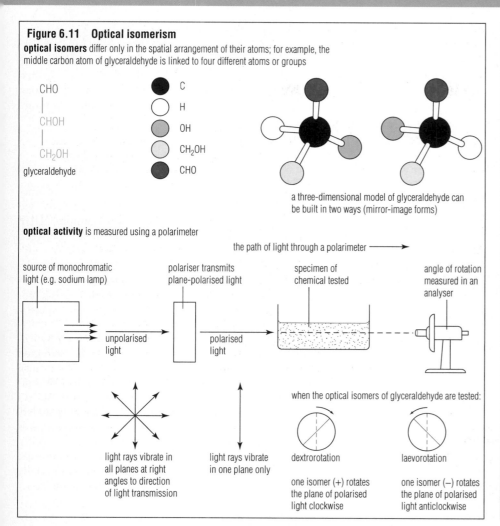

Figure 6.11 Optical isomerism

optical isomers differ only in the spatial arrangement of their atoms; for example, the middle carbon atom of glyceraldehyde is linked to four different atoms or groups

CHO
|
CHOH
|
CH₂OH

glyceraldehyde

C
H
OH
CH₂OH
CHO

a three-dimensional model of glyceraldehyde can be built in two ways (mirror-image forms)

optical activity is measured using a polarimeter

the path of light through a polarimeter ⟶

source of monochromatic light (e.g. sodium lamp)

polariser transmits plane-polarised light

specimen of chemical tested

angle of rotation measured in an analyser

unpolarised light

polarised light

light rays vibrate in all planes at right angles to direction of light transmission

light rays vibrate in one plane only

when the optical isomers of glyceraldehyde are tested:

dextrorotation

laevorotation

one isomer (+) rotates the plane of polarised light clockwise

one isomer (−) rotates the plane of polarised light anticlockwise

Figure 6.12 Pentose and triose sugars

Pentoses

Ribose and deoxyribose occur as straight-chain structures:

D(+) ribose

D(−) deoxyribose

They also occur in ring form:

β-D-ribose

β-D-deoxyribose

Trioses

glyceraldehyde

dihydroxyacetone

These are the structurally simplest carbohydrates that occur naturally.
They occur in phosphorylated form (the PO₄ group is often represented as —Ⓟ).
These two trioses are interconverted (isomerised) in the process of respiration.

---

sugar glyceraldehyde is an example (Figure 6.11). The two forms of glyceraldehyde have identical chemical properties, but in solution they rotate the plane of plane-polarised light in opposite directions; they are said to be optically active. The two forms of glyceraldehyde are **optical isomers**. Optical isomerism was discovered by Louis Pasteur in the middle of the nineteenth century (p. 12).

The optical isomer that rotates the plane of polarisation to the right (clockwise) is said to be **dextrorotatory** (represented in the name of the compound by a plus sign, +); the optical isomer that rotates the plane of polarisation to the left (anticlockwise) is **laevorotatory** (represented by a minus sign, −). Angles of rotation are measured in a **polarimeter** (Figure 6.11).

It is impossible to predict from the structure of a carbohydrate the direction in which it will rotate plane-polarised light. By convention, if the asymmetric carbon atom furthest from the functional group has its hydroxyl group in the same position as in (+)-glyceraldehyde the isomer is called the D-isomer; if it is in the same position as in (−)-glyceraldehyde it is the L-isomer. The D and L prefixes tell us nothing about the direction in which the isomer rotates plane-polarised light, however.

Optical isomerism and cell chemistry

Optical isomerism is a feature of monosaccharides and amino acids (p. 139). In general, cells produce and metabolise only one of two possible optical isomers of these metabolites. This is because each of the **enzymes** that catalyse the chemical changes involved can only catalyse the reaction of one of the two forms. (Enzyme action involves a close 'fit' between the enzyme and the molecule whose reaction it catalyses; unless their surfaces come close together, no reaction can occur, p. 183.) Optical isomerism of amino acids is discussed on p. 139.

Pentose and triose sugars

The only five-carbon sugars or **pentoses** commonly found in nature are ribose and deoxyribose (Figure 6.12). Both exist in the furanose ring form. Ribose occurs in coenzymes, in ATP and in ribonucleic acid (RNA). Deoxyribose occurs in deoxyribonucleic acid (DNA, p. 145).

The two naturally occurring three-carbon sugars (**trioses**) are glyceraldehyde (Figure 6.12) and dihydroxyacetone. Both these sugars are found in plant and animal cells, and they play an important part in carbohydrate metabolism (Figure 15.26, p. 318). They occur combined with a phosphate group, as in glyceraldehyde 3-phosphate (GALP).

Questions

1 Explain the differences between the molecular formula and the structural formula of a sugar (such as glucose).

2 What types of isomerism are shown by monosaccharide sugars?

3 Name the principal role of pentoses in living things.

4 What are the differences between
   a ribose and deoxyribose
   b glucose and fructose?

REDUCING AND NON-REDUCING SUGARS

All monosaccharides and some disaccharides are reducing sugars.

The test for a reducing sugar

2 cm³ of a solution of the sugar to be tested is added to an equal volume of Benedict's solution; the solutions are mixed, and heated with gentle agitation

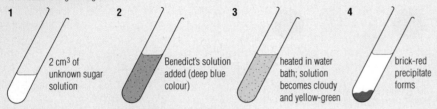

1. 2 cm³ of unknown sugar solution
2. Benedict's solution added (deep blue colour)
3. heated in water bath; solution becomes cloudy and yellow-green
4. brick-red precipitate forms

a colour change from blue to a cloudy yellow-green and the formation of a brick-red precipitate indicates that a reducing sugar is present

Explanation

Benedict's solution (blue) is an alkaline solution of copper(II) sulphate ($CuSO_4$). The aldehyde or ketone group of a monosaccharide sugar is able to reduce Cu^{2+} ions to Cu^+, itself being oxidised to a carboxyl (—COOH) group. A brick-red precipitate of copper(I) oxide is formed.

Disaccharides that are reducing sugars

Disaccharides formed by the reaction of the aldehyde group of one monosaccharide with a hydroxyl group of another contain a free aldehyde group, and will reduce Benedict's solution. For example:

galactose + glucose → lactose

a reducing sugar

Disaccharides that are non-reducing sugars

Disaccharides formed by the reaction of the aldehyde group of one monosaccharide with the aldehyde or ketone group of another are non-reducing sugars. For example:

glucose + fructose → sucrose

a non-reducing sugar

Testing for a non-reducing sugar

There is no specific test for non-reducing sugars. A disaccharide can, however, be hydrolysed to its constituent monosaccharides by boiling with dilute hydrochloric acid, and the products of hydrolysis will then reduce Benedict's solution (*Study Guide*, Chapter 6).

■ Disaccharides

The three commonly occurring disaccharides are maltose, lactose and sucrose (Figure 6.13). Disaccharides are formed by two monosaccharide units combining together with the elimination of a molecule of water:

glucose + glucose → maltose
glucose + galactose → lactose
glucose + fructose → sucrose

In cells the reactions are enzyme-catalysed and synthesis proceeds via nucleotide sugars; for example, UDP glucose (p. 177). The bond between monosaccharide units (residues) is a glycosidic link. Disaccharides may be hydrolysed to their constituent monosaccharides. In living cells this is accomplished in the presence of specific hydrolytic enzymes.

Maltose is a product of starch hydrolysis. Industrially, maltose is extracted from germinating barley (malt extract), and is used in brewing and in food manufacturing. **Lactose** is the sugar found in the milk of mammals. **Sucrose** is widely distributed in plants; it is the sugar that is translocated in the phloem (p. 337). Sucrose is also the 'sugar' humans use most often. It is extracted industrially from sugar cane (p. 118) and sugar beet.

■ Polysaccharides

Most of the carbon compounds that occur in living things can be conveniently

---

Figure 6.13   Disaccharides

maltose is a reducing sugar

CH₂OH ... OH ... HO ... OH ... O ... OH ... CH₂OH ... OH
maltose — α-1,4-glycosidic linkage
+ H—O—H water ⇌ (hydrolysis / condensation)
CH₂OH ... OH ... HO ... OH ... OH   D-glucose   +   CH₂OH ... OH ... HO ... OH ... OH   D-glucose

lactose is also a reducing sugar

HO ... CH₂OH ... O ... OH ... O ... CH₂OH ... OH ... OH
lactose ... OH
+ H—O—H water ⇌ (hydrolysis / condensation)
HO ... CH₂OH ... OH ... OH   β-D-galactose   +   CH₂OH ... OH ... HO ... OH ... OH   β-D-glucose

sucrose is a non-reducing sugar

CH₂OH ... OH ... HO ... α-1,2-glycosidic linkage ... O ... CH₂OH ... HO ... HOH₂C ... OH
sucrose
+ H—O—H water ⇌ (hydrolysis / condensation)
CH₂OH ... OH ... HO ... OH ... OH   α-D-glucose   +   HO ... O ... CH₂OH ... OH ... HOH₂C ... OH   β-D-fructose

in sucrose the aldehyde group of glucose and the ketone group of fructose are linked together in the glycosidic bond, and neither is free to participate in a reaction reducing copper(II) to copper(I)

(if rather arbitrarily) divided into the low-M_r compounds, the so-called 'small' molecules like sugars, amino acids and fatty acids, and the giant molecules or **macromolecules**. Macromolecules are loosely defined as substances of M_r 5000–10 000 and upwards, often much more. Macromolecules constitute about 90% of the dry mass of the cell.

The panel below illustrates the differences between the 'small' molecules and the macromolecules found in living things. Polysaccharides are one of the natural groups of macromolecules. They are long-chain carbohydrate molecules consisting of many condensed monosaccharide units (or residues). Examples are cellulose, glycogen and starch.

Cellulose

Cellulose is a straight-chain (that is, unbranched) polymer of between two and three thousand glucose units. Carbon atom 1 of one β-glucose links up with carbon atom 4 of the next, to produce a long chain. Linkages like these are known as β-1,4-glycosidic bonds. The shape and construction of the cellulose polymer allows the close packing of chains into fibres. These consist of bundles of about 2000 chains, held together by hydrogen bonds, and are known as **microfibrils** (see Figure 7.29, p. 167). Cellulose microfibrils are extremely strong and are able to resist tension stress.

Cellulose is hydrolysed chemically to glucose only with difficulty (heating with concentrated acid is required to achieve this in a laboratory experiment, for example). Certain fungi, bacteria and possibly protozoa produce a **cellulase** enzyme, which catalyses the hydrolysis, usually forming glucose or sugar acids.

Few animals produce cellulase. For example, herbivorous mammals, whose diets contain a high proportion of cellulose, harbour in their gut a vast population of symbiotic bacteria, which can digest the cellulose, thus making the glucose available to the host. Some insects that feed on wood, such as termites, have flagellate protozoa (p. 28) in their gut and it is these (or the symbiotic bacteria in the protozoa) that digest the cellulose of wood.

'Small' molecules *v.* the macromolecules (the structures of polypeptides, proteins and nucleic acid are shown on pp. 141 and 145)

'Small' molecules have a M_r in the range of 50–500.	'Giant' molecules (**macromolecules**) have a M_r of 10 000 up to millions. Macromolecules are polymers, made from many repeating units (monomers) by condensation reactions in which water is removed. These reactions require energy.
Examples	**Examples**

Sugars
monosaccharides such as the hexoses, e.g.

β-D-glucose

Polysaccharides
These may be branched-chain (e.g. glycogen) or straight-chain polymers (e.g. cellulose).
They are built from monosaccharides, mainly hexoses, linked by glycosidic bonds.
Usually only one type of sugar monomer is involved in each polysaccharide.
Some polysaccharides function as stores of potential energy (e.g. glycogen, starch).
Some polysaccharides have a structural role (e.g. cellulose).

in cellulose the bond is a β-1,4-glycosidic linkage

the monomer units within a polymer are known as residues

glucose residues in cellulose

Amino acids

Proteins
These are straight-chain polymers of the 20 different amino acids found in all cells, joined by peptide bonds.
The sequence of the amino acids is important in the properties and functioning of the protein.
Roles of proteins:
▶ structural (e.g. in membranes, chromosomes)
▶ enzymes (e.g. condensation enzyme)
▶ blood protein (e.g. antibodies, haemoglobin)

Organic bases
pyrimidines
purines

Nucleic acids
These are straight-chain polymers of sugar (pentoses), phosphate, and organic bases.
The sequence of organic bases is important in the properties and functioning of nucleic acids.
Role of nucleic acids:
▶ transmission and 'reading' of hereditary information

Fatty acids
$CH_3(CH_2)_nCOOH$
a fatty acid

glycerol

Lipids
These are much smaller molecules (M_r = 750–2500) and are **not** macromolecules; they associate together in large aggregates.

Starch

Starch is also a polymer of glucose. It is an important carbohydrate storage material in plants. It consists of a mixture of two polysaccharides, called **amylose** and **amylopectin**, the proportions of which vary with the source of the starch (Figure 6.14). Usually, however, amylopectin constitutes about 70% of starch.

Amylose is an unbranched chain of 200–1500 glucose residues linked by α-1,4-glycosidic bonds. The molecule takes the form of a helix.

The chemical difference between the α-1,4-glycosidic bonds of starch and the β-1,4-glycosidic bonds of cellulose accounts for the pronounced difference in physical properties of the two polysaccharides. Amylose is a powdery substance, while cellulose exists as tough microfibrils.

Amylopectin contains from 2000 to 200 000 glucose units per molecule. As in amylose, the glucose chains are wound in a helix. Here, however, the macromolecule consists of branched chains of glucose residues. Branch points (α-1,6-glycosidic bonds) occur about every 20 residues along the straight-chain sections.

Starch gives a characteristic blue-black colour with iodine solution (iodine dissolves in water in the presence of potassium iodide). The colour is the result of a complex formed between the amylose helix and the iodine molecules (*Study Guide*, Chapter 6).

Both amylose and amylopectin are only sparingly soluble in water; amylose is the slightly more soluble of the two. In the laboratory starch can be hydrolysed to glucose by heating with dilute acids. In plants and animals **amylases** catalyse the hydrolysis of starch.

Glycogen

Glycogen is also a branched-chain polymer of glucose. Chemically, glycogen is very similar to amylopectin, although the glycogen macromolecule is larger and much more highly branched than that of amylopectin. Glycogen is the carbohydrate storage product found in many animals, especially in the vertebrates, where it is stored in muscles and liver. In a TEM of a liver cell granules of glycogen are frequently observed in the cytoplasm. Glycogen is insoluble in water. It is readily hydrolysed by enzyme action to produce glucose.

Other polysaccharides and polysaccharide derivatives

Pectins and hemicelluloses

In addition to cellulose, plant cell walls contain many other polysaccharides, which contribute to the mechanical strength of the organism. Pectins are polysaccharides of galactose and galacturonic acid residues. Pectins become bound together by calcium ions (calcium pectate), and are important components of the first layers of a cell wall to be laid down (known as the middle lamella, p. 166).

Hemicelluloses are more of a mixture than are pectins, but their chief ingredients are pentose sugars and sugar acid residues.

An additional component, lignin (not a carbohydrate, p. 147), is of importance in woody tissues.

Inulin

Inulin, a polymer of fructose, is a reserve carbohydrate found in some groups of plants as an alternative to starch.

Chitin

Chitin is chemically related to cellulose. It occurs in the walls of fungal hyphae and in the exoskeletons of arthropods. The monomer from which the chitin polymer is built is a glucosamine (a glucose in which one —OH is replaced by —NHCOCH$_3$). These macromolecules are arranged in long, straight, parallel chains, but are very similar to cellulose in structure.

Mucopolysaccharides

Mucopolysaccharides (these are sometimes called glycosaminoglycans) consist of organic acids derived from sugar molecules condensed together with amino acids. They occur in vertebrate connective tissue, and as lubricants in various joints.

Figure 6.14 Starch

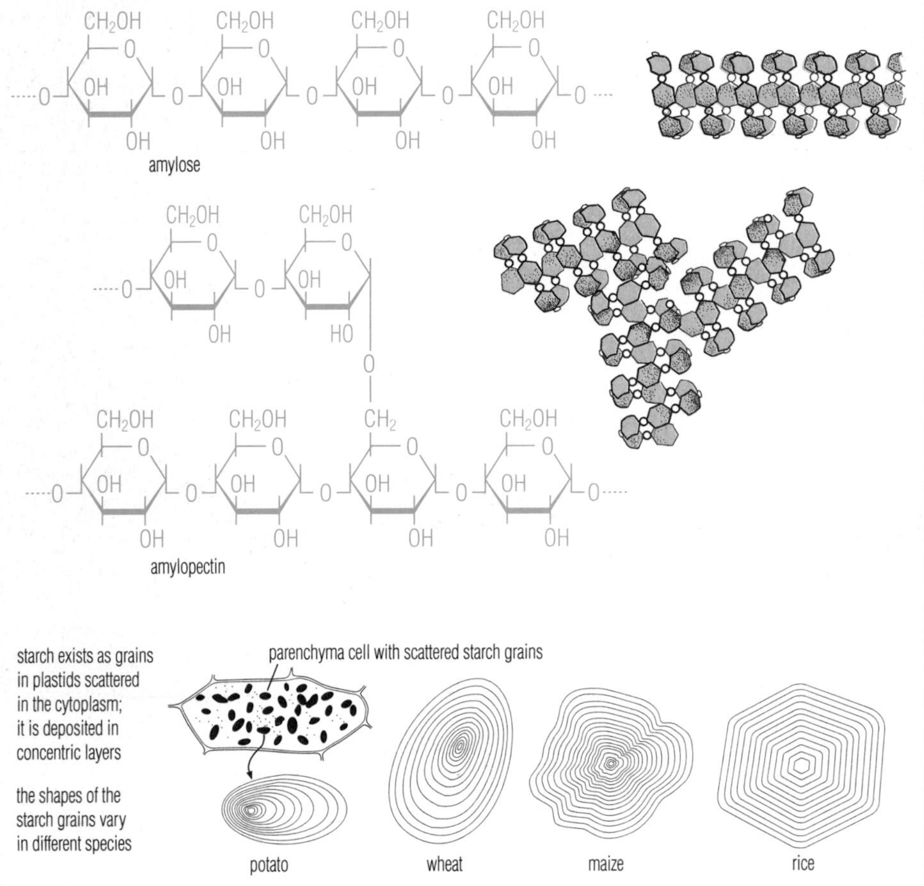

amylose

amylopectin

starch exists as grains in plastids scattered in the cytoplasm; it is deposited in concentric layers

the shapes of the starch grains vary in different species

parenchyma cell with scattered starch grains

potato   wheat   maize   rice

Questions

1 What is the difference between a condensation and a hydrolysis reaction? Give an example of each.

2 Outline the key steps by which malt extract is produced (p. 298), and by which sucrose is extracted from sugar cane (p. 118).

3 What are the key features of biological macromolecules?

4 What is the significance of hydrogen bonds in fibres of cellulose?

6.7 LIPIDS

The lipids are a group of substances that contain carbon, hydrogen and oxygen, as do carbohydrates, but in lipids the proportion of oxygen is much smaller. Lipids are also insoluble in water and are said to be **hydrophobic** ('water-hating'). Lipids are soluble in organic solvents such as ethanol and ether.

Lipids are a diverse group of chemicals. They can be classified as

▶ simple lipids – the animal fats, the vegetable oils and the waxes
▶ phospholipids and related compounds
▶ steroids.

The addition of water to a solution of lipid in ethanol produces a cloudy white emulsion of tiny lipid droplets; this is used as a test for lipids (*Study Guide*, Chapter 6).

■ The fats and oils

Fats and oils are among the simplest of the larger biological molecules. They are not macromolecules (p. 135), but because of their hydrophobic properties the molecules aggregate into globules. Globules of fat or oil are compact and relatively inert. They are the most abundant of the lipids found in nature.

Fats and oils are of the same basic chemical structure, the difference between them being their physical state at 20 °C. Fats are solid at this temperature, oils are liquid. Both fats and oils are esters formed by a condensation reaction between fatty acids and an alcohol called glycerol to form ester links (—COO—) (Figure 6.15).

Fatty acids consist of a long unbranched hydrocarbon chain, $CH_3(CH_2)_n$—, ending with a carboxylic acid (—COOH) group. The hydrocarbon chain is nonpolar, but the terminal carboxyl group is partially ionised and can form ionic bonds. The carboxyl group is therefore polar. Glycerol has three hydroxyl (—OH) groups, which may react with the carboxyl groups of three fatty acid molecules, eliminating three molecules of water. The product is a molecule of **triacylglycerol** (formerly called a triglyceride).

Saturated and unsaturated fats

The fatty acids from which triacylglycerols are made are long-chain monocarboxylic acids. Some are saturated fatty acids; others are unsaturated, that is, they have either one double bond (monounsaturated), or more than one (polyunsaturated), in the hydrocarbon chain (Figure 6.16).

Triacylglycerols made of fatty acids with shorter hydrocarbon chains and containing one or more double bonds (**unsaturated fats**) have low melting points. They are oils or soft fats at room temperature in temperate climates. Oils occur mainly in plants.

Triacylglycerols made from saturated fatty acids and having longer hydrocarbon chains (**saturated fats**) are solids at room temperature. The fats are characteristic of animals. Different species of animals produce fats of different melting points. Fat that is supple and flexible in our bodies at 30 °C might set solid and inflexible in the body of a fish living in the North Sea!

Figure 6.15   The formation of triacylglycerols, by a condensation reaction (water is eliminated) between glycerol (an alcohol) and monocarboxylic acids (fatty acids)

Figure 6.16   Saturated and unsaturated fatty acids

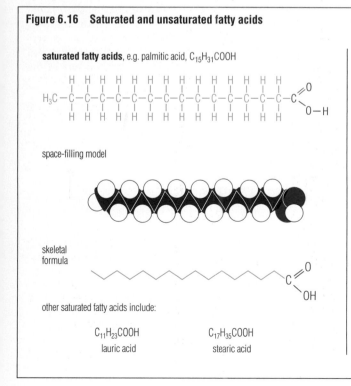

saturated fatty acids, e.g. palmitic acid, $C_{15}H_{31}COOH$

space-filling model

skeletal formula

other saturated fatty acids include:

$C_{11}H_{23}COOH$ $C_{17}H_{35}COOH$
lauric acid   stearic acid

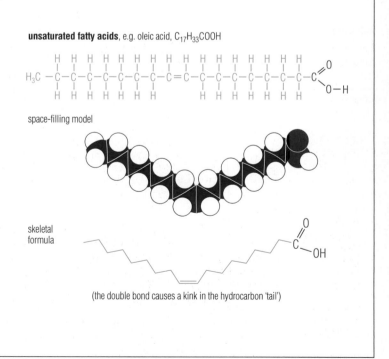

unsaturated fatty acids, e.g. oleic acid, $C_{17}H_{33}COOH$

space-filling model

skeletal formula

(the double bond causes a kink in the hydrocarbon 'tail')

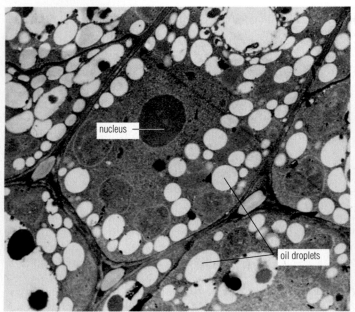 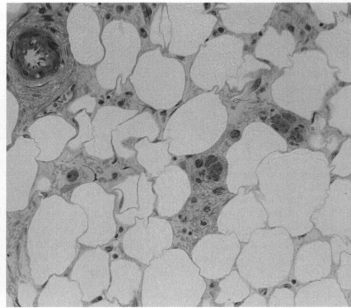

Figure 6.17   Stored fats and oils: TS of cells of sunflower cotyledon (left): droplets of oil occur in the cytoplasm; TS of adipose cells of a mammal (right)

Saturated fats and cholesterol

Saturated fats are more readily converted by the human body into the steroid cholesterol (also a lipid, see below). Cholesterol as a component of our diet has received much attention recently because of a suspected correlation between the cholesterol level in the blood and certain types of heart disease (p. 355). Dieticians recommend lowered intake of animal fat and the replacement of butter (containing animal fats) by margarines based on a higher concentration of unsaturated fats.

Roles of triacylglycerols in living things

Triacylglycerols are compact, insoluble, long-term energy stores. They can be stored at high concentrations, without requiring water as a solvent. While carbohydrates are the most direct source of energy in living things because they can be mobilised quickly, and used under aerobic or anaerobic conditions (p. 302), triacylglycerols release roughly twice as much energy per gram as do carbohydrates. The energy in fat cannot be released in the absence of oxygen, however, as in vertebrate muscle when operating in conditions of oxygen debt (p. 492).

Fat is a common food store in animals living in cold climates. In aquatic animals fats aid buoyancy. Hibernating animals store food as additional fat layers below the skin and around body organs. Oils are often a major food store in seeds (sunflower, for example, Figure 6.17) and in fruits such as maize and olives.

Triacylglycerols also release twice as much water as do carbohydrates when oxidised in respiration. Water produced from the oxidation of food is called **metabolic water**. In some situations metabolic water is of vital importance. These include the development of very young birds and reptiles while enclosed in their egg shells, and the daily metabolism of animals of habitats where water is scarce, such as that of the camel and kangaroo rat (p. 390).

■ Waxes

Waxes are esters of fatty acids, formed not with glycerol but with complex alcohols. They are used as waterproofing materials by both plants and animals. In plants waxes occur in the

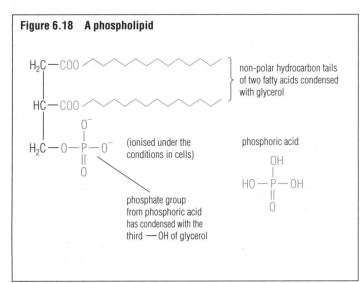

Figure 6.18   A phospholipid

H₂C — COO ~~~~~~~~ } non-polar hydrocarbon tails of two fatty acids condensed with glycerol

HC — COO ~~~~~~~~

H₂C — O — P — O⁻ (ionised under the conditions in cells)

phosphate group from phosphoric acid has condensed with the third — OH of glycerol

phosphoric acid

OH
|
HO — P — OH
||
O

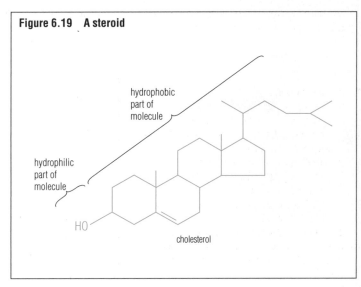

Figure 6.19   A steroid

hydrophobic part of molecule

hydrophilic part of molecule

HO

cholesterol

cuticle of the epidermis on leaves, stems, fruits and seeds. In animals they occur on the skin and on feathers. Wax is also a constituent of the honeycomb of bees, placed there by the worker bees.

■ Phospholipids

If one of the fatty acid groups in a triacylglycerol is replaced by a phosphate group then the molecule is known as a phospholipid (Figure 6.18). The phosphate group is ionised and therefore water-soluble. Phospholipids combine the hydrophilic ('water-loving') properties of the phosphate group with the hydrophobic properties of the hydrocarbon chains of the fatty acid parts of the molecule. At oil–water interfaces they become orientated with the hydrophobic groups in oil and the hydrophilic groups in water.

In living things, phospholipids have three roles: in membrane structure (p. 158), as a source of acetylcholine (p. 448), and in the transport of fat between gut and liver in mammalian digestion (p. 284).

■ Steroids

The 'skeleton' of a steroid molecule consists of complex rings of carbon atoms, quite different from that of a triacylglycerol. Steroids occur in both plants and animals.

In mammals, one of the most widespread steroids in the tissues is cholesterol (Figure 6.19). From cholesterol are made the bile salts, the sex hormones (p. 592) and the hormones of the adrenal cortex (p. 475). A steroid closely related to cholesterol occurs in the human skin and is converted to vitamin D by the ultra-violet radiation in sunlight (p. 272).

Questions

1 a How do fats differ from oils?
b What property is conferred on a lipid when it is combined with a phosphate group?
2 a Explain how fats can provide an organism with both energy and water.
b What might be the advantages and disadvantages of trying to drastically lower the level of cholesterol in the body?
3 a What is the chemical composition of a triacylglycerol?
b How do individual triacylglycerols differ?
4 Why does a white suspension form when a solution of a lipid in ethanol is mixed with water?

6.8 PROTEINS

About two-thirds of the total dry mass of a cell is composed of protein. Proteins are macromolecules; the M_r of most of the proteins found in cells lies between 20 000 and 10 million.

Proteins differ from carbohydrates and lipids in that they always contain the element nitrogen as well as carbon, hydrogen and oxygen. Sulphur is often present too. Some proteins combine with other substances to form complexes containing phosphorus, iron or certain trace elements.

■ Amino acids

Proteins are built up from a linear sequence of amino acids (Figure 6.20). As the name implies, each amino acid carries two functional groups, both of which are ionised in the cell:

▶ an amino group, $—NH_2$ (basic), and
▶ a carboxyl group, $—COOH$ (acidic).

A substance that is both acid and base is described as **amphoteric** (Figure 6.21). Most of the amino acids from which proteins are built have the two functional groups attached to the same carbon atom, known as the α-carbon atom (there are two exceptions, proline and hydroxyproline).

Many of the amino acids from which proteins are formed have a single amino group and a single carboxyl group and are therefore described as neutral amino acids. Some contain additional amino groups, however (the basic amino acids), and some additional carboxyl groups (the acidic amino acids). For example, arginine and histidine are basic amino acids, while aspartic acid and glutamic acid are acidic amino acids.

Amino acids differ in the nature of the fourth group attached to the α-carbon atom. More than 150 different types of amino acid are known to occur in cells, but only twenty or so are used to make proteins; these are listed in Table 9.7 (p. 207). Some of the amino acids in proteins may be modified chemically *after* the protein has been formed; an example is hydroxyproline in the protein collagen.

Optical isomerism

Like monosaccharides, almost all the amino acids from which the proteins of living things are made up have an asymmetric carbon atom (four different groups attached) in their molecules. The exception is the simplest amino acid, glycine. All the other amino acids therefore show optical isomerism (p. 133). Without significant exception, however, only the L-forms of amino acids are found in living things.

Figure 6.20   The structure of amino acids

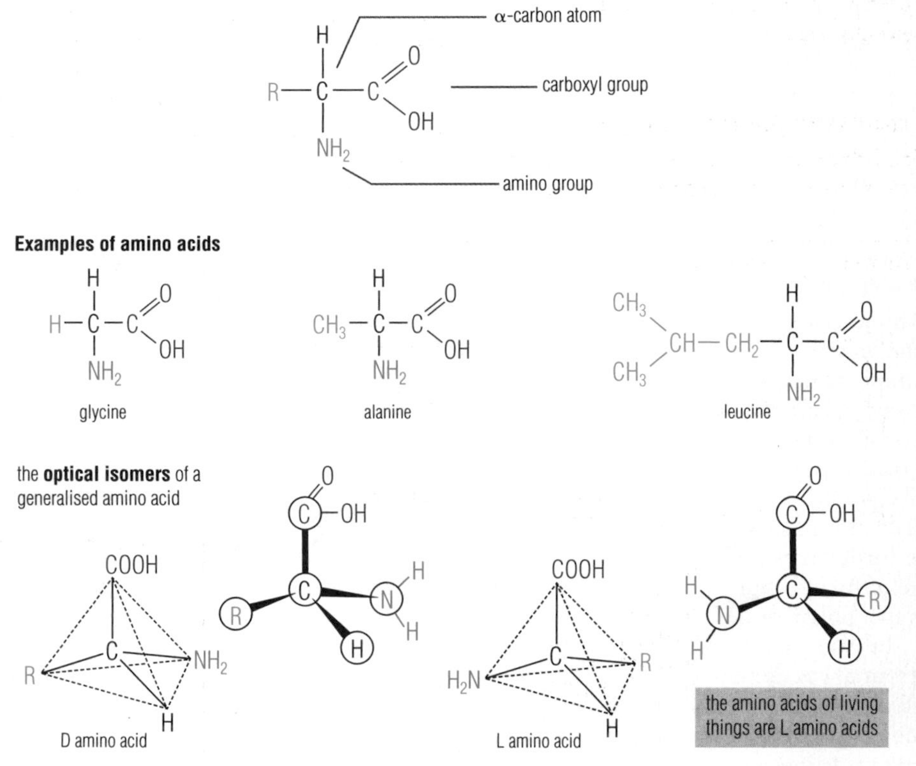

139

Amino acids as polar molecules

Amino acids are crystalline solids that dissolve in water but are insoluble in organic solvents. In neutral solution the amino acid molecule exists as a dipolar ion (**zwitterion**). The charge on the ion changes with pH. In an acid solution (low pH) the amino acid picks up H^+ ions and becomes positively charged. In alkaline solution (high pH) the amino acid donates H^+ ions to the medium and becomes negatively charged. The pH at which the amino acid is electrically neutral is called the **isoelectric point**.

Amino acids tend to stabilise the pH of a solution in which they are present, because they will remove excess H^+ or excess OH^- ions. They are acting as **buffers** (Figure 6.21). This buffering property is retained even when amino acids are incorporated into peptides and proteins, because of the presence of the additional amino groups and carboxyl groups of basic and acidic amino acid residues of the protein. Proteins play an important part in preventing changes in pH in cells and organisms.

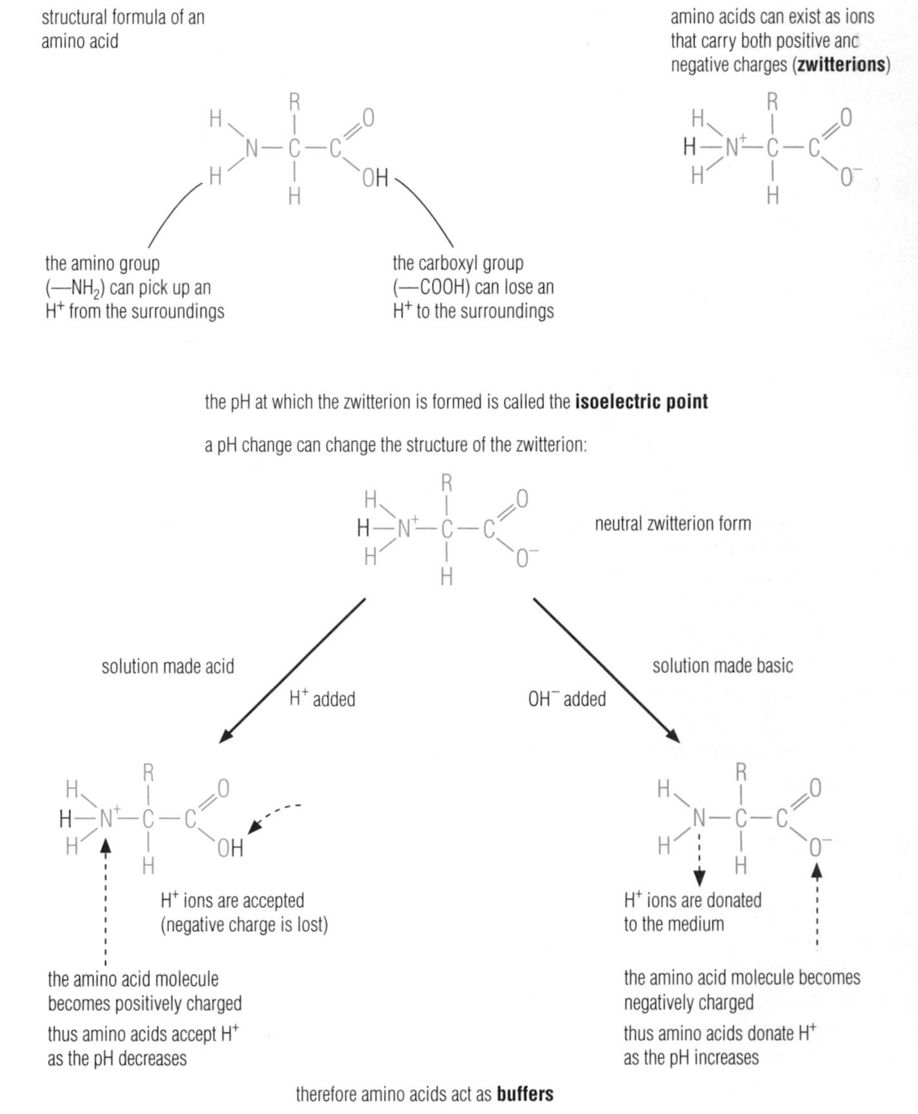

Buffers

In a buffer solution, the pH does not change significantly when a small amount of acid or alkali (base) is added to it. Some buffers contain a weak acid together with a soluble salt of a strong base with the same acid.

The pH of human blood is maintained between 7.35 and 7.45. This is achieved although the concentrations of carbon dioxide and carbonic acid in the blood vary greatly (p. 361). The buffer in blood consists of a mixture of phosphate ions, hydrogencarbonate ions and proteins.

■ Peptides

Amino acids can react with each other by condensation: the amino group of one amino acid molecule reacts with the carboxyl group of another with the elimination of water. The bond that is formed is called a **peptide bond**, and the resulting compound is a **dipeptide**. Three amino acid molecules will combine in this way to form a **tripeptide** (Figure 6.22), and so on. If many amino acids are joined in this way a **polypeptide** is formed.

In the presence of copper(II) ions (from a 1% copper sulphate solution) in alkaline solution (5% potassium hydroxide), a compound containing a peptide linkage forms a purple coloration, which

Figure 6.21   Amino acids are amphoteric

structural formula of an amino acid

the amino group (—NH₂) can pick up an H^+ from the surroundings

the carboxyl group (—COOH) can lose an H^+ to the surroundings

amino acids can exist as ions that carry both positive and negative charges (**zwitterions**)

the pH at which the zwitterion is formed is called the **isoelectric point**

a pH change can change the structure of the zwitterion:

neutral zwitterion form

solution made acid H^+ added

solution made basic OH^- added

H^+ ions are accepted (negative charge is lost)

the amino acid molecule becomes positively charged

thus amino acids accept H^+ as the pH decreases

H^+ ions are donated to the medium

the amino acid molecule becomes negatively charged

thus amino acids donate H^+ as the pH increases

therefore amino acids act as **buffers**

develops slowly, without heating. This is known as the **biuret test**, and is used to detect a peptide linkage (*Study Guide*, Chapter 6).

■ Polypeptides and proteins

The formation of proteins occurs on the ribosomes of cells (p. 208). Protein synthesis requires a supply of amino acids (the pool of amino acids in the cytoplasm), energy (obtained by reaction with ATP), and information (the linear sequence of bases in DNA, p. 204).

A polypeptide or protein chain is assembled, one amino acid residue at a time, leading to the formation of long-chain molecules. Common **polypeptides** of cells and organisms have twenty or more amino acid residues in their chains. For example, the melanocyte stimulating hormone (MSH), which controls skin

pigmentation in humans, has 22 amino acid residues. Glucagon (p. 505) has 29.

The hormone insulin contains 51 amino acid residues. (Insulin is produced by the β-cells of the islets of Langerhans in the pancreas (p. 505).) At or above this number a polypeptide chain is described as a **protein**, although there is no hard and fast dividing line between polypeptides and proteins. Normally, protein molecules are many hundreds of amino acid residues long, and the structure of some is further complicated because they contain several separate polypeptide chains linked together.

In 1953 the biochemist Fredrick Sanger, working at Cambridge, showed for the first time that a protein has a precisely defined **amino acid sequence** consisting of L-amino acids held by peptide linkages between amino groups and

Figure 6.22 The peptide bond

amino acids combine together, the amino group of one with the carboxyl group of the other

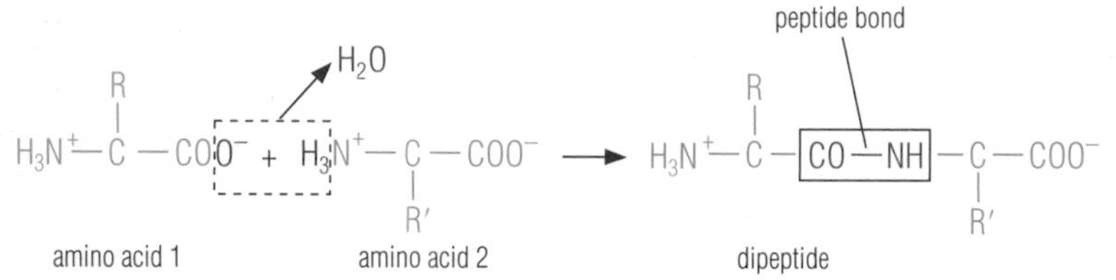

$$H_3N^+ - \overset{\overset{R}{|}}{C} - CO\boxed{O^- + H_3}N^+ - \overset{\overset{|}{}}{C} - COO^- \longrightarrow H_3N^+ - \overset{\overset{R}{|}}{C} - \boxed{CO - NH} - \overset{|}{C} - COO^-$$

amino acid 1   amino acid 2   dipeptide

peptide bond

two chemically different dipeptides can be formed from any two different amino acids; for example, glycine and alanine can react in two ways:

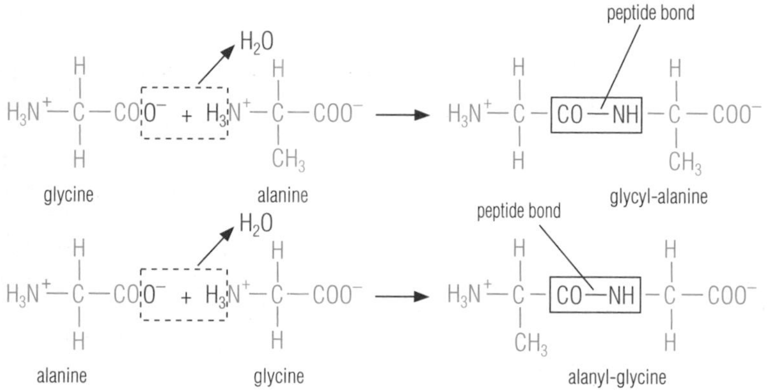

glycine   alanine   glycyl-alanine

alanine   glycine   alanyl-glycine

three amino acids combine together to form a tripeptide, e.g. glycyl-alanyl-cysteine:

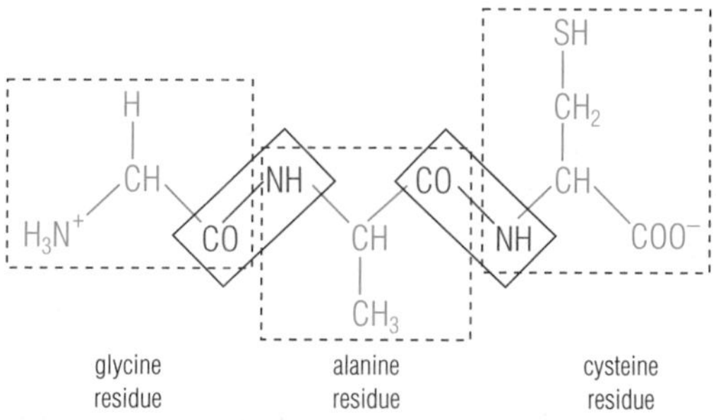

glycine residue   alanine residue   cysteine residue

there are six isomers of this tripeptide (gly-ala-cys, gly-cys-ala, cys-ala-gly, etc.), which are each chemically different

carboxyl groups. Sanger had elucidated the amino acid sequence of bovine insulin (Figure 6.23); the task took him ten years to complete.

Sanger's technique involved identifying the terminal amino acid residues by visible 'labelling' (he used a yellow compound). The 'labelled' insulin was completely hydrolysed to its constituent amino acids. The labelled amino acids were separated by paper chromatography (p. 249) and identified. This technique was combined with acid and enzymic hydrolysis of the protein chain into short-chain peptides in which the sequence of amino acids could be determined. For this work, Sanger was awarded his first Nobel prize in 1959.

■ Primary structure of a protein

The **primary structure** of a protein is the linear sequence of amino acids in its molecule. Proteins differ from each other in the variety, numbers and order of their constituent amino acids. In the living cell the sequence of amino acids in a polypeptide chain is specified by the DNA (p. 206).

Today, the task of determining the sequence of amino acids is carried out, at least in part, by machine. The process is much faster than that which Sanger applied to insulin. The primary structure of several thousand proteins is known.

Knowing the sequence of amino acids in a protein is important because this sequence determines practically all the properties of the protein. In a protein enzyme, about 90% of the amino acid residues are responsible for maintaining the three-dimensional structure. The rest are directly involved in enzyme catalytic action (Figure 8.20, p. 184).

Questions

1 a Write down generalised structural formulae for a monocarboxylic acid (p. 131) and an amino acid.
   b Why is an amino acid described as amphoteric?

2 What is meant by a condensation reaction? Show how it results in the formation of a peptide bond.

3 Living cells are not involved with the optical properties of isomers, yet L-amino acids occur naturally but D-amino acids do not. Why is this the case?

4 a Explain why the molecules of amino acids are described as polar.
   b How is it that amino acids are chemical buffers?

Figure 6.23 The amino acid sequence of bovine insulin: insulin is a small protein (a hormone) composed of 51 amino acid residues arranged as two polypeptide chains held together by disulphide bridges (for the meaning of the symbols see Table 9.7, p. 207)

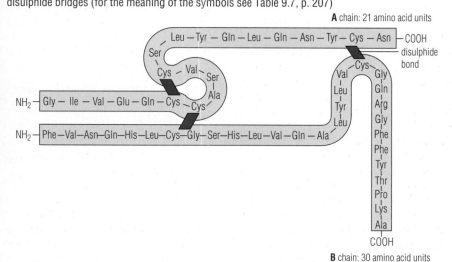

A chain: 21 amino acid units

Leu — Tyr — Gln — Leu — Gln — Asn — Tyr — Cys — Asn — COOH

disulphide bond

NH₂ — Gly — Ile — Val — Glu — Gln — Cys

NH₂ — Phe—Val—Asn—Gln—His—Leu—Cys—Gly—Ser—His—Leu—Val—Gln—Ala

B chain: 30 amino acid units

■ Secondary structure of a protein

Polypeptide chains may become folded or twisted in various ways. The most common ways are to coil to form a helix (α-helix) or to fold into sheets (β-sheets). These forms are referred to as the **secondary structure** of the protein.

When the polypeptide molecule coils up or folds up on itself, some of the atoms form so-called 'weak bonds': hydrogen bonds (p. 125), ionic bonds and van der Waals forces (Figure 6.24). Whilst none of these bonds is strong compared with covalent bonds, where many hundreds are formed they act to maintain the characteristic shape of the polypeptide or protein.

Some proteins contain **disulphide bonds**, formed by oxidation of sulphydryl groups of cysteine residues that can be brought adjacent to each other by the folding of the polypeptide chain. Bovine insulin contains three disulphide bonds (Figure 6.23).

α-helix

In many proteins part of the polypeptide chain takes the form of a helix (Figure 6.25). X-ray diffraction analysis (p. 202) shows that there are 3.6 amino acid residues per turn of the α-helix. The helical shape is maintained by hydrogen bonds.

Proteins contain different proportions of α-helix structures. A protein with a great deal of α-helical structure is keratin, which occurs in vertebrate skin and is also a component of hair, wool, nails, claws, beaks, feathers and horns.

β-sheets

The alternative secondary structure of protein is a sheet form, occurring as a flat zig-zag chain (Figure 6.25). This commonly occurs in insoluble structural proteins, but parts of soluble proteins often take this form too. Adjacent chains of β-sheets are stabilised by hydrogen bonding between NH and CO groups.

An example is the protein fibroin in silk, which consists almost entirely of the β-sheet form.

Parts of proteins are typically neither α-helices nor β-sheets. Sometimes these regions are represented as a 'random coil' but this is a misnomer; they are probably not random at all.

Collagen

Collagen is unique in that it forms a triple helical structure. The basic structure of collagen consists of three polypeptide chains, consisting largely of glycine and proline residues. It cannot form an α-helix because proline lacks the NH group necessary for internal hydrogen bonds. Three chains of this type, twisted together, achieve stability from the hydrogen bonds formed between the NH groups on the glycine residues in one chain and the CO groups of the proline in an adjacent chain.

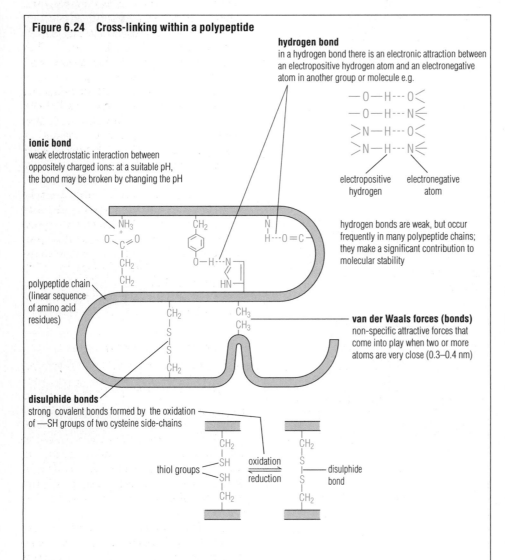

Figure 6.24   Cross-linking within a polypeptide

hydrogen bond
in a hydrogen bond there is an electronic attraction between an electropositive hydrogen atom and an electronegative atom in another group or molecule e.g.

electropositive hydrogen   electronegative atom

hydrogen bonds are weak, but occur frequently in many polypeptide chains; they make a significant contribution to molecular stability

ionic bond
weak electrostatic interaction between oppositely charged ions: at a suitable pH, the bond may be broken by changing the pH

polypeptide chain (linear sequence of amino acid residues)

van der Waals forces (bonds)
non-specific attractive forces that come into play when two or more atoms are very close (0.3–0.4 nm)

disulphide bonds
strong covalent bonds formed by the oxidation of —SH groups of two cysteine side-chains

thiol groups

oxidation ⇌ reduction

disulphide bond

Questions

1 a What three 'ingredients' are required for protein synthesis?
  b Where does protein synthesis occur in cells?
  c In what form does 'information' reach the site of protein synthesis (p. 208)?

2 Describe three types of bond that contribute to a protein's secondary structure.

3 In Figure 6.24, state which of the bonds shown contributes to
  a the protein's secondary structure
  b its tertiary structure
  c its solubility in water.

4 The possible number of different polypeptides (P) that can be made from a 'pool' of amino acids is given by:

$$P = A^n,$$

where A = the number of types of amino acid available, and n = the number of amino acid residues in the polypeptide molecule. Given the naturally occurring pool of twenty different types of amino acid, how many different polypeptides are possible if constructed from 5, 25 and 50 amino acid residues respectively?

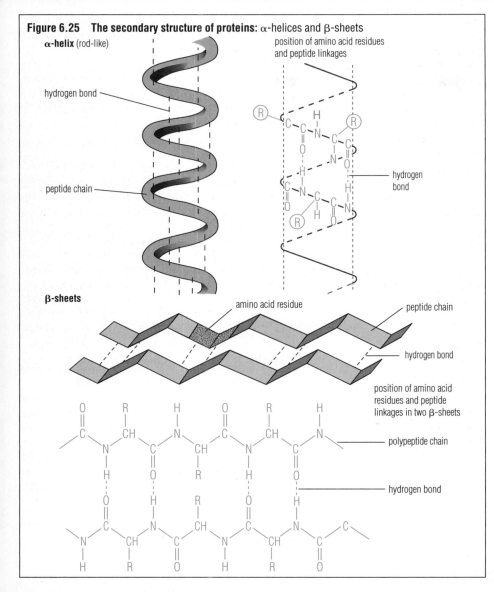

Figure 6.25 The secondary structure of proteins: α-helices and β-sheets

α-helix (rod-like)

hydrogen bond

peptide chain

position of amino acid residues and peptide linkages

hydrogen bond

β-sheets

amino acid residue

peptide chain

hydrogen bond

position of amino acid residues and peptide linkages in two β-sheets

polypeptide chain

hydrogen bond

Tertiary structure of a protein

Every protein molecule has a precise, compact, geometrical structure that is unique to that protein. For example, the enzyme lysozyme consists of 129 amino acid residues in the peptide chain, about 40% of which are in the α-helical form, 12% in the β-form and the rest in an irregular chain. The whole molecule is permanently folded into this complex shape, the **tertiary structure** of the protein. The shape is held by different types of bonding between adjacent parts of the chain (Figure 6.24). These include four disulphide bonds and numerous hydrogen bonds. Other groups in the peptide chain, especially those on the surface, form hydrogen bonds with water so that the molecule is water-soluble. Lysozyme (Figure 6.26) is a relatively small protein. Most proteins are very much larger, but the same principles of tertiary structure apply.

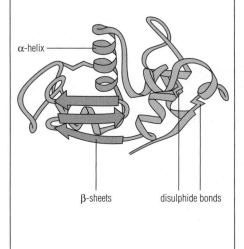

Figure 6.26 The tertiary structure of lysozyme, an antibacterial enzyme present in tears: a single polypeptide strand of 129 amino acid residues, part α-helix, part β-sheet, and part irregular sections

α-helix

β-sheets   disulphide bonds

Quaternary structure of proteins

Many complex proteins exist as aggregations of polypeptide chains held together by the various forms of bonding previously described. Their precise arrangement is described as the **quaternary structure of protein**. An example is haemoglobin, which consists of four separate polypeptide chains – two α-chains and two β-chains (Figure 17.28, p. 358). Haemoglobin also contains four haem groups within the molecule; proteins like this, which contain non-protein material in their molecules, are called **conjugated proteins**. The non-protein part is called the **prosthetic group**.

Roles of proteins

Some of the proteins of organisms are **fibrous proteins**, and these mostly perform structural functions. They include the collagen fibres of tendons, bones and connective tissues, myosin of muscle and keratin of hair. Many fibrous proteins are insoluble in water. They consist of long polypeptide chains with numerous cross-linkages.

Very many of the cell proteins are **globular proteins**, and these mostly function as enzymes, antibodies and hormones. Globular proteins consist of polypeptide chains tightly folded to form spherical shapes. They are soluble in water.

Much globular protein occurs on its own, but some is combined with non-protein material, forming conjugated proteins. Examples of conjugated proteins are the glycoproteins (protein + carbohydrate) of cell membranes (p. 159), and haemoglobin (p. 358). The functions and fates of proteins are described in Table 6.3 and Figure 6.27 (overleaf).

Metabolism of proteins

Organisms manufacture their own proteins from a pool of amino acids in the cytoplasm (p. 209). In addition plants and many microorganisms are able to manufacture their own amino acids (p. 333), whereas animals are unable to synthesise all twenty amino acids. An animal's chief source of amino acids is its diet, although humans can synthesise twelve of the twenty amino acids they need.

In living cells many proteins are constantly being broken down to the constituent amino acids and new protein is rebuilt. In humans, about 100 g of body protein are resynthesised each day.

Vertebrates do not store amino acids. Instead, excess amino acids are deaminated (that is, the —NH$_2$ part is removed and excreted, p. 508), and the remainder of the molecule is oxidised and used in respiration. If an animal is starved, the structural proteins of its body are drawn upon as a reserve and broken down to provide energy. If this goes on for any length of time the consequences are disastrous. Plants can store protein, however, and often do so in seeds and other structures.

■ Hydrolysis of proteins

The peptide bond can be broken by hydrolysis with dilute acids or by enzymes. An acid hydrolyses all the peptide bonds in a polypeptide quite indiscriminately, so breaking it down to its constituent amino acids. Some protease enzymes (p. 282) attack specific bonds, producing shorter peptide chains; others hydrolyse any peptide linkage present (Figure 6.28).

■ Denaturing of proteins

Denaturation is the loss of the three-dimensional structure of a protein molecule.

Table 6.3 The functions of proteins

Role	Examples
homeostatic	soluble proteins act as buffers, stabilising the pH wherever they occur in the body (p. 140)
structural	collagen of connective tissues (p. 215)
	keratin of skin, hair, etc. (p. 511)
hormonal	insulin, glucagon (regulation of glucose metabolism, p. 504)
enzymic	most enzymes are proteins, e.g. digestive enzymes of gut (p. 287), respiratory enzymes in all cells
transport	cell membrane protein and protein of membranes of cell organelles (involved in transport of metabolites and ions across membranes, p. 231)
	haemoglobin (oxygen transport in vertebrate blood, p. 358)
	myoglobin (oxygen transport in muscle, p. 360)
protection	antibodies (reacting with 'foreign' proteins and some other molecules, p. 366)
	fibrinogen, thrombin (in blood clotting mechanism, p. 364)
contractile	myosin, actin (muscle contraction mechanism, p. 491)
storage	casein in milk (p. 241)
	aleurone protein in seeds (p. 405)

Mild chemical treatment of proteins, such as an alteration in pH, changes the distribution of the charged groups along the polypeptide chain. This alters the bonds that maintain the secondary and tertiary structure of the protein. The protein molecules may begin to unfold, and the properties of the protein that depend upon shape are then lost. Reversal of the mildly denaturing conditions will allow the protein to return to its normal shape and so regain its properties.

Heating proteins usually denatures the protein irreversibly, leading to precipitation from solution. An example is the irreversible coagulation of egg white on boiling. Denaturation leading to coagulation may also be brought about by treatment with heavy metal ions or organic solvents.

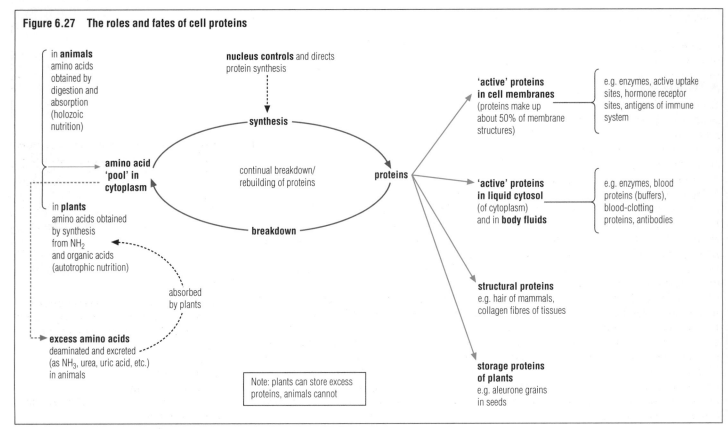

Figure 6.27   The roles and fates of cell proteins

in **animals**
amino acids obtained by digestion and absorption (holozoic nutrition)

nucleus controls and directs protein synthesis

synthesis

'active' proteins **in cell membranes** (proteins make up about 50% of membrane structures)

e.g. enzymes, active uptake sites, hormone receptor sites, antigens of immune system

amino acid 'pool' in cytoplasm

continual breakdown/ rebuilding of proteins

proteins

'active' proteins **in liquid cytosol** (of cytoplasm) and in **body fluids**

e.g. enzymes, blood proteins (buffers), blood-clotting proteins, antibodies

in **plants**
amino acids obtained by synthesis from NH$_2$ and organic acids (autotrophic nutrition)

breakdown

absorbed by plants

structural proteins
e.g. hair of mammals, collagen fibres of tissues

excess amino acids deaminated and excreted (as NH$_3$, urea, uric acid, etc.) in animals

Note: plants can store excess proteins, animals cannot

storage proteins **of plants** e.g. aleurone grains in seeds

Figure 6.28 Hydrolysis of polypeptides

hydrolysis of a dipeptide

H_3N^+—CH—CO—NH—CH—COOH $\longrightarrow$ H_3N^+—CH—COO$^-$ + H_3N^+—CH—COO$^-$

glycyl-alanine   glycine   alanine

hydrolysis by enzymes

the pancreatic enzyme chymotrypsin helps digestion of peptides by attacking peptide bonds where the amino acid on the carbon side of the —CONH— peptide bond is either phenylalanine or tyrosine:

—CH—CO—NH—CH—CO—NH—CH—CO—NH—CH—

phenyl-   alanine
alanine   OH$^-$ H$^+$
chymotrypsin

other proteases digest polypeptides at different points in the chain; trypsin hydrolyses any peptide bond where the amino acid at the carbon side of the —CONH— bond is either lysine or arginine:

trypsin will attack only these bonds

H_2N—Leu—Val—Glu—Phe—Ala—Gly—Arg—Ala—Ser—Trp—Lys—Gly—CO_2H

chymotrypsin will attack this bond

6.9 NUCLEIC ACIDS

Nucleic acids are the informational molecules of the cell; they constitute the genetic material of all living things as well as of viruses (Figure 6.29). Within the structure of nucleic acids are coded the 'instructions' that govern all cellular activities. This control is largely achieved by enzymes, which determine the kind and extent of the chemical reactions of life (Chapter 8), and enzyme synthesis is controlled by nucleic acids.

Two types of nucleic acid are found in living cells: deoxyribonucleic acid (DNA) and ribonucleic acid (RNA). Most of the DNA is in the nucleus, but while there is some RNA in the nucleus most is in cytoplasm, particularly in the ribosomes. (Some nucleic acid also occurs in chloroplasts and mitochondria, p. 163.)

Nucleic acids are long, thread-like macromolecules built up of nucleotides. Hydrolysis of a nucleotide yields three component substances: a pentose sugar, nitrogenous bases and phosphoric acid.

The nitrogenous bases found in nucleic acids are derived from one of two parent compounds, purine or pyrimidine (Figure 6.31, overleaf). Two purine-

Figure 6.29 The composition of nucleic acid

nucleic acid
nucleotides
(nucleoside)

nitrogenous base (purine or pyrimidine) — **pentose sugar** (deoxyribose or ribose) — phosphoric acid

OH
O=P—OH
HO

derived bases (adenine and guanine) and three pyrimidine-derived bases (thymine, cytosine and uracil) are found in nucleic acid. It has become a tradition in biology to abbreviate the names of the five bases to their first letters, A, G, T, C and U. DNA contains A, G, T and C; RNA contains A, G, U and C.

The sugars in nucleic acids are five-carbon sugars (pentoses), which exist in a five-sided ring form. The two pentose sugars occurring in nucleic acids, deoxyribose and ribose, are very similar (Figure 6.12). The 'deoxy' prefix indicates that this sugar lacks an oxygen atom that is present in ribose. Deoxyribose occurs in DNA, and ribose in RNA.

■ Nucleosides, nucleotides and nucleic acids

The combination of a pentose sugar with a base forms a compound known as a **nucleoside**. The combination of a nucleoside with a phosphate group forms a **nucleotide**. In Figure 6.30 there is a list of the five nucleotides from which **nucleic acids** are built.

A nucleic acid molecule is a string of nucleotides. The resulting compound, also called a polynucleotide, has a 'backbone' of —sugar–phosphate–sugar–phosphate— residues with a base residue attached to every sugar. Nucleic acid molecules are of great length; they have relative molecular masses of several millions.

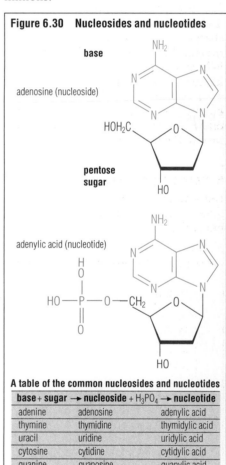

Figure 6.30 Nucleosides and nucleotides

base

adenosine (nucleoside)

pentose sugar

adenylic acid (nucleotide)

A table of the common nucleosides and nucleotides

base + sugar	→ nucleoside + H_3PO_4	→ nucleotide
adenine	adenosine	adenylic acid
thymine	thymidine	thymidylic acid
uracil	uridine	uridylic acid
cytosine	cytidine	cytidylic acid
guanine	guanosine	guanylic acid

Questions

1 For what general reasons are the membrane proteins important to the functioning of a cell?
2 What would be the effects of treating fresh insulin
   a with acids
   b with trypsin
   c with chymotrypsin?
3 Why are temperatures above 50 °C lethal to most living cells?

In a nucleic acid, the nucleotides are joined together by covalent bonds between the phosphate group of one nucleotide and the sugar of the next. It is the phosphate groups that give nucleic acids their acidic properties. The link, known as a **phosphodiester bridge**, occurs between carbon atom 3 (3') of one sugar and carbon atom 5 (5') of the next. (The 'dash' after the number means that we are referring to the sugar ring rather than that of the nitrogen-containing base.) Thus the nucleic acids have a direction. The base sequence of DNA and RNA is written in the 5' → 3' direction (Figure 6.32).

Overall, nucleic acids have a stable structure, an important property relating to their role in cells (p. 200). How the three-dimensional structure of DNA was elucidated, and how the significance of the sequence of bases became understood – an outstanding achievement of modern biology – is discussed in Chapter 9.

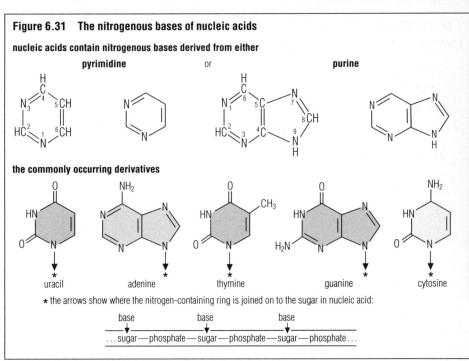

Figure 6.31   The nitrogenous bases of nucleic acids

nucleic acids contain nitrogenous bases derived from either

pyrimidine   or   purine

the commonly occurring derivatives

uracil   adenine   thymine   guanine   cytosine

★ the arrows show where the nitrogen-containing ring is joined on to the sugar in nucleic acid:

base   base   base

...sugar—phosphate—sugar—phosphate—sugar—phosphate...

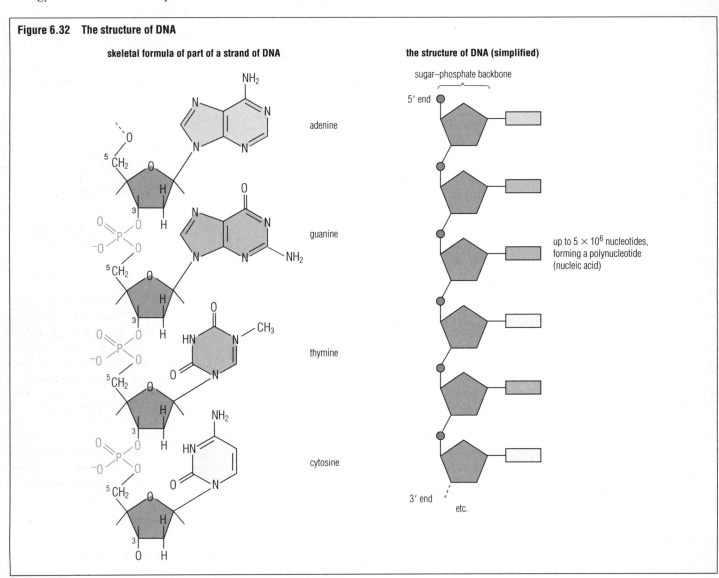

Figure 6.32   The structure of DNA

skeletal formula of part of a strand of DNA

the structure of DNA (simplified)

adenine

guanine

thymine

cytosine

sugar–phosphate backbone

5' end

up to 5×10^6 nucleotides, forming a polynucleotide (nucleic acid)

3' end

etc.

6.10 ADENOSINE TRIPHOSPHATE AND NUCLEOTIDE COENZYMES

In addition to being the building blocks of nucleic acids, certain nucleotides and their derivatives are important coenzymes. The role of coenzymes in enzyme action is discussed on p. 179.

■ Adenosine triphosphate

Adenosine triphosphate (ATP) consists of adenine linked to a pentose sugar (ribose), which in turn is linked to a string of three phosphate groups. ATP is found in every living cell.

The majority of energy-requiring (**endergonic**) reactions are powered by ATP. During reactions with other cell chemicals, hydrolysis of the third phosphate group of ATP yields energy that can be used to do useful work. ATP is therefore a reservoir of potential chemical energy. An ATP molecule breaks down to a molecule of adenosine diphosphate and inorganic phosphate. The structure of ATP is shown in Figure 8.9.

■ NAD and NADP

Nicotinamide adenine dinucleotide (NAD) and nicotinamide adenine dinucleotide phosphate (NADP) are coenzymes derived from nicotinic acid (a vitamin of the B complex). These molecules carry hydrogen ions and electrons involved in metabolism, NAD in respiration (p. 316) and NADP in photosynthesis (p. 259) (Figure 6.33).

■ Flavine nucleotides (FAD and FMN)

Flavine adenine dinucleotide (FAD) and flavine mononucleotide (FMN) are coenzymes derived from vitamin B_2. These substances are hydrogen-carrying molecules involved in oxidation and reduction reactions. They occur quite tightly associated with proteins, forming complexes known as flavoproteins (Figure 6.33).

■ Coenzyme A

Coenzyme A is involved in the addition and removal of two-carbon fragments from certain carbohydrates and lipids during respiration and in food storage reactions (Figure 6.33).

Questions

1 Briefly describe the differences between an organic base, a nucleoside, a nucleotide and a nucleic acid.

2 How does DNA differ from RNA, both in its structure and in its distribution in the cell?

3 What are the monomers in nucleic acid?

Figure 6.33 Nucleotide coenzymes involved in metabolism

1 nicotinamide adenine dinucleotide (NAD)

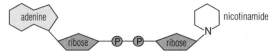

adenine — ribose — P — P — ribose — nicotinamide

the function of NAD is to carry protons and electrons in respiration; the complex structure can be simplified to the 'action end' (hydrogen transport) and the 'rest' (part that is identified by enzymes)

mode of action:

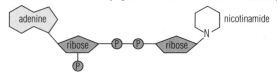

H—CONH₂ $2H^+, 2e^-$ H H—CONH₂ + H⁺

this reaction is usually written:

$$NAD^+ + 2H^+ + 2e^- \rightleftharpoons NADH + H^+$$
oxidised form   reduced form

2 nicotinamide adenine dinucleotide phosphate (NADP)
an almost identical molecule carrying out an identical role to that of NAD, but in photosynthesis

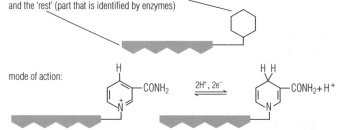

adenine — ribose — P — P — ribose — nicotinamide
P

3 flavine nucleotides

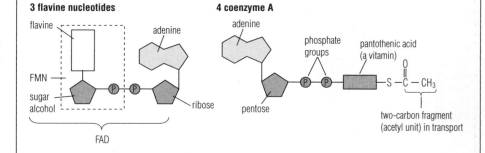

flavine
FMN
sugar alcohol — P — P — ribose — adenine
FAD

4 coenzyme A

adenine
phosphate groups
pantothenic acid (a vitamin)
pentose — P — P — S—C—CH₃
two-carbon fragment (acetyl unit) in transport

6.11 OTHER COMPOUNDS OF IMPORTANCE

There are other chemical substances of importance to cells and tissues. Some information concerning their chemistry is introduced where they arise in this book, in sufficient detail to understand their functions. These substances include the fat-soluble and water-soluble vitamins (p. 272), the mineral salts and trace elements (p. 273), the growth factors of plants (p. 422) and the hormones of animals (p. 472).

■ Lignin

Lignin occurs in the walls of xylem vessels, tracheids, fibres and sclereids. It accumulates between the cellulose microfibrils, and strengthens the resistance of walls to compression forces.

Lignin is neither a carbohydrate nor a protein. It is derived from tyrosine, an amino acid, which is oxidised to form phenolic compounds that polymerise. Lignin is digested by certain fungi and bacteria. It can be dissolved by sodium bisulphite solution, and this is exploited in the manufacture of paper pulp from wood.

FURTHER READING/WEB SITES

B Rockett and P Sutton (1996) *Chemistry for Biologists at Advanced Level.* John Murray
Royal Society of Chemistry: **www.chemsoc.org**

7 Cell structure

- The cell is the basic unit of structure and function; virtually all organisms consist of cells and cell products.
- Cells are extremely small and can be seen properly only when magnified and viewed through the lenses of a microscope.
- There is great variety in the shapes, sizes and structures of the cells of plants and animals. This variety reflects the varied functions for which cells are adapted.
- The fine structure of cells, known as cell ultrastructure, was discovered by the technique of electron microscopy.

- The cell is a three-dimensional sac of different organelles suspended in an aqueous medium and contained within a plasma membrane.
- The plasma membrane consists of a lipid bilayer in which proteins occur on, in and across the membrane. Lipids and proteins move about within the structure of the membrane.
- The cells of prokaryotes (bacteria and cyanobacteria) are extremely small when compared to eukaryotic cells, and are without a true nucleus or the organelles of a eukaryotic cell.

ANTHONY VAN LEEUWENHOEK (1632–1723): THE POWER OF THE SINGLE LENS

Life and times

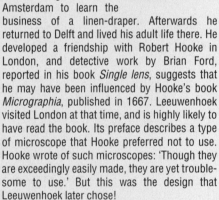

Leeuwenhoek (pronounced 'Laywenhook') was born in Delft, the son of a basket-maker. He had no formal training in science and a rather limited schooling before being sent to Amsterdam to learn the business of a linen-draper. Afterwards he returned to Delft and lived his adult life there. He developed a friendship with Robert Hooke in London, and detective work by Brian Ford, reported in his book *Single lens*, suggests that he may have been influenced by Hooke's book *Micrographia*, published in 1667. Leeuwenhoek visited London at that time, and is highly likely to have read the book. Its preface describes a type of microscope that Hooke preferred not to use. Hooke wrote of such microscopes: 'Though they are exceedingly easily made, they are yet troublesome to use.' But this was the design that Leeuwenhoek later chose!

In 1673 Leeuwenhoek's name first appeared in the annals of the Royal Society in London. He was elected a Fellow in 1680.

Contributions to biology

Leeuwenhoek was an enthusiastic microscopist. He produced his own microscopes (500 of them at least) of slips of metal, perforated by a single hole in which he sandwiched his hand-made lens. A series of simple screws moved the specimen across the field of view, and focused the image. The instrument was very small, yet magnified ×266 (the best of modern equipment is only four times as powerful). Leeuwenhoek's observations included:

- blood cells, spermatozoa and protozoa, and organelles such as cilia (p. 165)
- bacteria growing in infusions of plant materials (he did not dwell on the harmful effects of microorganisms, but rather on their structure and activities)

- the vinegar eelworm (a nematode) producing live young (rather than laying eggs)
- the cells of trees (he cut sections of 50 species of tree), and pits in plant cell walls
- the differences between 'spring' and 'summer' wood of trees (p. 410)
- that freshwater protozoa could survive being dried up; he took issue with Robert Hooke's acceptance of the idea of 'spontaneous generation', believing that where life appeared 'seeds' were present first
- the parasites of frogs and the liver fluke of sheep's liver
- that plant extracts and sulphur dioxide function as pesticides
- a practical study of conception in mammals.

Influences

Through the Royal Society of London, his results were published. In all, he sent 200 letters in a personal style, describing his experiments and observations. The resulting fame brought him visits by many microscopists. He remained secretive about his methods, however. He worked for himself and for the advancement of knowledge, and refused to teach others or to start a 'school'.

His simple microscope (and other designs of single lens microscope) have been dismissed as 'crude' and 'basic', following popularisation of the compound microscope since the early 1800s. The power of the single lens and its role in biological discovery have, however, been greatly underestimated.

an original Leeuwenhoek microscope

7.1 INTRODUCTION

In Chapter 1 it was established that living things are made of cells. The cell is the basic unit of structure and function; virtually all organisms consist of cells and cell products. This generalisation is known as the **cell theory**. The cell theory is usually credited to the European biologists Matthias Schleiden and Theodor Schwann in the period 1838–9, but cells were first observed very much earlier than this. For example, Leeuwenhoek made detailed observations of plant and animal cells with his simple microscope (see panel). Robert Hooke first used the term 'cell' in 1662.

■ Cell theory

Today the cell theory embraces four ideas:

- the cell is the building block of structures in living things
- the cell is derived from other cells by division
- the cell contains information that is used as instructions for growth, development and functioning (hereditary material, p. 192)
- the cell is the functioning unit of life; the chemical reactions of life (metabolism, p. 170) take place within cells.

Many biologists contributed to the development of the cell concept, which evolved gradually during the nineteenth century as a result of the steadily accelerating pace of developments in microscopy and biochemistry. Today the study of the structure of cells (**cytology**) is part of a major branch of biology known as **cell biology**. Cell biology embraces biochemistry (Chapter 8) as well as cytology, and is linked to many other aspects of biology – most importantly, to genetics (Chapters 9 and 28). The panel opposite identifies some steps in the foundations of modern cell biology.

■ Cell size

Cells are extremely small, and can be seen properly only when magnified and viewed through the lenses of a microscope. Cell dimensions are expressed in micrometres (μm); micrometres are sometimes called microns. Cells are in the size range of about 5–500 μm, but most are between 10 and 150 μm (Figure 7.1).

The Système International (SI) unit of length is the metre. The SI-derived units of length are

1 metre (m) = 1000 millimetres (mm)
1 mm (10^{-3} m) = 1000 micrometres (μm)
1 μm (10^{-6} m) = 1000 nanometres (nm)
1 nm (10^{-9} m) = 1000 picometres (pm).

The angstrom (Å) is 10^{-10} m, and is sometimes used to record the thickness of cell membranes (see section 7.4 on p. 157) and the sizes of certain macromolecules.

■ Two types of cellular organisation

The differences between prokaryotic and eukaryotic organisation were introduced on p. 19. This chapter is largely concerned with the structure of eukaryotic cells; the structural features of a prokaryotic cell are listed on p. 169.

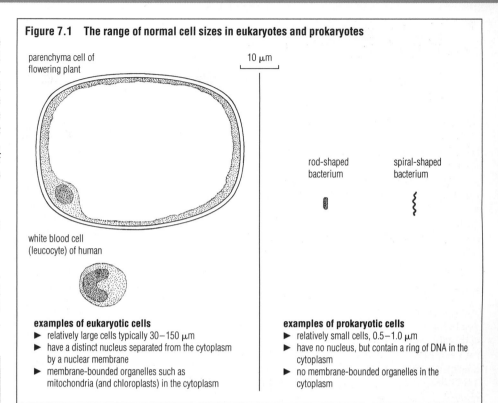

Figure 7.1　The range of normal cell sizes in eukaryotes and prokaryotes

parenchyma cell of flowering plant

10 μm

rod-shaped bacterium

spiral-shaped bacterium

white blood cell (leucocyte) of human

examples of eukaryotic cells
► relatively large cells typically 30–150 μm
► have a distinct nucleus separated from the cytoplasm by a nuclear membrane
► membrane-bounded organelles such as mitochondria (and chloroplasts) in the cytoplasm

examples of prokaryotic cells
► relatively small cells, 0.5–1.0 μm
► have no nucleus, but contain a ring of DNA in the cytoplasm
► no membrane-bounded organelles in the cytoplasm

Questions

1　Leeuwenhoek made many discoveries about cell structure, and yet is said to have founded no 'school'. What does this mean? What are the likely reasons for this?

2　What are the four features of the cell theory?

3　How did Leeuwenhoek contribute to the demise of the theory of 'spontaneous generation'?

KEY STEPS IN THE ORIGINS OF CELL BIOLOGY

Since antiquity	Magnifying properties of convex lens known	1839 Purkinje	Realised that the contents of cells ('cytoplasm + nucleus') were important, not just the walls
Early 17th century	The use of two convex lenses to make near objects look larger (the compound microscope) was practised in Europe	1838 Matthias Schleiden	Explained the derivation of plant tissues from cells
1632–1723	Antony van Leeuwenwoek (see opposite)	1839 Theodor Schwann	'Cells are organisms and entire animals and plants are aggregates of these organisms arranged according to definite laws'
1661 Marcello Malpighi	Used lenses, discovered capillaries, and may have described cells in writing of 'saccules' and 'globules'		
1662 Robert Hooke	Introduced the term 'cell' in describing the structure of cork; believed cell walls to be the important part of otherwise empty structures	1856 Rudolf Virchow	Established the idea that cells arise only by division of existing cells
1809 Jean de Monet de Lamarck	'No body can possess life if its containing parts are not of cellular tissue or formed by cellular tissues'	1862 Louis Pasteur	Disposed of the spontaneous generation theory of microbial appearance
1825–1945	With the invention of the immersion lens, the light microscope reached the limit of resolving power = 0.2 μm	1897 Eduard Buchner	Confirmed that fermentation requires only a cell extract, not whole cells
		1900	Rediscovery of work of Gregor Mendel, giving the theoretical basis of modern genetics
1828 Friedrich Wöhler	Synthesised urea, discrediting the view that organic compounds can be made only by living things	1930–46	Electron microscope developed, then used widely in cytology, revealing details of cell organelles (cell ultrastructure)
1831 Robert Brown	Reported his discovery of the cell nucleus	modern cell biology	Established that cells contain basic hereditary material (DNA) in the nucleus; roles of nucleic acid in the control of metabolism and of growth via protein synthesis understood

7.2 THE STRUCTURE OF EUKARYOTIC CELLS

There is great variety in the shapes, sizes and structures of cells of plants and animals. These differences can be observed by light microscopy.

When human cheek cells and the leaf cells of Canadian pondweed are examined by phase-contrast microscopy (p. 153), or are stained with dyes such as methylene blue (for cheek cells) or with iodine solution (for pondweed leaf cells) and are examined by the high power of the compound microscope, a range of structures is seen (Figure 7.2).

■ The plant cell

The plant cell is surrounded by a tough yet flexible and fully permeable external **cell wall**, laid down by the cytoplasm, and composed of cellulose (p. 135), pectins, hemicelluloses and some protein. The cytoplasm contains a more or less spherical **nucleus**. The nucleus controls and directs the activities of the cell, and the **cytoplasm** and its component structures are the sites of the chemical reactions of life. With the aid of particular stains the nucleus is seen to be surrounded by the **nuclear envelope**, or **nuclear membrane**, and to contain one or more compact, spherical **nucleoli** (singular, nucleolus) amidst an otherwise granular region of **chromatin**.

The cytoplasm is bounded by a **plasma membrane** (cell surface membrane or plasmalemma) and is pressed firmly against the cell wall by a large fluid-filled **vacuole**, which usually makes up the bulk of the plant cell. The vacuole is surrounded by a specialised membrane, the **tonoplast**, and is traversed by strands of cytoplasm. Within the cytoplasm are many tiny structures of various types called **organelles**. Certain types of organelles are visible by light microscopy, and in some plant cells these include tiny, discoid **chloroplasts**. Other organelles of the cytoplasm have been observed only by electron microscopy (p. 154). The cytoplasm may also contain stored food substances such as oil droplets. Starch grains occur in chloroplasts or leucoplasts (p. 164). Adjacent plant cell walls are cemented together by the **middle lamella** between them. There are cytoplasmic connections between cells, called **plasmodesmata** (singular, plasmodesma).

■ The animal cell

The animal cell is bounded by a plasma membrane, but it has no cell wall and contains no large permanent vacuoles. The nucleus is an identical structure to that in plants. An organelle called a **centrosome**, consisting of two **centrioles**, occurs beside the nucleus in animal cells. The cytoplasm may contain food granules, secretory granules and often temporary vacuoles or **vesicles**. It also contains organelles of various types, but no chloroplasts. The organelles of animal cells, like those of plant cells, are mostly too small to be seen by light microscopy.

■ Cell diversity

The shapes of cells vary widely (Figure 7.3). This variety often reflects the varied functions for which the cells are adapted, and is particularly evident in multicellular organisms. Here, groups of cells are usually highly specialised and carry out particular functions within the whole organism. The specialisation of cells is known as **division of labour** (Chapter 10).

■ Non-cellular organisation

In addition to the familiar unicellular or multicellular organisation of living things, there are a few multinucleate organs and organisms, without divisions into separate cells. These include those fungal hyphae that are **coenocytic** (p. 557) and vertebrate voluntary muscle fibres, known as **syncytia** (p. 218).

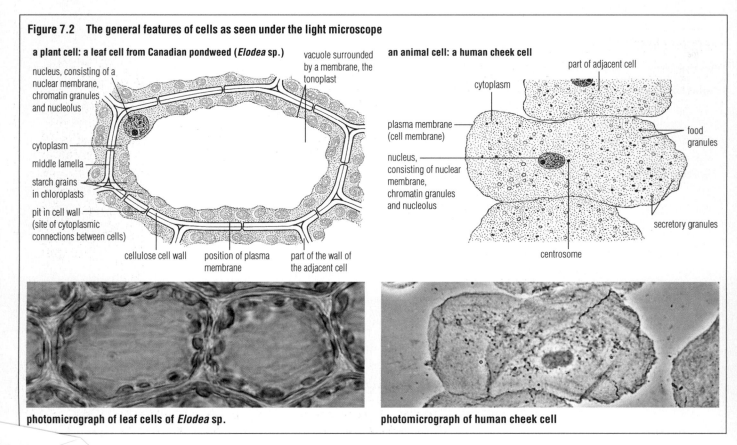

Figure 7.2   The general features of cells as seen under the light microscope

a plant cell: a leaf cell from Canadian pondweed (*Elodea* sp.)

nucleus, consisting of a nuclear membrane, chromatin granules and nucleolus

vacuole surrounded by a membrane, the tonoplast

cytoplasm

middle lamella

starch grains in chloroplasts

pit in cell wall (site of cytoplasmic connections between cells)

cellulose cell wall

position of plasma membrane

part of the wall of the adjacent cell

an animal cell: a human cheek cell

part of adjacent cell

cytoplasm

plasma membrane (cell membrane)

nucleus, consisting of nuclear membrane, chromatin granules and nucleolus

food granules

secretory granules

centrosome

photomicrograph of leaf cells of *Elodea* sp.

photomicrograph of human cheek cell

Figure 7.3 The diversity of animal and plant cells

Free-living unicellular organisms

Amoeba sp.: a genus of single-celled organisms living in soil and fresh water. Irregularly shaped; projecting from the body are pseudopodia concerned with locomotion and with food capture. Size approximately 400–500 μm.

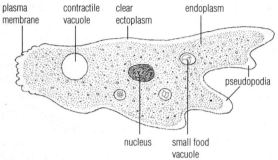

plasma membrane · contractile vacuole · clear ectoplasm · endoplasm · pseudopodia · nucleus · small food vacuole

Chlamydomonas sp.: a single-celled aquatic alga living in stagnant freshwater pools and ditches, often in such large numbers that the water is coloured green. The cell is motile, photosynthetic and light-sensitive. Size about 20×10 μm.

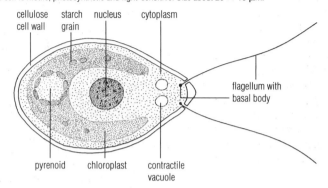

cellulose cell wall · starch grain · nucleus · cytoplasm · flagellum with basal body · pyrenoid · chloroplast · contractile vacuole

Cells of multicellular organisms
Mammalian cells

Ciliated epithelium cell: epithelia form a thin layer which covers surfaces of the body, internally and externally. Ciliated epithelium lines the air passages of the lungs, the oviducts and the spinal and brain canals.

This cell is attached to a basement membrane; arising from the free surface of the cell are cilia, each originating from a basal body. Size typically 10×3 μm.

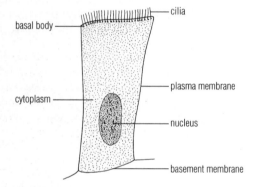

cilia · basal body · plasma membrane · cytoplasm · nucleus · basement membrane

Involuntary (unstriped) muscle cell: these muscle cells are present in layers in the wall of the gut, in blood vessels, in the iris, and in the dermis of the skin in mammals. Each cell is elongated and spindle-shaped, with very long myofibrils passing along its length. These can contract when stimulated. Size typically 100×7 μm.

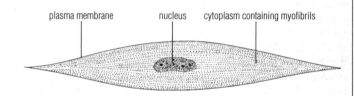

plasma membrane · nucleus · cytoplasm containing myofibrils

Flowering plant cells

Phloem sieve tube cell with companion cell: occurs in vascular tissue of angiosperms. The end walls are perforated by small pores. Sieve tubes are the site of solute (sugar and amino acids) transport. Mature sieve tubes are without a nucleus but are accompanied by companion cells. Size typically 30×140 μm.

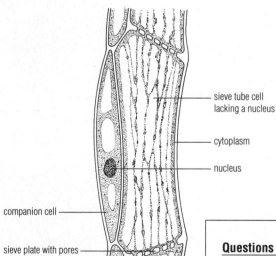

sieve tube cell lacking a nucleus · cytoplasm · nucleus · companion cell · sieve plate with pores · cellulose cell wall

Protoxylem vessel: occurs in the vascular tissue of angiosperms. Protoxylem vessels are the first-formed vessels, and the site of water movement. The spirally arranged wall thickenings consist of cellulose impregnated with lignin. Size typically 40×150 μm.

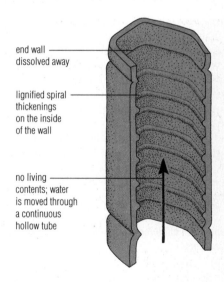

end wall dissolved away · lignified spiral thickenings on the inside of the wall · no living contents; water is moved through a continuous hollow tube

Questions

1 a What features do animal and plant cells have in common?
   b Make a table of the differences in structure between animal and plant cells.

2 Figure 7.3 illustrates some of the diversity in structure that occurs between different cells. What may be the advantages of such structural differences?

3 What are the difficulties in trying to describe a 'typical' plant or animal cell?

7.3 TECHNIQUES IN CELL BIOLOGY

Optical microscopes were the first powerful tools in the history of biology, and they remain one of the most useful. Structures that are too small to be seen by the unaided eye may be observed with the hand lens (simple microscope) or the compound microscope. In these instruments light rays passing from the specimen are transmitted through a single lens or system of lenses, and focused on the retina of the eye.

The subsequent invention and development of the electron microscope (p. 154) during the twentieth century has shown us the detailed structures of the cytoplasm and, in particular, the organelles. The separation of different organelles by the techniques of cell fractionation has led to extensive knowledge of cell biochemistry. Both the structure of organelles and their functions are now understood.

■ Optical microscopy

The **hand lens** consists of a biconvex lens in a supporting frame. It enables the eye to see an enlarged image of a small object held close to the eye. The hand lens is used to observe the external form of objects, rather than details of their internal structure. The power of magnification of the common hand lens is usually between ×10 and ×20.

The **compound microscope** (Figure 7.6) is an expensive precision instrument in which the magnifying powers of two convex lenses, the eyepiece and the objective lens, are used to produce a magnified image of very small objects. Light rays are transmitted through tiny objects or thin sections of larger objects. Refraction at the two lenses produces an enlarged, upside-down image. The light microscope is mostly used to investigate internal structure.

Preparation of biological material for optical microscopy

Biological specimens, living or preserved, can be observed with the compound microscope provided there is sufficient contrast for the structural differences of the specimen to be appreciated. 'Contrast' is the highlighting of tiny differences in the structure of cells. In biological material that is too transparent, contrast is normally achieved by staining the material.

Using vital dyes

Some dyes, known as **vital dyes** or stains, are taken up by living cells without killing them. These dyes colour particular structures within cells. For example, if yeast cells are treated with dilute Neutral Red dye or Congo Red dye (these dyes are also acid/base indicators, like litmus) and then fed to *Paramecium*, the stages of digestion of the yeast cells in the *Paramecium* cytoplasm can be observed.

Preserved and stained tissues

Much of what is known about the structure of cells was first learnt by observation of thin sections of tissues that had been preserved, sectioned, stained and permanently mounted on microscope slides (Figure 7.4). The steps in the process are described below.

▶ **Fixation:** preserving freshly killed tissues in as lifelike a condition as possible, by immersion in a chemical such as formalin. This also hardens the tissue so that sections can be cut.

▶ **Sectioning:** cutting microscopically thin slices of hardened tissue. The sections must be thin enough to let light through. The tissue is either embedded in wax or plastic and sectioned by microtome or supported in elder pith and sectioned using a hand-held razor.

▶ **Staining:** immersing sections in dyes that stain some structures better than others. For example, cytoplasm stains pink with eosin or green with the dye Light Green; nuclei stain blue with haematoxylin or red with safranin.

▶ **Dehydration:** immersion of stained sections in ethanol to remove water. This helps to make and keep the tissue transparent. Dehydration is carried out gradually, using a series of increasing concentrations of ethanol in water until 'absolute' (nearly pure) alcohol is reached.

▶ **Clearing:** alcohol in the specimens is replaced by a liquid such as xylene, which is miscible with the mounting medium. This completes the process of making the material transparent.

▶ **Mounting:** sections are mounted on a microscope slide in a medium (Canada balsam, for instance), which excludes air and protects indefinitely. A cover slip is added, and the mounting medium hardened.

Stained temporary mounts

The making of **temporary mounts** is a quicker process. It is especially appropriate for the examination of thin sections of plant organs (stem, leaf or root) cut by hand-held razor. Sections are dropped into water in a watch glass and stained. For example, lignified cell walls stain yellow with aniline sulphate and red with a mixture of phloroglucinol and concentrated hydrochloric acid, while stored starch grains stain blue-black with iodine solution. The stained sections are placed on a microscope slide and mounted in glycerol under a cover slip. Glycerol delays drying out of the preparation, which can safely be examined over a period of several hours or days, although it cannot be kept permanently (Figure 7.4).

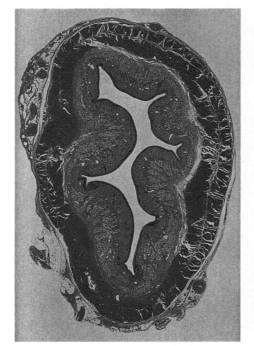 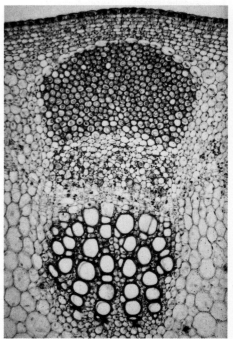

Figure 7.4 Photomicrographs of counterstained tissues in TS: permanent mounts showing (left) mammalian oesophagus (p. 280), and (right) a vascular bundle of a sunflower stem (p. 407)

Other techniques in light microscopy (see Figure 7.5)

Dark-ground microscopy

In dark-ground microscopy a special condenser allows only rays of light scattered by structures within the specimen to have a chance to enter the objective lens and form the image. The result is an image that appears bright against a dark background, with a high degree of contrast. Resolution is usually not as good as that obtained using normal (bright-field) transmitted light, but the technique does make it possible to view living, unstained cells.

Phase-contrast microscopy

The phase-contrast microscope has special fitments to the objective lens and sub-stage condenser, the effect of which is to exaggerate the structural differences between the cell components. As a consequence, structure within living, unstained cells becomes visible in high contrast and with good resolution. Phase-contrast microscopy avoids the need to kill cells or to add dyes to a specimen before it is observed microscopically.

Oil-immersion microscopy

In oil-immersion microscopy the light-gathering properties of the objective lens are enhanced by placing oil of the same refractive index as the glass microscope slide in the space between slide and objective lens. Normally the technique is used to view permanently mounted specimens.

The oil-immersion lens (×90 to ×100) gives higher magnification than the normal high-power objective lens. It is particularly appropriate for the study of bacteria and of objects of a size at or around the limit of the resolving power of the light microscope, such as the larger cell organelles.

Questions

1 What is meant by the terms 'vital dye' and 'permanent mount'? Give one example of each.
2 What is the intended effect of treating freshly killed tissue with a 'fixative'?
3 State the circumstances in which you would use
   a stained sections
   b phase-contrast microscopy
   c oil-immersion microscopy.

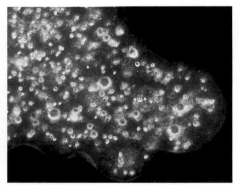
dark-ground microscopy (living specimen)

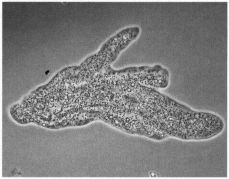
phase-contrast microscopy

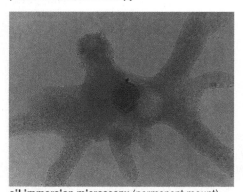
oil immersion microscopy (permanent mount)

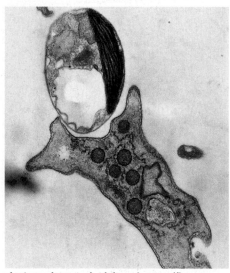
electronmicrograph of *Amoeba* engulfing an algal cell

Figure 7.5 ***Amoeba* observed by various microscopic techniques**

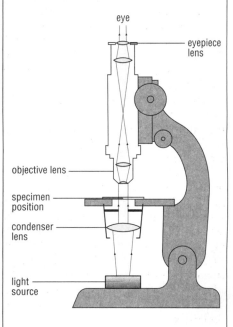
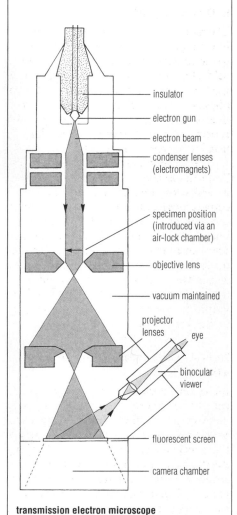

Figure 7.6 A comparison of the transmission electron microscope (TEM) and the optical (light) microscope

transmission electron microscope

Magnification and resolution

The **magnification** obtained with a compound microscope depends on which combination of lenses you use. For example, using a ×10 eyepiece and a ×10 objective lens (medium power), the image is magnified ×100. When a switch is made to the ×40 objective (high power) with the same eyepiece lens, the magnification becomes ×400. These are common orders of magnification used in laboratory work.

Magnification is defined as the number of times larger an image is than the specimen. In fact there is no limit to magnification. For example, if a magnified image is photographed, then further enlargement can be made photographically. This is what may happen with photomicrographs shown in books. Magnification is given by the formula

$$\text{magnification} = \frac{\text{size of image}}{\text{size of specimen}}$$

Suppose a particular animal cell of 50 μm diameter is photographed with a light microscope and the image is then enlarged photographically. If a print is made of the cell at 5 cm diameter (50 000 μm), the magnification is ×1000.

If a further enlargement is made, to show the same cell at 10 cm diameter (100 000 μm), then the image has been magnified ×2000.

In the second enlargement, the image size has been doubled, but *the detail will be no greater*. You will not be able to see, for example, details of cell membrane structure, however much the image is enlarged. This is because the layers making up a cell membrane are too thin to be seen as separate structures by light microscopy.

The ability to distinguish between two separate objects is known as **resolution**. If a microscope does not resolve separate small objects that are very close together then they will be seen as one object. Merely enlarging the image further will not separate them. Resolution is a property of lenses quite different from their ability to magnify – and more important (Figure 7.7). Resolution is limited by the wavelength of light. Light is composed of relatively long wavelengths, but shorter wavelengths give better resolution. For the light microscope, the resolving power is about 0.2 μm. This means two objects less than 0.2 μm apart will be seen as one object.

■ Electron microscopy

The electron microscope (EM) has powers of magnification and resolution that are greater than those of an optical (light) microscope. The resolving power (resolution) of a microscope is related to the wavelength of illumination being used: the shorter the wavelength, the greater the resolution. The wavelength of visible light is about 500 nm, whereas that of the beam of electrons used is 0.005 nm. At best the light microscope can distinguish two points which are 200 nm (0.2 μm) apart, whereas the transmission electron microscope can resolve points 1 nm apart when used on biological specimens.

In principle, an electron microscope operates on the same basis as a light microscope (Figure 7.6). Light rays are replaced by an electron beam. In transmission electron microscopy (TEM) a beam of electrons is passed through a section of material to produce the image. The electron beam is generated from a tungsten filament at incandescent heat (electron gun). The electrons, although moving at high speed, are of low energy. Electrons are easily deflected in collisions with atoms and molecules, including those of the gases of the air, so the interior of the microscope must be under vacuum. Because of the vacuum, specimens must be completely dehydrated, and living material cannot be examined; only ultra-thin sections or very small objects can be used. Cell membranes and other fine structures must first be made electron-opaque by the use of special stains.

The electron beam is focused by electromagnets and is projected on to a fluorescent screen for direct viewing, or on to a photographic plate for permanent recording. The resulting photograph is referred to as an **electronmicrograph** (see, for example, Figure 7.16).

Preparation of biological materials for the TEM

Stained, thin sections

Cells or tissues are killed and chemically 'fixed' in a lifelike condition, embedded in a resin medium and sectioned to less than 2 μm thick. The fragile specimens are floated off the knife point on to a supporting fine copper grid. Sections are stained with electron-dense materials, such as solutions of lead salts (for lipid-containing membranes) or uranyl salts (for proteins and nucleic acids). The dried grids with the stained sections are transferred to the EM.

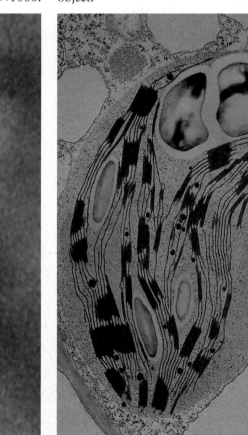

Figure 7.7 Magnification without resolution: chloroplasts at the same magnification (×25 000), by light microscopy (left, details completely unresolved), and by electron microscopy (right, showing internal structure – grana, stroma and starch grains – in detail)

Freeze-etching

This approach avoids the need to fix, section and stain biological matter before EM examination (Figures 7.8 and 7.9).

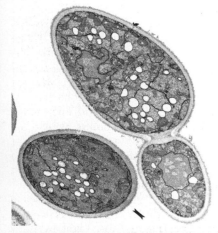

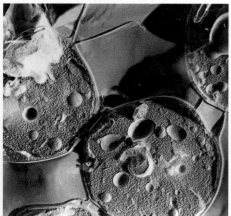

Figure 7.9 Yeast cells: a transmission electronmicrograph of a thin section (above); a freeze-etched preparation (×16 000 approximately) (below)

Questions

1. What magnification can be achieved with a ×10 objective and a ×5 eyepiece?
2. What are the differences between 'resolution' and 'magnification' in microscopy?

Scanning electron microscopy

In the scanning electron microscope the whole specimen is scanned by a beam of electrons and an image is created from the electrons reflected from the surface of the specimen. Scanning electronmicrographs (SEM) show depth of focus and a three-dimensional appreciation of the object. Another advantage is that larger specimens can be viewed than by TEM (Figure 7.10). The resolution is markedly better than that of the light microscope but is inferior to that obtained in TEM.

Figure 7.8 Freeze-etching a biological specimen for electron microscopy

1 biological specimen fast-frozen (e.g. using liquid nitrogen), and mounted in a vacuum chamber

2 the specimen is fractured along lines of weakness

(diagram labels: frozen specimen, support, about 0.1 cm, knife edge, line of previous fracture, potential fracture plane, exposed surface, vacuole, mitochondrion, nucleus, endoplasmic reticulum)

3 solid ice at the surface of the specimen evaporates away as water vapour (the surface is 'etched'); structures within the cell are exposed

4 a replica of the exposed surface is made by depositing a layer of carbon

(diagram labels: carbon, carbon arc, coating, frozen specimen, exposed etched surface, evaporated carbon for replica)

5 the carbon replica is shadowed with heavy metal in a vacuum chamber

(diagram label: platinum electrode)

6 the specimen is thawed, and the replica is floated off and examined in the TEM; the metal film is opaque to electrons, so it produces a hard shadow effect (see Figure 7.9 (bottom))

(diagram labels: shadowed metal, carbon replica)

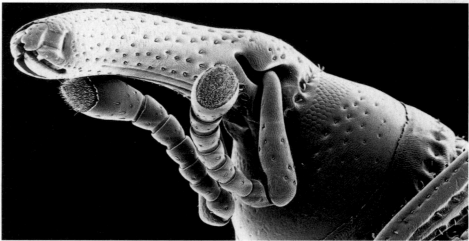

Figure 7.10 Scanning electronmicrograph: the head and the mouthparts of a weevil (×100)

Artefacts in microscopy

In microscopy the term 'artefact' refers to any feature observed in cells prepared for microscopy that is not present in life. Artefacts can be caused by the distortion of the components of the cell; for example, proteins in the cytoplasm and in cell membranes may coagulate and shrink. Alternatively, chemicals from outside the cell may appear in the cytoplasm as solid deposits. In both light microscopy and TEM, artefacts can be caused by the succession of chemical and mechanical treatments of biological material in the preparation of thin sections of stained cells.

Checking for artefacts

Phase-contrast microscopy permits the examination of unstained, living cells. In this way the existence (or absence) of structures observed in stained sections by light microscopy can be checked. Similarly, in EM the technique of freezing living material and 'etching' fracture surfaces allows a check on the existence of structures observed by TEM (compare the two micrographs in Figure 7.9).

■ Cell fractionation

Cell fractionation is the technique by which the tiny structures within cells, known as **organelles**, are isolated so that structure and function can be investigated outside the organism. The steps in cell fractionation are summarised in Figure 7.11. The isolation of chloroplasts is shown in Figure 12.26, p. 259.

Questions

1 When a tissue is 'fixed', the chemicals denature the proteins in the cell (by coagulating them, for example). How might this affect the interpretation of cell structure by light microscopy?

2 The electron microscope permits cells to be examined at high resolution and great magnification. What possible causes of error arise in the study of cell structure by TEM?

3 **a** Why must the interior of the EM be under vacuum?

 b What are the consequences of vacuum conditions for the preparation of living tissue for electron microscopy?

4 In the process of organelle extraction from leaf cells, the following steps are carried out:

 a pre-chilling the tissue

 b cutting and homogenising the tissue

 c filtering the homogenised tissue.

 What is the role of each step?

Figure 7.11   The processes of cell disruption and organelle isolation: all steps are carried out in the cold with chilled equipment, so minimising the self-digestion (autolysis) of cell organelles by the digestive enzymes released from cell vesicles damaged in cell disruption

1 the chilled tissue is cut up in cold isotonic buffer solution

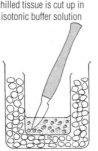

2 the tissue fragments are homogenised

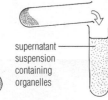

pestle homogeniser   or   blender

3 the suspension is filtered through layers of muslin

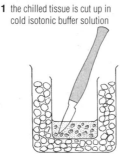

filtrate

4 the filtrate is centrifuged at low speed to remove part-opened cells, heavy starch grains and cell wall debris

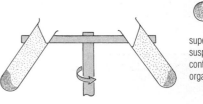

supernatant suspension containing organelles

5 the organelles are fractionated *either* by differential centrifugation

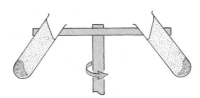

centrifuge at 500–600 *g* for 5–10 minutes

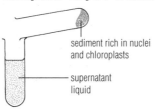

sediment rich in nuclei and chloroplasts

supernatant liquid

centrifuge at 10 000–20 000 *g* for 15–20 minutes

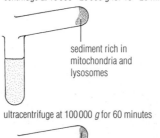

sediment rich in mitochondria and lysosomes

ultracentrifuge at 100 000 *g* for 60 minutes

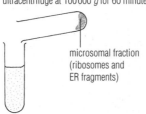

microsomal fraction (ribosomes and ER fragments)

or by density-gradient centrifugation

suspension of organelles added by pipette

least dense sucrose solution (dilute)

continuous density gradient of sucrose

most dense sucrose solution (concentrated)

centrifugation

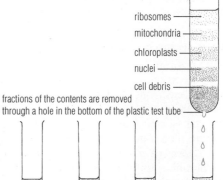

ribosomes
mitochondria
chloroplasts
nuclei
cell debris

fractions of the contents are removed through a hole in the bottom of the plastic test tube

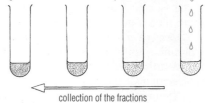

collection of the fractions (different types of organelle)

6 the organelle fractions are analysed chemically by chromatography; their enzymes are isolated and analysed, and their roles in metabolism are investigated

7.4 THE ULTRASTRUCTURE OF THE EUKARYOTIC CELL

The fine structure of cells, known as cell **ultrastructure**, was largely discovered by the techniques of electron microscopy and cell fractionation. Today the cell is visualised as a tiny, three-dimensional sac containing many membraneous organelles, of approximately half a dozen different types, together with other cell inclusions. Organelles are suspended within an aqueous medium and contained within a **plasma membrane**. The bulk of the substance of the cell is referred to as **cytoplasm**. The fluid that remains when all organelles are removed is referred to as **cytosol**. (At one time the term 'protoplasm' was used to describe the substance of cells, but today the word is not used.) Cytoplasm contains various components, including

▶ an aqueous solution of various essential ions (p. 273) and soluble organic compounds such as sugars and amino acids

▶ soluble proteins, present in the cytosol, many of which are enzymes

▶ the cell organelles such as the nucleus, mitochondria, ribosomes and endoplasmic reticulum

▶ a network of fine strands of globular proteins, known as microtubules and microfilaments (pp. 164–6), and collectively referred to as the **cytoskeleton**.

Figure 7.12 represents our current knowledge of cell ultrastructure. Most of the organelles and other structures of cells are common to animals and plants. For example, both animal and plant cells have a plasma membrane, a nucleus, mitochondria, endoplasmic reticulum, a Golgi apparatus, ribosomes and lysosomes.

Only plant cells have a cell wall containing cellulose. Plant cells usually contain a large permanent vacuole contained within a special membrane, the tonoplast. Green plant cells contain chloroplasts.

Only animal cells contain two centrioles (making up the centrosome), and often also temporary vacuoles or vesicles.

Figure 7.12   The ultrastructure of the eukaryotic cell

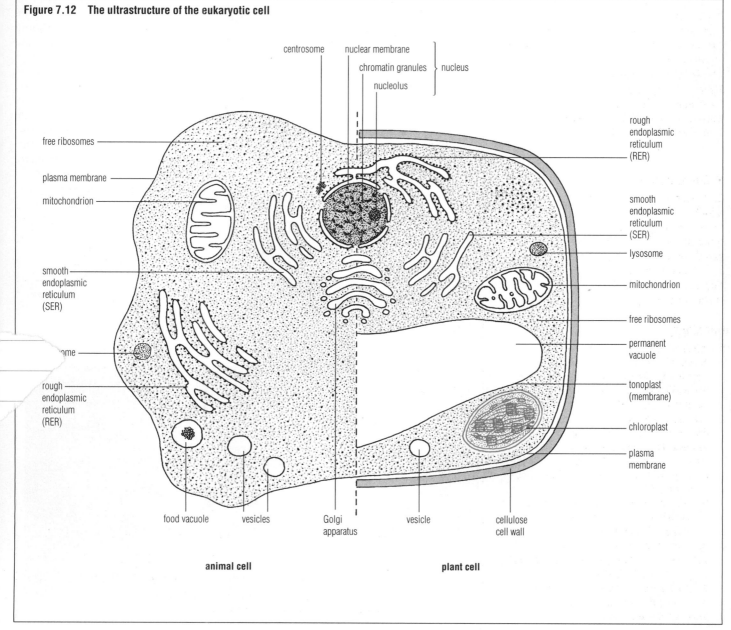

centrosome

nuclear membrane

chromatin granules ⎫ nucleus

nucleolus

free ribosomes

plasma membrane

mitochondrion

smooth endoplasmic reticulum (SER)

...ome

rough endoplasmic reticulum (RER)

food vacuole   vesicles   Golgi apparatus   vesicle   cellulose cell wall

animal cell

rough endoplasmic reticulum (RER)

smooth endoplasmic reticulum (SER)

lysosome

mitochondrion

free ribosomes

permanent vacuole

tonoplast (membrane)

chloroplast

plasma membrane

plant cell

■ The cell surface membrane (plasma membrane)

The membrane at the surface of the cell has been observed by electron microscopy. The presence of a thin selective cell barrier at the surface of the cell was inferred long before it was observed, however. Observations of the behaviour of cells under experimental conditions provided the evidence:

▶ The volumes of animal cells change when bathed in solutions of differing concentrations; cells swell or contract as if surrounded by a partially permeable membrane (p. 234).
▶ The uptake of dyes and other substances by cells is selective (p. 236).
▶ A cell loses its contents if its surface is ruptured.

The biochemical evidence for membrane structure includes the following:

▶ water-soluble compounds enter cells less readily than compounds that dissolve in lipids (non-polar compounds and hydrophobic substances)
▶ membranes contain a high proportion of lipids
▶ it has been estimated that lipids extracted from cell membranes are sufficient to form a layer two molecules thick (a bilayer), arranged with polar (hydrophilic) heads facing outwards and non-polar (hydrophobic) tails facing inwards (Figure 7.13)
▶ membranes contain more protein than lipid, but the protein is harder to isolate; initially, the suggestion was that the protein forms a thin layer coating the outer surfaces of the lipid bilayer (that is, the **unit membrane model** of membrane structure, Figure 7.13)
▶ work on the extraction of membrane protein has indicated that some protein is buried within the lipid bilayer
▶ freeze-etching studies of plasma membranes have shown that where the membrane is fractured along its midline, globular (spherical shaped) proteins occur on and buried in the lipid bilayer, rather than forming sheets of protein on the surface (Figure 7.14)
▶ studies of the reactions of membrane proteins with enzymes and antibodies have shown that some membrane proteins are exposed at one surface, some are exposed at both surfaces, and some are inaccessible
▶ in several types of cell the quantity of lipid has been found to be insufficient to form a complete continuous bilayer over the whole cell surface; the bilayer is interrupted by substantial quantities of other components, which are presumably proteins
▶ the tagging of membrane components with marker substances (fluorescent dyes) has shown that the molecules of the membrane are on the move; the membrane is a fluid, ever-changing barrier
▶ the lipid bilayer contains cholesterol molecules, which disturb the close packing of the phospholipid molecules around them, and appear to increase the flexibility of the membrane
▶ at the outer surface of the cell, antenna-like carbohydrate molecules form complexes with many membrane proteins (glycoproteins) and lipids (glycolipids).

Fluid mosaic model of membrane structure

In 1972 Singer and Nicolson proposed the fluid mosaic model of membrane structure (Figure 7.15). The fluid mosaic membrane is a dynamic structure in which much of the protein floats about, although some is anchored to organelles within the cell. Lipid also moves about.

This model explains the known physical and chemical properties of membranes; it is a starting point to understand the functioning of the cell. All the membranes of cells, including the tonoplast and those that make up the various organelles, are now considered to have the fluid mosaic construction.

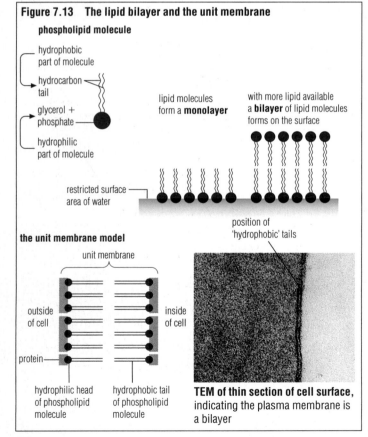

Figure 7.13   The lipid bilayer and the unit membrane

phospholipid molecule

- hydrophobic part of molecule
- hydrocarbon tail
- glycerol + phosphate
- hydrophilic part of molecule

lipid molecules form a **monolayer**

with more lipid available a **bilayer** of lipid molecules forms on the surface

restricted surface area of water

position of 'hydrophobic' tails

the unit membrane model

unit membrane

outside of cell

inside of cell

protein

hydrophilic head of phospholipid molecule

hydrophobic tail of phospholipid molecule

TEM of thin section of cell surface, indicating the plasma membrane is a bilayer

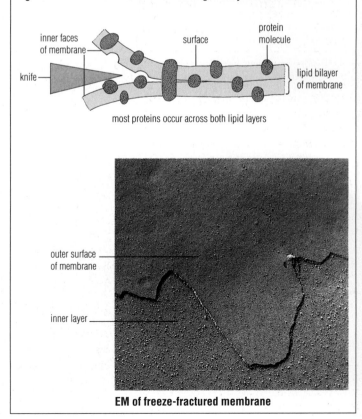

Figure 7.14   Plasma membrane investigated by freeze-fracture

inner faces of membrane

knife

surface

protein molecule

lipid bilayer of membrane

most proteins occur across both lipid layers

outer surface of membrane

inner layer

EM of freeze-fractured membrane

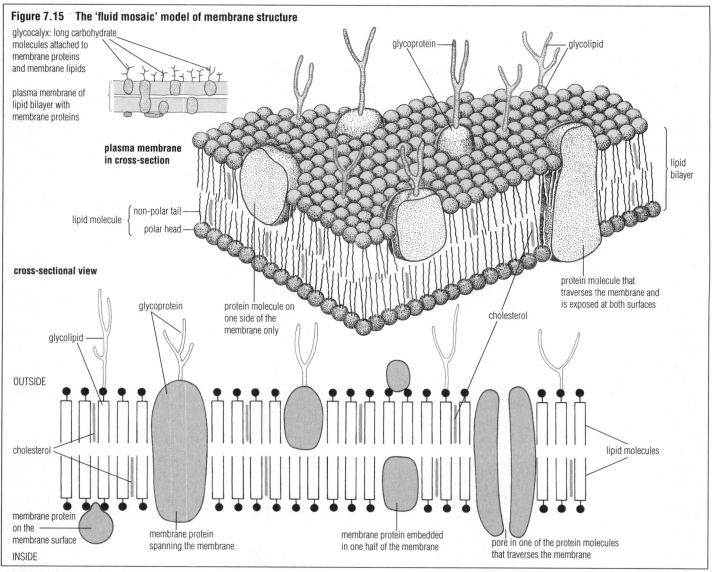

Figure 7.15   The 'fluid mosaic' model of membrane structure

glycocalyx: long carbohydrate molecules attached to membrane proteins and membrane lipids

plasma membrane of lipid bilayer with membrane proteins

glycoprotein

glycolipid

plasma membrane in cross-section

lipid bilayer

lipid molecule { non-polar tail / polar head

cross-sectional view

protein molecule that traverses the membrane and is exposed at both surfaces

glycolipid

glycoprotein

protein molecule on one side of the membrane only

cholesterol

OUTSIDE

cholesterol

lipid molecules

membrane protein on the membrane surface

INSIDE

membrane protein spanning the membrane

membrane protein embedded in one half of the membrane

pore in one of the protein molecules that traverses the membrane

■ Transport across the cell membrane

The cell surface is the frontier of the cell: water, energy, food materials, other nutrients and waste substances all have to cross this barrier. Meanwhile, the internal membranes are barriers between compartments within the cell. The internal membranes regulate movement of substances within the cell.

In addition to the transport of nutrients and respiratory gases, membranes are also receptors and message senders! The membrane receives and despatches specific messages such as nerve impulses, antibodies and hormones. The range of short-distance transport phenomena is summarised in Table 7.1.

Role of the glycocalyx

The glycocalyx (glycoproteins and glycolipids) of the fluid mosaic membrane varies in composition from one cell type to another. The roles of glycocalyx include

▶ cell–cell recognition, e.g. as antigens
▶ functioning as receptor sites for chemical signals, e.g. hormones
▶ binding cells together into tissues.

Questions

1   The fluid mosaic model of membranes is described as dynamic. What does 'dynamic' imply in this context?

2   What is the difference between a lipid bilayer and a 'double membrane'?

3   What is the chemical composition of the non-polar tail of a lipid (p. 137)?

Table 7.1   Membrane traffic: a summary of movement across cell membranes

'Traffic'		Related cell activity
diffusion of gases		respiration (Chapter 15), photosynthesis (Chapter 12)
osmosis		water movement (Chapter 11)
via membrane channels		ultrafiltration (kidney) (Chapter 18)
active transport by enzymic pumps		ion uptake (Chapter 11)
antibody attachment		defence (Chapter 24)
hormone attachment		growth, development and coordination (Chapter 21)
nerve impulse: direct electrical stimulation, chemical stimulation at synapse		response and coordination in animals (Chapter 21)
endocytosis (also exocytosis)	phagocytosis	white cells ingesting bacteria (Chapter 24)
	pinocytosis	feeding in protozoa (Chapter 14)

Figure 7.16 TEM of the nucleus of a pancreatic cell of a mammal (×25 000) (above); EM of a freeze-etched nuclear membrane showing the battery of nuclear pores (below)

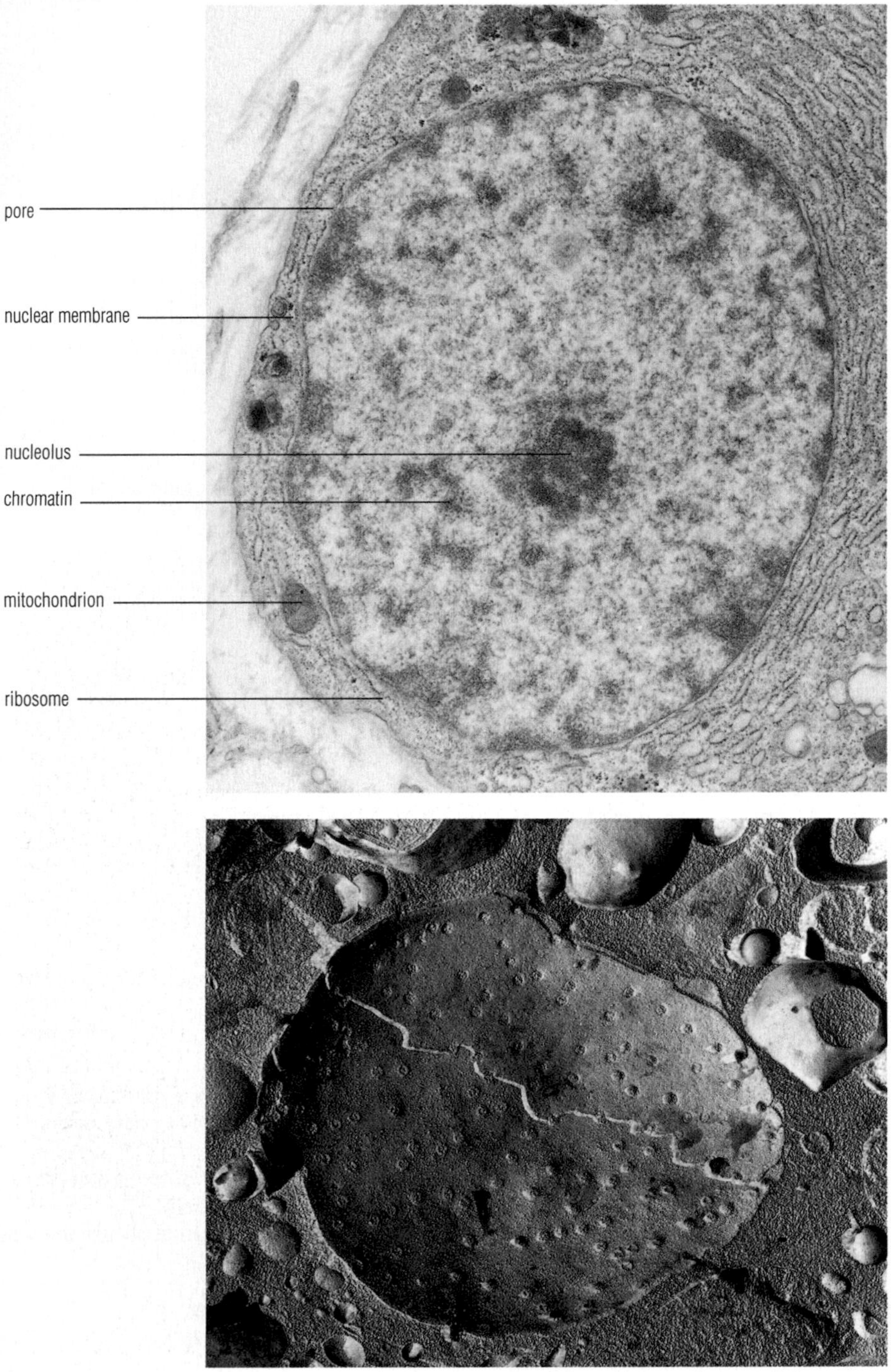

pore

nuclear membrane

nucleolus

chromatin

mitochondrion

ribosome

The nucleus is surrounded by a double membrane, which is perforated by very many pores (Figure 7.16). The pores are individually only 80–100 nm in diameter, yet they cover one-third of the membrane surface area.

The nucleus encloses the **chromosomes**. Chromosomes are visible only at times of cell division. For most of the cell cycle the chromosomes are dispersed as a diffuse network, called **chromatin**. Chromosomes and chromatin are normally visible only when the cells are stained with basic dyes. Staining with basic dyes also discloses that the nucleus contains one or more **nucleoli**. These small bodies occur in the nucleus at interphase.

The nucleus consists of protein (50% of the dry mass), DNA (20%) and RNA (20%). Most of the protein is histone, which together with the DNA makes up the chromosomes. The chromosomes contain the genetic (hereditary) information (Chapter 28) and they are replicated (reproduced) prior to nuclear and cell division (Chapter 9). The nucleus controls cell activity by regulating protein and enzyme synthesis.

■ Endoplasmic reticulum

Endoplasmic reticulum (ER) consists of a network of folded membranes forming sheets, tubes or flattened sacs in the cytoplasm (Figure 7.18). It originates from the outer membrane of the nucleus, to which it often remains attached. We can assume that ER is flexible and mobile since it occupies much of the cytoplasm of many cells, including those in which streaming movements of the cytoplasm are commonly observed.

The ER is the site of specific enzymically controlled reactions of cell biochemistry. The outer surface of some ER carries numerous ribosomes (see page opposite). The presence of ribosomes gives a granular appearance, and in this condition ER is described as **rough endoplasmic reticulum** (RER). RER is the site of the synthesis of proteins, which are packaged up in membraneous vesicles and either moved about the cell or despatched from it (pp. 208–9).

Smooth endoplasmic reticulum (SER) does not have this coating of ribosomes. SER is concerned with lipid metabolism.

■ The nucleus

The nucleus (Figure 7.16) is the largest organelle within the eukaryotic cell, so large that it is easily observed by light microscopy. It is a spherical or ovoid structure, 10–20 μm in diameter. Almost all cells contain one nucleus; the very few exceptions are

▶ mature phloem sieve tube cells (p. 227) and mature mammalian red blood cells (p. 216), which are without a nucleus

▶ the ciliated protozoan *Paramecium* (p. 29), and most cells of the club fungi (Basidiomycetes, p. 31), which have two nuclei

▶ certain mammalian white blood cells (p. 218), which have a much-lobed single nucleus.

Figure 7.17   Structure of the ribosome

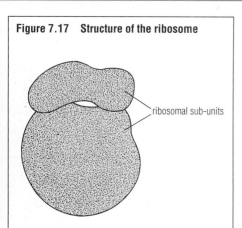

ribosomal sub-units

Figure 7.18   The endoplasmic reticulum (ER)

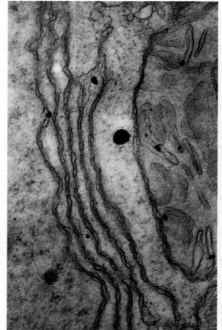

ribosome

rough endoplasmic reticulum

nucleus

ribosome

smooth endoplasmic reticulum

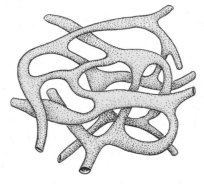

TEM of mammalian SER

TEM of mammalian RER

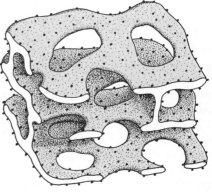

tubular form of ER

sheet form of ER

■ Ribosomes

Ribosomes are minute structures, 20 nm in diameter, each consisting of two sub-units (Figure 7.17). There are several thousand ribosomes per cell. Ribosomes are composed not of membrane, but of protein and RNA in roughly equal quantities. Ribosomes are the site of protein synthesis (p. 208). Here messenger RNA (mRNA) is 'read', and controls the sequence of amino acid residues in the proteins formed. When several ribosomes occur along a common strand of mRNA the whole structure is known as a **polysome**.

Ribosomes lying free in the cytoplasm are the site of synthesis of the proteins retained within the cell, whereas ribosomes bound to endoplasmic reticulum produce proteins that are subsequently secreted outside the cell.

Ribosomes occur in both prokaryotic and eukaryotic cells. The ribosomes of prokaryotes are distinctly smaller than those of eukaryotes. Ribosomes also occur inside mitochondria and chloroplasts of eukaryotic cells, but here they are smaller than the cytoplasmic ribosomes and actually resemble those of prokaryotes (p. 19). This observation has led to speculation about the origin of these two organelles within eukaryotic cells (pp. 168 and 666).

Questions

1   What is the most likely role of the pores in the nuclear membrane, given that the nucleus controls and directs the activities of the cell (p. 208)?

2   What is the relationship between histone, DNA, chromatin and chromosomes (p. 200)?

3   **a**   Where, in the cell, are the ribosomes?

    b   How does the function of ribosomes vary according to where they are found?

■ Golgi apparatus

The Golgi apparatus consists of a stack of flattened membraneous sacs. This organelle (it is named after the biologist who first reported it) is present in all cells, but is most prominent in those that are metabolically active. The Golgi apparatus is the site of synthesis of biochemicals; these are packaged into swellings at the margin of each sac, which become pinched off as vesicles (Figure 7.19).

The Golgi apparatus also collects proteins and lipids made in the ER, by the fusion of vesicles pinched off from the ER (RER or SER) with its own flattened sacs. In the Golgi apparatus additional substances are added and the products are repackaged into fresh vesicles, which are then cut off and moved to other parts of the cell. Here the secretions are discharged or deposited.

For example, certain of the digestive enzymes of pancreatic juice originate in the Golgi apparatus in pancreatic cells (p. 284). In the growth of a plant cell the Golgi apparatus forms the first layer of the new cell wall that develops between the two daughter cells as they divide (Figure 7.28).

■ Lysosomes

Lysosomes are small spherical vesicles, 0.2–0.5 μm in diameter or larger (particularly in plant cells). The lysosome is bounded by a single membrane, and contains a concentrated mixture of hydrolytic digestive enzymes (p. 187). If these digestive enzymes were not kept enclosed in a membraneous sac they would attack the other cell organelles.

Lysosomes originate either from the Golgi apparatus or directly from the endoplasmic reticulum. The enzymes they contain are used in the dissolution and digestion of redundant structures or damaged macromolecules from within or outside the cell. For example, when an animal cell ingests food into a food vacuole, lysosomes fuse with the vacuole and break down the contents, their enzymes digesting carbohydrates, fats, proteins and other components (Figure 7.20). Redundant cell organelles are enclosed in a membrane and broken down by lysosomes in a similar manner. The gland cells in some digestive organs package their digestive enzymes in lysosomes before releasing them outside the membrane. When a cell dies its own lysosomes release the enzymes that digest the remains of the cell in a process known as **autolysis**.

Figure 7.19   The Golgi apparatus

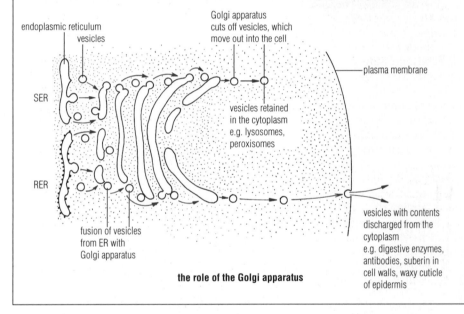

TEM of Golgi apparatus of *Euglena gracilis*

stereogram of Golgi apparatus at work

the role of the Golgi apparatus

Figure 7.20   Lysosomes

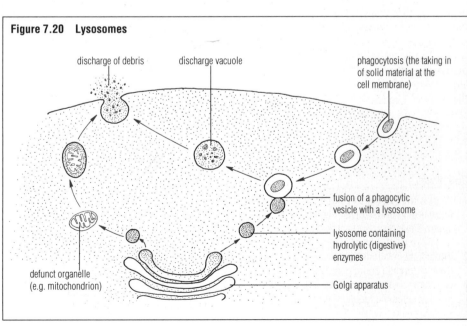

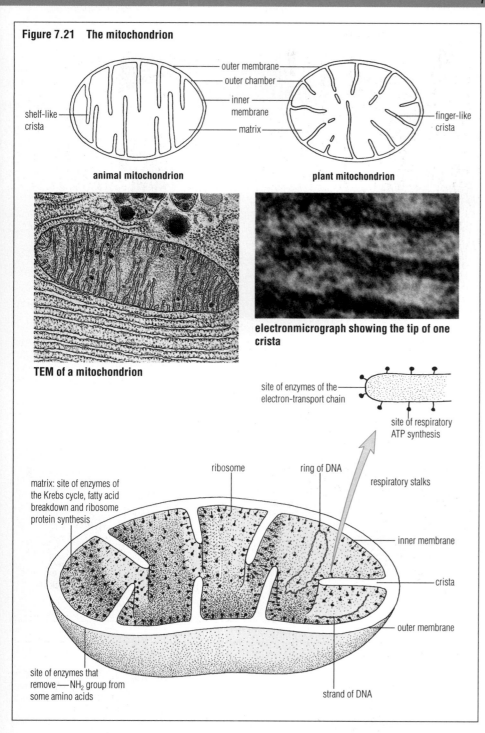

Figure 7.21   The mitochondrion

animal mitochondrion

- outer membrane
- outer chamber
- inner membrane
- matrix

shelf-like crista

plant mitochondrion

finger-like crista

TEM of a mitochondrion

electronmicrograph showing the tip of one crista

site of enzymes of the electron-transport chain

site of respiratory ATP synthesis

respiratory stalks

matrix: site of enzymes of the Krebs cycle, fatty acid breakdown and ribosome protein synthesis

ribosome

ring of DNA

inner membrane

crista

outer membrane

site of enzymes that remove —NH_2 group from some amino acids

strand of DNA

■ Vacuoles

Vacuoles are fluid-filled cavities bounded by a single membrane. They are formed either by the infolding and pinching-off of part of the cell membrane, or by enlargement of a vesicle cut off by the Golgi apparatus. Young plant cells usually contain several small vacuoles which, in the mature cell, have united to form a large permanent, central vacuole occupying over 80% of the cell volume (Figure 7.2). The plant vacuole is filled with a liquid known as **cell sap**, an aqueous solution of dissolved food materials, ions, waste products and pigments. The membrane around this type of vacuole is known as the **tonoplast** (p. 157).

The vacuoles of animal cells are usually very much smaller and less permanent. Small vacuoles are often called **vesicles**. They may contain engulfed solids or liquids.

■ Mitochondria

Mitochondria appear in electronmicrographs mostly as rod-shaped or cylindrical organelles, although occasionally they are more variable in shape. Mitochondria occur in all cells in very large numbers, possibly more than a thousand of them in a cell that is metabolically very active. They vary in size within the range 0.5–1.5 μm wide and 3.0–10.0 μm long. Each mitochondrion is bounded by a double membrane, the outer layer being a smooth continuous boundary. The inner membrane is extensively infolded to form partitions called **cristae**, which partially divide the interior (Figure 7.21). In the mitochondria of plants the cristae are commonly tubular and villus-like; in animal cells they are sheet-like plates.

The mitochondria are the sites of pyruvic acid oxidation by the Krebs cycle (p. 315), of ATP formation (p. 316) and of some protein synthesis and fatty acid metabolism.

■ Peroxisomes

Peroxisomes are spherical organelles bounded by a single membrane (Figure 7.22). Peroxisomes contain catalase, an enzyme that catalyses the decomposition of hydrogen peroxide (a potentially very harmful oxidising agent) to the harmless products water and oxygen. Hydrogen peroxide is a by-product of certain reactions of metabolism (including photorespiration in mesophyll cells, p. 267, and lactic acid oxidation in liver cells).

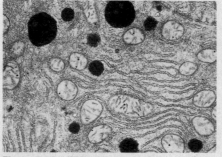

Figure 7.22   TEM of dark-staining peroxisomes from the liver cell of a rat

Questions

1 What is the functional relationship between the endoplasmic reticulum, Golgi apparatus and lysosomes?

2 List the possible functions of lysosomes.

3 a What functions take place in a mitochondrion?

  b Why would you expect actively metabolising cells to have large numbers of mitochondria?

4 In what ways do peroxisomes resemble lysosomes? In what ways do they differ?

■ Chloroplasts

Chloroplasts are members of a group of organelles known as **plastids**. Plastids normally contain pigments such as chlorophylls and carotenoids (p. 248). Photosynthesis occurs in chloroplasts in all green plants.

The chloroplast of *Chlamydomonas* is cup-shaped (p. 553), but those of most multicellular plants are biconvex discs 4–10 μm long and 2–3 μm wide. In land plants chloroplasts are normally restricted to the mesophyll cells of leaves (p. 255) and to the cells of the outer cortex of herbaceous (non-woody) aerial stems. Mesophyll tissue with chloroplasts is known as **chlorenchyma** (p. 223). There may be up to fifty chloroplasts in a mesophyll cell.

The chloroplast is bounded by a double membrane. The outer membrane is a smooth, continuous boundary; the inner gives rise to strands of branching membranes called lamellae extending throughout the organelle. The interior of the chloroplast is divided into the **grana** (singular, granum), which are surrounded by **stroma**. In the grana the lamellae are stacked in piles of flat, circular sacs called **thylakoids**, which contain photosynthetic pigments. In the stroma the thylakoids criss-cross loosely, suspended in an aqueous matrix containing ribosomes, lipid droplets and small starch grains (Figure 7.23).

Other members of the plastid family include **leucoplasts** (colourless plastids, common in storage organs such as roots and seeds) and **chromoplasts** (coloured plastids containing non-photosynthetic pigments, common in fruits and in flower petals and also in carrot root tissue).

■ Microtubules

Microtubules occur in most plant and animal cells. They are straight, unbranched hollow cylinders, 25 nm wide and usually quite short in length (Figure 7.24). They are made of protein, and are constantly being built up and broken down.

Microtubules are involved in the movement of cytoplasmic components within the cell. They also occur in centrioles, in the spindle (p. 195), in cilia and flagella and in the basal bodies. Microtubules appear to direct the passage of Golgi vesicles to deposition sites (Figure 7.26). Along with the microfilaments (p. 166) the microtubules constitute the cytoskeleton, which controls the shape and movements of the cell.

Figure 7.23   The chloroplast

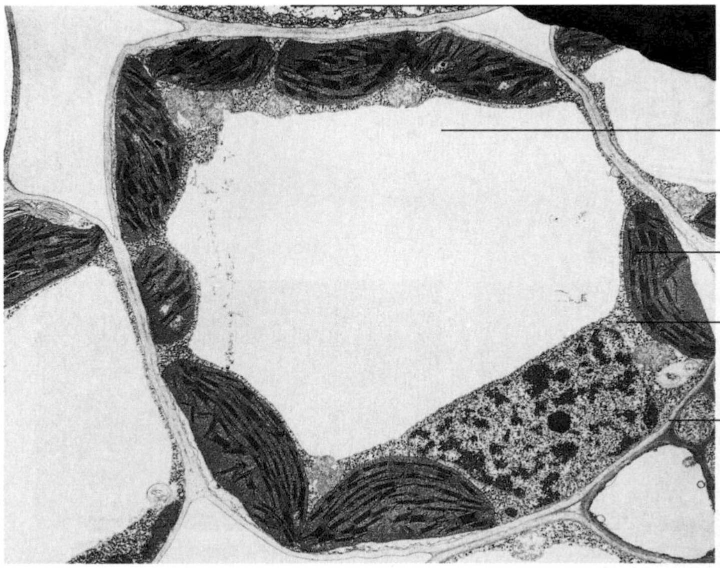

EM of chloroplasts in a plant cell

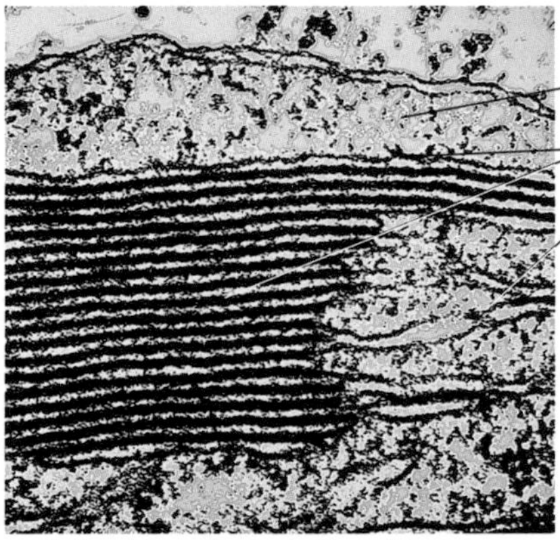

— vacuole
— chloroplast
— cytoplasm
— cell wall

part of a chloroplast greatly enlarged

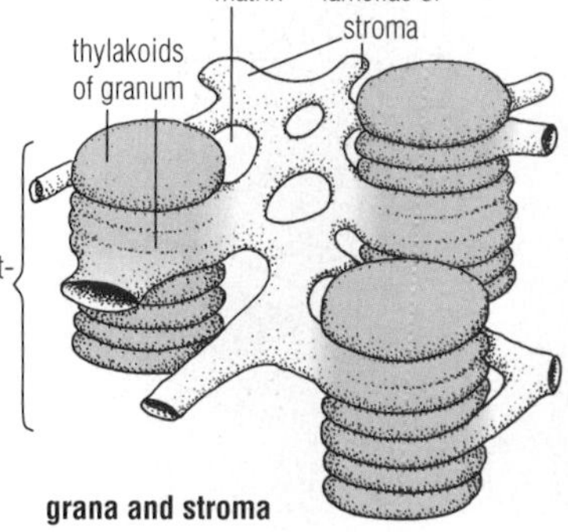

matrix
grana
stroma

stroma
(site of light-independent stage)
matrix   lamellae of stroma
thylakoids of granum
granum (site of light-dependent stage)

grana and stroma

Figure 7.24   Microtubules

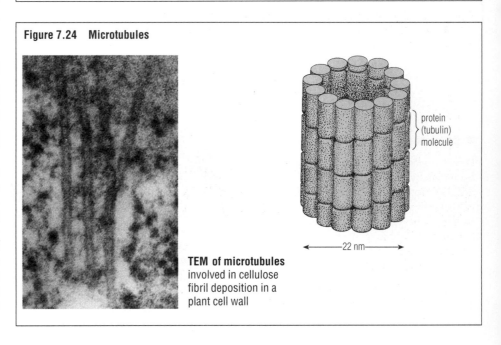

TEM of microtubules
involved in cellulose fibril deposition in a plant cell wall

protein (tubulin) molecule

←—22 nm—→

Figure 7.25   Structure of flagella and cilia

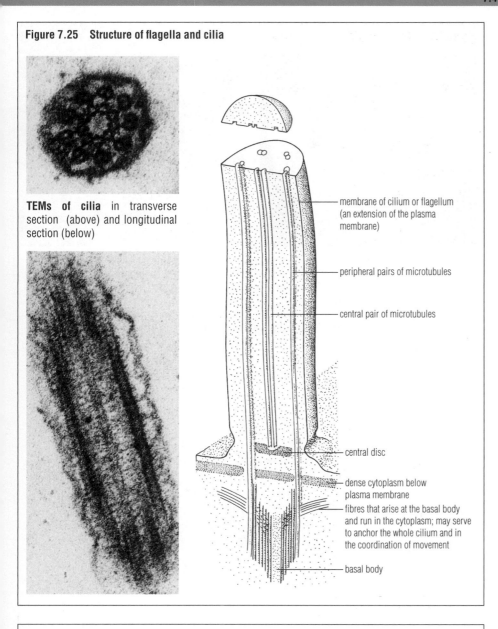

TEMs of cilia in transverse section (above) and longitudinal section (below)

membrane of cilium or flagellum (an extension of the plasma membrane)

peripheral pairs of microtubules

central pair of microtubules

central disc

dense cytoplasm below plasma membrane

fibres that arise at the basal body and run in the cytoplasm; may serve to anchor the whole cilium and in the coordination of movement

basal body

■ Flagella and cilia

Flagella and cilia are organelles that project from the surface of cells but are connected to a **basal body** just below the plasma membrane. Flagella occur singly or in small numbers, whereas cilia occur in large numbers on larger cells, and are typically shorter than flagella. Structurally, flagella and cilia are almost identical (Figure 7.25); both are able to move.

Flagella and cilia are enclosed by a plasma membrane. Internally, they consist of **microtubules** (p. 164) arranged in an outer ring of nine pairs surrounding one central pair. 'Side-arm' links join the outer pairs to the central pair and also to each other. The microtubules and side-arms are made of protein, and contain enzymes that release energy from ATP in the presence of magnesium ions. Flagella and cilia are believed to move by means of sliding movements of one member of a microtubule pair relative to the other, very like the movements generated by myosin and actin in muscle fibres (pp. 489–91). The movements of flagella and cilia produce forces that may be used in various ways:

▶ to move a cell (for example, a sperm, or a unicellular organism such as *Chlamydomonas* sp., p. 553)

▶ to propel fluids across cells (for example, the ciliated cells that move mucus along the bronchial lining, p. 312)

▶ to acquire food (for example, the feeding current generated by *Paramecium* in its oral groove, p. 290)

▶ to sense the environment (for example, by sensory hair cells, pp. 454–5).

■ Centrioles of the centrosome

The centrosome occurs in animal cells but not in plant cells. It consists of two hollow cylindrical bodies, the centrioles, lying at right angles to each other beside the nucleus. Each centriole is made of nine triplets of microtubules and is identical to the basal body that lies at the base of a flagellum or cilium. The centrioles separate and move to opposite ends of the nucleus before cell division.

Questions

1 **a** What form of energy conversion occurs frequently in chloroplasts (p. 260) but never in mitochondria?

   b Which part of the chloroplast is specially adapted for this energy change, and how?

2 State three functions of microtubules.

3 List four places where cilia are found, and state the functions in each situation.

Figure 7.26   The microtubule in axon transport: transport of other organelles along the outside of the microtubules allows swift exchange of substances between the nerve cell body and the synaptic terminal (p. 447)

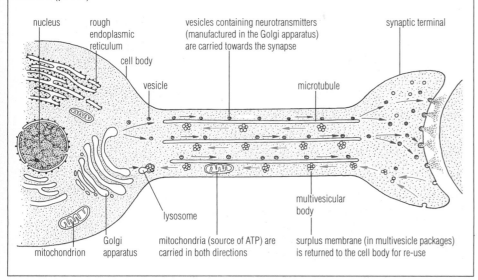

nucleus

rough endoplasmic reticulum

cell body

vesicles containing neurotransmitters (manufactured in the Golgi apparatus) are carried towards the synapse

synaptic terminal

vesicle

microtubule

lysosome

multivesicular body

mitochondrion

Golgi apparatus

mitochondria (source of ATP) are carried in both directions

surplus membrane (in multivesicle packages) is returned to the cell body for re-use

■ Microfilaments

Microfilaments are long but extremely thin fibres, about 5 nm wide, which make up a protein 'skeleton' within the cytoplasm of the cell. Microfilaments are involved in cytoplasmic movements, for example, during cytoplasmic streaming and cell movement, and the cytoplasmic constriction at cell division in animal cells.

■ Stored food granules

Stored food substances are non-living inclusions within cells, deposited as reserve substances or as a by-product of metabolism. They may occur within membraneous organelles, as do the starch grains in leucoplasts in plant cells. More often they occur freely within the cytoplasm. Examples in animal cells include fat and oil globules and glycogen granules. Within plant cells the inclusions may be oil globules, inulin crystals, aleurone (protein) grains and calcium oxalate crystals.

7.5 CELL WALLS

A cell wall is a structure external to the cell and is therefore not an organelle, although it is a product of various cell organelles.

The presence of a rigid external cell wall is a characteristic of plant cells. Plant cell walls consist of cellulose together with other substances, mainly other polysaccharides. The hyphae of fungi (p. 557) and the cell walls of prokaryotes (p. 551) have a similar structure but their chemical composition is quite different. Cell walls are secreted by the cell they enclose, and their formation and composition are closely tied in with the growth, development and functions of the cell they protect and support.

A new cell wall starts to form by the growth of cellulose microfibrils (Figure 7.27) from numerous small vesicles produced by the Golgi apparatus. When plant cells divide, these vesicles collect at the equator of the spindle, known as the **cell plate**, immediately after nuclear division (mitosis, p. 195; meiosis, p. 196). They then fuse together: their membranes form the new cell membrane, and their contents coalesce as the **middle lamella** (p. 136), a sticky, gel-like layer of calcium and magnesium pectate (Figure 7.28).

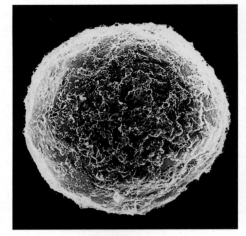
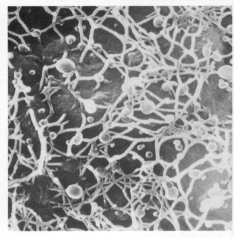

Figure 7.27   Cell wall growth studied in a moss leaf cell, the wall of which has been removed by treatment with cellulase (an enzyme that hydrolyses cellulose); progressive regrowth of cell wall microfibrils after four hours (left) and after ten hours (right)

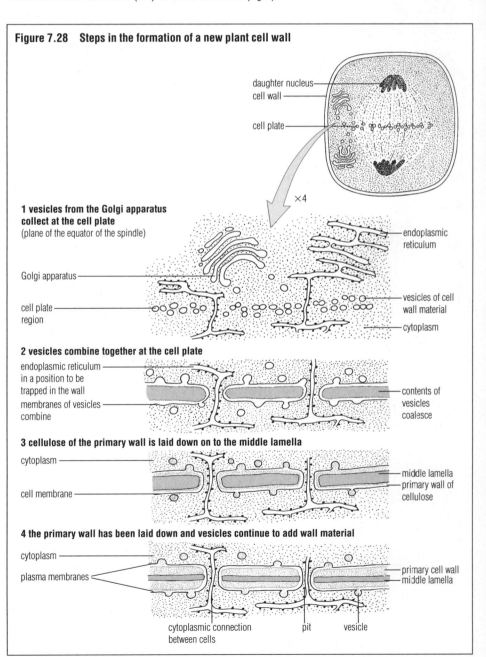

Figure 7.28   Steps in the formation of a new plant cell wall

daughter nucleus
cell wall
cell plate

×4

1 vesicles from the Golgi apparatus collect at the cell plate
(plane of the equator of the spindle)

endoplasmic reticulum
Golgi apparatus
cell plate region
vesicles of cell wall material
cytoplasm

2 vesicles combine together at the cell plate

endoplasmic reticulum in a position to be trapped in the wall
membranes of vesicles combine
contents of vesicles coalesce

3 cellulose of the primary wall is laid down on to the middle lamella

cytoplasm
cell membrane
middle lamella
primary wall of cellulose

4 the primary wall has been laid down and vesicles continue to add wall material

cytoplasm
plasma membranes
primary cell wall
middle lamella

cytoplasmic connection between cells   pit   vesicle

Figure 7.29 The cellulose microfibrils of the plant cell wall

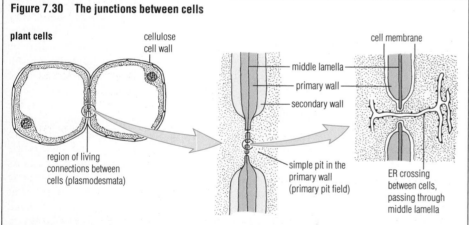

microfibril

(enlarged)

cellulose chains

the structure of microfibrils

SEM of a primary wall: the microfibrils run in all directions (×16 000)

SEM of a secondary wall: the microfibrils are very tightly packed; successive layers are arranged at different angles

■ Primary cell walls

The primary cell wall is deposited on both sides of the middle lamella, and is made from bundles of cellulose molecules known as microfibrils. Each microfibril bundle consists of approximately 2000 extremely long cellulose molecules (p. 135). The individual cellulose molecules are composed of about 3000 glucose residues condensed together (Figure 7.29). Microfibrils of cellulose are held together in the plant cell wall by a matrix of other polysaccharides known as pectates and hemicelluloses (p. 136).

In the primary wall, the microfibrils run in all directions. This arrangement allows for considerable stretching during cell growth.

Portions of endoplasmic reticulum are trapped across the middle lamella as the new wall is laid down, and these become the cytoplasmic connections between cells (the plasmodesmata). Here the wall is not thickened further, and depressions or thin areas known as pits are formed in the walls. Pits normally pair up between adjacent cells.

■ Secondary cell walls

Many more layers of cellulose are normally laid down on to the primary wall. These layers form the secondary wall. Secondary walls consist of tightly packed microfibrils, arranged in a particular direction. Often the cell elongates as the wall is forming, and the microfibrils are orientated in the direction of elongation of the cell.

The secondary wall layers become extremely thick in some cells and the whole wall may also be impregnated with **lignin** (p. 147), which gives it even greater strength. This is the case in fibres, tracheids and xylem vessels (p. 225). In cork tissue (tree bark, for example) the walls are impregnated with waxy **suberin**.

The outermost walls of the epidermal cells of stems and leaves are coated with the waterproofing substances known as **wax**. Waxes (p. 138) form an outer, highly water-repellent layer.

Other stem cells, and also root cells, lay down a substance called **cutin** in their walls. Cutin is a polymer built from unusual fatty acids. Cutin becomes bonded to the wall polysaccharides. Together, suberin and cutin make the aerial system of higher plants impervious to water, relatively unstretchable (p. 500) and highly resistant to invasion by bacteria or fungi.

■ The junctions between cells

We have seen that the plasma membranes of adjacent plant cells are separated by cell walls. The only connections between the cell protoplasts are normally the **plasmodesmata**.

The plasma membranes of adjacent animal cells are generally separated by a tiny fluid-filled space only 20 nm across. But in certain cases there are interactions between the membrane proteins of adjacent cells (see Figure 7.30, and overleaf).

Figure 7.30 The junctions between cells

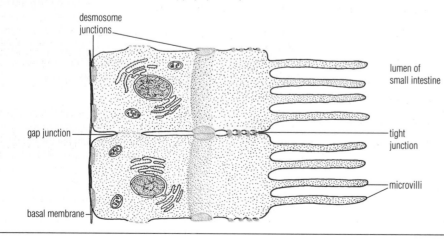

plant cells

cellulose cell wall

cell membrane

middle lamella

primary wall

secondary wall

region of living connections between cells (plasmodesmata)

simple pit in the primary wall (primary pit field)

ER crossing between cells, passing through middle lamella

animal cells: the distribution of junctions in the epithelial cells of mammalian small intestine; the adjacent cells are coupled mechanically (through spot desmosomes) and metabolically (through gap junctions), and the digestive fluids in the intestine are kept separate from the tissue fluid (by tight junctions)

desmosome junctions

lumen of small intestine

gap junction

tight junction

microvilli

basal membrane

Questions

1 How do Golgi vesicles contribute to
  a the middle lamella
  b the primary wall
  c the new cell membrane?

2 In plant cells, what is a pit? What is its probable function?

3 What is the relationship between a cellulose molecule, a cellulose microfibril and a cellulose-containing cell wall?

Three very common types of special junction between animal cells are described below:

▶ **Desmosomes** help to hold cells together. They are equivalent to spot welds in metal engineering. Dense fibrous material loops in and out of the desmosome region.

▶ **Tight junctions** are barriers that prevent or reduce fluid movements in the spaces between cells. At the tight junction the outer parts of adjacent membranes are fused.

▶ **Gap junctions** are tiny open channels in the plasma membrane through which small molecules and ions may pass. Gap junctions occur in a wide variety of cells, including certain muscle and nerve cells.

The structures of the junctions between cells are summarised in Figure 7.30 (p. 167).

7.6 APPLICATIONS OF CYTOLOGY

The techniques of cytology have important applications in biological research in fields as diverse as medicine, molecular genetics, biochemistry, agriculture, horticulture and drug manufacture.

■ Cell fractionation

So many different reactions take place simultaneously in a living cell that biochemical study requires the isolation of parts of cells in order to investigate pathways of metabolism. The ability to isolate cell components with minimal damage has underpinned the revolution in cell biology (Figure 7.11, p. 156). Isolated organelles are used in fundamental research; for example, isolated chloroplasts are used in the investigation of photosynthesis (p. 259). Isolated cell components are also used in applied and commercially focused research, such as the study of the effects of drugs on animal metabolism.

■ Tissue culture

Developments in cytology have also made possible the culturing of isolated cells. Theoretically, all cells have the capacity for independent existence provided they are supplied with the nutrients they require. Techniques of cell and tissue culture have been developed to a high level of reliability. Cells grown in glass culture chambers in the laboratory (*in vitro* culturing) are referred to as cultured cells. Cell cultures play an important part in fermenter technology, in the work of geneticists, in animal and plant breeding and in physiological and biochemical investigations into economically important plants and animals.

Animal tissue culture

Cells from young animals and cancer cells are most easily induced to grow *in vitro*. The culture medium has to be precisely controlled; conditions such as temperature and osmotic potential (p. 232) of the culture medium are normally critical. Mitosis occurs rapidly provided the cells are in contact with a surface such as the base of the culture vessel. Animal cells in tissue culture develop into mature cells of the same type as the cell from which the culture was started. One important application of the technique is in the design and testing of drugs.

Another recent successful development is the culturing of human stratified epithelium *in vitro* for use in grafts for the regeneration of the epidermis after severe burns and other forms of skin damage.

Plant tissue culture

Tissue culture of plant cells with intact walls has been successful, starting with fully differentiated cells dislodged from mature plant organs. With the correct balance of nutrients, such cells have been induced to become meristematic (p. 406). Whole new plants have been grown from a single mature, fully differentiated cell (p. 430).

Another important technique is the culturing of wall-free cells (protoplasts). The cell wall is removed by enzymic digestion using cellulase. Isolated plant protoplasts are used in genetic engineering in the service of horticulture, for example (p. 637).

Questions

1 In what ways are the cells of prokaryotes fundamentally different from those of eukaryotes?
2 When animal cells and cell-wall-free plant protoplasts are cultured in the laboratory, the water potential of the culture medium is critical to the cells' survival. Why is this (p. 234)?

7.7 PROKARYOTIC CELLS

The cells of prokaryotes are extremely small. Bacteria are generally between $0.5\,\mu m$ and $10\,\mu m$ long; rod-shaped bacteria of the genus *Bacillus* are usually $10.0\,\mu m$ long. The cells of most cyanobacteria are between $2\,\mu m$ and $5\,\mu m$ long (Figure 7.31). They have no distinct nucleus; within the cytoplasm is a ring of DNA, normally attached to the plasma membrane. Small ribosomes exist in the cytoplasm, but there are no large, complex membraneous organelles. The plasma membrane may be intucked and infolded, forming an increased surface area on which enzymes and pigments are held; in the cells of cyanobacteria, for example, photosynthetic pigments are housed on infolded membranes. Some prokaryotes have simple flagella, which do not consist of microtubules.

Table 7.2 summarises the differences between prokaryotic and eukaryotic cells.

■ A role for prokaryotes in the evolution of the eukaryotic cell?

The fossil record suggests that the first forms of life on Earth were very similar to present-day prokaryotes. Fossil prokaryotes are believed to be 3500 million years old (p. 666). Eukaryotic cells are believed to have arisen only 1000 million years ago.

It is probable that some of the organelles of the eukaryote cell arose through a symbiotic relationship between prokaryotes and eukaryotic cells. The chloroplast may have had its origin as a free-living photosynthetic bacterium or cyanobacterium that entered a eukaryotic cell and lived there mutualistically (p. 67). In a similar way the precursor of the mitochondrion may have been a free-living heterotrophic prokaryote with aerobic respiration. Both mitochondria and chloroplasts contain a ring of DNA, just as prokaryotes do, and the dimensions of these organelles are about the same as those of prokaryotes.

3 The DNA in mitochondria and chloroplasts of eukaryotes has been found to have different sequences of bases from the DNA of the eukaryote chromosomes (p. 666). What does this observation suggest about the origin of mitochondria and chloroplasts?
4 Name the three types of junction that occur between animal cells, and state their functions.

Figure 7.31   A rod-shaped bacillus (left) and a cyanobacterium (right)

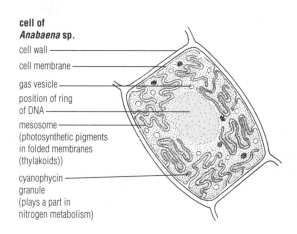

rod-shaped
bacterium

slime layer
cell wall
cell membrane
ring of DNA
food granules
ribosome

mesosome
(invaginations of the plasma
membrane; increased surface area,
site of enzymes and pigments)

pili (fimbriae), sometimes
used in conjugation
and DNA transfer

cell of
Anabaena sp.

cell wall
cell membrane
gas vesicle
position of ring
of DNA
mesosome
(photosynthetic pigments
in folded membranes
(thylakoids))
cyanophycin
granule
(plays a part in
nitrogen metabolism)

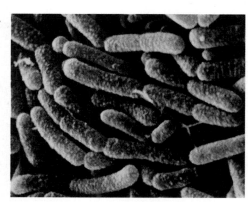

SEM of *Bacillus* sp.
(×10 000)

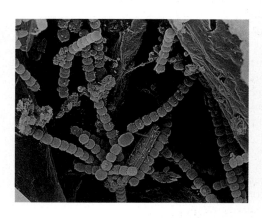

SEM of
Anabaena sp.
(×1000)

Table 7.2   The differences between prokaryotic and eukaryotic cells

Prokaryotic cells (bacteria, cyanobacteria)	Eukaryotic cells (green plants, fungi, animals)
Usually extremely small, 0.5–10 μm in diameter	Usually larger cells, 10–100 μm in diameter
Cell division preceded by DNA replication	Cell division preceded by mitosis or meiosis (DNA replicaton takes place earlier in the cycle)
Nucleus absent, circular DNA helix in cytoplasm; DNA is not supported with basic proteins (histones)	Nucleus present with distinct bounding membrane perforated by nuclear pores, and with DNA in linear chromosomes and combined with histone proteins
In some prokaryotes, a sexual system involves transfer of some DNA from one cell to another	Sexual system involves complete nuclear fusion between special sex cells (gametes), with equal contribution from both nuclei
Cell walls present, but chemically different from those of plants; they contain mucopeptides	Cell walls present in plants and fungi; they do not contain mucopeptides
Few organelles; membraneous structures absent or very simple and existing briefly in the cell	Many organelles, bounded by double membranes (chloroplasts, mitochondria, nucleus) and single membrane (Golgi apparatus, lysosomes, vacuole, endoplasmic reticulum)
Protein synthesised in small ribosomes	Protein synthesised in large ribosomes
Some cells have simple flagella; these are without microtubules, 20 nm in diameter	Some cells have complex cilia and flagella; these have 9 + 2 arrangement of microtubules, 200 nm in diameter
Some can fix nitrogen for use in amino acid synthesis	None have this facility

FURTHER READING/WEB SITES

The WWW Virtual Library – Cell Biology:
vl.bwh.harvard.edu/general_cell_biology.shtml

8 Cell biochemistry

- The chemical reactions of life are collectively known as metabolism. Metabolism consists of anabolic reactions in which complex compounds are synthesised, and catabolic reactions in which complex molecules are broken down.
- A sequence of metabolic reactions makes a metabolic pathway. The molecules involved in the steps in the pathway are known as metabolites or intermediates.
- Molecules contain potential chemical energy equal to the amount of energy needed for their synthesis. When molecules react, part of their potential chemical energy, known as free energy, may become available to do useful work.

- Free energy released in the aerobic respiration of glucose is used to drive energy-requiring processes such as muscle contraction and anabolic reactions such as protein synthesis, for example.
- Adenosine triphosphate (ATP) is a nucleotide found in all living cells. It is a reservoir of potential chemical energy that works in metabolism as an 'energy currency', linking energy-requiring and energy-yielding reactions.
- Enzymes are biological catalysts present in cells, most of them made of protein. They are highly specific in their action, and they allow the reactions of metabolism to occur very quickly, at the relatively low temperatures found in living things.
- Enzymes may be extracted from cells, and may be exploited in medical processes or biotechnological industries.

8.1 METABOLISM

Within an organism, thousands of chemical reactions are occurring simultaneously, and most are catalysed by specific enzymes (p. 178). The chemical reactions of life are collectively known as **metabolism**. The molecules involved in metabolic reactions are referred to as **metabolites**. Many metabolites are synthesised within the cells of an organism; others are taken in as food substances.

Metabolic reactions can be classified according to whether they involve breakdown or construction of organic molecules (Figure 8.1). In **anabolic reactions**, complex compounds are built up from simpler ones. Examples of anabolic processes are the synthesis of molecules like sugars, amino acids and fatty acids, and of the much larger macromolecules (p. 135), including the synthesis of proteins from amino acids (p. 139), photosynthesis (p. 246), the synthesis of starch, glycogen and cellulose from sugars, and the synthesis of fats.

Catabolic reactions are concerned with the breakdown of complex molecules. The breakdown of sugars and fats are examples of catabolism. Some of the energy released during these processes is used to drive anabolic reactions, but energy is also needed for movement, growth, repair and the disposal of waste, and for the maintenance of an internal environment.

■ Progress in biochemistry
Two German scientists, Otto Warburg and Hans Krebs, had important influences on the development of biochemistry. They used chemical methods to tackle biological problems, achieving spectacular results (see panels opposite).

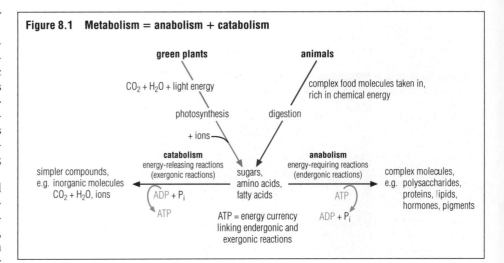

Figure 8.1   Metabolism = anabolism + catabolism

green plants

animals

$CO_2 + H_2O$ + light energy

complex food molecules taken in, rich in chemical energy

photosynthesis

digestion

+ ions

catabolism
energy-releasing reactions
(exergonic reactions)

simpler compounds, e.g. inorganic molecules $CO_2 + H_2O$, ions

sugars, amino acids, fatty acids

anabolism
energy-requiring reactions
(endergonic reactions)

complex molecules, e.g. polysaccharides, proteins, lipids, hormones, pigments

ADP + P_i

ATP

ATP

ATP = energy currency linking endergonic and exergonic reactions

ADP + P_i

■ Metabolic pathways: metabolites and intermediates

Often a single equation is used to summarise a major process of metabolism. The equation for glucose oxidation, called aerobic respiration, is an example:

$$C_6H_{12}O_6 + 6O_2 \rightarrow 6CO_2 + 6H_2O$$
$$+ \text{ENERGY}$$

Metabolic processes such as respiration do not consist of a single reaction, however. They take place in numerous steps occurring in a precise sequence. This happens because individual enzymes can only catalyse rather simple chemical transformations. One advantage of this is that energy release or input occurs in small amounts – that is, in safe and manageable quantities. Another advantage is that the substances formed at many of the steps may supply the building blocks required for other processes such as growth, maintenance and repair of cells.

A sequence of metabolic reactions, such as the steps of respiration, is known as a **metabolic pathway**. The individual molecules of a metabolic pathway, as

well as being known as metabolites, are also referred to as **intermediates**. Intermediates are compounds that are *en route* to forming cellular materials, and compounds in the course of being broken down with the release of energy. For example, some of the steps in glucose breakdown are

glucose
↓
triose phosphate
↓
pyruvic acid
↓
$CO_2 + H_2O$

Glucose, triose phosphate, pyruvate, carbon dioxide and water are all metabolites, but triose phosphate and pyruvate are intermediates of the pathway from glucose to carbon dioxide and water. (There are, of course, very many more intermediates in respiration.) Under the conditions in cells, organic acids such as pyruvic acid are partially ionised, so we refer to pyruvic acid as **pyruvate**.

OTTO WARBURG (1883–1970)

Life and times

Otto Warburg lived in Berlin from the age of thirteen (his father became professor of physics at the University there), apart from a brief period of military service during which he was wounded. Albert Einstein persuaded him to return to biological research.

Warburg had fluent command of spoken and written English, and regularly visited Britain. He was a fit man and loved horse riding; he rode every morning before going to work, and during his ride he did his most sustained thinking. He said he was a 'slow thinker'; when questions were put to him he would say 'I must think it over.' The answer would come next day!

Contributions to biochemistry

Warburg advanced biochemistry mainly through his own experimental work, using chemical methods to attack biological problems. Above all, he was a pioneer in the creation of new tools of investigation. His major field of interest was tissue respiration. In the **Warburg respirometer**, thin slices of still-respiring tissue were incubated in a small flask with a buffered nutrient solution. The uptake of oxygen was followed by monitoring the fall in pressure, measured in a U-shaped manometer attached to the flask.

Spectroscopic methods for the identification and analysis of cell constituents (metabolic intermediates) and enzymes were developed, along with other methods for the isolation of intermediates and enzymes. Warburg established the role of iron-containing cytochromes in respiration by measuring oxygen uptake in the presence of specific poisons (inhibitors, p. 184). Many vitamins were shown to be components of enzymes.

Warburg worked on aspects of photosynthesis for the last 50 years of his life, and also made a substantial contribution to cancer research. His work was developed by many others, and his students dominated biochemistry for a generation.

He was awarded a Nobel prize in 1931, but Warburg's later career was marred by a degree of isolation, arising from his increasingly intolerant attitude to the ideas of others.

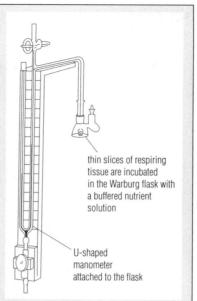

thin slices of respiring tissue are incubated in the Warburg flask with a buffered nutrient solution

U-shaped manometer attached to the flask

Warburg respirometer unit

In the process of Warburg constant volume manometry, the manometers are attached to a constant temperature tank so that the flasks are immersed in water. The water in the tank is circulated, and the respirometers are gently agitated between readings of the manometers. When used for photosynthesis research, fluorescent tubes are fitted below the glass base of the water bath in order to illuminate the Warburg flasks.

SIR HANS KREBS (1900–81)

Sir Hans Krebs

Hans Krebs was one of Warburg's students from 1928 to 1932. He improved the Warburg manometer and used it to show how, in the liver of many vertebrates, amino acids lose nitrogen (as —NH_2) and urea is formed by a process now known as the **ornithine cycle** (Figure 18.6, p. 377).

Krebs left Germany as a refugee in 1933, and worked in the Biochemical Laboratory at Cambridge. From 1935 to 1954 he was Professor of Biochemistry at Sheffield University. Here he established the pathway and role of the citric acid or tricarboxylic acid cycle (now known as the **Krebs cycle**); this is the chief energy-generating process within respiration, which occurs in mitochondria. He was elected a Fellow of the Royal Society in 1947, and shared the Nobel prize in 1953.

■ Orderliness in metabolism

The precise sequence followed by the steps of metabolism is related to the needs of organisms and cells. This orderliness of metabolism is due to three important features of structure and function of cells:

▶ The sequence of steps is **regulated by enzymes (biological catalysts)**. The production of enzymes is controlled by the nucleus (Chapter 9).

▶ Many **intermediates are located within specific cellular structures** known as **organelles** (p. 157). (It is sometimes said that metabolism is 'compartmentalised'.) Intermediates are therefore not necessarily available to react with other metabolites (Chapter 7).

▶ Movement of many metabolites into the cell and between the different metabolic compartments is by facilitated diffusion or by active transport using energy from respiration (Chapter 11).

■ The science of biochemistry

Biochemists aim to explain, in chemical terms, how living things work. As far as we know, nothing takes place in living matter that cannot ultimately be explained by the laws of physics and chemistry, even though the chemistry of life involves such a vast array of different chemical reactions. Biochemistry is concerned with the following:

▶ the **chemical composition of cells** and tissues (for a discussion of the range of chemicals found in living things, see Chapter 6)

▶ the **biological catalysts (enzymes)** that accelerate chemical changes within cells (see section 8.5); practically every chemical reaction occurring within the body requires a specific enzyme to catalyse it, and therefore there are about as many enzymes as metabolites

▶ **discovering all the metabolic pathways** by which chemical changes occur within the living cell (the techniques used are outlined in section 8.3)

▶ the energy requirements of metabolism; energy released in catabolism is retained in 'energy currency' molecules and made available 'on demand' to drive anabolic reactions, as well as other energy-requiring processes such as muscle contraction; the chief energy currency molecule is adenosine triphosphate (see section 8.4)

▶ **how chemical reactions are controlled and regulated** so that the cells, tissues and organs are able to maintain their organised structures (section 8.8); the way in which the DNA of the nucleus controls protein synthesis in the cytoplasm is discussed in Chapter 9.

Questions

1 Write out a single equation to summarise photosynthesis. In what way is this equation similar to that for respiration?

2 From your previous reading, give two examples of catabolic processes and two examples of anabolic processes.

8.2 ENERGY AND METABOLISM

Energy exists in many forms, including heat, light, sound, mechanical, chemical, electrical, magnetic and atomic energies. All these forms may be important in biology in one way or another, but it is heat, light and chemical energy that biochemistry is most concerned with. We define energy as the capacity to do work, and we measure it in **joules**. The quantities of energy involved in biological systems are such that the kilojoule is the unit commonly used in biology (1 kJ = 1000 J). For example, the daily energy requirement of an average-sized adult male of average activity is 7500 kJ; a gram of carbohydrate has an energy value of 17.2 kJ.

Energy can be stored, and stored energy is known as **potential energy**. For example, every molecule contains an amount of potential chemical energy equal to the amount of energy needed to synthesise that molecule. Living organisms made up of organic compounds represent a store of potential energy.

Energy that is no longer 'in store' but is 'on the move' is referred to as **kinetic energy** ('active' energy). When glucose is oxidised to carbon dioxide and water in aerobic respiration, energy is released. Some of the energy released in respiration is used to drive energy-requiring processes such as muscular contraction (often leading to movement), as well as anabolic reactions such as the synthesis of proteins.

■ Energy transformation

Energy can be transformed from one form to another. Energy transformation is referred to as **transduction**, and occurs in a transducer (Table 8.1). For example, electrical energy is transduced to light and heat energy in a lamp bulb. Light energy is transduced to chemical energy in a chloroplast during photosynthesis (p. 246). The chloroplast is a **transducer organelle**.

Energy cannot be created or destroyed. When energy is converted from one form to another, no energy is gained or lost (**First Law of Thermodynamics**, often called the Law of Conservation of Energy). Part of the total energy becomes converted into heat energy, however, and escapes (**Second Law of Thermodynamics**). In fact, all the energy reaching the Earth from the Sun is progressively transduced to heat and passes out into space through the actions of respiration, combustion and many other chemical reactions of the biosphere.

Figure 8.2 **The Sun as the source of energy for metabolism:** arrows indicate the direction of energy flow; the final fate of all energy is to be degraded to heat energy and lost into space – this is represented by the broken arrows

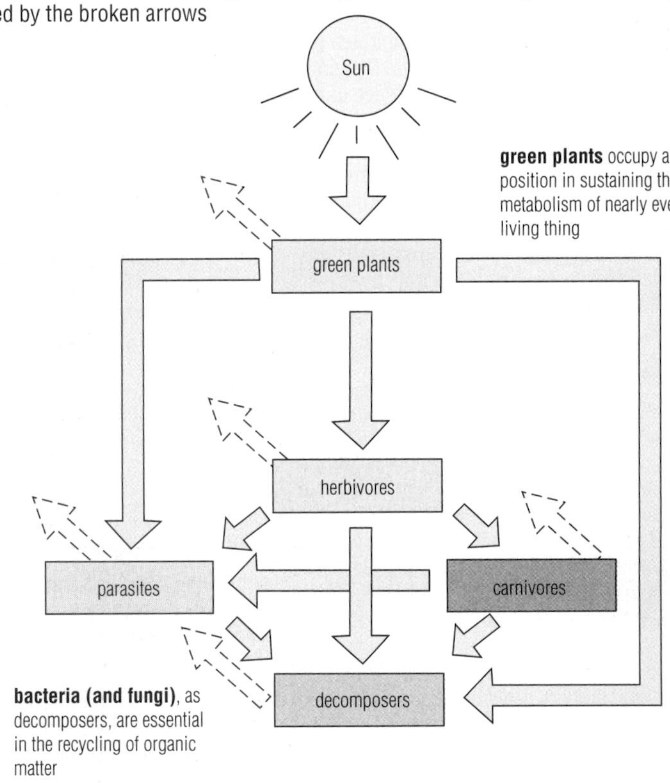

green plants occupy a central position in sustaining the metabolism of nearly every living thing

bacteria (and fungi), as decomposers, are essential in the recycling of organic matter

■ Transduction of sunlight via green plants

Practically all life, with the exception of the relatively few chemosynthetic organisms (p. 269), depends upon the radiant energy from the Sun. Yet it is only a tiny part of the Sun's energy that is absorbed by green plants in photosynthesis (p. 268). In this process, some of the Sun's radiant energy is turned into chemical energy stored in glucose, a product of photosynthesis. Other organisms depend ultimately on the photosynthetic process of green plants for their energy requirements (Figure 8.2).

Table 8.1   Energy transduction in biology

Energy change	Transducer
chemical → heat	all living cells
chemical → electrical	brain and nerve
chemical → mechanical	muscles
chemical → osmotic	kidney, and across membranes
chemical → radiant light	female glow-worm (see Figure 8.4)
light → chemical	chloroplasts
light → electrical	retina of eye
hydrostatic → electrical	semicircular canals of inner ear
mechanical → electrical	inner ear

■ Energy and chemical reactions

You have already learnt that chemical energy exists in the structure of molecules; every molecule contains a quantity of potential energy equal to the quantity of energy needed to synthesise it originally. The quantity of energy in a molecule changes when bonds in that molecule break and new bonds form – that is, when chemical reactions occur. Chemical energy does not exist in chemical bonds themselves, however, but rather in the whole structural arrangement of the molecule.

When molecules are broken down, only part of the potential chemical energy in the molecules is available to do useful work. This energy, which is known as **free energy** or **Gibbs energy**, is defined as the energy available to do work under conditions of constant temperature and pressure. Energy is released in any reaction in which the products of the reaction have less free energy than the original reactants. Reactions that release energy are known as **exergonic reactions**. On the other hand, reactions that require energy are called **endergonic reactions**. An endergonic reaction can only occur if it is coupled with an exergonic reaction so that the *net* reaction is exergonic (Figure 8.3).

Figure 8.3 Exergonic and endergonic reactions

an **exergonic reaction** is one in which the products have less potential energy than the reactants

for example:

A + B → AB

free energy emerges as heat energy, work etc.

when hydrogen gas is burnt in air:

$2H_2 + O_2 \rightarrow 2H_2O + HEAT\ ENERGY$

downhill reaction   energy given off

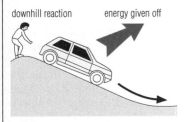

an **endergonic reaction** requires energy input, because the products contain more potential energy than the reactants

for example:

C + D → CD

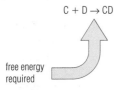

free energy required

in photosynthesis:

$2H_2O + LIGHT\ ENERGY \rightarrow 4(H) + O_2$

taken up in the production of carbohydrates etc.

energy put in   uphill reaction

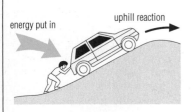

without free energy input, an endergonic reaction cannot proceed

Figure 8.4 Light emitted by the glow-worm (*Lampyris noctiluca*).

The female (flightless) uses light to attract a mate. Greenish-blue light is produced from chemical energy in the last three abdominal segments. Most of the energy involved is emitted as light; little heat is produced in the reactions

■ The equilibrium constant of a reversible reaction

A reversible reaction can be represented in the equation

$$A + B \rightleftharpoons C + D$$

in which molecules A and B react to form molecules C and D. As soon as molecules C and D start to accumulate, they will also react to form A and B. This is what the term 'reversible reaction' means; this is why there is a double arrow in the equation.

An example of such a reversible reaction occurring in living cells is

glutamate + pyruvate

$\updownarrow$ transamination (p. 209)

α-oxoglutarate + alanine

A reversible reaction has reached its **equilibrium point** when the rate of the forward reaction equals the rate of the reverse reaction. At equilibrium the reaction does not stop. The forward and backward reactions are taking place at equal rates; the concentrations of reactants and products remain steady.

The equilibrium point of a reversible reaction varies according to the reactants, the products and the conditions. For one reaction, say A + B → C + D, the equilibrium may not be reached until nearly all A and B has been converted to C and D. For another reaction, say E + F → G + H, equilibrium may be reached when only a little of E and F has been converted. The **equilibrium constant** K is the ratio of the concentration of the products to that of the reactants at equilibrium point, and is experimentally determined

$$K = \frac{[C] \times [D]}{[A] \times [B]}$$

where [C], [D], [A] and [B] are the molar concentrations of the reactants.

■ Free energy and the equilibrium point

A reaction goes forwards if it releases free energy, that is, if it is an exergonic reaction. The greater the drop in free energy, the more completely the reaction proceeds. For an exergonic reaction

$$A + B \rightleftharpoons C + D + ENERGY$$

the reaction will proceed, the products C and D accumulating at the expense of the reactants A and B, and K will be greater than 1.

If the energy difference between reactants and products is not very great there will be little tendency for the reaction to go in either direction; there will be approximately equal quantities of (A + B) and of (C + D) at equilibrium, and K will be close to 1.

■ Living things and disorder

We have learnt that it is not possible to use all the energy released in a reaction to perform work because some of the energy is transformed into heat and lost from the system (the Second Law of Thermodynamics). Energy transformations, such as chemical reactions, result in a reduction in the amount of usable energy (free energy) in the system. Left to themselves, systems and their surroundings move to a state of minimum energy. Since (other things being equal) an ordered state is of higher energy than a disordered state, this means that they tend to move to a state of maximum disorder.

Highly ordered systems such as cells and organisms reverse the tendency of steadily increasing disorder, during their lifetime. This is only temporary, however. Living organisms trap or capture energy from outside their system – ultimately, from the Sun. As a result, the organism's structures are highly organised.

Questions

1 In what respect can the mitochondrion (p. 316) be described as a transducer organelle?

2 What is the ultimate source of energy for metabolism of parasitic organisms?

3 What is the difference between an exergonic and an endergonic reaction? Give two examples of each.

4 In a reversible reaction at equilibrium, how does the rate of the forward reaction compare with that of the backward reaction?

8.3 INVESTIGATING METABOLISM

The unravelling of the biochemistry of metabolism is a challenging task. As a consequence, when individuals or a team of workers achieve a breakthrough in the understanding of a pathway, the pathway they have discovered is often named after them or their leader. For example, the Krebs cycle in aerobic respiration (p. 315) was named after Sir Hans Krebs, and the Calvin cycle in photosynthesis after Melvin Calvin (p. 262–3). Several biochemists have been awarded Nobel prizes as a result of their work with metabolic pathways.

Why do biochemical investigations present difficulties? The main reasons are as follows:

▶ Cells are minute structures, so tissue samples of more or less identical cells tend to be very small. Consequently, it is difficult to obtain large quantities of intermediates or enzymes for analysis. Biochemists must often work with extremely small samples in their analyses and investigations.

▶ To obtain metabolites and enzymes, tissue samples are normally first ground up in a blender. Disruption of cell structures brings the contents of damaged organelles together. Metabolites that in cells are held apart may react. As a result, abnormal chemical reactions can occur, unrepresentative of biochemical events in intact cells. To take an extreme example, lysosomes contain hydrolytic enzymes that in working cells digest redundant organelles (p. 162). The blending process may expose other organelles to these enzymes. To minimise the harmful effects of disruption of tissue samples, solutions and reagents are held at the temperature of melting ice (Figure 8.5). This slows down chemical change, but does not prevent abnormal reactions.

▶ Metabolic pathways are not independent of each other. It is frequently the case that two (or more) pathways have intermediates common to both. For example, glycerate 3-phosphate (GP) is an intermediate in aerobic respiration, in photosynthesis and in the synthesis of an amino acid (Figure 8.6). Under the conditions of an investigation, metabolites of linked pathways may be diverted along pathways not normally followed in the intact cell.

■ Analytical techniques of biochemistry

A biochemist develops a hypothesis about the possible or likely reactions in the cells of a tissue by which a pathway of biochemical events occurs. From the hypothesis, predictions are made about the intermediates and enzymes that may be present. The sequence of steps of that pathway is then confirmed (or rejected) by seeking answers to a series of questions using various analytical techniques.

1 Are the predicted intermediates of the pathway present in tissues? Soluble metabolites are extracted from samples of tissues taken from animals or plants or from cultures of unicellular organisms. The extract may require to be concentrated by evaporation of the solvent and tissue fluid. This is carried out at moderate to low temperatures and reduced pressure. The metabolites are then analysed by chromatography. This is a technique used to separate and identify unknown mixtures of closely related substances. Chromatography is carried out on a stationary medium, or phase: paper (paper chromatography), for example, or a thin film of dried solid (thin layer chromatography). A solvent travels through the stationary phase by capillarity, through a 'spot' containing a mixture of the unknown substances. Components of the spot are carried along with the solvent, but some are held back to varying extents by interaction with the stationary phase, and so a separation is effected. Identification of the separated spots is possible by running parallel chromatograms loaded with known substances thought to be present in the unknown mixture. The solvent front on the chromatogram is marked at the end of the run, and the Rf values of the spots calculated:

$$Rf = \frac{\text{distance moved by the solute spot}}{\text{distance moved by the solvent front}}$$

Rf values depend on the solvent used.

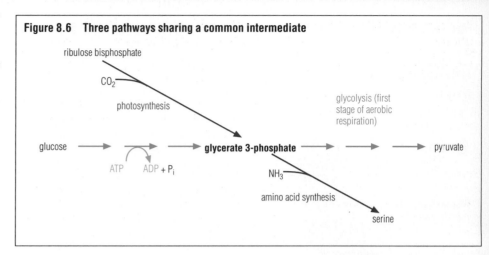

Figure 8.6   Three pathways sharing a common intermediate

ribulose bisphosphate

CO_2

photosynthesis

glucose → glycerate 3-phosphate → → → pyruvate

glycolysis (first stage of aerobic respiration)

ATP   ADP + P_i

NH_3

amino acid synthesis

serine

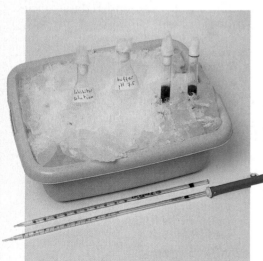

Figure 8.5   Biochemists work with small samples, held at the temperature of melting ice

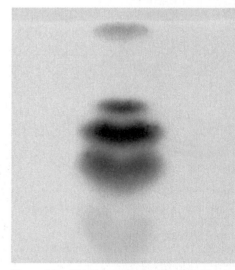

Figure 8.7   Thin layer chromatography: the solvent flow has separated the compounds

Thin layer chromatography is shown in Figure 8.7. Paper chromatography of chlorophyll pigments is shown in Figure 12.5, p. 249. Two-dimensional paper chromatography of the products of photosynthesis is shown in Figure 12.32, p. 262. Gas–liquid chromatography is shown in Figure 20.25, p. 432.

2 Are enzymes of the pathway present in the tissue? Enzymes (proteins) can be extracted from tissue that has been ground up. Many enzymes are in solution in the cytosol (other enzymes are integral parts of the structure of membranes, and may be more difficult to separate and purify for study).

Enzymes can be made to precipitate out from solution by the addition of a salt. The precipitated protein is usually dialysed (p. 388) to remove salt, re-dissolving in the process. In the laboratory individual proteins may be separated from the solution by various techniques, including ion-exchange chromatography and electrophoresis (Figure 8.8).

Individual enzymes, once isolated, catalyse the same metabolic reaction in a test tube (*in vitro*) that they catalyse in the cell (*in vivo*) and may therefore be studied in isolation from all the complex events going on in the cell. Tissues must be shown to contain all the enzymes necessary to catalyse a postulated pathway.

3 Does the addition of an intermediate enhance metabolism? If a pathway is deprived of its starting material (its substrate), all the intermediate stages will stop and the end-product will not accumulate (the tissue sample will have been starved). If one of the intermediate metabolites is now added to the system, the reactions 'downstream' of this metabolite will start up again and the end-product will be formed. This confirms that the added intermediate is an essential step in the pathway.

For example, if a tissue sample that has been ground up (homogenised) is deprived of the respiratory substrate, glucose, no pyruvate is formed (Figure 8.6). If triose phosphate is now added, pyruvate is again produced. This sort of investigation is called a 'feeding experiment'. Where carbon dioxide is one of the by-products of a pathway, feeding experiments can also be carried out in a Warburg apparatus (p. 171), where respiratory gas exchange is accurately measured.

4 Do intermediates become tagged with radioactive 'label' when labelled substrate is fed? Atoms of certain elements exist in more than one form, and these are known as isotopes of the element. Isotopes are classified as stable or unstable. Stable isotopes persist in nature because they do not undergo radioactive decay. Unstable isotopes are radioactive. This means they break down steadily, a process known as radioactive decay, often emitting α or β particles. The product of decay is a stable isotope; for example, $^{14}C \rightarrow {}^{14}N + \beta$ particle. The radiation emitted by radioactive isotopes fogs photographic film held in contact with the radioactive source in the dark, a technique known as autoradiography. The presence of radioactive isotopes can alternatively be detected by an instrument such as a Geiger counter.

The 'feeding' of radioactively labelled intermediates was used by Melvin Calvin to work out the carbon pathway in photosynthesis (p. 262) and has also been applied to investigation of protein synthesis (p. 208) and of glycolysis.

Figure 8.8 Electrophoresis

Protein separation

Proteins are first treated with a detergent known as SDS. This detergent 'unwinds' the protein molecule and attaches itself to the peptide bonds of the polypeptide chain, exposing negatively charged groups of SDS. The resulting negative charge on the denatured protein is proportional to the number of peptide bonds.

SDS-treated protein molecules all move towards the anode, but in passing through PAG the sieving effect brings about separation on the basis of their size: small molecules pass relatively quickly towards the anode whereas larger ones take longer.

Electrophoresis is a technique for separating molecules such as proteins or nucleic acid fragments on the basis of their net charge and mass. Commonly, gel electrophoresis is used, in which the stationary phase is polyacrylamide gel (PAG), and separation occurs when an electrical current is applied across the gel. The method requires only a tiny quantity of a mixture; for example, 10 µg is often sufficient.

DNA separation

DNA molecules are huge, and must first be cut into short fragments by the use of a restriction enzyme (p. 628).

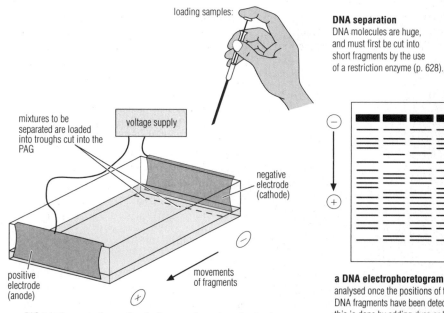

loading samples:

voltage supply

mixtures to be separated are loaded into troughs cut into the PAG

negative electrode (cathode)

positive electrode (anode)

movements of fragments

PAG is both a supporting medium for the separation and a molecular sieve. Separation of fragments depends on their charge and their molecular mass.

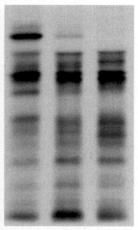

a **protein electrophoretogram** is analysed by staining the PAG

a **DNA electrophoretogram** can be analysed once the positions of the DNA fragments have been detected; this is done by adding dyes or by using radioactively labelled probes (p. 632)

8.4 ADENOSINE TRIPHOSPHATE AND METABOLISM

A large amount of energy is required to couple together glucose residues to make a cellulose molecule (p. 134). Similarly, it takes a great deal of energy to couple together the hundreds of amino acid residues needed to make a protein molecule (p. 140). These processes, like many others in metabolism, are endergonic (p. 172). Endergonic reactions are driven by coupling them to exergonic reactions. Adenosine triphosphate (ATP) has the function of linking exergonic reactions with endergonic reactions.

Adenosine triphosphate is a nucleotide (p. 147) consisting of an organic base (adenine), a five-carbon sugar (ribose) and a sequence of three phosphate groups linked together (Figure 8.9). ATP was first isolated from mammalian muscle; it has since been shown to occur in all types of cell, at concentrations varying between 0.5 and 2.5 mg cm^{-3}.

■ The ADP–ATP cycle and metabolism

ATP is formed from ADP and P$_i$ (inorganic phosphate) in respiration (p. 316), and in photosynthesis (p. 260). ATP is converted to ADP and P$_i$ in the processes of supplying essential free energy for the performance of mechanical work in muscle contractions and other cell movements, the active transport of molecules and ions across membranes, and the synthesis of macromolecules and other molecules from simple precursors in the cells (Figure 8.10).

■ How does ATP work in metabolism?

ATP is a reservoir of potential chemical energy. In metabolism, ATP acts as a common intermediate linking energy-requiring and energy-yielding reactions.

Hydrolysis of the terminal phosphate group of ATP is a highly exergonic reaction. In this reaction

$$ATP + H_2O \rightarrow ADP + PO_4^{3-}$$

adenosine diphosphate, orthophosphate (P$_i$)

the free energy released is approximately 30–33 kJ mol^{-1}. Hydrolysis of ADP to adenosine monophosphate (AMP) releases a similar amount of energy. These reactions will occur only in the presence of an appropriate catalyst (an enzyme, ATPase).

The hydrolysis of ATP to ADP is especially important in biological systems; ATP is also hydrolysed to AMP in metabolism on occasions. Hydrolysis of AMP to adenosine and P$_i$, which yields only a small amount of energy, is a biologically insignificant event, however.

■ Coupling of ATP to an endergonic reaction

In the building of glucose residues into polysaccharides, glucose phosphate is formed initially. Glucose phosphate is then added on to a growing polysaccharide chain by reaction with a nucleotide phosphate (such as ATP, but other nucleotide phosphates may be involved depending on the polysaccharide; see opposite page). The initial reaction between glucose and orthophosphate ion

glucose + orthophosphate → glucose phosphate

is an endergonic reaction, requiring free energy of about 16 kJ mol^{-1}. We have seen that hydrolysis of ATP is exergonic, releasing at least 30 kJ mol^{-1}:

$$ATP + H_2O \rightarrow ADP + P_i$$

The 'coupling together' of these two reactions produces another overall chemical reaction, which is strongly exergonic:

glucose + ATP → glucose + ADP phosphate

This reaction tends to proceed spontaneously, provided a suitable catalyst is present, with an energy release of more than 14 kJ mol^{-1} – a good illustration of the linking of ATP hydrolysis with an endergonic process.

■ Direct hydrolysis of ATP

In certain instances, direct hydrolysis of the terminal phosphate group of ATP is used to bring about changes. Muscle action is an example of a process driven by direct hydrolysis (p. 491).

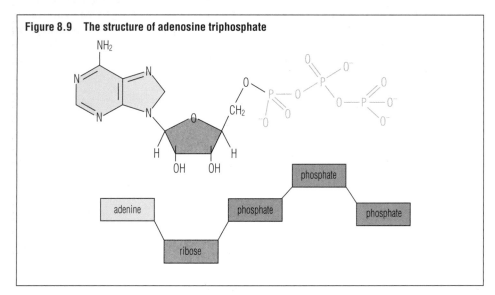

Figure 8.9 The structure of adenosine triphosphate

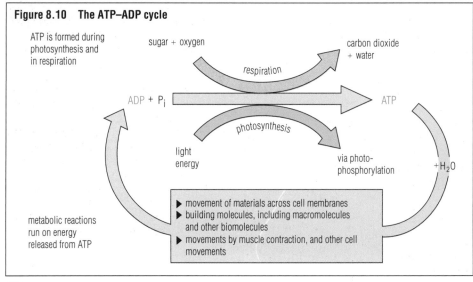

Figure 8.10 The ATP–ADP cycle

ATP is formed during photosynthesis and in respiration

sugar + oxygen

carbon dioxide + water

respiration

ADP + P$_i$ — ATP

photosynthesis

light energy

via photo-phosphorylation

+H$_2$O

metabolic reactions run on energy released from ATP

▶ movement of materials across cell membranes
▶ building molecules, including macromolecules and other biomolecules
▶ movements by muscle contraction, and other cell movements

THE FORMATION OF ATP

In 1961 Dr Peter Mitchell put forward the hypothesis that energy was held as a gradient in hydrogen ion concentration and used in ATP synthesis in chloroplasts (in photosynthesis) and mitochondria (in respiration). This idea is known as the **chemiosmotic theory** of ATP synthesis (see below). During the years that followed, evidence supporting the model was obtained by him and Dr Jennifer Moyle, and by other workers, and in October 1979 Mitchell was awarded the Nobel prize for chemistry in recognition of his part in developing the theory.

Mitchell's hypothesis proposed that ATP synthesis (by reversal of the hydrolysis reaction) is catalysed by the enzyme known as ATPase. A gradient of hydrogen ions (protons) is generated across the membrane of a mitochondrion (in respiration) or a chloroplast (in photosynthesis). The membrane is impermeable to hydrogen ions. The accumulation of hydrogen ions on one side of the membrane is a source of potential energy for ATP synthesis.

The enzyme catalysing the hydrolysis of ATP to ADP and P_i occurs in the membrane. The hydrogen ions are able to pass back across the membrane, through a channel in the ATPase. Flow of hydrogen ions through this channel causes reversal of ATP hydrolysis.

Mitchell's chemiosmotic theory

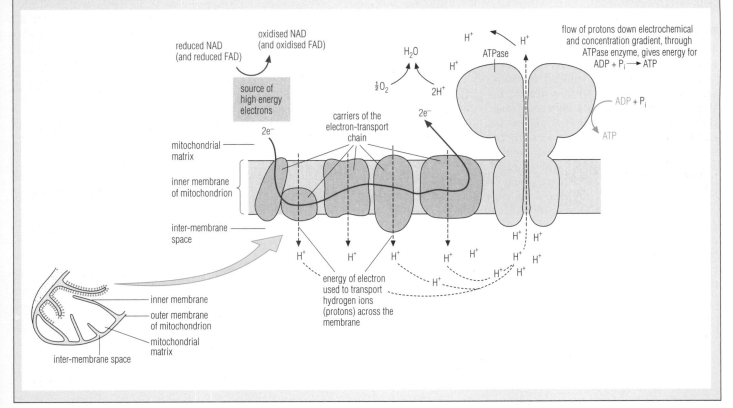

■ ATP and other phosphorylated nucleotides in metabolism

Apart from ATP, other phosphorylated nucleotides found in cells are guanosine triphosphate (GTP) and uridine triphosphate (UTP). They are used to 'drive' certain endergonic reactions: that is to say, their hydrolysis is coupled with an endergonic reaction so that the overall reaction is exergonic.

ATP is used in the syntheses of starch, fats and proteins. GTP is used in the synthesis of cellulose from glucose phosphate, and UTP in the synthesis of glycogen from glucose phosphate.

■ High-energy bonds?

Because of its functions in metabolism, ATP has been described as a 'high-energy compound' or been said to possess 'high-energy bonds'. Sometimes the so-called high-energy phosphate bond is represented by a 'squiggle' ($\sim$) in formulae and equations. But in fact there is nothing unusual about ATP as a chemical substance. It is a molecule composed of atoms held together by covalent bonds, and there is nothing abnormal about these bonds either. Energy is not actually stored in chemical bonds; there is no such thing as a high-energy phosphate bond. It is perhaps more helpful to think of ATP at work in metabolism as a dehydrating reagent: ATP will tend to pull water from other reactions and to be hydrolysed in the process.

Questions

1   Look at Figure 8.6, p. 174. Say why detecting the presence of glycerate 3-phosphate might not be good evidence that the glycolysis pathway proceeds as shown in Figure 15.21, p. 315.

2   Why may the disruption of cells cause abnormal chemical reactions to occur? What is the advantage of carrying out the disruption process at low temperatures?

3   The functional parts of ATP are the phosphate groups, but it is misleading to talk of high-energy phosphate bonds. So where is the energy in the ATP molecule?

4   ATP is a reservoir of potential chemical energy – an energy currency. What are the advantages to cells of this source of energy?

5   What do you understand by a 'coupled' reaction?

8.5 ENZYMES AND METABOLISM

The chemical reactions of metabolism occur quickly, even at the relatively low temperatures found in living things (from just above freezing to almost 40 °C). Thousands of simultaneous reactions occur in living cells, without interruption or interference from surrounding reactions. Life chemistry proceeds in an orderly and controlled manner under the apparently cramped conditions of the cell.

■ Enzymes as catalysts

These features of metabolism are made possible by the action of enzymes. Without enzymes, biochemical reactions would be too slow to sustain life at all. Enzymes are biological catalysts; most of them are proteins. A catalyst speeds up the rate of a chemical reaction without itself being changed at the end of the reaction. The properties of catalysts, and the special features of biological catalysts, are reviewed in the panel on the right. Enzymes differ from most inorganic catalysts in that they catalyse only one reaction, or only a small range of reactions. That is, unlike most inorganic catalysts, enzymes are highly **specific** in their actions.

THE PROPERTIES OF CATALYSTS

Inorganic catalysts
These are used in chemistry and industry.

▶ **They are effective in small amounts.** The number of moles of substrate converted by a mole of catalyst per minute is the turnover number, which for catalysts is generally between 100 and 3 000 000.
▶ **They are unchanged at the end of a reaction.** A catalyst is chemically involved in a reaction. It is consumed in one step and then regenerated in a subsequent step. It can be used repeatedly without undergoing permanent chemical change.
▶ **They speed up the rate at which equilibrium position is reached.** A catalyst speeds up the rates of both the forward and the backward reactions, and the rate at which equilibrium position is attained. A catalyst does not alter the position of equilibrium.
▶ **They lower the amount of energy required to activate the reacting molecules** (the activation energy for the reaction). Finely divided platinum, for example, provides sites on which substances are adsorbed in close proximity and can react easily.

Enzymes (biological catalysts)
These are (mostly) proteins that catalyse the chemical reactions of living systems.

▶ **Enzymes are extremely efficient.** The turnover numbers of enzymes are rarely less than 10 000, and are often higher. The enzyme carbonic anhydrase, found in red blood cells, can catalyse the conversion of 3.6×10^7 molecules of carbon dioxide to hydrogencarbonate per minute, per mole of enzyme.
▶ **They are extremely specific.** Most enzymes are specific to one particular type of substrate molecule, and usually one isomer of that substance. Other enzymes are specific to a group of similar substances, or to a particular type of chemical linkage.
▶ **They can be denatured.** Like most proteins, they are denatured by heat. Also certain products of reactions may irreversibly change proteins and denature the enzyme.
▶ **They are affected by pH.** Enzymes have an optimum pH, at which they act most efficiently.

■ The site of enzymes in cells

Enzymes formed and retained in the cell are known as **intracellular enzymes**, and may occur either in the cytoplasm or the nucleus. Cytoplasmic enzymes may occur either in the cytosol or inside the membraneous organelles (mitochondria or chloroplasts), or attached to the plasma membrane or membranes of cell organelles.

Certain other enzymes produced in the cell are packaged to be secreted from the cell, and work externally. Most digestive enzymes are of this sort. These are known as **extracellular enzymes**.

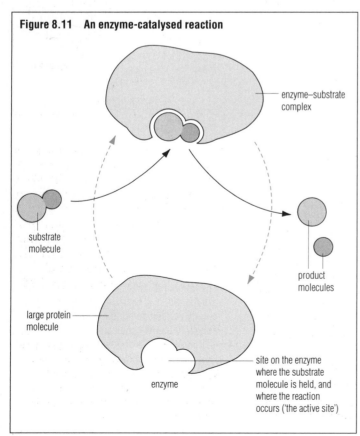

Figure 8.11   An enzyme-catalysed reaction

enzyme–substrate complex

substrate molecule

product molecules

large protein molecule

enzyme

site on the enzyme where the substrate molecule is held, and where the reaction occurs ('the active site')

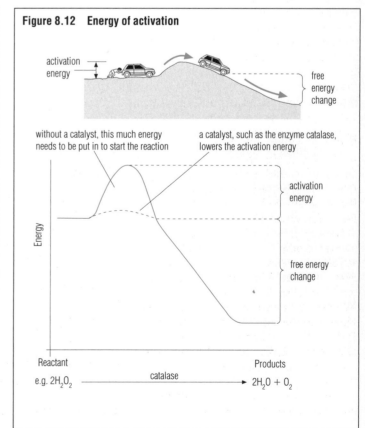

Figure 8.12   Energy of activation

activation energy

free energy change

without a catalyst, this much energy needs to be put in to start the reaction

a catalyst, such as the enzyme catalase, lowers the activation energy

activation energy

free energy change

Energy

Reactant
e.g. $2H_2O_2$

catalase

Products
$2H_2O + O_2$

■ How do enzymes work?

Enzymes are large molecules that work by reacting with another compound or compounds (the **substrate**) to form a short-lived enzyme–substrate complex. This complex is formed at a particular part of the enzyme molecule, usually at its surface, known as the **active site** (Figure 8.11). The complex breaks down to form the product(s), leaving an unchanged enzyme molecule, which is then available to catalyse another cycle. For example, one molecule of catalase (an intracellular enzyme catalysing the decomposition of hydrogen peroxide) can transform 5 600 000 substrate molecules per minute under optimal conditions. At the active site the enzyme works by reducing the amount of energy (the **energy of activation**) that substrate molecules need to have before they can undergo the chemical change concerned (Figure 8.12).

■ Enzymes as proteins

Most enzymes are made of protein, although a few consist of RNA. Of the protein-containing enzymes, some consist of protein only. Many other enzymes contain additional, non-protein portions and are therefore described as conjugated proteins; the protein part of these is called the **apoenzyme**. If the non-protein portion of an enzyme is readily detached and actually participates in reactions with two or more different enzymes, it is known as a **coenzyme**; the nucleotides NAD^+ and $NADP^+$ (p. 147) are examples. If the non-protein portion remains more or less permanently attached to one particular enzyme it is known as the **prosthetic group**. Haem, a ring-shaped organic molecule with an iron atom at its centre, forms the prosthetic group of haemoglobin, cytochrome (p. 358) and the enzymes catalase and peroxidase (which catalyse the decomposition of hydrogen peroxide). Chlorophyll is very similar to haem but has an atom of magnesium at its centre.

Sometimes a metal ion is necessary for the functioning of an enzyme. Some enzymes contain very tightly bound metal ions; an example is the zinc ion (Zn^{2+}) in carboxypeptidase, a digestive enzyme that hydrolyses the carboxyl terminal peptide bond in the polypeptide chain. The enzyme thrombokinase, which converts prothrombin to thrombin in the clotting of blood (p. 364), is activated by calcium ions (Ca^{2+}).

Figure 8.13 illustrates protein-only and conjugated protein enzymes.

Figure 8.13 Types of enzyme

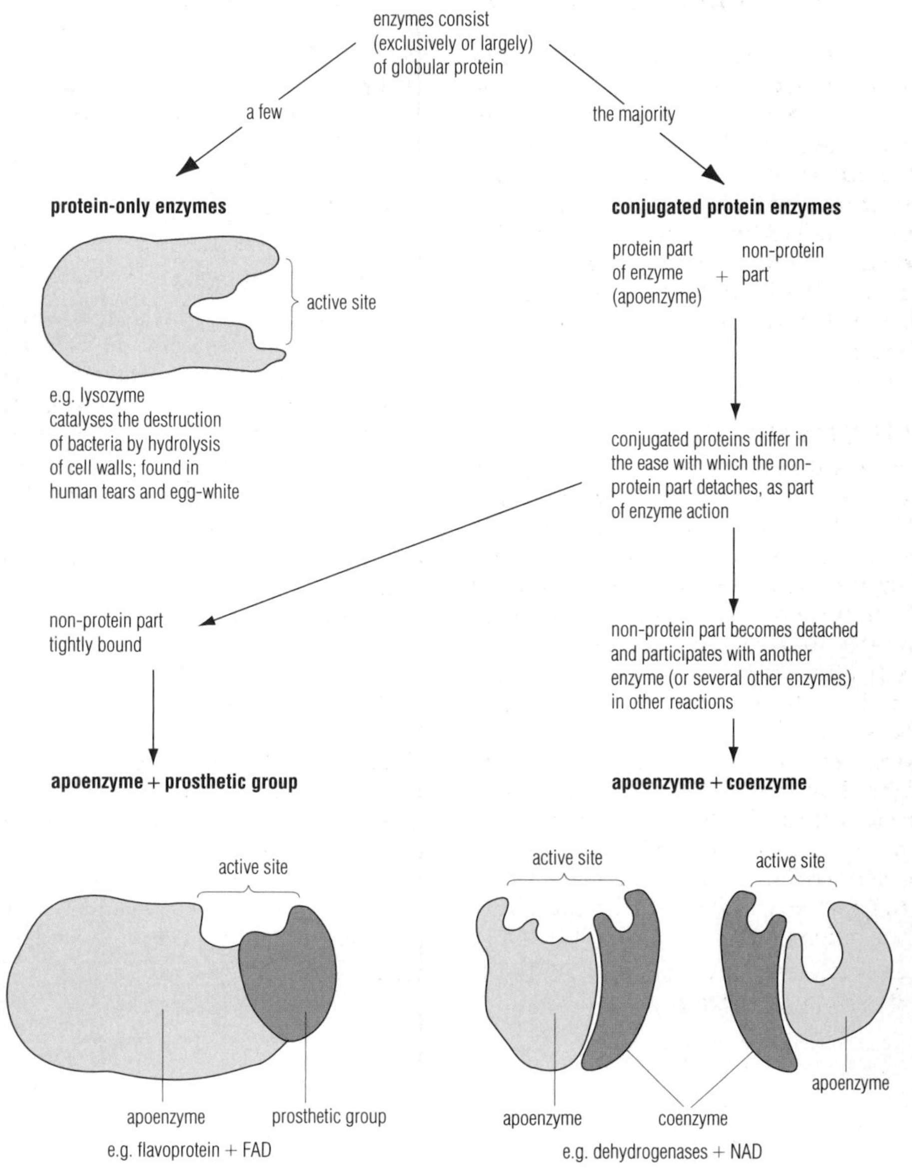

enzymes consist (exclusively or largely) of globular protein

a few → **protein-only enzymes**

active site

e.g. lysozyme catalyses the destruction of bacteria by hydrolysis of cell walls; found in human tears and egg-white

the majority → **conjugated protein enzymes**

protein part of enzyme (apoenzyme) + non-protein part

conjugated proteins differ in the ease with which the non-protein part detaches, as part of enzyme action

non-protein part tightly bound

non-protein part becomes detached and participates with another enzyme (or several other enzymes) in other reactions

apoenzyme + prosthetic group

active site

apoenzyme   prosthetic group

e.g. flavoprotein + FAD

apoenzyme + coenzyme

active site   active site

apoenzyme   coenzyme   apoenzyme

e.g. dehydrogenases + NAD

8.6 FACTORS AFFECTING ENZYME ACTION

Any factor that influences an enzyme will alter the rate of the reaction that is being catalysed. Enzyme reaction rates are investigated by measuring the amount of substrate changed, or the amount of product formed, in unit time. The properties of enzymes, and the ways in which their action is influenced by external conditions such as temperature and pH, reflect the fact that enzymes are proteins.

■ The effect of temperature

An investigation of the effects of temperature changes on enzyme action using amylase is shown in Figure 8.14. Several enzymes catalyse the breakdown of starch in plants, including α- and β-amylase and starch phosphorylase. (Starch phosphorylase degrades starch to glucose phosphate.)

'Amylase', as bought from a biological supplier or obtained as a crude extract from germinating seeds, is a mixture of enzymes. The mixture contains α- and β-amylase and maltase (p. 284).

Questions

1 In what ways do industrial inorganic catalysts differ from biological catalysts (enzymes)?
2 What is meant by the term 'active site'?
3 What are the main features that characterise the relationship between an enzyme and its substrate?

Amylase degrades starch to short chains of glucose residues about ten units long, known as dextrins, and eventually to maltose (a disaccharide) and glucose. Amylases catalyse the uptake of one molecule of water for each glucose–glucose bond cleaved, and so they are called hydrolytic enzymes.

In this study, samples of enzyme and substrate are brought to the temperature under investigation before they are mixed, and the rate of reaction is then determined at that temperature. With increased temperature there is increased kinetic energy of the molecules and hence greater reactivity. (That is, both the substrate molecules and the enzyme molecules are moving about more rapidly, and therefore have more chance of collisions.) In fact, most chemical reactions are speeded up by a rise in temperature; for most reactions, the rate roughly doubles for every 10 °C rise. So if we define the temperature coefficient (Q_{10}) of the reaction as

$$Q_{10} = \frac{\text{rate of reaction at } (x + 10) \text{ °C}}{\text{rate of reaction at } x \text{ °C}}$$

$Q_{10} \approx 2$ for most reactions.

■ Optimum temperature

Proteins are denatured by heat, however, and the rate of denaturation increases and becomes significant at higher temperatures. Heat denaturation really represents an irreversible destruction of the tertiary structure of an enzyme protein, changing its shape, and eventually destroying the active sites. As a result there is an 'apparent optimum temperature' for the action of amylase due to the dual effects of heat, firstly on the reaction and secondly on the stability of the enzyme protein (Figure 8.15).

■ The effect of pH

Small changes in the pH of the medium usually have dramatic effects on the efficiency of enzymes, and therefore upon the rate of reaction catalysed. Figure 8.16 shows how an investigation was set up to study the effect of pH change on the action of the digestive enzyme pepsin on heat-coagulated egg-white protein. The pH range studied was 1.5–4.5, using citric acid/sodium hydrogenphosphate buffer solution to adjust the pH. The samples were kept at body temperature throughout the experiment.

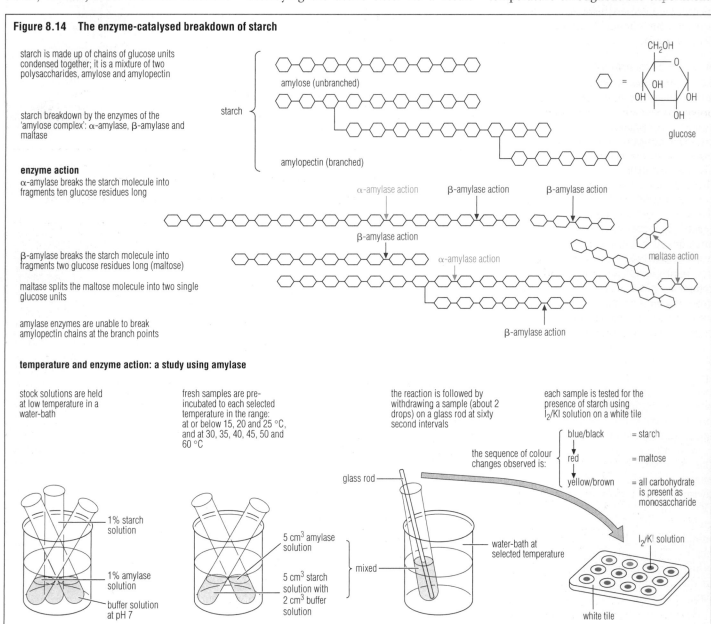

Figure 8.14   The enzyme-catalysed breakdown of starch

starch is made up of chains of glucose units condensed together; it is a mixture of two polysaccharides, amylose and amylopectin

starch breakdown by the enzymes of the 'amylose complex': α-amylase, β-amylase and maltase

amylose (unbranched)

amylopectin (branched)

starch

glucose

enzyme action

α-amylase breaks the starch molecule into fragments ten glucose residues long

β-amylase breaks the starch molecule into fragments two glucose residues long (maltose)

maltase splits the maltose molecule into two single glucose units

amylase enzymes are unable to break amylopectin chains at the branch points

α-amylase action   β-amylase action   β-amylase action

β-amylase action

α-amylase action

maltase action

β-amylase action

temperature and enzyme action: a study using amylase

stock solutions are held at low temperature in a water-bath

fresh samples are pre-incubated to each selected temperature in the range: at or below 15, 20 and 25 °C, and at 30, 35, 40, 45, 50 and 60 °C

the reaction is followed by withdrawing a sample (about 2 drops) on a glass rod at sixty second intervals

each sample is tested for the presence of starch using I₂/KI solution on a white tile

the sequence of colour changes observed is:

blue/black = starch

red = maltose

yellow/brown = all carbohydrate is present as monosaccharide

glass rod

1% starch solution

1% amylase solution

buffer solution at pH 7

5 cm³ amylase solution

5 cm³ starch solution with 2 cm³ buffer solution

mixed

water-bath at selected temperature

I₂/KI solution

white tile

The pepsin solution was added last to the medium, at which time the stopclock was started. The reaction was regarded as complete at the time at which the solution ceased to be at all cloudy (judged subjectively) – that is, the particles of egg-white had been completely digested to soluble peptides; this time was recorded. The rate of reaction in each tube was calculated, and the results are shown in Figure 8.16 (left-hand curve), together with those of a similar experiment using trypsin at higher pH values (right-hand curve).

The optimum pH is that at which the rate of reaction is at a maximum. At pH values only slightly above and below the optimum pH there is for many enzymes a marked fall-off in efficiency. Unlike the effects of heat on enzymes, however, the effects of pH are normally reversible, at least within limits. Restoring the pH to the optimum level usually restores the rate of reaction.

Amino acids, from which all proteins are made, contain both acidic ($-COO^-$) and basic ($-NH_3^+$) groups, and in a polypeptide chain there are many of these acidic and basic groups. The secondary and tertiary structure of the enzyme is maintained by bonds between these groups, and other interactions (Figure 8.20, p. 184). A change of pH in the environment of the protein alters the ionic charge of the acidic and basic groups, and causes the shape of the protein to change (the importance of protein shape is discussed in more detail overleaf). This change, and changes induced at the active site, may prevent the substrate molecule from making a very close fit to an active site. The enzyme-catalysed reaction fails to occur.

Questions

1 In studies of the effect of temperature on enzyme-catalysed reactions, why are the enzyme, substrate and buffer solutions pre-incubated to a particular temperature before they are mixed together?

2 Why is it correct to say an enzyme has an optimum pH but only an 'apparently optimum temperature'? What are the optimum pH values for pepsin and trypsin respectively?

3 How do you think the rate of an enzyme-catalysed reaction would be affected by an increase in the concentration of the enzyme? Explain your reasoning.

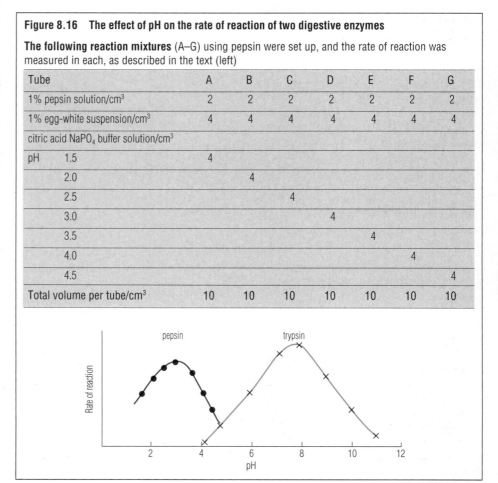

Figure 8.15   **Temperature and enzyme action:** an apparent optimum temperature

decrease in reaction rate due to denaturation of enzyme molecules

increase in reaction rate due to increased kinetic energy of substrate and enzyme molecules

actual rate of reaction as a result of the combined effects of these two influences

'apparent optimum temperature'

as temperature rises, more and more enzyme molecules are denatured and the rate of reaction appears to fall

rate of reaction

temperature/°C

enzyme in active state

active site

interpretation
at **lower temperatures** the enzyme is in an active state; the rate of reaction depends on the kinetic energy of the molecules

at **optimum temperature** the enzyme is still in an active state (shape)

the effect of **higher temperatures** is irreversibly to alter the shape and flexibility of the enzyme (protein) molecule and to change the chemical properties of the active site; reactions are no longer catalysed

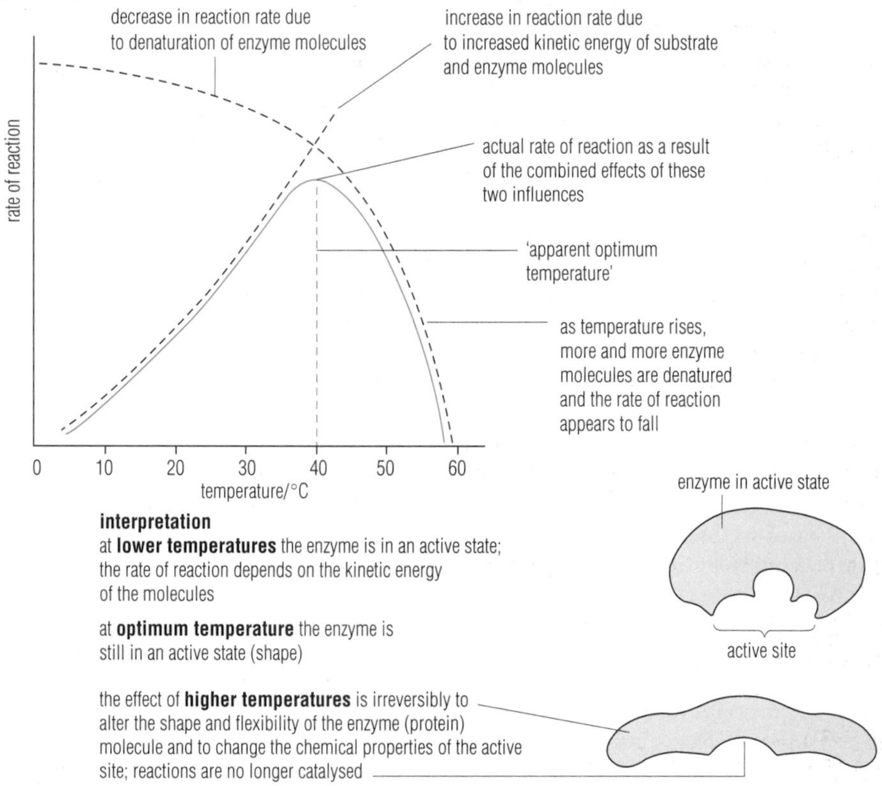

Figure 8.16   **The effect of pH on the rate of reaction of two digestive enzymes**

The following reaction mixtures (A–G) using pepsin were set up, and the rate of reaction was measured in each, as described in the text (left)

Tube		A	B	C	D	E	F	G
1% pepsin solution/cm³		2	2	2	2	2	2	2
1% egg-white suspension/cm³		4	4	4	4	4	4	4
citric acid NaPO₄ buffer solution/cm³								
pH	1.5	4						
	2.0		4					
	2.5			4				
	3.0				4			
	3.5					4		
	4.0						4	
	4.5							4
Total volume per tube/cm³		10	10	10	10	10	10	10

pepsin

trypsin

Rate of reaction

pH

■ The effects of substrate concentration

The effects of different concentrations of hydrogen peroxide on the rate of its enzyme-catalysed decomposition may be investigated as shown in Figure 8.17. At each substrate concentration it is the initial rate of reaction that is taken. This is because as soon as the reaction is under way the substrate concentration falls and products accumulate, causing the rate of reaction to fall off. So only the reaction rate at the start can be measured under the defined conditions.

The initial rate of reaction at each substrate concentration is plotted against substrate concentration. The resulting curve has two phases, even though the same amount of enzyme was present with each concentration of substrate. At lower substrate concentrations the rate of reaction increases in direct proportion to the substrate concentration, but at higher substrate concentration the rate of reaction becomes constant, and shows no increase.

The result can be interpreted in terms of the enzyme (E) forming a complex with the substrate (S):

$$E + S \rightleftharpoons ES \qquad (1)$$

This complex then breaks down to release the product (Pr):

$$ES \rightarrow E + Pr \qquad (2)$$

Reaction (1) is likely to be slightly faster than reaction (2). As the substrate concentration is increased the number of enzyme active sites that are initially 'engaged' would rise, up to a certain substrate concentration at which all sites are initially filled. Further increases in substrate concentration would not be able to speed up the initial rate of reaction. The hypothesis that enzymes work by forming a short-lived enzyme–substrate complex is thus supported by the shape of the curve obtained in the experiment.

8.7 THE MECHANISM OF ENZYME ACTION

The enzyme and substrate molecules are very different in size. Even if the substrate molecule is very large, such as a starch molecule, only one bond is in contact with the enzyme catalysing the reaction. Therefore any contact between enzyme and substrate can only be over a limited part of the surface of the enzyme. This region will have a certain 'shape', and will contain amino acid residues responsible for its specificity (making a shape that is complementary to that of the substrate), and also for its ability to catalyse some sort of chemical reaction.

In theory there could be many active sites on each protein molecule, but calculations from reaction rates and numbers of molecules catalysed show this is not the case. A few enzymes have more than one active site, but this is uncommon. Catalase is very unusual in having four active sites per molecule.

Evidence for an active site, which is only a small part of the whole enzyme protein, has come from X-ray diffraction studies (p. 203) of crystallised enzymes. It has been discovered that the surface of many enzyme proteins has an area that functions as a binding site for the substrate. The substrate molecule is precisely held in an optimum position to break or form a particular bond. It was not possible to study crystals of the

---

Figure 8.17 The effect of substrate concentration on an enzyme-catalysed reaction

The enzyme catalase catalyses the breakdown of hydrogen peroxide:

$$2H_2O_2 \xrightarrow{\text{catalase}} 2H_2O + O_2$$

Solutions containing different concentrations of substrate were prepared by diluting '20-volume' hydrogen peroxide solution. In the apparatus shown, using a fixed amount of enzyme solution for each experiment, the initial rate of oxygen production at a range of substrate concentrations was measured.

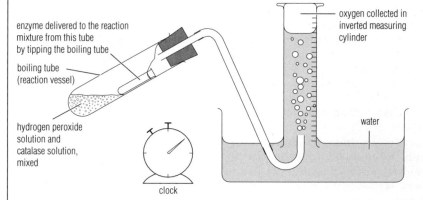

enzyme delivered to the reaction mixture from this tube by tipping the boiling tube

boiling tube (reaction vessel)

hydrogen peroxide solution and catalase solution, mixed

clock

oxygen collected in inverted measuring cylinder

water

only the initial rate of each reaction was used; this is found by measuring the gradient of a tangent to the first part of the curve

then the initial rate of reaction was plotted against the substrate concentration (see discussion in text)

interpretation

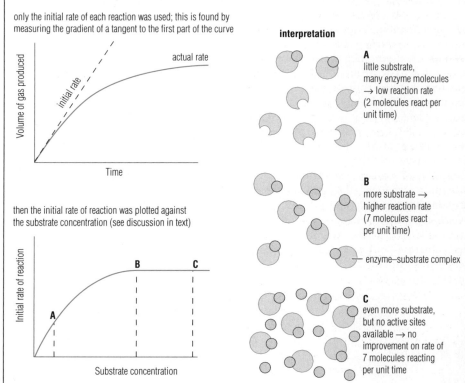

A little substrate, many enzyme molecules → low reaction rate (2 molecules react per unit time)

B more substrate → higher reaction rate (7 molecules react per unit time)

enzyme–substrate complex

C even more substrate, but no active sites available → no improvement on rate of 7 molecules reacting per unit time

enzyme–substrate complex, because the substrate was converted into the product too quickly. Rather, the crystalline complex studied was that of a structurally similar compound that combines with the active site but is not broken down there, such as a competitive inhibitor (p. 185). For many enzymes, however, no such compound is known.

In this model of an enzyme, the enormous polypeptide chain is visualised as consisting of many amino acid residues forming a sort of scaffolding that maintains the three-dimensional structure of the protein, a small part of this being an 'active site' (Figure 8.20).

■ Enzyme specificity: the 'lock and key' hypothesis

Enzymes are highly specific in the reactions catalysed. Some enzymes catalyse the transformation of one particular type of substrate molecule or, at most, a very restricted group of substrate molecules; some catalyse only one type of chemical change. This degree of specificity distinguishes enzymes from all other types of catalyst.

The specificity of enzymes is due to the configuration of the active site. This idea was originally developed as the 'lock and key' hypothesis (Figure 8.18): the substrate is the 'key' that fits exactly the enzyme 'lock'.

■ 'Induced fit' at the active site

Evidence from protein chemistry suggests that a small rearrangement of chemical groups occurs in both the enzyme and the substrate molecules when the enzyme–substrate complex is formed. It is suggested that the strain of these changes may play a part in the catalytic process; it may help induce a chemical change. This suggestion, which is called the 'induced fit hypothesis' (Figure 8.19), seems to offer a better model to explain enzyme action, given the available evidence. The modern view is that proteins are rather wobbly, flexible molecules and the change in shape on formation of an enzyme–substrate complex reflects this.

Questions

1 In an enzyme-catalysed reaction *in vitro*, why is the initial rate of reaction not maintained for long?
2 When all active sites of the enzyme are occupied, how does increasing the substrate concentration affect the rate of reaction?
3 Explain the difference between the 'lock and key' and 'induced fit' hypotheses of enzyme action.

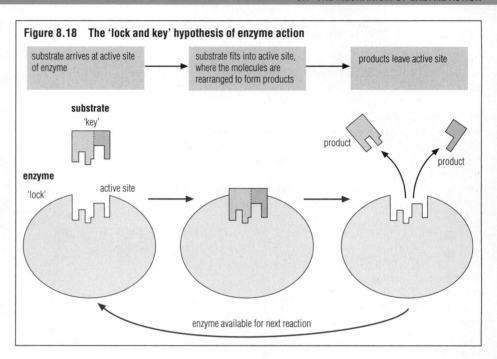

Figure 8.18   The 'lock and key' hypothesis of enzyme action

substrate arrives at active site of enzyme → substrate fits into active site, where the molecules are rearranged to form products → products leave active site

substrate 'key'

enzyme 'lock'   active site

product

product

enzyme available for next reaction

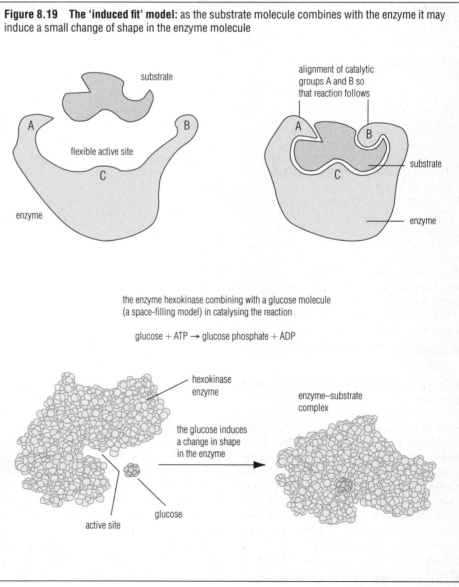

Figure 8.19   The 'induced fit' model: as the substrate molecule combines with the enzyme it may induce a small change of shape in the enzyme molecule

substrate

A   B

flexible active site

C

enzyme

alignment of catalytic groups A and B so that reaction follows

A   B

C

substrate

enzyme

the enzyme hexokinase combining with a glucose molecule (a space-filling model) in catalysing the reaction

glucose + ATP → glucose phosphate + ADP

hexokinase enzyme

enzyme–substrate complex

the glucose induces a change in shape in the enzyme

active site

glucose

Figure 8.20 The different roles of amino acid residues in an enzyme

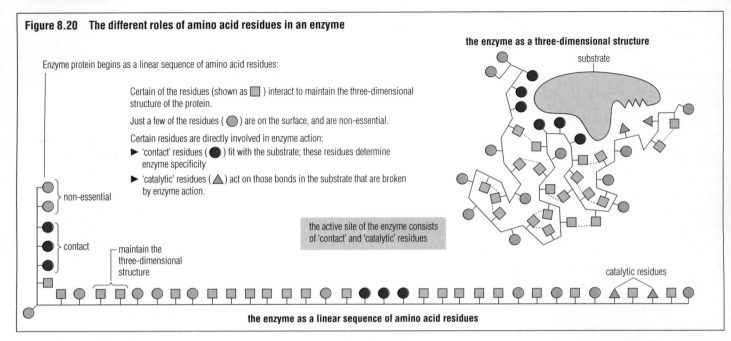

the enzyme as a three-dimensional structure

substrate

Enzyme protein begins as a linear sequence of amino acid residues:

Certain of the residues (shown as ▪) interact to maintain the three-dimensional structure of the protein.

Just a few of the residues (●) are on the surface, and are non-essential.

Certain residues are directly involved in enzyme action:
► 'contact' residues (●) fit with the substrate; these residues determine enzyme specificity
► 'catalytic' residues (▲) act on those bonds in the substrate that are broken by enzyme action.

the active site of the enzyme consists of 'contact' and 'catalytic' residues

non-essential

contact

maintain the three-dimensional structure

catalytic residues

the enzyme as a linear sequence of amino acid residues

■ What happens at the active site?

At a given temperature the molecules in a reaction mixture have a range of energies, but only a few molecules will have enough energy to react. This energy is called the **energy of activation**; it is an energy barrier to a reaction occurring. If the energy of activation is lowered, then more molecules can react (Figure 8.12).

The enzyme–substrate complex formed at the active site, albeit only momentarily, decreases the energy of activation of the reaction. How? Possibly because reactants are held close together in an exposed position, possibly because of strains or tensions created within substrate molecules, and possibly because of special chemical conditions at the active site (Figure 8.21).

■ Altering the rate of enzyme-catalysed reactions

The actions of enzymes may be enhanced or inhibited by various substances, some formed in the cell and others absorbed from the external environment. Studying the effects of these substances has sometimes yielded information on the structure and shape of enzymes and on the reactive groups of their active sites. Enzymologists may investigate the effects of inhibitors of various sorts in order to find out how enzymes work, but the study may also have commercial and medical aims too. Certain drugs and pesticides alter metabolism by enzyme inhibition.

Figure 8.21 How an enzyme may lower the activation energy

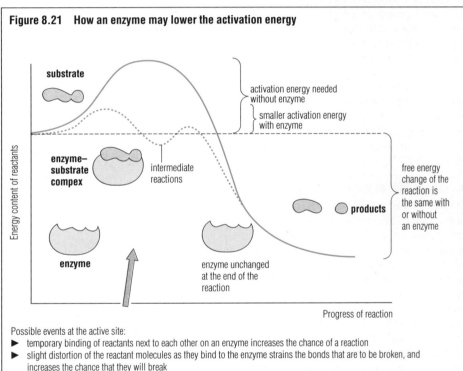

substrate

activation energy needed without enzyme

smaller activation energy with enzyme

enzyme–substrate compex

intermediate reactions

free energy change of the reaction is the same with or without an enzyme

products

enzyme

enzyme unchanged at the end of the reaction

Progress of reaction

Energy content of reactants

Possible events at the active site:
► temporary binding of reactants next to each other on an enzyme increases the chance of a reaction
► slight distortion of the reactant molecules as they bind to the enzyme strains the bonds that are to be broken, and increases the chance that they will break
► reactants are held by the enzyme in such a way that bonds are exposed to attack
► hydrophobic amino acids create a water-free zone in which non-polar reactants may react more easily
► acidic and basic amino acids in the enzyme facilitate the transfer of electrons to and from the reactants

Inhibitors

Irreversible inhibitors bind tightly and permanently to enzymes and destroy their catalytic properties. These drastic effects occur at very low concentrations of inhibitor, and we describe these substances as poisons.

Heavy-metal ions such as mercury, silver and lead are described as irreversible, **non-competitive inhibitors**.

These ions combine with sulphydryl (—SH) groups, forming strong covalent bonds. The sulphydryl groups are an integral part of protein structure. Only very rarely are they located in the active sites of enzymes (although the enzyme papain has an —SH group in the active site, inhibited by Hg^{2+}); in general, heavy-metal ions do not compete with substrate molecules for the active sites.

Reversible inhibitors bind less tightly to an enzyme, and some occur naturally in cells, where they appear to have a role in the regulation of metabolism (see Figure 8.22).

Some reversible inhibitors, known as **competitive inhibitors**, are substances with a close structural resemblance to the substrate of the enzyme. A competitive inhibitor competes with the substrate molecule for the active site of the enzyme. Because the inhibitor is not acted on by the enzyme it may then remain bound to it, excluding substrate molecules from the active site whilst it remains attached. This inhibition is classified as 'competitive' because, if the substrate concentration is raised to a sufficiently high level, the inhibitor is progressively replaced at active sites by the substrate.

An example of competitive inhibition is the action of malonate on succinate dehydrogenase (Figure 8.23). The enzyme ribulose bisphosphate carboxylase, which catalyses the reaction between carbon dioxide and the acceptor molecule in photosynthesis, is competitively inhibited by oxygen (p. 267).

Non-competitive reversible inhibitors, on the other hand, bear no resemblance to the substrate molecule. Nevertheless these inhibitors bind to the enzyme at or near the active site, reducing its rate of reaction and causing it to work less effectively. The inhibitor is not covalently bonded to the enzyme, however, and is not competing with the substrate for the active site. The cyanide ion is a non-competitive inhibitor of cytochrome oxidase, an enzyme involved in terminal oxidation in the hydrogen transport system of aerobic respiration (Figure 11.21, p. 239).

Investigating inhibitor effects by experiment

You can tell the difference between competitive and non-competitive inhibitors by their effects in enzyme-catalysed reactions in which the substrate concentration is increased (Figure 8.24). At low concentrations of substrate both inhibitors slow the rate of reaction. At higher substrate concentration the rate of reaction in the presence of a non-competitive inhibitor does not reach the maximum rate. This is because the non-competitive inhibitor keeps some of the enzyme 'out of action'. In effect, the enzyme concentration is lowered. Meanwhile, a competitive inhibitor is largely replaced at the active sites by substrate molecules, when the substrate is at high concentration.

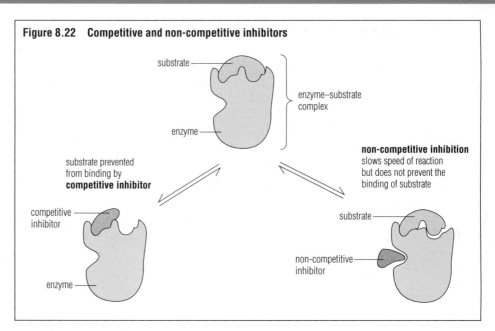

Figure 8.22   Competitive and non-competitive inhibitors

Figure 8.23   An example of competitive inhibition: succinate dehydrogenase catalyses the oxidation of succinate to fumarate (succinate and fumarate are intermediates of the Krebs cycle, p. 315); malonate is a competitive inhibitor of the enzyme

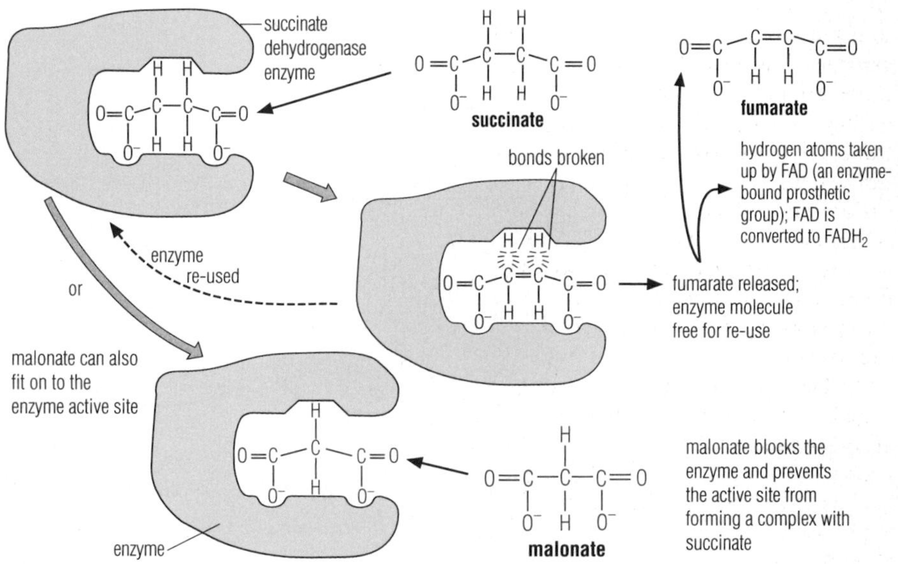

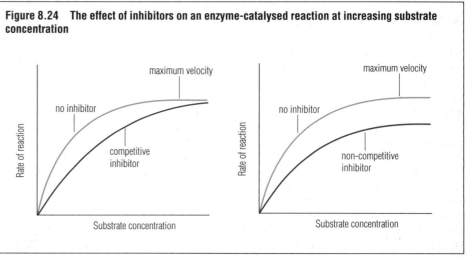

Figure 8.24   The effect of inhibitors on an enzyme-catalysed reaction at increasing substrate concentration

Figure 8.25   Allosteric activation and inhibition

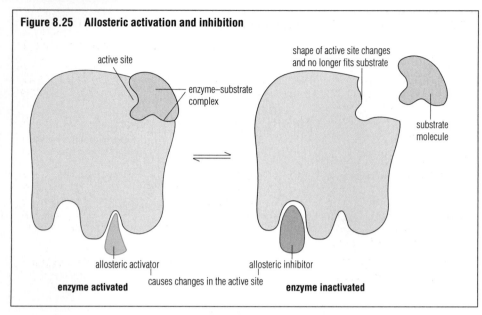

active site

enzyme–substrate complex

shape of active site changes and no longer fits substrate

substrate molecule

allosteric activator

enzyme activated

causes changes in the active site

allosteric inhibitor

enzyme inactivated

Figure 8.26   End-product regulation of a pathway: a case of allosteric inhibition

End-product inhibition occurs in the synthesis of certain amino acids.
In most cases of end-product inhibition, the enzyme catalysing the first step in the reaction sequence is inhibited by the end product. This prevents the accumulation of unnecessary intermediates, and is an example of negative feedback.

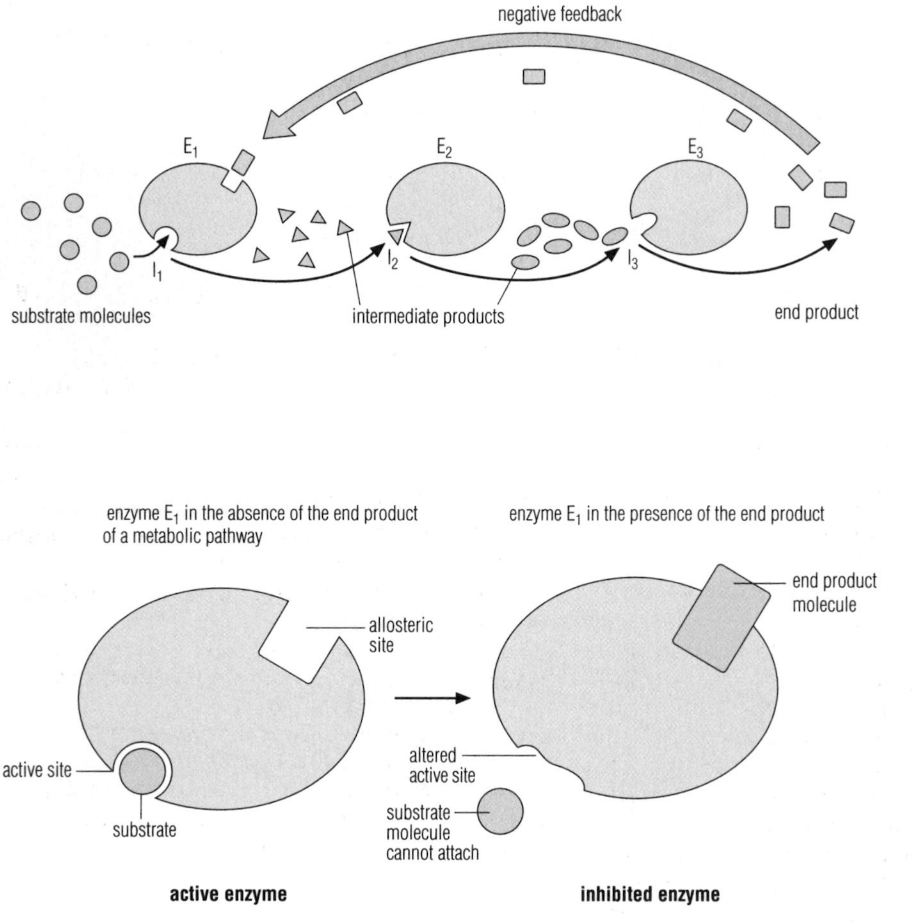

negative feedback

E_1

E_2

E_3

I_1

I_2

I_3

substrate molecules

intermediate products

end product

enzyme E_1 in the absence of the end product of a metabolic pathway

enzyme E_1 in the presence of the end product

allosteric site

end product molecule

active site

substrate

altered active site

substrate molecule cannot attach

active enzyme

inhibited enzyme

Allosteric effectors

Allosteric effectors are substances present in cells that reversibly bind with an enzyme away from the active site, yet cause a change in the structure of the active site. The activity of the enzyme is regulated in this way. Allosteric activators speed up catalysis; allosteric inhibitors slow down the reaction and provide a way of controlling enzymic activity in metabolism.

The enzyme phosphofructokinase catalyses the phosphorylation of fructose phosphate by ATP to give fructose bisphosphate. When ATP is at high concentration – for example, when the mitochondria are actively producing it – it inhibits this enzyme (ATP is an allosteric inhibitor) and glycolysis is inhibited. When ATP starts to be used up in the cell – as it is when muscle is contracting – its concentration falls and glycolysis is no longer inhibited. Allosteric modulation is illustrated in Figures 8.25 and 8.26.

8.8 ENZYMES AND THE CONTROL OF METABOLISM

Although many thousands of different reactions make up the metabolism of cells, the overall direction of metabolism is not chaotic. The biochemistry of the cell is regulated and controlled. In addition to catalysing the individual reactions in metabolism, enzymes play a part in the control of metabolism as a whole. How is this achieved?

■ Allosteric modulation of enzymes

Enzyme action may be allosterically activated or inhibited by the products of a metabolic sequence, or by other intermediates. Negative feedback control (Figure 8.26) of metabolism is made possible in this way.

■ The position of enzymes in the cell

Some enzymes are isolated in particular organelles, and even in specialised membranes within that organelle. Most of the 5000 or more cellular enzymes are not freely exposed to a general pool of metabolites. Substrate molecules are selectively transported (p. 236) across the membranes of the organelles to particular enzyme systems.

■ The specificity of enzymes

Enzymes are highly specific in the reactions they catalyse. Enzymes are also extremely efficient; that is, a tiny quantity of enzyme catalyses the reaction of a great deal of substrate. So, the production of relatively few molecules of a specific enzyme immediately enhances a particular reaction in metabolism. This brings about a qualitative change in metabolism, rather than enhancing metabolism in general.

■ The control of enzyme synthesis

Some of the proteins of the cell (including some enzyme proteins) are broken down to their constituent amino acids after a relatively brief 'life-span' (a short half-life). From the pool of amino acids in the cytoplasm new proteins are constantly produced. Protein synthesis is directly under the control of the messenger RNA in the cytoplasm (p. 208), and indirectly under the control of the nucleus. The nucleus controls and directs the growth and development of the cell and its metabolism: this is achieved by controlling which enzymes are present at any time, and therefore which chemical changes can occur. The factor that is important is therefore the turnover or half-life of the enzyme

■ The affinity of enzymes for their substrates

Different enzymes have different affinities (attractive powers) for their substrate molecules. Each enzyme has a particular substrate concentration that sustains maximum velocity of the reaction. In 1913 the biochemists Michaelis and Menten studied this aspect of enzyme reactions and introduced a constant, now known as the Michaelis constant. The **Michaelis constant** (K_m) is defined as the substrate concentration that sustains half maximum velocity. The K_m

▶ can be experimentally determined, at a specified pH and temperature
▶ is expressed in units of molarity
▶ is independent of enzyme concentration.

The actual values of K_m have been measured for very many enzymes. They fall in the range of 10^{-2} to 10^{-5} mol dm^{-3} of substrate. This is a very wide difference in concentration. It means that some enzymes are able to work at maximum velocity at incredibly low concentrations of substrate, whilst others only function effectively at much higher concentration.

THE NAMING AND CLASSIFICATION OF ENZYMES

In the history of biology enzymes have been named in several different ways. For example, enzymes have been named

▶ arbitrarily, at the time of discovery; e.g. the digestive enzymes trypsin, pepsin, rennin and ptyalin
▶ because of the property of the product formed; e.g. invertase (sucrase) forms invert sugar (a reference to optical properties)
▶ by adding the suffix -ase to the root of the name of the substrate; e.g. urease, which acts on urea
▶ by the type of reaction catalysed; e.g. dehydrogenases catalyse the removal of hydrogen.

Modern classification of enzymes

To avoid the continued haphazard naming of enzymes the International Union of Biochemistry introduced a classification in 1961. This scheme is based on the type of reaction the enzyme catalyses. Accordingly, enzymes are placed into six classes, and these six classes are further subdivided, first according to the substrate acted on. This systematic renaming of enzymes has not entirely caught on for everyday uses; the value of the scheme is that when enzymologists need to achieve absolute precision, the means to do so exist.

oxidoreductases catalyse biological oxidation and reduction by the transfer of hydrogen, oxygen or electrons from one molecule to another; e.g.

$$\text{ethanal} + \text{NADH}_2 \xrightarrow{\text{alcohol dehydrogenase}} \text{ethanol} + \text{NAD (p. 316)}$$

transferases catalyse the transfer of a chemical group from one substrate to another; e.g.

$$\text{glutamic acid} + \text{pyruvic acid} \xrightarrow{\text{aminotransferase}} \alpha\text{-ketoglutaric acid} + \text{alanine (p. 209)}$$

hydrolases catalyse the formation of two products from a larger substrate molecule by hydrolytic reaction; e.g.

$$\text{sucrose} \xrightarrow{\text{sucrase}} \text{fructose} + \text{glucose (p. 134)}$$

lyases catalyse non-hydrolytic addition or removal of parts of substrate molecules; e.g.

$$\text{pyruvic acid} \xrightarrow{\text{pyruvate decarboxylase}} \text{ethanal} + \text{CO}_2 \text{ (p. 315)}$$

isomerases catalyse internal rearrangements of substrate molecules (that is, isomerisation); e.g.

$$\text{glucose-1-phosphate} \xrightarrow{\text{phosphoglucomutase}} \text{glucose-6-phosphate (p. 177)}$$

ligases catalyse the joining together of two molecules with the simultaneous hydrolysis of ATP; e.g.

$$\text{amino acid} + \text{specific tRNA} + \text{ATP} \xrightarrow{\text{amino-acyl–tRNA synthetase}} \text{amino acid–tRNA complex} + \text{ADP} + \text{P}_i \text{ (p. 208)}$$

Take the case of two (or more) enzymes that catalyse the transformation of the same substrate molecule but in different reaction sequences. There are several instances of this in cells: see, for example, Figure 8.6 (and Figure 12.35, p. 264), where different enzyme systems compete for the substrate molecule glycerate 3-phosphate (GP). If the reserves of that substrate are low, and its supply is restricted, then the enzyme with the lowest K_m will claim more of the GP. One particular metabolic pathway will benefit at the expense of the others.

Knowing the K_m of enzymes is an important part of knowing about metabolism, quantitatively and qualitatively.

Questions

1 Why are several enzymes needed in a typical metabolic pathway?

2 Into what category of enzymes would you place pepsin?

3 How might a stomach cell increase its production of pepsin in response to the presence of food in the stomach?

4 In what ways might a cell reduce its production of a metabolite?

5 How does the effect on an enzyme of an irreversible inhibitor differ from that of an allosteric inhibitor?

8.9 ENZYME TECHNOLOGY

Enzymes can function outside the cells in which they are produced, and sometimes do so quite naturally. For example, in 1897 Edward Buchner obtained a cell-free extract from yeast and used it to ferment sugar to alcohol and carbon dioxide, exactly as intact yeast cells do. Today, many different enzymes have been routinely extracted from cells and used in industrial or medical processes. This area, enzyme technology, is one aspect of **biotechnology** (p. 112). Sometimes organisms, normally microorganisms, are engineered to produce a particular enzyme, so **genetic engineering** (p. 628) may be an aspect of enzyme technology, too. Examples of common industrial and medical uses of enzymes are given in Table 8.2.

■ Obtaining enzymes

Enzymes for industrial uses have been obtained from animals (for example, rennin from the stomachs of veal calves) and from flowering plants (for example, amylases from germinating barley). However, the majority of enzymes are obtained from microorganisms, mainly from bacteria and fungi. The microorganisms are selected strains, cultured in batches using liquid media (batch culture, p. 115). The steps to production of an extracellular enzyme (an enzyme secreted by a microorganism into its substrate) are illustrated in Figure 8.27. The **advantages** of this source of enzymes for industrial uses include the following:

▶ microorganisms have high growth rates and produce a higher proportion of enzyme in relation to body mass than most eukaryotic organisms, from a given quantity of substrate

▶ microorganisms can usually be cultured economically, using relatively low cost substrates for their nutrition, such as whey from the dairy industry, molasses from sugar refinement, or waste starch from flour milling

Table 8.2 Industrial and medical uses of enzymes: many of today's uses of enzymes are recent developments, but the use of enzyme preparations in industry has a very long history

Application	Enzyme	Use, and notes
1 Dairy industry	rennin from calves, lambs etc.	used in cheese manufacture to coagulate milk proteins; rennin of microbial origin is now commonly used, obtained from genetically engineered bacteria
2 Brewing industry	amylase from germinating barley, protease	breakdown of starch to glucose for fermentation by yeast in brewing, and breakdown of proteins to amino acids for benefit of yeast; enzymes of microbial origin are now used to prevent cloudiness during storage of beers (protease) and to produce 'low calorie' beers (amyloglucosidase)
3 Baking industry	protease	breakdown of proteins in flour for biscuit manufacture
	amylase	breakdown of some starch to glucose in flour for white bread, buns and rolls
4 Biological detergents	protease, lipase, amylase	removal of organic stains (e.g. food, saliva, blood) from clothes
	amylase	removal of starch residues in dishwashing machines
5 Leather tanning industry	protease in dog and pigeon faeces	treatment (called 'bating') of hides to remove hair and make the leather pliable; now replaced by enzymes obtained from slaughter houses and (most recently) of microbial origin
6 Textile industry	amylase	removal of the starch that is applied to the threads of some fabrics to protect from mechanical damage during weaving
7 Processed foods	amylase	manufacture of glucose syrup from starch
	glucose-isomerase	manufacture of fructose syrup from glucose, for 'low calorie' sweetening of manufactured foods (USA and Japan)
	trypsin	pre-digestion of some baby foods
8 Forestry and paper industry	ligninases from fungi	removal of lignin from pulverised wood, prior to use of wood cellulose in manufacturing processes
	amylase	partial breakdown of starch for sizing of paper – filling the gaps between the fibres to produce a smooth, 'quality' paper
9 Photographic industry	protease	digestion of gelatin of old film to allow recovery of silver (current 'film' and photographic 'papers' are actually plastics)
10 Medical/ pharmaceutical uses and analytical chemistry	trypsin	removal of blood clots, and in wound cleaning
	various enzymes	used in biosensors, e.g. for blood glucose etc.

Figure 8.27   The production of an extracellular enzyme

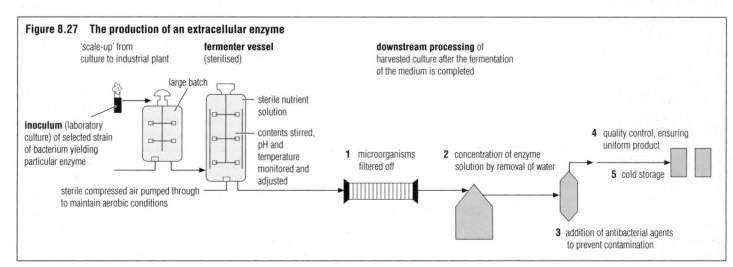

'scale-up' from culture to industrial plant

fermenter vessel (sterilised)

downstream processing of harvested culture after the fermentation of the medium is completed

large batch

inoculum (laboratory culture) of selected strain of bacterium yielding particular enzyme

sterile nutrient solution

contents stirred, pH and temperature monitored and adjusted

sterile compressed air pumped through to maintain aerobic conditions

1 microorganisms filtered off

2 concentration of enzyme solution by removal of water

3 addition of antibacterial agents to prevent contamination

4 quality control, ensuring uniform product

5 cold storage

Figure 8.28   Immobilisation of enzymes

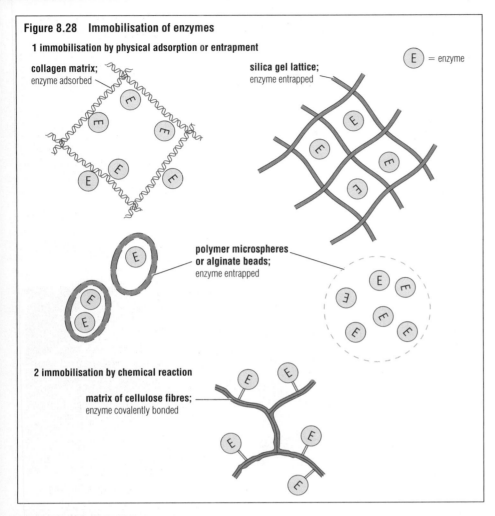

1 immobilisation by physical adsorption or entrapment

collagen matrix;
enzyme adsorbed

silica gel lattice;
enzyme entrapped

E = enzyme

polymer microspheres
or alginate beads;
enzyme entrapped

2 immobilisation by chemical reaction

matrix of cellulose fibres;
enzyme covalently bonded

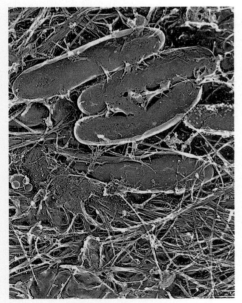

Figure 8.29   Immobilised bacterial cells in alginate beads. Alginate is a polysaccharide extracted from certain seaweeds. Bacteria rich in a required combination of enzymes are mixed with alginate solution and turned into pellets by extruding the mixture as droplets into calcium chloride solution. The pellets can be used to catalyse reactions, by packing them into a column and passing through a solution of the reactants

► bacteria, in particular, can be genetically engineered to contain novel genes and to have enhanced yields of particular proteins; for example, it is common for copies of genes obtained from higher plants or animals (including humans) or from other microorganisms to be added into a bacterial chromosome in order to produce a particular protein

► microorganisms occur naturally in widely different habitats, and some are able to survive and grow at particular extremes of temperature or pH; the enzymes of these organisms are fully functional at these extremes, making them particularly useful in biotechnology under certain industrial conditions.

■ Immobilised enzyme preparations

Isolated enzymes can be attached to, or retained within, an insoluble support material. The reaction mixture on which the enzyme acts is then passed over or through this framework. Enzymes held in these ways are called immobilised enzymes. The advantages of immobilisa-

tion are that the enzyme is available to be re-used for as long as enzymic activity is maintained, and the products of the reaction are not contaminated by the enzyme. This is particularly important when the enzyme was expensive to extract in the first place, and where the products might be degraded by the steps needed to remove enzyme from them. The methods of immobilisation may involve physical adsorption or chemical reaction with the inert framework materials, and are summarised in Figure 8.28.

■ Cell-free enzymes and immobilised whole cells

Most of the enzymes used on an industrial scale are normally secreted by microorganisms and naturally act upon their substrate outside the cell (extracellular enzymes). Cell-free enzymes like these are relatively easy to obtain in bulk. **Isolated free enzyme** is sometimes added to a reaction mixture in industrial processes, too, with no attempt made to reclaim enzyme on completion of the task for which it was selected (as in domestic detergents, for example; see overleaf). An advantage of this method of use may be a much higher level of enzyme activity than is normally shown by immobilised enzymes. A loss of enzymatic activity due to immobilisation may occur because the enzyme is attached near to its active site, or because the immobilising framework hinders the arrival of substrate molecules and the escape of product molecules from the active site.

Another approach in biotechnology is the use of immobilised whole cells (Figure 8.29). This is advantageous where the enzymes are expensive and difficult to extract from cells in an active form. It may be useful, too, where a sequence of interconnected reactions is being catalysed by a group of related enzymes.

However, the use of **whole-cell preparations** has the disadvantage that a substantial portion of the substrate is converted into bacterial biomass, whereas with cell-free enzymes there are no 'wasteful' side reactions possible because other enzymes are not present. Also, the optimum conditions to produce the product sought may not be the optimum conditions for the growth of the cell as a whole.

In fact there is now a rapid proliferation in the range of enzymes available, and in the uses to which they are being put. Undoubtedly, methods of use will develop apace, too.

■ Industrial uses of enzymes

Biological washing powders

The enzymes in washing powders have to be able to work in alkaline conditions and at temperatures normally between 30 and 90 °C. The genes coding for these proteins have mostly been obtained from heat-loving (thermophilic) bacteria such as occur naturally in hot springs or in compost heaps (p. 297). The first enzymes used by this industry were dry powders, and were found to cause allergic reactions in the factory workers during manufacture. This was overcome by encapsulating the enzymes. Today the marketing of detergents in liquid form also helps to avoid this problem.

Now, biological detergents are more important than ever before because they make for efficient cleaning at low temperatures (saving power in the washing process) and at low phosphate concentrations (reducing the dangers of the later eutrophication of rivers and coastal waters by sewage effluents, p. 101). Also, as a result of the wider use of enzymes in detergents, the products are milder in their effects on delicate fabrics and a wider range of materials may be cleaned by washing machines.

The enzymes used are **proteases** that degrade proteins, such as those from food residues, saliva, and blood stains in clothing, together with **amylases** to degrade starch residues, and **lipases** to remove grease. More recently, **cellulase** has been used to remove the microfibrils that form over the surface of cotton threads during machine washing. This brightens the colours and softens the cloth (Figure 8.30).

Fruit juice production

Fruit juice is extracted by crushing the fruit and pressing the pulp. Cellulose in cell walls is tough and resists mechanical damage. Degradation of the cell walls of fruit during commercial juice extraction is aided by the addition of microbial **cellulase** and **hemicellulase** enzymes. Much more juice is released by the use of these enzymes.

The raw juice always contains pectins (p. 136). These complex carbohydrates are present in the cellulose of the walls and in the middle lamella between cells, and exist in colloidal suspension in juice. The extracted juice is treated using **immobilised pectinase**. Enzymic conversion of pectins to short-chain sugars increases the volume of juice extracted. Removal of pectins also speeds up

---

Figure 8.30   Enzymes used in biological washing powders and their effects

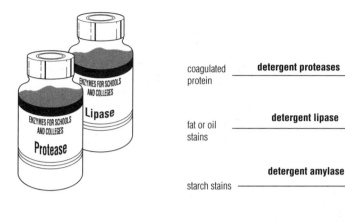

coagulated protein	→ **detergent proteases** →	soluble short-chain peptides
fat or oil stains	→ **detergent lipase** →	soluble fatty acids and glycerol
starch stains	→ **detergent amylase** →	soluble shorter-chain polysaccharides and sugars

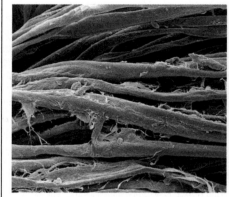

SEM of cotton fibres before and after treatment with a cellulase used for fabric conditioning

---

filtering of the juice during manufacture, and prevents it from becoming cloudy on cold storage.

Fructose syrup

Fructose corn syrup is a sweetener used widely in the USA in food and drinks. It is manufactured from starch obtained from corn (maize). The grains are milled to a starch slurry, and the enzyme amylase is added to produce a glucose syrup. This is then decolourised and concentrated. Finally, this syrup is passed down a column of immobilised **glucose isomerase** enzyme (an enzyme of the glycolysis pathway in all organisms) where the glucose is converted to fructose. Fructose has the same energy content as glucose, yet it is far sweeter to human taste buds than glucose (or sucrose), so far less is needed. Fructose corn syrup is used to sweeten manufactured foods without adding too many 'calories'.

Meat tenderisation

Meat is largely muscle protein, but is surrounded by connective tissue (Figure 22.16, p. 489), and it contains collagen fibres, which are tough and non-elastic. After death, natural 'tenderisation' occurs as the lysosomes of the dead cells release their hydrolytic enzymes and these bring about partial autolysis of muscle tissue, including the connective tissue and collagen. This natural tenderisation process is exploited when meat is 'hung' in cold storage before being cut into joints for sale.

An alternative approach is to use proteolytic enzymes; these are injected into the bloodstream prior to slaughter so that they are distributed to all parts of the muscle. Meat treated in this way can be eaten next day, so storage costs are saved. Animal welfare groups question the confidence expressed by advocates of this technique that the animal is caused no suffering, and so they object strongly to the practice. The enzymes are drained from the carcass after death, with the blood. Enzymes for meat tenderisation are obtained from the papaya or pawpaw plant (giving papain), from the pineapple, and from the fig. All three plants are rich in proteolytic enzymes.

Figure 8.31 The production of lactose-free milk

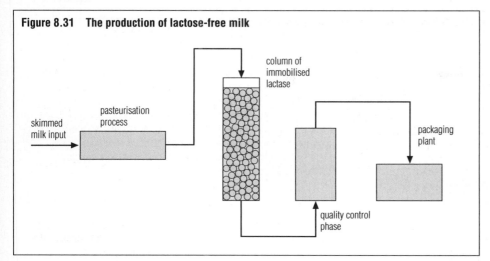

Figure 8.32 A hand-held biosensor for urea

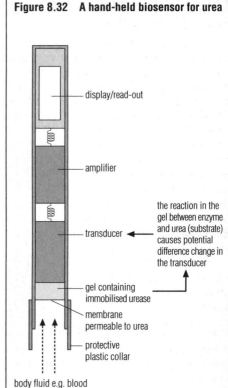

Lactose-free milk

Milk sugar, lactose, is normally hydrolysed to glucose and galactose in the ileum by the enzyme lactase, as it is absorbed into the bloodstream (Figure 13.18, p. 286). Some people do not have the necessary enzyme and cannot therefore digest lactose. Instead, this disaccharide remains in the gut, and causes most unpleasant reactions of diarrhoea, flatulence, and abdominal pain as it is digested by gut bacteria. Lactose-free milk is manufactured by passing milk through a column of immobilised lactase enzyme (Figure 8.31).

Biosensors

The specificity and sensitivity of enzymes is exploited in enzyme-based biosensor devices. Such devices are now being designed and used for the rapid and accurate detection of tiny traces of biologically important molecules.

In the biosensor designed to measure the quantities of urea in blood or urine samples (Figure 8.32), immobilised urease enzyme is housed in a supporting gel, protected by a partially permeable membrane. Urea in the sample is catalytically converted to carbon dioxide and ammonium ions. The ammonium ions are detected by a transducer, and the signal is amplified and displayed on a read-out. The amount of ammonium ions released (and therefore the strength of the electric current produced) in a fixed time is proportional to the concentration of urea in the solution.

The test strips used to measure glucose in urine samples are another example of the use of enzymes in the analysis of small quantities of biochemicals. These strips contain two enzymes – glucose oxidase and peroxidase – and a colourless hydrogen donor compound called chromogen. The strip is dipped into the urine sample. If glucose is present, then it is oxidised to gluconic acid and hydrogen peroxide. The second enzyme catalyses the reduction of hydrogen peroxide and the oxidation of chromogen. The product is water and the oxidised dye, which is coloured. The more glucose present in the urine, the more coloured dye is formed. The colour of the test strip is then compared to the printed scale to indicate the amount of glucose in the urine (Figure 8.33).

Figure 8.33 The urine glucose test: glucose is found in the urine of patients with diabetes (p. 506)

The principle involved in measuring glucose in urine using test

$$glucose + oxygen \xrightarrow{\text{glucose oxidase}} gluconic\ acid + H_2O_2$$

$$\underset{\substack{\text{reduced} \\ \text{chromogen} \\ \text{(colourless)}}}{DH_2 + H_2O_2} \xrightarrow{\text{peroxidase}} \underset{\substack{\text{chromogen} \\ \text{(coloured)}}}{2H_2O + D}$$

The process: test strip dipped into urine sample

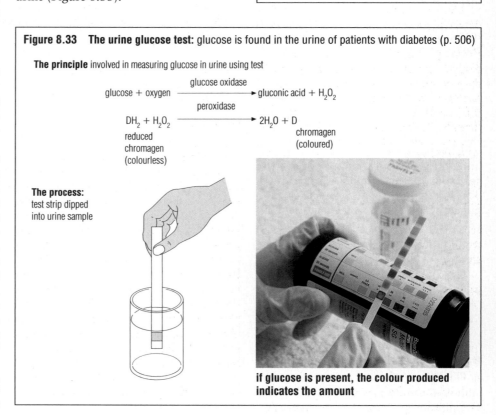

if glucose is present, the colour produced indicates the amount

FURTHER READING/WEB SITES

E J Wood, C A Smith and W R Pickering (1997) *Life Chemistry and Molecular Biology.* Portland Press
The Biochemical Society: **www.biochemsoc.org.uk**

9 The nucleus in division and interphase

- The nucleus contains information in its chromosomes. Chromosomes control cell activity, and are passed to new cells at cell division, and to new individuals in sexual reproduction. When cells divide the nucleus divides first.
- Mitosis is a division of the nucleus to produce two daughter cells containing chromosomes identical to the parent nucleus. Mitosis is associated with growth and with asexual reproduction.
- Meiosis is a division of the nucleus to produce four daughter nuclei each containing half the chromosome number of the parent nucleus. Meiosis is associated with sexual reproduction.

- Chromosomes contain a long, double-stranded DNA molecule in the form of a helix. Before the chromosome divides, its DNA strands separate and are copied (replicated) to form identical strands.
- The information in DNA is known as the genetic code, and takes the form of a sequence of bases. Three bases form a unit of information, coding for a particular amino acid in cell protein synthesis.
- Protein synthesis takes place in the cytoplasm using a copy of information from the nucleus.

9.1 INTRODUCTION

The nucleus is the largest organelle in the cell, and can normally be observed by light microscopy (see Figure 7.5, p. 153). By means of the transmission electron microscope (p. 154) it has been shown that the nucleus is bounded by a double membrane. This **nuclear membrane** is perforated by many tiny **pores**, about 50 nm in diameter. Pores are routes of communication between nucleus and cytoplasm.

Inside the nucleus are **chromosomes**. Chromosomes contain DNA. In non-dividing cells the chromosomes are very long and thin, and are dispersed as **chromatin**. Also present are one or more bodies known as **nucleoli**. A nucleolus has no membrane; it is an aggregation of RNA and protein.

The nucleus has a double function. First, it controls the activity of the cell throughout life. Second, it is the location within the cell of the hereditary material, which is passed from generation to generation during reproduction. Both of these functions depend upon 'information'. The information in the nucleus is contained within the chromosomes. It is the chromosomes that

► control cell activity
► are transmitted from cell to cell when cells divide
► are passed to the new individual when sex cells fuse together in sexual reproduction.

■ The number of chromosomes per nucleus

Although the nucleus is occupied by the chromosomes, these long structures become visible only during cell division (when they shorten and thicken), and then only when stained with special dyes or if observed by phase-contrast microscopy. This is when chromosomes can be counted.

The number of chromosomes in the cells varies between different species. But in any one species, the number of chromosomes per nucleus is normally constant (Table 9.1). Animal species tend to have many short chromosomes (4–6 μm in length), whereas plant species tend to have a few long chromosomes (up to 50 μm in length); there are many exceptions, however.

Table 9.1 Numbers of chromosomes in the nuclei of selected species

Crocus belansae (crocus)	6
Ranunculus acris (meadow buttercup)	14
Taraxacum officinale (common dandelion)	24
Homo sapiens (human)	46
Locusta migratoria (locust)	24 female, 23 male
Mus musculus (mouse)	40

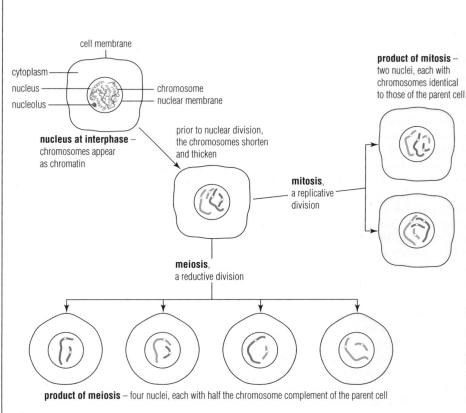

Figure 9.1 Types of nuclear division, and the changes in chromosome complement associated with them

cell membrane

cytoplasm

nucleus

nucleolus

chromosome

nuclear membrane

nucleus at interphase – chromosomes appear as chromatin

prior to nuclear division, the chromosomes shorten and thicken

mitosis, a replicative division

product of mitosis – two nuclei, each with chromosomes identical to those of the parent cell

meiosis, a reductive division

product of meiosis – four nuclei, each with half the chromosome complement of the parent cell

■ Nuclear divisions

When a cell divides, the nucleus divides first, followed by the cytoplasm. The period between nuclear divisions is known as **interphase**. During interphase the chromosomes take the form of greatly extended threads, and at this time they have a regulatory role in cell metabolism. Well before cell division, each chromosome makes a copy of itself (chromosome replication). When the nucleus divides the original and replicated chromosomes (they are temporarily called **chromatids**) are distributed equally between the daughter cells.

Nuclear divisions are of two types, depending on whether the number of chromosomes present in each of the daughter cells is the same as the number in the parent cell, or is half that number (Figure 9.1).

Mitosis

Mitosis is division of the nucleus to produce two daughter nuclei containing identical sets of chromosomes. For this reason mitosis is also called a **replicative division**. Mitosis takes place when an organism grows, and when tissues are repaired or replaced: for example, when red blood cells are produced in bone marrow, during cell division at plant meristems and during the production of skin cells to replace the cornified layer. Asexual reproduction can also occur as a result of mitotic division (p. 575).

Meiosis

Meiosis is division of the nucleus to produce four daughter nuclei, each containing half the number of chromosomes of the original nucleus. For this reason, meiosis is also called the **reductive division**. In sexual reproduction, when two sex cells (called gametes) fuse, the chromosome number of the new cell (the zygote) is doubled by the amalgamation of two nuclei. Meiosis normally precedes the formation of gametes, however, so the chromosome number remains constant from generation to generation. Meiosis is also associated with spore formation in certain plants (p. 563).

Questions

1 What structures of the interphase nucleus can be seen by electron microscopy (p. 154)?
2 What are the essential differences between mitosis and meiosis?
3 What events of the cell cycle occur in
   a the cytoplasm
   b the nucleus
   between cytokinesis and nuclear division?

■ The cell cycle

The cell cycle (Figure 9.2) is arbitrarily defined as the period from the formation of a cell by division to the point when that cell itself divides.

Most types of cell never divide again after they have grown and become specialised. This group includes the guard cells of the stomata of plant epidermis (p. 322), and many cells of the mammalian nervous system (p. 220).

Some unspecialised cells retain the ability to divide, and for these cells the cell cycle is shown below.

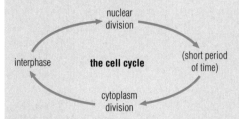

The length of this cell cycle is very variable. For example, cells of the epithelium lining the gut, and those forming the outer skin layer of mammals (Figure 10.4, p. 214), divide frequently and throughout the life of the organism. Cells of the mammalian bone marrow (Figure 10.10, p. 216) may divide about every 8–10 hours to produce new blood cells.

In other cells, division is seasonal. Cells that divide frequently at certain times of the year include cells of the plant growth regions (meristems, p. 222). For example, cells of a root-tip meristem (p. 408) may divide every 20–24 hours during periods of root growth. In this case, the length of the cell cycle depends upon external factors such as temperature and the supply of nutrients.

■ Interphase

Interphase is not a 'resting' phase, but a period during which the metabolic activity of the nucleus is intense. During interphase the nucleus is involved in protein synthesis, and the chromosomes themselves are copied or replicated.

Chromosome replication was detected when a radioactively labelled nucleotide (thymidine, p. 145) was made available to growing and dividing cells. Nucleotide was taken up only by the nuclei of cells during a restricted period of interphase when DNA synthesis was taking place.

Figure 9.2   The cell cycle

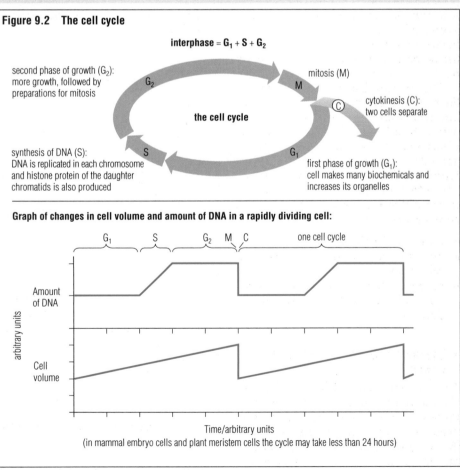

$$\text{interphase} = G_1 + S + G_2$$

second phase of growth (G_2): more growth, followed by preparations for mitosis

mitosis (M)

cytokinesis (C): two cells separate

the cell cycle

synthesis of DNA (S): DNA is replicated in each chromosome and histone protein of the daughter chromatids is also produced

first phase of growth (G_1): cell makes many biochemicals and increases its organelles

Graph of changes in cell volume and amount of DNA in a rapidly dividing cell:

G_1 S G_2 M C   one cell cycle

Amount of DNA

arbitrary units

Cell volume

Time/arbitrary units
(in mammal embryo cells and plant meristem cells the cycle may take less than 24 hours)

Figure 9.3   Mitosis in cells of the root tip of *Allium* sp. (onion): most species of *Allium* have 16 chromosomes (eight pairs) but, for simplicity, these drawings show four chromosomes only

1   Prophase

Chromosomes appear as long, thin threads; each thread has already replicated into two chromatids, but the chromatids are tightly held together and still appear joined at the centromere.

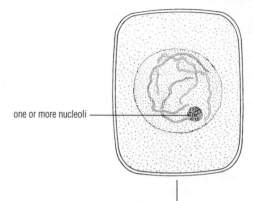

nuclear membrane
cell wall

chromatin granules

nucleolus

cytoplasm

one or more nucleoli

5   Interphase

Chromosomes now appear as chromatin.

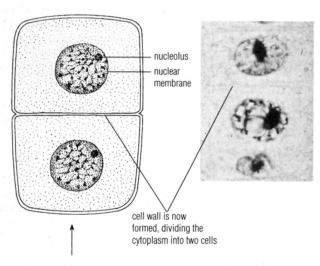

nucleolus
nuclear membrane

cell wall is now formed, dividing the cytoplasm into two cells

2   Metaphase

Nuclear membrane has broken down; each chromosome is attached to spindle fibres at the equator of the spindle by its centromere.

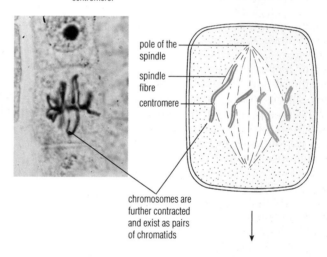

pole of the spindle

spindle fibre

centromere

chromosomes are further contracted and exist as pairs of chromatids

4   Telophase

Nucleolus and nuclear membrane reappear.

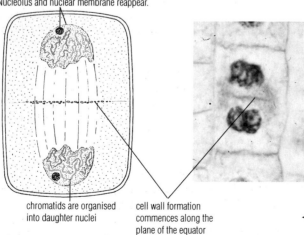

chromatids are organised into daughter nuclei

cell wall formation commences along the plane of the equator of the former spindle

3   Anaphase

Chromatids separate following division of the centromeres.

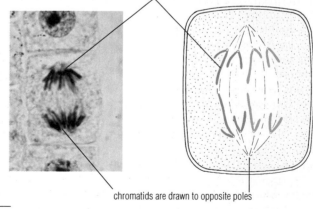

chromatids are drawn to opposite poles

9.2 MITOSIS

In mitosis the chromosomes, which have replicated by duplication during interphase (S-phase, Figure 9.2), are accurately separated and their chromatids distributed, one to each daughter nucleus. Once it starts, mitosis proceeds steadily, as a continuous process, although it is usually described in stages (four phases) to permit analysis of the changes taking place. The interphase part of the cell cycle is the period of intense biochemical activity, in which the information of the chromosomes is used to control and direct the activities of the cell.

The stages of mitosis, as observed in a plant cell, are shown in Figure 9.3. For simplicity and clarity the drawings show a cell containing only four chromosomes. In animals the process is essentially the same, although in animal cells a centrosome consisting of two centrioles (p. 165) is present outside the nuclear membrane. The centrioles appear to be involved in spindle formation. The centrosome is absent in plant cells.

Prophase

The chromosomes first become visible as long, thin threads that increasingly shorten and condense by a process of coiling. (Remember, each chromosome has already replicated.) The two parts of a chromosome, the chromatids, are held attached at one point, the **centromere**. Since the chromosomes have replicated, the DNA content has doubled.

Throughout most of prophase the joined chromatids appear as a single chromosome even though there are actually two identical, helically coiled strands lying side by side. Only as the shortening and thickening process continues does it become possible to observe the chromatids as separate structures. During prophase the nucleoli (p. 160) of the nucleus gradually disappear.

Metaphase

In this phase the nuclear membrane normally breaks down. The contents of cytoplasm and nucleus intermingle. In animal cells (only) the centrioles of the centrosome move to opposite points in the cell, called **poles**. Then, in all cells at this stage of mitosis, a series of microtubules (known as the **spindle**) is laid down between the poles. These fine protein fibres converge on the poles, and are spaced out at the equator region.

Each undivided centromere becomes attached to a spindle fibre at various places and then moves to the equatorial plate. Now the shortening and thickening of the chromosomes is at its maximum, and the individual chromatids and the centromeres are clearly visible.

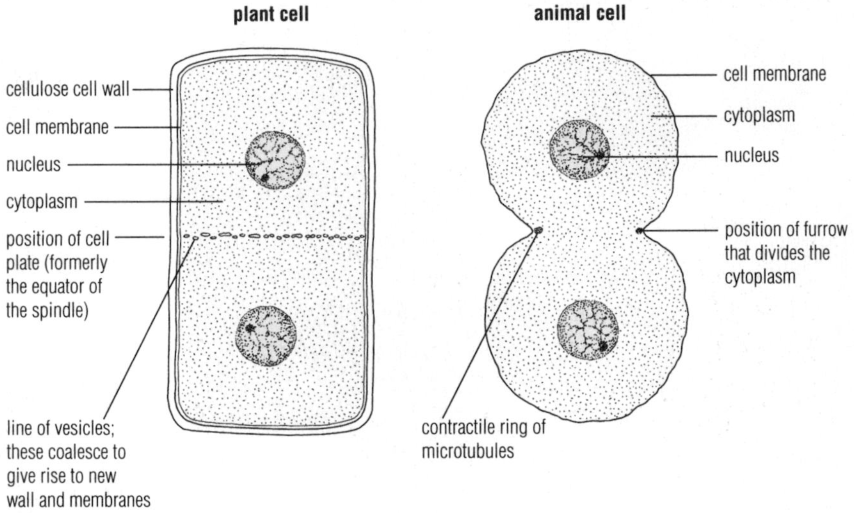

Anaphase

This is the briefest stage in mitosis. The centromeres divide, and the chromatids move to each pole, centromeres leading. This movement appears to be by disassembly of the protein of the microtubules (p. 164) that make up the spindle. A part of the centromere known as the **kinetochore** reacts with microtubule protein in this way.

Telophase

The two groups of chromatids (now daughter chromosomes) converge on the poles of the spindle apparatus. Each group becomes surrounded by a nuclear membrane. The chromosomes uncoil, the nucleoli re-form, and the two nuclei take on the granular appearance of interphase.

■ Division of the cytoplasm

Division of the cytoplasm, or **cytokinesis**, usually follows nuclear division (Figure 9.4). In cytokinesis the cell organelles such as mitochondria and chloroplasts become evenly distributed between the daughter cells.

In animal cells the cell membrane begins to intuck (invaginate) at the midpoint of the cell (at the equator of the spindle). This constriction of the cell circumference forms a furrow in the cytoplasm which then extends to divide the cytoplasm into two cells. Microtubules are involved in the constriction process.

In plant cells a new wall, known initially as the **cell plate**, forms at the midpoint of the cell (at the equator of the spindle). The cell plate forms by the coalescence of tiny vesicles. It extends outwards to the existing cell wall, and separates the two daughter cells.

■ The significance of mitosis

Each daughter cell produced by mitosis has a full set of chromosomes, identical to those of the parent cell. No variation in genetic information arises by mitosis.

In growth and development it is important that new cells all carry the same information (the same chromosomes) as the existing cells of the organism. Similarly, when damaged tissues are repaired the new cells must be exact copies of the cells being replaced. Mitosis ensures this is so.

In asexual reproduction, of which various forms exist (pp. 562, 580), the offspring produced are identical to the parents since they are produced as a result of mitotic cell division. As a consequence, the offspring have all the advantages of the parents in mastering the same habitat – and any disadvantages, too. The offspring produced by asexual reproduction are often described as **clones**.

Questions

1 What is the major feature of each of the four phases of mitosis?
2 Name one tissue in the human body, and one in a flowering plant, where you would expect to find cells dividing by mitosis.
3 What is the chief role of microtubules in cells (p. 164)?

Figure 9.4 Cytokinesis in plant and animal cells

plant cell

cellulose cell wall
cell membrane
nucleus
cytoplasm
position of cell plate (formerly the equator of the spindle)
line of vesicles; these coalesce to give rise to new wall and membranes

animal cell

cell membrane
cytoplasm
nucleus
position of furrow that divides the cytoplasm
contractile ring of microtubules

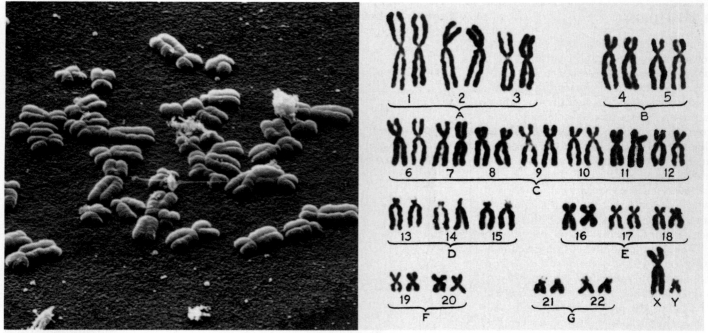

Figure 9.5   Human chromosomes, observed at the metaphase stage: (left) SEM of chromosomes from a burst cell, and (right) photomicrograph of homologous pairs of human chromosomes in decreasing order of length

9.3 HOMOLOGOUS CHROMOSOMES

We have seen that the cell contains a fixed number of chromosomes in the nucleus, and that nuclear divisions during growth result in the exact duplication of these chromosomes in the daughter nuclei (mitosis).

Examination of the chromosomes at certain stages of nuclear division discloses another important feature. Chromosomes occur in pairs! (Figure 9.5.) These pairs of chromosomes are known as **homologous pairs**. Members of a homologous pair are identical in length and in the position of the centromere, and can be identified by their characteristic shape. This becomes apparent during meiosis, when homologous chromosomes pair up (Figure 9.6).

Homologous pairs of chromosomes arise in the first place because one member of the pair has come from the male parent and the other from the female parent. Homologous pairs are then maintained by the exact replication that takes place in each mitotic division.

A cell with a full set of chromosomes (all homologous pairs) in its nucleus is said to be in the **diploid** condition. The sex cells (gametes) produced by meiosis contain a single set of chromosomes. This condition is known as **haploid**. When haploid gametes fuse the resulting cell (zygote) has the diploid condition restored.

■ Chromosome abnormalities

Abnormalities sometimes cause variations in the number of the chromosomes in a cell (or sometimes changes to the structure of individual chromosomes). Such an event, known as a chromosome **mutation**, is likely to be of profound significance for the organism. Examples of such changes are discussed in Chapter 28, p. 625).

Figure 9.6   Homologous chromosomes about to pair

Chromosomes are made up of a linear series of genes. Individual genes are too small to be seen.

The centromere, a small constriction on the chromosome is not a gene.

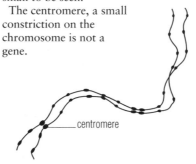

centromere

Early in nuclear division, a disc shaped protein molecule attaches to the centromere. When the spindle forms, a spindle fibre attaches to that protein, Then centromeres of sister chromatids split apart and the chromatids are pulled to opposite poles of the cell by the attached spindle fibre.

9.4 MEIOSIS

Meiosis involves two successive divisions of the nucleus. Both the divisions superficially resemble mitosis; the first division (I), however, separates homologous chromosomes and the second division (II) separates chromatids. Four cells are formed as a result of meiosis, each with half the chromosome number of the original cell. The reduction in chromosome number is not a wholly random process; one of each homologous pair is present in each nucleus formed. But which member of each pair enters which gamete is a matter of chance.

The four cells produced by meiosis are often found together, at least temporarily, and are then referred to as a **tetrad**; the tetrads of pollen grains (Figure 26.16, p. 571) are an example.

■ The process of meiosis

Meiosis is slower and more complex than mitosis but like mitosis, once started, it proceeds steadily as a continuous process of nuclear division (Figure 9.7).

Meiosis is preceded by an interphase stage during which the chromosomes of the nucleus are replicated. When meiosis begins the chromosomes make their appearance as long threads, shortening and thickening by coiling. In meiosis the behaviour of chromosomes during the prophase I is especially complex.

Figure 9.7   Meiosis in an animal cell (for simplicity, the process is illustrated for a cell containing four chromosomes only): in a plant cell the process is essentially the same, except that the centrosome is absent and cytokinesis is by formation of a cell plate

1 Cell at interphase

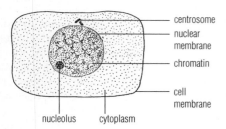

centrosome
nuclear membrane
chromatin
cell membrane
nucleolus   cytoplasm

2 Early prophase I   Four chromosomes have condensed to long, thin threads.

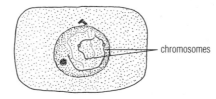

chromosomes

3 Mid-prophase I   Homologous chromosomes have paired to form bivalents; the nucleolus is breaking down.

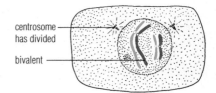

centrosome has divided
bivalent

4 Late prophase I   Homologous chromosomes now begin to repel each other; individual chromatids, with chiasmata, become visible.

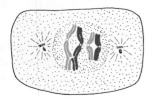

5 Metaphase I   The spindle has formed; bivalents are attached to spindle fibres at the equator of the spindle.

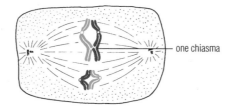

one chiasma

6 Anaphase I   Chromosomes move apart to the poles of the spindle, centromeres first; each chromosome consists of two chromatids united at the centromere.

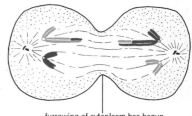

furrowing of cytoplasm has begun

7 Telophase I   The cell has divided; the nuclear membrane re-forms around the chromosomes.

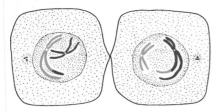

8 Prophase II   The chromosomes shorten and thicken again; the nuclear membrane breaks down.

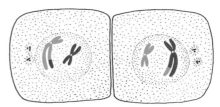

9 Metaphase II   The spindle re-forms; the chromosomes move to the equator and become attached to spindle fibres.

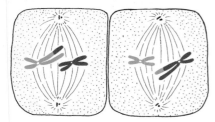

10 Anaphase II   The centromeres divide and separate, pulling the sister chromatids to opposite poles.

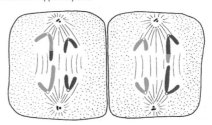

11 Telophase II   The nuclear membrane re-forms; the chromosomes disperse and appear as chromatin; the cytoplasm furrows; and the nucleoli reappear.

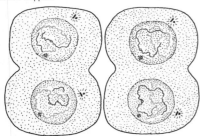

12 Interphase

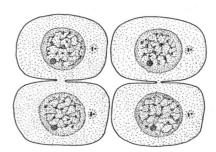

Figure 9.8   The changes to a homologous pair of chromosomes during meiosis

1 two chromosomes (homologous chromosomes); these occur together, in the nucleus of one cell

2 homologous chromosomes pairing

3 bivalent continues to shorten and thicken (prophase I)

4 chromatids become visible; homologous chromosomes begin to repel each other; the positions of the chiasmata become visible (late prophase I)

5 the chromosomes part, but there is still attraction between sister chromatids; chiasmata hold the bivalent together (metaphase I)

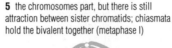

chiasma   chiasma

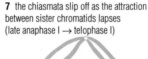

6 chromosomes separated; each now consists of a pair of chromatids united at the centromere (anaphase I)

7 the chiasmata slip off as the attraction between sister chromatids lapses (late anaphase I → telophase I)

Stages **8**, **9** and **10** illustrate the changes to only one of the two chromosomes shown in stages **1** to **7**.

8 chromosome at the equatorial plate (metaphase II)

10 the chromosomes unwind and elongate (telophase II)

9 the centromere divides and the chromosomes move apart (anaphase II)

The two chromosomes are now separated, lying in the nuclei of two different cells.

Prophase I

The chromosomes appear in the light microscope as single thin threads with many bead-like thickenings along their lengths. They are observed to come together in specific pairs, point by point all along their length. Paired chromosomes (homologues) are joined by an extremely thin protein framework extending their whole length. The product of pairing of homologous chromosomes is termed a **bivalent**.

Homologous chromosomes, paired together as bivalents, continue to shorten and thicken. At a certain stage in prophase I the chromosomes appear double-stranded, as the sister chromatids of each chromosome separate and become visible (Figure 9.8).

As the bivalents shorten and open out to reveal their chromatids it is evident that some breakage and rejoining of non-sister chromatids has taken place earlier when they were closely paired. When non-sister chromatids from homologous chromosomes break and rejoin they do so at exactly corresponding sites, so that a cross-shaped structure called a **chiasma** is formed at one or more places along a bivalent (Figure 9.9). At this stage of meiosis there is still attraction between sister chromatids throughout their full length, and this attraction of sister chromatids actually keeps the bivalents together.

The chromatids are now at their shortest and thickest. The disappearance of the nucleoli and the nuclear membrane marks the end prophase I.

Metaphase I

The spindle forms, and the bivalents attach to it and then move to the equatorial plate. Centromeres become attached to individual spindle fibres.

Anaphase I

Attraction of sister chromatids lapses and homologous chromosomes move to opposite poles. Each centromere drags behind it two chromatids, still united at the centromere.

Telophase I

This step usually blends directly into the second prophase, for the cell often fails to divide into two after the first nuclear division.

Prophase II

If the nuclear membranes were re-formed after telophase I, then they are broken down in prophase II.

Figure 9.9 The chiasma has a mechanical function in meiosis (as well as a genetic role)

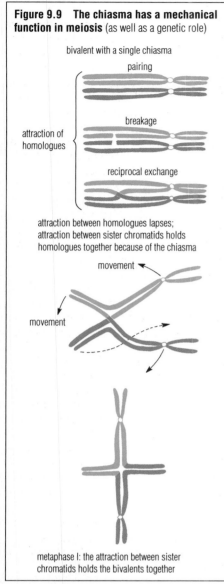

bivalent with a single chiasma

pairing

attraction of homologues

breakage

reciprocal exchange

attraction between homologues lapses; attraction between sister chromatids holds homologues together because of the chiasma

movement

movement

metaphase I: the attraction between sister chromatids holds the bivalents together

Metaphase II

Two new spindles are formed at right angles to the original one. The chromosomes, each consisting of two chromatids held together at the centromeres, attach to the spindle fibres by their centromeres, and then move to the equatorial plate.

Anaphase II

The centromeres divide and separate, pulling the sister chromatids to opposite poles.

Telophase II

Nuclear membranes form around the four groups of chromatids and four nuclei are formed. Once again the chromosomes become dispersed and appear as chromatin granules. Nucleoli reappear.

The process of meiosis is now complete, and is followed by cytokinesis.

Table 9.2 A comparison of mitosis and meiosis

	Mitosis	Meiosis First division	Meiosis Second division
THE PROCESS			
Prophase	Chromosomes become visible, finally as two chromatids joined at the centromere; nuclear membrane breaks down; nucleolus disappears; spindle apparatus forms	Chromosomes become visible; homologous chromosomes pair to form bivalents which eventually appear four-stranded, each chromosome consisting of two chromatids*; nucleolus disappears; spindle apparatus forms	Chromosomes reappear as two chromatids joined at the centromere; spindle apparatus forms
Metaphase	Chromosomes at equator of spindle; centromeres attach to spindle fibres	Bivalents at equator of spindle; centromeres attach to spindle fibres	Chromosomes at equator of spindle; centromeres attach to spindle fibres
Anaphase	Centromeres divide; chromatids move to opposite poles	Chromosomes (each of two sister chromatids joined at their centromeres) move to opposite poles	Centromeres divide; chromatids move to opposite poles
Telophase	Spindle apparatus breaks down; nuclear membrane re-forms; cytoplasm divides	Spindle apparatus breaks down; nuclear membrane may re-form; cytoplasmic division may occur or may be delayed until the end of meiosis	Spindle apparatus breaks down; nuclear membrane re-forms around each nucleus; cytoplasm divides
THE PRODUCTS			
	Two identical cells, each with the diploid chromosome number	Two non-identical nuclei, each with the haploid chromosome number (i.e. one of each homologous pair per nucleus)	Four non-identical cells, each with the haploid chromosome number
THE CONSEQUENCES AND THE SIGNIFICANCE			
	Mitosis produces identical diploid cells. Mitosis permits growth within multicellular organisms. Asexual reproduction and vegetative propagation occur as a result of mitotic division.	Meiosis produces haploid cells from a diploid cell. Meiosis contributes to genetic variability by: ▶ reducing the chromosome number by half, permitting fertilisation, and the combination of genes from two parents ▶ permitting the random assortment of maternal and paternal homologous chromosomes ▶ recombination of segments of individual maternal and paternal homologous chromosomes during crossing over*.	

* See also Chapter 28, where the Law of Heredity and genetic principles are discussed

Questions

1 Explain why chromosomes occur in homologous pairs in diploid cells.
2 What is the fate of homologous chromosomes when a diploid nucleus becomes haploid?
3 What are the essential differences between prophase of mitosis and prophase I of meiosis?
4 What events in meiosis result in the gametes having different chromosomal content?
5 What is a chiasma?

9.5 WHAT ARE CHROMOSOMES COMPOSED OF?

The chromosomes of eukaryotic cells are composed of DNA, protein and a small amount of RNA. DNA makes up only about one-third of the mass of the chromosome; there is approximately twice as much protein as DNA (Table 9.3).

Table 9.3 The chemical composition of the eukaryotic chromosome/%

DNA	27
protein	67
RNA	6

■ Chromosome protein

About one-half of the protein present in the chromosome is **histone**. This protein has a high concentration of amino acid residues with additional basic ($-NH_2$) groups, such as lysine and arginine. Histone binds strongly to DNA, taking the form of beaded sub-units on the DNA chain called **nucleosomes**. Thus histone plays an important part in the packaging of the DNA.

Non-histone proteins in the chromosomes contain more of the acidic amino acid residues (those with additional $-COOH$ groups) than does histone. Many of these proteins have turned out to be enzymes that catalyse the transcription (p. 208) and replication (p. 205) of nucleic acids, and are known as polymerase enzymes. The remainder of the non-histone protein is the protein scaffold to the chromosome structure (see below).

■ Chromosome DNA

The total quantity of DNA in a eukaryotic chromosome is phenomenal. DNA occurs as a single strand in the form of a double helix (p. 204) running the length of the chromosome, one double helix on each chromatid. For example, human chromosome number 1 (Figure 9.5) is 10 μm long at metaphase (the period when it is at its shortest and most compact), yet it contains about 7 cm of DNA!

The DNA strand is looped around and attached to the histone 'beads', and is arranged along a protein framework or scaffold. During mitosis and meiosis the whole structure is further condensed and compacted together. Throughout interphase the DNA takes on a more open structure, and is accessible to enzyme action and to transcription (by messenger RNA, p. 208).

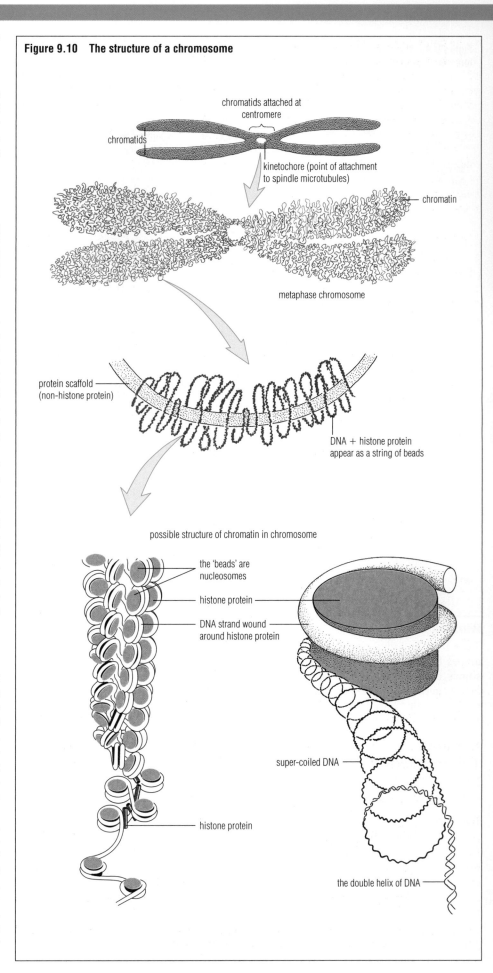

Figure 9.10   The structure of a chromosome

chromatids attached at centromere

chromatids

kinetochore (point of attachment to spindle microtubules)

chromatin

metaphase chromosome

protein scaffold (non-histone protein)

DNA + histone protein appear as a string of beads

possible structure of chromatin in chromosome

the 'beads' are nucleosomes

histone protein

DNA strand wound around histone protein

super-coiled DNA

histone protein

the double helix of DNA

■ The centromere of the chromosome

From the S-phase of the cell cycle (Figure 9.2), prior to mitosis and meiosis, chromosomes consist of sister chromatids, attached at their centromeres. The centromere is the last part of the chromosome to separate into daughter chromatids in nuclear division, and consists of a length of DNA that is never transcribed for protein synthesis. Chromatids are attached at sites on either side of the kinetochore (Figure 9.10).

■ Is DNA the hereditary material?

We know DNA is the genetic material as a result of the work of biochemists, bacteriologists and virologists. For example, biochemical analysis of nuclei has shown that

▶ all the diploid cells of a given organism have the same amount of DNA at corresponding stages of the cell cycle (whereas the amounts of other substances, such as protein, carbohydrate and fats, vary widely between different types of cell in the organism)

▶ haploid gametes (eggs, sperms) have only half the amount of DNA that is in the diploid cells they are derived from

▶ the more complex an organism (that is, the more different kinds of protein it can synthesise), the greater is the content of DNA per nucleus (Table 9.4).

Table 9.4 The DNA content of cells

	DNA per cell/pg*
bacteria	0.05
fungi	0.10
green plants	2.5
molluscs	1.2
fish	2.0
birds	2.5
bird sperm	1.24
mammals	6.0

* pg = picogram;   1 pg = 1×10^{-12};
all figures are approximate

■ An experiment with pneumonia-causing bacteria

The role of DNA was first established in bacteria. These prokaryotes lack true chromosomes. Figure 9.11 is a summary of an experiment with a pneumonia-causing bacterium, *Streptococcus pneumoniae*, carried out in England in 1928. Pneumonia was a major cause of death at that time, and medical researchers were looking for an effective vaccine against the pathogen.

One strain was virulent; if it was injected into mice, they developed pneumonia and died. Colonies of this strain have a smooth appearance, and it is referred to as the 'S' strain. Bacteria of the S strain can form a polysaccharide capsule around the cell wall, effectively protecting the cell from the mouse's defence mechanism. The ability to produce a polysaccharide capsule is genetically determined.

The other strain used in these experiments did not cause pneumonia. No polysaccharide capsule was formed, and the strain was not protected from the mouse's defence mechanism. Colonies of this strain had a rough appearance, and the strain is referred to as the 'R' strain.

The researcher found that heat treatment of the virulent S strain rendered it harmless, exactly as he expected. (Heat killed the cell by denaturing the proteins.) The surprise came when a mixture of live R strain and heat-treated S strain was administered: the mice died! Furthermore, the dead mice were found to contain live capsule-forming *Streptococcus*. Apparently the genetic message for capsule formation had passed from dead to live cells, and the latter had become virulent.

It was concluded that some substance had passed from the remains of the dead S cells into the living R cells. Only very much later was this substance shown, by a different team of researchers, to be DNA. The mechanism by which the genetic message for capsule formation had passed from dead S cells to live R cells became known as **transformation**.

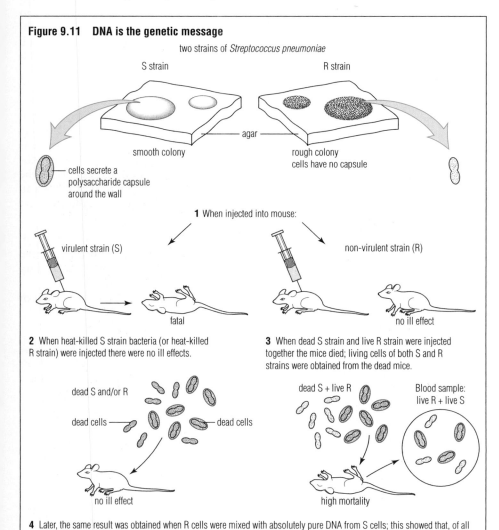

Figure 9.11   DNA is the genetic message

two strains of *Streptococcus pneumoniae*

S strain / R strain

agar

smooth colony

rough colony
cells have no capsule

cells secrete a polysaccharide capsule around the wall

1 When injected into mouse:

virulent strain (S)

non-virulent strain (R)

fatal

no ill effect

2 When heat-killed S strain bacteria (or heat-killed R strain) were injected there were no ill effects.

3 When dead S strain and live R strain were injected together the mice died; living cells of both S and R strains were obtained from the dead mice.

dead S and/or R

dead cells — dead cells

no ill effect

dead S + live R

Blood sample: live R + live S

high mortality

4 Later, the same result was obtained when R cells were mixed with absolutely pure DNA from S cells; this showed that, of all the compounds present in the S cell, it was the DNA that carried the genetic information.

Questions

1 What are the roles of the various forms of protein in chromosomes?

2 What is the significance of the presence of an external polysaccharide capsule in *Streptococcus pneumoniae* to
   a the bacterium,
   b the experimenter?

■ The Hershey–Chase experiment with bacteriophage

Confirmation that DNA is the genetic 'information' came from an experiment with a bacteriophage virus (a type of virus that attacks a bacterium). Viruses consist of a protein coat surrounding a nucleic acid core (p. 548). Viruses gain entry to a host cell and then (and only then) can take over the nuclear and cytoplasmic machinery. The infected cell is 'switched on' to the replication of the virus components (protein and nucleic acid), which are then assembled into new virus. Eventually the cell breaks down (lysis) and the virus particles escape. The 'life cycle' of a bacteriophage is shown in Figure 25.2 (p. 549).

The experiment by Alfred Hershey and Martha Chase in 1957 used the bacteriophage known as T_2 (Figure 9.12). The host bacterium of T_2 is *Escherichia coli* (p. 552). It is now known that T-phage viruses have a quite complex 'head and tail' structure. The external wall of the 'head' and the 'tail' consist of protein. The head contains a long DNA molecule, coiled up. When T_2 bacteriophages infect *E. coli* the protein coat of the virus remains outside the host cell, but the DNA enters it. Soon after, the host cell

becomes a 'factory' producing new T-phage viruses. The virus consists of nucleic acid and protein only, so which of these two components carries the genetic information? To us, it appears that it is DNA, since it is this component of the virus that enters the host cell, but at the time of the Hershey–Chase experiment this had not been established.

In the experiment, one sample of the T_2 virus was labelled with radioactive phosphorus-32 (phosphorus occurs in nucleic acid but not in protein). Consequently, this sample of the virus was labelled in the nucleic acid core.

The second sample of the T_2 virus was labelled with radioactive sulphur-35 (sulphur occurs in protein but not in nucleic acid). This sample of the T_2 virus was labelled in the protein coat.

Two identical cultures of *E. coli* were infected, one with the nucleic acid-labelled virus, and the other with the protein-labelled virus. Soon after the bacteria were infected the viruses replicated. Only from the bacteria infected with virus labelled with ^{32}P were radioactive viruses subsequently produced. This experiment established that DNA carries the genetic information controlling the production of new viruses.

9.6 DNA: THE DOUBLE HELIX MODEL

The basic unit of structure of nucleic acids, the nucleotide, was introduced in Chapter 6, p. 145. Deoxyribonucleic acid (DNA) is a polymer of repeating nucleotides. Individual nucleotides consist of three components: phosphoric acid, a pentose sugar (deoxyribose or ribose) and an organic base (a purine or a pyrimidine derivative) (p. 146). Nucleotides themselves are combined together to form the long strand of nucleic acid (Figure 6.32, p. 146).

Interest in the chemical composition and the three-dimensional structure of DNA grew steadily from the time it was realised that nucleic acid might be the genetic material, early in the twentieth century. It fell to Francis Crick and James Watson, working at the Cavendish Laboratory in Cambridge, to propose the now accepted model of DNA (see panel).

Crick and Watson used the chemical and physical information available to them. Some of this evidence was published data, the full significance of which had been overlooked. Their main evidence for this model was as follows.

X-ray diffraction patterns

When X-rays are passed through crystallised DNA they are scattered in a way that represents the arrangement of atoms within DNA. By analysis of the X-ray diffraction pattern obtained, details of the three-dimensional structure of DNA can be deduced. This evidence indicates that DNA

► is a long, thin molecule, diameter 2 nm
► is coiled in the form of a helix, the helix making one complete twist every 3.4 nm
► has ten bases to each complete twist of the helix
► consists of two strands.

This crucial experimental evidence came in 1953 from the work of M H K Wilkins and Rosalind Franklin, working at King's College, London (Figure 9.13).

Figure 9.12 The Hershey–Chase experiment: this experiment resolved whether genetic information in a virus (bacteriophage, T_2 parasitic upon the bacterium *E. coli*) resides in the protein coats of the virus particles or in the DNA at their centre

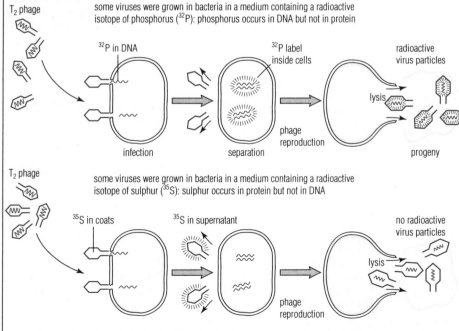

T_2 phage

some viruses were grown in bacteria in a medium containing a radioactive isotope of phosphorus (^{32}P): phosphorus occurs in DNA but not in protein

^{32}P in DNA
^{32}P label inside cells
radioactive virus particles

lysis

infection   separation   phage reproduction   progeny

T_2 phage

some viruses were grown in bacteria in a medium containing a radioactive isotope of sulphur (^{35}S): sulphur occurs in protein but not in DNA

^{35}S in coats ^{35}S in supernatant   no radioactive virus particles

lysis

phage reproduction

The experiment demonstrated that the DNA part of the virus carries the genetic information for phage reproduction.

Question

What would have been the outcome of the Hershey–Chase experiment if protein had been the carrier of genetic information?

FRANCIS CRICK AND THE FOUNDATION OF MOLECULAR BIOLOGY

Life and times

Francis Crick was born in Northampton, eldest son of a shoe manufacturer. His secondary schooling was at Northampton Grammar School and, later, at Mill Hill School. He progressed to University College, London and gained Class II Honours in physics with subsidiary maths, in 1937. He commenced research in physics but was rescued by war from a dull project.

He was drawn into Admiralty research, and distinguished himself in marine mine design. Afterwards came the opportunity for a fresh start. 'Only gradually did I realise that [my] lack of qualifications could be an advantage. By the time most scientists have reached the age of 30 they are trapped by their own expertise.' One of his interests was the 'borderline between the living and the non-living.' But he wrote:

I knew very little biology, except in a very general way, till I was over thirty, since my first degree was in physics. It took me a little time to adjust to the rather different way of thinking necessary in biology. It was almost as if one had to be born again.

Meanwhile, the Medical Research Council (MRC) had decided to support the entry of physics into biological research, and in 1945 Crick took up an MRC studentship at the Cavendish Laboratory, Cambridge, to study protein structure under Max Perutz.

Jim Watson was brought up in a fairly poor family living among the steel mills of Chicago and, having gained an interest in ornithology from his father, had read zoology at the University of Chicago. Subsequently he had research experience in viruses at the University of Illinois, before reaching the Cavendish Laboratory to work on a studentship on the biochemistry of viruses in 1951. Here he met Crick, with whom he shared an optimistic enthusiasm that it should be possible to understand the nature of the gene (p. 210) in molecular terms.

Contribution to biology

Within two years of meeting, Crick and Watson had achieved their understanding of the nature of the gene in chemical terms. They brought together the diverse contributions of many workers, and from all this evidence they deduced the likely structure of DNA and built a model. This was of two separate chains, wound around each other (double helix). The backbone of the chains was formed by —sugar–phosphate— residues. Organic bases attached to the sugars were directed inwards and held by hydrogen bonds.

The model suggested that DNA could uncoil, and that the separated strands could act as templates in replication. Genetic information was coded by the sequence of bases along the chain. Copies of the DNA (in the form of RNA) were made (transcription) and these directed the sequence of amino acids in the proteins produced by the cell (translation).

Francis Crick

Influences

This, the greatest biological advance of the century, was made by combining the concepts of physics and biology. It created a new branch of biology: molecular biology.

Few scientists become household names in their own lifetime, but the youthful discoverers of the structure of DNA are an exception. The importance of their discovery thrust genetics and molecular biology to the forefront of biological and medical research.

Base ratio analysis

Edwin Chargaff was a Czech-American who began working at Columbia University, New York, in 1935. Chargaff investigated the organic base composition of nucleic acids: some of his results are shown in Table 9.5. (The organic bases adenine (A), guanine (G), thymine (T) and cytosine (C) are shown in Figure 6.31, p. 146.) He showed that a single organism contains many different kinds of RNA, but its DNA is essentially one kind, characteristic of the organism.

Chargaff's analysis of DNA established the following rules:

▶ the number of purine bases (A + G) present = the number of pyrimidine bases (T + C)
▶ the number of adenine bases = the number of thymine bases
▶ the number of guanine bases = the number of cytosine bases.

These observations are best interpreted as adenine being always paired with thymine (A = T), and guanine with cytosine (G = C). (You will notice in Table 9.5 the close correspondence in the percentages of adenine and thymine, and those of guanine and cytosine, in each species.) Chargaff's rules are a key clue to the structure of DNA.

Figure 9.13   Wilkins and Franklin's study of the X-ray diffraction pattern of DNA: the pattern is produced by directing a narrow X-ray beam through a crystalline fibre of a DNA salt, and allowing the scattered rays to fall on photographic film; the film is developed and the pattern is analysed mathematically

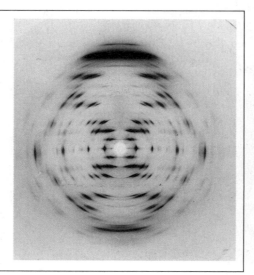

Table 9.5   The organic base of composition of DNA from a range of organisms/molar proportions %

Source of DNA	Adenine	Guanine	Thymine	Cytosine
human	30.9	19.9	29.4	19.8
sheep	29.3	21.4	28.3	21.0
hen	28.8	20.5	29.2	21.5
turtle	29.7	22.0	27.9	21.3
salmon	29.7	20.8	29.1	20.4
sea urchin	32.8	17.7	32.1	17.3
locust	29.3	20.5	29.3	20.7
wheat	27.3	22.7	27.1	22.8
yeast	31.3	18.7	32.9	17.1
Escherichia coli (a bacterium)	24.7	26.0	23.6	25.7

■ Watson and Crick's model

Watson and Crick were eventually successful in building a model of the DNA molecule in the form of a double helix: two polynucleotide chains, coiled around the same axis and interlocking. The —sugar–phosphate— chains formed 'backbones' to the molecule, and were on the outside. The base pairs were together, on the inside, held together by hydrogen bonds between specific pairs of bases (Figure 9.14).

Because of base pairing, the sequence of bases in one strand would determine the sequence of bases in the other. The two strands are said to be complementary. The bases of the two strands fit together only if the sugar molecules they are attached to point in opposite directions. The strands of DNA are said to be 'antiparallel'.

Figure 9.14　The DNA molecule of the chromosome: a summary

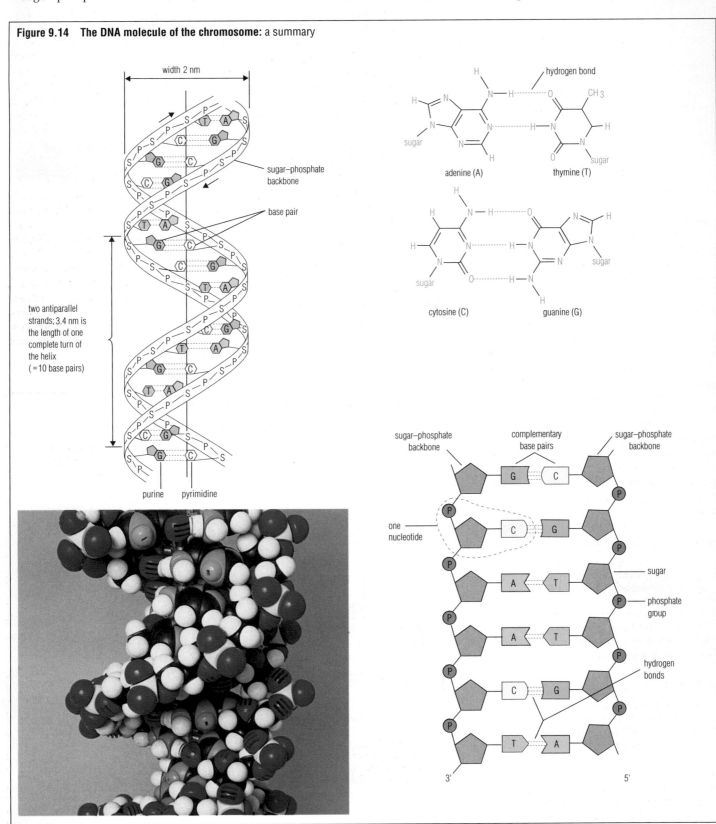

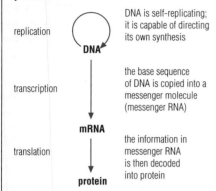

Figure 9.15   The central dogma of molecular biology: a one-way flow of coded information from DNA during protein synthesis

The double helix DNA and the coding of information

Watson and Crick published their model in the scientific research journal *Nature*, in 1953. Towards the end of this account came the sentence 'It has not escaped our notice that the specific pairing [of bases] we have postulated immediately suggests a possible copying mechanism for the genetic material.' An understatement!

The model they built suggests that DNA may replicate by the uncoiling of the helix, both strands acting as templates for the formation of complementary DNA. This would occur during the replication of chromosomes, prior to mitosis and meiosis.

The model also suggests that genetic information could be coded by the sequence of bases along the DNA chain. Crick then proposed a 'central dogma', summarising the way in which DNA both replicated itself and dictated conditions in the cell (Figure 9.15).

■ The public recognition

In 1962 Watson and Crick, together with Maurice Wilkins, were awarded the Nobel prize for medicine for elucidating the structure of DNA. Rosalind Franklin, whose evidence from X-ray crystallography played such an important part in the hypothesis, had died in 1958, aged only 37.

Questions

1 Which five nucleotides occur in nucleic acids (p. 145)? Which nucleotides occur in DNA?
2 What type of evidence of molecular structure can be obtained from X-ray diffraction studies of macromolecules?
3 What is meant by antiparallel strands?
4 Explain how experimental evidence rules out the pairing of cytosine and thymine, and of adenine and guanine, in the DNA helix.

9.7 REPLICATION OF DNA

DNA must replicate (make copies of itself) if genetic information is to be available for transmission to daughter cells, and from generation to generation in reproduction.

As we have seen, the Crick and Watson model of DNA immediately suggested a method by which replication of DNA could occur. The hydrogen bonds between complementary base pairs must break, and the two strands unwind ('unzip'). Each single strand of DNA would then act as a mould or **template**. Complementary nucleotides would assemble alongside their respective partners, and the sugar–phosphate backbone form by condensation reactions between nucleotides. Once this had occurred along the whole length of the DNA molecule two identical DNA double helices would have formed.

This model of replication is called **semi-conservative**, since each daughter molecule formed contains (conserves) one of the original (parental) strands, and has alongside it one newly synthesised strand (Figure 9.16).

■ DNA replication *in vitro*

Replication of a single strand of DNA was carried out in a test tube (*in vitro*) within three years of the publication of Watson and Crick's paper, using components and an enzyme obtained from the bacterium *Escherichia coli*. The enzyme was named **DNA polymerase**.

Later it was discovered this enzyme catalyses the addition of nucleotides to the strand running in the $5' \rightarrow 3'$ direction. A second enzyme, **DNA ligase**, catalyses the final assembly of the strand running in the $3' \rightarrow 5'$ direction (Figure 9.17, overleaf).

■ Evidence for semi-conservative replication

Meselson and Stahl (1958), also working with *E. coli*, obtained experimental results that were possible only if DNA replicated semi-conservatively. Nucleic acid of the chromosomes of growing and dividing bacteria was labelled with the 'heavy' isotope of nitrogen, ^{15}N. Then these bacteria were transferred to a medium in which the nitrogen available was the normal isotope, ^{14}N. The changes in concentration of ^{15}N and ^{14}N in the chromosomes of the succeeding generations were measured (Figure 9.18, overleaf).

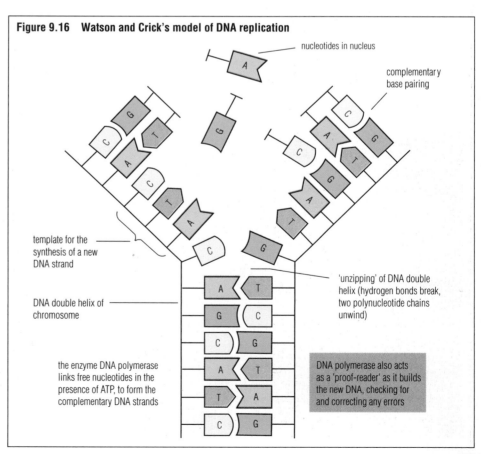

Figure 9.16   Watson and Crick's model of DNA replication

nucleotides in nucleus

complementary base pairing

template for the synthesis of a new DNA strand

DNA double helix of chromosome

the enzyme DNA polymerase links free nucleotides in the presence of ATP, to form the complementary DNA strands

'unzipping' of DNA double helix (hydrogen bonds break, two polynucleotide chains unwind)

DNA polymerase also acts as a 'proof-reader' as it builds the new DNA, checking for and correcting any errors

Figure 9.17 Replication of DNA: how it happens

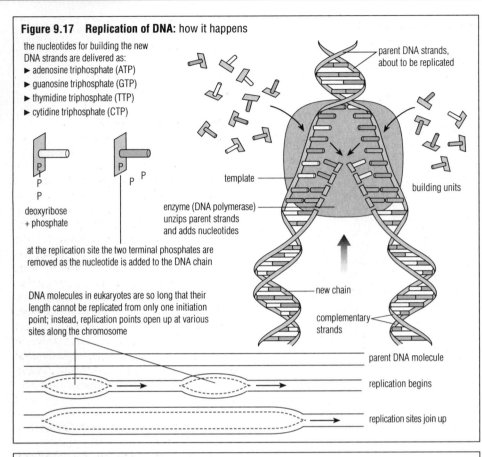

the nucleotides for building the new DNA strands are delivered as:
► adenosine triphosphate (ATP)
► guanosine triphosphate (GTP)
► thymidine triphosphate (TTP)
► cytidine triphosphate (CTP)

deoxyribose + phosphate

at the replication site the two terminal phosphates are removed as the nucleotide is added to the DNA chain

DNA molecules in eukaryotes are so long that their length cannot be replicated from only one initiation point; instead, replication points open up at various sites along the chromosome

parent DNA strands, about to be replicated

template

building units

enzyme (DNA polymerase) unzips parent strands and adds nucleotides

new chain

complementary strands

parent DNA molecule

replication begins

replication sites join up

Figure 9.18 The Meselson–Stahl experiment

Stock of the bacterium *Escherichia coli* was grown in a nutrient medium in which the nitrogen is the ordinary (light) isotope, ^{14}N.

When DNA is extracted from these cells and centrifuged on a salt density gradient, the DNA separates out at the point at which its density equals that of the salt solution.

position of 'light' DNA

density →

^{14}N DNA — DNA with two 'light' strands

Cultures of these cells were grown for many generations in a medium containing the 'heavy' nitrogen isotope, ^{15}N; all the DNA became labelled with ^{15}N.

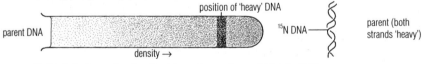

position of 'heavy' DNA

parent DNA

density →

^{15}N DNA — parent (both strands 'heavy')

The labelled cells were transferred back to a culture medium containing the 'light' nitrogen isotope (^{14}N), and allowed to grow.

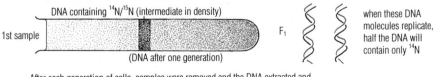

DNA containing $^{14}N/^{15}N$ (intermediate in density)

1st sample

(DNA after one generation)

F_1

when these DNA molecules replicate, half the DNA will contain only ^{14}N

After each generation of cells, samples were removed and the DNA extracted and analysed.

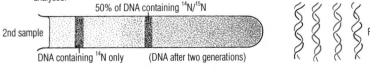

50% of DNA containing $^{14}N/^{15}N$

2nd sample

DNA containing ^{14}N only   (DNA after two generations)

F_2

The results can be explained if DNA replicates semi-conservatively (i.e. each strand of an existing double helix serves as a template for the synthesis of a new strand).

9.8 THE GENETIC CODE

The role of DNA is to instruct the cell to make specific proteins. Information in DNA determines the sequence of amino acids of the proteins which the cell manufactures ceaselessly. This information exists in the sequence of bases in DNA, and is known as the **genetic code**.

Proteins are made from some twenty different types of amino acid. An average protein contains several hundred amino acids, condensed together in a specific sequence. DNA consists of only four different bases. Thus there are only four 'letters' to the base code 'alphabet' with which to code for all twenty amino acids (Table 9.6).

It was immediately clear that combinations of three bases would be most likely to code for the amino acids. This is known as a 'triplet' code.

An experiment in which a single base was added to or deleted from a nucleic acid supported these assumptions. This experiment, known as Crick's frame-shift experiment and outlined in Table 9.6, confirmed that the code was a triplet code. Triplets became known as **codons**. These experiments also helped to establish the code as a non-overlapping one. For example, the sequence GUACGAGGA is read as GUA, CGA, GGA and not as either GUA, UAC, ACG. . . or as GUA, ACG, GAG. . . .

■ Deciphering the code

The code was eventually broken by the preparation of lengths of nucleic acid (messenger RNA, in effect) in which one triplet code (codon) was repeated many times.

Initially, the synthetic RNA was made from single nucleotides, producing structures such as –UUU–UUU–UUU– (which codes for the amino acid phenylalanine), –CCC–CCC–CCC– (which codes for proline) and –AAA–AAA–AAA– (which codes for lysine).

Subsequently, synthetic messenger RNAs with various known sequences of nucleotides were produced, enabling the complete code to be deciphered. These synthetic nucleic acids were added to cell-free protein-synthesising systems containing all twenty amino acids and other metabolites, and the polypeptides formed were analysed.

■ The triplet code

The genetic code is given in Table 9.7. This is a listing of the codons of mRNA.

Table 9.6 Codes for synthesising amino acids: how many bases are needed?

DNA contains four bases, **A**denine, **T**hymine, **G**uanine, and **C**ytosine, the alphabet of a code.

if 1 base = 1 amino acid (singlet code), 4 amino acids can be coded } possibilities

if 2 bases = 1 amino acid (doublet code), 4^2 (=16) amino acids can be coded } too few

if 3 bases = 1 amino acid (triplet code), 4^3 (= 64) amino acids can be coded (with only about 20 amino acids to code, some triplets must be redundant)

if 4 bases = 1 amino acid, 4^4 (= 256) amino acids can be coded (far more possibilities than are needed)

Singlet code	Doublet code				Triplet code			
A	AA	AG	AC	AT	AAA	AAG	AAC	AAT
G	GA	GG	GC	GT	AGA	AGG	AGC	AGT
C	CA	CG	CC	CT	ACA	ACG	ACC	ACT
T	TA	TG	TC	TT	ATA	ATG	ATC	ATT
					GAA	GAG	GAC	GAT
					GGA	GGG	GGC	GGT
					GCA	GCG	GCC	GCT
					GTA	GTG	GTC	GTT
					CAA	CAG	CAC	CAT
					CGA	CGG	CGC	CGT
					CCA	CCG	CCC	CCT
					CTA	CTG	CTC	CTT
					TAA	TAG	TAC	TAT
					TGA	TGG	TGC	TGT
					TCA	TCG	TCC	TCT
					TTA	TTG	TTC	TTT

Francis Crick's 'frame-shift' experiment

In 1961 Crick provided experimental evidence for the triplet hypothesis. Working with the bacteria *E. coli*, he made additions or deletions of one or more nucleotides to specific genes. The changes (a type of **gene mutation** (p. 627)) caused the genes to be misread, and resulted in abnormal phenotypes. But when **three** bases were added or subtracted the resulting phenotypes were normal or almost normal.

(It is also the sequence of bases in the non-transcribing strand of DNA, and it is the complementary base sequence to that in the coding strand of DNA, p. 208.)

In the genetic code there are more codons than there are amino acids, and so the code is described as **degenerate**. Only two amino acids are coded for by a single triplet code: methionine (AUG) and tryptophan (UGG). A few others have two codons. Most amino acids are coded for by several triplet codes, up to a maximum of six. In effect the code contains more 'words' than are required, and the spare capacity in the language is used rather than left unquoted. Some of the words code for a 'full stop', however, rather than for another amino acid. These are the protein chain-terminating codes (UAA, UAG, UGA).

Another important discovery is that, with scarcely any exceptions, the code is universal. This means that any particular triplet code represents the same amino acid whether in bacteria, cyanobacteria, green plants, fungi or animals.

Questions

1 What sequence of changes is catalysed by RNA polymerase?
2 What is the likely significance of the discovery that the DNA code is universal (p. 651)?
3 Write the sequence of amino acids in a polypeptide coded for by UCCUACCACCAAAAA.
4 What kind of experimental results would you expect to see if the Meselson–Stahl experiment was carried on for three generations?

Table 9.7 The 20 amino acids found in proteins and the genetic code

Amino acid	Abbreviation
alanine	Ala
arginine	Arg
asparagine	Asn
aspartic acid	Asp
cysteine	Cys
glutamine	Gln
glutamic acid	Glu
glycine	Gly
histidine	His
isoleucine	Ile
leucine	Leu
lysine	Lys
methionine	Met
phenylalanine	Phe
proline	Pro
serine	Ser
threonine	Thr
tryptophan	Trp
tyrosine	Tyr
valine	Val

The genetic code (the code is given here in terms of the mRNA codons)

first base	second base U	second base C	second base A	second base G	third base
U	UUU UUC Phe / UUA UUG Leu	UCU UCC UCA UCG Ser	UAU UAC Tyr / UAA UAG stop	UGU UGC Cys / UGA stop / UGG Trp	U C A G
C	CUU CUC CUA CUG Leu	CCU CCC CCA CCG Pro	CAU CAC His / CAA CAG Gln	CGU CGC CGA CGG Arg	U C A G
A	AUU AUC AUA Ile / AUG Met	ACU ACC ACA ACG Thr	AAU AAC Asn / AAA AAG Lys	AGU AGC Ser / AGA AGG Arg	U C A G
G	GUU GUC GUA GUG Val	GCU GCC GCA GCG Ala	GAU GAC Asp / GAA GAG Glu	GGU GGC GGA GGG Gly	U C A G

207

9.9 PROTEIN SYNTHESIS

Protein synthesis requires a supply of amino acids (the pool of amino acids in the cytoplasm), energy (obtained by reactions with ATP) and information (the linear sequence of bases of DNA, which determines the primary structure of protein, p. 141). When these are brought together protein is synthesised in three steps: **transcription**, **activation** and **translation** (Figure 9.19).

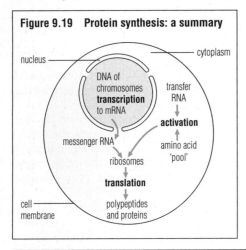

Figure 9.19   Protein synthesis: a summary

■ Step one: transcription

Transcription (Figure 9.20) is the process of transferring part of the coded information of DNA of chromosomes in the nucleus to the ribosomes in the cytoplasm. It involves a nucleic acid molecule, made of RNA (p. 145) and known as **messenger RNA** (mRNA). This is a single-strand molecule. mRNA is manufactured in the nucleus from one strand of the double helix, referred to as the coding strand. The enzyme catalysing the transcription is RNA polymerase. RNA polymerase attaches to the double helix at a 'start' signal (normally the codon for the amino acid methionine). First the hydrogen bonds are broken in the region of the DNA to be copied (the gene or genes that are to be 'expressed'), and the DNA strands unwind. One DNA strand is then copied by base pairing of ribonucleotides, and the ribonucleotides are condensed together to make a strand of mRNA. The base sequence of the mRNA is complementary to the coding strand of DNA. Once formed, the mRNA passes out into the cytoplasm and becomes attached to a ribosome.

■ Step two: amino acid activation

Amino acids are activated for protein synthesis by combining with a short length of RNA, known as **transfer RNA** (tRNA). The activation processes involve ATP (Figure 9.21). There are more than twenty different tRNA molecules, one for each of the twenty or so amino acids coded for in cell proteins. All tRNA molecules have a clover leaf shape, but they differ in the sequence of bases, known as the **anticodon**, which is exposed on one of the 'clover leaves'. This anticodon is complementary to the codon of mRNA. The enzyme catalysing the formation of the amino acid–tRNA complex 'recognises' one type of amino acid and the corresponding tRNA.

■ Step three: translation

Translation (Figure 9.22) occurs in the ribosomes (p. 161): these are tiny, near-spherical two-component structures, consisting of protein and RNA. Ribosomes move along the length of a mRNA strand, 'reading' the codons from the 'start' codon. In the ribosome, complementary anticodons of amino

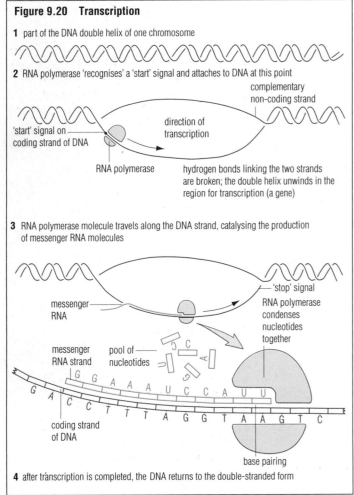

Figure 9.20   Transcription

1 part of the DNA double helix of one chromosome

2 RNA polymerase 'recognises' a 'start' signal and attaches to DNA at this point

complementary non-coding strand

'start' signal on coding strand of DNA

direction of transcription

RNA polymerase

hydrogen bonds linking the two strands are broken; the double helix unwinds in the region for transcription (a gene)

3 RNA polymerase molecule travels along the DNA strand, catalysing the production of messenger RNA molecules

messenger RNA

'stop' signal
RNA polymerase condenses nucleotides together

messenger RNA strand

pool of nucleotides

G G A A A U C C A U U
G A C C T T T A G G T A A G T C

coding strand of DNA

base pairing

4 after transcription is completed, the DNA returns to the double-stranded form

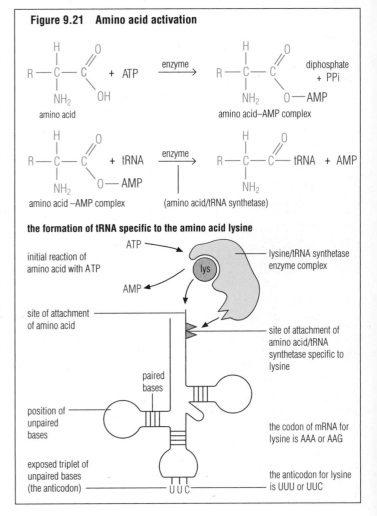

Figure 9.21   Amino acid activation

amino acid + ATP $\xrightarrow{\text{enzyme}}$ amino acid–AMP complex + diphosphate + PPi

amino acid –AMP complex + tRNA $\xrightarrow{\text{enzyme}}$ (amino acid/tRNA synthetase) + AMP

the formation of tRNA specific to the amino acid lysine

initial reaction of amino acid with ATP

ATP

AMP

lys

lysine/tRNA synthetase enzyme complex

site of attachment of amino acid

site of attachment of amino acid/tRNA synthetase specific to lysine

paired bases

position of unpaired bases

the codon of mRNA for lysine is AAA or AAG

exposed triplet of unpaired bases (the anticodon)

U U C

the anticodon for lysine is UUU or UUC

acid–tRNA complexes are held in place by hydrogen bonds. The amino acids are then joined by a peptide bond. Thus the ribosome is acting as a supporting framework, holding mRNA and two amino acid–tRNA complexes together, and enabling an enzyme to catalyse the formation of a polypeptide bond between adjacent amino acids. The ribosome moves on, three bases at a time, and the cycle of events is repeated. The tRNAs are released into the cytoplasm and are re-used in amino acid activation. The whole process continues until a 'stop' codon is reached.

A polypeptide chain is assembled, one amino acid residue at a time, and when completed is discharged into the cytoplasm. The sequence of amino acid residues is the primary structure of that protein. The polypeptide chain then takes up a three-dimensional structure, forming the secondary, tertiary and quaternary structure of the protein (pp. 142–3). Once the protein has taken up a particular shape, the shape is maintained by various intramolecular forces (p. 142).

■ The fate of the proteins formed

Proteins that are used in the cell are formed on ribosomes in the cytosol. (Where many ribosomes move along a single mRNA strand, the whole structure is called a polyribosome or **polysome**, Figure 9.22). On the other hand, proteins for export from the cell, such as digestive enzymes secreted outside the cell membrane, are formed on ribosomes attached to endoplasmic reticulum (p. 160).

Certain of the proteins formed become structural proteins, such as collagen fibres (p. 142); other proteins are enzymes (p. 179). The range of roles of proteins in cells and organisms is listed in Table 6.3 (p. 144).

The cycle of formation and breakdown of cell protein

Most of the enzymes that are important in metabolic regulation are short-lived; their concentration and activity may change rapidly. Damaged and defective proteins are also quickly broken down.

Other proteins have a longer life. Proteins, on breakdown, contribute to the pool of amino acids in the cells from which new proteins are constructed.

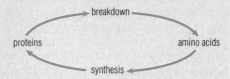

The source of amino acids in plants and mammals

Mammals have heterotrophic nutrition (p. 270). The pool of amino acids in their cells is added to mainly by the digestion of dietary proteins, although mammals can synthesise some of the amino acids required. Furthermore, mammals cannot store proteins to any significant extent, apart from the protein in structures such as muscle. Consequently, the excess protein taken in is deaminated and the ammonia released is excreted either directly or after conversion to urea (p. 377).

Green plants have autotrophic nutrition (p. 246), and manufacture the elaborated nutrients they require. The element nitrogen is absorbed mainly as nitrate ions. Green plants then manufacture their own amino acids, first by reducing NO_3^- to $-NH_2$, and then forming the amino acid glutamic acid. All other essential amino acids are formed from keto acids by **transamination**; that is, by transferring the amino group.

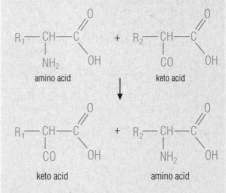

For example,
  glutamic acid + pyruvic acid →
    α-ketoglutaric acid + alanine
Plants are able to store proteins in certain circumstances, such as in seeds and food storage tissues (p. 404).

Question

What are the key differences between transcription and translation? Where in the cell do they occur?

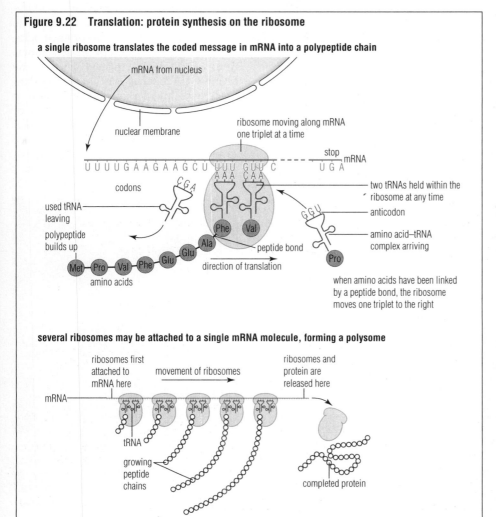

Figure 9.22   Translation: protein synthesis on the ribosome

a single ribosome translates the coded message in mRNA into a polypeptide chain

mRNA from nucleus

nuclear membrane

ribosome moving along mRNA one triplet at a time

stop

mRNA

U U U U G A A G A A G C U U U U G U U C   U G A

codons

C G A

A A A   C A A

used tRNA leaving

polypeptide builds up

two tRNAs held within the ribosome at any time

anticodon

amino acid–tRNA complex arriving

Phe   Val

G G U

Ala

Glu   Glu

Pro

Met   Pro   Val   Phe

peptide bond

direction of translation

amino acids

when amino acids have been linked by a peptide bond, the ribosome moves one triplet to the right

several ribosomes may be attached to a single mRNA molecule, forming a polysome

ribosomes first attached to mRNA here

movement of ribosomes

ribosomes and protein are released here

mRNA

tRNA

growing peptide chains

completed protein

9.10 WHAT IS A GENE?

Earlier we referred to lengths of DNA that are being transcribed as **genes**. A gene is also defined as a hereditary factor (p. 608). More precisely, a gene is a specific region of a chromosome that is capable of determining the development of a specific trait or characteristic of an organism. Thus a gene is a specific length of the DNA helix, about a thousand bases long, which codes for a polypeptide or RNA product.

■ An experiment using *Neurospora*

The idea that one gene is a length of DNA that codes for one protein (or, at least, a polypeptide) was established in 1940 by the biochemical geneticists George Beadle and Edward Tatum at Stamford University. The organism they chose for their experiments was the fungus *Neurospora crassa* (p. 559), a fungus whose spores are ubiquitous, and which is a pest organism in bakeries because it can damage flour stocks. The reasons for their choice were as follows:

▶ in the laboratory, *Neurospora* can be grown on a defined 'minimal' medium containing only sucrose, inorganic salts and the growth factor biotin

▶ the life cycle is short, and the organism is relatively simple biochemically

▶ the diploid nucleus contains seven pairs of chromosomes on which gene positions are easily mapped

▶ the haploid condition lasts for most of the life cycle, so recessive genes (p. 608) are immediately detected

▶ the ascospores (p. 559) occur in a linear series in the ascus and can be dissected out to isolate reproductive progeny.

Neurospora was exposed to X-rays in order to cause random mutations (p. 627). Exposed spores were then cultured separately in order to detect mutant strains unable to synthesise nutrients needed for growth (that is, mutants that could no longer grow on minimal medium). Then, by adding extra growth factors or amino acids, one by one, it was determined which substance a mutant fungus could no longer synthesise. For example, one mutant grew only when arginine was added. Beadle and Tatum established that this mutant had lost the gene coding for a single enzyme, the enzyme catalysing the synthesis of arginine. Only when arginine was supplied in the medium could this mutant grow (Figure 9.23). They had shown that one gene was responsible for the production of a single enzyme.

■ Sickle-cell anaemia

Sickle-cell anaemia is a genetically controlled condition in which the normal haemoglobin of red blood cells (haemoglobin A) is replaced by another haemoglobin (haemoglobin S). Haemoglobin S, when deoxygenated, forms long fibrous molecules which distort and damage the red cells (Figure 9.24). Sickle-cell red blood cells are less efficient than normal cells in the delivery of oxygen to tissues, and when distorted they may block the fine capillaries. The disease is relatively rare. Rather more common is sickle-cell trait, in which only some of the red blood cells become sickle-shaped.

The cause of this condition is a mutation in one base of one codon in the gene for one component protein of haemoglobin (p. 358). As a result of the gene mutation, the mRNA formed carries the codon GUA (codes for valine) rather than GAA (which codes for glutamic acid) at one point in the chain. The discovery of the cause of sickle-cell anaemia provided further evidence that a gene is a sequence of DNA, coding for a linear sequence of amino acids making up a single protein.

Figure 9.23   Beadle and Tatum's experiments with *Neurospora*

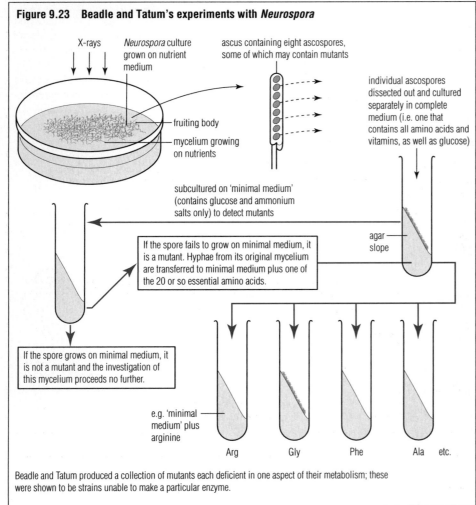

X-rays

Neurospora culture grown on nutrient medium

ascus containing eight ascospores, some of which may contain mutants

fruiting body

mycelium growing on nutrients

individual ascospores dissected out and cultured separately in complete medium (i.e. one that contains all amino acids and vitamins, as well as glucose)

subcultured on 'minimal medium' (contains glucose and ammonium salts only) to detect mutants

agar slope

If the spore fails to grow on minimal medium, it is a mutant. Hyphae from its original mycelium are transferred to minimal medium plus one of the 20 or so essential amino acids.

If the spore grows on minimal medium, it is not a mutant and the investigation of this mycelium proceeds no further.

e.g. 'minimal medium' plus arginine

Arg   Gly   Phe   Ala   etc.

Beadle and Tatum produced a collection of mutants each deficient in one aspect of their metabolism; these were shown to be strains unable to make a particular enzyme.

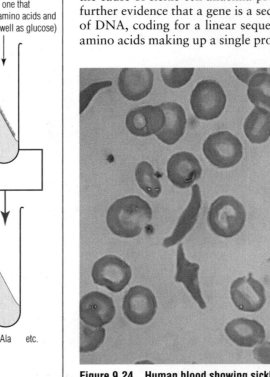

Figure 9.24   Human blood showing sickle-cell trait (×3000)

■ How the genes are controlled

A cell nucleus contains more coded instructions than are required at any one moment. For example, in a multicellular organism every nucleus contains the coded information relating to the development and maintenance of all mature tissues and organs of the body. During development the genetic information in the nucleus is used in a selective way. Genes are expressed only in cells they relate to, when they are needed; for example, a gene for pepsin production is present in all the cells of the body but it is active only in cells of the gastric glands in the stomach. How is this gene regulation brought about?

Much of what is known of gene regulation comes from the study of prokaryotes. For example, François Jacob and Jacques Monod used the fact that in a bacterium such as *Escherichia coli* many of the enzymes are synthesised continuously, but other enzymes are synthesised only in the presence of an inducing substance. They demonstrated that, in *E. coli*, enzymes needed to metabolise glucose are continuously produced, but when the culture medium is changed to one containing lactose but no glucose there is an initial delay in growth. This is because the formation of the lactose-metabolising enzyme is induced by the presence of lactose (and ceases within minutes of *E. coli* being returned to a glucose-containing medium).

Jacob and Monod discovered that, in addition to the genes for the enzymes of lactose metabolism, the bacterial DNA also contained a regulatory gene and a length called an operator site (Figure 9.25). When lactose is absent a repressor protein made by the regulatory gene binds with the operator site and blocks transcription of the genes for lactose metabolism. When lactose is present it binds to the repressor molecule and prevents the blocking of the operator site: transcription then proceeds, and the genes for lactose metabolism are 'switched on'.

In eukaryotes the mechanism discovered by Jacob and Monod in *E. coli* does not operate. In fact, in eukaryotes gene regulation is less well understood than in prokaryotes.

■ Genetic engineering

The ability to manipulate genetic material is known as recombinant DNA technology, or **genetic engineering** (explored fully in Chapter 29). By this technology, genes can be manipulated both inside and outside cells (referred to as *in vivo*, meaning 'in life', and *in vitro*, or 'in glassware', respectively).

Questions

1 Why is *Neurospora* useful as a research organism in biochemical genetics?
2 Why is it necessary for the eukaryotic cell to regulate the expression of certain genes?
3 How does knowledge of the cause of sickle-cell trait support the concept of a gene as a linear sequence of bases?

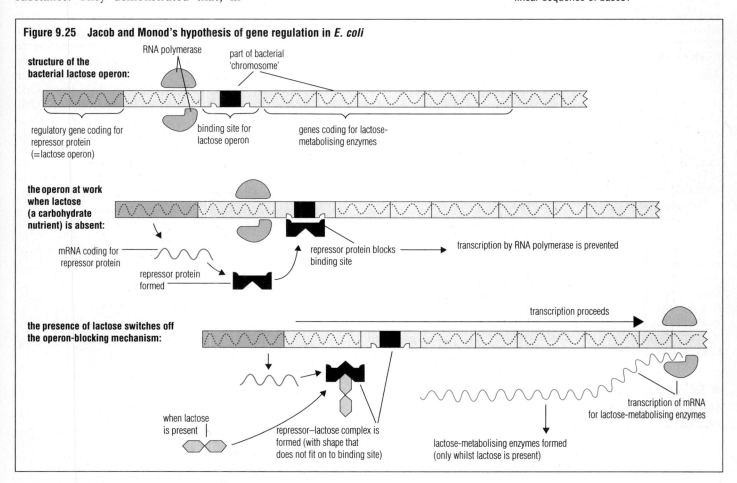

Figure 9.25   Jacob and Monod's hypothesis of gene regulation in *E. coli*

structure of the bacterial lactose operon:

RNA polymerase

part of bacterial 'chromosome'

regulatory gene coding for repressor protein (=lactose operon)

binding site for lactose operon

genes coding for lactose-metabolising enzymes

the operon at work when lactose (a carbohydrate nutrient) is absent:

mRNA coding for repressor protein

repressor protein formed

repressor protein blocks binding site

transcription by RNA polymerase is prevented

the presence of lactose switches off the operon-blocking mechanism:

transcription proceeds

when lactose is present

repressor–lactose complex is formed (with shape that does not fit on to binding site)

lactose-metabolising enzymes formed (only whilst lactose is present)

transcription of mRNA for lactose-metabolising enzymes

FURTHER READING/WEB SITES

F Crick (1988) *What Mad Pursuit*. Weidenfeld and Nicolson
Browse: **mcb.harvard.edu/BioLinks.html**

10 Cell specialisation

- Histology is the study of the microscopic structure of tissues and organs. A tissue is a group of cells of similar structure that perform a particular function. An organ is a collection of tissues and has a specific role.
- Specialised tissues of animals are classified as:
  ▶ epithelia – single or multiple sheets of cells covering internal and external surfaces
  ▶ connective tissue – binding and support tissues
  ▶ muscle tissue – contractile tissue made of elongated cells or fibres
  ▶ nervous tissue – nerve cells that conduct nerve impulses.
- Specialised tissues of plants are classified as:
  ▶ parenchyma and collenchyma – packing and support tissues
  ▶ sclerenchyma (fibres) – reinforced support and strengthening tissues of stems and roots
  ▶ xylem – water-conducting tissue found in the vascular bundles
  ▶ phloem – food-conducting tissue in the vascular bundles.

10.1 MULTICELLULAR ORGANISATION AND DIVISION OF LABOUR

The new cells formed after mitosis and cell division normally enlarge and become specialised for a particular function. Multicellular organisms consist of many groups of specialised cells, making up their tissues and organs.

A **tissue** is a group of cells of more or less similar structure that performs a particular function. Examples of tissues are muscle tissue of mammals (Figure 10.12, p. 219), and parenchyma tissue of flowering plants (Figure 10.19, p. 223). The cells of some tissues are all of one type; for example, squamous epithelium in mammals consists of a single layer of identical cells (Figure 10.3). Other tissues are mixtures of different types of cell; phloem tissue in flowering plants, for example, consists of sieve tube elements, companion cells, parenchyma cells and sometimes fibres.

An **organ** is a collection of tissues. It has a definite form and structure, and it performs one or more specific functions. Examples of organs are the leaf of a flowering plant (p. 255) and the heart of a vertebrate animal, such as a mammal. Organs of animals are part of larger functional units, known as **systems**. For example, the heart is part of the vascular system, which consists of a heart and all the blood vessels.

■ Differentiation

Differentiation is the process by which unspecialised structures become modified and specialised for the performance of specific functions. This chapter is concerned with the specialisation of cells to form the tissues of multicellular organisms. Differentiation begins as soon as cells have been formed by cell division.

Differentiation involves changes in both the shape and the physiology of a cell, which enable it to carry out a specialised function. In some nerve cells, for instance, long fibres growing out from the cell allow conduction of impulses over a distance, while modifications of the chemical and physical changes in its cytoplasm help to establish and conduct the nerve impulses (p. 445).

Differentiation can also be seen as a **division of labour**. When cells become specialised for one function they often lose the ability to carry out other functions. The advantage of division of labour is that specialised cells perform specific tasks more effectively than can unspecialised cells. A disadvantage could be that the more profoundly a cell becomes specialised, the less readily it can change its function if an unusual need arises. Sometimes the changes to cells are drastic and irreversible, such as the total loss of living contents in mature xylem vessels of flowering plants, or the loss of the nucleus of red blood cells.

■ Histology

The study of the structure of tissues and organs is known as **histology**. Histology is carried out largely by microscopy (p. 152). For example, a series of very thin sections of tissue or organ are cut both longitudinally (LS) and transversely (TS). These sections are then suitably stained and examined microscopically.

Some cells are best examined intact. One way of obtaining isolated and intact cells is by **maceration**; that is, the process of dissolving away the softer substances of an organ, leaving only the harder structures present (Figure 10.25, p. 226).

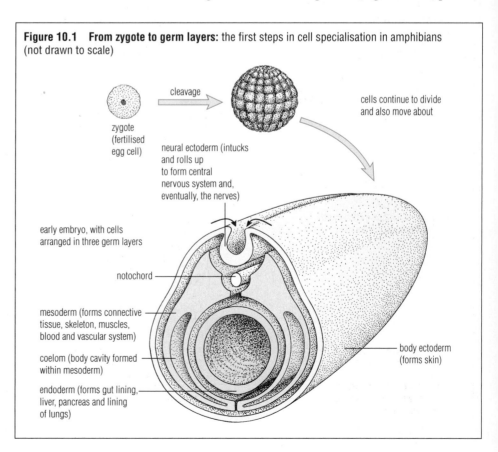

Figure 10.1   From zygote to germ layers: the first steps in cell specialisation in amphibians (not drawn to scale)

cleavage

zygote (fertilised egg cell)

cells continue to divide and also move about

neural ectoderm (intucks and rolls up to form central nervous system and, eventually, the nerves)

early embryo, with cells arranged in three germ layers

notochord

mesoderm (forms connective tissue, skeleton, muscles, blood and vascular system)

coelom (body cavity formed within mesoderm)

endoderm (forms gut lining, liver, pancreas and lining of lungs)

body ectoderm (forms skin)

10.2 SPECIALISATION OF ANIMAL CELLS

Most animals are formed from a fertilised egg cell, known as a **zygote**. The early development of an organism from the zygote (embryonic development, p. 413) is fundamentally similar in all animal species. Development of the embryo is conveniently divided into three phases: cleavage, gastrulation and organogeny. The early phases of embryonic development in amphibians, which show the main steps in the development of a vertebrate particularly clearly, are illustrated in Figure 10.1.

During **cleavage** the zygote formed at fertilisation undergoes repeated mitotic cell division. The new cells become organised into a tiny hollow ball of cells.

The cells continue to divide but also move about and become rearranged as special layers, known as **germ layers**, within the enlarging embryo. This phase is known as **gastrulation**. Three germ layers are formed:

▶ the **ectoderm**, which subsequently forms the nervous system and skin
▶ the **mesoderm**, which forms connective tissue, skeletal tissues, muscles, and the vascular and renal systems
▶ the **endoderm**, which forms the gut lining, the liver and pancreas, and the lining of the lungs.

The cells of the germ layers undergo further cell division and specialisation, and form the organs of the embryo. This phase is called **organogeny**. The organs are made of four basic types of tissue, known as **epithelia**, **connective tissues**, **nerves** and **muscle** (Figure 10.2). These tissues all form from cells of the germ layers in the early growth and development of the mammal.

10.3 EPITHELIA

An **epithelium** is a tissue consisting of one or a few layers of cells, and is found on the internal and external surfaces of organs.

Unlike most tissues, epithelia consist of fairly simple cells. Those epithelia that are only one cell thick are called **simple epithelia**; those that are several cells thick are called **compound** or **stratified epithelia**. Epithelial cells are attached to the underlying tissues by a **basement membrane**, made of a network of white, wavy, non-elastic collagen fibres. The junctions between epithelial cells tend to be strongly held together by desmosomes (Figure 7.30, p. 167).

Many epithelia protect the underlying tissues against mechanical injury such as abrasion or pressure. Other epithelia protect against water loss and dehydration, against infection by microorganisms, and sometimes also from radiant energy. Epithelia are involved in the physiological processes that take place across surfaces, including respiratory gas exchange, excretion, elimination of waste and absorption. As a result, the free surface of epithelial cells may be adapted for absorption or secretion, or the cells may bear cilia that beat and so move liquids over their surface. Many epithelia are sensitive layers, with nerve endings penetrating between the cells.

Epithelia may be classified in various ways: by the shape of the cells, or by the number of cell layers present. Functionally there are three types of epithelium.

■ Squamous epithelium

Squamous (or pavement) epithelium (Figure 10.3, overleaf) consists of thin, smooth and strong sheets of cells. Squamous epithelium occurs in blood vessels, and lines the chambers of the heart and also the inside of the body cavity (the peritoneum). The thinness of squamous epithelium facilitates diffusion across it, while its smoothness facilitates the passage of fluids and lubricates the movement between adjacent surfaces. (The epithelium that lines tubes such as blood vessels is called an **endothelium**, and is derived from the embryonic mesoderm.)

■ Columnar and cubical epithelia

Columnar and cubical epithelia (Figure 10.4, overleaf) have either a secretory or an absorptive function or both, and always occur close to blood vessels. They often contain mucus-secreting goblet cells. Columnar and cubical epithelia line ducts and glands such as the salivary glands and sweat glands. Ciliated epithelia line the oviducts and respiratory surfaces, where the cilia cause movements of liquid such as the mucus stream that clears away dust and bacteria from the lungs (p. 309). Columnar epithelium lines the villi of the small intestine, and has an outer border of microvilli (p. 167). The types of junction between the epithelial cells of the small intestine are shown in Figure 7.30 (p. 167).

■ Stratified epithelium

Stratified epithelium (Figure 10.5, overleaf) consists of several layers of cells, which form a tough, impervious barrier. The cells of the outermost layers are dead, and are continuously replaced from below. The innermost layer of cells, known as the **generative layer**, is in an active state of mitotic cell division. Above the generative layer the cells become

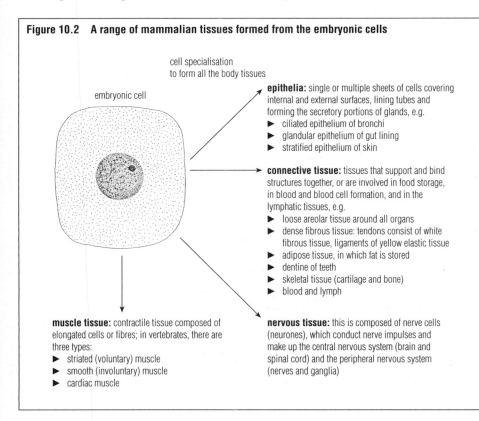

Figure 10.2   A range of mammalian tissues formed from the embryonic cells

cell specialisation to form all the body tissues

embryonic cell

epithelia: single or multiple sheets of cells covering internal and external surfaces, lining tubes and forming the secretory portions of glands, e.g.
▶ ciliated epithelium of bronchi
▶ glandular epithelium of gut lining
▶ stratified epithelium of skin

connective tissue: tissues that support and bind structures together, or are involved in food storage, in blood and blood cell formation, and in the lymphatic tissues, e.g.
▶ loose areolar tissue around all organs
▶ dense fibrous tissue: tendons consist of white fibrous tissue, ligaments of yellow elastic tissue
▶ adipose tissue, in which fat is stored
▶ dentine of teeth
▶ skeletal tissue (cartilage and bone)
▶ blood and lymph

muscle tissue: contractile tissue composed of elongated cells or fibres; in vertebrates, there are three types:
▶ striated (voluntary) muscle
▶ smooth (involuntary) muscle
▶ cardiac muscle

nervous tissue: this is composed of nerve cells (neurones), which conduct nerve impulses and make up the central nervous system (brain and spinal cord) and the peripheral nervous system (nerves and ganglia)

Figure 10.3 Squamous (or pavement) epithelium

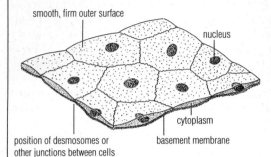

smooth, firm outer surface
nucleus
position of desmosomes or other junctions between cells
cytoplasm
basement membrane

squamous (or pavement) epithelium

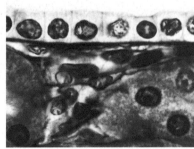

photomicrograph of squamous epithelium lining the thorax

Figure 10.4 Columnar and cubical epithelia

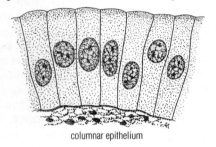

columnar epithelium

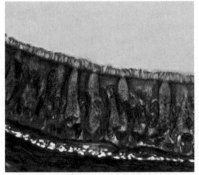

photomicrograph of ciliated columnar epithelium

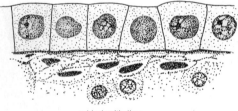

cubical epithelium

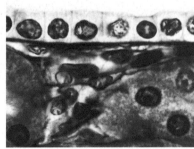

photomicrograph of cubical epithelium of the renal tubule

Figure 10.5 Stratified epithelium: this is a multilayered (compound) epithelium

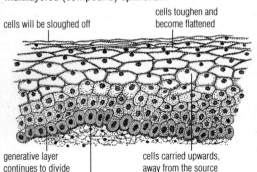

cells will be sloughed off
cells toughen and become flattened
generative layer continues to divide mitotically
cells carried upwards, away from the source of nutrients
basement membrane

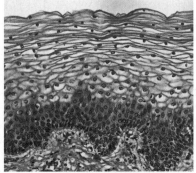

photomicrograph of stratified epithelium of the cervix

progressively isolated from the oxygen and nutrients supplied by the blood. The cells die and eventually slough off.

This multilayered structure provides protection from abrasion. A stratified epithelium occurs in the buccal (mouth) cavity and in the vagina, for example. In the skin (epidermis) additional protection is afforded by the dying cells, the proteins of which become modified into a hard, waterproof layer. The protein of this layer (**keratin**) provides a tough surface, referred to as 'cornified'. The greater the friction that a given area of skin experiences, the thicker the cornified layer becomes. For example, after regular hard physical activity such as rowing or digging, the epidermis of the hands becomes noticeably toughened up.

Questions

1 Give one example each of an organ, a tissue and an undifferentiated cell to be found in
  a a flowering plant (p. 406)
  b a mammal.
2 What is the connection between 'differentiation' and 'division of labour'?
3 Give examples of mammalian epithelia that respectively
  a bear cilia
  b protect the body against mechanical injury
  c facilitate respiratory gas exchange.
4 What do you think might be the function of the ciliated epithelium of the oviduct?

10.4 CONNECTIVE TISSUES

Connective tissues are the supporting tissues of the body. They contain large amounts of intercellular material or **matrix**, as well as specialised cells. Connective tissues develop from the embryonic mesoderm. The matrix is normally secreted by the connective tissue cells, and forms the bulk of the tissue.

There are many types of connective tissue with widely differing properties and functions (Figures 10.6 and 10.7). The structure in relation to function of two examples of connective tissue, namely bone and blood, is detailed next.

■ Skeletal tissues

The skeletal tissues of mammals consist of cartilage and bone. In both these types of connective tissue the matrix is comparatively hard.

Cartilage

In cartilage, the matrix is of a protein called **chondrin**. Chondrin is secreted by cells (**chondrocytes**) that become embedded in the matrix in small clusters. Blood vessels do not penetrate cartilage; oxygen and nutrients reach these cells by diffusion, and waste products diffuse away.

Cartilage tissue is firm but flexible, resists strain and is elastic. The common form of cartilage in mammals, known as **hyaline cartilage** (Figure 10.8), forms the entire skeleton of the embryo. During development, the cartilage skeleton is replaced by bone. In adults, hyaline cartilage is restricted to the surfaces of bones at joints (p. 486), the flexible connections between ribs and sternum (p. 484), and the incomplete rings that hold open the trachea and bronchi (p. 308).

Other forms of cartilage occur in mammals, some of it permeated by numerous elastic fibres. **Elastic cartilage** occurs in the pinna of the outer ear and in the epiglottis (p. 278). Some cartilage contains collagen fibres, which give substantial additional strength and rigidity. This **fibro-cartilage** occurs in the intervertebral discs of the spinal column, forming tough 'shock-absorbing' pads (p. 484).

Bone

Bone makes up the greater part of the skeleton in many vertebrates, and approximately 15% of the body mass of the adult mammal.

Bone is a very strong and rigid tissue, composed of approximately 70% mineral and 30% organic matter. The organic components of bone are numerous bone cells known as **osteocytes**, together with a matrix of mucopolysaccharide substance containing collagen fibres. The inorganic components are very hard deposits of minerals such as calcium phosphate (85%) and calcium carbonate (10%), together with magnesium and calcium fluorides (5%).

Osteocytes secrete and maintain the bone matrix. This is impervious to tissue fluid, but osteocytes are supplied with nutrients from blood vessels that permeate the bone tissue. These blood vessels run through the matrix in canals (**Haversian canals**). Slender cytoplasmic threads (processes) extend from immature osteocytes through tiny tubes (**canaliculi**) in the matrix. These cytoplasmic processes are later withdrawn, and the canals fill with tissue fluid. Canaliculi are the routes by which nutrients, waste materials and respiratory gases move between blood plasma and bone cells.

Figure 10.6 Types of connective tissue, classified by role or function

| binding together organs within body cavity, skin to sub-dermal tissue (loose connective tissue Figure 10.7) | food storage adipose tissue | protection and provision of supporting framework | producing blood and lymph |

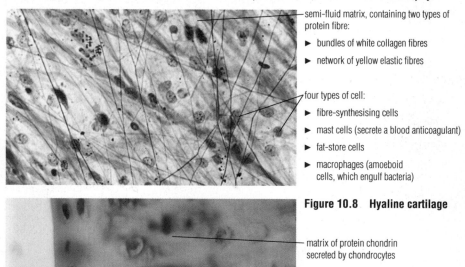

Figure 10.7 Loose connective tissue: also known as areolar tissue, this tissue binds tissues and organs together. The transparent matrix contains flexible fibres, some inelastic and others elastic, so that this tissue has strength and resilience. The cells present lay down the components of the connective tissue, help protect against disease, and assist in the responses of the tissue to mechanical injury

— semi-fluid matrix, containing two types of protein fibre:
► bundles of white collagen fibres
► network of yellow elastic fibres

— four types of cell:
► fibre-synthesising cells
► mast cells (secrete a blood anticoagulant)
► fat-store cells
► macrophages (amoeboid cells, which engulf bacteria)

Figure 10.8 Hyaline cartilage

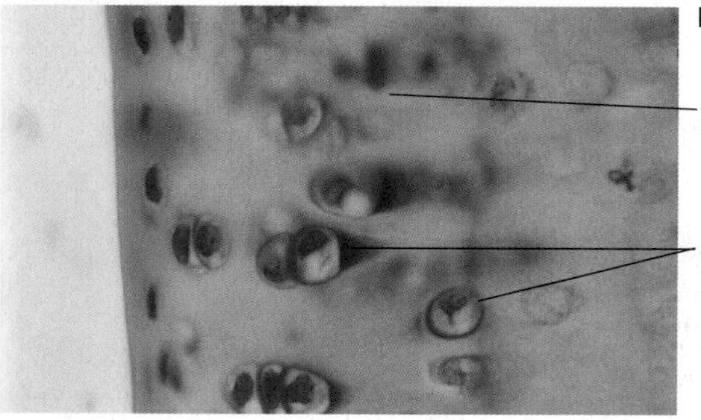

— matrix of protein chondrin secreted by chondrocytes

— chondrocytes

There are two types of bone tissue: compact bone and spongy bone.

Compact bone

Compact bone is very dense. It makes up the greater part of the bones of the body. Compact bone consists of osteocytes and matrix arranged in **Haversian systems**, that is, in cylinders of bone surrounding central Haversian canals. Each cylinder consists of concentric rings of osteocytes and the calcified matrix. An artery and a vein run through each Haversian canal, and capillaries pass via canaliculi to the osteocytes in spaces known as **lacunae** (Figure 10.9). Additionally, occasional large blood vessels in larger canals connect the blood vessels of the Haversian canals to the exterior of the bone.

At the outer and inner surfaces of a bone there are no Haversian systems, but bone tissue is laid down as a solid layer. The exterior of the bone is covered with a tough fibrous membrane of dense connective tissue called **periosteum**.

Spongy bone

Spongy or **cancellous bone** occurs within larger bones, and is always surrounded by compact bone. Spongy bone consists of thin bars or sheets of bone called **trabeculae**, interspersed with large spaces occupied by bone marrow (see below). The trabeculae contain osteocytes, which are more or less irregularly dispersed in the matrix (Figure 10.9, overleaf). The matrix contains a rather smaller proportion of inorganic material than does the matrix of compact bone. The trabeculae develop along the lines of stress within bone.

The spaces within the spongy bone at the head of the long bones contain **red bone marrow tissue**. This very soft tissue is less dense than bone, and is the site of red blood cell formation (p. 216). **Yellow marrow tissue**, consisting principally of fat, fills the spaces within the spongy bone of the shaft.

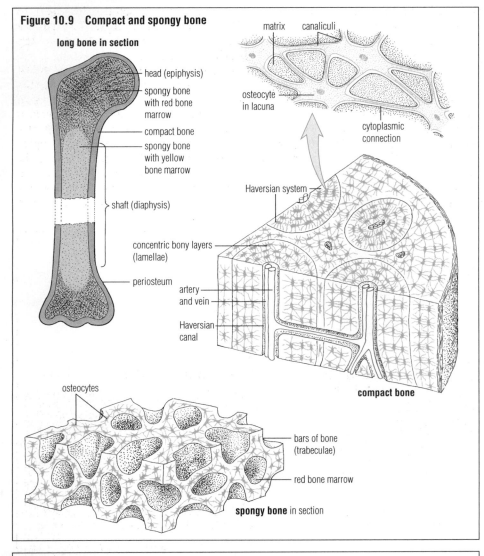

Figure 10.9 Compact and spongy bone

long bone in section

- head (epiphysis)
- spongy bone with red bone marrow
- compact bone
- spongy bone with yellow bone marrow
- shaft (diaphysis)
- concentric bony layers (lamellae)
- periosteum
- artery and vein
- Haversian canal

matrix   canaliculi

- osteocyte in lacuna
- cytoplasmic connection

Haversian system

compact bone

osteocytes

- bars of bone (trabeculae)
- red bone marrow

spongy bone in section

Figure 10.10 Ossification and growth of a mammalian long bone

Bones increase in length by deposition of cartilage at the ends of the diaphysis (shaft), at the epiphysial plate. This additional cartilage is then ossified. The epiphysial plate remains active until the individual reaches its full adult size.

- cartilage
- compact bone forms under the periosteum at the margin of the diaphysis
- spongy (cancellous) bone with yellow marrow forms in the centre of the diaphysis
- spongy (cancellous) bone with red marrow forms in the epiphysis
- epiphysial plate
- compact bone formation in the epiphysis
- centre of shaft breaks down and is replaced by yellow bone marrow
- articular cartilage

mature bone

stages in ossification

Formation of bone

Most bones of the body are formed by ossification (that is, the laying down of the inorganic components of bone) of pre-existing cartilage by the bone-forming cells called **osteoblasts** (Figure 10.10). (Osteoblasts are actively dividing bone cells; osteocytes have stopped dividing.) Examples of bones of cartilaginous origin are the limbs, girdles and vertebral column.

The bones of the skull and the clavicles (collar bones) are not formed by the ossification of cartilage. These bones, called **dermal** or **membrane bones**, are formed by direct ossification of the dermis of the skin.

Mature bones are in a state of dynamic equilibrium, continuously being reabsorbed and reconstructed by the action of bone cells. This process enables bones to grow, and to adapt structurally to changes in mechanical use. It also allows repair of fractures. The movements of calcium and phosphate ions between the bone matrix and blood plasma are controlled by hormones (p. 472).

Questions

1 Note the structure (Figure 10.7) and function (Figure 10.6) of loose connective tissue. In what ways does the structure of this tissue facilitate its functions?
2 What feature of the matrix of bone makes this tissue harder and more rigid than cartilage?
3 What is the physiological importance of red bone marrow?
4 Where does cartilage occur in
   a an embryo
   b an adolescent
   c a mature human?

■ Blood

Blood is a fluid connective tissue contained within a closed system of tubes (arteries, veins and capillaries, p. 351) through which it is circulated by the pumping action of the heart (p. 348). It is the chief transport system of the body (p. 358), and has an important role in the body's defence against disease (p. 364). Blood makes up 6–10% of the body mass in mammals.

Blood is an unusual connective tissue in that its specialised cells are surrounded by a fluid matrix (called **plasma**) that is not secreted by blood cells themselves. Plasma makes up 55% of the blood by volume, and surrounds the erythrocytes (red cells), leucocytes (white cells) and cell fragments called thrombocytes (platelets). Cells make up 45% of the blood by volume (Figure 10.11).

Plasma

Plasma is a straw-coloured, slightly alkaline fluid, consisting largely of water. Its composition is carefully regulated by the body, and most components are maintained at a constant concentration. The functions of plasma are

- ▶ to carry nutrients to the cells
- ▶ to remove excretory products from the cells
- ▶ to distribute hormones from endocrine glands to target organs
- ▶ to distribute heat energy from warm parts of the body to the cooler parts.

These functions contribute to the homeostasis of the body (p. 502). The plasma also helps to protect the body from pathogens by forming clots at wound sites and distributing antibodies.

Blood cells

Erythrocytes (red blood cells)

Erythrocytes are small, circular, biconcave discs, 7–8 µm in diameter (Figure 10.11). They are formed in red bone marrow tissue opposite. When first formed, an erythrocyte has a nucleus but little haemoglobin. As the cell matures the haemoglobin content increases to 90% of the cell's dry mass. Towards the end of this maturation process the nucleus is squeezed out.

Erythrocytes carry oxygen, combined as oxyhaemoglobin (p. 358), from the region of the body at high oxygen tension (the lungs) to regions of low oxygen tension (respiring tissues). Erythrocytes also carry carbon dioxide as hydrogencarbonate (p. 361) from regions of the body at high carbon dioxide tension (respiring tissues) to the region of low carbon dioxide tension (the lungs). Their

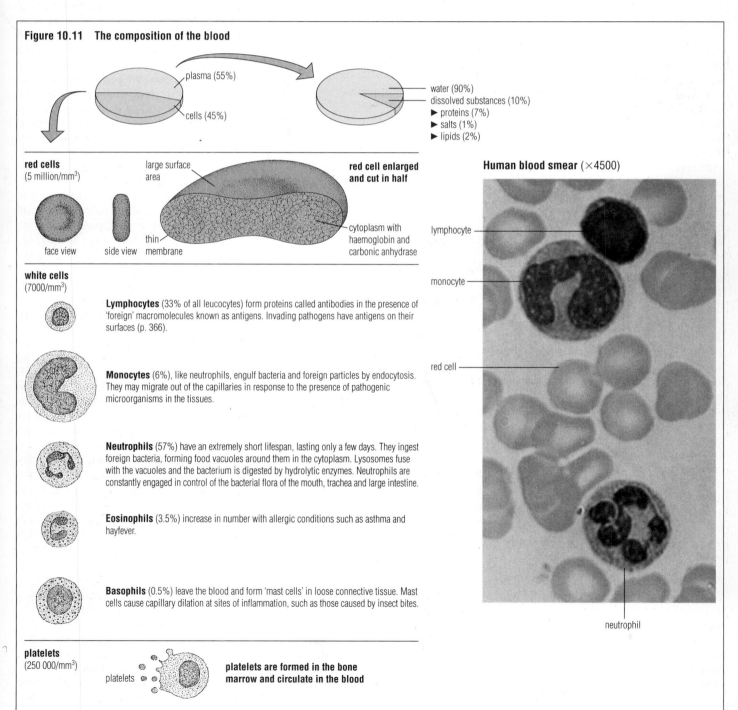

Figure 10.11   The composition of the blood

plasma (55%)

cells (45%)

water (90%)
dissolved substances (10%)
▶ proteins (7%)
▶ salts (1%)
▶ lipids (2%)

red cells
(5 million/mm³)

large surface area

red cell enlarged and cut in half

face view   side view   thin membrane

cytoplasm with haemoglobin and carbonic anhydrase

Human blood smear (×4500)

lymphocyte

monocyte

red cell

white cells
(7000/mm³)

Lymphocytes (33% of all leucocytes) form proteins called antibodies in the presence of 'foreign' macromolecules known as antigens. Invading pathogens have antigens on their surfaces (p. 366).

Monocytes (6%), like neutrophils, engulf bacteria and foreign particles by endocytosis. They may migrate out of the capillaries in response to the presence of pathogenic microorganisms in the tissues.

Neutrophils (57%) have an extremely short lifespan, lasting only a few days. They ingest foreign bacteria, forming food vacuoles around them in the cytoplasm. Lysosomes fuse with the vacuoles and the bacterium is digested by hydrolytic enzymes. Neutrophils are constantly engaged in control of the bacterial flora of the mouth, trachea and large intestine.

Eosinophils (3.5%) increase in number with allergic conditions such as asthma and hayfever.

Basophils (0.5%) leave the blood and form 'mast cells' in loose connective tissue. Mast cells cause capillary dilation at sites of inflammation, such as those caused by insect bites.

neutrophil

platelets
(250 000/mm³)

platelets

platelets are formed in the bone marrow and circulate in the blood

cytoplasm contains the enzyme carbonic anhydrase, which catalyses the formation and breakdown of the hydrogencarbonate ion (p. 361).

There are vast numbers of erythrocytes in the adult body. Each cubic millimetre of blood, which is a drop the size of a pin head, contains about five million. The lifespan of an erythrocyte is about 120 days. Consequently one 120th part of the body's total (about 200 000 million cells) is replaced every day. This is approximately three million every second! This prodigious metabolic activity occurs in the liver and the spleen (where erythrocytes are broken down) and in the red bone marrow (where they are formed). Any malfunction of the liver or the bone marrow will rapidly result in an altered blood cell count. In the breakdown process most of the iron is retained for re-use, held as an iron-containing protein. The remainder of the haemoglobin molecule is converted into the yellow-green bile pigments, bilirubin and biliverdin, which are excreted by the liver in the bile (p. 507). The amino acids of the protein portion of the haemoglobin become part of the body's pool of amino acids, available for fresh protein synthesis.

Leucocytes (white blood cells)

There are relatively few leucocytes in the blood: about one leucocyte for every 700 erythrocytes. There are five different types of leucocyte (Figure 10.11). Lymphocytes and monocytes have non-granular cytoplasm and are known as **agranulocytes**. Neutrophils, eosinophils and basophils have granular cytoplasm and are known as **granulocytes**. They are all concerned with the body's protection against infection (see Chapter 24). Leucocytes are short-lived cells. They are formed from bone marrow cells, but in certain cases they mature in the lymph nodes, the thymus gland or the skin. On average, white blood cells spend only 10% of their lives in the blood, and so they are also correctly referred to as 'white cells'.

Platelets

Platelets are cell fragments produced by large cells in the bone marrow, and there are between 150 000 and 400 000 platelets in each cubic millimetre of blood. They are disc-shaped and very small (only 2 μm in diameter). Platelets have no nucleus, but are rich in hydrolytic enzymes. They have an important role in blood clotting.

Sites of red cell formation in the growing mammal

In the very young fetus the first-formed blood is made in blood vessels. Later in development, before the bones and bone marrow are fully established, erythrocytes are produced in the liver, kidney, spleen and muscles. In children erythrocytes are made in the marrow of all the bones, but in adults most erythrocyte formation is in the red bone marrow of the skull, ribs, sternum and vertebrae. Blood cells originate from bone marrow cells known as **stem cells**. Bone marrow is capable of producing erythrocytes three or four times as fast as they are broken down. After severe bleeding (or blood donation) the erythrocyte count quickly returns to normal. When the rate of loss exceeds that of production the body suffers from a condition known as anaemia. When the body acclimatises to reduced oxygen tension at high altitude the number of erythrocytes increases (p. 362).

10.5 MUSCLE

Muscle tissue consists of elongated cells or fibres, held together by connective tissue. Muscle cells and fibres are highly specialised in that they are able to shorten to a half or even a third of their 'resting' length. There are three types of muscle tissue in a mammal's body: voluntary (or skeletal) muscle, involuntary (or smooth) muscle and cardiac muscle. All three types of muscle tissue are derived from the embryonic mesoderm. Muscle tissue makes up about 40% of the mass of the adult body.

■ Voluntary muscle

Voluntary muscle mainly occurs attached to the skeleton in the trunk, limbs and head, either directly on to the bone or indirectly, via a tendon.

Voluntary muscle consists of thousands of elongated cylindrical multinucleated muscle fibres, lying parallel to one another. (This unusual multinucleate structure is called a **syncytium**, or **coenocytic** arrangement.) Voluntary muscle is also referred to as **striped** or **striated muscle**. This is because, under the microscope, the muscle fibres have alternate light and dark bands or striations, along their entire length. Individual muscle fibres vary widely in length, from one to 50 mm long, and are between

10 and 100 μm in diameter. Muscle fibres do not extend the whole length of the muscle; there are many more in the mid-region of the muscle than near the ends.

Groups of fibres are bound into bundles by a thin film of connective tissue, and the whole muscle is bounded by a sheath of connective tissue. Blood vessels run longitudinally beside the fibres with many capillary connections (Figure 10.12).

Each multinucleate muscle fibre is contained by a plasma membrane, called the **sarcolemma**, surrounding cytoplasm known as **sarcoplasm**. The sarcoplasm contains hundreds or even thousands of **myofibrils**, each having characteristic cross-striations. Sarcoplasm also contains numerous mitochondria and smooth endoplasmic reticulum. The physiology and biochemistry of contraction of voluntary muscle are described on pp. 488–92.

Voluntary muscle is under control of the **voluntary nervous system** via motor nerves from the brain and spinal cord (p. 461). Every voluntary muscle cell is served by a motor nerve fibre, ending in a **motor end plate**. When stimulated, muscle fibres contract quickly and powerfully with a short **refractory period** (the period following stimulation during which a nerve or muscle fibre is incapable of responding), but they fatigue relatively quickly.

■ Involuntary muscle

Involuntary muscle occurs in sheets or bundles in the walls of the intestine (p. 283), in the tubes and ducts of the respiratory (p. 308), urinary (p. 388) and genital (pp. 589–90) systems, in the walls of blood vessels (p. 351) and in the iris and ciliary body of the eye (p. 456). It is **unstriped** (**unstriated**) when seen under the microscope, in contrast to voluntary muscle of the skeleton.

An involuntary muscle cell is spindle-shaped, about 0.02–0.5 mm long and 5–10 μm in diameter – much shorter and thinner than a voluntary muscle fibre. The cytoplasm contains a single nucleus and inconspicuous myofibrils, but there are no cross-striations (Figure 10.13).

Questions

1 Blood plasma components are maintained at a fairly constant composition. What is the importance of maintaining a constant concentration of hydrogen ions in the blood (p. 180)?

2 There is particular risk for the body if radioactive isotopes are taken up and built into bone structure. What are these dangers (p. 627)?

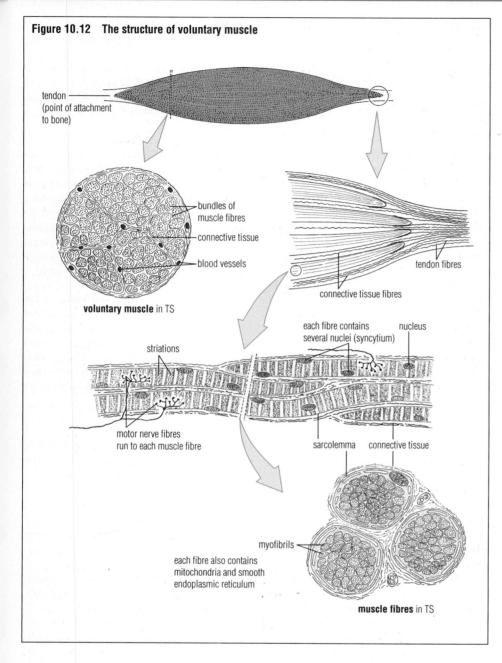

Figure 10.12   The structure of voluntary muscle

tendon
(point of attachment
to bone)

bundles of
muscle fibres

connective tissue

blood vessels

tendon fibres

connective tissue fibres

voluntary muscle in TS

striations

each fibre contains
several nuclei (syncytium)

nucleus

motor nerve fibres
run to each muscle fibre

sarcolemma   connective tissue

myofibrils

each fibre also contains
mitochondria and smooth
endoplasmic reticulum

muscle fibres in TS

Involuntary muscle is under the control of the **autonomic nervous system** (p. 462), and is served by a general nerve network, known as a **plexus**, rather than each cell or muscle fibre being served directly. Involuntary muscles (and also cardiac muscle, below) are capable of sustained and rhythmical contractions without noticeable fatigue.

■ Cardiac or heart muscle

Cardiac muscle consists of cylindrical branching columns of short coenocytic (p. 150) fibres, forming a three-dimensional network. Each cardiac muscle fibre, like a voluntary muscle fibre, is surrounded by a sarcolemma. Abundant mitochondria are present. Capillaries in the muscle tissue provide a rich blood supply.

Cardiac muscle occurs only in the walls of the four chambers of the heart. The fibres are connected by special electrical junctions called **intercalated discs** (Figure 10.14). The impulses that cause contraction arise within the heart muscle, and it is described as **myogenic**. All the muscle fibres contract in unison, creating the efficient pumping action of the heart. Cardiac muscle tissue contracts rhythmically from the moment of its formation until its death. The heart is also supplied (innervated) by the vagus nerve and other autonomic nerves (p. 350).

Questions

1   How does cardiac muscle differ from both voluntary muscle and smooth muscle in its structure and its function?
2   What is the special importance for muscle tissue of the presence of mitochondria and blood capillaries?

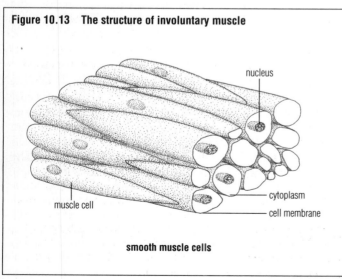

Figure 10.13   The structure of involuntary muscle

nucleus

muscle cell

cytoplasm

cell membrane

smooth muscle cells

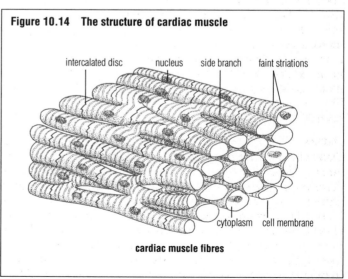

Figure 10.14   The structure of cardiac muscle

intercalated disc   nucleus   side branch   faint striations

cytoplasm   cell membrane

cardiac muscle fibres

10.6 NERVOUS TISSUE

The nervous system consists of the brain and spinal cord, known as the **central nervous system** (CNS, p. 460), and the nerves and associated ganglia known as the **peripheral nervous system** (PNS, p. 460). Nervous tissue is made up of cells called **neurones**, which are adapted for the transmission of electrical impulses, together with associated **neuroglia cells**.

Neurones are the pathways of communication between the brain and the body. Electrical impulses pass along the neurones from organs that receive stimuli (the **receptors**), such as a skin sense cell (p. 450), to organs that effect change (the **effectors**), such as a muscle or gland. Impulses are also generated spontaneously in the brain and are conducted to effectors via the peripheral nerves.

Neurones differ considerably in structure, but all neurones have three things in common. There is a **cell body** containing a nucleus. Impulses are brought in towards the cell body either by a fine cytoplasmic fibre called a **dendron** or by many such fibres (**dendrites**); a single fibre called the **axon** takes impulses away from the cell. Neurones can be classified by the number of their fibres (Figure 10.15) or by their function.

The cytoplasm of a neurone cell body is densely packed with ribosomes, Golgi apparatus, rough endoplasmic reticulum and mitochondria. Microtubules run from the cell body out along the axons and dendrons. Neurones are nourished by nutrients flowing along these processes and by nutrients received from the surrounding neuroglia cells.

■ Types of neurone

The functional classification of neurones is based on the direction in which they transmit impulses and on the positions they occupy in the nervous system. On this basis, there are three types of neurone (Figure 10.16).

Afferent neurones

An afferent (or sensory) neurone conducts impulses along a dendron from a receptor organ to the central nervous system. Cell bodies of sensory neurones occur in clusters called **ganglia** next to the spinal cord (Figure 21.29, p. 461). An axon carries the impulse into the central nervous tissue.

Intermediary or relay neurones

Intermediary or relay neurones occur within the central nervous system (brain or spinal cord) or in sympathetic ganglia (p. 462). They receive impulses from sensory neurones or from other intermediary neurones. These impulses they relay either to motor neurones or to other intermediary neurones.

Efferent neurones

Efferent (or motor) neurones transmit impulses along axons from the central nervous system to the muscles and glands.

■ Synapses

Neurones rarely have direct electrical contact with each other at their extremities; the electrical impulse cannot pass from one neurone to another. The junction points between the tips of connecting nerve fibres or between fibres and cell body are called **synapses**. At the synapse the nerve ending has a tiny swollen or bulbous tip (Figure 21.10, p. 448). The impulse is conducted across the synapse by chemical means (Figure 21.11, p. 449).

■ Other cells of the nervous system

In addition to neurones, nervous tissue contains cells known as **neuroglia cells**. These are ten times more numerous than neurones, and are found throughout the central nervous system. Neuroglia mechanically support the neurones, and they supply nourishment to neurone fibres. Some neuroglia are involved in the memory process by storing information in the form of RNA (p. 145). Certain cells of the neuroglia form the fibre sheaths (see opposite page).

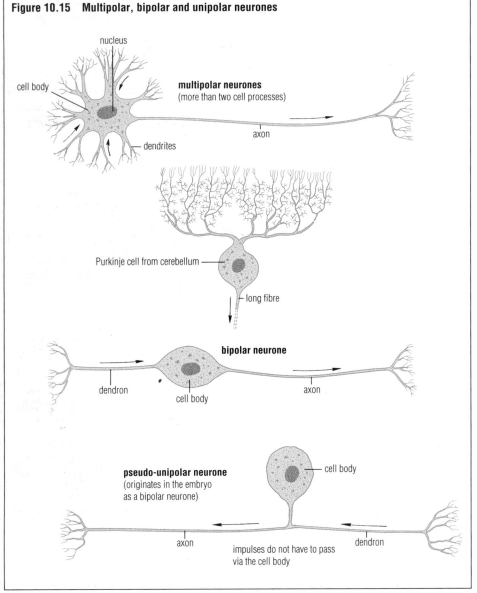

Figure 10.15 Multipolar, bipolar and unipolar neurones

nucleus

cell body

multipolar neurones
(more than two cell processes)

axon

dendrites

Purkinje cell from cerebellum

long fibre

bipolar neurone

dendron

cell body

axon

pseudo-unipolar neurone
(originates in the embryo as a bipolar neurone)

cell body

axon

impulses do not have to pass via the cell body

dendron

■ Nerves

When nerve fibres (dendrons and/or axons) occur together, wrapped up in connective tissue and protected by special sheath cells, they are identified as **nerves**. The nerves of the peripheral nervous system are bundles of nerve fibres enclosed in connective tissue. Most cell bodies of neurones are confined to the central nervous system (p. 460) but where they occur together in nerves they form a bulge called a **ganglion**.

Some nerves consist only of dendrons running from sense organs to the central nervous system. These nerves are known as **sensory nerves**; the optic nerve (p. 456) is an example.

Some nerves contain only axons running to effector organs such as muscles. These are known as **motor nerves**; the oculomotor nerve (which controls the muscles that move the eye) is an example.

Some nerves, such as the vagus nerve (pp. 288 and 350), are mixtures of sensory and motor nerve fibres and are known as **mixed nerves**.

■ Schwann cells

Nerve fibres are surrounded by specialised cells that form an electrically insulating and protective sheath. These cells are known as **Schwann cells**.

Where Schwann cells wrap round the fibres many times they form a **myelin sheath**. These fibres are said to be **myelinated**. A myelin sheath consists of layers of fatty membrane rich in a phospholipid called **lecithin** (p. 139). At intervals along each fibre, at the point where the Schwann cell ends, the fibre is unprotected. These points are called **nodes of Ranvier** (Figure 10.17); they are significant in the transport of nerve impulses (Figure 21.8, p. 447). Cranial and spinal nerves are myelinated.

A fibre that is simply enclosed or partly enclosed in a Schwann cell is said to be **unmyelinated**. Unmyelinated fibres have no nodes of Ranvier. Autonomic nerves are unmyelinated.

Questions

1 a What is the difference between a sensory neurone and a sensory nerve?
  b How might you try to detect whether a nerve had a sensory or a motor function?
2 a What features do all neurones have in common?
  b How do these features vary in the three types of neurone?

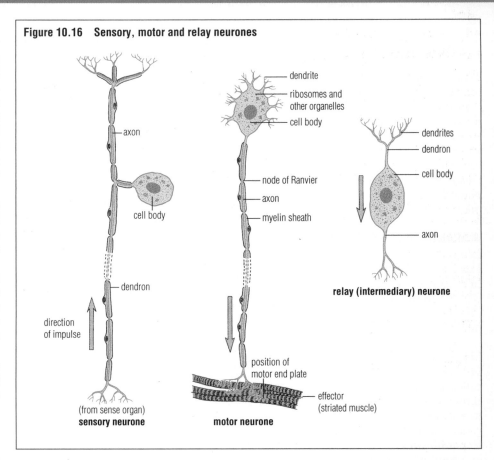

Figure 10.16 Sensory, motor and relay neurones

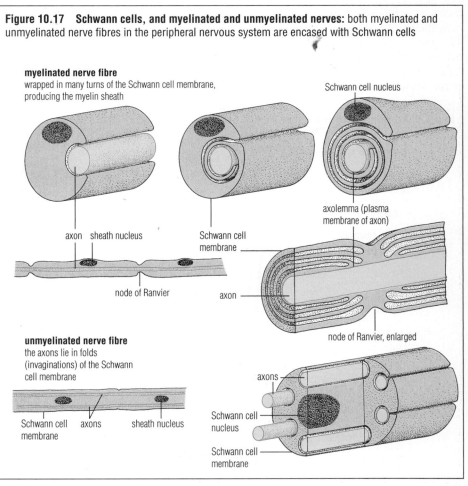

Figure 10.17 Schwann cells, and myelinated and unmyelinated nerves: both myelinated and unmyelinated nerve fibres in the peripheral nervous system are encased with Schwann cells

10.7 SPECIALISATION OF FLOWERING PLANT CELLS

The embryo of the flowering plant is formed in the seed, which develops from a fertilised ovule (p. 573). The embryo, a many-celled and more-or-less spherical structure, grows buried in the endo-sperm, the food store of the embryo. As the embryo develops, it becomes flat-tened, and forms an embryonic stem or **plumule**, embryonic root or **radicle** and one or two seed-leaves or **cotyledons**.

■ Meristematic cells and meristems

Initially, all the cells of an embryo are capable of further cell division and are said to be **meristematic**. Meristematic cells of flowering plants are small and thin-walled, with dense cytoplasm. The nucleus occupies a large part of the cell.

As the embryo grows into an independent seedling, embryonic cells become restricted to specialised regions, called **primary meristems**. These persist throughout the life of the plant. Primary meristems occur at the tips of stems and roots (apical meristems), and are responsible for growth in length of stem and root. In addition, a lateral meristem forms from the **procambial strand**, which runs the length of the stem and the root. This meristem is responsible for the for-mation of the vascular tissue in the stem (and, later, for the growth in girth of the stem) (Figure 10.18).

Table 10.1 The primary tissues of flowering plants

Primary tissue	Where it is found in herbaceous plants	Chief functions
Simple tissues (of one type of cell only):		
Parenchyma	Forms the greater part of cortex and pith	Little structural specialisation; site of most physiological and biochemical processes of the plant; turgidity within the parenchyma, counteracted by the epidermis, supports the plant
Modified forms of parenchyma:		
epidermis	Outermost layer, one cell thick; with or without stomata	Reduces water loss; protection, forming a continuous 'skin'; also contributes to support
mesophyll	In leaves, between the upper and lower epidermis	Those mesophyll cells that contain chloroplasts are the site of photosynthesis
endodermis	Around the stele (central vascular tissue) of roots	Selective barrier to metabolites entering or leaving conducting tissue
Collenchyma	Often immediately below the epidermis in young stems, in leaves at tip and in mid-rib region	Elastic, flexible support in leaves and young stems
Sclerenchyma	Outer cortex in mature stems	Support
fibres	In and around vascular bundles, in and around xylem	
sclereids	Singly or in groups in stems, leaves, fruits and seeds	Support and mechanical protection
Compound tissues (of different types of cell):		
Xylem		
vessels	Vascular system of stem and root	Transport of water and mineral salts
tracheids		
Phloem		
sieve tubes	Vascular system as above	Transport of sugars and amino acids
companion cells		Work in association with sieve tubes

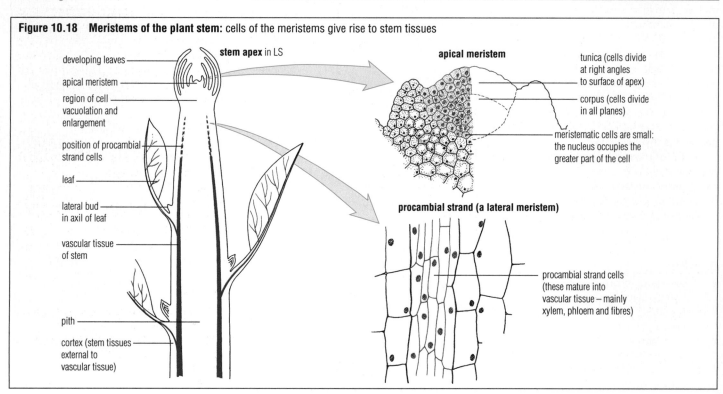

Figure 10.18 Meristems of the plant stem: cells of the meristems give rise to stem tissues

■ Primary tissues of the flowering plant

The structure of plant cells and tissues is studied by examining plant organs sectioned in different planes – for example, stem tissues viewed in transverse (TS) and longitudinal sections (LS). Within the stem, some tissues occur as vascular tissues (vascular bundles), some as cortex (tissues external to the vascular tissue) and some as the pith (tissues internal to the vascular tissues).

The first-formed plant tissues, found in seedlings and in the tissues of a herbaceous (non-woody) plant, are called **primary tissues** (Table 10.1). Primary plant tissues are formed from cells cut off from the primary meristems (Figure 10.18). The primary plant tissues are either **simple plant tissues** consisting of one type of cell only (parenchyma, collenchyma and sclerenchyma are simple plant tissues) or **compound plant tissues** consisting of more than one type of tissue (xylem and phloem are compound plant tissues). All these plant tissues are discussed in the sections that follow.

10.8 PARENCHYMA

■ Structure and occurrence

Parenchyma (Figure 10.19) consists of living cells, with relatively thin walls containing cellulose, pectin and hemicellulose. Parenchyma cells are roughly round when seen in transverse section, but in longitudinal section they usually appear elongated. They may become quite large in size, and prominent air spaces commonly occur between the cells. In the stem, parenchyma develops from cells cut off by a region of the apical meristem known as the corpus region (Figure 10.18).

Parenchyma tissues make up the bulk of the herbaceous plant. They form the packing tissues of non-woody stems and roots, and may also occur in the vascular bundles. Green leaves mostly consist of a specialised form of parenchyma.

■ Role

Parenchyma cells show relatively little structural specialisation but are the main sites of the physiological and biochemical processes of the plant, including photosynthesis, starch storage, ion accumulation and protein synthesis.

Turgid parenchyma cells are incompressible and exert pressure on the epidermis, creating support for the herbaceous stems, the root and the leaves (p. 500).

Parenchyma cells may retain the ability to divide even when mature, and they play an important part in secondary growth and in wound repair.

■ Specialised forms of parenchyma

Parenchyma cells become modified and more specialised in certain parts of the plant. Examples include the epidermis, mesophyll cells of leaves, and the endodermis and pericycle (p. 327) of the root.

Epidermis

Epidermis is a continuous, compact layer of cells on the surface of the plant body. The layer is one cell thick. The epidermis of aerial parts of plants contains stomata (p. 322), and all the epidermal cells of the stems and leaves have a waxy cuticle on their outer walls. The epidermis often bears unicellular or multicellular hairs (root hairs in the case of the epidermis covering young roots, p. 326).

The epidermis forms a thin 'skin' which gives mechanical support and protection against certain fungi and bacteria. It is concerned with gaseous exchange by the aerial parts of the plant (p. 253), with the absorption of water and solutes by the roots (pp. 326 and 330) and with the restriction of transpiration (p. 325).

Mesophyll tissue of leaves

Mesophyll is the parenchymatous tissue of the leaf. The cells are packed with chloroplasts (pp. 164 and 255). Mesophyll cells are the 'power house' of the photosynthetic leaf, an organ specialised for photosynthesis (p. 255). Photosynthetic cells in green stems are of similar structure, but are called **chlorenchyma**.

Questions

1 Where do meristematic cells occur? How do they contribute to growth?
2 What might be the functions of a parenchyma cell in the outer layers of
   a a stem
   b a root?

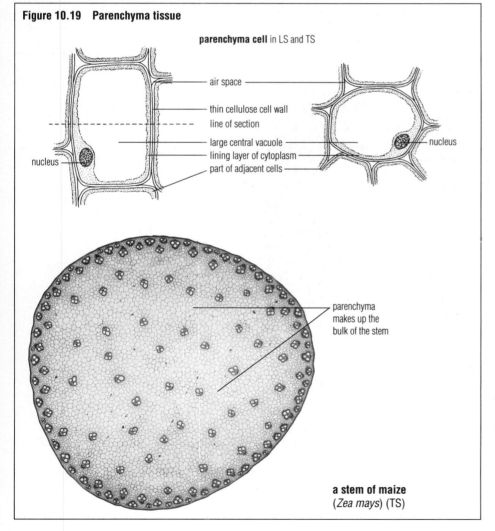

Figure 10.19 Parenchyma tissue

parenchyma cell in LS and TS

air space
thin cellulose cell wall
line of section
large central vacuole
lining layer of cytoplasm
part of adjacent cells
nucleus
nucleus

parenchyma makes up the bulk of the stem

a stem of maize
(*Zea mays*) (TS)

10.9 COLLENCHYMA

■ Structure and occurrence

Like parenchyma, collenchyma develops from cells cut off by the corpus region of the apical meristem (Figure 10.18) and consists of living cells. They are mostly quite narrow and elongated in shape compared with those of parenchyma. In collenchyma there are additional cell wall deposits at the corners of the cells. These deposits of cellulose, pectin and hemicellulose provide additional mechanical strength, and result in the characteristic unevenly thickened walls of collenchyma (Figure 10.20).

■ Role

Collenchyma is a strengthening tissue that is living, and which consequently can continue to grow and change in shape. It is a flexible tissue, and can be stretched. Collenchyma functions as the supporting tissue of young and growing stems; it stretches with the growth and movement of the stem. It often occurs in complete cylinders of tissue immediately below the epidermis, or as discrete ridges of tissue (Figure 10.20). Collenchyma is also the supporting tissue of leaves.

10.10 SCLERENCHYMA

■ Structure and occurrence

Sclerenchyma exists as long, narrow, pointed cells called **fibres**, and also as shorter, circular or more irregularly shaped cells called **sclereids**. Fibres occur in and around the vascular tissue, and may also develop below the epidermis as a cylinder of supporting tissue in some older stems. Sclereids may occur singly or in groups in stems, leaves, fruits and seeds.

Fibres and sclereids are thick-walled. The 'cells' consist of walls alone, for the cell contents die when the walls become thickened and impermeable, cutting off nutrients and water. The walls are composed of layers of cellulose impregnated with lignin (p. 147). Lignin, when combined with cellulose, resists stretching (has high tensile strength) and buckling (has high compressional strength). Sclerenchyma cells are commonly connected by pits, and in these regions the walls are unlignified (Figure 10.21).

■ Role

Sclerenchyma is a supporting tissue that develops in organs when their growth in length is completed.

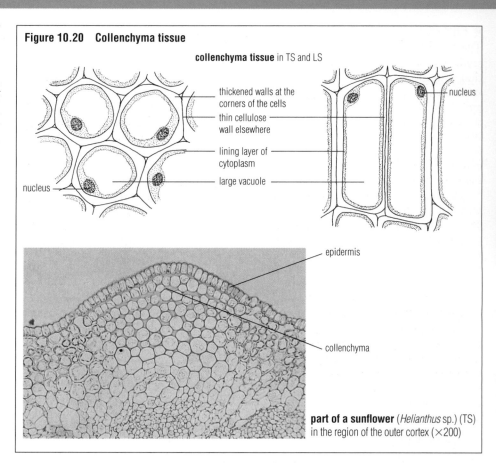

Figure 10.20   Collenchyma tissue

collenchyma tissue in TS and LS

thickened walls at the corners of the cells
thin cellulose wall elsewhere
lining layer of cytoplasm
large vacuole
nucleus
nucleus

epidermis

collenchyma

part of a sunflower (*Helianthus* sp.) (TS) in the region of the outer cortex (×200)

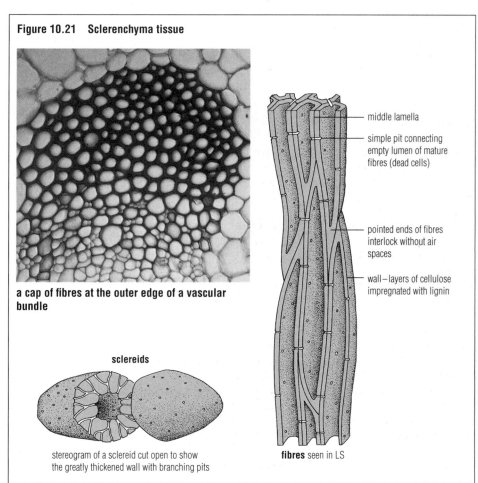

Figure 10.21   Sclerenchyma tissue

a cap of fibres at the outer edge of a vascular bundle

middle lamella
simple pit connecting empty lumen of mature fibres (dead cells)
pointed ends of fibres interlock without air spaces
wall – layers of cellulose impregnated with lignin

sclereids

stereogram of a sclereid cut open to show the greatly thickened wall with branching pits

fibres seen in LS

10.11 XYLEM

Xylem is a compound tissue consisting of xylem vessels and tracheids, often occurring with xylem parenchyma and fibres. Neither xylem vessels nor tracheids are living cells when mature.

■ Xylem vessels

Structure and occurrence

Xylem tissue develops from procambial strand cells (Figure 10.18), on the inner side of the strand in the developing stem (Figure 19.18, p. 408). Xylem vessels have thickened walls, impregnated with lignin.

The walls of the first-formed xylem vessels (called **protoxylem**) are thickened by lignified cellulose either as rings (such vessels are called **annular vessels**, Figure 10.22) or as spirally arranged bands (**spiral vessels**). In the growing stem the protoxylem develops and matures among other actively elongating cells and it is therefore subject to mechanical stress. The vessels are stretched, and many are eventually destroyed.

Xylem formed later is called **metaxylem**. This matures after elongation of the stem is completed. The walls of metaxylem vessels are massively thickened on the inner surface with cellulose impregnated with lignin, but with irregular areas of unthickened wall (**reticulate vessels**) or with organised pit areas (**pitted vessels**) (Figure 10.23).

Vessels become tubes

The end walls of developing vessels partially or almost totally disappear during the maturation of xylem, leaving a structure known as the **perforation plate**. Once all the transverse walls have been broken down, the vessels effectively become continuous tubes. In pitted vessels specially thin, unlignified areas (pits) occur in the side walls. Pits facilitate lateral movements of water, since the lignified walls themselves are impervious.

Role

Xylem has two main functions: conduction of water and mineral salts through the plant, and mechanical support. The strength of lignified cell walls enables xylem vessels to resist compression and tension. Xylem vessels are the characteristic conducting elements of angiosperm stems; comparatively few tracheids are present.

The water column in xylem vessels is under tension because water is drawn up the stem as a result of the reduction in pressure in the leaves when evaporation occurs from the mesophyll cells (p. 321). The thickening in xylem vessels helps to resist the tendency for the vessels to collapse under tension.

Figure 10.22   Stages in the formation of an annular xylem vessel (seen in LS)

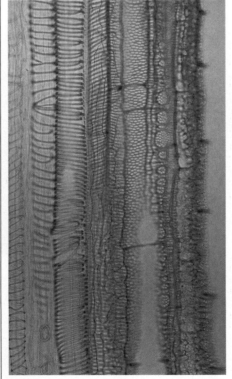

Figure 10.23   Xylem vessels

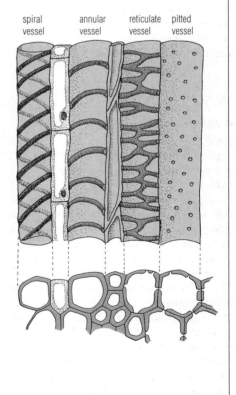

vessels in stem of marrow (*Cucurbita* sp.) (LS)

Questions

1   How does water escape laterally from xylem vessels?

2   Protoxylem vessels permit further extension growth in the stem. How is this possible?

■ Tracheids

Structure and occurrence

Tracheids are a form of xylem occurring among the vessels in many dicotyledons (p. 34). Conifers such as the Scots pine and ferns (p. 33) do not contain vessels; their role is taken by tracheids.

Tracheids are long, narrow cells with pointed ends. They are non-living when mature, and their walls are thickened by lignified bands (Figure 10.24).

Role

Tracheids are involved in mechanical support, and may also be the site of water movement. There are no perforation plates in tracheid walls; water passes from one tracheid to the adjacent one through connecting pits, in which a thin membrane of primary wall materials separates the hollow cell centres (lumen). By contrast, in xylem vessels, water moves directly and relatively freely past perforated end walls.

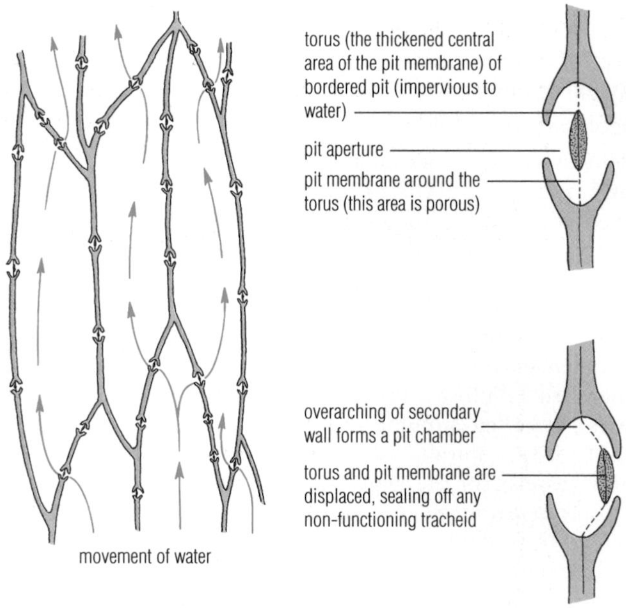

■ Secondary growth of flowering plants

Secondary growth occurs in trees and shrubs, and in certain herbaceous plants. Secondary growth results in additional xylem (secondary xylem) and phloem (secondary phloem) being formed in rows. Secondary xylem and phloem are more regular than their primary counterparts. Secondary growth leading to secondary thickening of the stem is introduced on p. 410.

■ Secondary xylem

Secondary xylem persists from year to year once formed, until the death and decay of the woody plant, although it often ceases to function in water transport after the season in which it was formed.

Secondary xylem is of great economic importance as a fuel (firewood), as the raw material in papermaking (wood pulp) and as commercial timber. It is therefore of much interest to plant anatomists and to botanists employed by the forestry industry. It is generally studied by microscopic examination of thin sections of wood or by maceration (p. 212), which permits the examination of individual lignified vessels, fibres and tracheids from a wood sample (Figure 10.25).

Figure 10.24 Tracheids

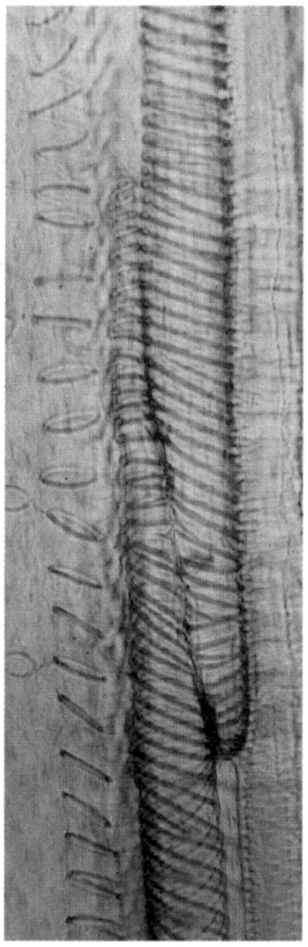

tracheids of Scots pine (*Pinus sylvestris*) (left) seen in LS

torus (the thickened central area of the pit membrane) of bordered pit (impervious to water)

pit aperture

pit membrane around the torus (this area is porous)

overarching of secondary wall forms a pit chamber

torus and pit membrane are displaced, sealing off any non-functioning tracheid

movement of water

Figure 10.25 Secondary xylem

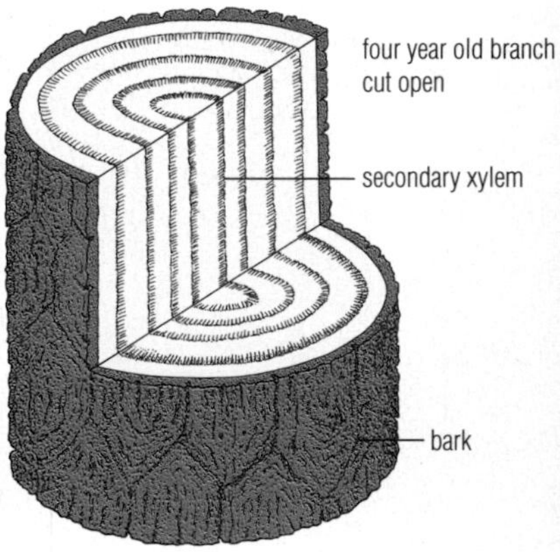

four year old branch cut open

secondary xylem

bark

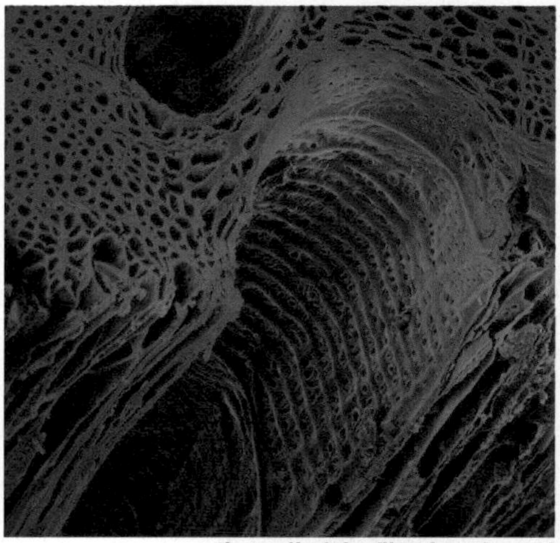

the wall of the fibre has rings of thickening on the inside

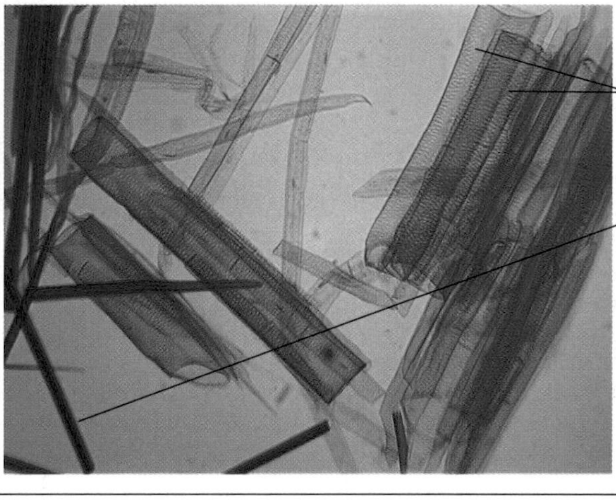

xylem vessels

fibre

macerated apple wood (maceration involves dissolving out the middle lamellae of tissue)

10.12 PHLOEM

Phloem is a compound tissue containing sieve tubes and companion cells, often with parenchyma and sometimes with fibres too.

■ Structure and occurrence

Phloem develops from procambial strand cells on the outer side of the strand in the growing stem. The sieve tube nucleus degenerates before the cell is mature, yet the sieve tube cell remains living, dependent on an adjacent companion cell. There are direct connections between sieve tubes and companion cells via plasmodesmata (p. 340). The walls of sieve tubes and companion cells are composed largely of cellulose. As the sieve tube cells mature the transverse walls develop a pattern of pores, and are then known as **sieve plates**. Thus dissolved substances are able to flow along the sieve tubes (Figure 10.26).

The first-formed phloem tissue, or **protophloem**, becomes crushed during the development of the later-formed phloem (**metaphloem**).

■ Role

Sieve tubes are the site of solute transport. Sucrose is transported from the leaves to other parts of the plant, and amino acids are transported all over the plant. Phloem transport is discussed on p. 340.

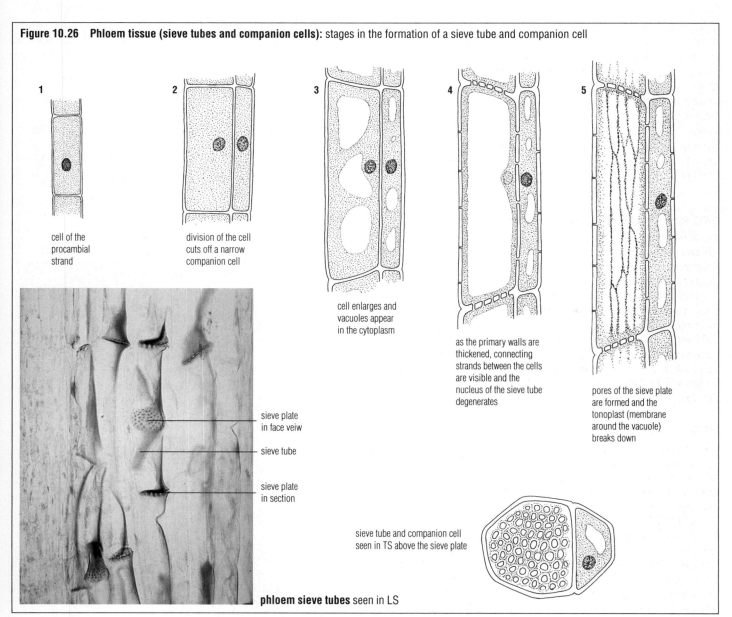

Figure 10.26   Phloem tissue (sieve tubes and companion cells): stages in the formation of a sieve tube and companion cell

1   cell of the procambial strand

2   division of the cell cuts off a narrow companion cell

3   cell enlarges and vacuoles appear in the cytoplasm

4   as the primary walls are thickened, connecting strands between the cells are visible and the nucleus of the sieve tube degenerates

5   pores of the sieve plate are formed and the tonoplast (membrane around the vacuole) breaks down

sieve plate in face veiw

sieve tube

sieve plate in section

sieve tube and companion cell seen in TS above the sieve plate

phloem sieve tubes seen in LS

Questions

1   How can vessels be responsible for carrying water even when they contain no nucleus or cytoplasm?

2   State the tissue and the direction in which you would expect each of the following to be moving in a flowering plant: sucrose, amino acids, water and salts.

FURTHER READING/WEB SITES

W H Freeman and Brian Bracegirdle (1976) *An Advanced Atlas of Histology*. Heinemann
C J Clegg and G Cox (1978) *Anatomy and Activities of Plants*. John Murray

11 Movement in and out of cells

- Substances enter and leave cells across the plasma membrane, a 'fluid mosaic' of a lipid bilayer with globular proteins in it. Movement of substances occurs by diffusion, active transport or bulk transport.
- Diffusion occurs due to the continuous, random movements of particles in liquids and gases.

- Osmosis is a special case of diffusion, involving water and a partially permeable membrane.
- Active transport is a selective movement of mainly useful substances, requiring energy from respiration.
- Bulk transport is possible due to the dynamic, mobile structure of the cell membrane, and involves the formation of tiny vacuoles of substances to be moved.

11.1 THE TRAFFIC ACROSS CELL MEMBRANES

Traffic into and out of living cells occurs on the grand scale! Cells exchange respiratory gases, absorb nutrients and vitamins, and take in and lose water. Cells also dispose of excretory products. Some cells secrete substances – certain enzymes and hormones, for example. Plant cells secrete the entire substance of their cell walls. Many animal cells receive nerve impulses, mostly in the form of chemical transmitter substances. Plant and animal cells take in hormones and growth substances. Energy enters the cell not only as chemical energy in molecules but also as heat energy and light. Much energy leaves cells as heat energy.

The extent of the traffic between the cell and its external environment is summarised in Figure 11.1.

■ The plasma membrane

All cells are separated from their environment by a plasma membrane. The plasma membrane facilitates transport of substances into and out of the cell, but it prevents the loss of important molecules from the cell and prevents the entry of many toxic substances, too. What is this barrier, the plasma membrane, made of?

The structure of the plasma membrane and its appearance in electronmicrographs (TEM and SEM) are shown in Figures 7.13 and 7.14 (p. 158). The main structural component of the cell membrane is a double layer of phospholipid molecules (p. 158). Lying on, within and across the lipid bilayer are large protein molecules. In fact, the membrane is made up, very roughly, of equal amounts of protein and lipid. The membrane has a fluid structure, and our understanding of its structure is summarised in the fluid mosaic model (Figure 7.15, p. 159).

■ The external environment of the cell

The immediate environment of the living cell is always aquatic, and therefore all exchanges between cell and environment occur in an aqueous medium. Extracellular liquid bathes all living cells in all organisms, whether they are aquatic or terrestrial, animal or plant.

For example, in mammals the outermost layer of the body consists of dead cells. Below these dead cells of the outer skin all the living cells are bathed in tissue fluid (p. 353).

In flowering plants all cells are contained within a cellulose cell wall. The pores and free spaces of the cell wall contain an extremely dilute solution of various ions and some metabolites as extracellular fluid.

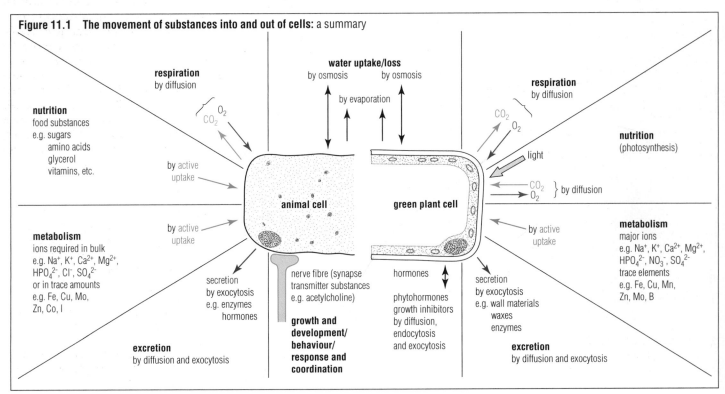

Figure 11.1 The movement of substances into and out of cells: a summary

■ How do substances move across membranes?

Water can cross the plasma membrane and so can some solutes, but not others (p. 158). Consequently, the cell membrane is described as **partially permeable**. The most important property of the plasma membrane of cells is its partial permeability. Because of this property, the cell membrane plays a key part in determining the composition of the cytoplasm by controlling what material gets in and out.

Cells absorb from, or exchange matter with, their environment by five processes: diffusion, osmosis, active transport, phagocytosis and pinocytosis. These processes are examined in this chapter.

■ Movement within the cell

Substances in the cytosol of cytoplasm move about by diffusion. Sometimes they are also carried about the cell by streaming movements of the cytoplasm. But free movement of substances is impeded in eukaryotic cells because their contents are compartmentalised in various membrane-bound organelles (p. 157). Substances are not distributed uniformly between the different organelles. Membranes are selective and control the movement of substances between the organelles and the cytosol. In effect, the partially permeable membranes of the organelles contain specialised micro-environments within the cell.

■ Investigating membrane permeability

The permeability of membranes may be studied using beetroot tissue because the cells in it contain in their vacuoles a water-soluble and intensely red pigment (an anthocyanin derivative). When the plasma membranes of these cells are damaged (by treating them with heat or with some chemical, for example) the red pigment escapes from the cells. It can be detected by eye and its concentration measured using a colorimeter.

An experiment to investigate the effect of temperature on the permeability of the plasma membrane is shown in Figure 11.2. It is found that the effect of heat on membrane proteins, as with most proteins, is to denature them irreversibly (p. 144). Above a certain temperature the integrity of the plasma membrane is so disrupted that the contents of the vacuole stream out. In this way the 'thermal death point' of cells can be estimated. The method can be adapted to find which chemicals are harmful to cell membranes, and at what concentrations they cause damage (*Study Guide*, Chapter 11).

Figure 11.2   Investigating membrane permeability

cutting cylinders of beetroot tissue

washing the cylinders

Experimental method

Preparation of washed beetroot tissue cylinders
Ten cylinders, each 3 cm long × 0.5 cm diameter, were cut, washed and placed in distilled water.

Heat treatment
A water-bath was heated to 70 °C. A cylinder of tissue was submerged in the water for 1 minute, then withdrawn and placed in 15 cm³ of water in a labelled test tube at room temperature for 15 minutes. The tissue cylinder was then removed and discarded.

The water-bath was then cooled to 65 °C, and a second cylinder of tissue was heat-treated, allowed to stand in a tube of fresh distilled water for 15 minutes and discarded. The process was repeated using heat treatment with water that was 5 °C cooler each time, down to a final temperature of 25 °C. The distilled water in the test tubes became coloured by the pigment that had escaped from the tissues.

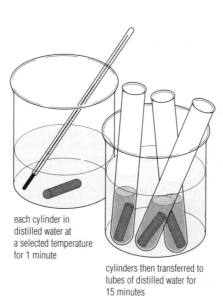

each cylinder in distilled water at a selected temperature for 1 minute

cylinders then transferred to tubes of distilled water for 15 minutes

Measuring pigment loss from tissue
The concentration of pigment in the solutions in the test tubes was related to the intensity of the colour of the solutions, which could be measured accurately with a colorimeter. The colour of the filter used for the determination was the colour that was complementary to that of the sample.

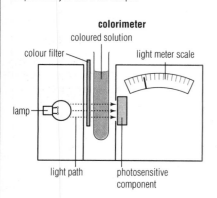

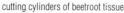

colorimeter

results

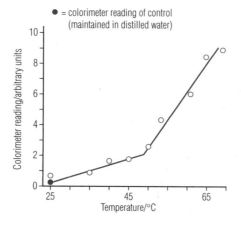

● = colorimeter reading of control (maintained in distilled water)

Questions

1 Which solutes are found to pass most readily through the plasma membrane? What does their passage suggest about the chemical composition of the membrane (p. 158)?

2 What is meant when we say the plasma membrane is partially permeable?

3 **a** Why is it necessary to wash beetroot tissue cylinders in running water before using them to investigate membrane permeability?

   b What is the approximate thermal death point of the cells of beetroot tissue shown in Figure 11.2?

11.2 DIFFUSION

When pollen grains are mounted in water and viewed at high magnification through a microscope they are seen to make continuous, erratic movements. This movement was first reported in 1827 by the Scottish botanist Robert Brown while studying pollen grains under the microscope. He observed minute particles within pollen grains in constant random motion. This movement of suspended and visible particles is now called Brownian motion. In **Brownian motion** the small particles that we can see are being bombarded on all sides by randomly moving molecules, which are far too small for us to see. Einstein (1879–1955) suggested that Brownian motion provided evidence for the kinetic theory of matter.

According to that theory, all matter is composed of tiny invisible particles, whether atoms, molecules or ions. The continuous random movement of particles in liquids and gases causes the transport and mixing that we call **diffusion**.

Diffusion is the **movement of particles from a region of their high concentration to a region of their low concentration** (down a concentration gradient). Where there is a difference of concentration in a gas, or in the solute in a liquid, net movement will occur to areas of lower concentration, resulting in the molecules becoming evenly distributed. The motion of particles that causes diffusion continues ceaselessly, but once particles are evenly distributed there is no *net* movement in any particular direction (Figure 11.3).

The net diffusion of different types of particle can take place in opposing directions. Diffusion of particles in one direction does not hinder the movements of other particles in an opposite direction. Two-way diffusion occurs constantly at lung surfaces, for example, when carbon dioxide diffuses out of the blood at the same time that oxygen diffuses in. Diffusion takes place across membranes so long as the membrane is fully permeable to the diffusing molecules.

■ Fick's Law

The factors affecting the rate of diffusion across a membrane (Table 11.1) are summarised in Fick's Law:

rate of diffusion $\propto$

$$\frac{\text{surface area} \times \text{difference in concentration}}{\text{membrane thickness}}$$

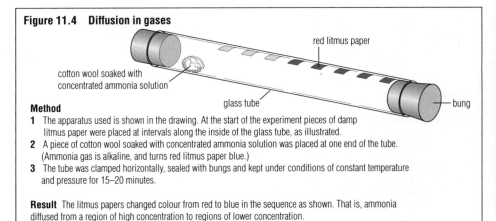

Figure 11.3   Diffusion in liquids

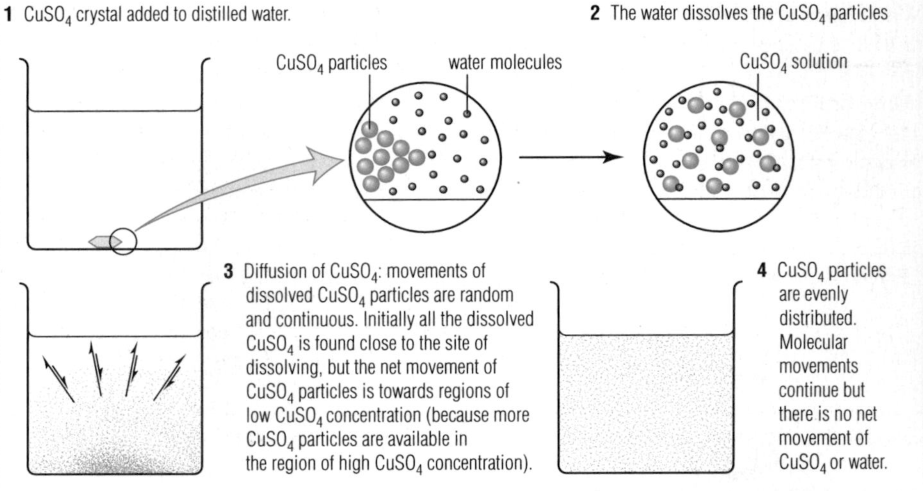

1 $CuSO_4$ crystal added to distilled water.

2 The water dissolves the $CuSO_4$ particles

$CuSO_4$ particles   water molecules $CuSO_4$ solution

3 Diffusion of $CuSO_4$: movements of dissolved $CuSO_4$ particles are random and continuous. Initially all the dissolved $CuSO_4$ is found close to the site of dissolving, but the net movement of $CuSO_4$ particles is towards regions of low $CuSO_4$ concentration (because more $CuSO_4$ particles are available in the region of high $CuSO_4$ concentration).

4 $CuSO_4$ particles are evenly distributed. Molecular movements continue but there is no net movement of $CuSO_4$ or water.

Figure 11.4   Diffusion in gases

red litmus paper

cotton wool soaked with concentrated ammonia solution

glass tube   bung

Method
1   The apparatus used is shown in the drawing. At the start of the experiment pieces of damp litmus paper were placed at intervals along the inside of the glass tube, as illustrated.
2   A piece of cotton wool soaked with concentrated ammonia solution was placed at one end of the tube. (Ammonia gas is alkaline, and turns red litmus paper blue.)
3   The tube was clamped horizontally, sealed with bungs and kept under conditions of constant temperature and pressure for 15–20 minutes.

Result   The litmus papers changed colour from red to blue in the sequence as shown. That is, ammonia diffused from a region of high concentration to regions of lower concentration.

Table 11.1   The rate of diffusion

Factors affecting the rate of diffusion	The qualities needed to achieve rapid diffusion across a surface
1 The concentration gradient: the greater the difference in concentration between two regions, the greater the amount that diffuses in a given time	A fresh supply of substance for diffusion needs to reach the surface, and substances that have crossed it need to be transported away
2 The distance over which diffusion occurs: the shorter the distance over which diffusion occurs, the greater the rate of diffusion across it	The structure or surface needs to be thin
3 The area across which diffusion occurs: the larger the area, the greater the rate of diffusion	The surface area needs to be large
4 The structure through which diffusion occurs: pores in a barrier may enhance diffusion	A greater number of pores and a larger size of pore may enhance diffusion (cf. the effect of pore perimeter and pore frequency on diffusion through stomata of leaf epidermis, Chapter 16)
5 The size and type of diffusing molecule: smaller molecules diffuse faster than larger molecules	Thus oxygen may diffuse more quickly than does carbon dioxide
molecules soluble in the substance(s) of a barrier diffuse faster through it	Fat-soluble substances diffuse faster across cell membranes than do water-soluble substances

■ Facilitated diffusion

The lipid bilayer is a barrier to many larger molecules like glucose, to polar molecules like amino acids and fatty acids, and to charged particles like ions. Nevertheless, these substances often pass across membranes by diffusion, assisted by special proteins, in a process known as **facilitated diffusion**. These assisting protein molecules are large and completely traverse the lipid bilayer. They are hydrophilic, with a water-filled channel running through them. Facilitated diffusion proteins are of two types.

Channel proteins are fixed-shape molecules, and the sites for the diffusion of many charged ions. These channels are selective about which ions can pass through. For example, there are different channels for potassium ions and for sodium ions in nerve fibres. Many channel proteins are 'gated', in that they allow the passage of ions only under particular conditions. In nerve cells the signal for an ion channel to 'open' may be a particular change in the potential across the membrane (p. 445).

Carrier proteins are molecules thought to undergo rapid shape changes when in contact with a particular solute molecule. The outcome is assisted diffusion across a membrane. Movements of glucose into red cells occur in this way, as does the entry of ADP into mitochondria and the exit to the cytosol of ATP (Figure 11.5). Neither channel proteins nor carrier proteins require energy from metabolism for their actions, and movement of the solute occurs down a concentration gradient.

Figure 11.5   Movement of ATP and ADP into and out of the mitochondrion

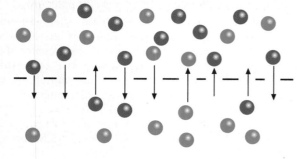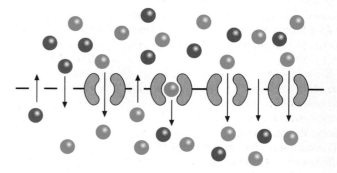

Passive diffusion
▶ the rate depends on the concentration gradient
▶ similar molecules will diffuse at similar rates
▶ diffusion can occur in either direction
▶ equilibrium is reached when concentrations are equal
▶ energy from ATP is not needed

Facilitated diffusion
▶ specific molecules diffuse faster than others
▶ occurs via special channel or carrier proteins
▶ diffusion is faster in one direction
▶ equilibrium is reached when concentrations are equal
▶ energy from ATP is not needed

Figure 11.6   Diffusion of oxygen enhanced by oxyhaemoglobin formation: the combination of oxygen with haemoglobin (Hb) in the red cells to make oxyhaemoglobin (HbO) effectively reduces the concentration of free oxygen in the plasma, so maintaining a steep diffusion gradient

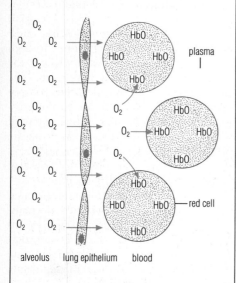

■ Diffusion in living things

Examples of diffusion in operation in living things are numerous. In fact, we can assume that movement occurs by diffusion wherever the membrane has pores that are large enough for the solute molecule to pass or is fully permeable to the solute, provided there is a concentration gradient for diffusion (see Figure 11.6, for example).

The absorption of oxygen into organisms and into cells and the movement of carbon dioxide in the reverse direction occur by diffusion. The movement of glucose from the gut of an insect into the body cells is by diffusion; this diffusion continues because once the absorbed glucose arrives inside the cells it is converted into less-soluble carbohydrates, thus maintaining the concentration gradient of glucose.

On the other hand, diffusion is not the process by which many substances enter cells. Usually a substance is absorbed into cells across a membrane that is apparently impermeable to that substance. Moreover, most absorption takes place into cells or organelles that already contain much more of the solute concerned than exists outside – that is, movement is largely *against* a concentration gradient. In these cases absorption is by entirely different mechanisms that are described as 'active transport'. Active transport is discussed in section 11.4.

Questions

1   How do molecular size and differences in concentration affect the rate of diffusion?
2   Carbon dioxide diffuses across a membrane until the concentration on both sides is the same. What then happens to the carbon dioxide molecules?
3   a   In the experiment shown in Figure 11.4, why is it necessary for the litmus paper to be damp?
    b   Why must the tube be sealed from the air and held under constant temperature conditions?
    c   The movements of molecules are random, yet the initial movement of ammonia was down the tube in the direction of the litmus paper. Why was this so?

11.3 OSMOSIS

In the experiment shown in Figure 11.7 a bag made of a partially permeable membrane, containing sucrose solution, was lowered into pure water in a beaker. After some minutes the bag became stretched and turgid, due to a net inflow of water. Water had moved from a region of high concentration of free water molecules (the molecules of pure water in the beaker) to a region of low concentration of free water molecules (the water molecules of the concentrated sucrose solution). A special case of diffusion, known as **osmosis**, is demonstrated here.

Osmosis in living organisms is the diffusion of water molecules from a region of their high concentration to a region of their low concentration through a partially permeable membrane.

Water molecules in pure water are all free to move about at random and thus diffuse into the partially permeable bag. In the concentrated sucrose solution most water molecules are associated with sucrose molecules. We can think of individual sucrose molecules as hydrated by a surrounding 'cloud' of water molecules clinging round them. Consequently, in the concentrated sucrose solution many water molecules move much more slowly than in pure water, and most of them are retained with the hydrated sucrose molecules.

■ Water movement and the concept of water potential

We have already seen that water molecules possess kinetic energy in that, when together in liquid (or gaseous) form, they move about very rapidly in random directions. The **water potential** of a solution is the term given to the tendency for water molecules to enter or leave that solution by osmosis. Water potential is a term derived from thermodynamics (p. 172), and is a measure of the free kinetic energy of the water molecules in the solution.

Pure water has the highest water potential, which by definition is set somewhat surprisingly as zero. The effect of dissolving solute molecules into pure water is to reduce the concentration of water molecules and therefore to lower the water potential. Consequently, solutions (at atmospheric pressure) have *negative* values of water potential. Water diffuses from a region of high water potential (less negative or zero value) to a region of lower water potential (more negative value).

■ Components of water potential

Two important factors that determine the water potential of solutions in and around living cells are the presence of dissolved solutes (giving rise to a **solute potential**) and the mechanical pressure acting on water (**pressure potential**).

The concept of water potential allows the effects of factors acting on water in a system (such as a cell) to be brought together under a single term. The following equation (in which water potential is represented as the Greek letter ψ, psi) is used:

$$\text{water potential} = \text{solute potential} + \text{pressure potential}$$

$$\psi = \psi_s + \psi_p$$

Solute potential of a solution

We have already seen (Figure 11.7) that when an aqueous solution is separated from pure water by a partially permeable membrane, a net movement of water into the solution occurs by osmosis. This is because the presence of solute molecules in water lowers the water potential, making it more negative. Water molecules tend to diffuse in from the high concentration in pure water where the water potential is at a maximum. The solute potential is the negative component of water potential due to the presence of solute molecules. Solute potential was previously referred to as 'osmotic pressure' or 'osmotic potential', and these terms are sometimes still used in animal physiology and medicine.

The solute potential of a concentrated solution can be demonstrated dramatically in an apparatus known as an **osmometer**. In this, the pressure that must be applied to stop water entering the solution by osmosis is the solute potential of that solution.

> **UNITS OF PRESSURE**
>
> In the past, various units have been used to express pressure, such as the 'atmosphere', the bar and mm of mercury. Now the pascal (Pa) and its multiple the kilopascal (kPa) are used. The pascal is defined as a pressure of one newton per square metre ($1\ N\ m^{-2}$).
>
> 1 atm = 101.325 kPa
> 1 mm mercury = 133.3 Pa
> 1 bar = 100.005 kPa

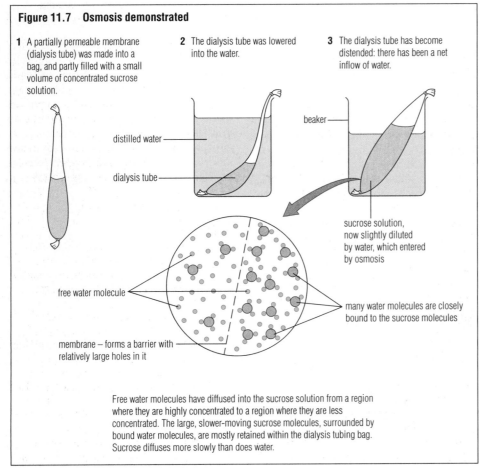

Figure 11.7   Osmosis demonstrated

1 A partially permeable membrane (dialysis tube) was made into a bag, and partly filled with a small volume of concentrated sucrose solution.

2 The dialysis tube was lowered into the water.

3 The dialysis tube has become distended: there has been a net inflow of water.

distilled water

dialysis tube

beaker

sucrose solution, now slightly diluted by water, which entered by osmosis

free water molecule

many water molecules are closely bound to the sucrose molecules

membrane – forms a barrier with relatively large holes in it

Free water molecules have diffused into the sucrose solution from a region where they are highly concentrated to a region where they are less concentrated. The large, slower-moving sucrose molecules, surrounded by bound water molecules, are mostly retained within the dialysis tubing bag. Sucrose diffuses more slowly than does water.

Figure 11.8   Solute potential demonstrated

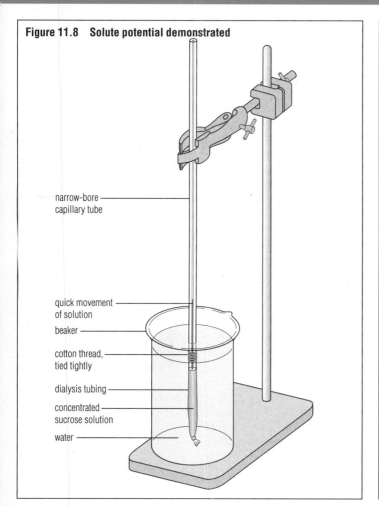

narrow-bore capillary tube

quick movement of solution

beaker

cotton thread, tied tightly

dialysis tubing

concentrated sucrose solution

water

Figure 11.9   The effect of mechanical pressure on water potential

1 A length of dialysis tubing containing sucrose solution was knotted at each end to make a model cell; the model was lowered into water.

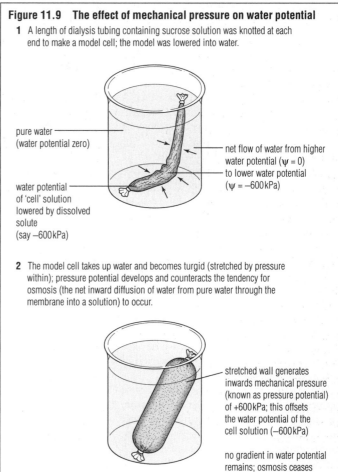

pure water (water potential zero)

net flow of water from higher water potential ($\psi = 0$) to lower water potential ($\psi = -600\,kPa$)

water potential of 'cell' solution lowered by dissolved solute (say $-600\,kPa$)

2 The model cell takes up water and becomes turgid (stretched by pressure within); pressure potential develops and counteracts the tendency for osmosis (the net inward diffusion of water from pure water through the membrane into a solution) to occur.

stretched wall generates inwards mechanical pressure (known as pressure potential) of $+600\,kPa$; this offsets the water potential of the cell solution ($-600\,kPa$)

no gradient in water potential remains; osmosis ceases

The osmometer shown in Figure 11.8 consists of a bag made from narrow dialysis tubing securely tied to a length of narrow-bore capillary tube, and immersed in pure water. The height of water raised in the long capillary tube demonstrates the pressure that can be built up by osmosis.

Pressure potential

If a pressure greater than atmospheric pressure is applied to pure water or to a solution inside a partially permeable bag, it increases the water potential. Hydrostatic pressure to which water is subjected is called **pressure potential**. Pressure potential is usually positive, but in the xylem of a transpiring plant (p. 321) the water column is under tension and the pressure potential is negative. In most cells (such as a turgid plant cell, p. 234), the pressure potential is positive. Pressure potential is represented by the symbol ψ_p. Mechanical pressure in cell water relations was previously known as 'turgor pressure' or 'wall pressure', but these terms have been dropped by plant physiologists.

The practical effect of the pressure on water potential of a solution is illustrated in Figure 11.9, using a model cell consisting of a small quantity of sugar solution enclosed in a bag made of partially permeable dialysis tube. This is submerged in pure water: osmosis (a net inward diffusion of water) occurs, and the model cell rapidly becomes stretched. The bag, now acting like a plant cell wall, generates mechanical pressure against the cell contents as it inflates. Eventually the mechanical pressure becomes sufficiently high to counterbalance further water uptake by osmosis. At this point, water molecules are moving inwards through the membrane at the same rate that water molecules are moving out through it.

■ Water molecules across the lipid bilayer

Water diffuses in and out of cells rapidly, yet the lipid bilayer is theoretically impermeable to water. How does it get across?

Because water is a small molecule, it may diffuse across the lipid bilayer to some extent. For example, where the fluid mosaic membrane is rich in phospholipids with unsaturated carbon tails,

the membrane is especially leaky. Why? Look at the shape of these molecules (Figure 6.16, p. 137). How closely will they fit together?

The main routes of water diffusion, however, are protein-lined pores. A membrane protein that forms highly selective water-channels in cell membranes has recently been discovered. It occurs widely, and in large quantities, in many cell membranes. The gene coding for this protein has also been identified.

Questions

1 If a concentrated solution of sucrose is separated from a dilute solution of sucrose by a partially permeable membrane, which solution

  a has a higher concentration of water molecules?

  b has a lower water potential?

  c will experience a net gain of water molecules?

2 Water moves from a higher to a lower water potential. If solution A has a ψ of $-800\,kPa$ and solution B has a ψ of $-650\,kPa$,

  a which solution has a higher ψ?

  b does more water diffuse from solution B to solution A than in the opposite direction?

■ Cells and water potential

The plasma membrane of living cells is a partially permeable membrane, and osmosis is the principal mechanism by which water enters and leaves plant and animal cells. Whether the net direction of water movement is into or out of cells depends on whether the water potential of the cell solution is more negative or less negative than the water potential of the external solution. We will examine the consequences of these alternative conditions for an animal cell (red cell) and a plant cell (parenchyma cell) in turn, but first we shall look at the effect on net water movement of an environment with the same water potential as that of the cell solution.

When the external water potential is the same as that of the cell (whether it is an animal or a plant cell), there is no net movement of water into or out of it. Two systems at the same water potential (Figure 11.10) are often said to be **isotonic** (the term is most commonly used in animal physiology and medicine).

When the external water potential is less negative than that of the cell (that is, the cell is surrounded by a solution with a less negative water potential – a more dilute solution – than that of the cell cytoplasm or vacuole, referred to as a **hypotonic** solution), there is net water movement into the cells. The cytoplasm becomes distended by water uptake; if this distension causes pressure against a cell wall (plant cell) the cell is described as **turgid**. A pressure potential develops. Plant cells, surrounded by a cellulose cell wall, are protected by their wall from rupture; the wall is like a corset. Animal cells such as red blood cells have no external wall to protect them from rupture. Their plasma membrane is unable to resist the pressure potential that develops, and these cells may swell until they burst. Within the bodies of animals the fluid bathing the cells, known as tissue fluid, is maintained at the same water potential as the cell solution (that is, it is kept isotonic with it). The danger that cells may take up excessive water is thus avoided (Figure 11.11).

When external water potential is more negative than that of the cell (that is, the cell is surrounded by a solution of more negative water potential – a more concentrated solution – than that of the cell cytoplasm, referred to as a **hypertonic** solution) there is net movement of water out of the cell (Figure 11.12). Cells and tissues in this condition become **flaccid** (the opposite of turgid).

Figure 11.10   When the external water potential and that of the cell are the same . . .

When cells are in an aqueous environment of the same water potential as that of the cell (isotonic solution) there is no net water movement.

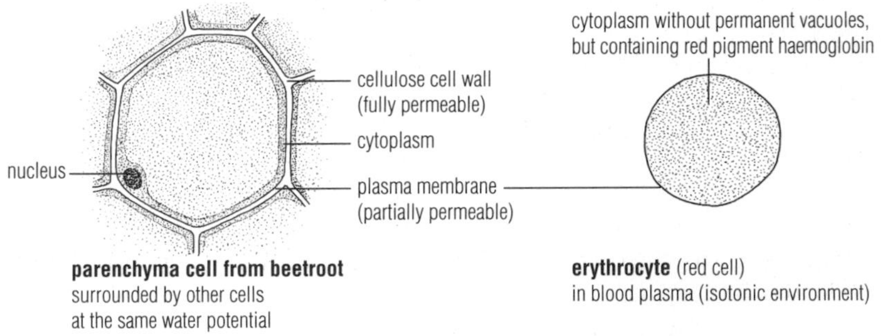

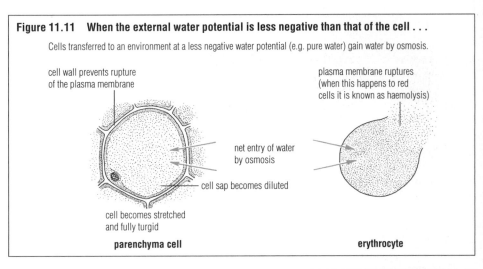

cytoplasm without permanent vacuoles, but containing red pigment haemoglobin

cellulose cell wall (fully permeable)

cytoplasm

nucleus

plasma membrane (partially permeable)

parenchyma cell from beetroot surrounded by other cells at the same water potential

erythrocyte (red cell) in blood plasma (isotonic environment)

Figure 11.11   When the external water potential is less negative than that of the cell . . .

Cells transferred to an environment at a less negative water potential (e.g. pure water) gain water by osmosis.

cell wall prevents rupture of the plasma membrane

plasma membrane ruptures (when this happens to red cells it is known as haemolysis)

net entry of water by osmosis

cell sap becomes diluted

cell becomes stretched and fully turgid

parenchyma cell

erythrocyte

Figure 11.12   When the external water potential is more negative than that of the cell . . .

Cells transferred to an environment at a more negative water potential (e.g. concentrated sugar solution) lose water by osmosis.

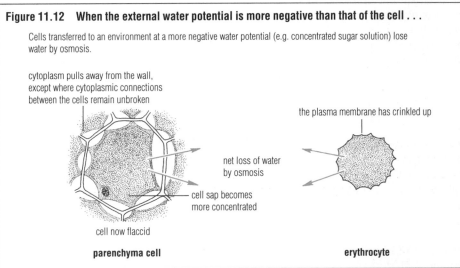

cytoplasm pulls away from the wall, except where cytoplasmic connections between the cells remain unbroken

the plasma membrane has crinkled up

net loss of water by osmosis

cell sap becomes more concentrated

cell now flaccid

parenchyma cell

erythrocyte

When a red cell loses water by osmosis the shrinking of the cell volume leads to crinkling of the plasma membrane, called **crenation**.

In plant cells in which water loss by osmosis has become severe, the cytoplasm and plasma membrane pull away from the cell walls, and the cells are then said to have become **plasmolysed**. The plasmolysed condition is normally seen under laboratory conditions; it may be extremely rare in nature. Cells are not harmed by plasmolysis, at least initially; they are able to take up water again when it becomes available. Plasmolysis may, however, break the plasmodesmata (p. 167) that connect adjacent cells through holes in the cell walls.

■ Water relations of plant cells

When a cell at **incipient plasmolysis** (that is, when the plasma membrane has just begun to shrink away from the cell wall) is placed in pure water, water is taken up by osmosis. The effects of water uptake by such a flaccid plant cell on the water potential, the solute potential and the pressure potential are shown below in Figure 11.13.

In a plasmolysed cell, the water potential is due solely to the solute potential of the cell solution, because the pressure potential of a flaccid cell is zero. As water uptake progresses the cell solution and cell sap (vacuole solution) become more dilute and the solute potential becomes less negative.

Pressure potential is generated when sufficient water has been taken up to cause the cell contents to press against the cell wall. With continuing water uptake, the cell continues to inflate and its wall is stretched. The pressure potential rises progressively to reach a value that completely offsets the solute potential. Water uptake ceases when the water potential of the cell rises to zero. Its water potential is then the same as that of pure water.

■ Water relations of animal cells

Animal cells do not have a cell wall to protect them; instead, in most animals the osmotic concentration of cells and tissue fluid is regulated. This process is referred to as **osmoregulation** (p. 372). The kidney of the mammal (p. 380) has an important osmoregulatory function, for example.

Amoeba, along with many other protozoans, normally has a water potential more negative than the pond water in which it lives (many animal physiologists prefer to explain osmosis in terms of concentration difference; they would say that *Amoeba* has a higher osmotic pressure than the pond water). The difference in water potential leads to a net inflowing of water into the cytoplasm of the *Amoeba*, which uses its contractile vacuole to pump out the excess water that accumulates inside its single cell. The pressure potential (called turgor or hydrostatic pressure by some animal physiologists) of this organism is thus not permitted to rise (Figure 11.14). *Amoeba* may also adjust the quantity of dissolved substances in the cytoplasm (by active pumping of ions, for example) so that water entering by osmosis is reduced to a volume that the contractile vacuole can pump out efficiently (Figure 18.7, p. 378).

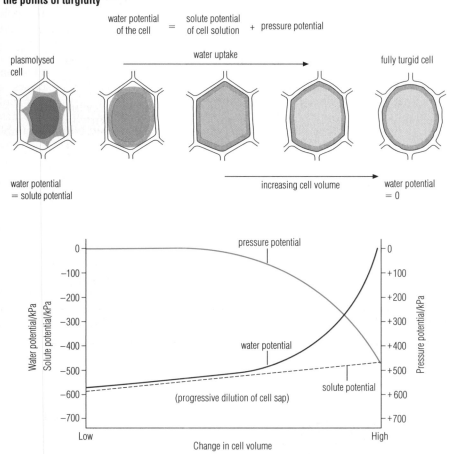

Figure 11.13 The water potential of a flaccid (plasmolysed) plant cell in water as it hydrates to the points of turgidity

Figure 11.14 The water potential of *Amoeba*, and the role of the contractile vacuole

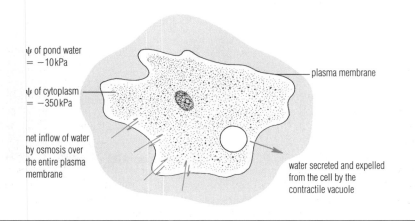

Questions

1 What is the outcome of immersing an animal cell in a solution of substantially lower water potential than that of the cell? What would be the effect on a plant cell?

2 Explain what water movement is taking place, and why, at the point in Figure 11.13 when the increase in cell volume reaches three-quarters of its maximum.

3 a What is the normal response of *Amoeba* if water intake by osmosis increases?

   b What will happen to an *Amoeba* if the action of the contractile vacuole is inhibited by poison?

4 What might be the effect on root cells if an excess of chemical fertiliser is added to the soil?

11.4 ACTIVE TRANSPORT ACROSS MEMBRANES

Transport of materials across membranes requires energy. We have already seen that the energy source for diffusion is the intrinsic *kinetic energy* of molecules and ions. In 'active transport' it is *metabolic energy* produced by the cell (and stored as ATP) that transports substances across cell membranes.

■ Evidence for active transport

Indirect evidence for the transport of useful substances such as ions and other metabolites across membranes comes from measurements of the concentrations of essential ions and metabolites inside and outside cells.

Many essential substances occur at much higher concentrations inside cells than outside. In fact, many metabolically active substances are found to be hoarded in cells where they have been absorbed from very dilute external solutions, and pumped into cytoplasm already containing significant amounts of these substances. The direction of their accumulation is the exact opposite to that to be expected if they had accumulated by diffusion. Figure 11.15 shows the concentrations of certain ions inside and outside mammalian skeletal muscle and *Valonia*, a marine green alga having giant multi-nucleate cells. It shows marked differences between the concentrations of five metabolically important ions in the interior of the cells and in the external solution. For example, compare the concentrations of sodium (Na^+) and potassium (K^+) ions inside and outside both types of cell.

■ How active transport occurs

Active transport requires metabolic energy. The mechanism by which this energy is used to transport ions or metabolites across a membrane is referred to as a 'pump'. **Biological pumps** take the form of carrier molecules in membranes. These are large protein molecules buried in the membrane or located on one side of the lipid bilayer, but able to move within and across it.

> ### ACTIVE TRANSPORT OF METABOLITES: A SUMMARY
> The evidence suggests that
> ▶ salts are absorbed as ions
> ▶ both ions and covalent substances (p. 128) are taken in or excluded selectively
> ▶ the maintenance of the marked differences in concentration between the interior and the exterior of a cell requires a constant expenditure of energy by the cell; energy is made available from respiration and is used both to maintain the leakproof property of the cell membranes and to transport substances across them.

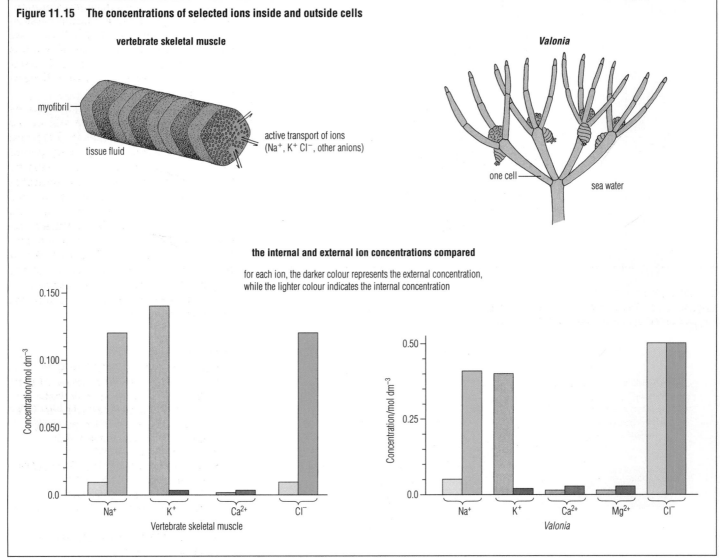

Figure 11.15   The concentrations of selected ions inside and outside cells

vertebrate skeletal muscle

myofibril

tissue fluid

active transport of ions (Na^+, K^+ Cl^-, other anions)

Valonia

one cell

sea water

the internal and external ion concentrations compared

for each ion, the darker colour represents the external concentration, while the lighter colour indicates the internal concentration

Concentration/mol dm^{-3}

Na⁺   K⁺   Ca²⁺   Cl⁻
Vertebrate skeletal muscle

Na⁺   K⁺   Ca²⁺   Mg²⁺   Cl⁻
Valonia

Figure 11.16 A primary pump for active transport: how it may operate

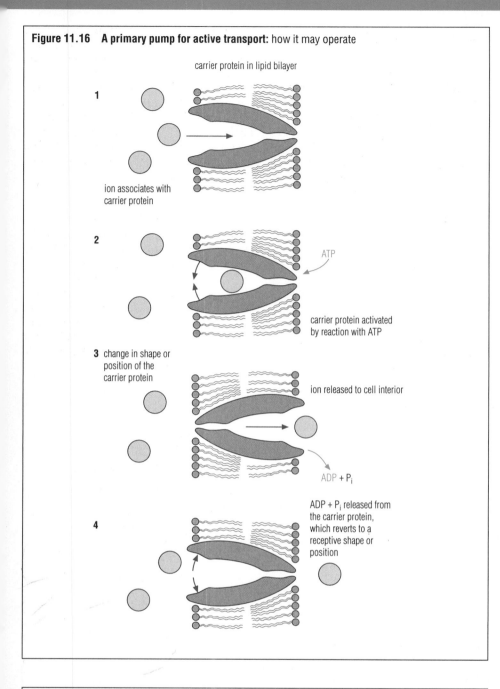

carrier protein in lipid bilayer

1

ion associates with carrier protein

2

ATP

carrier protein activated by reaction with ATP

3 change in shape or position of the carrier protein

ion released to cell interior

ADP + P$_i$

ADP + P$_i$ released from the carrier protein, which reverts to a receptive shape or position

4

Biological pumps are of two types, known as primary and secondary pumps. In a **primary pump** the energy is directly involved in the movement of the metabolite or ion across the membrane (Figure 11.16).

Some primary pumps at work

One example of a primary pump is an ion pump that is present in many cell membranes, in which the carrier molecule is a protein of the plasma membrane. After reaction with ATP, it pumps sodium ions in one direction across the membrane (Figure 11.17).

An ion pump of this type is of great importance in animal cells, since these are not protected by a rigid cell wall. If such a cell contains an excessive quantity of ions there is a danger that it may take up so much water by osmosis that it swells up and bursts (lyses). The sodium pump is an essential mechanism for keeping down the sodium ion concentration of a cell.

Another form of ion pump simultaneously transports sodium and potassium ions (in opposite directions). It too uses energy from ATP. This type of pump is almost universal in animal cell membranes, and is common in plant cell membranes also.

A pump that transports sucrose occurs in the epithelial cells of the villi of the small intestine of mammals (p. 285), and is another example of a primary pump. The protein of this pump is an enzyme that completes digestion of the disaccharide sucrose. The formation of the enzyme–substrate complex on the outside of the cell membrane triggers the hydrolysis of sucrose to its component monosaccharides (p. 132), and also a change in the shape or position of the enzyme. The products of sucrose hydrolysis are then released onto the cytoplasm side of the membrane.

Figure 11.17 The sodium pump of animal cells: a sodium pump, consisting of a special protein in the plasma membrane, is universal to all animals

sodium pump (many of these proteins occur in a cell membrane)

ADP + P$_i$

ATP

plasma membrane

sodium ion diffuses to pump

trapped solute

Na$^+$

H$_2$O

Questions

1 From the data in Figure 11.15, name
   a an ion that living cells accumulate, despite being in low concentration externally
   b an ion that living cells exclude, even if the external concentration is relatively high.
2 Why is ATP described as a source of 'metabolic energy' (p. 176)?
3 State a role in cell metabolism for each of the following ions: K$^+$, Ca^{2+} and NO$_3^-$.
4 a Suggest reasons why the relative concentrations of K$^+$ and Na$^+$ ions in cells are not identical.
   b Which of these two ions appears to be pumped out of the cell?

■ Secondary pumps

In some cells the presence of sodium ions is essential for the movement of organic molecules such as glucose, amino acids and other organic substances across membranes. These organic molecules accumulate against a concentration gradient. To help transport the organic molecule, sodium ions have to be present at a higher concentration on the side of the membrane from which the organic molecules are moved.

These observations led to the hypothesis that a 'secondary pump' exists. In a secondary pump, active transport of an organic molecule is coupled to and driven by the flow of a second solute (an ion, normally sodium) down a concentration gradient that has been created by the action of associated primary pumps (driven by metabolic energy provided by ATP). This mechanism is illustrated here by two examples.

The first example concerns the movement of glucose across the epithelium cells of the small intestine into the blood (Figure 11.18). By means of a primary pump, sodium ions are pumped out of the epithelial cells of the villus (p. 283). Sodium ions are then at higher concentration outside the cells, and a gradient in sodium ions exists across the cell membrane.

Glucose molecules cross the cell membrane from the lumen of the gut while bound to a carrier protein molecule, which also binds with sodium ions. The sodium ions diffuse down their own concentration gradient, across the carrier protein (which is believed to traverse the cell membrane). They transport the glucose molecules with them. On the other side of the membrane sodium ions and glucose dissociate from the carrier protein, and pass into the cytosol of the epithelial cell.

Finally, glucose molecules combine with a carrier protein in the membrane of the base of the epithelial cell and are transported across into the blood capillary network of the villus by facilitated diffusion (p. 231).

The mechanism of the loading of sucrose into phloem sieve tubes provides a further example (Figure 11.19). Sucrose made in the leaf mesophyll cells (p. 255) is transported into sieve tubes (p. 337) by special **transfer**

Figure 11.18   Transport of glucose across the epithelium of the villus

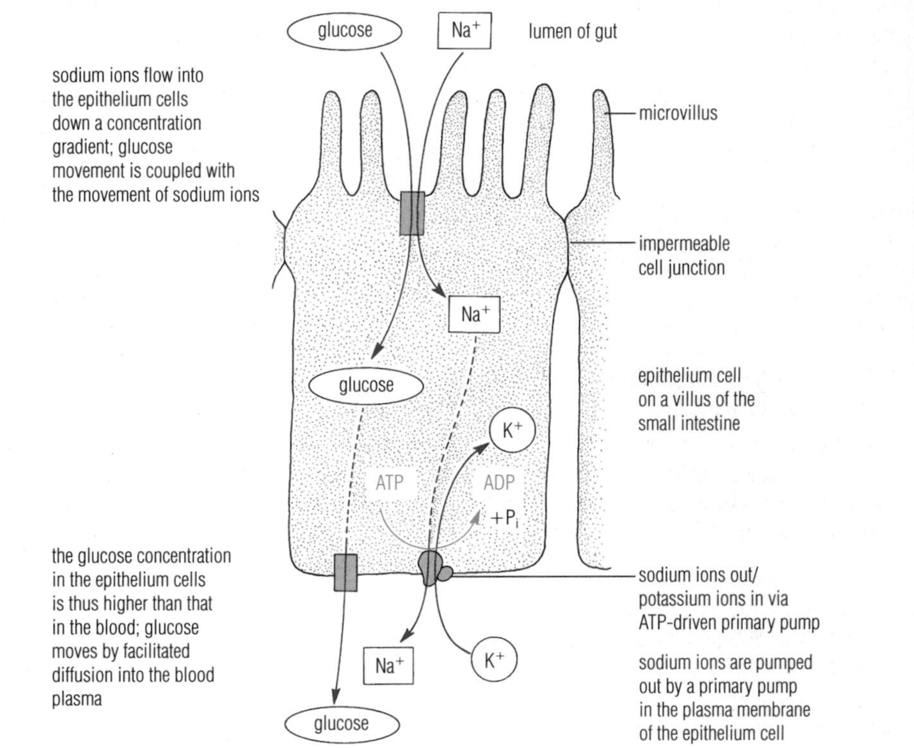

Figure 11.19   Transport of glucose from transfer cells to phloem sieve tube cells

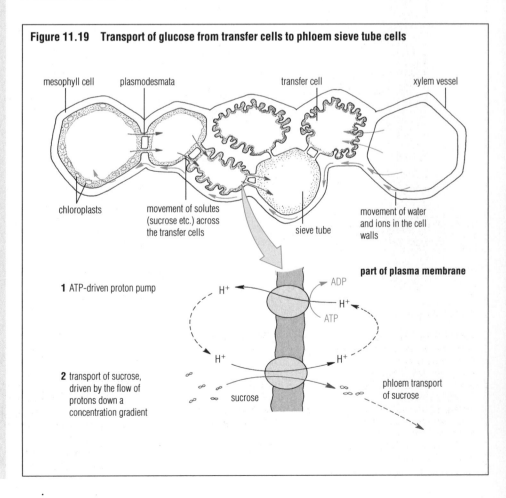

cells. The transfer cells have highly convoluted cell walls, which appear to increase the surface area of the cell membrane available for exchange of substances. The loading of sucrose into sieve tubes, across the plasma membrane of the transfer cell, is the work of a secondary pump.

An ATP-driven primary pump exports hydrogen ions (protons) from the sieve tube into the cytosol of the transfer cell. The re-entry of hydrogen ions into the sieve tube is coupled with the transport of sucrose.

■ Evidence that respiration is involved in active uptake

The rate of respiration of living tissue can be measured using a Warburg manometer, using samples of tissues (Chapter 8, p. 171). The use of the Warburg manometer to measure oxygen uptake and carbon dioxide output of tissue is outlined below. The flask to contain the tissue in the Warburg apparatus is arranged with a side arm, from which solutions of ions or respiratory inhibitors can be added during measurements of respiration of the tissue.

An experiment to investigate respiration rate and uptake of ions by plant tissue discs is outlined in Figure 11.20. The results of the investigation are shown in Figure 11.21. This experiment could also be carried out with animal tissue; the results would be similar for all tissues.

In the experiment in which potassium (K^+) and chloride (Cl^-) ions were available to the plant tissue discs, the tissue responded by active uptake of ions, and its respiration rate also increased. This additional respiration has been described as 'salt respiration' because of its association with ion accumulation. Salt respiration was shown to be linked to ion uptake by the introduction of a specific inhibitor of respiration. The cyanide ion is a specific inhibitor of enzymes that contain metal, particularly the electron-carrier molecules known as cytochromes (p. 316). When cyanide was added, ion accumulation and the extra respiration known as salt respiration stopped instantly.

The link between respiration and ion uptake is most likely to occur through the supply of ATP from respiration and its use in priming the membrane pumps. (It is during oxidation and reduction of the cytochromes of the electron-transport pathway that most ATP is generated in respiration.)

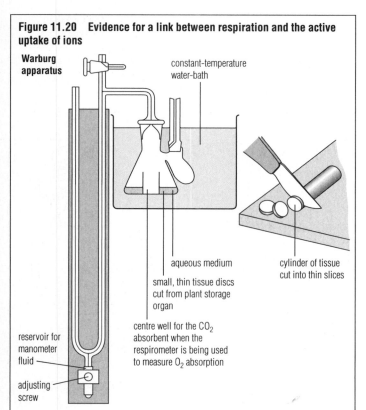

Figure 11.20 Evidence for a link between respiration and the active uptake of ions

Warburg apparatus

constant-temperature water-bath

aqueous medium

cylinder of tissue cut into thin slices

small, thin tissue discs cut from plant storage organ

centre well for the CO_2 absorbent when the respirometer is being used to measure O_2 absorption

reservoir for manometer fluid

adjusting screw

The investigations

sample 1: placed in a Warburg apparatus for **oxygen uptake measurement** at regular intervals over a five-hour period, involving in turn
▶ tissue in pure water
▶ with dilute KCl solution added
▶ with dilute solution of KCN (a respiratory inhibitor) added

sample 2: placed under identical conditions for **ion absorption measurement** at regular intervals (as determined from the disappearance of K^+ and Cl^- ions from the medium) during periods
▶ when KCl was available
▶ when KCN was also present

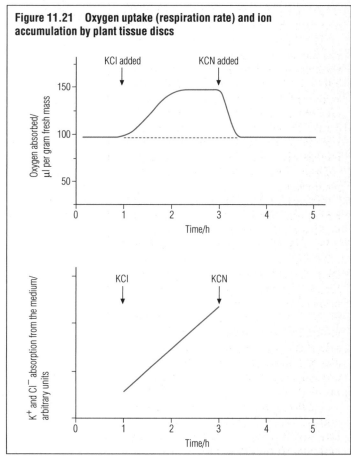

Figure 11.21 Oxygen uptake (respiration rate) and ion accumulation by plant tissue discs

KCl added   KCN added

Oxygen absorbed/ μl per gram fresh mass

Time/h

KCl   KCN

K^+ and Cl^- absorption from the medium/ arbitrary units

Time/h

Questions

1 What are the key differences between a primary and a secondary pump mechanism for active uptake of substances into cells?
2 What is an ionophore protein (p. 231)? What happens when it is activated?
3 Why are thin slices of tissue used in experiments on the metabolism of cells rather than blocks or cubes of tissue?

11.5 BULK TRANSPORT ACROSS THE PLASMA MEMBRANE

Another mechanism by which substances enter the cell is known as **endocytosis**. This is a form of bulk transport of materials. We have seen that the plasma membrane of cells has a dynamic, mobile structure (represented by the fluid mosaic model of membrane structure, Figure 7.15, p. 159). The flexibility of the cell membrane is an important factor in the bulk transport of materials into the cells.

Endocytosis occurs by invagination (in-tucking) of the plasma membrane, which thus forms small vesicles or food vacuoles that become detached and enter the cytosol. Alternatively, vacuoles are formed from tiny cups of membrane at the cell surface. When the materials engulfed in this manner are largely solids the process is called **phagocytosis** (meaning 'cell eating'). When the materials taken in are liquids or suspensions of finely divided solids the process is known as **pinocytosis** ('cell drinking'). Pinocytic vesicles are often formed at the end of a pinocytic channel, itself formed by invagination of the plasma membrane (Figure 11.22).

Vesicle and food vacuole formation are active processes, which require energy from respiration. Materials to be absorbed into the cytosol from a vacuole or vesicle must first cross the membrane before truly entering the cytosol. For example, the contents of food vacuoles are only digested after lysosomes (p. 162) have fused with the food vacuole and delivered hydrolytic enzymes and acids.

Pinocytosis occurs in many protozoans, in certain cells of the liver and kidneys and in some plant cells. Many cells carry out phagocytosis, but the process is particularly associated with the following three specialised functions:

► It is the main feeding mechanism of *Amoeba* (p. 291) and other members of the protozoan phylum Rhizopoda.
► In the response of mammals to an invasion by foreign bacteria, certain of their white blood cells (neutrophils and monocytes, p. 364) can recognise the invading organisms, and engulf and destroy them by phagocytosis.
► Quantitatively, the most important example of phagocytosis is the disposal in mammals of senescent or damaged cells and cell debris; for example, human phagocytic cells typically remove approximately 3×10^{11} erythrocytes daily (p. 218).

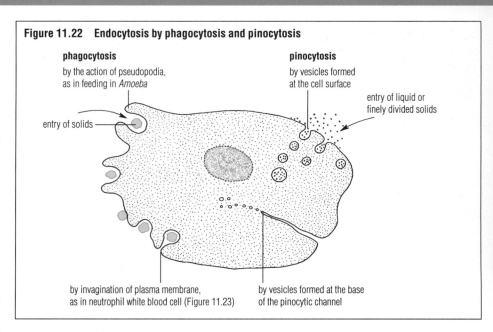

Figure 11.22   Endocytosis by phagocytosis and pinocytosis

phagocytosis
by the action of pseudopodia, as in feeding in *Amoeba*

entry of solids

pinocytosis
by vesicles formed at the cell surface

entry of liquid or finely divided solids

by invagination of plasma membrane, as in neutrophil white blood cell (Figure 11.23)

by vesicles formed at the base of the pinocytic channel

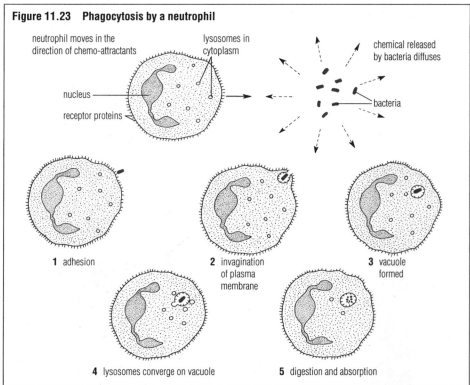

Figure 11.23   Phagocytosis by a neutrophil

neutrophil moves in the direction of chemo-attractants

lysosomes in cytoplasm

chemical released by bacteria diffuses

nucleus

receptor proteins

bacteria

1 adhesion
2 invagination of plasma membrane
3 vacuole formed
4 lysosomes converge on vacuole
5 digestion and absorption

■ The stages in phagocytosis

Phagocytosis by a neutrophil (Figure 11.23) involves the following steps:

1 The engulfing cells detect chemo-attractive molecules (usually small peptide molecules) released by the target matter, and respond by moving towards it.

2 The material to be engulfed becomes attached to the phagocytic cells by means of some 'lock and key' mechanisms involving receptor proteins on the cell surface membrane.

3 Local contractile processes of each cell's cytoskeleton (microfilaments, p. 167) become active in the cytoplasm. These consist of the proteins actin and myosin; these proteins react with ATP, and help to form the pseudopodia that result in a new food vacuole around the foreign material.

4 The engulfed particles are digested by enzymes delivered to the vacuole by lysosomes (p. 162). There is a characteristic burst of respiratory activity in the neutrophil following vacuole formation.

11.6 SECRETION FROM CELLS BY BULK TRANSPORT

Substances are secreted from cells by the reverse of endocytosis. Multicellular organisms contain several different types of very specialised cell whose main function is to manufacture and export products that are put to use elsewhere in the organism or outside it. The general name for these processes is **exocytosis**.

Examples of secretory cells include the cells that produce enzymes in the pancreas (p. 284), those that secrete hydrochloric acid in the gastric glands (p. 281), those that secrete hormones in the various ductless glands (p. 472) and the sweat-secreting cells of the sweat glands of human skin (p. 511).

Some secretory cells produce solutions of ions and relatively simple molecules. Others secrete suspensions of large and complex molecules such as proteins, lipids and polysaccharides; these chemicals are manufactured in the Golgi apparatus or endoplasmic reticulum, and packaged in vesicles (p. 162). These are transported across the cytoplasm to the surface of the cell and then released into the environment around the cell by reverse pinocytosis or reverse phagocytosis.

■ An example of exocytosis: milk secretion

The milk secreted by mammals is an aqueous colloidal suspension of proteins, lipids, sugar (mainly lactose), ions and water, together with tiny quantities of other substances. The proteins are produced in the rough endoplasmic reticulum (p. 160) of the secretory cells of mammary glands. Lipids are formed in the smooth endoplasmic reticulum of the same cells. Proteins and lipids are discharged from the cells by reverse phagocytosis.

Lactose is produced by the actions of a soluble enzyme dissolved in the cytoplasm. It is then discharged from the microvilli by reverse pinocytosis.

Sodium and potassium ions are pumped out across the cell membrane by ion pumps (Figure 11.16). Water follows the ions by osmosis (Figure 11.24).

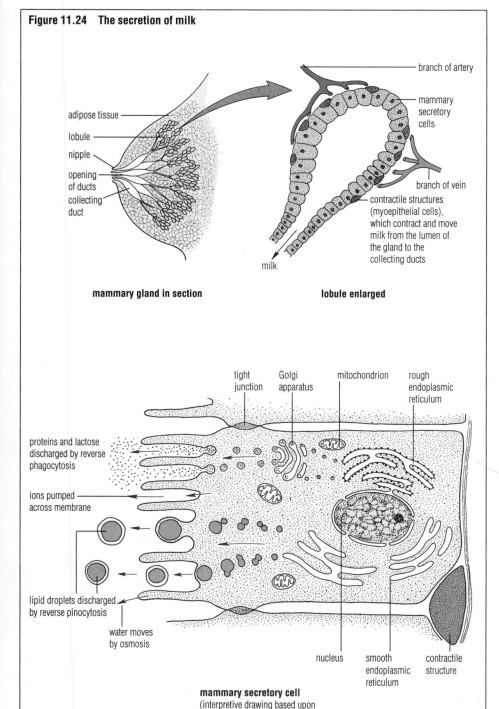

Figure 11.24   The secretion of milk

adipose tissue
lobule
nipple
opening of ducts
collecting duct
milk

mammary gland in section

branch of artery
mammary secretory cells
branch of vein
contractile structures (myoepithelial cells), which contract and move milk from the lumen of the gland to the collecting ducts

lobule enlarged

tight junction
Golgi apparatus
mitochondrion
rough endoplasmic reticulum

proteins and lactose discharged by reverse phagocytosis
ions pumped across membrane
lipid droplets discharged by reverse pinocytosis
water moves by osmosis

nucleus
smooth endoplasmic reticulum
contractile structure

mammary secretory cell
(interpretive drawing based upon an electronmicrograph)

Questions

1 By what processes do oxygen, water, glucose and bacteria respectively enter a living neutrophil white blood cell?
2 What are the steps involved when a cell takes substances from outside the cell by formation of vesicles?
3 In the production of milk, what are the roles of
   a the endoplasmic reticulum (p. 160)?
   b ribosomes (p. 161)?

FURTHER READING/WEB SITES

Browse: **vl.bwh.harvard.edu**

EXAMINATION QUESTIONS SECTION 2

Answer the following questions using this section (chapters 6 to 11) and your further reading.

1 a i State **two** advantages of using an electron microscope to study cells. (2)
ⅱ Outline how a sample of tissue may be prepared for the electron microscope. (4)

This is a diagram of a mitochondrion, as seen using an electron microscope.

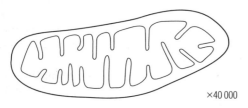

×40 000

b Calculate the **actual** length of the mitochondrion. Show your working. (2)

Mitochondria are not found in prokaryotic cells.

c State **four** other ways in which prokaryotic cells differ from eukaryotic cells. (4)

UoCLES AS/A Modular: Biology Foundation February 1997 Section A Q1

2 The diagram represents a phospholipid molecule.

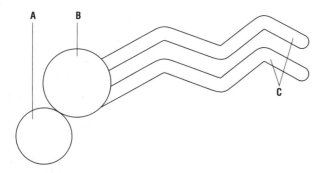

a i Name the parts of the molecule **A, B** and **C.** (1)
ⅱ Explain how the phospholipid molecules form a double layer in a cell membrane. (2)
b Cell membranes also contain protein molecules. Give **two** functions of these protein molecules. (2)

NEAB AS/A Biology Processes of Life Module Test (BYO1) March 1998 Section A Q1

3 The disaccharide sucrose is a non-reducing sugar. If this carbohydrate is hydrolysed, it forms a molecule of glucose and a molecule of fructose when the glycosidic link is broken.
a State what is meant by
i a *non-reducing sugar*; (1)
ⅱ a *glycosidic link*. (2)
b Copy and complete the table that follows to show the difference between
i the structure, and
ⅱ the function
of starch (amylose) and cellulose. (5)

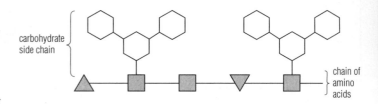

	Amylose	Cellulose
i structure		
ii function		

UoCLES AS/A Modular: Biology Foundation February 1997 Section A Q2

4 Below is a diagram of the molecular structure of part of a glycoprotein found at the cell surface membrane.

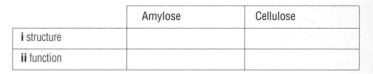

a State how the following are linked in a glycoprotein:
i adjacent amino acids; (1)
ⅱ adjacent molecules of the sugar in the carbohydrate side chain. (1)
b State **two** functions of glycoproteins in cell surface membranes. (2)
c Explain why cell surface membranes are described as having a *fluid mosaic* structure. (4)

UoCLES AS/A Modular: Biology Foundation June 1998 Section A Q1

5 The diagram shows a metabolic pathway in which substrate **A** is converted, with the aid of enzymes, to the end-product **D.**

a Giving an explanation for your answers, suggest what would happen to the rate of production of the end-product **D** if:
i the concentration of substrate **A** were reduced; (1)
ⅱ the concentration of **enzyme 1** were increased but the concentrations of the other enzymes remained constant; (1)
ⅲ the temperature rose from 15 °C to 25 °C. (1)
b Suggest how molecule **D** could act as an *end-product inhibitor*. (2)

AEB A Biology Paper 1 June 1998 Q3

6 Catalase is an enzyme which breaks down hydrogen peroxide into oxygen and water. The activity of catalase can be measured by soaking small discs of filter paper in a solution containing the enzyme. The discs are immediately submerged in a dilute solution of hydrogen peroxide. The filter paper discs sink at first but float to the surface as oxygen bubbles are produced. The reciprocal of the time taken for the discs to rise to the surface indicates the rate of reaction.

An experiment was carried out to investigate the effect of substrate concentration on the activity of catalase. A filter paper disc was soaked in a solution containing catalase, and then submerged in a buffer solution containing hydrogen peroxide. The time taken for the disc to rise to the surface was recorded. This experiment was repeated using a range of concentrations of hydrogen peroxide.

The results are shown in the graph below.

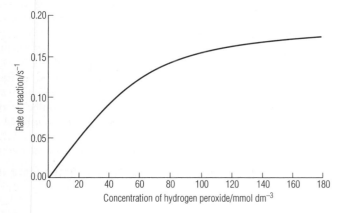

a State why a buffer solution was used in this experiment. (1)

b i Describe the relationship between the rate of reaction and the concentration of hydrogen peroxide as shown by the graph. (3)

ii Explain this relationship between substrate concentration and the rate of reaction. (3)

c Describe how a solution containing 160 mmol of hydrogen peroxide per dm³ would be diluted to prepare a solution containing 80 mmol of hydrogen peroxide per dm³. (2)

d Describe how this experiment could be modified to investigate the effect of temperature on the activity of catalase. (4)

London AS/A Biology and Human Biology Module Test B/HB1 January 1998 Q7

7 The microbial enzymes α-amylase, amyloglucosidase and glucose isomerase are used in the industrial production of high fructose corn sweeteners (HFCS) from corn starch. The diagram below summarises the stages of the process.

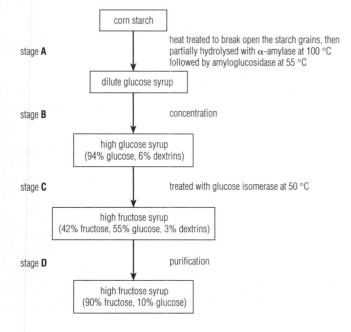

a The corn starch is partially hydrolysed by the enzymes α-amylase and amyloglucosidase at stage **A**.

i Name the type of bond which is hydrolysed by these enzymes. (1)

ii Suggest an explanation, apart from breaking open the starch grains, for the high temperatures used during hydrolysis in stage **A**. (2)

b Both glucose and fructose are hexoses with the same general formula. The enzyme, glucose isomerase, catalyses the re-arrangement of the atoms in the glucose molecule so that fructose is formed.
State the general formula of a hexose. (1)

c The conversion of the high glucose syrup into high fructose syrup at stage **C** is carried out in a reactor vessel, containing immobilised glucose isomerase.

i Explain what is meant by *enzyme immobilisation*. (2)

ii Explain **two** advantages of using immobilised enzymes in industrial processes. (4)

d Suggest **one** reason why fructose is used in preference to glucose as a sweetener in drinks and confectionery. (1)

London A Biology and Human Biology Module Test B/HB4A Microorganisms and Biotechnology June 1998 Q6

8 a Distinguish between *tissue* and *organ*, giving an example of each in an animal. (4)

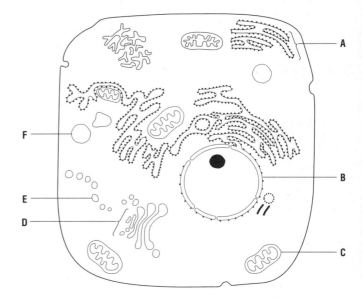

The diagram shows the ultrastructure of an animal cell (×9000).

b Name the structures **A** to **D**. (2)

c Calculate the actual diameter of the structure labelled **F**. Show your working. (2)

d Describe the functional relationship between structures **A**, **D** and **E**. (3)

OCR (Cambridge Modular) AS/A Sciences: Biology Foundation March 1999 Q1

9 a Name the type of bond that holds together the two strands of nucleotides in a DNA molecule. (1)

Genetic drugs are short sequences of nucleotides. They act by binding to selected sites on DNA or mRNA molecules and preventing the synthesis of disease-related proteins. There are two types.

Triplex drugs are made from DNA nucleotides and bind to the DNA forming a three-stranded helix.

Antisense drugs are made from RNA nucleotides and bind to mRNA.

b Name the process in protein synthesis that will be inhibited by:
  i triplex drugs; (1)
  ii antisense drugs. (1)

c The table shows the sequence of bases on part of a molecule of mRNA.

Base sequence on coding strand of DNA								
Base sequence on mRNA	A	C	G	U	U	A	G	C	U
Base sequence on antisense drug								

Copy and complete the table to show:
  i the base sequence on the corresponding part of the coding strand of a molecule of DNA; (1)
  ii the base sequence on the antisense drug that binds to this mRNA. (1)

AEB A Biology Paper 1 June 1998 Q5

10 The diagram shows three adjacent plant cells.

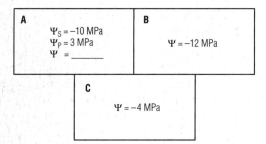

Copy the diagram.
a **i** Calculate the water potential of cell **A**. Write your answer in cell **A** in your diagram. (1)
  ii Show, by means of arrows on your diagram, the direction of water movement between these cells. (1)
  iii Explain why the water potential of a sucrose solution has a negative value. (2)
b An artificial membrane can be made which consists only of a lipid bilayer. The diagram compares the permeability of such an artificial membrane with a biological cell membrane.

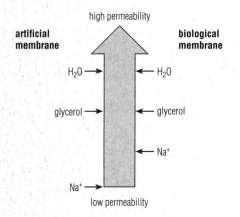

  i Explain why the permeability of both membranes to glycerol is the same. (1)
  ii Explain why the permeability of the two membranes to sodium ions differs. (2)

AEB A Biology Paper 1 June 1998 Q1

11 The diagram shows the sequence of bases in a short length of mRNA.

A U G G C C U C G A U A A C G G C C A C C A U G

a **i** What is the maximum number of amino acids in the polypeptide for which this piece of mRNA could code? (1)
  ii How many different types of tRNA molecule would be used to produce a polypeptide from this piece of mRNA? (1)
  iii Give the DNA sequence which would be complementary to the first five bases in this piece of mRNA. (1)
b Name the process by which mRNA is formed in the nucleus. (1)
c Give **two** ways in which the structure of a molecule of tRNA differs from the structure of a molecule of mRNA. (2)

AQA (NEAB) AS/A Biology: Continuity of Life (BY02) Module Test March 1999 Q5

12 State the property of water and explain its importance in each of the following cases:
a the cooling of a crop plant, (2)
b the survival of fish in ice-covered lakes, (2)
c insects, such as pondskaters, which can walk on water, (2)
d the transport of glucose and ions in a mammal. (2)

OCR (Cambridge Modular) AS/A Sciences: Biology Foundation June 1999 Q5

13 The graph shows the percentage change in length of cylinders of potato which had been placed in sucrose solutions of different concentrations for 12 hours.

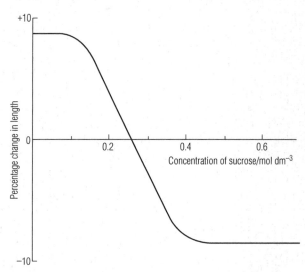

a **i** In terms of water potential, explain the change in length which occurred when the cylinder of potato was placed in a sucrose solution of concentration 0.3 mol dm^{-3}. (2)
  ii Explain the shape of the curve between sucrose concentrations of 0.4 and 0.6 mol dm^{-3}. (1)
b Potato tubers store starch. As they start to grow or sprout, some of this starch is converted to sugars. Make a copy of the graph, and on it sketch a curve to represent the changes in length you would expect if the investigation had been carried out with sprouting potatoes. (2)

AQA (AEB) A Biology: Module Paper 4 June 1999 Q1

3

The Maintenance of the Organism

A dragonfly, caught in a spider's web, is wrapped in silk prior to being digested; trapped prey is first immobilised with poisons injected by the spider's bite

12 Autotrophic nutrition

- Autotrophic organisms make their own organic nutrients (e.g. sugars, amino acids) from an external supply of relatively simple inorganic nutrients (e.g. carbon dioxide, water, mineral ions) and energy.
- In green plants the energy source is light, and the process is known as photosynthesis. The leaf of a flowering plant is an organ specialised for photosynthesis.
- Light energy is absorbed by the green pigment molecule chlorophyll, housed in the chloroplasts, especially in the mesophyll cells of leaves.

- Photosynthesis is a complex set of reactions that can be divided into two stages:
  1 in a light-dependent stage, light energy is used to split water, forming ATP and reduced NADP, and oxygen (waste product) is given off
  2 in a light-independent stage, carbon dioxide is built up into sugars using the products of the light-dependent stage.
- The autotrophic nutrition of green plants sustains the nutrition of other organisms, directly or indirectly, as shown by food chains. Photosynthesis also maintains the composition of the atmosphere, providing oxygen and removing carbon dioxide.

12.1 INTRODUCTION

Nutrition is the process by which organisms obtain energy to maintain life functions, and matter to create and maintain structure. Both energy and matter are obtained from **nutrients**. The various methods by which organisms obtain nutrients are summarised on p. 290.

In **autotrophic nutrition** organisms make their own organic nutrients from an external supply of relatively simple inorganic raw materials (inorganic nutrients) and energy. If the energy source is sunlight, the process is known as **photosynthesis**. This is the form of nutrition of green plants, algae and photosynthetic bacteria (cyanobacteria). It is one of the most important of all biochemical processes because almost all living things depend upon photosynthesis, directly or indirectly, for their organic nutrients. This chapter concentrates on the process of photosynthesis.

Other autotrophic organisms obtain energy from specific chemical reactions, rather than from light. This process, known as **chemosynthesis**, is the mode of nutrition of certain bacteria. Chemosynthetic organisms do not contain chlorophyll, and are not green in colour. Chemosynthesis is discussed in section 12.10 (p. 269).

■ Photosynthesis in outline

In green plants photosynthesis takes place in the cells containing chloroplasts. Most chloroplasts occur in the mesophyll cells of green leaves, but some are also found in parenchyma cells below the epidermis of herbaceous stems. In the chloroplasts, the energy of sunlight is trapped in the pigment **chlorophyll**. Light energy absorbed by chlorophyll is converted into the chemical energy of specific organic compounds (mainly sugars and starch, but also some amino acids and

PHOTOSYNTHESIS: A SUMMARY

The process in the chloroplast can be summarised by the equation:

carbon dioxide	+	water	+	light energy	$\xrightarrow{\text{chlorophyll in}}$ chloroplast	organic compounds e.g. sugars	+	oxygen
(raw materials)				*(energy source)*		*(products)*		*(waste product)*
$6CO_2$	+	$6H_2O$	+	light	$\xrightarrow{\text{chlorophyll in}}$ chloroplast	$C_6H_{12}O_6$	+	$6O_2$

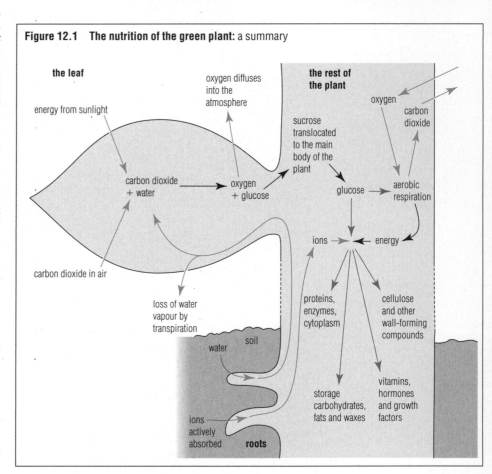

Figure 12.1   The nutrition of the green plant: a summary

lipids), all of which possess much greater energy than the starting compounds

carbon dioxide and water. Oxygen is the waste product (Figure 12.1).

■ The significance of photosynthesis

Photosynthesis and plant metabolism

The whole of the metabolism of the green plant is sustained by the products of photosynthesis. Sugar and other metabolites are transported from the chloroplasts to all parts of the plant. These products are used in the living cells as the building blocks for all the other substances the plant requires. Synthesis of some metabolites requires certain ions, which are absorbed from the soil; for example, nitrate ions are required for the conversion of sugars to amino acids (the building blocks of proteins). Nitrate ions and other ions required in substantial quantities are known as **macronutrients**. Other ions, such as those of magnesium and iron, are required in trace amounts only, and are called **micronutrients**. The macro- and micronutrients essential to plant metabolism are listed in Table 16.1 (p. 328).

The products of photosynthesis also supply the energy needed to carry out chemical change. Energy is held in ATP made during respiration of sugars (p. 316) and by photophosphorylation during the light-dependent reaction of photosynthesis (p. 260).

Photosynthesis and food chains

The survival of life depends upon photosynthesis. This is because all organisms get their nutrients from green plants, either directly or indirectly (with the exception of the few chemoautotrophs, p. 269).

The feeding relationships between organisms are represented in a **food chain** (p. 47). Food chains commence with phytoplankton or green plants, or parts of them, whether living or dead. All other organisms (**heterotrophic organisms**) depend directly or indirectly upon the products of autotrophic nutrition. Green plants themselves depend on the action of the organisms of decay (**saprotrophs**) to obtain recycled inorganic nutrients as ions for their nutrition. This **cycling of nutrients** and the **flow of energy** are shown in Figure 3.48 (p. 72).

Photosynthesis and the composition of air

The concentration of carbon dioxide in the atmosphere is currently 0.035% by volume. Carbon dioxide is added to the air by the respiration of animals, plants and microorganisms, and by the combustion of fossil fuels. The release of carbon dioxide from combustion is now on the increase, contributing to a possible warming of our planet (the 'greenhouse effect', p. 84). Nevertheless, photosynthesis takes place on so great a scale that it re-uses on a daily basis almost as much carbon dioxide as is released into the atmosphere. This is illustrated by the **carbon cycle** (Figure 12.2).

The seasonal variation in the level of carbon dioxide in the atmosphere is greater than the net annual increase (Figure 4.13, p. 84). Actually, the yearly increase is half that expected if all the carbon dioxide produced by burning fossil fuel were to remain in our atmosphere. Phytoplankton of the oceans has an important role in the global carbon cycle, converting carbon dioxide that dissolves in the water to organic carbon. A large proportion of the phytoplankton of the spring bloom (Figure 4.14) sinks to the depths and remains submerged for hundreds of years. The oceans are an important 'sink' for carbon (as are the tropical rain forests, p. 81).

Photosynthesis is the only natural process that releases oxygen into the atmosphere. In the early history of the Earth the element oxygen occurred in various compounds, including oxides in the rocks and water, but not as a gas in the atmosphere. After the evolution of photosynthetic organisms oxygen gas began to accumulate, and to date the atmospheric concentration of oxygen has reached about 21%

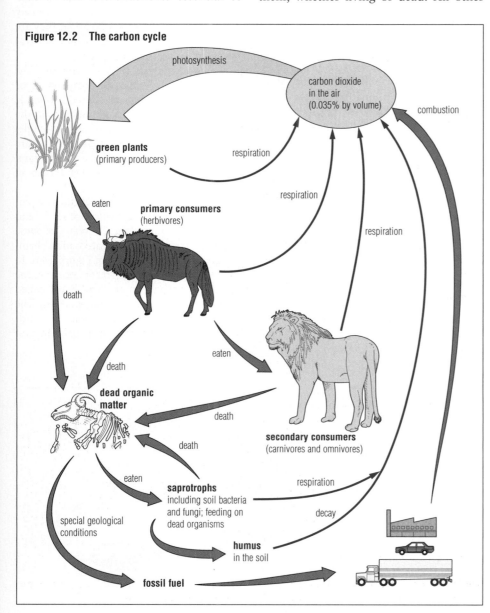

Figure 12.2   The carbon cycle

photosynthesis

carbon dioxide in the air (0.035% by volume)

combustion

green plants (primary producers)

respiration

respiration

respiration

eaten

primary consumers (herbivores)

death

eaten

death

dead organic matter

death

death

secondary consumers (carnivores and omnivores)

eaten

respiration

saprotrophs including soil bacteria and fungi; feeding on dead organisms

decay

special geological conditions

humus in the soil

fossil fuel

Questions

1   What is the waste product of photosynthesis?
2   In what form is the carbohydrate made by photosynthesis
   a   transported away from the leaf?
   b   used in metabolism?

Figure 12.3   Testing a green leaf for starch

1 a leaf is detached from a healthy green plant that has been exposed to full light for several hours

2 the leaf is dipped into very hot water for 5 seconds; this kills the cells, making them fully permeable

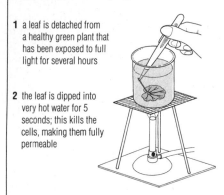

3 the leaf is rolled and inserted into a test tube of boiling ethanol solution and kept immersed for 5 minutes; this dissolves out much of the chlorophyll

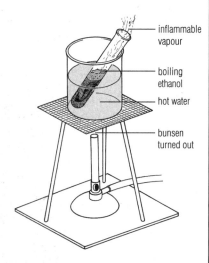

— inflammable vapour

— boiling ethanol

— hot water

— bunsen turned out

4 the leaf is withdrawn and dipped into hot water; this softens the leaf tissue

5 the leaf is spread out on a white tile and iodine solution is applied over its surface; a blue-black colour develops where starch is present

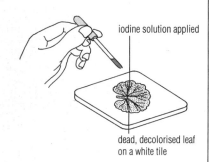

iodine solution applied

dead, decolorised leaf on a white tile

6 provided the leaf was initially free from starch, the presence of starch at the end of the experiment is acceptable evidence that photosynthesis has occurred

12.2 THE CONDITIONS FOR PHOTOSYNTHESIS

From the definition of photosynthesis we can expect that the process needs light, chlorophyll in chloroplasts, carbon dioxide and water. Since photosynthesis is a bio-chemical process involving many enzymes, a suitable temperature is also necessary. These are essential conditions for photo-synthesis in a plant cell.

■ Testing whether photosynthesis has occurred

It is relatively easy to detect whether photosynthesis has taken place in the labo-ratory by testing for the accumulation of starch in illuminated leaves, using the blue-black colour that develops when iodine molecules associate with a form of starch known as amylose (p. 136). Here we are assuming that the starch has been formed from sugar produced in chloroplasts in the light (Figure 12.3).

Alternatively, in the case of aquatic plants in particular, we can demonstrate that oxygen is evolved in the light. Many water plants have large air spaces in the stems and leaves. These act as gas reservoirs for photosynthesis and respiration, and they make the plant organs buoyant. When the plant is held inverted below the water and the stem is cut across, gas bubbles out from the cut stem. This gas, which contains gases released by the plant cells into the air spaces, can be collected by downward dis-placement of water, as shown in Figure 1.13 (p. 13). When tested with a glowing splint, it is found to contain a higher pro-portion of oxygen than that in air.

■ Chlorophyll

The requirement of chlorophyll for photosynthesis is demonstrated in plants with variegated leaves (Figure 12.4). In a variegated leaf there are areas of mesophyll that are unable to manufacture chlorophyll. These patches are yellow or cream-coloured, in contrast to the surrounding green leaf tissue. The cells without chloro-phyll have lost the genetic instructions (genes) for chlorophyll production. Com-monly occurring plants with variegated leaves include varieties of privet, laurel, geranium and ivy.

What are the pigments in chlorophyll?

The photosynthetic pigments in a leaf can be separated by chromatography. Chro-matography is carried out on a stationary, absorptive medium (paper in the case of paper chromatography, powdered solid in column chromatography, or a thin film in thin-layer chromatography). A solvent travels through the stationary phase by cap-illarity, and this is the propelling force. The substances being separated are **differen-tially adsorbed** (that is, temporarily attached to the stationary phase) as they are propelled along. This brings about separa-tion of the chlorophyll pigments.

For the analysis of leaf pigments, the pigments can be extracted into ethanol or propanone (acetone) or another organic solvent, and the individual pigments in the solution can then be separated by single-directional paper chromatography (Figure 12.5). This has to be carried out with care, because the extracted pigments bleach in direct sunlight.

Chlorophyll is in fact a mixture of pig-ments: **chlorophyll** *a*, **chlorophyll** *b* and certain **carotenoids**. The chemistry of these pigments has been investigated; their chem-ical structures are shown in Figure 12.6. In the intact plant these pigments occur in the chloroplasts, held in the structures called grana (stacks of membranes, p. 164). The chlorophyll in the membrane is sand-wiched between alternating protein and lipid layers.

Figure 12.4   The need for chlorophyll

1 a potted plant with variegated leaves is held in the dark for 48 hours to destarch the leaves

2 a leaf from the plant is detached and tested for starch

3 the plant is exposed to full light for 8–12 hours

4 another variegated leaf is detached; the outline of the leaf is drawn, and the areas of chlorophyll-free tissue are drawn on the diagram

5 the leaf is tested for starch

6 the blue-black colour due to starch appears only in the tissue that had contained chlorophyll

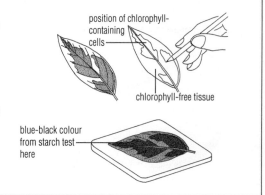

position of chlorophyll-containing cells

chlorophyll-free tissue

blue-black colour from starch test here

Figure 12.5 The extraction and separation of photosynthetic pigments (safety precautions and the essential practical details of this experiment are given in the *Study Guide,* Chapter 12)

1 **The extraction** is made either from fresh green leaves (e.g. nettle), dipped into boiling water to kill the cells and make them permeable, or from dried powdered grass leaves; the material is ground up with aqueous propanone (acetone) and sand.

2 **Chromatographic separation** can be carried out on a narrow strip of filter paper, suspended in a boiling tube (separation will be complete in about 30–60 minutes).

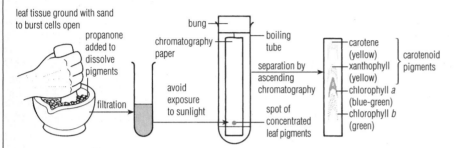

leaf tissue ground with sand to burst cells open

propanone added to dissolve pigments

filtration

chromatography paper

bung

boiling tube

separation by ascending chromatography

avoid exposure to sunlight

spot of concentrated leaf pigments

carotene (yellow)
xanthophyll (yellow) } carotenoid pigments

chlorophyll *a* (blue-green)

chlorophyll *b* (green)

Loading the chromatogram: the chromatogram is loaded at a point just above the line of contact with the solvent; a tiny spot of pigment extract is spotted and allowed to dry, and that spot is neatly respotted. Loading is repeated until a dark green spot is produced.

Running the chromatogram: a solvent mixture that causes the individual pigments to be separated during the run is selected; the loaded chromatogram is lowered into the solvent and the solvent travels up the paper. The run is stopped when the solvent front nears the top of the paper.

3 **Identification of the pigments:** the solvent front at the end of the run is marked and the Rf values of the pigments can be calculated according to the formula

$$Rf = \frac{\text{distance moved by the pigment}}{\text{distance moved by the solvent front}}$$

Figure 12.6 The chemistry of chlorophyll pigments

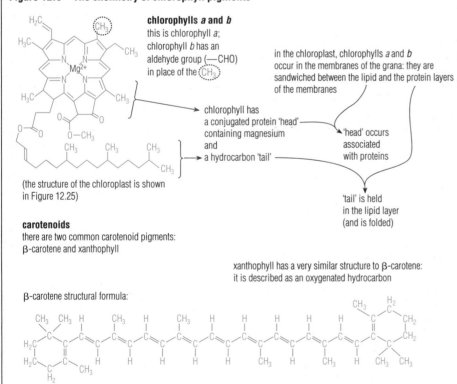

chlorophylls *a* and *b*
this is chlorophyll *a*;
chlorophyll *b* has an aldehyde group (—CHO) in place of the CH₃

in the chloroplast, chlorophylls *a* and *b* occur in the membranes of the grana: they are sandwiched between the lipid and the protein layers of the membranes

chlorophyll has a conjugated protein 'head' containing magnesium and a hydrocarbon 'tail'

'head' occurs associated with proteins

(the structure of the chloroplast is shown in Figure 12.25)

'tail' is held in the lipid layer (and is folded)

carotenoids
there are two common carotenoid pigments: β-carotene and xanthophyll

xanthophyll has a very similar structure to β-carotene: it is described as an oxygenated hydrocarbon

β-carotene structural formula:

vertebrate animals oxidise the molecules of β-carotene taken in with food, each molecule yielding two molecules of vitamin A

Abnormal colouring in leaves

Leaves may also lose chlorophyll as a result of certain diseases. For example, when leaf cells are invaded by viruses or fungi, and the membranes of organelles such as chloroplasts are destroyed, chlorophylls are broken down too. Another cause of discoloration is deficiency of certain minerals and ions (particularly magnesium, iron and nitrate, p. 329).

Light is essential for chlorophyll formation in flowering plants; plants grown in the dark are yellow and etiolated (p. 420). During ageing and senescence of leaves many substances are broken down, including the chlorophylls (Figure 12.7). In tree leaves, before abscission occurs (p. 439), chlorophylls, proteins and various cofactors of enzymes are turned into simpler, soluble substances, and are translocated to cells of stems and roots, before the leaves are cut off.

Figure 12.7 Deciduous leaves in summer (above), and prior to abscission (below)

■ Light

The example of a controlled experiment shown in Figure 1.13 (p. 13) is confirmation that an aquatic plant requires light both to photosynthesise and to produce oxygen.

The requirement for light in photosynthesis can be demonstrated in terrestrial plants by investigating the effect of a period of light or darkness on starch accumulation in an attached starch-free leaf. (To obtain starch-free leaves the whole plant is held in total darkness for 48 hours. In this period starch previously stored in the leaf is converted to sucrose, translocated to storage sites in the stem, root and seeds, and redeposited as starch. The leaves are said to have been 'destarched'.) Starch is made only in those parts of a leaf that receive light (Figure 12.8).

Figure 12.8   The effect of light on the accumulation of starch in a green leaf

1 the plant is kept in the dark for 48 hours; this is called 'destarching'

2 a leaf is detached and tested for starch; this establishes that 'destarching' has taken place

3 a lightproof stencil is fixed to one leaf and the plant is exposed to full sunlight for 6 hours; part of the leaf is illuminated and part is in the dark

4 the leaf with the stencil is detached, the stencil is removed and the leaf is tested for starch

5 the result, known as a starch print, shows that starch is present only in the part of the leaf that has been exposed to the light

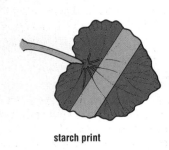

starch print

What is light?

Light is a form of the electromagnetic radiation produced by the Sun. Visible light forms only a small fraction of the total electromagnetic radiation reaching the Earth. Of the visible spectrum, certain wavelengths are absorbed by green plants, including the red and blue parts of the spectrum. Other wavelengths, particularly green light of course, are mostly reflected or transmitted (Figure 12.9).

The true nature of light is not fully understood. Like other forms of energy, it is described in terms of its detectable effects on matter.

The **wave theory** of light holds that light travels in waves of different lengths, and the amount of energy varies inversely with the wavelength of the light. In the visible spectrum, blue light has a relatively short wavelength and therefore has considerable energy, whereas red light has the longest wavelength and less energy.

The **quantum theory** of light holds that light consists of particles called quanta or photons. Quanta of red light have less energy than quanta of blue light.

The wave theory explains certain characteristics of light better than does the quantum theory, and vice versa. Although these concepts are quite different they are both useful to us.

Light and chlorophyll

In solution, chlorophyll absorbs light energy but the energy cannot be used to carry out photosynthesis. This is because the chlorophyll has been separated from the proteins and enzyme machinery of the chloroplast. When the pigments are in the chloroplast the light energy trapped is immediately passed to surrounding proteins as chemical energy, which is used in photosynthesis. A solution of chlorophyll, on the other hand, re-emits the energy of absorbed light as light of longer wavelength (red **fluorescence**); by comparison, intact chloroplasts emit relatively little red fluorescence. The mechanism of the light-dependent stage of photosynthesis is discussed on p. 259.

Figure 12.9   The light absorbed by chlorophyll

the spectrum of electromagnetic radiation

gamma rays	10^{-10}
X-rays	10^{-8}
ultra-violet	10^{-6}
visible	10^{-4}
infra-red (heat)	10^{-2}
microwaves	10^0
radio-waves	10^2 cm

the spectrum of visible light

violet	400 nm
blue	
	500 nm
green	
	600 nm
yellow	
orange	
	700 nm
red	
	750 nm

chlorophyll pigment extracted from the leaf appears green because all wavelengths of light except green are absorbed

visible light → prism → absorbed / green transmitted / absorbed

Figure 12.10   Chlorophyll: the absorption spectrum and the action spectrum

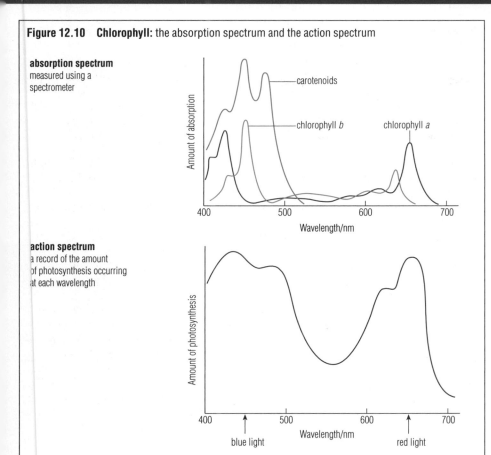

absorption spectrum
measured using a
spectrometer

action spectrum
a record of the amount
of photosynthesis occurring
at each wavelength

Absorption spectra and action spectra

The **absorption spectra** of the photosynthetic pigments are obtained by measuring the absorption of light of different wavelengths by solutions of each pigment (first separated by column chromatography and then washed from the column one by one). The results are plotted as a graph showing the amount of light absorbed by the solution over the wavelength range of visible light.

The **action spectrum** for photosynthesis is obtained by projecting light of these same wavelengths in turn, for a unit of time, on to aquatic green pondweed. An apparatus is used in which the gas evolved (largely oxygen) can be collected and its volume measured. The result is a graph showing rate of oxygen production (as a measure of the rate of photosynthesis) at different wavelengths.

The absorption spectrum for chlorophyll pigments and the action spectrum for photosynthesis (Figure 12.10) show that the wavelengths of light absorbed by chlorophyll pigments, namely red and blue light, are very similar to the wavelengths that cause photosynthesis. The absorption and action spectra match quite well. So the wavelengths optimally absorbed are the ones that provide most energy for photosynthesis; both blue and red light are used by green plants as the energy source for photosynthesis.

Discovery of the role of red and blue light in photosynthesis

An investigation of photosynthesis by Engelmann, published in 1880, used a species of motile aquatic aerobic bacterium. Where these bacteria are found to accumulate he knew that oxygen was also present. Filamentous algae (p. 554) are made up of rows of photosynthetic cells. Engelmann split sunlight into its constituent colours by means of a prism, and projected them on to cells so that the different colours of light were received by different parts of the filament. The aerobic bacteria in the water collected around the filaments, mainly in the areas where the chloroplasts were receiving red and blue light (Figure 12.11)

Figure 12.11   Engelmann's investigation

The filamentous algae used included *Spirogyra*, which has a ribbon-shaped chloroplast, and *Cladophora*, the cells of which are uniformly packed with chloroplasts. In the water were placed motile aerobic bacteria of the genus *Pseudomonas*. These bacteria migrate towards regions of higher oxygen concentration. The algae filaments were illuminated with light spots, either white or of selected wavelengths.

Spirogyra

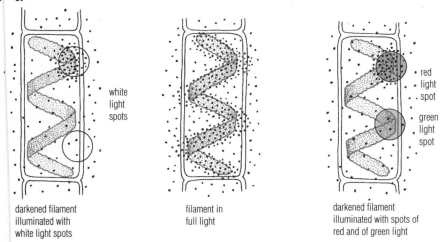

darkened filament
illuminated with
white light spots

filament in
full light

darkened filament
illuminated with spots of
red and of green light

Oxygen was released only from chloroplasts that had been illuminated with white or red light.

Cladophora
A filament was illuminated with light split into the visible spectrum. Bacteria collected (and therefore oxygen was released) only where the filament received red or blue light.

Questions

1 What is the difference between an absorption spectrum and an action spectrum in the context of photosynthesis?

2 It is safe to expose chlorophyll in green plants to sunlight, but extracted chlorophyll solution must be protected from sunlight (Figure 12.5, p. 249). Explain why this is so.

■ A suitable temperature

A rise in temperature increases the rate of a chemical reaction, but as the temperature rises the rate of heat-denaturation of protein also increases. Consequently enzyme-catalysed reactions have an apparent optimum temperature above which denaturation dominates and the rate of reaction decreases (p. 180).

Photosynthesis has an optimum temperature that varies with the species. In most terrestrial plants of temperate regions the optimum temperature is about 20–30 °C; above 40 °C the process is terminated by enzyme denaturation. Photosynthetic algae of the natural hot-water springs that occur in volcanic areas and the seaweeds of the cold seas of the northern hemisphere have altogether different optima, of course.

Figure 12.12   The need for carbon dioxide

1 two potted plants are held in the dark for 48 hours in order to destarch the leaves

2 a leaf from each plant is detached and tested for starch

3 plant A is enclosed with an open dish of soda lime, which absorbs carbon dioxide from the air around the plant;
plant B is enclosed with a dish of sodium hydrogencarbonate solution, which gives off carbon dioxide gas into the air around the plant

4 both plants are placed in full sunlight for 6–12 hours

5 the enclosing bags are removed, and a leaf from each plant is detached and tested for starch

6 the results:
leaf from plant B (supplied with carbon dioxide) gives a positive test for starch;
leaf from plant A (without an external carbon dioxide supply) gives a negative test for starch

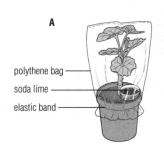

polythene bag
soda lime
elastic band

saturated sodium hydrogencarbonate solution

Figure 12.13   The effect of light intensity on the net movement of carbon dioxide

To prepare hydrogencarbonate indicator
Hydrogencarbonate indicator stock solution is prepared by dissolving 0.2 g thymol blue and 0.1 g cresol red in 20 cm^3 ethanol (dye solution). Then 0.84 g sodium hydrogencarbonate is dissolved in 900 cm^3 distilled (or de-ionised) water. The dye solution is added to this solution and the volume made up to 1 litre.

As required, working solution is made by diluting a sample of stock solution 10 times (adding nine times its own volume of distilled water), and equilibrating this solution with atmospheric air (bubbling through air from outside the laboratory) for 10 minutes. The solution turns red (a small sample in a test tube will appear red–orange).

Colour changes
Changes in carbon dioxide concentration in the air above hydrogencarbonate indicator solution bring about a more or less immediate change in the colour of the indicator. Increasing acidity (more CO_2 added) turns the indicator solution yellow. Decreasing acidity (carbon dioxide lost from the solution) turns the indicator solution cherry-red, then purple.

leaves of terrestrial plants enclosed with hydrogencarbonate indicator

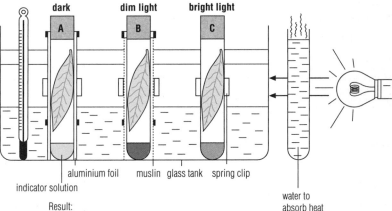

dark   dim light   bright light

A   B   C

aluminium foil   muslin   glass tank   spring clip
indicator solution

water to absorb heat

Result:
▶ the indicator solution in A (dark) goes yellow
▶ the indicator solution in B (dim light) remains cherry-red
▶ the indicator solution in C (bright) goes purple

The results are interpreted as due to the following balances between photosynthesis and respiration:

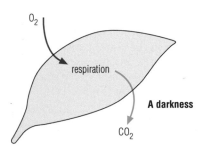

O_2
respiration
A darkness
CO_2

no photosynthesis; CO_2 concentration rises

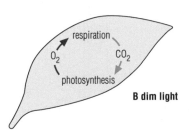

respiration
O_2 CO_2
photosynthesis
B dim light

rates of respiration and photosynthesis equal; no exchange of gases with air

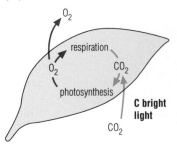

O_2
respiration
O_2 CO_2
photosynthesis
C bright light
CO_2

photosynthesis faster than respiration; CO_2 concentration falls

■ Carbon dioxide

The necessity for carbon dioxide in photosynthesis may be demonstrated by enclosing a previously destarched green plant in air from which atmospheric carbon dioxide has been removed by soda lime granules. In this experiment it is found that leaves deprived of carbon dioxide do not form starch in the light, whereas the leaves of a control plant with access to atmospheric carbon dioxide do form starch (Figure 12.12). Of course, carbon dioxide cannot be totally excluded from the cells of a green leaf since carbon dioxide is being formed and released in respiration. In daylight photosynthesis masks the respiratory production of carbon dioxide because respiratory carbon dioxide is immediately re-used in the chloroplasts.

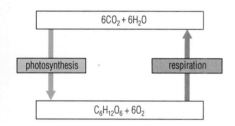

Net movement of carbon dioxide

Carbon dioxide is a soluble gas; it dissolves in water to form carbonic acid. Carbon dioxide entering a leaf cell first dissolves in the surface film of water, and the reverse path is taken by carbon dioxide leaving the cell.

The net movement of carbon dioxide between green leaves and the surrounding air can be detected using sodium hydrogencarbonate indicator solution. This is a pH indicator that is cherry-red when in equilibrium with the atmosphere, which contains 0.035% of carbon dioxide. If carbon dioxide is subsequently removed from the air the indicator becomes purple, but if carbon dioxide accumulates in the air the indicator becomes yellow (Figure 12.13).

Using this indicator solution in an enclosed space with healthy green plant leaves in bright light (active photosynthesis masking respiration) we detect that carbon dioxide is absorbed from the air. In the dark (respiration occurs but not photosynthesis) there is a net output of carbon dioxide. In dim light there is no net movement of carbon dioxide (this is the compensation point, Figure 12.14).

Plants and atmospheric carbon dioxide

Studies of the diurnal pattern of carbon dioxide concentration in the air around crop plants provide evidence supporting the view that green plants export carbon dioxide in the dark but are net importers of carbon dioxide in full light. The results of one study are given in Figure 12.15.

■ Water

The role of water in photosynthesis cannot be investigated by depriving a plant of water because this would kill it. Therefore an alternative method is used. For example, an illuminated green aquatic plant, when supplied with water containing the isotope oxygen-18, gives off oxygen gas containing ^{18}O (Figure 12.30, p. 261).

Questions

1 Design an experiment to test whether temperature has an effect on photosynthesis.

2 a In Figure 12.12, why is the control plant B enclosed in a plastic bag?

  b It is assumed that starch found in a leaf is evidence of photosynthesis in the leaf. Where else might the starch have come from?

Figure 12.14   Light intensity and the compensation point

▶ in darkness, a green plant cannot photosynthesise, but respiration continues

▶ as light intensity increases in the daylight, so does the rate of photosynthesis

▶ eventually, the compensation point is reached when all the CO_2 produced in respiration by the plant is re-used in photosynthesis, and there is no net loss or gain in CO_2

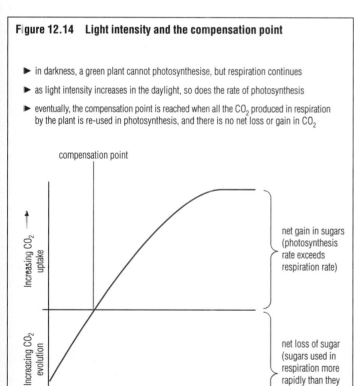

Figure 12.15   Green plants and atmospheric carbon dioxide concentration

measurements of carbon dioxide concentration in the air among a grass crop; the crop was about 85 cm tall, air was sampled at 40 cm above the soil, and there was daylight from 0500 to 2100 hours

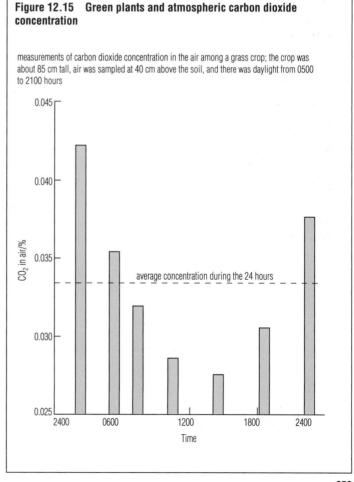

12.3 MEASURING THE RATE OF PHOTOSYNTHESIS

The rate of photosynthesis can be measured with the apparatus shown in Figure 12.16. The effects of light intensity (varied by varying the distance of the lamp from the plant), carbon dioxide concentration (varied by using sodium hydrogencarbonate solutions of different concentrations) and temperature can be investigated.

The experiment requires a freshly cut shoot of pondweed (*Elodea* sp.) from which a vigorous stream of gas bubbles is rising. Either the shoot is weighted down at the stem apex so that the cut base is uppermost, or the tip of the shoot is cut away so that gas can escape from the upper part of the stem. The gas evolved (oxygen-enriched air, not pure oxygen) in unit time is collected and the volume measured.

The apparatus shown in Figure 12.16 may be used to investigate the effects of varying light on the rate of photosynthesis whilst keeping other factors – for example, temperature and carbon dioxide concentration – constant (see Figure 12.21, p. 257).

12.4 THE LEAF AS A FACTORY FOR PHOTOSYNTHESIS

The leaf is an organ specialised for photosynthesis (Table 12.1, p. 256). Leaves themselves are arranged on the stem to receive maximum illumination.

■ Leaf mosaics

Looking down upon a leafy shoot (or looking upwards from below), we see an almost continuous expanse of leaf surface in which leaves overlap each other to a very slight extent (Figure 12.17). Leaf arrangements that minimise overlap are known as leaf mosaics.

Question

In the apparatus shown in Figure 12.16, why would it be essential to have a thermometer in the water surrounding the pondweed during the experiment?

Figure 12.17 Leaf arrangements in which there is a minimum of overlap

Some mosaics arise from the regular arrangement of leaf attachments to the stem (phyllotaxy)

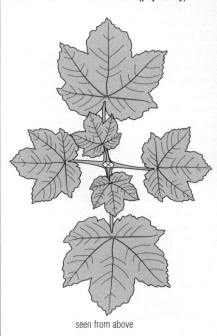

seen from above

The sycamore (*Acer pseudoplatinus*) has simple leaves (blade not divided into leaflets); the leaves are attached in pairs and each succeeding pair occurs at right angles (opposite and decussate arrangement). The older leaves have the longest leaf stalks, which cause the blades to project beyond the shade of the leaf above.

Some mosaics arise from the growth of leaf stalks

seen from above

The common lime (*Tilia* × *vulgaris*) has alternately arranged leaves that largely avoid overlapping by growth and twisting of the leaf stalks.

Figure 12.16 Measuring the rate of photosynthesis in pondweed

- syringe
- burette clip
- gas burette
- lamp
- dilute solution of sodium hydrogencarbonate
- gas-collecting bulb
- inverted shoot of *Elodea*

2 unscrew the clip to draw gas from the bulb into the stem

3 measure the gas column

1 allow gas to collect for a fixed period of time

■ Leaf structure

The leaf, an organ specialised for photosynthesis (Figure 12.18), consists of a **lamina** (leaf blade) connected to the stem by a **petiole** (leaf stalk). The lamina is a thin structure in which many cells are held in well-illuminated positions. The whole leaf is supported by a system of branching **veins** that form a fine network throughout the lamina (Figure 12.19). A vein is a vascular bundle surrounded by a few parenchyma cells normally without chloroplasts, known as the **bundle sheath**. This is why veins normally appear lighter in colour than the rest of the leaf.

The leaf is surrounded by a tough, continuous, protective epidermis, which also provides support by counteracting the hydrostatic pressure of all the turgid cells of the leaf (p. 500), binding them together. The epidermis contains pores called **stomata**, which allow gaseous exchange (p. 322).

the net venation of a poplar leaf (*Populus nigra*)

the parallel venation of a grass leaf

Figure 12.19 The veins of leaves: most dicotyledons have a network of veins in the leaf (above), whereas most monocotyledons have a system of parallel veins (below)

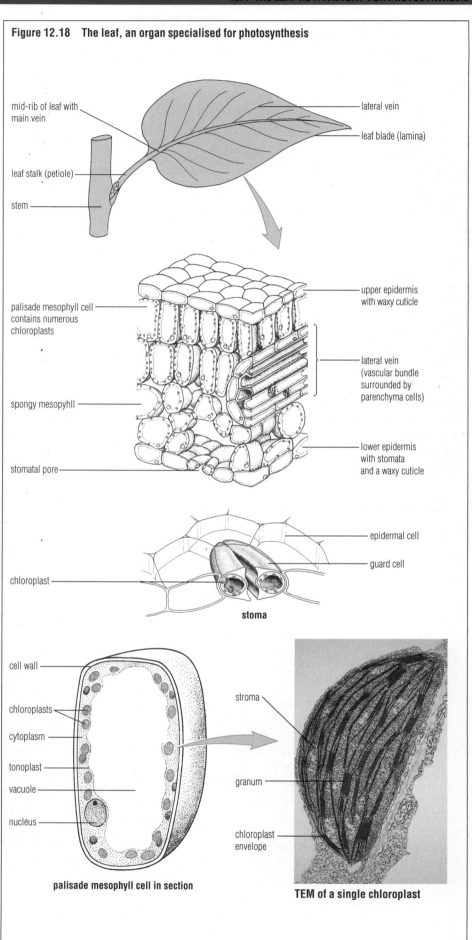

Figure 12.18 The leaf, an organ specialised for photosynthesis

mid-rib of leaf with main vein

lateral vein

leaf blade (lamina)

leaf stalk (petiole)

stem

palisade mesophyll cell contains numerous chloroplasts

upper epidermis with waxy cuticle

spongy mesopyhll

lateral vein (vascular bundle surrounded by parenchyma cells)

lower epidermis with stomata and a waxy cuticle

stomatal pore

epidermal cell

guard cell

chloroplast

stoma

cell wall

chloroplasts

cytoplasm

tonoplast

vacuole

nucleus

stroma

granum

chloroplast envelope

palisade mesophyll cell in section

TEM of a single chloroplast

Figure 12.20   The functioning leaf: the leaf blade in section, showing the site of photosynthesis in relation to the intake and transport of the raw materials and the output of products

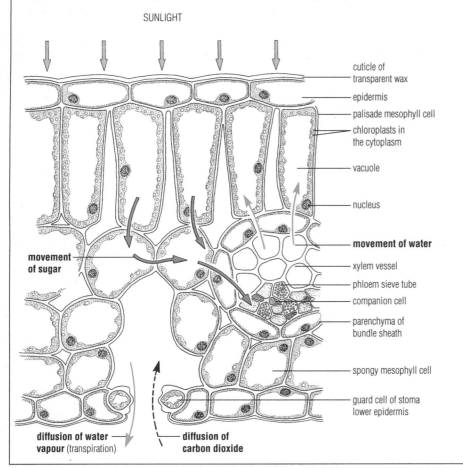

SUNLIGHT

- cuticle of transparent wax
- epidermis
- palisade mesophyll cell
- chloroplasts in the cytoplasm
- vacuole
- nucleus
- **movement of water**
- xylem vessel
- phloem sieve tube
- companion cell
- parenchyma of bundle sheath
- spongy mesophyll cell
- guard cell of stoma lower epidermis

movement of sugar

diffusion of water vapour (transpiration)

diffusion of carbon dioxide

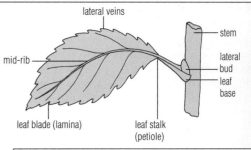

lateral veins

mid-rib

leaf blade (lamina)

leaf stalk (petiole)

stem

lateral bud

leaf base

Mesophyll is the parenchymatous tissue containing chloroplasts. The palisade (upper) mesophyll cells, which contain most chloroplasts, can be thought of as the power house of the photosynthetic leaf. They receive water from the xylem vessels of the vein network. Carbon dioxide diffuses into the surface film of water around the mesophyll cells. Water inevitably evaporates from all internal surfaces and water vapour diffuses out of the stomata (transpiration, p. 321). The sugar formed in photosynthesis is transported to the phloem sieve tubes of the vein network and is translocated away (p. 337). Figure 12.20 shows this process.

Questions

1   Outline three ways in which the structure of a green leaf facilitates photosynthesis.
2   List the principal functions of leaf veins.

Table 12.1   The structural adaptation of the green leaf for photosynthesis: a summary

Feature	Structure	Function/role
External shape and structure	There is great diversity of leaf shape and form; leaf morphology is outlined on pp. 32–4	
	Most plants have numerous leaves; most leaves are thin with a large surface area and a flat leaf blade	Allows maximum absorption of incident light energy Facilitates inward diffusion of carbon dioxide gas to the mesophyll cells
Epidermis	Modified parenchyma cells, with thick external walls and with a waxy cuticle on the outer surface	Support and protection of the leaf tissue: ▶ support provided by containing and counteracting the hydrostatic pressure of the turgid cells of the leaf, binding them together
	Epidermis has epidermal hairs (extensions of cells of epidermis), particularly on young leaves	▶ protects leaf tissue from invasion by predators and fungal parasites and from excessive water loss by evaporation
Stomata	Pores in the leaf epidermis between two guard cells (p. 322); changes in the turgidity of the guard cells cause the pores to open or close	Regulate entry of carbon dioxide gas and the loss of water vapour from the leaf
Palisade mesophyll	Modified parenchyma cells with numerous chloroplasts in the cytoplasm; cells are elongated in shape and packed closely, but with air spaces also	'Power house' of the leaf; chloroplasts are the site of the transduction of light energy to chemical energy Sugars and other substances are produced in the chloroplasts and oxygen is released
Spongy mesophyll and air spaces	Modified parenchyma cells with some chloroplasts; cells mostly more or less spherical or elongated, with large interconnecting air spaces between them	Facilitates gaseous exchange by diffusion
Veins, containing vascular bundles	Network of small vascular bundles continuous with those of stem and root	Supports the leaf tissues as a large, thin, flexible organ
	Vascular bundles contain ▶ xylem vessels ▶ phloem sieve tubes	Deliver water to the leaf cells Translocate away the organic products of photosynthesis

12.5 THE INTERACTION OF FACTORS

The rate of photosynthesis (measured as the volume of oxygen produced in unit time) can be investigated using the apparatus shown in Figure 12.16 (p. 254). The effect of varying the light intensity on the rate of photosynthesis is shown in graph A, Figure 12.21. Here the rate was determined at different intensities of light, ranging from low to high light intensity. The graph shows that the rate of photosynthesis rose steadily as the light intensity increased, but then it levelled off (point X). What stopped the rate of photosynthesis from continuing to increase?

The answer to this question can be found by repeating the experiment at different concentrations of carbon dioxide (Graph B in Figure 12.21). At a higher concentration of carbon dioxide (0.13% CO_2), the rate of photosynthesis continued to increase with increasing light intensity until the rate levelled of again (point Y). However, the increased carbon dioxide concentration sustained a maximum rate of photosynthesis almost double that of the lower concentration of carbon dioxide.

■ **Interpretation of the experiment**

Photosynthesis is a biochemical process involving a series of interconnected reactions. All these reactions contribute to the overall rate of the process, and they depend on several essential conditions being favourable. At any one time, the rate of photosynthesis will be limited by the slowest of these reactions and the essential condition (factor) for it to occur. In fact, the overall rate of photosynthesis will be limited by the factor that is in shortest supply. This factor, whichever one it is, is known as the **limiting factor**.

For example, a limited supply of either carbon dioxide or light could limit the overall rate of photosynthesis, since both are essential for photosynthesis. If a plant has a plentiful supply of carbon dioxide, but the light is set at a very low intensity, then light is limiting. If the light intensity is raised, the plant immediately photosynthesises faster. Had the carbon dioxide concentration been raised instead, the rate would have been unchanged, since carbon dioxide was not the rate-limiting factor at the time.

Returning to the curves in the graphs in Figure 12.21, up to points X and Y we can see that light is the factor limiting the overall rate of photosynthesis, but above these points, the concentration of carbon dioxide becomes the limiting factor.

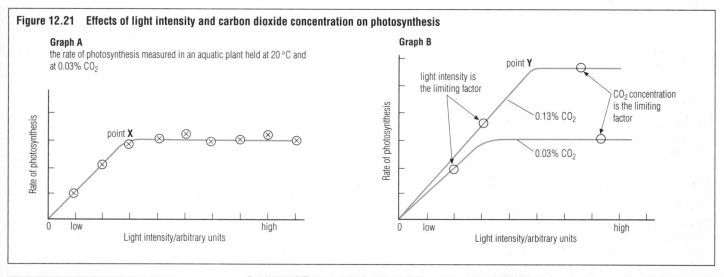

Figure 12.21   Effects of light intensity and carbon dioxide concentration on photosynthesis

Graph A
the rate of photosynthesis measured in an aquatic plant held at 20 °C and at 0.03% CO_2

Graph B

point **Y**

light intensity is the limiting factor

0.13% CO_2

CO_2 concentration is the limiting factor

0.03% CO_2

Limiting factors in the plant's environment

With plants in their natural habitats, the factor limiting the rate of photosynthesis will change with the time of day and the conditions. For example, at first light in the early morning, light will be the limiting factor. Later, in maximum sunlight, the carbon dioxide concentration may limit the rate of photosynthesis.

With commercial crops, grown intensively in glasshouses where environmental conditions are controlled, the limiting effects of light intensity or carbon dioxide concentration can be reduced or removed (by additional lighting, or by enriching the atmosphere with carbon dioxide gas), thereby improving productivity (Figure 12.22).

Figure 12.22   A commercial greenhouse with controlled environmental conditions (intensity and duration of light, and atmospheric CO_2 concentration)

The effect of temperature on the rate of photosynthesis

The effect of temperature on the rate of photosynthesis, under controlled laboratory conditions, depends upon the light intensity. The effect of increasing temperature (for example, in the range from 10 °C to 30 °C) actually has little effect under conditions of low light. Under higher intensities, the rise in temperature increases the rate of photosynthesis significantly. This relationship is shown in Figure 12.23.

The importance of this discovery lies in the fact that temperature has virtually no effect on a photochemical reaction (for example, when we take a picture with a camera, or when light energy alters a chlorophyll molecule) but speeds up a chemical or enzymic reaction (for example, when a photographic film is developed, or carbon dioxide is converted to sugar).

From the graph in Figure 12.23, at low light intensity (when light is the limiting factor) a temperature rise has no effect on the rate of photosynthesis. But at higher light intensities (when carbon dioxide is the limiting factor) a temperature rise increases the rate. This is interpreted as showing that photosynthesis is made up of two stages: a light-dependent stage (which, like all photochemical events, is temperature indifferent) and a light-independent stage (which, like all biochemical steps catalysed by enzymes, is temperature sensitive) (Figure 12.24). This idea has also been confirmed by investigations of the biochemical reactions of photosynthesis in isolated chloroplasts.

The site of photosynthesis

The structure of the chloroplast is introduced in Figure 7.23 (p. 164). The grana are the sites of the light-dependent stage, and the stroma is the site of the light-independent stage (Figure 12.25).

Question

a Why is it that we try to control the temperature when we develop a photographic film but not when we take the picture?

b What is the main effect of a small rise in temperature on the rate of photosynthesis when light and carbon dioxide concentration respectively are limiting factors?

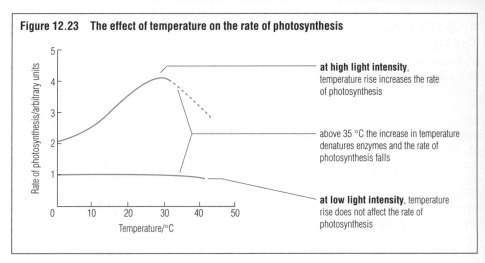

Figure 12.23   The effect of temperature on the rate of photosynthesis

at high light intensity, temperature rise increases the rate of photosynthesis

above 35 °C the increase in temperature denatures enzymes and the rate of photosynthesis falls

at low light intensity, temperature rise does not affect the rate of photosynthesis

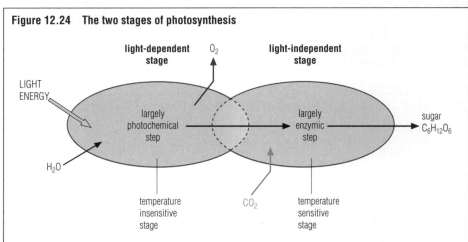

Figure 12.24   The two stages of photosynthesis

light-dependent stage

O_2

light-independent stage

LIGHT ENERGY

largely photochemical step

largely enzymic step

sugar $C_6H_{12}O_6$

H_2O

CO_2

temperature insensitive stage

temperature sensitive stage

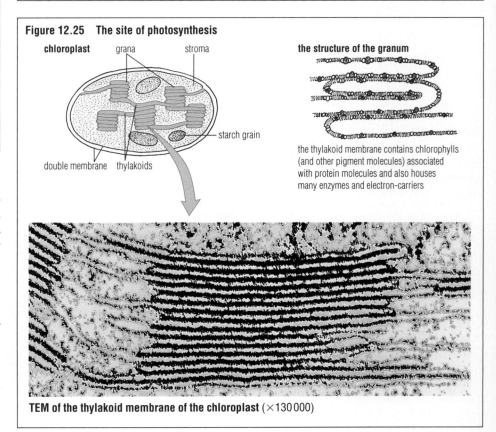

Figure 12.25   The site of photosynthesis

chloroplast   grana   stroma

the structure of the granum

double membrane   thylakoids

starch grain

the thylakoid membrane contains chlorophylls (and other pigment molecules) associated with protein molecules and also houses many enzymes and electron-carriers

TEM of the thylakoid membrane of the chloroplast (×130 000)

Figure 12.26   Isolating intact chloroplasts by differential centrifugation

1 chilled green leaf tissue is ground up in a homogeniser (blender)

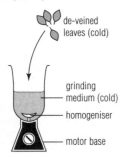

de-veined leaves (cold)

grinding medium (cold)

homogeniser

motor base

2 the homogenate is filtered through muslin (cheesecloth) to remove the larger debris (part-ground cells)

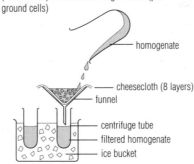

homogenate

cheesecloth (8 layers)

funnel

centrifuge tube

filtered homogenate

ice bucket

3 the filtered homogenate is centrifuged at low speed

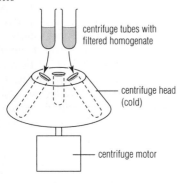

centrifuge tubes with filtered homogenate

centrifuge head (cold)

centrifuge motor

4 the heavier particles and the larger organelles are precipitated; the supernatant is decanted into a fresh tube and the pellet of debris is discarded

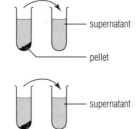

supernatant

pellet

supernatant

pellet

5 the supernatant liquid is recentrifuged at high speed

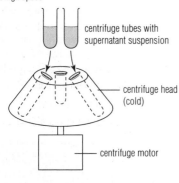

centrifuge tubes with supernatant suspension

centrifuge head (cold)

centrifuge motor

6 the high-speed centrifugation precipitates the lower-mass organelles as a pellet

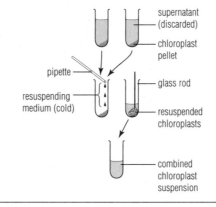

supernatant (discarded)

chloroplast pellet

pipette

glass rod

resuspending medium (cold)

resuspended chloroplasts

combined chloroplast suspension

Figure 12.27   The light-dependent stage: a summary

LIGHT ENERGY

granum

O_2

H_2O

NADP

photolysis and reduction

$NADPH_2$

LIGHT ENERGY

ADP + P_i

photophos-phorylation

ATP

12.6 THE LIGHT-DEPENDENT STAGE

Chloroplasts are obtained from green leaves by breaking up the cells in a blender. The solutions used are ice-cold so as to prevent the action of hydrolytic enzymes released from damaged lysosomes in the cell cytosol. The solutions are isotonic with the cytosol so that no abrupt osmotic changes occur and disrupt delicate membranes. The organelles of slightly different masses are then separated by centrifugation of the suspension at progressively higher speeds (Figure 12.26). Once the chloroplast fraction has been sedimented into a pellet, it is re-suspended in solution and used for investigations of the reactions of photosynthesis.

■ Developments with isolated chloroplasts

Working with isolated chloroplast suspensions, it has been established that the light-dependent stage of photosynthesis occurs in the grana and consists of the following two processes (Figure 12.27).

▶ **The photochemical splitting of water and the formation of reduced NADP.** Chloroplasts contain the naturally occurring electron acceptor $NADP^+$. This is reduced in the light to $NADPH_2$ (this is a 'shorthand' representation of $NADPH + H^+$). The transfer of electrons and hydrogen ions from water to NADP is an endergonic reaction (p. 172). It is achieved by chlorophyll molecules that have been momentarily 'excited' by light energy.

▶ **The production of ATP from ADP and P_i.** This photophosphorylation reaction is also driven by energy from light trapped in the chlorophyll.

Reduced NADP and ATP are used up in the light-independent stage in the stroma, by which sugar is synthesised from carbon dioxide (p. 262).

Question

Consider the process of isolating organelles from intact cells (Figure 12.26).

a Why is the process carried out at or about the freezing point of water?

b What type of cell structure is certain to pass through the muslin filter?

c Chloroplasts are re-suspended in a medium that is isotonic with the cell contents. Why?

■ Mechanism of the light-dependent stage

1 Photosystems trap light energy

Light energy is trapped by photosynthetic pigments arranged in centres called **photosystems**. In each photosystem, several hundred chlorophyll molecules, plus accessory pigments (carotene and xanthophyll), harvest light energy and 'funnel' the energy to a single chlorophyll molecule, known as the **reaction centre**. (Figure 12.28) The different pigments of the photosystem absorb light energy of different wavelengths, making each centre more efficient.

There are two types of photosystem present in the thylakoid membranes:

▶ **Photosystem I** has a reaction centre of a chlorophyll *a* molecule with maximum light absorption at 700 nm wavelength. This reaction centre is also referred to as P700.

▶ **Photosystem II** has a reaction centre of a chlorophyll *a* molecule with maximum light absorption at 680 nm wavelength. This reaction centre is also referred to as P680.

The chlorophyll molecule of the reaction centre becomes 'excited' by the light energy received. As a result, a succession of high energy electrons is released from this chlorophyll molecule, and these electrons bring about the biochemical changes of the light-dependent reaction. The spaces vacated by high energy (excited) electrons are continuously refilled by non-excited or 'ground state' electrons. We will examine this sequence of reactions next.

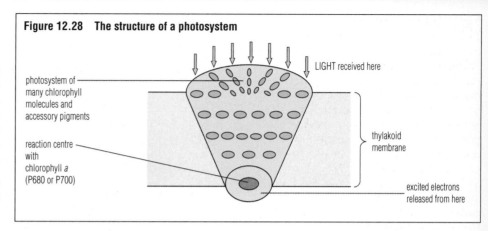

Figure 12.28   The structure of a photosystem

photosystem of many chlorophyll molecules and accessory pigments

reaction centre with chlorophyll *a* (P680 or P700)

LIGHT received here

thylakoid membrane

excited electrons released from here

2 Photophosphorylation and reduction of NADP

First, the excited electrons from photosystem II (P680) are picked up by the first of a chain of electron carriers, and passed along this sequence of carriers. As the electrons are moved along, the energy released is 'coupled' to reactions that synthesise ATP from ADP and P_i. This process is known as **photophosphorylation** because it is driven by light energy. As a result, the excitation level of these electrons falls back to 'ground state' and they come to fill the vacancies in the reaction centre of photosystem I (P700). In effect, these electrons have been transferred from photosystem II to photosystem I.

Meanwhile, the 'holes' in each reaction centre P680 also have to be filled by ground state electrons. These come from water molecules. In fact, it is the positively charged vacancies in P680 that are powerful enough to cause the splitting of water (**photolysis**). The reaction, which is catalysed by an enzyme, releases the required ground state electrons, as well as hydrogen ions and oxygen atoms. The oxygen atoms combine to form molecular oxygen, the waste product of photosynthesis. The hydrogen ions are used in the reduction of NADP (see below).

Next, the excited electrons from photosystem I (P700) are taken up by a different electron acceptor. Two at a time, they are then passed to NADP, which, with the addition of hydrogen ions from photolysis, is reduced to form NADPH + H^+ (= $NADPH_2$).

By this sequence of reactions, repeated again and again at very great speed throughout every second of daylight, the products of the light-dependent stage (ATP and reduced NADP) are formed (Figure 12.29). They do not accumulate, however, as they are immediately used in the fixation of carbon dioxide in the surrounding stroma (light-independent stage, p. 262). Then ADP, P_i and NADP diffuse back into the grana for re-use.

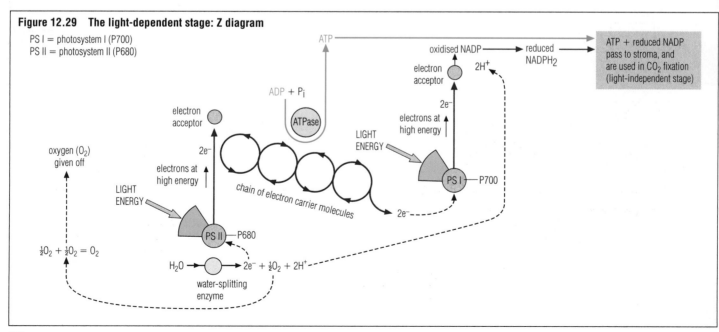

Figure 12.29   The light-dependent stage: Z diagram

PS I = photosystem I (P700)
PS II = photosystem II (P680)

ATP

ADP + P_i

ATPase

oxidised NADP → reduced $NADPH_2$

ATP + reduced NADP pass to stroma, and are used in CO_2 fixation (light-independent stage)

electron acceptor

$2H^+$

electron acceptor

$2e^-$

$2e^-$

electrons at high energy

LIGHT ENERGY

PS I — P700

oxygen (O_2) given off

$2e^-$

electrons at high energy

LIGHT ENERGY

chain of electron carrier molecules

$2e^-$

PS II — P680

$\frac{1}{2}O_2 + \frac{1}{2}O_2 = O_2$

H_2O → $2e^- + \frac{1}{2}O_2 + 2H^+$

water-splitting enzyme

Non-cyclic and cyclic photophosphorylation

The high energy electrons that provide the energy for ATP synthesis in photosynthesis originated from water, and fill the vacancies in P680. They are then moved on from P680 to P700 (during which ATP synthesis occurs), and finally are used to reduce NADP. Therefore, the photophosphorylation reaction in which they are involved is known as **non-cyclic** photophosphorylation.

If, for some reason, photosystems I and II cannot 'fire' together (as when carbon dioxide concentration is limiting the rate of photosynthesis) then the excited electrons from P700 fall back to their point of origin. They do this via electron carriers to which ATP synthesis is coupled. This type of photophosphorylation is called **cyclic** because the electrons have returned to the photosystem from which they originated.

Oxygen from photosynthesis comes from water

The use of water containing the isotope oxygen-18 has confirmed that the oxygen released in photosynthesis comes from water. This is a non-radioactive isotope, so it is detected by difference in mass from oxygen-16, using a mass spectrometer (Figure 12.30).

Figure 12.30   Photolysis of water

Illuminated suspensions of *Chlorella* cells were provided with *either* carbon dioxide enriched with $C^{18}O_2$, *or* water enriched with $H_2^{18}O$.
The oxygen evolved was collected and was analysed by the **mass spectrometer** to determine the oxygen-18 concentration.

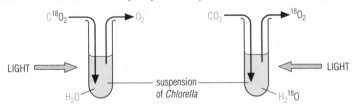

The results demonstrated that the oxygen evolved in photosynthesis came from water:

$$4H_2O \xrightarrow[\text{chlorophyll}]{\text{LIGHT ENERGY}} 4(H^+ + e^-) + 4 \bullet OH \longrightarrow 2H_2O + O_2$$

hydrogen acceptor   reduced hydrogen acceptor

This experiment established that the 'traditional' balanced equation for photosynthesis:

$$6CO_2 + 6H_2O + \text{LIGHT ENERGY} \xrightarrow[\text{chloroplast}]{\text{chlorophyll pigments in}} C_6H_{12}O_6 + 6O_2$$

is incorrect in suggesting that some or all of the oxygen evolved comes from carbon dioxide. This is because twelve oxygen atoms are produced, but the six water molecules on the left-hand side of the equation contain only six.
The equation for photosynthesis is less misleading when written as:

$$6CO_2 + 12H_2O + \text{LIGHT ENERGY} \longrightarrow C_6H_{12}O_6 + 6O_2 + 6H_2O$$

Photosystems I and II in the thylakoid membrane

The photosystems have been observed as particles in the thylakoid membrane, occurring alongside the electron carriers and other enzymes of the light-dependent step. These include the stalked particles, ATP synthetase (ATPase), the enzyme by which ATP is formed. ATP formation is catalysed when the hydrogen ions that build up within the thylakoid space flow out through a channel in each ATPase molecule, down their electrochemical gradient (Figure 12.31).

Figure 12.31   Photophosphorylation: the mechanism

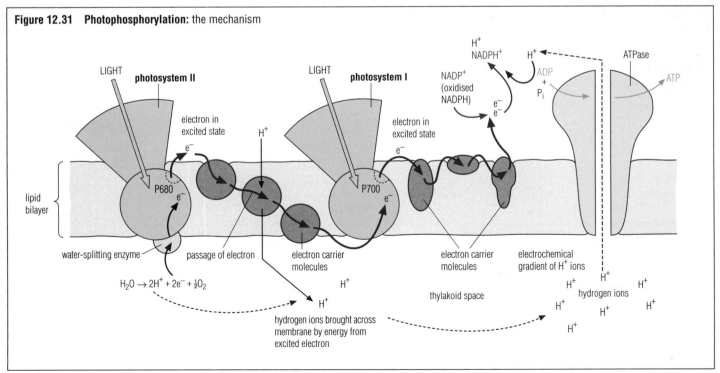

12.7 THE LIGHT-INDEPENDENT STAGE

In the light-independent stage, carbon dioxide is converted to carbohydrate. The light-independent stage occurs in the stroma of the chloroplast and involves many reactions, each catalysed by a different enzyme. The light-independent stage requires the products of the light-dependent stage (ATP and NADPH$_2$), but does not itself require light; this is why it is referred to as a 'dark' reaction. Of course the light-dependent stage and the light-independent stage occur simultaneously in chloroplasts within the leaf during daylight, each dependent upon the other.

■ How the light-independent stage was investigated

The sequence of steps in the light-independent stage (sometimes referred to as 'the path of carbon in photosynthesis') was investigated using radioactively labelled carbon dioxide ($^{14}CO_2$) 'fed' to an illuminated culture of the unicellular green alga *Chlorella*. This technique was pioneered between 1946 and 1953 by a team at the University of California, led by Melvin Calvin. (Radioactive isotopes had become available for research in 1945, a by-product of war work upon the atomic bomb.) The team also exploited the then relatively recent invention of paper chromatography. Calvin's work with carbon-14 and *Chlorella* is an example of a 'feeding experiment' (p. 175); his method is described in Figure 12.32. The sequence of reactions of the light-independent stage was deduced from the changing distribution of radioactivity between the metabolites with time.

Calvin acquired his interest in photosynthesis whilst a postgraduate student in Manchester in the 1930s. On return to the USA he worked almost exclusively on this aspect of plant metabolism. He was awarded a Nobel Prize in 1961.

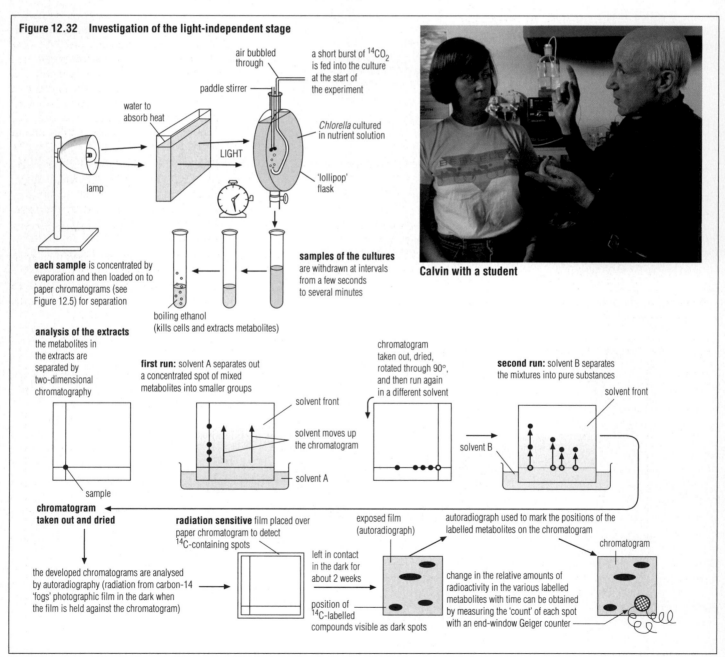

Figure 12.32   Investigation of the light-independent stage

a short burst of $^{14}CO_2$ is fed into the culture at the start of the experiment

air bubbled through

paddle stirrer

water to absorb heat

Chlorella cultured in nutrient solution

LIGHT

'lollipop' flask

lamp

Calvin with a student

each sample is concentrated by evaporation and then loaded on to paper chromatograms (see Figure 12.5) for separation

samples of the cultures are withdrawn at intervals from a few seconds to several minutes

boiling ethanol (kills cells and extracts metabolites)

analysis of the extracts the metabolites in the extracts are separated by two-dimensional chromatography

first run: solvent A separates out a concentrated spot of mixed metabolites into smaller groups

solvent front

solvent moves up the chromatogram

solvent A

sample

chromatogram taken out, dried, rotated through 90°, and then run again in a different solvent

second run: solvent B separates the mixtures into pure substances

solvent front

solvent B

chromatogram taken out and dried

radiation sensitive film placed over paper chromatogram to detect ^{14}C-containing spots

left in contact in the dark for about 2 weeks

position of ^{14}C-labelled compounds visible as dark spots

exposed film (autoradiograph)

autoradiograph used to mark the positions of the labelled metabolites on the chromatogram

chromatogram

the developed chromatograms are analysed by autoradiography (radiation from carbon-14 'fogs' photographic film in the dark when the film is held against the chromatogram)

change in the relative amounts of radioactivity in the various labelled metabolites with time can be obtained by measuring the 'count' of each spot with an end-window Geiger counter

■ The path of carbon in photosynthesis

The first product of the light-independent stage is glycerate 3-phosphate (GP). Some of this is metabolised to produce the molecule that first accepts carbon dioxide. The remainder of the initial products are metabolised (using ATP and $NADPH_2$) into carbohydrates such as sugars, sugar phosphates and starch, and later into lipids, amino acids such as alanine, and organic acids such as malate (Figure 12.33).

At first, a two-carbon acceptor molecule for carbon dioxide was sought, because the first product of carbon dioxide fixation was known to be a three-carbon compound. But the acceptor molecule for carbon dioxide eventually proved to be the five-carbon molecule ribulose bisphosphate. As carbon dioxide adds to the five-carbon acceptor (ribulose bisphosphate), the six-carbon product at once splits into two three-carbon GP molecules – hence the initial confusion. The enzyme involved is called **ribulose bisphosphate carboxylase**.

The reactions of the light-independent stage, known as the **Calvin cycle**, can be divided into four phases (Figure 12.34):

► carbon dioxide fixation, the **carboxylation phase**
► reduction of glycerate 3-phosphate (GP) to triose phosphate, the **reduction phase**
► the re-formation of acceptor molecules, the **regeneration phase**
► the metabolism of triose phosphate to carbohydrates, lipids and amino acids, the **product synthesis phase.**

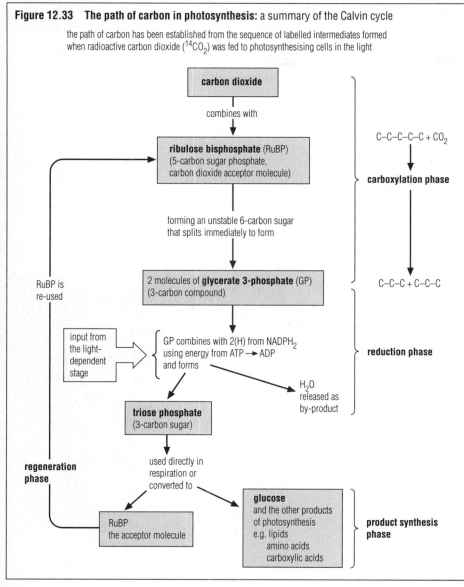

Figure 12.33   The path of carbon in photosynthesis: a summary of the Calvin cycle

the path of carbon has been established from the sequence of labelled intermediates formed when radioactive carbon dioxide ($^{14}CO_2$) was fed to photosynthesising cells in the light

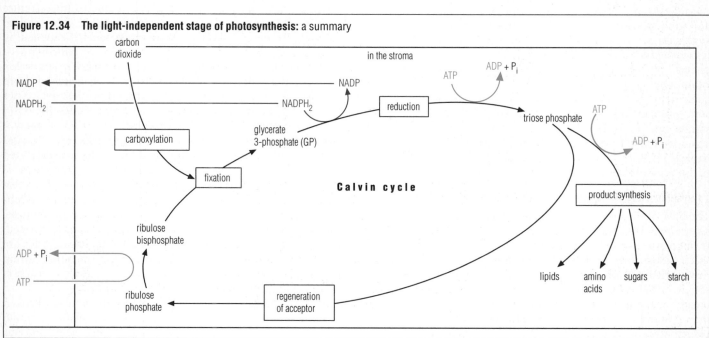

Figure 12.34   The light-independent stage of photosynthesis: a summary

Figure 12.35   The product synthesis phase of photosynthesis

key

→ glycolysis + Krebs cycle

→ photosynthesis (light-independent stage)

→ formation of products

Sugar and starch are not the only products of photosynthesis. In Calvin's experiments, radioactive carbon accumulated in amino acids and lipids as well. The reactions of glycolysis and the Krebs cycle are the pathways by which GP and triose phosphate are converted to all these products.

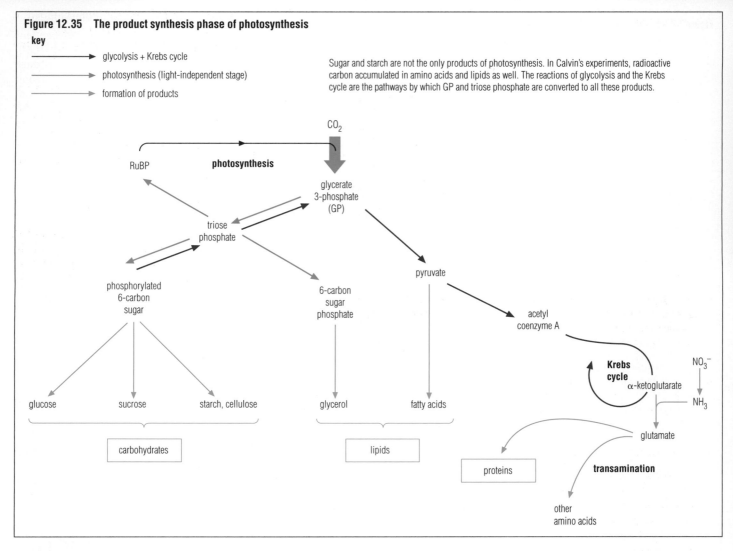

■ Metabolism of the products of photosynthesis

The first carbohydrate products of photosynthesis are glycerate 3-phosphate and triose phosphate. These three-carbon sugars do not accumulate, but rather are immediately converted to glucose phosphates, glucose and other substances. From these products, and using the inorganic ions (p. 328) absorbed from the soil, the plant manufactures all the substances needed for life (Figure 12.35). As a result of this biosynthesis, the plant is entirely self-sufficient. What are the steps in this self-sufficiency?

1 Transport of sugars and storage of starch

The bulk of the products of photosynthesis are exported from the leaf. Sugar is translocated (p. 337) as sucrose, a disaccharide formed from glucose and fructose. Much of this sugar is stored as starch in the cells of the stem or roots, mostly in the parenchyma ground tissue there. Starch is formed from glucose phosphates.

2 Respiration of sugar

Stored starch is drawn upon for a supply of the sugar needed for respiration. Respiration involves glycolysis (p. 315), in which triose phosphate and glycerate 3-phosphate are intermediates. Respiration occurs in every living cell, but regions of the plant undergoing growth and development – such as stem and root tips, young leaves, fruits and seeds – have the highest respiration rates. Energy from respiration is required to drive the endergonic reactions of biosynthesis, for example.

3 Synthesis of lipids

Lipids are formed from glycerol and fatty acids (p. 137). Glycerol is formed directly from triose phosphate, and fatty acids are built up from acetyl coenzyme A. Both of these are intermediates of the carbon pathway in photosynthesis. These reactions occur in the cytoplasm, where the enzymes for lipid synthesis also occur. The breakdown of fatty acids occurs in mitochondria.

4 Synthesis of proteins

Proteins are built of a linear series of some twenty different amino acids (p. 139). Plants are able to manufacture all of these amino acids. The carbon skeleton of an amino acid is supplied by organic acids that are intermediates of respiration (largely from the Krebs cycle). The amino group is synthesised from nitrates absorbed from the soil. After absorption by the root hairs (p. 330), nitrates are reduced to nitrites, a reaction catalysed by nitrate reductase. Nitrites are then reduced to ammonia, catalysed by nitrite reductase. The ammonia is combined with the organic acid α-ketoglutarate, an intermediate of the Krebs cycle p. 315, catalysed by transaminase. The amino acid glutamate is formed. From glutamate all other amino acids are formed by transferring the amino group from other organic acids, a reaction known as transamination.

Proteins of plants are regularly broken down, moved about the plant as amino acids, and rebuilt into new proteins

A SUMMARY OF PHOTOSYNTHESIS

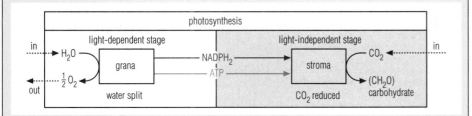

Light-dependent stage	Light-independent stage
► occurs in the thylakoid membranes of the grana	► occurs in the stroma
► largely a photochemical change, requiring light energy	► a series of biochemical changes, each reaction catalysed by an enzyme
► light energy is converted to chemical energy in the form of ATP and NADPH$_2$; water is split into hydrogen and oxygen; hydrogen is combined in NADPH$_2$; oxygen gas is released as a waste product	► carbon dioxide is converted to compounds such as carbohydrates (with the chemical energy of ATP and NADPH$_2$); the reactions of the light-independent stage are known as the Calvin cycle
► chlorophylls are grouped together in units of about 300 molecules (known as photosystems); two types exist, photosystems I and II	► carbon dioxide is combined with ribulose bisphosphate (the acceptor substance) and the product splits instantly into two molecules of glycerate 3-phosphate (GP, the first product of photosynthesis)
► light energy absorbed by the photosystems causes electrons from chlorophyll to be raised to a high energy level and to pass to NADPH$_2$; ATP is generated; water is split and provides the electrons to the photosystem and the hydrogen for NADPH$_2$ production:	► GP is reduced to a three-carbon sugar, triose phosphate; then, in a series of reactions, the acceptor molecule is regenerated and sugars, starch and other substances are formed from triose phosphate:

$$2H_2O + 2NADP \xrightarrow[\text{chlorophyll}]{\text{light}} O_2 + 2NADPH_2$$

$$ADP + P_i \xrightarrow[\text{chlorophyll}]{\text{light}} ATP \text{ (considerable, but variable amount)}$$

$$\begin{array}{cc} 3ATP & 3ADP + 3P_i \\ CO_2 \searrow\nearrow & \searrow\nearrow (CH_2O) + H_2O \\ 2NADPH_2 & 2NADP \end{array}$$

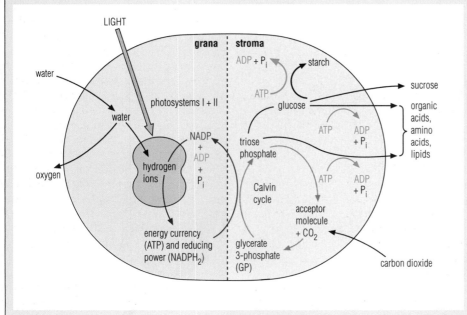

where required. In the young growing plant, proteins are concentrated at the growing points and in the leaves. In the mature plant, the bulk of protein exists in the leaves, present mainly as the enzyme ribulose bisphosphate carboxylase (Figure 16.20, p. 331). As the plant switches to seed formation, however, proteins are transported to the seeds where they are stored for the later benefit of the embryo, at germination (Figure 19.13, p. 405).

5 Other syntheses

Synthesis of wall compounds, vitamins, plant growth regulators, pigments and all other essential substances occurs in the plant, as required, using the products of photosynthesis.

Questions

1 a What types of molecule make up a photosystem?
   b Why are the reaction centres of photosystems I and II referred to as P700 and P680?
2 In non-cyclic photophosphorylation, what is the ultimate fate of electrons displaced from the reaction centre of photosystem II?
3 Why does the traditional equation for photosynthesis imply that some of the oxygen must come from carbon dioxide?
4 Why is it that the light-independent stage of photosynthesis can occur only in the light in the intact green plant?
5 What reaction do you think might be catalysed by the enzyme ribulose bisphosphate carboxylase?
6 What are the significant differences between
   a the starting materials and the end products of photosynthesis?
   b the light-dependent stage and the light-independent stage of photosynthesis?
7 List the steps by which a carbon atom in carbon dioxide gets into a protein.

12.8 PHOTOSYNTHESIS IN C₄ PLANTS

The fixation of carbon dioxide in the light-independent stage of photosynthesis (section 12.7, p. 262) is referred to as the C₃ **pathway** because the first product, glycerate 3-phosphate (GP), is a three-carbon compound. 'C₃' is a shorthand name for the C-fixing reactions of the light-independent stage found in the majority of plants in temperate climates.

In many tropical and subtropical plants the initial products of carbon dioxide fixation are both GP and malic acid (as malate in the physiological conditions of the cell). Malate is a four-carbon (C₄) compound, so the shorthand name for the plants that form malate *and* GP during photosynthesis is C₄ **plants**. These plants form GP and malate simultaneously, and do so in the light.

■ The biochemistry of C₄ photosynthesis

Leaves of C₄ plants have a distinctive sheath of cells around the vascular bundles of the leaf, known as **bundle sheath cells** (Figure 12.36). These cells are packed with chloroplasts. This distinctive arrangement of chloroplast-containing cells around the vascular bundles is referred to as **Krantz anatomy**.

In the leaves of C₄ plants, malate is formed in the outer mesophyll cells and is exported to the inner bundle sheath cells, immediately around the vascular bundles. Here carbon dioxide is regenerated during the process of malate breakdown to pyruvate. Carbon dioxide concentration is thus enhanced in the bundle sheath cells, where it is fixed photosynthetically via the normal (C₃) pathway. Enhanced production of carbohydrate results (because photorespiration is inhibited by higher carbon dioxide levels, see opposite page). C₄ plants occur where the climate is hot (tropical/subtropical). Under these conditions C₄ plants achieve faster rates of photosynthesis than can C₃ plants under the same conditions.

■ The biogeography of C₄ plants

Photosynthetic strategy is an important factor in determining where a plant lives (Figure 12.37).

Most C₄ species live in hot conditions because they require a high rate of respiration to generate an excess of the carbon dioxide acceptor PEP (phosphoenolpyruvate, Figure 12.38). Moreover, many of the enzymes of C₄ plants have optimum temperatures well above 25°C.

Figure 12.36   Maize (*Zea mays*) as a 'C₄ plant'

Maize is a cultivated grass from tropical/subtropical regions. Varieties of maize have been bred (genetically modified) to grow successfully in temperate regions too. Now, maize fruits – on or off the 'cob' – are enjoyed by humans across the world.

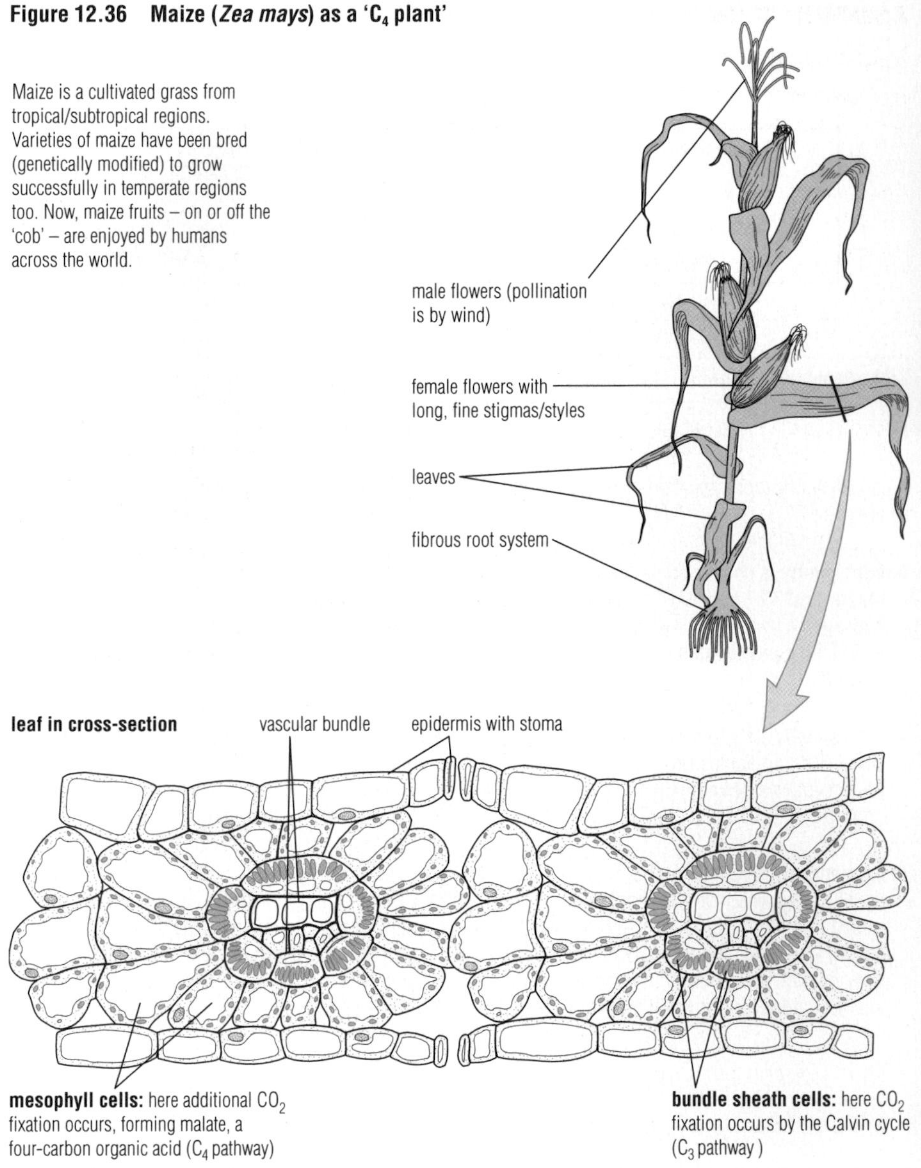

male flowers (pollination is by wind)

female flowers with long, fine stigmas/styles

leaves

fibrous root system

leaf in cross-section   vascular bundle   epidermis with stoma

mesophyll cells: here additional CO₂ fixation occurs, forming malate, a four-carbon organic acid (C₄ pathway)

bundle sheath cells: here CO₂ fixation occurs by the Calvin cycle (C₃ pathway)

Figure 12.37   The biogeography of C₄ plants

C₄ plants do best in the tropics; the percentage of grasses using the C₄ mechanism in North America diminishes towards the north

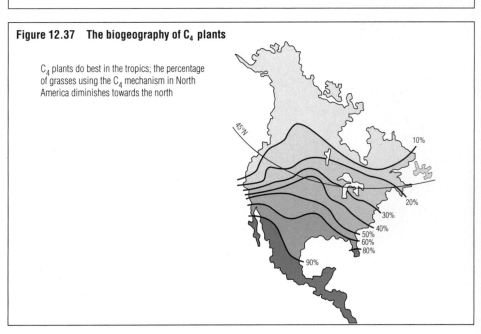

45°N

10%

20%

30%

40%

50%
60%
80%

90%

Figure 12.38   C₄ plants

C₄ plants are mainly tropical and subtropical in origin; examples include:

maize (*Zea mays*)
sugar cane (*Saccharum officinale*)
sorghum (*Sorghum bicolor*)

loves-lies-bleeding (*Amaranthus caudatus*)
globe amaranth (*Gomphrena globosa*)
sun plant (*Portulaca grandifolia*)

TS of a young maize leaf
there are two distinct layers of green cells around the vascular bundles: outer mesophyll cells and inner bundle sheath cells

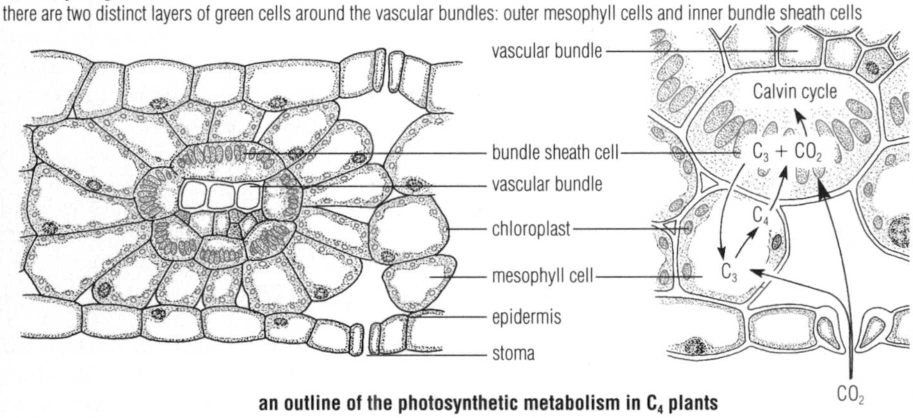

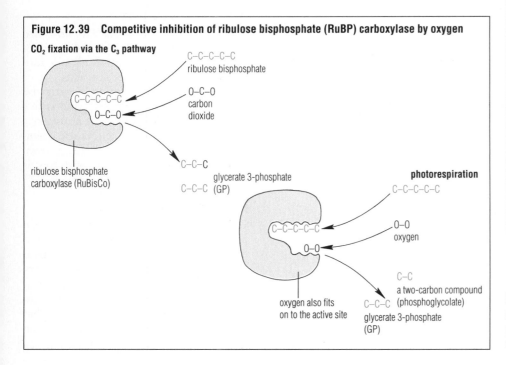

an outline of the photosynthetic metabolism in C₄ plants

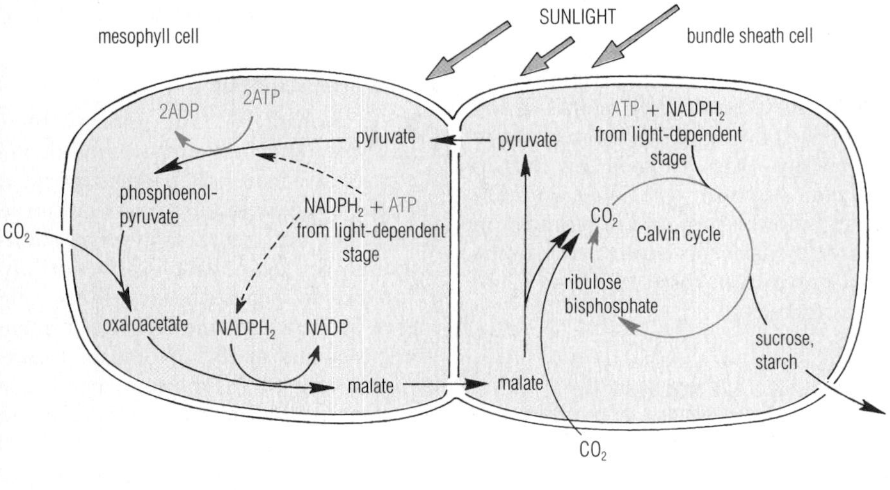

Figure 12.39   Competitive inhibition of ribulose bisphosphate (RuBP) carboxylase by oxygen

CO₂ fixation via the C₃ pathway

■ Oxygen concentration and photorespiration

In the light-independent stage of photosynthesis the enzyme ribulose bisphosphate (RuBP) carboxylase catalyses the reaction between carbon dioxide and RuBP to form two molecules of GP (p. 263). This is the carboxylation phase of the Calvin cycle.

$$CO_2 + RuBP \xrightarrow{\text{RuBP carboxylase}} 2GP$$

Oxygen may displace carbon dioxide at the active site of the enzyme, however, and the products are then GP and a two-carbon compound, phosphoglycolate. Oxygen is thus a competitive inhibitor (p. 185) of the enzyme RuBP carboxylase; an increase in oxygen concentration favours the uptake of oxygen rather than carbon dioxide (Figure 12.39).

Phosphoglycolate is metabolised to GP in the mesophyll cells in a series of reactions known as **photorespiration**. Only one molecule of GP is produced from two molecules of phosphoglycolate. Overall photorespiration is a wasteful process, and may reduce the potential yield from photosynthesis in C₃ plants by up to 50%.

An increase in the carbon dioxide concentration at the chloroplasts favours the carboxylation reaction of RuBP carboxylase as compared with the oxygen interaction. The mechanism evolved by C₄ plants raises the carbon dioxide concentration at the site of the carboxylation reaction, and is therefore a mechanism preventing photorespiration. C₄ plants do not show photorespiration.

It is likely that photosynthesis evolved in an atmosphere of low oxygen concentration and high carbon dioxide concentration (p. 666). Currently the atmosphere contains about 21% of oxygen (due to the success of photosynthesis!). Photorespiration may be an increasing threat to C₃ plants. So far, it appears that no new and effective type of carboxylase, not susceptible to oxygen, has yet evolved.

Question

Suggest why the rate of photosynthesis in C₄ plants is not enhanced by higher atmospheric carbon dioxide concentrations, yet in C₃ plants it is.

12.9 ECONOMIC ASPECTS OF PHOTOSYNTHESIS

■ The productivity of photosynthesis

Only about half the energy emitted by the Sun and reaching our outer atmosphere gets through to the soil level and plant life (the remainder is reflected into space as light or heat energy). Of the radiation reaching green plants, about 45% is in the visible wavelength range (400–700 nm) and can be used by the plant in photosynthesis. Much of the light energy reaching the plant is reflected from the leaves or transmitted through them, however. Of the energy absorbed by the stems and leaves of plants, much is lost in the evaporation of water.

A small quantity of the energy reaching the green leaf is absorbed by the photosynthetic pigments and used in photosynthesis; only one-quarter of this light energy ends up as chemical energy in molecules like glucose. The remainder is lost as heat energy in the various reactions of the light-dependent and light-independent stages. Finally, much of the energy in glucose is lost as heat energy in cellular respiration and other reactions of metabolism. The remainder is retained in new materials, either new structures or stored food, and represents the net primary productivity of the plant.

The value given in Figure 12.40 for photosynthetic productivity of about 5.5% applies to a fully grown crop plant. This value is achieved for only a short period in the growth cycle of the crop plant. For part of the year agricultural land is uncultivated. In fact, for a crop of temperate zones (such as wheat), primary productivity is typically 0.5–1.5%, whereas for a crop plant of the tropics and subtropics (such as sugar cane, a C_4 plant) primary productivity is typically 1.0–2.5%.

The efficiency with which wild plants incorporate sunlight energy into stored glucose molecules in semi-natural and natural ecosystems is even lower. The primary productivity of natural ecosystems is shown in Figure 3.52 (p. 74).

Photosynthesis in relation to human energy demands

Professor David Hall of King's College, London, has calculated that each year the relatively 'inefficient' green plants of the Earth's biosphere fix about ten times more energy than is used in human energy consumption worldwide, and that the energy presently stored in biomass on the Earth's surface is equivalent to the total of our proven fossil fuel reserves of oil, coal and gas.

With the potential for greater productivity by green plants via photosynthesis it is hard to believe in a genuinely unsurmountable energy crisis for humans. There may be many ways we can make better use of photosynthesis.

Energy farming

'Energy farming' is a term used to mean growing plant materials for their value as fuel for humans. Cane sugar as an energy crop is discussed on p. 118.

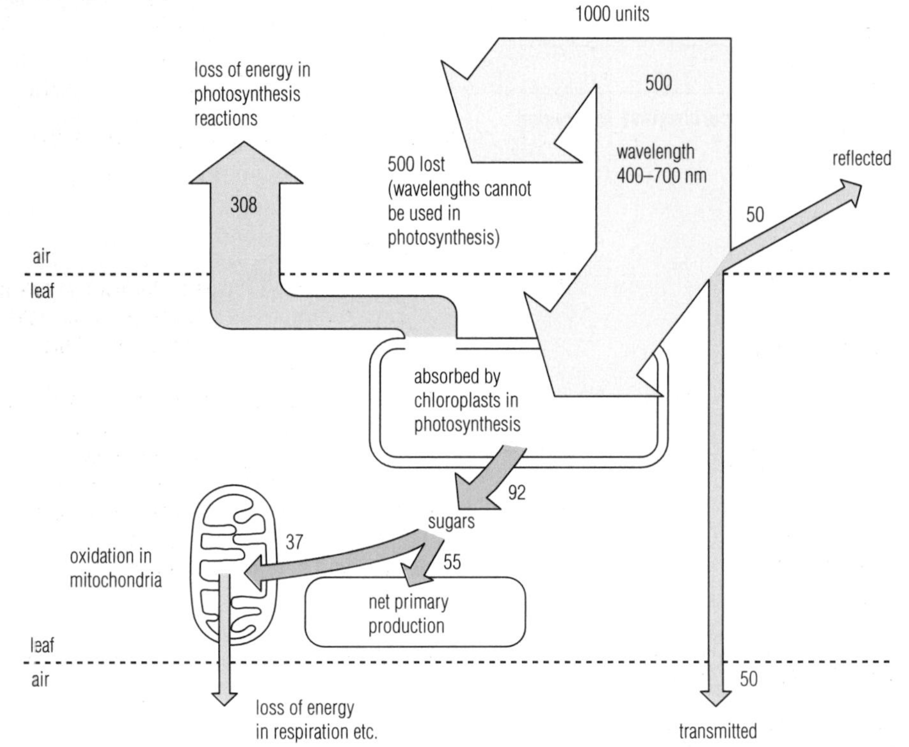

12.10 AUTOTROPHIC PROKARYOTES

The prokaryotes include the bacteria and the cyanobacteria (formerly called the blue-green algae). Certain species of the prokaryotes are autotrophic.

■ The cyanobacteria

The cyanobacteria are photosynthetic organisms. They use carbon dioxide and water as raw materials and the energy of sunlight to manufacture sugar and other substances, very much as in green plants, although the cyanobacteria do not have chloroplasts. The photosynthetic pigments of the cyanobacteria are in membrane systems in the cytoplasm formed by intucking of the plasma membrane. The autotrophic nutrition of the cyanobacterium *Anabaena* is illustrated in Figure 16.26 (p. 333).

■ Autotrophic bacteria

The autotrophic bacteria fall into two distinct groups according to the source of energy used to build carbon dioxide into carbohydrates and other nutrients.

Some of the autotrophic bacteria use light energy for the purpose, and so we classify this as **photosynthetic nutrition**.

Other autotrophic bacteria use chemical energy in the form of energy from specific chemical reactions, and this is known as **chemosynthetic nutrition**.

Photosynthetic bacteria

Photosynthetic bacteria, like the green plants and cyanobacteria, require light energy. They trap light in a pigment called bacteriochlorophyll, which is very similar to the photosynthetic pigment of higher plants. These bacteria differ from other photosynthetic organisms in that the source of hydrogen for the reduction of carbon dioxide is not water.

Figure 12.40   The fate of the light energy that reaches the green leaf of a crop plant (taken arbitrarily to total 1000 units per unit time); note that less than 6% of the incoming energy ends up as net primary production

Some use hydrogen sulphide

The group of photosynthetic bacteria known as the purple sulphur bacteria (Thiorhodaceae) use hydrogen from hydrogen sulphide:

$$CO_2 + 2H_2S \rightarrow (CH_2O) + H_2O + 2S$$

Sulphur bacteria occur at the bottom of ponds and rock pools, where hydrogen sulphide is produced as a by-product of the decay of organic matter by anaerobic bacteria. Light energy for photosynthesis in these bacteria is absorbed by bacteriochlorophyll. The bacteriochlorophyll of the purple sulphur bacteria has a slightly different absorption spectrum from that of the green plant chlorophylls. As a consequence, these bacteria can absorb energy from the light that passes through the aquatic plants growing above them.

Some use organic compounds

The brown or red bacteria (Athiorhodaceae) use relatively simple organic compounds, such as ethanoic acid, as the source of hydrogen.

Chemosynthetic bacteria

The chemosynthetic bacteria synthesise complex organic compounds from inorganic substances (carbon dioxide and water) but they obtain the necessary energy from other chemical reactions that they are able to catalyse. The metabolism of many chemosynthetic bacteria is of significant economic importance.

The nitrifying bacteria

Nitrifying bacteria are chemosynthetic autotrophs that use energy from two exergonic chemical reactions (p. 172) in order to drive the synthesis of carbohydrates.

$$2NH_3 + 3O_2$$

by *Nitrosomonas* sp.

$$2NHO_2 + 2H_2O + ENERGY$$

$$2NHO_2 + O_2$$

by *Nitrobacter* sp.

$$2HNO_3 + ENERGY$$

These reactions are key steps in the natural cycling of combined nitrogen. The nitrogen cycle is presented in Figure 16.28 and discussed on p. 334.

Iron bacteria

The tiny rod-shaped bacterium *Thiobacillus ferro-oxidans* occurs naturally deep down in ore dumps, away

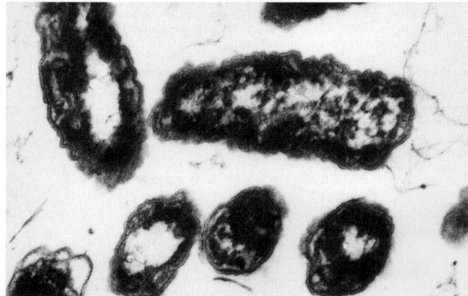

Figure 12.41   Autotrophic bacteria: EM of freeze-etched *Nitrobacter agilis* (above); TEM of *Thiobacillus ferro-oxidans* (below)

from light. It requires an acid medium in order to carry out the oxidation of iron ions.

$$CuFeS_2 + 2Fe_2(SO_4)_3 + 2H_2O + 3O_2 \rightarrow$$

$$CuSO_4 + 5FeSO_4 + 2H_2SO_4 + ENERGY$$

The energy released in this reaction is used in part to fix carbon dioxide and water to carbohydrate. The metabolism of this bacterium is exploited in microbial mining (p. 113).

Questions

1 What factors contribute to the fact that only a small proportion of light energy reaching the leaf surface is built into the matter of the plant's structure?

2 In addition to carbon dioxide and water and an external source of energy, what other materials are likely to be required by autotrophic bacteria for their nutrition?

3 What use do nitrifying bacteria make of the energy released from the oxidation of nitrogen-containing compounds?

FURTHER READING/WEB SITES

The WWW Virtual Library – Botany:
www.ou.edu/cas/botany-micro/www-vl

13 Holozoic nutrition of mammals

- In holozoic nutrition, food is taken into a digestive system, such as the gut of the mammal, broken down (digested) to smaller molecules, and the required nutrients absorbed into body tissues.
- The food eaten is the source of the proteins, lipids, carbohydrates, vitamins, minerals and water required to provide the nutrients essential for growth, repair and respiration by the body tissues.
- The gut is a long muscular tube, locally specialised to carry out particular steps in digestion and absorption. Associated with the gut are various glands that produce watery secretions, some of which contain digestive enzymes.

- Digestion is brought about both mechanically, by teeth and the churning action of the muscular walls of the gut, and by the digestive enzymes secreted on to the gut contents or held in the membrane of the gut lining to assist in digestion and absorption.
- The products of digestion are mostly absorbed into the villi of the small intestine. Amino acids and sugars pass into blood capillaries, while fatty acids pass mostly into the lacteals of the lymphatic system.
- The body tissues take up (assimilate) and use the nutrients that circulate in the blood. The concentrations of nutrients in the blood circulation are regulated largely by the liver.

13.1 INTRODUCTION: HETEROTROPHIC NUTRITION

A heterotroph is an organism incapable of manufacturing its own food; it depends upon other organisms as a source of nutrients. All animals are heterotrophs, as are fungi and many bacteria. Heterotrophic nutrition literally means 'other feeding', in contrast to autotrophic nutrition or 'self-feeding'

(Chapter 12). Heterotrophs depend upon autotrophs either directly or indirectly for their food supply, a fact that is made evident by studying food chains and food webs (Figure 3.4, p. 46).

So heterotrophic organisms feed on complex, ready-made organic food to obtain the nutrients they require. There are three forms of heterotrophic nutrition, each a distinctive strategy for obtaining essential nutrients.

■ Holozoic nutrition

In holozoic nutrition food is taken into a specialised digestive system (a pouch or tube) in the organism. Here the food is broken down to smaller molecules, which are then absorbed. Holozoic nutrition is characteristic of free-living animals; human nutrition is an example.

■ Saprotrophic nutrition

Saprotrophic organisms live in decaying organic matter (the dead remains of plants and animals or their products), which they use for food. Digestion by saprotrophs is extracellular and the products of digestion are absorbed. The nutrition of the fungus *Rhizopus* (Figure 14.16, p. 296) is an example.

■ Parasitic nutrition

A parasite lives in close association with the organism on which it feeds, known as its host, for all or most of its life. The relationship is advantageous to the parasite, but harmful to the host. An example of a parasite is the tapeworm *Taenia solium* in the human small intestine (see Figure 14.22, p. 299).

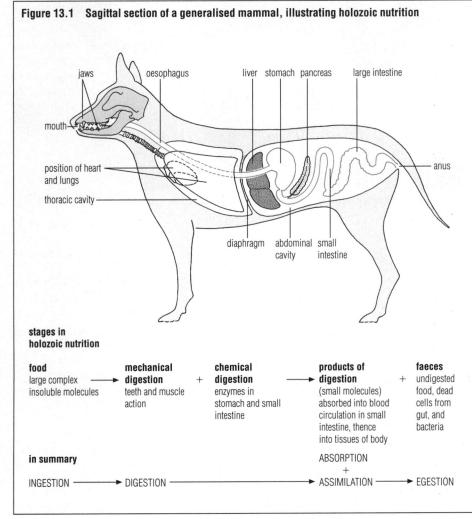

Figure 13.1 Sagittal section of a generalised mammal, illustrating holozoic nutrition

jaws oesophagus liver stomach pancreas large intestine

mouth

position of heart and lungs

thoracic cavity

anus

diaphragm abdominal cavity small intestine

stages in holozoic nutrition

food	mechanical	chemical	products of	faeces
large complex insoluble molecules	digestion teeth and muscle action	digestion enzymes in stomach and small intestine	digestion (small molecules) absorbed into blood circulation in small intestine, thence into tissues of body	undigested food, dead cells from gut, and bacteria

in summary

INGESTION ⟶ DIGESTION ⟶ ABSORPTION + ASSIMILATION ⟶ EGESTION

13.2 HOLOZOIC NUTRITION OF MAMMALS

Holozoic nutrition is illustrated by reference to a generalised mammal in Figure 13.1. Food, which may have to be hunted or searched for, is taken into the **alimentary canal**, known simply as the **gut**. Digestive cells secrete enzymes into the lumen of the gut, where digestion occurs. The soluble products of digestion are selectively **absorbed** into the blood system and then **assimilated** into the cells. The undigested remains of food (faeces) are eliminated from the digestive system.

Table 13.1 Estimated average requirement for energy and reference nutrient intakes for selected nutrients (per day)

Age range /years	Estimated average requirement for energy/MJ	Protein /g	Calcium /mg	Iron /mg	Zinc /mg	Vitamin A /mg	Thiamin /mg	Vitamin B_6 /mg	Folic acid /μg	Vitamin C /mg
4–6	7.16	19.7	450	6.1	6.5	500	0.7	0.9	100	30
7–10	8.24	28.3	550	8.7	7.0	500	0.7	1.0	150	30
Males										
11–14	9.27	42.1	1000	11.3	9.0	600	0.9	1.2	200	35
15–18	11.51	55.2	1000	11.3	9.5	700	1.1	1.5	200	40
19–50	10.60	55.5	700	8.7	9.5	700	1.1	1.5	200	40
50+	10.60	53.3	700	8.7	9.5	700	0.9	1.4	200	40
Females										
11–14	7.92	41.2	800	14.8	9.0	600	0.7	1.0	200	35
15–18	8.83	45.0	800	14.8	7.0	600	0.8	1.2	200	40
19–50	8.10	45.0	700	14.8	7.0	600	0.8	1.2	200	40
50+	8.00	46.5	700	8.7	7.0	600	0.8	1.2	200	40
Pregnant	+0.80	+6.0				+100	+0.1		+100	+10
Lactating (0–4 months)	+2.20	+11.0	+550		+6.0	+350	+0.2		+60	+30

data taken from Ministry of Agriculture, Fisheries and Food, *'Manual of Nutrition'* tenth edition (1995)

Figure 13.2 A calorimeter used to investigate the energy content of foods

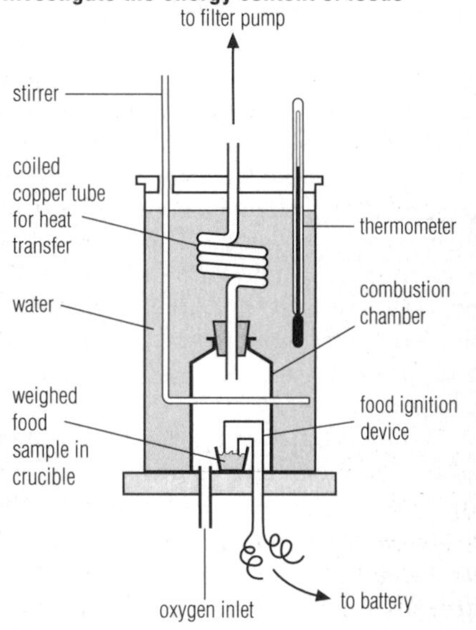

■ Herbivores, carnivores and omnivores

Holozoic animals may be classified according to whether their diet is made up of plants, animals or both:

- **herbivores** feed exclusively on plant materials; examples include the rabbit and sheep
- **carnivores** feed on other animals; examples include the cat and the hedgehog
- **omnivores** feed on both plant and animal material; examples include humans, the pig and the badger.

■ Diet

The food we choose to eat makes up our diet. A balanced diet consists of the essential nutrients in the correct proportions. Malnutrition arises when an organism is under- or over-supplied with any of these nutrients. The essential nutrients for the balanced diet of a mammal consist of the following components:

- metabolic fuel; this is normally taken in as carbohydrates and fats (lipids), but proteins provide 10–15% of the energy required by humans in developed countries
- combined nitrogen to build the proteins needed, taken in as proteins or amino acids
- vitamins
- water
- minerals
- roughage, as dietary fibre.

Carbohydrates and fats: the supply of energy

The major sources of carbohydrate are the sugars and starch, together with cellulose (dietary fibre), found in foods of plant origin. Glycogen stored in animal tissues (muscle and liver) breaks down to glucose after the death of an animal. Fats are obtained as animal fats and plant oils. The main role in diet of both carbohydrates and fats is to provide energy. Energy can also be obtained from proteins in the diet. The chemistry of carbohydrates and lipids is discussed in Chapter 6, pp. 131 and 137.

The energy required by the body varies with the sex, age, body size and activity of the animal. The human body's requirement for energy is determined by

- the background rate of metabolism, known as **basal metabolism**; this is our rate of energy metabolism, resting but awake, at a comfortable external temperature, after having fasted for twelve hours, and is the basic rate against which to compare energy demands of various activities
- how physically active we are in work and leisure; additional energy is required for physical exercise
- the growth and repair work undertaken in the body; for example, the rapid growth that occurs in adolescents makes an additional demand of about 8 kJ per hour.

The recommended daily requirements for energy and nutrients for different human groups – now referred to as the 'estimated average requirement for energy' and the 'reference nutrient intakes for selected nutrients (per day)' – are listed in Table 13.1.

The energy value of carbohydrate, fat and protein can be estimated by burning a sample of known mass in oxygen, in a calorimeter (Figure 13.2). The energy available is released as heat, which raises the temperature of the water surrounding the combustion chamber. The amount of energy in food is expressed in joules; 4.18 J of heat energy are needed to raise 1 g water through 1 °C. On this basis, the energy provided by the individual nutrients is as follows:

- carbohydrate $16\,kJ\,g^{-1}$
- fat $37\,kJ\,g^{-1}$
- protein $17\,kJ\,g^{-1}$

(The corresponding figure for alcohol is $29\,kJ\,g^{-1}$). Thus the approximate energy values of individual items of food can be calculated, provided that the proportions of carbohydrate, fat and protein they contain are known.

The amount of energy in food used to be expressed as 'calories' (in fact, kilocalories) and this term persists in popular literature. One 'calorie' = 4.18 J.

Questions

1 a What features do holozoic, saprotrophic and parasitic nutrition have in common?
  b How do they differ from each other?
2 Work out the likely food web (p. 46) for a human eating cornflakes with milk, bacon and egg, toast and marmalade.
3 Potato crisps typically contain 6% protein, 36% fat and 50% carbohydrate. How much energy does a packet (30 g) of crisps provide?

Proteins and amino acids

Dietary proteins may come from plant and animal sources. Humans obtain most of their plant proteins from cereals, nuts and seeds including peas and beans. Animal protein comes from meat, fish, milk, cheese and eggs. The main role of dietary proteins is as a source of twenty amino acids from which the organism's own proteins are made. The mammal can also respire protein (after deamination, pp. 377 and 508) and this is the fate of the excess protein eaten. Amino acids and proteins are discussed further in Chapter 6.

Whilst plants can synthesise all amino acids they need, because of their ability to convert any one amino acid into another (transamination, p. 209), mammals are unable to meet their full requirements in this way. For example, humans cannot synthesise nine of the required amino acids, so these nine amino acids are

essential ingredients of the diet. Consequently, the 'quality' of dietary protein depends on the essential amino acids it contains. In general, plant proteins are of a lower nutritional quality than animal proteins, but the overall proportions of amino acids in any plant or animal food differ from those needed by humans. In mixed diets, however, there is complementation between different proteins. In fact the concept of protein quality is only relevant when total protein intake is at or below the level required. When intake is greater than requirements, there is no problem. Protein requirements for humans are about 40 g per day; the average daily intake in the UK is about 90–100 g.

Vitamins

Vitamins are organic compounds that are required in small amounts for the maintenance of normal health. Most vitamins cannot be made in the body, and so must

be provided in the diet. Vitamins affect a variety of body functions, although they do not provide energy or function as building materials. Most vitamins are coenzymes (p. 179). Systematic study of the role of vitamins in diet was initiated by Fredrick Gowland Hopkins in 1910 (see panel opposite).

Certain of the vitamins (A, D, E and K) are soluble in lipid and may accumulate in the body in fatty tissues. Vitamin A can be stored in relatively large amounts in the body, and excessive intake of vitamin A can lead to levels in the liver that are so high that they are toxic.

The other vitamins (C and B complex) are water-soluble and are stored to a lesser extent. Since vitamins cannot be manufactured in the human body, their persistent absence from an individual's diet may lead to specific vitamin-deficiency diseases. Table 13.2 lists the

Table 13.2   Vitamins in human diet

Vitamin	Major food source	Function and deficiency symptoms
A (retinol) (fat-soluble)	fish-liver oil, animal liver, dairy products (margarine contains added vitamin A); all green vegetables contain carotene, which is converted to retinol in the body	needed for growth and function of surface tissues (epithelia), especially those secreting mucus (deficiency leads to dry skin, mucous membrane degeneration)
		essential for vision (deficiency leads to night blindness, and eventually to complete blindness in children)
		required for tissue differentiation; has a crucial role in embryogenesis
B vitamins (water-soluble) thiamine (B$_1$)	widely distributed in plant and animal food (meat, wholemeal bread, vegetables); white bread is fortified with thiamine	essential for release of energy from carbohydrate (coenzyme in decarboxylation) (deficiency leads to beri-beri disease, a form of nerve degeneration, where diet is rich in carbohydrate but low in thiamine, as with diets consisting mostly of 'polished' rice)
riboflavine (B$_2$)	widely distributed in foods including milk	essential for releasing energy from food (coenzyme in electron transport); specific deficiency signs rarely seen
nicotinic acid (niacin, B$_3$)	meat, yeast extract, potatoes, wholemeal bread, coffee; the amino acid tryptophan can be converted to nicotinic acid	essential for the release of energy from food (NAD$^+$ and NADP$^+$) (deficiency causes a skin disease called pellagra, and diarrhoea)
pantothenic acid (B$_5$)	in most foods	forms part of coenzyme A molecule
pyridoxine (B$_6$)	meat, fish, eggs, whole cereals, some vegetables	essential for the metabolism of amino acids
B$_{12}$ (several cobalt-containing compounds)	liver, yeast extract; none in vegetable foods	essential for nucleic acid synthesis in rapidly dividing cells (deficiency leads to pernicious anaemia)
folic acid	liver, white fish, raw leafy green vegetables (readily lost in cooking)	essential for nucleic acid synthesis (deficiency leads to anaemia during pregnancy)
biotin	liver, yeast extract, vegetables; made by bacteria in the intestine	essential for metabolism of fats
C (ascorbic acid) (water soluble)	potatoes, green vegetables and citrus fruits; humans are one of few animals unable to make vitamin C	essential for maintenance of connective tissue (deficiency leads to bleeding from small vessels, and to scurvy – bleeding from gums, wounds failing to heal)
D (calciferol) (fat soluble)	fish-liver oil, butter, margarine, egg yolks; made by the action of sunlight on the skin	essential for maintaining levels of calcium and phosphorus in the blood by enhancing absorption in the intestine, and regulating exchange between blood and bones (deficiency leads to rickets – failure of growing bones to calcify, leading to bow-legs and knock-knees in children – and in adults painful bones, which may fracture easily)
E (tocopherol) (fat soluble)	major source is plant oils	conclusive evidence of its essentiality in human nutrition, associated with intake of polyunsaturated fatty acids
K (phylloquinone) (fat soluble)	in dark green leafy vegetables; made by bacteria in the intestine	essential for the normal clotting of blood (involved in the manufacture of prothrombin in the liver) (deficiency leads to delay in the blood-clotting mechanism)

HOPKINS AND THE VITAMIN HYPOTHESIS

Frederick Gowland Hopkins (1861–1947) believed that the chemical reactions in living cells, although complex, are understandable in terms of the laws of chemistry and physics. He was given a research post at Trinity College, Cambridge, in 1910, and there began his work on what were then called 'accessory food factors'. In fact, Hopkins' work was the first general scientific study of vitamins.

In an investigation of the nutrition of a group of young rats, Hopkins found that they failed to grow on a diet made up of pure protein, carbohydrate, fat, salts and water. A control group grew normally, however, when supplied with the same diet with the addition of milk. When Hopkins added 2 cm³ of milk to the daily diet of each rat of the first group, they began to put on weight and grow. (Milk, the sole item of diet for most mammals for the first part of their lives, provides an almost complete diet: carbohydrates, fats, proteins, minerals and a variety of vitamins.)

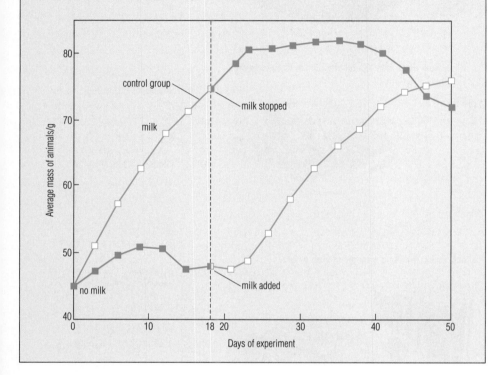

essential vitamins, the foods that contain them, their roles in metabolism and symptoms of their deficiency.

Water

Water makes up approximately 70% of the total body mass. Water is lost in excretion (p. 377) and sweating (p. 513), as vapour during gaseous exchange and as a component of faeces. Between 2.5 and 3.0 dm³ of water per day are lost in these ways. Water intake must balance loss; humans can survive for about a month without food, but only a few days without water.

Water is gained by the body in drinking (but alcoholic wines and spirits have a dehydrating effect), as a major component of the food eaten and as metabolic water (p. 138) made available when lipids – and to a lesser extent carbohydrates – are oxidised.

Minerals

The principal mineral ions required in the human diet and their roles in metabolism are listed in Table 13.3.

Questions

1. How might the experiment shown in the panel on the left be redesigned to demonstrate that the vitamins in milk are responsible for the observed improvement in growth in young mammals?

2. Why must excess amino acid be deaminated before the products of protein digestion are respired to release energy (pp. 377 and 508)?

Table 13.3 Minerals required in the human diet

Mineral	Food source	Functions
Major minerals		
iron, Fe^{2+}, Fe^{3+}	most foods, particularly in meats	in haemoglobin in blood, muscle protein myoglobin, many enzymes, cytochromes
calcium, Ca^{2+}	milk, cheese, bread	as calcium phosphate in bones and teeth, in muscles for contraction, in neurones for nerve function, and for normal clotting of blood
phosphorus, PO_4^{3-} etc.	nearly all foods	as calcium phosphate in bones and teeth, in all cells as a component of nucleic acids, and in ATP (utilisation of energy)
magnesium, Mg^{2+}	widespread in food of vegetable origin because it is an essential constituent of chlorophyll	in bones, and required by enzymes of energy utilisation in cells
sodium and chlorine, Na^+, Cl^-	at low level in all foods, but added to most prepared foods	involved in maintaining water balance of body; sodium essential for nerve and muscle function (there is a danger of excessive intake)
potassium, K^+	vegetables have higher K^+/Na^+ ratio than meat	complementary action with sodium in the functioning of cells
Trace elements (required in minute amounts only)		
copper	shellfish and liver	associated with certain enzymes
chromium	common in foods	involved in the metabolism of glucose
fluorine	drinking water, tea	structure of bones and teeth, resistance of teeth to decay
iodine	seafoods	essential constituent of the hormones produced by the thyroid
manganese	tea, nuts, spices, whole cereals	associated with certain enzymes
selenium	meat, fish, cereals	needed for enzymes in the red blood cells
zinc	meat and dairy products	in healing wounds, activity of many enzymes

Roughage

Roughage, or dietary fibre, is made up of the indigestible cellulose walls of plant material. These non-starch polysaccharides consist of cellulose, hemicellulose, pectins and lignin. They provide bulk to the gut contents in the colon by retaining water, and stimulate movement by peristaltic activity (p. 280) leading to defecation. The presence of adequate dietary fibre in the diet helps to prevent intestinal disorders.

Daily intake of food

It is a useful (although rather tedious) task to analyse a typical diet of people in your community. This requires reference to published tables of the energy, protein, mineral and vitamin contents of foods (see Further Reading at the end of this chapter) and the table of recommended daily amounts of nutrients (Table 13.1). In Figure 13.3 the food values of alternative snack meals are contrasted.

In developed countries, a tradition of eating 'three good meals a day' has for many people been replaced by more frequent eating of 'snacks' and of light and 'low-calorie' pre-prepared meals, and by eating at 'fast food' outlets. These changes are associated with affluence, and with the idea of 'eating out' as a leisure activity. Other developments influencing diets include the availability of convenience foods, the fact that the adults of many households are out at work, and a changing pattern of work and leisure hours.

The effects of these changes on the individual diet may become a matter of concern, particularly as far as they result in inappropriate diets for vulnerable ('at risk') groups within the population, including adults with stressful jobs and little recreational activity, young children and growing adolescents. A barrier to wise eating is, of course, poverty imposed by low income.

Questions

1. Consider the snacks analysed in Figure 13.3.
   a. The 'pub snack' is low in fat yet has a high energy content. Which of the food items are providing this energy?
   b. What are the sources of carbohydrate in the convenience food snack?
   c. What are the sources of calcium in the wholefood meal?
2. Which vitamin might be in short supply in a strictly vegetarian (vegan) diet in which no animal products are present?

Figure 13.3   Alternative snack meals analysed

a typical 'high street' convenience food snack

lemonade
white bread roll
'quarterpounder' hamburger
pickle
french fries
peas

585 g of food and drink
providing 2.9 MJ of energy

a 'wholefood' lunch with careful attention to balance in diet

milk
wholemeal roll with ham
pickle
mixed salad
apple

560 g of food and drink
providing 2.1 MJ of energy

a 'pub snack' including a couple of pints of beer

2 pints beer
brown bread roll with ham
pickle
mixed salad

1380 g of food and drink
providing 2.8 MJ of energy

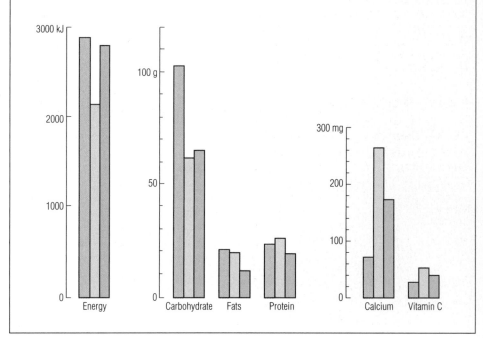

13.3 ISSUES IN HUMAN NUTRITION

■ Malnutrition in the less-developed world

About two-thirds of the people in the world receive barely enough food to supply the energy they need (Figure 13.4). Insidious malnutrition is widespread in many regions of the globe. People rarely die of their undernourishment, however. Instead, they may succumb to very common and curable diseases to which their resistance has been lowered. This is because the immune system is vulnerable to undernourishment. Measles, for example, is a major cause of death in malnourished children, but not in those who are well fed.

When the body is deprived of a sufficient supply of energy, reserves of glycogen in the liver and other tissues are used up before fat stores around the body are mobilised. With continuing starvation and the absence of remaining fat reserves, muscle protein starts to be used as a source of energy. This is the final phase of starvation.

Protein deficiency is a special form of malnutrition. It arises as described above, or when the diet contains enough energy but not enough protein. This happens when starchy foods like maize, yam or cassava are all that is eaten, for example, because protein-rich seeds of legumes (such as peas, beans, soya beans and chick peas), or cereals like wheat, or meat protein are not available. The condition, known as protein energy malnutrition, produces a range of symptoms in individuals (kwashiorkor or marasmus). Whichever symptoms develop, stunted growth and reduced resistance to infection follow.

■ Causes of world hunger

The roots of world hunger lie in poverty. There are enormous disparities in the disposable wealth of peoples and communities of the world today. Most hungry people have insufficient resources to obtain the food they need. Meanwhile, people in developed countries often explain world hunger as a result of overpopulation, and of natural disasters such as floods and droughts.

Overpopulation does contribute to malnutrition. In many communities of the less-developed world the birth rate is high, and many people compete for

too little food. The chronically poor, however, often see their children as a valuable workforce and their only support in their old age. Children are not seen as additional mouths to feed, even though the density of population may prevent the region from becoming self-supporting.

On a world scale, food production is successful; there have been spectacular improvements in the production of cereals (wheat, rice, millet and so on). The crisis exists because of the uneven distribution of food. Excess cereals are produced in North America and in Europe, whereas the deficiency occurs in the developing countries largely in the southern hemisphere. Furthermore, many developing countries give over land to grow food items such as tea, coffee and cocoa for export to developed, industrialised countries.

Food items are exported to earn the international currency to pay for manufactured imports, and to pay interest on existing debts.

Does 'weather' contribute to famine? Sometimes freak weather conditions, such as the severe hurricanes of tropical and subtropical regions, do threaten agricultural production. On the other hand, so-called 'natural disasters', such as the very severe flooding experienced in tropical lowlands, may result from human actions. For example, there is some evidence that deforestation of rain forest causes flooding after heavy hill rains, and leads indirectly to desertification (Figure 4.11, p. 83). Another example is the loss of good farming land and of the food produced there as a result of a steady drift to the cities and towns by farmers and peasants.

Figure 13.4   Issues in malnutrition by undernourishment

1   world population in relation to energy intake

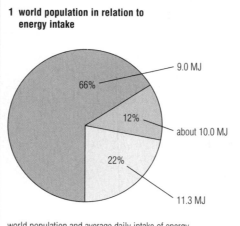

world population and average daily intake of energy

2   geographical distribution of population and cereal production

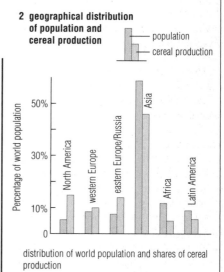

distribution of world population and shares of cereal production

3   symptoms of protein energy malnutrition

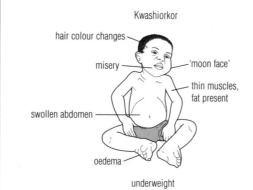

Kwashiorkor

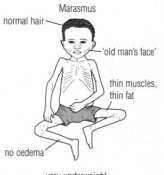

Marasmus

■ Malnutrition in the developed world

Obesity

Obesity is the most prevalent form of malnutrition in affluent societies. Many overweight people consume too many nutrients in relation to their energy expenditure. Obesity is a medical and public health problem for people in all age-groups in developed countries. Obese people suffer an excess mortality proportional to the extent of excess weight. They are vulnerable to diabetes (p. 506), hypertension (p. 354) and coronary heart disease (p. 356).

There are problems in defining what is a normal body weight and therefore in defining what constitutes being 'overweight' and 'obese'. The most widely accepted means of assessing obesity is the **body mass index**:

$$\text{body mass index (BMI)} = \frac{\text{body mass in kg}}{\text{height (in m)}^2}$$

A BMI of 25+ indicates that a person is overweight, and one of 30+ indicates obesity.

'Slimmer's diseases'

Some body fat is essential for health, but a few individuals develop an aversion to food, and take steps to reduce their weight. The conditions known as anorexia nervosa and bulimia nervosa are thought to be on the increase, particularly among young Caucasian females from middle or upper social classes. In anorexia, deliberate dieting, and sometimes deliberate vomiting, lead to serious weight loss and even the loss of consecutive menstrual cycles. Patients have an obsessive fear of gaining weight or becoming fat; they see themselves as much fatter than they actually are.

In bulimia, periods of excessive eating ('binge eating') are followed by self-induced vomiting and use of laxatives to achieve weight control. Here, patients do not necessarily lose excessive weight and their menstrual cycles remain normal.

These two conditions are believed to have more to do with anxiety about maturation and sexuality than with diet, in many cases.

Cholesterol

Cholesterol is a lipid used to synthesise steroid hormones in the adrenal glands and gonads (p. 592), and is an essential component of the plasma membranes of cells (p. 159). Cholesterol is taken in the diet in meat, dairy products (in butter fat), and eggs, but also manufactured in the liver if insufficient is eaten. Cholesterol is transported around the body in the plasma combined with proteins as lipoprotein (p. 508). A high level of blood cholesterol is a cause of coronary heart disease because lipoproteins pass through the lining of arteries and lead to fatty/fibrous deposits, causing atherosclerosis (p. 354). If the diet is low in saturated fats, if obesity is avoided and if the total fat intake is kept low, the need to transport fats in the plasma is minimised and the danger of atherosclerosis is reduced.

Alcohol

The favourite (and often acceptable) mood-altering drug of many human societies is alcohol. A small amount of alcohol may be beneficial rather than harmful, but persistent over-consumption increases the risk of harm to almost every body function (Figure 13.5). Alcohol abuse is a significant underlying cause of serious illness and death in the adult population.

Fruit and vegetables

These components of diet have been recommended for the roughage content they provide, but it is now known that diets rich in fresh fruit and lightly cooked or uncooked vegetables (five helpings a day) substantially reduce the risks of several forms of cancer, including cancers of the breast, large intestine, stomach and prostate. Diets high in fibre also affect the rate of absorption of sugar, and are correlated with lower incidence of diabetes (p. 506).

Figure 13.5 Alcoholic drinks: the risks

Beer, wine and spirits are often referred to as 'alcohol'. Strictly speaking, these are **alcoholic drinks**, which contain anything from 2% to 40% of ethanol. Some contain sugar, too.

A large glass of wine, half a pint of cider, a glass of sherry and a single whisky all contain the same amount of ethanol – about 8 g. This is often referred to as '**1 unit of alcohol**'.

Safe upper limits per week:
► man – 21 units, plus 2 or 3 days without alcohol
► woman – 14 units, plus 2 or 3 days without alcohol
► pregnant woman – a zero intake

The risks of excessive drinking

► short-term effects on the brain, and long-term brain and nerve damage

► increased risk of oesophageal and pancreatic cancer

► deficiency of many nutrients if the drinker loses interest in food

► obesity, with its attendant health problems (alcoholic drinks have high energy values)

► damage to the immune system, resulting in greater susceptibility to infection

► impaired sexual and reproductive function in both men and women

► high blood pressure – but the good news is that moderate alcohol consumption (up to 2 units a day) seems to lower blood pressure

► weaker and less regular heartbeat – but alcohol does not increase the risk of having a heart attack

► inflammation of the stomach, causing severe pain

► liver damage – enlarged liver, fatty liver, jaundice, maybe cirrhosis

---

Questions

1 Children of both affluent and poor parents in countries such as the UK may be obese. What are the likely reasons in each case?

2 Cholesterol is a normal component of our diet, but is also synthesised in the liver. Why is this substance required in our bodies (p. 592)?

13.4 THE STRUCTURE OF THE GUT

Food cannot be said to have entered a person's body until it has been digested and absorbed into the blood. Digestion and absorption occur in the gut, which is a long, hollow muscular tube (Figure 13.6). Each region of the gut is locally specialised to carry out particular steps in the overall process of mechanical and chemical digestion and absorption. The gut is also organised to allow movement of its contents in one direction only. The main regions of the gut are the mouth and buccal cavity, pharynx (throat), oesophagus (gullet), stomach, small intestine (duodenum, jejunum and ileum), caecum and appendix (which form a cul-de-sac), large intestine (colon and rectum) and anus. The total length of the human gut is 8–9 metres, of which two-thirds is the small intestine.

The gut is suspended from the dorsal wall of the abdominal cavity by tough connective tissue called **mesentery**. In the mesentery run the blood vessels, nerves and lymphatics that serve the gut. The gut wall consists of four distinct layers known as the **serosa** (the outer layer), **muscularis mucosa** (external muscle), **submucosa** and **mucosa**. The mucosa surrounds the hollow **lumen** where the food passes and digestion occurs.

The gut wall contains glands whose linings are formed by intuckings (invaginations) of the epithelium lining the surface of the lumen. The glands produce copious quantities of watery secretions, some of which contain digestive enzymes, as detailed in Table 13.4 (p. 287).

Figure 13.6 The gut, the site of digestion and absorption

nasal cavity
buccal cavity
mouth
epiglottis
(cartilage at opening to the larynx)
oesophagus
trachea (windpipe)
palate
(bone and cartilage platform)
pharynx (throat)
glottis
(opening of the larynx between the vocal folds)

diaphragm
gall bladder
liver
pancreas
duodenum
(first part of small intestine; the main site of digestion)
ileum
(completion of digestion and absorption of food)
caecum
appendix
stomach
(food storage, digestion of proteins)
pyloric sphincter
pancreatic duct
colon
(absorption of water)
rectum
anal sphincter muscle
anus

Secretion by the glands is coordinated with the presence of food in that region of the gut (p. 288). Digestive juices also contain mucus. Mucus, which is also secreted by the epithelium of the mucosa (Figure 13.7), protects the tissues of the gut from attack by the digestive juices (**autodigestion**, see 'Peptic ulcers', p. 282), and it lubricates the linings of the canal as the contents are moved along.

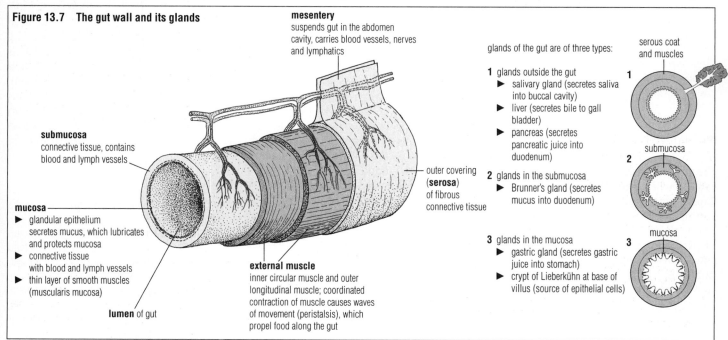

Figure 13.7 The gut wall and its glands

mesentery
suspends gut in the abdomen cavity, carries blood vessels, nerves and lymphatics

submucosa
connective tissue, contains blood and lymph vessels

mucosa
▶ glandular epithelium secretes mucus, which lubricates and protects mucosa
▶ connective tissue with blood and lymph vessels
▶ thin layer of smooth muscles (muscularis mucosa)

lumen of gut

external muscle
inner circular muscle and outer longitudinal muscle; coordinated contraction of muscle causes waves of movement (peristalsis), which propel food along the gut

outer covering (**serosa**) of fibrous connective tissue

glands of the gut are of three types:

1 glands outside the gut
▶ salivary gland (secretes saliva into buccal cavity)
▶ liver (secretes bile to gall bladder)
▶ pancreas (secretes pancreatic juice into duodenum)

2 glands in the submucosa
▶ Brunner's gland (secretes mucus into duodenum)

3 glands in the mucosa
▶ gastric gland (secretes gastric juice into stomach)
▶ crypt of Lieberkühn at base of villus (source of epithelial cells)

serous coat and muscles
1

submucosa
2

mucosa
3

Figure 13.8 The buccal cavity and pharynx, structure and function: the movements of lips, lower jaw, tongue, soft palate, epiglottis and larynx in swallowing are shown by dotted lines

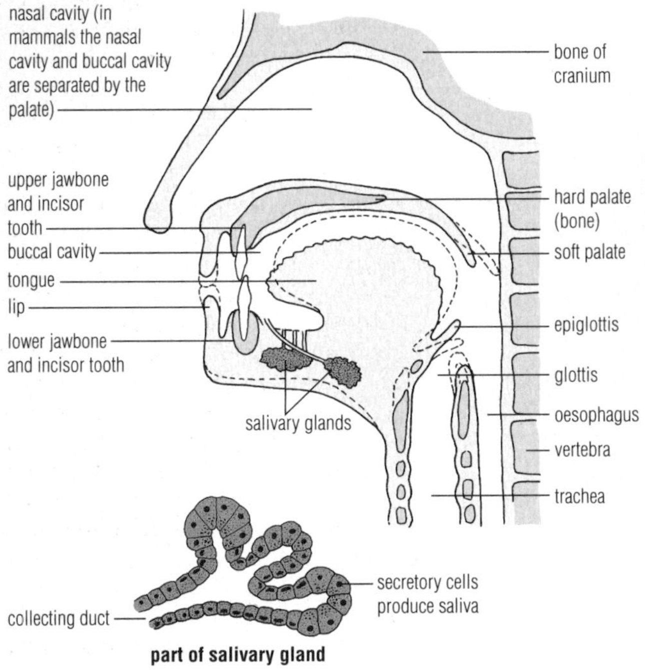

nasal cavity (in mammals the nasal cavity and buccal cavity are separated by the palate)

bone of cranium

upper jawbone and incisor tooth

hard palate (bone)

buccal cavity

soft palate

tongue

lip

epiglottis

lower jawbone and incisor tooth

glottis

oesophagus

salivary glands

vertebra

trachea

collecting duct

secretory cells produce saliva

part of salivary gland

Figure 13.9 The structure of human teeth

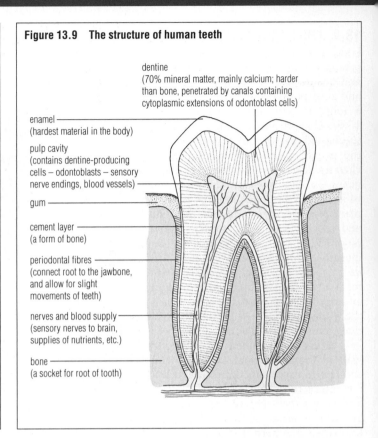

dentine (70% mineral matter, mainly calcium; harder than bone, penetrated by canals containing cytoplasmic extensions of odontoblast cells)

enamel (hardest material in the body)

pulp cavity (contains dentine-producing cells – odontoblasts – sensory nerve endings, blood vessels)

gum

cement layer (a form of bone)

periodontal fibres (connect root to the jawbone, and allow for slight movements of teeth)

nerves and blood supply (sensory nerves to brain, supplies of nutrients, etc.)

bone (a socket for root of tooth)

Peritoneum

The outer wall of the gut, the mesenteries and the inner surface of the abdominal cavity are lined by an epithelium called the **peritoneum**. This smooth, moist layer of cells reduces friction by lubricating the surfaces of the abdominal cavity.

Peritonitis is the inflammation of the peritoneum. Inflammation may be caused by contaminating pathogenic bacteria from outside the body (introduced by accidental injury or surgical wounds) or by bacteria or secretions released into the abdomen by perforation of the gut wall (for example, from a ruptured appendix).

◼ The mouth and buccal cavity

The buccal cavity (Figure 13.8) contains the muscular tongue, and is supported by jaws in which the teeth are set in sockets. Lips, tongue and teeth work together to capture and receive food, to move food about the mouth, and to cut, grind and chew food into smaller pieces. The tongue also possesses taste buds (p. 452) and so has a role as a sensory organ; the human tongue is important in speech, too.

The buccal cavity is lined by a stratified epithelium (p. 213) continuous with the external skin (this is because the buccal cavity of the embryo mammal is formed by intucking of the ectoderm). Inside the cavity the epithelial cells become flattened; they are not cornified, however, and the dead outer layer is therefore continually worn away by friction with food during mastication, and replaced by cells from deeper layers.

Saliva

About 1.5 dm³ of saliva is secreted from the salivary glands into the mouth each day. Saliva is a watery solution of salts including sodium hydrogencarbonate, mucus, the digestive enzyme amylase and the antibacterial enzyme lysozyme. The solution is very slightly alkaline. Secretion of saliva is controlled by the nervous system and occurs in response to the sight, smell, taste or thoughts about food.

Saliva is an essential feature in the functionings of the mouth and throat. It keeps the surface lubricated for speech. It moistens and binds chewed food, enabling it to be formed into a bolus, and also lubricates the bolus for ease of swallowing. Salivary amylase catalyses the hydrolysis of starch to maltose, and lysozyme reduces bacterial contamination.

Mastication in the buccal cavity

The teeth of the articulated lower jaw are worked against the teeth of the fixed upper jaw in order to cut and crush food. The chewing action of the teeth (known as **mastication**) reduces food to small particles, increasing the surface area on which enzymes can act. Mastication also facilitates swallowing.

Mastication is made possible in mammals because of the presence of a **palate** separating the air path (nasal cavity) from the mouth. As a consequence, food can be retained in the mouth rather than swallowed whole, between breaths. Since food is retained in the mouth for cutting, crushing, grinding or shearing, according to diet, mammals have evolved different types of teeth (**heterodont dentition**), each type being specialised for a particular function. This condition contrasts with the **homodont dentition** of fish and reptiles. The structure of human teeth is illustrated in Figure 13.9.

Questions

1 What steps in digestion are commenced in the buccal cavity?

2 Reptiles and amphibians swallow food whole, immediately it is caught, but in mammals food is retained in the mouth whilst it is cut up and chewed. How is this possible?

3 Part of the jawbone consists of compact bone and part of spongy bone (Figure 13.10). What is the difference in structure between these two types of bone (p. 215)?

The structure of teeth

The part of a tooth projecting above the gum, the crown, is covered by hard enamel. The neck of the tooth is enclosed by the gum, and the root is embedded in the jawbone. The bulk of a tooth consists of tough, hard dentine around a central pulp cavity. Radiating out through the dentine run fine channels containing cytoplasmic extensions of the dentine-forming cells (**odontoblasts**). The bodies of these cells lie in the pulp cavity, which also contains sensory nerve endings and blood vessels. Covering the root in its socket is a bone-like cement. The tooth is held in the socket by numerous **periodon-** **tal fibres** which both anchor and suspend it, permitting slight movement in response to the pressures of biting, chewing and grinding.

The permanent set of teeth of the adult mammal consists of incisors, canines, premolars and molars. A mammal forms two complete sets of teeth during its life; the permanent set is preceded by a deciduous or milk set. The number of teeth of each type in a dentition may be represented in a **dental formula**: the dental formulae of the deciduous set and the permanent set of humans are shown in Figure 13.10.

Dental disease

Dental disease occurs as a result of the actions of bacteria. The mouth has a natural microflora of bacteria, which includes certain pathogenic species. Many species of bacteria can be harmful if allowed to accumulate in 'stagnation sites' between over-crowded teeth or in the **plaque** that can build up on teeth. Plaque forms from a mixture of food debris, saliva and bacteria which, undisturbed, progressively hardens as it attaches to the teeth. Plaque may harbour two types of harmful microorganism.

Firstly, bacteria that feed on gum tissue cause inflammation of the gum and bleeding as they feed (peridontal disease). Subsequently the gum tissue becomes white and dead as the bacteria pass on to and destroy the periodontal fibres. Ultimately, the tooth is lost.

Secondly, bacteria that convert sugars in the diet to organic acids cause dental caries. These acids slowly erode the enamel. Once acids and bacteria reach the dentine, decay occurs more quickly and the pulp cavity is eventually attacked, causing the severe pain of toothache.

Dental disease may be prevented by a combination of factors. The following action is important:

► choosing a diet avoiding sugary snacks and sweets (which are quickly converted to acids); sucrose specifically encourages the growth of plaque-forming bacteria

► regular and careful brushing of teeth in order to remove debris between teeth, to prevent plaque build-up, and to allow free flow of saliva with its neutralising and antibacterial properties; the regular use of dental floss will remove plaque on and between the teeth before it can accumulate

► having regular dental check-ups and treatment for any disease before serious harm occurs

► an appropriate intake of fluoride which, when included in the diet in small quantities, decreases the incidence of dental caries; drinking water, and toothpaste, are the means most frequently chosen to deliver effective amounts of this ion, which helps mineralisation of the teeth.

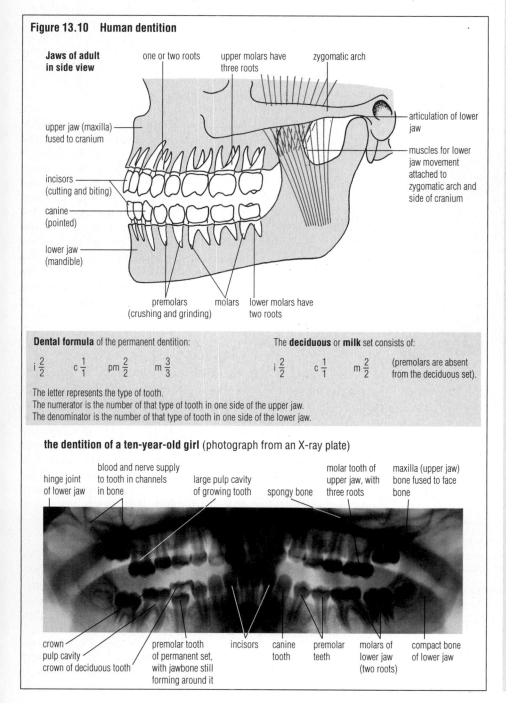

Figure 13.10 Human dentition

Jaws of adult in side view

one or two roots — upper molars have three roots — zygomatic arch

upper jaw (maxilla) fused to cranium

articulation of lower jaw

muscles for lower jaw movement attached to zygomatic arch and side of cranium

incisors (cutting and biting)

canine (pointed)

lower jaw (mandible)

premolars (crushing and grinding) — molars — lower molars have two roots

Dental formula of the permanent dentition:

$$i\,\frac{2}{2} \qquad c\,\frac{1}{1} \qquad pm\,\frac{2}{2} \qquad m\,\frac{3}{3}$$

The **deciduous** or **milk** set consists of:

$$i\,\frac{2}{2} \qquad c\,\frac{1}{1} \qquad m\,\frac{2}{2} \qquad \text{(premolars are absent from the deciduous set).}$$

The letter represents the type of tooth.
The numerator is the number of that type of tooth in one side of the upper jaw.
The denominator is the number of that type of tooth in one side of the lower jaw.

the dentition of a ten-year-old girl (photograph from an X-ray plate)

hinge joint of lower jaw — blood and nerve supply to tooth in channels in bone — large pulp cavity of growing tooth — spongy bone — molar tooth of upper jaw, with three roots — maxilla (upper jaw) bone fused to face bone

crown — pulp cavity — crown of deciduous tooth — premolar tooth of permanent set, with jawbone still forming around it — incisors — canine tooth — premolar teeth — molars of lower jaw (two roots) — compact bone of lower jaw

Swallowing

Food is masticated and mixed with saliva. The tongue presses the food against the hard palate, and forms a **bolus** (a ball) of food. The bolus is then pushed to the back of the mouth, ejected into the pharynx and moved to the upper part of the oesophagus. The soft palate rises, closing off the nasal cavity, the glottis is constricted by a sphincter muscle, and the epiglottis acts as a chute, directing food over the top of the trachea and down the oesophagus (Figure 13.8).

Swallowing is a deliberate or voluntary action, but once the food is in the oesophagus it is moved by reflex action.

■ The oesophagus

The oesophagus is a straight, narrow, thick-walled muscular tube about 25 cm long, leading from the pharynx to the stomach. The glandular epithelium of the oesophagus pours copious quantities of mucus into the lumen, lubricating the passage of the bolus. The wall of the upper part of the oesophagus contains voluntary muscle (p. 218), the middle part both voluntary and involuntary muscle, and the lower part involuntary muscle only.

The bolus is propelled down the oesophagus towards the stomach by a wave of muscular contraction and relaxation, known as **peristalsis** (Figure 13.11). Circular muscles of the oesophagus wall in front of the bolus relax, whilst those immediately behind it contract.

■ The stomach

The stomach is a J-shaped muscular bag located high in the abdominal cavity, on the left-hand side just below the diaphragm. The lining of the muscular wall lies in folds when the stomach is empty, but can distend to about 5 dm³ after a large meal. The stomach is a site of partial digestion, and a mixed meal will remain in the stomach for up to four hours. An extremely fatty meal will remain longer. The exit from the stomach into the duodenum is controlled by a circular sphincter muscle, the **pyloric sphincter** (Figure 13.12).

The functioning of the stomach

The wall of the stomach has three layers of smooth muscle (compared with the generalised structure shown in Figure 13.7, the stomach has an additional layer of oblique muscle). The stomach epithelium is liberally supplied with mucus-secreting cells, and it also contains millions of tiny pits called gastric glands. These glands secrete between 400 and 800 cm³ gastric juice per meal. Secretion is coordinated with the arrival of the food. The incoming food is thoroughly mixed with the gastric juice by the repeated churning action of the stomach wall, and the contents of the stomach quite quickly become a semi-liquid known as **chyme**. Very little digested food is absorbed from the stomach into the blood supply, but some water and ions are taken up at this point, and the drugs aspirin and alcohol are readily absorbed here. Periodic relaxation of the pyloric sphincter muscle permits successive small quantities of chyme to pass into the duodenum, until the stomach is empty.

Gastric juice

There are four types of cell in the gastric glands, each with a specific role in the production of gastric juice.

Oxyntic cells secrete 0.15M hydrochloric acid at about pH 1, in sufficient quantities to create an acid environment of pH 1.5–2.0 for the bolus on arrival in the stomach. This is the optimum pH for the protein-digesting enzymes present in the gastric juice. It is an extremely acid environment.

The blood supply to these cells delivers hydrogen ions at 4×10^{-8}M concentration. The oxyntic cells concentrate hydrogen ions about three million times to deliver their acidic product. This is an energy-demanding process, dependent upon a substantial supply of ATP. Oxyntic cells contain vast numbers of mitochondria.

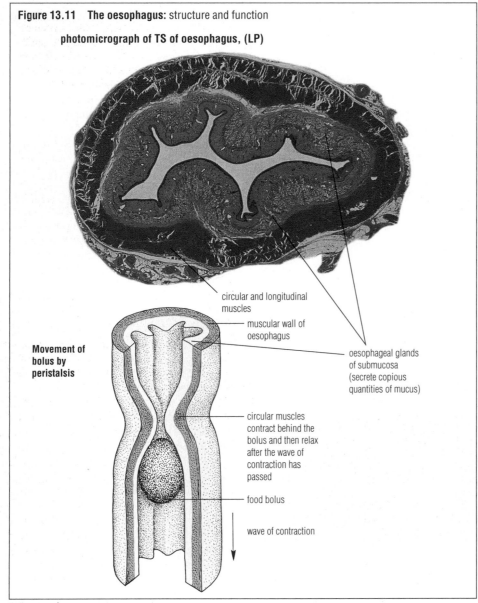

Figure 13.11 The oesophagus: structure and function

photomicrograph of TS of oesophagus, (LP)

circular and longitudinal muscles

muscular wall of oesophagus

oesophageal glands of submucosa (secrete copious quantities of mucus)

Movement of bolus by peristalsis

circular muscles contract behind the bolus and then relax after the wave of contraction has passed

food bolus

wave of contraction

Figure 13.12   The human stomach: its wall and glands

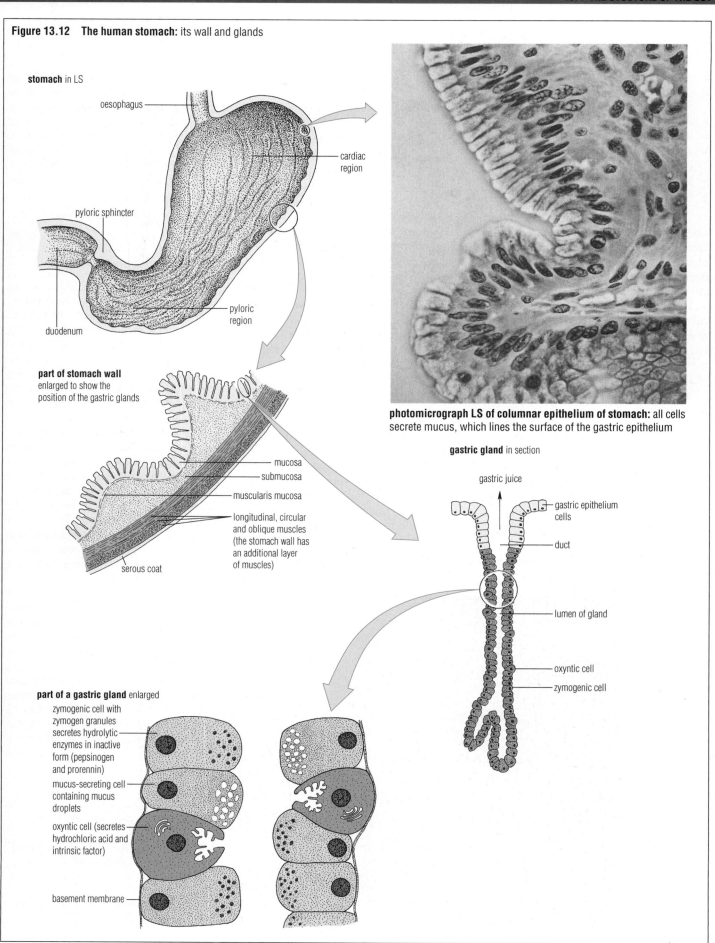

stomach in LS

oesophagus

cardiac region

pyloric sphincter

pyloric region

duodenum

part of stomach wall enlarged to show the position of the gastric glands

mucosa

submucosa

muscularis mucosa

longitudinal, circular and oblique muscles (the stomach wall has an additional layer of muscles)

serous coat

photomicrograph LS of columnar epithelium of stomach: all cells secrete mucus, which lines the surface of the gastric epithelium

gastric gland in section

gastric juice

gastric epithelium cells

duct

lumen of gland

oxyntic cell

zymogenic cell

part of a gastric gland enlarged

zymogenic cell with zymogen granules secretes hydrolytic enzymes in inactive form (pepsinogen and prorennin)

mucus-secreting cell containing mucus droplets

oxyntic cell (secretes hydrochloric acid and intrinsic factor)

basement membrane

The **roles of hydrochloric acid** in the gastric juice, in addition to establishing an optimum pH, are

▶ to kill bacteria in the ingested food
▶ to denature proteins and soften fibrous connective tissues in the food
▶ to activate pepsin (see below)
▶ to activate rennin (see below)
▶ to render calcium and iron salts suitable for absorption in the intestine
▶ to begin the hydrolysis of sucrose
▶ to split nucleoproteins to nucleic acid and protein.

Oxyntic cells also produce **intrinsic factor**, a substance essential for the absorption of vitamin B_{12}. Failure of intrinsic factor secretion results in pernicious anaemia.

Zymogenic cells (also known as chief cells) synthesise and secrete **pepsinogen**, the precursor of the protein-digesting enzyme **pepsin**. In young mammals at the milk-fed stage, **prorennin** – the precursor of rennin – is also produced and secreted.

The hydrochloric acid in the gastric juice initiates conversion of pepsinogen to pepsin. This occurs by the removal of a few terminal amino acid residues at one end of the protein chain. The first-formed pepsin then quickly completes the conversion of the remainder of the pepsinogen, in a process known as **autocatalysis**. If pepsin were to be produced in active form in the zymogenic cells it would digest the cell protein. As it is, pepsin becomes active only in the stomach where it begins the digestion of proteins in the chyme, hydrolysing them to shorter-chain polypeptides. For example:

Val–Tyr–Ala–Ala–Gly–Phe–Gly–Arg
       ↑   ↑
    pepsin   pepsin

Pepsin breaks the protein chain at various points, but especially at bonds with tyrosine and phenylalanine. (There are about seven different pepsins in human gastric juice.)

Hydrochloric acid is responsible for the conversion of prorennin to rennin. Rennin, when present, commences the digestion of milk proteins such as caseinogen by coagulating them. In the coagulated form they are then digested by the action of pepsin (Table 13.4, p.287).

Mucus-secreting cells in the gastric glands and in the stomach epithelium secrete copious quantities of mucus. Mucus forms a barrier between the stomach lining and the gastric juice. It protects the stomach wall and glands from self-digestion (autolysis) by pepsin and hydrochloric acid, which could give rise to stomach ulcers.

The cells of the gastric glands and the stomach epithelium survive only a few days, despite the protection provided. There is a very rapid rate of replacement of these cells throughout life.

Peptic ulcers

Normally the gastric mucosa is protected from the digestive action of pepsin by mucus. Any local failure of the mucus supply results in erosion of gut wall due to the actions of pepsin and hydrochloric acid. A peptic ulcer may result, either in the stomach (gastric ulcer) or in the duodenum (duodenal ulcer). A pathogenic bacterium, *Helicobacter pylori*, is one cause of gastric ulcers.

The discovery of stomach function

Stomach digestion and its control have been investigated by various experimental techniques.

1 A gastric fistula (that is, an opening from the stomach to the exterior) permits the stomach contents to be sampled while a meal is being digested. A fistula may be formed accidentally, for example due to a gunshot wound that heals abnormally, or may be opened by surgery in order to investigate stomach secretion, using experimental animals.

2 Similar investigations with a surgically constructed stomach pouch (part of the stomach wall sealed off from the bulk of the stomach, and opening to the exterior) permits the gastric juice formed to be sampled, uncontaminated by food.

3 The interior of the stomach may be viewed by passing down a gastroscope with an optical fibre link to a video camera and display screen.

4 The contents of the stomach may be sampled at intervals during digestion by passing a plastic tube down into the stomach. Such a tube is passed through the nose and has to be swallowed.

5 X-ray examination of the gut, following an X-ray-opaque 'barium meal', may facilitate the detection of abnormalities in the structure or functioning of the gut (Figure 13.13).

Figure 13.13 **The passage of a 'barium meal':** barium sulphate is insoluble and opaque to X-rays, and shows up as pale (unexposed) areas on the X-ray plates

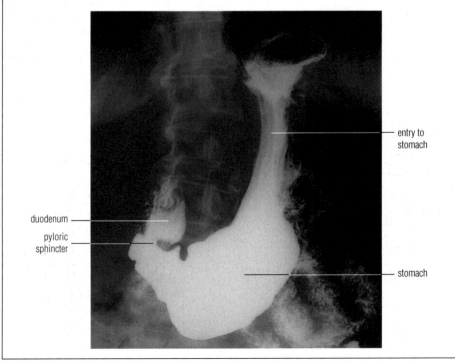

entry to stomach

duodenum

pyloric sphincter

stomach

Questions

1 What is the role of the epiglottis during swallowing?
2 Waves of muscular contraction (peristalsis) move across the stomach wall after food arrives there. What part does this peristalsis play in digestion?
3 Contrast the roles of mucus secreted in the salivary glands, the oesophagus and the stomach.

■ The small intestine

The human small intestine is 5–6 metres long. The first region, called the **duodenum**, occupies only 30 cm. The wall of the small intestine is relatively thin, but in the duodenum the submucosa layer is thrown into inwardly projecting folds, which are just visible in the X-ray photograph of Figure 13.13. In the rest of the small intestine, the **ileum**, these inner foldings of the wall decrease and disappear (Figure 13.15). (In humans a zone intermediate between the duodenum and the ileum, the **jejunum**, is identified.)

From the stomach the chyme passes into the duodenum, where the bulk of digestion is completed. Into the duodenum runs the bile from the gall bladder and pancreatic juice from the pancreas (Figure 13.14).

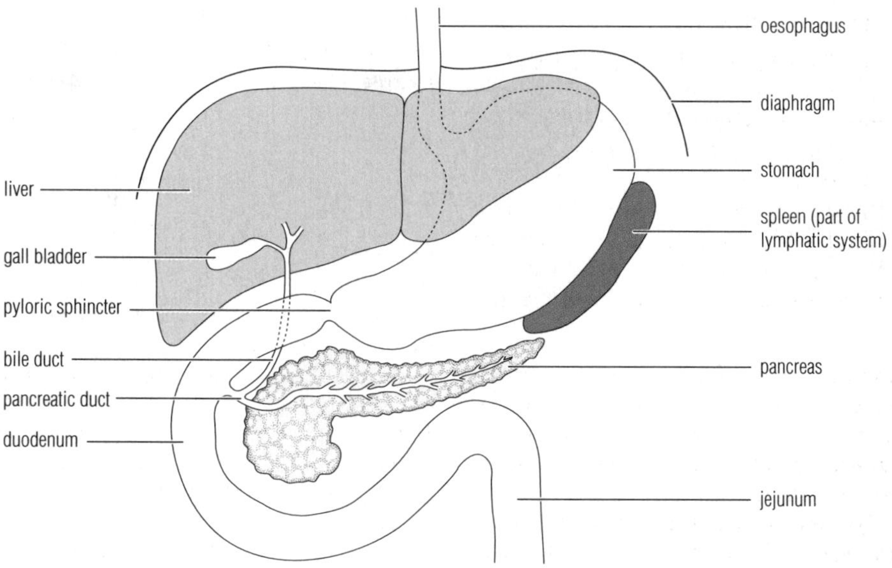

Villi

Throughout the small intestine the mucosa is shaped into a vast number of projections called **villi**, some finger-like, some leaf-like. Villi increase the surface area for absorption very considerably. Each villus is richly supplied with blood capillaries and lymph vessels. Smooth muscle of the wall extends into the villi. Contraction and relaxation of this muscle moves the villi among the passing food.

Also throughout the small intestine are intestinal glands (**crypts of Lieberkühn**) at the bases of the villi. (In the duodenum there are in addition small, rounded **Brunner's glands**).

The epithelium of the villi consists of unstratified, columnar epithelial cells, together with numerous **goblet cells**,

which secrete mucus. The outer surface of the epithelial cells carries a **brush border** of about 2000 **microvilli** per cell (Figure 13.15). These cells of the epithelium are short-lived. The epithelial cells, exposed to the mechanical movement of chyme and to the chemical action of digestive enzymes, including proteases secreted by the pancreas, are rapidly destroyed. After only one or two days *in situ* they are worn away and sloughed off into the lumen of the intestine. They are discharged with the digestive enzymes they contain, which are then added to the digesting food.

Dislodged epithelial cells are replaced by cells from deep in the intestinal gland. The cells at the base of the gland repeat-

edly divide mitotically. As new cells form they move up and out on to the villi, taking from two to four days to move into place.

The small intestine is thus a long, very flexible tube, its surface area enormously extended by the villi and microvilli, which facilitate digestion, and also absorption by the intestinal cells of the products of digestion. Each villus has an extremely extensive lymph and blood supply, which assists in the transport of nutrients to the rest of the body. The small intestine is strongly self-protective, by means of a copious production of mucus and a mechanism for the rapid replacement of cells damaged by contact with food and digestive juices.

Figure 13.14 The duodenum, gall bladder and pancreas

oesophagus

diaphragm

stomach

spleen (part of lymphatic system)

liver

gall bladder

pyloric sphincter

bile duct

pancreatic duct

duodenum

pancreas

jejunum

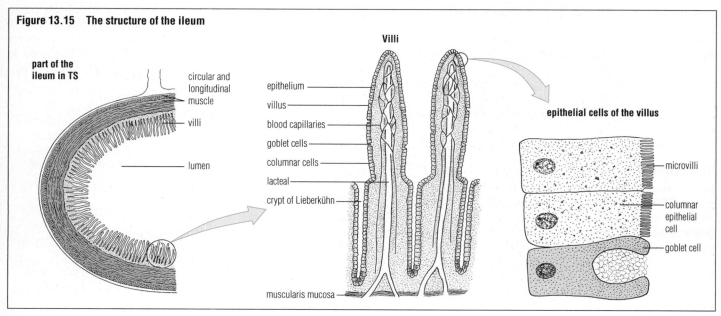

Figure 13.15 The structure of the ileum

part of the ileum in TS

circular and longitudinal muscle

villi

lumen

Villi

epithelium

villus

blood capillaries

goblet cells

columnar cells

lacteal

crypt of Lieberkühn

muscularis mucosa

epithelial cells of the villus

microvilli

columnar epithelial cell

goblet cell

Digestion in the small intestine

In the duodenum digestion is aided by the action of the digestive juices of the pancreas, together with that of bile from the gall bladder. Secretions from Brunner's glands of the duodenum coat the mucosa with mucus and alkaline salts. This coating functions as a first line of defence against the stomach enzymes, since in the upper part of the duodenum the stomach acidity has not been fully neutralised by the mixing in of bile with the chyme.

Bile

Bile is produced in liver cells (p. 507), collects in the gall bladder (in humans and some other mammals) and passes intermittently into the duodenum via the bile duct. Bile is a yellow-green, mucous fluid containing bile salts, bile pigments, cholesterol and salts. Bile lowers the surface tension of fat globules, causing large drops to break up into tiny droplets (**emulsification**), enormously increasing surface area and assisting subsequent hydrolysis of fat by lipase (Figure 13.16).

Bile is strongly alkaline, being rich in various alkaline salts, and, with pancreatic juice, plays a major part in neutralising the acidity of the chyme.

Pancreatic juice

The pancreatic juice is an alkaline fluid, pH about 7.5–8.8. It contains several enzymes. Pancreatic **amylase** catalyses the hydrolysis of starch to maltose. Pancreatic **lipase** catalyses the hydrolysis of fats to fatty acids and glycerol. Pancreatic juice also contains protein-digesting enzymes in an inactive form, known as **trypsinogen** and **chymotrypsinogen**. Trypsinogen is activated to **trypsin** by **enteropeptidase** secreted by epithelial cells of the villi. Chymotrypsinogen is activated to **chymotrypsin** by trypsin. Both enzymes, once activated, hydrolyse proteins to polypeptides.

Pancreatic juice also contains a **nuclease** that catalyses the hydrolysis of nucleic acid to nucleotides, and a variety of **peptidases** that catalyse the release of free amino acids from polypeptides.

'Intestinal juice'

In the small intestine digestion is completed by **enzymes from the pancreas** (detailed above), by enzymes from the **disintegration of epithelial cells** after these have been shed into the lumen of the gut, and by **enzymes bound to the microvilli of the epithelium** lining the villi. The intestinal glands (the crypts of Lieberkühn and also the Brunner's

glands) secrete mucus and alkaline salts, but do *not* secrete enzymes (except enteropeptidase, mentioned earlier).

As we have seen, the cells of the intestinal glands migrate out of the crypts to become epithelial cells lining the villi. It is in these cells that absorption and digestion of food is completed (Figure 13.17, p. 285). Here the remaining carbohydrate and protein is finally broken down within the cell membrane of the microvilli of the epithelial cells. Short-chain peptides are digested to amino acids (by di- and tri-peptidases), maltose to glucose (by maltase), sucrose to glucose and fructose (by sucrase) and lactose to glucose and galactose (by lactase). These steps in digestion are combined with transport across the membrane.

Peristalsis moves the mixture of food and enzymes along the small intestine.

The presence of the food stimulates the secretion of the intestinal juice, and the chyme now takes the form of a watery emulsion. The products of digestion in this fluid are now ready for absorption.

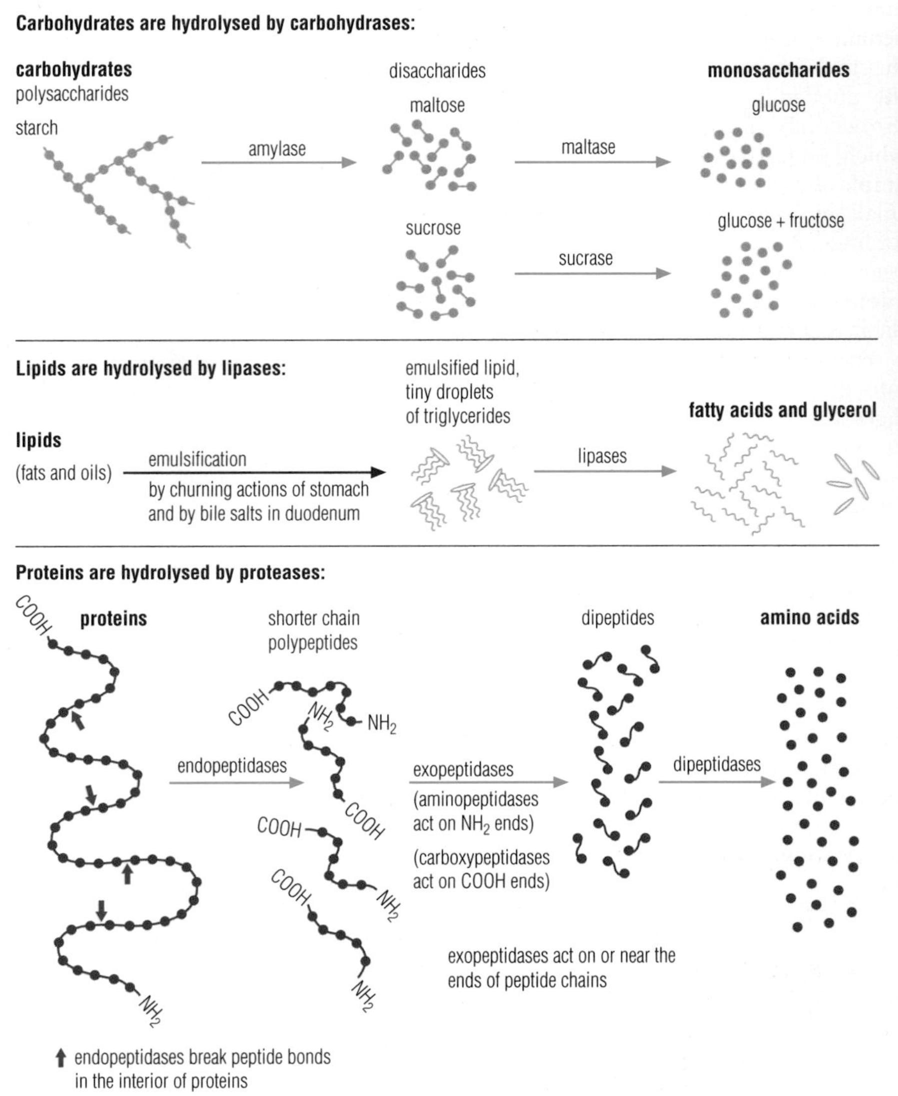

Questions

1 Bile is not a digestive enzyme. What is it? What is its role in digestion and absorption?

2 How is digesting food moved in the small intestine?

3 What are the final digestion products of, respectively, starch, sucrose, protein and lipid?

4 a What changes in pH occur during digestion of our food?

  b What would be the effect on the enzymes of the pancreatic juice if the pH of food in the duodenum remained the same as that in the stomach?

Figure 13.16   The steps to digestion of carbohydrates, lipids and proteins

Carbohydrates are hydrolysed by carbohydrases:

Lipids are hydrolysed by lipases:

Proteins are hydrolysed by proteases:

↑ endopeptidases break peptide bonds in the interior of proteins

Figure 13.17   The structure and function of the villus

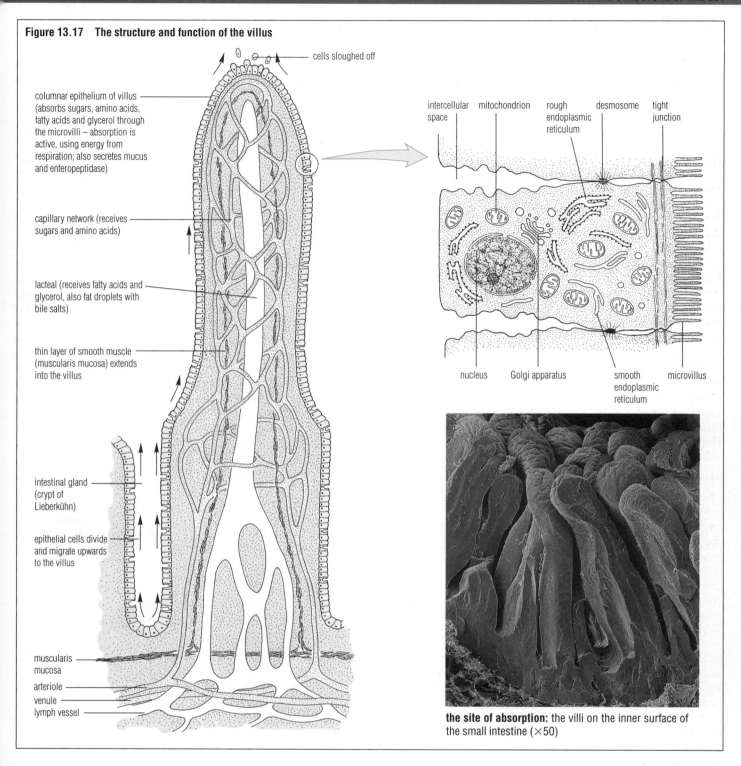

cells sloughed off

columnar epithelium of villus (absorbs sugars, amino acids, fatty acids and glycerol through the microvilli – absorption is active, using energy from respiration; also secretes mucus and enteropeptidase)

capillary network (receives sugars and amino acids)

lacteal (receives fatty acids and glycerol, also fat droplets with bile salts)

thin layer of smooth muscle (muscularis mucosa) extends into the villus

intestinal gland (crypt of Lieberkühn)

epithelial cells divide and migrate upwards to the villus

muscularis mucosa

arteriole

venule

lymph vessel

intercellular space   mitochondrion   rough endoplasmic reticulum   desmosome   tight junction

nucleus   Golgi apparatus   smooth endoplasmic reticulum   microvillus

the site of absorption: the villi on the inner surface of the small intestine (×50)

■ Absorption of digested food

For the products of digestion to be absorbed into the body, the particles must by hydrolysed to molecules small enough to be transported across the cell membrane. These small molecules are absorbed as they make contact with the epithelial cells.

The process of absorption takes place in the small intestine: it starts in the duodenum and is completed in the ileum. The small intestine has a huge internal surface area over which nutrients can be absorbed, due to its great length, the vast numbers of villi projecting into the lumen and the thousands of microvilli in each epithelial cell (Figure 13.17). Epithelial cells expend energy in the active transport processes (p. 236). Within the villi are the lacteals (lymph capillaries, part of the lymphatic system, p. 354) and blood capillaries, which pick up the absorbed food substances and transport them all over the body (Figure 13.19).

Questions

1   Why is it essential for digestive enzymes such as proteases to be secreted in an inactive state?

2   **a**   Why is digestion an essential stage in feeding?

   b   Name two essential constituents of food that are not digested.

Absorption of carbohydrate

Carbohydrate is absorbed in the form of monosaccharide sugar. Carbohydrate digestion is largely completed in the lumen of the gut, but some digestion of the disaccharides maltose, sucrose and lactose is completed in the absorption process; the enzymes maltase, sucrase and lactase are components of the membrane protein of the microvilli. As soon as the disaccharide forms a complex with the enzyme, hydrolysis occurs, and movement of the enzyme across the fluid mosaic membrane (p. 159) discharges the monosaccharide molecules into the cell interior (Figure 13.18).

The uptake of sugar across the cell membrane is an active process, probably linked with the transport of sodium ions (p. 238). All sugars are taken up into the blood capillaries and transported to the liver via the hepatic portal vein.

Absorption of protein

Protein is absorbed as di- and tri-peptides. Uptake is an active process involving a carrier protein molecule in the membrane of the microvilli; there are different carrier proteins for many of the peptides. Peptides transported across the membrane as such are then hydrolysed to free amino acids in the cytoplasm. Also located on the surface of the microvilli are many aminopeptidases, which complete digestion of short-chain peptides to the constituent amino acids as they transport them across the membrane (Figure 13.18).

Amino acids enter the blood capillary network of the villi and are transported to the liver via the hepatic portal vein.

Absorption of lipid

The products of lipid digestion are mostly absorbed into the lymphatic system, although some fatty acids and glycerol enter the blood capillaries and go directly to the liver. Most of the products of lipid digestion enter the lacteal vessel and are transported via the thoracic duct into the bloodstream in the region of the neck (Figure 13.19).

In addition, extremely small droplets of fats, when emulsified in the presence of the bile salts sodium glycocholate and sodium taurocholate, are sufficiently water-soluble to be taken up into the blood capillaries and carried to the liver as lipid. The bile salts are reclaimed by the liver cells and returned to the gall bladder in the bile for re-use in fat absorption.

Some water and inorganic salts, as well as all the vitamins, are also absorbed in the small intestine.

■ Assimilation of absorbed food

Sugars and amino acids are transported to the liver via the hepatic portal vein, where they may be further processed (Figure 13.19). The liver, the largest organ in the abdomen, plays a key part in the assimilation process. Most liver functions are concerned with the maintenance of a constant internal environment (homeostasis, p. 507).

Blood sugar level is maintained at a constant threshold level by the liver (p. 504). Glucose from the blood is used by cells and tissues to fuel tissue respiration. Excess glucose is converted to glycogen and stored in the liver and in the muscles round the body. The liver turns the glycogen back to glucose as and when the level of glucose in the blood falls below the threshold level. Any further glucose surplus to immediate requirements is metabolised to yield fats, which are then stored in adipose tissue.

Figure 13.18 Digestion and absorption in epithelial cells of the villi

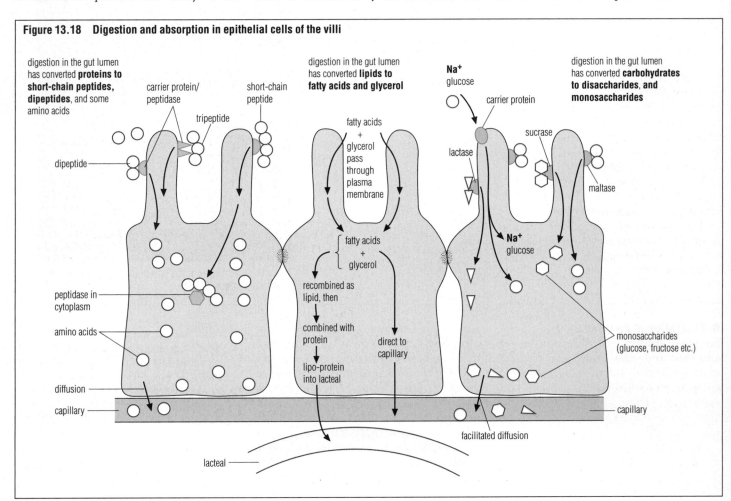

Table 13.4 Digestive juices of the human gut: a summary

Secretion	Site of production	Site of action	Trigger for secretion	Enzymes and other secretions	Substrate	Products	Comments/Other actions of the digestive juice
saliva (slightly alkaline)	salivary glands	mouth cavity	expectation and reflex stimulation	amylase mucus hydrogencarbonate ions lysozyme	starch	maltose	makes food slippery; buffer action provides almost neutral medium for action of salivary amylase; attacks bacteria
gastric juice (distinctly acid)	stomach wall (gastric glands)	stomach	reflex stimulation, stretching of stomach wall, and hormone (gastrin)	water mucus HCl	pepsinogen	pepsin	softens food; prevents gastric juices from damaging the stomach wall; stops the action of salivary amylase and allows pepsin to work; kills many germs
				pepsin	protein	polypeptides	*(see below)
				rennin	caseinogen (soluble)	casein (insoluble)	
bile (alkaline)	liver (via gall bladder)	duodenum	reflex action and hormones	water bile pigments hydrogencarbonate ions bile salts	fats	fat droplets	waste material – excreted with faeces or absorbed and re-excreted later; buffer action stops pepsin, helps intestinal enzymes
pancreatic juice (alkaline)	pancreas	duodenum	hormones	water hydrogencarbonate ions			
				amylase	starch	maltose	buffer action as above
				trypsin	protein	polypeptides	*endopeptidases break bonds between amino acid residues in the interior of proteins; exopeptidases break bonds attaching terminal amino acid residues.
				peptidases	polypeptides	amino acids	
				lipase	fats	fatty acids, glycerol	
				nucleases	nucleic acids	nucleotides	
intestinal juice	duodenal glands	small intestine	contact with intestinal lining	water mucus enteropeptidase			protects intestinal mucosa activates trypsinogen forming trypsin; trypsin then activates more trypsinogen and chymotrypsinogen
	surface of villi	microvilli		peptidase	short-chain peptides	amino acids	*(see above)
				maltase	maltose	glucose	
				sucrase	sucrose	glucose, fructose	
				lactase	lactose	glucose, galactose	

Amino acids are used in the synthesis of new protein by being built up to make enzymes or incorporated into the cytoplasm, including the cell membranes. Proteins are used in cell growth and cell repair. Excess amino acids cannot be stored; they are deaminated in the liver (p. 508).

Fats largely by-pass the liver by entering the lymphatic system, which empties into large veins entering the heart (Figure 13.19). Fat stored in the subcutaneous adipose tissue is the major energy reserve of the body. Fats and lipids required in metabolism are converted back to fatty acids and glycerol and respired or metabolised to other substances (p. 318). Fat stored around the heart and kidneys and in the mesenteries of the abdomen has a protective function, cushioning the organs against physical damage.

■ The large intestine

The human large intestine consists of the caecum and appendix, colon and rectum, and is about 1.5 metres long. In nonruminant herbivorous mammals, such as the horse and rabbit, the **caecum** and **appendix** are relatively long structures and have an important role in cellulose

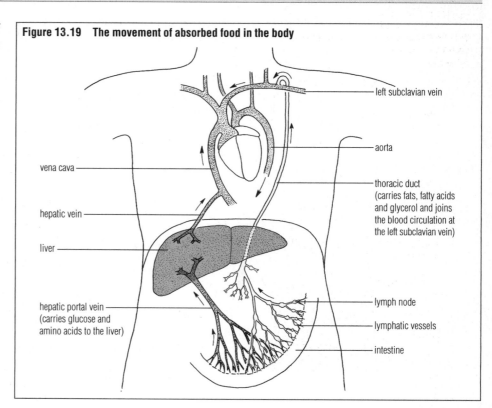

Figure 13.19   The movement of absorbed food in the body

left subclavian vein

aorta

thoracic duct (carries fats, fatty acids and glycerol and joins the blood circulation at the left subclavian vein)

vena cava

hepatic vein

liver

hepatic portal vein (carries glucose and amino acids to the liver)

lymph node

lymphatic vessels

intestine

digestion. In humans the caecum and appendix are short, and are perhaps inherited from cellulose-digesting pre-

human ancestors (p. 668). The appendix does have lymph nodes and may help to control intestinal pathogens.

287

The slurry entering the colon is an aqueous residue of undigested substances, such as cellulose and other fibrous plant materials, and water from the various copious digestive juices added, together with mucus secretions of the epithelium cells of the gut and water taken in as a component of the meal. It also contains dead mucosal cells and a huge microflora of bacteria and certain protozoa. The major component of this population of microorganisms is the bacterium *Escherichia coli* (p. 28). The bacterial flora of the colon synthesise biotin (a B vitamin) and vitamin K (Table 13.2, p. 272).

In the **colon**, water and mineral salts are absorbed. What remains in the large intestine is now referred to as **faeces**. Bacteria comprise about 50% of the faeces, whatever the diet. The remaining undigested solid material and dead epithelial cells nourish the large microflora of the colon. Decomposition products of bile pigments, added to the digesting food in the duodenum, colour the faeces uniformly.

The **rectum** is a short muscular tube that passes through the pelvis (p. 484) to terminate at the **anus**. It is closed by two sphincter muscles, the inner one consisting of smooth muscle under control of the autonomic nervous system and the outer one of striated muscle under voluntary nervous control. Discharge of faeces from the body is referred to as **defecation**. In babies defecation is a reflex action in response to distension of the rectum by faeces.

■ Nervous and hormonal control of secretion

Most digestive juice is secreted only when there is food in the gut. Coordination of secretion with the presence of food involves the nervous and endocrine systems.

1 Saliva

Food in the mouth in contact with the taste bud receptors of the tongue (p. 452) initiates a reflex action (p. 443). Sensory neurones carry impulses from these receptors to the brain. From the brain, impulses travel along motor neurones of the autonomic nervous system to the salivary glands (effector organs), stimulating the secretion of saliva. In addition we may similarly respond to the sight or smell of food that has been good to eat in the past. Alternatively, we may only need to imagine an excellent food item to find ourselves salivating! These latter cases are examples of conditioned reflexes; our nervous system has become conditioned to initiate salivation, so we respond in much the same way as Pavlov's dogs responded to the bell (p. 521).

2 Control of secretion of gastric juice

Mammals often feed intermittently, with periods when the stomach holds little or no food. The sight, taste or smell of food causes the brain to send nerve impulses via a branch of the vagus nerve (p. 464) to the stomach. This causes secretion of gastric juice rich in pepsinogen. The effect of the nervous stimulation persists for about an hour.

When a meal enters the stomach, the stretching effect on the walls stimulates further secretion of gastric juice. This is referred to as mechanical control. But the main effect of the presence of a meal is mediated by the hormone **gastrin**. The presence of food in the stomach causes certain cells in the stomach epithelium to secrete gastrin into the blood stream. Gastrin circulates in the body and on reaching the gastric glands causes further secretion of gastric juice (Figure 13.20).

---

3 Control of digestion in the small intestine

The stimulus of acid chyme entering the duodenum triggers the secretion of intestinal juice locally, and also the secretion of two hormones, **secretin** and **cholecystokinin**. These hormones circulate in the blood. On reaching the target organs they stimulate

▶ the secretion of pancreatic juice by the pancreas
▶ release of bile by the gall bladder
▶ the formation of pancreatic juice by the pancreas
▶ the formation of bile by the liver.

A third hormone, **enterogastrone**, is released at the same time. This inhibits the stomach lining from further release of acid.

Questions

1 What is absorbed into the bloodstream in
   a the small intestine
   b the colon
   from a meal containing a baked potato with butter and salt, a mixed fresh-vegetable salad and a meringue (egg white and sugar)?
2 What components of faeces are true excretory products (i.e. are waste products of the metabolism of cells)?

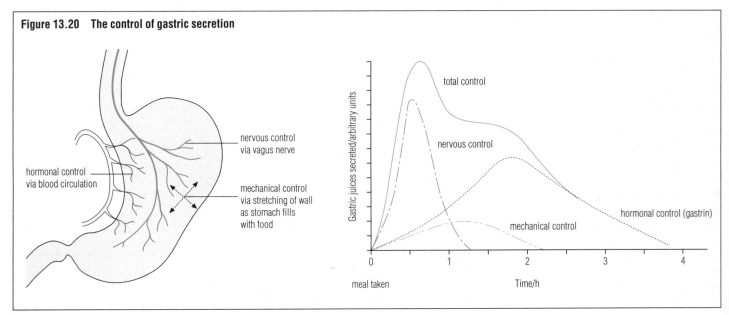

Figure 13.20   The control of gastric secretion

hormonal control via blood circulation

nervous control via vagus nerve

mechanical control via stretching of wall as stomach fills with food

Gastric juices secreted/arbitrary units

total control

nervous control

mechanical control

hormonal control (gastrin)

meal taken   Time/h

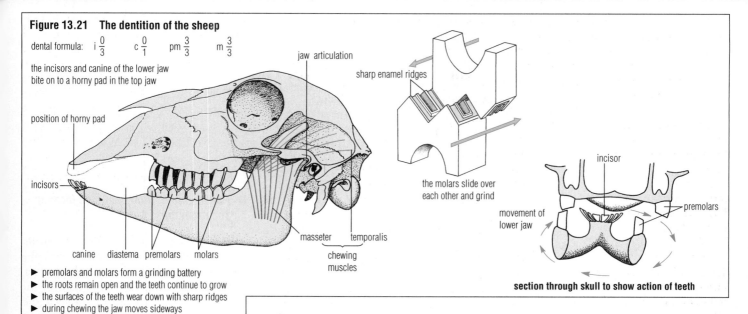

Figure 13.21   The dentition of the sheep

dental formula: i $\frac{0}{3}$ c $\frac{0}{1}$ pm $\frac{3}{3}$ m $\frac{3}{3}$

the incisors and canine of the lower jaw
bite on to a horny pad in the top jaw

position of horny pad

incisors

canine   diastema   premolars   molars

jaw articulation

sharp enamel ridges

the molars slide over
each other and grind

masseter   temporalis

chewing
muscles

movement of
lower jaw

incisor

premolars

section through skull to show action of teeth

▶ premolars and molars form a grinding battery
▶ the roots remain open and the teeth continue to grow
▶ the surfaces of the teeth wear down with sharp ridges
▶ during chewing the jaw moves sideways

■ Nutrition of herbivorous mammals

Herbivorous mammals crop vegetable matter using sharp-edged, opposing incisors (in certain animals, including cows and sheep, the incisors of the lower jaw cut against a horny pad on the upper jaw). Canine teeth are either absent or indistinguishable from incisors. A toothless space (**diastema**) separates the incisors from the premolars. Here the tongue operates, moving the freshly cut vegetable matter to the grinding surfaces of the premolar and molar teeth (known as the cheek teeth). The articulation of the lower jaw with the cranium allows a circular grinding action in a horizontal plane. The cheek teeth interlock, very much as the bottom of letter W fits the top of letter M (Figure 13.21). The grinding surfaces become worn down, exposing sharp-edged enamel ridges which further increase the efficiency of the grinding mill. The wear on a herbivore's teeth is considerable. As a consequence, the teeth of herbivores have open, unrestricted roots and continue to grow throughout life.

Mammals and cellulose digestion

Herbivorous mammals do not secrete cellulase enzymes. Rather, they exploit bacteria that are able to secrete cellulase enzymes. Such bacteria live in a specific region of the gut. In non-ruminant mammals, such as the rabbit and the horse, they live in the caecum (p. 287). In ruminant mammals such as the cow and the sheep the bacteria are housed in a specialised stomach which has four chambers (Figure 13.22).

Figure 13.22   The digestion of cellulose (plant cell walls) by ruminants

Ruminants such as cows and sheep have a four-chambered 'stomach', with three chambers derived from the lower part of the oesophagus (rumen → reticulum → omasum) and one chamber (the abomasum) that is the true stomach.

Cellulose digestion takes place in six stages:

1 In the mouth, grass cropped by the incisors/horny pad is ground up by the premolars and molars, swallowed with copious saliva and passed to the rumen.

2 The rumen functions as a fermentation vat. Cud is mixed with anaerobic cellulolytic bacteria, which break down cellulose to glucose, which in turn is fermented to organic acids. These fatty acids are absorbed into the blood through the rumen wall and are the major source of energy for ruminants. The fermentation produces carbon dioxide and methane, which are belched out. The rumen bacteria also form proteins from inorganic nitrogen (ammonium salts), and B-complex vitamins are also synthesised. The rumen also contains protozoa (ciliates) that feed on the bacteria.

3 The fermented grass passes to the reticulum and is formed into balls ('cud'), which are regurgitated to the mouth for further chewing. The cud is then swallowed and passed to the omasum.

4 In the omasum much water is re-absorbed from the cud (a cow secretes 100–190 dm^3 of digestive juices each day). The firmed-up remainder of the cud passes to the abomasum.

5 In the abomasum, normal gastric secretions begin to digest the proteins of grass and also of the bacteria and ciliates, including the proteins synthesised in the rumen (see **2**).

6 Chyme passes to the duodenum and then to the small intestine, where digestion is completed and the products of digestion are absorbed.

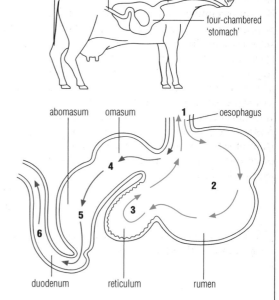

to small intestine   oesophagus

four-chambered
'stomach'

abomasum   omasum   oesophagus

duodenum   reticulum   rumen

FURTHER READING/WEB SITES

Food and diet: MAFF (1995) *Manual of Nutrition*. HMSO (10th edition). HMSO
A E Bender and D A Bender (1985) *Food Tables*. OUP

14 Variety in heterotrophic nutrition

- Heterotrophic organisms feed on ready-made organic compounds to obtain the nutrients they require, either by taking in food and digesting it (holozoic nutrition), or by digesting food outside the body (saprotrophic nutrition), or by living in or on a specific living food source for most of their life cycle (parasitic nutrition).
- Animals with holozoic nutrition feed on small particles, fluids or soft tissues, or larger particles.
- The majority of saprotrophs are fungi or bacteria, and their breaking down of dead organic matter in the environment is of great ecological significance.
- Parasites have to enter or hold onto their host tissues and resist attack by the host defence systems. Once established, they have a ready food supply, but require a mechanism of reaching and infecting new hosts.

14.1 INTRODUCTION

Heterotrophic organisms feed on complex, ready-made organic foods to obtain the nutrients they require. They depend for their nutrients on other organisms, dead or alive, or on the by-products of other organisms, such as their cell walls, food stores or faeces. There are three forms of heterotrophic nutrition: **holozoic**, **saprotrophic** and **parasitic**, as outlined in a general classification of nutrition shown in Figure 14.1.

■ Holozoic nutrition

In holozoic nutrition, illustrated by reference to humans and other mammals in Chapter 13, food is taken into the digestive system and broken down, and the useful products absorbed. There is a great variety of feeding mechanisms among holozoic animals. Methods of feeding have been classified on the relative size of food particles ingested (a classification introduced by the marine zoologist C M Yonge).

Animals that feed on relatively small particles such as unicellular or tiny multi-cellular organisms suspended in an aquatic environment are called **microphagous feeders**. Animals that feed on fluids or dissolved substances are called **fluid feeders**. Animals that feed on relatively large particles are called **macrophagous feeders**.

14.2 EXAMPLES OF MICROPHAGOUS FEEDERS

■ *Amoeba*: food vacuoles formed by pseudopodia

Amoeba proteus lives on the bottom of shallow lakes and ponds and feeds on microscopic organisms such as smaller protozoa and algae. Prey is enclosed in a food vacuole (with a little of the pond water) by the action of pseudopodia flowing around the food item. Lysosomes, which contain hydrolytic enzymes, fuse with the vacuole and the prey is digested there. Digestion is first acidic and then goes into an alkaline phase. The products of digestion are absorbed into the cytoplasm. Undigested remains are discharged at the surface of the *Amoeba* (Figure 14.2).

■ *Paramecium*: food vacuoles formed at cytostome

Paramecium caudatum lives in fresh-water ponds that contain decaying organic matter and numerous bacteria. Bacteria are the main component of the diet. The whole surface of *Paramecium* is covered by cilia. Rapid and coordinated beating movements of these cilia drive the slipper-shaped protozoan through the water, rotating it as it moves. Cilia of the oral groove generate a feeding current in the pond water and bacteria are selectively accumulated at the **cytostome** at the base of the oral groove. Here, food vacuoles are formed and circulate along a pathway through the cytoplasm. Digestive enzymes are delivered by lysosomes, which fuse with the food vacuoles. Investigations with prey impregnated with coloured pH indicators show that food vacuoles are first made acidic (pH 3), killing the prey, then alkaline (pH 8). The products of digestion are absorbed into the cytoplasm. Undigested remains are discarded at a definite point towards the posterior of the animal (Figure 14.3).

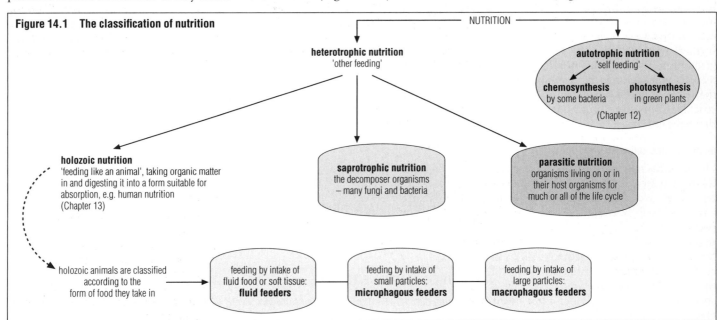

Figure 14.1 The classification of nutrition

NUTRITION

heterotrophic nutrition
'other feeding'

autotrophic nutrition
'self feeding'

chemosynthesis
by some bacteria

photosynthesis
in green plants

(Chapter 12)

holozoic nutrition
'feeding like an animal', taking organic matter in and digesting it into a form suitable for absorption, e.g. human nutrition
(Chapter 13)

saprotrophic nutrition
the decomposer organisms – many fungi and bacteria

parasitic nutrition
organisms living on or in their host organisms for much or all of the life cycle

holozoic animals are classified according to the form of food they take in

feeding by intake of fluid food or soft tissue: **fluid feeders**

feeding by intake of small particles: **microphagous feeders**

feeding by intake of large particles: **macrophagous feeders**

Figure 14.2 Nutrition in *Amoeba*

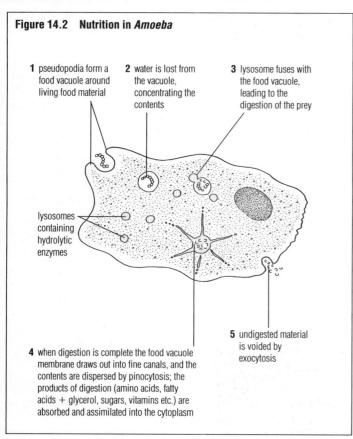

1 pseudopodia form a food vacuole around living food material

2 water is lost from the vacuole, concentrating the contents

3 lysosome fuses with the food vacuole, leading to the digestion of the prey

lysosomes containing hydrolytic enzymes

5 undigested material is voided by exocytosis

4 when digestion is complete the food vacuole membrane draws out into fine canals, and the contents are dispersed by pinocytosis; the products of digestion (amino acids, fatty acids + glycerol, sugars, vitamins etc.) are absorbed and assimilated into the cytoplasm

Figure 14.3 Nutrition in *Paramecium*

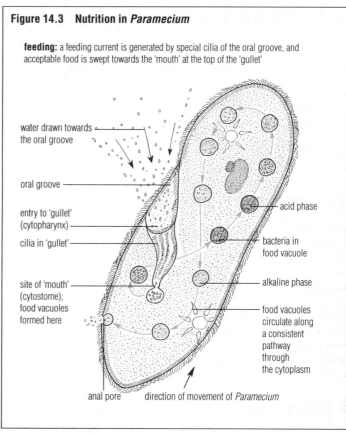

feeding: a feeding current is generated by special cilia of the oral groove, and acceptable food is swept towards the 'mouth' at the top of the 'gullet'

water drawn towards the oral groove

oral groove

entry to 'gullet' (cytopharynx)

cilia in 'gullet'

site of 'mouth' (cytostome); food vacuoles formed here

acid phase

bacteria in food vacuole

alkaline phase

food vacuoles circulate along a consistent pathway through the cytoplasm

anal pore   direction of movement of *Paramecium*

Figure 14.4 Nutrition in *Mytilus*

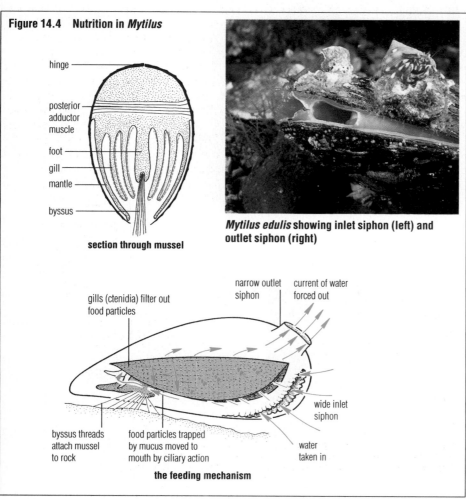

hinge

posterior adductor muscle

foot

gill

mantle

byssus

section through mussel

Mytilus edulis showing inlet siphon (left) and outlet siphon (right)

narrow outlet siphon

current of water forced out

gills (ctenidia) filter out food particles

byssus threads attach mussel to rock

food particles trapped by mucus moved to mouth by ciliary action

wide inlet siphon

water taken in

the feeding mechanism

■ *Mytilus*: filter feeding by cilia

Mytilus edulis, the common mussel, lives on the middle shore (Figure 3.8, p. 48), attached by special threads (**byssus threads**) to rocks and breakwater structures. A shell of two valves provides protection; the valves close at low tide, retaining sea water over the gills. *Mytilus* is a **filter feeder**; it extracts plankton from the respiratory water current. The water is drawn in over the gills by the action of cilia that line the gill surfaces. Plankton organisms are trapped in a stream of mucus secreted on the gills, and are directed towards the mouth where they are ingested (Figure 14.4).

Questions

1 Lysosomes are 'packages' of hydrolytic enzymes.

  a What contains the packages?

  b Where do lysosomes originate in the eukaryotic cell (p. 162)?

2 In what ways are the feeding methods of *Amoeba* and *Paramecium* similar? In what ways are they different?

14.3 EXAMPLES OF FLUID AND SOFT-TISSUE FEEDERS

■ Aphids: sucking sap from the phloem

Insects have no internal jaws, but the limbs of the head segments have evolved as external mouthparts. The aphids ('greenfly' and 'blackfly' are examples) are insects that possess piercing mouthparts adapted for sucking juices of plants. When not in use the mouthparts (known as the **proboscis**) are held horizontally below the thorax. The proboscis consists of long, thin **mandibles** and **maxillae** sheathed in the **labium**. In use, the tip of the labium guides the proboscis to a suitable feeding site (Figure 16.34 illustrates how aphids are exploited in phloem research). Then the saw-edged mandibles make an incision and the slender, needle-like maxillae follow into the wound. The maxillae (secondary jaws) house two canals; saliva from the insect is pumped down one (the saliva canal), and food is drawn up the other (the food canal) (Figure 14.5).

■ Worker honeybees: feeding on nectar and pollen

In the honeybee's mouthparts the labium is formed into a long sucking tube ('tongue') sheathed by other mouthparts. The tongue is used for sucking nectar from flowers (p. 570). Nectar is drawn into the honey sac, from where it can be regurgitated for the benefit of the bee community, back in the hive. When not in use the mouthparts are folded back under the head, leaving the mandibles free for chewing pollen, manipulating wax and attacking intruders.

The other item of a bee's diet is pollen. This adheres to the hairy body, and can be collected by combing actions of the legs. Pollen is carried back to the hive as pellets attached to the 'pollen baskets' of the hind legs (Figure 14.6).

■ Butterflies: sucking nectar from flowers

In butterflies the mouthparts form a sucking tube called the proboscis, by which nectar is obtained from flowers (Figure 14.7). The proboscis consists of two grooved maxillae hooked together to form a feeding channel. When the proboscis is not in use, contraction of muscles inside each half of the tube coils up the proboscis below the head.

■ Mosquitoes: sucking blood

Female mosquitoes are blood-sucking insects (Figure 14.8). They are responsible for the transmission of debilitating diseases such as malaria (p. 535), yellow fever and elephantiasis in tropical countries. The mouthparts consist of needle-like mandibles and maxillae (piercing organs), together with a grooved **labrum**. The labrum, with the tongue-like **hypopharynx** that lies behind it, forms the food channel through which blood is taken up. (The hypopharynx is not a 'true' mouthpart but it becomes of importance in the feeding processes of the mosquito and certain other liquid-feeding insects.) A tube within the hypopharynx delivers saliva, a secretion

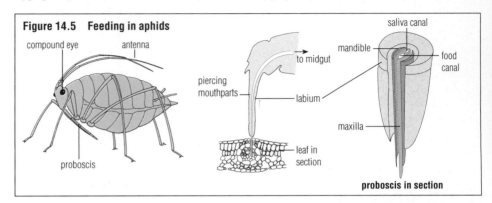

Figure 14.5 Feeding in aphids

compound eye — antenna — piercing mouthparts — labium — leaf in section — saliva canal — mandible — food canal — maxilla — proboscis — to midgut

proboscis in section

Figure 14.6 Feeding in worker honeybees

worker bees collect nectar in the honey sac and pollen in the pollen baskets on the hind legs before returning to the hive

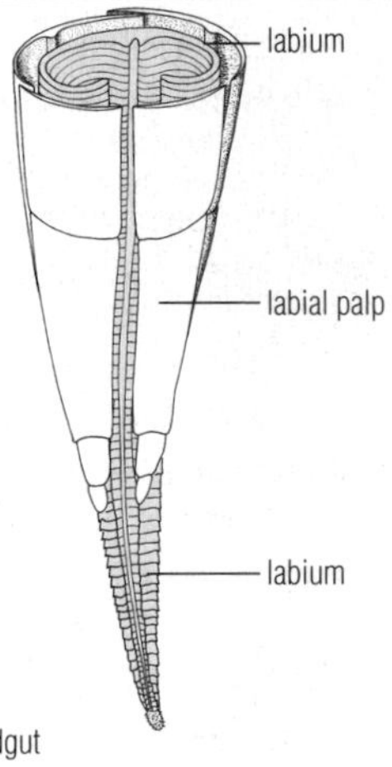

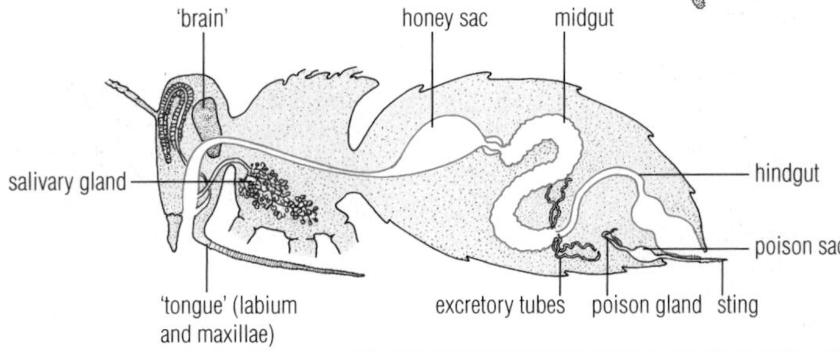

labium — labial palp — labium — 'brain' — salivary gland — 'tongue' (labium and maxillae) — excretory tubes — poison gland — sting — honey sac — midgut — hindgut — poison sac

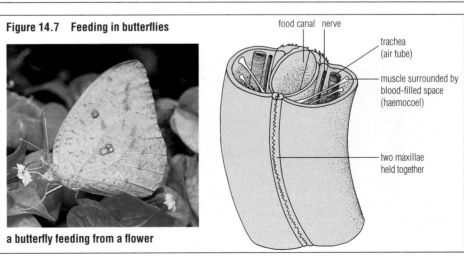

Figure 14.7 Feeding in butterflies

food canal — nerve — trachea (air tube) — muscle surrounded by blood-filled space (haemocoel) — two maxillae held together

a butterfly feeding from a flower

Figure 14.8 Feeding in the mosquito

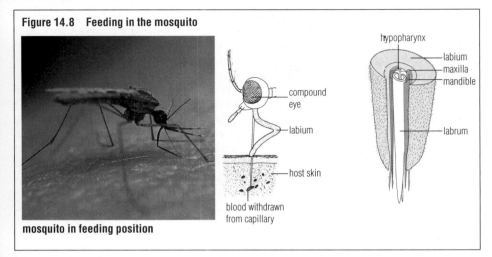

mosquito in feeding position

Labels: hypopharynx, labium, maxilla, mandible, labrum, compound eye, labium, host skin, blood withdrawn from capillary

Figure 14.9 Feeding in the garden spider (*Araneus* sp.)

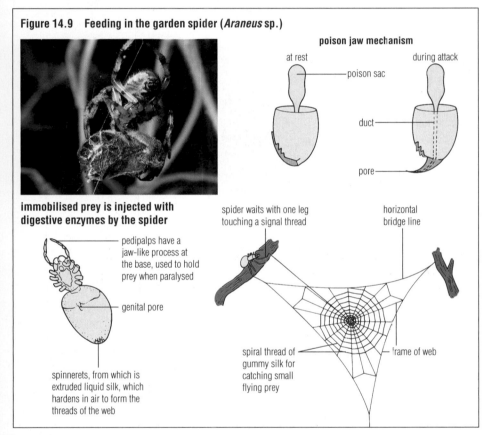

immobilised prey is injected with digestive enzymes by the spider

poison jaw mechanism

at rest — poison sac — during attack

duct

pore

pedipalps have a jaw-like process at the base, used to hold prey when paralysed

genital pore

spinnerets, from which is extruded liquid silk, which hardens in air to form the threads of the web

spider waits with one leg touching a signal thread

horizontal bridge line

spiral thread of gummy silk for catching small flying prey

frame of web

Figure 14.10 Feeding in the housefly

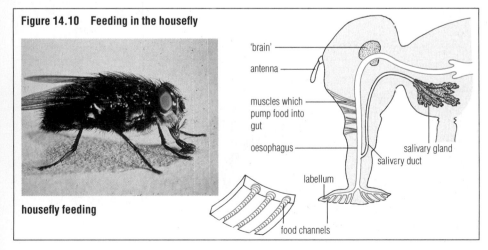

housefly feeding

Labels: 'brain', antenna, muscles which pump food into gut, oesophagus, salivary gland, salivary duct, labellum, food channels

---

that prevents the host blood from clotting (p. 364). All these structures are carried in a groove in the labium. The male mosquito lacks the piercing organs and it feeds on plant juices. In Britain the common blood-sucking 'mosquito' is *Culex*, which rarely bites humans.

■ Spiders: liquidising the contents of prey

The garden spider (*Araneus*) feeds on insects that become trapped in its web (Figure 14.9). Silk glands in the abdomen discharge liquid silk (a protein) via spinnerets. The silk quickly hardens. The silk of the supporting framework of the web is dry, but that forming the spiral web is glue-like. Flying insects become trapped in the web. The spider injects the struggling prey with poison, and wraps it in silk. It then injects the prey with digestive enzymes, and sucks up its liquidised body structures. Finally it cuts loose the empty body of the prey.

■ Housefly: mopping up liquid food

The fly (*Musca domestica*) has sucking mouthparts, or proboscis, formed from a modified labium (Figure 14.10). At the tip of the proboscis are two lobes (**labella**), each containing numerous channels. When the fly feeds it places its proboscis on the food. Liquids such as milk are sucked up into the gut by muscle action. Solid food is first digested and made soluble by saliva secreted from the salivary glands. The housefly's catholic taste in organic matter (human food, rotting organic matter including human faeces), together with its habit of walking over its food, accounts for its role in the transmission of infectious diseases of the bowel, paratyphoid and typhoid fevers, diarrhoea of babies, 'food poisoning' due to *Salmonella* and poliomyelitis.

Questions

1 How is the feeding mechanism of aphids exploited in research into phloem transport?
2 a In what ways are the feeding habits of bees and butterflies advantageous to the plants they visit?
   b How do flowers favoured by bees differ from those normally chosen by butterflies?
3 State which of the above methods of feeding might be classified as
   a potentially harmful to humans
   b potentially useful to humans
   c neither harmful or beneficial.
4 In what ways are the feeding methods of aphids and mosquitoes similar? In what ways are they different?

14.4 EXAMPLES OF MACROPHAGOUS FEEDERS

■ *Hydra*: trapping mobile aquatic prey

Hydra lives in freshwater ponds and streams, attached to weeds. It feeds on small aquatic animals such as *Cyclops* and *Daphnia* (p. 38). Prey moving close to the waving tentacles may become trapped and immobilised by the action of poison cells (**nematoblasts**) on the tentacles. Coordinated movements of the tentacles cause the prey to be pushed, intact and whole, through the mouth into the **enteron** (Figure 14.11). Digestion begins in the enteron through the action of digestive enzymes secreted by cells of the endoderm, initially working on the external surface of the prey. Digestion is completed intracellularly when particles of the prey become dislodged and are taken into food vacuoles in endoderm cells. Undigested food is discharged through the mouth.

■ The garden snail: rasping and scraping plant food

The garden snail (*Helix aspersa*) is herbivorous, and feeds on a variety of plant material. The mouth opens into a buccal cavity containing a special rasping organ, the **radula**, and an opposing tough jaw plate. The radula is made of hardened but flexible protein and chitin, and bears backwardly pointing teeth. The radula is reinforced by a cartilage-like support (**odontophore**), and the radula and its support are moved by an array of muscles. The whole is referred to as the **buccal mass**.

In feeding, the buccal mass is extended through the mouth and used to rasp and scrape plant material into shreds (Figure 14.12). Food particles, consisting of partly shredded cells, are taken into the buccal cavity and then passed down the oesophagus to the rest of the gut, where they are digested and the products absorbed.

■ The grasshopper: biting and chewing plant food

Grasshoppers feed largely on leafy vegetation, which needs to be cut and ground up before digestion. In the grasshopper the maxillae hold the food while it is cut up by the mandibles before being passed into the mouth.

In front of the mouthparts the labrum, an 'upper lip' plate, helps grip food for chewing. Behind the labrum are the paired mandibles, the cutting apparatus.

Behind the mandibles is a pair of maxillae, the palps of which are sensory, and concerned with sensing the acceptability of food. The labium, a 'lower lip', performs a similar function to the maxillae. The labial palps assist in moving the ground-up food into the mouth.

In the grasshopper and related insects, mechanical digestion continues in the gizzard. It is followed by enzymic digestion in the gut by enzymes secreted by the gastric caeca and the stomach (Figure 14.13).

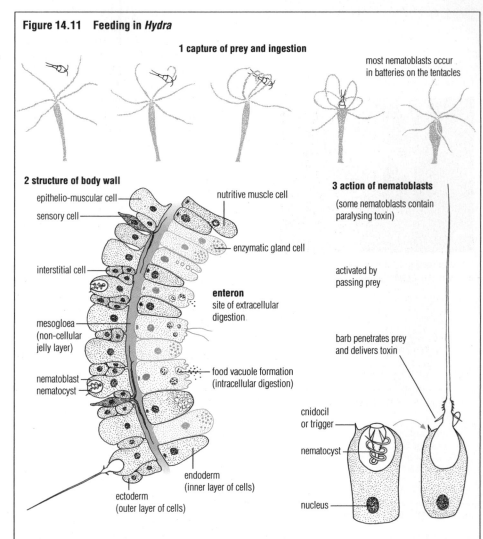

Figure 14.11   Feeding in *Hydra*

1 capture of prey and ingestion

most nematoblasts occur in batteries on the tentacles

2 structure of body wall

epithelio-muscular cell
sensory cell
interstitial cell
mesogloea (non-cellular jelly layer)
nematoblast
nematocyst
ectoderm (outer layer of cells)
nutritive muscle cell
enzymatic gland cell
enteron
site of extracellular digestion
food vacuole formation (intracellular digestion)
endoderm (inner layer of cells)

3 action of nematoblasts
(some nematoblasts contain paralysing toxin)

activated by passing prey

barb penetrates prey and delivers toxin

cnidocil or trigger
nematocyst
nucleus

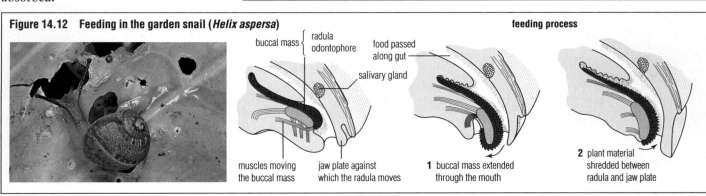

Figure 14.12   Feeding in the garden snail (*Helix aspersa*)

feeding process

buccal mass
radula
odontophore
food passed along gut
salivary gland

muscles moving the buccal mass
jaw plate against which the radula moves
1 buccal mass extended through the mouth
2 plant material shredded between radula and jaw plate

Figure 14.13   Feeding in the field grasshopper (*Chorthippus* sp.)

the adult grasshopper, showing the mouthparts

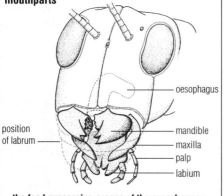

position of labrum —

— oesophagus

— mandible
— maxilla
— palp
— labium

the food-processing organs of the grasshopper

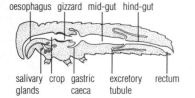

oesophagus gizzard mid-gut hind-gut

salivary crop gastric excretory rectum
glands caeca tubule

The earthworm: feeding on dead organic matter in soil

The earthworm is a terrestrial, soil-living animal. Earthworms feed on organic matter, and on the saprotrophic fungi and bacteria in decaying matter found in soil. Some earthworms eat vegetable matter such as dead leaves, which they drag down from the surface of the soil at night. Food is drawn into the gut by the action of the muscular pharynx. Mucus is secreted, lubricating the movement of soil and food through the long, tubular gut. The regions of the gut and their roles in digestion are shown in Figure 14.14. Most digestion occurs in the intestine, the surface of which is increased by a dorsal longitudinal fold (**typhlosole**) that protrudes into the gut. After digestion, food is absorbed into the blood system. The bulk of the soil passes through the worm and in some species is deposited as worm casts at the soil surface.

The frog: prey capture by protrusible tongue

The adult frog (*Rana temporaria*) is a carnivore and feeds on a wide range of small animals. (Tadpoles, however, are herbivorous at most stages of their development.) The tongue is attached at the front of the mouth, immediately behind the lower jaw. When a suitable prey passes close to the frog the mouth is opened and the tongue flicked out by muscular contraction. If the prey becomes attached by the sticky secretion that covers the tongue then it is brought

Figure 14.15   Frog catching prey: adult frog (*Rana temporaria*) catches a stationary insect with its sticky tongue

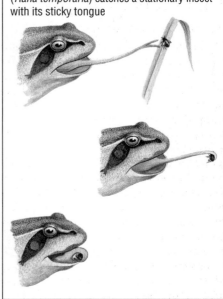

to the mouth when the tongue returns (Figure 14.15). This movement is very rapid, and is best observed by high-speed photography. The frog's teeth are for holding prey rather than for chewing. The prey is quickly swallowed.

Figure 14.14   Feeding by the earthworm (*Lumbriscus* sp.)

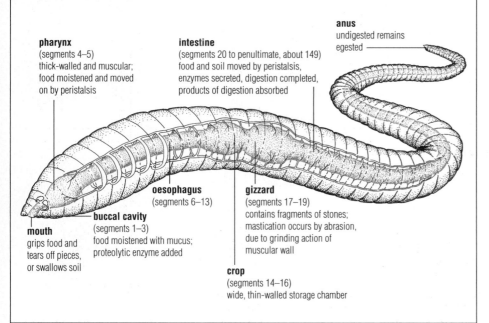

pharynx
(segments 4–5)
thick-walled and muscular; food moistened and moved on by peristalsis

intestine
(segments 20 to penultimate, about 149)
food and soil moved by peristalsis, enzymes secreted, digestion completed, products of digestion absorbed

anus
undigested remains egested

oesophagus
(segments 6–13)

mouth
grips food and tears off pieces, or swallows soil

buccal cavity
(segments 1–3)
food moistened with mucus; proteolytic enzyme added

gizzard
(segments 17–19)
contains fragments of stones; mastication occurs by abrasion, due to grinding action of muscular wall

crop
(segments 14–16)
wide, thin-walled storage chamber

Questions

1   Digestion in *Hydra* is completed in food vacuoles in endoderm cells lining the enteron. How might the products of digestion reach the other cells of *Hydra*?

2   In the garden snail, how does the shredding of plant cells by a radula working against the jaw plate aid the processes of digestion in the gut?

3   What is the value to the grasshopper of the sensory palps that form part of the mouthparts?

4   How are the feeding and burrowing actions of earthworms advantageous in the formation of soils (p. 60)?

5   In what respects is digestion in *Amoeba* and *Hydra* similar?

6   In an aquarium, pond snails help to keep the inside of the glass free from algae. How might their feeding apparatus enable them to do this?

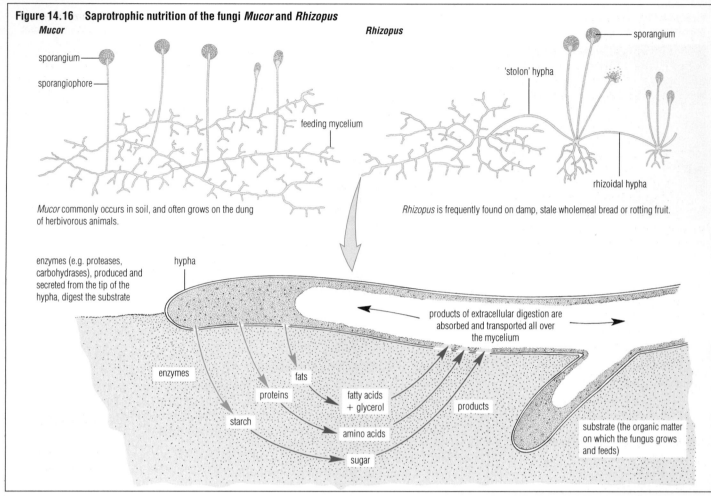

Figure 14.16 Saprotrophic nutrition of the fungi *Mucor* and *Rhizopus*

Mucor

Rhizopus

sporangium

sporangiophore

feeding mycelium

sporangium

'stolon' hypha

rhizoidal hypha

Mucor commonly occurs in soil, and often grows on the dung of herbivorous animals.

Rhizopus is frequently found on damp, stale wholemeal bread or rotting fruit.

enzymes (e.g. proteases, carbohydrases), produced and secreted from the tip of the hypha, digest the substrate

hypha

products of extracellular digestion are absorbed and transported all over the mycelium

enzymes

fats

proteins

fatty acids + glycerol

products

starch

amino acids

substrate (the organic matter on which the fungus grows and feeds)

sugar

14.5 SAPROTROPHIC NUTRITION

Saprotrophs feed on the dead and decaying matter on which they live and grow. Saprotrophs secrete enzymes onto the food source, and digestion is external to the organism (extracellular). The products of digestion are re-absorbed. The vast majority of bacteria and fungi are saprotrophic. The feeding mechanism of the saprotrophic fungi *Mucor* and *Rhizopus* is illustrated in Figure 14.16. The nutrients and the conditions of the physical environment required by *Mucor* and *Rhizopus* are defined in Figure 14.17.

■ Ecological significance of saprotrophic nutrition

Saprotrophs are of great ecological importance. Organic materials of dead organisms and their waste products are broken down during saprotrophic feeding and the component chemical elements are eventually released for re-use by autotrophs. The general role of saprotrophs (in the form of detritivores) in the cycling of nutrients in the biosphere is summarised in Figure 3.48 (p. 72).

Figure 14.17 Nutrients required by saprotrophic fungi: for successful growth the fungus requires nutrients (○) and favourable conditions (★)

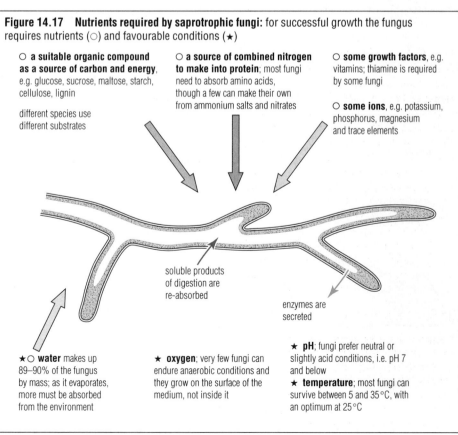

○ **a suitable organic compound as a source of carbon and energy**, e.g. glucose, sucrose, maltose, starch, cellulose, lignin

different species use different substrates

○ **a source of combined nitrogen to make into protein**; most fungi need to absorb amino acids, though a few can make their own from ammonium salts and nitrates

○ **some growth factors**, e.g. vitamins; thiamine is required by some fungi

○ **some ions**, e.g. potassium, phosphorus, magnesium and trace elements

soluble products of digestion are re-absorbed

enzymes are secreted

★ **water** makes up 89–90% of the fungus by mass; as it evaporates, more must be absorbed from the environment

★ **oxygen**; very few fungi can endure anaerobic conditions and they grow on the surface of the medium, not inside it

★ **pH**; fungi prefer neutral or slightly acid conditions, i.e. pH 7 and below

★ **temperature**; most fungi can survive between 5 and 35 °C, with an optimum at 25 °C

■ Recycling by coprophilous fungi

The dung of herbivores (such as rabbits, horses and cows) contains a natural flora of fungi, bacteria and their spores. When fresh dung is incubated in moist conditions fungal fruiting bodies appear. These fruiting bodies disperse spores towards the light, often by a violent or explosive mechanism. Successful spore dispersal requires that the spores reach surrounding vegetation that may be eaten by other herbivores. The succession of fungi that feed on herbivorous dung is illustrated in Figure 14.18.

The first organisms to flourish on fresh dung are fast-growing bacteria and fungi that absorb the small quantities of sugar, amino acid and fatty acid present. These substances are leftovers from digestion that have not been absorbed in the herbivore's gut. The remaining organisms in the succession specialise in digesting the increasingly insoluble and resistant organic materials.

Organisms that feed on dung are described as **coprophilous**.

■ Recycling in garden compost

A succession of microorganisms flourishes in disposed organic matter. In nature, soil fertility is maintained by the slow decomposition of dead plant and animal remains at or near the soil surface. In the garden the process can be speeded up by maintaining a large, compact heap of moist, finely chopped plant material within an insulated container (Figure 14.19). In heaped garden waste (compost) the heat generated in the rotting materials by the respiration of the microorganisms present has three important effects. The raised temperature speeds up the decay processes, it selectively favours organisms adapted to survive at higher temperatures, and it may even be high enough to kill the weed seeds present in the rotting vegetation. Well-rotted compost can be dug into the soil, where further decay releases the ions essential for plant growth (p. 328).

Discussion point

Mown meadow grass cut to make silage and freshly cut lawn mowings that are put to compost are of very similar composition, yet silage (p. 108) and compost are totally different products. What are the special conditions that are maintained in a well-managed silage heap that prevent the rotting of grass?

Figure 14.18   A typical succession of fruiting bodies of coprophilous fungi on herbivorous dung

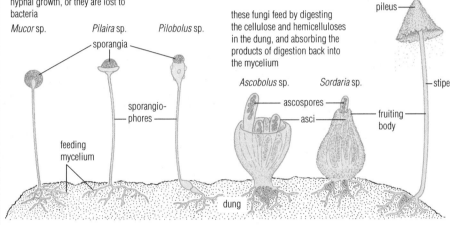

the first fruiting bodies to appear on the surface of the dung are members of the plant-like and conjugation fungi (Phycomycetes), such as *Mucor* sp., *Pilaira* sp. and *Pilobolus* sp.; they appear in one to seven days, and last for up to fourteen days

fruiting bodies of sac fungi (Ascomycetes), such as *Ascobolus* sp. and *Sordaria* sp., appear after five to six days and may persist for four to six weeks

finally the fruiting bodies of the club fungi (Basidiomycetes), such as *Coprinus* sp., appear normally after one or two weeks; they persist until the spores have been dispersed

these fungi absorb any remaining sugars and amino acids and digest starch, fats and proteins in the dung; these relatively short-lived food sources must be exploited quickly by hyphal growth, or they are lost to bacteria

these fungi feed by digesting the lignified (woody) remains of fibres and xylem vessels in the dung

these fungi feed by digesting the cellulose and hemicelluloses in the dung, and absorbing the products of digestion back into the mycelium

Figure 14.19   Garden compost

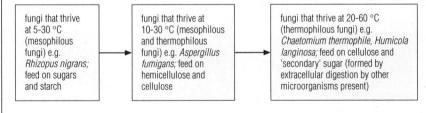

a sequence of microorganisms is involved in aerobic decay within compost

fungi that thrive at 5-30 °C (mesophilous fungi) e.g. *Rhizopus nigrans*; feed on sugars and starch

→ fungi that thrive at 10-30 °C (mesophilous and thermophilous fungi) e.g. *Aspergillus fumigans*; feed on hemicellulose and cellulose

→ fungi that thrive at 20-60 °C (thermophilous fungi) e.g. *Chaetomium thermophile*, *Humicola langinosa*; feed on cellulose and 'secondary' sugar (formed by extracellular digestion by other microorganisms present)

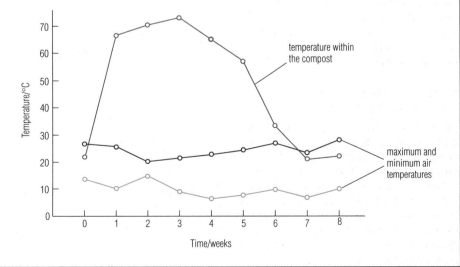

the variation of temperature within a compost heap

temperature within the compost

maximum and minimum air temperatures

■ Yeast, an economically important saprotroph

Yeasts (Figure 14.20) are saprotrophic unicellular fungi, commonly occurring on the surface of plants and animals, on animal mucous membranes and in soil and dung. Yeasts feed on sugars, and they produce ethanol as a waste product of anaerobic respiration. The ethanol formed is toxic to living cells, including the yeast cells. Consequently the final concentration of ethanol rarely exceeds 10%, although certain mutant strains may produce a final concentration as high as 12%. In these cases the yeast cells are killed off by 10% ethanol but the enzymes they have secreted into the medium remain active, and the enzyme-catalysed conversion of sugar to ethanol continues.

Yeasts are of economic importance in traditional industries of brewing (Figure 14.21), wine-making and baking, and in the manufacture of 'meat' and vegetable extracts. Yeasts are also of increasing importance in biotechnology – for example, they are essential in the production of ethanol from sugar cane ('gasohol', p. 118). Yeasts have also been genetically engineered (p. 635) to produce quantities of human enzymes, and also a vaccine against hepatitis.

Figure 14.20 Yeast

a thin section of a yeast cell

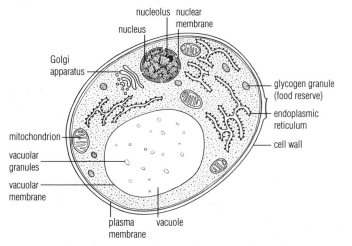

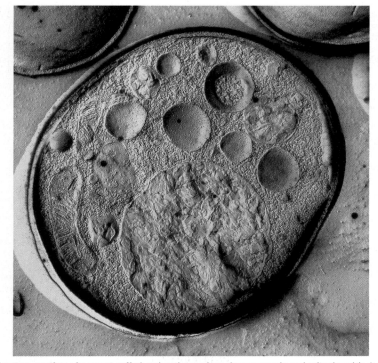

freeze-fracture section of a yeast cell showing the nucleus, large vacuole and mitochondria (×15 000)

Figure 14.21 The role of yeast in brewing

Malting
Barley grains are soaked for 2–3 days, drained, and incubated for 10 days at 13–17 °C. Roots and shoots start to grow. Starch reserves are mobilised as sugar. (This process is necessary because yeast does not produce amylase and therefore cannot hydrolyse starch to sugar.) The temperature is then raised to between 40 and 70 °C to halt germination.

Cracking
Grains are lightly roasted to 80 °C and then passed between rollers, which crack them open.

Mashing
Hot water (62–68 °C) is used to wash the sugars and amino acids from the grains. Left behind from this process are brewer's grains, used in the feeding of milking herds of cows.

Boiling
The liquor from the mashing process, known as wort, is boiled for several hours to concentrate it. Dried hops are added for flavour and for their antimicrobial properties. Cooling follows.

Fermentation
Yeast (*Saccaromyces cerevisiae* or *S. carlsbergensis*) is now added, and converts sugar to ethanol and carbon dioxide over a period of 2–5 days. As the yeast grows it forms a thick head, which is skimmed off the surface. This is sold as a cake or as a slurry (e.g. in yeast extract manufacture). The carbon dioxide produced is solidified at low temperature and under high pressure and sold as 'dry ice'.

fermentation occurs in a large, deep tank

Conditioning
The beer should be 4–8% ethanol (10% in barley wine). It is stored in barrels to allow final fermentation and for clearing of the yeast.

Under modern quality control and marketing procedures it may be filtered, pasteurised, standardised (brought to a consistent colour and flavour) and canned.

14.6 PARASITIC NUTRITION

Parasites (p. 66) are organisms that live on (**ectoparasite**) or in (**endoparasite**) another organism (the **host**) for all or the greater part of the life cycle, and obtain nourishment at the expense of the host. A wide range of organisms obtains nutrients in this way. An example of an ectoparasite of humans is the body louse (Figure 3.38, p. 66) which feeds by piercing the skin and sucking blood. The louse has limbs adapted for secure attachment to the body of the host, and mouthparts adapted to bite through the stratified epithelium of the epidermis (p. 511).

■ The pork tapeworm (*Taenia solium*)

The pork tapeworm is an endoparasite of the human gut. The adult tapeworm is a ribbon-like platyhelminth (p. 35) up to ten metres long. Most of the tapeworm's body consists of a linear series of segments known as **proglottids**. The anterior part is a 'head' or **scolex**, a small muscular knob bearing suckers and a double row of curved hooks and thus well adapted for attachment to the intestinal wall. Behind the scolex is a narrow neck region where the proglottids are formed. The body of the tapeworm defends itself from the digestive enzymes of the host by the secretion of large quantities of mucus and by the local inactivation of enzymes by inhibitor substances at the surface of the proglottids.

The life cycle of the pork tapeworm is illustrated in Figure 14.22. The secondary host, the pig, may become infected if it feeds in drainage channels contaminated by human faeces, for example. Humans may become infected by eating undercooked infected pigmeat.

Each proglottid contains a hermaphrodite set of reproductive organs; it is assumed that self-fertilisation occurs, since the infected human host does not support more than one tapeworm. After fertilisation most internal structures of the proglottid break down, leaving an enlarged uterus packed with tiny hooked embryos inside egg capsules.

Larval and adult stages of *Taenia* obtain nutrients from the part of the host body in which they shelter. The proglottids have no feeding or digestive organs (although guts are found in free-living species of the platyhelminths); the surface of the proglottids is specialised for the active, selective absorption of the digested food available in the human small intestine.

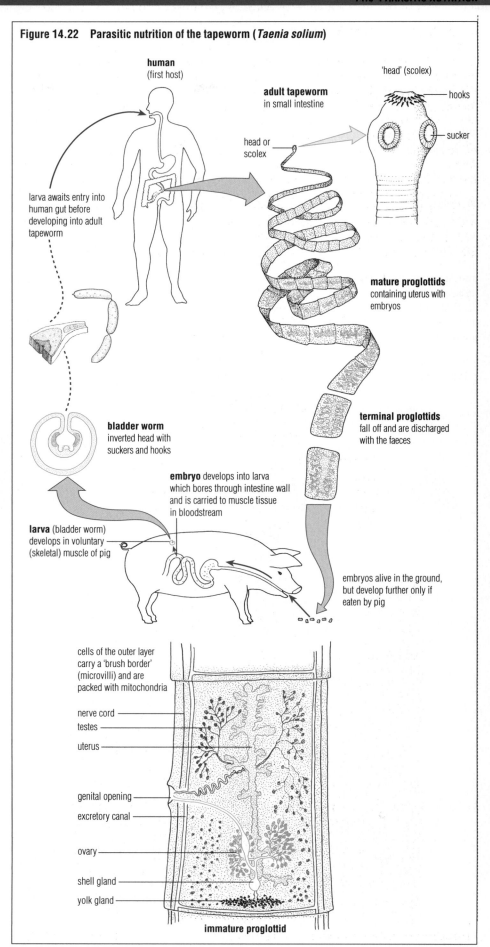

Figure 14.22   Parasitic nutrition of the tapeworm (*Taenia solium*)

human (first host)

adult tapeworm in small intestine

'head' (scolex)

hooks

sucker

head or scolex

larva awaits entry into human gut before developing into adult tapeworm

mature proglottids containing uterus with embryos

terminal proglottids fall off and are discharged with the faeces

bladder worm inverted head with suckers and hooks

embryo develops into larva which bores through intestine wall and is carried to muscle tissue in bloodstream

larva (bladder worm) develops in voluntary (skeletal) muscle of pig

embryos alive in the ground, but develop further only if eaten by pig

cells of the outer layer carry a 'brush border' (microvilli) and are packed with mitochondria

nerve cord

testes

uterus

genital opening

excretory canal

ovary

shell gland

yolk gland

immature proglottid

The symptoms caused by the presence of a pork tapeworm in the human bowel are usually few, though there may be some abdominal discomfort. A danger arises from careless personal hygiene or ineffective disposal of sewage, however, since the tapeworm segments can re-infect their host (or another human) directly, without passing through the second host. Multiple infections can cause illness and debilitation.

■ Potato blight (*Phytophthora infestans*)

Phytophthora infestans is a fungus-like member of the Protoctista (p. 28). It is a parasite of the potato plant and so has important economic implications for human communities. Potato blight was first recorded in the British Isles in 1845. In that year, and for several years following, it destroyed much of the potato crop in Ireland, where the tubers formed the staple food for most of the rural population; the result was the notorious Irish Potato Famine. Since then, in the warmer wetter areas of Britain the disease has severely damaged the potato crop in some years, although today much effort goes into breeding resistant potatoes, and in the design and application of effective fungicides.

Phytophthora overwinters in infected tubers and spreads into the new shoots as they form in the spring (Figure 14.23). The mycelium (p. 557) feeds on the shoot tissues and spreads throughout the whole plant, killing the tissues, and then produces sporangia at the surface of infected leaves. Sporangia are carried in the wind and, if conditions are warm and humid, they develop on other potato leaves that they land on. *Phytophthora* cannot live saprotrophically on dead plant material; it can develop only on a living potato plant. It is therefore described as an obligate parasite (see opposite).

Hyphae of *Phytophthora* penetrate host cells by the secretion of a cellulase enzyme at the hyphal tip. Once inside the cell the hypha swells into a **haustorium**, a special organ for the digestion and absorption of host cell cytoplasm.

Questions

1 Where would you expect to be able to find living populations of *Rhizopus* and *Mucor* respectively?

2 Why does a heap of fresh lawn mowings heat up?

3 Outline two different ways in which the nutrition of yeast is exploited in human food production.

4 Both yeasts and bacteria are single-celled organisms. What are the differences in structure between yeasts and bacteria?

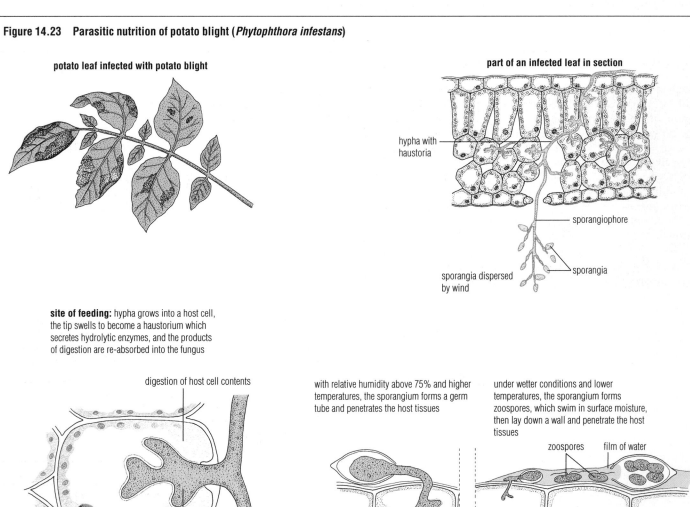

Figure 14.23   Parasitic nutrition of potato blight (*Phytophthora infestans*)

potato leaf infected with potato blight

part of an infected leaf in section

hypha with haustoria

sporangiophore

sporangia dispersed by wind

sporangia

site of feeding: hypha grows into a host cell, the tip swells to become a haustorium which secretes hydrolytic enzymes, and the products of digestion are re-absorbed into the fungus

digestion of host cell contents

the host cell finally dies   haustorium

with relative humidity above 75% and higher temperatures, the sporangium forms a germ tube and penetrates the host tissues

under wetter conditions and lower temperatures, the sporangium forms zoospores, which swim in surface moisture, then lay down a wall and penetrate the host tissues

zoospores   film of water

■ Points about parasitism

Adaptations to parasitism

Many of the differences between free-living organisms and related parasitic species are recognised as adaptations to parasitism. These features include

► mechanisms for attachment to and penetration of host tissues
► degeneration of redundant body systems
► mechanisms aiding dispersal of parasites to new hosts
► mechanisms resisting attack by the host defence systems.

Many of the parasites described here and in Chapters 3, 24 and 25 illustrate several if not all of these adaptations.

Obligate *v.* facultative parasites

Obligate parasites are unable to survive and reproduce in the absence of a host. The tapeworm and the potato blight are examples of obligate parasites.

Some parasites can live independently in the absence of a host, and are described as **facultative parasites** (they have the facility to be parasitic if the opportunity arises). Several fungal parasites are like this; one example is the bootlace fungus (*Armillaria mellea*), which lives saprotrophically on rotting tree stumps. This fungus may send out long black cords (hyphae growing together) that can infect, parasitise and kill living trees (Figure 14.24).

Virulent and benign parasites

Some parasites, such as the Dutch elm disease fungus (p. 66), quickly kill their hosts. Others live with their hosts causing few harmful effects; the protozoan *Monocystis* lives on the developing sperms in the seminal vesicles of earthworms, and appears to cause no harm to its host.

Those parasites that cause dramatic and terminal effects on their hosts have the problem of reaching a new host soon after successfully establishing. Since a major problem for parasites is successful transmission to new hosts, a parasite that kills its host soon after establishing itself may be seen as a relatively recently evolved relationship. By a similar argument, cases of mutualism (p. 67) may have evolved from parasitic relationships in which the harm has been replaced by shared benefits.

Parasitism and trophic levels

Ecologists identify organisms as belonging to particular trophic levels (p. 47). On this classification parasites are at trophic level 2 (primary consumers), trophic level 3 (secondary consumers) or above.

Figure 14.24 Facultative parasitism of *Armillaria mellea* (Bootlace fungus)

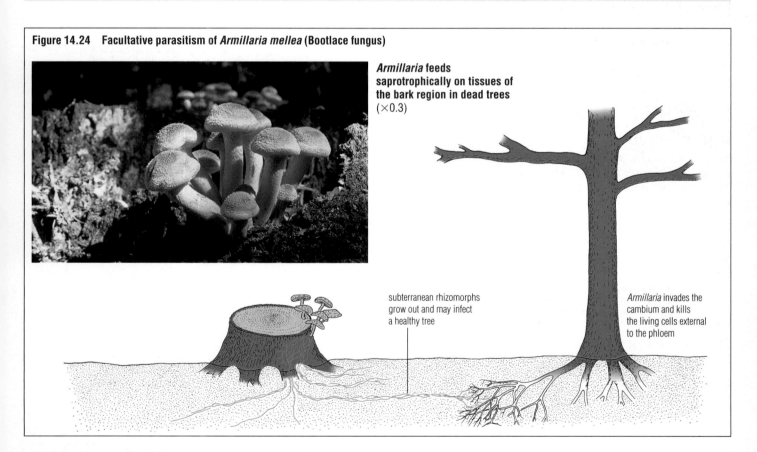

Armillaria feeds saprotrophically on tissues of the bark region in dead trees (×0.3)

subterranean rhizomorphs grow out and may infect a healthy tree

Armillaria invades the cambium and kills the living cells external to the phloem

---

Questions

1 In what ways does saprotrophic nutrition differ from parasitic nutrition?

2 A range of adaptations of successful parasites facilitates their nutrition and reproduction, and therefore their survival. List these features and give an example of each from the parasites described here and in Chapters 3, 24 and 25.

FURTHER READING/WEB SITES

London Zoo: **www.londonzoo.co.uk**

15 Respiration and gaseous exchange

- The energy required by living cells is released by the breakdown of sugar in respiration. As a result of respiration, gaseous exchange occurs between cells and the environment.
- The site of gaseous exchange is called the respiratory surface. In small organisms the respiratory surface is the body surface where this is permeable to gases. These organisms (and individual cells) have a very large surface area:volume ratio, essential for efficient diffusion.
- Many larger animals have evolved special organs of gaseous exchange, such as gills or lungs, and some have ventilation mechanisms for moving air or water over the respiratory surface.

- In many animals, a blood circulation system transports dissolved gases between respiring cells and the respiratory surface.
- Breakdown of sugars in cells, called tissue respiration, consists of a series of oxidation–reduction reactions, in which sugar is typically oxidised to carbon dioxide, and oxygen is reduced to water.
- Alternatively, sugar may be partially broken down to ethanol or lactic acid in a process known as fermentation.

15.1 INTRODUCTION

Living things require energy for the activities of life. Energy is released in cells by the breakdown and oxidation of sugars and other substances by the process known as **respiration**. The term applies to the chemical reactions by which energy is released (sometimes called tissue respiration). As a consequence of respiration gases are exchanged between respiring cells and the environment, the process of **gaseous exchange**. Respiration occurs in the living cell, and provides the chemical energy source (ATP, p. 176) that drives the cellular machinery.

■ Aerobic respiration

Respiration requiring oxygen is referred to as **aerobic respiration**. The waste products of aerobic respiration are carbon dioxide and water. Aerobic respiration may be represented by the equation below, which is a summary of the reaction only (it gives the incorrect impression that respiration is a single-step process):

$$\text{glucose + oxygen}$$
$$\downarrow$$
$$\text{carbon dioxide + water + ENERGY}$$

Or, using chemical formulae:

$$C_6H_{12}O_6 + 6O_2$$
$$\downarrow$$
$$6CO_2 + 6H_2O + \text{ENERGY}$$

All plants and animals require oxygen and respire aerobically. They are described as **aerobes**. Some aerobes (and sometimes certain tissues within aerobes) have the ability to survive and satisfy their energy requirements for a short time in the absence of oxygen. Such organisms (or tissues) are described as **facultatively anaerobic**, since they have the facility to do this if necessary. Vertebrate skeletal muscle is facultatively anaerobic (p. 492).

■ Anaerobic respiration

The maintenance of a supply of energy in the absence of oxygen is known as **anaerobic respiration**. Flowering plants and fungi such as yeasts respire anaerobically by **alcoholic fermentation**, in which the waste products are ethanol and carbon dioxide:

$$\text{glucose}$$
$$\downarrow$$
$$\text{ethanol + carbon dioxide + ENERGY}$$

Or:

$$C_6H_{12}O_6$$
$$\downarrow$$
$$2CH_3CH_2OH + 2CO_2 + \text{ENERGY}$$

In **lactic acid formation** the sole waste product is lactic acid. Vertebrate skeletal muscle tissue may respire anaerobically, forming lactic acid:

$$\text{glucose} \rightarrow \text{lactic acid + ENERGY}$$

Or:

$$C_6H_{12}O_6$$
$$\downarrow$$
$$CH_3CHOHCOOH + \text{ENERGY}$$

Some organisms respire only in the absence of oxygen and are known as **obligate anaerobes**. The bacterium causing tetanus ('lock-jaw'), *Clostridium tetani*, is an example.

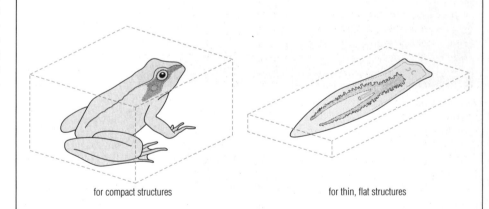

Figure 15.1   Size and shape, the surface area: volume ratio

for compact structures

for thin, flat structures

'Organism'	Compact			Flat and thin		
	small	medium	large	small	medium	large
Dimensions/cm	1×1×1	2×2×2	4×4×4	2×1×0.5	8×2×0.5	16×8×0.5
Volume/cm³	1	8	64	1	8	64
Surface area/cm²	6	24	96	7	42	280
Surface area:volume ratio	6.0	3.0	1.5	7.0	5.25	4.4

Figure 15.2   Gaseous exchange in elder (*Sambucus nigra*)

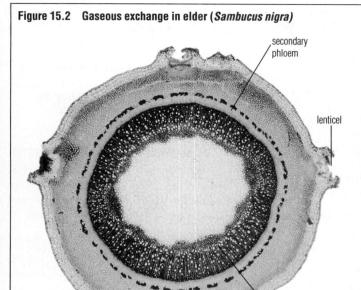

secondary
phloem

lenticel

secondary xylem

TS of elder stem

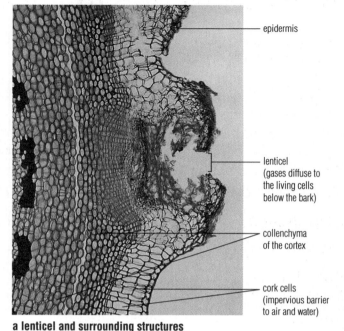

epidermis

lenticel
(gases diffuse to
the living cells
below the bark)

collenchyma
of the cortex

cork cells
(impervious barrier
to air and water)

a lenticel and surrounding structures

A lenticel is a small area in the bark where the cork-forming layer is more active than elsewhere, and its cells divide to form loosely arranged cells in place of the more usual cork cells. The lenticel contains large and rounded cells. The air spaces of the lenticel are continuous with the air spaces of the cortex.

15.2 GASEOUS EXCHANGE

In respiring tissues, concentration gradients of oxygen and the waste products (mainly carbon dioxide) are generated between the cells and the environment. We have already seen that molecules move from regions of high concentration to regions of low concentration by **diffusion**. For effective diffusion to occur, a large concentration difference is required across an extensive surface area where the barrier is thin and permeable. Gases enter and leave cells in aqueous solution. Diffusion as a process is discussed in Chapter 11, and the factors affecting the rate of diffusion are summarised in Table 11.1 (p. 230).

The site of gaseous exchange is referred to as the **respiratory surface**. In some organisms this is the whole external surface, but in very many organisms a specialised respiratory surface has evolved. Why do you think this is so?

■ Gaseous exchange in small animals

In small organisms the respiratory surface is the general body surface. These organisms (and individual cells) have a large surface area:volume ratio (that is, the surface area available for gas exchange per unit of volume, Figure 15.1), and the external surface is fully permeable to the gases. Gases diffuse in and out of the organism over the whole plasma membrane. The external surface of organisms such as *Amoeba* (and *Chlorella*) (Figure 1.9, p. 7) is an efficient respiratory surface. This is also the case in many small, multicellular animals that are relatively inactive and sluggish; examples include cnidarians such as *Hydra* (p. 34) and sea anemones. Some other multicellular organisms may become quite large, but because they remain sluggish they require less oxygen than active organisms of a similar size, and so require no specialised respiratory surface. An example is the sponge *Leucosolenia* (Figure 1.10, p. 8). This organism has numerous channels through which a feeding current of water is drawn, which also facilitates gaseous exchange.

In a large multicellular organism, the shape may determine whether unaided diffusion can meet its needs. A flattened or thin shape has a greater surface area than a compact structure of the same volume (Figure 15.1). The aquatic flatworm *Planaria* (p. 35) reaches a length of 10–12 mm, yet has evolved no specialised respiratory surface; the whole external surface of this multicellular organism is an efficient surface for gaseous exchange. The flatworms, as their name implies, have flattened bodies, and so the distances between respiring cells and respiratory surface are relatively short.

■ Gaseous exchange in plants

The shape and structure of plants facilitate gaseous exchange by diffusion. A terrestrial flowering plant has many air spaces between the cells of stem, leaf and root. These air spaces are continuous; oxygen diffuses into the air space through stomata (the pores in stems and leaves) and carbon dioxide and water vapour diffuse out. Oxygen also enters the root tissue by diffusion from the soil air, after dissolving in the surface film of moisture. Diffusion of carbon dioxide and oxygen in the green leaf is discussed on p. 256, and diffusion of water vapour on p. 323.

Woody flowering plants (trees and shrubs) have an external, impervious bark. Here, gaseous exchange occurs through small, powdery patches in the stem, called **lenticels** (Figure 15.2).

Questions

1   List three characteristics of an efficient respiratory surface, giving the reason why each is effective.
2   What is the role of ATP in living organisms (p. 176)?
3   Green leaves in the light appear not to respire, in that they give out oxygen and take in carbon dioxide. Explain this apparent anomaly.
4   By what pathway and process does oxygen from the atmosphere reach the cambium cells of the tree trunk?

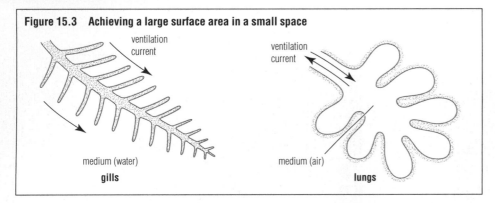

Figure 15.3   Achieving a large surface area in a small space

ventilation current

medium (water)

gills

ventilation current

medium (air)

lungs

most vertebrates (in blood cells) and in many non-vertebrates (either in cells or in the plasma).

15.3 GASEOUS EXCHANGE IN SOME NON-VERTEBRATES

■ The earthworm (*Lumbricus terrestris*)

The earthworm (Figure 15.4) inhabits burrows in damp soil, and emerges to feed (and to reproduce) in the darkness; there is less danger of desiccation at night.

Gaseous exchange occurs through the skin. Here the blood capillaries bring the blood, which contains the respiratory pigment haemoglobin, close to the external environment. The epidermis of the skin consists of columnar epithelial cells (p. 213) and mucus-secreting cells. The epidermis is covered by a protective layer (a thin cuticle covered by mucus), and is also the main respiratory surface. The cuticle is secreted by the epidermal cells, and consists of a fine network of collagen fibres.

Water loss during gaseous exchange is a potential threat to earthworms. In hot, dry periods they burrow deep into the soil and become totally inactive.

■ Gaseous exchange in large organisms

Large organisms have a low surface area:volume ratio. Often, if they are animals, they have a higher rate of metabolism than many unicellular organisms, and so require more oxygen and produce more carbon dioxide per unit of body volume (Figure 17.1, p. 342). Moreover, the surface of the body may have become toughened and impervious; in some it is enclosed in a protective shell. Most such organisms have evolved special gaseous exchange mechanisms.

▶ **Gills** or **lungs** are relatively compact organs in which the gaseous exchange surface is greatly increased by intuck-ings, leaf-like plates or chambers with folded linings (Figure 15.3).

▶ **A ventilation mechanism** moves a fresh supply of air or water over the respiratory surface maintaining a relatively high level of oxygen and a relatively low level of carbon dioxide in the medium at the site of gaseous exchange.

▶ **An internal transport system**, such as a blood circulation system, maintains the concentration difference across the gaseous exchange surface and facilitates the movement of dissolved gases between respiring cells and the respiratory surface.

▶ The presence of a **respiratory pigment** in the blood increases its oxygen-carrying capacity. The most commonly occurring respiratory pigment is haemoglobin (p. 343), found in

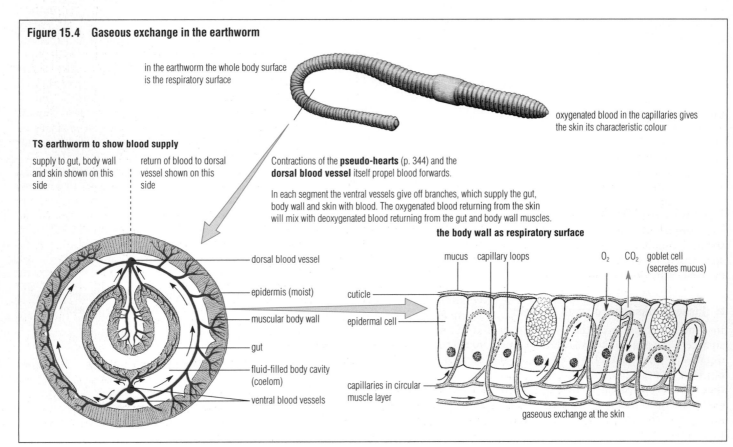

Figure 15.4   Gaseous exchange in the earthworm

in the earthworm the whole body surface is the respiratory surface

oxygenated blood in the capillaries gives the skin its characteristic colour

TS earthworm to show blood supply

supply to gut, body wall and skin shown on this side

return of blood to dorsal vessel shown on this side

Contractions of the **pseudo-hearts** (p. 344) and the **dorsal blood vessel** itself propel blood forwards.

In each segment the ventral vessels give off branches, which supply the gut, body wall and skin with blood. The oxygenated blood returning from the skin will mix with deoxygenated blood returning from the gut and body wall muscles.

the body wall as respiratory surface

mucus   capillary loops

O_2 CO_2 goblet cell (secretes mucus)

dorsal blood vessel

epidermis (moist)

muscular body wall

gut

fluid-filled body cavity (coelom)

ventral blood vessels

cuticle

epidermal cell

capillaries in circular muscle layer

gaseous exchange at the skin

Figure 15.5 Gaseous exchange in insects

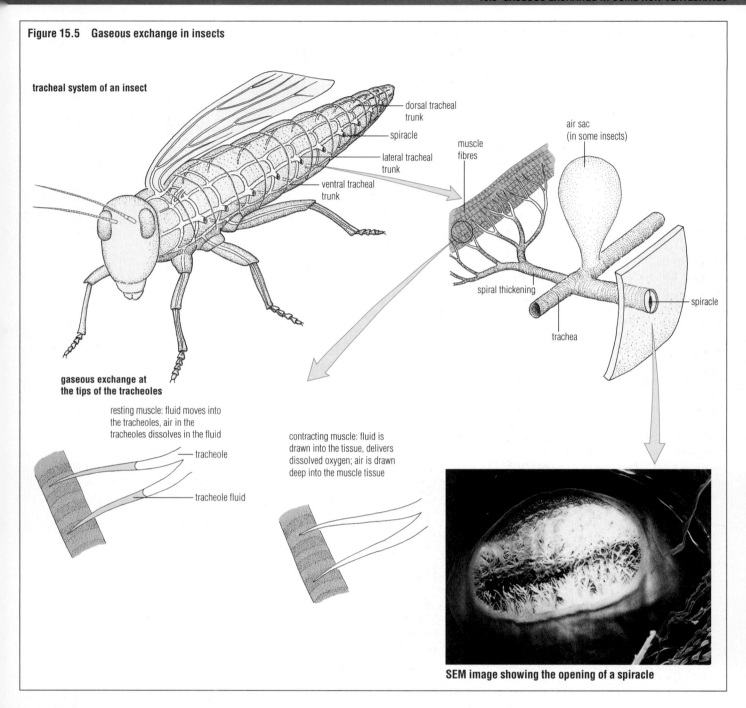

tracheal system of an insect

dorsal tracheal trunk

spiracle

lateral tracheal trunk

ventral tracheal trunk

muscle fibres

air sac (in some insects)

spiral thickening

trachea

spiracle

gaseous exchange at the tips of the tracheoles

resting muscle: fluid moves into the tracheoles, air in the tracheoles dissolves in the fluid

tracheole

tracheole fluid

contracting muscle: fluid is drawn into the tissue, delivers dissolved oxygen; air is drawn deep into the muscle tissue

SEM image showing the opening of a spiracle

■ Insects: a tracheal system

The body of an insect is protected by an external skeleton (exoskeleton), which is impervious to oxygen. Air reaches the individual cells and tissues via an elaborate system of branching tubes called **tracheae.** The tracheae open to the exterior via controlled valves called **spiracles;** most adult insects have two pairs of spiracles in the thorax and eight pairs in the abdomen. Large tracheal tubes run along and across the body. Smaller tracheae branch profusely from these, finally forming tiny **tracheoles.** Tracheoles end among the cells of the organs, as shown in Figure 15.5.

Oxygen diffuses into the cells at the fine endings of the tracheal system. Here the tracheoles have especially thin, permeable walls, and are in intimate contact with the cells. In the resting condition body fluid diffuses into the tracheoles. During activity, the tracheolar end cells rapidly pump ions into their cytoplasm, so making them hypertonic. As water is withdrawn from the tracheoles by osmosis, air moves closer to the cells and diffusion is faster.

A ventilation mechanism

Grasshoppers and many other insects also have **air sacs** in their tracheal system. These air sacs function as bellows during vigorous body movements, speeding up the movement of gases to and from the tissues. Insects also have a body fluid, which is referred to as blood although the system contains no respiratory pigment and plays no part in oxygen transport. **Carbon dioxide** is transported in the blood, however, and diffuses out through the exoskeleton.

Questions

1 How do earthworms avoid loss of water vapour from their respiratory surfaces during unfavourable conditions?

2 How is gaseous exchange carried out in insects?

15.4 GASEOUS EXCHANGE IN VERTEBRATES

■ Fish

Fish obtain oxygen from water by means of gills. Gill surfaces are large, especially in active fish. There is a high rate of water flow over the gills, which have a fine structure that allows water and blood to be in close contact. Efficient gaseous exchange is achieved by the stream of water flowing over the gills and the blood flow through the gills being in opposite directions (a countercurrent flow) – it is often so efficient that 80–90% of the oxygen available may enter the blood.

Bony fish such as the trout have four pairs of gills, each supported by a bony arch. The gills consist of two rows of gill filaments, arranged in a V-shape. Each filament bears numerous thin-walled gill plates. In the gill plates the vascular network is arranged so that the blood that is about to leave the filament encounters water from which no oxygen has been removed. As water flows on over the plates it encounters blood of decreasing oxygen content. The counter-current mechanism ensures that as blood leaves the gill plates it has a partial pressure of oxygen almost as great as that of the incoming water (Figure 15.6).

Figure 15.6   Gaseous exchange in a bony fish

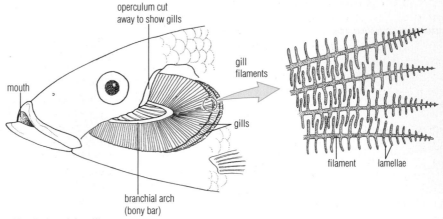

Ventilation of the gills

Water is moved over the gills in a continuous unidirectional flow by maintaining a lower pressure in the opercular cavity than in the bucco-pharynx.

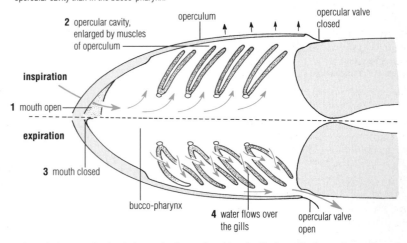

1 Inspiration: water is taken in, by opening the mouth and lowering the floor of the bucco-pharynx, thus reducing pressure below that of exterior and drawing in water.
2 At the same time the opercular cavity is enlarged, causing the pressure to fall below that of the bucco-pharynx.
3 Expiration: the mouth closes and the floor of the bucco-pharynx is raised; water is forced over the gills.
4 The opercular muscles relax, closing the opercular cavity and allowing water to pass out at the opercular valve.

Countercurrent flow of oxygenated water and deoxygenated blood in the gill filaments and lamellae

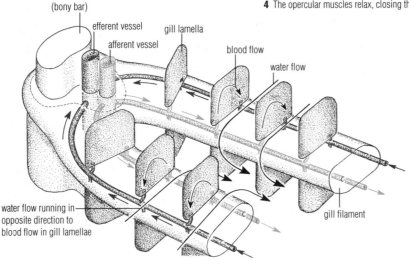

In a countercurrent system the blood constantly meets water with a relatively higher concentration of dissolved oxygen in it. Because a concentration gradient is maintained across the whole gill surface, bony fish extract 80% of the oxygen in water.
(In parallel flow, less than 50% of the available oxygen would be absorbed.)

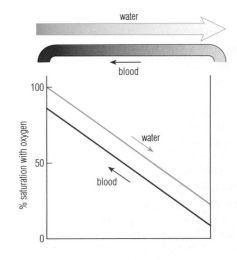

The gills are protected from the exterior by a movable bony flap, the **operculum**. The flap acts both as a valve permitting exit of water and as a pump drawing water past the gill filaments. The pumping actions of the mouth (buccal) chamber, together with actions of the operculum, deliver an almost continuous flow of water over the gills.

Oxygen from water

Water and air are contrasting media for gaseous exchange. Aquatic organisms have a restricted oxygen concentration to draw on; oxygen makes up about 21% of the volume of the air, whereas in natural water dissolved oxygen is, at best, only 1% by volume. In water, oxygen diffuses more slowly because water is far more dense and viscous than air.

An increase in the temperature of the water, or an increase in the concentration of dissolved salts, lowers the amount of oxygen that can dissolve. In polluted water the oxygen supply is especially problematical. Dissolved or suspended organic matter is decomposed by aerobic bacteria, which can cause the temporary removal of all dissolved oxygen. The result is often the death of many aquatic organisms from asphyxia (see the discussion of biological oxygen demand on p. 101).

For organisms of a similar size and level of metabolic activity, a much greater volume of water must normally pass across the respiratory surface of an aquatic animal than the volume of air that must reach the respiratory surface of a land animal.

■ Birds

Birds maintain a high and constant body temperature with a high metabolic rate and an enhanced requirement for energy, particularly during flapping flight. They require an efficient gaseous exchange system to maintain their metabolic rate and active life style.

The respiratory system of birds is a unique system of air sacs and lungs (Figure 15.7). The lungs consist of fine, highly vascular tubes, open at both ends. Here air is continuously ventilated (one-way flow) from the extensive system of air sacs to the exterior. Gaseous exchange is very efficient; virtually no residual gas remains in the lung tubes.

The air sacs function as bellows. During inhalation all sacs fill with air, the posterior ones with fresh air and the anterior ones with air from the lungs. During exhalation air flows out of the air sacs: it flows from the posterior sacs to the lungs, and from the anterior sacs to the exterior via the bronchus. Ventilation of the air sacs is driven by contraction and relaxation of intercostal and abdominal muscles. During vigorous flight contractions of the pectoral muscles enhance ventilation.

Questions

1 Which physical properties of water influence the gaseous exchange process in fish?

2 Which features of the gills of bony fish make them efficient organs for gaseous exchange?

3 In what ways is ventilation in fish similar to that in birds?

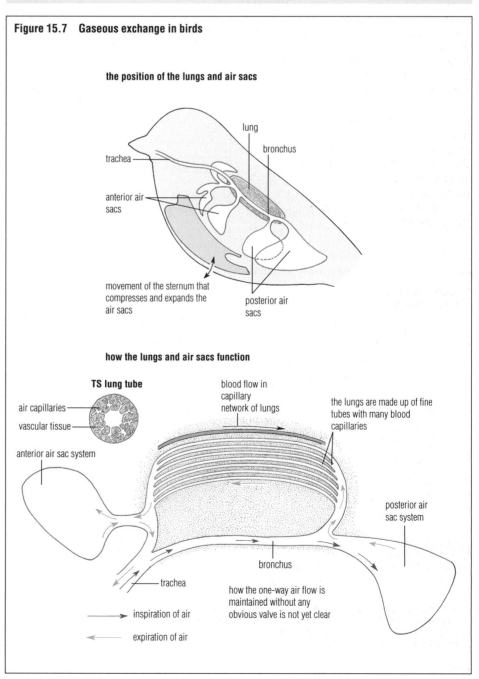

Figure 15.7   Gaseous exchange in birds

the position of the lungs and air sacs

trachea · lung · bronchus · anterior air sacs · posterior air sacs · movement of the sternum that compresses and expands the air sacs

how the lungs and air sacs function

TS lung tube · air capillaries · vascular tissue · anterior air sac system · blood flow in capillary network of lungs · the lungs are made up of fine tubes with many blood capillaries · posterior air sac system · bronchus · trachea · how the one-way air flow is maintained without any obvious valve is not yet clear

→ inspiration of air
← expiration of air

■ Mammals

Gaseous exchange in mammals, including humans, occurs in a pair of **lungs**. Lungs are enclosed (with the heart) in an air-tight compartment, the **thorax**, delineated by the thorax wall and the diaphragm.

The system of ribs forms a protective cage around these organs. In humans the **rib-cage** consists of twelve pairs of bony ribs, each pair articulating with a thoracic vertebra (p. 484), and able to move obliquely upwards and outwards. The first ten pairs of ribs are joined by cartilage to the breast bone (sternum); the remaining two pairs are not. Intercostal muscles, attached between the ribs, move the rib-cage. The thoracic and abdominal cavities are separated by the **diaphragm**; this is a sheet of muscle shaped like a flattened dome, attached to the body wall at the base of the rib-cage (Figure 15.8).

The inner surface of the thoracic cavity and the outer surfaces of the lungs are lined with a smooth membrane, the **pleural membrane**. The space between the membranes, the pleural cavity, contains lubricating fluid (**pleural fluid**).

Ventilation of the lungs

Lungs are delicate, compact and highly elastic organs. Air is drawn into the lungs when pressure there is lower than atmospheric pressure. Air is forced out of the lungs when pressure there is higher than atmospheric pressure. Pressure changes in the lungs are brought about by changes in the volume of the thorax. The volume of the thorax changes as a result of movements of ribs and diaphragm, caused by muscle contraction (Figure 15.9). Air flow in the mammal is tidal; air enters and leaves the lungs along the same route. In effect the lungs are a cul-de-sac, and always contain residual air that cannot be expelled.

In addition to the normal breathing movements detailed in Figure 15.9, 'forced breathing' may occur under conditions of vigorous exercise. Additional muscles then come into use, especially the abdominal wall muscles.

Air reaches the lungs via the mouth and nostrils by passing through the **larynx** (voice box) and **trachea** to the lungs. Entry into the larynx is via a slit-like opening, the **glottis**, which, with the **epiglottis**, prevents entry of food into the air channels.

The trachea lies beside and in front of the oesophagus (p. 277). A hard mass of food passing down the oesophagus might interrupt the air supply to the lungs.

Figure 15.8 The structure of the human thorax

nasal cavity
epiglottis
glottis
oesophagus
larynx
trachea
intercostal muscles
rib
position of the heart
left lung
bronchus
pleural membranes
pleural cavity containing pleural fluid
diaphragm
abdominal cavity

Figure 15.9 Ventilation of the lungs

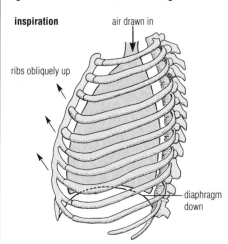

inspiration air drawn in

ribs obliquely up

diaphragm down

1 diaphragm muscles contract; the diaphragm flattens

2 the external intercostal muscles contract; ribs and sternum move up and out

The volume of the thorax (and therefore that of the lungs) is increased; the pressure in the lungs is reduced below atmospheric pressure, and air flows in.

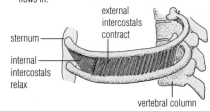

external intercostals contract
sternum
internal intercostals relax
vertebral column

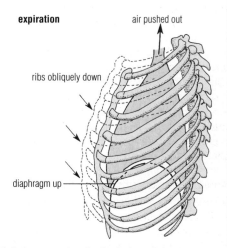

expiration air pushed out

ribs obliquely down

diaphragm up

1 diaphragm muscles relax; the diaphragm becomes more dome-shaped

2 the external intercostal muscles relax, the viscera return to a relaxed position; ribs and sternum move down and in

The volume of the thorax is decreased; the pressure in the lungs is increased above atmospheric pressure, and air flows out.

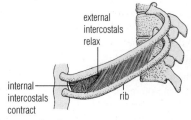

external intercostals relax
internal intercostals contract
rib

how the intercostal muscles move the ribs

Figure 15.10 The site of gaseous exchange in humans

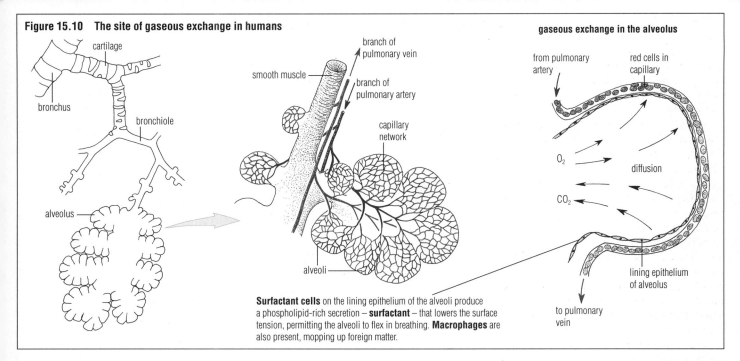

gaseous exchange in the alveolus

Surfactant cells on the lining epithelium of the alveoli produce a phospholipid-rich secretion – **surfactant** – that lowers the surface tension, permitting the alveoli to flex in breathing. **Macrophages** are also present, mopping up foreign matter.

The trachea is protected against closure by a series of closely packed C-shaped rings of strong, flexible cartilage, arranged horizontally in its wall. These cartilage rings also keep the trachea open as the air pressure there falls below atmospheric pressure during inspiration.

The trachea splits into two **bronchi**, each serving one lung. Within the lungs, the bronchi themselves divide into smaller and smaller **bronchioles**. The walls of the bronchi and the larger bronchioles are also strengthened and protected from collapse by cartilage rings in their walls.

Alveolar structure and gaseous exchange

The bulk of the lung tissue consists of millions of microscopic air sacs or **alveoli** in grape-like clusters, each served by a tiny bronchiole (Figure 15.10). Networks of capillaries surround the alveoli; they are supplied with blood by the pulmonary artery and are ultimately drained by the pulmonary vein. Elastic connective tissue separates the clusters of alveoli. The 700 million alveoli in the lungs of an adult human provide an area for gaseous exchange of 100–150 m² of lung surface (approximately equal to the area of a 'doubles' tennis court!).

The alveolar walls consist of squamous epithelium (p. 213) only 10 μm thick. Oxygen dissolves in the film of water on the surface of the wall. Dissolved oxygen diffuses across the epithelial cells and the capillary endothelial cells into the blood plasma. In the plasma, oxygen diffuses into the red cells (erythrocytes) and combines with haemoglobin to form oxyhaemoglobin. Carbon dioxide diffuses from the red cells and from the plasma into the alveoli.

The passage of blood through the lungs is slow. This is because the capillaries are extremely narrow, only just wide enough for red cells to squeeze through. But this slow passage of blood through the lungs facilitates gaseous exchange: there is time for oxygen to combine with haemoglobin and for carbon dioxide to come out of solution.

The oxygenated blood leaving the lungs by the pulmonary vein is under relatively low pressure. It returns to the right side of the heart before being pumped to the rest of the body (p. 351). The chemical mechanisms of transport of respiratory gases in the red cells are discussed in Chapter 17 (p. 358).

Change in composition of breathed air

The air that is breathed out (expired air) is the product of mixing of the incoming atmospheric air with the residual volume of air in lungs, which cannot be expired, and this is reflected in its composition (Table 15.1).

Table 15.1 The composition of air in lungs/%

	Inspired air	Alveolar air	Expired air
oxygen	20.95	13.80	16.40
carbon dioxide	0.04	5.50	4.00
nitrogen	79.01	80.70	79.60

Protection of the lungs

Hairs in the nostrils trap and filter out larger dust particles from the incoming air stream. A second line of defence occurs in the nasal passages, trachea and bronchi, for these tubes are lined with a ciliated epithelium with numerous goblet cells. The mucus produced by the goblet cells moistens the incoming air and traps the finest dust particles. The ciliated epithelium beats the mucus stream up into the buccal cavity, where it is swallowed.

Superficial blood vessels in the nasal cavity warm the incoming air. Odours in the air are detected by the olfactory cells in the olfactory mucous membrane in the nasal cavities (p. 452).

Surfactant cells, lodged in the alveoli (Figure 15.11), produce a phospholipid-rich secretion known as **surfactant**, which lines the inner surface of the alveoli. It lowers the surface tension, allowing the alveoli to flex as the pressure of the thorax rises and falls.

Also present here are **macrophages** (p. 310). These cells of the body's defence mechanism ingest any bacteria and dust that enter the bronchioles and reach the alveolar lining.

Questions

1 In what ways does lung structure facilitate gaseous exchange?
2 List the steps and detail the pathway by which air is drawn into the alveoli of the lungs.

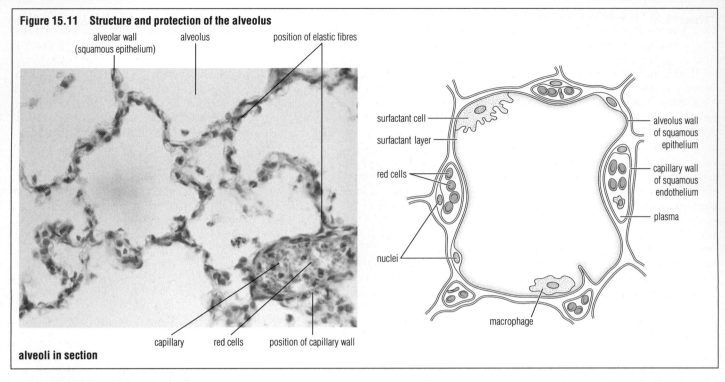

Figure 15.11   Structure and protection of the alveolus

alveolar wall (squamous epithelium)

alveolus

position of elastic fibres

surfactant cell

surfactant layer

red cells

nuclei

alveolus wall of squamous epithelium

capillary wall of squamous endothelium

plasma

macrophage

capillary

red cells

position of capillary wall

alveoli in section

15.5 INVESTIGATING HUMAN BREATHING

The pattern of change in lung volume during human breathing can be analysed using a recording **spirometer**, which requires the cooperation of a healthy volunteer! The spirometer is a precision instrument that consists of a chamber of about 6 dm³ capacity suspended over water. The chamber can be filled with either air or oxygen. The lid of the chamber is arranged to rise and fall as the subject breathes in and out through the mouthpiece. A record of volume change may be made on a kymograph (the original arrangement), as shown in Figure 15.12; alternatively, movements can be registered via a chart recorder using an electronic arm, or by means of a computer interfaced via a position transducer (see the *Study Guide*, Chapter 15).

Using the spirometer the capacity of the lungs (Figure 15.13) can be measured. With the spirometer chamber filled with oxygen, and with the carbon dioxide breathed out by the experimental subject taken up by a suitable absorbent, the consumption of oxygen during respiration can be estimated.

■ Lung capacity

During normal, rhythmical breathing a person takes in and expels approximately 450 cm³ of air. This volume is referred to as the **tidal volume**. The tidal volume is only about 10% of the total capacity of the lungs. With an extra deep breath, an additional 3 dm³ of air (**inspiratory reserve volume**) can be taken in. If as much air as possible is expired, an additional 1.5 dm³ of air can be forced out (**expiratory reserve volume**). The total of these extreme changes is referred to as the **vital capacity** of the lungs. The vital capacity is between 4 and 6 dm³ of air. However hard we breathe out, approximately 1.5 dm³ of air remains in the lungs (**residual volume**).

Air taken into the lungs mixes with the residual air. About 350 cm³ of the inspired air reaches the alveoli to mix with the gas there. The remainder occupies the 'dead space': that is, the trachea, the bronchi and the bronchioles where no significant gaseous exchange is possible.

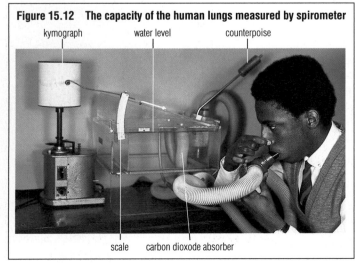

Figure 15.12   The capacity of the human lungs measured by spirometer

kymograph

water level

counterpoise

scale

carbon dioxode absorber

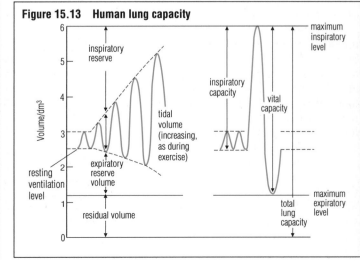

Figure 15.13   Human lung capacity

Volume/dm³

inspiratory reserve

tidal volume (increasing, as during exercise)

resting ventilation level

expiratory reserve volume

residual volume

maximum inspiratory level

inspiratory capacity

vital capacity

total lung capacity

maximum expiratory level

■ Control of the breathing mechanism

Our breathing rate varies. For example, when resting we normally take 15–20 breaths per minute.

Ventilation of the thorax is controlled by the **respiratory centre** in the part of the hindbrain known as the **medulla oblongata** (p. 466). The respiratory centre consists of a group of neurones and is organised in two parts, an **inspiratory centre** and an **expiratory centre**. The neurones of the respiratory centre are connected by cranial nerves (p. 464) to the diaphragm and rib muscles.

Breathing occurs automatically by involuntary reflex action (p. 462). The rate of breathing is continually adjusted to meet the body's immediate needs. During periods of increased activity mammals ventilate the lungs more deeply and more rapidly. Faster and deeper breathing results in an increased supply of oxygen, and also eliminates from the body the additional carbon dioxide produced by increased respiration. The breathing rate slows down when the demand for oxygen falls during periods of rest.

Adjustment of the intrinsic rhythm is an example of a homeostatic mechanism of regulation by feedback control (Chapter 23, p. 503).

Chemoreceptors in the medulla (and to a lesser extent in the carotid arteries and aorta, Figure 17.13, p. 350) are sensitive to changes in the carbon dioxide concentration, and to a much lesser extent that of oxygen. A rise in the carbon dioxide concentration (and a fall in the oxygen concentration) results in impulses being sent to the inspiratory centre. From here nerve impulses are sent to the diaphragm and intercostal muscles, causing an increased rate of contraction and faster and deeper inspiration. As the lungs expand, their **stretch receptors** are stimulated and send impulses to the expiratory centre. The inspiratory centre is switched off and expiration follows, with the result that the stretch receptors are no longer stimulated. The expiratory centre in its turn becomes inactive and inspiration can begin again (Figure 15.14).

The responses of the intercostal muscles and diaphragm muscles can also be consciously overridden by impulses from the higher centres of the brain, as in singing, clarinet playing and talking, for example.

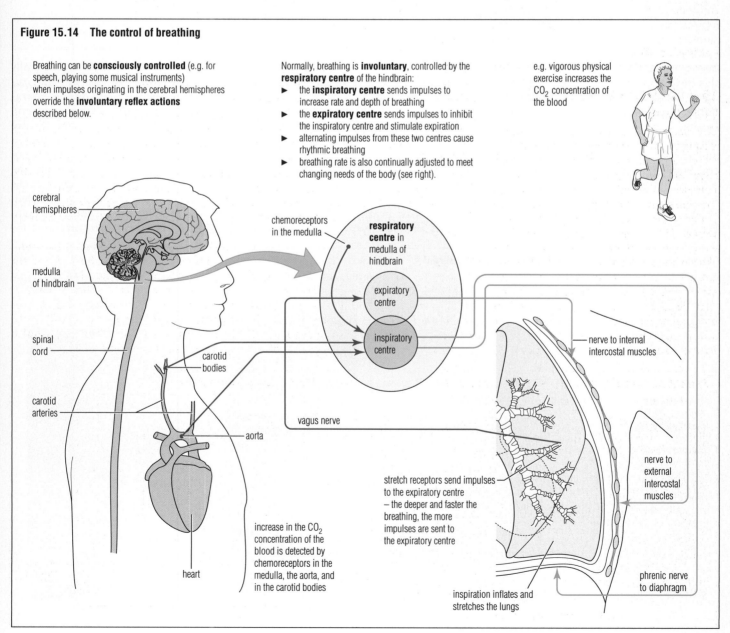

Figure 15.14   The control of breathing

Breathing can be **consciously controlled** (e.g. for speech, playing some musical instruments) when impulses originating in the cerebral hemispheres override the **involuntary reflex actions** described below.

Normally, breathing is **involuntary**, controlled by the **respiratory centre** of the hindbrain:
► the **inspiratory centre** sends impulses to increase rate and depth of breathing
► the **expiratory centre** sends impulses to inhibit the inspiratory centre and stimulate expiration
► alternating impulses from these two centres cause rhythmic breathing
► breathing rate is also continually adjusted to meet changing needs of the body (see right).

e.g. vigorous physical exercise increases the CO_2 concentration of the blood

cerebral hemispheres

medulla of hindbrain

spinal cord

carotid bodies

carotid arteries

aorta

heart

vagus nerve

chemoreceptors in the medulla

respiratory centre in medulla of hindbrain

expiratory centre

inspiratory centre

nerve to internal intercostal muscles

nerve to external intercostal muscles

stretch receptors send impulses to the expiratory centre – the deeper and faster the breathing, the more impulses are sent to the expiratory centre

increase in the CO_2 concentration of the blood is detected by chemoreceptors in the medulla, the aorta, and in the carotid bodies

inspiration inflates and stretches the lungs

phrenic nerve to diaphragm

■ Respiratory disease

Mainly due to smoking

Smoking tobacco greatly increases the risk of illness, disability and death from bronchitis, emphysema and lung cancer, as well as heart disease and diseases of the vascular system. This is because when cigarette smoke is inhaled into the lungs the secretion of viscous mucus by goblet cells is stimulated, but the movements of cilia of the epithelium lining the airways (Figure 15.15) are inhibited. Mucus in which is trapped dust and carcinogenic chemicals accumulates in the bronchioles, and the smallest bronchioles are blocked off. Irritation of the airways by the smoke results, made worse by the cilia having lost their protective function. However, a person who gives up smoking can, in time, achieve a life expectancy very similar to that of a non-smoker (Figure 15.16).

Chronic bronchitis is a very serious disease in the UK (about 1 in 2000 die of this disease every year). It has a gradual onset but is of long duration (typically lasting about three months). The bronchi become inflamed, mainly due to the retention of the tar in cigarette smoke, and excess mucus ultimately has to be coughed up. There is accumulation of phlegm and recurring attacks of coughing. As the disease progresses the bronchioles narrow, and breathing may become difficult as a result. Cigarette

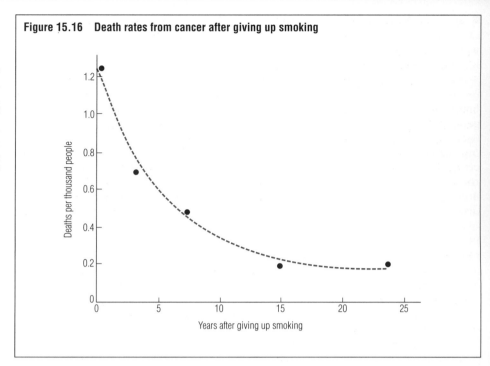

Figure 15.16   Death rates from cancer after giving up smoking

smoking is the most important cause of chronic bronchitis, but other forms of air pollution, and some respiratory infections may be the cause in some cases.

Emphysema is caused by chemicals in cigarette smoke, and by industrial dust or other air pollutants, which harm the cells that regulate the structure and condition of the elastic fibres in the lung connective tissue. As a result, naturally occurring enzymes that break down the elastic material of the alveoli walls are not checked by natural inhibitors. The walls of the alveoli lose their elasticity and can no longer elastically recoil during expiration. Lung tissue breaks down and alveoli merge to form larger air sacs (Figure 15.17). Also, dust particles that collect in the lungs cause the numbers of phagocytic cells present to increase; in the process of phagocytosis, these cells release an enzyme that digests the delicate structures of the alveoli. The combined effect is that the effective area for

Figure 15.15   Ciliated epithelium of the trachea and bronchi

ciliated epithelia line many respiratory surfaces; they often contain mucus-secreting goblet cells, and the cilia move a stream of mucus

this type of ciliated columnar epithelium occurs in the trachea and is called **'pseudostratified'** for here the columnar cells are of different lengths, and their nuclei occur at different levels

cilia

columnar epithelium cell

nucleus

cytoplasm

basement membrane

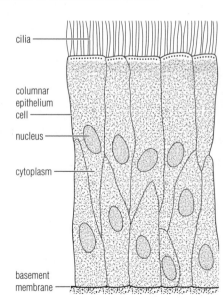

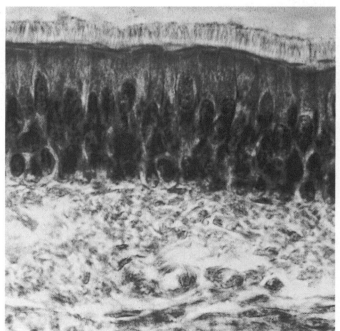

pseudostratified ciliated columnar epithelium of the trachea

gaseous exchange is decreased, and the walls of the expanded sacs become thickened with fibrous connective tissue. Gaseous exchange becomes exceedingly inefficient, and the patient is progressively transformed from a state of mere breathlessness to being a complete invalid. Emphysema can be prevented by stopping smoking, but the damage done cannot be cured or reversed.

Carbon monoxide poisoning results from absorption of carbon monoxide in tobacco smoke because this gas combines irreversibly with haemoglobin, inhibiting the formation of oxyhaemoglobin (p. 358). Consequently, oxygen transport by the blood is impaired. The drug **nicotine** is also present in tobacco smoke. Its effects are a temporary increase in the heart rate and constriction of the peripheral blood vessels, leading to a temporary increase in blood pressure.

Lung cancer is another disease that may result from persistent exposure to cigarette smoke. The basal cells of the ciliated epithelium, which replace epithelial cells as they wear out, start to divide abnormally when persistently exposed to smoke. The result is a thickening of the epithelium of the bronchioles. A carcinoma (a type of cancer) may develop among the mass of irregular cells formed, possibly triggered by carcinogenic chemicals in the smoke (Figure 15.18). If these cells break free, the cancer may spread through the lungs and on to other organs. Lung cancer has taken over from tuberculosis as a major killer disease. Heavy cigarette smokers are twenty times more likely to develop this disease than are non-smokers.

Asthma

Asthma is a disease involving our immune systems (as are the conditions of hay fever and eczema). In an asthma attack, the airways in the lungs become narrow due to contraction of the smooth muscle in the walls of the bronchioles. Breathing becomes difficult. Extra mucus, produced as part of the asthmatic condition, exaggerates the symptoms. But breathing difficulties are mainly due to an acute inflammatory response (p. 364). Release of histamine causes many of the symptoms, including narrowing of the airways and attraction of phagocytic cells. A chronic long-term effect of asthma is the build up of fibrous tissue in place of the normal epithelium of the alveoli (p. 309). The gas exchange surface is reduced.

Asthma is increasing in incidence. Some argue that we are underexposed to disease-causing organisms as affluence influences our life styles. Perhaps our immune system, under-challenged, has become hypersensitive and over-reacts to certain conditions we may experience in our lungs. This is one hypothesis, but asthma attacks are mainly triggered by irritants like pollen, dust from pets, droppings from house dust mites, by certain viruses, and by oxides of nitrogen present in vehicle exhaust fumes. People diagnosed as susceptible to an asthma attack carry an inhaler to treat themselves (Figure 15.19).

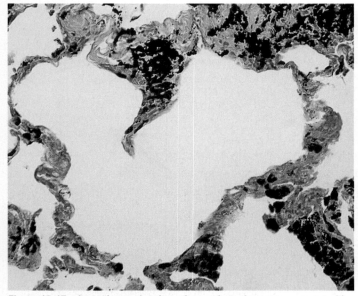

Figure 15.17 Lung tissue showing advanced emphysema: compare with healthy alveoli shown in Figure 15.11, p. 310

Figure 15.19 Asthma patient using an inhaler

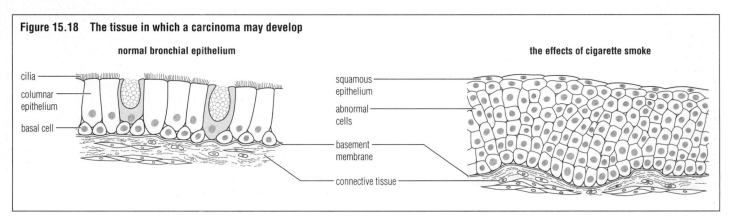

Figure 15.18 The tissue in which a carcinoma may develop

normal bronchial epithelium

the effects of cigarette smoke

cilia
columnar epithelium
basal cell

squamous epithelium
abnormal cells
basement membrane
connective tissue

15.6 TISSUE RESPIRATION

In tissue respiration the chemical energy of organic molecules such as glucose is made available for use in the living cell. Much of the energy released in respiration is lost in the form of heat energy, but cells are able to retain significant amounts of chemical energy in **adenosine triphosphate** (ATP, p. 176). ATP, found in all cells, is the universal '**energy currency**' in living systems. ATP is a relatively small molecule, and is soluble. It diffuses from the site of energy conversion to the sites where energy is required – for example, in muscles for movement (p. 491), in membranes for active transport (pp. 237 and 340) and in ribosomes for protein synthesis (p. 208). ATP can provide the energy for a wide variety of activities at short notice (see Figure 8.10, p. 176).

■ Tissue respiration as a series of redox reactions

Tissue respiration involves a series of oxidation–reduction (**redox**) reactions in which respiratory substrates are oxidised to carbon dioxide, and oxygen is reduced to water. In order to understand cellular respiration it is necessary to understand the nature of oxidation and reduction (see panel, right).

The amount of energy in a molecule depends upon the degree of oxidation. An oxidised substance has less energy than a reduced one. For example, methane has more energy than carbon dioxide; fats have more energy than carbohydrates. The amount of energy in a molecule also depends partly on its size; larger molecules contain more energy than small molecules. Thus a molecule of glucose, $C_6H_{12}O_6$, contains more energy than does a molecule of methane, CH_4.

In summary, glucose (the major respiratory substrate) is a relatively large molecule containing six carbon atoms, all in a reduced state. During respiration glucose undergoes a series of enzyme-catalysed oxidation reactions. These reactions can be grouped into three major phases (Figure 15.20):

▶ **glycolysis**, in which glucose is converted to pyruvate
▶ the **Krebs cycle**, in which pyruvate is converted to carbon dioxide and water
▶ the **electron-transport system**, in which hydrogen removed in oxidation is converted to water, and ATP is synthesised.

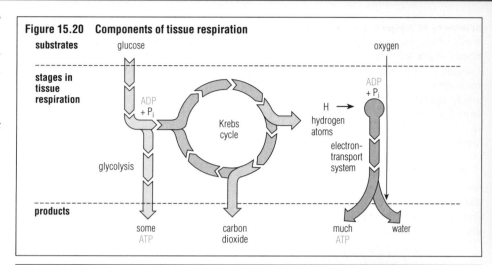

Figure 15.20   Components of tissue respiration

substrates — glucose — oxygen

stages in tissue respiration — glycolysis — $ADP + P_i$ — Krebs cycle — H → hydrogen atoms — $ADP + P_i$ — electron-transport system

products — some ATP — carbon dioxide — much ATP — water

SIX POINTS ABOUT REDOX REACTIONS

1 One type of **oxidation** is **the addition of oxygen**, e.g.

$$2Ca + O_2 \longrightarrow 2CaO$$
calcium metal   oxygen   quicklime (calcium oxide)

Another type is **the removal of hydrogen**, e.g.

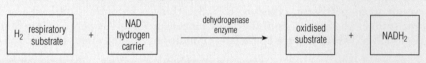

H_2 respiratory substrate + NAD hydrogen carrier —dehydrogenase enzyme→ oxidised substrate + $NADH_2$

2 **Oxidation** is associated with a loss of electrons, e.g.

calcium atom + oxygen atom → calcium ion + oxygen ion (= calcium oxide)

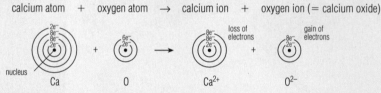

nucleus   Ca   O   Ca²⁺ (loss of electrons)   O²⁻ (gain of electrons)

The calcium atom has lost two electrons; it has become oxidised.
The oxygen has been reduced to calcium oxide; it has gained two electrons.

Mnemonic = **o**xidation **i**s **l**oss (of electrons), **r**eduction **i**s **g**ain (**oilrig**)

3 **Redox reactions**   When one substance is oxidised, another substance is reduced; for example, in a furnace or smelter the overall reaction might be:

metal oxide + carbon → metal + carbon monoxide
an ore   (as charcoal or coke)

For instance: CuO + C → Cu + CO
The carbon has been oxidised as the metal oxide has been reduced; this is a **redox** (**red**uction–**ox**idation) reaction. Oxidation–reduction reactions are not confined to those involving oxygen and hydrogen.

4 **Oxidising and reducing agents**   An iron(II) salt can be oxidised to the iron(III) salt by chlorine gas:

$$2Fe^{2+} + Cl_2 \rightarrow 2Fe^{3+} + 2Cl^-$$
salt   salt

The electron removed from iron(II) has been accepted by the chlorine gas, which is an **oxidising agent**. Oxidising agents accept electrons, and reducing agents donate electrons. Iron(II) is a **reducing agent**.

5 **Redox substances vary in their tendency to attract electrons**   In the following sequence of oxidising and reducing agents there is an increasing tendency to attract electrons:

A   B   C   D   E   F

A is the strongest reducing agent in the sequence. F is the strongest oxidising agent. An electron 'loaded' on at C will tend to pass from C to D to E to F.

6 **Cellular respiration is a series of redox reactions!**

Figure 15.21 Glycolysis: a summary

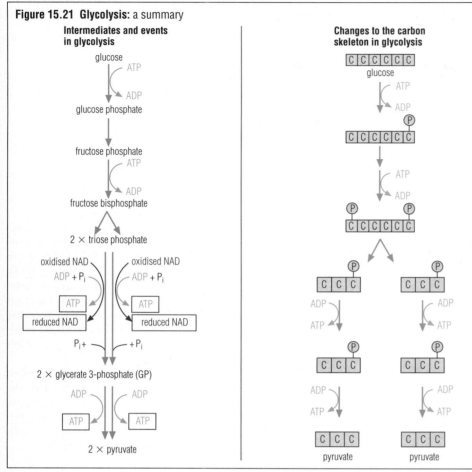

Intermediates and events in glycolysis

Changes to the carbon skeleton in glycolysis

Glycolysis

Glycolysis, the series of reactions in which six-carbon sugar is broken down to two molecules of the three-carbon pyruvate ion, is outlined in Figure 15.21. Glycolysis occurs in the cytosol.

Glucose is first **phosphorylated** by reaction with ATP, which activates it. Conversion to fructose phosphate follows, and a further phosphate group is added, forming fructose bisphosphate. Note that two molecules of ATP are consumed per molecule of glucose, *at this stage*. Next the phosphorylated 6C-sugar is split (lysis) into two 3C-sugar phosphates, called triose phosphate. Finally, the triose phosphate molecules are **oxidised** to pyruvate by removal of hydrogen, producing reduced NAD. Two molecules of ATP are formed for each triose phosphate oxidised. This means that there are four ATPs formed during oxidation, and a *net gain* of two ATPs in glycolysis, per molecule of glucose.

The reactions of glycolysis are common to both aerobic and anaerobic (p. 316) respiration, but the fate of the reduced NAD that is formed differs. In aerobic respiration it is oxidised in the mitochondria with the formation of ATP, during the third phase of glycolysis (Figure 15.21).

The Krebs cycle

The reactions of this cycle (Figure 15.22), discovered by Hans Krebs (p. 171), occur within the mitochondrion.

Pyruvate, the product of glycolysis, enters the mitochondrion and is first decarboxylated by removal of carbon dioxide. The remaining two-carbon fragment reacts with the sulphydryl group of coenzyme A (CoA—SH) to form acetyl-coenzyme A (acetylCoA); in the same step, hydrogen atoms are transferred to the hydrogen acceptor NAD, forming NADH$_2$. AcetylCoA reacts with oxaloacetate (the ion of a four-carbon organic acid) in the presence of an enzyme. The acetyl group splits off, citrate (six-carbon) is formed and CoA—SH is released for re-use. Then, in a cyclic series of reactions, citrate is converted back to the oxaloacetate ion, which is then available to repeat the cycle. At certain points in the cycle two molecules of carbon dioxide are given off, a pair of hydrogen atoms is removed and a molecule of ATP is formed. The hydrogen atoms are taken up by NAD, with the exception of one pair that is attached directly to flavine adenine dinucleotide (FAD, p. 147).

Figure 15.22 The Krebs cycle: a summary

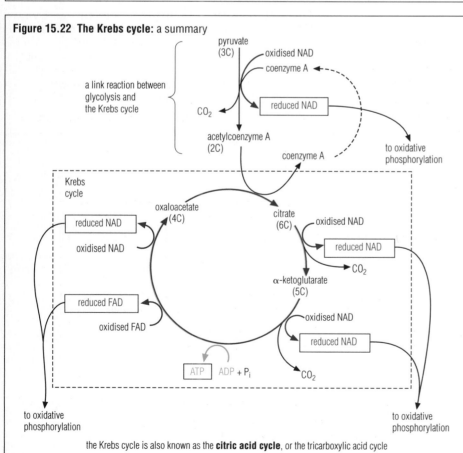

a link reaction between glycolysis and the Krebs cycle

the Krebs cycle is also known as the **citric acid cycle**, or the tricarboxylic acid cycle

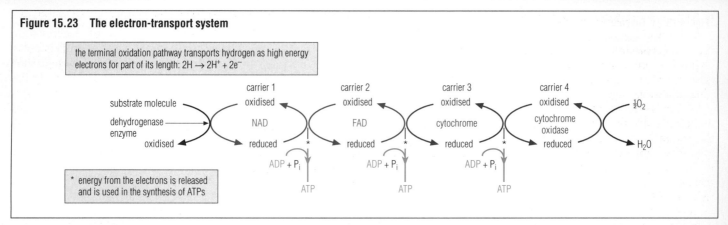

Figure 15.23   The electron-transport system

the terminal oxidation pathway transports hydrogen as high energy electrons for part of its length: $2H \rightarrow 2H^+ + 2e^-$

* energy from the electrons is released and is used in the synthesis of ATPs

■ Electron-transport pathway

The pairs of hydrogen atoms removed from respiratory intermediates during the oxidation of glucose are accepted by a **hydrogen acceptor**, normally **NAD**. These dehydrogenation reactions are catalysed by enzymes known as **dehydrogenases**. The hydrogen atoms in $NADH_2$ are ultimately oxidised to water by oxygen in a process involving several intermediate steps. The process is known as the electron-transport pathway, or sometimes the hydrogen-transport pathway. At three points of transfer in the chain, energy is released and is subsequently used in the synthesis of ATP from ADP and phosphate in the presence of an ATP-synthetase enzyme (Figure 15.23).

The mitochondrion is the site of the electron-transport pathway. The inner membrane is infolded in a series of **cristae**, which project into the fluid-filled interior or **matrix** (Figure 15.24). The enzymes of the Krebs cycle are in the matrix, and the carriers and enzymes of

the electron-transport chain that oxidise $NADH_2$ and synthesise ATP are proteins attached to the inner mitochondrial membrane. Glycolysis occurs in the cytosol, and reduced NAD formed there diffuses into the mitochondrion before it can be reoxidised.

15.7 ANAEROBIC RESPIRATION

In tissues in the absence of oxygen, glycolysis can continue initially but the Krebs cycle and the electron-transport pathway are blocked. Consequently, sugars are only partially broken down to the intermediates pyruvate and reduced hydrogen acceptor ($NADH_2$), which would then accumulate. Tissue would run out of NAD (which would all be converted to $NADH_2$) in the quantities required to maintain glycolysis. Glycolysis, too, would become blocked unless an alternative acceptor for the hydrogen atoms became available.

In **anaerobic respiration** the product of glycolysis, **pyruvate** itself, may become the hydrogen acceptor. This is the case in lactic acid fermentation (Figure 15.25). Lactic acid fermentation occurs in vertebrate muscle when, through vigorous activity, the oxygen supply becomes insufficient to maintain oxidative phosphorylation of respiratory substrates (p. 492).

Alternatively **ethanal** (acetaldehyde), formed by decarboxylation of pyruvate, is the hydrogen acceptor. This is the case in alcoholic fermentation. Alcoholic fermentation occurs in higher plant cells and yeasts under anaerobic conditions.

Questions

1 During which steps in aerobic respiration is ATP generated from ADP and P_i?
2 In the electron-transport chain, is NAD or cytochrome a stronger reducing agent?
3 What type of reaction is catalysed by
   a a carboxylase,
   b a dehydrogenase enzyme?

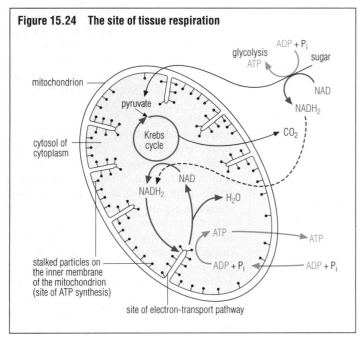

Figure 15.24   The site of tissue respiration

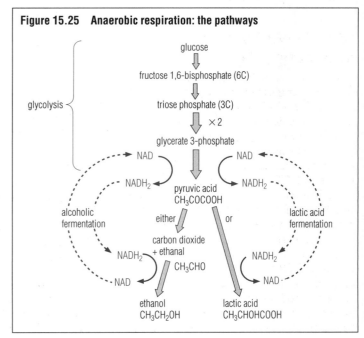

Figure 15.25   Anaerobic respiration: the pathways

■ Fate of the products of anaerobic respiration

A considerable amount of energy remains trapped in ethanol and lactic acid. This energy cannot be tapped as long as anaerobic conditions last.

In animals much of the energy locked up in lactic acid (which exists as lactate in the conditions of the cell) may be liberated later. For example, lactate produced in resipring tissues, including skeletal muscle, is converted back to pyruvate and reduced NAD. Some of this pyruvate is respired (Krebs Cycle and electron transport system) to form ATP. This ATP may be used to metabolise the remainder of the lactate to glucose via pyruvate. The glucose formed may be stored as glycogen, although it may ultimately be completely oxidised to carbon dioxide and water. These metabolic changes can go on in most tissues of the body; the liver is not the sole organ for processing excess lactate. Further, much of the lactate present in the body during periods of physical excercise is used immediately as a respiratory substrate. Heart muscle, for example, readily respires lactate from the onset of strenuous activity.

Ethanol is toxic, and so **plants** can only respire anaerobically for relatively short periods; for example, plant roots do so under temporarily waterlogged conditions. Respiration must then revert to being aerobic if the organism is to survive. On the return of aerobic conditions plants can and do metabolise ethanol to acetyl-CoA, which is then involved in reactions of established pathways. On the other hand, **yeasts** are unable to metabolise the ethanol they produce in fermentation.

15.8 ENERGY YIELD OF RESPIRATION

The energy available from the complete oxidation of glucose (in a calorimeter, for example) is $-2880\,kJ\,mol^{-1}$. In respiration much of this is not available to do useful work because a large proportion is lost as **heat energy** during the individual chemical reactions. Only the energy that is trapped in ATP is available for metabolism. The free energy liberated by the hydrolysis of ATP is $-30.7\,kJ\,mol^{-1}$. From these figures the efficiency of respiration can be estimated.

In aerobic respiration 36 to 38 molecules of ATP are formed for every molecule of glucose used (Table 15.2), according to how efficiently the hydrogen atoms removed during glycolysis are transported into the mitochondrion.

Table 15.2 The yield of ATP in aerobic and anaerobic respiration: the yield of a respiratory pathway may be measured in ATPs formed per molecule of glucose respired

Stage/step	Aerobic respiration	Anaerobic respiration
1 Glycolysis		
Net ATP formed directly	2	2
ATP formed when reduced NAD is oxidised (1 reduced NAD = 3 ATPs)	6	–
2 Krebs cycle (2 'turns' per glucose molecule respired)		
In the pyruvate → acetylCoA step, ATP formed when reduced NAD is oxidised (1 reduced NAD = 3 ATPs)	6	–
Net ATP formed directly	2	–
3 Electron-transport pathway		
ATP formed when reduced NAD is oxidised (1 reduced NAD = 3 ATPs)	18	–
ATP formed when reduced FAD is oxidised (1 reduced FAD = 2 ATPs)	4	–
Totals	38	2

Therefore the maximum efficiency of aerobic respiration is

$$\frac{38 \times -30.7}{-2880} \times 100 = 40.5\%$$

This figure compares favourably with the estimated efficiency of the petrol engine (approximately 25%).

In anaerobic respiration the bulk of the energy of glucose is not made available in the cell, since the products of fermentation are still in a highly reduced state. The net production of two molecules of ATP from glycolysis is the total yield from anaerobic respiration of a molecule of glucose. For this reason the yield of energy from anaerobic respiration contrasts unfavourably with that from aerobic respiration; it is given by

$$\frac{2 \times -30.7}{-2880} \times 100 = 2.1\%$$

In animals, much of the energy in lactate is released after subsequent metabolism, however, just as the energy in ethanol becomes available to plants once aerobic respiration is re-established.

Questions

1 How may the absence of oxygen switch off the Krebs cycle and the electron pathway?
2 Write an equation for the reaction between pyruvic acid and $NADH_2$.
3 Which intermediates of the Krebs cycle provide the carbon skeletons of specific amino acids?
4 Suggest likely
   a plant species, and
   b mammal tissues
   in which respiration of fat may occur (Figure 15.26, overleaf).

15.9 RESPIRATION AND FURTHER METABOLISM

The obvious connection between respiration and metabolism is 'ATP'. Respiration is the major source of the ATPs required to drive metabolism. Additionally, respiration and metabolism are interdependent. Metabolism is the source of a range of possible respiratory substrates, and respiration is the source of many of the intermediates required in metabolism for the synthesis of the other molecules. So respiration and the rest of metabolism are intimately connected.

■ Alternative respiratory substrates

Both fats and proteins can be used as alternatives to sugar in respiration. The pathways by which these respiratory substrates feed into aerobic respiration are shown in Figure 15.26 (overleaf).

Fats

The mammalian liver (p. 508) and the seeds of many flowering plants (p. 404) contain stores of fats or oils that may be used in aerobic respiration. Lipase catalyses the hydrolysis of fats and oils to glycerol and fatty acids.

In respiration, glycerol is phosphorylated with ATP, dehydrogenated with NAD and converted to triose phosphate. Triose phosphate is fed into the glycolysis pathway and respired. There is a net yield of 19 molecules of ATP from the oxidation of triose phosphate and of the $NADH_2$ formed.

Fatty acids contain a long hydrocarbon chain. This is oxidised by the successive removal of two-carbon fragments, in the form of acetylcoenzyme A. This process, known as **B oxidation**, occurs in the matrix of the mitochondria. The acetylcoenzyme A is then oxidised to carbon dioxide and water by the Krebs cycle and the electron-transport pathway, and the coenzyme A is available for re-use. Fatty acids are an important source of energy. For example, the oxidation of stearic acid $(CH_3(CH_2)_{16}COOH, p. 319)$ yields about 145 ATP molecules.

Proteins

Proteins are also used as a respiratory substrate by animals, especially carnivorous animals.

Proteins are first hydrolysed to amino acids. Then individual amino acids are deaminated (that is, the amino group, $—NH_2$, is split off and is excreted as ammonia, urea or uric acid, p. 508). The residual carbon compound, a keto acid, then enters the respiration pathway as pyruvic acid, acetylcoenzyme A or a Krebs cycle acid such as oxaloacetic acid or α-ketoglutaric acid.

■ Intermediates from respiration

Aerobic respiration serves anabolic reactions by providing a pool of intermediates and starting substances from which most of the complex biochemicals of living cells are manufactured. The respiratory pathways as starting points for synthesis of biochemicals are summarised in Figure 15.26.

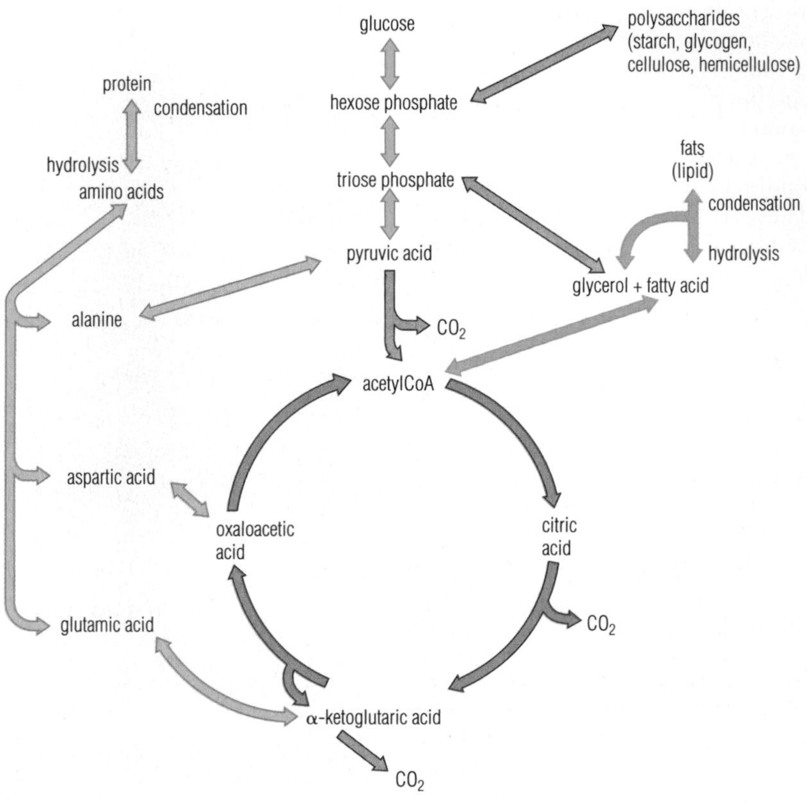

15.10 MEASURING RESPIRATION

The rate of respiration in small animals and germinating seeds may be measured in a simple **respirometer** in which the carbon dioxide given out is absorbed by soda lime, and the oxygen uptake is detected by displacement of manometric fluid in a glass U-tube. Since this manometer may also respond to change in temperature or pressure, a differential respirometer with a thermobarometer attached to the manometer improves accuracy (Figure 15.27). Respiration of tissue samples can be accurately measured in the Warburg respirometer (Figure 11.20, p. 239).

Measurement of the rate of respiration (oxygen uptake in a given time) allows investigation of factors that affect respiration. There are many such factors,

Figure 15.26   Tissue respiration and its connections with the rest of metabolism

polysaccharides (starch, glycogen, cellulose, hemicellulose)

glucose

protein — condensation — hexose phosphate

hydrolysis

amino acids — triose phosphate

fats (lipid)

condensation

hydrolysis

pyruvic acid

glycerol + fatty acid

alanine

CO_2

acetylCoA

aspartic acid

oxaloacetic acid

citric acid

CO_2

glutamic acid

α-ketoglutaric acid

CO_2

CO_2

Figure 15.27   A simple method of measuring respiration

The respiring organism absorbs O_2 and gives out CO_2. The CO_2 is absorbed by soda lime, so the only volume change results from oxygen uptake and will cause the manometer fluid to rise on the right. The syringe is depressed from time to time to keep the level constant. The volume of O_2 absorbed in a given time will be shown by the reading on the syringe.

a differential respirometer

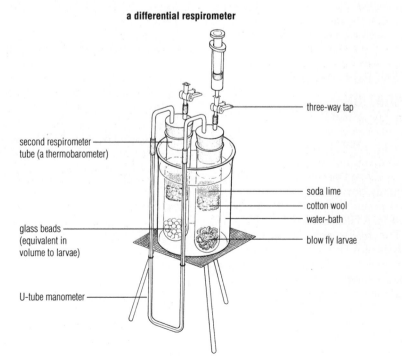

three-way tap

second respirometer tube (a thermobarometer)

soda lime
cotton wool
water-bath

glass beads (equivalent in volume to larvae)

blow fly larvae

U-tube manometer

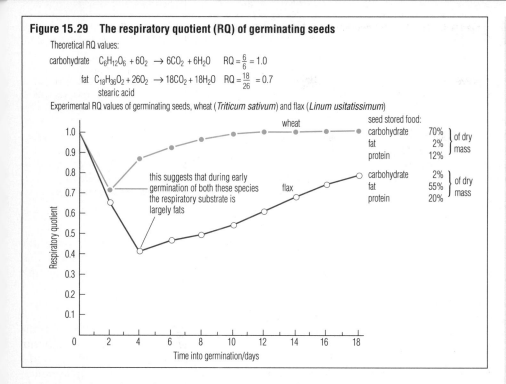

Figure 15.29 The respiratory quotient (RQ) of germinating seeds

Theoretical RQ values:

carbohydrate $C_6H_{12}O_6 + 6O_2 \rightarrow 6CO_2 + 6H_2O$ $RQ = \frac{6}{6} = 1.0$

fat $C_{18}H_{36}O_2 + 26O_2 \rightarrow 18CO_2 + 18H_2O$ $RQ = \frac{18}{26} = 0.7$
stearic acid

Experimental RQ values of germinating seeds, wheat (*Triticum sativum*) and flax (*Linum usitatissimum*)

seed stored food:
wheat
carbohydrate 70% of dry
fat 2% mass
protein 12%

this suggests that during early germination of both these species the respiratory substrate is largely fats

flax
carbohydrate 2% of dry
fat 55% mass
protein 20%

both internal (such as the age and stage of development of the organism) and external (such as the temperature, Figure 15.28).

■ Respiratory quotient

The **respiratory quotient** is the ratio of the volumes of carbon dioxide and oxygen exchanged in a given time:

$$RQ = \frac{\text{carbon dioxide evolved}}{\text{oxygen consumed}}$$

The value of measuring RQ of a tissue is that it can suggest which respiratory substrates are being used. Respiration of carbohydrates gives an RQ of 1.0, whereas that of fats gives an RQ of 0.7 (because fats are more highly reduced than sugars and more oxygen is required for their oxidation). In fact, organisms frequently respire a mixture of substrates. If anaerobic respiration is taking place the RQ will be less than 1, even when carbohydrate is the substrate. Thus the RQ value is not conclusive evidence of which respiratory substrate is being catabolised (Figure 15.29).

■ Control of respiration

The rate of respiration is directly related to the demand for energy in cells and tissues. If the amount of ATP present is large, the rate of respiration slows. This is because a high concentration of ATP is an inhibitor of an enzyme needed for phosphorylation of sugar at the start of glycolysis. In the presence of excess ATP, the supply of pyruvate for the Krebs cycle dries up, and ATP formation slows and stops. When the demand for ATP increases again, inhibition of glycolysis is lifted and ATP formation recommences.

FURTHER READING/WEB SITES

C J Clegg (1998) *Illustrated Advanced Biology: Mammals Structure and Function*. John Murray
Browse:
vl.bwh.harvard.edu/metabolism.shtml

■ Measuring the human body's demands for energy

Respiration sustains the chemical changes of the body, since it provides the energy for metabolism. The rate of metabolism can be estimated by measuring the heat energy produced in a given time, in a special calorimeter room, or by measuring gaseous exchange during different activities.

Studies of human metabolism have established that the body's requirements for energy are determined by at least three factors, outlined below.

Basal metabolism This is the background rate of metabolism, the energy required to maintain the body's constant internal environment (heartbeat, nerve-impulses, breathing requirements, temperature, chemical changes, and physical posture). Metabolism during sleep approximately equals basal metabolism. The energy requirement of basal metabolism of a woman is typically 250 kJ h^{-1}, and for a man it is 315 kJ h^{-1}.

The degree of physical activity Additional energy is required for exercise and physical work, whether in employment or leisure pursuit. For example, the energy expenditure above basal metabolic rate during various activities is as follows:

a gentle stroll   585.8 kJ h^{-1}
a brisk walk   1004.2 kJ h^{-1}
intense sustained
 physical activity
 (e.g. sawing wood)   1882.8 kJ h^{-1}

In the absence of physical activity the demand for energy (*above* basal metabolism) is very low; for example:

keyboard skills   167.4 kJ h^{-1}
concentrated brain work   0.0 kJ h^{-1}

Growth and repair work The processes of growth and development and the repair of damaged tissues undertaken by the body generate further demands for energy and materials. For example, the rapid growth that occurs between the ages of 11 and 16 years demands an additional 7.7 kJ h^{-1}, whereas the normal replacement processes of a person who has completed growing are met within the basal metabolism requirements.

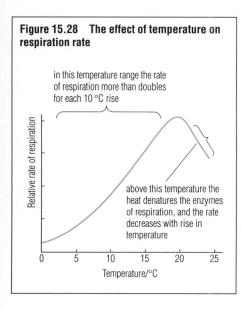

Figure 15.28 The effect of temperature on respiration rate

in this temperature range the rate of respiration more than doubles for each 10 °C rise

above this temperature the heat denatures the enzymes of respiration, and the rate decreases with rise in temperature

16 Uptake and transport within plants

- Movement of water across cells occurs mostly in the spaces in the cellulose walls (apoplast), but may also occur through the cytoplasm (symplast).
- Transpiration is the evaporation of water from the aerial parts of plants. It generates a force that pulls water up the stem from the roots. Water movement occurs though xylem vessels, which are hollow tubes.
- Stomata are the pores, mainly in the leaf epidermis, through which diffusion of gases (CO_2 and O_2) and water vapour occurs. Stomata are opened and closed by the plant.

- Water and ion uptake from the soil occurs through the roots, mainly at the root hairs. Water uptake by a cell occurs by osmosis, whereas ions are selectively absorbed, using energy from respiration.
- Combined nitrogen is one of the inorganic nutrients essential for survival of all living things. It is constantly recycled. Recycling involves soil, atmosphere and water, plus the actions of microorganisms.
- Translocation is the movement of organic molecules in the phloem tissues, e.g. from leaves where the sugars are formed to growing points and storage sites. Phloem transport is an active process of living cells.

16.1 INTRODUCTION

Within a multicellular organism, materials are transported to and from individual cells, in all parts of the organism. As a consequence, efficient means of internal transport are essential. In Chapter 11 the mechanisms for transport are discussed. Substances may move by diffusion (p. 230), by osmosis (a special case of diffusion, p. 232), by active transport (p. 236), and by various forms of mass flow, including endocytosis and exocytosis (p. 240) and cytoplasmic streaming (p. 341).

■ Transport within the flowering plant

Transport within higher plants is of nutrients (carbon dioxide, water and essential ions), oxygen for respiration, elaborated foods (mainly sugar and amino acids) and 'hormones'. These substances are transported in different ways: water and most ions are transported in the xylem vessels (p. 225), carbon dioxide for photosynthesis and oxygen for respiration diffuse in the system of air spaces that permeates plant structures, and elaborated foods are transported in the phloem tissues (p. 227). The transport of plant hormones is discussed in Chapter 20 (p. 422).

16.2 THE MOVEMENT OF WATER IN PLANTS

There are distinct routes by which water moves through the tissues of flowering plants (Figure 16.1).

Apoplast

The **apoplast** pathway consists of the spaces within the cellulose cell walls. The volume of these spaces, known as free space, is considerable. It includes the water-filled spaces of dead cells and the hollow tubes of the xylem vessels, as well as the intermolecular spaces in the cell walls. The apoplast route totally avoids the contents of living cells. Water travelling via the apoplast meets little resistance. Up to 90% of the water that travels through the plant goes by this route.

Symplast

The **symplast** is a pathway through the living contents of the cells, the cytoplasm. The contents of cells are linked by the cytoplasmic connections (plasmodesmata, p. 167) that cross the cell walls through pits. Cytoplasm presents a considerable resistance to the flow of water through cells, due to the various organelles and membranes. Nevertheless, a significant quantity of water travels by the symplast.

The vacuolar route

Water also moves into and out of cell vacuoles by osmosis, across partially permeable membranes (plasma membrane and tonoplast). The vacuolar route is not a significant pathway of water transport, but it is the route by which individual cells absorb water from the apoplast.

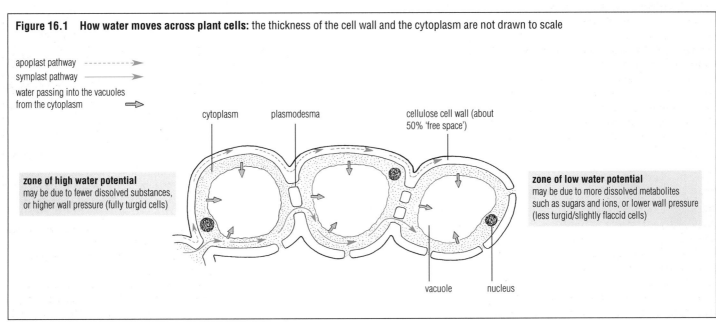

Figure 16.1   How water moves across plant cells: the thickness of the cell wall and the cytoplasm are not drawn to scale

apoplast pathway

symplast pathway

water passing into the vacuoles from the cytoplasm

cytoplasm   plasmodesma   cellulose cell wall (about 50% 'free space')

zone of high water potential
may be due to fewer dissolved substances, or higher wall pressure (fully turgid cells)

zone of low water potential
may be due to more dissolved metabolites such as sugars and ions, or lower wall pressure (less turgid/slightly flaccid cells)

vacuole   nucleus

The movement of water through plants

To the casual observer, the flow of water from the soil to the tops of tall trees appears to defy the laws of hydraulics! The heights of some species and the quantities of water that flow up to the canopy are phenomenal. For example, the giant redwoods of California (*Sequoiadendron giganteum*) and the blue gum trees of Australia (*Eucalyptus* sp.) grow to heights of more than 100 metres. These species transport many litres of water up the trunks in just a few hours on a hot day.

Transpiration

To understand how water moves to the top of a tall tree we have to appreciate that only a minute fraction of the water taken up to the aerial system (leaves, stems, flowers and buds) is retained there or used in photosynthesis and growth. The bulk of the water moving through the plant (more than 99%, in fact) evaporates from the surfaces of cells inside the leaves and escapes from the leaves as water vapour. The evaporation of water from the aerial parts of plants is known as **transpiration**. The process of transpiration is illustrated in Figure 16.2, and the pathway of water from soil solution (p. 62) to vapour in the atmosphere via root, stem and leaf in Figure 16.3.

Cohesion of the water column

The heat energy for evaporation of water vapour from the leaves is provided by the Sun. A continuous column of water extends from the external surface of the mesophyll cells throughout the free spaces of the plant cell walls, to the water inside xylem vessels. Because of its unique cohesive properties (its tensile strength, p. 126), water is drawn up the xylem to replace the water that has evaporated from the walls of the mesophyll cells in the leaves. Transpiration maintains a water potential gradient by which water moves from the soil into the root hairs, and across the cortex of the root to the central vascular tissue, too (Figure 16.12, p. 326).

The role of transpiration

Transpiration has a cooling effect, countering the heating effect of the Sun on the whole shoot system of the plant. This benefit apart, transpiration is merely an unhelpful consequence of the structure and functioning of the aerial parts of the plant, rather than a process with a clear use. The plant functions as a wick, drying the soil. The stream of water travelling up the xylem vessels does, however, serve to conduct ions and mineral salts from roots to stem, leaves and growing points.

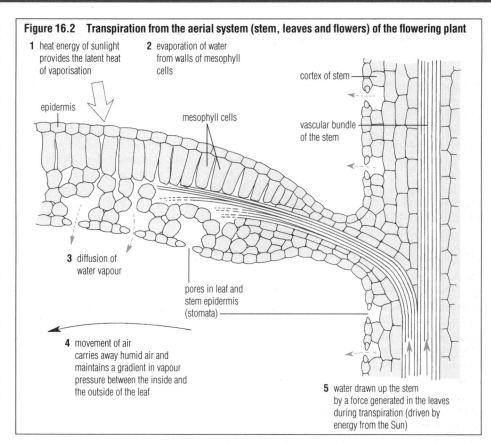

Figure 16.2 Transpiration from the aerial system (stem, leaves and flowers) of the flowering plant

1 heat energy of sunlight provides the latent heat of vaporisation

2 evaporation of water from walls of mesophyll cells

epidermis

mesophyll cells

cortex of stem

vascular bundle of the stem

3 diffusion of water vapour

pores in leaf and stem epidermis (stomata)

4 movement of air carries away humid air and maintains a gradient in vapour pressure between the inside and the outside of the leaf

5 water drawn up the stem by a force generated in the leaves during transpiration (driven by energy from the Sun)

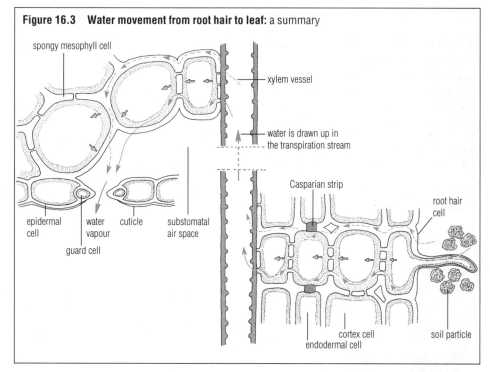

Figure 16.3 Water movement from root hair to leaf: a summary

spongy mesophyll cell

xylem vessel

water is drawn up in the transpiration stream

Casparian strip

root hair cell

epidermal cell

water vapour

cuticle

substomatal air space

guard cell

cortex cell

endodermal cell

soil particle

Questions

1 What are the differences between the apoplast and the symplast of cells?

2 What factors affect the water potential of a plant cell (p. 232)?

3 In hot, dry weather the plant acts as a wick, drying out the soil. What advantages, if any, does transpiration confer on the green plant?

■ Stomata

Stomata are found in the epidermis of leaves, stems and parts of the flowers of flowering plants. Each stoma consists of a pore surrounded by two **guard cells**. The surrounding epidermal cells abutting the stoma are called **subsidiary cells** (Figure 16.4). The guard cells are firmly joined at both ends, but are able to separate in the mid-region of their length (Figure 16.8).

Most stomata occur in the leaves. In the dorsi-ventrally flattened leaves typical of most dicotyledons (p. 34), stomata are present mainly in the lower epidermis (Figure 16.5). There are no stomata in the root epidermis.

The role and importance of stomata

The open stomata of a leaf represent an extremely small area of the total leaf. Despite this the epidermis is not a serious barrier to the inward diffusion of carbon dioxide (facing page). But in conditions of water deficit when the life of the plant may be threatened by desiccation the stomata close automatically. Water loss is virtually prevented until the deficit can be made good.

The opening and closing of stomata

The stomata of most species tend to open in the light and close in the dark, but not all species conform to this pattern. Some species have stomata that open well before dawn and close during the hours of daylight (Figure 16.6), and a few species have an entirely different regime of opening and closing (see Crassulacean plants, p. 375).

The opening and closing of stomata is caused by changes of turgor pressure in the guard cells (Figure 16.8). The guard cells take up water from surrounding cells, and become turgid. The shape of the guard cell walls, and the directions in which the microfibrils of cellulose in the cell walls are laid down, all play a part in causing the pore between the guard cells to open as they become turgid (Figure 16.8).

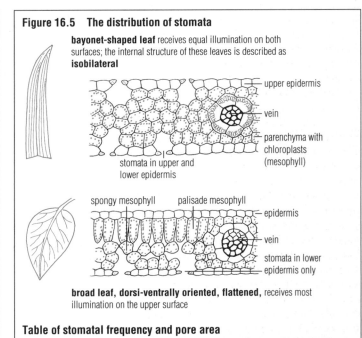

Figure 16.6 Stomatal aperture in the potato (outer circle) and maize (middle circle): the innermost circle shows the hours of darkness

---

Figure 16.4 The structure of stomata

stoma in TS

ordinary epidermal cells

cuticle

subsidiary cell, contains no chloroplasts

pore

guard cell, contains chloroplasts (the guard cells are the only cells in the epidermis to contain chloroplasts)

surface view of stomata

Figure 16.5 The distribution of stomata

bayonet-shaped leaf receives equal illumination on both surfaces; the internal structure of these leaves is described as **isobilateral**

upper epidermis

vein

parenchyma with chloroplasts (mesophyll)

stomata in upper and lower epidermis

spongy mesophyll   palisade mesophyll

epidermis

vein

stomata in lower epidermis only

broad leaf, dorsi-ventrally oriented, flattened, receives most illumination on the upper surface

Table of stomatal frequency and pore area

Species	Frequency/mm^{-2}		Pore area/% of leaf area
	upper surface	lower surface	
Monocotyledons (leaves mostly isobilateral for part or all of the leaf)			
onion (cylindrical leaves)	175	175	2.0
wheat	50	40	0.6
maize	98	108	0.7
Dicotyledons (leaves are dorsi-ventrally flattened)			
oak	0	340	0.8
sunflower	120	175	1.1
tobacco	50	190	0.8
broad bean	65	75	1.0

The opening and closing of stomata can be investigated by means of a porometer, which measures the resistance to air flow through the leaf. Alternatively, indirect measurement can be made by investigating the time taken for dry cobalt chloride paper to turn from blue to pink when held securely to the surface of leaves (*Study Guide*, Chapter 16).

Diffusion through pores

Open stomata occupy 1–2% of the total leaf area, but allow 50–60% of the diffusion that would occur if no barrier covered the mesophyll cells.

This is so because the rate of diffusion of gases through small pores is proportional to the perimeter of the pore, not to its area. The smaller the pore, the greater is the proportion of molecules that can diffuse near or at its perimeter.

Once through the pores, the molecules accumulate over them in a zone of still air. Hemispherical diffusion shells form over the pores (Figure 16.7). At the edge of a pore the concentration gradient (which is the cause of diffusion) is greatest.

As a stomatal pore starts to close there is no initial change in the dimensions of its perimeter, and therefore little effect on the rate of diffusion. Diffusion through the pore only ceases when it closes completely.

Consequently, very small open pores, well spaced, avoid overlap of the diffusion shells and provide little check on diffusion. The stomata of sunflower leaves, for example, could allow diffusion of much more carbon dioxide and water vapour than normally passes to and from the mesophyll cells, even at peak times. Nevertheless, when water is scarce the stomata are invaluable, since they then close and prevent excessive water loss.

Questions

1 Which cells of the plant epidermis contain chloroplasts?
2 Small changes in the size of a stomatal pore have little effect on diffusion. Why is this so?
3 How might you demonstrate that stomata open because of turgor pressure?
4 When do Crassulacean plants open their stomata (pp. 375–6)?

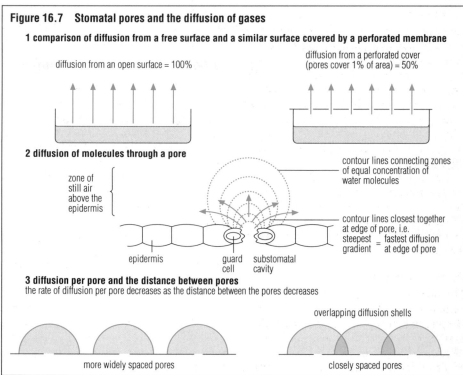

Figure 16.7 Stomatal pores and the diffusion of gases

1 comparison of diffusion from a free surface and a similar surface covered by a perforated membrane

diffusion from an open surface = 100%

diffusion from a perforated cover (pores cover 1% of area) = 50%

2 diffusion of molecules through a pore

zone of still air above the epidermis

contour lines connecting zones of equal concentration of water molecules

contour lines closest together at edge of pore, i.e. steepest gradient = fastest diffusion at edge of pore

epidermis   guard cell   substomatal cavity

3 diffusion per pore and the distance between pores
the rate of diffusion per pore decreases as the distance between the pores decreases

overlapping diffusion shells

more widely spaced pores   closely spaced pores

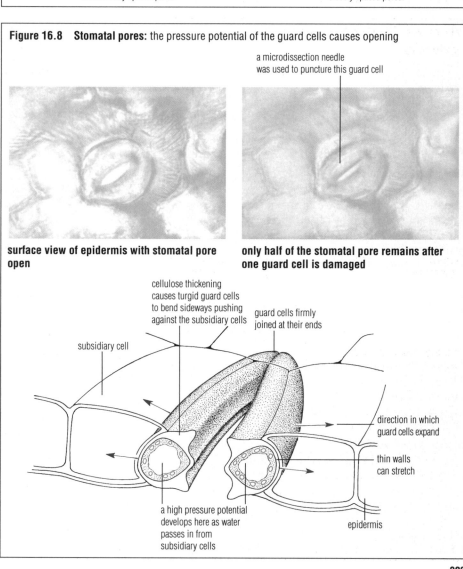

Figure 16.8 Stomatal pores: the pressure potential of the guard cells causes opening

a microdissection needle was used to puncture this guard cell

surface view of epidermis with stomatal pore open

only half of the stomatal pore remains after one guard cell is damaged

cellulose thickening causes turgid guard cells to bend sideways pushing against the subsidiary cells

guard cells firmly joined at their ends

subsidiary cell

direction in which guard cells expand

thin walls can stretch

a high pressure potential develops here as water passes in from subsidiary cells

epidermis

Stomata: the mechanism for opening and closing

The guard cells are the only epidermal cells to contain chloroplasts. In the light, photosynthesis occurs and sugars are formed. An early explanation of the opening mechanism was that sugar manufactured in guard cells lowered the water potential, and caused water to enter by osmosis from the surrounding epidermal cells. This is not an adequate explanation, for the following reasons:

► the response of stomata is too quick to be due to the slow accumulation of sugar

► in several plants, the stomata open before daybreak

► a marked accumulation of potassium ions and malate has been observed in guard cells prior to opening of the pore

► a plant growth regulator (an inhibitor of growth, p. 431) causes stomatal closure.

A recent hypothesis of stomatal opening suggests that the accumulation of potassium and malate ions in the guard cells lowers the water potential and causes stomatal opening (Figure 16.9).

Figure 16.9　The opening and closing of stomata: a hypothesis

Opening

1 a potassium ion pump in the cell membranes actively transports K$^+$ ions from subsidiary cells to guard cells

2 carbon dioxide concentration falls (due to photosynthesis); the pH rises; the starch/malate equilibrium shifts to the right (due to pH sensitive enzymes):

starch $\rightleftharpoons$ malate

3 K$^+$ and malate ions accumulate in the vacuoles of guard cells; water flows into guard cells, causing high hydrostatic pressure in these cells

Closing

1 K$^+$ ion pump in the cell membranes of the guard cells transfers K$^+$ ions to the subsidiary cells

2 carbon dioxide concentration rises (due to respiration only); the pH falls; the starch/malate equilibrium shifts to the left:

starch $\rightleftharpoons$ malate

3 ions are lost from the guard cells; water flows from the guard cells to the subsidiary cells, leading to lowered hydrostatic pressure in the guard cells

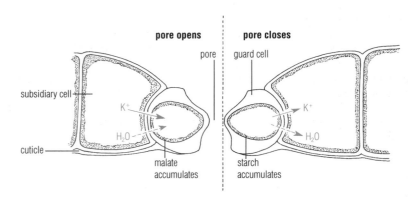

■ The rate of transpiration

Measuring transpiration rate

It is common practice to use a potometer (Figure 16.10) to measure transpiration (the loss of water vapour from the aerial system). In fact, the potometer measures water uptake by a leafy shoot. As the shoot transpires, the water vapour it has lost is replaced by liquid water drawn in from the potometer via the xylem of the stem. Water uptake is assumed to be the same as the water lost by the shoot due to transpiration.

Factors affecting transpiration rate

The potometer may be used to investigate the effects of environmental factors on the rate of transpiration (the rate of evaporation from the leaves).

Humidity (vapour pressure)

The humidity of the atmosphere affects the gradient of water vapour between the substomatal cavity and the air outside the leaf, and hence the rate of diffusion of water vapour. Low humidity (low vapour pressure) outside the leaf favours transpiration because it increases the gradient. Most

Figure 16.10　A potometer to measure transpiration

the movement of water along the capillary tube is timed, using a stopclock; between readings the water can be returned to the start by letting in water from the resevoir

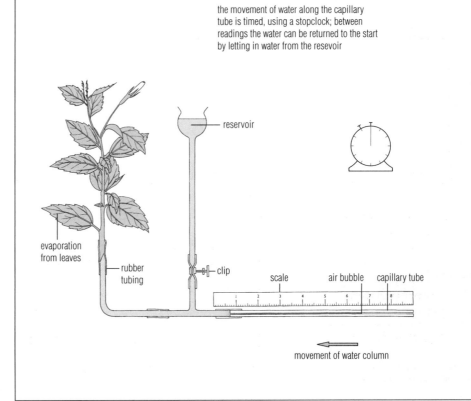

plants adapted to very arid conditions show structures that reduce the loss of water vapour. These features are known as xeromorphic (Table 18.2, p. 375).

Temperature

High temperature speeds up transpiration by providing the latent heat (enthalpy) of vaporisation. At the same time high temperature also lowers the relative humidity of air. Both of these changes increase the gradient in water vapour concentration between the substomatal cavity and the external atmosphere.

Air movements

Moving air sweeps away the water vapour in the air outside the stomata (Figure 16.2), speeding up diffusion of water vapour by making a steeper gradient between the substomatal cavity and the air. Consequently windy conditions lead to an increased rate of transpiration. Some plants adapted to dry conditions (xerophytes) have rolled or folded leaves, or have developed epidermal hairs – all measures that help retain moist air around the stomatal apertures.

Atmospheric pressure

Water vapour pressure decreases as the atmospheric pressure decreases with increasing altitude. The lower the atmospheric pressure, the greater is the rate of evaporation of water from leaves. Most alpine plants show xeromorphic features, which help reduce water loss.

Light

Light affects transpiration indirectly. Stomata tend to be open in the light, and so transpiration is more likely to occur in the light than in the dark. Sunlight is also the chief source of heat energy for the plant, enhancing evaporation.

Water supply

Shortage of water in the soil leads to wilting of the plant and closure of the stomata. As the soil dries out the remaining soil water becomes a more concentrated solution. When the water potential of the soil water is lower (more negative) than that of the cells of the root the **permanent wilting point** has been reached, and plants die quickly.

Features of leaf structure

We have seen that leaf shape and the position of stomata influence water vapour loss in xerophytes. In these plants the cuticle is extremely thick, and there is negligible loss of water vapour through the cuticle (cuticular transpiration). On the other hand, shade plants with a large leaf surface area and a thin cuticle may lose much water vapour via the cuticle.

Water stress

Simultaneous measurements of water uptake and water loss by transpiration show that the supply of water into the plant often falls behind the rate of water loss at times of peak transpiration. This places the plant tissues of roots, stem and leaves under some degree of water shortage, referred to as stress.

Evidence of this stress comes from measurement of the diameter of a tree trunk over a 24-hour period (*Study Guide*, Chapter 16). In a large tree there is an easily detectable shrinkage in the diameter of the trunk during the day, which recovers during the night when transpiration has stopped and water uptake has made good the losses (Figure 16.11).

If the water stress is sufficiently severe the plant may release a growth inhibitor (abscisic acid, p. 431), which triggers stomatal closure, probably by activating the metabolic ion pump mechanism by which potassium ions are moved from the guard cells to the subsidiary cells (Figure 16.9).

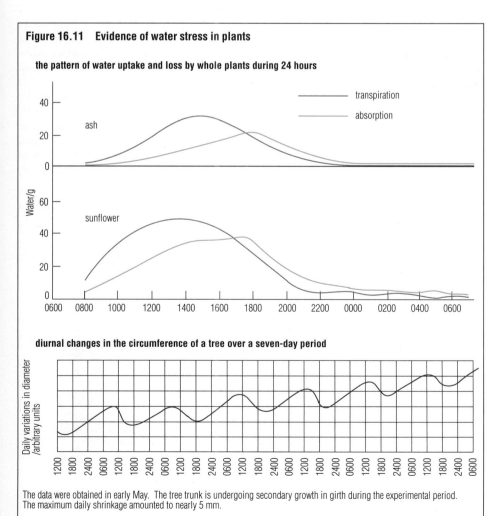

Figure 16.11 Evidence of water stress in plants

the pattern of water uptake and loss by whole plants during 24 hours

diurnal changes in the circumference of a tree over a seven-day period

The data were obtained in early May. The tree trunk is undergoing secondary growth in girth during the experimental period. The maximum daily shrinkage amounted to nearly 5 mm.

Questions

1 a How do changes in turgor of the guard cells tend to cause both the opening and the closing of the stomatal pore?

   b What conditions might lead to stomatal closure?

2 a What does the potometer measure?

   b How could you measure loss of water vapour from a leafy shoot?

3 What adaptations favourable to survival in arid conditions are commonly shown by plants?

4 In Figure 16.11 the pattern of water uptake and loss by intact plants during a 24-hour period is illustrated.

   a For about how long in a 24-hour period were the plants losing water faster than it was absorbed?

   b What are the likely explanations for the slow-down in transpiration after 14.00 hours?

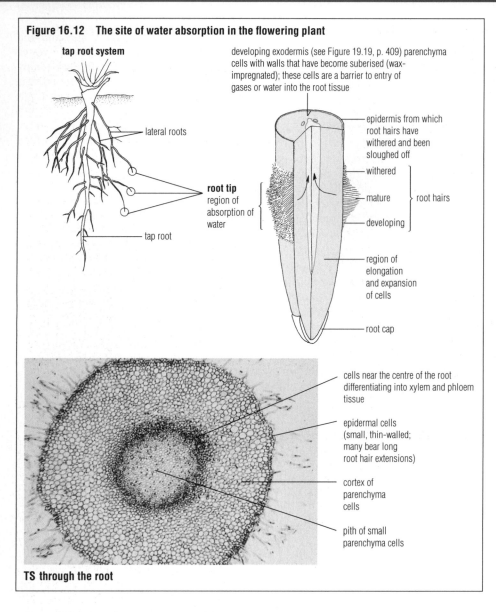

Figure 16.12 The site of water absorption in the flowering plant

tap root system

lateral roots

root tip
region of absorption of water

tap root

developing exodermis (see Figure 19.19, p. 409) parenchyma cells with walls that have become suberised (wax-impregnated); these cells are a barrier to entry of gases or water into the root tissue

epidermis from which root hairs have withered and been sloughed off

withered

mature — root hairs

developing

region of elongation and expansion of cells

root cap

cells near the centre of the root differentiating into xylem and phloem tissue

epidermal cells (small, thin-walled; many bear long root hair extensions)

cortex of parenchyma cells

pith of small parenchyma cells

TS through the root

■ Water uptake by the roots

The root system grows in the soil and provides anchorage for the aerial system of the plant. The uptake of water (and mineral salts, as ions) by roots from the soil is more or less restricted to the younger, most recently formed cells of the region of extension growth and the region of root hairs (Figure 16.12). Root hairs are short-lived, and older parts of the root are rendered less permeable by the deposition of suberin in the outer cortical cells (the exodermis, p. 409).

Most plants form root hairs, which increase the surface area for absorption. Root hairs are tiny extensions of individual cells of the root epidermis. They are formed in the region of the root quite close to the root tip, immediately after the cells formed at the root apex have completed extension growth (the region of elongation). The root hairs grow out between the soil particles and make intimate contact with the film of soil water on them (Figure 16.13).

Water movement across the root cortex

We have seen that water can move through plant tissues from vacuole to vacuole by osmosis, and through the cytoplasm of the cells via the plasmodesmata (the symplast), but that the bulk of water passes through the 'free spaces' of the cell walls (the apoplast) (Figure 16.1).

Water converges onto the central vascular tissue enclosed by the endodermis. Entry into this central region of the root, the stele (p. 409), is through the cells of the endodermis. Endodermal cells are

Figure 16.13 Root hairs, the absorbing surface of plants

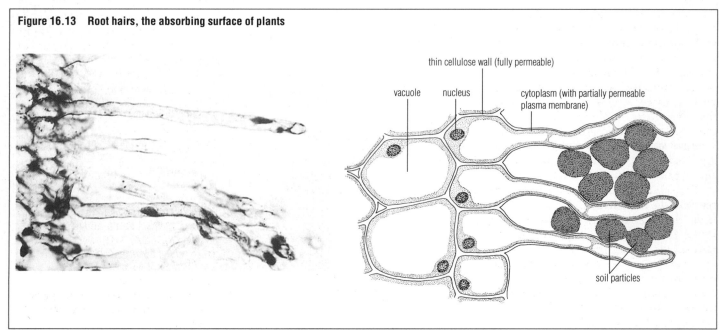

thin cellulose wall (fully permeable)

vacuole nucleus

cytoplasm (with partially permeable plasma membrane)

soil particles

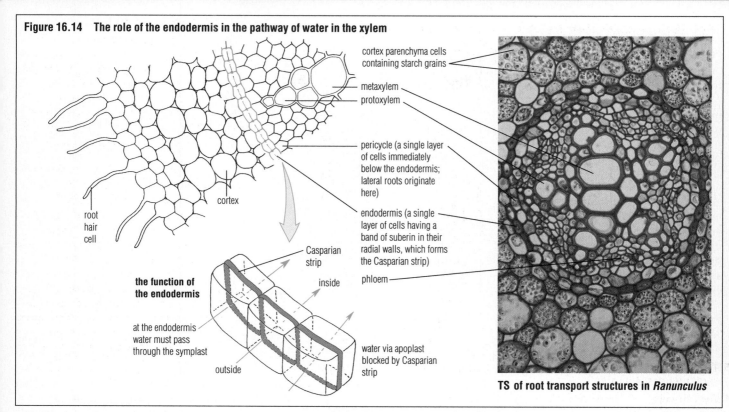

Figure 16.14 The role of the endodermis in the pathway of water in the xylem

root hair cell

cortex

cortex parenchyma cells containing starch grains

metaxylem

protoxylem

pericycle (a single layer of cells immediately below the endodermis; lateral roots originate here)

endodermis (a single layer of cells having a band of suberin in their radial walls, which forms the Casparian strip)

phloem

the function of the endodermis

Casparian strip

inside

at the endodermis water must pass through the symplast

outside

water via apoplast blocked by Casparian strip

TS of root transport structures in *Ranunculus*

subtly different from other cells of the root in having a strip of waxy substance consisting of suberin, forming the **Casparian strip**, in their radial walls. The Casparian strip prevents movement of water via the cell walls (apoplast), and so all water (and also solutes) moving across the root to the central vascular tissues must pass through the cytoplasm of the endodermal cells (Figure 16.14).

Role of the Casparian strip

The bulk of water lost from the leaves by evaporation (transpiration) is drawn up in the apoplast from the soil by a gradient in water potential (Figure 16.15). Without the Casparian strip the plant would function as an indiscriminate wick, drawing up the soil solution into the aerial system. The result would be that ion concentrations in the aerial system would increase steadily without any control or regulation for as long as transpiration continued.

The effect of the Casparian strip is to regulate the ions that the plant draws in from the soil solution to its stem and leaves. The cytoplasm in the cells of the endodermis is selective about the type and number of ions absorbed. Only the ions absorbed by the cytoplasm of the endodermis are available to be carried up to the aerial system in the xylem stream (see section 16.3 overleaf).

Figure 16.15 The water potential (ψ) of soil, plant and air

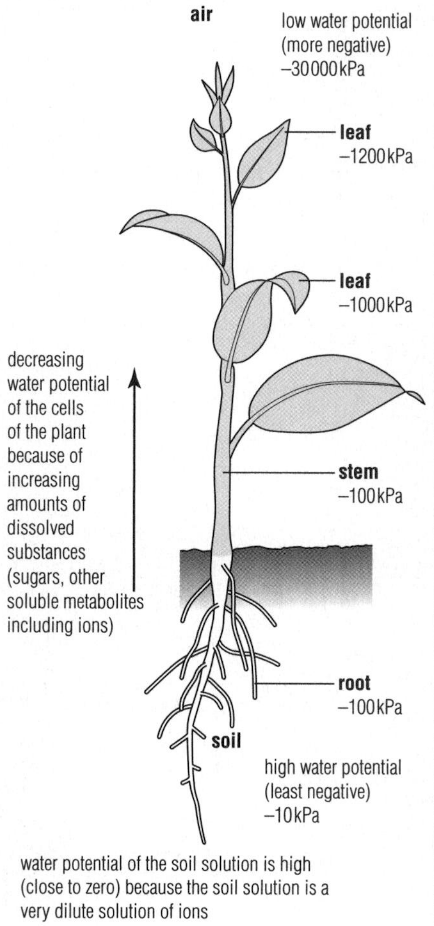

air

low water potential (more negative) −30 000 kPa

leaf −1200 kPa

leaf −1000 kPa

decreasing water potential of the cells of the plant because of increasing amounts of dissolved substances (sugars, other soluble metabolites including ions)

stem −100 kPa

root −100 kPa

soil

high water potential (least negative) −10 kPa

water potential of the soil solution is high (close to zero) because the soil solution is a very dilute solution of ions

Root pressure

If the stem of a plant is cut at its base, water exudes from the cut stump. This exudation suggests that water, as well as being drawn up the stem by tension generated in the leaves, is pushed up by a force generated in the root. This force is called **root pressure** (see *Study Guide*, Chapter 16).

Root pressure results from the vacuolar pathway of water through the endodermis, which causes the centre of the root to function as a simple osmometer. The soil solution is normally very dilute. It has a higher (less negative) water potential than the water inside the root stele. As a result water crosses into the stele by osmosis, and creates the root pressure. On its own, however, root pressure is never strong enough to drive water to the tops of tall trees at the speeds achieved during a normal growing season.

Questions

1. Why are only limited amounts of water (or ions) absorbed into roots once a suberised exodermis has been formed?
2. What features of root hairs facilitate absorption from the soil?
3. What is the consequence of the Casparian strip for the apoplast pathway of water movement?

16.3 ION UPTAKE IN PLANTS

Plants manufacture sugar and other elaborated foods by photosynthesis. Some of the products of photosynthesis are respired to release energy for metabolism, and some are used as the building blocks for the synthesis of the other compounds required for the growth of the plant. Ions are needed to make these essential components.

The mineral salts required in substantial quantities, the macronutrients, include the metal ions potassium, sodium and calcium, and the non-metal ions nitrate and phosphate. Other inorganic nutrients, required in much smaller amounts, are the micronutrients or trace elements, such as manganese and copper. The essential mineral salts, their roles and functions, are listed in Table 16.1.

■ Discovering the mineral nutrient requirements of plants

The growing of terrestrial plants with roots in an aerated solution is known as water culture or hydroponics (p. 331).

Figure 16.16   Water culture in plants

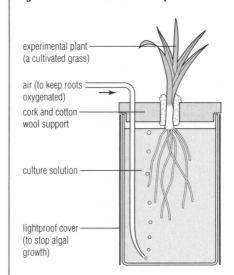

experimental plant (a cultivated grass)

air (to keep roots oxygenated)

cork and cotton wool support

culture solution

lightproof cover (to stop algal growth)

The culture solution contains a mixture of soluble salts, giving nutrient concentrations as follows:

Element	Concentration/ mg dm^{-3}
Macronutrients:	
K	195
N (as NO_3^-)	210
Ca	200
S	48
P	41
Mg	24
Fe (as Fe^{3+})	5.6
Micronutrients:	
Mn	0.55
Zn	0.065
Cu	0.064
B	0.37
Mo	0.019
Co	0.0006
Cl	3.55

This technique can be used to discover which ions are essential for normal, healthy growth (Figure 16.16).

A 'balanced' solution (complete culture solution) is one that provides all the necessary ions at appropriate concentrations. Solutions are prepared from extremely pure chemicals dissolved in deionised water. Carefully cleaned glassware is essential. Many grass species are suitable for studies of ion deficiency; they depend almost exclusively on the external supply because their seeds are small, and contain only small reserves of ions.

Table 16.1   Macro- and micronutrients of plants

Substance	Major functions	synthesis of metabolites	transport	used in metabolism
			Sites and venues	
Macronutrients: nitrogen as nitrate, NO_3^-	amino acid synthesis for proteins, nucleic acid synthesis	in roots, NO_3^- is reduced to NH_2— and combined with organic acids to form amino acids	as amino acids in the phloem	in all tissues, particularly where growth occurs and enzymes are synthesised
sulphur as sulphate SO_4^{2-}	sulphur-containing amino acids, vitamins and co-factors thiamine, biotin, coenzyme A	in all living tissues, especially in roots (amino acid formation)	as amino acids in the phloem, as SO_4^{2-} in xylem	in all tissues, including stem and root tips, young leaves
phosphorus as phosphate, PO_4^{3-}	ATP, nucleic acids, phospholipids	in all living tissues	as inorganic PO_4^{3-} in xylem, as sugar phosphate etc. in the phloem	in all tissues
potassium as potassium ion, K^+	anion/cation balance across cell membranes, active transport across membranes, enzyme activity		in phloem and xylem	in all tissues including stomata, leaves, phloem sieve tubes
calcium as calcium ion, Ca^{2+}	cell wall synthesis, enzyme activity	in all living tissues, especially stem and root apices	in xylem	in meristematic and differentiating tissues
iron as iron(II) ion, Fe^{2+}	cytochromes, chlorophyll synthesis, enzyme activity	in all living tissues	in xylem	in all tissues
magnesium as magnesium ion, Mg^{2+}	chlorophyll, activation of ATPase enzyme	in all living tissues, especially leaves	in xylem	in leaves, and all tissues
Micronutrients: *manganese* as Mn^{2+}	activation of carboxylase enzymes			
zinc as Zn^{2+}	component of carbonic anhydrase enzyme			
copper as Cu^{2+}	component of some oxidase enzymes, and as component of photosynthetic photosystem			
molybdenum as Mo^{3+}	activates nitrate reductase in nitrogen fixation			
boron as BO_3^{3-} or $B_4O_7^{2-}$	essential for meristem activity			
chlorine as Cl^-	component of the oxygen evolution step in photosynthesis			

Table 16.2 Ion-deficiency symptoms of plants. The effects of deficiencies of specific ions on plant growth can be studied using the water culture method. Alternative culture solutions are used, each having one selected element absent. Plants grown in deficient media are compared with control plants grown in complete culture solution. Knowledge of the deficiency symptoms commonly shown by plants lacking an adequate supply of an essential nutrient is important in horticulture and agriculture

Deficient element	Typical symptoms shown by deficient plants
Nitrogen	Nitrogen deficiency leads to reduced protein synthesis, and hence reduced growth of all organs, and to yellowing (chlorosis) of leaves. Chlorosis first appears in the older leaves. This happens because of the mobilisation and circulation of ions in plants (p. 330)  **a healthy leaf**  **a leaf showing nitrogen deficiency**
Sulphur	Sulphur-deficiency symptoms are very similar to nitrogen-deficiency symptoms, because most proteins contain numerous sulphur-containing amino acids
Phosphorus	Severe shortage of phosphorus affects all aspects of growth, because of the role of ATP in all aspects of metabolism; poor root growth and the failure to absorb all other minerals as ions are the most immediately influential outcomes of phosphorus deficiency
Potassium	The symptoms of potassium deficiency are similar to those of phosphorus deficiency; the leaves are dark green, and older leaves have a mottled appearance (necrotic spots)
Calcium	Calcium in the plant is immobilised in cell walls already formed; deficiency is noticed in the growing points, and in young leaves; root growth is also severely restricted
Magnesium	Existing magnesium in the plant is mobilised and transported to newly formed leaves; symptoms of magnesium deficiency appear in the oldest leaves first
Iron	Iron(II) ions are not particularly mobile in the plant and are not transferred from one organ to another; deficiency of iron leads to chlorosis of young leaves

Figure 16.17 Cations and anions available in the soil (not drawn to scale)

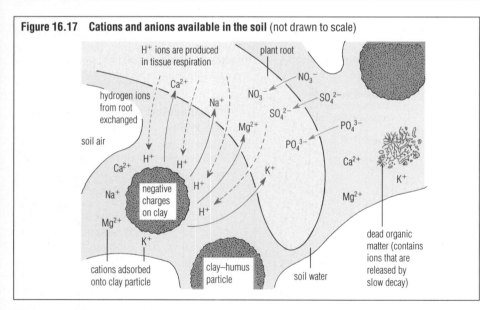

The soil as the source of nutrients

The mineral ions required by terrestrial plants are normally obtained from the soil. The components of soil are shown in Figure 3.26 (p. 61). The mineral nutrients occur in various forms in soil (Figure 16.17).

▶ **Mineral ions in solution in soil water** These are freely available to plants (as well as to the fungi and microorganisms of the soil), but are equally free to be washed away (leached) from the soil by excess rainfall or by flooding.

▶ **Mineral ions adsorbed on to clay particles** Clay particles are extremely small, have a vast surface area (collectively), and retain a reserve of charged ions adsorbed onto their surfaces. Ions are held by opposite charges on the clay surface. These ions can become available to plants but are not free to be leached away.

▶ **Mineral ions in humus** Humus is a black gum-like substance, derived from decayed plant and animal remains. It occurs around the mineral skeleton of the soil. Further decay of humus releases ions progressively.

▶ **Mineral ions stored in the mineral skeleton** The mineral skeleton of soil (for example, the sand and clay particles) is a long-term store of essential ions, but these are not available to plants until released and made soluble by weathering processes.

Questions

1 Why are de-ionised water and exceptionally clean glassware essential in water-culture studies of the mineral ion requirements of plants?

2 Why do plants deficient in nitrogen and magnesium tend to have very yellow leaves?

3 Symptoms of deficiency in calcium show up in young leaves, but symptoms of nitrogen deficiency appear in the older leaves. Why is this?

4 In what ways are the elements nitrogen and sulphur carried and distributed in plants?

■ The process of ion uptake

Ions required by plants are absorbed from the soil by the tips of roots in the same region of the root as water is absorbed (Figure 16.18). The processes of ion absorption and water absorption are entirely different, however. Water uptake occurs by mass flow, and is related to the rate of water loss from the leaves. Ion uptake is largely an active, selective process (p. 236). It is related to the requirement of the plant for particular ions.

Diffusion and active uptake

Ions are absorbed from the soil either by diffusion or by active uptake. Diffusion occurs where the internal concentration of an ion is lower than the external concentration. Most ions absorbed by plant cells are, however, already at a higher concentration inside the cell than outside (see Figure 11.15, p. 236).

Plant cells accumulate essential ions and retain them. Energy from respiration is used to 'pump' in ions, and to maintain the cell membrane as a barrier to outward diffusion of ions and other valuable metabolites. We believe that protein

carrier molecules at the external surface of the plasma membrane are activated by reaction with ATP. A protein–ion complex is formed. The carrier–ion complex then changes shape, and discharges the ion inside the cell (Figure 11.16, p. 237).

Ions may be taken up into cells of the epidermis, cortex and endodermis. Ions are absorbed into the cytoplasm and vacuole from the dilute soil solution that occupies the free space of the cell walls.

■ Transport of ions in the stem

Ions are transported up the stem in the stream of water carried in the xylem (Figure 16.19). The ions are actively absorbed into the growing cells in the leaf and stem. Meanwhile the bulk of the water brought up the stem is lost by evaporation.

■ Mobilisation and circulation of ions in plants

When essential inorganic nutrients are in short supply, a plant may mobilise existing ions. Mobilisation may occur from the oldest or first-formed leaves, taking nutrients to new growing points. Nutrients moved about the plant in this way include nitrogen (Figure 16.20), potassium, sulphur and phosphorus. Nitrogen and sulphur are translocated to the new growing points mostly in the form of amino acids, and phosphorus as sugar phosphate. As a consequence, deficiencies of these elements show up first in the oldest parts of a plant.

Other ions are built into compounds from which they cannot easily be mobilised. For example, calcium and iron are virtually immobile in the plant. Deficiency of calcium and iron shows up first at the growing points and in the newest leaves (Table 16.2).

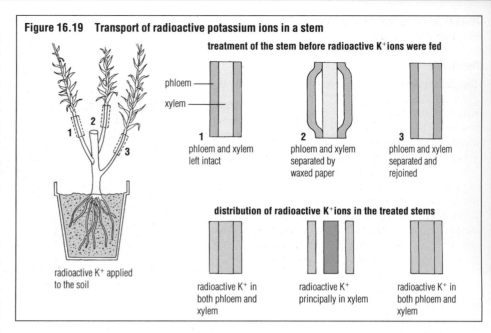

Figure 16.19   Transport of radioactive potassium ions in a stem

treatment of the stem before radioactive K⁺ ions were fed

phloem
xylem

1 phloem and xylem left intact

2 phloem and xylem separated by waxed paper

3 phloem and xylem separated and rejoined

radioactive K⁺ applied to the soil

distribution of radioactive K⁺ ions in the treated stems

radioactive K⁺ in both phloem and xylem

radioactive K⁺ principally in xylem

radioactive K⁺ in both phloem and xylem

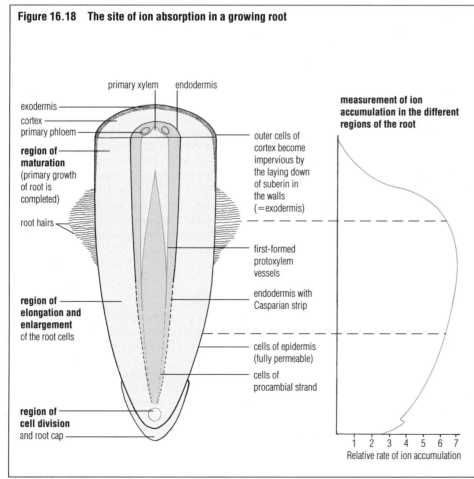

Figure 16.18   The site of ion absorption in a growing root

exodermis
cortex
primary phloem
primary xylem
endodermis

region of maturation (primary growth of root is completed)

root hairs

region of elongation and enlargement of the root cells

region of cell division and root cap

outer cells of cortex become impervious by the laying down of suberin in the walls (=exodermis)

first-formed protoxylem vessels

endodermis with Casparian strip

cells of epidermis (fully permeable)

cells of procambial strand

measurement of ion accumulation in the different regions of the root

1 2 3 4 5 6 7
Relative rate of ion accumulation

Maintaining soil fertility for commercial crops

Fertilisers

The harvesting of crops removes mineral elements, and cropping interrupts the natural recycling of nutrients. Loss of nutrients can be offset by the addition of manure or compost (natural organic waste materials) or artificial fertilisers (chemicals). A fertiliser is a material that supplies nutrients to plants, either macronutrients or micronutrients, sometimes both. Mostly, fertilisers are used to supply the elements nitrogen, phosphorus and potassium in order to promote growth rates and total yields of the crop.

The use of fertilisers has contributed to increased productivity and has permitted new, poorer areas of land to be used for agriculture. The aim is to target the delivery of fertiliser to ensure the correct balance of ions at the peak periods of uptake and growth. For example, in the investigation shown in Figure 16.20, the period from day 40 to day 60 after germination was the ideal time for 'top dressing' with nitrogen-containing fertiliser.

Over-use of fertilisers has led to environmental problems. Excessive application of organic fertilisers may prove as harmful as the application of excess soluble nitrogen fertilisers. In both cases ions intended for the crop may be washed into drainage channels and rivers, and lead to eutrophication (p. 100) of natural waters.

Figure 16.21 Hydroponic culture of cucumbers: the plants are rooted in pumice chips and individually supplied with nutrients in solution via a fine plastic pipe; run-off from the pots is collected and recycled, after the nutrient composition has been adjusted

Cultivating plants without soil

Crops may be grown without soil, provided that a solution containing all the macro- and micronutrients is made available to the roots. The process of growing plants in this way is referred to as **hydroponic culture** (Figure 16.21). This system has been successful in producing high-yielding crops such as tomatoes and cucumbers in glasshouses in temperate climates. It is also useful in arid regions, such as the edge of a desert to which fresh water is pumped. In these situations the light and temperature regimes are good for growth, and the plants have access to the water and nutrients they need.

Questions

1 'Diffusion plays a limited part in ion uptake by plant roots.' What evidence is there for this statement (see p. 236)?
2 Which cells of the root are able to absorb ions from the soil solution?
3 In a green leaf of wheat (Figure 16.20), most combined nitrogen exists as ribulose bisphosphate carboxylase (Chapter 12, p. 267).
   a What is the role of this enzyme?
   b What is the fate of this enzyme during grain formation?

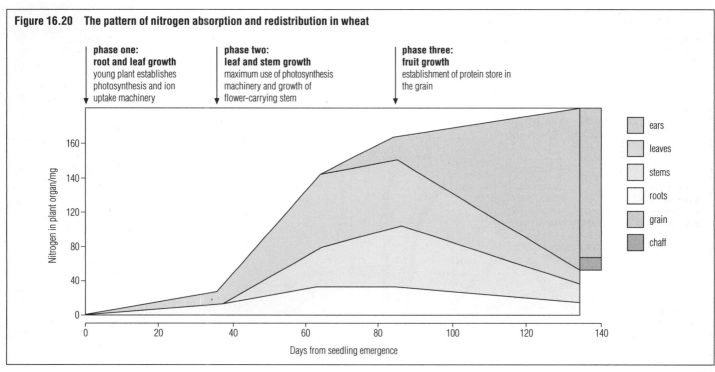

Figure 16.20 The pattern of nitrogen absorption and redistribution in wheat

Crop rotation

Crop rotation is the practice of growing different crops on the same ground in successive years, rather than repeatedly growing the same crop, year after year (Figure 16.22). The advantages of crop rotation include

▶ the prevention of the build-up in the soil of a particular group of weeds or parasites associated with a single crop

▶ variation in the demands made upon soil reserves of both macro- and micronutrients from year to year

▶ the chance to include a leguminous crop (see p. 334) in the rotation so that combined nitrogen levels can be built up by the actions of nitrogen-fixing bacteria in root nodules.

Today the use of traditional rotation patterns is limited. Cereals are often grown for many years on the same field without loss of yields. This has been made possible because of improved varieties of crops with increased yields, and by the effective use of expensive fertilisers and pesticides.

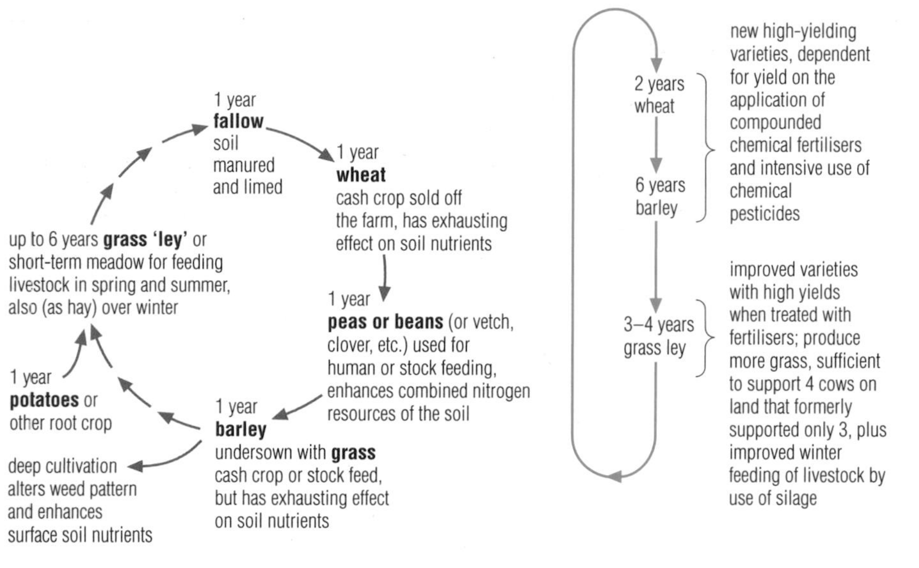

16.4 BIOGEOCHEMICAL CYCLING

The recycling of inorganic nutrients is essential for the survival of all living things, for otherwise the available resources, which are limited, would become exhausted. The cycling processes by which essential elements are used and released are called **biogeochemical cycles**; one example is the carbon cycle (Figure 12.2, p. 247).

The cycling processes involve interactions between soil, the atmosphere, the oceans and freshwaters, and living things. Microorganisms in particular play an important part in decay and the release of nutrients from dead organisms and their waste products. The cycling of the elements sulphur and phosphorus is illustrated in Figures 16.23 and 16.24, and the cycling of nitrogen is examined in detail below.

■ The nitrogen cycle

Although the element nitrogen makes up approximately 80% of the Earth's atmosphere, nitrogen compounds available for use by living things are rather scarce. This is because atmospheric nitrogen consists of stable dinitrogen molecules (N_2), and the amount of energy needed to break the bond of dinitrogen is relatively high.

Many prokaryotes can break the dinitrogen bond and form organic nitrogen

---

Figure 16.22 Crop rotation

1 year **fallow** soil manured and limed

1 year **wheat** cash crop sold off the farm, has exhausting effect on soil nutrients

up to 6 years **grass 'ley'** or short-term meadow for feeding livestock in spring and summer, also (as hay) over winter

1 year **peas or beans** (or vetch, clover, etc.) used for human or stock feeding, enhances combined nitrogen resources of the soil

1 year **potatoes** or other root crop

1 year **barley** undersown with **grass** cash crop or stock feed, but has exhausting effect on soil nutrients

deep cultivation alters weed pattern and enhances surface soil nutrients

2 years wheat

6 years barley
new high-yielding varieties, dependent for yield on the application of compounded chemical fertilisers and intensive use of chemical pesticides

3–4 years grass ley
improved varieties with high yields when treated with fertilisers; produce more grass, sufficient to support 4 cows on land that formerly supported only 3, plus improved winter feeding of livestock by use of silage

---

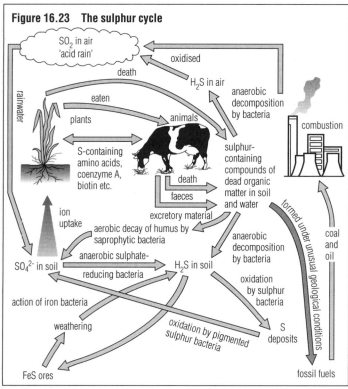

Figure 16.23   The sulphur cycle

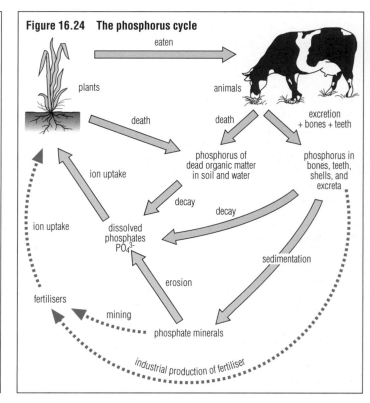

Figure 16.24   The phosphorus cycle

compounds. Much combined nitrogen available to plants is supplied by the actions of free-living and mutualistic (p. 67) prokaryotes able to fix atmospheric nitrogen. These bacteria and cyanobacteria are able to reduce atmospheric nitrogen to ammonium ions and combine these with organic acids of the respiratory pathway to produce the amino acids required for protein synthesis (Figure 16.25). On death or predation of these organisms, the combined nitrogen enters the food chains (Figure 16.28).

■ Nitrogen-fixing cyanobacteria

The free-living, filamentous cyanobacterium *Anabaena* contains many vegetative cells, which are photosynthetic, and a few thick-walled cells called **heterocysts**, which contain no photosynthetic pigments. In *Anabaena*, nitrogen fixation is restricted to the heterocysts. The locations of photosynthesis and nitrogen fixation in the cells of the filament are illustrated in Figure 16.26.

At the start of the long history of life on Earth, the cyanobacteria were some of the earliest inhabitants. Despite being an ancient group, however, cyanobacteria are widespread and abundant all over the world today.

Cyanobacteria have acted as 'green manure' on those parts of the Earth's surface moist enough for survival. For example, the cyanobacterial mats of paddy fields of the tropical and subtropical region maintained soil fertility and facilitated the survival of human communities long before humans as farmers were aware of fertilisers, or of plant nutrition. It has been calculated that about $625\,g$ of nitrogen are fixed annually per km^2 of the paddy fields of Asia and the Indian subcontinent. Combined nitrogen from the cyanobacteria becomes available mostly when they die and decay.

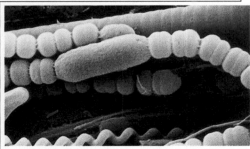

part of an *Anabaena* filament with heterocyst ($\times 1500$)

Figure 16.25   The pathway of gaseous nitrogen fixation by prokaryotes

N_2 → (nitrogenase) → NH_3 → (organic acids) → amino acids → protein

protein → enzymes

protein → cytoplasm and membranes

energy from ATP

reducing power (●H) from the respiration of glucose

The enzyme nitrogenase contains iron and molybdenum, and the reaction requires magnesium ions. Nitrogenase is destroyed or inactivated by oxygen; cells able to fix nitrogen have to protect the enzyme from oxygen.

Figure 16.26   *Anabaena* and nitrogen fixation

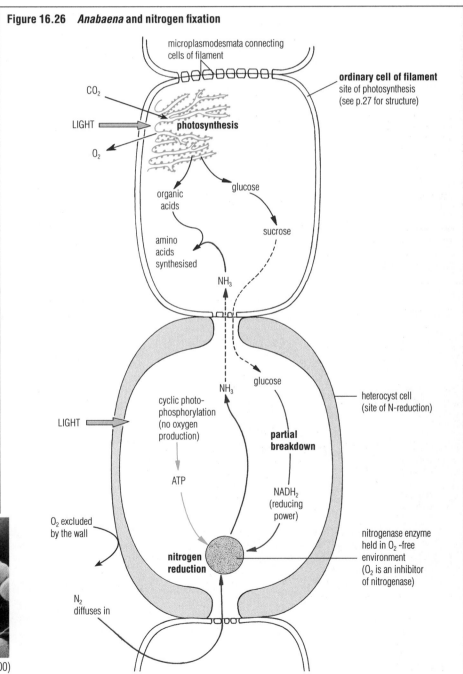

microplasmodesmata connecting cells of filament

ordinary cell of filament
site of photosynthesis (see p.27 for structure)

CO_2

LIGHT

photosynthesis

O_2

organic acids

glucose

amino acids synthesised

sucrose

NH_3

NH_3

glucose

heterocyst cell (site of N-reduction)

cyclic photophosphorylation (no oxygen production)

LIGHT

partial breakdown

ATP

NADH₂ (reducing power)

O_2 excluded by the wall

nitrogenase enzyme held in O_2-free environment (O_2 is an inhibitor of nitrogenase)

nitrogen reduction

N_2 diffuses in

Figure 16.27 The formation and functioning of leguminous root nodules

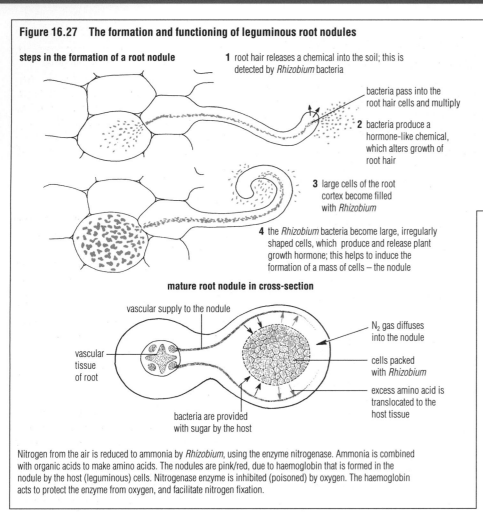

steps in the formation of a root nodule

1 root hair releases a chemical into the soil; this is detected by *Rhizobium* bacteria

bacteria pass into the root hair cells and multiply

2 bacteria produce a hormone-like chemical, which alters growth of root hair

3 large cells of the root cortex become filled with *Rhizobium*

4 the *Rhizobium* bacteria become large, irregularly shaped cells, which produce and release plant growth hormone; this helps to induce the formation of a mass of cells – the nodule

mature root nodule in cross-section

vascular supply to the nodule

vascular tissue of root

N₂ gas diffuses into the nodule

cells packed with *Rhizobium*

excess amino acid is translocated to the host tissue

bacteria are provided with sugar by the host

Nitrogen from the air is reduced to ammonia by *Rhizobium*, using the enzyme nitrogenase. Ammonia is combined with organic acids to make amino acids. The nodules are pink/red, due to haemoglobin that is formed in the nodule by the host (leguminous) cells. Nitrogenase enzyme is inhibited (poisoned) by oxygen. The haemoglobin acts to protect the enzyme from oxygen, and facilitate nitrogen fixation.

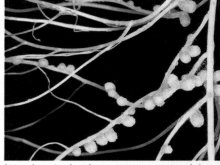

lateral roots showing numerous root nodules

Rhizobium in the nodules of leguminous plants

Bacteria of the genus *Rhizobium* are also nitrogen-fixing organisms. They are free-living in the soil, but they also occur in the nodules that form on the roots of members of the Leguminosae family of flowering plants (Figure 16.27). Their relationship with the roots is another example of mutualism (p. 67).

The presence of the nodules enables leguminous plants to grow successfully even where soil nitrates are scarce. On the death of the plant, soil fertility is further improved because the nodules break down and bacteria and ammonium compounds are released into the soil. This is why leguminous plants are included in crop rotations.

Figure 16.28 The nitrogen cycle

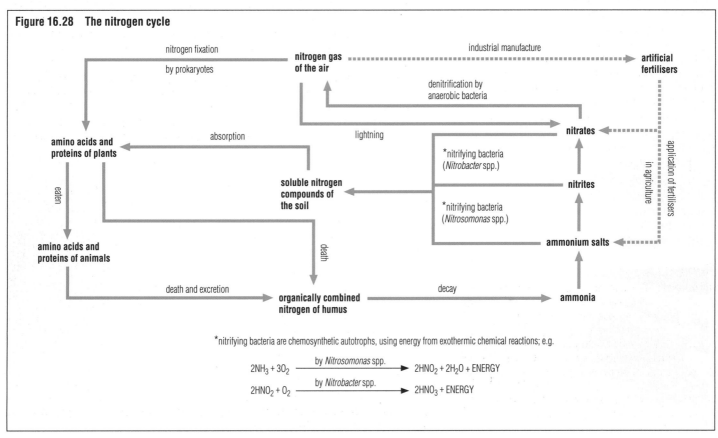

*nitrifying bacteria are chemosynthetic autotrophs, using energy from exothermic chemical reactions; e.g.

$2NH_3 + 3O_2$ — by *Nitrosomonas* spp. → $2HNO_2 + 2H_2O + ENERGY$

$2HNO_2 + O_2$ — by *Nitrobacter* spp. → $2HNO_3 + ENERGY$

Other steps in the nitrogen cycle

The other naturally occurring processes that fix nitrogen are the actions of ultra-violet light in the upper atmosphere, and of lightning. Additionally, in the internal-combustion engine nitrogen in the air in the cylinders forms nitrogen oxides under the influence of intense pressures, high temperatures and the presence of oxygen. Altogether these sources contribute no more than 10% of combined nitrogen.

Humans have developed an industrial process, the Haber–Bosch process, by which dinitrogen is catalytically reduced to ammonia using hydrogen gas under conditions of relatively high temperature and pressure. Ammonia is the raw material for the production of nitrogenous fertilisers; about 15% of the total global combined nitrogen input in agriculture is manufactured in this way.

Nitrates and ammonium salts absorbed by plants are converted to proteins. Animals obtain combined nitrogen from plants by eating them and digesting the proteins. On the death and decay of organisms, ammonia is released as a result of bacterial decomposition. Ammonia reacts with soil chemicals to form ammonium salts, which are then converted to nitrates by chemosynthetic bacteria such as *Nitrosomonas* and *Nitrobacter*. Nitrates, being water-soluble, are either rapidly absorbed by plants or microorganisms, or washed out of the soil by excess soil water.

In the anaerobic conditions of a water-logged soil chemosynthetic denitrifying bacteria flourish, reducing nitrates and ammonium salts to dinitrogen gas. These denitrifying bacteria occur naturally in soil, sewage and compost heaps, and their metabolism is responsible for a considerable loss of combined nitrogen to the atmosphere.

The nitrogen cycle is illustrated in Figure 16.28.

Questions

1 **a** What advantages arise from the inclusion of leguminous plants in a crop rotation?
  b Which leguminous plants are frequently grown in grass leys?

2 What is the benefit to nitrifying bacteria in converting ammonia to nitrites and nitrates?

3 What do we mean by the term 'green manure'?

4 In *Anabaena*, what are the respective sources of nitrogen, hydrogen and energy for the production of amino acids?

16.5 SPECIAL METHODS OF OBTAINING NUTRIENTS

The supply of nutrients (particularly of nitrogen) is a factor that may limit the growth of plants in certain habitats. Shortage of nutrients may arise when

▶ the soil pH is so low that ions exist only in insoluble forms
▶ high rainfall leaches out the soluble ions as they become available
▶ ions are not released at a season when plants most require them.

Some plants have evolved features that help overcome these problems.

■ Carnivorous plants

Carnivorous plants occur in habitats where most nutrients are in short supply and where available nitrates are virtually non-existent. For example, round-leaved sundew (*Drosera rotundifolia*, Figure 16.29) is common in acid bogs on mountains and hills where, because of low temperatures and the low pH of the peaty soil, the decay of plant and animal remains is very slow. High rainfall washes away soluble ions as they are released.

Leaves of carnivorous plants are adapted to trap small animals, which are detained and digested. Glands on the leaf surface secrete protease enzymes, which catalyse the digestion of proteins of the animal's body to amino acids. These amino acids, and probably also ions such as phosphate and potassium, are absorbed. Finally the trap re-opens and the non-digested remains are blown away. The carnivorous plant thus obtains combined nitrogen and some other nutrient ions from animals, but at the same time it produces sugars by photosynthesis.

The leaves of the sundew are reddish, densely glandular and fringed with long hairs. When an insect touches the surface it becomes stuck to the glandular hairs, which then bend over to cover the insect and digest it. The leaf partially folds up, too.

■ Mycorrhizae

The roots of many plants are intimately associated with species of fungi to mutual advantage, in a relationship known as a **mycorrhiza**. Mycorrhizae are a form of symbiosis known as **mutualism**. The fungus causes little or no damage to the host root tissue; the relationship is a nutritional alliance. Mycorrhizae are of two types, depending on the arrangement of the fungal hyphae in the host tissue.

Ectotrophic mycorrhizae

In **ectotrophic mycorrhizae** the fungal hyphae enclose the roots in a dense sheath and penetrate between the host cells to a limited extent. Mycorrhizal roots are much branched, but individually short and stunted compared with non-infected roots. Forest and woodland trees have ectotrophic mycorrhizae. The fungal fruiting bodies, various woodland mushrooms, are common in temperate woodland in autumn.

The fungal hyphae permeate the soil for distances far beyond the reach of the tree roots. The fungus obtains the sugars it requires from the tree. As a consequence, it is in a good position to

Figure 16.29   Sundew (*Drosera rotundifolia*): growing in a peat bog (left), and a fly trapped on a single leaf, held by the secretions of glandular hairs (right)

Figure 16.30   The ectotrophic mycorrhizal relationship

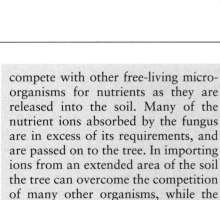

lateral root

normal root tips with root hairs

mycorrhizal root

mushroom fruiting body

mycorrhizae

mycorrhizal root in section

hyphae between the cells of the cortex (exchange of sugars and ions between the fungus and the tree occurs here)

phloem

xylem

ectotrophic mycorrhizal sheath

fungal hypha

cortex

hyphae in contact with the soil particles and soil solution (absorption of ions occurs)

experimental investigation of the nutritional alliance of ectotrophic mycorrhizae

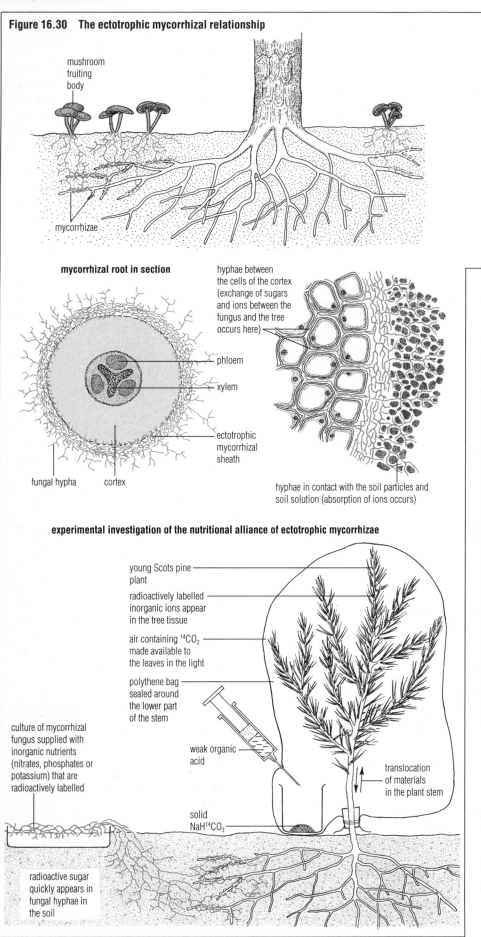

young Scots pine plant

radioactively labelled inorganic ions appear in the tree tissue

air containing $^{14}CO_2$ made available to the leaves in the light

polythene bag sealed around the lower part of the stem

weak organic acid

translocation of materials in the plant stem

culture of mycorrhizal fungus supplied with inorganic nutrients (nitrates, phosphates or potassium) that are radioactively labelled

solid $NaH^{14}CO_3$

radioactive sugar quickly appears in fungal hyphae in the soil

compete with other free-living micro-organisms for nutrients as they are released into the soil. Many of the nutrient ions absorbed by the fungus are in excess of its requirements, and are passed on to the tree. In importing ions from an extended area of the soil the tree can overcome the competition of many other organisms, while the fungal symbiont benefits by obtaining its carbon and energy supplies from the tree. The experimental identification of this nutritional alliance between the fungus and host is illustrated in Figure 16.30.

Endotrophic mycorrhizae

The fungi of **endotrophic mycorrhizae** form only a loose network of hyphae on the root surface but develop extensively within the root tissues, penetrating the host cells.

In the endotrophic mycorrhizae of orchids and heathers, the fungus feeds saprophytically in the soil on dead organic matter. Meanwhile the plant's cells digest the hyphae that have penetrated the root cells, releasing food from the fungus (Figure 16.31).

Questions

1 What factors in the habitat of a carnivorous plant such as *Drosera* specifically inhibit its anabolism? How may these adverse factors be overcome by the plant?

2 **a** What are the benefits to ectotrophic mycorrhizae of their mutualistic association with the roots of trees?

   b How does the tree benefit?

Figure 16.31 The endotrophic mycorrhizal relationship: orchid root tissue in section, showing the hyphae of endotrophic mycorrhizae

fungal hyphae

hyphae within the cortex cells of the host

dead organic matter in the soil

Figure 16.32 Sieve tube element with companion cell

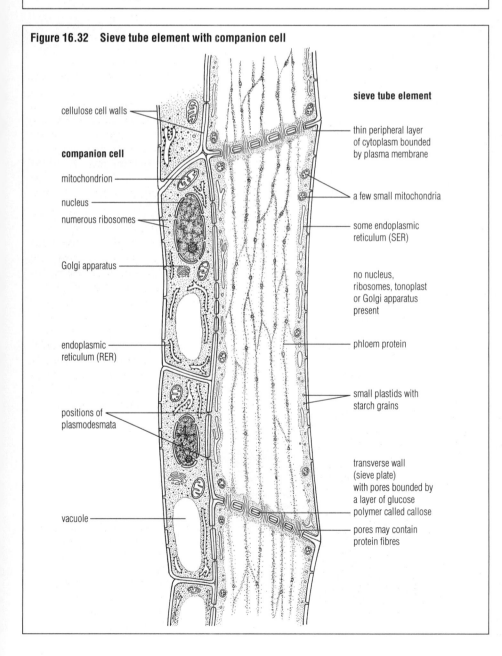

cellulose cell walls

companion cell

mitochondrion

nucleus

numerous ribosomes

Golgi apparatus

endoplasmic reticulum (RER)

positions of plasmodesmata

vacuole

sieve tube element

thin peripheral layer of cytoplasm bounded by plasma membrane

a few small mitochondria

some endoplasmic reticulum (SER)

no nucleus, ribosomes, tonoplast or Golgi apparatus present

phloem protein

small plastids with starch grains

transverse wall (sieve plate) with pores bounded by a layer of glucose polymer called callose

pores may contain protein fibres

16.6 MOVEMENT OF ELABORATED FOOD

Sugar is manufactured in the leaves by photosynthesis and transported to other parts of the plant where it is used in growth, development and repair, or to and from sites where it is stored as starch. Amino acids are moved from the cells where they are synthesised (often at the root tips) to places in the plant where proteins are being synthesised. Sugars and amino acids are transported in the phloem tissue, and movement in phloem is known as **translocation**.

■ The structure of phloem

Phloem is a living tissue consisting of sieve tubes and companion cells (Figure 16.32). These cells often occur with some parenchyma and fibres (p. 407).

Sieve tubes consist of narrow, elongated cells about 150–1000 µm long and 10–15 µm in diameter. These cells, known as sieve tube elements, are joined end to end to form a system of tubes that runs throughout the plant. The end walls, the sieve plates, are perforated by pores. Each sieve tube cell loses its nucleus at maturity (and also its Golgi apparatus, ribosomes, tonoplast and most of its mitochondria). The lateral walls become thickened by additional cellulose. The formation of phloem from the cells of the procambial strand is described on p. 227.

Companion cells are small, with thin cellulose walls. Each has a nucleus, and the cytoplasm is rich in mitochondria and the other organelles common to cells that are metabolically active. Companion cells are connected to sieve tube elements by plasmodesmata (p. 167). We may assume that companion cells service and maintain the sieve tube elements in various ways.

■ Phloem as the site of translocation

The earliest evidence that phloem is the site of sugar transport came from (destructive) bark-ringing experiments, and from laborious analyses of the changes in sugar content of xylem, phloem and leaf tissue over 24-hour periods. More recently, plant physiologists have exploited the techniques of radioactive tracing with labelled metabolites, and direct sampling of the contents of individual sieve tubes using the mouth parts of aphids as micropipettes.

Figure 16.33 Bark ringing: the result of carrying out ringing during the growing season

removal of all tissues external to the xylem, i.e. the cambium, phloem, cortex parenchyma, cork cambium and cork (phelloderm)

sugar (mainly sucrose) being transported down the phloem accumulates as starch above the ring

Figure 16.35 Changes in sugar concentration in cotton plants: the variable marked 'radiation' is a measure of the input of energy from the Sun

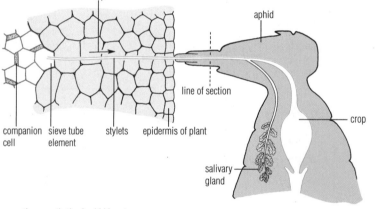

Figure 16.34 Aphid mouthparts as a micropipette

aphid with proboscis inserted into stem tissue

sap

aphid

line of section

companion cell

sieve tube element

stylets

epidermis of plant

crop

salivary gland

the anaesthetised aphid is cut away

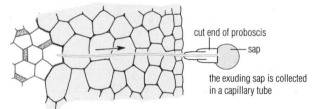

cut end of proboscis

sap

the exuding sap is collected in a capillary tube

phloem sap exuding from the cut end of the proboscis

Bark ringing

The removal of a ring of bark from around the trunk of a woody plant takes away all tissues external to the xylem (Figure 16.33). When the ring was cut in the summer, sugar from the leaves (destined for the roots) accumulated in the outer stem, above the ring. The treated trees did not wilt, so it was clear that the xylem functioned unharmed. The complete ringing of the main trunk of a tree led to the death of the plant because the roots were deprived of sugars; phloem tissue cannot regenerate across a wide gap. Bark ringing is destructive, and is not practised now. It may occur naturally in severe winter weather when rabbits gnaw away bark for food.

Analysis of sugar concentration in green plants

Measurements of the concentration of sugar in the leaves, bark and woody tissues of cotton plants throughout a 24-hour period (Figure 16.35) showed that the sugar build-up in the leaves in the light was followed by sugar accumulation in the phloem but not in the xylem. The data were obtained by analysis of batches of plants harvested at intervals. Large samples of cotton plants were analysed, in order to overcome variations between individual plants.

Use of the mouthparts of aphids as micropipettes

Aphids have hollow, needle-like mouthparts, which they insert into sieve tubes of the herbaceous stem to feed on the sap (p. 292). To sample the phloem sap, the aphid is anaesthetised by a stream of carbon dioxide gas and the proboscis cut off, leaving the mouthparts *in situ*. The sap exudes from the tiny tube and can be collected by capillarity into a fine glass tube (Figure 16.34). It may be analysed qualitatively by chromatography, and quantitatively by chemical analysis.

The use of radiocarbon as a tracer

Carbon dioxide labelled with radioactive carbon ($^{14}CO_2$), fed to green leaves in the light, is turned into radioactively labelled sugars (p. 175) and the sugars are transported about the plant (Figure 16.36). The movements of the labelled sugars can be followed by, for example, sectioning tissues and locating the radioactivity by autoradiography.

In autoradiography, the dried tissues (or organ, such as a whole leaf) are firmly held against photographic film in the dark for an extended period. (The more radioactive isotope that is present, the shorter is the exposure time required.) Then, when the film is developed, the presence of radioactivity in parts of the tissue shows up as 'fogging' of the negatives.

Figure 16.36   Using radiocarbon to investigate phloem transport

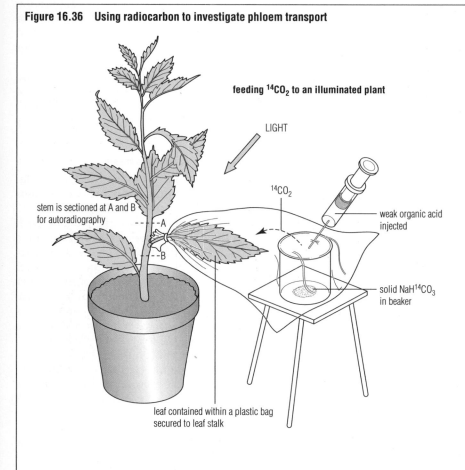

feeding $^{14}CO_2$ to an illuminated plant

LIGHT

$^{14}CO_2$

weak organic acid injected

stem is sectioned at A and B for autoradiography

solid $NaH^{14}CO_3$ in beaker

leaf contained within a plastic bag secured to leaf stalk

location of ^{14}C-labelled sugars in stem section by autoradiography

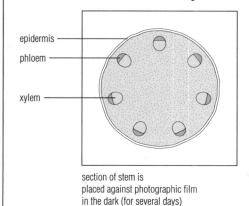

epidermis

phloem

xylem

section of stem is placed against photographic film in the dark (for several days)

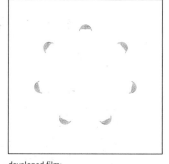

developed film: emulsion is fogged by the presence of radioactivity in the phloem

Questions

1 What are the differences between translocation and transpiration?

2 Where in the stem of the herbaceous plant are phloem sieve tubes and xylem vessels respectively to be found (p. 407)?

3 In the cotton plant stem, which component of vascular tissue has a low concentration of sugar?

4 What are the steps to obtaining an autoradiograph like that shown in Figure 16.36?

Figure 16.37   A model demonstrating pressure flow: the hydrostatic pressure developed in bulb A forces the sugar solution into B; mass flow continues until the concentrations of sugar in A and B are equal

flow of solution
(≡ phloem)

concentrated sugar solution (very low water potential)

partially permeable reservoirs (rigid)

water, or dilute sugar solution

water movement from high to low water potential by osmosis

water movement by hydrostatic pressure

A

B

water
(≡ xylem)

water

Figure 16.38   Pressure flow hypothesis of phloem transport: a summary

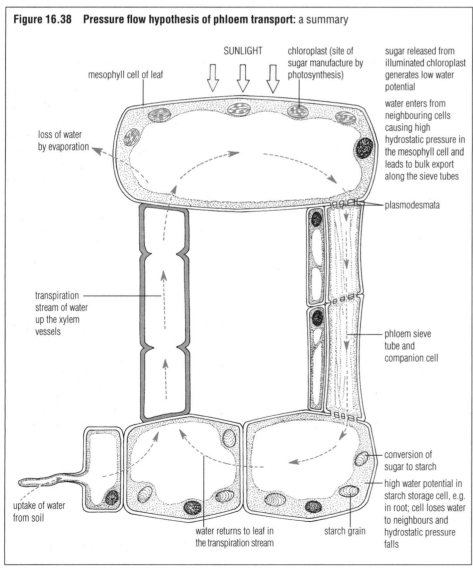

SUNLIGHT

mesophyll cell of leaf

chloroplast (site of sugar manufacture by photosynthesis)

sugar released from illuminated chloroplast generates low water potential

water enters from neighbouring cells causing high hydrostatic pressure in the mesophyll cell and leads to bulk export along the sieve tubes

loss of water by evaporation

plasmodesmata

transpiration stream of water up the xylem vessels

phloem sieve tube and companion cell

conversion of sugar to starch

high water potential in starch storage cell, e.g. in root; cell loses water to neighbours and hydrostatic pressure falls

uptake of water from soil

water returns to leaf in the transpiration stream

starch grain

■ The mechanism of phloem transport

Sugars are loaded into the phloem sieve tubes via transfer cells (Figure 11.19, p. 238). A secondary pump mechanism is involved. Transfer cells are specialised parenchyma cells with an expanded internal surface area, which may facilitate the expenditure of energy in the movement of solutes across cell membranes.

Solutes move in the phloem sieve tubes at 25–100 cm h^{-1}, which is faster than diffusion alone would allow. Movement in sieve tubes is dependent on living phloem tissue having aerobic conditions. It is slowed down and stopped by high temperatures, and by the application of trichloromethane (chloroform) or cyanide. This evidence indicates that energy from respiration is essential for translocation.

How is energy involved? As yet no single theory of the mechanism of phloem transport is firmly established; translocation is the subject of uncertainty and controversy.

The pressure flow hypothesis

The principle of pressure flow of a solution down a gradient of hydrostatic pressure is illustrated in Figure 16.37.

The sieve tubes serving the mesophyll cells of an illuminated leaf will contain a high concentration of sucrose, and therefore will have a low (more negative) water potential. As a consequence, water flows into the phloem from the xylem and a high hydrostatic pressure is created. The sieve tubes in a leaf are a 'source' region for pressure flow.

In the cells of the root, sugars are turned to starch. Here there is a low concentration of sucrose, and therefore the cells will have a relatively high (less negative) water potential. As a consequence, water flows out into neighbouring cells by osmosis, and a low hydrostatic pressure develops. Root cells are a 'sink' region.

The gradient in hydrostatic pressure between source and sink drives a mass flow of water and dissolved solutes. If sugar continues to be formed in the source tissue, and if sucrose is converted to starch or respired at the sink tissue, then the gradient will be maintained and flow will continue (Figure 16.38).

There is some evidence in support of the pressure flow hypothesis:

▶ the contents of the phloem sieve tubes are under marked pressure, and sieve tube sap exudes when the phloem tissue is cut or damaged

Figure 16.39 Alternative hypotheses of phloem transport

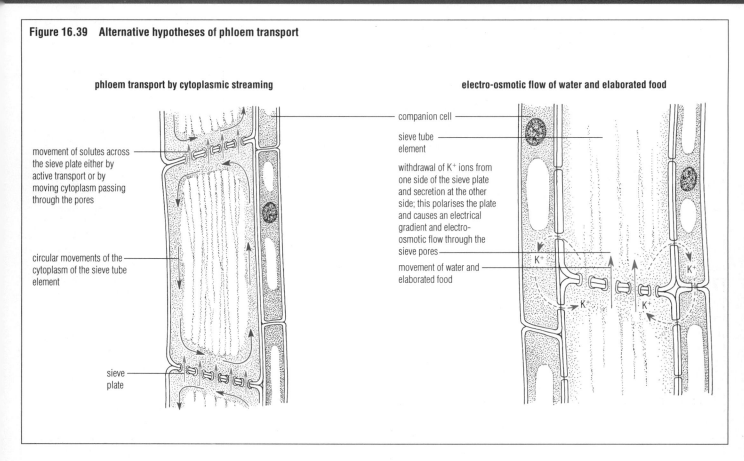

phloem transport by cytoplasmic streaming

movement of solutes across the sieve plate either by active transport or by moving cytoplasm passing through the pores

circular movements of the cytoplasm of the sieve tube element

sieve plate

electro-osmotic flow of water and elaborated food

companion cell

sieve tube element

withdrawal of K$^+$ ions from one side of the sieve plate and secretion at the other side; this polarises the plate and causes an electrical gradient and electro-osmotic flow through the sieve pores

movement of water and elaborated food

K$^+$

▶ appropriate gradients in concentration of sucrose between the sink regions and source regions have been found in numerous plants.

The evidence against the pressure flow hypothesis includes the following:

▶ sugars and amino acids have been observed to move at different speeds and in different directions in the same vascular bundles; in these cases phloem translocation may not occur in the direction of the 'deepest' sink, but in different directions according to the metabolites being transported

▶ the sieve plate is an impediment to mass flow; the end walls might be expected to have been lost in the course of evolution.

Cytoplasmic streaming

The living cytoplasm of many cells (viewed by phase-contrast microscopy, for example, p. 153) is constantly on the move, sweeping cell organelles about the cell. The streaming of cytoplasm in individual sieve tube elements could be responsible for bidirectional movements along individual sieve tubes, provided there was some mechanism to transport solutes across the sieve plates (Figure 16.39). Evidence against this hypothesis includes the following:

▶ streaming has been observed in immature sieve tube tissue only

▶ streaming movements are not fast enough to account for the observed rates of phloem transport.

Electro-osmosis

The energy from metabolism required for phloem transport may be invested by the companion cells in withdrawing ions, such as potassium ions, from the sieve tube on one side of a sieve plate and secreting them on the other side, creating a potential difference across the plate.

Polar water molecules would then be swept along with the stream of ions drawn through the pores by the potential difference. This stream would carry the solutes present in the sieve tubes with it (Figure 16.39). The value of this theory is that it actually requires two factors:

▶ living cells dependent on energy from respiration for translocation to occur

▶ a sieve plate structure across which the potential difference can be developed. Wherever food-transporting cells have evolved in the plant kingdom they have sieve plates in their end walls. An example is in the trumpet hyphae of large seaweeds such as oarweed (Laminaria).

Against this hypothesis is the lack of evidence for it. Neither does it offer any explanation of how movement occurs in the sieve element between the plates.

Questions

1 Why are conditions of high temperature and exposure to trichloromethane or cyanide harmful to respiring cells?

2 What are the important differences between a 'source' and a 'sink' region of the plant as regards the pressure flow hypothesis?

3 How might a potential difference arise across the sieve plates of a sieve tube system?

FURTHER READING/WEB SITES

A W Galston (1994) *Life Processes of Plants*. Scientific American Library/W H Freeman and Co
Science and Plants for Schools: **www.rmplc.co.uk/orgs/saps**

17 Transport within animals

- In small animals, internal transport of nutrients and waste products is by diffusion and active transport alone. In animals with a blood circulatory system, the organs may be surrounded by the blood (as in the open circulation of insects) or the blood may be enclosed in a system of vessels (as in the closed circulation of vertebrates).
- In closed blood circulatory systems, blood is either pumped once through the heart in each complete circulation of the body (as in the single circulatory system of fish) or twice in each complete circulation (as in the double circulatory system of mammals).
- The four-chambered heart of mammals is divided into right and left halves. The right side pumps blood to the lungs (pulmonary circulation) and the left side pumps to the rest of the body (systemic

circulation). From the heart, blood is delivered to capillaries in the tissues by arteries and returns to the heart in veins.
- The mammalian blood circulation transports respiratory gases (particularly the red cells), waste products, nutrients and hormones around the body. Blood proteins of the plasma help maintain a constant pH. Heat is distributed between the organs.
- The blood circulation system (particularly the white cells) helps combat disease in several ways: in the immediate response of the body to damage (inflammation); by clotting when a blood escape occurs; by phagocytosis of invading cells; and through its part in the immune response to foreign substances.

17.1 INTRODUCTION

Throughout an organism's life, materials are constantly being moved to and from all parts of the body. A complex internal transport system has evolved in many animals, especially in those that are larger and more compactly built. In many smaller animals, however, internal transport is by diffusion (and active transport from cell to cell): this is only possible where the distances between parts of the organism, and from any part of the body to the surface, are extremely short.

■ What determines the need for a transport system?

The larger an animal, the more likely it is to require an internal transport system. But it is not possible accurately to define a size of animal at which a circulatory system becomes essential. Other factors must be taken into account as well.

For example, shape is important. In a flattened and leaf-like animal the surface area:volume ratio is relatively high (Figure 15.1, p. 302), and the distances over which diffusion occurs between the exterior and the innermost cells are much shorter than in a more compact organism.

Another factor of importance is how active the organism is. The more active the animal, the higher is its level of metabolic activity and the greater its need to deliver oxygen and nutrients round the body. For example, an animal that maintains a constant body temperature (an **endotherm**, p. 510) is able to remain active over a wide range of external temperatures. It requires a lot more energy than does an animal of a similar size having a body temperature that varies with the environment (an **ectotherm**). This is shown by the data in Figure 17.1, in which the metabolic rate (energy expenditure) of unicellular organisms and

larger animals is expressed as a function of size. We see that the larger the animal the greater is its metabolic rate – the most important factor affecting the rate of metabolism of an animal is its size!

■ Animals without a blood vascular system

There are many smaller animals in which the supply of oxygen to the cells can be maintained by diffusion alone. The Platyhelminthes and the Cnidaria are examples of phyla in which there is no blood vascular system (Figure 17.3).

The jellyfish *Aurelia* (p. 34) may grow up to 15 cm in diameter, but certain giant jellyfish reach a diameter of 1.5 metres. Approximately 90% of the bulk of these larger cnidarians is non-living jelly, the **mesogloea**. A living cellular layer covers the external surface and lines the gut and canals. These cells are never very far from a supply of oxygen or food. In fact the system of radiating canals functions as a transport system of a sort, circulating sea water (the source of oxygen) and digested food all round the body (Figure 17.2).

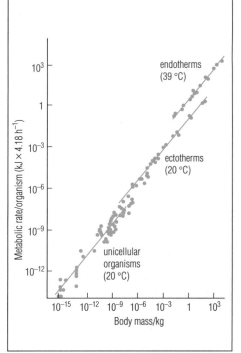

Figure 17.1 Metabolic rate of animals as a function of size, in endotherms, ectotherms and unicellular organisms

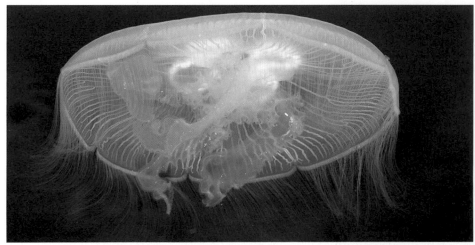

Figure 17.2   The jellyfish *Aurelia*: in jellyfish, ciliary currents carry sea water with food particles around the canal system that radiates throughout the body

Figure 17.3 Small animals without a transport system

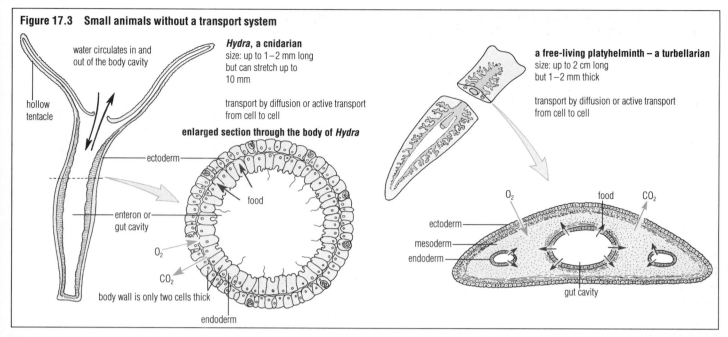

water circulates in and out of the body cavity

hollow tentacle

Hydra, a cnidarian
size: up to 1–2 mm long
but can stretch up to
10 mm

transport by diffusion or active transport
from cell to cell

enlarged section through the body of *Hydra*

ectoderm

enteron or
gut cavity

O_2

CO_2

food

body wall is only two cells thick

endoderm

a free-living platyhelminth – a turbellarian
size: up to 2 cm long
but 1–2 mm thick

transport by diffusion or active transport
from cell to cell

O_2 food CO_2

ectoderm

mesoderm

endoderm

gut cavity

Figure 17.4 Open and closed circulatory systems compared

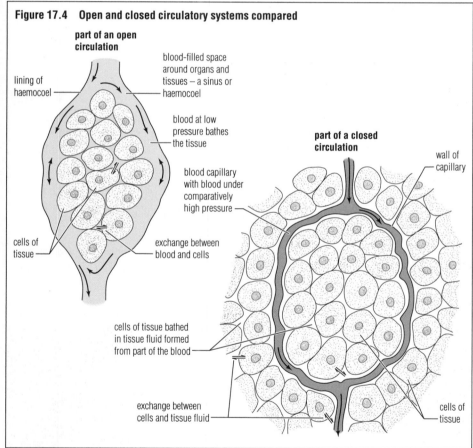

part of an open circulation

lining of haemocoel

cells of tissue

blood-filled space around organs and tissues – a sinus or haemocoel

blood at low pressure bathes the tissue

blood capillary with blood under comparatively high pressure

exchange between blood and cells

cells of tissue bathed in tissue fluid formed from part of the blood

exchange between cells and tissue fluid

part of a closed circulation

wall of capillary

cells of tissue

■ Animals with a blood vascular system

The internal transport systems of larger animals normally consist of a circulating fluid called **blood**, which is pumped by one or more muscular hearts either through a system of spaces (an open circulation) or through a system of tubes (a closed circulatory system) (Figure 17.4).

Typically, the blood contains a **respiratory pigment** that greatly increases the oxygen-carrying capacity of the blood. Not all animals have a respiratory pigment, however; insects lack one and instead pipe air directly to all respiring tissues (Figure 15.5, p. 305). The range of respiratory pigments found in animals is listed in Table 17.1. These pigments are chemically very similar, but differ in the metal they contain. Some pigments occur in solution in the plasma, others in blood cells. Haemoglobin (p. 358) occurs not only in all vertebrate groups but also in some non-vertebrates.

Questions

1 In a small animal such as a flatworm (platyhelminth, Figure 17.3), where adequate oxygen is delivered to cells by diffusion, which metabolites may be taken into cells by active uptake?

2 Examine the graph in Figure 17.1. Suggest reasons why
   a metabolic rate increases with increasing body mass
   b an endotherm's metabolic rate is higher than that of an ectotherm of the same size.

Table 17.1 Respiratory pigments of the animal kingdom

Pigment	Site in blood	Groups found in
chlorocruorin (green, iron-containing)	plasma	polychaete annelids
haemoerythrin (brown, iron-containing)	cells	annelids
haemocyanin (blue, copper-containing)	plasma	gastropod and cephalopod molluscs, crustaceans
haemoglobin (red, iron-containing)	plasma	annelids, molluscs
	cells	all vertebrate classes

■ An open circulatory system

In an open circulatory system blood is pumped at relatively low pressure from the heart into spaces within the body (Figure 17.5). These spaces are called **sinuses** or, collectively, a **haemocoel**. In an open circulation, blood bathes the organs directly, but there is little control over the direction of circulation. Blood slowly returns to the collecting vessel. Here valves and waves of contraction of the muscular wall move the blood forward to the head region where the open circulation is recommenced. Arthropods and molluscs have this type of circulation (insects possess a tracheal system as well, p. 305).

■ A closed circulatory system

In a closed circulatory system the blood circulates in a continuous system of tubes, the blood vessels (Figure 17.6). Blood is pumped by a muscular heart at relatively high pressure, and a rapid flow rate results. Organs are not in direct contact with the blood, but they are bathed by fluid leaking out from the narrow, thin-walled parts of the system, the capillaries. This fluid, the **tissue fluid**, is the medium in which exchange takes place between blood and the body tissues, before it returns to the blood vessels.

Single and double circulations

Closed circulatory systems are of two types, depending on whether the blood passes through the heart once or twice in each circulation of the body (Figure 17.7).

Fish have a single circulation. The fish's blood is pumped from the heart, through the gills, and then on to the rest of the body before returning to the heart. Blood goes once through the heart in each circulation of the body.

Birds and mammals have a double circulation. Here, blood is first pumped from the heart to the lungs via the **pulmonary circulation**, and returns to the heart again; from there it is pumped to the body tissues via the **systemic circulation**. One side of the heart serves the pulmonary circulation and pumps deoxygenated blood. The other side of the heart serves the systemic system and pumps oxygenated blood. The two sides of the heart are structurally separated from each other.

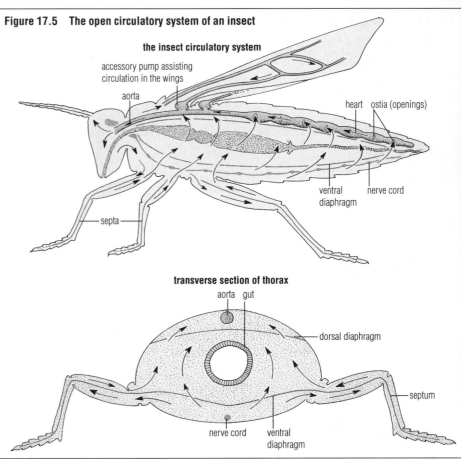

Figure 17.5 The open circulatory system of an insect

Figure 17.6 The closed circulatory system of the earthworm

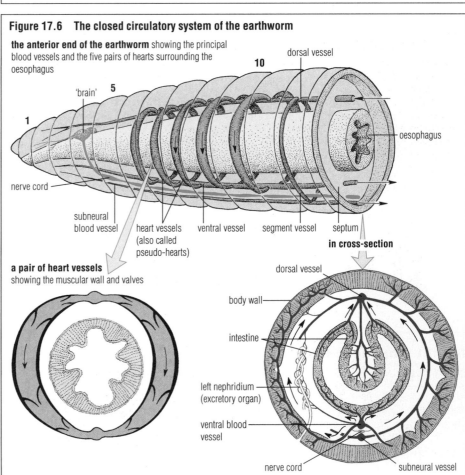

Figure 17.7 Single and double circulatory systems: arrows indicate the direction of the blood flow

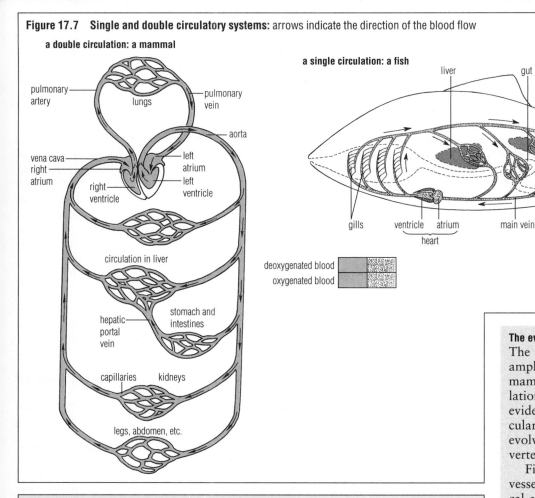

a double circulation: a mammal

a single circulation: a fish

deoxygenated blood
oxygenated blood

COELOM OR HAEMOCOEL: WHAT IS THE DIFFERENCE?

The internal organs of an annelid (such as an earthworm) are in a cavity called the **coelom**. The internal organs of an arthropod (such as an insect) are in a cavity called a **haemocoel**. The difference between these cavities lies in the way the embryo develops in the two groups.

Initially, a block of mesoderm containing a small cavity, a coelom, occurs on either side of the gut. **In annelids**, the mesoderm expands into and almost obliterates the blastocoel. In fact the blastocoel is reduced to being the cavity of the dorsal and ventral blood vessels. The coelom becomes the main body cavity. By contrast, **in arthropods**, the coelom remains small and the blastocoel becomes the main body cavity. It forms an open blood system.

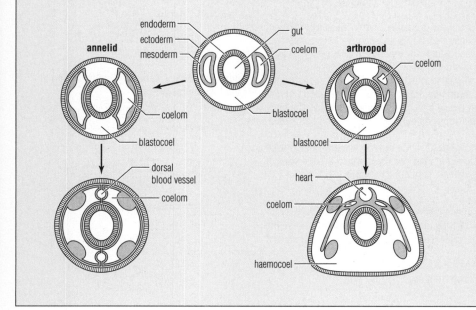

The evolution of the vertebrate circulation

The five classes of vertebrates (fish, amphibians, reptiles, birds and mammals) have distinct types of circulation. There is, however, anatomical evidence to suggest that the blood vascular systems of vertebrates may have evolved from a common fish-like vertebrate ancestor.

Fish exhibit an arrangement of vessels rather similar to the hypothetical ancestral circulation. Other vertebrates show increasing divergence from this pattern, associated with the development of lungs in place of gills, and of a double circulation in place of the single circulation of fish. In amphibians the heart is partially divided and there is some double circulation of blood. In reptiles division of the heart into two 'sides' is still incomplete. Only birds and mammals have a true double circulation.

Questions

1 In the open circulation there is 'little control over the circulation'.
   a What does this mean?
   b Why is this a disadvantage?
2 A closed circulatory system does not allow blood to bathe the organs of the body. How is exchange between blood and body organs arranged?
3 What advantage arises from returning oxygenated blood to the heart before it circulates around the body?
4 Suggest two major developments in body systems that became necessary when terrestrial vertebrates evolved from a fish-like ancestor.

345

17.2 THE HUMAN HEART

The human heart, discussed here as an example of the mammalian heart, is a hollow, muscular organ about the size of a clenched fist. It is shaped like an inverted cone. The heart lies in the thoracic cavity between the lungs, in a position protected by the sternum. The heart wall consists of **cardiac muscle** (Figure 10.14, p. 219 and Figure 17.8), a special type of muscle that is unique to the heart. **Coronary arteries** arise from the base of the aorta and pass over the surface of the heart (Figure 17.8), supplying the capillaries of the cardiac muscle with oxygenated blood. Deoxygenated blood is collected by the **coronary veins**, which drain into the right atrium.

The heart is surrounded by and tightly contained in a membranous sac, the **pericardium**. Fluid secreted by the pericardium reduces friction between the beating heart and the surrounding stationary tissues. The strongly non-elastic nature of the pericardium prevents the heart from becoming overfilled by blood and therefore overstretched. The pericardium is also attached to the diaphragm and firmly anchors the heart in position.

■ The chambers of the heart

The cavity of the heart is divided into four chambers (Figure 17.8); the chambers of the right side of the heart are completely separate from the chambers of the left side. The two upper chambers are thin-walled, and are called **atria**. The two lower chambers are thick-walled, and are called **ventricles**. The right atrium receives deoxygenated blood from the body. The right ventricle receives this deoxygenated blood from the right atrium and pumps it to the lungs. The left atrium receives oxygenated blood from the lungs. The left ventricle receives this oxygenated blood from the left atrium and pumps it to the body.

The muscular wall of the left ventricle is about three times as thick as that of the right ventricle. This difference is related to the greater distances that blood is pumped by the left ventricle, and the huge resistance that it has to overcome in serving the whole body (the right ventricle serves the lungs only). Despite the difference in thickness of the walls of the two ventricles, the volumes of the left and right ventricles must be the same to ensure balance of blood delivered to and returning from the lungs with that delivered to the rest of the body.

■ Heart valves and one-way flow

The direction of blood flow through the heart is maintained by valves. Massive valves, the **atrio-ventricular valves**, separate the upper chambers from the lower. The edges of the atrio-ventricular valves are supported by non-elastic strands, the **chordae tendinae**, which are anchored to the muscular wall of the ventricles below. The atrio-ventricular valve of the right side of the heart is called the **tricuspid valve**; that of the left side is the **bicuspid** or **mitral valve**. Pocket-shaped valves, the **semilunar valves**, lie between the ventricles and the pulmonary artery and aorta. Each of these valves allows blood to flow through it in one direction only.

Questions

1 What are the significant differences of structure and function between voluntary (skeletal) muscle and cardiac muscle (p. 218)?
2 In what ways does the pericardium protect the organ it surrounds?
3 What are the main consequences for the blood circulation of a 'hole in the heart' between the left and right ventricles?
4 What is the role of the chordae tendinae when the ventricle walls contract?

WILLIAM HARVEY

Life and times

William Harvey (1578–1657) was the eldest of seven sons in a close-knit Kent farming family. His education took him to Canterbury Grammar School and then to Caius College, Cambridge, and he went on to study medicine in Padua, the greatest medical school at that time.

He then settled in London, practised medicine, and became physician at St Bartholomew's Hospital and, later, physician to Charles I. After the Civil War he continued his research. *On the Motions of the Heart and Blood*, a masterly book of 100 pages and the first ever published on physiology, appeared in 1628.

Contributions to biology

Modern biology is based on observation and experiment. Prior to Harvey's work, ideas from ancient Greece and Rome had dominated the thoughts of scholars as far as any enquiry was concerned. Respect for these classical studies took the place of investigation. The writings of Galen (AD 131–201) were the textbooks for medicine up to the seventeenth century.

Harvey's discoveries were based upon observation of hearts in many different animals, and upon comparisons between working hearts and human-made pumps. Harvey measured the blood output by the heart and calculated the quantities of blood that would have to be manufactured by the body each second if blood was consumed in the tissues and replaced by the liver, as Galen had stated, rather than being circulated as Harvey believed.

Harvey also demonstrated the flow of blood towards the heart in the veins in the arm. He showed that when an arm is lightly ligatured the veins show up with small swellings, which are the sites of valves. Blood can be pushed past these points towards the heart but not backwards.

Influences

Harvey was an enthusiastic, cautious and skilful experimenter whose observations led to the conclusion '. . . therefore blood must circulate'. Harvey never directly observed the connection between the arteries and veins, however. (Capillaries were first observed and described by the Italian physiologist Marcello Malpighi (1628–94).)

Harvey also worked on embryology, and contributed to the downfall of the concept of spontaneous generation with his firm statement 'All living things develop from eggs'.

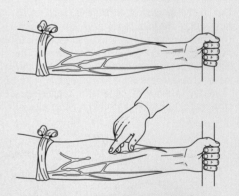

interpretation of Harvey's demonstration of the blood circulation

the swellings in the veins are the sites of valves

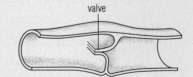

Figure 17.8   The mammalian heart

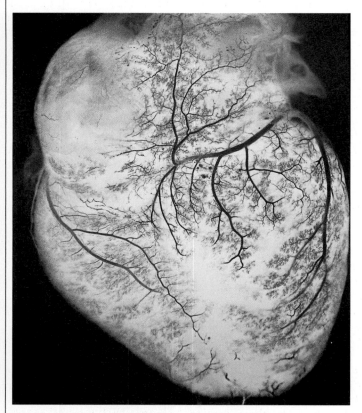

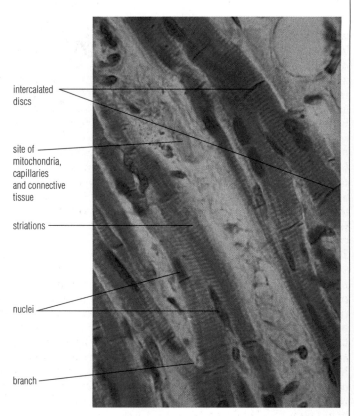

the external appearance of the human heart (ventral view): vessels supplying the heart muscle (coronary arteries) have been injected with dye to emphasise the capillary networks

intercalated discs

site of mitochondria, capillaries and connective tissue

striations

nuclei

branch

photomicrograph of cardiac muscle, LS (×800)

the heart in longitudinal section

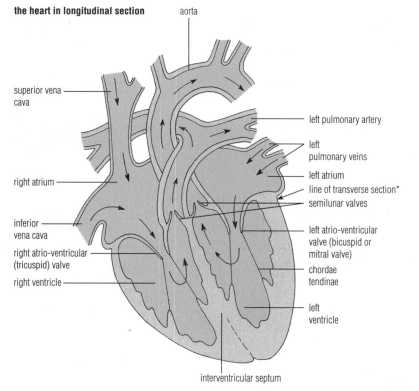

aorta

superior vena cava

left pulmonary artery

left pulmonary veins

left atrium

line of transverse section*

semilunar valves

right atrium

inferior vena cava

left atrio-ventricular valve (bicuspid or mitral valve)

right atrio-ventricular (tricuspid) valve

chordae tendinae

right ventricle

left ventricle

interventricular septum

the heart in transverse section, cut between the atria and the ventricles in the region of the valves*

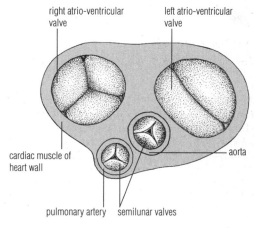

right atrio-ventricular valve

left atrio-ventricular valve

cardiac muscle of heart wall

aorta

pulmonary artery

semilunar valves

■ The heart as a pump

The **cardiac cycle** (Figure 17.9) is the sequence of events of a heartbeat by which blood is pumped round the body. The pumping action of the heart consists of alternate contractions (called **systole**) and relaxations (**diastole**). The time taken for one cardiac cycle of contraction and relaxation is 0.8 seconds when the heart is beating at a rate of 75 beats per minute. Contraction of cardiac muscle is followed by relaxation and elastic recoil of the heart, due to the presence of connective tissue fibres in the muscle, which are temporarily distorted in the contraction. As the cardiac muscle relaxes, they revert to their original shape.

The cardiac cycle

Stage 1 The right and left ventricles relax simultaneously (ventricular diastole); as they do so, blood flows into them from the atria. The tricuspid and bicuspid valves (the atrio-ventricular valves) are open and the right and left atria contract (atrial systole). Contraction of the atrial wall also has the effect of sealing off the venae cavae and pulmonary veins, preventing backflow of blood into the veins as the blood pressure in the atria rises.

Stage 2 The right and left ventricles contract simultaneously (ventricular systole) and the right and left atria relax (atrial diastole). Blood is pumped from the ventricles into the pulmonary artery and aorta, and the semilunar valves are opened. Meanwhile, the tricuspid valve and the bicuspid valve are closed by the raised ventricular pressure.

Stage 3 After ventricular systole there is a short period of simultaneous atrial and ventricular relaxation (diastole).

In ventricular diastole the high pressure developed in the aorta and pulmonary artery and the falling pressure in the ventricles cause a slight backflow of blood, closing the semilunar valves and preventing further backflow.

Relaxation of the atrial wall and the contraction of the ventricles initiates the refilling of the atria by blood under relatively low pressure. Venous blood flows into the right atrium, and oxygenated blood flows from the pulmonary vein into the left atrium. The atria gradually become distended.

During ventricular diastole, pressure in the ventricle falls. The increasing volume of blood in the atria commences to enter the ventricles passively, pushing open the atrio-ventricular valves.

Figure 17.9 **The events of the cardiac cycle,** illustrated by reference to the left side only (simultaneous events take place in the right side of the heart)

the cardiac cycle

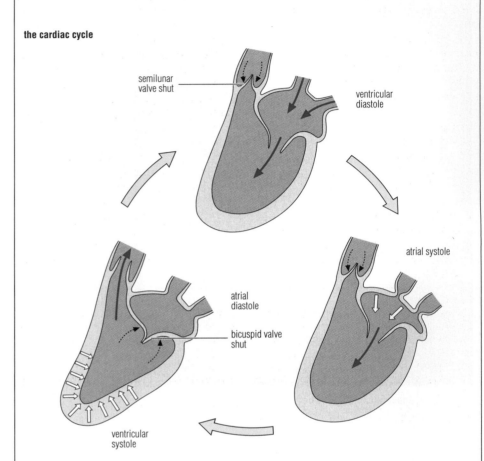

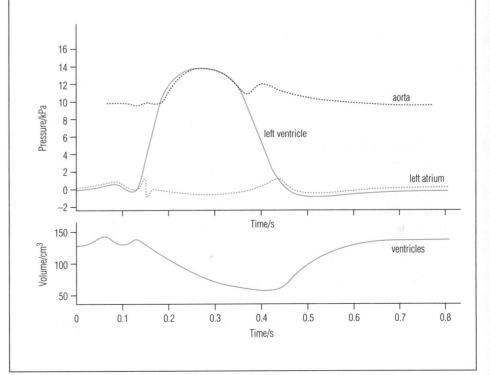

The control of heartbeat

Cardiac muscle continues to contract rhythmically even after the heart has been surgically removed from the body, provided that it is maintained in a favourable medium. So the origin of the heartbeat is not a nerve impulse (it is not **neurogenic**) but rather is an inherent property of the cardiac muscle (it is **myogenic**).

The pacemaker

Excitation originates in a tiny structure in the wall of the right atrium, known as the **sinoatrial node (SAN)** or **pacemaker** (Figure 17.10). Muscle tissue conducts the excitation to both atria. At the base of the right atrium a second node, the **atrio-ventricular node (AVN)**, passes on the excitation via bundles of exceptionally long muscle fibres, the **Purkinje fibres**

(collectively called the **bundles of His**), to all parts of both ventricles. In this way contraction of the ventricles is initiated. As the ventricles contract the atria relax. Subsequently the ventricles relax and the cycle is complete.

After being stimulated there is a brief period when muscle is insensitive to further stimulation (the refractory period, p. 446). Cardiac muscle has a relatively long refractory period. Because of this, cardiac muscle is able to beat forcefully, without fatigue and without developing a permanently contracted state known as **tetanus**.

A symptom in some forms of heart disease is failure of the pacemaker. In many cases, normal heart rhythm can be maintained by a battery-powered device, implanted in the thorax, that delivers small, regular electrical pulses to the heart.

Figure 17.10 The pacemaker, and myogenic stimulation

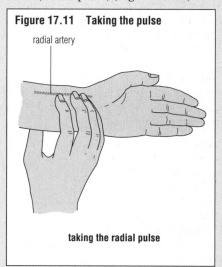

HEART SOUNDS

The valves of the heart close passively whenever there is a tendency for blood to flow in the reverse direction. The relative pressures in the atria, ventricles and arteries determine the opening and closing of the valves, since blood flows from a region of high pressure to a region of lower pressure.

The sounds of the valves of the heart closing can be heard by using a stethoscope (or by holding the ear very close to someone's chest). For example, the first sound is associated with the simultaneous closure of the atrio-ventricular valves. This sound is like the syllable 'lub' spoken very softly. Associated with closure of the semi-lunar valves in ventricular diastole comes the second heart sound of 'dup'. As the heart pumps, the repeating sequence 'lub dup (pause) lub dup' is heard.

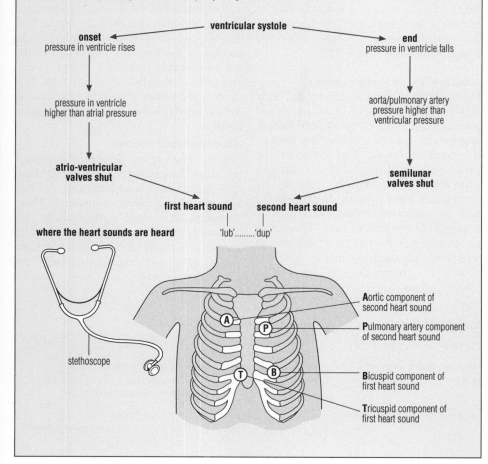

Pulse

Ventricular contraction (systole) forces a wave of blood through the arteries. The expansion of the arteries can be felt as a **pulse**, particularly where the artery is near the skin surface, and passes over a bone. The pulse is traditionally taken above the wrist (radial pulse) (Figure 17.11).

Figure 17.11 Taking the pulse

Questions

1 When heart muscle ceases to contract the heart reverts to its relaxed position. What causes this change in shape?
2 What is the purpose of a doctor being trained to know the normal sounds a heart makes as it beats?
3 a What do we mean by saying the origin of the heartbeat is 'myogenic'?
  b How is it known that the origin of the heartbeat is myogenic?

Regulation of cardiac output

The volume of blood pumped from the heart depends upon:

▶ how fast the heart beats (= **heart rate**);
▶ how much blood is pumped out per beat (= **stroke volume**).

The **cardiac output** = stroke volume × heart rate.

How the pacemaker is regulated

The pacemaker is regulated according to the needs of the body. For example, heart rate is increased during physical activity, from the 75 beat per minute of the 'at rest' heart, up to 200 beats a minute in strenuous exercise. This regulation of the pacemaker is brought about by reflex action, by nerves of the autonomic nervous system (p. 426). The pacemaker receives impulses from the **cardiovascular centre** (an inhibitory centre and an accelerator centre) in the medulla of the hind brain, via two (antagonistic) nerves (Figure 17.13).

▶ A **sympathetic nerve**, part of the sympathetic system, speeds up the rate of heartbeat (by releasing noradrenalin);
▶ A branch of the **vagus nerve**, part of the parasympathetic nervous system, slows down the rate of heartbeat (by releasing acetylcholine).

Figure 17.13   The regulation of heartbeat by the nervous system

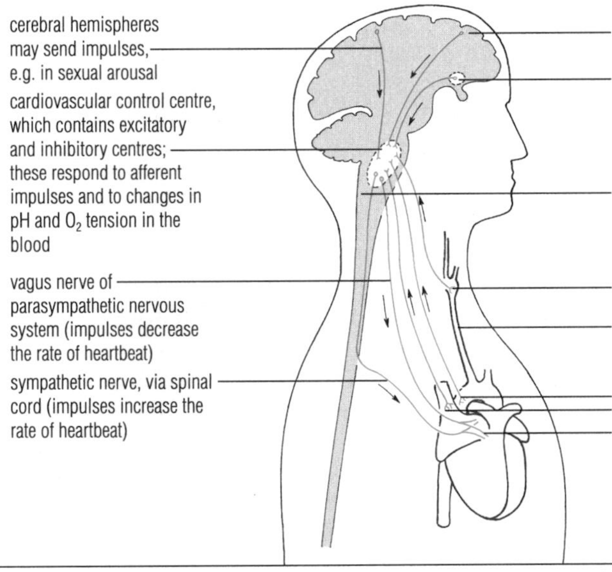

cerebral hemispheres may send impulses, e.g. in sexual arousal

cardiovascular control centre, which contains excitatory and inhibitory centres; these respond to afferent impulses and to changes in pH and O_2 tension in the blood

vagus nerve of parasympathetic nervous system (impulses decrease the rate of heartbeat)

sympathetic nerve, via spinal cord (impulses increase the rate of heartbeat)

cerebral cortex

hypothalamus may send impulses, e.g. in alarm or rage

medulla oblongata

baroreceptors of carotid sinus

carotid artery

baroreceptors of aortic arch
baroreceptors of vena cava
nerve fibres supply SA node, AV node and heart muscle

Figure 17.12   Physical exercise and blood flow back to the heart

to the heart

vein

valve open

blood flow in the veins is assisted by alternating contraction and relaxation of skeletal muscle

muscle contracted

valve closed

muscle now relaxed

backflow is prevented by valves, and muscle movements squeeze the blood along

vigorous exercise speeds up the return of blood to the heart

The cardiovascular centre receives impulses from **baroreceptors** (stretch receptors) located in the walls of the aorta, the carotid arteries, and the right atrium, when changes in blood pressure occur at these positions. When blood pressure is high in the arteries, the rate of heartbeat is lowered by impulses from the cardiovascular centre via the vagus nerve. When blood pressure is low, the rate of heartbeat is increased by impulses from the cardiovascular centre via the sympathetic nerve (Table 17.2).

Stroke volume of the heart can change

During intense physical activity the body may restrict the flow of blood to the gut (so that blood flow to the skeletal muscle is enhanced) and then blood flow back to the heart increases. This stretches the cardiac muscle fibres as they are about to contract. Cardiac muscle responds by contracting more strongly during systole, and an increased volume of blood is pumped out per unit of time. This direct relationship between the degree of stretching of cardiac muscle and the power of contraction is known as **Starling's Law**.

As the volume of blood flow back to the heart is increased, the heart is slightly stretched and distended. By this process (from regular exercise) a permanent increase in output is maintained, even when resting. So the 'trained' heart at rest has a greater stroke volume and can therefore pump the same quantity of blood as an untrained heart, but at a lower heart rate. Of course, the increased stroke volume stretches the aorta and carotids which then send impulses to the cardiac inhibitory centre of the cardiovascular centre. From here, impulses slow down the heart rate. This is an automatic fail-safe mechanism which prevents the heart from damaging itself.

Changes in blood composition

During periods of persisting strenuous exercise, the oxygen tension of the blood falls, the carbon dioxide tension rises and the pH of the blood is lowered. These changes in blood composition are detected in the cardiovascular control centre; as a result, impulses that stimulate heart rate are transmitted via the sympathetic nerve.

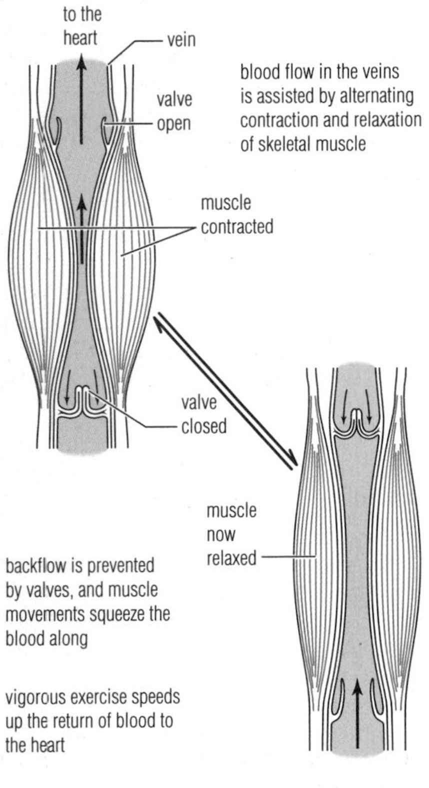

Table 17.2   Modulation of the heart by the autonomic system

Stretch receptors stimulated	Source of stimulation	Outcome in pacemaker/ventricle wall
vena cava baroreceptors	strenuous muscular activity increasing blood return to heart	sympathetic nerve fibres from accelerator centre liberate noradrenaline, causing increased heart rate
aorta and carotid baroreceptors	increased cardiac output (volume of blood × number of beats)	parasympathetic nerve fibres from inhibitory centre liberate acetylcholine, which reduces heart rate

Figure 17.14   The layout of the mammalian circulation, and the positions of the major vessels in the human blood circulation

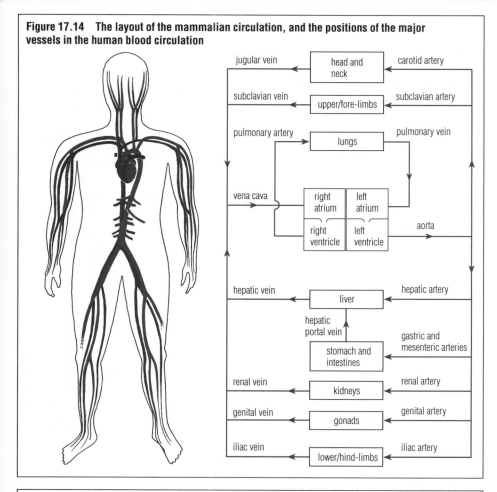

At times of stress, **adrenaline** is secreted from the medulla of the adrenal glands (the 'flight or fight' response, p. 475). Adrenaline speeds up the heartbeat. **Thyroxine**, produced by the thyroid gland, raises basal metabolic rate, causing enhanced metabolic activity. So, an indirect effect of thyroxine is increased cardiac output.

17.3 ARTERIES, VEINS AND CAPILLARIES

■ Layout of the blood circulation

The role of the circulatory system is to transport blood to all parts of the body. Each organ has a major **artery** supplying it with blood from the heart and a major **vein** that returns blood to the heart. The general plan of the mammalian circulation is shown in Figure 17.14.

■ The structure of arteries and veins

The walls of arteries and veins consist of three layers. There is an innermost layer of squamous endothelium (this is a form of squamous epithelium, p. 213, that lines the heart and the blood and lymph vessels, and forms the walls of the capillaries), a middle layer of smooth muscle and elastic fibres (**tunica media**) and an outer layer of fibrous connective tissue (**tunica externa**). The walls of arteries are much thicker than those of veins, enabling the arteries to withstand the high pressure of blood caused by ventricular contractions (Figures 17.15 and 17.16). The diameter of an artery and the thickness of its walls steadily decrease as its distance from the heart increases.

The walls of the veins have substantially fewer elastic and muscular fibres than those of the arteries, but their thickness increases as they approach the heart. Blood is under low pressure in the veins, and backflow of the blood is a possibility. Backflow is prevented, however, by valves in the veins.

Figure 17.15   Arteries and veins: structure in relation to function

the wall structure of blood vessels

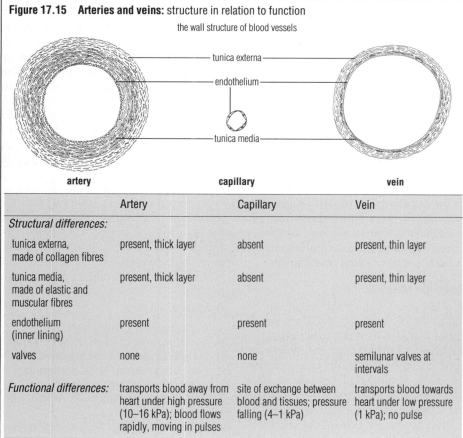

	Artery	Capillary	Vein
Structural differences:			
tunica externa, made of collagen fibres	present, thick layer	absent	present, thin layer
tunica media, made of elastic and muscular fibres	present, thick layer	absent	present, thin layer
endothelium (inner lining)	present	present	present
valves	none	none	semilunar valves at intervals
Functional differences:	transports blood away from heart under high pressure (10–16 kPa); blood flows rapidly, moving in pulses	site of exchange between blood and tissues; pressure falling (4–1 kPa)	transports blood towards heart under low pressure (1 kPa); no pulse

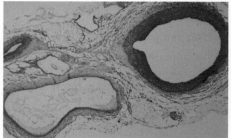

Figure 17.16   Photomicrograph showing transverse sections through an artery (top right) and a vein (bottom left)

External pressure exerted by movements of surrounding tissues (due to contractions of muscles) helps drive the blood in the direction of the heart (Figure 17.12). Standing motionless for a prolonged period may reduce the volume of blood returning to the heart, and even lead to fainting through reduced blood flow to the brain.

■ The capillary network

Arteries divide into smaller vessels called **arterioles**, which further divide into narrow, thin-walled tubes called **capillaries**. The walls of capillaries consist of endothelium only (Figure 17.15). Capillaries branch profusely, and in the resulting capillary networks (capillary beds) the blood comes close to the cells of the body. Capillary networks are numerous; no cell is very far from a capillary (Figure 17.18, facing page).

The body controls the quantity of blood flowing through capillary networks. Regulation is by three features of the circulatory system.

▶ Localised increases in metabolites (such as carbon dioxide, hydrogen ions or potassium ions) and localised deficiencies (lack of oxygen) cause arteriole dilation, and increased blood flow in the vicinity. Thus local conditions ensure the blood supply is regulated to meet the tissues' metabolic needs.

▶ Arterioles supplying a capillary network have **pre-capillary sphincter muscles**. When these contract the flow of blood into the capillary network is reduced or temporarily prevented (vasoconstriction).

▶ Many capillary networks, especially in the feet, hands and stomach, have **bypass vessels** or **shunts** (also equipped with sphincter muscles), which allow the short-circuiting of individual networks (Figure 23.19, p. 512).

Capillaries join together to form small vessels called **venules**. Venules join together to form veins.

■ The control of blood pressure

Blood pressure is regulated in the heart by the control of the cardiac output (p. 350), and in the body circulation by regulation of the diameter of arterioles supplying the capillary networks that serve the tissues of the body (the peripheral circulation).

The diameter of the arterioles determines the resistance to circulation of

Figure 17.17　The control of blood pressure in the capillaries

1　by the autonomic nervous system

- hypothalamus
- cardiovascular centre in medulla
- vasodilator impulses increase blood flow in arterioles and capillaries
- pre-capillary sphincter
- arteriole
- in resting muscle, pre-capillary sphincters close; most blood by-passes capillary bed
- venule
- capillary
- in active muscle, build-up of metabolites causes pre-capillary sphincters to dilate; more blood flows through capillary bed

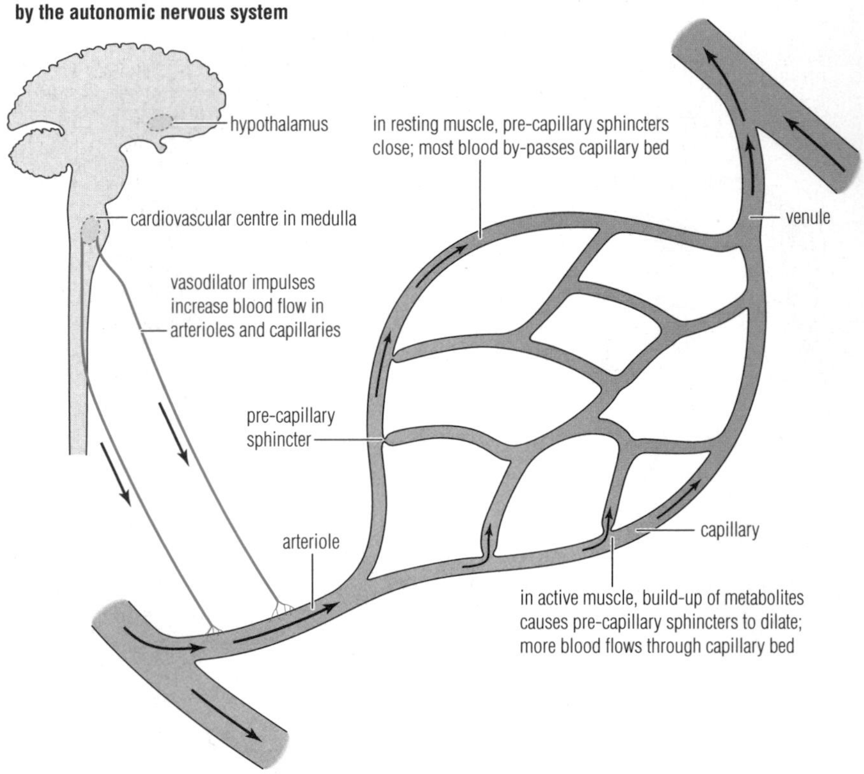

2　by endocrine secretions of the endothelium
endothelium cells (the capillary wall) sense changes in blood pressure, partial pressure of oxygen and blood flow; in response, hormones that target the arterioles are released, either **endothelin**, causing vasoconstriction, or **nitric oxide**, causing vasodilation

blood. When the arterioles are constricted (vasoconstriction) the blood pressure rises; when they are dilated (vasodilation) the blood pressure falls. Smooth muscle in the walls of the arterioles is innervated by neurones of the autonomic nervous system running from the cardiovascular centre (p. 350) and from the hypothalamus (Figure 17.17). Increased frequency of impulses in the sympathetic fibres mostly causes vasoconstriction (so reducing the blood flow in the capillaries). Reduced frequency of impulses in these fibres causes vasodilation. Some arterioles receive vasodilation fibres; increased frequency of impulses in these fibres also causes vasodilation.

Role of the endothelium

The flattened cells that make up the lining layer of blood vessels (and the entire wall of the capillaries) are anchored to a basement membrane and attached to each other by tight junctions (p. 167). Endothelial cells have many functions. Their cytoplasm contains pinocytotic vesicles, which transport substances from one side of the cell to the other. In addition, endothelial cells have

the ability to sense changes in the blood (for example, changes in pressure, O_2 partial pressure, blood flow) and to respond by the production and release of regulatory substances. For example, **endothelin** has the effect of vasoconstriction when it reaches the smooth muscles of the arterioles, causing heightened blood pressure. **Nitric oxide**, by contrast, brings about vasodilation when in contact with smooth muscle, causing lowered blood pressure. The nitric oxide from endothelial cells also prevents the formation of blood clots by suppressing the activation of platelets. The ability to secrete nitric oxide is diminished in diseased blood vessels, which are therefore more likely to produce a thrombus (p. 355).

Questions

1　During vigorous activity the heart beats quickly. What mechanisms in the body cause the raised heart rate in this situation?

2　What is adrenaline (p. 475)? What are its general effects on body activity?

3　What is the hepatic portal vein? What organs does it connect?

4　Summarise the role of the hindbrain (medulla oblongata) in controlling heartbeat.

■ Tissue fluid: exchange between blood and cells

The capillaries are the sites of exchange between the blood and the cells of the body (Figure 17.18). The blood is contained in a closed system, but fluid derived from plasma escapes through the walls of the capillaries. This fluid, called **tissue fluid**, bathes the cells.

Tissue fluid has a composition similar to that of plasma, but lacks its proteins. Tissue fluid is an aqueous solution of oxygen, glucose, amino acids, fatty acids, hormones and inorganic ions. These are the components of blood required by the cells of the body. In addition, tissue fluid picks up carbon dioxide and other excretory substances as it surrounds the cells and supplies nutrients to them.

Blood pressure and diffusion are the factors responsible for the movements of water and solutes into and out of the capillaries (Figure 17.19). At the arterial end of a capillary bed, the diffusion gradient for solutes such as oxygen, glucose and ions favours movement from the capillaries to the tissue fluid. These are substances that are being used up during cell metabolism. On the other hand, the diffusion gradient for solutes such as carbon dioxide and other waste products of metabolism favours the movement of these substances from the tissue fluid to the capillaries.

The diffusion gradient for water at the arterial end is from tissue fluid to blood, because the plasma proteins reduce the water potential of blood. Nevertheless, because the blood pressure is high, water is forced out through the capillary wall, carrying solutes with it. This is ultrafiltration, as in the kidney glomerulus (p. 381).

At the venous end of the capillary bed, the blood pressure is lower and water diffuses into the capillaries by osmosis. The solute potential of the plasma proteins causes a net inflow of water.

The net effect of these forces is that water, oxygen and the solutes needed by cells pass from the capillaries to the tissue fluid, and water, carbon dioxide and waste products pass from the tissue fluid into the capillaries (Figure 17.19).

The remainder of the tissue fluid eventually passes back into the blood circulation as lymph via the lymphatic system.

■ Lymph

The lymphatic system (Figure 17.20, overleaf) is really a part of the vascular system. Whilst 90 % of the tissue fluid finds its way back into the capillaries, the remainder enters the lymphatic system, and becomes known as **lymph**. The lymph capillaries are very similar to veins, but with many more valves along their length. Lymph capillaries join together to form larger vessels, the **lymphatics**. The movement of lymph along the lymphatics is largely due to the movements of the surrounding skeletal muscles as they contract and relax during movement, squeezing the connective tissue in which the vessels lie. The walls of the lymphatics contain valve-like pores that allow the entry of cell debris and bacteria. Before reaching the blood, lymph passes through one or more **lymph nodes**, where these suspended solids are removed by cells of the lymph node by the process of phagocytosis (p. 364).

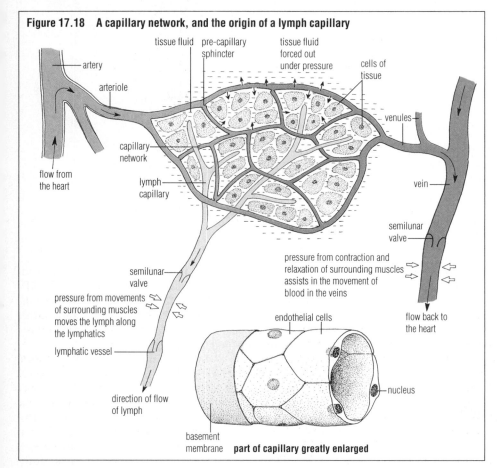

Figure 17.18   A capillary network, and the origin of a lymph capillary

artery
arteriole
tissue fluid
pre-capillary sphincter
tissue fluid forced out under pressure
cells of tissue
capillary network
flow from the heart
lymph capillary
venules
vein
semilunar valve
semilunar valve
pressure from contraction and relaxation of surrounding muscles assists in the movement of blood in the veins
pressure from movements of surrounding muscles moves the lymph along the lymphatics
lymphatic vessel
endothelial cells
flow back to the heart
nucleus
direction of flow of lymph
basement membrane
part of capillary greatly enlarged

Figure 17.19   Exchange at the capillaries

water and solutes
waste products and water
tissue fluid
osmosis
glucose, amino acids
ions
O₂
carbon dioxide and other waste products
water
blood
blood pressure (hydrostatic pressure)
water potential
diffusion and filtration
diffusion
osmosis
blood pressure
arterial end
capillary
venous end

Questions

1 When the body becomes chilled in a cold environment, what is the sequence of changes that reduces blood supply to the capillaries of the skin (p. 512)?

2 a How is tissue fluid formed from the blood?
  b Which components of the blood are normally not found in tissue fluid?

3 Which components of tissue fluid are likely to be taken up by cells? Which components are produced by cells?

Lymph drains into the blood circulation via thoracic ducts that join the veins in the neck. The right thoracic duct drains the right side of the head and thorax and the right arm. The left thoracic duct drains the remainder of the body. The lacteals of the small intestine (p. 285) are connected to the left thoracic duct.

Oedema

When an excess of tissue fluid is formed and is not re-absorbed, the organs and tissues of the body swell up with the increased fluid. This condition is known as **oedema**. Typically, oedema is caused by an increase in capillary blood pressure, or by failure of the excretion of sodium and chloride ions in the kidneys (p. 388).

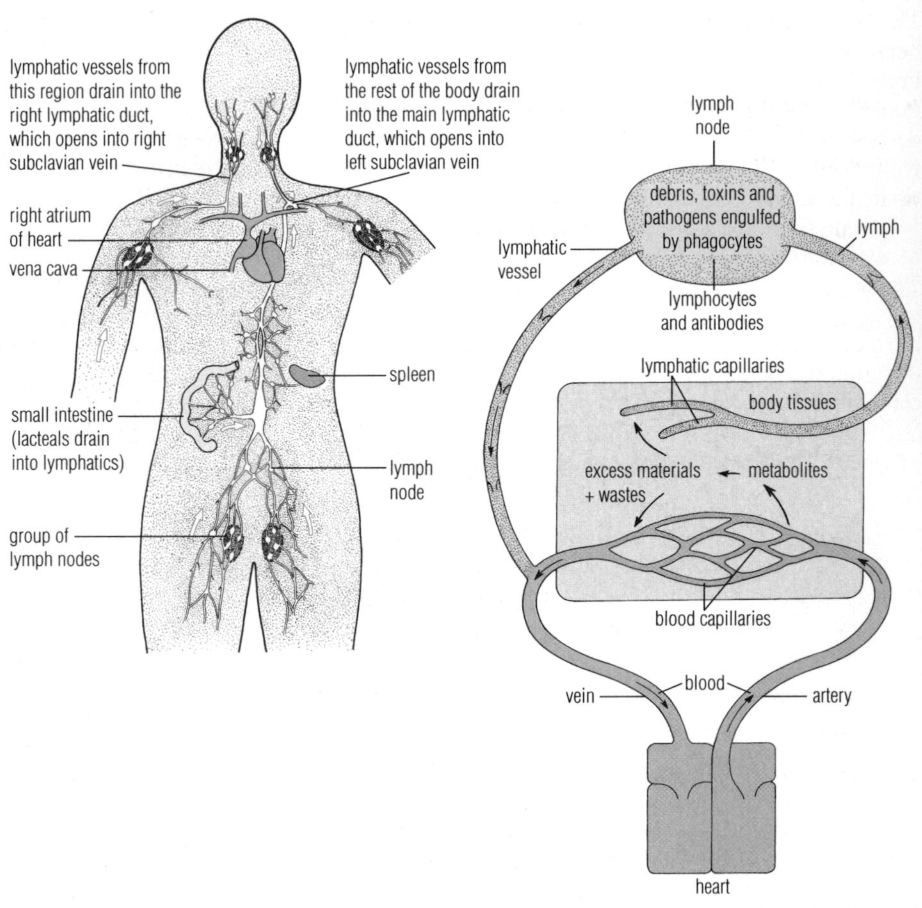

17.4 THE CARDIOVASCULAR SYSTEM IN HEALTH AND DISEASE

■ Effects of exercise

Short-term effects

In periods of intense physical activity the role of the cardiovascular system is to maintain an increased supply of oxygen and nutrients (for example, glucose and fatty acids) to the muscles, and to remove the CO_2 and heat produced. If the activity is anticipated, then impulses from the cerebral cortex to the vascular control centre in the medulla (Figure 17.13, p. 350) and the effect of adrenaline released stimulate an increased heart rate. During heavy physical activity, the output of blood from the left ventricle rises from about $5 \, dm^3 \, min^{-1}$ to $15–20 \, dm^3 \, min^{-1}$ in untrained individuals, and to $30 \, dm^3 \, min^{-1}$ in athletes, due to increased rate of contraction (heart rate) and more complete emptying of the ventricles (stroke volume). Blood pressure rises steeply. Veins draining the muscles dilate but arterioles supplying organs not immediately involved (such as gut, liver, kidney) constrict, increasing the return of blood to the heart. As activity persists and heat builds up, vasodilation of skin arterioles also occurs, and heat loss is increased.

Long-term effects

If a person trains regularly using extended periods of intense physical activity under aerobic muscle conditions (p. 492), the strength of the heart muscle and the volume of the heart chambers

Figure 17.20 The lymphatic system: layout and role

increase. The heart comes to pump the same volume of blood as that of an untrained person but at a lower heart rate. All aspects of the heart improve from 'endurance' training of this sort, down to the number and size of the mitochondria present between the cardiac muscle fibres. On the other hand, brief periods of excessively heavy activity creating anaerobic conditions in the muscles have no such advantageous after-effects.

■ Hypertension (persistently raised blood pressure)

Blood pressure is usually measured with the sphygmomanometer (Figure 17.21). An inflatable cuff around the upper arm is pressurised until the pulse in the brachial artery of the lower arm can no longer be heard. Pressure in the cuff is then decreased, first until sound is just heard (systolic pressure) and then further, until the sound is 'muffled' (diastolic pressure). Normal systolic and diastolic pressures are about 15.8 and 10.5 kPa respectively (medical workers usually give these values as 120 and 70–80 mmHg). Blood pressure depends on age, but a value of 18.6/12.0 kPa (140/90 mmHg) is typically at the threshold of hypertension, and values above 21.3/12.6 kPa (160/95 mmHg) are dangerously high.

Possible **causes** of hypertension include atherosclerosis (which reduces the elasticity of the arterial wall and narrows the lumen), kidney disease (when this triggers the release of a vasoconstricting substance within the blood) and excessive secretion by the adrenal glands (resulting in high levels of aldosterone (p. 388) and causing the retention of salt and water by the kidneys).

The **dangers** from hypertension are to the heart, brain and kidneys. The heart uses more energy in pumping, leading to angina pectoris (chest pains owing to inadequate blood supply to the heart muscle) and increased chance of myocardial infarction (p. 356). In the brain, the arteries are less well protected by surrounding tissues than in other organs and are in greater danger of rupture and haemorrhage. In the kidneys the arterioles supplying the kidney tubules thicken and develop a reduced lumen, and the blood supply to the cells of the tubules may be dangerously reduced.

The **treatment** of hypertension includes

▶ overweight patients being placed on a diet to lose weight, because blood pressure often falls with weight loss

▶ diet adjustment to reduce salt intake, as this helps to reduce blood volume

▶ patients who smoke being strongly advised to stop smoking, as nicotine is a vasoconstricting drug that elevates blood pressure

▶ prescription of a beta-blocking drug (such as atenolol), since this blocks off the β-receptors for adrenaline at the sympathetic terminals on the pacemaker of the cardiac muscle (Figure 17.22); stimulation by the sympathetic nerve increases heart rate and force of contraction, thereby raising blood pressure.

■ Atherosclerosis

Diseases of blood vessels are responsible for more human deaths in developed countries (that is, in societies where most people live well beyond the age of fifty) than any other single cause. Most of these deaths are due to **atherosclerosis**.

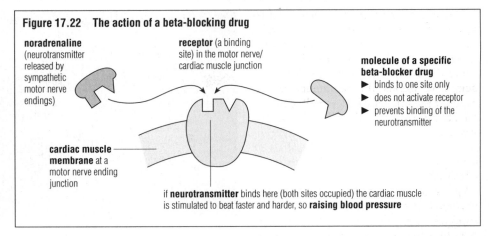

Figure 17.22 The action of a beta-blocking drug

noradrenaline (neurotransmitter released by sympathetic motor nerve endings)

receptor (a binding site) in the motor nerve/cardiac muscle junction

molecule of a specific beta-blocker drug
▶ binds to one site only
▶ does not activate receptor
▶ prevents binding of the neurotransmitter

cardiac muscle membrane at a motor nerve ending junction

if **neurotransmitter** binds here (both sites occupied) the cardiac muscle is stimulated to beat faster and harder, so **raising blood pressure**

Atherosclerosis is a condition of progressive degeneration of artery walls. A healthy artery has a pale, smooth, glistening lining. In an unhealthy artery there are yellow fatty streaks under the endothelium. A fat deposit, called an **atheroma**, is built up from cholesterol taken up from the blood as low-density lipoproteins (p. 508). Dense white fibrous tissue may also be laid down, forming irregular raised patches that impede blood flow. The smooth lining breaks down and blood comes into contact with the fatty and fibrous deposits. The blood platelets, source of triggering substances in the clotting mechanism, stick to the roughened surface. A blood clot, or **thrombus**, is likely to form (Figure 17.23).

Aneurysm

Atherosclerosis may lead to weakening of the artery wall, giving rise to stretching and a bulge. This kind of local dilation of an artery is known as an **aneurysm**. An aneurysm may burst at any time.

Embolisms and thromboses

A blood clot formed at an atherosclerotic site may break away and be carried around the circulation (an **embolus**). An embolus may come to lodge in and obstruct a small artery. The obstruction is known as an **embolism**.

Blood clots formed at an atherosclerotic site may accumulate and cause blockage of the artery there. A **thrombosis** is a blockage caused at the site of the atheromatous lesion.

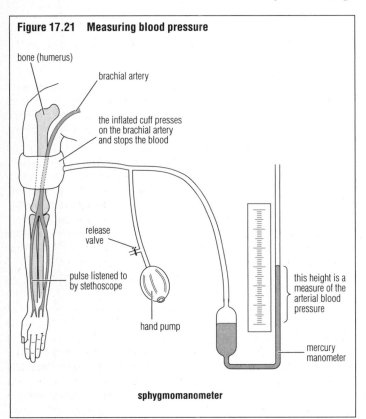

Figure 17.21 Measuring blood pressure

bone (humerus)

brachial artery

the inflated cuff presses on the brachial artery and stops the blood

release valve

pulse listened to by stethoscope

hand pump

this height is a measure of the arterial blood pressure

mercury manometer

sphygmomanometer

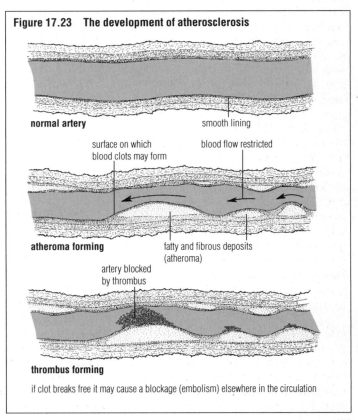

Figure 17.23 The development of atherosclerosis

normal artery

smooth lining

surface on which blood clots may form

blood flow restricted

atheroma forming

fatty and fibrous deposits (atheroma)

artery blocked by thrombus

thrombus forming

if clot breaks free it may cause a blockage (embolism) elsewhere in the circulation

If an obstruction occurs in the coronary artery, a region of heart muscle is deprived of an adequate blood supply (Figure 17.24). This leads to acute chest pains, known as **angina**. If this condition is not treated in time the heart muscle supplied by the blocked vessel may die ('heart attack' or **myocardial infarction**) because of the lack of adequate oxygen. An outcome of the demise of patches of heart muscle is the increasing likelihood of periods of irregular heartbeat rhythm, or of the heartbeat being replaced by uncoordinated twitching movements – a condition known as ventricular fibrillation. In these cases, the heart ceases to be an effective pump. If the heart becomes ineffective other vital organs, such as the brain, may be deprived of adequate oxygenated blood and the heart attack may be fatal. Abnormal heart function is detected by electrocardiography (Figure 17.26).

Coronary arteries damaged by atherosclerosis can be surgically by-passed with parts of a vein or artery taken from elsewhere in the body (Figure 17.25).

If a thrombosis or embolism blocks an artery supplying part of the brain, a 'stroke' is the outcome (Figure 17.27). The brain depends on blood for a continuous supply of oxygen and glucose. Interruption of blood flow for only a few minutes may cause death of vital neurones, and if the interruption persists, all brain cells in the affected area die. The result is some degree of impairment of the body function that is controlled by the damaged region of the brain. The degree to which lost facilities are recovered is normally limited; the brain cells destroyed cannot be replaced.

Vulnerability factors

The factors that predispose people to atherosclerosis and therefore to coronary heart disease and strokes are listed in the panel.

Questions

1 How do the conditions within an artery affected by atherosclerosis favour the formation of blood clots (pp. 355 and 364)?
2 **a** Why is blood pressure essential?
  b What do people usually mean if they say they 'have blood pressure'?
3 You cannot do much about a hereditary predisposition to atherosclerosis, but how could your life style lessen the risk?

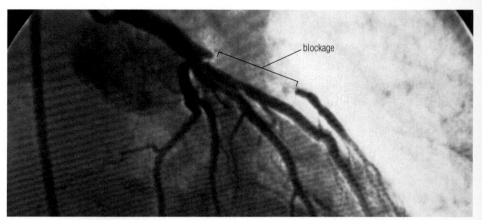

Figure 17.24   A damaged coronary artery: the photograph (an angiogram) shows part of an adult human heart that has a large blockage in the coronary artery

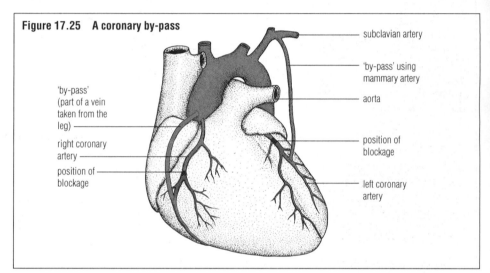

Figure 17.25   A coronary by-pass

subclavian artery

'by-pass' using mammary artery

aorta

position of blockage

left coronary artery

'by-pass' (part of a vein taken from the leg)

right coronary artery

position of blockage

FACTORS THAT PREDISPOSE TO CORONARY HEART DISEASE

▶ **age:** risk of coronary atherosclerosis increases with age, but the evidence is that the condition may start to develop very early in life
▶ **sex:** women before menopause (p. 601) rarely suffer from heart disease (oestrogen appears to protect women); coronary disease mainly afflicts adult males and manifests itself in middle age (45–55 years)
▶ **diet:** individuals with hereditary high levels of blood cholesterol (p. 276) are more prone to develop coronary disease; in patients with diseased arteries, the risk of a coronary attack may be reduced by a change in diet to lower blood cholesterol levels, and indeed for most people a diet with a low fat content may be helpful from an early age
▶ **weight:** obesity is a major factor in increasing the stress on heart and arteries
▶ **heredity:** coronary heart disease is much more common in some families than in others
▶ **smoking:** heavy cigarette smokers are much more likely to develop atherosclerosis
▶ **lack of exercise:** regular exercise aids a healthy circulation, and the physically inactive are more at risk from heart disease

▶ **other diseases:** diabetes mellitus and high blood pressure increase the risk of coronary disease
▶ **socio-economic disadvantage:** as nations become prosperous coronary disease becomes less of a problem for the well-educated managerial classes, and preferentially affects those in the lower social levels; poor housing and little education are strong indicators of high coronary mortality
▶ **psychological and personality factors:** sleep disturbances, hypochondria and the continuing experience of emotional exhaustion are predictors of angina (acute chest pains), infarction and death from heart attack; the annual rate of all coronary disease in Type A people, who have a high potential for 'hostility' and 'anger' emotions*, far exceeds that of Type B people

*Type A personalities are competitive, aggressive and rushed; they invest intense efforts and feel the pressures of time and the burdens of responsibility.
Type B personalities are easy-going and relaxed.

Figure 17.26   Electrocardiography

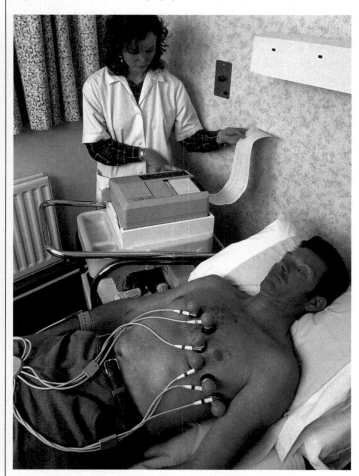

in electrocardiography electrodes are attached to the patient's chest, and the electrical activity detected is displayed as an electrocardiogram by means of the chart recorder; an electrocardiogram is taken to detect abnormalities that may confirm a suspected heart attack, for example

a normal electrocardiogram (ECG) analysed

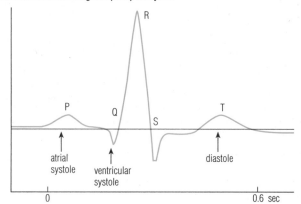

a normal ECG trace

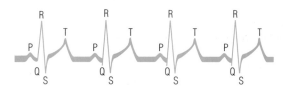

Abnormal traces showing

1 tachycardia

heart rate is over 100 beats/minute

2 ventricular fibrillation

uncontrolled contraction of the ventricles – little blood is pumped

3 heart block

ventricles not always stimulated

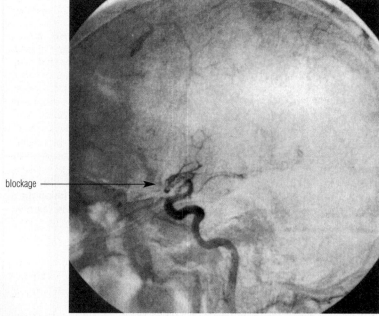

blockage

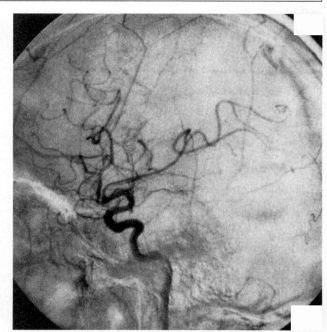

Figure 17.27   Treatment of a stroke: the blockage of a cerebral artery (left) has been cleared (right) by injection of a compound called tissue plasminogen activator; removal of a blockage must be achieved promptly if the patient is to make a full recovery

17.5 BLOOD AS A TRANSPORTING MEDIUM

The composition of the blood is discussed in Chapter 10 (p. 216).

The blood circulatory system is the **internal communication system** of the body by which a **constant internal environment** is maintained (homeostasis). The blood system is also an essential part of the body's defence against disease. Merely to list the functions of the blood circulatory system is to understate its key part in the survival of the organism. The transport functions of blood and its role in the body's defence against disease are discussed here. Homeostasis, the maintenance of a controlled internal environment, is discussed in Chapter 23 (p. 502). In summary, the blood circulation system facilitates the following:

▶ **tissue respiration**, by transporting oxygen to all tissues, and carbon dioxide back to the lungs
▶ **hydration**, by transporting water to all tissues
▶ **nutrition**, by transporting soluble organic and inorganic nutrients as raw materials for growth, replacement and repair, and as respiratory substrates to supply energy
▶ **excretion**, by transporting waste products to the kidneys, lungs, sweat glands and liver
▶ **temperature regulation**, by the distribution of heat between the organs that produce excess and all other tissues of the body, and by vasoconstriction and vasodilation in the skin
▶ **maintenance of a constant pH**, via the maintenance and circulation of the plasma proteins
▶ **growth, development and coordination**, by the transport of hormones and other regulatory substances from the endocrine glands to all parts, including the target organs.

■ Transport of oxygen

The red blood cells contain haemoglobin, the molecule that transports oxygen. Typically, the human body requires between $250\,cm^3$ (at rest) and $1000\,cm^3$ (in strenuous exercise) of oxygen per minute. At body temperature, about $0.3\,cm^3$ of oxygen can be dissolved in $100\,cm^3$ of blood plasma, but the total oxygen-transporting facility of blood is about $20\,cm^3$ per $100\,cm^3$. Thus some 98% of the oxygen transported is carried by haemoglobin.

Haemoglobin and oxygen transport

Haemoglobin is a compact molecule made up of four interlocking sub-units. The sub-units are identical conjugated proteins, each consisting of a globular protein (globin) attached to a haem unit. The haem units comprise a porphyrin ring containing an atom of iron(II) at the centre (Figure 17.28). The haem groups in haemoglobin are responsible for its red colour, and are the site of oxygen transport. An iron atom combines with a molecule of oxygen, but without oxidation of the iron(II). In fact, the oxygen molecules fit into pockets in the haemoglobin called **binding sites**. Up to four oxygen molecules may be carried by each haemoglobin molecule. The reaction between haemoglobin and oxygen to form oxyhaemoglobin is reversible:

$$\underset{\text{haemoglobin}}{Hb} \times 4O_2 \rightleftharpoons \underset{\text{oxyhaemoglobin}}{HbO_8}$$

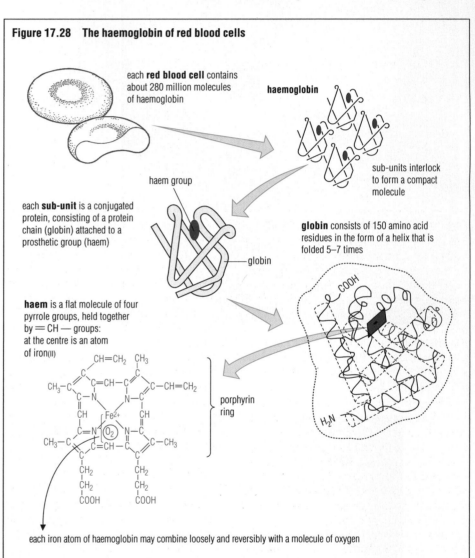

Figure 17.28   The haemoglobin of red blood cells

each **red blood cell** contains about 280 million molecules of haemoglobin

haemoglobin

sub-units interlock to form a compact molecule

haem group

each **sub-unit** is a conjugated protein, consisting of a protein chain (globin) attached to a prosthetic group (haem)

globin

globin consists of 150 amino acid residues in the form of a helix that is folded 5–7 times

haem is a flat molecule of four pyrrole groups, held together by $=CH-$ groups: at the centre is an atom of iron(II)

porphyrin ring

each iron atom of haemoglobin may combine loosely and reversibly with a molecule of oxygen

Oxygen tension (partial pressure)

In a mixture of gases, each component gas exerts a pressure (its **partial pressure** or **tension**) in proportion to its molar percentage in the mixture. Table 17.3 gives the partial pressures of the component gases in dry air at sea level (atmospheric pressure = $101.3\,kPa$).

Table 17.3   Partial pressures of the components of air

Component gases	Percentage composition	Partial pressure/kPa
oxygen	21	21.3
carbon dioxide	0.035	negligible
nitrogen	79	80.0

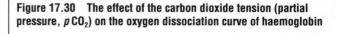

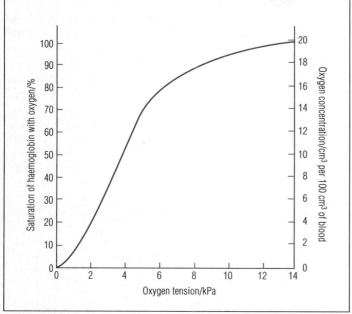

Figure 17.29 The oxygen dissociation curve of human haemoglobin at body temperature: this data is obtained experimentally, and the scale on the x-axis covers the range of oxygen tensions (partial pressures, pO_2) in the human body

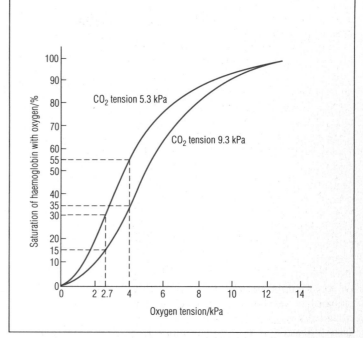

Figure 17.30 The effect of the carbon dioxide tension (partial pressure, $p\,CO_2$) on the oxygen dissociation curve of haemoglobin

Oxygen tension and oxyhaemoglobin formation

The affinity of haemoglobin for oxygen is measured experimentally by determining the percentage saturation with oxygen of samples of blood after exposure to air mixtures containing different partial pressures of oxygen. The result, called an **oxygen dissociation curve**, is shown in Figure 17.29.

The oxygen dissociation curve is S-shaped. Why this shape, rather than, say, a straight line?

The answer is that the S-shape is due to the way in which oxygen binds with haemoglobin. The first molecule of oxygen to bind does so with difficulty, distorting the shape of the haemoglobin molecule in the process. But as a result of this initial distortion the subsequent three molecules of oxygen are taken up progressively more quickly.

In the body, too, the amount of oxygen held by haemoglobin depends upon the partial pressure. In the lungs air is saturated with water vapour, and so the partial pressures of the component gases are different from those in the air outside the body (Table 17.4).

The oxygen tension in the lungs results in about 95% saturation of haemoglobin. The oxyhaemoglobin is carried in the blood to respiring tissues, where the oxygen tension is much lower (0.0–4.0 kPa). At this partial pressure of oxygen, oxyhaemoglobin readily breaks down. Oxygen is thus made available in the tissue fluid.

In summary, haemoglobin has a high affinity for oxygen where the oxygen tension is high (the lungs), but a low affinity for oxygen where the oxygen tension is low (respiring tissues). This property makes haemoglobin an efficient respiratory pigment.

The effect of carbon dioxide

The carbon dioxide tension of blood arriving at respiring tissues is about 5.26 kPa; it rises to about 6.05 kPa as the blood passes through the capillary network. The effect of a rising carbon dioxide tension is to decrease the affinity of haemoglobin for oxygen (Figure 17.30). Thus the carbon dioxide released in respiring tissue accelerates the delivery of oxygen. This phenomenon is called the **Bohr effect**, after the physiologist who discovered it.

Carboxyhaemoglobin

The iron of haemoglobin is not oxidised to iron(III) in oxyhaemoglobin formation. This oxidation can occur, however. For example, carbon monoxide converts iron(II) to iron(III) in its reaction with haemoglobin, with the formation of carboxyhaemoglobin. In this form haemoglobin cannot carry oxygen. Very little haemoglobin occurs in this form except in cases of carbon monoxide poisoning and in the blood of smokers, where about 10% is in the form of carboxyhaemoglobin. Carboxyhaemoglobin does not dissociate; the reaction between carbon monoxide and haemoglobin is irreversible. Smokers lose the use of the haemoglobin that combines with carbon monoxide.

Questions

1 a What change in the percentage saturation of haemoglobin occurs if the oxygen tension drops from 4 to 2.7 kPa when the partial pressure of carbon dioxide is 5.3 kPa?
   b What will be the change if the partial pressure of carbon dioxide is 9.3 kPa?
2 What properties of haemoglobin make it an efficient means of transporting oxygen?
3 Why might smoking impair an athlete's performance?

Table 17.4 Partial pressures of the components of alveolar air

Component gases	Percentage composition	Partial pressure/kPa
oxygen	13.1	13.3
carbon dioxide	5.2	5.3
nitrogen	75.5	76.4
water vapour	6.2	6.3

Myoglobin and haemoglobin

Myoglobin is a conjugated protein chemically similar to haemoglobin. It consists of a single haem–globin unit only. Myoglobin occurs largely but not exclusively in voluntary (skeletal) muscle tissue, and gives 'meat' its characteristic red colour. The important property of myoglobin is that it has a greater affinity for oxygen than that of haemoglobin. As a consequence, myoglobin functions as an oxygen store in muscle tissue. When the demand for oxygen temporarily exceeds the rate of supply, as it does in periods of prolonged and extreme muscular exertion, the oxygen of oxymyoglobin is released (Figure 17.31).

If muscular activity persists after the myoglobin-based oxygen supply is also exhausted (the muscle has moved into 'oxygen debt', p. 492), then the muscle must respire by lactic acid fermentation. Once fatiguing muscular activity ends, myoglobin reserves of oxygen are replenished from oxyhaemoglobin of the blood. The lactate that accumulated is metabolised in the muscle or in other tissues (p. 317).

There are very high levels of myoglobin in the muscles of diving animals (p. 363) and in the flight muscles of flighted birds (p. 497).

Figure 17.31 Myoglobin and the supply of oxygen to voluntary muscle

in the muscles, oxygen from oxyhaemoglobin is passed to and retained by myoglobin

oxyhaemoglobin in blood

lungs

muscle

haem group

heart
left side pumps oxygenated blood to the tissues

muscle myoglobin, one unit

Graph: Saturation of pigment with oxygen/% (y-axis) versus Oxygen tension/kPa (x-axis), showing curves for myoglobin and haemoglobin.

(p. 492) (p. 317) (p. 363) (p. 497)

Haemoglobin in other animals

The chemical compositions of haemoglobins are not the same in all animals and, as a consequence, the oxygen-carrying capacities also vary. A comparison of the dissociation curves of oxyhaemoglobins of two animals with that of humans (Figure 17.32) demonstrates this. For ease of comparison of the properties of haemoglobins from different species, the terms 'loading tension' and 'unloading tension' are used. The loading tension is the oxygen tension at which 95% of haemoglobin is oxygenated. The unloading tension is the oxygen tension at which 50% of the haemoglobin releases its oxygen.

The lugworm *Arenicola* has a low metabolic rate. It lives in the sand of the shore in the intertidal zone (p. 48). The lugworm pumps sea water through its burrow, giving access to the limited amount of dissolved oxygen present. The loading tension of haemoglobin in the lugworm is at or about the partial pressure of oxygen in sea water, however, so this animal is able to retain the oxygen available and deliver it to the respiring cells.

By comparison, pigeons have access to the 21% of oxygen in air, and have the additional advantage of continuous ventilation of their lungs (p. 307). In these circumstances the high loading tension of pigeon haemoglobin is no problem. The very high metabolic rate of the bird requires, however, that a great deal of oxygen is made available to all tissues. Here, the high unloading tension ensures that oxygen is readily available to the respiring cells.

Figure 17.32 A comparative approach to oxygen transport

Arenicola marina, the marine lugworm

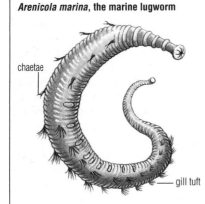

chaetae

gill tuft

dissociation curves of haemoglobin

Graph: Saturation of haemoglobin with oxygen/% (y-axis) versus Oxygen tension/kPa (x-axis), showing curves for Arenicola, human and pigeon.

Arenicola obtains oxygen from sea water, which it pumps through its burrow; the sea water is relatively low in dissolved oxygen

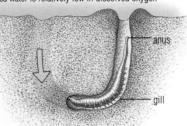

anus

gill

***Arenicola* in its burrow**

the pigeon (***Columba palumbus***), a typical bird

pigeon in flight: the bird's lung/air sac system is very efficient in obtaining oxygen from the air

body movements and general level of metabolic activity of the bird make great demands upon the oxygen supply to the tissues

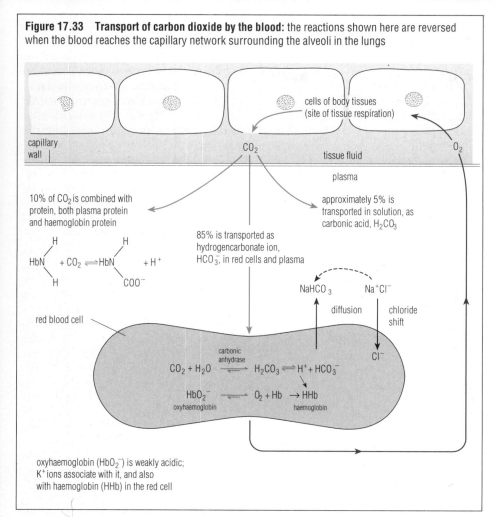

Figure 17.33 **Transport of carbon dioxide by the blood:** the reactions shown here are reversed when the blood reaches the capillary network surrounding the alveoli in the lungs

cells of body tissues
(site of tissue respiration)

capillary wall

CO_2 tissue fluid O_2

plasma

10% of CO_2 is combined with protein, both plasma protein and haemoglobin protein

$$HbN \begin{matrix} H \\ \\ H \end{matrix} + CO_2 \rightleftharpoons HbN \begin{matrix} H \\ \\ COO^- \end{matrix} + H^+$$

approximately 5% is transported in solution, as carbonic acid, H_2CO_3

85% is transported as hydrogencarbonate ion, HCO_3^-, in red cells and plasma

$NaHCO_3$ Na^+Cl^-

diffusion   chloride shift

red blood cell

Cl^-

$$CO_2 + H_2O \xrightarrow{\text{carbonic anhydrase}} H_2CO_3 \rightleftharpoons H^+ + HCO_3^-$$

$$\underset{\text{oxyhaemoglobin}}{HbO_2^-} \longrightarrow O_2 + Hb \rightarrow \underset{\text{haemoglobin}}{HHb}$$

oxyhaemoglobin (HbO_2^-) is weakly acidic; K^+ ions associate with it, and also with haemoglobin (HHb) in the red cell

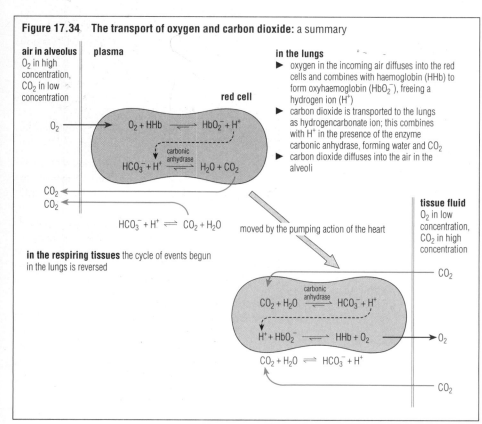

Figure 17.34 **The transport of oxygen and carbon dioxide:** a summary

air in alveolus
O_2 in high concentration, CO_2 in low concentration

plasma

in the lungs
▶ oxygen in the incoming air diffuses into the red cells and combines with haemoglobin (HHb) to form oxyhaemoglobin (HbO_2^-), freeing a hydrogen ion (H^+)
▶ carbon dioxide is transported to the lungs as hydrogencarbonate ion; this combines with H^+ in the presence of the enzyme carbonic anhydrase, forming water and CO_2
▶ carbon dioxide diffuses into the air in the alveoli

red cell

O_2

$$O_2 + HHb \rightleftharpoons HbO_2^- + H^+$$

$$HCO_3^- + H^+ \xrightleftharpoons{\text{carbonic anhydrase}} H_2O + CO_2$$

CO_2
CO_2

$$HCO_3^- + H^+ \rightleftharpoons CO_2 + H_2O$$

moved by the pumping action of the heart

tissue fluid
O_2 in low concentration, CO_2 in high concentration

in the respiring tissues the cycle of events begun in the lungs is reversed

CO_2

$$CO_2 + H_2O \xrightleftharpoons{\text{carbonic anhydrase}} HCO_3^- + H^+$$

$$H^+ + HbO_2^- \rightleftharpoons HHb + O_2$$

O_2

$$CO_2 + H_2O \rightleftharpoons HCO_3^- + H^+$$

CO_2

■ Transport of carbon dioxide

Carbon dioxide is transported both in the plasma and in red cells, from tissues all over the body to the lungs. A little carbon dioxide reacts with water to form carbonic acid, and is transported as such, in the plasma. A slightly larger quantity reacts with blood proteins (Figure 17.33). The red cells, however, are rich in the enzyme carbonic anhydrase, which accelerates the formation of carbonic acid. Consequently, most of the carbon dioxide diffuses into the red cells. Carbonic acid dissociates into hydrogencarbonate ions and hydrogen ions, as it is formed:

$$CO_2 + H_2O \rightleftharpoons H_2CO_3 \rightleftharpoons H^+ + HCO_3^-$$

The hydrogen ions are buffered by the haemoglobin, preventing the blood from becoming acidic (and causing oxyhaemoglobin to dissociate into oxygen and haemoglobin). Oxygen diffuses into the respiring cells (Figure 17.34). At the same time, hydrogencarbonate ions diffuse out into the plasma, and chloride ions into the red cells, maintaining electrical neutrality. The membranes of the red blood cells contain a sodium pump (p. 237), which pumps sodium ions out of these cells. Sodium ions associate with hydrogencarbonate ions in the plasma.

■ Transport of other substances

The blood transports water and soluble food substances (sugars, amino acids and fatty acids), together with various other metabolites and hormones, all dissolved in the plasma. These substances are delivered to the cells via the tissue fluid, formed as the plasma escapes through the walls of the capillaries. The tissue fluid also picks up waste products to be excreted, and is returned to the blood directly via the capillaries, or indirectly via the lymphatic system.

The blood also plays a part in temperature regulation by the transport of heat all over the body, and by vasoconstriction and vasodilation in the skin.

Questions

1 Outline in tabulated form the part played by the blood in the transport of respiratory gases.
2 Name the key enzyme involved in carbon dioxide transport in the blood. What component of the blood contains this enzyme?
3 The haemoglobin of *Arenicola* can take up oxygen at the very low tension in which it occurs in sea water. What does this imply about the oxygen tension in the animal's cells?
4 What important properties of enzymes are well illustrated by carbonic anhydrase (p. 178)?

17.6 SOME UNUSUAL CIRCULATORY SYSTEMS

■ Circulation in the human fetus

The fetal circulation is similar to that of the adult, but with some important differences. The reason for these differences is that the functions performed in the adult by the lungs, liver, kidneys and gut are largely performed in the fetus by the placenta. The fetus is supplied with nutrients and oxygen through the placenta, and waste materials are returned to the maternal circulation by the same route.

There are several special features of the fetal circulation (Figure 17.35).

▶ An umbilical artery and an umbilical vein connect the fetal circulation to the placenta. Blood is pumped to the placenta by the fetal heart.

▶ A hole in the wall of the heart, the foramen ovale, connects the right and left atria. Most of the deoxygenated blood returning to the right side of the heart is passed to the left atrium and pumped via the left ventricle to the aorta and on to the placenta.

▶ A short vessel, the ductus arteriosus, connects the pulmonary artery to the aorta. Most of the blood leaving the heart by the pulmonary artery is re-directed to the aorta. This occurs because pressure in the pulmonary circulation is higher than in the aorta.

▶ A vessel, the ductus venosus, carries blood from the placenta and fetal gut directly to the posterior vena cava, largely by-passing the fetal liver. The liver in the fetus manufactures blood, rather than having the regulatory role it has in the adult mammal.

▶ Fetal haemoglobin combines more readily with oxygen than adult haemoglobin does (Figure 17.36). This is essential for the fetus to be able to extract oxygen from the maternal blood supply.

Changes at parturition

At parturition (birth) the gaseous exchange mechanism must switch from placenta to lungs, nutrition from placenta to gut, and excretion from placenta to kidneys. A baby emerges from a watery environment at 37 °C to an air temperature that is variable. Evaporation causes rapid cooling. The stimulation of the skin by a cooling environment causes the baby to cry. This leads to inflation of the lungs with air, and gaseous exchange commences. The resistance within the pulmonary circulation falls dramatically and the resistance in the aorta rises as the gut, liver and kidneys start to function for themselves.

The placenta then ceases to function. The umbilical cord is ligatured, and rapidly withers away, leaving a scar, the umbilicus.

The foramen ovale and the ductus arteriosus close off within the first few days. The ductus venosus becomes sealed off by muscular contraction.

■ Mammals at high altitude

The percentage of oxygen by volume in air does not vary significantly between sea level and high altitudes, but the atmospheric pressure falls with increasing altitude. The quantity of oxygen in a sample of air is thus determined by the pressure, since it is the pressure that determines how many moles of gas are compressed into a unit volume of the air. At sea level the atmospheric pressure is 101.3 kPa, the oxygen makes up almost 21% of the air, and the oxygen tension is 21.2 kPa. Table 17.5 indicates how oxygen tension changes with altitude.

At high altitude it is difficult to load haemoglobin with the oxygen. The result is a lowered percentage saturation of haemoglobin with oxygen. This is detected in the circulation system by chemoreceptors (p. 311). The body's immediate response is to take extra deep breaths; as a result, more carbon dioxide is lost from the body. This causes a small but significant rise in pH of the blood. An outcome is that the chemoreceptors are rendered ineffective, and ventilation is hampered.

The body cannot adapt immediately to high altitude, but with time

▶ a more alkaline urine is secreted, the pH of the blood is returned to normal, and the carbon dioxide chemoreceptors become sensitive to the carbon dioxide present; normal ventilation rates are then maintained

▶ bone marrow produces more erythrocytes to raise the oxygen-carrying capacity of the blood.

Animals that have evolved at high altitude have a form of haemoglobin that loads more easily at lower partial pressures of oxygen; the loading tension of the haemoglobin of the llama (a high-altitude mammal) is much lower than that of the haemoglobin of lowland mammals (Figure 17.37).

Figure 17.35   The blood circulation of the mammalian fetus

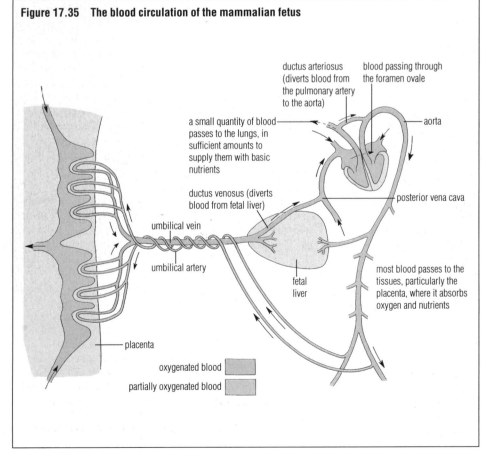

ductus arteriosus (diverts blood from the pulmonary artery to the aorta)

blood passing through the foramen ovale

a small quantity of blood passes to the lungs, in sufficient amounts to supply them with basic nutrients

aorta

ductus venosus (diverts blood from fetal liver)

posterior vena cava

umbilical vein

umbilical artery

fetal liver

most blood passes to the tissues, particularly the placenta, where it absorbs oxygen and nutrients

placenta

oxygenated blood

partially oxygenated blood

Figure 17.36   The oxygen dissociation curves of fetal and adult haemoglobin in humans

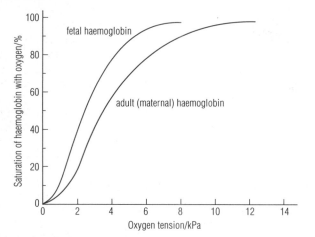

Table 17.5   Change in oxygen tension with altitude

Altitude/m above sea level	Atmospheric pressure/kPa	Oxygen content/%	Oxygen tension/kPa
0	101.3	20.9	21.2
2500	74.7	20.9	15.7
5000	54.0	20.9	11.3
7000	38.5	20.9	8.1
10000	26.4	20.9	5.5

Figure 17.37   High altitude and the transport of oxygen

Acclimatisation of mammals to breathing at high altitude

	Altitude/m	Red blood cell count/ $\times 10^{12}$ dm^3
human	0 (sea level)	5.0
	5000+ as a temporary visitor	5.95
	as a resident	7.37
rabbit	0	4.55
	5000+	7.00
sheep	0	10.5
	4500+	12.05

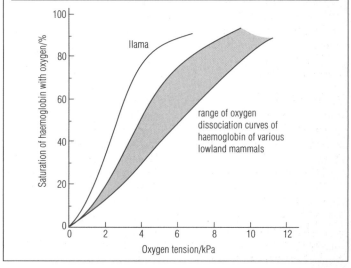

Diving mammals

Seals (Figure 17.38), walruses, whales and dolphins, and related air-breathing mammals, are able to endure long periods under water, sometimes at considerable depths. Maximum duration of dive varies with the species; it may be 15 minutes to one hour, or even longer.

Adaptations of the circulatory system in diving mammals

Compared with the circulation systems of terrestrial mammals

▶ blood makes up a greater proportion of the body mass
▶ there is an increased concentration of red cells in the blood
▶ there is a higher concentration of haemoglobin in the red cells
▶ during a dive the heartbeat slows down automatically, with blood pressure maintained by constriction of the arteries
▶ blood is distributed to the vital organs during a dive by constriction of the veins that drain the less immediately important organs, such as the kidneys
▶ there is a higher concentration of myoglobin in the muscles (an additional oxygen store)

▶ a higher concentration of lactic acid in muscle tissue can be tolerated, due to a reduced sensitivity to lowered pH; as the concentration of carbon dioxide rises in the body, breathing and heartbeat are not automatically stimulated.

Figure 17.38   A diving mammal: the common seal (*Phoca vitulina*)

Questions

1 List three main changes in the blood circulation of the mammal at birth.
2 Why is the oxygen tension lower in air at high altitude, even though the percentage of oxygen is the same as at sea level?
3 What changes occur when humans become adapted to high altitudes?

363

17.7 DEFENCE AGAINST DISEASE

The blood circulation plays a key part in the resistance of the body to infection. Any break in the body surfaces, internal or external, may lead to an invasion of tissues by pathogenic microorganisms. When a surface of the body is breached, the body's responses are as follows:

▶ **inflammation** and the phagocytosis of invading bacteria
▶ **clotting** of escaping blood
▶ the **immune response** to infection.

■ Inflammation

Inflammation is a rapid localised response to tissue damage caused by cuts, blows or bites, or to injected chemicals such as those in a nettle (*Urtica*) sting or an insect sting, or due to presence of microorganisms. Alternatively, over-exposure to sunlight may produce a similar reaction. The symptoms are localised redness, pain, heat and swelling.

Inflammation is a part of the body's internal defence system. The steps to the inflammatory response are described below.

1 Vasodilation of arterioles and increased permeability of blood vessels: these are triggered by the escape of histamines and other substances from damaged cells. Mast cells (large cells in connective tissue) and white cells called basophils in the blood (Figure 10.11, p. 217), together with blood platelets attached to damaged blood vessels, are the main sources of the histamines. The increase in the diameter of blood vessels means **more blood** is retained in the area, raising the temperature locally. Increased permeability of the capillaries allows more **tissue fluid** to escape leading to localised swelling (oedema). **White cells**, blood-clot forming substances (Figure 17.41), and **complement proteins** (special plasma proteins) also accumulate. Another of the substances present is **interferon**, secreted by macrophages after exposure to foreign antigens (p. 366). The effect of interferon is to make the surrounding cells resistant to attack by viruses.

2 Removal of microorganisms and their toxins: invading microorganisms are engulfed by white cells, mainly neutrophils (Figure 17.39). These white cells move about in the damaged tissues by amoeboid movement, and detect microorganisms that have become coated with **opsonins** (Figure 17.40).

Opsonins are either **antibodies** produced in the bloodstream (p. 366) or particular complement proteins activated by the presence of the bacteria. Once the opsonins are attached, the microorganisms are especially vulnerable to phagocytosis and destruction by digestion. The increased blood supply inactivates and removes toxins, and the debris from damaged tissues is removed by macrophages, the body's main 'rubbish-collecting' cells.

■ Clotting of blood

When blood escapes from the blood vessels at a wound the process of clot formation is set in train, sealing off the gap. Clotting is rapid, normally preventing serious blood loss and, at the same time, the wholesale invasion of the tissues by bacteria. But it is important that the mechanism of clot formation is not activated in the intact circulation. Blood clots (embolisms) formed inside arteries and released into the circulation may come to obstruct small arterioles elsewhere in the vascular system, cutting off the blood supply to tissues (thrombosis, p. 355). Under normal conditions, clotting is prevented by a substance called **heparin**, present in low concentration in the plasma, and secreted by mast cells in connective tissue and in the liver. Also **nitric oxide** secreted by the capillary endothelial cells (p. 353) helps prevent clot formation by suppressing activation of blood platelets.

Coagulation is initiated by the release of factors from the damaged cells at the wound, and by the exposure of blood to collagen fibres in the damaged vessel walls. Collagen fibres are not normally in contact with the blood.

The process of blood clot formation is complex, involving many steps (Figure 17.41). The advantage of the complexity is as a fail-safe mechanism. Most conditions required for the formation of a clot are not normally met, so that clotting of blood within the intact blood circulation is not likely. In some people there is an excessive tendency to form clots in the circulation, but this tendency may be suppressed by anticoagulant drugs. Others suffer from a condition known as haemophilia, in which the blood contains too low a concentration of an essential clotting factor. Haemophilia is caused by a single recessive gene on the X chromosome (p. 620).

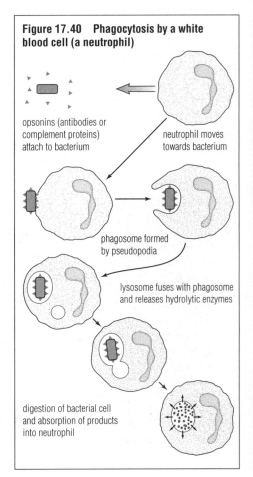

Figure 17.39 The phagocytic cells of the body's defence mechanism

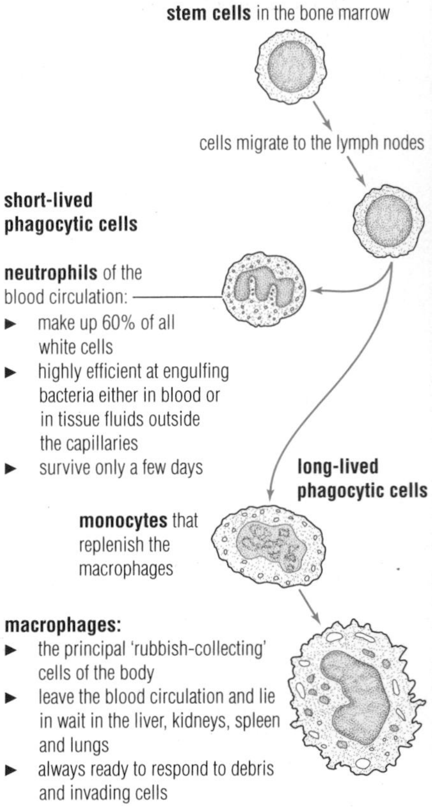

stem cells in the bone marrow

cells migrate to the lymph nodes

short-lived phagocytic cells

neutrophils of the blood circulation:
▶ make up 60% of all white cells
▶ highly efficient at engulfing bacteria either in blood or in tissue fluids outside the capillaries
▶ survive only a few days

long-lived phagocytic cells

monocytes that replenish the macrophages

macrophages:
▶ the principal 'rubbish-collecting' cells of the body
▶ leave the blood circulation and lie in wait in the liver, kidneys, spleen and lungs
▶ always ready to respond to debris and invading cells

Figure 17.40 Phagocytosis by a white blood cell (a neutrophil)

opsonins (antibodies or complement proteins) attach to bacterium

neutrophil moves towards bacterium

phagosome formed by pseudopodia

lysosome fuses with phagosome and releases hydrolytic enzymes

digestion of bacterial cell and absorption of products into neutrophil

Figure 17.41   The clotting of blood

steps in the formation of a blood clot

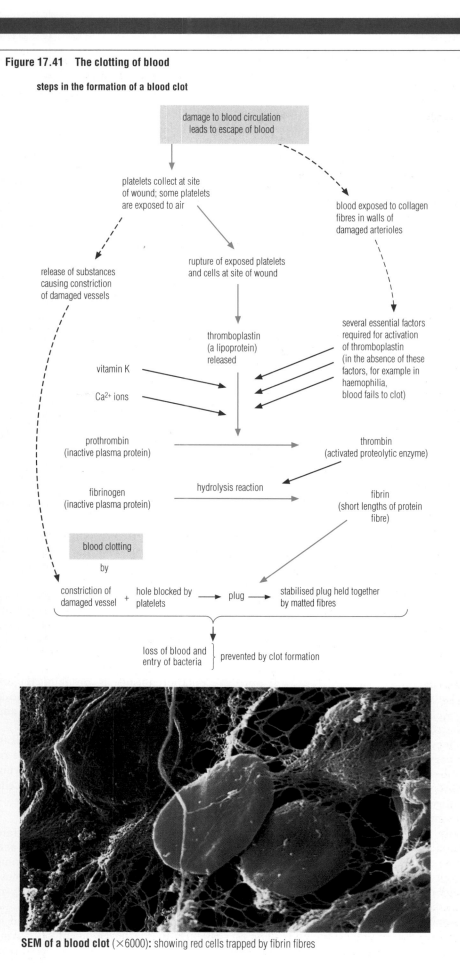

SEM of a blood clot (×6000): showing red cells trapped by fibrin fibres

Sealing of wounds

Platelets alone may become sticky and block a small wound, but the normal response to a break in blood vessels is due to the coagulation of a soluble blood protein, fibrinogen, forming insoluble fibrin fibres. Then the fibrin fibres form a compact network in which blood cells become entrapped and the whole plug is firmly stuck to the surrounding tissue as a hardening scab.

What triggers fibrin formation at the site of damage? The first significant event is the collection of blood platelets at the damage site. The collecting platelets release a substance called thromboplastin. Thromboplastin, along with calcium ions and vitamin K, causes the plasma protein prothrombin to be converted to the enzyme thrombin at the site of the damaged vessel. Thrombin then converts fibrinogen into insoluble fibrin fibres.

Eventually, the inflammation site reverts to more normal conditions. In the next phase of recovery, the repair process, cells called fibroblasts migrate in and synthesise collagen fibres. These form new tissues together with dividing epidermal cells. Capillaries grow into the new tissue, bringing nutrients and oxygen. The tissue differentiates into skin beneath the scab, which eventually falls off.

■ Phagocytosis elsewhere in the body

The white cells, neutrophils and monocytes, are able to accumulate anywhere in the tissues or circulatory system where there is an invasion of microorganisms.

Role of monocytes

Monocytes are formed from stem cells (p. 217) that have migrated to the lymph nodes. Monocytes are long-lasting cells, and develop into macrophage cells when these are required in the body. **Macrophages** are the principal detritus-collecting cells of the body. They leave the blood circulation by amoeboid movement, and lie in wait at strategic sites around the body, ready for the arrival of foreign matter or detritus, which they then 'mop up'.

Questions

1 Despite the complex fail-safe control system in blood clotting, clots may form within the intact circulation system. What conditions favour this (p. 355)?

2 Phagocytic cells are found in the airways of the lungs.
   a What is their role?
   b Where have they come from?
   c How did they get there?

■ Immunity and the immune system

Immunity is the ability of an organism to resist infection. The immune response is based upon the recognition of a 'foreign' or non-self chemical substance produced by an invading organism, followed by the production of chemicals and cells that effectively repel the invasion. The invading foreign substance recognised as 'non-self' is called an **antigen**. The chemical substance produced by the threatened organism in response to the intrusion is known as an **antibody**. An antibody may neutralise the antigen, or it may harm or cause the destruction of the organism producing it.

The ability to recognise foreign cells and to take steps to reject them is tied in with the evolution of a specialised white blood cell type, called the **lymphocyte**. Lymphocytes are found particularly in vertebrates; there are none in animals on the evolutionary scale of the level of arthropods, for example. Many non-vertebrates, including arthropods, can tolerate the exchange of tissues between organisms and the presence of foreign cells in their bodies.

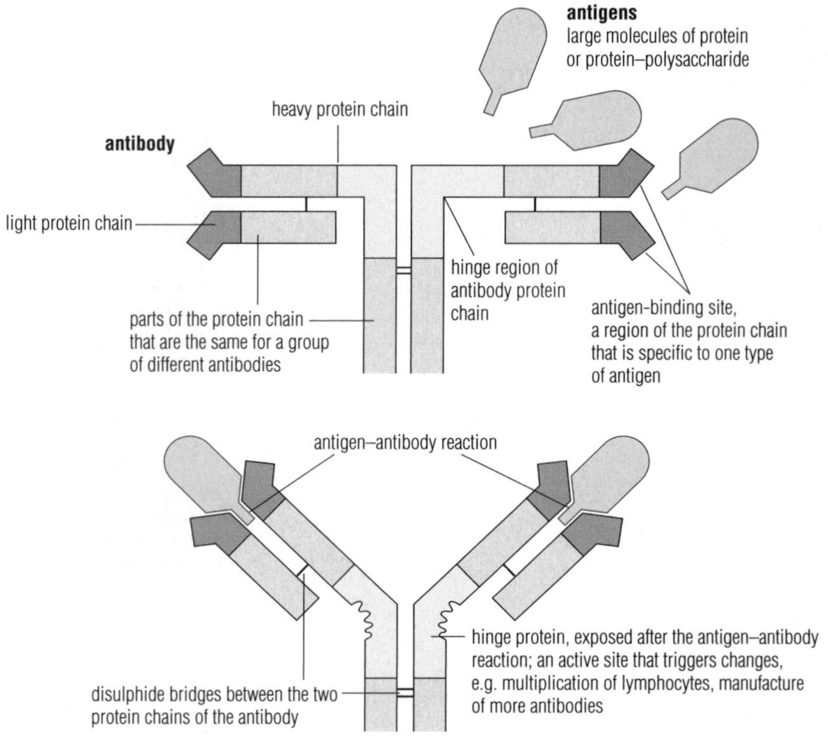

The chemical nature of antigens and antibodies

Antigens are large organic molecules, usually made of protein, or of protein and polysaccharides, although a few antigens are large polysaccharide molecules. Antigens may occur on the walls or external membranes of invading organisms, or may be secreted or injected by an invading microorganism. Antigens are recognised by an invaded organism as 'non-self'; that is, as 'foreign' matter.

The body responds to the presence of an antigen by the production of an antibody. Antibodies are proteins known as **immunoglobulins**. Each antibody is specific to only one type of antigen. The generalised structure of an immunoglobulin is shown in Figure 17.42. The specificity of the antibody resides in the sequence of amino acids in the parts of the immunoglobulin where the antigen actually keys in, and where the antibody response is switched on. In a lymphocyte, part of each antibody molecule is exposed on the external membrane as the receptor site for the appropriate antigen.

Antibodies are first and foremost binding proteins; they do not kill the invading cells themselves. But by clustering on the surface of a foreign cell, antibodies trigger a response that leads to the death of the invader (Figure 17.43).

Figure 17.42   The structure and functioning of an antibody

The development of lymphocytes

Lymphocytes originate in the bone marrow from stem cells similar to those that give rise to other blood cells. Unlike the other blood cells, however, the lymphocytes migrate to the lymph nodes, where they mature and continue to multiply in very large numbers throughout life. Lymphocytes migrate to the lymph nodes by one of two routes, giving rise to two types of lymphocyte (known as T and B lymphocytes) (Figure 17.43).

B lymphocytes and the antibody-mediated immune response

The B lymphocytes, or 'bone-derived lymphocytes', migrate directly from bone marrow to lymph nodes. Like the T lymphocytes they operate from there, and recognise antigens in the same way (see below). They respond differently, however: B lymphocytes respond to antigen by proliferating to form larger, plasma cells. These have a lifespan of only a few days. They respond to the antigens in their environment by switching on their protein synthesis machinery to produce and secrete vast quantities of antibody specific to the antigen. Some of the cells are subsequently retained in the lymph nodes as 'memory cells', so that if the infection returns to the body there is an almost immediate immune response. This is known as **acquired immunity**.

Types of antigen–antibody interaction

Antibodies may work to destroy antigens in several different ways:

▶ neutralising toxins released by bacteria, rendering them harmless so that they can be secreted from the body in the kidney tubules

▶ precipitating antigen molecules by causing them to form insoluble clumps; in this form the antigens may be engulfed by phagocytic cells, digested and rendered harmless

▶ attaching to the cells of invading microorganisms as opsonins (Figure 17.40, p. 364) so that phagocytosis and destruction by neutrophils and other white cells can follow

▶ working with complement proteins in the plasma, attaching to the walls of invading microorganisms, identifying them for the attention of T killer cells (see below).

T lymphocytes and the cell-mediated response

The first-formed T lymphocytes migrate from the bone marrow of a fetus and reach the lymph nodes via the thymus gland. Thus they are known as 'thymus-derived lymphocytes' (hence the shortened name of T lymphocytes).

The influence of the thymus gland is essential to the development of T lymphocytes, for it is the thymus gland that weeds out and destroys lymphocytes that

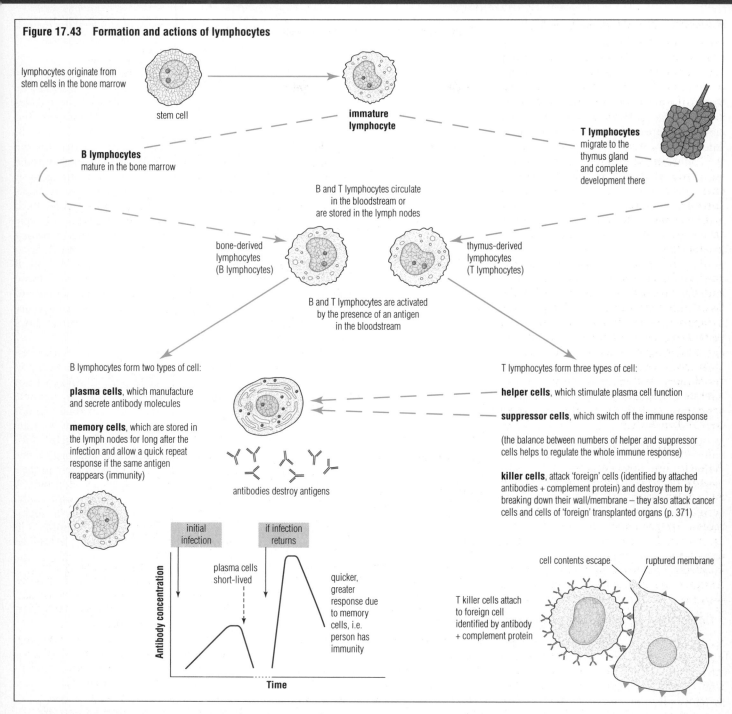

Figure 17.43   Formation and actions of lymphocytes

lymphocytes originate from stem cells in the bone marrow

stem cell

immature lymphocyte

T lymphocytes migrate to the thymus gland and complete development there

B lymphocytes mature in the bone marrow

B and T lymphocytes circulate in the bloodstream or are stored in the lymph nodes

bone-derived lymphocytes (B lymphocytes)

thymus-derived lymphocytes (T lymphocytes)

B and T lymphocytes are activated by the presence of an antigen in the bloodstream

B lymphocytes form two types of cell:

plasma cells, which manufacture and secrete antibody molecules

memory cells, which are stored in the lymph nodes for long after the infection and allow a quick repeat response if the same antigen reappears (immunity)

antibodies destroy antigens

T lymphocytes form three types of cell:

helper cells, which stimulate plasma cell function

suppressor cells, which switch off the immune response

(the balance between numbers of helper and suppressor cells helps to regulate the whole immune response)

killer cells, attack 'foreign' cells (identified by attached antibodies + complement protein) and destroy them by breaking down their wall/membrane – they also attack cancer cells and cells of 'foreign' transplanted organs (p. 371)

initial infection

if infection returns

plasma cells short-lived

quicker, greater response due to memory cells, i.e. person has immunity

Antibody concentration

Time

cell contents escape

ruptured membrane

T killer cells attach to foreign cell identified by antibody + complement protein

would otherwise react to the body's own cells. Thus a normal immune system cannot produce antibodies against the body's own tissues.

B and T lymphocytes look alike but behave quite differently in response to the presence of an antigen. Three types of T lymphocytes form: helper cells, suppressor cells and killer cells.

▶ **Helper cells** stimulate plasma cells into antibody production and activate the phagocytic cells of the bloodstream. When a person is infected by human immuno-deficiency virus (HIV) it is the helper cells that are

attacked. This is why AIDS patients are ultimately vulnerable to a wide range of infections.

▶ **Suppressor cells** inhibit the activities of plasma cells and phagocytic cells.

▶ **Killer cells** destroy body cells identified by the opsonins and complement proteins attached to their plasma membranes, or by the antigens present on the membranes. For example, cells attacked by a virus typically carry virus antigen on their plasma membrane and may be destroyed by killer cells before replication of the virus is completed. Incidentally, it is the killer

cells of our immune system that identify the cells of skin grafts and of transplanted organs as foreign and set about slowly rejecting them (p. 371).

Questions

1 a   What do we mean by immunity?
   b   Outline in tabulated form the parts played by the blood in protection of the body.
2 a   What are the differences between an antigen and an antibody?
   b   Where are antigens and antibodies found in the body?
3 Distinguish between
   a   granulocytes (p. 213) and lymphocytes
   b   T lymphocytes and B lymphocytes.

Types of immunity

Passive immunity involves antibodies that have been acquired from elsewhere. For example, antibodies are acquired by a fetus from the maternal circulation, across the placenta. Then again, the first-formed milk (colostrum) supplied to the newly born mammal may be rich in antibodies.

Today, many different antibodies can be prepared from the blood of human donors. For example, antibodies against measles, mumps, hepatitis, tetanus, chicken pox and rabies are available. These antibodies can be injected to confer some immunity, but this immunity is not permanent.

Active immunity arises when an organism manufactures its own antibodies. This is a common outcome of an infection, and afterwards the body is able to respond without delay to subsequent infections. Active immunity may be very long-term or permanent, as in the cases of 'diseases we get only once', such as mumps and measles.

Immunisation or **vaccination** is an artificial way of achieving the same protection. Pharmacologists have been able to manufacture vaccines from living but harmless (attenuated) pathogens, from dead pathogens or from the toxic chemicals that the pathogen releases, which have been rendered harmless (toxoids). Exposed to the vaccine, the body makes antibodies, and more-or-less permanent protection from infection is achieved.

17.8 HUMAN BLOOD GROUPS

Human blood groups, a form of the immune system, were discovered during the early attempts at giving blood transfusions between individuals. If a person suffers a minor haemorrhage and up to a litre of blood is lost, the body can normally recover. When a massive haemorrhage occurs, however, and the body loses 40% or more of its red blood cells and plasma, a transfusion of blood is essential if the patient is to survive.

If a patient is to receive blood, the blood must be compatible with that of the patient. Transfusions of blood bring about an immune response if the donor's blood and the recipient's blood are incompatible. When blood of an incompatible type is transfused the red blood cells of the donated blood clump together (agglutinate), and block up the fine capillaries. Ultimately, the cell membranes of the 'foreign' red blood cells are broken down (haemolysis) and the haemoglobin is discharged into the plasma. Free haemoglobin blocks the ultrafiltration mechanism of the kidneys (p. 382). Today, transfusions of incompatible blood are extremely rare. Before 1900, prior to the discovery of the ABO system, blood transfusions often proved fatal.

We now know that human blood can be categorised into a variety of blood groups. About twenty blood group systems have been discovered for human blood; the ABO system and the rhesus system are the most important. Blood groups are a special example of the expression of the antigen–antibody response.

■ The ABO blood system

The ABO system reflects the presence or absence of two mucopolysaccharides (p. 136), known as **agglutinogens** and referred to as agglutinogens A and B. These agglutinogens are carried on the external surface of the plasma membrane of the red blood cells. Agglutinogens A and B function as antigens. The red cells of any individual may possess one, both or neither of these antigens. People whose cells carry agglutinogen A belong to blood group A, those with B belong to blood group B, those with A and B belong to blood group AB, and those with neither agglutinogen belong to blood group O. These blood groups are inherited according to Mendelian laws (pp. 608, 612 and 622).

For each agglutinogen there is a corresponding **agglutinin**. Agglutinins occur in solution in the plasma. Agglutinins are not produced as a result of an immune response reaction, and individuals do not produce the agglutinin that reacts with the agglutinogen on their own red cells. They do, however, produce the other agglutinins of the ABO system and carry them in their blood plasma. The plasma of group A blood contains 'b' agglutinin (anti-B) only. The plasma of group B contains 'a' agglutinin (anti-A) only. Neither agglutinin is in the plasma of group AB blood, but both 'a' and 'b' agglutinins are present in the plasma of group O blood (Figure 17.44).

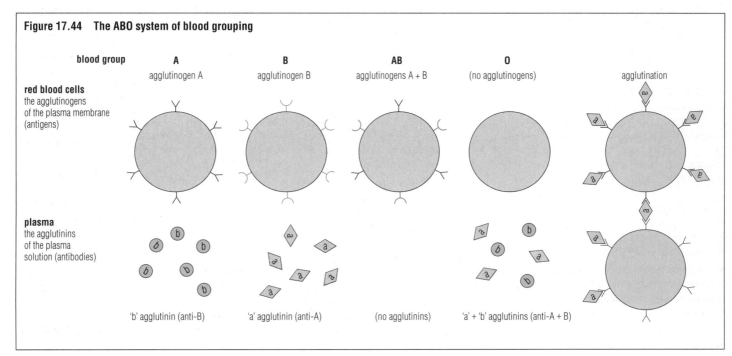

Figure 17.44   The ABO system of blood grouping

blood group	**A**	**B**	**AB**	**O**	
	agglutinogen A	agglutinogen B	agglutinogens A + B	(no agglutinogens)	agglutination

red blood cells
the agglutinogens of the plasma membrane (antigens)

plasma
the agglutinins of the plasma solution (antibodies)

| 'b' agglutinin (anti-B) | 'a' agglutinin (anti-A) | (no agglutinins) | 'a' + 'b' agglutinins (anti-A + B) | |

Blood transfusions

When blood from a donor is added to the blood of the recipient, it is necessary to avoid bringing together corresponding agglutinogen and agglutinin. For example, agglutinogen A and 'a' agglutinin (anti-A) must not be mixed since agglutination will occur; these bloods are incompatible.

In reality, if only a small quantity of blood is to be introduced, there is a margin of safety in certain combinations. In an incompatible transfusion the recipient's red cells are not clumped, because the donor's agglutinins are quickly diluted in the plasma, but the donor's cells are inevitably clumped by the recipient's agglutinins.

Thus blood of group O, which contains no agglutinogens on the red blood cells, can be given in small amounts to all other blood groups. People with group O blood are said to be universal donors. Patients of blood group O themselves can safely receive only group O blood (Table 17.6).

Similarly, individuals of blood group AB, which have no agglutinins in their plasma, can receive small amounts of either group A or group B blood. A patient of group AB is said to be a universal recipient.

Table 17.6 Blood transfusions: donors and recipients

Group	Can donate blood to	Can receive blood from
A	A and AB	A and O
B	B and AB	B and O
AB	AB	all groups
O	all groups	O

■ The rhesus system

The rhesus factor was discovered from work on rhesus monkeys. The 'factor' is actually due to a group of antigenic substances on the membrane of the red blood cells, of which the main one is D-antigen. Approximately 75% of the human population have D-antigen on their red cells and are said to be rhesus positive (Rh+). The rest are without D-antigen and are rhesus negative (Rh−).

The blood does not contain any preformed plasma antibodies associated with the rhesus factor. But a Rh− person does form a plasma antibody, called anti-D, if the blood is exposed to Rh+ cells. The first reaction in the blood of a Rh− person to whom a transfusion of Rh+

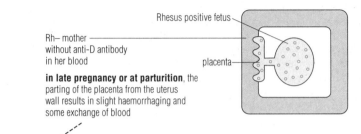

Figure 17.45 The cause and prevention of haemolytic disease of the newborn

Rhesus factor (Rh+) is an inherited dominant characteristic. A Rh+ person may be homozygous (DD) or heterozygous (Dd). A Rh− person can only be homozygous (dd).

first pregnancy of Rhesus negative mother (dd) with Rhesus positive fetus (Dd):

Rhesus positive fetus

Rh− mother without anti-D antibody in her blood

placenta

normally no exchange of blood occurs across the placenta during gestation

in late pregnancy or at parturition, the parting of the placenta from the uterus wall results in slight haemorrhaging and some exchange of blood

as a result, the mother's blood would become sensitised to Rh+ and form anti-D antibodies

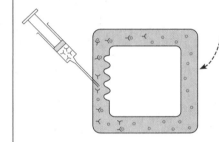

this is avoided by treatment of mother immediately after childbirth with anti-D antibodies to kill all Rh+ cells that have entered her body to prevent her own anti-D production

Haemolytic disease of the newborn occurs where a Rh− mother has been allowed to become sensitised to Rh+ blood (and so carry anti-D antibodies in her blood) as a result of the first pregnancy. In this case, in **a second pregnancy** with a Rh+ fetus, anti-D antibodies may cross the placenta from maternal to fetal circulation. Anti-D antibodies cause haemolysis of the fetal red cells.

blood is given is the slow, steady production and accumulation of anti-D in the plasma. Otherwise there are no external signs of the mismatch. A second transfusion of Rh+ blood into the now-sensitised Rh− patient will result in haemolysis of the donor red blood cells. Consequently, a sensitised Rh− patient must receive only Rh− blood.

A Rh+ person does not produce any rhesus antibodies and can be given Rh− or Rh+ blood in complete safety.

The rhesus factor and pregnancy

If a rhesus negative mother is carrying a rhesus positive fetus, a complication can arise concerning the rhesus system.

The placenta facilitates the exchange of nutrients and excretory materials without the exchange of maternal and fetal blood (p. 598). But in the later stages of pregnancy, the stresses of muscular contraction at the uterus/placenta interface can cause a limited exchange of blood. As a result the mother's blood circulation will become sensitised to the fetus's Rh+ blood cells, and she will produce

anti-D in plasma. Maternal anti-D can cross the placental membranes, and consequently in a future pregnancy a Rh+ fetus will be at risk. In about 10% of the subsequent (second) pregnancies in which the fetus is again Rh+, this reaction causes the dangerous condition known as haemolytic disease of the newborn (Figure 17.45) in which the 'rhesus baby' is usually born prematurely, anaemic and jaundiced (p. 509). The blood of such an infant has to be completely replaced by a transfusion of healthy blood.

A Rh− mother who has a Rh+ baby is given an injection of anti-D antibodies to destroy any Rh+ cells that have entered her circulation. The anti-D antibodies do not persist long and her own lymphocytes have not had time to become sensitised.

Questions

1 Why are people with group O blood called universal donors?

2 How might the blood of a rhesus negative person become sensitised and react to rhesus positive blood?

17.9 CURRENT ISSUES IN IMMUNOLOGY

■ AIDS, the acquired immune deficiency syndrome

AIDS is a disease caused by a virus called human immune deficiency virus (HIV). This virus was first identified in 1983. HIV is tiny, less than 0.1 μm in diameter. It belongs to a group of viruses known as **retroviruses**. Retroviruses are so called because, within a host cell, their RNA is translated into DNA (the reverse of the 'central dogma', p. 205) and incorporated into the DNA of the host nucleus. The acquired immune deficiency syndrome that may develop from HIV infection is a variety of infections that result from the abnormal destruction of certain T lymphocytes, known as CD4 cells (helper cells, Figure 17.43, p. 367).

In an infection by HIV the protein of the virus attaches to the protein on the surface of the CD4 cells. Once inside, the virus disintegrates, releasing RNA and an enzyme called reverse transcriptase. The enzyme causes the cell to translate the viral RNA to DNA. The DNA enters the nucleus and is incorporated into the cell's own DNA. Thus a gene representing the HIV becomes a permanent part of the infected person's CD4 lymphocyte cells.

As in other viral infections, an immune response is mounted by the body in response to infection by HIV. However, in contrast to viruses causing measles or mumps, which are generally eliminated by the immune system, HIV avoids clearance by two strategies. Firstly, it infects cells of the immune system itself, so impairing its ability to respond (Figure 17.46). Secondly, it avoids recognition by the immune system by latently infecting cells. This means that cells infected with the virus produce no viral proteins, so the immune system cannot detect the cells and therefore does not destroy them. In addition, the virus changes parts of its proteins that are recognised, so keeping itself 'invisible' to the immune system most of the time. For this reason the development of a vaccine for AIDS is very difficult (p. 17).

The way in which HIV causes AIDS is still not very well understood. It is much debated whether most infected cells are destroyed by the immune system or by the virus itself. Populations of CD4 cells gradually decline to such low levels that the immune system cannot control replication of HIV, or resist infection by other pathogens. This leads to the onset of AIDS. Death usually results from opportunist infections such as pneumonia.

Figure 17.46   HIV and the AIDS disease of the white blood cells

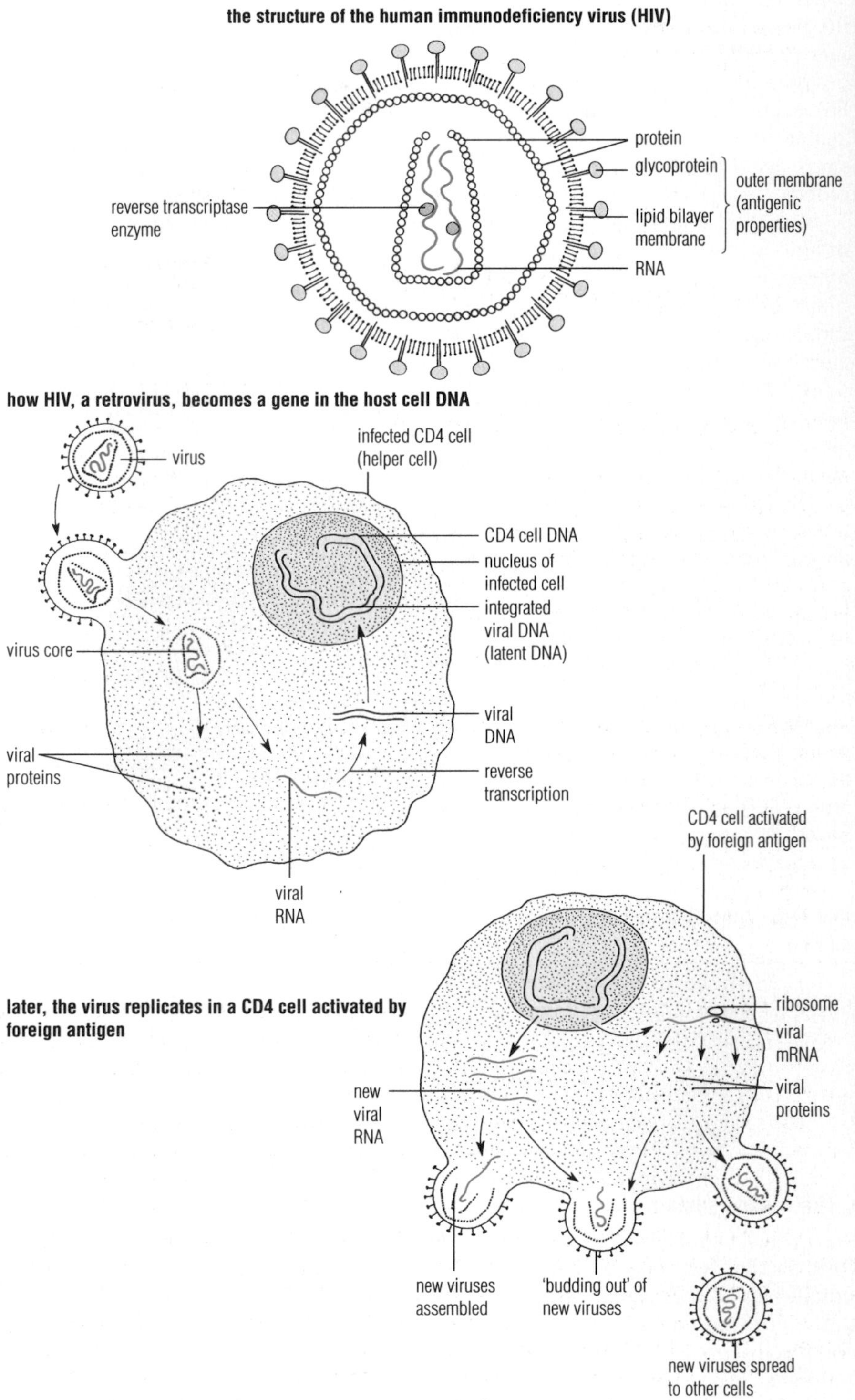

the structure of the human immunodeficiency virus (HIV)

how HIV, a retrovirus, becomes a gene in the host cell DNA

later, the virus replicates in a CD4 cell activated by foreign antigen

Questions

1 What distinguishes a retrovirus from other viruses?
2 What is the effect of an immunosuppressant drug?
3 An extensive burn might be treated by a skin graft taken from another site on the patient's body. Why is this graft not rejected?

Figure 17.47   Formation of monoclonal antibodies

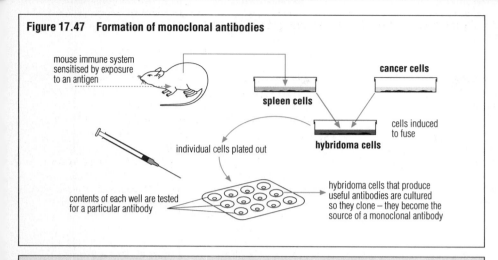

mouse immune system sensitised by exposure to an antigen

spleen cells

cancer cells

cells induced to fuse

hybridoma cells

individual cells plated out

contents of each well are tested for a particular antibody

hybridoma cells that produce useful antibodies are cultured so they clone – they become the source of a monoclonal antibody

DEFENCE AGAINST DISEASE IN MAMMALS: A SUMMARY

Maintenance of a barrier to infection
▶ the physical barrier of the skin
▶ the acid/enzyme barrier in the stomach
▶ the mucus of the trachea, bronchi and lungs
▶ the acidity of the vagina
▶ lysozymes (digestive enzymes) in nose, saliva, urine
▶ clotting of blood when the circulation is breached

Protection by commensal flora of microorganisms (these compete with pathogens for resources)
▶ on skin, in gut, in vagina

Phagocytic actions
▶ of neutrophils (and monocytes) of blood and tissue fluid
▶ of macrophages all over the body

Chemical means of inhibiting invasion forces
▶ interferon to combat viral infection

Immune system, a specific response to particular antigens
▶ B cell production of antibodies
▶ T cell destruction of foreign or infected cells
▶ passive immunity
   1 from antibodies received *in utero* and from first-formed milk
   2 from antibodies obtained from humans or other animals and injected
▶ acquired immunity
   1 from antibodies manufactured by the body during infections
   2 via immunisation (vaccination)

■ Monoclonal antibodies

Monoclonal antibodies are pure antibodies of one type, each made from only one type of B cell, which clones itself, making identical copies. The problem with B lymphocyte plasma cells that secrete antibodies is that they have a short life span and cannot divide. This problem has been overcome by fusing a B cell with a cancer cell (which goes on dividing indefinitely) forming an 'immortal' cell called a **hybridoma** (Figure 17.47). These form a clone of cells that secrete a particular antibody. Monoclonal antibodies are used for pregnancy testing (p. 600), for preventing rejection of transplants (above) and for tissue typing for transplants. A likely future use is in the treatment of disease by targeting particular cells.

■ Autoimmune disease

The body's immune mechanism is programmed to recognise its own cells and proteins: it does not normally produce T cells against itself. Abnormally, however, antibodies are produced against the body's tissue antigens – for instance, in diseases such as rheumatoid arthritis, rheumatic fever, multiple sclerosis and one form of diabetes. These are a few examples of such **autoimmune diseases**, many of which are rather rare.

■ Allergy

An **allergy** is an antigen–antibody reaction that occurs in certain individuals upon exposure to substances that are innocuous to other individuals under similar conditions. The common forms of allergy are hay fever, asthma, urticaria, childhood eczema and various gastrointestinal disturbances. In these conditions the body reacts to antigens found on pollen grains, fungal spores, dust, certain food constituents, the bites or stings of 'minibeasts' or plants, and substances on or in fur or feathers. The production of specific antibodies is stimulated and the resulting antigen–antibody reaction leads to the release of histamine within the body. The histamine causes inflammation, skin rash and (in asthma) constriction of the bronchioles. At the same time, the number of certain white blood cells called **eosinophils** increases. They relieve the symptoms by removing histamine from the circulation.

■ Organ transplantation and the immune response

Modern surgery makes possible the transplantation of tissues and organs. In the most straightforward case, tissues are grafted from one area of the body to another, as in many skin grafts. However, in most cases organs or tissues are taken from recently dead donors or living volunteers. The organs or tissues inevitably carry antigens on their cells, which the recipient's body recognises as 'foreign', and it therefore produces antibodies against them. The transplanted organ may be attacked and destroyed by T lymphocytes; it is said to be 'rejected'.

Today, tissues of donor and recipient are typed and matched as far as possible in order to minimise the degree of rejection. Even when the individuals are extremely closely related, however, there are very many antigens in the grafted organ that are incompatible with the recipient.

Rejection has to be minimised by the use of **immunosuppressant** drugs; that is, drugs that block antibody production in the recipient. The patient is then susceptible to all kinds of infections because he or she cannot respond to the antigens of invading pathogenic microorganisms.

A better way to overcome rejection would be to take out the lymphocytes (killer cells) specifically targeted to destroy the cells carrying antigens of the implanted tissue or organ. In some cases, this can already be done using monoclonal antibodies (see below) that recognise these killer cells and suppress only them. Rejection of kidney transplants has been overcome in this way, for example.

FURTHER READING/WEB SITES

The heart: an online exploration: **sln.fi.edu/biosci/biosci.html**
Browse: **mcb.harvard.edu/BioLinks.html**

18 Osmoregulation and excretion

- Excretion is the removal from the body of the waste products of metabolism, together with toxins and substances present in excess. Maintaining the correct balance of water and solutes in the body is known as osmoregulation.
- In plants, nitrogenous compounds are resources for protein synthesis, rather than excretory products. A wall protects the plant cell from mechanical damage due to excess water uptake; plants are vulnerable to having too little water, rather than too much.
- Many plants show structural adaptations to their environment, particularly to the supply of water available to them. This is most evident in water plants (hydrophytes), saltmarsh plants (halophytes) and plants adapted to arid environments (xerophytes).

- In animals, excretion is mainly concerned with disposal of nitrogenous waste. Animal cells are not contained within a cell wall, and they are potentially threatened by both too little water and too much.
- Whilst many freshwater animals safely excrete nitrogenous waste as ammonia, insects convert excess nitrogen to uric acid, which is disposed of as a solid, thereby reducing water loss in excretion.
- In mammals, excretion and osmoregulation are carried out by the kidneys, which consist of millions of tubules called nephrons. Nephrons work by ultrafiltration, selective re-absorption and secretion, and are able to form urine more concentrated than the blood, so minimising water loss in excretion.

18.1 INTRODUCTION

The chemical reactions of metabolism produce a variety of by-products, some of which are toxic if allowed to accumulate in the organism. **Excretion** is the process by which the organism removes such substances. Also excreted are metabolites present in excessive concentrations. Excretion plays an important part in the processes by which the internal environment is regulated to maintain more or less constant conditions (**homeostasis**, p. 502).

What constitutes a waste product depends on the organism and its environment. Green plants in the light excrete oxygen, whereas animals excrete nitrogenous waste formed by deamination of excess amino acids. Green plants do not excrete nitrogenous waste; on the contrary, they re-use any breakdown products of nitrogen metabolism (p. 209) (Table 18.1).

Excretion is an entirely different process from either egestion or secretion.

Egestion is the elimination of undigested matter in the diet, which is discharged from the organism without being taken into the cells and metabolised. Egestion is one step in holozoic nutrition (p. 270).

Secretion is the discharge of materials that have been formed by the organism for use inside or outside the body. Examples of secretion include the addition of digestive enzymes to food materials, the release of growth regulators or hormones within the plant or animal, the ejection of defensive chemicals when under attack, and the pouring of sweat on to the skin as a means of temperature regulation.

Both egested materials and secretions may also contain substances that are being excreted from the body. Examples include the salts in sweat, and the bile pigments.

Table 18.1 Summary of excreted substances

Excreted substance	How formed	How and where excreted
oxygen	in photosynthesis in green plants	leaves of green plants, e.g. by diffusion through the stomata of terrestrial plants
carbon dioxide	in respiration of living things	from the gas exchange surface (e.g. lungs or gills), and by diffusion from the surface of cells
		in terrestrial flowering plants by diffusion through intercellular spaces and stomata
water	in aerobic respiration	by evaporation or osmosis at the cell surface, also retained as an essential component of all cells
ions	in synthesis of metabolites, pigments, enzymes, hormones, etc.	ions in excess, those that cannot be stored and those at toxic concentrations may be excreted, e.g. from sweat glands, kidney or chloride secretory cells of gills, or combined with organic compounds and deposited in dead cells of woody plants
bile pigments	in breakdown of haemoglobin in liver	in bile in the gall bladder, and subsequently egested with faeces
organic waste molecules, including organic acids such as oxalic and pectic acids	in highways and byways of plant metabolism, including synthesis of substances that are harmful to predators and herbivores	deposited in cells of stem or leaf
nitrogenous compounds, e.g. ammonia, urea, uric acid	in metabolism of proteins and nucleic acids	ammonia is lost by diffusion from small, aquatic organisms; urea is excreted by organs of nitrogenous excretion, e.g. kidney

Osmoregulation is defined as the mechanism by which the balance of water and dissolved solutes is regulated. An appropriate quantity of water is essential to maintain the volume of cells and organisms within relatively narrow limits. The dissolved solute composition of cells and body fluids determines the solute potential, and therefore affects water content. As part of osmoregulation, organisms regulate their content of both non-electrolytes (sugars and amino acids) and electrolytes (inorganic ions such as K^+, Na^+, H^+, Cl^-, NO_3^- and HCO_3^-), as well as their water content. Osmoregulation and excretion are closely connected processes. By excretion excess water and dissolved solutes are lost from the body.

■ The importance of osmoregulation and excretion

The concentrations of a wide range of metabolites affect the operation of essential life processes. For example, appropriate concentrations of solutes are essential to the activity of enzymes, to the rates of various metabolic processes, to the formation and actions of hormones, in nerve action via impulse transmission, and for muscular contraction. Many aspects of metabolism are dependent upon the presence of specific metabolites and ions at particular concentrations.

The mechanisms of obtaining, retaining and eliminating water and certain solutes (osmoregulation and excretion) in organisms of the sea, fresh water and dry land vary significantly. We shall examine examples of these structural and functional adaptations in plants and animals.

18.2 EXCRETION AND OSMOREGULATION IN PLANTS

■ Excretion

Green plants excrete oxygen in the light. Carbon dioxide is excreted in the dark, when no longer re-absorbed by the process of photosynthesis. Plants may store organic waste substances in mature or moribund cells, such as cells of the heartwood or the bark. For example, the organic acids oxalic and pectic acid may be combined with various excess cations and stored as large crystalline deposits in this way.

Excretion differs in plants and animals

Important differences between plant and animal metabolism make the process of excretion in plants of less significance than excretion in animals.

Firstly, a plant has a lower metabolic rate than that of an animal of the same size or mass, largely because plants are stationary, and metabolic waste accumulates more slowly.

Secondly, as primary producers, plants synthesise their organic requirements as the necessary raw materials

GUTTATION

guttation in barley

Water-secreting glands

Abnormally, under conditions of high humidity and therefore low transpiration, some species exude water from points known as **hydathodes** at the edges of leaves. This process is known as **guttation**.

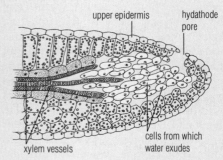

hydathode in section

become available; for example, soluble nitrogenous compounds such as ammonia and nitrate are resources for protein production rather than unwanted substances to be excreted. Animals obtain proteins as part of their diet, and dispose of any excess by deamination and excretion, for they cannot store proteins.

Thirdly, much of the structure of plants is based on carbohydrates rather than on proteins; structural proteins play a less important part in the life of plants than they do in animals. Carbon dioxide and water are the end-products of carbohydrate metabolism, and these substances are easily excreted by gaseous diffusion.

■ Osmoregulation

The cytoplasm and plasma membrane of the plant cell are surrounded by a cellulose cell wall. This cell wall protects the cytoplasm from mechanical damage that might result from excess water absorption. When a plant cell is fully stretched by excessive water uptake the cellulose wall generates a pressure potential that opposes the solute potential of the cell sap and terminates further net water uptake (the water potential of the cell is zero, p. 232).

In fact, plants are normally more vulnerable to having too little water than too much. Desiccation is a real danger to plants, not only because they contain more water than animals do but also because plants are rooted to a spot that may dry out. Plants show adaptations that prevent excessive water loss (for example, xerophytes, p. 375).

■ Plants and water conservation

Plant structure – the arrangement of tissues in stem, leaf and root – is essentially similar in all flowering plants. Individual species show some **modifications to their structures**, however, and most of these modifications are adaptations to environmental conditions. Some of the most striking modifications relate to the conditions of water supply. In fact, flowering plants are conveniently classified on the basis of structure in relation to the prevailing water supply:

▶ **hydrophytes** – water plants e.g. water lily (*Nymphaea alba*), rice (*Oryza sativa*)

▶ **halophytes** – saltmarsh plants e.g. glasswort (*Salicornia* spp.), cordgrass (*Spartina angelica*)

▶ **xerophytes** – plants of arid conditions e.g. marram grass (*Ammophila arenaria*), sorghum (*Sorghum bicolor*)

▶ **mesophytes** – plants of areas with adequate water supply.

Questions

1 Distinguish between excretion, egestion, osmoregulation and secretion by means of definitions and examples.

2 In which of the following processes is osmosis involved, and how?
   a movement of water into guard cells in the light
   b movement of water through the xylem
   c entry of water into cytoplasm and vacuole
   d passage of water across a cell of the endodermis
   e entry of ions into root hair cells.

3 Suggest reasons why excretion is less critical to plants than to many animals.

Hydrophytes

Hydrophytes grow submerged or partially submerged in water. Water is a supportive medium, which normally presents no problems of water supply (the problems of saltmarsh plants are examined below). They have little or no lignified supporting tissues (fibres or tracheids are wholly or almost absent from the submerged parts of hydrophytes). Xylem tissue, in which water is transported, is poorly developed throughout the plant. The stems and leaves have little or no cuticle (a layer of wax that reduces evaporation of water from the surface of terrestrial plants). Characteristically, stems and leaves have large continuous air spaces, forming a reservoir of oxygen and carbon dioxide, which also provides buoyancy to the plant tissues when submerged (Figure 18.1).

Rice as a hydrophyte

Rice (*Oryza sativa*) is an annual, tropical swamp plant, now grown in subtropical regions too. This cultivated grass produces a grain that has been a source of food for human communities for about 5000 years. It is the staple cereal for more people than any other crop.

Rice is unique among major food crops in that it grows in standing water. It almost certainly evolved in conditions similar to those of the lands of South East Asia today, subject to heavy monsoon rains that cause seasonal flooding. It can be grown on dry land, rather like wheat, but yields from flooded rice are much greater than dryland rice. The adaptations shown by rice include those described below:

► the young seedlings are able to survive drought stress that occurs before the monsoon rains begin

► when the growing plants become submerged by rising floodwater they are able to elongate rapidly (10 cm or more in a day), keeping part of the stem and leaves in the air

► the stem and roots contain aerenchyma (parenchyma with huge air spaces between the cells, Figure 18.2) that facilitates diffusion of oxygen and carbon dioxide in submerged regions of the plant

► roots are maintained in the ground, but additional roots develop at the nodes of the stem (points of leaf attachment), and nutrient ions are absorbed from the floodwater

► flowering is timed (photoperiodism, p. 434) to occur when floodwaters have normally reached their peak

Figure 18.1   The hydrophyte water lily (*Nymphaea alba*): this plant is native to lakes, ponds and ditches that are naturally eutrophic waters

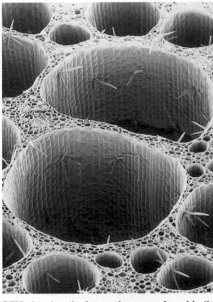

floating leaves, with long stalks (petioles) and almost circular blades (laminae)

large, showy, floating flower

stout rhizome rooted in the mud of the pond

SEM showing the large air spaces found in the stem and petioles of many water plants (×30)

Figure 18.2   Rice as a swamp plant

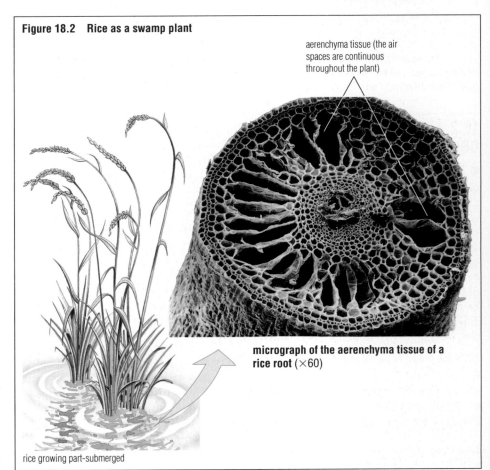

aerenchyma tissue (the air spaces are continuous throughout the plant)

micrograph of the aerenchyma tissue of a rice root (×60)

rice growing part-submerged

► many varieties of rice are able to tolerate total submergence of the fully grown crop for several days, caused by flash flooding; the cells and tissues of the plant are able to tolerate the ethanol that accumulates due to the anaerobic respiration these conditions cause (p. 316).

Figure 18.3 Profile of a saltmarsh, and the features of some typical halophytes

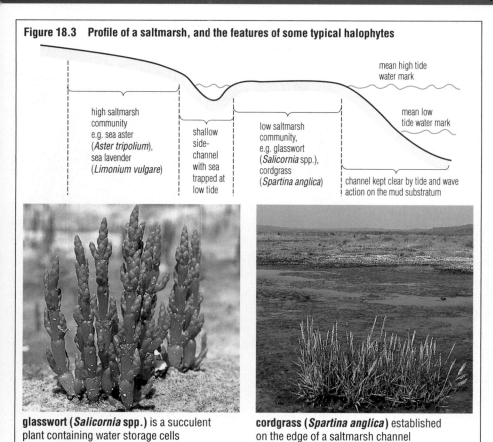

high saltmarsh community e.g. sea aster (*Aster tripolium*), sea lavender (*Limonium vulgare*)

shallow side-channel with sea trapped at low tide

low saltmarsh community, e.g. glasswort (*Salicornia* spp.), cordgrass (*Spartina anglica*)

mean high tide water mark

mean low tide water mark

channel kept clear by tide and wave action on the mud substratum

glasswort (*Salicornia* spp.) is a succulent plant containing water storage cells

cordgrass (*Spartina anglica*) established on the edge of a saltmarsh channel

The cells of both plants are able to retain salts as ions. Their water potential is low, and may be lower (more negative) than that of sea water. Stem and leaf tissues in both plants contain large air spaces, by which gases are made available to the submerged parts of the plants, including the root cells.

Table 18.2 Xeromorphic features: a summary

Features that minimise water loss	Effect	Example
Reducing transpiration		
thick cuticle to leaf and stem	reduce cuticular transpiration	ivy (*Hedera helix*) sea holly (*Eryngium maritimum*), *Pinus* sp.
layers of hairs on the epidermis	traps moist air and reduces the diffusion gradient	lamb's ear (*Stachys lanata*)
massing of leaves into a rosette at ground level	traps moist air between leaves (moisture greatest at ground level)	dandelion (*Taraxacum officinale*), daisy (*Bellis perennis*)
reduction in the numbers of stomata	reduces the pores through which transpiration can occur	*Nerium* sp.
stomata in pits or depressions, or at the bottoms of grooves	moist air trapped outside the stomata, reducing diffusion	*Hakea* sp., *Pinus* sp.
leaves reduced to scales on a photosynthetic stem	reduces surface area	broom (*Cytisus scoparius*)
closure of stomata in the light	dark CO_2 fixation	CAM plants
Storage of water		
succulent stem and/or leaves	water storage cells	prickly pear cactus (*Opuntia* sp.), *Bryophyllum* sp.
Survival of desiccation		
leaf rolled up or folded when flaccid	reduction of area from which transpiration can occur	marram grass (*Ammophila arenaria*), heathers (*Erica* sp.)
deep and extensive root system	tapping lower water table	*Acacia* shrub
superficial root system	absorbing overnight condensation	most cacti

Halophytes

Halophytes are the plants found in areas of high salinity, such as estuaries and saltmarshes. Many saltmarsh plants grow rooted in soil irrigated by sea water and may be inundated by sea water at certain high tides, at least at the equinoxes (Figure 18.3).

In the saltmarsh environments the salinity is variable. At high tides salinity is that of sea water. Once the tide has turned, however, the salt concentration may change quickly. For example, salinity may exceed that of sea water when high temperatures and winds cause evaporation, and the remaining soil water is concentrated. Alternatively, when there is heavy rainfall, or when the flow of river water dilutes the sea water, the concentration of salts will fall well below that in sea water.

Many plants of the salt-tolerant communities of saltmarshes and estuaries store water in succulent tissues. Often the concentration of salts in their cells is high, higher than that of sea water; they can thus take up water from the sea water by osmosis.

Halophytic plants also tend to be rooted in waterlogged, anaerobic mud. The extensive air spaces throughout the stems and roots of these plants make air available to all cells, and give buoyancy to the stems and leaves when the plants are submerged at the highest tides.

Xerophytes

Xerophytes are found in habitats where water is scarce. Plants that survive successfully under these conditions show features that minimise water loss due to transpiration, adaptations that are referred to as **xeromorphic features** (Table 18.2). These include a thickened cuticle, and a reduced number of stomata. The remaining stomata are arranged so that moist air is trapped outside the pores (by a dense mat of epidermal hairs, for example). As a result the diffusion gradient between the interior of the leaf and its environment is much reduced, and the outward diffusion of water vapour is slowed.

Some xerophytes survive drought by storing water in succulent leaves or stems, or by growing roots that tap unusual water sources. In others, loss of water causes rolling or folding of the leaf, protecting the stomata still further (Figure 18.4). Another adaptation to water scarcity by green plants is the physiological mechanism of crassulaceans, known as crassulacean acid metabolism (CAM).

Figure 18.4 Leaf of marram grass (*Ammophila arenaria*)

Marram grass (*Ammophila arenaria*) colonises sand dunes around the coast of western Europe. It endures extremely dry conditions.
The outer surface of the leaf has a thick cuticle and no pores (stomata). Stomata are restricted to furrows on the inner surface of the leaf.
When water loss from the leaf exceeds uptake, the hinge cells become flaccid and cause the leaf to roll up. When rolled up, the ratio of its external surface to its volume is small.

tissue map of part of the lamina

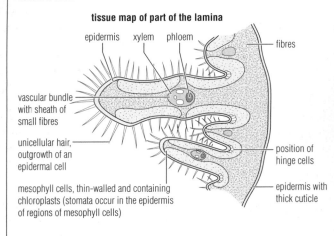

epidermis   xylem   phloem

fibres

vascular bundle with sheath of small fibres

unicellular hair, outgrowth of an epidermal cell

position of hinge cells

mesophyll cells, thin-walled and containing chloroplasts (stomata occur in the epidermis of regions of mesophyll cells)

epidermis with thick cuticle

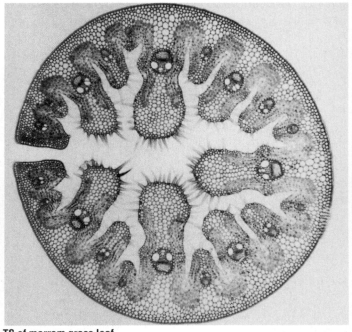

TS of marram grass leaf

CAM plants have stomata open in the dark and fix carbon dioxide, forming malic acid (hence 'acid metabolism'). In the light the stomata close, which prevents loss of water vapour but also prevents the entry of carbon dioxide. However, malic acid is broken down again in the day, releasing carbon dioxide inside the leaf, and photosynthesis takes place unhindered.

Sorghum as a xerophyte

Sorghum (*Sorghum vulgare*) is a cultivated grass of semi-arid climates (Figure 18.5). In many African countries sorghum grains are used for making an unleavened bread (bread without yeast to 'raise' it). Sorghum has a deep, extensive root system, and the leaves show xeromorphic features (Table 18.2).

Mesophytes

Mesophytes flourish in habitats with adequate water supply. Most native plants of temperate and tropical zones, and most of our crop plants, are mesophytes. They are adapted to grow best in well-drained soils with their aerial system exposed to moderately dry air. Loss of water vapour from the leaves may be substantial, particularly in the early part of the day, but excessive loss of water is prevented by closure of the stomata. Any deficit is made good by the water uptake that continues at night (Figure 16.11, p. 325). Structures of some typical mesophytes are illustrated in Figure 19.19 (stem and root), and Figure 12.18 (leaf).

Mesophytes may be able to survive unfavourable seasons (periods of physiological drought, for example, due to water supplies being frozen). Many woody perennial mesophytes of temperate zones shed their leaves before winter sets in, successfully reducing water loss in periods when supplies are severely restricted (Figure 20.35, p. 439). Herbaceous perennials lose their aerial system during the unfavourable season, surviving as underground organs such as bulbs, corms, and rhizomes (p. 578). Most annual mesophytes survive the winter as dormant seeds (p. 438).

Questions

1 In very warm, dry conditions mesophytes may experience mild wilting by mid-day.
   a Why does this wilting occur?
   b Why is loss of water vapour terminated only when the stomata are closed (p. 322)?
2 Describe three significant features of structure that effectively reduce water loss from the aerial parts of xerophytes.
3 CAM plants open their stomata only in the dark.
   a How does this modification aid survival in arid conditions?
   b How does it still permit these plants to obtain carbon dioxide for photosynthesis in light?
4 What adaptive features would you expect to find in sections taken respectively through a stem and a leaf of a hydrophyte plant?

Figure 18.5   Sorghum (*Sorghum vulgare*)

18.3 EXCRETION AND OSMOREGULATION IN ANIMALS

■ Excretion

Excretion in animals is mainly concerned with nitrogenous waste, and to a lesser extent with other waste products of metabolism. The three compounds of importance in nitrogenous excretion of animals are ammonia, urea and uric acid.

Ammonia (NH_3)

Ammonia is an extremely soluble gas of low molecular mass. It diffuses rapidly, even when dissolved in water. Ammonia is a highly toxic substance; organisms cannot tolerate an accumulation of ammonia.

The toxicity of ammonia is reduced by dilution with relatively large amounts of water. Animals of freshwater habitats continuously take in water by osmosis, and so are able to excrete ammonia in dilute solution without suffering any dehydration. It is not surprising, then, that excretion of ammonia is typical of organisms of aquatic habitats, at either larval or adult stage, or at both. Ammonia diffuses across exchange surfaces such as gills in an aquatic medium.

Urea ($CO(NH_2)_2$)

Urea is less soluble in water than ammonia, and much less toxic. Consequently, less water is needed for the safe elimination of urea from the body than for that of ammonia. Urea excretion is seen in organisms of aquatic and terrestrial habitats.

Urea is synthesised from ammonia and carbon dioxide by reaction with the amino acid ornithine. In a metabolic pathway of several steps the intermediate amino acid arginine is formed; arginine is then split by enzyme action to urea and ornithine. This cyclic process for the production of urea is known as the **ornithine cycle** (Figure 18.6).

Uric acid

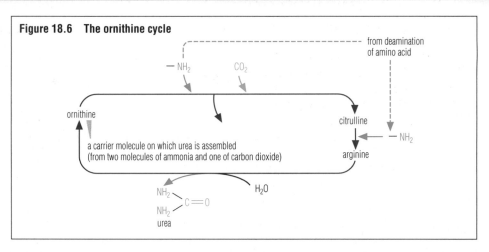

Figure 18.6 The ornithine cycle

Uric acid is a much larger molecule than urea. It is a purine, similar in structure to adenine and guanine (p. 146). Energy from metabolism is required in the formation of uric acid, but once formed it is virtually insoluble and therefore non-toxic to the organism. Uric acid is produced as a colloidal suspension. Further water absorption leads to crystals of urates being formed. These are discharged as a thick paste or as solid pellets.

Uric acid excretion occurs in organisms that develop in an enclosed egg (where water is severely limited), or which normally experience a very dry terrestrial environment as adult organisms.

■ Osmoregulation

Animals in fresh water (and other aquatic environments of water potential less negative than that of the cytosol) experience a net inflow of water by osmosis. An unchecked inflow of water first distends the cells, then disrupts and destroys them, for animal cells are not contained in a cellulose cell wall as plant cells are. Animals are potentially threatened both by too little water *and* by too much!

Osmoregulation in aquatic animals

Osmoregulation problems may be avoided by adjustment of the animal's fluids to become isotonic (p. 234) with the surrounding aquatic environment. In this situation there is no net flow of water. Most marine non-vertebrate animals have body fluids that are isotonic with sea water.

Aquatic animals with body fluids that are hypertonic to their surrounding medium have to remove water from the body; examples include the freshwater protozoa and freshwater bony fish. In the eliminating process some essential ions may be lost as well. In response, the animal actively absorbs selected ions from the medium to maintain its ionic concentration.

Conversely, aquatic animals such as marine bony fish have body fluids that are hypotonic to their medium, and experience a net outflowing of water. The organism must acquire water to protect its cells. This is normally achieved by taking in the external solution and selectively excreting the salts as ions, so that only the water is retained.

Osmoregulation in terrestrial animals

Terrestrial animals, like other terrestrial organisms, must conserve water. Many species, such as the insects (Figure 18.8, p. 378), have evolved mechanisms that prevent persistent loss of water from their bodies. Some animals are adapted for survival in arid regimes, or during periods of prolonged drought; examples include the kangaroo rat (*Dipodomys spectabilis*) (Figure 18.24, p. 390) and the camel (Figure 18.25). Some animals, however, respond to the ever-present danger of desiccation by restricting themselves to environments that are highly moist, as most amphibians do.

Questions

1 The properties of uric acid make this substance especially useful in excretion by certain animals. Why?

2 Why is 'too much water' rarely a problem to plant cells, but potentially hazardous to animal cells?

3 Certain birds live their lives almost entirely at sea. What process must these birds be able to carry out to avoid desiccation when drinking sea water?

4 The cnidarian *Hydra* has no contractile vacuoles in its cells. How does the active secretion of salts into the enteron assist it in osmoregulation?

Osmoregulation and excretion in some non-vertebrates

Protozoa

We have seen that protozoa are small, unicellular animals with a relatively large surface area:volume ratio. Protozoa live in aquatic habitats, and their excretory products diffuse out through the plasma membrane. The nitrogenous excretory substance, ammonia, is produced at low concentrations, and is rapidly diluted in the surrounding water.

Marine protozoa are isotonic with sea water; there is no net water movement, and in most there are no osmoregulatory structures.

Freshwater protozoans are hypertonic to their surroundings, and a net inflow of water occurs. One or more contractile vacuoles actively remove water from the cell (Figure 18.7). The liquid expelled by the vacuole is hypotonic to the cell fluid. Meanwhile the membrane of the proto-zoan actively transports ions to maintain the ionic concentration and ion balance of the cytoplasm.

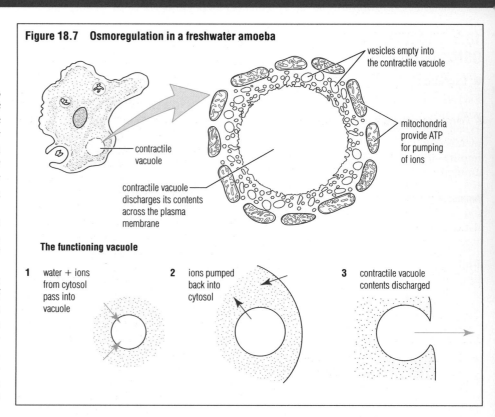

Figure 18.7   Osmoregulation in a freshwater amoeba

vesicles empty into the contractile vacuole

mitochondria provide ATP for pumping of ions

contractile vacuole

contractile vacuole discharges its contents across the plasma membrane

The functioning vacuole

1 water + ions from cytosol pass into vacuole

2 ions pumped back into cytosol

3 contractile vacuole contents discharged

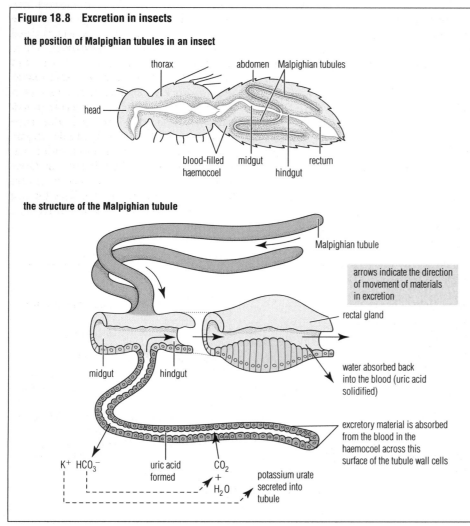

Figure 18.8   Excretion in insects

the position of Malpighian tubules in an insect

thorax

abdomen   Malpighian tubules

head

blood-filled haemocoel   midgut   hindgut   rectum

the structure of the Malpighian tubule

Malpighian tubule

arrows indicate the direction of movement of materials in excretion

rectal gland

midgut   hindgut

water absorbed back into the blood (uric acid solidified)

excretory material is absorbed from the blood in the haemocoel across this surface of the tubule wall cells

K^+ HCO_3^- uric acid formed CO_2 + H_2O potassium urate secreted into tubule

Insects

Insects (phylum Arthropoda) are adapted to survive and prosper as terrestrial organisms. Their ability to conserve water in the process of excretion is an important factor in their success on dry land. The cells of insects form uric acid as the nitrogenous excretory material. Uric acid is removed from the blood (as potas-sium urate) at the **Malpighian tubules**. These closed tubes lie in the haemocoel and empty into the alimentary canal at the junction of the midgut and the hindgut (Figure 18.8).

The wall of the upper part of the Malpighian tubule secretes urate into the lumen of the tubule. Carbon dioxide and water diffuse into the lumen too. In the lower part of the tubule the contents of the lumen react together, forming uric acid, potassium hydrogencarbonate and water. Uric acid is moved down into the gut and the potassium hydrogencarbon-ate is actively transported back into the blood. Meanwhile, in the rectum much more water is withdrawn from the faeces and transported back into the haemo-coel; the uric acid present becomes solid pellets, which leave the body with the faeces.

This method of nitrogenous excretion is an important aspect of water conserva-tion in insects. In addition, the hard external skeleton is coated by a layer of impervious wax. The system of tracheae by which oxygen is piped to all body tissues has openings, known as spiracles (p. 305). Valves control movements of gases through the spiracles, and reduce loss of water vapour from the tracheae.

■ Osmoregulation and excretion in vertebrates

The excretory and osmoregulatory organ of vertebrates is the **kidney** (which in non-mammal groups is variously supplemented by other organs, such as the gills and rectal glands of fish, and the salt gland in some birds). A pair of kidneys occurs in the dorsal wall in the body cavity, the coelom, with a supply of blood from the dorsal aorta via renal arteries, and drained by renal veins. The functional unit of the kidney is a tiny elongated tubule, the **renal tubule** or **nephron**. A kidney consists of a large number of nephrons, which eventually empty into a single duct, the **ureter**, running to the exterior of the body.

At one end of the nephron, blood supply and tubule are in close association in a structure called the **Malpighian body**. This consists of a cup-shaped chamber, the **renal capsule**, which encloses a knot of capillaries known as the **glomerulus**. In the glomerulus blood pressure forces part of the plasma out of the thin capillary walls. The capillary walls prevent the blood cells and protein molecules from passing through, so the filtered fluid is like tissue fluid. The filtrate passes from the renal capsule down the remainder of the tubule, and on this journey the useful metabolites are selectively re-absorbed. Additional waste substances may also be actively secreted into the ultrafiltrate from the cells of the wall of the tubule. By these processes the excreted liquid, urine, is formed.

Fish

Fish are thought to have originated in the sea, but it is believed that these animals adapted to fresh water early in their evolutionary history. Some fish subsequently remained in fresh waters, giving rise to our modern freshwater bony fish; freshwater fish have body fluids that are hypertonic to their environment.

Other fish then re-invaded the sea, however, and have given rise to modern marine fish, both cartilaginous and bony. The body fluids of marine fish are normally hypotonic to sea water (unlike those of most marine non-vertebrates, which evolved, adapted and remained in the sea).

The external surface of the fish is more or less impervious to water and to dissolved substances, so you might think that the external solution is unimportant. In fact, however, the respiratory current of water through mouth and pharynx brings the aquatic environment into close contact with thin-walled gills and the blood they contain. The ingestion of food brings some of the environment into contact with the body of the fish, too.

Freshwater bony fish

The body fluids of freshwater fish are hypertonic to their aquatic environment. Water flows into the body by osmosis through the gill surfaces. The kidneys of these fish contain many large Malpighian bodies, with large glomeruli. Here a high rate of filtration produces a large volume of glomerular filtrate. As this solution passes down the tubule, ions and useful metabolites are re-absorbed. The fish discharges copious quantities of very dilute urine. The urine also contains the nitrogenous excretion compound, ammonia, in very dilute concentrations (Figure 18.9).

Some loss of ions occurs in the process of discharging urine, and these are replaced by ion uptake elsewhere in the body. For example, cells in the gills accumulate chloride ions into the blood, against a concentration gradient, using energy from respiration (an example of active transport, p. 236).

Marine bony fish

The body fluids of marine bony fish are hypotonic to sea water. At the gill surfaces the fish loses water by osmosis and also gains ions from the sea water by diffusion, thus increasing the concentration of the body fluids. If there were no compensatory mechanisms, these processes would quickly lead to dehydration.

In fact, marine bony fish swallow sea water to replace the water lost. The excess ions, mainly chloride ions in the blood from ingested sea water and gained by diffusion, are actively secreted by means of special excretory cells in the gills (and sodium ions follow passively). In this way the concentration and volume of body fluids is maintained (Figure 18.9).

In marine bony fish the kidneys are small and have very few glomeruli; the nitrogenous excretory products formed are urea and trimethylamine oxide. These are much less toxic than ammonia, and require much less water for safe excretion. In the kidney tubules there is a low filtration rate, and a relatively small quantity of urine is produced. The urine is isotonic with the body fluids. (Some marine bony fish have no glomeruli at all, and do not filter their blood. In these animals urine is formed by secretion from the cells of the renal tubule. This type of kidney is described as aglomerular.)

Figure 18.9 Excretion and osmoregulation in bony fish (teleosts)

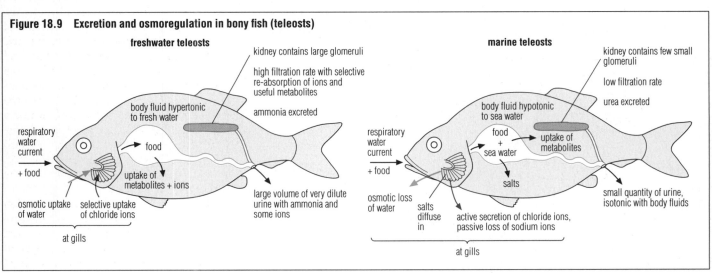

freshwater teleosts
- kidney contains large glomeruli
- high filtration rate with selective re-absorption of ions and useful metabolites
- ammonia excreted
- body fluid hypertonic to fresh water
- respiratory water current + food
- food
- uptake of metabolites + ions
- osmotic uptake of water
- selective uptake of chloride ions
- at gills
- large volume of very dilute urine with ammonia and some ions

marine teleosts
- kidney contains few small glomeruli
- low filtration rate
- urea excreted
- body fluid hypotonic to sea water
- respiratory water current + food
- food + sea water
- uptake of metabolites
- salts
- osmotic loss of water
- salts diffuse in
- active secretion of chloride ions, passive loss of sodium ions
- at gills
- small quantity of urine, isotonic with body fluids

18.4 THE HUMAN KIDNEY

The human kidneys are each the size of a clenched fist – about 10 cm long, 6 cm wide and 3 cm deep. They lie in the abdominal cavity, attached to the dorsal wall on either side of the vertebral column, behind the liver. Each kidney is enclosed in a fibrous coat, covered by the peritoneum (p. 278), and usually surrounded by fat. The left kidney lies very slightly higher than the right.

In longitudinal section the kidney can be seen to have two distinct regions, an outer **cortex** and an inner **medulla**. The human kidney contains at least a million **nephrons**, arranged so that part of each is in the cortex and part in the medulla (Figure 18.10). Kidneys are served by renal arteries and drained by renal veins. Within the kidney the renal artery branches progressively into tiny arterioles, each serving a single nephron.

■ The nephron

Each nephron is a minute tubule about 3 cm long divided into six distinct regions, each with a specific function.

The nephron begins as a group of blood capillaries (the **glomerulus**) surrounded by a cup-like capsule (the **renal capsule**). The renal capsule is formed by the invagination of the blind end of the nephron. This first region of the nephron is known as the **Malpighian body**.

The Malpighian body leads into the remainder of the tubule. First comes the **proximal convoluted tubule**, then the descending and ascending limbs of the **loop of Henle**, next the **distal convoluted tubule**, and finally the **collecting duct**.

Table 18.3 The composition of fluids in the kidney (concentrations are given in g per 100 cm³ of fluid)

Component	Blood plasma entering the glomerulus	Filtrate in renal capsule	Urine in collecting duct
water	90–93	97–99	96
blood proteins	7–9	some*	0.0
glucose	0.10	0.10	0.0
urea	0.03	0.03	2.0
other N-containing compounds	0.003	0.003	0.24
ions:			
sodium	0.32	0.32	0.30–0.35
chloride	0.37	0.37	0.60
others (Ca^{2+}, Mg^{2+}, K^+, PO_4^{3-}, SO_4^{2-})	0.038	0.038	0.475
pH	7.35–7.45		4.7–6.0 (average 5.0)

*Some of the smallest blood protein molecules only

The blood supply to the nephron begins as an afferent arteriole serving the glomerulus. From the glomerulus blood is carried by the efferent arteriole to two other capillary structures. One is a capillary network serving the proximal and distal convoluted tubules. The other is a single straight capillary running beside the limbs of the loop of Henle. These capillary structures are known as the **vasa recta** (Figure 18.11).

Functionally, the nephron is indistinguishable from its blood supply; nephron and capillaries form an integrated unit.

Osmoregulation and excretion in the nephron

The composition of body fluids is influenced more by output by the kidneys than by intake via the gut, and therefore the kidneys are important guardians of the internal environment. By means of the regulated production of urine from the blood the nephrons carry out the dual processes of excretion and osmoregulation. Excretion occurs by the lowering of the blood levels of urea and of any other toxic compounds present. Osmoregulation occurs by the regulation of the water and salt concentrations of the blood.

Approximately 1.0–1.5 litres of urine are normally produced each day. However, copious quantities of very dilute urine are produced when large quantities of water are taken in through food and drink. The concentration of urine increases (and the volume decreases) when significant amounts of water are lost through sweating.

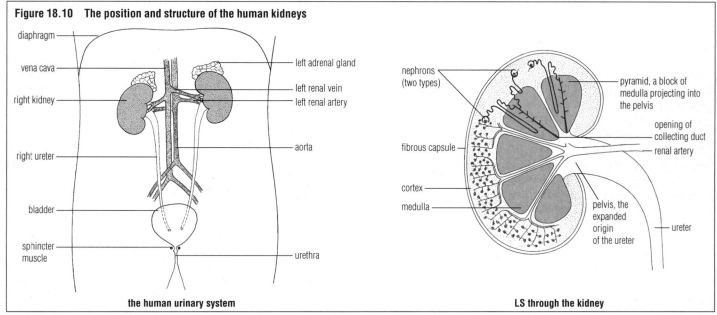

Figure 18.10   The position and structure of the human kidneys

the human urinary system

LS through the kidney

The principles of urine formation

The daily output of urine contains about 50 g solutes, the chief components being urea (30 g) and sodium chloride (15 g). Also present is the surplus of other metabolites and inorganic ions from catabolism. The pH of the blood is kept constant at 7.4; that of urine varies according to whether an excess of hydrogen or of hydroxide ions has been absorbed in the food (Table 18.3).

The operation of the nephron involves several distinct mechanisms, listed below and discussed in more detail on the next few pages.

1 Ultrafiltration

High blood pressure in the glomerulus forces water and other small molecules in the plasma through the walls of the capillaries and of the renal capsule into the tubule lumen, a process known as **ultrafiltration**.

2 Selective re-absorption

Those components of the filtrate in the nephron that are useful to the body are re-absorbed in the blood system by active transport across the cells of the tubule wall. These re-absorbed substances include amino acids and glucose, as well as other compounds, to maintain the water and salt composition of the body fluids.

3 Secretion

Other substances, not required by the body, are added to the filtrate from the blood by active transport from the cells of the tubule wall.

4 Differential permeability

Parts of the walls of the tubule are impermeable to water, ions or urea; these substances thus cannot diffuse or be transported back into the blood in those regions. The permeability of the walls of the collecting duct and the distal convoluted tubule is controlled by hormones.

5 Urine storage

The urine passes to the **bladder** via long tubes, the **ureters** (in humans the ureters are about 25 cm long), and is stored there. The bladder is a distensible sac with a wall of smooth muscle.

The bladder can hold 400–500 cm³ of urine. Once 200 cm³ or so has collected stretch receptors are stimulated, and this stimulation initiates the desire to discharge urine (micturate). The urine passes to the exterior through the urethra (p. 388). Sphincter muscles close the exit from the bladder.

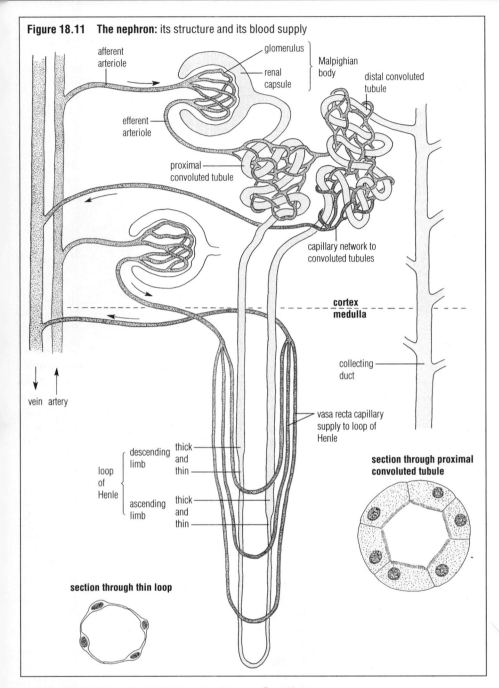

Figure 18.11   The nephron: its structure and its blood supply

afferent arteriole
glomerulus
renal capsule
Malpighian body
distal convoluted tubule
efferent arteriole
proximal convoluted tubule
capillary network to convoluted tubules
cortex
medulla
collecting duct
vasa recta capillary supply to loop of Henle
vein   artery
loop of Henle
descending limb — thick and thin
ascending limb — thick and thin
section through proximal convoluted tubule
section through thin loop

Cortical and juxtamedullary nephrons

There are two types of nephron, differing both in their position in the kidney and in the length of the loop of Henle.

Cortical nephrons occur largely in the cortex and have short loops of Henle, just reaching the medulla.

Juxtamedullary nephrons are in the cortex close to the junction with the medulla. They have long loops of Henle that extend deep into the medulla.

Most nephrons in human kidneys (80–90%) are cortical. The proportions of cortical and juxtamedullary nephrons in other mammals vary (p. 384).

Questions

1 a Which region of the kidney contains the Malpighian bodies, and which region the loops of Henle and collecting ducts?
   b Urine is carried first to the bladder and thence onwards to the exterior of the body. What is the name of the tube in each case?
2 Most nephrons of human kidneys are cortical.
   a What is a juxtamedullary nephron?
   b Under what environmental conditions are animals most likely to have evolved juxtamedullary nephrons?

■ Urine formation

Ultrafiltration in the renal capsule

Ultrafiltration occurs from the capillaries of the glomerulus into the lumen of the renal capsule (Figure 18.12), due to the pressure of blood in the kidney and the sieve-like quality of the walls of the glomerular capillaries and the renal capsule (Figure 18.13). The sieve action permits many smaller-sized molecules to pass through, but retains in the blood almost all the blood proteins (and blood cells).

The sieve

The capillaries of the glomerulus are more permeable than capillaries elsewhere in the body. This permeability is due to the presence of numerous pores or **fenestrations** (0.1 μm in diameter) between the endothelium cells that make up the capillary wall. These cells are attached to the basement membrane, through which the filtrate next passes. This basement membrane is a continuous layer. As a consequence, whilst much of the plasma passes through, all blood proteins and cells are retained within the arterioles.

The final part of the sieve mechanism consists of the walls of the renal capsule. The cells of these walls, instead of forming a continuous sheet, are organised into an irregular supporting network. The filtrate, having passed between the endothelium cells and through the basement membrane, now passes into the capsule via large, slit-like pores between these processes.

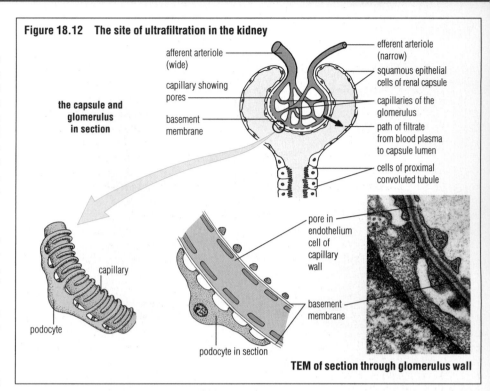

Figure 18.12   The site of ultrafiltration in the kidney

the capsule and glomerulus in section

afferent arteriole (wide)

capillary showing pores

basement membrane

efferent arteriole (narrow)

squamous epithelial cells of renal capsule

capillaries of the glomerulus

path of filtrate from blood plasma to capsule lumen

cells of proximal convoluted tubule

capillary

podocyte

pore in endothelium cell of capillary wall

basement membrane

podocyte in section

TEM of section through glomerulus wall

Blood supply and filtrate production in the kidneys

A volume of blood equivalent to that in the whole body passes through the kidneys every four or five minutes. This means that every minute the kidneys process 1200 cm³ of blood. Of this, 650 cm³ is plasma. About 125 cm³ of this plasma (one-fifth of the volume) is filtered out into the capsules. In all, about 180 litres of filtrate is formed by the kidneys each day, but obviously far less is lost from the body as urine! As the filtrate flows along the remainder of the tubule its composition is changed, and much is re-absorbed. The quantity of urine finally formed is 1–1.5 litres per day, or about 1 cm³ per minute.

Blood pressure in the glomerulus

The blood pressure in the glomerulus is high because the diameter of the efferent arteriole is much smaller than that of the afferent arteriole. A hydrostatic pressure is generated, which forces molecules of M_r below about 68 000 into the renal capsule. These molecules include glucose, amino acids, vitamins, some hormones, urea, other simple nitrogenous compounds, ions and small proteins, and also some of the water. Retained in the blood are red and white cells, platelets, most plasma proteins and the rest of the water. In Table 18.3 (p. 380) the composition of blood and the filtrate are compared.

Movement of filtrate into the capsule is opposed by the hydrostatic pressure of filtrate already there, and the osmotic potential of the blood plasma in the glomerulus. However, the net filtration pressure is greater:

$$\text{net filtration pressure} = \begin{pmatrix}\text{hydrostatic pressure of blood in glomerulus}\end{pmatrix} - \begin{pmatrix}\text{hydrostatic pressure of capsular filtrate} + \text{plasma osmotic potential}\end{pmatrix}$$

$$= 8.0 \text{ kPa} - (2.7 \text{ kPa} + 4.0 \text{ kPa})$$

Therefore, net filtration pressure is 1.3 kPa.

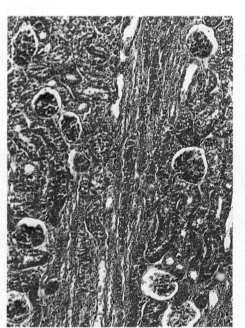
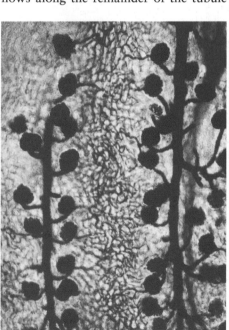

Figure 18.13   Photomicrographs of cortex of kidney showing renal capsules (left), and injected to show blood supply (right) (×50)

Re-absorption in the proximal convoluted tubule

The proximal convoluted tubule is the longest and widest section of the whole nephron, approximately 14 mm long and 60 µm wide. The wall is one cell thick. The cells of the tubule are packed with **mitochondria**, and the surface facing into the tubule has a **brush border** of microvilli (Figure 18.14). Individual tubule cells are imperviously 'welded' together by tight junctions (p. 167) just below the microvilli. Behind the tight junctions are channels of intercellular space. At the base of the cells runs the **capillary network** of the convoluted tubules, derived from the efferent arteriole (Figure 18.11, p. 381).

Active re-absorption of the glomerular filtrate occurs in the proximal convoluted tubule, and a large part of the filtrate is re-absorbed into the capillary network here. For example, glucose and amino acids are actively re-absorbed from the filtrate into the proximal convoluted tubule cells, and are then actively transported (pumped) into the fluid of the intercellular spaces by carrier molecules in the plasma membrane. From here

these metabolites then diffuse into the blood in the capillary network.

Sodium ions pass from the filtrate into the proximal convoluted tubule wall cells by both diffusion and by an active transport mechanism with glucose by means of a secondary pump (p. 238). There is also some exchange of sodium ions outside the cell with hydrogen ions inside. Movement of sodium ions from the wall cells into the fluid of the intercellular spaces is by active transport using energy from ATP. This movement of dissolved substances from filtrate into the blood capillary network brings about the osmotic movement of water in the same direction. In this way about 70% of the water of the filtrate is re-absorbed from the proximal convoluted tubule into the blood of the capillary.

About 50% of the urea in the filtrate diffuses back into the blood here, because of the concentration differences generated by the water re-absorption between the filtrate, the proximal convoluted tubule cells and the blood.

Here in the proximal tubule the cells actively excrete poisonous substances

from the blood into the filtrate, along with some nitrogen-containing substances such as creatinine. Any blood proteins forced into the filtrate by extra high blood pressure in the glomerulus are taken out of the filtrate here by pinocytosis at the base of the microvilli. As a result of these processes a much reduced volume of filtrate, isotonic with the body fluids, passes into the loop of Henle.

Questions

1. a List the main components of the blood that are likely to be filtered in the nephron.
   b Which of these are likely to be re-absorbed in the kidney tubule?
   c What are the energy sources for these two processes?
2. What is the role of the basement membrane in ultrafiltration in the renal capsule?
3. The cells of the proximal convoluted tubule have a 'brush border'. What does this mean, and what advantage does it confer?

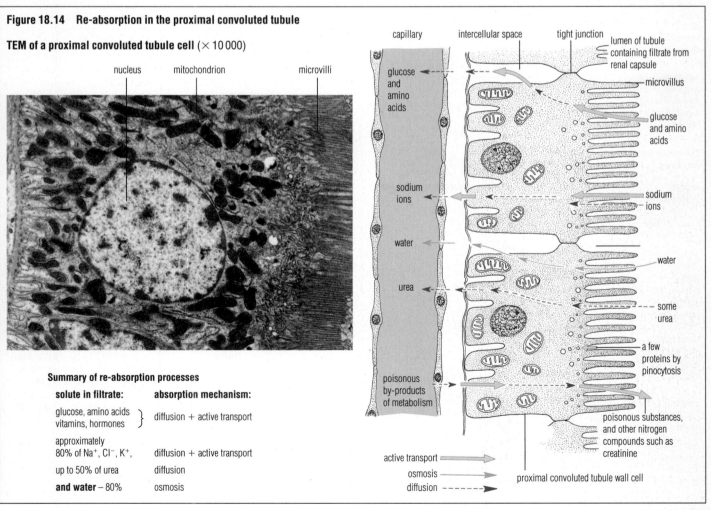

Figure 18.14   Re-absorption in the proximal convoluted tubule

TEM of a proximal convoluted tubule cell (× 10 000)

nucleus   mitochondrion   microvilli

capillary   intercellular space   tight junction

lumen of tubule containing filtrate from renal capsule

microvillus

glucose and amino acids

glucose and amino acids

sodium ions

sodium ions

water

water

urea

some urea

a few proteins by pinocytosis

poisonous by-products of metabolism

poisonous substances, and other nitrogen compounds such as creatinine

proximal convoluted tubule wall cell

active transport
osmosis
diffusion

Summary of re-absorption processes

solute in filtrate:	absorption mechanism:
glucose, amino acids vitamins, hormones }	diffusion + active transport
approximately 80% of Na⁺, Cl⁻, K⁺,	diffusion + active transport
up to 50% of urea	diffusion
and water – 80%	osmosis

The role of the loop of Henle

Urea is always expelled from a mammal's body in solution, so some water loss during excretion is inevitable. Mammals are capable, however, of forming urine that is more concentrated than blood plasma, thus maximising the amount of water retained. The nephron of mammals is able to do this because it possesses a loop of Henle. The loop of Henle has a key role in water conservation.

The ability to conserve water by producing hypertonic urine has enabled mammals to colonise dry habitats. The lengths of the loops of Henle and of the collecting ducts, and the general thickness of the medulla region of the kidneys, increase progressively in animals best adapted to these drier habitats. Figure 18.15 illustrates the correlation between the position and dimensions of the loop of Henle and the relative concentration of urine produced in three mammals living in very different habitats.

The structure of the loop of Henle

The loop of Henle has two regions – a **descending limb** and an **ascending limb** – together with a parallel blood capillary system, the **vasa recta**. Parts of both limbs are thicker than the rest. The different regions of the tubule have different functions. The greater part of these structures lies in the medulla region, along with the collecting ducts.

Countercurrent systems in the medulla

You may have already learnt about countercurrent mechanisms. For example, one exists in the gills of fish, and facilitates gas exchange between the blood of the fish and water flowing over the gills (p. 306). In the gills, adjacent streams of oxygenated water and deoxygenated blood flow in opposite directions, facilitating the exchange of respiratory gases.

There are two countercurrent systems in the medulla of the kidneys concerned with the exchange of solutes and water.

1 A countercurrent multiplier (Figure 18.16, and see opposite) occurs in the loop of Henle. The effect of the countercurrent exchange mechanism of the loop of Henle is to build up the concentration of ions across the medulla. An increasing concentration of ions in the medulla facilitates the later withdrawal of water, mostly from the collecting ducts.

2 The vasa recta, the blood vessels serving this region of the kidney, have a dual role to deliver nutrients to the cells of the medulla, and to carry away water re-absorbed from the filtrate.

Figure 18.15 Kidney structure and water conservation: a comparative study

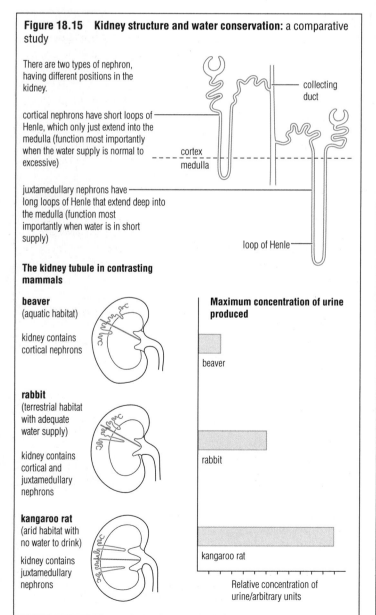

There are two types of nephron, having different positions in the kidney.

cortical nephrons have short loops of Henle, which only just extend into the medulla (function most importantly when the water supply is normal to excessive)

juxtamedullary nephrons have long loops of Henle that extend deep into the medulla (function most importantly when water is in short supply)

collecting duct

cortex
medulla

loop of Henle

The kidney tubule in contrasting mammals

beaver
(aquatic habitat)

kidney contains cortical nephrons

rabbit
(terrestrial habitat with adequate water supply)

kidney contains cortical and juxtamedullary nephrons

kangaroo rat
(arid habitat with no water to drink)

kidney contains juxtamedullary nephrons

Maximum concentration of urine produced

beaver

rabbit

kangaroo rat

Relative concentration of urine/arbitrary units

Figure 18.16 The countercurrent multiplier hypothesis of the loop of Henle: outcome of this process is an increasing gradient of salt concentration in the medulla; this enables excess water to be withdrawn by osmosis from the urine in the collecting duct

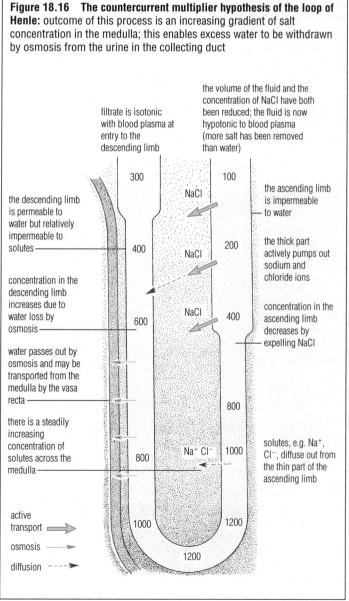

filtrate is isotonic with blood plasma at entry to the descending limb

the volume of the fluid and the concentration of NaCl have both been reduced; the fluid is now hypotonic to blood plasma (more salt has been removed than water)

the descending limb is permeable to water but relatively impermeable to solutes

the ascending limb is impermeable to water

the thick part actively pumps out sodium and chloride ions

concentration in the descending limb increases due to water loss by osmosis

concentration in the ascending limb decreases by expelling NaCl

water passes out by osmosis and may be transported from the medulla by the vasa recta

there is a steadily increasing concentration of solutes across the medulla

solutes, e.g. Na$^+$, Cl$^-$, diffuse out from the thin part of the ascending limb

300   100
NaCl

400   200
NaCl

600   400
NaCl

800
Na$^+$ Cl$^-$ 1000

1000   1200

1200

active transport →
osmosis →
diffusion --→

The functioning loop of Henle

The wall of the descending limb of the loop of Henle is fully permeable to water but only slightly permeable to sodium and chloride ions. The wall of the ascending limb is largely impermeable to water. The cells of the ascending limb actively transport sodium and chloride ions from the filtrate into the medulla. Water is unable to move out with the ions, however. The ions slowly diffuse into the descending limb, the rate of movement allowing the concentration of sodium and chloride ions to build up in the medulla. As a consequence, water leaves the descending limb by osmosis. At the bend of the loop of Henle the concentration of the medulla intercellular fluid reaches a maximum. The collecting ducts also pass through this region of the medulla (Figure 18.11, p. 381). More water may be drawn from the tubular fluid in the collecting duct into the medulla, by the osmotic gradient created.

The role of the vasa recta

You will, no doubt, realise that all the reactions taking place in the kidney tubules require a great deal of energy. The cells in the medulla will therefore need a good blood supply to deliver oxygen and nutrients and take away the waste products of their metabolism.

If the food and oxygen were to be delivered by a normal capillary bed (p. 353), the blood flow would also carry away all the sodium and chloride ions that had built up to a high concentration in the medulla.

The vasa recta, however, are not in the form of a normal capillary bed. The fine vessels run parallel to the loops of Henle rather than forming a diffuse branching network. Consequently, the blood in the vasa recta comes into equilibrium with the solutes in each part of the medulla through which the vessels pass. The blood gains ions during the descending journey and loses them again from the ascending loop.

In this way the vasa recta supply essential materials to the cells of the medulla without affecting the osmotic gradient there.

Questions

1 From the data in Figure 18.16, why can we say the loop of Henle is implicated in water retention?

2 What is the essential feature of a countercurrent flow system?

3 Why is the countercurrent flow in the loop of Henle called a multiplier system?

Quantifying water relations in the medulla

In medical physiology it remains the practice to describe the osmotic properties of solutions in terms of concentration (rather than in terms of the reduced free energy of water molecules in a solution, p. 232).

The concentration of a solute is often expressed as the number of moles in a litre. Where the molar concentration of ions or molecules is very low the millimole, mM, is used as a unit (1 M = 1000 mM). In osmosis all the dissolved substances contribute to the osmotic concentration, and the sum of the millimolar concentrations of all the molecules and ions is referred to as the milliOsmolar concentration (mOsm).

The concentration of the blood plasma is 300 mOsm, whereas that of human urine is commonly about 950 mOsm (but ranges from 50 to 1400 mOsm). In order to be able to extract a large proportion of water by osmosis and generate such concentrated urine, the osmotic concentration in the medulla must be about four times that of the blood plasma (or 1200 mOsm). Only at such a **concentration difference** is sufficient water withdrawn from the filtrate as the fluid passes from the descending limb of the loop of Henle and the collecting duct.

The cells of the ascending limb of the loop of Henle that transport sodium chloride from the filtrate out into the solution in the medulla are capable of creating a gradient of about 200 mOsm. This gradient is sometimes referred to as the 'single effect'. (The concentration of the medulla solution may also be increased by the presence of urea that has escaped from the collecting ducts under particular conditions, p. 387.) How is the required gradient of 1200 mOsm routinely generated?

It is generated by multiplying the 'single effect' many times over between the two streams of fluid moving in opposite directions. This is what is meant by the term '**countercurrent multiplier**' (Figure 18.16). As sodium and chloride ions are pumped out of the ascending limb the surrounding fluid found between the cells (interstitial fluid) becomes hypertonic. The descending limb is permeable, and so water passes out into the medulla and sodium and chloride ions diffuse into the fluid of the descending limb. The fluid in the descending limb becomes more concentrated. As the fluid moves onwards the sequence of changes is repeated again and again, producing a higher and higher concentration in the deep medulla and at the bend in the loop.

In the ascending limb the concentration of sodium and chloride ions becomes progressively less, because it is from here that they are being pumped out. Finally, the fluid leaving the ascending loop is actually slightly hypotonic to the blood plasma, but reduced in volume. More salt has been lost than water. The final step in the production of a concentrated urine is completed in the collecting duct.

INVESTIGATING TUBULE FUNCTION

By means of microdissection equipment, it has been possible to sample the filtrate as it leaves the renal capsule in working tubules in frogs and rats. By this technique the precise concentrations of a range of metabolites in the filtrate has been estimated, and compared with concentrations in the blood plasma.

The transport of metabolites across short excised (cut out) lengths of tubule has also been investigated by physiologists, in order to investigate the role in re-absorption of metabolites of different regions of the nephron tubule.

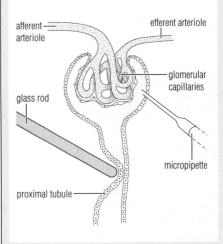

sampling the filtrate in the renal capsule

The proximal convoluted tubule is closed by pressure on it with a fine glass rod. A micropipette inserted into the capsule is used to withdraw a sample of filtrate.

Regulation in the distal convoluted tubule

The role of the cells of the wall of the distal convoluted tubule is the fine control of the blood composition. These cells have a brush border of microvilli lining the lumen of the tubule, and the cell contents include many mitochondria.

The pH of blood is maintained at a constant value of 7.4. Abrupt change in blood pH is prevented by the presence in the plasma of proteins, while hydrogen-carbonate and phosphate ions act together as buffers (p. 140). These substances are able to counteract the effects of a sudden and excessive release of hydrogen or hydroxide ions in the course of metabolism. In the distal convoluted tubule longer-term adjustments in the ion balance of the blood are made. If the blood pH is tending to fall below 7.4, the cells of the distal convoluted tubule secrete hydrogen ions into the urine; if the pH is tending to rise, hydrogencarbonate ions are secreted (Figure 18.17).

The cells of the distal tubule also regulate the plasma concentrations of ions such as sodium, chloride and calcium. Where necessary these ions are removed from the urine into the blood by active transport across the tubule wall.

Water re-absorption in the collecting duct

The permeability to water of the walls of the collecting duct is controlled by **anti-diuretic hormone** (ADH), secreted by the pituitary gland. When the walls are made

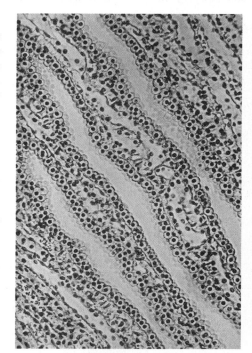
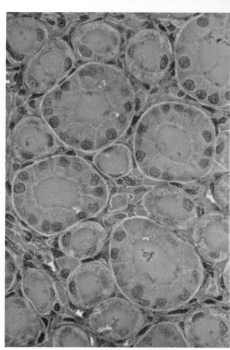

Figure 18.18   Medulla of kidney in LS (left, ×250), and in TS (right, ×500)

permeable water is re-absorbed from the urine in the collecting duct by osmosis, as a result of the high salt concentration in the medulla, through which the collecting duct runs (Figure 18.18). When the walls are impermeable, however, no water is lost from the urine as it travels down the collecting duct. Control by ADH secretion leads to the production of hypotonic or hypertonic urine, according to the body's varying demand for water (Figure 18.19).

ADH and osmoregulation

When the intake of water exceeds the body's normal requirements, the urine produced is copious and dilute. This condition is known as **diuresis**. Little or no ADH is secreted into the blood in these conditions.

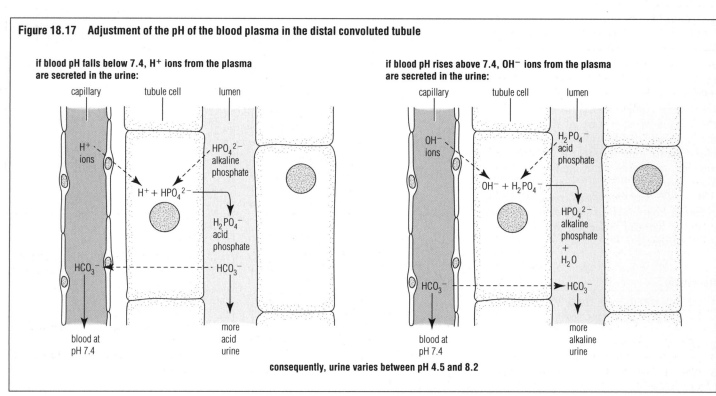

Figure 18.17   Adjustment of the pH of the blood plasma in the distal convoluted tubule

if blood pH falls below 7.4, H$^+$ ions from the plasma are secreted in the urine:

capillary   tubule cell   lumen

H$^+$ ions

HPO$_4^{2-}$ alkaline phosphate

H$^+$ + HPO$_4^{2-}$

H$_2$PO$_4^-$ acid phosphate

HCO$_3^-$ HCO$_3^-$

blood at pH 7.4

more acid urine

if blood pH rises above 7.4, OH$^-$ ions from the plasma are secreted in the urine:

capillary   tubule cell   lumen

OH$^-$ ions

H$_2$PO$_4^-$ acid phosphate

OH$^-$ + H$_2$PO$_4^-$

HPO$_4^{2-}$ alkaline phosphate + H$_2$O

HCO$_3^-$ HCO$_3^-$

blood at pH 7.4

more alkaline urine

consequently, urine varies between pH 4.5 and 8.2

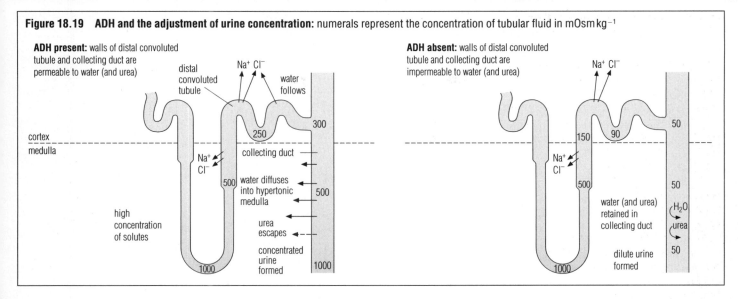

Figure 18.19 **ADH and the adjustment of urine concentration:** numerals represent the concentration of tubular fluid in mOsm kg⁻¹

ADH present: walls of distal convoluted tubule and collecting duct are permeable to water (and urea)

ADH absent: walls of distal convoluted tubule and collecting duct are impermeable to water (and urea)

On the other hand, when little water is ingested, when heavy sweating occurs (see the discussion of temperature regulation mechanism on p. 513) or if a large quantity of salt is absorbed in the diet, then the solute potential of the body fluids becomes more negative (medical physiologists still say the osmotic pressure rises, p. 232). The change in solute potential is detected by a group of cells in the hypothalamus (p. 466) known as **osmoreceptors**. As a result an impulse is generated, which passes to the posterior pituitary gland at the base of the brain (p. 474), triggering the release of ADH into the bloodstream. In the target organ, the kidney, ADH causes increased permeability of the walls of the collecting duct (and of the distal convoluted tubule). More water is withdrawn from the filtrate and urine, into the cortex and medulla, and it passes back into the blood capillaries. A reduced volume of hypertonic urine is discharged by the kidney (Figure 18.19).

A second effect of ADH on the nephron is to increase the permeability of the collecting duct to urea. In the presence of ADH more urea diffuses out into the medulla. This urea in the medulla lowers the solute potential of the kidney tissues still further. As a consequence more water moves by osmosis from the descending tubule of the loop of Henle and from the collecting duct itself into the medulla, and is carried away in the capillary network and vasa recta. More water is retained in the body and the urine is further concentrated.

When a large volume of water is ingested or very little sweating occurs, or there is extremely low salt intake in the diet, the solute potential of the blood

Figure 18.20 **Antidiuretic hormone (ADH)**

ADH is released either when the solute potential becomes more negative or when the blood pressure falls

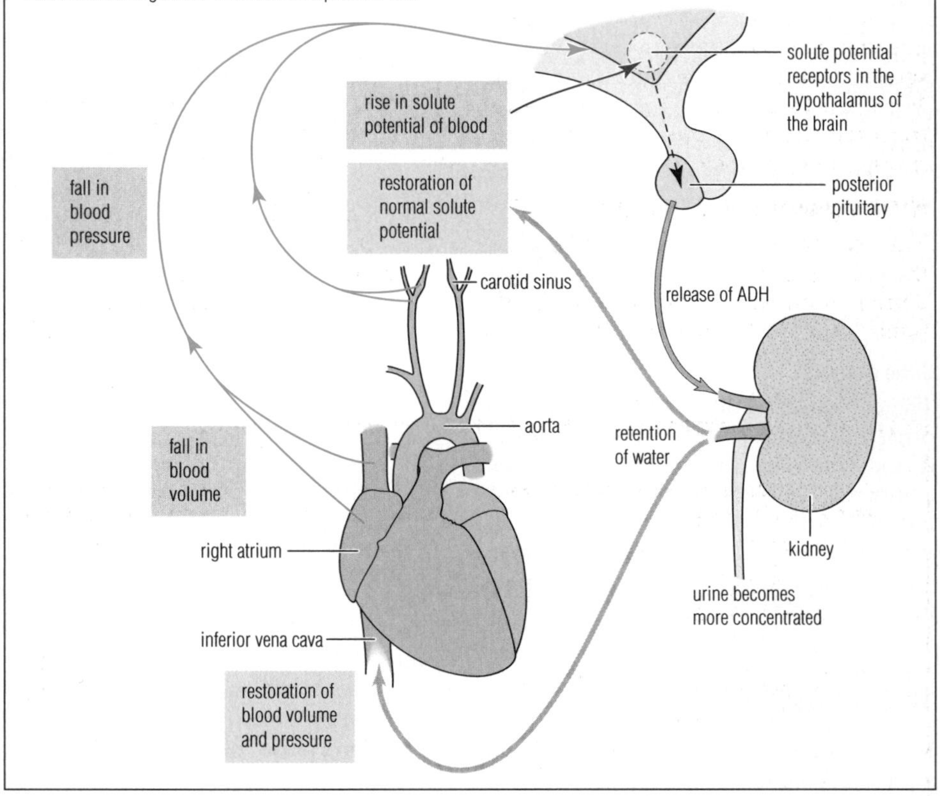

becomes less negative. This condition is detected by the osmoreceptor cells of the hypothalamus. The result is an inhibition of ADH release. Without ADH in the bloodstream the walls of the collecting duct (and of the distal convoluted tubule) become less permeable to water. Less water is absorbed into the medulla as the urine passes down and a large volume of hypotonic urine is produced (Figures 18.20 and 18.21).

Questions

1 Why is it important that excess of hydrogen and hydroxide ions is removed from the blood circulation by the action of the distal convoluted tubule?

2 a What is the effect of ADH on the permeability of the walls of the collecting duct?

b In what circumstances in the body is ADH released?

Aldosterone

The hormone aldosterone, secreted by the cortex of the adrenal gland, has a secondary effect on the water content of the body. The primary role of aldosterone is the maintenance of a constant level of sodium in the plasma. The presence of aldosterone in the blood triggers the active uptake of sodium from the filtrate in the nephron into the tissue fluid of the cortex (known as interstitial fluid). From here the sodium ions diffuse into the capillaries. This movement of sodium ions is accompanied by osmotic movement of water. The adrenal glands are discussed on p. 475.

Emptying of the bladder

Urine formed by the kidneys is temporarily stored in the bladder (Figure 18.22). Emptying of the bladder, known as **micturition**, normally occurs well before the limit capacity is reached. Micturition takes place by contraction of the bladder wall and relaxation of the sphincter muscles at the neck of the bladder. The bladder wall consists of smooth muscle, which is under the control of the autonomic nervous system. Micturition is a reflex in response to stretching of the walls. The reflex becomes suppressed by voluntary nervous action.

Urine analysis

The composition of a sample of urine is often tested in medical diagnostic procedures. This is because urine of abnormal composition is a relatively easily detected indicator of disease or malfunction of many of the body systems. For example, when glucose appears in the urine it may be as a symptom of diabetes (p. 506).

18.5 KIDNEY FAILURE

Kidney failure may be an outcome of a variety of conditions including

▶ bacterial infection
▶ external mechanical injury
▶ high blood pressure.

In the event of rapid onset kidney failure (known as acute renal failure) the tubules are no longer able to excrete water, sodium ions or urea, and these substances accumulate in the blood. If the glomeruli are damaged, plasma proteins may escape into the filtrate and appear in the urine. But it is rare for symptoms of

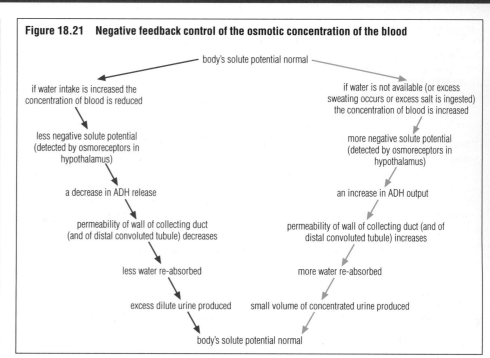

Figure 18.21 Negative feedback control of the osmotic concentration of the blood

body's solute potential normal

if water intake is increased the concentration of blood is reduced → less negative solute potential (detected by osmoreceptors in hypothalamus) → a decrease in ADH release → permeability of wall of collecting duct (and of distal convoluted tubule) decreases → less water re-absorbed → excess dilute urine produced

if water is not available (or excess sweating occurs or excess salt is ingested) the concentration of blood is increased → more negative solute potential (detected by osmoreceptors in hypothalamus) → an increase in ADH output → permeability of wall of collecting duct (and of distal convoluted tubule) increases → more water re-absorbed → small volume of concentrated urine produced

body's solute potential normal

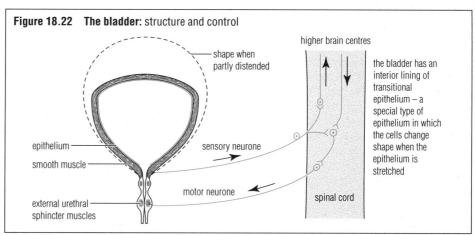

Figure 18.22 The bladder: structure and control

shape when partly distended

epithelium

smooth muscle

external urethral sphincter muscles

sensory neurone

motor neurone

higher brain centres

the bladder has an interior lining of transitional epithelium – a special type of epithelium in which the cells change shape when the epithelium is stretched

spinal cord

kidney failure to arise whilst 50% of the nephrons remain functional. This is because the human kidney has a large reserve of renal function (each kidney contains about a million nephrons).

In the early stages of acute renal failure, a careful choice of diet (regulation of fluid and salt consumption and restriction on protein intake) can prevent the accumulation of waste products in the blood.

■ Haemodialysis

If kidney failure worsens, the first line of defence may be **haemodialysis**. Haemodialysis is carried out in an 'artificial kidney', often referred to as a 'dialysis machine' (Figure 18.23).

In dialysis the blood is withdrawn from the body from an opening in a vein, an anticoagulant is added to prevent clotting, and the blood is then circulated through the dialysis apparatus before

being returned to the body. During dialysis, blood is pumped repeatedly through a tube of partially permeable membrane (cellophane) bathed in a fluid of carefully adjusted composition (the dialysate). The dialysis membrane has minute pores, which allow solutes to diffuse through, but not blood cells or protein molecules. Through such a membrane, substances such as salts and glucose, which are needed by the body, also diffuse out. Consequently the dialysing fluid is made up to have the same concentration as normal tissue fluid, so that the outward and inward diffusion of, for example, glucose are in equilibrium. Excess salts, urea and toxic substances are absent from the dialysate, however, and so diffuse out of the blood.

Dialysis goes on for 6–10 hours, during which the composition of the dialysate is adjusted to restore the blood to its correct composition.

Haemodialysis enables a patient with kidney failure to be kept healthy more or less indefinitely, provided that treatment is given three times each week. Strict regulation of diet and fluid intake has to be maintained.

An alternative arrangement is to use the abdominal cavity. The peritoneum that lines it functions as the dialysis membrane. The dialysis fluid (1–2 litres) is introduced into the abdomen and left there for a number of hours before being siphoned out and replaced. This is repeated 3–4 times in one day. Because of the large area of peritoneum and its rich blood supply, the exchange that occurs in a dialysis machine occurs here, too. The procedure is cheaper and simpler than using a dialysis machine, but carries a slight risk of causing peritonitis.

■ **Transplantation**

In a young patient with kidney failure, transplantation of a healthy kidney that will be accepted by the body and function normally is preferable to dialysis as the long-term solution. The failed kidney is removed. Then the donated kidney is placed in the pelvis region of the abdomen (Figure 18.23), and linked to the internal iliac artery and vein (p. 351). The transplanted ureter is inserted directly into the bladder.

The problems with transplantation are immunological: the transplanted kidney tends to be rejected by the immune response of the recipient (p. 371). In order to minimise the danger of rejection, the tissue cell types of recipient and donor are matched as closely as possible. At transplantation the antibody-producing cells are suppressed, and subsequently some immunosuppressive drugs have to be administered permanently. Unfortunately these drugs reduce the patient's resistance to infections.

Organs for transplant become available from patients who have died from, say, a brain tumour, a condition that does not affect the organs sought for transplant, or injury in an accident. Kidneys must be removed within an hour of death and chilled to slow down deterioration, and must be transplanted within about 12 hours.

Healthy people have an excess of kidney function and can survive with a single kidney. Consequently it is possible for a living person to donate a kidney to another. (When the donor is the identical twin, rejection is very unlikely.)

In order to obtain a donor organ, permission is required from the donor (if living) and from close relatives. Many people carry a donor card to facilitate the use of their organs after their death. Although some 2000 kidney transplant operations are routinely carried out each year in the United Kingdom, very many more people await the availability of a suitable kidney.

Kidney transplantation raises ethical issues for the medical profession, since the supply of kidneys falls far short of the numbers needed. Should some people be given priority over others in the waiting lists, for example?

Figure 18.23   Kidney failure may be rectified by haemodialysis or transplantation

An opening (fistula) is formed in the patient's arm by surgery. Blood flows from the artery directly into the vein, which develops a thickened wall as a result of the increased pressure and volume of blood flowing. Dialysis needles are inserted into the fistula for haemodialysis treatment.

In the dialysis process blood flows through a very long cellophane tube-type structure, surrounded by fresh dialysis fluid; diffusion of small molecules (e.g. sugars, amino acids) and ions is prevented by adjusting their concentrations in the dialysate to be the same as in blood.

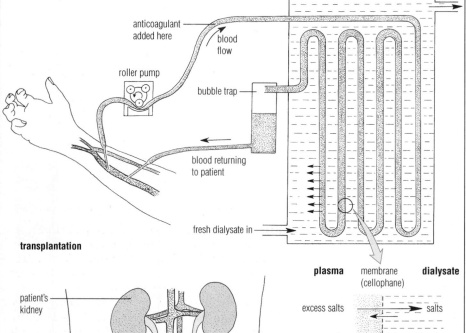

the dialysis process

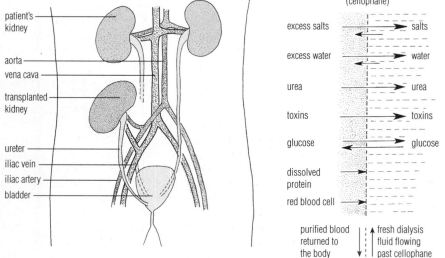

transplantation

Questions

1 What advantage arises from arranging for blood and dialysate to flow in opposite directions in the kidney machine?

2 How is it possible for the dialysate composition to be used to withdraw excess water from blood?

3 Why is an anticoagulant required in the process of withdrawing blood for dialysis?

4 Why is an identical twin the best source of organs for transplantation?

18.6 CASE STUDIES IN ANIMAL OSMOREGULATION AND EXCRETION

■ Mammals of arid regions

Typically, animals of arid regions are adapted to survive with little or no liquid water intake. This group of animals includes the **kangaroo rat** (*Dipodomys* sp., Figure 18.24), which lives in hot dry deserts, but hides in a burrow during the day. The kangaroo rat is able to survive without access to any drinking water at all. Extremely concentrated urine is excreted (Figure 18.15, p. 384), and no sweat is produced.

The **camel** is also adapted to life in an arid climate. The camel can go for long periods without drinking, although there is no water storage facility in its body. (Of course, it obtains metabolic water, p. 138, from the oxidation of its food reserves.) The camel is able to tolerate elevation of the body temperature, and to endure an unusually high degree of dehydration (Figure 18.25). Two species of camel occur in various desert regions of the world: the dromedary (one-humped), *Camelus dromedarius*, and the Bactrian (two-humped), *Camelus bactrianus*. The food reserves of fat are stored in the hump(s).

Figure 18.24   The water relations of the kangaroo rat, a desert-adapted mammal

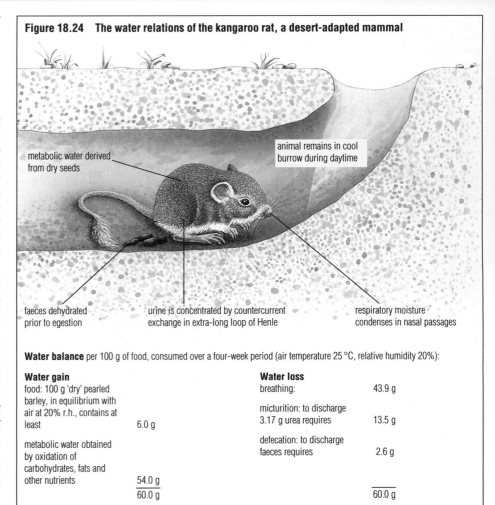

metabolic water derived from dry seeds

animal remains in cool burrow during daytime

faeces dehydrated prior to egestion

urine is concentrated by countercurrent exchange in extra-long loop of Henle

respiratory moisture condenses in nasal passages

Water balance per 100 g of food, consumed over a four-week period (air temperature 25 °C, relative humidity 20%):

Water gain		Water loss	
food: 100 g 'dry' pearled barley, in equilibrium with air at 20% r.h., contains at least	6.0 g	breathing:	43.9 g
		micturition: to discharge 3.17 g urea requires	13.5 g
metabolic water obtained by oxidation of carbohydrates, fats and other nutrients	54.0 g	defecation: to discharge faeces requires	2.6 g
	60.0 g		60.0 g

Figure 18.25   Camels and survival in the desert

The structural and physiological adaptations shown by camels to survival in desert conditions include:

▶ **the conservation of water by heat storage**
When unlimited water is available, body temperature fluctuates (day–night) by only about 3 °C, but in the exposed desert it varies by up to 7 °C. By allowing body temperature to rise in the heat of the day, water that would otherwise evaporate is saved. 'Day heat' stored in the body is lost at night.

▶ **water vapour conservation in breathing**
Dried mucus lining the nostrils absorbs the water vapour in exhaled air. Inhaled air picks up water vapour from the mucus. Hence water vapour is shunted between nostrils and lungs, rather than lost.

▶ **toleration of water depletion**
In going without drinking for up to a week, 25% of body mass may be lost. Blood volume is maintained by drawing liquid from the tissues. Urea, held in the blood, causes this transference of water. The ovoid red cells of the blood travel easily through the capillaries during times of dehydration.

▶ **adaptations of feet, nostrils and mouth**
The feet are structured for movement on shifting sand. The nostrils are slit-like (closeable) and guarded by dense hairs. The eyes are shielded from wind-blown sand by long, dense eye lashes. The lips are split and very mobile, allowing selective browsing of foliage protected by thorns.

Bactrian camel (*Camelus bactrianus*)

head (nostrils and mouth) of a dromedary camel (*Camelus dromedarius*)

Figure 18.26 The life cycle of the eel, a fish that migrates between marine and freshwater habitats

Mature eels migrate from freshwater rivers, out across the Atlantic Ocean. They breed in the Sargasso Sea. The young eels return to Europe, and move up rivers to sites where they feed and grow.

Figure 18.27 The role of the salt glands in the survival of sea birds

Sea birds with access to sea water and sea food are only able to survive because of their salt glands.

Land mammals quickly dehydrate without access to fresh water.

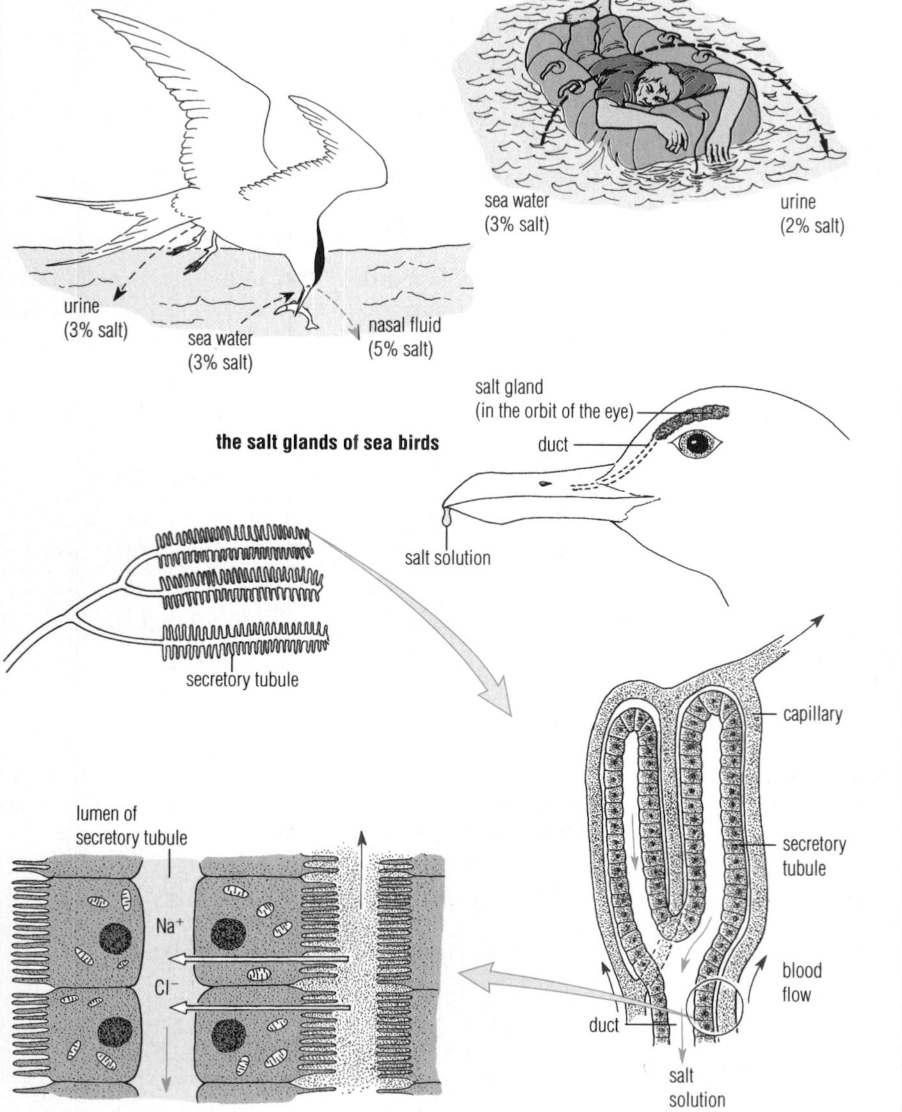

sea water (3% salt)

urine (2% salt)

urine (3% salt)

sea water (3% salt)

nasal fluid (5% salt)

the salt glands of sea birds

salt gland (in the orbit of the eye)

duct

salt solution

secretory tubule

lumen of secretory tubule

Na⁺

Cl⁻

capillary

secretory tubule

duct

blood flow

salt solution

Question

In general, what kinds of mechanisms are used by animals in order to maintain a constant internal environment?

■ Migratory fish

Most animals cannot tolerate pronounced changes in external salinity; aquatic animals do not normally move between the sea and fresh water, and vice versa. Exceptions are the salmon and eels (Figure 18.26) that migrate between sea water and fresh water more than once in their life cycles.

Such fish (known as **euryhaline fish**) have cells in the gills that move ions in or out of their body fluids. In fresh water, when much salt is lost in copious urine production, salts are pumped into the body. Salts are pumped out in the marine phases of the life cycles when sea water is taken in to compensate for water lost by osmosis. This mechanism is thought to be part of how they adapt.

■ Birds

Birds have a high metabolic rate, an efficient gaseous exchange system and a relatively high body temperature. As a consequence, water vapour loss in the expired air can be high, and birds require a supply of water. Conservation of water is achieved in nitrogenous excretion; birds produce uric acid, which is discharged in a semiliquid or firm paste form, needing little or no water to remove it.

Some birds have adapted to a marine environment. Birds that feed at sea ingest large quantities of salt, and some have access only to sea water. At the same time, birds have limited capacity to concentrate urine (at most, their urine is only two or three times more concentrated than their plasma). As a result the kidney of birds exposed to high salt intakes cannot excrete the excess salt without draining valuable body water. Hence the need for a **salt gland** (Figure 18.27), which produces a salt secretion considerably hypertonic to plasma. In many sea birds, salt is removed from the blood by salt glands, which are situated in the orbits of the eyes and open into the nasal cavity. With the aid of salt glands, marine birds can thus maintain their body fluid composition and supply tissues with the necessary water.

FURTHER READING/WEB SITES

C J Clegg (1998) *Illustrated Advanced Biology: Mammals Structure and Function*. John Murray

EXAMINATION QUESTIONS SECTION 3

Answer the following questions using this section (chapters 12 to 18) and your further reading.

1 a The process of photosynthesis can be subdivided into two stages, one dependent on light, the other independent of light.

   i Copy and complete the table to show the substances used in, and the end products of, each of these stages. (Do not include solar energy.) (2)

	Light dependent stage	Light independent stage
Substances used	1. Water 2. Inorganic phosphate 3. ADP 4. NADP	1. Reduced NADP 2. ATP 3. _____ 4. _____
End products	1. ATP 2. _____ 3. _____	1. NADP 2. ADP 3. Inorganic phosphate 4. Carbohydrate

   ii What are the functions of reduced NADP and ATP in the light independent stage of photosynthesis? (2)

b The graph shows the effect of temperature on the rate of photosynthesis.

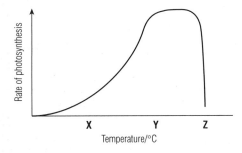

   i Explain why increasing the temperature from **X** °C to **Y** °C increases the rate of photosynthesis. (2)

   ii Explain why increasing the temperature from **Y** °C to **Z** °C decreases the rate of photosynthesis. (2)

AQA (NEAB) AS/A Biology: Processes of Life (BY01) Module Test June 1999 Q6

2 a The diagram summarises the light independent reaction of photosynthesis.

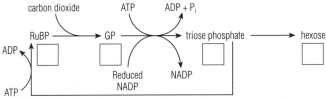

   i Copy the diagram and complete the **four** boxes to show the number of carbon atoms in a molecule of each substance. (1)

   ii Where in the chloroplast does the light independent reaction take place? (1)

   iii Explain why the amount of GP increases after a photosynthesising plant has been in darkness for a short time. (2)

b Describe the role of water in the light dependent reaction of photosynthesis. (2)

AQA (AEB) A Human Biology: Paper 1 June 1999 Q2

3 The diagram shows chemical pathways involved in respiration and photosynthesis.

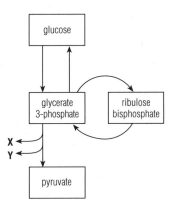

a Name the process that produces pyruvate from glucose. (1)

b Name the compounds labelled **X** and **Y**. (2)

c **i** In which part of a chloroplast is glycerate 3-phosphate converted into ribulose bisphosphate? (1)

   ii Describe the role of ribulose bisphosphate in photosynthesis. (1)

AQA (NEAB) A Biology: Paper 1 June 1999 Q3

4 A student investigated the effect of temperature on the rate of respiration of maggots. This was measured by calculating the rate of oxygen uptake, using a respirometer. The apparatus was set up as shown in the diagram.

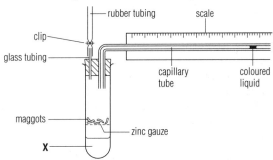

a **i** Name a suitable chemical to be used at **X**. (1)

   ii Explain the purpose of chemical **X**. (1)

The apparatus was placed in a water bath at 25 °C with the clip open. After 10 min the clip was closed and the position of the liquid in the capillary tube was recorded at 5 min intervals. The results obtained are shown in the graph.

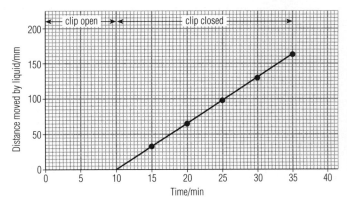

b Explain
  i why the apparatus was left for 10 minutes before closing the clip; (2)
  ii why the liquid moved after the clip was closed. (1)
The capillary tube had a cross-sectional area of 2.5 mm².
c Calculate the rate of oxygen uptake, in mm³ per minute, between 20 and 30 minutes. Show your working. (2)
d Explain how the results would be expected to differ if the investigation was carried out at 15 °C. (2)
Another student carried out the same investigation under exactly the same conditions, except that a different mass of maggots was used.
e Explain what the students needed to do in order to be able to compare their results. (1)
The student carried out the investigation with some green leaves to compare their rate of respiration with that of the maggots.
f State **two** precautions that the student should have taken in order to make a valid comparison (2)

UoCLES AS/A Modular: Central Concepts in Biology June 1998 Section A Q3

5 The diagram shows some of the processes that occur within a mitochondrion and some of the substances that enter and leave it.

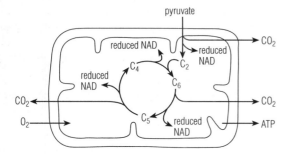

a Name the process that
  i produces pyruvate in the cytoplasm;
  ii produces reduced NAD and carbon dioxide in the matrix of the mitochondrion. (2)
b Describe how ATP is produced from reduced NAD within the mitochondrion. (3)

NEAB A Biology Paper 1 June 1997 Section A Q3

6 **a** **i** Give **one** similarity between the way in which oxygen from the atmosphere reaches a muscle in an insect and the way it reaches a mesophyll cell in a leaf. (1)
  ii Give **one** difference in the way in which carbon dioxide is removed from a muscle in an insect and the way in which it is removed from a muscle in a fish. (1)
The diagram shows the way in which water flows over the gills of a fish. The graph shows the changes in pressure in the buccal cavity and in the opercular cavity during a ventilation cycle.

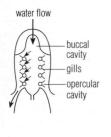

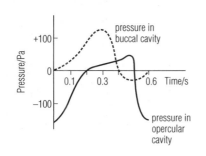

b Use the graph to calculate the rate of ventilation in cycles per minute. (1)
c For most of this ventilation cycle, water will be flowing in one direction over the gills. Explain the evidence from the graph that supports this. (2)
d Explain how the fish increases pressure in the buccal cavity. (2)

NEAB A Biology Paper 1 June 1997 Section B Q17

7 An experiment was carried out with cells of carrot tissue to determine the effect of temperature on the absorption of potassium ions.

Slices of carrot tissue were immersed in a potassium chloride solution of known concentration. The changes in concentration of potassium ions in the solution were determined at intervals for 6 hours. From these measurements, the mass of potassium ions taken in by the carrot cells was found. The experiment was carried out at 2 °C and 20 °C. The solutions were aerated continuously.

The results are shown in the graph below. Absorption of potassium ions is given as micrograms of potassium per gram of fresh mass of carrot tissue ($\mu g\ g^{-1}$).

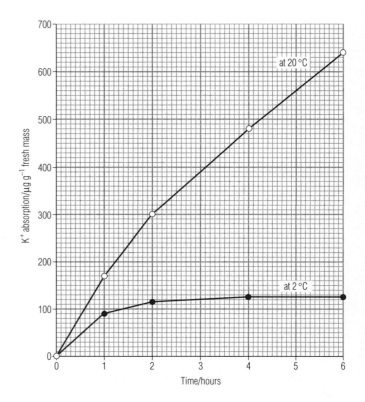

a During the first hour, some of the potassium ions enter the cells by diffusion. State **two** conditions which are necessary for a substance to enter a cell by diffusion. (2)
b **i** Calculate the mean rate of absorption of potassium ions at 20 °C, between 2 and 6 hours. Show your working. (3)
  ii Compare the rate of absorption of potassium ions at 2 °C and 20 °C during this experiment. (3)
  iii Suggest an explanation for the differences in the rates of absorption of potassium ions at the two temperatures. (3)

London AS/A Biology and Human Biology Module Test B/HB1 June 1998 Q7

8 The diagram shows part of the nitrogen cycle between the atmosphere, the plants and the soil in a field of clover (a legume) and grass.

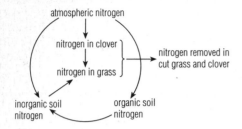

a Name
   i the process by which atmospheric nitrogen is converted into organic nitrogen; (1)
   ii an organism capable of carrying out this process. (1)
b Outline how atmospheric nitrogen may be converted into inorganic nitrogen without the aid of microorganisms. (2)
c Outline the sequence of processes involved in the conversion of organic to inorganic nitrogen in the soil. (4)
d In terms of *the cycling of nitrogen* in this field, state
   i **two** consequences of the farmer cutting and removing the grass and clover; (2)
   ii the effect of the soil becoming waterlogged. (2)

UoCLES AS/A Modular: Central Concepts in Biology November 1997 Section A Q1

9 The secretion of digestive juices into the mammalian gut is controlled by the endocrine and nervous systems. Copy and complete the table about the secretion of digestive juices. (4)

Stimulus that triggers secretion	Effect	Digestive juice secreted
_____	parasympathetic nerve stimulates salivary gland	Saliva
Contact of food with stomach lining	_____ secreted	Gastric juice
Contact of food with duodenum lining	Cholecystokinin secreted	_____
Contact of food with duodenum lining	_____ secreted	Alkaline fluid from pancreas

AQA (NEAB) A Biology: Paper 1 June 1999 Q5

10 a Explain what is meant by the term *translocation*, in plants. (2)
   b Outline the role of companion cells in translocation. (3)
   c Outline **one** experimental method that has been used to show that translocation takes place in phloem. (2)

OCR (Cambridge Modular) AS/A Sciences: Transport, Regulation and Control November 1999 Q4

11 During the cardiac cycle the pressure in the different chambers of the heart varies in a regular pattern.
   a State in which chamber of the heart
      i the greatest pressure in generated; (1)
      ii the greatest change in pressure occurs. (1)
The diagram shows the changes in blood pressure during one cardiac cycle.

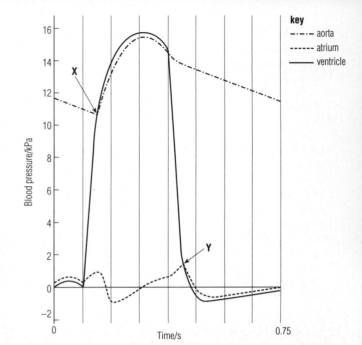

b With reference to the diagram
   i explain what is happening at **X** and **Y**; (3, 3)
   ii calculate the heart rate in beats per minute. Show your working. (2)

OCR (Cambridge Modular) AS/A Sciences: Transport, Regulation and Control November 1999 Q1

12 The diagram shows two oxyhaemoglobin dissociation curves.

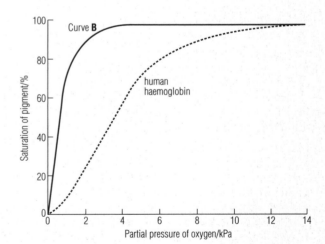

a Use the graph to explain why human haemoglobin:
   i is saturated with oxygen in the lungs; (1)
   ii releases oxygen when it reaches the tissues. (1)
b Explain what causes human oxyhaemoglobin to give up a greater proportion of the oxygen it carries during vigorous exercise. (2)
c Lugworms live in deep sand on sea shores. When the tide is in, they obtain oxygen by pumping water over their gills. At low tide there is no water to pump over the gills. Curve **B** on the graph shows the dissociation curve of lugworm oxyhaemoglobin. Explain the advantage of haemoglobin of this type to the lugworm. (2)

AEB A Biology Paper 1 June 1998 Q11

4 The Responding Organism

Analysis of flapping flight in the little owl (*Athena nocturna*) by multiflash photography

19 Growth and development

- Growth consists of a permanent, irreversible increase in size, whilst development is a change in shape, form and complexity of an organism.
- Growth of a population of unicellular organisms consists of repeated division of individual cells. There is a limited amount of differentiation in unicellular organisms.
- Measurement of growth in numbers of unicellular organisms, plotted against time, typically shows an S-shaped curve.
- Measurement of growth of multicellular organisms, by recording changes in their length, height or mass in unit time, may be analysed in different ways to provide insights into growth and development processes.

- Germination and growth of a seed leads to the formation of new tissues, including the vascular tissues, by primary growth.
- In vertebrates, an embryo is formed from a fertilised egg. During growth the embryo forms the tissues and organs of the adult.
- In non-vertebrates, growth and development of the fertilised egg to adult typically involve juvenile stages, called larvae, markedly different in structure from the adult.

19.1 INTRODUCTION

The changes that take place in an organism between fertilised egg stage and adult structure are referred to as growth and development. Growth and development are distinctive characteristics of living things, yet defining them is not altogether straightforward. For example, when an organism grows it normally increases in size, yet increase in size alone may be an inadequate definition of growth. Thus the change in size of a cell absorbing water by osmosis (p. 232) is easily and quickly reversible, but this volume change is not growth. Conversely, the dramatic increase in the number of cells as a vertebrate zygote undergoes cleavage (p. 413) is a case of growth and development without any increase in volume or mass of cytoplasm. Our definitions must allow for these points.

Growth in an organism consists of a permanent and more or less irreversible increase in size, commonly accompanied by an increase in solid matter, dry mass and amount of cytoplasm. Growth is a quantitative concept concerned with the increasing mass of an organism.

Development refers to the changes in shape, form and degree of complexity that accompany growth. Development, which is a qualitative concept, involves **differentiation**, the process by which unspecialised structures become modified and specialised for the performance of specific functions.

Growth and development commonly go hand in hand, and are often referred to collectively as **morphogenesis** (from Greek words meaning, literally, 'the origin of form').

■ Microorganisms and multicellular organisms

It is not easy to examine the growth pattern of individual unicellular organisms because of their small size. Instead, we study the growth of populations of microorganisms. In these cases, growth refers to the numbers of individuals.

There are similarities and differences between growth in unicellular and multicellular organisms, which we will consider next.

Figure 19.1 Exponential growth

Cell division:

Number of cells:

original cell	1
1st generation	$2 = 2^1$
2nd generation	$4 = 2^2$
3rd generation	$8 = 2^3$
4th generation	$16 = 2^4$
division continues . . . nth generation	$= 2^n$

When the number of cells doubles at each division (or in unit time) a graph of cell numbers plotted against time gives a steeply rising (exponential) curve.

A useful check that growth is exponential is to plot the $\log_{10}$ of cell numbers (or use semi-log graph paper), since the log of exponential growth is a straight line.

Generations (number of cell divisions)	0	1	2	3	4	5	6	7	8	8	10
Numbers of cells formed	1	2	4	8	16	32	64	128	256	512	1024
$\log_{10}$ of cell numbers	0.000	0.301	0.602	0.903	1.204	1.505	1.806	2.107	2.408	2.709	3.010

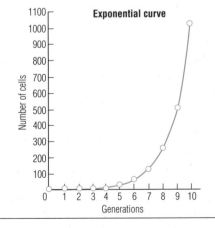

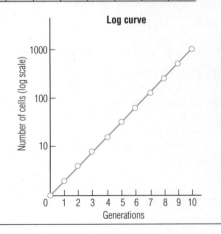

Figure 19.2 Growth of a population of bacteria

Hours	No. viable cells/cm³ medium	Total no. cells*/cm³ medium
0	20 000	20 000
2	21 900	27 200
4	496 000	540 000
6	5 430 000	6 400 000
8	81 900 000	105 760 000
12	83 400 000	126 300 000
24	80 500 000	127 600 000
36	1 120 000	127 900 000

*Including dead and dying cells

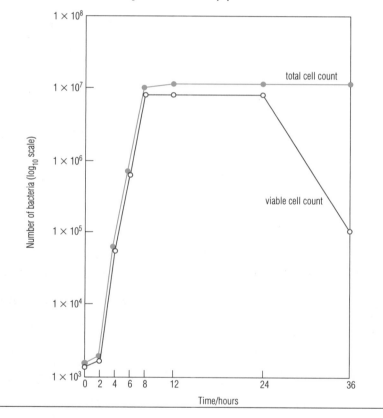

growth curve for the population of *E. coli*

Growth of unicellular organisms

The growth of an individual cell may be limited by the distance over which the nucleus can exert its control. Whether or not this is so, when a cell reaches a particular size it may divide, forming two new, smaller cells. Cell division is followed by growth of the daughter cells to full size (Figure 1.8, p. 6). There is a limited amount of differentiation in unicellular organisms.

Growth in a population of unicellular organisms consists of repeated divisions of individual cells. When a microorganism divides into approximately equal parts, the division is known as **binary fission**. Divisions are often more or less synchronised to occur simultaneously in all cells of the population. Binary fission is a form of asexual reproduction (p. 556) in unicellular organisms, resulting in growth of the population.

Growth of multicellular organisms

Growth of a multicellular organism involves division of cells, the expansion of cells and some movements of cells about the organism. Within a multicellular organism groups of cells typically become specialised, and perform a particular function. **Specialisation** involves biochemical as well as structural changes. For example, developing red cells acquire large quantities of haemoglobin and carbonic anhydrase in their cytoplasm and lose their nuclei. Specialisation of cells may lead to greater efficiency in functions such as transport, movement, support, digestion or food production, and defence of the organism. Greater size and complexity is a consequence of the organism becoming multicellular.

■ 'Nature *v.* nurture' in growth and development

The processes of growth and development are influenced by many factors, some internal to the organism and some environmental or external. From observation of living things it may be difficult to determine whether a particular feature of growth such as body height is determined largely by internal factors, namely the inherited instructions each organism receives in its genes (p. 210), or by some environmental factor such as the food supply available to the organism. This is because the effects of internal and external factors interact.

The interaction of internal and external factors in human growth and development are sometimes referred to in literature as 'nature *v.* nurture'. The distinction is one of 'genetic *v.* environmental' influences.

19.2 THE MEASUREMENT OF GROWTH

■ Measuring the growth of unicellular organisms

In a study of the growth of a culture of unicellular organisms it is population growth that is measured by counting the numbers of individuals present in the culture at specific time intervals. Starting with the inoculation of a fresh culture, there is normally a brief period of adaptation by the cells to the new conditions. Once adapted, cells may divide every 20 or 30 minutes whilst conditions (temperature, adequate nutrients and the absence of accumulated waste products) remain favourable. Thus during early growth, the number of cells doubles at each division (Figure 19.1). This is the **exponential phase** of population growth. Subsequent growth is progressively inhibited by exhaustion of nutrient resources, and by the accumulation of toxic waste products. As time proceeds some cells die. Consequently the total number of cells present is higher than the number of viable (living) cells present (Figure 19.2).

Estimating population sizes

The total number of cells in a population of microorganisms is estimated using a small representative sample of the culture. The numbers of cells present may be counted directly, using an instrument such as a haemocytometer (Figure 19.3), or indirectly, by measuring the change in turbidity (absorbance) of the culture as the density of the culture increases.

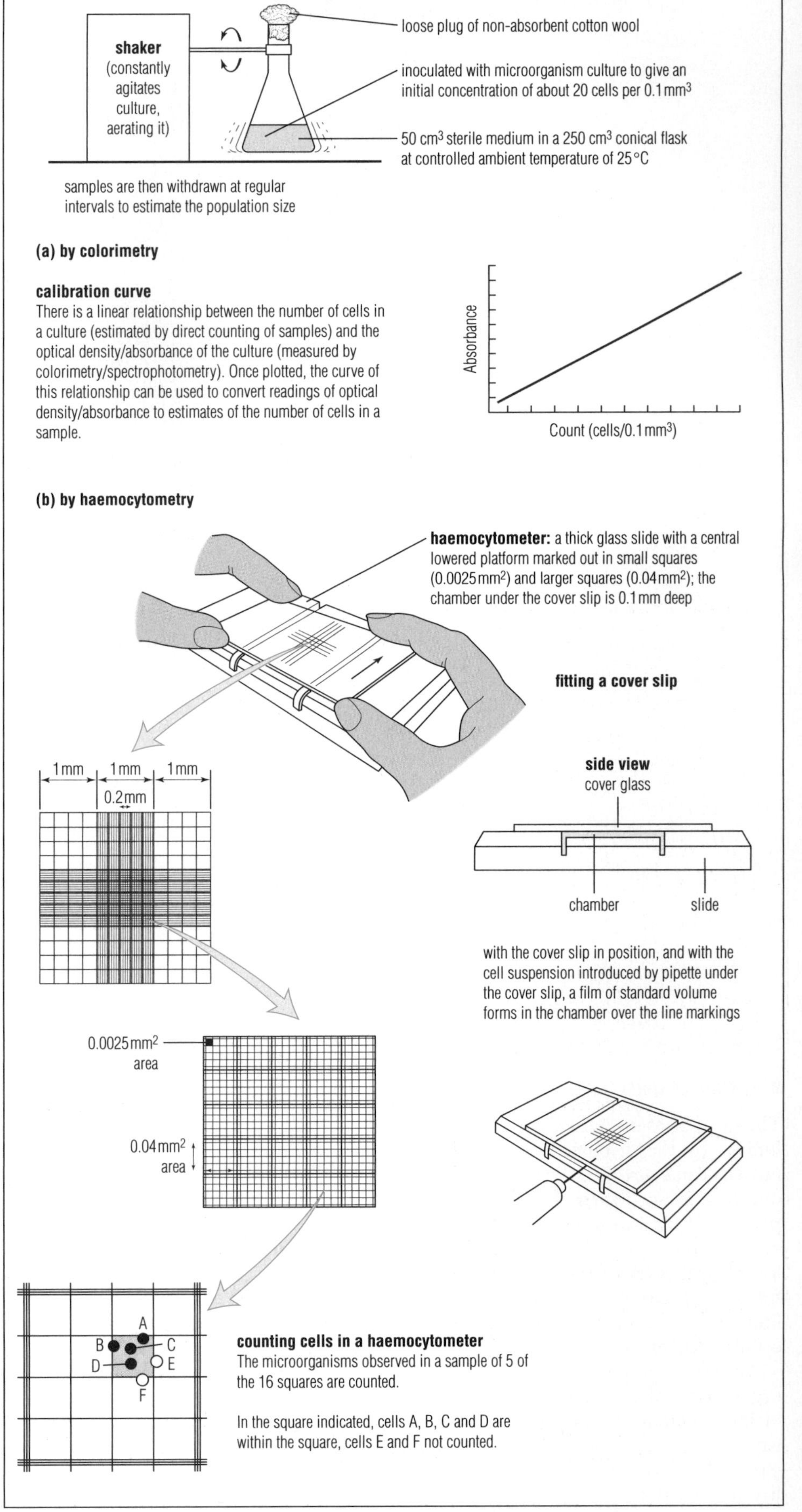

■ Measuring the growth of a multicellular organism

The growth of a multicellular animal or plant is normally measured by recording changes in length or height or mass in unit time. Growth measurements of this type are illustrated in Figures 19.4 and 19.5. The growth of parts of large organisms, such as plant fruits or leaves, may similarly be measured by changes in length, area or mass (Figure 19.6, p. 401).

Fresh mass *v.* dry mass

Measurements of the fresh mass of an organism are relatively easily and quickly carried out. The organism is not injured, and elaborate preparation of the samples may be unnecessary. Fresh mass measurements permit repeated measurements on the same organism or, preferably, a large group of organisms in order to rule out minor individual differences.

The problem with fresh mass measurements is the likelihood of significant variations in water content. Water makes up 70–90% of the fresh mass of living organisms, and so measurements of fresh mass may give inconsistent readings due to fluctuations in water content.

Dry mass measurements of organisms determine the masses obtained after all water has been removed by drying. Dry mass values, collected over a period of growth, provide a more accurate estimate of growth than fresh mass measurements.

Dry mass measurement, however, requires the destruction of the sample of organisms in the process. Consequently, a large population of individuals is needed, all of the same age and of approximately the same size. Measurements are made on samples drawn from the population in order to rule out the effects of individual differences. The dry mass of the batch is obtained by killing the organisms and heating the tissues to 110 °C in an oven to drive off water. The sample is then allowed to cool in a desiccator, weighed and then repeatedly heated and cooled, and the mass measured until a constant dry mass is obtained.

Figure 19.3 Measurement of population growth in a culture of microorganisms

Figure 19.4   Growth of a maize plant (*Zea mays*)

the cycle of growth and development

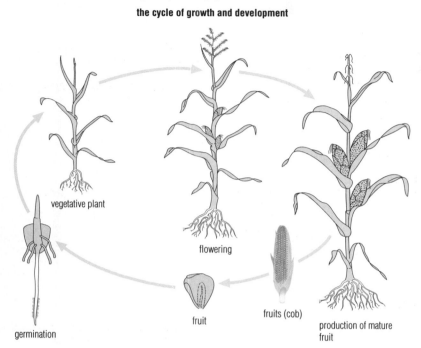

Amount of protein

An increase in amount of living substance, the cytoplasm, with time may be a better indicator of growth. In this analysis, the protein content of tissues can be estimated, and may be taken as representing the amount of cytoplasm. Approximately one-sixth of the mass of a protein is due to nitrogen in combined form (the rest is due to the other elements present in amino acid residues).

Growth can therefore be monitored by regularly measuring the content of combined nitrogen over the period of growth of samples of a large population of organisms at a similar stage of development. The analysis is carried out by heating a weighed sample with concentrated sulphuric acid, thus converting all the combined nitrogen in the sample to ammonia (NH_3), which can then be determined by titration.

Results: growth measurements by height of plant

Days after planting	Height/cm	Growth rate (the increase in height during each 10-day period)	Relative growth rate (the increase in height as a % of the growth at time of measurement)
10	2	2	100
20	7	5	71
30	20	13	65
40	40	20	50
50	75	35	46
60	110	35	31
70	140	30	21
80	150	10	7
90	155	5	3
100	160	5	3
110	160	0	0
120	160	0	0

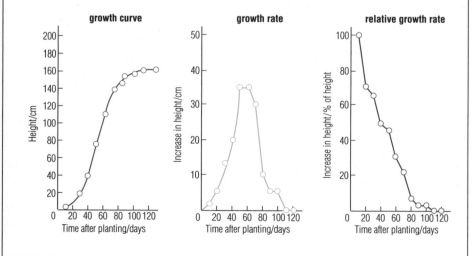

Questions

1 **a** How would you define growth in a multicellular organism, such as a flowering plant or human?
  b How does development differ from growth?
2 What changes do you think may occur in cells adapting to a new culture medium?
3 What advantage may be gained from choosing to plot the log of cell number counted in culture over a period of growth?
4 How might a bacteriologist estimate the changes in the numbers of viable cells in a culture over a period of time (p. 99)?
5 When measuring growth in a living organism, what are the advantages and the disadvantages of taking fresh mass measurements?

Analysing data on growth

The data obtained from investigations of growth in successive intervals of time (growth rate) can be analysed in different ways, providing insight into the nature of growth processes.

Size v. time

The plot of size *v.* time, the direct method of presenting growth data, results in a **growth curve** such as that shown in Figure 19.6. This curve is sometimes referred to as the 'actual' growth curve. The shape of the growth curve of colonies of microorganisms is usually sigmoid (S-shaped). A very similar pattern of growth against time is detected in the growth of individual organs and in the growth of many large, multicellular plants and animals. There are distinct phases to the normal S-shaped growth curve (Figure 19.7).

▶ **The lag phase** Initially growth is slow, with little or no cell division or cell growth occurring. In this period the organisms are adapting to available resources. For example, hydrolytic enzymes specific to the food source are produced in the cytoplasm (enzyme synthesis, pp. 187 and 211).

▶ **The logarithmic phase** Growth proceeds at an exponential rate of increase. This is a time of no apparent constraints on growth; supplies of nutrients are, at the very least, adequate, and waste products have not accumulated.

▶ **The linear growth phase** Growth now proceeds at a steady, relatively constant rate. Growth is limited to some extent by internal or external factors, or by both. This third phase is not always apparent, however. For example, it is virtually absent from the growth curve of the bacteria shown in Figure 19.2, but it can be identified in the human growth curve in Figure 19.5.

▶ **The stationary phase** This is a plateau on the curve. The organism or the population is maintained but growth has ceased.

Growth rate

A **growth rate curve** is obtained by finding the increase in growth between measurements and plotting this increase in growth against time. The result is often a bell-shaped curve (the middle curve in Figure 19.4). It is sometimes referred to as an absolute growth rate curve.

This form of presentation emphasises changes in the rate of growth with time. We see that the rate of growth steadily increases until a maximum rate is reached, and then falls off.

Relative growth rate

The **relative growth rate** gives an alternative insight on growth. To produce this curve the increase in growth between successive readings is expressed as a percentage of the growth that has already occurred, and plotted against the time.

An example of a relative growth rate curve is shown on the right in Figure 19.4.

The rationale for this plot is the observation that growth is an internal process, largely dependent upon the amount of growth that has already been achieved. Growth rate of the maize plant is shown to be a maximum at the beginning, gradually reducing with time.

Figure 19.5   Human growth: the table below shows the mean mass in kilograms of males and females in the UK at yearly intervals, together with the growth rates and percentage growth rates calculated at yearly intervals

Age/years	Boys Mass/kg	Growth rate/kg y^{-1}	Growth/%	Girls Mass/kg	Growth rate/kg y^{-1}	Growth/%
0	3.40		100	3.36		100
1	10.07	6.67	66.2	9.75	6.39	65.5
2	12.56	2.49	19.8	12.29	2.54	20.7
3	14.61	2.05	14.0	14.42	2.13	14.8
4	16.51	1.90	11.5	16.42	2.00	12.2
5	18.89	2.38	12.6	18.58	2.16	11.6
6	21.90	3.01	13.7	21.09	2.51	11.9
7	24.54	2.64	10.8	23.68	2.59	10.9
8	27.26	2.72	10.0	26.35	2.67	10.1
9	29.94	2.68	8.9	28.94	2.59	8.9
10	32.61	2.67	8.2	31.89	2.95	9.2
11	35.20	2.59	7.4	35.74	3.85	10.8
12	38.28	3.08	8.0	39.74	4.00	10.1
13	42.18	3.90	9.2	44.95	5.21	11.6
14	48.81	6.63	13.6	49.17	4.22	8.6
15	54.48	5.67	10.4	51.48	2.31	4.5
16	58.30	3.82	6.5	53.07	1.59	3.0
17	61.78	3.48	5.6	54.02	0.95	1.8
18	63.05	1.27	2.0	54.39	0.37	0.7

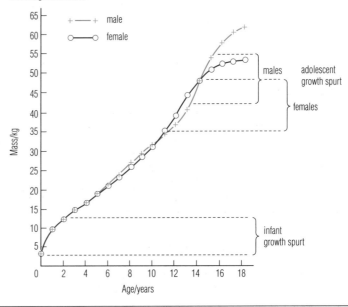

human growth curve

Figure 19.6 Growth of a cucumber fruit

Data from the measurement of five comparable cucumbers (the graph shows mean values and their range)

Day	1	2	3	4	5	6	7	8	9	10	11	12	13	14	15	16
Length/cm	9.7	12.0	14.4	17.1	20.4	24.0	26.8	29.1	31.0	32.8	34.5	36.1	37.6	38.2	38.6	38.6
Gain/cm		2.3	2.4	2.7	3.3	3.6	2.8	2.3	1.9	1.8	1.7	1.6	1.5	0.6	0.4	0.0

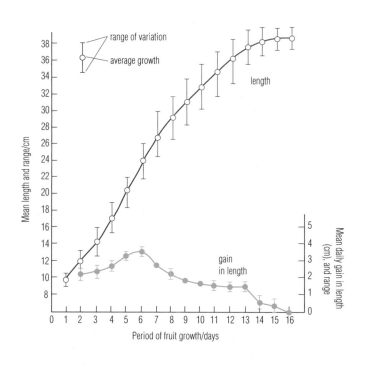

cucumber fruits at various stages in growth

Figure 19.7 The sigmoid curve of growth

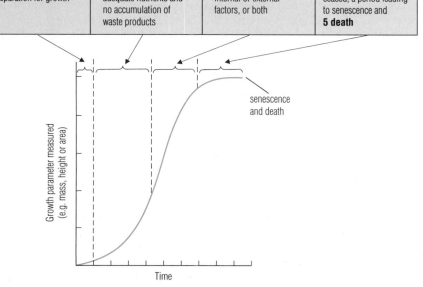

1 lag phase
little or no cell multiplication or growth; a period of adaptation or preparation for growth

2 log phase
exponential growth; a period of no constraint on growth, with adequate nutrients and no accumulation of waste products

3 linear growth phase
a decelerating phase; a period when the growth rate becomes limited by internal or external factors, or both

4 stationary phase
the organism (or population) is maintained but new growth has ceased; a period leading to senescence and **5 death**

senescence and death

Questions

1 A casual glance at the relative growth rate curve in Figure 19.4 might suggest that the organism is shrinking rather than growing. What is the correct interpretation?

2 In what ways may the increase in amount of protein with time be a better indication of the rate of growth than, for example, an increase in mass with time?

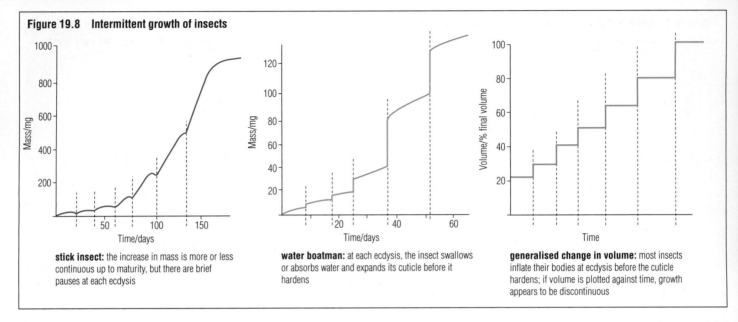

Figure 19.8   Intermittent growth of insects

stick insect: the increase in mass is more or less continuous up to maturity, but there are brief pauses at each ecdysis

water boatman: at each ecdysis, the insect swallows or absorbs water and expands its cuticle before it hardens

generalised change in volume: most insects inflate their bodies at ecdysis before the cuticle hardens; if volume is plotted against time, growth appears to be discontinuous

19.3 DIFFERENT GROWTH CURVES

The growth of very many organisms follows the sigmoid curve of growth shown in Figure 19.7. In some organisms, however, the growth curves are significantly different. Two examples are discussed here.

■ Human growth

In the growth of a human after birth there are two phases of rapid growth, one in early infancy and the other in adolescence. Between these there is a period of steady growth, and so the human growth curve appears as two sigmoid curves joined (Figure 19.5, p. 400).

■ 'Intermittent' growth of insects

In members of the phylum Arthropoda, which includes the insects, a hard, inflexible external skeleton interferes with a smooth growth rate. As a consequence, arthropods must moult periodically during growth, throwing off the old cuticle and growing a new, larger one.

Moulting is known as **ecdysis** (p. 476). The old cuticle softens, and its recoverable resources are dissolved and absorbed into the new exoskeleton, which is forming below. The old cuticle then splits, and is discarded. At the moment of moulting the insect enlarges its surface area by swallowing and retaining air or liquid until the new cuticle has hardened, so that there is room for growth within the new cuticle. Certain blood-sucking insects enlarge the body by ingesting a huge meal of blood after moulting. The meal is subsequently digested, the nutrients incorporated into body tissue, and the excess fluid and other water excreted.

From the plots in Figure 19.8, it appears that the external skeleton interferes with the rate of growth. But dry mass measurements of insect growth, made on large samples of individuals drawn from a uniform population, show that true growth is not profoundly interrupted by ecdysis. These studies provide a normal sigmoid growth curve for insects.

19.4 PATTERNS OF GROWTH

■ Allometric v. isometric growth

The steady increase in mass or height of an organism may give a misleading impression of the development of particular organs and tissues. Some organs grow at rates that differ from the growth rate of the whole organism, a phenomenon known as **allometric growth.** Many organisms show allometric growth to some extent.

An example of allometric growth is shown in Figure 19.9. Note, for example, the rapid growth during early childhood of the lymph tissue followed by partial loss during adolescence. Lymph tissue provides the young organism with natural immunity and the necessary defence mechanisms against diseases. By contrast, the last organ system to be developed is the reproductive system. The human head and the brain similarly develop comparatively rapidly early in life, whereas the full development of the legs is delayed. All primate infants remain for a long time in the care of their parents, protected and fed. Parental training and 'education' is a feature of the young primate's early life.

In **isometric growth**, on the other hand, an organ grows at the same rate as the rest of the organism. During isometric growth proportions of the parts remain the same. This type of growth is shown in insects such as the locust (apart from its wings) illustrated in Figure 19.10.

■ Limited v. unlimited growth

Limited growth is illustrated by the pattern of growth of an annual plant. Organisms with limited growth grow up to a predetermined size within the normal range for the organism; growth is then complete. The growth of the organs of an annual plant, such as its fruits and seeds, also show a pattern of limited growth (Figure 19.6, p. 401). Once the fruit has reached full size it does not grow further, but plays a specific role in seed dispersal.

Unlimited growth is illustrated by a colony of sedentary animals such as *Obelia*, by corals and sponges (p. 34) and by perennial woody plants (trees, Figure 19.11), which add to their mass, height or length during each growing season. Growth continues more or less indefinitely, until the size of the organism causes its demise, if an earlier attack by parasite or predator has not destroyed its structures.

Figure 19.9   Allometric growth of human organs and tissues: some parts grow at different rates to the whole organism

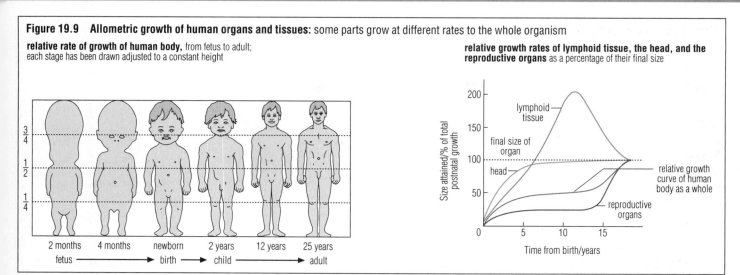

relative rate of growth of human body, from fetus to adult; each stage has been drawn adjusted to a constant height

2 months   4 months   newborn   2 years   12 years   25 years

fetus ——————→ birth —→ child ——————→ adult

relative growth rates of lymphoid tissue, the head, and the reproductive organs as a percentage of their final size

Size attained/% of total postnatal growth

200
150
final size of organ — 100
50
0

lymphoid tissue
head
relative growth curve of human body as a whole
reproductive organs

0   5   10   15
Time from birth/years

Figure 19.10   Isometric growth of a locust: the growth of parts occurs at the same rate as the whole organism

locust from first nymphal stage to adult

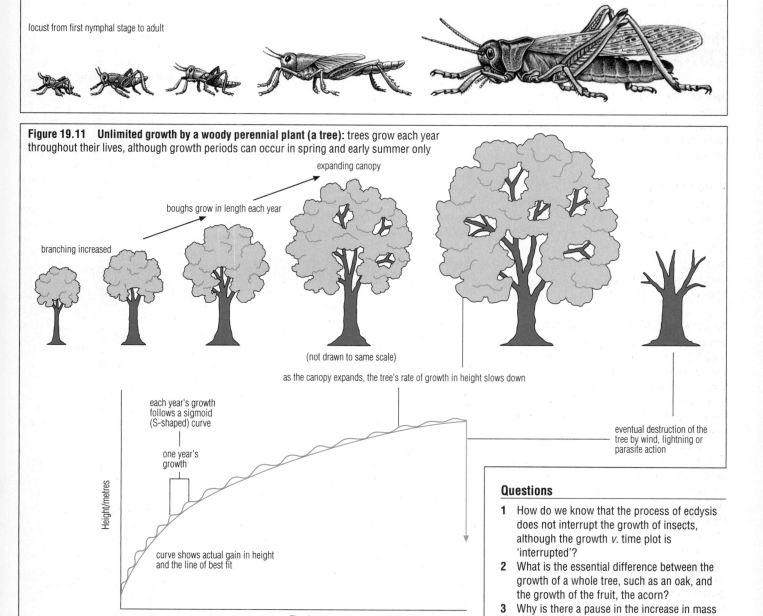

Figure 19.11   Unlimited growth by a woody perennial plant (a tree): trees grow each year throughout their lives, although growth periods can occur in spring and early summer only

expanding canopy

boughs grow in length each year

branching increased

(not drawn to same scale)

as the canopy expands, the tree's rate of growth in height slows down

each year's growth follows a sigmoid (S-shaped) curve

one year's growth

Height/metres

curve shows actual gain in height and the line of best fit

Time/years

eventual destruction of the tree by wind, lightning or parasite action

Questions

1   How do we know that the process of ecdysis does not interrupt the growth of insects, although the growth *v.* time plot is 'interrupted'?

2   What is the essential difference between the growth of a whole tree, such as an oak, and the growth of the fruit, the acorn?

3   Why is there a pause in the increase in mass of the stick insect at each ecdysis?

19.5 GROWTH AND DEVELOPMENT IN THE FLOWERING PLANT

■ The seed

The seed develops from the fertilised ovule and contains an embryonic plant and a food store (Figure 19.12). The embryo consists of an embryonic shoot (**plumule**), root (**radicle**) and one or two seed-leaves (**cotyledons**). The base of the plumule, situated just above the attachment of the cotyledons, is called the **epicotyl**. The base of the radicle, just below the attachment of the cotyledons, is called the **hypocotyl**. The food store is contained either in tissues surrounding the embryo, known as the **endosperm** (endospermic seeds, such as maize), or within the cotyledons (non-endospermic seeds, such as the broad bean and sunflower). The whole seed is enclosed and protected by a tough seed coat, the **testa**, derived from the wall of the ovule.

Germination: the necessary conditions

The essential environmental factors for germination are an adequate supply of water, a suitable temperature and an appropriate partial pressure of oxygen. (Certain species also require either the presence or the absence of light, p. 439.) The essential internal conditions required for germination are the maturity of the embryo and the overcoming of dormancy (p. 438).

Water uptake

Germination can commence after the uptake of water by the seed. Water is absorbed through the micropyle and the testa. Water absorption is **imbibition**, the process of uptake of water by the dry colloidal substances of the seed, which include lipids and the dry cell wall substances of the dehydrated seed tissue (*Study Guide,* Chapter 19). The resultant swelling of the seed may rupture the testa. Subsequent movement of water in the hydrated seed tissue is by osmosis.

Water is essential to the vacuolation of the growing cells and for the activation of the enzymes that catalyse the biochemical reactions of germination. Turgor of plant cells is a force for cell expansion when the walls are in a condition permitting stretching. Water is a reagent in the hydrolysis of stored food substances, and is required for the translocation of hydrolysed food reserves – the sugars, amino acids and fatty acids – to the sites of growth in the embryo.

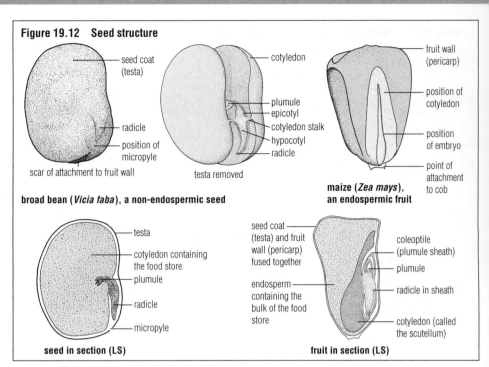

Figure 19.12   Seed structure

broad bean (*Vicia faba*), a non-endospermic seed

maize (*Zea mays*), an endospermic fruit

seed in section (LS)

fruit in section (LS)

Biodeterioration of seeds in storage

If moist seeds are stored in bulk at room temperature they may start to germinate, and can rapidly heat up due to the waste heat produced as food reserves are respired. The heat accumulates in the bulk of the seeds, and as the temperature rises the proteins are denatured. As a result, germination slows down again. The seeds are worthless, since their food store is lost and their cell proteins destroyed.

Successful, safe seed storage is an important skill for humans the world over. Damage to stored seeds is a major problem, for example, in third world countries today. Dried seeds are an extremely nutritious and compact form of food, very attractive to animals, fungi and bacteria as well as to humans.

Temperature

The optimum temperature for germination is the optimum for the enzymes involved in mobilisation of food reserves, provided that other factors are non-limiting. This temperature varies from species to species; wheat seeds germinate in the temperature range 1–35 °C, and maize in the range 5–45 °C.

Oxygen

Respiration makes available the energy for metabolism and growth (p. 302). Germinating seeds respire very rapidly, and require oxygen for aerobic respiration. Seeds will not germinate in the total absence of oxygen. Diffusion of oxygen through the testa may be slow, however, and in the early stages of germination seeds may rely on some anaerobic respiration, at least until the testa has ruptured.

Mobilisation of stored food

The stored foods of seeds consist of carbohydrates, lipids and proteins (Table 19.1). Starch forms the major food reserve of most seeds, but in the sunflower and some other seeds oil makes up about half of the food stored. In pea and bean seeds protein is an additional important reserve food.

Food is stored in seeds in an insoluble form, and must be hydrolysed to soluble substances early in germination, following the hydration of the seed. Stored food is hydrolysed to produce the substrates for respiration (sugars) and the building blocks for synthesis (substances such as sugars, amino acids and fatty acids).

Table 19.1   Major storage compounds in seeds /% dry mass

Species	Starch	Protein	Lipid
barley (*Hordeum vulgare*)	76	12	3
maize (*Zea mays*)	75	12	9
wheat (*Triticum aestivum*)	75	12	2
pea (*Pisum sativum*)	56	24	6
soybean (*Glycine max*)	26	37	17
peanut (*Arachis hypogea*)	12	31	48
rape (*Brassica napus*)	19	21	48

Figure 19.13   The mobilisation of food stores in seeds

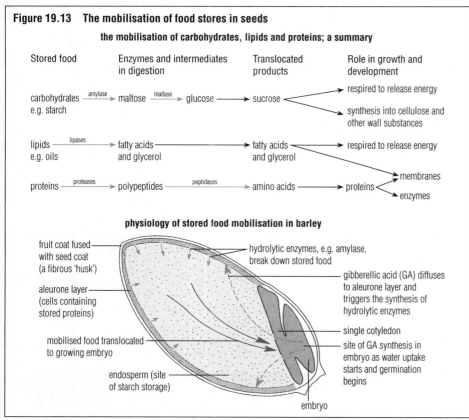

the mobilisation of carbohydrates, lipids and proteins; a summary

physiology of stored food mobilisation in barley

Figure 19.14   Epigeal and hypogeal germination

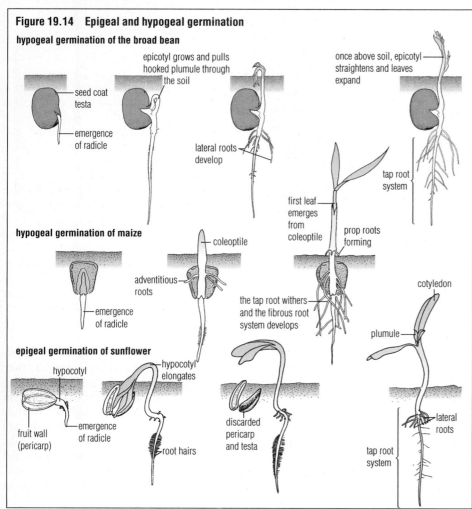

hypogeal germination of the broad bean

hypogeal germination of maize

epigeal germination of sunflower

Some hydrolytic enzymes already exist in the seed and await only water uptake and an appropriate temperature to become activated. Other enzymes are produced in response to hormones released from the developing embryo. Release of hydrolytic enzymes is triggered by gibberellic acid (GA, p. 429), which is produced in the embryo and diffuses to the food store. GA works directly on the nuclei of the food storage cells, activating the genes that code for hydrolytic enzymes (Figure 19.13). Since GA is formed by the embryo only after sufficient water has been absorbed, mobilisation of food reserves is linked to germination (Figure 19.14).

Epigeal and hypogeal germination

After the seed coat has ruptured, the first organ to emerge is normally the radicle, followed somewhat later by the plumule. Germination is classified according to the position of the cotyledons in germination (Figure 19.14).

In **epigeal germination** the cotyledons are carried up and out of the soil by the growth of the hypocotyl. The cotyledons appear as simple green leaves. During growth through the soil the hypocotyl remains hooked and the plumule tip lies between the cotyledons, protected from abrasion damage by the soil.

In **hypogeal germination** of dicotyledons the cotyledons remain below ground. The plumule is carried aloft by the growth of the epicotyl. The hooked shape taken on by the plumule as it is drawn up through the soil protects the tip from damage by soil abrasion.

In monocotyledons such as maize, the plumule grows through the soil, with the first leaves protected by the coleoptile (p. 422).

Questions

1   a   In the germination of seeds, why are oxygen and water required?

    b   Under anaerobic conditions, what end-products of respiration would you expect to accumulate? Why are they particularly injurious if not rapidly removed?

2   a   Why does the embryonic seed require a food store?

    b   What is the difference between an endospermic and non-endospermic seed?

3   a   What is the value to the seedling of establishing a root system as the first step in growth of the new plant?

    b   What is the essential difference between epigeal and hypogeal germination?

■ Meristems and plant growth

We have seen that, initially, many cells of the developing embryo are capable of dividing repeatedly. Cells in this condition are described as meristematic. A **meristem** is a group of plant cells that retain the ability to divide by mitosis. Once the very young embryo stage is past, growth in multicellular plants is confined to the meristems. Subsequent growth of the plant occurs by cell division at the meristems. There are three types of meristem, which are illustrated in Figure 19.15.

▶ **Apical meristems** These occur at the tips of the stem and root and are responsible for the **primary growth** of the plant. Growth at the apical meristems leads to increase in length of stem and root.

▶ **Lateral meristems** These occur as cylinders towards the outer part of the stem, and are responsible for the **secondary growth** of the plant, resulting in thickening.

▶ **Intercalary meristems** These occur at the nodes of grasses. (The **node** is the stem at the point of attachment of a leaf.) Intercalary meristems allow an increase in length in positions other than the tip; grass stems continue to grow even when the apical meristem has been cut off. The apical meristem of grasses produces the inflorescence, normally at a later stage in the growing season.

Primary growth of the plant

The seedling undergoes primary growth to form the herbaceous (non-woody) green plant. Primary growth is initiated by divisions of meristematic cells at the apex of stem and root, the apical meristems. These cells, at and just behind the extreme tip of a growing stem, make up the zone of cell division. The structure of the meristem of a stem is shown in Figure 10.18 (p. 222).

The primary growth of the stem

Cell division

The apical meristem of the stem is composed of an outer 'coat' or **tunica** of two or three rows of regularly arranged meristematic cells, surrounding a central mass of irregularly arranged meristematic cells, the **corpus**. During mitosis in cells of the tunica, the plane of division is at right angles to the surface. Cells of the corpus, on the other hand, divide haphazardly in any plane.

Figure 19.15   Meristems: a summary

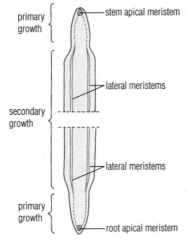

apical meristems occur at the tips of stems and roots; they contribute to primary growth of the stem and root (increase in length of the plant)

lateral meristems occur away from the tips of stems and roots near to the periphery of the stem; they contribute to secondary growth (an increase in girth)

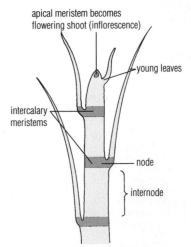

intercalary meristems are parts of the apical meristem that have become separated from the apex during growth; examples are those in the nodes in the stems of grasses

intercalary meristems allow grasses to continue growth despite grazing or cutting

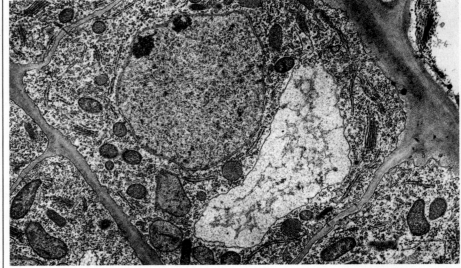

TEM of a cell detached from a meristem: enlargement has begun with the formation of a permanent vacuole (seen here in the cytoplasm, next to the nucleus) (×3000)

Cell enlargement

As new cells are formed by divisions at the meristems the outer, older cells are left behind. These cells rapidly increase in size, and form the zone of cell expansion in the growing stem. Cells expand by osmotic uptake of water, leading to the formation of vacuoles in the cytoplasm, and the stretching and growth of the primary cell wall (Figure 19.16). The final shape of the mature plant cell – whether, for example, it is approximately spherical or extremely thin and elongated – is largely determined by the directions in which the cellulose microfibrils were laid down.

Cell differentiation

Further down the stem, well back from the apical meristem, the cells become specialised. This is known as the zone of differentiation.

Certain cells left behind by the meristem do not vacuolate. Instead they remain meristematic and form long strands, running parallel to the stem. These are the procambial strands (p. 222). Procambial strand cells are longer than other meristematic cells. They eventually differentiate to form the vascular tissue. The stages of the formation of a vascular bundle are shown in Figure 19.17.

Figure 19.16   Expansion of a growing plant cell

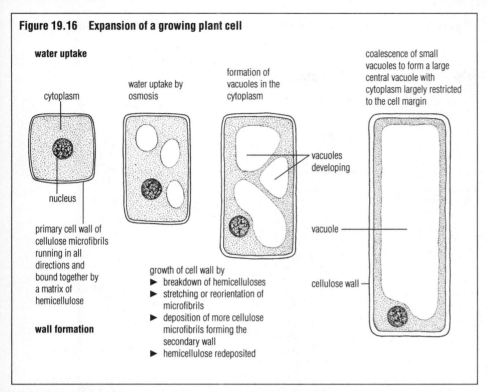

water uptake

cytoplasm

nucleus

primary cell wall of cellulose microfibrils running in all directions and bound together by a matrix of hemicellulose

wall formation

water uptake by osmosis

formation of vacuoles in the cytoplasm

vacuoles developing

growth of cell wall by
► breakdown of hemicelluloses
► stretching or reorientation of microfibrils
► deposition of more cellulose microfibrils forming the secondary wall
► hemicellulose redeposited

coalescence of small vacuoles to form a large central vacuole with cytoplasm largely restricted to the cell margin

vacuole

cellulose wall

Figure 19.17   The procambial strand and vascular tissue

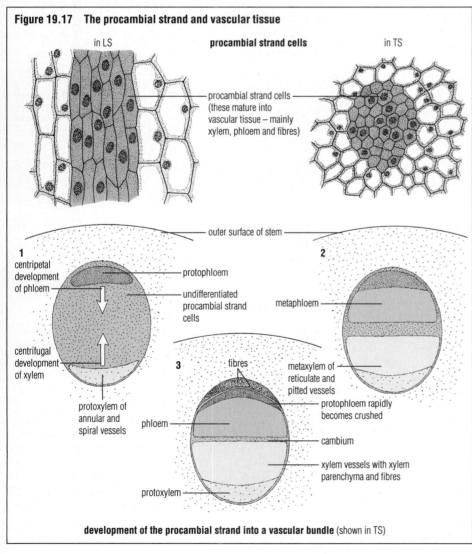

in LS   **procambial strand cells**   in TS

procambial strand cells (these mature into vascular tissue – mainly xylem, phloem and fibres)

outer surface of stem

1
centripetal development of phloem

protophloem

undifferentiated procambial strand cells

centrifugal development of xylem

protoxylem of annular and spiral vessels

2

metaphloem

3   fibres

metaxylem of reticulate and pitted vessels

protophloem rapidly becomes crushed

phloem

cambium

xylem vessels with xylem parenchyma and fibres

protoxylem

development of the procambial strand into a vascular bundle (shown in TS)

Formation of vascular tissue

The innermost cells of the procambial strands of the stem form the protoxylem, consisting of annular and spiral xylem vessels (p. 225). Cells at the outer side of the procambial strands form the protophloem. Phloem tissue consists of sieve tube elements and companion cells (p. 227).

Subsequently, metaxylem and metaphloem tissues are formed. The first-formed metaxylem vessels appear next to the protoxylem, and the remainder develop across the procambial strand towards the outside of the stem (centrifugally). Metaxylem consists of reticulate and pitted xylem vessels (p. 225). Cells of the procambial strand may also differentiate into fibres, a form of sclerenchyma (p. 224).

The first-formed metaphloem tissue appears next to the protophloem, and the remainder develops across the outer part of the procambial strand, towards the inside of the stem (centripetally). The formation and enlargement of the more robust metaxylem and metaphloem tends to crush the first-formed vascular tissue, particularly the protophloem.

The differentiation of phloem involves the formation of phloem sieve tubes and accompanying companion cells (p. 227). These tissues are living cells when mature, with little additional thickening and no lignification in their walls.

The other cells cut off from the corpus give rise to the cells of the ground tissues of the stem, largely parenchyma (p. 223) and collenchyma (p. 224).

Questions

1   State where in a flowering plant you would find
   a   a lateral meristem
   b   an intercalary meristem.
2   The apical meristem occurs in a zone of cell division. What is the next zone in a growing stem or root?
3   What are the significant differences in structure between protoxylem and metaxylem (p. 225)?
4   What are the roles of parenchyma and collenchyma in the herbaceous stem (pp. 223–4)?
5   Describe the changes that take place in a cell that eventually differentiates into
   a   a sieve tube element (p. 227)
   b   a vessel element (p. 225).

The primary growth of the root

The apical meristem of the root is a mass of irregularly arranged cells, which divide and cut off new cells in all directions. The cells cut off ahead of this meristem form the **root cap**. The root cap is a mass of undifferentiated cells. These cells are treated as expendable by the plant; they become worn away and are sloughed off as the root grows and pushes through the soil.

The bulk of the cells cut off by the root apical meristem gives rise to the central procambial strand, and to the tissues of the cortex. Protoxylem and protophloem develop from the procambial strand cells followed by metaxylem and metaphloem. The positions of xylem and phloem development in the root contrast with those in the stem (Figure 19.18).

The central vascular tissue of the root is referred to as the **stele**. The stele is surrounded by a single layer of cells, the **endodermis**, and immediately within this layer is a layer of cells, the **pericycle** (Figure 19.19).

Origin of leaves and lateral buds

The leaves and lateral buds in the stem originate in the apical meristem. The tunica of the stem apical meristem gives rise to tiny swellings of cells, known as the leaf **primordia**, at the surface of the stem tip. The leaf primordia subsequently grow and differentiate into leaves and lateral buds. The point of attachment of the leaf and lateral bud on the stem is the **node**. The stem between the nodes is referred to as the **internode**.

Cambium

After primary growth is completed and the mature primary tissues of stem and root are formed, some procambial strand cells remain in the vascular bundles of the stem and in the stele of the root. These meristematic cells lie between the metaxylem and metaphloem, and are known as **cambium**. Cambium cells are capable of further growth, leading to secondary growth of roots and stems (p. 410).

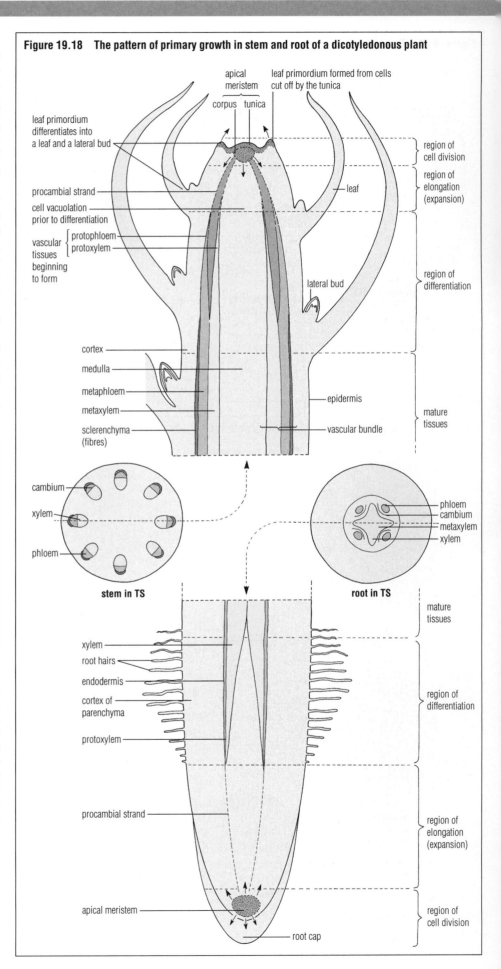

Figure 19.18   The pattern of primary growth in stem and root of a dicotyledonous plant

Figure 19.19 Photomicrographs of mature primary growth of the stem and root of *Ranunculus* sp. (buttercup)
the stem

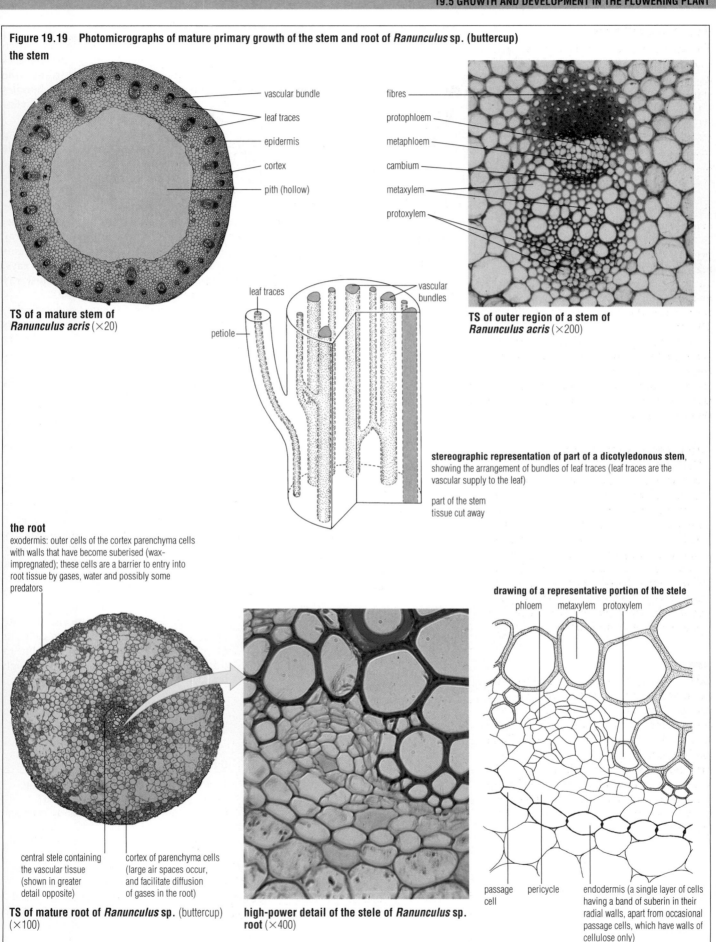

TS of a mature stem of
Ranunculus acris (×20)

vascular bundle
leaf traces
epidermis
cortex
pith (hollow)

fibres
protophloem
metaphloem
cambium
metaxylem
protoxylem

TS of outer region of a stem of
Ranunculus acris (×200)

leaf traces
petiole
vascular bundles

stereographic representation of part of a dicotyledonous stem,
showing the arrangement of bundles of leaf traces (leaf traces are the
vascular supply to the leaf)

part of the stem
tissue cut away

the root
exodermis: outer cells of the cortex parenchyma cells
with walls that have become suberised (wax-
impregnated); these cells are a barrier to entry into
root tissue by gases, water and possibly some
predators

central stele containing
the vascular tissue
(shown in greater
detail opposite)

cortex of parenchyma cells
(large air spaces occur,
and facilitate diffusion
of gases in the root)

TS of mature root of *Ranunculus* sp. (buttercup)
(×100)

high-power detail of the stele of *Ranunculus* sp.
root (×400)

drawing of a representative portion of the stele
phloem   metaxylem   protoxylem

passage
cell
pericycle
endodermis (a single layer of cells
having a band of suberin in their
radial walls, apart from occasional
passage cells, which have walls of
cellulose only)

The origin of lateral roots

Lateral roots originate at a considerable distance from the apical meristem, in contrast to stem structures such as leaves and lateral buds. Lateral roots arise from the pericycle layer within the central stele opposite a protoxylem group. These new cells add to the volume of the root, causing a bump on its surface early in their development. The new lateral root grows out through the cortex of the root before reaching the soil (Figure 19.20).

Secondary growth of plants

Secondary growth takes place in woody perennial plants (trees and shrubs), and normally results in the formation of a large amount of secondary xylem called **wood** and of an external cork layer known as **bark**. The primary structure of the plant is profoundly modified by secondary growth.

Secondary growth is the result of cell division of lateral meristems, and ultimately leads to an increase in girth of the plant. Two lateral meristems are involved: a **vascular cambium** ring, and the **cork cambium**.

The ring of vascular cambium

Secondary growth commences when parenchyma cells between the bundles become meristematic. As a result, the cambium in the vascular bundles becomes connected into a complete cylinder around the stem (Figure 19.21).

Then the cells of the cambium ring divide, cutting off new cells to the outside and the inside of the ring. Many more cells are cut off to the inside than towards the outside. The inner cells differentiate into **secondary xylem** tissue and the outer cells into **secondary phloem** tissue. Smaller cells in the cambium ring are **ray initials**, and cut off cells that form chains of living cells, known as **medullary rays**, connecting cortex with pith. Most ray cells are similar to parenchyma.

The secondary xylem develops relatively thick, lignified walls, and persists from year to year. The phloem cells are much less rigid, and under the pressure generated in the stem from secondary growth, the older phloem becomes crushed (Figure 19.22).

The cork cambium

A second lateral meristem, the cork cambium or **phellogen**, is formed by tangential division of parenchyma cells more or less immediately below the epidermis. The formation of a cork cambium is essential in a stem undergoing secondary thickening; strains at the circumference of the stem, resulting from secondary thickening in the interior, ultimately cause the epidermis to split. The cork cambium, however, replaces the epidermis, forming a protective, impervious bark at the surface of the woody stem.

The bark contains **lenticels**, discrete groups of powdery parenchyma cells cut off by the phellogen. Gases are able to diffuse into and out of the stem through lenticels (p. 303).

Figure 19.20   The origin of lateral roots

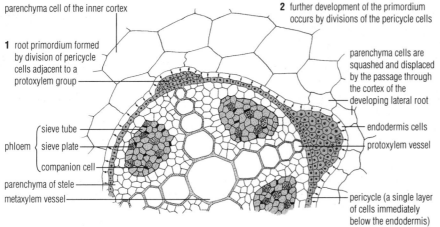

parenchyma cell of the inner cortex

2 further development of the primordium occurs by divisions of the pericycle cells

1 root primordium formed by division of pericycle cells adjacent to a protoxylem group

parenchyma cells are squashed and displaced by the passage through the cortex of the developing lateral root

endodermis cells

phloem { sieve tube / sieve plate / companion cell }

protoxylem vessel

parenchyma of stele

metaxylem vessel

pericycle (a single layer of cells immediately below the endodermis)

early stages in development of lateral roots in the pericycle

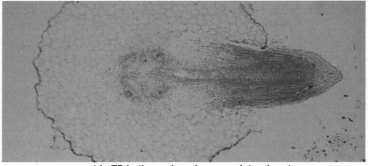

root in TS in the region of a young lateral root

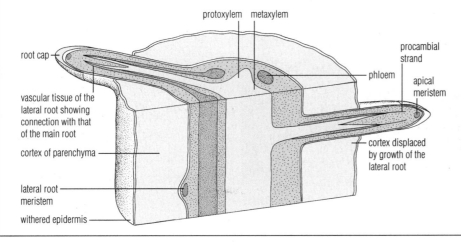

protoxylem   metaxylem

root cap

procambial strand

phloem

apical meristem

vascular tissue of the lateral root showing connection with that of the main root

cortex of parenchyma

cortex displaced by growth of the lateral root

lateral root meristem

withered epidermis

Questions

1 What is the difference between a vascular bundle and a leaf trace (Figure 19.19)?

2 How does the structure of the apical meristem of a root differ from that of the stem?

3 **a** In addition to transport of water, ions and elaborated food, what is the other major role of the vascular bundles of the stem (p. 500)?

   b How do the structure and composition of the vascular bundle assist in this function?

4 In what ways does the origin of a lateral root differ from the origin of a lateral stem?

Figure 19.21 The ring of vascular cambium

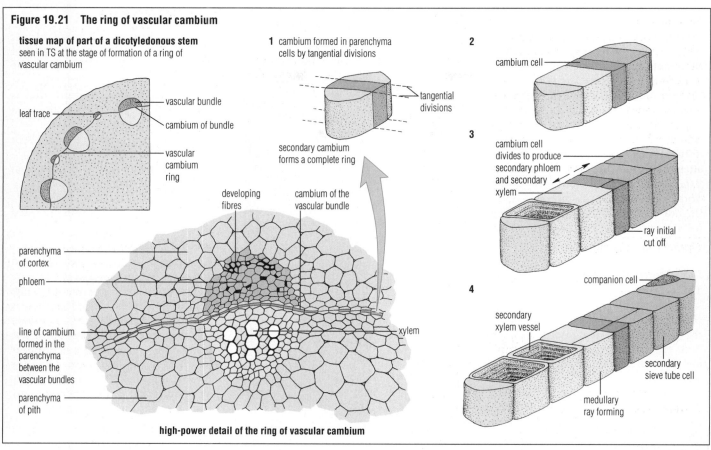

tissue map of part of a dicotyledonous stem
seen in TS at the stage of formation of a ring of vascular cambium

leaf trace
vascular bundle
cambium of bundle
vascular cambium ring

parenchyma of cortex
phloem

developing fibres
cambium of the vascular bundle

line of cambium formed in the parenchyma between the vascular bundles
parenchyma of pith

xylem

high-power detail of the ring of vascular cambium

1 cambium formed in parenchyma cells by tangential divisions

tangential divisions

secondary cambium forms a complete ring

2 cambium cell

3 cambium cell divides to produce secondary phloem and secondary xylem

ray initial cut off

4 companion cell

secondary xylem vessel

secondary sieve tube cell

medullary ray forming

Figure 19.22 Stages in the first year of secondary growth in a dicotyledonous stem and root

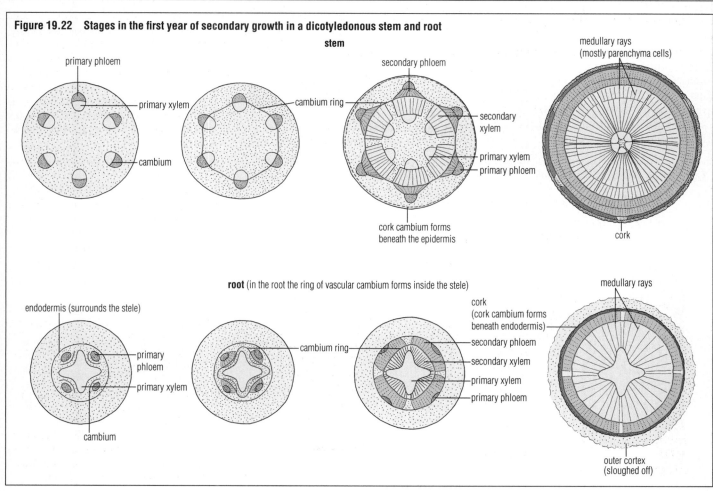

stem

primary phloem
primary xylem
cambium

cambium ring

secondary phloem
secondary xylem
primary xylem
primary phloem

cork cambium forms beneath the epidermis

medullary rays (mostly parenchyma cells)

cork

root (in the root the ring of vascular cambium forms inside the stele)

endodermis (surrounds the stele)
primary phloem
primary xylem
cambium

cambium ring

cork (cork cambium forms beneath endodermis)
secondary phloem
secondary xylem
primary xylem
primary phloem

medullary rays

outer cortex (sloughed off)

Annual rings

In trees of northern temperate climates at least, activity of the vascular cambium (and of the apical meristem) is normally restricted to spring and early summer each year. Spring-formed vessels are larger and thinner-walled than the later-formed vessels, which are also surrounded by more fibres. As a consequence there are alternate bands of 'early' wood and 'late' wood, forming concentric rings. These are known as **annual rings** (Figure 19.24). When a tree is felled the age can normally be determined by counting the annual rings exposed on the cut surface.

Dendrochronology

The ring of new growth that a tree lays down each year is wide in years when the conditions are favourable to the growth of the tree and narrow in poor years. Consequently, the prevailing climate imposes a pattern in the annual growth rings.

This study of the pattern, known as **dendrochronology**, provides a climate record that can be traced back through archaeological time. Sufficient oak timbers from trees of different ages whose growth periods have overlapped have been found in ancient buildings or buried in archaeological sites, bogs or river gravel, and analysed to provide a developing master chronological record. This record is a means of dating other samples of oak timber of uncertain history.

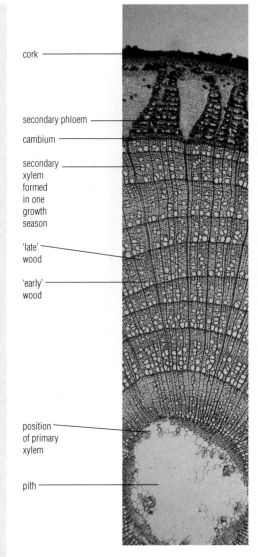

cork

secondary phloem

cambium

secondary xylem formed in one growth season

'late' wood

'early' wood

position of primary xylem

pith

Figure 19.24 Annual growth rings: TS of part of the stem of a four-year-old lime tree (*Tilia vulgaris*) (×10)

Features of timbers

Obvious features of a piece of cut wood are its annual growth rings, and lateral branches absorbed into the wood by subsequent growth, the 'knots'. In polished furniture that is produced from selected woods, such as oak, the presence of large and prominent medullary rays often appears as a feature.

The properties of timbers obtained from different species of tree are of importance to foresters and the timber trade. Some of these features are listed in Figure 19.23.

Anatomy of monocotyledons

The grasses, including the cultivated grasses such as maize and wheat, are examples of monocotyledonous plants (p. 33). In the monocotyledons there is a different arrangement of tissues from that found in dicotyledons. For example, the vascular bundles appear scattered across the stem (see, for example, Figure 10.19, p. 223). Very few monocotyledons undergo any form of secondary thickening.

Questions

1 What are the key differences between primary growth and secondary growth?
2 In secondary thickening, what is
   a an annual growth ring?
   b cork tissue?
3 Why can we say, in theory at least, that a growing tree trunk of many years standing is really an elongated cone shape, rather than a cylinder?

Figure 19.23 Features of cut timber

diagram of a tree cut in transverse section (TS), radial longitudinal section (RLS) and tangential longitudinal section (TLS)

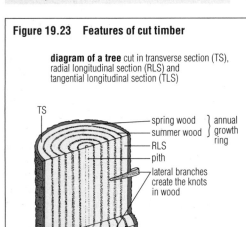

TS

spring wood
summer wood } annual growth ring
RLS
pith
lateral branches create the knots in wood

TLS

bark

Commercial timbers

Timber	Special features	Common uses
Hardwoods: (little furniture is now made from solid hardwood; for economy, a veneer of hardwood is used)		
oak	strong and hard, resists nails, has to be secured at joints with pegs	used in buildings, as beams, durable fences, high-quality furniture
ash	toughest native wood	used for hammer, axe and spade handles
beech	hard strong wood that can be worked in any direction	used for handles, legs of chairs, hand tools
walnut	attractively figured wood, mechanically strong	used in furniture, carved and turned bowls, gunstocks
holly	exceptionally hard and heavy wood, always very low in water content	used for carving, walking sticks; burned as firewood, with a high heating power
plane	grows well in the polluted air of cities, even where root space is limited	more valuable as a street plant than as timber
Softwoods:		
Scots pine	relatively fast-growing	used for telegraph poles, railway sleepers, joists and rafters in buildings
Sitka spruce		manufacture of chipboard, insulation boards and hardboard, and in papermaking

19.6 GROWTH AND DEVELOPMENT IN MULTICELLULAR ANIMALS

In sexual reproduction (p. 582), a new organism originates from the zygote formed by fusion of male and female gametes. Many multicellular animals start development from a fertilised egg in this way. The subsequent growth and development of the zygote produces a new multicellular body. In the early stages of development an embryo is formed. We have already seen, from the flowering plant (p. 573), that development at the embryo stage involves cell division, maturation and specialisation of cells. In animals, embryogenesis also involves movement and repositioning of certain groups of cells.

Most of the early stages in the development of a fertilised egg to a multicellular embryo are common to all animals, whether vertebrate or non-vertebrate. Embryonic development is a continuous process, but is conveniently divided into three major stages. These are

▶ **cleavage**, the division of the zygote into a mass of daughter cells

▶ **gastrulation**, the rearrangement of cells into distinct layers; the embryo at this stage is called a **gastrula**

▶ **organogeny**, the process of tissue and organ formation.

Whilst the early stages of embryogenesis show striking uniformity, at least up to and including the gastrula stage, the differences that do appear are due to the presence of differing quantities of the embryo's food reserves, known as **yolk**, in the eggs of different species. Typically, the eggs of birds and reptiles have the largest quantities of yolk and their early development diverges most strikingly from those with little yolk. Of course, by the later stages of an animal's development many variations can be seen, typical of the particular species.

■ Juvenile forms, and metamorphosis

Development involves a progressive increase in complexity, as the fertilised egg develops into the multicellular adult. In the process of development in some organisms an intermediate form, often referred to as a larva, is formed. Larvae are juvenile forms that follow on from the egg stage in the life cycles. The larvae later become adults as a result of subsequent development.

In most non-vertebrate phyla the larva is totally unlike its adult. For example, barnacles have a nauplius larva, molluscs have a trochophore or veliger larva, and annelids have a trochophore larva (see Figure 19.29, p. 415). Even in the crustaceans and holometabolous insects (see below), the larvae are quite different from the adults. In these cases, when the time comes for the larva to change into adult the process involves something of an upheaval, known as **metamorphosis**. The hormonal control of metamorphosis in insects is discussed on p. 476.

The larval stage of some non-vertebrate species is very similar to the adult form, however, at least in superficial appearance. The larvae of grasshoppers and locusts are good examples of this.

The existence of larval forms and of metamorphosis is not restricted to non-vertebrate animals. The development of the aquatic tadpole into a terrestrial frog is an example from among the vertebrates. But relatively speaking, there are vastly more non-vertebrate than vertebrate species with larvae. The importance of larval stages in the life cycles of non-vertebrates is discussed in section 19.8, p. 415. We shall use the development of the frog to illustrate some principles of growth and development in vertebrates.

19.7 GROWTH AND DEVELOPMENT IN VERTEBRATES

The main sequence of events in development of a multicellular embryo in amphibians, up to the point where formation of the major organs is under way, is illustrated in Figures 19.25 and 19.26. Development of amphibians is not typical of the development of other groups of vertebrates, but all vertebrates go through similar stages. The further development of the mammalian (human) embryo, and the formation of its embryonic membranes, are discussed in Chapter 27. The way in which cells of the vertebrate embryo differentiate into the four basic types of tissue, namely epithelia, connective tissues, nerves and muscles, is described in Chapter 10.

■ Fertilisation and cleavage

Development of the amphibian embryo is normally initiated by penetration of the egg membrane by the head of the sperm. This event triggers the second meiotic division (p. 591), and a haploid female nucleus is formed. The female nucleus then fuses with the male nucleus from the sperm. At this time the cytoplasm of the amphibian egg cell is often visibly asymmetric or 'polarised', because most stored food (yolk) is retained in the cytoplasm of one half of the cell, referred to as the **vegetal pole**. The other half is called the **animal pole**.

Fusion of sperm and egg nuclei triggers repeated mitotic cell divisions. Thousands of cells, called **blastomeres**, are produced and form a spherical mass, the **morula**. The embryo does not increase in mass at this stage; the cells become smaller and smaller with each division. The cells then form into a hollow sphere, the **blastula**. The fluid-filled space in the blastula is called the **blastocoel**.

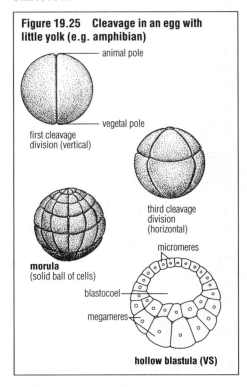

Figure 19.25 Cleavage in an egg with little yolk (e.g. amphibian)

- animal pole
- vegetal pole
- first cleavage division (vertical)
- third cleavage division (horizontal)
- **morula** (solid ball of cells)
- micromeres
- blastocoel
- megameres
- **hollow blastula (VS)**

The divisions of cleavage are not in the least haphazard. In the amphibian embryo two vertical divisions at right angles to each other precede one horizontal division, followed by alternate vertical and horizontal divisions. (In other vertebrate classes the extent of the divisions and the speed of cleavage depends on how much yolk is stored in the unfertilised egg.) When little or no yolk is present the blastomeres are small and equal in size.

A frog's egg contains sufficient yolk to affect cleavage divisions. Only a few large cells called **macromeres** are formed in the vegetal region, whereas at the animal pole many small **micromeres** are formed. A bird's egg contains so much yolk that cleavage is entirely restricted to the animal pole.

■ Gastrulation

In gastrulation (Figure 19.26) cell division continues, but is overshadowed by a series of coordinated movements of cells in which the blastula, a hollow ball of cells, is converted into a complex embryo of several layers of cells. In the frog, gastrulation first involves the formation of a circular opening, the **blastopore**. Along the dorsal lip of the blastopore cells invaginate into the blastocoel in an organised way, forming an entirely enclosed two-layered cup-shaped structure, the **gastrula**. The new cavity is called the **archenteron** or 'primitive gut'. At the end of gastrulation the enlarged embryo elongates, and becomes tubular.

A large quantity of yolk results in a different method of gastrulation. In birds and mammals there is no invagination. The cells forming the inner layer migrate into position.

■ Organogeny

The cells of the gastrula continue to divide. At the same time, blocks of these cells move within the developing embryo, to form a gastrula consisting of three germ layers. This condition is known as **triploblastic**.

Development of the triploblastic gastrula

The innermost layer of cells in the gastrula gives rise to the **endoderm**, the outer layer of cells to the **ectoderm**, and the cell masses between these layers to the **mesoderm**. Cell differentiation then commences, and the cells of the three layers develop into the tissues and organs of the embryo (Figures 19.27 and 19.28).

Fate maps

Injecting different coloured vital dyes (p. 152) into the cells of a blastula has disclosed that the fates of embryonic cells are determined at a very early stage in development. Experimental embryologists have been able to construct what are called 'fate maps' of the cells of a blastula, showing the prospective position and roles of the tissues to be formed from individual cells in the development of an adult frog (or other vertebrate).

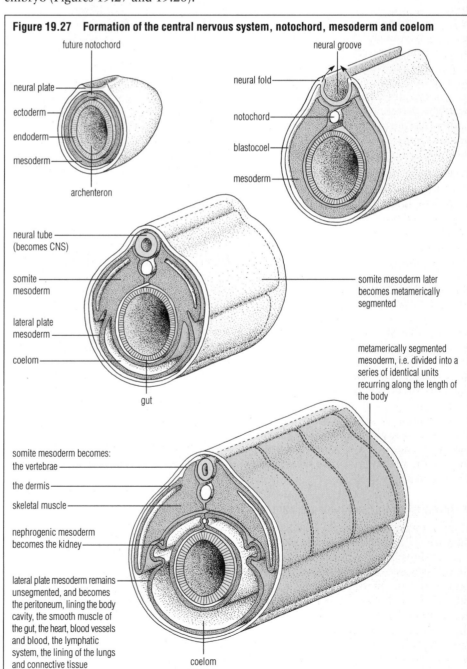

Figure 19.26 Stages in gastrulation as seen in a frog embryo, a vertebrate embryo with a moderate yolk store

blastula
- animal pole
- blastocoel
- vegetal pole

migration of cells begins
- blastopore begins to form

- formation of new cavity
- ectoderm
- endoderm
- blastopore
- mesoderm begins to tuck in

gastrula
- archenteron cavity (gut)
- remains of blastocoel
- blastopore
- mesoderm pushes between ectoderm and endoderm

Figure 19.27 Formation of the central nervous system, notochord, mesoderm and coelom

- future notochord
- neural plate
- ectoderm
- endoderm
- mesoderm
- archenteron

- neural groove
- neural fold
- notochord
- blastocoel
- mesoderm

- neural tube (becomes CNS)
- somite mesoderm
- lateral plate mesoderm
- coelom
- gut
- somite mesoderm later becomes metamerically segmented

metamerically segmented mesoderm, i.e. divided into a series of identical units recurring along the length of the body

somite mesoderm becomes:
- the vertebrae
- the dermis
- skeletal muscle
- nephrogenic mesoderm becomes the kidney
- lateral plate mesoderm remains unsegmented, and becomes the peritoneum, lining the body cavity, the smooth muscle of the gut, the heart, blood vessels and blood, the lymphatic system, the lining of the lungs and connective tissue
- coelom

Role of the ectoderm

The tissues of the central nervous system (p. 460) are formed from cells of part of the upper ectoderm, known as the **neural plate**. The neural plate invaginates to form the **neural tube**, thereby separating itself from the rest of the ectoderm. Ultimately the anterior part of the neural tube forms the brain and the cranial nerves, and the remainder the spinal cord and the peripheral nerves (p. 460). The rest of the ectoderm gives rise to the epidermis of the skin and the associated structures (hair, nails and so forth in mammals).

Role of mesoderm and endoderm

Within the embryo a group of cells breaks free from the mesoderm and forms a rod of tissue, the **notochord**. The notochord initially supports the embryo, but is ultimately replaced by the vertebral column. The remainder of the mesoderm and the endoderm form tubular masses that run the length of the embryo. The endoderm forms the gut, including its lining and glands.

The rest of the mesoderm layer subdivides into three blocks, known as the somite mesoderm, the nephrogenic mesoderm and the lateral plate mesoderm. The somite mesoderm and the nephrogenic mesoderm form a series of paired blocks running the length of the embryo. This is known as **metameric segmentation**. The nephrogenic mesoderm forms the kidneys, and the somite mesoderm which forms on each side of the notochord gives rise to the dermis of the skin, skeletal muscle, vertebrae and body wall muscles.

The lateral plate mesoderm forms around the endoderm. It is not segmented. The coelom, the main body cavity in the adult vertebrate, persists here. The lateral plate mesoderm develops into smooth muscle, heart, blood vessels and blood, lymphatic system, the lining of the lungs, and connective tissue of the vertebrate body.

19.8 GROWTH AND DEVELOPMENT IN NON-VERTEBRATES

In the growth and development of non-vertebrates from fertilised egg to adult form the organism normally passes through one or more juvenile or intermediate stages, known as larvae. Larval forms occur in most non-vertebrate phyla, including the insects (arthropods) with their caterpillar, maggot or nymph larvae. The common features of larval forms are as follows:

► they are markedly different in structure from the adult that they eventually become (Figure 19.29)
► they lead an existence fully independent from the adult; in particular, they feed in a different way
► they are not normally a reproductive phase and do not reproduce sexually (the larvae of many parasites do reproduce asexually, however).

Parasitic species may exploit their larval stages in the transmission to new hosts, as in the liver fluke (a trematode platyhelminth, see below).

■ The roles of the larval stage in the life cycle

The larva as a dispersal phase

Organisms that in adult form move about little or not at all are described as **sessile**. Examples of sessile non-vertebrates that we have already encountered are the barnacle (*Semibalanus*, p. 55) and the mussel (*Mytilus*, pp. 37 and 291). Both these species are found anchored in the intertidal zone of the shore. Their eggs hatch into tiny **motile** larvae, which move about and feed in surface waters of the sea, as members of the zooplankton. They are dispersed by the currents, but eventually they settle down and complete development to the adult forms.

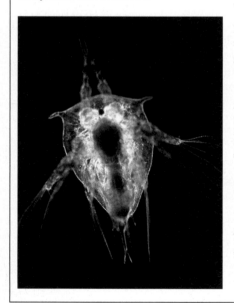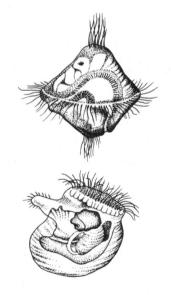

Figure 19.29 The nauplius larva of *Semibalanus balanoides* (left) and the trochophore larva of *Arenicola* (annelid) (top right) and the veliger larva of *Mytilus* (bottom right); these larvae are totally unlike their adult forms

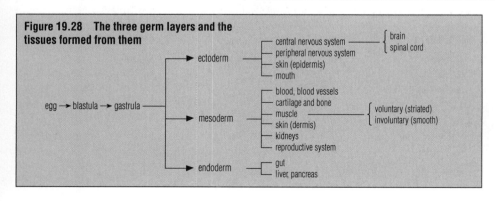

Figure 19.28 The three germ layers and the tissues formed from them

egg → blastula → gastrula
- ectoderm
  - central nervous system — { brain, spinal cord }
  - peripheral nervous system
  - skin (epidermis)
  - mouth
- mesoderm
  - blood, blood vessels
  - cartilage and bone
  - muscle — { voluntary (striated), involuntary (smooth) }
  - skin (dermis)
  - kidneys
  - reproductive system
- endoderm
  - gut
  - liver, pancreas

Questions

1 What are the essential differences between cleavage and gastrulation?
2 The presence of yolk appears to influence the way in which an embryo develops. What is the value of yolk to the embryo?
3 What are the three generative layers of a triploblastic gastrula?
4 What is meant by 'metameric segmentation'? What form does metameric segmentation take in annelids and insects respectively?

Figure 19.30   Larval and adult forms of the great water-beetle, _Dytiscus marginalis_

The larva as a feeding phase

The larva is an important feeding and growth phase in the life cycle. For example, the aquatic larva of the great water-beetle, _Dytiscus marginalis_ (Figure 19.30) increases its mass by 52 times in the larval period of about 50 days. The maggots of some meat-eating flies increase in mass more than 450 times in about 70 hours.

Commonly, both larva and adult are feeding stages, although they generally have different diets. In fact, they normally occupy different niches (p. 65). Consequently, there is greatly reduced intraspecific competition for food. For example, with most butterfly and moth species a caterpillar is the relatively long-lived feeding stage. Most caterpillars have well-developed jaws for cutting off bits of vegetation, whereas the adult feeds on liquids (nectar) via fine tubular mouthparts (p. 292).

In extreme cases the larva may be the only stage in the life cycle able to feed. For example, in mayflies (_Ephemera_) the aquatic nymph is the feeding stage; the adult mayfly is unable to feed.

The larva as asexual reproduction phase (in parasitic species)

In parasitic organisms it is essential that new host organisms are colonised. The chances of colonising new hosts are increased when larval stages in the parasite's life cycle reproduce asexually, so vastly increasing their numbers. An example is the redia larva in the life cycle of the liver fluke, _Fasciola hepatica_ (p. 35). Here the intermediate host is the snail. The infected snail harbours several successive larval stages, illustrated and named in Figure 19.31. These stages lead up to the cercaria larvae that ultimately escape from the snail and become the metacercarian larvae, encysted on blades of grass, awaiting chance ingestion by a sheep, the primary host.

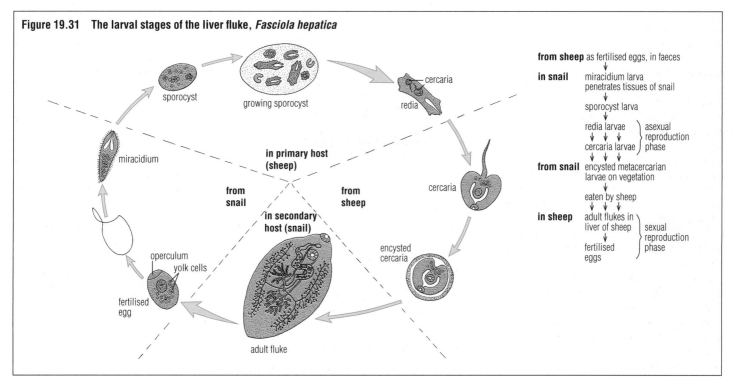

Figure 19.31   The larval stages of the liver fluke, _Fasciola hepatica_

19.9 GROWTH AND DEVELOPMENT IN INSECTS

The apparently 'intermittent' growth of insects (phylum Arthropoda), caused by the presence and repeated moulting (ecdysis) of the external skeleton, is described on p. 402.

■ Two types of insect life cycle

In the life cycle of many species of insect the egg develops into a larva that is quite different from the adult form. The individual undergoes an abrupt change of form, known as metamorphosis, between the final larval form and the adult (imago). The butterflies and moths, beetles and flies are examples of insects with complete metamorphosis in the life cycle. Such insects are referred to as **holometabolous insects** (Figure 19.33).

In other species of insect the egg develops via a series of **nymphs**, each essentially a miniature adult but lacking wings. Moulting and expansion growth occur at each nymphal stage. The locust (Figures 19.32 and 19.33), grasshopper and cockroach are examples of insects with this kind of incomplete metamorphosis. These insects are referred to as **hemimetabolous insects**. Each ecdysis results in a larval form that more closely resembles the adult, but the final moult still reveals the greatest degree of change.

■ Diet and life cycle

The diet of the locust, like that of the grasshopper, is leafy vegetation. This material is chopped and ground by the external mouthparts before digestion in the gut (p. 294). The gut contains carbohydrases, lipases and proteases, and the products of digestion – sugars, fatty acids and amino acids – are absorbed and assimilated. The nymphs of the locust, as well as appearing as miniature adults, have the same diet and range of enzymes. The differences in diet of the larva (caterpillar) and the adult butterfly are reflected in the digestive enzymes present in their guts. The caterpillar produces carbohydrases, lipases and proteases, and the products of digestion are used in growth, as well as in respiration. A butterfly typically produces sucrase (a carbohydrase) only. Sucrose in the nectar it draws up from flowers is hydrolysed to glucose, which is respired for energy. Butterflies need energy mainly for flight and reproduction.

Figure 19.32   A plague of locusts: when locust larvae are overcrowded they become highly excited and form into huge swarms; the swarms devour most vegetation in their path

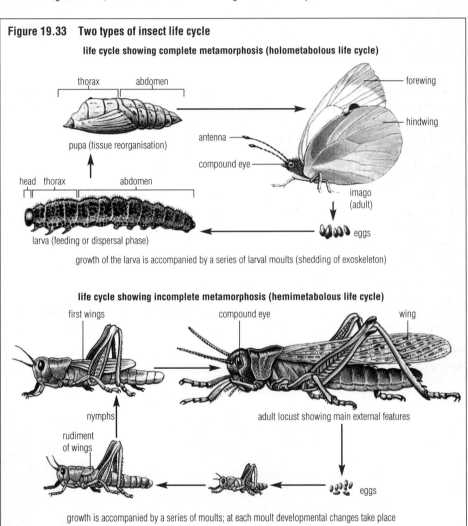

Figure 19.33   Two types of insect life cycle

Questions

1   What do we mean by
   a   niche
   b   intraspecific competition (p. 64)?
   Give examples of adult and larval forms which occupy different niches.
2   In insect life cycles, what is the difference between a nymph and a caterpillar or grub?
3   What is the value to an organism of having a larval stage in its life cycle?
4   **a**   What is meant by metamorphosis?
   b   To what extent is metamorphosis in the life cycle of the butterfly comparable with metamorphosis in the frog?

FURTHER READING/WEB SITES

Science and Plants for Schools: **www.rmplc.co.uk/orgs/saps**

20 Response and coordination in plants

- Plant movements are mostly slow growth movements, but plants are highly sensitive to environmental cues and stimuli, and detect and respond to small changes.
- Plant stems are sensitive to direction of light (positive phototropism), and stems and roots are sensitive to the pull of gravity (geotropism).
- A plant growth regulator (hormone) known as auxin (indoleacetic acid) is involved in plant growth and growth responses.

- Other plant hormones active in the regulation of growth and development are gibberellins, cytokinins, abscisic acid and ethene.
- Phytochrome is a reactive pigment molecule present in plants. The phytochrome molecule changes shape in red and far-red light.
- Flowering, dormancy of seeds and buds, and leaf fall are plant processes that may be regulated by environmental factors in which phytochrome is involved in some way.

20.1 SENSITIVITY

Living things exist in a changing environment. Some changes are favourable and advantageous to the organism, others are unfavourable or harmful, and many are of little significance. The ability of living things to detect changes and make appropriate responses is essential to their survival. This ability to detect change and to respond is referred to as **sensitivity**. Sensitivity is one of the characteristics of all living things. In fact, organisms are sensitive to change in both their internal and their external environments, as we shall see.

■ Plant and animal sensitivity compared

The sensitivity of larger animals is obvious to the most casual observer, as in animals hunting for food, escaping when under attack, and coordinating breeding activity with the season when food for them and their young is plentiful. Sensitivity in animals often involves complex sense organs and an elaborate nervous system, and frequently hormones (chemical messengers) too. Animal responses are usually quick movements, involving muscles, bones and joints. Animals tend to adapt to

new situations by modifying their behaviour. The sensitivity of animals is discussed in Chapter 21, p. 442.

Most plants are anchored organisms, growing in one place and remaining there even if the environment is not wholly favourable. As a result, plant responses are less evident than those of animals, but no less important to the plant. Yet plants have no nervous system and no muscle tissue. Consequently plant sensitivity may lead to responses less dramatic than those of animals. Rapid movements by plants are extremely rare (but see Figure 20.3); plant responses are generally quite slow growth or turgor movements. Plants tend to adapt to new situations by modifying their growth.

■ Observing sensitivity and movement in plants

The continuous movements of plants

Time-lapse photography makes it possible to monitor the continuous movements of plants. The technique involves making a record (a moving picture, for example) from individual photographs taken at intervals with a constant time period allowed to elapse between photo-

graphs. A time-lapse film of a shoot of a germinating seedling breaking through the soil can convey the extent of movements in the slow but important changes of seed germination (Figure 20.1).

As a plant grows the stem tip is not carried up in a straight line but follows a helical path, in response to internal stimulation. This movement, known as **nutation**, occurs because at any time one part of the apical meristem is growing more quickly than the rest, and the region of more rapid growth moves slowly around the apex. Nutation may be detected by time-lapse photography. Alternatively, the apparatus shown in Figure 20.2 allows the recording of nutation movement with simple laboratory equipment.

In climbing plants the tips of stems (or of tendrils) show exaggerated nutation movements. For example, the parasitic plant dodder (*Cuscuta europaea*) twines its stem around that of its host, the stem tip describing a path of about 10 cm diameter during the early growth of the seedling. The exaggerated nutation movements of climbing plant stems increase their chances of making contact with a supporting structure.

Figure 20.1 Time-lapse photography of sunflower (*Helianthus annuus*) germination at the point of emergence of the aerial system

Unusual quick movements of plants

Not all plant movements are slow growth movements. For example, the so-called sensitive plant (*Mimosa pudica*) folds up its leaves if any part of the plant is touched or damaged. During daylight, the compound leaves are expanded and held out from the stem. When darkness falls, the whole leaf folds up. The leaf similarly responds, very quickly, if any leaflet is touched or damaged (Figure 20.3). These leaf movements are due to sudden loss of water from the mass of parenchyma cells (a turgor change, p. 234) in pad-like swellings, called **pulvini**, found at the base of each leaflet, mid-rib and petiole. The effect of the stimulus is conducted quickly to all parts of the plant from a single damaged leaflet, as if by a (non-existent) nervous system.

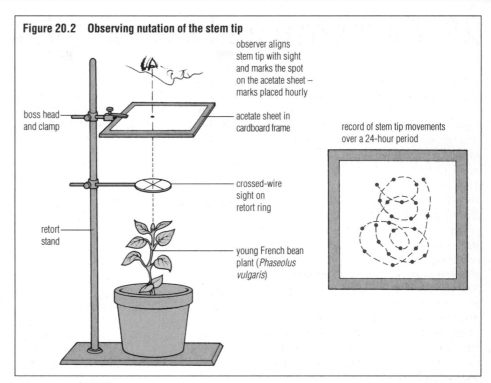

Figure 20.2   Observing nutation of the stem tip

observer aligns stem tip with sight and marks the spot on the acetate sheet – marks placed hourly

boss head and clamp

acetate sheet in cardboard frame

record of stem tip movements over a 24-hour period

crossed-wire sight on retort ring

retort stand

young French bean plant (*Phaseolus vulgaris*)

Figure 20.3   **The response of the sensitive plant (*Mimosa pudica*) to touch:** mimosa is a tropical weed of the New World, which has been introduced as a house plant in certain temperate countries; here a leaflet of the leaf in the foreground has just been touched

Action potentials in plants

The speed of conduction of the effect of the stimulus from the tip of the leaf of *Mimosa* to the leaf base and to other leaves has been measured. It reaches a maximum of about 3 cm per second. Amazingly, an electrical change (an action potential, p. 445) has been shown to be associated with this response. The phenomenon is not restricted to *Mimosa*, either. Apparently the action potentials of plants travel through ordinary cells, and cross between cells via the plasmodesmata (p. 167).

Questions

1   What are the key differences between animal and plant sensitivity?
2   What observations and measurements of movement by the sensitive plant would you need to carry out in order to formulate an explanation of the underlying mechanisms?
3   What might be the biological advantages to the sensitive plant of its response to touch? (Your speculation is as likely to be right as anybody else's.)

Plants and light

The response of green plants to light is complex. In total darkness plant stems grow thin and weak, and with tiny, undeveloped leaves. Chlorophyll (p. 248) is not formed, and the aerial system (stem and leaves) appears pale yellow. This growth is at the expense of stored food. The growing point with terminal bud is curved over and hooked, as in a young stem growing up through the soil. The condition of plants grown in the dark is described as **etiolated**. By contrast, the shoots in full light are short, with straight, thick, sturdy stems and large, dark green expanded leaves.

Since sunlight is essential for photosynthesis, by which the green plant manufactures nutrients (Chapter 12, p. 246), it is easy to assume that light is essential for plant growth also. In fact light appears to **inhibit stem growth** in length but **enhances leaf expansion**. Light is **essential for chlorophyll formation**.

Phototropism

When germinating seedlings receive light from one direction only (unilateral illumination), their stems grow towards the light source (Figure 20.4). A growth response of a plant organ to an external stimulus in which the direction of the stimulus determines the direction of the response is called a **tropism**. When light is the stimulus, as here, the response is known as **phototropism**. The stem is said to be **positively phototropic** because the tip of the stem grows towards the light. Tropisms are summarised in Table 20.1 (p. 427).

The region of the plant stem that curves towards the light source is the zone of elongation (by cell enlargement, pp. 407–8). This can be shown by marking the stems of growing seedlings at regular intervals, and then placing some of the marked seedlings to grow in normal illumination and others in unilateral illumination (Figure 20.5). The results show that the plant's response to unilateral illumination is a growth response.

Positive phototropism becomes apparent under the unusual circumstances of exposure of a young green plant to continuous unilateral light. In this circumstance the plant stem is strongly phototropic in the seedling stage. Once the growing plant has well-established

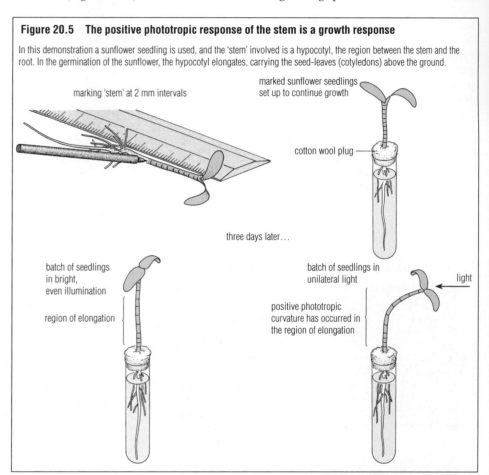

Figure 20.5 The positive phototropic response of the stem is a growth response

In this demonstration a sunflower seedling is used, and the 'stem' involved is a hypocotyl, the region between the stem and the root. In the germination of the sunflower, the hypocotyl elongates, carrying the seed-leaves (cotyledons) above the ground.

marking 'stem' at 2 mm intervals

marked sunflower seedlings set up to continue growth

cotton wool plug

three days later...

batch of seedlings in bright, even illumination

region of elongation

batch of seedlings in unilateral light

light

positive phototropic curvature has occurred in the region of elongation

1 seedlings illuminated from one side have stems that curve towards the light source (positive phototropism)

2 seedlings in total darkness have greatly elongated, spindly stems and small, yellow leaves

3 seedlings illuminated from all sides have short, stout stems and dark green leaves

Figure 20.4 Garden pea seedlings grown in light, in the dark and in unilateral illumination

Figure 20.6 'Suntracking' or the heliotropism of leaves and capitulum of the sunflower plant

Figure 20.7   The response of plant growth to gravity: plants secured to the slowly revolving clinostat plate receive equal gravity stimulation on all sides

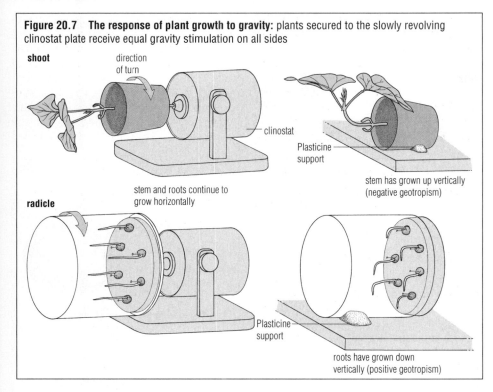

shoot

direction of turn

clinostat

Plasticine support

stem and roots continue to grow horizontally

stem has grown up vertically (negative geotropism)

radicle

Plasticine support

roots have grown down vertically (positive geotropism)

Figure 20.8   Photonasty of the dandelion (*Taraxacum officinale*) capitulum

(left) Flower head open in the light, and (right) closed in the dark

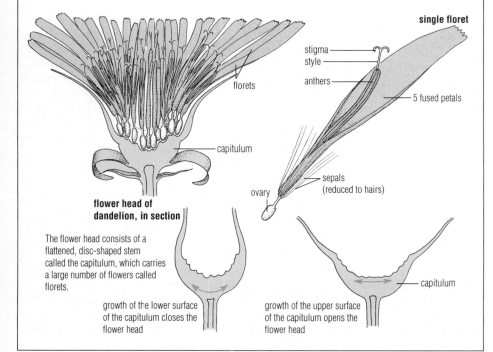

single floret

stigma
style
anthers

5 fused petals

florets

capitulum

sepals (reduced to hairs)

ovary

capitulum

flower head of dandelion, in section

The flower head consists of a flattened, disc-shaped stem called the capitulum, which carries a large number of flowers called florets.

growth of the lower surface of the capitulum closes the flower head

growth of the upper surface of the capitulum opens the flower head

leaves, however, the phototropic response normally devolves to the leaves (Figure 20.6). This response of leaves is described as 'suntracking' or **heliotropism**. The sensitivity of young stems to unilateral light does not normally persist in the fully grown plant. Once past the seedling stage, stems growing in natural conditions respond to the stimulus of gravity rather than to unilateral light.

Plants and gravity

When growing plants are placed horizontally they exhibit a growth response to the stimulus of one-sided gravity (**geotropism** or **gravitropism**). The tip of the stem grows away from the pull of gravity (**negative geotropism**), and the root tips grow towards it (**positive geotropism**). The plant responses to unilateral gravity can be recognised as growth responses because the resulting curvatures occur in the region of elongation (cell enlargement) of stem and root. As a control, plants held horizontally on a slowly revolving turntable (a **clinostat**) experience gravity equally on all sides. Stem and root tips of these plants exhibit no growth curvature, but continue to grow horizontally (Figure 20.7).

Lateral roots and stems usually grow at an angle to the direction of gravity, a response referred to as **plagiotropism**.

'Sleep movements' of plants

The so-called sleep movements of certain flowers and leaves are examples of **nastic movements** or **nasties**. In nastic movements the direction of the response of the plant organ is not determined by the direction of the stimulus. For example, a dandelion (*Taraxacum officinale*) flower head opens up in the light and closes up at dusk or in low light (Figure 20.8). This is referred to as **photonasty**. Flowers of crocus and tulip make similar responses, stimulated by changes in temperature (**thermonasty**) as well as changes of light. Opening and closure are due to localised growth in a particular part of the flower or inflorescence.

Questions

1   Compile a checklist of the effects of light on plant growth and development that you now know about.
2   What everyday observations would you use to show a sceptical friend that light inhibits plant growth?
3   How would you attempt to show that dandelion flowers are light-sensitive, rather than temperature-sensitive?

TROPISMS, NASTIES AND TAXES: A SUMMARY

Plant movements made in response to external stimuli fall into three categories.

1 Tropisms
(tropic movements)
A growth movement of a plant organ in response to an external stimulus in which the direction of the stimulus determines the direction of the response, e.g. positive phototropism (Figure 20.5)

2 Nasties
(nastic movements)
A growth movement of a plant organ in response to an external stimulus in which the direction of the response is not determined by the direction of the stimulus, e.g. photonasty (Figure 20.8)

3 Taxes
(tactic movements)
A movement by a motile organism (or motile gamete) in response to an external stimulus, in which the direction of the stimulus determines the direction of the response, e.g. positive chemotaxis of gametes (Figure 26.3, p. 564)

Other important examples of plant sensitivity are detailed elsewhere:

4 Stomatal movements The opening or closing of stomata (the pores in the epidermis of the aerial system of plants) is controlled by internal conditions in the leaf and coordinated with light and darkness (Chapter 16, p. 322)

5 Environmental coordination of germination, bud break, flowering and leaf fall These stages in the growth and development of plants are activated by conditions in the plants' environment (p. 434)

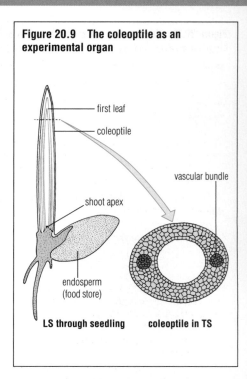

Figure 20.9 The coleoptile as an experimental organ

first leaf
coleoptile
vascular bundle
shoot apex
endosperm (food store)

LS through seedling   coleoptile in TS

20.2 A HORMONE BASIS FOR PLANT SENSITIVITY?

Among the internal factors that play a part in plant growth and sensitivity are substances known as plant growth hormones or, better, as **plant growth substances**. In learning about these substances, you should bear in mind certain points from the outset:

▶ there are five major types of compound, naturally occurring in plants, that we now classify as plant growth substances (p. 433)

▶ these substances tend to interact with each other in the control of growth, rather than working in isolation (p. 432)

▶ plant growth substances occur in very low concentrations in plant tissues, making their extraction and investigation difficult (pp. 424 and 432)

▶ some, at least, of the plant growth substances are also produced by bacteria or fungi present in the environment of plants, making it advisable to carry out growth experiments in sterile conditions, if possible (p. 428)

▶ plant growth substances cannot be directly compared with animal hormones because they function differently (p. 430)

▶ the precise role of each growth substance in plant growth and sensitivity is not fully understood.

To appreciate the current uncertainties within this fascinating and economically important topic we shall first consider how the various plant growth regulators were discovered, and the aspect of growth or sensitivity in which they were first implicated. You will find that, although some of the earliest explanations of 'plant hormone' control of

growth and sensitivity are now doubted, there is little agreement about exactly how plant sensitivity is mediated at cell and growth substance levels. Lots more has yet to be discovered.

■ The discovery of auxin

Several rather simple but ingenious experiments pointed to the possibility of the existence of plant growth regulators. Charles Darwin (p. 11) published the first evidence in 1881 in his book *The Power of Movements in Plants*. Darwin investigated the response of grass coleoptiles to unilateral light. (The structure of a coleoptile is shown in Figure 20.9.) His results apparently indicated that the tip of the coleoptile perceived the stimulus of unilateral light but that the growth response occurred lower down the organ. He concluded that some 'influence' is transmitted from the tip to the growth region.

These studies and subsequent investigations, including the work of Boysen-Jensen and Paal (Figure 20.10), promoted the idea that the plant response to external stimuli takes place in three phases:

▶ **perception** – a unilateral stimulus, such as light, is perceived (at the tip of the coleoptile, for instance)

▶ **transduction** – conduction of the 'influence', in the form of a growth hormone, to the site of response

▶ **response** – usually due to differential growth rates in the region of extension growth.

This idea was in line with the idea of animal hormones developed by an animal physiologist, Ernest Starling, and published in 1905, although plant hormones are not exactly analogous with animal hormones.

■ Went and the bioassay of auxin

Fritz Went, a Dutchman, continued with this line of enquiry in 1928. He showed that an excised oat coleoptile tip, when placed on an agar gel block, appeared to generate growth-promoting properties in the agar. When the agar block was placed on one side of the coleoptile stump, the result was a curvature of the coleoptile. An untreated agar block produced no curvature.

Went interpreted this result as due to a growth hormone having passed from the coleoptile tip into the agar block, and then to the stump. Went named this hormone **auxin**. It was assumed that auxin worked by stimulating extension growth in the cells. Went's results also suggested that the degree of curvature of the coleoptile was related to the concentration of auxin over a wide range. This latter observation was the basis of his method (a biological assay or **bioassay**) of measuring the concentration of hormone obtained from the coleoptile tips.

A major difficulty in the study of plant growth regulators is that the quantities of these substances present in plants are very small indeed, perhaps a fraction of a millionth of a gram (μg). Available chemical tests may be too insensitive to measure quantities as small as these. In a bioassay the concentration of an unknown sample of auxin is estimated from the degree of curvature it causes in a coleoptile curvature test, using the calibration curve (Figure 20.11).

Figure 20.10 Steps in the discovery of plant growth substances

Charles Darwin (1880) **investigated the growth of grass coleoptiles in unilateral light (phototropism).**

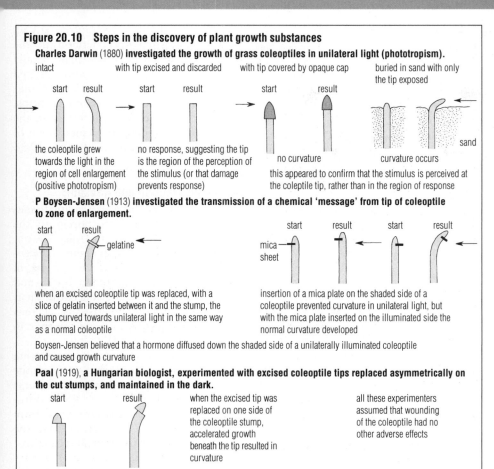

intact

the coleoptile grew towards the light in the region of cell enlargement (positive phototropism)

with tip excised and discarded

no response, suggesting the tip is the region of the perception of the stimulus (or that damage prevents response)

with tip covered by opaque cap

no curvature

buried in sand with only the tip exposed

curvature occurs

this appeared to confirm that the stimulus is perceived at the coleoptile tip, rather than in the region of response

P Boysen-Jensen (1913) **investigated the transmission of a chemical 'message' from tip of coleoptile to zone of enlargement.**

when an excised coleoptile tip was replaced, with a slice of gelatin inserted between it and the stump, the stump curved towards unilateral light in the same way as a normal coleoptile

mica sheet

insertion of a mica plate on the shaded side of a coleoptile prevented curvature in unilateral light, but with the mica plate inserted on the illuminated side the normal curvature developed

Boysen-Jensen believed that a hormone diffused down the shaded side of a unilaterally illuminated coleoptile and caused growth curvature

Paal (1919), **a Hungarian biologist, experimented with excised coleoptile tips replaced asymmetrically on the cut stumps, and maintained in the dark.**

when the excised tip was replaced on one side of the coleoptile stump, accelerated growth beneath the tip resulted in curvature

all these experimenters assumed that wounding of the coleoptile had no other adverse effects

Paal believed that a hormone had diffused from the tip and had accelerated growth

Questions

1 On the basis of the curvature of the oat coleoptile shown here, what was the concentration of auxin (IAA) in the agar block? (Refer to Figure 20.11.)

2 If you were to repeat Darwin's experiment as represented in Figure 20.10, what difference to the procedure would you propose to enable you to be more confident of the validity of the results?

3 Why is it necessary to use a biological assay (bioassay) for auxin rather than some chemical test of concentration?

Figure 20.11 The oat coleoptile curvature test for auxin: operations involving coleoptiles and agar are carried out in the dark and at 100% relative humidity

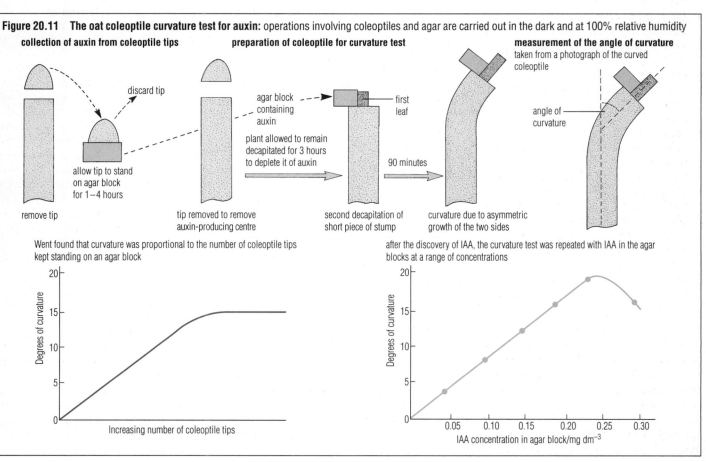

collection of auxin from coleoptile tips

discard tip

remove tip

allow tip to stand on agar block for 1–4 hours

preparation of coleoptile for curvature test

agar block containing auxin

plant allowed to remain decapitated for 3 hours to deplete it of auxin

tip removed to remove auxin-producing centre

first leaf

90 minutes

second decapitation of short piece of stump

measurement of the angle of curvature taken from a photograph of the curved coleoptile

angle of curvature

curvature due to asymmetric growth of the two sides

Went found that curvature was proportional to the number of coleoptile tips kept standing on an agar block

Degrees of curvature vs *Increasing number of coleoptile tips*

after the discovery of IAA, the curvature test was repeated with IAA in the agar blocks at a range of concentrations

Degrees of curvature vs *IAA concentration in agar block/mg dm^{-3}*

■ The chemistry of auxin

Auxin activity has subsequently been demonstrated in the organs of various higher plants. Auxin seems to be important in a wide range of organisms, and not only in higher plants. The media in which various fungi have been cultured show auxin activity at a high level.

The concentration of auxin in higher plant tissues is very low, too low to allow biochemists to isolate a sample for traditional methods of chemical analysis. In 1934 F Kogl, working in Holland, isolated a substance from human urine that was extremely active in causing curvature in the oat coleoptile test. This substance proved to be the organic acid called indoleacetic acid (IAA, Figure 20.12).

Figure 20.12   The chemical formula of indoleacetic acid (IAA)

$$HC \overset{\overset{\displaystyle H}{|}}{\underset{\underset{\displaystyle H}{|}}{C}} \quad C \quad C-CH_2-COOH$$

■ Is 'auxin' indoleacetic acid?

The 'auxin' of Went's experiments was then assumed to be IAA. This compound, like natural auxin, is active in extremely small amounts. But it was impossible to obtain sufficient IAA from coleoptile tissue to carry out chemical tests to confirm this belief. 'Auxin' is present in tissues at such low concentration that about 20 000 *tonnes* of coleoptile tip would be required to yield one gram of IAA!

By 1972, however, with the availability of a new analytical technique sensitive to extremely low concentrations (high-resolution mass spectrometry), it was possible to establish that the 'auxin' molecule diffusing from coleoptiles was indeed the chemical substance IAA. Since the initial discovery of IAA as an auxin, this substance has also been detected in most plant species. IAA is regarded as the principal auxin in plants. Consequently, from now on in this discussion we will refer to naturally occurring auxin as IAA.

There are some other substances present in plant tissue that also show 'auxin' properties. Most of these substances are closely related to IAA. One example is indole-3-acetonitrile (IAN), which is particularly common in the tissues of members of the plant family Cruciferae (wallflowers and related plants). Several synthetic 'auxins', manufactured by agricultural chemicals firms, are used as herbicides (p. 91) or in the artificial propagation of plants. Some of these are introduced in Figure 20.36 (p. 440).

■ IAA distribution and transport

IAA is synthesised in meristematic tissues (Figure 20.13) and transported to the regions of growth in the plant. Consequently IAA is found in highest concentration at the tips of stems and roots, in young growing leaves, and in flowers and fruits, and in decreasing concentration at a distance from these meristems. There is substantially less IAA in root tips than in stem tips (Figure 20.14).

Transport of IAA is polar, moving through tissues from the apex of the shoot or root, and not at all in the reverse direction. Movement is not due to diffusion but to active transport of IAA from cell to cell. This movement is at the expense of energy from metabolism.

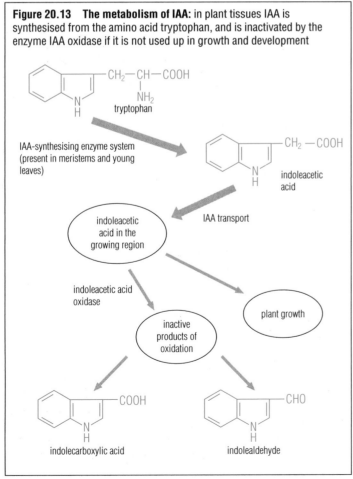

Figure 20.13   The metabolism of IAA: in plant tissues IAA is synthesised from the amino acid tryptophan, and is inactivated by the enzyme IAA oxidase if it is not used up in growth and development

---

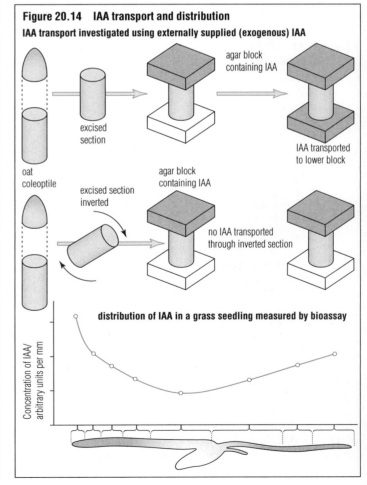

Figure 20.14   IAA transport and distribution

IAA transport investigated using externally supplied (exogenous) IAA

distribution of IAA in a grass seedling measured by bioassay

Concentration of IAA/ arbitrary units per mm

20.3 IAA AND TROPISMS: SOME CLASSICAL EXPERIMENTS

Experiments on the response of the grass coleoptile to unilateral light (Figure 20.10, p. 423) initiated the study of tropisms and led to the subsequent discovery of 'auxin' (p. 422). Further development of these experimental methods gave evidence for the hypothesis that 'auxin' distribution was responsible for the tropic responses of stems and roots. Some of these classical experiments are described in this section.

■ Phototropism

Went (1928) allowed unilateral light to fall upon an excised oat (*Avena*) coleoptile tip in contact with two agar blocks, arranged to receive the auxin from illuminated and darkened sides separately (Figure 20.15). He found that more auxin collected in the agar on the darker side than on the illuminated side. This uneven distribution of auxin was thought to explain the asymmetric growth of the coleoptile, with enhanced growth on the darkened side. Went believed auxin was inactivated on the illuminated side and that light also induced the transverse transport of auxin to the darkened side. The phototropic curvature observed in unilateral light was thus thought to be a result of differential auxin distribution induced by light (but see p. 428). This suggestion was also independently put forward by Cholodny at about the same time, and so it is referred to as the Cholodny–Went hypothesis.

Figure 20.15 'Auxin' and phototropic curvature of coleoptiles: this observation was made independently by Cholodny and Went

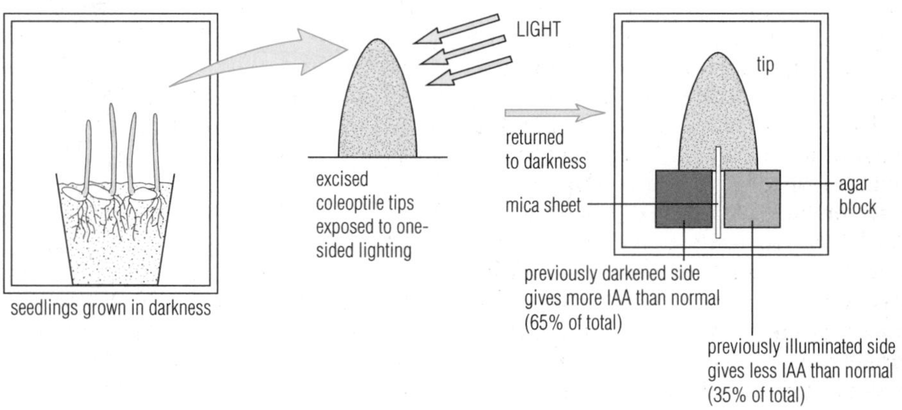

LIGHT
tip
returned to darkness
mica sheet
agar block
excised coleoptile tips exposed to one-sided lighting
seedlings grown in darkness
previously darkened side gives more IAA than normal (65% of total)
previously illuminated side gives less IAA than normal (35% of total)

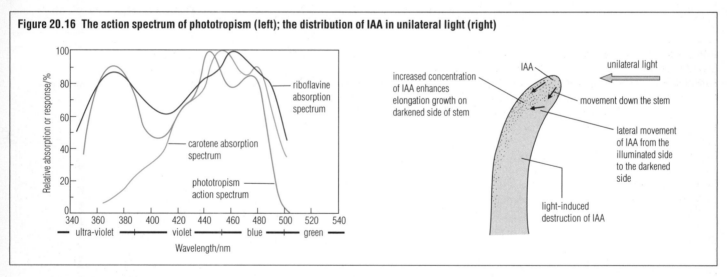

Figure 20.16 The action spectrum of phototropism (left); the distribution of IAA in unilateral light (right)

riboflavine absorption spectrum
carotene absorption spectrum
phototropism action spectrum
Relative absorption or response/%
Wavelength/nm
ultra-violet — violet — blue — green

increased concentration of IAA enhances elongation growth on darkened side of stem
IAA
unilateral light
movement down the stem
lateral movement of IAA from the illuminated side to the darkened side
light-induced destruction of IAA

Phototropism, the action spectrum

Research to identify the photoreceptor for phototropism began by attempts to match the action spectrum for phototropism with the absorption spectrum of pigments present in the light-sensitive tissues. Blue light is most effective in causing phototropism, and red is inactive. This suggests that a yellow pigment is the photoreceptor.

In fact two pigments may be implicated. It is found that the absorption spectrum of carotene matches the action spectrum in the visible range, whereas the absorption spectrum of riboflavine matches the action spectrum in the ultra-violet (Figure 20.16). Whilst it is probable that both substances are involved, it is not evident how, when activated by light, these pigments bring about the asymmetric distribution of IAA in stem or coleoptile tip (or of an inhibitor of growth, should one prove to be involved in the phototropic response).

Questions

1 IAA is synthesised from an amino acid but is not itself described by chemists as an amino acid. Why not?

2 In experiments on phototropism the coleoptiles are grown in the dark except for brief periods of exposure to unilateral light. Why is it necessary to keep coleoptiles in the dark?

3 Why, do you think, does a plant have an enzyme system for destroying IAA?

4 According to the graph in Figure 20.14, which regions of the seedling have respectively the lowest and the highest concentrations of IAA?

5 How might a plant physiologist obtain an action spectrum for phototropism?

■ Geotropism

From work on coleoptile tips there is evidence that geotropism, like phototropism, may be due to a redistribution of IAA (the Cholodny–Went hypothesis). Greater amounts of IAA have been obtained from the lower side of horizontally placed coleoptile tips. When radioactively labelled IAA is applied to cut and intact coleoptile tips held horizontally, the labelled compound is found to accumulate on the lower side.

When a seedling is placed horizontally, the stem tip grows up (negative geotropism) and the root tip grows down (positive geotropism), due to differential growth rates of the upper and lower sides of stem and root. In this position, IAA may collect on the lower side, at least in the stem tip. If this is the case, why does a root tip respond by growing down yet the stem tip responds by growing up? The answer may be that a raised concentration of IAA stimulates coleoptile (and stem) growth but inhibits root growth (Figure 20.17). Root growth is stimulated by lower IAA concentrations than are required to stimulate stem growth.

There is doubt whether the movement of IAA is responsible for geotropic responses of stems and roots. For one thing, the concentrations of IAA measured in upper and lower surfaces of horizontally held organs are in general too small to account for the growth curvatures observed. An alternative explanation of geotropism may involve growth inhibitors, including abscisic acid. Inhibitors are present in the tissues of stems and roots responding to the stimulus of gravity. The simple auxin hypothesis of geotropism has been the subject of considerable criticism.

Statolith theory

How could the stimulus of gravity trigger changes in growth in the upper or lower surfaces of stems and roots? This really remains a mystery. A theory has been advanced, however, and there is some evidence to support it.

According to this theory, the gravity stimulus is perceived in cells containing large starch grains near the stem or root apex. These starch grains are called **statoliths**. Statoliths change their position in the cell when the plant is moved from a vertical to a horizontal position.

The suggestion is that when the statoliths fall on to endoplasmic reticulum in a different part of the cell they bring about the release of calcium ions there. The effect of the raised concentration of

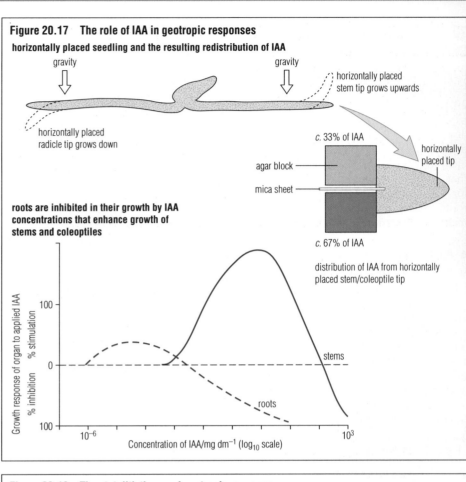

Figure 20.17   The role of IAA in geotropic responses

horizontally placed seedling and the resulting redistribution of IAA

roots are inhibited in their growth by IAA concentrations that enhance growth of stems and coleoptiles

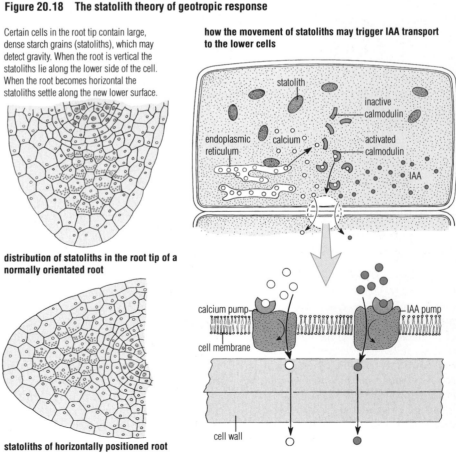

Figure 20.18   The statolith theory of geotropic response

Certain cells in the root tip contain large, dense starch grains (statoliths), which may detect gravity. When the root is vertical the statoliths lie along the lower side of the cell. When the root becomes horizontal the statoliths settle along the new lower surface.

distribution of statoliths in the root tip of a normally orientated root

statoliths of horizontally positioned root

how the movement of statoliths may trigger IAA transport to the lower cells

calcium ions is to activate calmodulin, a small protein that is present in cells and which is known to initiate the actions of several enzyme systems. These enzymes then activate growth of the cell. Activated calmodulin may switch on the cell membrane pumps for calcium ions and for IAA (or some growth inhibitor). Calcium ions and growth-regulatory substance would be passed across the tissue of the stem or root from cell to cell, extending the region of enhanced or inhibited growth (Figure 20.18).

20.4 THE MANIFOLD ROLES OF IAA

■ Apical dominance

In most species of plant the apical bud of the main shoot grows much more vigorously than the lateral buds below it. The apical bud appears to suppress growth of the lateral buds behind it (Figure 20.19). The degree of dominance of the apical bud is, however, very variable between species. For example, in tall, unbranched species apical dominance is strict, and extends over the entire stem. Many varieties of sunflower are like this. In shorter, much-branched plants apical dominance may extend only as far as the first few lateral buds. The tomato is a plant of this kind.

Apical dominance is imposed by the young growing leaves of the apical bud. These synthesise IAA at concentrations that suppress growth of buds below. If the apical bud is removed one or more lateral buds take over growth. If the apical bud stump is treated with IAA the growth of the lateral buds is suppressed. In some species the plant hormone gibberellin (p. 429) augments IAA action.

■ Other roles of IAA

There is good evidence that IAA is transported backwards from the apex of stem and root and regulates cell enlargement and the differentiation of the vascular tissue. The roles of IAA in stem, leaf, fruit and roots are summarised in Table 20.2.

Table 20.1 Examples of tropic and nastic responses (tropisms and nasties are defined in the panel on p. 422)

Stimulus	Tropisms	Nasties
light	**phototropism**: coleoptiles, young stems and some leaves are positively phototropic; certain roots, e.g. climbing roots of ivy (*Hedera helix*), are negatively phototropic	**photonasty**: composite flowering head (capitulum) of dandelion (*Taraxacum officinale*) closes in low light intensity and re-opens in full light
temperature		**thermonasty**: petals of tulip (*Tulipa*) flowers close at low temperature and during rapid fall in temperature; they re-open on the return of warmth
gravity	**geotropism**: coleoptiles and stems are negatively geotropic (or gravitropic); roots are positively geotropic (rhizomes and dicotyledonous leaves grow at 90° to gravity = *diageotropism*; lateral roots and lateral stems grow at some other angle to gravity = *plagiotropism*)	
chemicals including air	**chemotropism**: pollen tubes grow towards a chemical produced at the micropyle of the ovule **aerotropism**: pollen tubes are negatively aerotropic	
touch/solid surface	**thigmotropism**: roots may be positively thigmotropic*; tendrils and other climbing organs of climbing plants are positively thigmotropic	**thigmonasty**: leaves of the sensitive plant (*Mimosa pudica*) fold down when touched

*For a review of this response of roots see P W Freeland, *J. Biol. Ed.* **7**(6) (1973), 23–32

Table 20.2 The manifold roles of IAA

Organ	Effect
stem	promotes cell enlargement in region behind the apex
	promotes cell division in the lateral cambium, leading to secondary growth
leaves	inhibits leaf abscission
lateral buds	inhibits growth
fruits	promotes growth
	inhibits fruit abscission
root	promotes cell enlargement in region behind apex at low concentration only – inhibits at higher concentration
	initiates adventitious root formation in cuttings

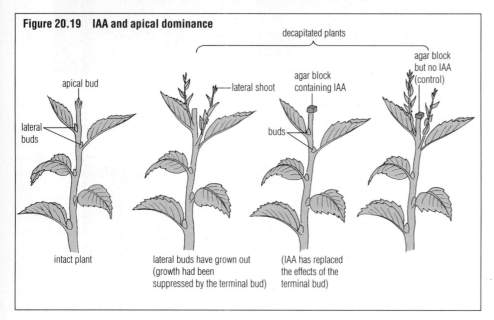

Figure 20.19 IAA and apical dominance

decapitated plants

apical bud

lateral shoot

agar block containing IAA

agar block but no IAA (control)

lateral buds

buds

intact plant

lateral buds have grown out (growth had been suppressed by the terminal bud)

(IAA has replaced the effects of the terminal bud)

Questions

1 In planning an experiment similar to that shown in Figure 20.19, what steps and procedures should be incorporated to ensure the results are significant?

2 A horizontally placed seedling is thought to acquire increased concentrations of IAA on the lower side of stem and root, yet the root tip grows down and the stem tip up. Why?

3 List plants where apical dominance is strictly adhered to, and others where only the first two or three lateral buds are suppressed.
   Use diagrams to contrast the branching pattern of these plants.

■ The role of IAA in tropism is controversial

Plant physiologists have critically re-examined the early studies by which 'auxin' was discovered and its role in tropism proposed. Some oppose the 'classical' story because of the following points:

▶ In the early experiments there was no attempt to maintain sterile conditions. We now know, however, that certain microorganisms release IAA into their surroundings. The IAA obtained from excised coleoptiles may have been produced by contaminating bacteria or fungi, rather than by the plant cells. Alternatively, the IAA could have been a product of the breakdown of proteins in the damaged stem tip.

▶ Much experimental evidence for plant growth regulation by IAA has been obtained using coleoptiles. The coleoptile is an unusual organ, found only in members of the grass family of monocotyledonous plants (Figure 20.9). The coleoptile is different in structure and behaviour from young roots or stems. It may have been unwise to build a general hypothesis on the regulation of growth in stems and roots on evidence of the responses of this untypical organ. For example, under some conditions the effect of cutting into coleoptiles is sufficient to promote changes in growth.

▶ Experimental results were more variable than the published accounts suggest. Many early experiments were not subject to proper statistical analysis. Those who are critical of the classical story point to contradictory experimental data submerged in what they described as the 'simplistic accounts usually favoured by plant physiology textbooks'.

▶ Many published IAA concentration gradients between illuminated and darkened sides (phototropism) and between upper and lower sides (geotropism) of coleoptiles are too small to have caused the spectacular growth curvatures observed. The levels of applied (exogenous) IAA that do result in pronounced growth curvatures are much higher than those found in intact plants (Figure 20.20, right).

▶ An inhibitor of IAA transport applied around the coleoptile tip reduces IAA export from tip to extension region by over 90%, but the extension growth of the coleoptile is unaffected (Figure 20.20, left).

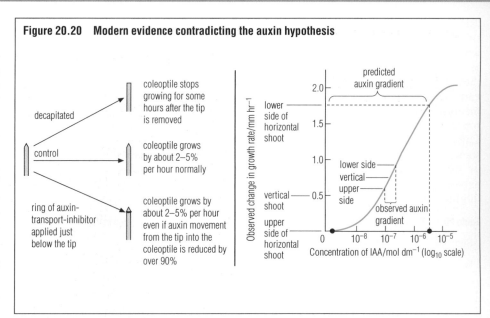

Figure 20.20   Modern evidence contradicting the auxin hypothesis

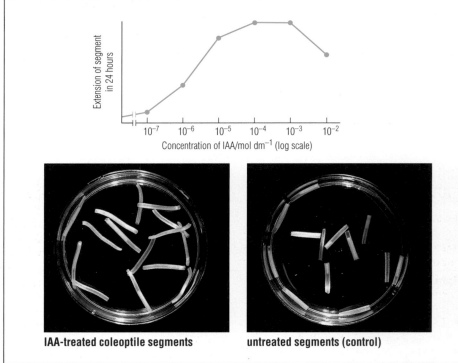

Figure 20.21   IAA promotes extension growth in sterile coleoptile segments: the relationships between the external concentration of IAA and the extension of oat (*Avena*) coleoptile segments, grown in presence of an antibiotic to prevent bacterial contamination

IAA-treated coleoptile segments   **untreated segments (control)**

Critical views on the roles of IAA in tropism can be read in Firn (1990) 'Phototropism', *J.Biol.Ed.* **20**(3), 153–7.

Meanwhile, other plant physiologists maintain their support for the idea that differences in IAA concentration cause the tropic responses.

■ The status of IAA as a plant growth substance

There is no doubt that IAA occurs naturally, and that it is a strong promoter of plant extension growth. In experiments conducted under sterile conditions, stem, root and coleoptile segments bathed in IAA solutions consistently show increases in length in the presence of an external concentration of IAA ranging from 10^{-7} to 10^{-3} M (Figure 20.21).

There is evidence that IAA does play a part in plant growth regulation, from the stimulation of extension growth to the maintenance of apical dominance. IAA is also implicated in other aspects of the regulation and coordination of plant growth.

Figure 20.22   Gibberellic acid and its effect on growth in height: the effect of GA on the growth of dwarf variety pea plants

several different forms of GA occur in plants; the red parts of the molecule are where there are differences between the various forms of GA

the GA was applied in alcoholic solution to the first leaf above ground during germination

μg GA applied in alcoholic solution

10.24   5.12   2.56   1.28   0.64   0.32   0.16   0.08   0.04   0.02   0.01   plant treated with alcoholic solution only   untreated plant

control plants

20.5 THE RANGE OF PLANT GROWTH SUBSTANCES

The successes of early searches for plant hormones comparable to those of animals led plant physiologists in many countries to seek to link aspects of plant growth and development with specific chemicals. Today five groups of substances, all occurring naturally in plants, have been identified as having some dramatic effects, even in low concentrations, on growth. These substances, which tend to interact in their effects as plant growth regulators, are discussed in this section.

■ Gibberellins

Gibberellins were first detected in the 1920s by Japanese plant pathologists working on a disease of rice known as 'foolish seedling' disease. Infected rice plants grew abnormally tall, with spindly stems. They became so weak they toppled over. From these plants a fungus, *Gibberella fujikori,* was obtained and cultured in liquid medium, entirely independently of the host plant. The used culture medium, with the fungal mycelium filtered out, itself induced abnormal growth when applied to healthy rice plants.

The substance produced by the fungus that caused rice stems to 'bolt' was eventually purified and chemically analysed. It was named gibberellic acid (Figure 20.22). Many higher plants have been found to contain gibberellic acid (GA).

The roles of GA in healthy tissues

GA is formed in chloroplasts, in the embryo of the seed and in young leaves, although it is absent from genetically dwarf varieties. Several different forms of the gibberellic acid molecule have been discovered and, consequently, these plant growth regulators are often referred to as gibberellins, rather than GA.

Gibberellins are involved in extension growth of the stem, causing elongation of internodes. When applied to intact plant stems their greatest effects are on dwarf varieties. Dwarf pea plants (genetically dwarf) become tall, for example.

The highest concentrations of gibberellins occur in fruits and seeds as they form, and in seeds as they germinate. The role of gibberellins in the mobilisation of stored food in germinating barley fruits is illustrated in Figure 19.13 (p. 405). There are other aspects of growth and development in higher plants, including dormancy, germination and the onset of flowering, in which gibberellins are also implicated (pp. 434–7).

Gibberellins for commercial use are produced in fermenters (p. 113), using fungal cultures. Gibberellin is used in the brewing industry to improve the malting of barley by stimulating amylase synthesis, thus enhancing the conversion of starch to sugars.

■ Cytokinins

Cytokinins were discovered during tissue culture experiments in which mature parenchyma tissue was isolated and cultured under sterile conditions, either on agar or in liquid media. This research sought to establish the essential growth requirements to make a single mature cell grow into a whole new plant, rather as a plant embryo grows from a single cell.

Experiments showed that mature cells did become meristematic and grew again when liquid coconut milk or DNA autoclaved in acid solution were added to the medium. Coconut milk, which is a liquid form of endosperm (p. 404), formed a natural source of a range of possible nutrients, vitamins and plant growth substances. Tissue supplied with an inappropriate range and balance of nutrients remained as a lump of undifferentiated parenchyma. On analysis, culture media that sustained the differentiation of shoots or roots were found to contain, among other compounds, a new group of plant growth substances, now named **cytokinins**. These are chemicals derived from the nucleic acid base adenine (p. 146). Cytokinins are produced in dividing cells throughout the plant. In mature plants cytokinins are produced in the root tips and travel to the shoots in the transpiration stream.

■ Plant and animal hormones compared

The concept of a 'hormone' was first developed by animal physiologists, and the term was applied to the chemical messengers being discovered in animals. Animal hormones are described in detail in Chapter 21. We now recognise that plant hormones are different from animal hormones in some important ways.

► Animal hormones are produced in specific organs or glands in specialised cells or tissues. Plant growth regulators may be produced in restricted areas of plants, but they are manufactured in relatiively unspecialised cells.

► Animal hormones are transported to all parts of the body in the bloodstream, but they have their effects in quite restricted target organs or tissues. Plant growth regulators are not always transported widely in the plants, and some are formed at the site where they are active.

► The effects of animal hormones are mostly specific to a particular tissue or organ in the body, and do not influence other pats of the body or more than one process. Plant growth regulators lack specificity, and they tend to influence different tissues and organs, sometimes in contrasting ways.

Today, plant hormones, or phytohormones, are often referred to as 'plant growth substances' in order to avoid direct comparison with animal hormones. Plant growth substances play an important part in regulating growth and development. They are not the only method of regulation, however. Other factors, such as the balance of nutrients reaching a meristem or developing organ, play a part in the regulation of growth and development.

Figure 20.23   Cytokinins, and their role in growth and development

cytokinins are mainly derived from the nucleic acid adenine

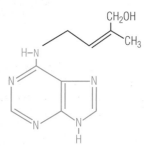

a natural cytokinin
(the adenine residue is in blue)

explants of parenchyma tissue from stems of tobacco (*Nicotiana* sp.) were cultured on an agar preparation containing nutrients

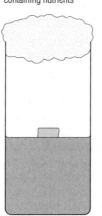

no growth

nutrients in agar with 0.2 mg dm^{-3} cytokinin

cell enlargement

nutrients in agar with 2 mg dm^{-3} IAA

discs of sterile parenchyma tissue from tobacco stem placed in each tube

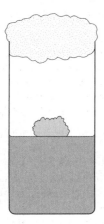

cell division and cell enlargement but producing only a mass of undifferentiated cells

nutrients in agar with 2 mg dm^{-3} IAA and 0.2 mg dm^{-3} cytokinin

root formation

nutrients in agar with 2 mg dm^{-3} IAA and 0.02 mg dm^{-3} cytokinin

shoot formation

nutrients in agar with 0.02 mg dm^{-3} IAA and 1.0 mg dm^{-3} cytokinin

Roles of cytokinins

Cytokinins promote cell divisions, working with other growth-promoting substances to do so. This was established in experiments conducted with small pieces of living plant tissue known as **explants**. The explants were cultured in sterile conditions with a supply of sugars, minerals and vitamins. The resulting organ differentiation in the explants depended on the balance between IAA, GA and cytokinin present. The results are summarised in Figure 20.23.

Cytokinins do not influence growth in length of the stem, but are involved in the breaking of dormancy in buds and seeds (p. 438), and in the release of lateral buds from dominance by the apical bud (p. 427). Added to mature leaves, they delay the onset of senescence.

■ Abscisic acid

The discovery of growth inhibitors in plants came from the investigation of

a young plant grown from a tissue explant and cultured on agar with the necessary nutrients; the plant has established leaves, stem and roots

Figure 20.24   Abscisic acid and fruit fall in the cotton plant

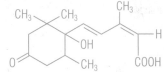
abscisic acid (ABA)

the concentration of abscisic acid in cotton fruit builds up from 0 to 4–5 µg per fruit in the 20 days prior to fruit fall

a cotton plant (*Gossypium* sp.) and its open boll

dormancy of buds and seeds and of the abscission of fruits and leaves. An extract with the properties of a growth inhibitor, first made from dormant buds of the sycamore tree (*Acer pseudoplatanus*), was found to contain an active ingredient, named **abscisic acid** (ABA). ABA, now detected and measured by gas–liquid chromatography (Figure 20.25, overleaf), has been isolated from several different plants.

The roles of ABA in plants

ABA is formed in mature leaves and in ripe fruits and seeds. ABA accumulates in leaves of deciduous trees in late summer, just prior to the winter buds becoming dormant. It also occurs in dormant seeds (p. 438), and in deciduous tropical plants at the onset of the dry season. Investigations of fruit fall in the cotton plant (*Gossypium* sp.) showed that ABA accumulates in cotton plant tissues as the fruit ripens (Figure 20.24).

ABA is also involved in the closure of stomata under condition of physiological stress induced by prolonged drought. Plants in this condition produce ABA in chloroplasts, and this inhibitor reaches the guard cells. Here ABA suppresses the ability of the guard cells to retain potassium ions, leading to prolonged stomatal closure (p. 324).

Questions

1   How might the breakdown products of protein contribute to the appearance of IAA in damaged tissues (p. 428)?
2   How could contaminating microorganisms cause the distortion of results of experiments involving parts of plants and the effects of plant growth regulators (p. 428)?
3   From Figure 20.22 (p. 429), what appears to be the optimum concentration of GA in promoting growth in dwarf pea plants?
4   What is the effect of GA on the stored proteins of germinating seeds (p. 405)?
5   What role does adenine have in metabolism, other than being involved in cytokinin synthesis?
6   a   Explain why the term 'plant growth substance' is now preferred to the term 'plant hormone'?
     b   Suggest illustrations of each of the three differences between animal and plant 'hormones' discussed above, using examples from a plant and the endocrine system of a mammal (p. 472).

Figure 20.25 Gas–liquid chromatography

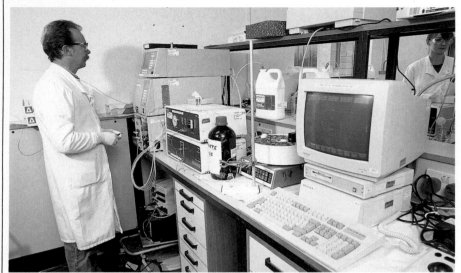

scientists working in a plant biochemistry research laboratory; the advanced analytical equipment in regular use includes the gas chromatograph

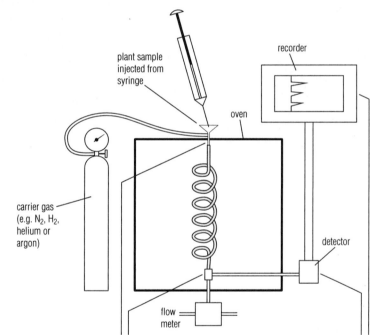

Typically, a column contains a high-boiling alkane supported on a porous inert solid such as alumina. Carrier gas is passed through, and a sample is injected into the flow at the top of the column.

Components of the sample pass through at different rates and are detected in sequence, causing signals to pass to the recorder.

Gas chromatogram

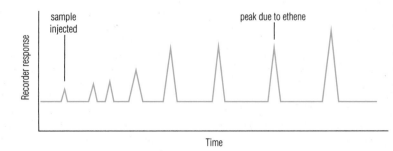

■ Ethene

Ethene is a relatively simple organic molecule that exists as a gas at normal temperatures. It is thus different from the other growth regulators, IAA, GA, cytokinins and ABA. You may know of ethene as the compound from crude oil that is polymerised to form polythene.

In fact, ethene is produced naturally in trace amounts in most of the organs of higher plants. It is detected by the technique of gas–liquid chromatography (Figure 20.25), making possible the measurement of very small quantities.

The role of ethene in development

It has been known for many years that fruits produce ethene as they ripen. Unripe fruits, such as apples, oranges, lemons and bananas, ripen more quickly if they are exposed to ethene. The speeding up of ripening (Figure 20.26) has been of important commercial value to horticulture, particularly where a crop is transported unripe, and can be ripened to coincide with a period of demand for the fruit.

Ethene, in addition to promoting fruit ripening, has an inhibitory effect on stem growth, particularly during periods of physiological stress. It contributes to the breaking of bud dormancy.

■ The interactions of plant growth regulators

So far the individual plant growth regulators have been introduced in turn, and their roles illustrated independently.

In plants, however, many aspects of development and response are influenced by two or more plant growth regulators. The control of growth processes can only occasionally be attributed to a single plant growth substance.

Plant growth substances may interact to reinforce each other's effects, a process known as **synergism**; for example, GA and IAA are synergistic in stem elongation. Alternatively, plant growth substances may oppose each other's effects, known as **antagonism**; for example, cytokinins are antagonistic to maintenance of apical dominance of the terminal bud by IAA.

Ethene produced in leaves is an essential regulator of abscission (p. 439). In the earliest stage of abscission, IAA is antagonistic to ethene, but later IAA becomes synergistic to ethene's action.

The interactions of plant growth substances in growth and sensitivity are summarised in Table 20.3.

Table 20.3 The interactions of phytohormones in growth and development

Process	Cellular changes	Roles of: IAA	GA	Cytokinins	ABA	Ethene
In stems:						
stem extension growth (growth in length)	cell division at apex	–	promotes	promotes	inhibitory during physiological stress	
	cell enlargement and differentiation	promotes	promotes with IAA	–	antagonistic to IAA action	promotes bud dormancy
cambial activity (growth in girth)	as above	promotes	promotes with IAA	–	antagonistic to IAA action	–
bud formation	as above	promotes slightly	promotes slightly	promotes	–	–
apical dominance of terminal bud	as above	inhibits lateral buds	augments IAA action	antagonistic to IAA action	promotes bud dormancy	–
tropic responses of stems	cell enlargement	promotes	promotes	–		may inhibit IAA action
In roots:						
root extension growth	cell division at apex cell enlargment and differentiation	promotes at very low level, inhibits at higher levels	–	–	–	–
root initiation	cell division in meristematic cells cell enlargement and differentiation	promotes when applied to cuttings	inhibits	–	–	–
tropic response of roots	cell enlargement	promotes	promotes	–		may inhibit IAA action
In leaves:						
leaf growth	cell enlargement	–	promotes	promotes	–	–
stomatal opening/closing	movement of K⁺ ion in/out of guard cell	–	–	promotes opening of stomatal pore	promotes closing of stomatal pore	–
ageing (senescence) of leaves	loss of proteins and other substance from cells	delays	delays	delays	promotes	promotes

– means that the effects are unknown or absent

Figure 20.26   Ethene and fruit ripening

ethene production by plant tissues is variable; it is highest in apical meristems (about 40 µl h⁻¹ kg⁻¹ fresh mass), and in many ripening fruit (up to 500 µl h⁻¹ kg⁻¹ fresh mass)

a demonstration of the use of ripe fruit as a source of ethene to ripen green tomatoes

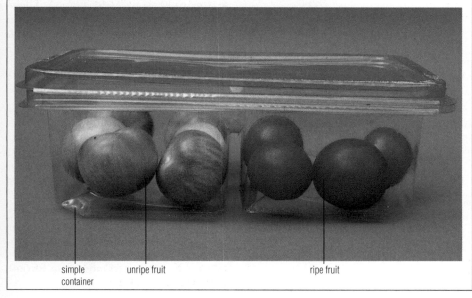

simple container   unripe fruit   ripe fruit

Questions

1   Why does a ripe tomato, enclosed with unripe fruit but not in contact with them, facilitate their ripening?

2   Using Table 20.3, give two examples of phytohormones working in support of each other (synergism) and two examples of phytohormones counteracting each other's influence (antagonism).

3   If you removed the apical bud from a shoot and applied either IAA, cytokinin or abscisic acid to the cut surface, what would you expect to happen in each case a few days later?

4   Design an experiment to test the hypothesis that a volatile substance emitted by the ripe fruit enhances the ripening process in unripe tomatoes.

20.6 GERMINATION, FLOWERING, DORMANCY AND LEAF FALL

As you know, stages in plant development such as germination, flowering and dormancy are geared to the seasons. As a result, the organism is active at times when conditions are favourable and dormant when they are unfavourable. How is plant development coordinated with environmental opportunities?

■ Flowering

There is a great deal of variation in the flowering period of plants. For example, flowering occurs with differing frequencies in the life cycles of annual, biennial and perennial plants. Different plant species flower at different times in the year, some in spring, some in summer and some in autumn. In temperate regions very few flower in winter.

Whenever it does occur, the flowering process shows certain common features.

Vegetative development, the first requirement

After germination (p. 404) a young plant grows vegetatively, forming an aerial system of leaves and stem, and a root system below ground. Vegetative growth continues until a certain minimum size is reached. Only then can the plant switch to reproductive development – a condition of 'ripeness to flower' has been achieved. Eventually one or more growing points of the stem converts from leaf formation to flower production.

Table 20.4 Examples of short-day and long-day plants

Short-day plants	Long-day plants
chrysanthemum (*Chrysanthemum* spp.)	barley (winter) (*Hordeum vulgare*)
cocklebur (*Xanthium strumarium*)	Canary grass (*Phalaris arundinacea*)
coffee (*Coffea arabica*)	clover, red (*Trifolium pratense*)
kalanchoë (*Kalanchoë blossfeldiana*)	dill (*Anethum graveolens*)
morning glory (*Ipomoea hederacea*)	henbane, annual (*Hyoscyamus niger*)
rice (*Oryza sativa*)	oat (*Avena sativa*)
soya bean (*Glyceria soja*)	rose mallow (*Hibiscus syriacus*)
strawberry (*Fragaria chiloensis*)	rye grass, Italian (*Lolium italicum*)
sugar cane (*Saccharum officinarium*)	spinach (*Spinacia oleracea*)
tobacco (*Nicotiana tabacum* var. 'Maryland Mammoth')	wheat (winter) (*Triticum aestivum*)

Day-neutral plants
antirrhinum (snapdragon) (*Antirrhinum magus*)
balsam (*Impatiens balsamina*)
bean, string (*Phaseolus vulgaris*)
cucumber (*Cucumis sativus*)
tomato (*Lycopersicon esculentum*)

Onset of flowering

There are several species where only ripeness to flower is required for flowering to occur. For example, some varieties of tomato grow vegetatively until the thirteenth node and then switch to flower formation (Figure 20.27). Many weed species also flower immediately a certain size has been reached.

Most plants flower only at a particular time of the year, suggesting that some seasonal change triggers the switch from vegetative growth to flower production. The most obvious seasonal changes are those of day-length and temperature. We shall see that one or other of these environmental signals cues the onset of flowering in many plants.

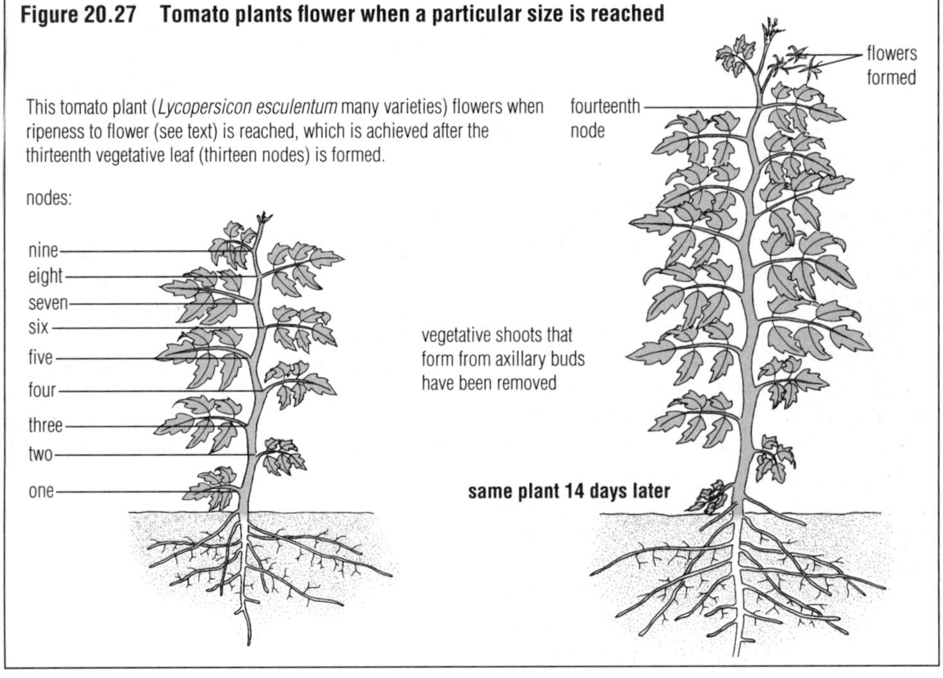

Day-length control of flowering (photoperiodism)

In temperate parts of the world, such as Britain, daylight in midwinter lasts for only eight hours but in midsummer for about seventeen hours. Plants that germinate in spring reach ripeness to flower as days are lengthening. After midsummer there is a steady shortening of day-length through the autumn. Some plants of temperate zones require long summer days to trigger flowering. These are known as **long-day plants**. Long-day plants make up the majority of summer-flowering annuals of temperate regions, and of most biennial plants as well.

In other plants flowering is triggered by the shorter days, and these plants are known as **short-day plants**. Short-day plants make up the bulk of autumn-flowering plants of temperate regions. In the tropics the day-length changes little from season to season. Most plants of low-latitude zones north or south of the Equator are short-day plants (Table 20.4).

Photoperiodic day-length is relative
Photoperiodism was discovered by two American agricultural botanists, W W Garner and H A Allard, in 1920.

The concept of photoperiodism refers not to absolute length of day and night, but whether the day is longer or shorter than a critical length.

A short-day plant will not flower if the daily illumination exceeds a certain critical number of hours. For example, *Chrysanthemum* will not flower if exposed to more than fifteen

Figure 20.27 Tomato plants flower when a particular size is reached

This tomato plant (*Lycopersicon esculentum* many varieties) flowers when ripeness to flower (see text) is reached, which is achieved after the thirteenth vegetative leaf (thirteen nodes) is formed.

nodes:
nine, eight, seven, six, five, four, three, two, one

vegetative shoots that form from axillary buds have been removed

same plant 14 days later

flowers formed
fourteenth node

hours of daylight (that is, it requires a minimum of nine hours of darkness).

A long-day plant will not flower unless it receives more than a certain number of hours of illumination. For example, the oat (*Avena*) requires at least nine hours illumination per day, and will flower most quickly if maintained in uninterrupted illumination.

Thus *Chrysanthemum*, a short-day plant, is able to flower in day-lengths longer than that needed by the oat, a long-day plant (Figure 20.28).

Low-temperature control of flowering (vernalisation)

In some species, the young plant or seedling requires a period of cold treatment before flowering will occur. Biennial plants such as carrot and parsnip grow vegetatively in their first year, and require overwinter cold treatment before they can flower the following year. Some varieties of cereals, such as winter wheat, flower in summer only if planted the preceding autumn and subjected to low winter temperatures. 'Spring' varieties of the same species are planted in spring, flower a few weeks later and can still be harvested by the early autumn (Figure 20.29).

The hormone control of flowering

When a plant has been exposed to a regime of light and dark for long enough to promote flowering, then it is said to be **photoperiodically induced**. Structural changes in the photoperiodically induced plant are not immediately visible, but flowering is now inevitable, given suitable growing conditions. What is the mechanism of this change from vegetative growth to flower formation?

The number of cycles of light/darkness needed for photoperiodic induction is quite limited in most plants, ranging from one cycle (only 24 hours) in the cocklebur to 8–30 cycles in the chrysanthemum, depending upon the species. Thus the amount of light energy needed to cue flowering is relatively small. This is in keeping with a hypothesis that flowering is switched on after the production of a small quantity of messenger chemical, a plant growth regulator, rather than as a result of the accumulation of metabolites in substantial quantities as in photosynthesis.

The hypothesis that flowering is controlled by a specific phytohormone was the idea of the Russian physiologist,

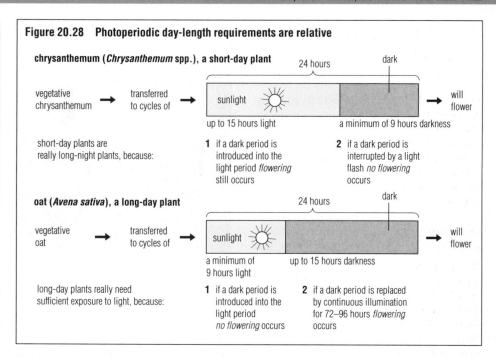

Figure 20.28 Photoperiodic day-length requirements are relative

chrysanthemum (*Chrysanthemum* spp.), a short-day plant

short-day plants are really long-night plants, because:

1 if a dark period is introduced into the light period *flowering still occurs*
2 if a dark period is interrupted by a light flash *no flowering occurs*

oat (*Avena sativa*), a long-day plant

long-day plants really need sufficient exposure to light, because:

1 if a dark period is introduced into the light period *no flowering occurs*
2 if a dark period is replaced by continuous illumination for 72–96 hours *flowering occurs*

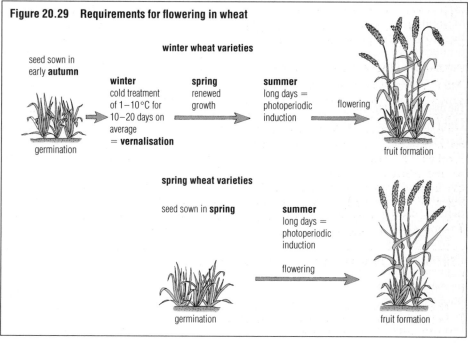

Figure 20.29 Requirements for flowering in wheat

winter wheat varieties

seed sown in early **autumn**

germination → **winter** cold treatment of 1–10 °C for 10–20 days on average = **vernalisation** → **spring** renewed growth → **summer** long days = photoperiodic induction → flowering → fruit formation

spring wheat varieties

seed sown in **spring**

germination → **summer** long days = photoperiodic induction → flowering → fruit formation

Chailakhyan, in 1936. The hypothetical hormone was given the name '**florigen**', and there is a certain amount of evidence for its existence.

The leaves of the plant are the organs that detect the exposure to correct photoperiods, but response occurs at the growing points. The evidence for a flowering hormone produced in the leaves and transported in the stems to the growing points is based on several experiments, summarised in Figure 20.30 (overleaf).

For example, only a small part of one leaf needs to be exposed to the correct photoperiods for the plant to flower.

Questions

1 List some annual, biennial and perennial (woody and herbaceous) plants growing in your area, and note down when they commonly flower.
2 Why can we say that short-day plants are really long-night plants?
3 What do winter wheat plants require to induce flowering that is not required by spring wheat varieties?

If the leaves of an experimental plant are removed immediately after or soon after photoperiodic induction, then flowering will not occur for a period of time. Eventually the growing points do receive the 'cue' to flower in sufficient quantity because subsequent removal of leaves does not prevent flowering.

Experiments have been carried out with grafted plants, in which a photoperiodically induced stem or leaf is grafted on to a non-induced plant and kept under a light/dark regime unfavourable for the switch to flowering. Flowering nevertheless took place. Grafting of vernalised plants on to non-vernalised plants has similarly led to flowering in both plants.

The hypothesis that attempts to explain the dependence of flowering on a specific photoperiod suggests that when leaves are exposed to appropriate periods of light and darkness, they produce a chemical (a hormone), which is translocated to the growing points. When this hormone arrives in sufficient quantities at the growing points, buds are induced to produce flowers instead of (or as well as) leaves. The hormone ('florigen') has not, however, been isolated and identified.

The discovery of phytochrome

Light influences many aspects of plant growth and development, quite apart from its role in photosynthesis. The effect of light on growth and development is known as **photomorphogenesis**. Both the duration of the illumination and the quality (wavelength) of the incident light are important.

The first stage in any light-driven process is the absorption of light by a pigment. The existence of a special pigment system involved in photomorphogenesis (including photoperiodism) was discovered by analysis of the action spectrum of the light that induced changes in plant growth and development. An **action spectrum** is a graph of the response of a plant to varying wavelengths of light; for example, the action spectrum for chlorophyll in photosynthesis is illustrated in Figure 12.10 (p. 251). Coincidence of the action and absorption spectra is evidence that the pigment concerned is the photoreceptor for the observed plant response.

The pigment now known as **phytochrome** was discovered during studies of certain varieties of lettuce seed that require light in order to germinate. The germination of these varieties is stimulated by red light (660 nm wavelength), and inhibited by far-red light (730 nm).

Figure 20.30   Evidence for a flowering hormone in cocklebur (*Xanthium* sp.)

Xanthium growing as a weed species in a cotton crop

Xanthium in flower

flowering occurs when leaves are exposed to the correct regime of light/dark, even if only part of one leaf is photoinduced

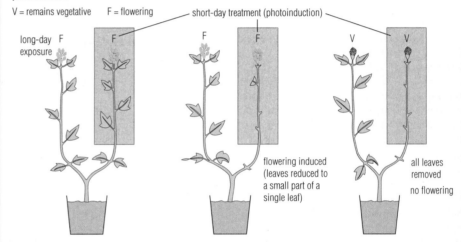

flowering 'hormone' can be transmitted between plants across a graft union

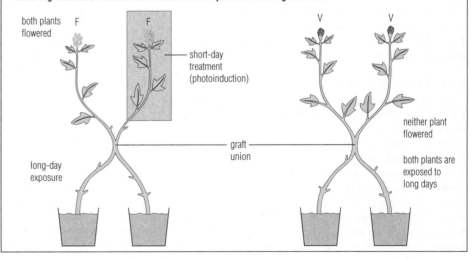

The action spectrum of photoperiodism was similar. For example, in the short-day plants soybean and cocklebur, a brief period of illumination in the long night prevented flowering. This illumination was effective if red light was used, but its effect was cancelled if the plants were exposed to far-red light afterwards.

The initial identification of a pigment system that could absorb light at red and far-red wavelengths was difficult because of the compound's low concentration in

Figure 20.31 Phytochrome, absorption and action spectra

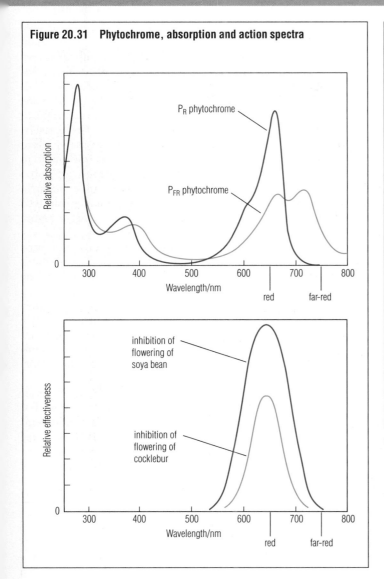

Figure 20.32 Phytochrome and flowering, a tentative hypothesis

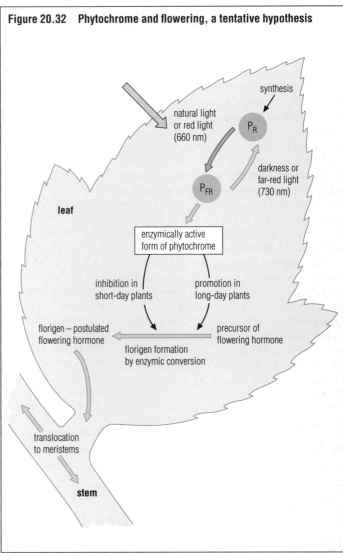

cells and because of the masking effects of chlorophyll. Isolation of the pigment, which was named phytochrome, was first achieved in 1959.

The chemistry of phytochrome

Phytochrome turned out to be a large, conjugated protein ($M_r = 120\,000$). The pigment part of the molecule is chemically a close relative of phycocyanin (p. 27), found in cyanobacteria and red algae. The protein part of phytochrome contains a high proportion of acidic, basic and sulphur-containing amino acids. Consequently, the phytochrome molecule is highly charged and highly reactive, capable of many rearrangements leading to changes in shape, and possibly to chemical reactions between neighbouring molecules. Phytochrome occurs attached to cell membranes.

Phytochrome exists in two interconvertible forms. One (P_R) is a blue pigment with its maximum absorption peak in the red part of the spectrum at 660 nm (Figure 20.31). The other form (P_{FR}) is blue-green, with maximum absorption in the far-red region, at the edge of visibility for the human eye. When P_R is exposed to light or to red light it is converted to P_{FR}. In the dark, or when exposed to far-red light, P_{FR} is converted back to P_R.

How is phytochrome involved in flowering?

The answer to this question is not clear.

Phytochrome is potentially a very reactive molecule. It may work by causing the formation of new enzymes or by activating existing ones. It is the P_{FR} form of phytochrome that is enzymatically active. P_{FR} promotes flowering in long-day plants but it apparently inhibits flowering in short-day plants (Figure 20.32).

Other hormones influence flowering in certain species under particular conditions. For example, GA promotes flowering in certain long-day plants, and inhibits flowering in certain short-day plants. ABA stimulates flowering in certain short-day plants, and inhibits it in certain long-day plants. Neither GA nor ABA is a candidate for the elusive 'florigen', however.

Questions

1 How can it be shown that it is the leaves of a particular species that detect the photoperiods, not the growing points themselves?

2 What are the essential conditions for the photoperiodic induction of flowering in the short-day plant *Chrysanthemum*?

3 What reasons can you suggest to explain why 'florigen' has not been isolated? (Your ideas may be just as likely to be close to the truth as those of an expert in this issue.)

■ The dormancy state

Dormancy is a period when active growth is suspended.

Dormancy may occur because external conditions are unfavourable for growth. This type of dormancy is said to be **imposed**. For example, several species of annual weeds, including groundsel (*Senecio vulgaris*), shepherd's purse (*Capsella bursa-pastoris*) and chickweed (*Cerastium* spp.) have no regular dormant phase in the life cycle, although seeds fail to germinate in the very coldest weather. On the other hand, when growth is arrested for a period, whether or not external conditions are suitable, dormancy is said to be **innate**.

Most plants of temperate regions pass through a phase of innate dormancy at some stage in the life cycle. These periods of dormancy mostly coincide with an unfavourable season. In higher plants innate dormancy involves seeds, underground organs such as rhizomes, corms or tubers, and the winter buds of woody twigs.

Bud dormancy

In the trees and shrubs of temperate regions, the growing regions of stems are protected by buds. The structure of a winter bud is shown in Figure 20.33.

Winter buds are formed in summer while growing conditions are favourable. Although formed in the growing season, these buds do not normally break until after winter. Bud dormancy at this stage is imposed by the surrounding leaves, for if the plant is unexpectedly defoliated – by insect attack or by humans, for example – the buds will grow and produce a second wave of foliage in the same season.

By late autumn innate dormancy of the buds is established, triggered, it seems, by the short days of autumn. This is a long-night requirement (as in short-day plants), for if the dark periods are interrupted by light then innate dormancy is delayed.

For bud dormancy to be broken, the woody plant requires quite prolonged exposure to low temperatures. Consequently dormancy diminishes in the buds as winter proceeds. The buds of twigs brought indoors in October, November, and early December will not break. Those brought in from January onwards will do so increasingly quickly, particularly after a cold spell.

Seed dormancy

Many seeds do not germinate as soon as they are formed and dispersed. This is particularly the case with seeds of wild plants. Most seeds have a dormant period and may germinate only after specific conditions have been met. For example, at the time of dispersal (p. 574) seed development may be incomplete, and the embryo may be structurally or physiologically immature. Alternatively, or in addition, the seed coat may prevent germination through impermeability to water or oxygen, or because it is too tough for the embryo to grow through. Examples of embryo dormancy and seed coat dormancy are illustrated in Figure 20.34.

Under natural conditions the seed matures in time, and the thick seed coat is weakened by abrasion in the soil and by the actions of soil microorganisms.

If seeds germinated as soon as they were shed in late summer or early autumn, they would produce plants that might succumb to harsh winter conditions before they could reproduce. Dormancy helps to avoid this hazard.

The dormancy of seeds of many species is overcome by low temperatures. For this to be effective the seeds require temperatures of 0–5°C for two to three weeks.

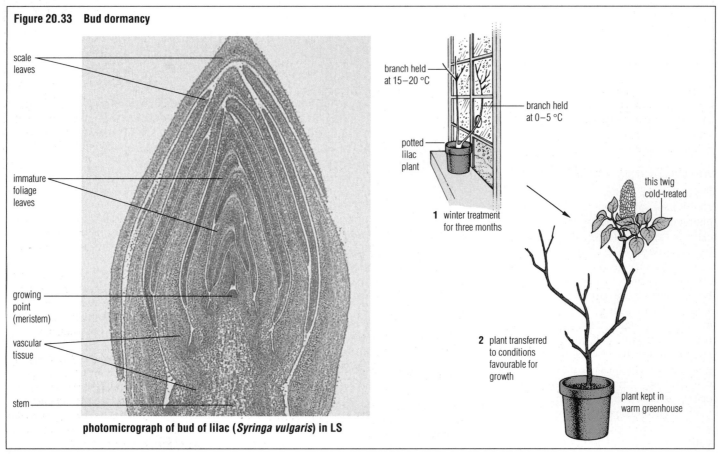

Figure 20.33 Bud dormancy

scale leaves

immature foliage leaves

growing point (meristem)

vascular tissue

stem

photomicrograph of bud of lilac (*Syringa vulgaris*) in LS

branch held at 15–20 °C

branch held at 0–5 °C

potted lilac plant

1 winter treatment for three months

this twig cold-treated

2 plant transferred to conditions favourable for growth

plant kept in warm greenhouse

Figure 20.34   Seed dormancy: seed coat dormancy and embryo dormancy

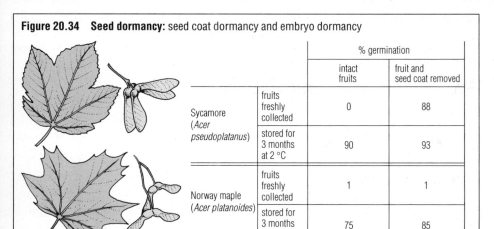

		% germination	
		intact fruits	fruit and seed coat removed
Sycamore (*Acer pseudoplatanus*)	fruits freshly collected	0	88
	stored for 3 months at 2 °C	90	93
Norway maple (*Acer platanoides*)	fruits freshly collected	1	1
	stored for 3 months at 2 °C	75	85

'Hormone' control of germination

The role of plant growth substances in dormancy has been studied by observing the effects of external application of plant growth substances and by investigating the changing concentrations of these substances in buds and seeds as they develop. The results indicate that no single substance controls dormancy, but rather that dormancy is controlled by interactions between growth inhibitors (such as ABA) and growth promoters (such as GA, cytokinins and ethene).

Figure 20.35   Leaf fall in a woody plant

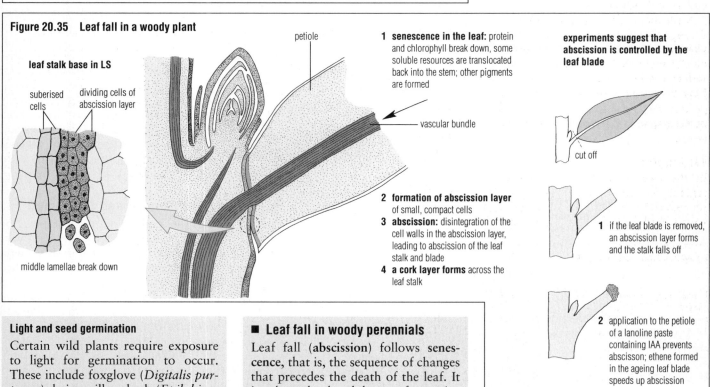

leaf stalk base in LS

suberised cells

dividing cells of abscission layer

middle lamellae break down

petiole

vascular bundle

1 **senescence in the leaf:** protein and chlorophyll break down, some soluble resources are translocated back into the stem; other pigments are formed

2 **formation of abscission layer** of small, compact cells
3 **abscission:** disintegration of the cell walls in the abscission layer, leading to abscission of the leaf stalk and blade
4 **a cork layer forms** across the leaf stalk

experiments suggest that abscission is controlled by the leaf blade

cut off

1 if the leaf blade is removed, an abscission layer forms and the stalk falls off

2 application to the petiole of a lanoline paste containing IAA prevents abscisson; ethene formed in the ageing leaf blade speeds up abscission

Light and seed germination

Certain wild plants require exposure to light for germination to occur. These include foxglove (*Digitalis purpurea*), hairy willow herb (*Epilobium hirsutum*) and dock (*Rumex crispus*). In a few species, germination is inhibited by light.

Work on the light-induced breaking of dormancy in the variety of cultivated lettuce known as 'Grand Rapids' (no longer available) led to the discovery of the pigment phytochrome. Exposure of seeds to white light triggered germination. So too did exposure to red light (660 nm), but this was reversed by exposure to far-red light (730 nm). Seeds exposed for a time to light of these two wavelengths alternately germinated only when red light was the last illumination given. Phytochrome (p. 436) has been extracted from lettuce seeds.

■ Leaf fall in woody perennials

Leaf fall (**abscission**) follows **senescence**, that is, the sequence of changes that precedes the death of the leaf. It involves the breakdown of proteins and chlorophyll and the synthesis of certain other pigments. Senescence is triggered by short days (and is consequently delayed in leaves around a street lamp) and the low temperatures of autumn (Figure 20.35).

Leaf fall requires the development of an **abscission layer** at the base of the petiole (leaf stalk). The abscission layer is a compact zone of small cells. Then the middle lamella of the cells of the abscission layer becomes gelatinous, resulting in abscission. A layer of cork forms beneath the abscission layer, sealing off the leaf scar.

Abscission is inhibited by IAA and promoted by ethene. Ageing leaves no longer produce IAA but are the site of ethene production.

Questions

1 What are the advantages of abscission and leaf fall in the autumn to trees in temperate regions?

2 For what reasons have cultivated crop plants been bred to germinate as soon as they are planted, given favourable conditions?

3 What physical properties of cork cells make them an effective barrier at the leaf scar (p. 410)?

4 How do the conditions required for flowering in Figure 20.33 contrast with those in Figure 20.29? How do you interpret this difference?

5 Examine Figure 20.34. How does the data given support the idea of dormancy being due to immaturity of the embryo in the Norway maple but not in the sycamore?

20.7 APPLICATIONS OF PLANT GROWTH SUBSTANCES IN INDUSTRY

■ Synthetic alternatives to auxin

Substances that alter the rate of plant growth or the pattern of plant development have proved to be useful in the destruction of weeds and in the improvement of crop productivity. After the chemical structure of indoleacetic acid (IAA) was determined, a search for synthetic alternatives was initiated. There are many compounds similar to IAA that have comparable effects upon growth and development, and several have proved more effective than IAA (Figure 20.36). The fact that the mechanism of plant growth regulator action is not fully understood has not prevented these substances from becoming useful tools in agriculture and horticulture.

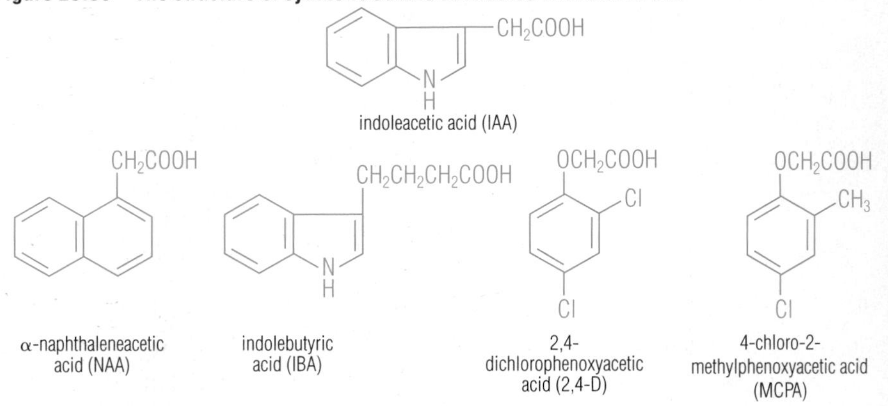

■ Hormones in the rooting of cuttings

A length of plant stem cut from the parent plant and placed in moist soil normally grows roots at its base, and the 'cutting' takes root and develops into a new plant. Propagation via cuttings is common practice in horticulture and in plant breeding research. Exposure to IAA enhances the initiation of roots, but artificial auxins can be used instead (Figure 20.37). In particular, α-naphthalene-acetic acid (NAA) is more useful than IAA because it is more easily synthesised, is less costly, and is not inactivated by IAA oxidase enzyme in the plant and therefore persists.

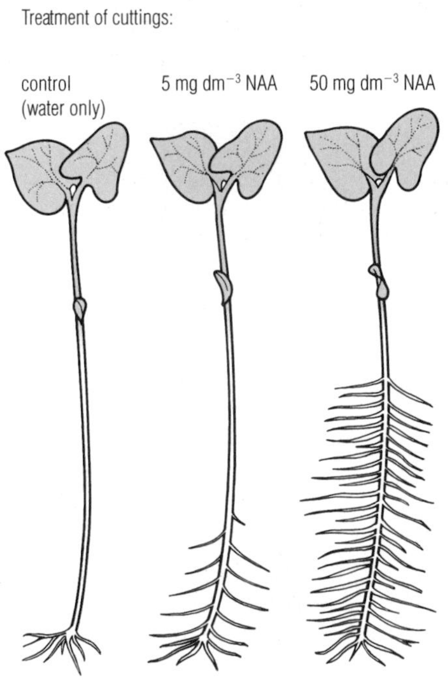

■ Delay of pre-harvest fruit drop

Fruit falls from plants when an abscission layer forms at the base of the fruit stalk, in much the same way as in leaf fall (Figure 20.35). As with leaves, the application of an auxin delays abscission.

Commercial growers often use sprays of 2,4-dichlorophenoxyacetic acid (2,4-D) on apple, pear and citrus crops. 2,4-D is more effective than exogenous IAA in delaying fruit fall, even when applied at lower concentrations (Figure 20.38).

■ Artificial auxins as herbicides

Research into hormone weedkillers or herbicides was commenced in 1940, and was initially shrouded in military secrecy. Two of the synthetic auxins, 2,4-D and the substance known as MCPA (Figure 20.36), proved to be powerful herbicides. At very low concentrations they affect

Figure 20.36 The structure of synthetic auxins contrasted with that of IAA

indoleacetic acid (IAA)

α-naphthaleneacetic acid (NAA)

indolebutyric acid (IBA)

2,4-dichlorophenoxyacetic acid (2,4-D)

4-chloro-2-methylphenoxyacetic acid (MCPA)

Figure 20.37 Stimulation of root initiation in bean cuttings using NAA

Treatment of cuttings:

control (water only)

5 mg dm^{-3} NAA

50 mg dm^{-3} NAA

growth in a similar way to IAA. At slightly higher concentrations they are toxic to broad-leaved plants (dicotyledons). Cultivated grasses such as wheat and barley are unaffected.

Herbicides of this type are taken up by the leaves and are transported to the growing points. Herbicides and insecticides that are taken up and translocated within the plant system are said to be **systemic**. These herbicides disrupt growth of meristems, causing death from excessive growth. Because they are translocated within plants they kill the roots also. At the concentration applied, these substances are comparatively harmless to animals, but they may cause serious

ecological problems if used excessively or indiscriminately (p. 91).

Herbicides are used to destroy broad-leaved weeds and so increase productivity in the cereal crops treated (p. 92). Of course, they cannot be used on broad-leaved crops such as the legumes. Many other selective herbicides have been developed; Dalapon (2,2-dichloropropionic acid), for example, is a systemic herbicide that is selectively toxic to monocotyledons, such as the grasses.

■ Parthenocarpic fruit development

After the pollination of a flower, the ovules develop into seeds and the ovary develops into a fruit. For fruit growth, IAA is required. IAA is normally supplied from the fertilised ovules as they develop into seeds (Figure 20.39). If ovules are not fertilised the seeds fail to develop and the young fruit falls off the plant.

Parthenocarpy

Fruit development in the absence of a fertilised ovule is referred to as **parthenocarpic** fruit development. Parthenocarpic plants do not need pollination and fertilisation to form seeds and fruits. There are instances of parthenocarpy in nature (the dandelion is one), and this process has been exploited by selective propagation to produce seedless fruit including grapes, bananas, pineapples and cucumbers. In the seedless fruit the ovule's placenta (p. 568) is the likely source of the auxin.

Plant growth regulators are used in fruit production in many ways. For example, 2,4-D is used to induce flowering in pineapples. GA is used with grapes at fruit-set to elongate the clusters, increasing the numbers and sizes of berries. Fruit such as bananas that are picked unripe for safe transit are induced to ripen by exposure to ethene gas.

Figure 20.38 Ripening orange fruit: premature drop of unripe fruits can be prevented when trees are sprayed with a solution of 2,4-D (25 ppm)

Figure 20.39 The seed as the source of IAA for fruit development: if the 'seeds' (achenes) of strawberry are removed after pollination they can be replaced by externally supplied (exogenous) auxin

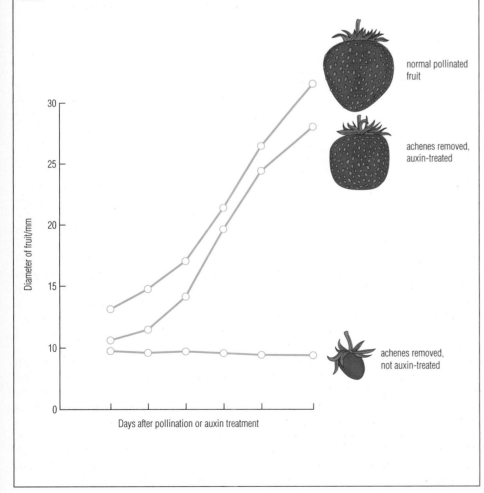

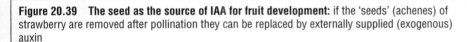

normal pollinated fruit

achenes removed, auxin-treated

achenes removed, not auxin-treated

Diameter of fruit/mm

Days after pollination or auxin treatment

■ Tumour formation

When a cut surface in a plant such as a sunflower is treated with lanolin paste containing a high concentration of IAA, a tumour arises. Cells around the wound proliferate abnormally. Cells enlarge and a centre of meristematic activity is formed. A disorganised mass of cells results.

The common soil bacterium *Agrobacterium tumefaciens* can infect plants at a point of injury and cause a tumour growth similar to that induced by a high concentration of IAA. What causes this tumour?

Research has shown the bacterial cell implants into the infected cells a gene (p. 636) that codes for enzymes that synthesise the plant growth hormones IAA and cytokinin. Consequently these plant growth regulators are formed in the infected, wounded cells at the command of the bacterium! As the levels of these hormones rise in the wound, the cells grow out of control.

Because of the ability of *Agrobacterium* to introduce genes into plant cells, this bacterium is exploited in plant breeding to produce genetically engineered crops. We will return to this phenomenon again in Chapter 29.

Questions

1 Why are some synthetic plant growth regulators more useful to us than naturally occurring substances?
2 What possible advantage to a fruit tree is the tendency for early fruit drop?
3 How may a systemic herbicide of broad-leaved plants improve productivity in cereal crops?

FURTHER READING/WEB SITES

A W Galston (1994) *Life Processes of Plants*, *Scientific American Library/* W H Freeman and Co
Plant hormones: **www.plant-hormones.bbsrs.ac.uk**

21 Response and coordination in animals

- Sensitivity and immediate response in animals occur via a system of receptors (sense organs) and effectors (muscles and glands) connected by nerve cells. Longer term coordination is via hormones.
- The fibres of neurones (nerve cells) conduct impulses as temporary reversals of the potential difference that exists between the inside and outside of the fibre.
- At the tiny gaps (synapses) at the junctions between neurones, transmission of the impulse is by diffusion of a chemical transmitter substance.
- Sense organs contain cells that are sensitive to changes in their environment. They generate impulses in the nerve fibres that serve them, in response to such stimuli.

- The nervous system of vertebrates consists of a central nervous system (brain and spinal cord) and a peripheral nervous system. Both systems may be involved in communication between receptors and effectors.
- The endocrine system of ductless glands produces and releases hormones, either at particular stages of growth, or under particular physiological conditions.
- Hormones are distributed by the circulation system throughout the body but induce a response only in target organs or tissues.
- Hormones are either steroids or proteins, and both stimulate the activity of particular enzymes in target cells, but by different mechanisms

21.1 INTRODUCTION

Sensitivity is the ability to detect change and to respond to it. Changes that are detected and lead to responses are called **stimuli** (singular 'stimulus'). All living things respond to stimuli of one kind or another. Many stimuli arise externally to an organism, but others arise from an organism's internal environment.

An essential feature of living things is their ability to coordinate internal activities. The various physiological processes, including respiration, growth and nutrition, are related to other processes and to the body's general needs. They are also related to changes external to the body. The processes by which the activities of an organism are made to function as an integrated whole are referred to as **coordination**.

The processes of coordination and response to stimuli of plants and animals are quite different. Animal responses tend to be quick movements; animals often respond to change by adjusting their behaviour. In contrast, plant responses are normally slow growth movements. The differences between plant and animal sensitivity are reviewed on p. 418.

■ Animal sensitivity

Protozoa and small animals

Both the detection of change and the response to it may occur within a single cell, sometimes with the involvement of specialised parts of cells, the organelles (p. 157). This is the case in protozoa (Figure 21.1) and in the cells of smaller multicellular animals. Here, sensitivity is a property of the cytoplasm of one cell.

Large multicellular animals

In multicellular animals such as the ~~nimals~~, the detection and response to

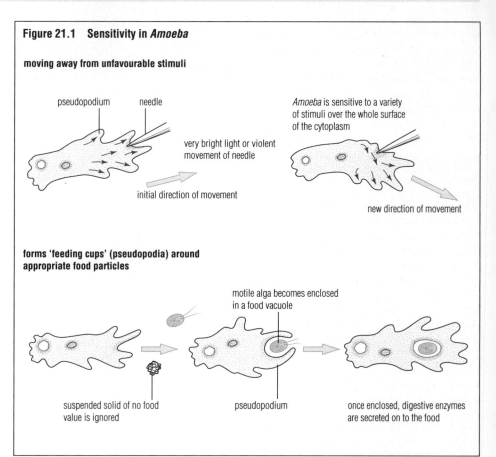

Figure 21.1   Sensitivity in *Amoeba*

moving away from unfavourable stimuli

pseudopodium   needle

very bright light or violent movement of needle

initial direction of movement

Amoeba is sensitive to a variety of stimuli over the whole surface of the cytoplasm

new direction of movement

forms 'feeding cups' (pseudopodia) around appropriate food particles

motile alga becomes enclosed in a food vacuole

suspended solid of no food value is ignored

pseudopodium

once enclosed, digestive enzymes are secreted on to the food

external environmental changes, and the regulation of the internal environment of the body, are brought about by two multicellular systems: the nervous system and the endocrine system. The actions of these two systems are themselves coordinated, but we need to examine them separately before we consider how they interact.

■ The nervous system

Neurones

The nervous system contains cells called **neurones** (see Figure 10.15, p. 220). Neurones take various forms, but each neurone has a cell body and a number of long extensions, the nerve fibres, specialised for transmitting information in the form of electrochemical impulses. Nerve impulses are transmitted in a few milliseconds, so nervous coordination is extremely fast and responses are virtually immediate. Impulses can be transmitted over considerable distances within the body, travelling along nerve fibres to the particular points in the body served by these fibres. Consequently the effects of impulses are directed and localised. This means that nervous control is precise rather than diffuse in its effects.

Receptors and effectors

Associated with the neurones of the nervous system is a system of receptors and effectors (Figure 21.2).

Receptors are the sense organs that detect change. Some receptors are large elaborate organs; examples include the eye and the ear. Others, such as the skin mechanoreceptors (p. 450), are structurally small and simple organs that can detect internal or external changes. At a receptor a stimulus (a form of energy such as light, sound waves or mechanical pressure) is converted into an impulse (an electrochemical signal). A weak stimulus promotes one or a few impulses per second. A strong stimulus promotes very many impulses per second. Impulses are generated by neurones and transmitted (normally via other neurones) to some effector.

Effectors produce the body's response to stimulation. The effectors are muscle tissues responsible for body movement and position, and glands capable of secreting various substances.

Neurones form reflex arcs

We have already seen that neurones are of three distinct types, known as sensory, motor and relay neurones (Figure 10.16, p. 221). Many of the body's neurones are assembled into pathways of impulse transmission called **reflex arcs** (Figure 21.3). The transmission of impulses from a stimulated receptor to an effector organ, via a reflex arc, produces a **reflex action**. Examples of reflex action include the automatic blink you make when a fast-moving object passes close to your eyes.

Reflex arcs are a fundamental component of nervous coordination. Reflex actions allow rapid responses of the body without involvement of the higher centres of the brain. Nevertheless, in the vertebrates at least, impulses having their origin in a reflex arc are also sent to the brain. Indeed, many reflex actions can be overruled by the higher centres, if we choose. Within the human brain, more than 10^{11} neurones receive and transmit impulses. Furthermore, much of a vertebrate's behaviour may be initiated by the brain, rather than being an automatic response to stimulation mediated through a reflex arc. The brain is the seat of consciousness, and it permits us to perform highly complex operations. We shall return to the functioning brain later.

■ The endocrine system

The endocrine system, a second system of coordination, consists of glands that manufacture chemical messengers called **hormones**. Hormones are secreted into the blood and transported to all parts of the body. Certain cells and tissues within the **'target' organ** respond, usually by a change in their metabolism. As a result, hormone regulation is generally quite slow, and the effects are diffuse. The endocrine system of the mammal (human) is discussed on p. 472.

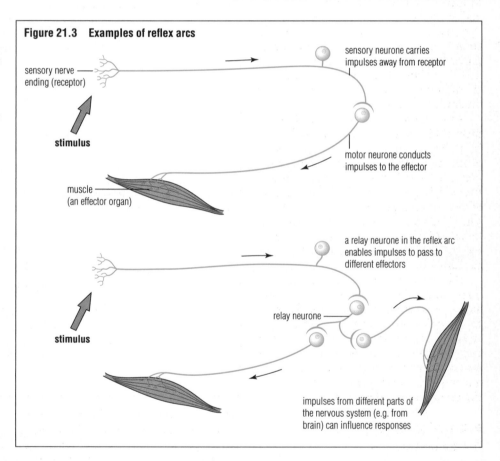

Figure 21.2 Coordination by the mammalian nervous system

external stimuli   internal stimuli

receptors (sense organs)

coordination of responses by the brain (part of the nervous system)

impulses transmitted by the neurones of the nervous system

effector organs
e.g.

muscles   glands

responses

'feedback' information

Figure 21.3 Examples of reflex arcs

sensory nerve ending (receptor)

stimulus

muscle (an effector organ)

sensory neurone carries impulses away from receptor

motor neurone conducts impulses to the effector

a relay neurone in the reflex arc enables impulses to pass to different effectors

relay neurone

stimulus

impulses from different parts of the nervous system (e.g. from brain) can influence responses

Questions

1 **a** Tabulate the essential differences between animal and plant sensitivity (p. 418).
   b Can we say plants are less sensitive than animals?

2 When an *Amoeba* passes close to a solid object and ignores it, but forms a food cup of pseudopodia around an adjacent flagellate, to what stimuli might it be responding?

3 Suggest a sequence of cells that might be involved in the eye-blink reflex arc.

21.2 THE TRANSMISSION OF AN IMPULSE

Nerve fibres of mammals may be very long, but they are also extremely thin; the axons of most neurones are no more than 20 μm in diameter. It is therefore difficult to use such tiny structures to investigate what an impulse is and how nerve impulses are transmitted.

On the other hand, the nervous system of certain non-vertebrates contains some axons (known as 'giant fibres') that are as much as a millimetre in diameter. These giant nerve fibres are associated with rapid transmission of nerve impulses (p. 445). Experimental investigations were therefore first conducted by using 'giant' axons. This approach was developed simultaneously in the USA and the UK in the late 1930s.

The giant nerve fibres of squid are large enough for a microelectrode to be inserted. A microelectrode is a very fine, hollow glass needle, filled with a salt solution, which conducts electricity. It is inserted into the interior of the axon. A second or reference electrode, a tiny metal plate, is placed on the surface of the axon at some distance from the first. Both electrodes are connected to a cathode ray oscilloscope, via preamplifiers, which increase the current substantially. The nerve fibre is electrically stimulated to generate a nerve impulse (Figure 21.4).

Figure 21.4   Investigating the transmission of an impulse

Squids swim by jet propulsion. Water taken into the mantle cavity is expelled at speed through the funnel. Impulses from the ganglion travel along the giant nerve fibres, and cause contraction of the muscles of the mantle.

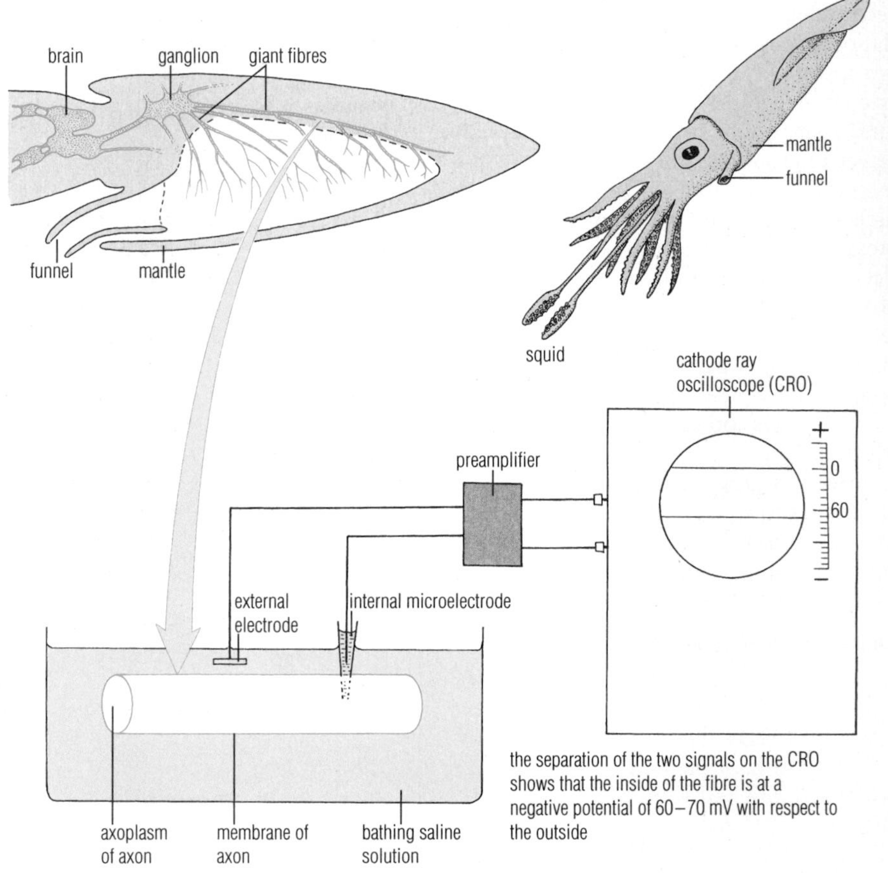

the separation of the two signals on the CRO shows that the inside of the fibre is at a negative potential of 60–70 mV with respect to the outside

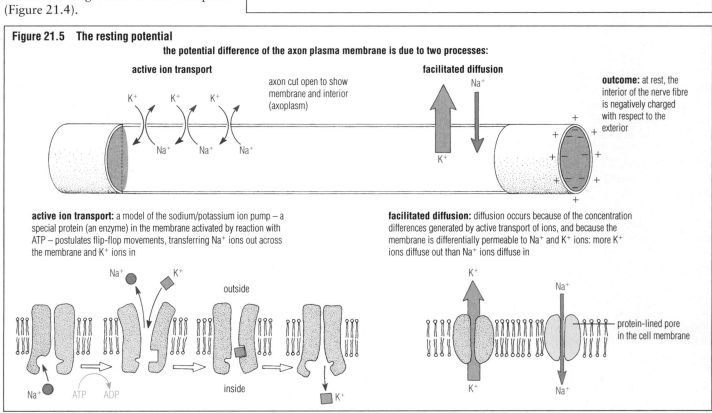

Figure 21.5   The resting potential

the potential difference of the axon plasma membrane is due to two processes:

active ion transport

facilitated diffusion

axon cut open to show membrane and interior (axoplasm)

outcome: at rest, the interior of the nerve fibre is negatively charged with respect to the exterior

active ion transport: a model of the sodium/potassium ion pump – a special protein (an enzyme) in the membrane activated by reaction with ATP – postulates flip-flop movements, transferring Na⁺ ions out across the membrane and K⁺ ions in

facilitated diffusion: diffusion occurs because of the concentration differences generated by active transport of ions, and because the membrane is differentially permeable to Na⁺ and K⁺ ions: more K⁺ ions diffuse out than Na⁺ ions diffuse in

protein-lined pore in the cell membrane

■ The resting potential of an axon

These investigations established that a potential difference is maintained between the inside and the outside of an axon. The inside is negatively charged with respect to the outside, and the membrane is said to be **polarised**. The potential difference (the **resting potential**) is due to a difference in the concentration of ions on either side of the membrane. How does this difference arise?

In the plasma membrane of animal cells, including neurones, there are numerous **sodium/potassium pumps** (p. 237). Each sodium/potassium pump, driven by energy from ATP, actively transports two potassium ions into the cell to every three sodium ions transported out. Consequently, the cytoplasm has a high concentration of potassium ions and a low concentration of sodium ions, in contrast to the exterior, where there is a low concentration of potassium ions and a high concentration of sodium ions. These gradients in concentration across the membrane are known as **electrochemical gradients**.

Because of their electrochemical gradients, sodium ions tend to diffuse back in and potassium ions tend to diffuse out. The **channel proteins** (p. 275) through which this diffusion occurs are much more permeable to potassium ions than sodium ions (Figure 21.5). Because of this, and also because there are immobilised, negatively charged protein ions retained inside the cell, the outside of the cell contains many more positive ions than the cytoplasm inside it. In a neurone, these processes create a **resting potential difference** of up to $-70\,mV$ (millivolts) inside compared to outside.

The resting potential is the normal condition at the plasma membranes of animal cells. In most animal cells (described as non-excitable cells) this potential difference remains constant. But in a few animal cells (known as **excitable cells**, and including neurones, sensory cells and muscle cells), the potential difference across the plasma membrane may change. This change is brought about only momentarily, when the excitable cells are stimulated, and an impulse is promoted. Note that the ATP used in setting up the resting potential provides the energy for the generation of an impulse.

■ The impulse

An impulse or **action potential** is a temporary and local reversal of the resting potential, arising when an axon is stimulated. It takes only a few milliseconds to happen (Figure 21.6). During an action potential, the membrane potential falls until the inside of the membrane (the axoplasm) becomes positively charged with respect to the exterior. In fact the potential changes from about $-70\,mV$ to about $+40\,mV$ at the point of stimulation. The membrane at this point is said to be **depolarised**.

This change in potential across a membrane comes about because some of the sodium and potassium channels have **voltage-sensitive gates**. This means their channels open and close in response to a change in the membrane potential difference. When the gates are closed there is little ion movement, but when the gates are open ions flow through by diffusion. One type of gated channel protein is permeable to sodium ions and the other type is permeable to potassium ions. During the resting potential all gates are closed.

A stimulus depolarises a neurone's plasma membrane by causing a local increase in permeability of the membrane to sodium ions. This localised depolarisation opens the gated sodium channels, allowing a large number of sodium ions to flow in down their electrochemical diffusion gradient. Consequently, the interior of the membrane of the neurone fibre becomes progressively more positive with respect to the outside of the membrane. Then, almost instantly, the sodium gates close. At the same time, gated potassium channels open and potassium ions flow out down their electrochemical gradient. The interior of the membrane starts to become less positive again, and the process of re-establishing the resting potential has begun.

An impulse in the form of this reversal of charge runs the length of the neurone fibre as a wave of depolarisation. The impulse is propagated (self-generated) by the effect of the sodium ions entering. They create an area of positive charge, causing a local current to be set up with the negatively charged resting region immediately ahead (Figure 21.6). The impulse lasts for only two milliseconds at each point along the fibre before the resting potential is re-established.

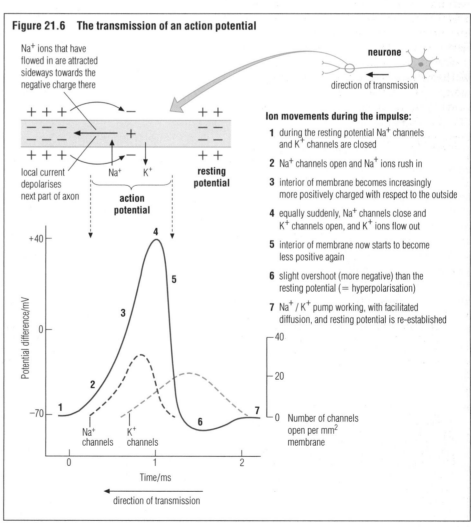

Figure 21.6   The transmission of an action potential

Na$^+$ ions that have flowed in are attracted sideways towards the negative charge there

local current depolarises next part of axon

neurone

direction of transmission

action potential

resting potential

Ion movements during the impulse:

1 during the resting potential Na$^+$ channels and K$^+$ channels are closed

2 Na$^+$ channels open and Na$^+$ ions rush in

3 interior of membrane becomes increasingly more positively charged with respect to the outside

4 equally suddenly, Na$^+$ channels close and K$^+$ channels open, and K$^+$ ions flow out

5 interior of membrane now starts to become less positive again

6 slight overshoot (more negative) than the resting potential (= hyperpolarisation)

7 Na$^+$/K$^+$ pump working, with facilitated diffusion, and resting potential is re-established

Potential difference/mV

Na$^+$ channels

K$^+$ channels

Number of channels open per mm^2 membrane

Time/ms

direction of transmission

■ The refractory period

For a brief period following the passage of an action potential the axon is no longer excitable. This phase is called the **refractory period**, and it lasts 5–10 milliseconds. Because of the length of the refractory period the maximum frequency of impulses is between 500 and 1000 per second.

Initially the block on the conduction of a second impulse is absolute; that is to say, no stimulus, however strong, will generate an impulse. This **absolute refractory period** is extremely brief, normally only 1 millisecond in duration. At this stage there is a huge excess of sodium ions inside the nerve fibre, and the membrane is temporarily impermeable to the passage of ions.

Subsequently, in the **relative refractory period**, the resting potential is progressively restored. The first stage of restoration is due to the outward diffusion of potassium ions. Next, the continuing actions of the sodium/potassium ion pumps and the differing rates of diffusion of sodium and potassium ions re-establish the potential difference across the membrane. During this period it becomes increasingly possible for an action potential to be generated.

■ The threshold of stimulation

A stimulus must be at or above a certain minimum strength, known as the **threshold of stimulation**, in order to initiate the transmission of an action potential (Figure 21.7). Thus a stimulus evokes either a full response or no response at all. When a stimulus is too weak, the influx of sodium ions into the neurone is slight, and normal polarity is very quickly re-established. Such a stimulus is said to be **sub-threshold**.

A stimulus at or above the threshold strength initiates one or more impulses. With very strong stimulation the frequency of action potentials passing along an axon increases, up to the maximum rate permitted by the refractory period. The size of the action potentials is more or less constant, however; it does not depend upon the stimulus strength.

■ The speed of transmission of impulses

Nerves are concerned with the conduction of information vital for survival; the faster the action potential travels, the more efficient is the nervous system. Two structural features of axons influence the speed of conduction of an action potential.

The diameter of the axon

One feature is the diameter of the axon. You will remember that the earliest studies on resting and action potentials were carried out on giant nerve fibres, up to 1 mm in diameter. In fact nerve fibres are of various diameters, typically from about 0.5 μm up to almost 1000 μm (1 mm) in diameter.

Speed of transmission of the action potential along the axon membrane depends on the resistance offered by the axoplasm within. This resistance is related to the diameter of the axon; the narrower the axon, the greater is its resistance and the lower the velocity of conduction of the action potential. The larger the diameter, the faster the axon will conduct action potentials.

In axons of small diameter (about 50 μm or less) an action potential is conducted at about $0.5\,\mathrm{m\,s^{-1}}$. By contrast, giant axons (500–1000 μm in diameter) conduct action potentials at about $100\,\mathrm{m\,s^{-1}}$.

The tiny nerve fibres of non-vertebrate animals are non-myelinated and therefore conduct impulses quite slowly. Where non-vertebrate animals have rapid conduction of impulses it is achieved by having very large axons. For example, the fast-conducting giant nerve fibres of the squid enable the animal to 'jet propel' itself at high speed away from potential danger (after a very brief delay), when the need arises. Normal movements of the squid are coordinated more slowly, involving the ordinary neurones of the nervous system. Another example of giant fibres playing an important part in 'escape' movements is in the earthworm (p. 480).

Myelin sheaths

The second factor that determines the speed of transmission of an impulse along an axon is the presence (or absence) of a **myelin sheath**. The structure and arrangement of the membranes of Schwann cells that provide the myelin sheath are shown in Figure 10.17 (p. 221). In vertebrates, most of the neurones have myelinated axons. Unmyelinated axons are typical of the nervous systems of non-vertebrates.

Myelin consists largely of lipid, and has a high electrical resistance. At points at intervals of about 1–2 mm along the myelinated axon, the axon membrane is exposed. These points, known as **nodes of Ranvier**, are the junctions between the sheath cells (Schwann cells).

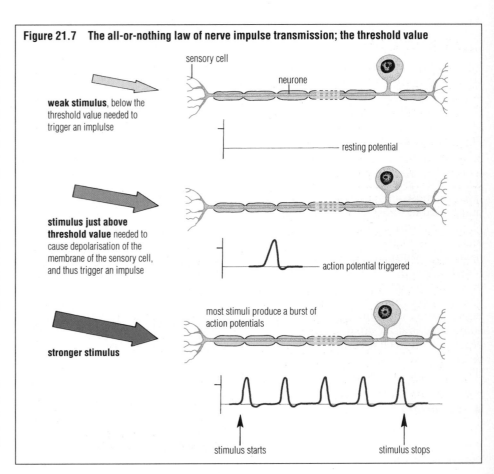

Figure 21.7   The all-or-nothing law of nerve impulse transmission; the threshold value

sensory cell

neurone

weak stimulus, below the threshold value needed to trigger an implulse

resting potential

stimulus just above threshold value needed to cause depolarisation of the membrane of the sensory cell, and thus trigger an impulse

action potential triggered

most stimuli produce a burst of action potentials

stronger stimulus

stimulus starts   stimulus stops

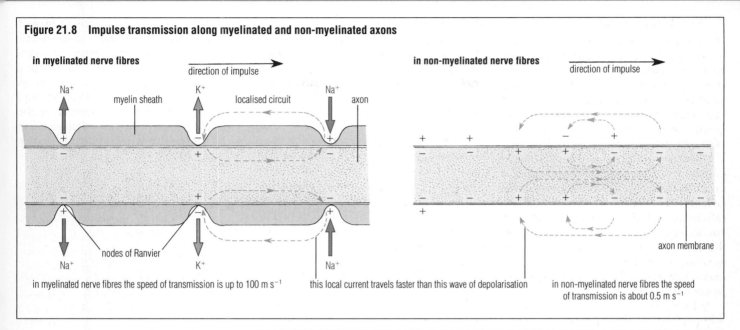

Figure 21.8   Impulse transmission along myelinated and non-myelinated axons

in myelinated nerve fibres

direction of impulse

Na⁺   K⁺   localised circuit   Na⁺

myelin sheath   axon

nodes of Ranvier

Na⁺   K⁺   Na⁺

in myelinated nerve fibres the speed of transmission is up to 100 m s⁻¹

this local current travels faster than this wave of depolarisation

in non-myelinated nerve fibres

direction of impulse

axon membrane

in non-myelinated nerve fibres the speed of transmission is about 0.5 m s⁻¹

In myelinated axons the action potential jumps from node to node (because depolarisation is restricted to the nodes) at high speed. This is referred to as **saltatory conduction** (because the impulse 'leaps' along). By contrast, the step-by-step depolarisation along the surface on an unmyelinated axon, referred to as **continuous conduction**, is much slower (Figure 21.8).

21.3 THE SYNAPSE

An impulse is transmitted from one neurone to the next at a structure called a **synapse**. Synapses occur between the swollen tips (**synaptic knobs**) of terminals of the axon of one neurone, the **presynaptic neurone**, and a dendrite or cell body of another neurone, the **postsynaptic neurone**. At the synapse the neurones are in extremely close contact, but with a tiny gap between them (known as the **synaptic cleft**), normally about 20 nm wide. The appearance of synapses is shown in Figure 21.9, and the structure of a synapse in section is shown in Figure 21.10 (overleaf).

Impulses are also conducted at the points of contact between the terminal fibres of motor neurones and muscle fibres. Such a junction is a special type of synapse, called a **motor end plate**. A motor end plate is structurally different from a synapse (Figure 22.16, p. 489), but functionally they are really the same.

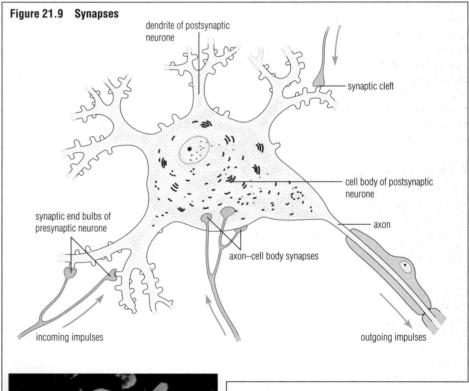

Figure 21.9   Synapses

dendrite of postsynaptic neurone

synaptic cleft

cell body of postsynaptic neurone

axon

synaptic end bulbs of presynaptic neurone

axon–cell body synapses

incoming impulses

outgoing impulses

SEM of synaptic junction between axon endings and a nerve cell body

Questions

1   Why do myelinated axons conduct faster than non-myelinated fibres of the same diameter?

2   Why is an axon unable to conduct an impulse for a brief period after an action potential has been conducted?

3   What is the difference in the response of a stimulated nerve fibre to a weak and a strong stimulus?

4   A row of dominoes is set up, standing on their ends a short distance apart. When the end domino is pushed over it causes the whole row to fall. Say how each stage of this effect is a model for the production and transmission of an action potential.

■ Transmission across the synapse

Transmission of an action potential across a synapse is normally by the diffusion of a chemical, known as a **neurotransmitter substance**. The most commonly occurring transmitter substance in mammals is the chemical **acetylcholine (ACh)**. Other important transmitter substances are **noradrenaline** and **adrenaline**. Neurotransmitters occur stored in tiny vesicles in the synaptic knob (Figure 21.10). Because of the minute quantities of chemical substances involved at any one time, the technical problems in identifying these chemicals are very great. Nevertheless there is evidence that γ-aminobutyric acid, dopamine and glutamic acid are also important neurotransmitters at certain synapses.

Steps in chemical transmission

The sequence of events at a synapse appears to be as follows (Figure 21.11):

1 The arrival of an impulse at the synaptic knob opens calcium ion channels in the presynaptic membrane briefly, and calcium ions flow in from the synaptic cleft.

2 The calcium ions induce a few vesicles containing transmitter substance to fuse with the presynaptic membrane and release their contents into the synaptic cleft.

3 Once released, the transmitter diffuses across the synaptic cleft within a fraction of a millisecond. Here it binds with a **receptor protein** on the membrane of the postsynaptic neurone. In this membrane, there are receptor sites specific for each transmitter substance. Each receptor protein controls a channel in the membrane which, when open, allows one or more types of ion (sodium, potassium or chloride, for example) to pass.

4 Typically, when the transmitter substance binds to the receptor it causes the opening of sodium ion channels, and sodium ions flow in (an **excitatory synapse**). The entry of sodium ions depolarises the postsynaptic membrane. If depolarisation reaches the threshold level, an action potential is generated in the postsynaptic neurone and travels down the axon to the next synapse or to an end plate.

5 Meanwhile, the action of the neurotransmitter does not persist. Removal of transmitter substance from the synaptic cleft (by enzyme action) prevents the continuous 'firing' of impulses in the postsynaptic neurone. For example, the enzyme cholinesterase hydrolyses ACh to choline and ethanoic acid, which are inactive as transmitters.

6 The inactivated forms of transmitter substance then re-enter the presynaptic knob and are resynthesised as transmitter substance, which is retained in vesicles awaiting re-use.

Summation in the postsynaptic neurone

Often the release of neurotransmitter in response to a single action potential is not sufficient to develop an action potential in a postsynaptic neurone. Instead, an excitatory postsynaptic potential builds up as packages of transmitter substance from individual synaptic vesicles arrive. Normally several vesicles have to be released before the threshold potential required to propagate an action potential is achieved. This addition effect in the postsynaptic membrane is known as **summation**.

From the SEM in Figure 21.9 it can be seen that many presynaptic fibres converge on a single postsynaptic neurone. A single relay neurone may have many thousands of synaptic knobs terminating at its dendrites and cell body. The summation process is described as **spatial** when impulses from several different axons contribute to the total.

Temporal summation occurs when several impulses from a single axon follow on in quick succession and cause an action potential to arise in the postsynaptic neurone.

Excitatory and inhibitory synapses

The account of synaptic transmission given on this page relates to an excitatory synapse. That is, the incoming impulses excite the postsynaptic neurone and generate an impulse.

Some synapses have the reverse effect. These are **inhibitory synapses**. Here, release of transmitter into the

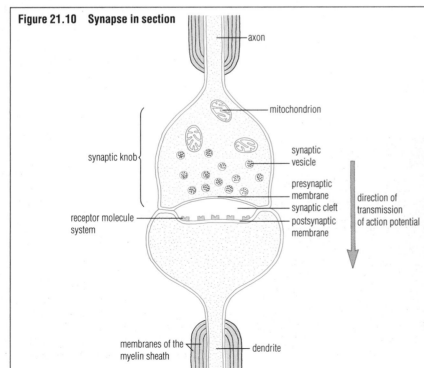

Figure 21.10 Synapse in section

- axon
- mitochondrion
- synaptic knob
- synaptic vesicle
- presynaptic membrane
- synaptic cleft
- postsynaptic membrane
- receptor molecule system
- direction of transmission of action potential
- membranes of the myelin sheath
- dendrite

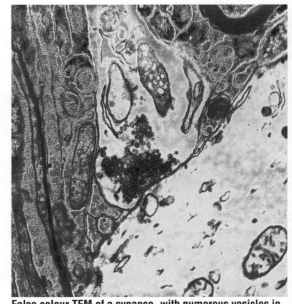

False colour TEM of a synapse, with numerous vesicles in the synaptic knob

Figure 21.11 Chemical transmission at the synapse

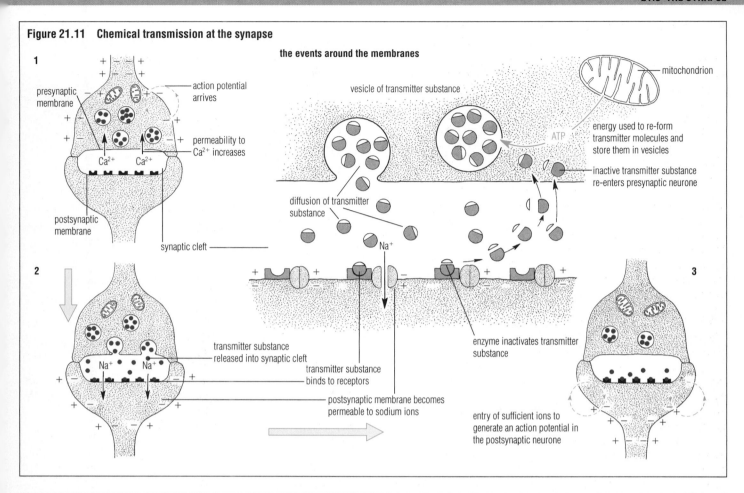

the events around the membranes

1
- action potential arrives
- presynaptic membrane
- permeability to Ca²⁺ increases
- Ca²⁺   Ca²⁺
- postsynaptic membrane
- synaptic cleft

vesicle of transmitter substance

mitochondrion

energy used to re-form transmitter molecules and store them in vesicles

ATP

inactive transmitter substance re-enters presynaptic neurone

diffusion of transmitter substance

Na⁺

2
- transmitter substance released into synaptic cleft
- transmitter substance binds to receptors
- postsynaptic membrane becomes permeable to sodium ions

Na⁺   Na⁺

enzyme inactivates transmitter substance

3
- entry of sufficient ions to generate an action potential in the postsynaptic neurone

---

synaptic cleft leads to opening of ion channels in the postsynaptic membrane through which chloride ions enter and potassium ions leave. As a result, the interior of the postsynaptic neurone becomes more negative (to −80 mV, for example). This increases the threshold, and the postsynaptic neurone is less easily stimulated.

A postsynaptic neurone may have both excitatory and inhibitory synapses. In this case, an impulse will not be generated unless the combined effects of the excitatory impulses and inhibitory impulses exceed the threshold.

Electrical synapses

At just a few synapses the action potential is transmitted directly across the synaptic cleft. In this type of synapse, known as an electrical synapse, the gap is very much smaller than where transmission is by a chemical. Electrical transmission is very quick. One example of this phenomenon is at the retinal ganglion cells in the mammalian eye (p. 458). Electrical synapses are not affected by drugs and, unlike chemical synapses, are not susceptible to fatigue.

■ The role of the synapse

The role of the synapse is to convey action potentials between neurones. We have already seen that the action potential information may lead to either excitation or inhibition of an impulse in the postsynaptic neurone. The evolution of the synapse may have wider implications, however, such as those listed below.

1 A postsynaptic neurone may have a large number of synapses with presynaptic neurones from many different sources. Some of these synapses will be excitatory and so reinforce each other. Others will be inhibitory and tend to cancel out the excitation. This enables the 'neurone plus synapse' system to collect information from different sense organs and compute an appropriate pattern of response.

2 The synapses effectively filter out infrequent and low-level stimuli of no significance. In effect, the synapse may remove 'background noise' from the body's communication system.

3 The synapses protect the response system from overstimulation, by becoming fatigued. The continuous transmission of an action potential sooner or later exhausts the immediate supply of transmitter substance.

4 The presence of millions of neurones in the brain with vast numbers of synapses may be the physical basis of learning and memory (p. 465).

Questions

1 By what process do ions (such as sodium, potassium and chloride ions) cross the postsynaptic membrane when a neurotransmitter substance has caused ion-specific channels to be opened?

2 What do you think is the function of organelles such as mitochondria and Golgi apparatus in the synaptic knobs of neurones?

3 What might be the effect of a drug or poison that inhibits the breakdown of neurotransmitter substances once these have become attached to receptor molecules of the postsynaptic membrane (p. 450)?

■ Drugs and the synapse

Synaptic conduction can be altered by drugs. Several different types of chemical interfere at the synapse, either amplifying or inhibiting transmission of impulses.

1 Amplification at the synapse may be due to substances mimicking the actions of natural transmitters. The drug LSD, for instance, mimics natural brain transmitters and causes hallucinations. Amplification may also be due to reduction in the threshold for excitation of the postsynaptic membranes, resulting in facilitation. The stimulants caffeine and nicotine work in this way.

Other substances block the enzymic breakdown of transmitter substances once they are attached to the receptor proteins of the postsynaptic membrane. They consequently prolong the effect of the neurotransmitters. For example, some military nerve gases, and also the poison strychnine, enhance synaptic transmission in this way, so that a person poisoned by these substances is racked by uncontrollable muscular contractions.

2 Inhibition of synaptic transmission may be due to chemicals that block transmission across synapses. The drug atropine is thought to act by preventing an action potential being generated by ACh when it attaches to its receptor protein on the postsynaptic membrane. Curare, a poison obtained from South American plants of the genera *Strychnos* and *Chondrodendron*, similarly blocks nerve muscle junctions and causes paralysis. Both of these substances have applications in medicine.

Other substances activate inhibitor synapses in the brain, making the brain neurones resistant to excitation. Some of these are used as 'painkillers' (analgesics) and 'tranquillisers' (such as Librium, Valium and Mogadon).

Research into the action of painkilling opiate drugs such as heroin and morphine has shown they mimic the activity of naturally occurring painkillers (known as endorphins).

Persistent, abnormally high blood pressure may be treated by drugs called beta blockers. These interfere with certain neurotransmitter receptors in the synapses of the autonomic nervous system serving the heart muscle (p. 354).

21.4 THE SENSORY SYSTEM

All cells are sensitive to change in the environment, but sense cells are specialised to detect stimuli and to respond

Table 21.1 The sense organs of mammals

Sense data (form of energy)	Type of receptor	Location in body
Mechanoreceptors, responding to mechanical stimulation:		
light touch	touch receptors e.g. Meissner's corpuscle Merkel's disc hair root plexus	mostly in dermis of skin; some free endings extend into epidermis
touch and pressure	touch and pressure receptors e.g. Pacinian corpuscle	dermis of skin
movement and position	stretch receptors e.g. muscle spindle; proprioceptors	skeletal muscle
sound waves and gravity	sensory hair cells	cochlea utriculus semicircular canals } ear
blood pressure	baroreceptors	aorta and carotid artery
Thermoreceptors, responding to thermal stimulation:		
temperature change at skin	free endings	dermis of skin
internally	cells in hypothalamus	brain
Chemoreceptors, responding to chemical stimulation:		
chemicals in air	sense cells of olfactory epithelium	nose
taste	taste buds	tongue
blood oxygen, CO_2, H^+	carotid body	carotid artery
osmotic concentration of blood	osmoregulatory centre in hypothalamus	brain
Photoreceptors, responding to electromagnetic stimulation:		
light	rod and cone cells of retina	eye

by producing an action potential. A stimulus is some form of energy, mechanical, chemical, thermal or light (photic). The sense cell converts (transduces, p. 172) the energy of the stimulus into the electrochemical energy of an action potential (p. 445), which is then conducted to other parts of the nervous system.

Some free nerve endings (dendrites) may be relatively unspecific and produce sensations of pain in response to, for example, pressure, heat or cold. Most sensory cells are, however, especially sensitive to only one type of stimulation, either heat, light, touch or chemicals, for instance. Some of these may be further specialised to respond maximally only to one class of chemical or one wave-band of light.

Sense organs contain large numbers of sensory cells that respond to one of the types of stimulus. Table 21.1 is a classification of mammalian sense organs.

■ Mechanoreceptors of the skin

The localisation of mechanoreceptors of the skin of a mammal is shown in Figure 23.15 (p. 511). The role and functioning

of one of them, the pressure receptor known as a Pacinian corpuscle (Figure 21.12), will serve to illustrate how a simple sense cell converts stimulation into an impulse.

Pacinian corpuscles occur deep in the skin all over the body, and also in the capsules around joints (p. 486). Each is a tiny structure consisting of a nerve ending wrapped in a many-layered capsule of connective tissue. Impulses generated in the Pacinian capsule are conducted away along a myelinated sensory fibre.

The generation of impulses by a sense organ

When strong pressure is applied to the skin, the fibrous connective tissue capsule of a Pacinian corpuscle is deformed. The result is that the nerve ending is mechanically stretched. At rest, the nerve ending maintains a resting potential of $-70\,\text{mV}$, typical of a neurone. When stimulated, the permeability of the membrane is temporarily changed. Sodium ions flow in, and the membrane starts to be depolarised (it becomes less negatively charged internally). This kind of localised

Figure 21.12 The Pacinian corpuscle (touch and pressure receptor)

Pacinian corpuscle seen in section

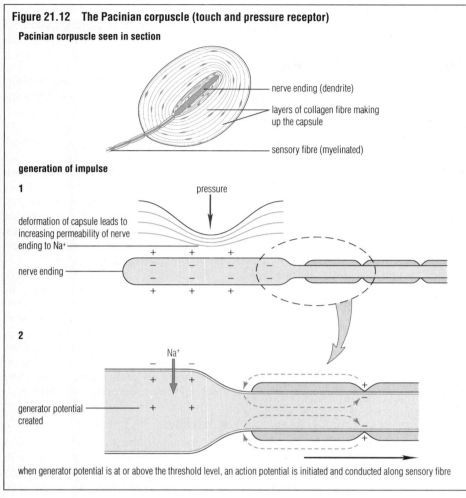

- nerve ending (dendrite)
- layers of collagen fibre making up the capsule
- sensory fibre (myelinated)

generation of impulse

1

pressure

deformation of capsule leads to increasing permeability of nerve ending to Na⁺

nerve ending

2

Na⁺

generator potential created

when generator potential is at or above the threshold level, an action potential is initiated and conducted along sensory fibre

Figure 21.13 Stimulus, generator potential and action potential

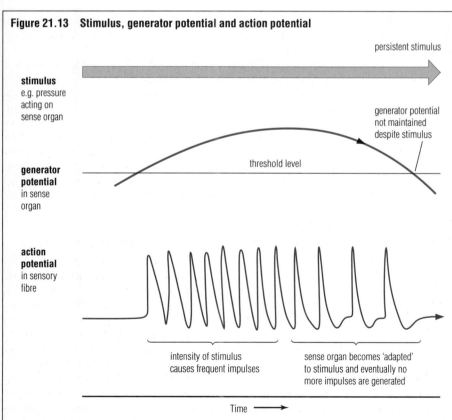

persistent stimulus

stimulus
e.g. pressure acting on sense organ

generator potential not maintained despite stimulus

generator potential
in sense organ

threshold level

action potential
in sensory fibre

intensity of stimulus causes frequent impulses

sense organ becomes 'adapted' to stimulus and eventually no more impulses are generated

Time →

depolarisation in a sensory cell is described as a **generator potential**.

The functioning of this sense organ has been investigated by applying micro-electrodes to the nerve ending in the capsule, and to the sensory fibre leading from the sense organ. The potential differences detected can be amplified and displayed on the cathode ray oscilloscope (Figure 21.4), so that they can be compared. By this means it was established that the stronger the stimulus applied to the sense organ, the greater is the resulting generator potential. If the generator potential reaches or exceeds a fixed threshold value it causes an action potential (p. 445) to flow along the sensory axon (Figure 21.13).

The sensitivity of a sense organ

The strength of the stimulus determines the frequency of impulses generated. The stronger the stimulus, the more frequent are the impulses that flow, at least initially. If the external pressure is kept more or less constant, however, the generator potential is not necessarily maintained. Sooner or later the permeability of the membrane to sodium ions changes. As a result, the frequency of impulses slows and stops. The sense organ is said to have **adapted**. Some sense organs adapt very quickly. The sense receptors in the nose and the fine touch receptors of the skin are examples. Pain receptors adapt hardly at all, however, and stretch receptors of skeletal muscles (p. 493) do so very slowly.

Sense organ adaptation is a familiar phenomenon. A pungent odour in a confined space may be very unpleasant at first, but you quickly cease to be aware of it; only the comments of a new arrival may remind you that the smell is still there! In the same way, you cease to notice the touch of clothing on your skin.

Remember, continuous chemical transmission of an impulse across the synapse is also subject to 'fatigue'. The supply of transmitter substance may be so depleted that action potentials flow less frequently or stop altogether.

Questions

1. What could be the value to the body of Pacinian corpuscles that occur around joints?
2. What is the biochemical explanation of 'fatigue' in the transmission of impulses at a synapse?
3. What advantages may result from the convergence of axon endings from several neurones on one postsynaptic neurone?

21.5 TASTE AND SMELL

The sensations of taste and smell are detected in the mouth and nose. In mammals taste and smell play a role in food choice, the detection of harmful substances, the detection of danger (whether an abiotic danger such as fire, or a biotic danger such as a predator), the establishment of territory and finding a mate. The human sense of smell is relatively modest compared with that of many other mammals.

'Taste' is detected by special **taste receptor cells**, located on the tongue and roof of the mouth. Humans have four types of taste cell, each responding maximally to the chemicals that produce sensations sweet, sour, bitter and salt. (Most specialised sense cells do not respond solely to one specific type of stimulus; rather they respond to a range of stimuli but give a maximal response to only one of them.) For example, acids give us the sensation of sourness, while salts such as sodium chloride, sodium nitrate and potassium chloride all taste 'salty'. Taste receptors are grouped into **taste buds**, each bud responding maximally to one of the four sensations. For example, taste buds at the tip of the tongue respond maximally to sweet sensations (Figure 21.14).

The sense of smell results from vapours drawn into the nasal passage. These substances stimulate the **olfactory cells** in the roof of the nasal cavity. In contrast to 'taste', several thousands of different odours can be detected by a relatively small number of different types of receptor cell, possibly only seven (the detectors of 'primary' odours). These receptors are stimulated by very low concentrations of vapour. For example, the unpleasant sulphur-containing substance mercaptan, which is added to the public gas supply to aid the detection of leaks, is detectable at $0.4\,ng\,l^{-1}$. Continuous exposure to most odours, however unpleasant, usually leads to diminished sensitivity (known as 'adaptation').

The 'flavour' of food is produced largely by the sense of smell, in response to the vapours that reach the nose from food in the mouth.

21.6 HEARING AND BALANCE

The ear (Figure 21.15) performs two distinct and major sensory functions, those of hearing and balance. There are three regions of the ear: the outer, middle and inner ear. Each region has a distinctive role.

The **outer ear** is a flap of elastic cartilage covered by skin (the **pinna**) and a tube (ear canal) leading to the ear drum. The **middle ear** is an air-filled chamber cut off from the outer ear by the ear drum, and from the fluid-filled **inner ear** by the oval and round windows. The Eustachian tube, which connects the middle ear to the pharynx, is normally closed. It opens briefly during swallowing. When opened, the Eustachian tube ensures that air pressure is equal on both sides of the ear drum.

The smallest bones in the body occur in the middle ear. These bones, known as the **hammer** (**malleus**), **anvil** (**incus**) and **stirrup** (**stapes**), together traverse the middle ear, from ear drum to oval window. They are held in place by ligaments and muscles.

The inner ear is a fluid-filled cavity of membranous canals, surrounded by extremely hard bone. The inner ear consists of the **cochlea** (concerned with transduction of sound waves to nerve impulses), the **utriculus**, **sacculus** and the three **semicircular canals**, which lie at right angles to each other. These latter structures are concerned with balance.

■ Hearing

The pinna of the outer ear acts as a funnel, channelling sound waves towards the ear drum. Dogs, rabbits and many other mammals have mobile pinnae, and

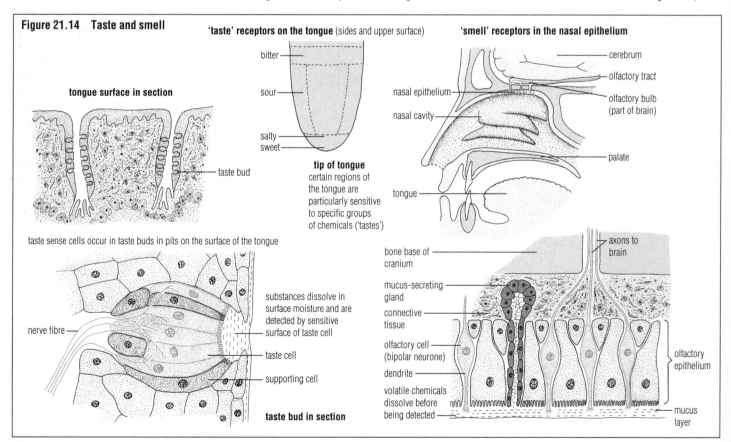

Figure 21.14 Taste and smell

'taste' receptors on the tongue (sides and upper surface)

'smell' receptors in the nasal epithelium

tongue surface in section

bitter
sour
salty
sweet

tip of tongue
certain regions of the tongue are particularly sensitive to specific groups of chemicals ('tastes')

taste bud

cerebrum
olfactory tract
olfactory bulb (part of brain)
nasal epithelium
nasal cavity
palate
tongue

taste sense cells occur in taste buds in pits on the surface of the tongue

nerve fibre

substances dissolve in surface moisture and are detected by sensitive surface of taste cell
taste cell
supporting cell

taste bud in section

bone base of cranium
axons to brain
mucus-secreting gland
connective tissue
olfactory cell (bipolar neurone)
dendrite
volatile chemicals dissolve before being detected
olfactory epithelium
mucus layer

Figure 21.15 The structure of the ear

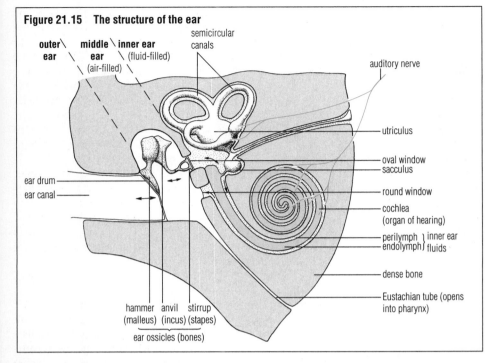

Figure 21.16 The cochlea and hearing

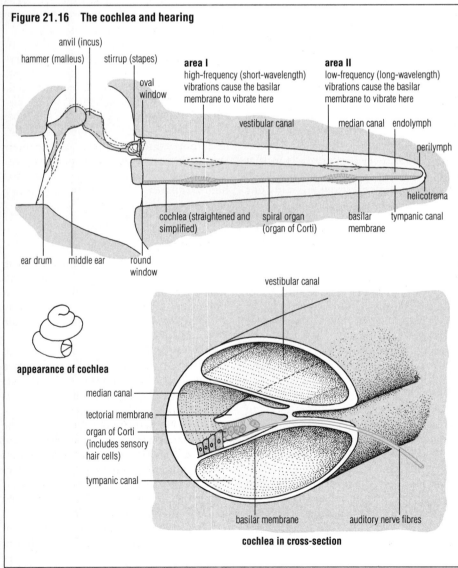

area I
high-frequency (short-wavelength) vibrations cause the basilar membrane to vibrate here

area II
low-frequency (long-wavelength) vibrations cause the basilar membrane to vibrate here

appearance of cochlea

cochlea in cross-section

use them to locate and focus on the source of a sound.

The **ear drum** is a thin, strong sheet of elastic connective tissue that vibrates in response to sound wave pressure. Its vibrations are transmitted to the oval window by the middle ear bones. These form a lever system, increasing the effective pressure on the oval window 20 times.

The cochlea (Figure 21.16) consists of a spirally coiled, fluid-filled tube, which is divided longitudinally into three compartments, separated by membranes. The coiled cochlea transduces mechanical vibrations to nerve impulses. When these are relayed to the brain they give us the sensation of sound.

The upper and lower compartments of the cochlea contain fluid called **perilymph**. These compartments communicate at the tip of the cochlea through a tiny hole called the **helicotrema**. The upper compartment, known as the **vestibular canal**, is in contact with the oval window. The lower compartment, the **tympanic canal**, is in contact with the round window.

Between these two is the middle compartment, known as the **median canal**, which is filled by **endolymph**. The median canal consists of the **basilar membrane**, and projecting into the canal is an inflexible **tectorial membrane**, parallel with the basilar membrane, and running the full length of the cochlea. Immediately beneath the tectorial membrane the basilar membrane supports the **organ of Corti**. This consists of sensory cells that connect with the auditory nerve, and which have sensory hairs projecting upwards, making contact with the tectorial membrane. The organ of Corti contains about 25 000 hair cells in total.

We do not understand the precise mechanism by which sound waves are translated into the sense of hearing, but some factors that may be significant are discussed overleaf.

Questions

1. How are the sensations of taste and smell detected?
2. a Why do you get a popping sensation in your ears when you climb a mountain or go up in an aeroplane?
   b How do you attempt to adjust the pressure on both sides of the ear drum?
3. What are the steps by which sound waves at the ear drum cause impulses to be generated in the cells of the organ of Corti?
4. In what form does the mechanical energy of sound leave the inner ear?

The mechanism of hearing

Vibrations from the ear drum and ossicles enter the cochlea via the oval window. Since fluid cannot be compressed, movements of the oval window cause similar movements of the perilymph and endolymph within the cochlea, and are eventually transmitted to the round window. Pressure waves in the perilymph and endolymph cause the hair cells of the organ of Corti, attached to the basilar membrane, to rub or pull against the tectorial membrane. The resulting movement of the sensory hair may cause the sensory cell to become depolarised. When a hair cell is stimulated in this way a generator potential develops in it, producing an action potential, which is transmitted to the brain via a branch of the auditory nerve.

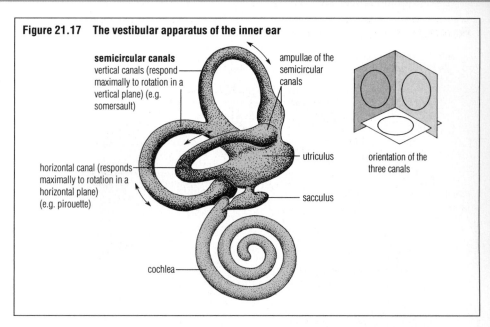

Figure 21.17   The vestibular apparatus of the inner ear

semicircular canals
vertical canals (respond maximally to rotation in a vertical plane) (e.g. somersault)

ampullae of the semicircular canals

horizontal canal (responds maximally to rotation in a horizontal plane) (e.g. pirouette)

utriculus

orientation of the three canals

sacculus

cochlea

How does the ear discriminate between sounds?

Sound has at least three qualities that human ears can normally differentiate. These are differences in pitch, amplitude and tone.

The **pitch** of a sound, its frequency or wavelength, is how we refer to high and low notes. High notes are produced by sound waves of high frequency (short wavelength), whereas low notes are produced by sound waves of lower frequency (longer wavelength). Frequency of sound is measured in hertz (Hz; 1 Hz = 1 vibration per second). The human ear is sensitive in the range 40–16 000 Hz, but is most sensitive in the range 800–8500 Hz.

The **amplitude** of a sound, its intensity, is how we refer to its loudness. Loudness is expressed in decibels (dB). (The dB scale is logarithmic, so that a 20 dB sound is 100 times louder than one of 10 dB.) A whisper is at 30 dB, normal talking at 60 dB, and painfully loud sounds at 120 dB or more.

The **tone** of a sound is its quality or timbre. Tone is determined by the number of different frequencies that make up a sound.

The mechanism by which the cochlea discriminates between pitch (high or low sounds), amplitude (intensity or loudness) and tone (quality or timbre) is not known with certainty. The following explanation represents a general consensus.

High-frequency or high-pitched sounds (short wavelength) cause the basilar membrane to vibrate near the base of the cochlea (area I in Figure 21.16). Here the basilar membrane is narrower and under higher tension, and vibrates at higher frequencies. Low-frequency or low-pitched sounds (long wavelength) cause the basilar membrane to vibrate near the apex of the cochlea (area II in Figure 21.16). Here the basilar membrane is broad and under lower tension, and it vibrates at lower frequencies. It is assumed that localised vibration of the basilar membrane, triggered by vibrations at a particular frequency, causes the sensory cells in that region of the cochlea to be stimulated. Whilst a pure sound stimulates only one region of the basilar membrane, a sound of several frequencies will stimulate many regions of the membrane. In this way the ear may detect the quality or tone of sound.

The intensity of sound is determined by the amplitude of the sound waves striking the ear drum. Loud sounds cause greater displacement of the basilar membrane than do softer sounds. Consequently, impulses may be fired at greater frequency in the auditory nerve.

Hearing defects

The ear's processing of sound is an extremely delicate and easily damaged mechanism, vulnerable to drugs, mechanical damage and exposure to intense sounds.

Conductive hearing loss occurs when the outer and middle ear fail to conduct vibrations to the cochlea, perhaps because of wax blockage of the ear canal or a punctured ear drum following mechanical injury or infection. Serious conductive loss that cannot be cured can be overcome with a hearing aid, which amplifies sound waves before they enter the ear.

Perceptive hearing loss is due to defects of the cochlea or the auditory nerve, or in the brain. These types of defect may be inherited, or acquired from infections or mechanical injury, or caused by exposure to very high levels of noise (such as that from amplified music at a pop concert or from a personal stereo, or from heavy machinery). There is a progressive loss of sensitivity to high-frequency sounds with increasing age.

■ Balance

Above the cochlea, but connected to it, there lies a system of fluid-filled canals and sacs referred to as the **vestibular apparatus** (Figure 21.17). These structures are concerned with the equilibrium (balance) of the body.

The vestibular apparatus consists of the three semicircular canals, each with a swelling called an **ampulla**, which detects movements of the head, and the utriculus and sacculus, which detect changes in posture. The sensory receptors within the vestibular apparatus have hair-like extensions embedded in dense structures that are supported in fluid endolymph. Movements cause deflection of the hairs and, if the hairs are sufficiently stimulated, the generation of an action potential.

The ability of the body to maintain equilibrium is controlled by the brain, and is based on sense data from all over

Figure 21.18   The semicircular canals and ampullae

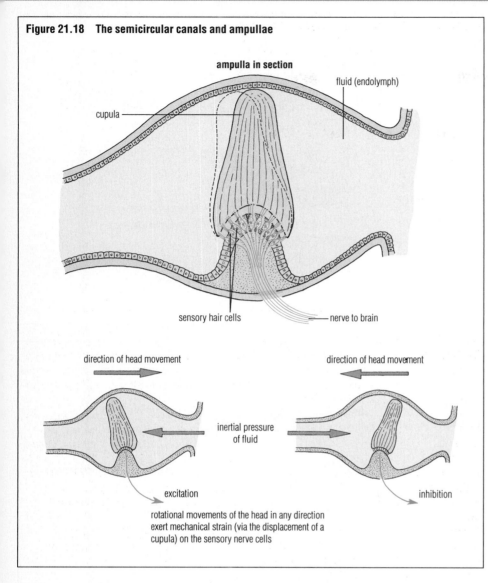

ampulla in section

fluid (endolymph)

cupula

sensory hair cells

nerve to brain

direction of head movement

direction of head movement

inertial pressure
of fluid

excitation

inhibition

rotational movements of the head in any direction
exert mechanical strain (via the displacement of a
cupula) on the sensory nerve cells

Figure 21.19   Utriculus and sacculus, and posture

fluid (endolymph)

the otoliths increase the
density of the otolithic
membrane; movement of
the membrane as a whole
distorts and stimulates the
hair cells

otolithic membrane

sensory hair cells
(macula)

nerve to brain

- the macula of the utriculus is on the floor (horizontal)
- the macula of the sacculus is on the side walls (vertical)

the body, not just from the inner ear. For example, proprioceptors in muscles and joints (p. 493) respond to changes in position and movements of body structures. In addition, an essential part is played by visual information about the position of the body in relation to the environment, received via the eyes. (Just try to keep your balance whilst standing on one leg, first with eyes open, then with eyes closed!)

■ Detecting movements of the head

The semicircular canals of the inner ear provide sense data on head movements due to rotation. Inside the ampulla on each semicircular canal there is a raised structure containing sensory hair cells. The hairs are embedded in a gelatinous mass, the **cupula**. When the head moves quickly the inertia of the endolymph within the semicircular canals deflects the dense, gelatinous cupula in one or more ampullae. Sensory hairs are stimulated and impulses are sent to the brain. Since the semicircular canals are mutually perpendicular, movements in any plane can be detected (Figure 21.18).

■ Posture

Sensory data relating to posture (slow movements) come from structures called **maculae** in the walls of the utriculus and sacculus. The macula consists of hair (receptor) cells that have their hairs embedded in a thick glycoprotein layer covered by calcium carbonate crystals, called the **otolithic membrane** (Figure 21.19). This structure is more dense than the endolymph in which it is suspended. As the head tilts or moves, the otolithic membrane moves, pulling on the sensitive hairs and making them bend. The sense cells are stimulated to varying degrees, causing action potentials to be sent to the brain.

Questions

1 a   What is the difference between pitch and intensity of sound?
   b   How may the cochlea differentiate between them?
2   Where in the body do we detect the sense data that enable us to keep a physically balanced position?
3   How does the role of the semicircular canals differ from that of the utriculus and sacculus?

21.7 SIGHT

The mammalian eyes, spherical structures about 2.5 cm in diameter, fit into bony sockets (orbits) in the skull (Figure 21.20). The adjacent tear glands secrete a watery solution of salts, together with some mucus and the bactericidal enzyme lysozyme. This solution moistens, lubricates and protects the eyes, and is drained away by a duct into the nasal passage.

The eye (Figure 21.21) is essentially a three-layered structure. The outer **sclera** contains and protects the eye. At the front of the eye this layer, known as the **cornea**, is transparent. The cornea plays a key part in the focusing of an image on the retina.

The middle layer of the eye, known as the **choroid**, is black, and prevents reflection of light in the interior of the eyeball. It contains a capillary network. Towards the front of the eye the choroid layer forms the ciliary body and iris. The **ciliary body** contains a circular muscle, and is attached to the lens by means of the **suspensory ligament**. The lens is a transparent, elastic, biconvex structure. Immediately in front is the **iris**. This circular disc of tissue contains radial and circular muscle fibres, and has a central hole, the **pupil**. The iris contains the pigment responsible for 'eye colour'. In front of the lens is a chamber containing transparent liquid, the **aqueous humour**. Behind the lens the eye is filled with transparent, jelly-like material, the **vitreous humour**.

Inside the choroid layer is the **retina**, which consists of neurones and photoreceptor cells. The photoreceptor cells are of two types, known as **rods** and **cones**. These synapse with bipolar neurones, which themselves synapse with ganglion neurones (electrical synapses, p. 449). Nerve fibres of the ganglion neurones lead from the retina to form the optic nerve. The point where the nerve fibres meet and become the optic nerve is the **blind spot** (see the *Study Guide*, Chapter 21). The optical axis is a theoretical line through the centre of the lens; the fovea, the place of most acute vision, is on the optical axis. Cones are concentrated at and around the fovea. In all, the human retina contains about 7 million cones and 10–20 million rods.

■ Accommodation; focusing of the image

Light rays from an object on which the eyes focus are bent (refracted), mostly by the cornea, to form a sharp image on the retina (Figure 21.22). The image is

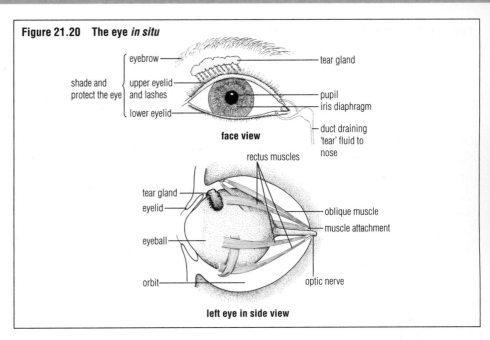

Figure 21.20   The eye *in situ*

shade and protect the eye
- eyebrow
- upper eyelid and lashes
- lower eyelid

tear gland
pupil
iris diaphragm
duct draining 'tear' fluid to nose

face view

rectus muscles
tear gland
eyelid
eyeball
orbit
oblique muscle
muscle attachment
optic nerve

left eye in side view

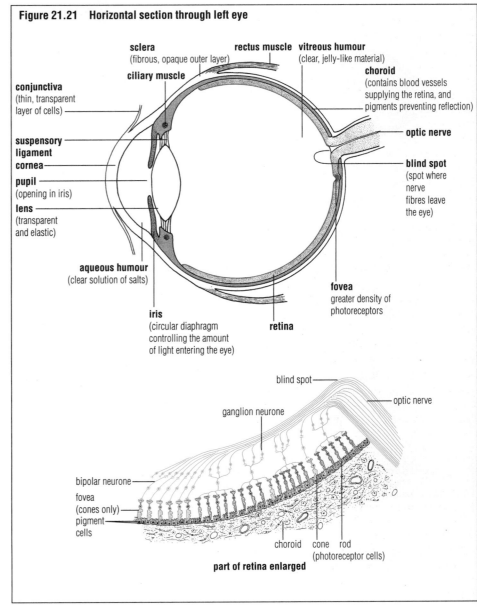

Figure 21.21   Horizontal section through left eye

sclera (fibrous, opaque outer layer)
ciliary muscle
rectus muscle
vitreous humour (clear, jelly-like material)
choroid (contains blood vessels supplying the retina, and pigments preventing reflection)

conjunctiva (thin, transparent layer of cells)
suspensory ligament
cornea
pupil (opening in iris)
lens (transparent and elastic)

optic nerve
blind spot (spot where nerve fibres leave the eye)

aqueous humour (clear solution of salts)
iris (circular diaphragm controlling the amount of light entering the eye)
retina
fovea greater density of photoreceptors

blind spot
optic nerve
ganglion neurone
bipolar neurone
fovea (cones only)
pigment cells
choroid
cone   rod (photoreceptor cells)

part of retina enlarged

Figure 21.22 Focusing an image on the retina (changes depicted in this series of drawings are greatly exaggerated); greatest refraction occurs at cornea

lens pulled less convex (flatter) by fluid pressure

ciliary muscle relaxed

rays focused onto retina

suspensory ligament taut

weakly divergent rays of light from distant object

for distant objects

lens reverts to natural (more convex) shape

strongly diverging rays of light from a near object

ciliary muscle contracted

rays focused onto retina

suspensory ligament not stretched

for near objects

cornea

ciliary muscle

suspensory ligament

lens

iris

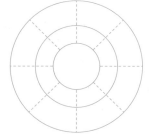

ciliary muscle

suspensory ligament

lens

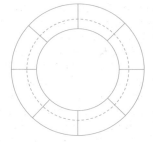

inverted and reversed by the lens, yet things are 'seen' upright because of interpretative actions of the brain.

Most refraction occurs at the cornea, because of its high refractive index in comparison with air, and also because of its curved surface. The degree of refraction by the cornea cannot be varied, however, and depends upon the angle at which light strikes it. (This, in turn, depends on the distance of an the object from the cornea.) The function of the lens is to adjust the degree of refraction to produce a sharp image on the retina, whatever the distance of an object from the eye. By changing shape, the lens focuses light rays from near or distant objects.

Lens tissue is elastic, and variation in the curvature of the lens surface is brought about by a change in the degree of contraction of the circular ciliary muscle to which the lens is attached. When the ciliary muscle relaxes there is increased tension on the suspensory ligament because of the outward pressure of the humours on the sclera, and the lens is pulled into a thinner, less sharply convex shape. This shape focuses light from distant objects on the retina. When the ciliary muscle contracts there is decreased tension on the suspensory ligament and the lens, being elastic, resumes its more convex shape. In this state, light from close objects is focused. At the same time the size of the pupil is reduced; accommodation in the eye involves both the regulation of pupil size (as shown in Figure 21.23) and the focusing processes by which an image is formed. Reduction in pupil size increases the depth of the visual field.

Figure 21.23 The iris diaphragm mechanism

in bright light

▶ circular muscle fibres of iris contract
▶ radial muscle fibres relax
▶ pupil diameter decreases
▶ less light enters the eye (preventing damage to the retina)
▶ depth of focus is increased

in dim light

▶ radial muscle fibres of iris contract
▶ circular muscle fibres relax
▶ pupil diameter increases
▶ more light enters the eye (enabling vision at low light intensity)
▶ depth of focus is decreased

A reflex action
Change of pupil size is a reflex action, controlled by the autonomic nervous system. The radial muscle fibres are served by the sympathetic nervous system, the circular muscle fibres are served by the parasympathetic nervous system.

Atropine and pupil size
The circular muscle fibres of the eye can be prevented from contracting by the drug atropine, which acts by blocking the neurotransmitter acetylcholine. Opticians sometimes apply atropine to the eyes in tiny quantities prior to examination of the retina using an opthalmoscope.

Questions

1 What is the principal function of the lens? How does the eye change the shape of the lens?
2 Why does the greatest degree of refraction of light occur at the cornea?
3 How does variation of pupil size also influence accommodation?
4 What are the structural differences between the blind spot and fovea of the eye?

■ The light-sensitive cells of the retina

The eye is sensitive to electromagnetic radiation in the wavelength range 380–760 nm (p. 250). Light falling on the retina causes structural change in the photopigment in the rods and cones (Figure 21.24).

Rod cells

Rods are involved in vision at low light intensity. In the rod cells the photosensitive pigment is **rhodopsin** (visual purple). When exposed to light rhodopsin splits into opsin (a protein) and retinal (a derivative of vitamin A). The split occurs because on absorbing light the retinal molecule changes shape about a double bond (a *cis–trans* isomerisation) and no longer binds tightly to the opsin molecule. Opsin alters the permeability of rod cells to sodium ions, causing a generator potential to arise. As a result, an action potential is generated in the ganglion cell and propagated to the brain along a nerve fibre (Figure 21.26). The pattern of nerve impulses conveys information to the brain about the changing visual field. Rhodopsin re-forms in the absence of further light stimulation.

At high light intensity rhodopsin is 'bleached', because it is broken down faster than it is re-formed. This is why our eyes have to adapt to the dark when we move from bright light. In total darkness it takes 30 minutes for all the rhodopsin to be re-formed.

Cone cells

Cones are sensitive to high light intensities, and to differences in wavelength. Mammals with cones in their retinas are able to distinguish colours. Here the photosensitive pigment is **iodopsin**. There are three forms of iodopsin, each responding to light of a different wavelength; in effect, there are red-, green- and blue-sensitive cones. Photodecomposition of iodopsin produces a generator potential, but in cones the photosensitive pigment quickly re-forms.

The mechanism of colour vision is not known; the **trichromatic theory** is a working hypothesis. According to this there are three types of cone; each has a different form of iodopsin that is maximally sensitive to either blue, green or red light. Colours are detected by the relative degree of stimulation of the three types of cone. White light stimulates all three types of cone equally.

■ The role of the brain in vision

Observations of the three-dimensional world about us are reduced to a two-dimensional image on the surface of the retina (Figure 21.25). As a consequence, action potentials generated in the rods and cones are carried by neurones of the optic nerves to the visual cortex of the brain (Figure 21.37, p. 467). Each eye views left and right sides of the visual field. The brain, however, receives action potentials from the left and right visual fields separately. The visual cortex, located at the rear and base of the cerebral hemispheres, combines the images into a single impression – our sight – by a process known as **perception**. Perception involves the interpretation of sense data from the retina, in terms of our existing experience and expectations, and is thus in large part a subjective process. Activities concerning the subjective nature of perception are described in the *Study Guide*, Chapter 21.

Figure 21.24 The retina and vision

Rod cells are distributed evenly throughout the retina. Many rod cells share a single bipolar neurone connection to the brain (synaptic convergence). Rod cells therefore give poor visual acuity (sharpness of vision).

Cone cells are concentrated in and around the fovea. Each cone cell has its own bipolar connection to the brain. Cone cells therefore give good visual acuity.

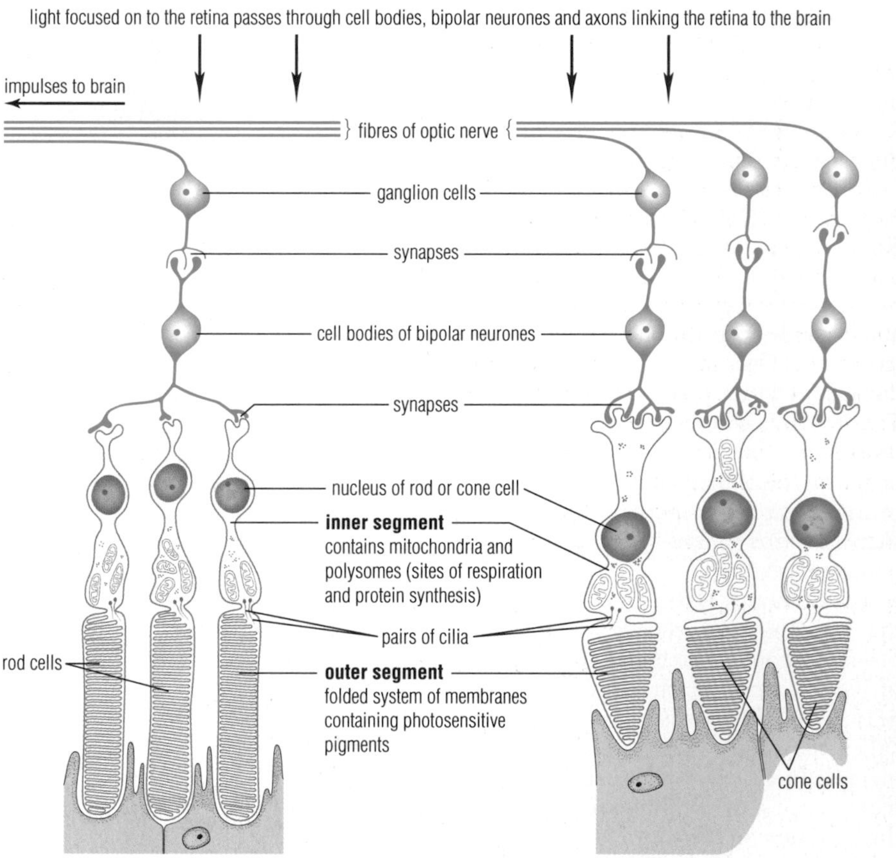

light focused on to the retina passes through cell bodies, bipolar neurones and axons linking the retina to the brain

impulses to brain

} fibres of optic nerve {

ganglion cells

synapses

cell bodies of bipolar neurones

synapses

nucleus of rod or cone cell

inner segment
contains mitochondria and polysomes (sites of respiration and protein synthesis)

pairs of cilia

rod cells

outer segment
folded system of membranes containing photosensitive pigments

cone cells

Figure 21.25 The eyes and visual cortex

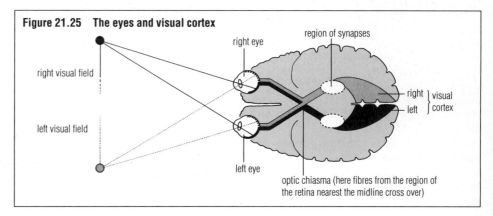

right eye

region of synapses

right visual field

left visual field

left eye

right } visual
left } cortex

optic chiasma (here fibres from the region of the retina nearest the midline cross over)

The mechanism of photoreception in rod cells

Rod cells each have an outer and an inner segment with distinctively different roles. Mitochondria are housed in the inner segment, and provide the ATP that drives the battery of sodium pumps in the plasma membrane of this segment. These continuously pump out sodium ions into the tissue fluid around the rod cells.

The visual pigment of rod cells, **rhodopsin**, is housed in disc-shaped membranes in the outer segment. In the dark (rhodopsin in the *cis* form) the plasma membrane of the outer segment allows sodium ions to diffuse back in. These ions reduce the negative charge inside the rod cell to about 40 mV. In this state, the outer segment releases a special transmitter substance (glutamate) into the tissue fluid. The effect of this is to inhibit transmission of an action potential in the ganglion cell.

When light of low intensity falls on the rod cell it changes the rhodopsin in the outer segment to the *trans* form. This indirectly causes the sodium channels to close, decreasing the permeability of the plasma membrane to sodium ions. Meanwhile, the inner segment continues to pump out sodium ions, so the interior of the rod cell becomes even more negative than usual (hyperpolarised). Hyperpolarisation stops the release of the special transmitter substance. As a result, in the ganglion cell that synapses with the rod cell an action potential is formed and passes down the optic nerve to the brain.

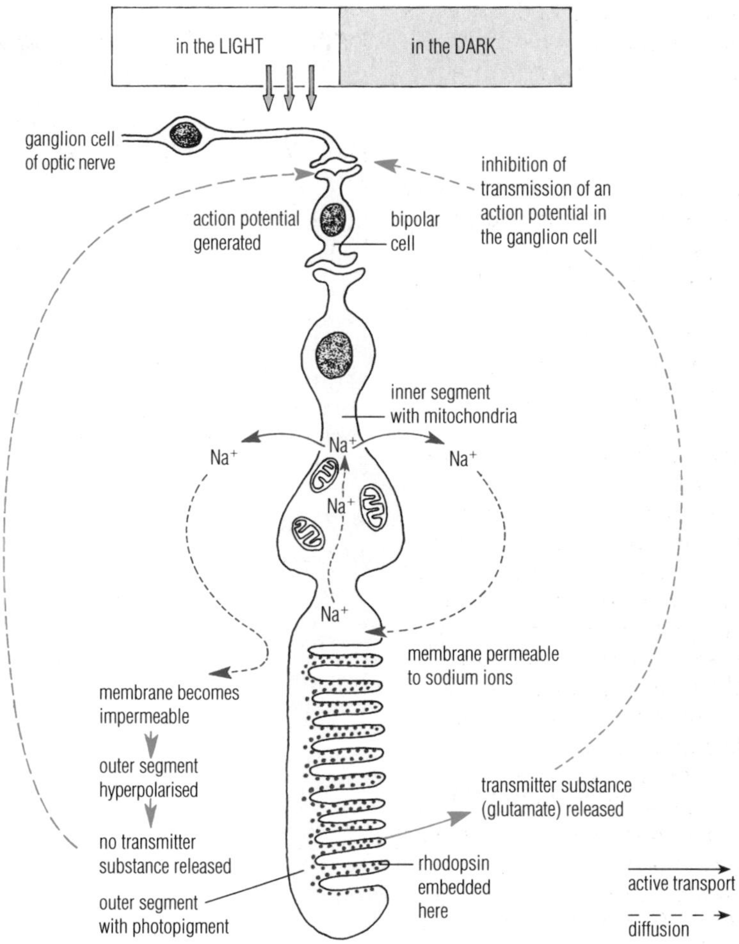

Stereoscopic vision

In an animal with stereoscopic vision, the visual fields of both eyes overlap, so that images of an object are focused on the retinas of both eyes. Consequently, the eyes simultaneously produce slightly different images. The brain resolves these into a three-dimensional impression.

Predators, such as members of the cat family and birds of prey such as owls (p. 95), have well-developed stereoscopic vision; so too do tree-climbing primates. By contrast, prey species tend to have laterally placed eyes. Their field of vision is at or near 360°, but they have minimal stereoscopic vision (Figure 21.27).

Figure 21.26 The working rod cell

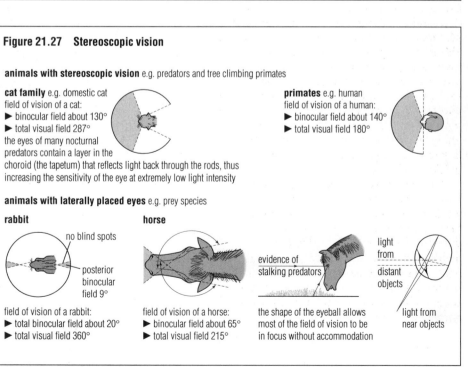

Figure 21.27 Stereoscopic vision

Colour blindness

Colour blindness is believed to be due to failure of one or more of the three classes of primary colour cones. Deficiencies in the cones most receptive to red and green light gives rise to red–green colour blindness. Colour blindness is a sex-linked characteristic. It is an inherited condition that affects males very much more often than females because the gene for colour blindness is a recessive gene carried on the X chromosome (p. 622).

Questions

1 Make a table contrasting rods and cones under the headings: shape; relative numbers; distribution in retina; visual acuity; light sensitivity; ability to discriminate light of different wavelengths.
2 What are the advantages to a grazing animal of having eyes at the side of the head rather than at the front?

21.8 THE NERVOUS SYSTEM OF MAMMALS

The mammalian nervous system (Figure 21.28) is dual in nature, and forms the central control and quick communication system of the body. The **central nervous system** (CNS), consisting of the brain and spinal cord, coordinates and controls the activities of the animal. The **peripheral nervous system** (PNS), the nerves and ganglia, forms the connecting link between the organs and the CNS. A **nerve** is an elongated structure consisting of a bundle of nerve fibres enclosed within a sheath of connective tissue. A **ganglion** is a collection of nerve cell bodies, also enclosed in connective tissue, lying along a nerve.

■ Peripheral nerves and reflex actions

Many body functions and actions are controlled by reflex action (see Figure 21.3, p. 443).

The reflex action is a rapid, automatic response resulting from nervous impulses initiated by a stimulus. A reflex action involves the central nervous system (the spinal cord or the brain) and sensory and motor neurones. It does not necessarily involve conscious awareness. For example, the opening and closing of the pupil in the eye (Figure 21.23, p. 457) involve the brain, but it is an automatic response of which we are not aware, and over which we have no control.

Another example of a human reflex action is the knee-jerk reflex (Figure 21.30). This reflex action involves the spinal cord.

■ The spinal cord

The spinal cord runs dorsally the length of the body. It is a cylindrical structure with a tiny central canal. This canal contains cerebrospinal fluid, and is continuous with the ventricles of the brain (p. 463). The central part of the cord contains cell bodies, synapses and non-myelinated relay neurones. It is known as the **grey matter**. The outer part of the cord contains myelinated fibres running longitudinally, and is known as the **white matter**.

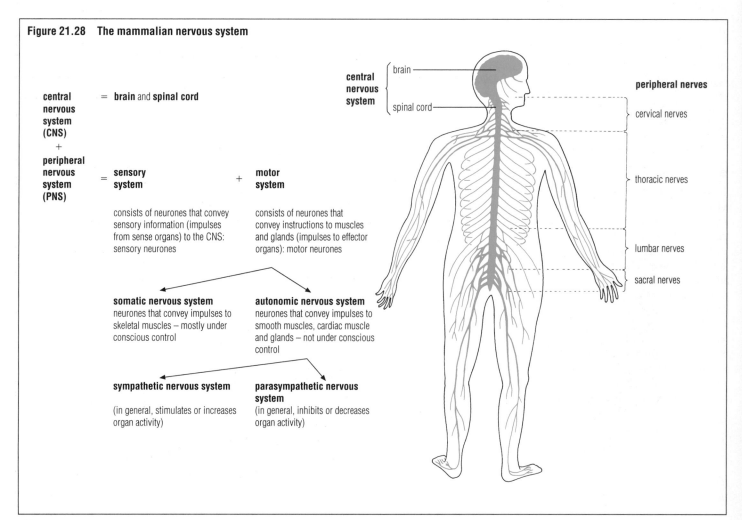

Figure 21.28   The mammalian nervous system

The spinal cord is enclosed and protected by the vertebrae (p. 484). Between each two vertebrae a pair of spinal nerves arises, one nerve on each side. There are 31 pairs of spinal nerves in humans. Each spinal nerve joins the spinal cord at two points, or roots: the sensory fibres enter through the dorsal root; motor fibres leave through the ventral root.

The functions of the spinal cord are to relay impulses in and out at any particular point along the cord, and to relay impulses up and down the body, including to and from the brain. The structure and functioning of the spinal cord are illustrated in Figures 21.29 and 21.30.

■ Impulses to and from the brain

With any reflex action, such as that illustrated in Figure 21.30, there is a pathway for impulses to be sent to the brain via ascending nerve fibres that originate at synapses in the grey matter of the spinal cord. The brain may store this information (in which case it contributes to memory), or may correlate it with sense data from, say, the eyes. It may also be compared with earlier stored experience.

As a result, impulses may be despatched from the brain via excitatory descending nerve fibres to enhance the response. Alternatively the response may be overridden by impulses from the brain along inhibitory nerve fibres. These various alternatives are all part of the basis of complex behaviour (p. 519). For example, when the type of response is modified by past experience the reflex is said to be **conditioned** (p. 521).

Figure 21.30 The spinal cord and the human knee-jerk reflex action

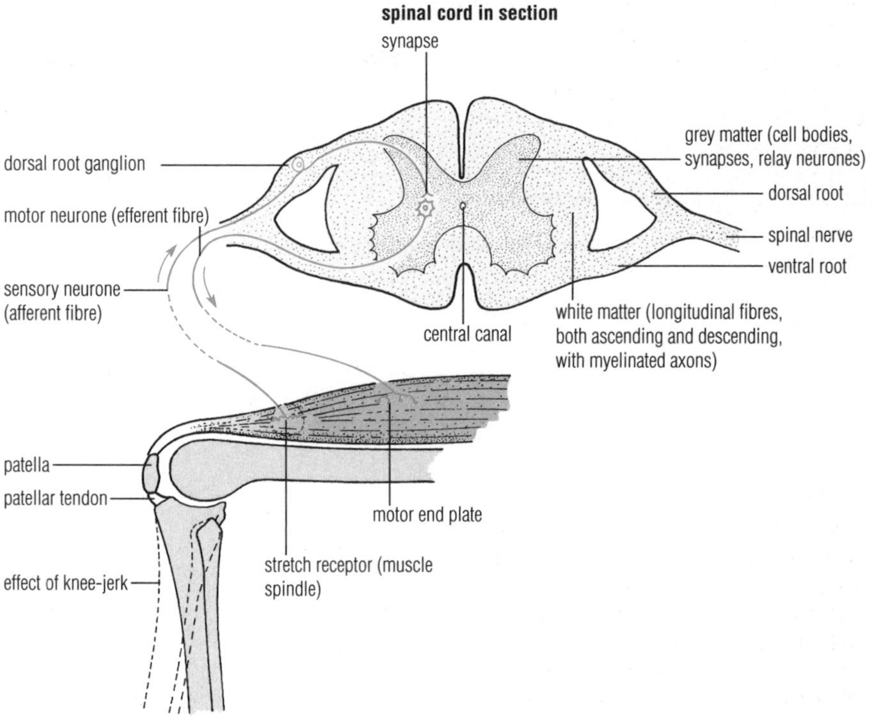

spinal cord in section

synapse

dorsal root ganglion

motor neurone (efferent fibre)

sensory neurone (afferent fibre)

grey matter (cell bodies, synapses, relay neurones)

dorsal root

spinal nerve

ventral root

central canal

white matter (longitudinal fibres, both ascending and descending, with myelinated axons)

patella

patellar tendon

motor end plate

effect of knee-jerk

stretch receptor (muscle spindle)

The knee-jerk as a monosynaptic reflex
The simplest type of reflex involves two neurones and one synapse – for example, when a muscle is stretched. One way of stretching a muscle is to apply a sharp tap with a rubber hammer to the patellar tendon. Stretch receptors in the muscle are stimulated. An impulse is conducted via a sensory neurone and crosses a synapse in the grey matter of the spinal cord to a motor neurone. The impulse is then conducted via the motor neurone to the extensor muscle of the leg, producing a contraction (via the motor end plate) and causing a knee-jerk.

Figure 21.29 A polysynaptic reflex

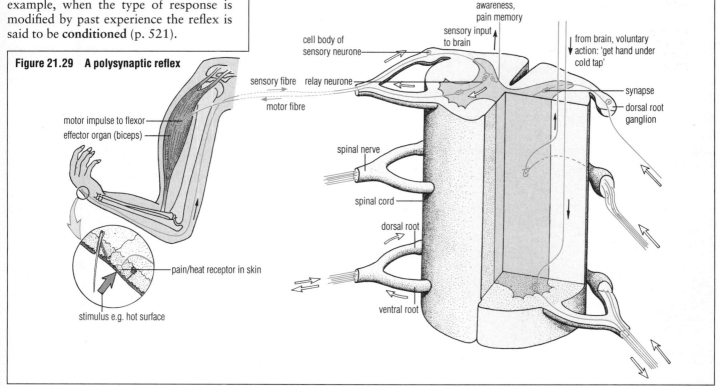

motor impulse to flexor

effector organ (biceps)

pain/heat receptor in skin

stimulus e.g. hot surface

cell body of sensory neurone

awareness, pain memory

sensory input to brain

from brain, voluntary action: 'get hand under cold tap'

sensory fibre relay neurone

motor fibre

synapse

dorsal root ganglion

spinal nerve

spinal cord

dorsal root

ventral root

■ The autonomic nervous system

The autonomic nervous system is concerned with control of the internal environment, and is not involved with the skeletal muscles. Nerve fibres of the autonomic nervous system run from the central nervous system to the various internal organs such as the heart, lungs, intestines and glands. Control by the autonomic nervous system is involuntary (conditioning, p. 521), although certain parts of the system, such as the anal and bladder sphincter muscles, can be put under voluntary control.

The autonomic nervous system consists of two components, the **sympathetic** and **parasympathetic nervous systems**. Both sympathetic and parasympathetic nerves contain a preganglionic neurone and a postganglionic neurone. In the sympathetic nervous system the synapse between the two neurones is near the spinal cord, whereas in the parasympathetic nervous system the synapse is in the effector organ (Figure 21.31).

Sympathetic and parasympathetic nerves have opposite effects on the organs they innervate. We say their effects are **antagonistic**. For example:

Sympathetic nervous system

▶ increases cardiac output
▶ increases blood pressure
▶ increases ventilation rate.

Parasympathetic nervous system

▶ decreases cardiac output
▶ reduces blood pressure
▶ decreases ventilation rate.

These antagonistic effects result from different chemical transmitters released at the nerve endings (Figure 21.32). The preganglionic fibres of both nervous systems all produce acetylcholine, but the nerve endings of postganglionic fibres of the sympathetic nerve endings mostly secrete noradrenaline whereas the parasympathetic nerve endings mostly release acetylcholine.

Overall control of the autonomic nervous system comes from centres in the hindbrain, such as the cardiac and vasomotor centre (Chapter 17, p. 349), the respiratory centre (Chapter 15, p. 311) and the thermoregulatory centre (Chapter 23, p. 514).

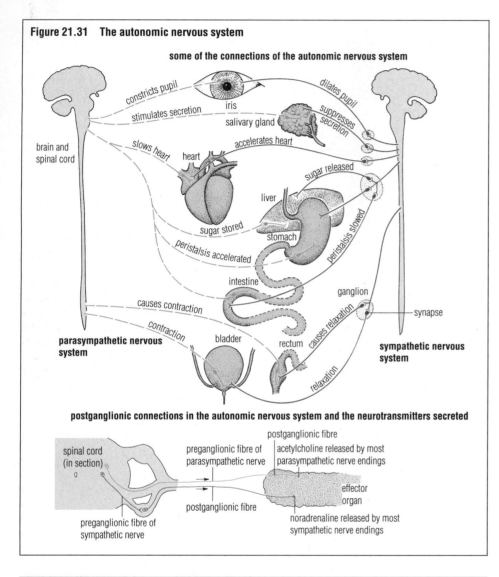

Figure 21.31   The autonomic nervous system

some of the connections of the autonomic nervous system

constricts pupil
dilates pupil
iris
stimulates secretion
suppresses secretion
salivary gland
slows heart
accelerates heart
brain and spinal cord
heart
sugar released
liver
sugar stored
stomach
peristalsis slowed
peristalsis accelerated
intestine
ganglion
synapse
causes contraction
causes relaxation
contraction
bladder
rectum
relaxation
parasympathetic nervous system
sympathetic nervous system

postganglionic connections in the autonomic nervous system and the neurotransmitters secreted

spinal cord (in section)
preganglionic fibre of parasympathetic nerve
postganglionic fibre
acetylcholine released by most parasympathetic nerve endings
effector organ
postganglionic fibre
preganglionic fibre of sympathetic nerve
noradrenaline released by most sympathetic nerve endings

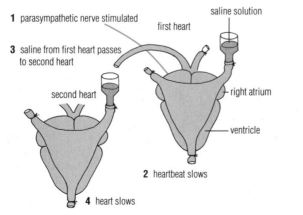

Figure 21.32   Autonomic nerves exert their effects through the release of chemicals (a discovery made by Otto Loewi in 1920)

A living heart was removed from a frog with its parasympathetic nerve fibres intact. Electrical stimulation of the sympathetic nerve speeded up the heartbeat, whereas electrical stimulation of the parasympathetic nerve slowed it down.

The same response was shown by a second heart if it was bathed in a saline solution from the first heart.

The nature of the chemical messengers was discovered later.

1 parasympathetic nerve stimulated
first heart
saline solution
3 saline from first heart passes to second heart
second heart
right atrium
ventricle
2 heartbeat slows
4 heart slows

Questions

1 What is meant by saying that a response is 'involuntary'?
2 State what nervous pathways are likely to be involved in the following activities:
   a stepping back from the kerb on seeing a car approaching
   b coughing when a crumb 'goes down the wrong way'.
3 What do you think might be the advantage to humans of the knee-jerk reflex?
4 Figure 21.29 is headed 'A polysynaptic reflex'. That means that it involves many synapses. How many synapses are shown?
5 Where is the control centre for the autonomic nervous system?

21.9 THE BRAIN

We have already seen that the vertebrate central nervous system develops in the embryo from the neural tube (p. 414). Initially this is a simple tube with a central canal containing cerebrospinal fluid. During development, the anterior part first enlarges to form the **primary brain vesicle**. This vesicle then subdivides as it grows, forming three irregular vesicles, the forebrain, midbrain and hindbrain. The various parts of the mature brain develop from these vesicles by selective thickening and folding processes in the walls and roof of the brain. As the brain develops the central canal becomes four small, fluid-filled cavities known as **ventricles**. The ventricles are connected together and to the spinal canal by narrow ducts.

White and grey matter are present in the brain, as in the spinal cord. Grey matter, which makes up the interior of the brain, consists of the cell bodies and synapses of groups of neurones, organised into groups called 'nuclei' or centres. (Later on in development of the human brain, grey matter forms an additional layer on parts of the forebrain and hindbrain; see below.) White matter consists of myelinated nerve fibres in bundles, and occurs towards the outside of the brain. Blood capillaries are also present in brain tissue.

■ Development of the human brain

In the human brain the fundamental division into fore-, mid- and hind-brain becomes obscured by the differentiated growth of parts of the embryonic brain. In fact, the structures of the adult brain develop from five regions of the brain, formed by subdivisions of the primary brain vesicles during development (Figure 21.33). These five regions and their origins are shown in Table 21.2.

The **forebrain** gives rise to two regions, the telencephalon and the diencephalon. The roof of the telencephalon grows out at both sides to form the cerebral hemispheres. These enlarge so greatly that they come to overlie the rest of the brain. The floor of the telencephalon is occupied by the basal ganglia, and the tip of the telencephalon forms the olfactory lobes.

Outgrowths from the sides of the diencephalon form the optic vesicles, which develop into the retina of the eyes. An outgrowth from the roof of the diencephalon forms a tiny gland, the pineal

body, and the remainder of the roof forms the anterior choroid plexus. An outgrowth of the floor of the diencephalon contributes to the formation of the master endocrine gland, the pituitary gland (by fusion with part of the roof of the pharynx). Internally, the sides of the diencephalon thicken to form the thalamus, whilst the floor itself becomes the hypothalamus.

The **midbrain** becomes the mesencephalon, a small region connecting the forebrain and hindbrain. The roof here forms a structure called the four-lobed corpora quadrigemina.

The **hindbrain** also gives rise to two regions, the metencephalon and myelencephalon. The roof of the metencephalon enlarges to form the cerebellum, and the floor forms the pons. The medulla oblongata forms from the floor and side walls of the myelencephalon. It is not sharply demarcated from the spinal cord with which it merges.

Table 21.2 The structures of the human brain, and the regions of the embryonic brain from which they develop

Primary brain vesicles of embryo	Forebrain		Midbrain	Hindbrain	
Five regions of adult brain formed by subdivision	telencephalon	diencephalon	mesencephalon	metencephalon	myelencephalon
Structures of adult human brain	cerebral hemispheres	pineal body	four-lobed corpora quadrigemina	cerebellum	
	basal ganglia	thalamus hypothalamus		pons	medulla
	olfactory lobes	pituitary gland			
		anterior choroid plexus*			posterior choroid plexus*

* The choroid plexus is a region where the cells are well supplied with blood capillaries and where the cerebrospinal fluid of the ventricles is secreted.

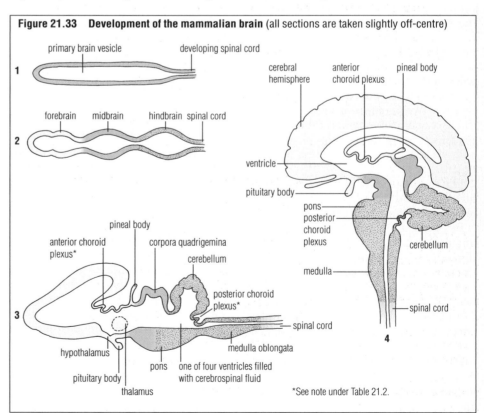

Figure 21.33   Development of the mammalian brain (all sections are taken slightly off-centre)

1   primary brain vesicle   developing spinal cord

2   forebrain   midbrain   hindbrain   spinal cord

3   anterior choroid plexus*   pineal body   corpora quadrigemina   cerebellum   posterior choroid plexus*   spinal cord   medulla oblongata   hypothalamus   pituitary body   pons   thalamus   one of four ventricles filled with cerebrospinal fluid

4   cerebral hemisphere   anterior choroid plexus   pineal body   ventricle   pituitary body   pons   posterior choroid plexus   medulla   cerebellum   spinal cord

*See note under Table 21.2.

Table 21.3 Cranial nerves and the structures innervated*

Cranial nerve	Innervation		Ventral view of brain
I Olfactory	sensory nerve:	smell, from the nose	
II Optic	sensory nerve:	sight, from the retina	
III Oculomotor	motor nerve:	eye movements, to the eyeball muscles	
IV Trochlear	motor nerve:	eye movements, to the eyeball muscles	
V Trigeminal	mixed nerve:	sensory, from the face (touch and pain)	
		motor, to the jaw (movement)	
VI Abducens	motor nerve:	eye movements, to the eyeball muscles	
VII Facial	mixed nerve:	sensory, from the tongue (taste)	
		motor, to the facial muscles (expression)	
VIII Auditory	sensory nerve:	hearing, from the cochlea	
		balance, from the semicircular canals	
IX Glossopharyngeal	mixed nerve:	sensory, from the tongue (taste)	
		motor, to the pharynx (swallowing)	
X Vagus	mixed nerve:	sensory, from internal organs	
		motor, to the same internal organs	
XI Accessory	motor nerve:	head movements, to muscles of neck	
XII Hypoglossal	motor nerve:	tongue movements, to muscles of tongue	

* Nerves for all other functions enter the brain via the spinal cord.

The role of the human brain

Humans (*Homo sapiens*) are the 'thinking species'. This facility resides in our brains. But the fundamental role of the brain is to coordinate and control the activities of the body. The brain controls all body functions except those under the control of simple spinal reflexes. How is this achieved?

1 The brain receives impulses from sensory receptors.

The embryonic regions of the brain (possibly equivalent to the regions of the primitive vertebrate brain) connect with the principal sense organs. That is, the forebrain is concerned with smell, the midbrain with sight and the hindbrain with balance. The adult human brain, when fully developed, receives impulses from the whole body via the spinal cord and the twelve pairs of cranial nerves. These cranial nerves connect the brain to the sense organs and muscles of the head and, via the vagus nerve, to the internal organs.

2 In the association centres the brain integrates and correlates the incoming information from different sense organs.

By integration we mean that the impulses coming in simultaneously from different sense organs are interpreted. Impulses are correlated with 'experiences' within stored memory. The brain has the ability to rehearse the possible consequences of different responses. Integration of stimuli is an essential prelude to coordination of body activities and functions.

3 The brain initiates activity by sending impulses to the effector organs (muscles and glands) causing them to function.

Impulses from the association and motor centres coordinate bodily activities so that the complete range of mechanisms and chemical reactions work efficiently together.

4 The brain stores information and builds up the memory bank of past experiences in a form that is normally accessible for future reference.

5 The brain also initiates impulses from its own self-contained activity, as when some idea suddenly occurs, or a problem is resolved, or another experience is remembered for its relevance.

The brain is the seat of the 'personality' and the emotions. It is the brain that enables us to imagine, to create, to plan, to calculate and to predict. Various forms of abstract reasoning are conducted in the brain.

How we know about brain function

The earliest known reference to the human 'brain' is on an Egyptian papyrus, dated to approximately 1800 BC, recording the treatment of a sword wound to the head.

In the long story of the discovery of the structure and functions of the human brain, detective work on people with severe but discrete head wounds has played an important part. These investigations required accurate observation of the precise site of injury (mostly by subsequent post-mortem examination of the cranium), followed by correlation with records of impairment of body function during life. The importance of this sort of enquiry is illustrated by the discovery of the role in sight of the visual cortex from the experiences of soldiers surviving bullet wounds in the rear of the skull. Today, similar discoveries come from studying the effects of localised lesions caused by strokes and tumours.

Other techniques have contributed also. For example, certain patients have given their permission for parts of the brain exposed during essential surgery to be electrically stimulated and the effects in the whole body to be noted; such studies led to the discovery of the role of parts of the motor cortex (Figure 21.38, p. 467).

Physiological experiments have been conducted on mammals and members of other vertebrate classes involving, for example, the removal of parts of a healthy brain or the severing of connections within the brain. The resulting altered behaviour gave insights into the roles of parts of the brain. Many of these approaches would be unacceptable today; we are now more reluctant to conduct inessential surgery that may cause pain.

Electroencephalography (Figure 21.34) is the technique of detecting and recording the electrical activity of the brain – during sleeping and when awake and alert, or with open and closed eyes, for example. Comparison of the electrical outputs of healthy, diseased and damaged brains gives clues to brain function. For example, altered patterns of electrical activity give evidence of brain damage or dysfunction, as in epilepsy.

Figure 21.34 Electroencephalography A patient is wired with electrodes (right) connected to an electroencephalogram machine. The 'brain waves' from the sleeping patient are being recorded during the various phases of sleep. In the electroencephalogram (below) the traces are due to brain activity (1 and 2), eye movement (3 and 4), heart activity (5) and muscles of the throat/neck (6 and 7). The trace includes a period of sleep known as 'rapid eye movement' sleep or REM

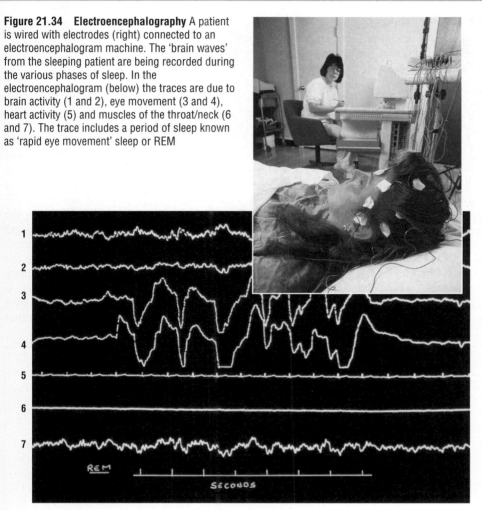

■ The functions of the parts of the human brain

The human brain (Figure 21.35) is a large organ, of mass approximately 1.3 kg. The brain contains 10^{11}–10^{12} neurones and at least the same number of non-excitable cells, the **neuroglial cells** (p. 220). Each neurone is in synaptic contact with about a thousand other neurones (Figure 21.36, overleaf).

The brain is surrounded by the bones of the cranium, and further protection comes from suspension of the brain in cerebrospinal fluid, which provides a shock-absorbing environment. This protects cells and blood vessels from damage due to jolting or jarring movements. The cerebrospinal fluid is contained within membranous coverings that surround the whole central nervous system. These membranes (there are three) are known as the **meninges**. Rarely, the membranes may become inflamed by bacterial or viral infection, a condition known as meningitis.

Questions

1 What does it mean to say that a particular cranial nerve is sensory, motor or mixed?
2 Explain what we mean by the term 'cephalisation' (p. 469).
3 What does an 'association centre' consist of? In what activities of the brain is it involved?

Figure 21.35 The human brain in sagittal section (in a bilaterally symmetrical animal, the sagittal plane is a vertical section dividing the body into right and left halves). This sagittal section passes *between* the cerebral hemispheres

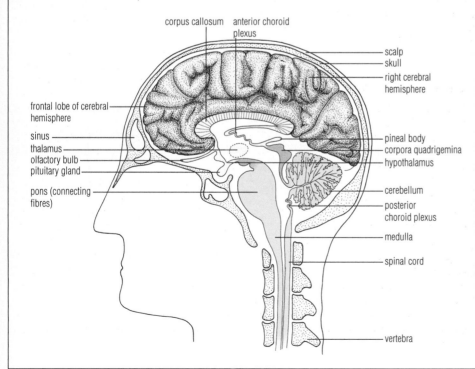

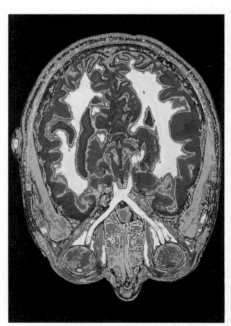

a computer tomography (CT) scan (computer simulation based on X-ray data) of a whole, **healthy brain,** showing both halves of the brain enclosed within the cranium

The hindbrain

The **medulla oblongata** contains regulatory centres for the autonomic nervous system. These are concerned with the automatic adjustment of heart rate and blood pressure (Chapter 17, pp. 350 and 352) and of the ventilation rate of the lungs (Chapter 15, p. 311), and with swallowing, salivation, vomiting, coughing and sneezing. The medulla receives sensory information from internal organs via the vagus nerve (see Table 21.3) and is influenced by impulses from the hypothalamus.

In the medulla the ascending and descending tracts of nerve fibres connecting the spinal cord with the parts of the brain cross over, so that the left side of the brain controls the right side of the body, and vice versa.

The **pons** is a 'bridge' of white fibres connecting the medulla with the upper parts of the brain. The pons also contains centres relaying impulses to the cerebellum.

The two hemispheres of the **cerebellum** have a folded external surface layer consisting of grey matter (non-myelinated neurones). The cerebellum is concerned with the control of involuntary muscle movements of posture and balance and of precise, voluntary manipulations involved in hand work, speech and writing. The cerebellum coordinates these movements, rather than initiating them.

The midbrain

The midbrain controls certain auditory reflexes such as the movements of the head to locate sounds, and visual reflexes involved in focusing on objects. It also houses the **reticular activating systems**, groups of neurones that arouse the brain and maintain consciousness. Here neurones also filter impulses, preventing overloading of the forebrain with too much information. Of the great number of stimuli we receive at any time, we are consciously aware of only a few. When the reticular activating system is inactive the body and brain pass into a state of sleep. Regular periods of sleep are essential to avoid fatigue of the nervous system.

The forebrain

The **thalamus** is a 'clearing house' for sensory impulses received from other parts of the brain and relayed on to appropriate parts of the sensory areas of the cortex of the cerebral hemispheres. It is also involved in the perception of both pleasure and pain.

The **hypothalamus** is the main control centre for the autonomic nervous system. Centres in the hypothalamus regulate the body activities concerned with maintaining a constant internal environment (homeostasis, p. 502). The hypothalamus also controls feeding and drinking reflexes, and aggressive and reproductive behaviour.

The hypothalamus is exceptionally well supplied with blood vessels, and it is here that body temperature and the levels of metabolites (such as sugars, amino acids and ions) and hormones in the blood are monitored. The hypothalamus works with the pituitary gland, which is attached to it, in controlling the release of most hormones (p. 474). Thus the hypothalamus is also part of the important linkage between the nervous and endocrine systems.

The **pineal gland**, also located in this part of the brain, is concerned with seasonal changes in physiology and behaviour in mammals other than humans. It is also involved in the process by which mammals measure time.

The **cerebral hemispheres** (cerebrum) form the bulk of the human brain, and consist of densely packed nerve cells and myelinated nerve fibres. The ratio of the size of the cerebral hemispheres to the size of the whole nervous system is larger in the human than in any other mammal. The cerebral hemispheres coordinate the body's voluntary activities and many involuntary ones. Each hemisphere is divided into four lobes, each lobe being named after the bone of the cranium that

---

Figure 21.36 Neural connections

Each neurone has a large number of terminal synaptic knobs.
A great variety of connections can be made. The ultimate response is the summation of all the excitatory (and inhibitory) stimuli received.

divergent connections

the impulse from a single presynaptic neurone makes synapses with a number of other neurones

impulses from one pathway are relayed to many pathways – information travels in various directions at the same time, thus allowing a variety of responses to a single stimulus

convergent connections

impulses are received from several presynaptic neurones

support strong excitation or inhibition, or allow one action in response to different stimuli

reverberating connections

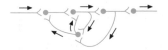

branches from postsynaptic neurones feed back to the presynaptic neurone, reinforcing its activity

once 'fired' the signal may last for some time – for a few seconds or for many hours

parallel after-discharge connections

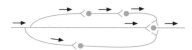

the presynaptic neurone stimulates a group of postsynaptic neurones, which then stimulate a common postsynaptic neurone, thus producing a strong, precise response

the postsynaptic neurone sends out a stream of impulses, but without feedback

covers it (Figure 21.37). The hemispheres have a vastly extended surface achieved by folding (convolution) with deep grooves. This surface is covered by a layer of grey matter about 3 mm deep, known as the **cerebral cortex**, and is densely packed with non-myelinated neurones known as pyramidal cells. These have a mass of dendrites and an almost unimaginable number of synaptic connections. Sensory, motor and (pyramidal) relay neurones are interconnected in this way, and a tremendous number of behavioural possibilities is one of the consequences. The bulk of the cerebral hemispheres consists of white matter – myelinated neurones, which connect the cerebral hemispheres with the midbrain, hindbrain and spinal cord. Beneath, the left and right hemispheres are connected by a tract of fibres, the **corpus callosum**.

The areas of the cortex with special sensory and motor functions have been mapped out (Figure 21.38). The **sensory areas** receive impulses from receptors all over the body. The **motor areas** are where motor impulses serving the whole body, via the spinal cord, originate. There are also **association areas** that interpret, integrate and store information.

The functions of the bulk of the front part of the cerebral hemispheres are less well understood. They appear to be concerned with personality, imagination and intelligence. An interior part of the cerebrum forms the **limbic system**, the seat of emotional aspects of behaviour and of the memory function. At the base of the cerebrum are the **basal ganglia**, a mass of grey matter concerned with muscular control and with connections to the remainder of the brain.

Figure 21.37   The cerebral hemispheres in humans

- lateral fissure
- central fissure
- parietal lobe
- **anterior**
- **posterior**
- occipital lobe
- frontal lobe
- cerebellum
- temporal lobe

the surface area of the cerebral hemisphere is increased by numerous infoldings (convolutions)

each cerebral hemisphere is divided into four lobes

in section
non-myelinated pyramidal neurones make up the grey matter of the cerebral cortex

- cerebral cortex
- thalamus
- ventricle
- hypothalamus
- basal ganglia (neurones linked to the reticular formation of the midbrain)

Figure 21.38   Location of function in the cerebral hemispheres

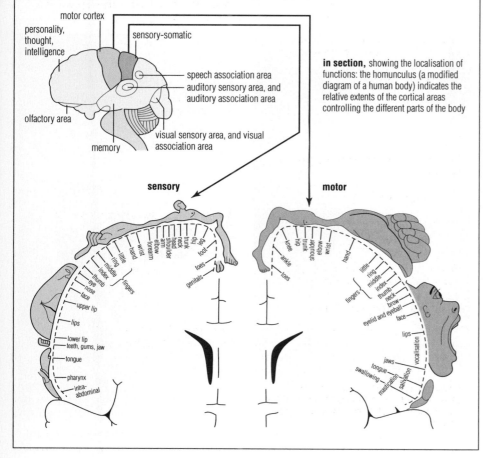

- motor cortex
- personality, thought, intelligence
- sensory-somatic
- speech association area
- auditory sensory area, and auditory association area
- olfactory area
- memory
- visual sensory area, and visual association area

in section, showing the localisation of functions: the homunculus (a modified diagram of a human body) indicates the relative extents of the cortical areas controlling the different parts of the body

sensory

motor

The blood–brain barrier

Substances in the blood are prevented from automatically entering the brain by the 'blood–brain barrier'. This is formed by the endothelial cells that line the blood vessels of the brain. These cells function as 'gatekeepers'. They allow essential nutrients in but keep out substances that would interfere with the functioning of the nerve cells. Without this barrier the brain might go into uncontrolled activity, resulting in a seizure or 'fit' if some potent substance has been introduced into the blood, perhaps in food or drink.

Questions

1 What is the importance of the layers of tissues that surround the brain inside the cranium?

2 Why is death immediate if brain and spinal cord are severed below the medulla?

3 Look at the homunculus diagram in Figure 21.38. Say what it implies about
   a the sensitivity of the face
   b the dexterity of the fingers.

4 By what two methods does the hypothalamus control basic bodily functions? Give examples of each.

5 Make a labelled diagram of a synaptic system between four neurones that you would expect to produce excitation of a postsynaptic fibre on most occasions. Explain how it would work.

21.10 THE NERVOUS SYSTEM IN NON-VERTEBRATES

■ The nerve net of *Hydra*

Hydra is a freshwater cnidarian (p. 34). The structure of *Hydra* and the arrangement of cells in its two-layered body wall are shown in Figure 14.11 (p. 294). This tiny animal was first observed and described by Leeuwenhoek (p. 148).

The bulk of the cells in the inner (endoderm) and outer (ectoderm) layers of the body wall are muscle tail cells. Between these layers is the structureless jelly called the mesogloea. The nervous system of *Hydra* consists of a network of neurones lying in the ectoderm and mesogloea (Figure 21.39) (there is no CNS). The receptors are sense cells, largely restricted to the ectoderm, and concentrated on the tentacles and around the mouth. The effectors include the muscle tails of ectoderm and endoderm cells. The muscle tails of the ectoderm are arranged longitudinally, and their contraction causes the body to shorten. The muscle tails of the endoderm are transversely arranged, and contraction of these makes the body long and thin.

When a sensory cell is stimulated impulses are transmitted slowly through the neurone network, radiating out in all directions. The distance the impulses travel depends upon the strength of the stimulation. Conduction is slow, possibly because of the number of synapses to be crossed. The first impulse to arrive at a synapse does not usually stimulate the adjacent neurone, but each impulse helps to 'charge' the postsynaptic membrane so that subsequent impulses pass. This system produces only localised responses at first, but if stimulation is prolonged the whole body may become involved. For example, a normal stimulus of trapped food (Figure 14.11, p. 294) initiates first a local bending of tentacles towards the mouth. Then other tentacles bend, and the mouth opens and engulfs its prey.

■ Sea anemone

Some cnidarians, including the sea anemone, contain specialised tracts of neurones around the mouth and in the body wall, additional to the irregular network of neurones. These elongated relay neurones function as through conduction tracts, which allow the animal to move its body (to contract, say) very quickly when stimulated violently by a predator, for example.

Figure 21.39   The nerve net of *Hydra*, with the sensory cells and muscle tails it connects

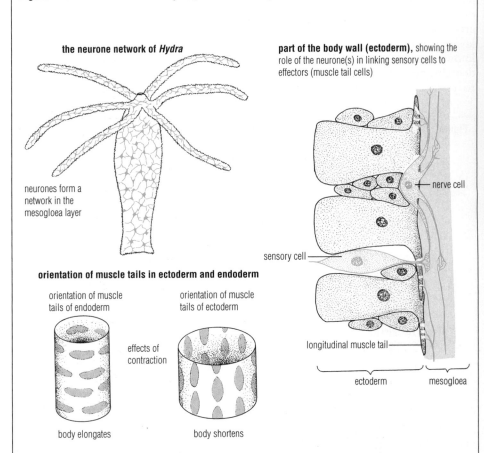

the neurone network of *Hydra*

neurones form a network in the mesogloea layer

part of the body wall (ectoderm), showing the role of the neurone(s) in linking sensory cells to effectors (muscle tail cells)

nerve cell

sensory cell

longitudinal muscle tail

ectoderm

mesogloea

orientation of muscle tails in ectoderm and endoderm

orientation of muscle tails of endoderm

orientation of muscle tails of ectoderm

effects of contraction

body elongates

body shortens

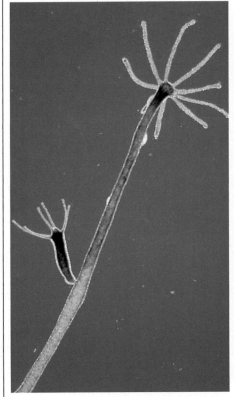

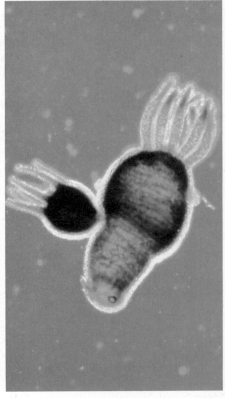

Hydra extended and contracted by contraction of the muscle tails of the endoderm (left) and ectoderm (right)

Figure 21.40   The central and peripheral nervous system of *Lumbricus*

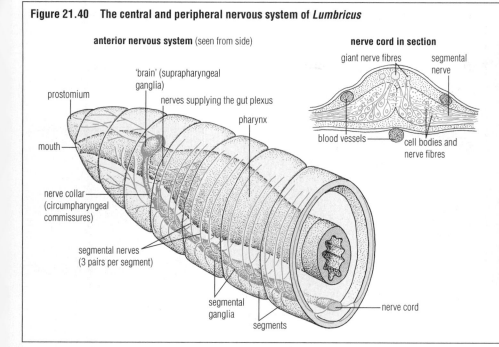

anterior nervous system (seen from side)

- prostomium
- 'brain' (suprapharyngeal ganglia)
- nerves supplying the gut plexus
- pharynx
- mouth
- nerve collar (circumpharyngeal commissures)
- segmental nerves (3 pairs per segment)
- segmental ganglia
- segments
- nerve cord

nerve cord in section

- giant nerve fibres
- segmental nerve
- blood vessels
- cell bodies and nerve fibres

***Lumbricus* may forage above ground at night**

■ Nerve cord and simple 'brain' of *Lumbricus*

The earthworm *Lumbricus* is a soft-bodied segmented worm, an annelid (p. 36), adapted to subterranean life. Locomotion through the soil is by waves of contraction and relaxation of the circular and longitudinal muscles of the body wall. During movement, parts of the body are anchored to the sides of its burrow by the temporary extension of tough bristles (chaetae), four pairs per segment (p. 480).

The central nervous system of the earthworm is composed of a ventral nerve cord running the length of the body, with a ganglion in each segment. At the anterior end, a 'brain' (supra-pharyngeal ganglion) occurs above the pharynx, and is connected to another large ganglion (subpharyngeal ganglion) by two nerves (circumpharyngeal commissures). From each segmental ganglion of the ventral nerve cord three pairs of nerves per segment serve (innervate) the body wall and gut in that segment (Figure 21.40).

The sense organs of the earthworm are relatively simple structures, consisting of individual cells or small groups of cells. The earthworm, which lives in the soil – a relatively uniform environment – generally moves away from the light and towards moisture. If touched or attacked it withdraws instantly into its burrow. It is unresponsive to sound, but can detect vibrations transmitted through solid objects. In very dry weather and in very

cold periods it retreats deep into the soil and survives tightly coiled up.

The light sensors of the earthworm are single-celled structures. They contain a simple lens and a nerve fibril network. Light-sensitive cells occur in the epidermal layer of the skin in the dorsal surface of the body, but are absent from the ventral surface. They are most abundant at the ends of the body, particularly the anterior end.

Many of the earthworm's sensory cells (Figure 21.41) have tiny hair-like extensions that project through the cuticle of the skin. Some sensory cells are sensitive to touch and pressure, others to tempera-

Figure 21.41   The sensory organs of the earthworm

Sensory organs of the earthworm are mostly concentrated at the anterior end and consist of single cells or groups of cells sensitive to particular stimuli, e.g. epidermal sense organs sensitive to touch, pressure, chemical stimuli and temperature change.

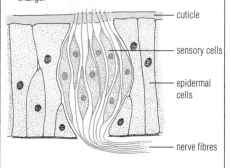

- cuticle
- sensory cells
- epidermal cells
- nerve fibres

ture or to chemical stimuli. The internal environment is supplied with sensory receptors (**proprioceptors**) in the muscles and other organs. In the mouth and pharynx are chemical receptors (taste).

The earthworm illustrates an early stage in the evolutionary trend towards the development of a head ('cephalisation') in mobile animals. The anterior end is the first part of the animal to encounter environmental stimuli. Cephalisation involves the concentration of sense organs and nervous tissue at the anterior end. The many incoming sensory fibres contribute to the enlargement of ganglia at the anterior of the animal. This constitutes the evolutionary pressure giving rise to a 'brain'.

The role of the 'brain' of the worm is to control the total body response to stimuli such as light and touch. The sub-pharyngeal ganglion controls eating and burrowing. Neither of these anterior ganglia control the ordinary locomotory movements of the body, however. These are discussed in Chapter 22.

Questions

1 Suggest two reasons why conduction through the nerve net of *Hydra* is slow.
2 What are the roles of the CNS and segmental nervous system in the activities of the earthworm?
3 What is the particular value of giant nerve fibres in the nervous system of the earthworm?

■ Insects: specialisation of nerve cord and sense organs

The insects are a class of arthropods (p. 38) that is of special interest because of the vast numbers of species that have evolved (p. 18). The arthropod body plan can be seen as an elaboration of the segmented body of annelids, although arthropods show a higher degree of cephalisation than do the annelids. The outer surface of the arthropod is a rigid, many-layered cuticle, and the segments of the body bear jointed appendages.

The nervous system of arthropods consists of a 'brain' in the form of a dorsal ganglion, connected by a nerve collar around the gut to the first ganglion of the ventral nerve cord. The nerve cord is double, with segmental ganglia. Probably because of the heavy external skeleton, the arthropods have a great variety of sensory organs connecting their central nervous system with the external environment.

Within the arthropods the pattern of repeating segments in which every segment performs all or most functions is obscured by associations of groups of segments, specialised to perform functions for the entire animal. The number of ganglia in the ventral nerve cord has become reduced by fusions (Figure 21.42).

Figure 21.42 The central nervous systems of arthropods

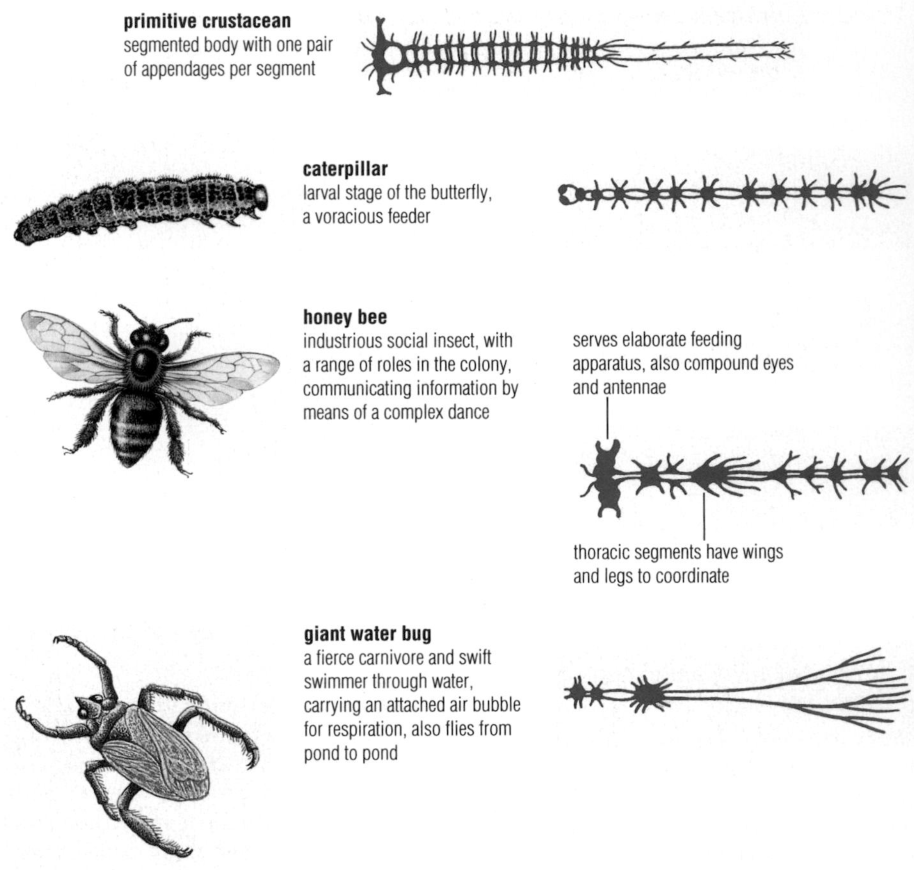

primitive crustacean
segmented body with one pair of appendages per segment

caterpillar
larval stage of the butterfly, a voracious feeder

honey bee
industrious social insect, with a range of roles in the colony, communicating information by means of a complex dance

serves elaborate feeding apparatus, also compound eyes and antennae

thoracic segments have wings and legs to coordinate

giant water bug
a fierce carnivore and swift swimmer through water, carrying an attached air bubble for respiration, also flies from pond to pond

The nervous system of the grasshopper

The grasshopper is a typical insect in that the body is divided into three regions (head, thorax and abdomen) and it has two pairs of wings and three pairs of legs attached to the thorax (Figure 21.43).

The nervous system consists of an anterior dorsal 'brain' and ventral ganglionated nerve cord. Pairs of peripheral nerves innervate the organs of head, thorax and abdomen. The dorsal bilobed ganglion relays sensory information from the anterior sense organs to the rest of the body and plays a role in regulating and coordinating movements, rather than initiating detailed activity in the body (in vertebrates the brain initiates and controls movements). The anterior ganglia of the ventral nerve cord similarly control the movements of the mouthparts, as well as being concerned with balance. The segmental ganglia in the grasshopper control many local reflex actions.

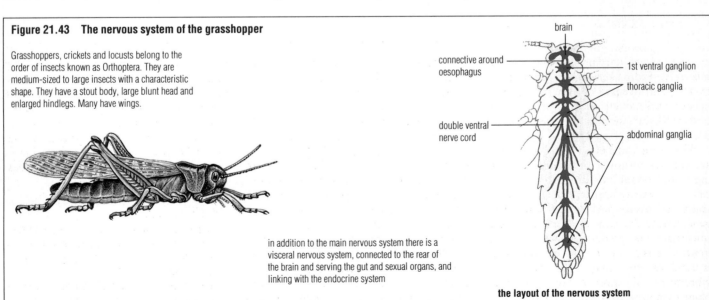

Figure 21.43 The nervous system of the grasshopper

Grasshoppers, crickets and locusts belong to the order of insects known as Orthoptera. They are medium-sized to large insects with a characteristic shape. They have a stout body, large blunt head and enlarged hindlegs. Many have wings.

in addition to the main nervous system there is a visceral nervous system, connected to the rear of the brain and serving the gut and sexual organs, and linking with the endocrine system

brain

connective around oesophagus

double ventral nerve cord

1st ventral ganglion

thoracic ganglia

abdominal ganglia

the layout of the nervous system

Figure 21.44 **The sense organs of insects:** besides those shown, other external sense organs include cells sensitive to temperature changes and (in some insects) auditory organs for 'hearing', while tiny internal receptors (proprioceptors) exist in muscles and other organs all over the body

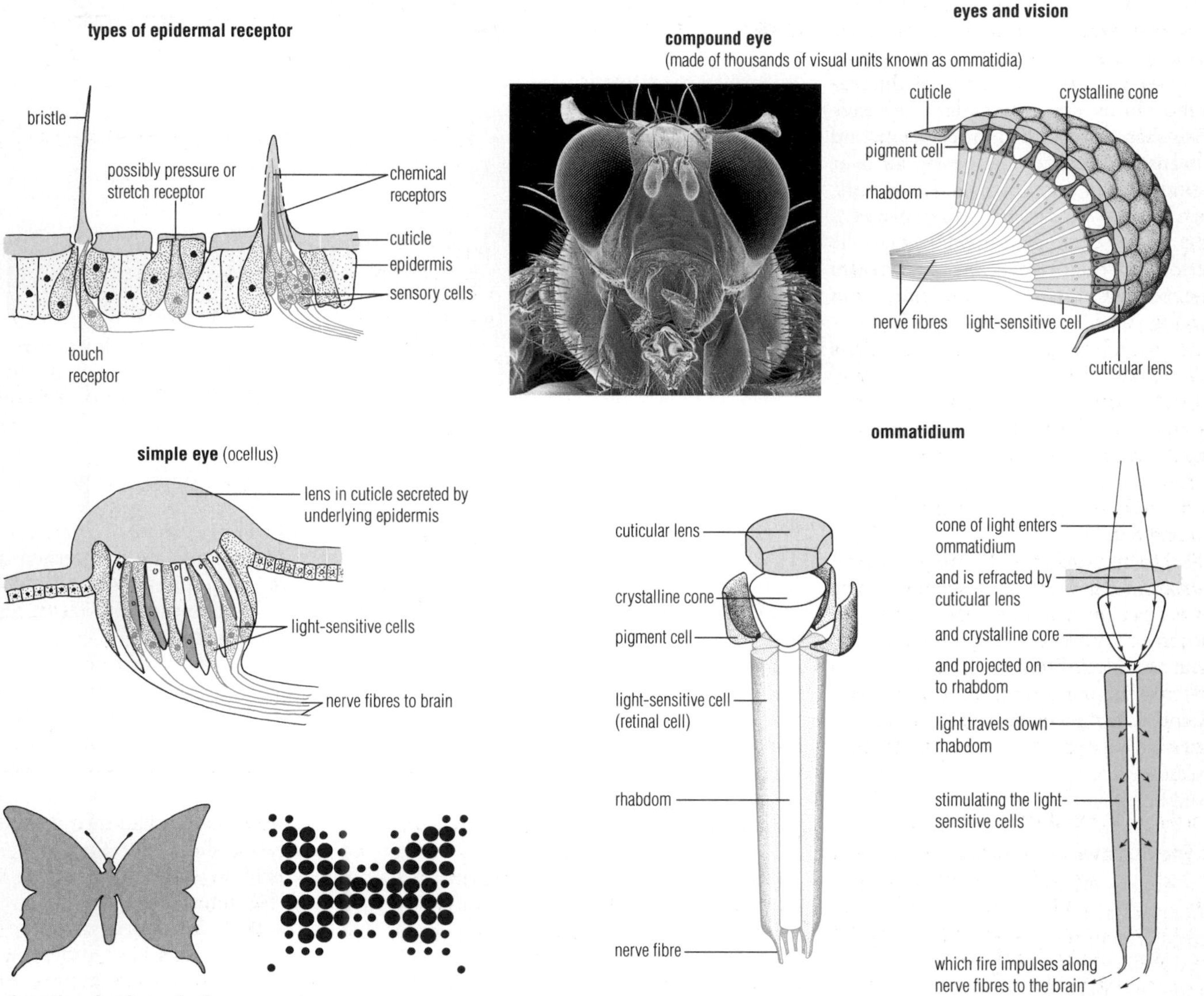

types of epidermal receptor

eyes and vision

compound eye
(made of thousands of visual units known as ommatidia)

simple eye (ocellus)

ommatidium

formation of an image by the compound eye
(left) outline of a butterfly,
(right) how it might be registered by the eye of a dragonfly at a distance of about 30 cm

dismantled

in operation

Insects have extremely well-developed sense organs. In the grasshopper there are two large compound eyes and also simple eyes or **ocelli**. In the simple eye no image is formed, but the ocellus is sensitive to degrees of light intensity.

The compound eye consists of thousands of identical visual units, the **ommatidia**. An ommatidium (Figure 21.44), with its pigment cells extended, is a self-contained unit. Each ommatidium points in a slightly different direction from the rest. It does not produce the sort of sharp image that is generated in the human eye, but it does detect shapes and movement. Movements of an object are recorded in many ommatidia, each slightly differently from the rest. Movement will be perceived as a change in pattern, many hundreds of times over.

In many insects, the ommatidia are able to regulate the intensity of light reaching the light-sensitive cells. This occurs by movements of the pigment within the pigment cells, either up around the lens or down at the base of the cells. The result appears to be functionally equivalent to the movements of the iris in the mammalian eye.

Other sense organs consist of a receptor cell or cells underlying a modified portion of the cuticle, and responding to a wide range of different stimuli. The antennae are sensitive to chemicals and to touch. Chemical stimuli are very important to insects. They can detect a mere trace of chemicals present in the air, and by these stimuli they locate and select food, identify their own species and members of their colony, and also their mates. Touch receptors exist as tactile hairs or triggers and as thin domed areas of cuticle above a receptor cell.

Questions

1 Movement in insects presents more complex control problems than in annelids. How is this reflected in the central nervous systems of these two groups?
2 Why might contraction of the pigment cells in a compound eye result in blurred vision?

471

21.11 THE ENDOCRINE SYSTEM

The endocrine system controls body activity by releasing chemical messengers called **hormones** into the bloodstream. Hormones are produced in small quantities in the cells of ductless or **endocrine glands**. Hormones are secreted by exocytosis directly into the bloodstream, and are transported indiscriminately all over the body. They act, however, on specific sites (referred to as the **target organs**). They cause changes to the metabolic reaction of their target organs before being broken down and excreted from the body.

The structure of the endocrine glands can be contrasted with that of the many ducted glands (exocrine glands), such as the salivary glands and sweat glands. These deliver their secretions through tubular ducts, whereas the endocrine glands release their secretions directly into the bloodstream.

The body contains two distinct communication and control systems: the nervous system, discussed above, and the endocrine system. These work in fundamentally different ways (Table 21.4), for the regulation of body activity and for achieving homeostasis (p. 502). These two systems do not work in isolation, however; their interactions are discussed below.

■ The discovery of hormone action

Hormone production has been investigated by surgical removal of a suspected endocrine organ from a healthy animal. The experimenter then demonstrates that the changes that follow can be reversed by grafting the organ back, or by administering an extract of the organ. By this method Dr A Berthold (1849) of Göttingen established that the presence of at least one testicle in a young cockerel was essential for the growth of the comb, wattles and spurs of the adult male cockerel. This demonstration can be said to have initiated endocrinology, but the identification of the male hormone involved in this demonstration was not achieved until the 1930s.

The term 'hormone' was first used by W M Bayliss and E H Starling of University College, London, in 1905, in describing their work on the secretion of digestive juice by the pancreas in dogs. They established that pancreatic juice was secreted when food entered the duodenum from the stomach (p. 283), even when all nerves to the pancreas were cut. They argued that a hormone they named 'secretin' was carried in the blood from the gut wall to the pancreas, stimulating the secretion of pancreatic juice.

■ The endocrine system of humans

The distribution of the endocrine glands in the human body is shown in Figure 21.45. The endocrine system is a complex array of very different glands. In this chapter we shall discuss the activities of the hypothalamus, the pituitary, thyroid and adrenal glands and the cells in the pancreas that secrete hormones, in order to gain insights into the mechanisms of hormone action. The roles of other endocrine glands are discussed elsewhere.

The coordination of nervous and endocrine control

The nervous system and the endocrine systems do not work in isolation, but interact with each other in regulating body functions. This is made possible in at least two important ways.

Firstly, one part of the brain, the hypothalamus, is also an endocrine gland (see p. 474). The hypothalamus is connected to the pituitary gland, which lies immediately below it, just outside the brain. A major role of the pituitary gland is to control many of the other endocrine glands. In effect, the pituitary is a master gland.

Secondly, the level of all hormones in the blood is monitored in the hypothalamus. This is an example of negative feedback control (p. 503). When specific hormones are at a low level the hypothalamus triggers stimulatory activity by the pituitary. When specific hormones are at a high level in the blood the hypothalamus triggers inhibitory activity in the pituitary.

The chemistry of hormones

Chemical analysis of hormones has shown that the majority can be divided into two groups, namely the steroid hormones and the amine (or peptide) group of hormones.

Steroid hormones are produced by endocrine glands of mesodermal origin (p. 414). The actions of many of the steroid hormones are concerned with long-term responses of the body. Examples include the oestrogens, testosterone and aldosterone.

Amine or **peptide hormones** are produced by endocrine glands of ecto- or endodermal origin; examples include antidiuretic hormone (ADH), thyroid-stimulating hormone (TSH) and insulin. Hormones in this group tend to bring about a response within a few minutes and their actions are of relatively short duration; they are usually held in store in endocrine cells, available for immediate release.

It has been discovered that the two groups of hormones have different, distinctive mechanisms of action within target cells.

Mode of action of hormones

Hormones are carried in the bloodstream, and so they reach every cell in the body. The effects of hormones are to alter the metabolic reactions only in certain cells, called **target cells**. These cells possess specific receptor molecules on their surfaces; if a cell lacks an appropriate receptor site it will not respond to the hormone.

In the case of amine or peptide hormones, the receptor/hormone complex may activate existing enzymes stored in the cytosol. In the case of steroid hormones the hormone may activate a specific gene in the nucleus, causing protein (enzyme) synthesis.

Activated enzymes then alter target cell metabolism, but in different ways (see Figure 21.46).

Table 21.4 The endocrine and nervous systems compared

Nervous system	Endocrine system
▶ communication is by means of electrochemical impulses transmitted via nerve fibres	▶ communication is by means of chemical messengers (hormones) transmitted in the bloodstream
▶ impulses targeted on specific cells reached by nerve fibres, e.g. muscle cells, gland cells, or other neurones (nerve cells)	▶ hormones 'broadcast' to the target cells in any location in the body
▶ causes muscles to contract or glands to secrete	▶ causes changes in metabolic activity
▶ impulses produce their effects within a few milliseconds	▶ hormones mostly have their effects over many minutes or over several hours, or even longer
▶ the effects of nervous stimulation are short-lived and reversible	▶ the effects of endocrine stimulation are fairly long-lasting

Figure 21.45 The human endocrine system: a summary

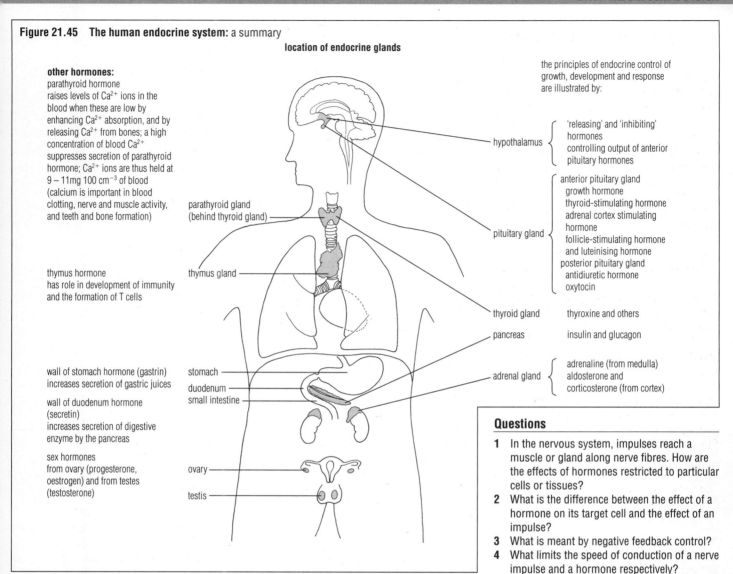

location of endocrine glands

other hormones:

parathyroid hormone
raises levels of Ca²⁺ ions in the blood when these are low by enhancing Ca²⁺ absorption, and by releasing Ca²⁺ from bones; a high concentration of blood Ca²⁺ suppresses secretion of parathyroid hormone; Ca²⁺ ions are thus held at $9 - 11$mg 100 cm⁻³ of blood (calcium is important in blood clotting, nerve and muscle activity, and teeth and bone formation)

thymus hormone
has role in development of immunity and the formation of T cells

wall of stomach hormone (gastrin) increases secretion of gastric juices

wall of duodenum hormone (secretin) increases secretion of digestive enzyme by the pancreas

sex hormones
from ovary (progesterone, oestrogen) and from testes (testosterone)

parathyroid gland (behind thyroid gland)

thymus gland

stomach
duodenum
small intestine

ovary

testis

the principles of endocrine control of growth, development and response are illustrated by:

hypothalamus — 'releasing' and 'inhibiting' hormones controlling output of anterior pituitary hormones

pituitary gland — anterior pituitary gland
growth hormone
thyroid-stimulating hormone
adrenal cortex stimulating hormone
follicle-stimulating hormone and luteinising hormone
posterior pituitary gland
antidiuretic hormone
oxytocin

thyroid gland — thyroxine and others

pancreas — insulin and glucagon

adrenal gland — adrenaline (from medulla)
aldosterone and corticosterone (from cortex)

Questions

1 In the nervous system, impulses reach a muscle or gland along nerve fibres. How are the effects of hormones restricted to particular cells or tissues?
2 What is the difference between the effect of a hormone on its target cell and the effect of an impulse?
3 What is meant by negative feedback control?
4 What limits the speed of conduction of a nerve impulse and a hormone respectively?

Figure 21.46 How hormones may act on target cells yet leave other cells unaffected

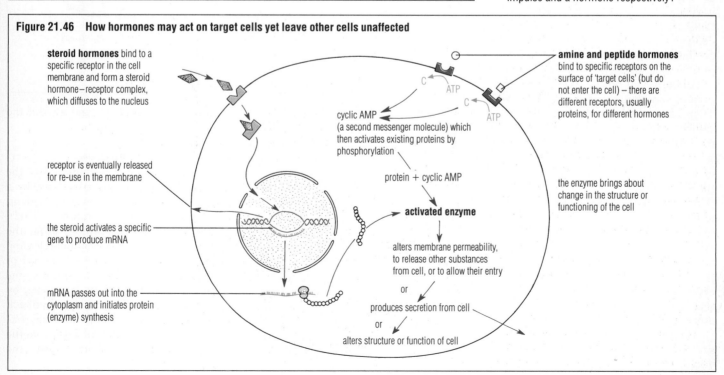

steroid hormones bind to a specific receptor in the cell membrane and form a steroid hormone–receptor complex, which diffuses to the nucleus

receptor is eventually released for re-use in the membrane

the steroid activates a specific gene to produce mRNA

mRNA passes out into the cytoplasm and initiates protein (enzyme) synthesis

cyclic AMP
(a second messenger molecule) which then activates existing proteins by phosphorylation

protein + cyclic AMP

activated enzyme

alters membrane permeability, to release other substances from cell, or to allow their entry

or

produces secretion from cell

or

alters structure or function of cell

amine and peptide hormones bind to specific receptors on the surface of 'target cells' (but do not enter the cell) – there are different receptors, usually proteins, for different hormones

the enzyme brings about change in the structure or functioning of the cell

Figure 21.47   The hypothalamus and pituitary gland

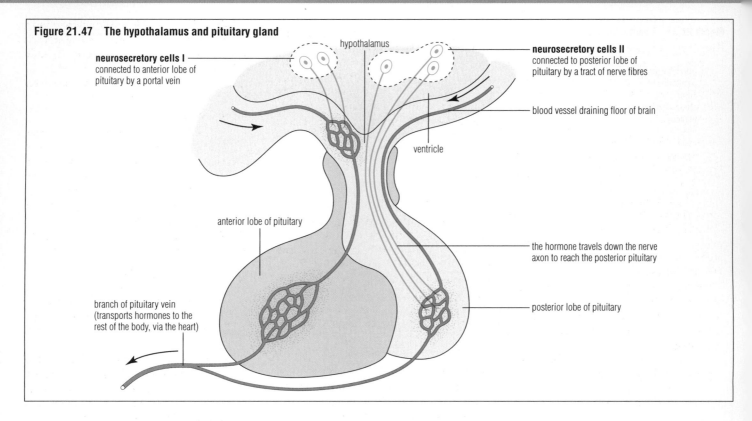

Hypothalamus and pituitary gland

In the hypothalamus the levels of hormones (and of other substances of the blood) are monitored. The hypothalamus also controls indirectly many important functions such as body temperature (p. 509), hunger, thirst and sleep. Additionally, the hypothalamus secretes two groups of hormones that control the activity of the anterior lobe of the pituitary gland. One of the groups of hypothalamic hormones stimulates, the other depresses, the release of other hormones produced and stored in the anterior pituitary gland.

All hormones produced in the anterior pituitary regulate other endocrine glands. Thyroid-stimulating hormone (TSH) controls the secretion of hormones by the thyroid gland, and adrenocorticotrophic hormone (ACTH) regulates the secretion of the hormones by the adrenal cortex (p. 475). Follicle stimulating hormone (FSH) stimulates spermatogenesis in the testes and growth of follicles in the ovaries, and luteinising hormone (LH) affects the hormonal functions of ovary and testes (pp. 589 and 593). In addition the anterior pituitary secretes somatotrophin, a growth hormone that also regulates the production of other hormones by the liver.

The posterior pituitary gland is the site for storage of two hormones produced by the hypothalamus in its

Figure 21.48   A neurosecretory cell of the hypothalamus

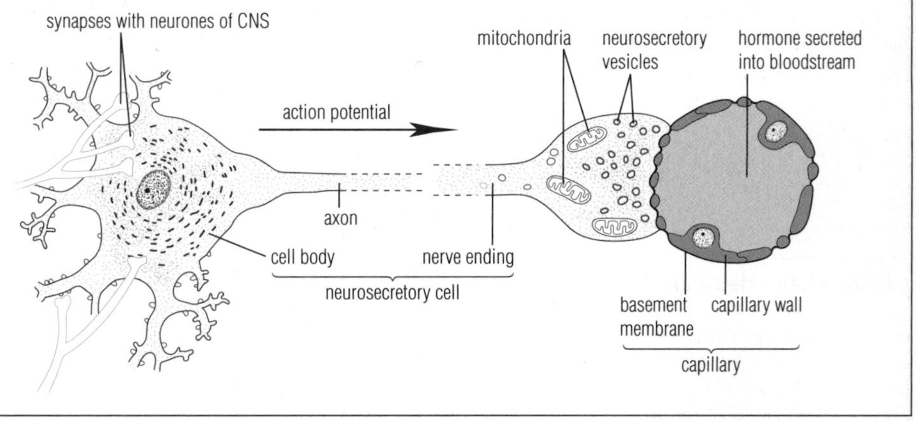

neurosecretory centre II (Figures 21.47 and 21.48). Antidiuretic hormone (ADH) controls re-absorption of water by the kidney tubule during urine formation. Its effect is to prevent dehydration of the body (p. 386). The other hormone, oxytocin, has a role in childbirth (p. 602).

The thyroid gland

The thyroid gland lies in the neck, overlying the trachea and near the larynx. This endocrine gland secretes two thyroxine hormones, which it manufactures from the amino acid tyrosine, with the addition of iodine that has been taken in with the diet. The thyroxine hormones affect all body cells. They cause an increase in the rate of tissue metabolism (basal metabolic rate, p. 271), and also in the rate of glucose metabolism.

Control of the secretion of thyroxines

The secretion of thyroxines by the thyroid gland is directly stimulated by a hormone from the pituitary gland (thyroid-stimulating hormone, TSH), and indirectly by a hormone from the hypothalamus (thyrotrophic-releasing factor, TRF; Figure 21.49). The level of thyroxine in the blood controls the further secretion of both hormones by the mechanism of negative feedback control (p. 503). That is, excess thyroxine inhibits the secretion of TRF from the hypothalamus and TSH from the pituitary gland.

Calcitonin

The thyroid gland also secretes a hormone called calcitonin when calcium ions are at too high a level in the blood. This hormone causes a reduction in the soluble calcium ions in circulation.

Disorders of the thyroid gland

Overactivity of the thyroid gland (**hyperthyroidism**) gives the symptoms of increased metabolic activity (raised body temperature, for example) and raised heart rate. Results include restlessness, irritability and loss of body mass. The patient may go on to develop protruding eyes and a swollen thyroid gland (known as a goitre). This condition may be caused by a failure of the negative feedback control mechanism, or by a thyroid tumour. It may be cured by surgical removal of part of the thyroid gland.

Underactivity of the thyroid gland (**hypothyroidism**) in adults gives the symptoms of lowered basal metabolism and increased body mass (myxoedema).

The patient becomes less alert. In children the condition is known as cretinism. Here the patient shows retarded mental, physical and sexual development. The condition can be corrected to some extent by taking thyroid tablets. Treatment is most effective if started early in life or during fetal development.

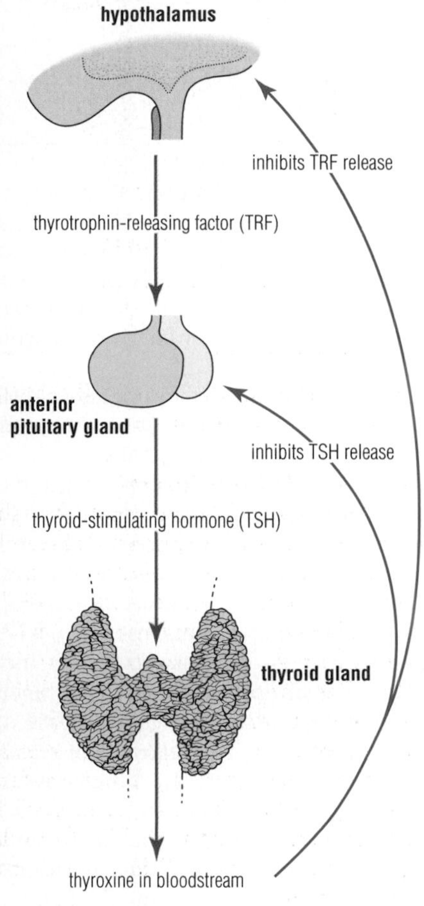

Adrenal glands

The adrenal glands lie immediately above the kidneys (Figure 18.10, p. 380). Each adrenal gland has two parts, an outer adrenal cortex and an inner adrenal medulla. The medulla is stimulated by the sympathetic nervous system (p. 462) but the cortex is stimulated by hormones from the pituitary. The size of the adrenal glands may vary according to the degree of stress experienced; in people suffering anxiety states they become enlarged.

The adrenal cortex

The adrenal cortex produces a number of steroid hormones, known collectively as corticoids. These hormones are produced from cholesterol (p. 139), which is synthesised here or absorbed from the blood (dietary cholesterol, p. 276). Steroid hormones are lipid-soluble, and are able to pass through cell membranes. They work by activating specific genes in the nucleus (Figure 21.46).

Corticoid hormones are slow-acting and have lasting effects. They can be classified into two groups.

Glucocorticoids, also known as cortisols, are concerned with glucose metabolism and are produced in periods of anxiety, fever and disease. Perception of these body states is by the brain and results in hormone production in the hypothalamus, acting via the anterior lobe of the pituitary. The result is an enhanced cortisol output. These hormones affect both carbohydrate and protein metabolism. Glucose and glycogen formation are enhanced at the expense of that of fats and proteins. At the same time the pool of amino acids is increased at the expense of proteins.

Overproduction of cortisol results in **Cushing's syndrome**, associated with obesity, muscular wasting, high blood pressure and diabetes. Destruction of the cortex causes **Addison's disease**, in which low blood sugar, reduced blood pressure and fatigue are symptoms.

Mineralocorticoids are concerned with water retention by control of inorganic ion distribution. These hormones include aldosterone, which increases the reabsorption of sodium and chloride ions by the kidney tubule (p. 388).

The adrenal medulla

The medulla of the adrenal gland consists of neurosecretory cells, capillaries and nerves. This area of the adrenal gland produces two hormones, adrenaline and noradrenaline. These substances augment the action of the sympathetic nervous system (p. 461), and produce the 'flight or fight' response, preparing the body for exertion and high physical and mental performance. The effects of the hormones are as follows:

► glycogen in the liver is converted to glucose, so increasing blood sugar levels and sustaining muscular contraction

► bronchioles become dilated and smooth muscles of the gut relaxed; thus the diaphragm is lowered, more air is inhaled and more oxygen is made available to the body

► the amplitude (stroke volume) and rate of heartbeat and blood pressure increase, which increases the delivery of blood to the tissues

► there is vasoconstriction of blood vessels to the gut and reproductive organs and inhibition of peristalsis, thus increasing the blood supply to voluntary muscles

► a decreased sensory threshold, increased mental awareness and dilation of the pupils of the eyes give rise to heightened sensitivity and more rapid response, together with increased range of vision

► hair erector muscles contract, making the hair stand on end, so giving the impression in furry mammals of increased size.

Figure 21.49 The control of thyroid activity, a case of negative feedback control

hypothalamus

inhibits TRF release

thyrotrophin-releasing factor (TRF)

anterior pituitary gland

inhibits TSH release

thyroid-stimulating hormone (TSH)

thyroid gland

thyroxine in bloodstream

Questions

1 What is a portal vein? Where in the body are portal veins found (p. 351)?
2 How do the anterior and posterior pituitary differ in relationship to the hypothalamus? How do they differ in their function?
3 Explain how negative feedback works with respect to secretion of
   a thyroxine
   b oestrogen (p. 593).

■ The pancreas

The bulk of the cells in the pancreas produce digestive enzymes, which are discharged via the pancreatic duct (it is therefore largely an exocrine gland) into the duodenum. There are, however, patches of pancreatic cells, known as **islets of Langerhans**, which are endocrine glands. These consist of two distinct types of cell: **α-cells**, which produce the hormone glucagon, and **β-cells**, producing the hormone insulin. Both hormones have a role in regulating blood sugar levels, but their effects are **antagonistic**. The structure of the pancreas and the regulation of blood sugar are discussed in Chapter 23, p. 504.

Insulin is released in response to a rise in blood glucose levels. Its effect is to lower blood glucose levels (it is the only hormone to do so), and so failure to produce insulin by the body leads to a very serious disease (insulin-dependent diabetes). If the blood glucose level rises above a certain level the kidney is unable to re-absorb it. Glucose appears in the urine, an indicator to the condition known as diabetes mellitus. This disease can be controlled by injections of insulin and a regulated diet.

Glucagon is secreted in response to a fall in the blood glucose level. It stimulates the liver to convert glycogen to glucose and enhances mobilisation of fatty acids from adipose tissue.

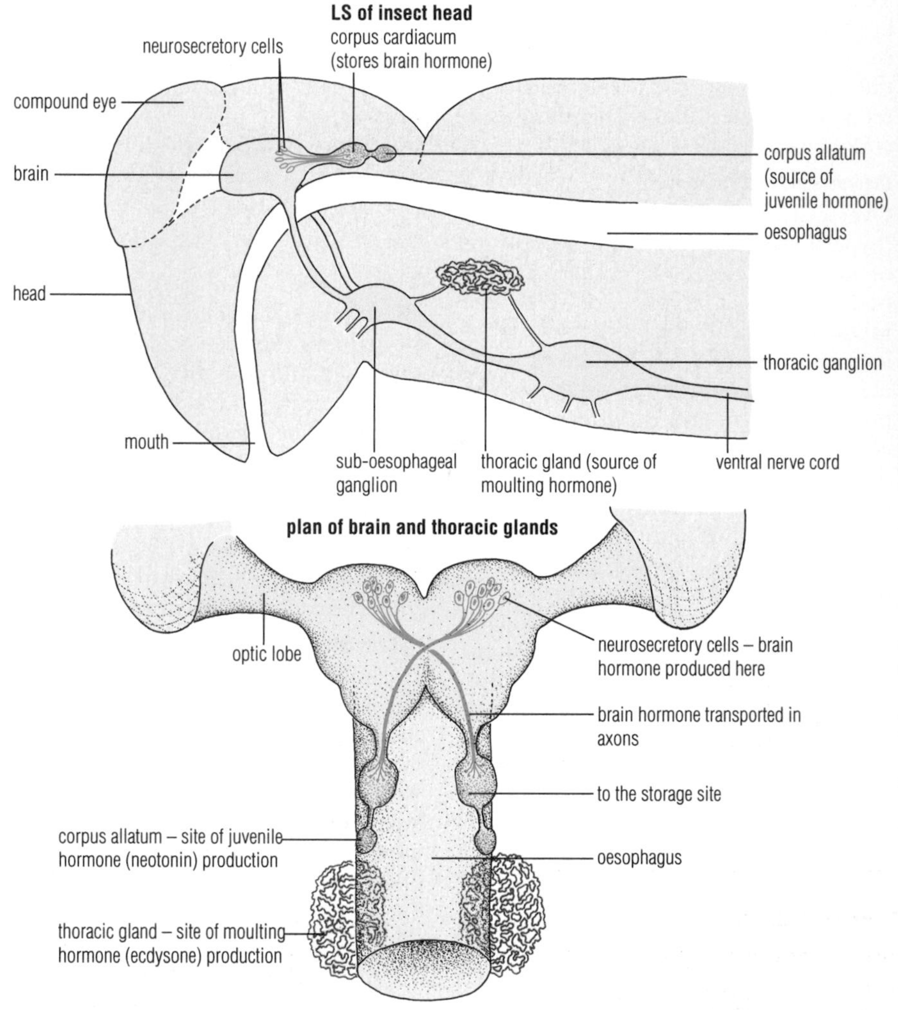

21.12 THE ROLE OF HORMONES IN NON-VERTEBRATES

Research into the hormones of non-vertebrate species has lagged behind investigations of the hormones of mammals. This is to be expected, given the fact that most non-vertebrate species are tiny animals. Remember, the main source of experimental evidence comes from the techniques of surgical removal (and transplanting) of hormone-producing glands in otherwise healthy organisms. Many non-vertebrates are just too small for biologists to carry out these operations with confidence.

Nevertheless, there is evidence that hormones are present in the species of many non-vertebrate phyla. In particular, the existence of insect hormones and their role in the growth and development of insect larvae into adult forms has been known since the pioneering work of Sir Vincent Wigglesworth in the period 1935–70. The great economic importance of insects, together with the relative

Figure 21.50 The sites of hormone production in an insect

LS of insect head

neurosecretory cells / corpus cardiacum (stores brain hormone)
compound eye
brain
head
mouth
corpus allatum (source of juvenile hormone)
oesophagus
thoracic ganglion
sub-oesophageal ganglion
thoracic gland (source of moulting hormone)
ventral nerve cord

plan of brain and thoracic glands

optic lobe
neurosecretory cells – brain hormone produced here
brain hormone transported in axons
to the storage site
corpus allatum – site of juvenile hormone (neotonin) production
oesophagus
thoracic gland – site of moulting hormone (ecdysone) production

ease of rearing many insect species in laboratories, have contributed to progress with this class of arthropods.

We have already seen that insect life cycles differ: in some, a complete metamorphosis occurs between larval and adult stages, while in others larval stages are miniature adults and metamorphosis is incomplete (p. 417). All insects have an external skeleton, which restricts growth, and which is periodically replaced during growth and development. The replacement of the external skeleton (ecdysis) makes growth of insects appear 'intermittent' (p. 402).

■ Growth and metamorphosis in insect larvae

Growth of the larval form and the process of metamorphosis is controlled by hormones. A balance of three hormones is responsible for regulating the events of ecdysis and metamorphosis. These hormones have been named according to where they originate or what they do.

The '**brain hormone**' originates in the 'brain'. 'Brain hormone' is produced in neurosecretory cells on the dorsal surface of the 'brain' and stored in the corpus cardiacum (Figure 21.50).

The hormone that initiates ecdysis (moulting) is called **ecdysone**. Ecdysone is stored in the thoracic glands.

The hormone that ensures retention of juvenile characters during early moults is called **juvenile hormone**. Juvenile hormone is stored in the corpus allatum.

In the endocrine control of development of hemimetabolous insects (p. 417), external factors such as changes in food supply, or in the abiotic environment (light, temperature), trigger release of 'brain hormone'. This causes the release of ecdysone and (initially) a high concentration of juvenile hormone. The result is a larval moult. During the final moult, only ecdysone is released and the adult is produced (Figures 21.51 and 21.52).

Figure 21.51   The sequence of secretion of insect hormones in moulting and metamorphosis

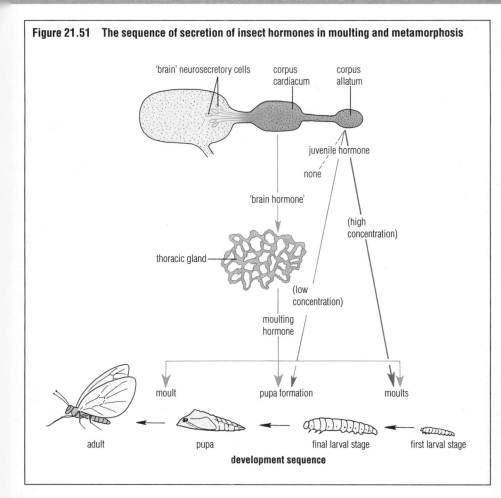

Figure 21.52   Moulting (above) and emergence from a pupa (below) in butterflies

■ Pheromones

Not all hormones produced by animals affect the individual that produces them. Chemicals that operate between members of the same species are social hormones or **pheromones**. A pheromone is a volatile substance produced and discharged by an organism, which induces a physiological response in another organism of the same species. The term 'pheromone' was first applied to such chemical messengers in 1959. We now recognise that pheromones are produced by many species of insect.

Most pheromones are highly potent, and insects are extremely sensitive to smell. The olfactory receptors of insects are usually situated on the antennae. Some pheromones enhance the chance of contact between the sexes. These are examples of **signalling pheromones**, used to induce a behaviour response. For example, the female silk moth (*Bombyx mori*) produces a scent (pheromone) that attracts male silk moths over distances of 4 km or more! The use of sex pheromones is not restricted to insects. Several species of wild mammal are believed to use them, although this is hard to verify experimentally.

Social insects such as ants make use of signalling pheromones to locate food sources and warn of danger. The alternative type of pheromone, **priming pheromone**, which acts upon individuals to modify development, is also illustrated by social insects. Worker bees are females maintained in a sterile state by a pheromone ('queen factor') produced by a fertile queen. This substance spreads among the workers whilst the queen is present, preventing maturation of the workers' ovaries. As a colony enlarges the pheromone is diluted, and if the colony becomes too large a second queen may arise.

Questions

1   Suggest what might happen if the corpus allatum was removed from an early larval stage of an insect.

2   Pheromones are used as a way of controlling insect pests. Suggest how an apple grower might use pheromones to reduce the incidence of codling moths, which damage the crop.

FURTHER READING/WEB SITES

The WWW Virtual Library – Neuroscience: **neuro.med.cornell.edu/VL**

22 Support and locomotion

- Plants' support comes from the pressure of the contents of turgid cells against cell walls. Some walls are strengthened with lignin. Animal cells lack walls, and are supported by means of a skeleton.
- The movement of whole organisms is known as locomotion. Many plants are stationary, but animals typically are supported in a way that allows them to move about.
- Locomotion may be brought about by the movements of cilia and flagella, by cytoplasmic streaming (amoeboid movement) or by contraction of muscles anchored to skeletal material across joints.

- Animal skeletons are either internal (a system of bones) or external, or hydrolastic skeletons of fluid held under pressure in compartments surrounded by muscles.
- Insects, which have jointed external skeletons with jointed limbs, are relatively small animals that, as a group, have mastered movement both on land and in the air.
- Muscle is a tissue built of protein filaments. Contraction is brought about by a 'ratchet mechanism' that causes the filaments to slide past each other, shortening the length of the whole muscle.

22.1 INTRODUCTION

Living things support themselves and maintain positions in which they can carry out the essential processes of life (p. 2). The support system in plants usually brings leaves, flowers and fruits into the optimum position for photosynthesis, pollination and seed dispersal. In animals the support systems are often adapted to methods of locomotion and feeding. In all organisms, the support system maintains the shape of the body that is essential for consistent and coordinated functioning of its parts.

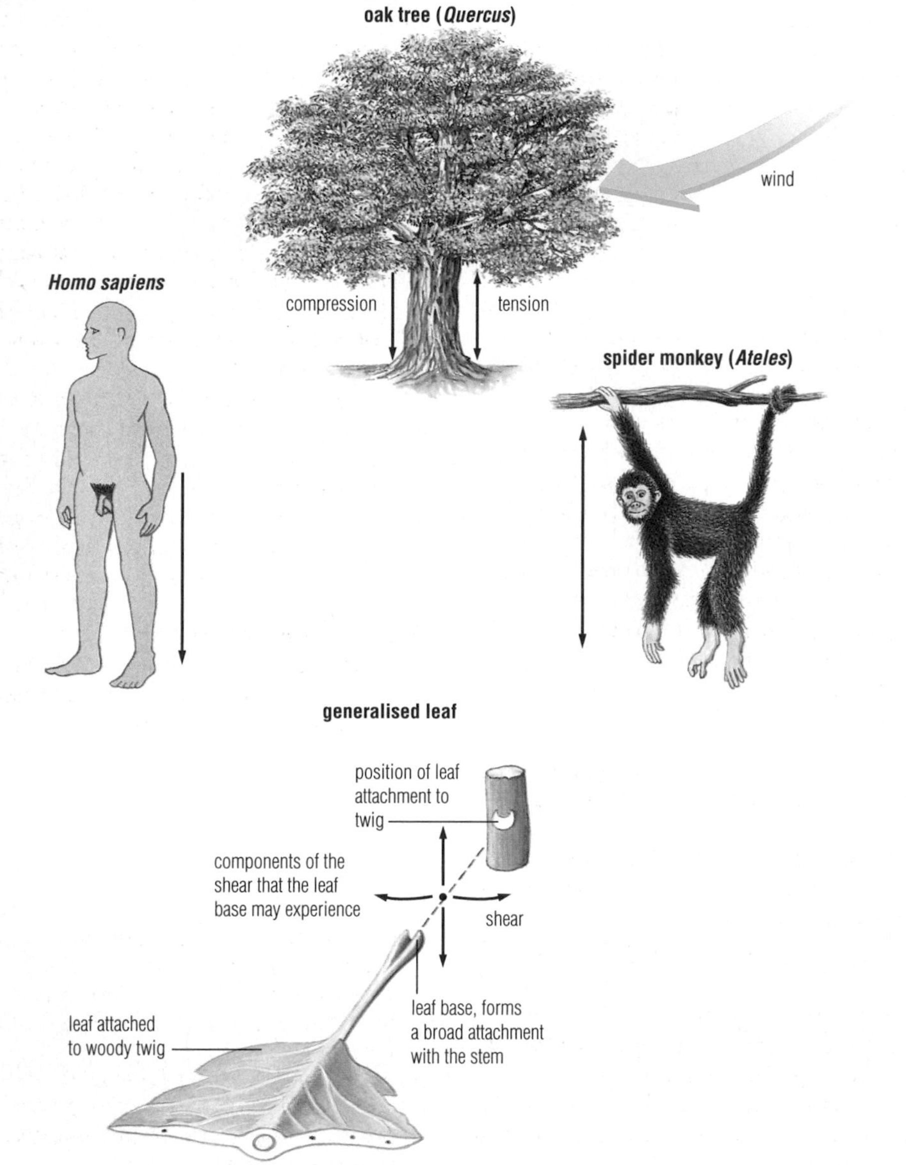

Plant cells are surrounded by a tough **cell wall**, which confers rigidity and support in turgid cells (p. 500), and provides some protection. In most multicellular terrestrial plants (trees and herbaceous plants, for example) the walls of some cells have become impregnated with the hardening substance **lignin** (p. 147). Lignified cells such as sclerenchyma (fibres) help support stems and roots rigidly.

Animal cells have no cell wall and little inherent rigidity. Many multicellular animals are supported and protected by a **skeleton**. Skeletons are often strengthened by inorganic salts such as calcium carbonate (common in shells) and calcium phosphate (in bone). Animal tissues and organs are surrounded by, and attached to or suspended from, the skeleton; it is also the skeleton that gives the animal its permanent shape and form.

■ Mechanical stress acting on living things

When mechanical force is applied to a body it comes under stress (Figure 22.1). The stress an organism experiences arises from supporting the weight of the body tissues, and from the forces generated by movement of the body. Environmental forces such as wind and waves may also stress an organism's body. Stress may take the form of

- ▶ **compression**, as when parts are being pushed together (put under positive pressure)
- ▶ **tension**, as when parts are pulled apart (stretched)
- ▶ **shear**, as when parts of a body slide past each other.

The bodies of living things must be strong enough to resist the stresses they experience, without the strengthening materials interfering with other activities.

Figure 22.1 Mechanical stress that organisms resist

oak tree (*Quercus*)

wind

compression   tension

Homo sapiens

spider monkey (*Ateles*)

generalised leaf

position of leaf attachment to twig

components of the shear that the leaf base may experience

shear

leaf attached to woody twig

leaf base, forms a broad attachment with the stem

■ Support systems and locomotion

Movement is a characteristic of all living things. There is movement within plant and animal cells – for example, during cytoplasmic streaming (p. 229) and when chromatids are pulled apart by spindle fibres during nuclear division (p. 195). Movement occurs in the tissues and organs of multicellular organisms; examples include the beating of the heart in vertebrates, and the movement of elaborated foods in the sieve tubes of flowering plants (p. 341).

The movement of whole organisms from place to place is a somewhat different phenomenon, known as **locomotion**. Whilst most plants are stationary organisms, very many animals support themselves in a way that allows them to move about. Animals move through the air, over land or in water. These contrasting media make different demands and confer different advantages with respect to support and locomotion.

Water is a supportive medium but dense and viscous, and resists movement. Air is much less dense, but a great deal of effort is needed to generate sufficient lift to support the animal off the ground. The ground provides terrestrial animals with a surface for support and propulsion, but most of them hold their bodies clear of the ground to minimise friction from the surface.

The purposes of animal locomotion include the search for food, avoidance of predators and other dangers, the search for a mate and in reproduction, migratory movements and the search for a more favourable environment. In many motile multicellular animals the support required is provided by the skeleton.

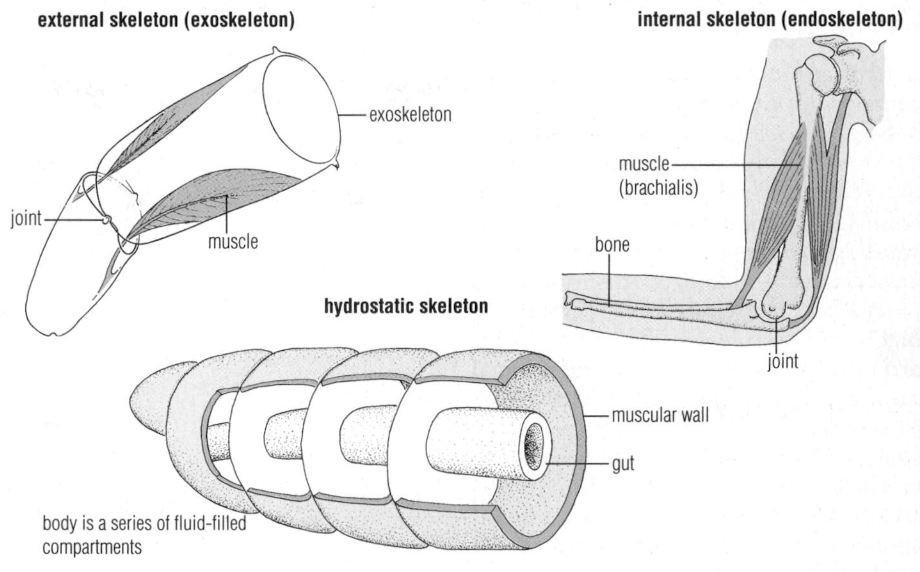

Skeletons of animals

Three major roles are fulfilled by the skeletons of animals.

1 Animal skeletons support the body, and enable the animal to resist compression forces. Within the body many organs are supported, attached to the skeleton.

The dense nature of water provides aquatic animals with a great deal of external support, not available to terrestrial and flying animals. Consequently, the largest aquatic animals are very much larger than any terrestrial animals in existence. Terrestrial animals require more rigid skeletons than those of aquatic animals, especially if the body is held off the ground.

Figure 22.2 Animal skeletons

external skeleton (exoskeleton)

exoskeleton

joint

muscle

internal skeleton (endoskeleton)

muscle (brachialis)

bone

joint

hydrostatic skeleton

muscular wall

gut

body is a series of fluid-filled compartments

2 Skeletons act as levers and facilitate locomotion. In many animals, locomotion is the result of the interaction of nervous, muscular and skeletal systems. Muscles for locomotion are attached to movable parts of the skeleton, across joints. By contraction of these muscles the body can act on its environment, exerting forces and moving itself.

Movement is not restricted to organisms or parts of organisms containing a hard skeleton. For example, the human tongue is a muscular organ that can make movement in many planes and directions, yet has no rigid skeleton. Similarly the trunk of the elephant functions like a fifth limb, yet has no rigid skeleton either. Further, movement occurs in many animals that lack a skeleton, as shown in amoeboid movement and movement by the actions of cilia and flagella (p. 500).

3 A skeleton may provide protection to delicate internal organs. In humans the brain is protected within the bony cranium, the eyes and ears are partially or totally surrounded by bone, and the heart and lungs are contained within the rib-cage. In animals with external skeletons virtually all soft tissues are protected.

Types of skeleton

There are three distinctly different types of skeleton (Figure 22.2).

A vertebrate has a rigid **internal skeleton** (**endoskeleton**) consisting of many component parts, made of cartilage or

bone. Movable parts of the skeleton articulate at joints. Many of the soft tissues of the body surround the bones, protected only by the skin. Support and movement in the endoskeleton of mammals is examined on pp. 484–95.

An insect (and, in fact, any member of the phylum Arthropoda, p. 37) has a firm or hard jointed **external skeleton** (**exoskeleton**) which encloses most of the body. Another form of exoskeleton is the hard shell of many molluscs. Here the role of the skeleton is almost exclusively for protection.

The earthworm illustrates a third type of skeleton, a **hydrostatic skeleton**, in which fluid is held under pressure in compartments surrounded by muscles. Since the liquid cannot escape and is of constant volume, it forms an incompressible 'skeleton' around which muscles can contract.

Support in the flowering plant

Support in flowering plants is introduced in Chapter 19 (p. 410), and discussed later in this chapter (p. 500).

Questions

1 For what reasons is it important that an organism maintains its characteristic shape?
2 What sort of forces would you expect the roots, stem and branches of a tree to experience?
3 Where in the human body are hyaline cartilage and compact bone respectively to be found (pp. 215–16)? What are the chief mechanical properties of these two tissues?

22.2 LOCOMOTION WITH A HYDROSTATIC SKELETON

A hydrostatic skeleton consists of fluid contained in a limited space, enclosed by muscles (normally circular and longitudinal muscles) of the body wall. This type of skeleton is most effective when the fluid is held in independent, more or less watertight, compartments, because the musculature is able to work in different parts of the body in turn. The result is a highly flexible, extensible body. This arrangement of circular and longitudinal muscles surrounding a hydrostatic skeleton occurs in the nematodes (p. 36), in some molluscs and in the annelids. In vertebrates the same principle (muscular contraction exerted against semi-liquid contents in a tube) is employed in the movement of food along the gut (peristalsis, p. 280).

■ Locomotion in the earthworm

All the segmented worms (phylum Annelida, p. 36) lack any rigid skeleton, but their fluid-filled coelom (the body cavity between gut and external wall) functions as a hydrostatic skeleton. In the earthworm the fluid is held in compartments separated by muscular segmental septa.

The body wall of the earthworm has both longitudinal and circular muscles, and these function antagonistically to power locomotory movements. Contractions of the circular muscles extend the segments. Contractions of the longitudinal muscles shorten the segments. Locomotion is made possible because the fully shortened segments can be secured to the ground by the chaetae, when these are fully extended. The chaetae anchor the stationary parts of the worm, resisting the backward pull exerted by the longitudinal muscles of the posterior segments as they contract (Figure 22.3).

When an earthworm moves forwards the circular muscles at the anterior end contract (Figure 22.4, stage A). The coelomic fluid, the volume of which is constant, is subjected to compression, which causes the relaxed longitudinal muscles in the same segments to lengthen. The anterior part of the body is narrowed in diameter, and extended forwards.

Following this extension movement the longitudinal muscles at the anterior end contract and the circular muscles relax, and the body now shortens and bulges here. Meanwhile the rear of the earthworm is brought forwards by a wave of contraction of the longitudinal muscles passing down the body (Figure 22.4, stages B and C). Successive waves of this peristaltic muscle contraction pass down the worm, causing locomotion. The chaetae are alternately extended to grip the burrow wall, and then retracted as that part of the body is moved forwards.

Nervous control of locomotion in the earthworm

Ordinary locomotory movements are controlled by reflexes between the segments, involving stretch receptors in the longitudinal muscles, and sensory and motor neurones running between sense organs and muscles in neighbouring segments. The segmental ganglia and the relay neurones permit communication between segments (Figure 22.5).

Giant nerve fibres are present in the ventral nerve cord, running the length of the worm. Giant fibres innervate the longitudinal muscles of the body wall. Stimulation of the giant nerve fibres results in the speedy transmission of impulses (p. 446) the full length of the worm. These impulses cause almost simultaneous contraction of the longitudinal muscles in each segment. With the posterior end firmly anchored in the soil the earthworm contracts suddenly, and may well escape danger by withdrawing rapidly into its burrow.

Figure 22.3 The machinery for locomotion in the earthworm

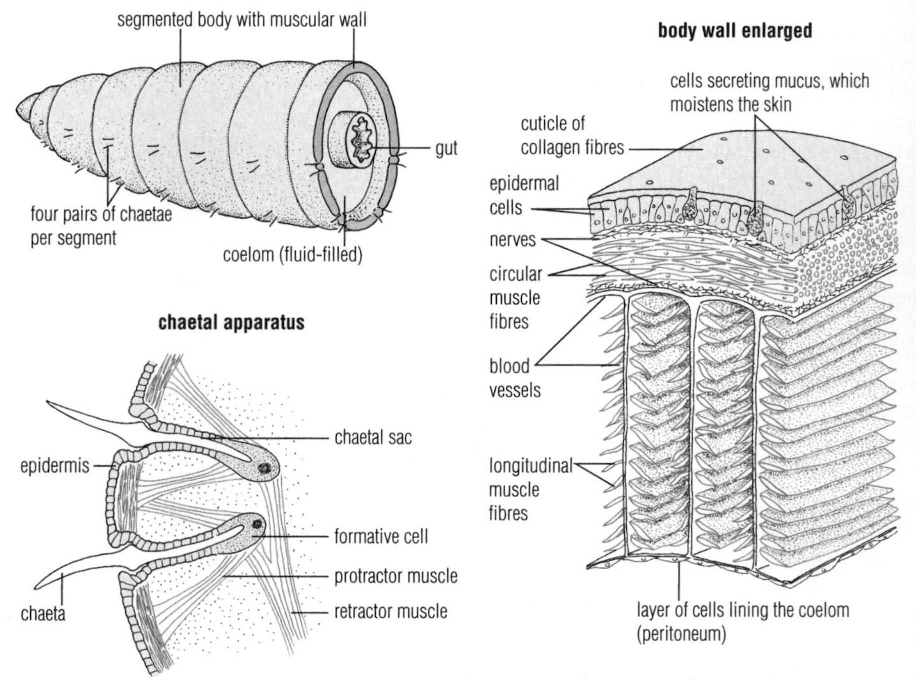

segmented body with muscular wall

four pairs of chaetae per segment

gut

coelom (fluid-filled)

chaetal apparatus

epidermis

chaeta

chaetal sac

formative cell

protractor muscle

retractor muscle

body wall enlarged

cells secreting mucus, which moistens the skin

cuticle of collagen fibres

epidermal cells

nerves

circular muscle fibres

blood vessels

longitudinal muscle fibres

layer of cells lining the coelom (peritoneum)

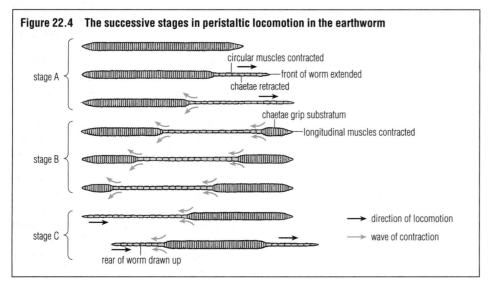

Figure 22.4 The successive stages in peristaltic locomotion in the earthworm

stage A

circular muscles contracted

front of worm extended

chaetae retracted

stage B

chaetae grip substratum

longitudinal muscles contracted

stage C

rear of worm drawn up

direction of locomotion

wave of contraction

Figure 22.5   Nervous control of locomotion in the earthworm

principal neurone connections between segments (horizontal longitudinal section)

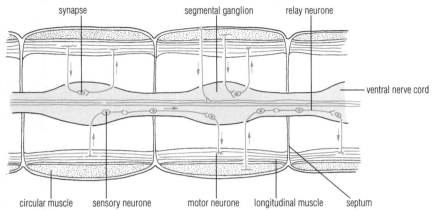

synapse · segmental ganglion · relay neurone

ventral nerve cord

circular muscle · sensory neurone · motor neurone · longitudinal muscle · septum

movement

1 contraction of longitudinal muscles here pulls on stretch receptors in the next segment

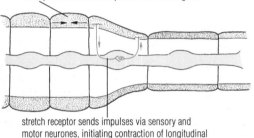

stretch receptor sends impulses via sensory and motor neurones, initiating contraction of longitudinal muscle

2 contraction in this segment stimulates stretch receptors

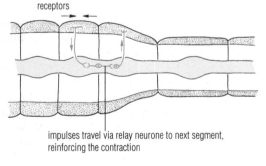

impulses travel via relay neurone to next segment, reinforcing the contraction

3 impulses via giant fibres stimulate muscles in all segments to contract simultaneously

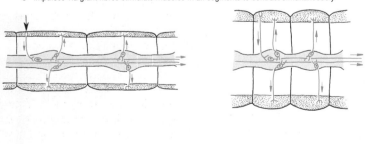

22.3 LOCOMOTION WITH AN EXOSKELETON

An exoskeleton is a protective and supportive structure, external to the body, but secreted by the outermost layer of cells. Exoskeletons of two types are found in non-vertebrates.

A **rigid exoskeleton**, commonly known as a shell, is typical of the molluscs. The shell covers a large part of the body, and the remainder can be drawn into the shell when necessary. In bivalve molluscs the shell has a joint. The mollusc shell is secreted by the epidermis of a part of the body known as the **mantle**.

A **jointed exoskeleton** is found in the arthropods. The Arthropoda is the largest phylum within the animal kingdom, and the insects make up the largest class of animals in the phylum (p. 38). We will focus on insects in discussing locomotion with an exoskeleton.

■ Efficiency of an exoskeleton

Given a limited quantity of material from which to build a skeleton (as in a very small animal), it turns out that a hollow tubular structure surrounding the body (that is, an exoskeleton) will provide greater support than a solid cylindrical rod within it (an endoskeleton) made from the same quantity of material. If we try to design increasingly larger structures whilst keeping the amount of skeleton-building material constant, the strength of the tubular exoskeleton decreases very rapidly as it is increased in diameter, in comparison with the same material used as a solid rod. To prevent this loss of strength, as the animal grows in size the quantity of material going into an external skeleton must be increased disproportionately. An exoskeleton loses its advantage in large animals because if it were to provide adequate support it would have to be impossibly cumbersome. In other words an exoskeleton is only superior if the animal remains small!

Other features of exoskeletons are significant too. For example, a tough external coat may provide some protection from predators. If waterproof, it will reduce water loss by evaporation. Against this must be set the disruption to a smooth and steady growth process that an exoskeleton imposes. The exoskeleton has to be periodically shed (ecdysis, p. 476), leaving the animal especially vulnerable as the new skeleton hardens.

■ Structure of the insect exoskeleton

The body of the insect is covered by a cuticle (Figure 22.6) made of a complex mixture of nitrogen-containing poly-saccharides of which chitin (p. 136) is a major component, together with protein. There is a waxy outer layer. The cuticle covers the whole body. It is formed into discrete overlapping plates with soft and flexible regions (joints) between the plates, facilitating movement. Chitin is hard and tough, but light. Its weight, therefore, does not prevent rapid movements. Indeed most insects are skilful and swift in their movements. The cuticle is secreted by the epidermis. During moulting (p. 402) enzymes from the epidermis dissolve the old cuticle as the new one is formed.

■ Locomotion in insects

Legs and walking

The exoskeleton of an insect's leg consists of a series of hollow cylinders held together by joints. Joints consist of soft, pliable membranes. The first section of the insect's leg joins to the body by a form of ball-and-socket joint. All the other joints are hinged structures. Movement at the joints is caused by contraction of muscles attached to the inside of the exoskeleton, across the joints, in antagonistic pairs (known as **flexor** and **extensor** muscles).

Insects have three pairs of legs (Figure 22.7). During walking movements the body is supported on a tripod of three legs. The other three legs pull or push the

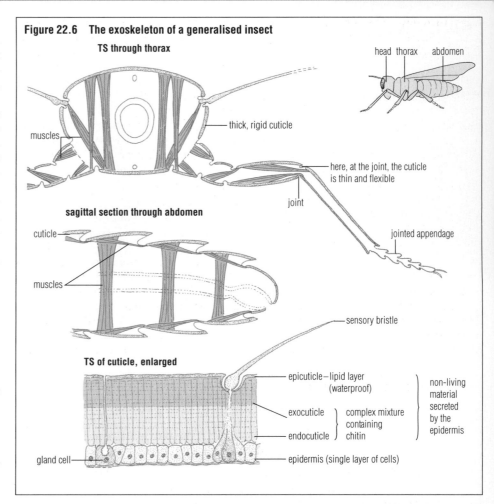

Figure 22.6   The exoskeleton of a generalised insect

body forwards, before or after they are lifted. Many insects have a pair of claws and an adhesive pad at the end of each leg. These insects can climb up vertical surfaces and walk upside down, too.

Insects use their legs for walking over many different types of surface. The rear legs of some insects are adapted for jumping. Many insects use the front pair of legs for holding food or manipulating

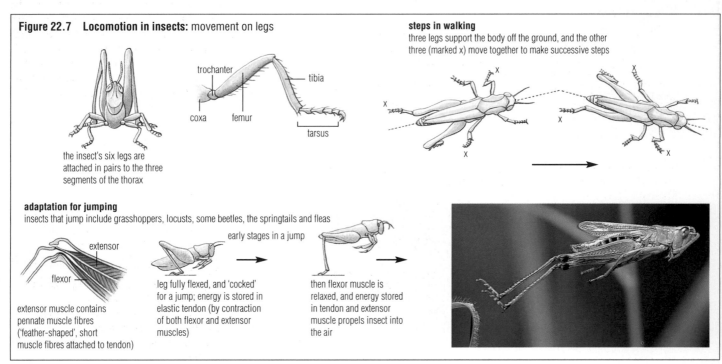

Figure 22.7   Locomotion in insects: movement on legs

the insect's six legs are attached in pairs to the three segments of the thorax

trochanter
tibia
coxa   femur
tarsus

steps in walking
three legs support the body off the ground, and the other three (marked x) move together to make successive steps

adaptation for jumping
insects that jump include grasshoppers, locusts, some beetles, the springtails and fleas

extensor
flexor

extensor muscle contains pennate muscle fibres ('feather-shaped', short muscle fibres attached to tendon)

early stages in a jump

leg fully flexed, and 'cocked' for a jump; energy is stored in elastic tendon (by contraction of both flexor and extensor muscles)

then flexor muscle is relaxed, and energy stored in tendon and extensor muscle propels insect into the air

other materials. The three pairs of legs are an efficient 'undercarriage' when a flying insect lands.

Flight in insects

Insects are the only non-vertebrates that can fly. Most insects are powerful flying machines. The power of flight has played a large part in the success (numerical dominance) of insects in the animal kingdom. The earliest insects were wingless, and a few present-day groups have lost the power of flight. Most insects have two pairs of wings and three pairs of legs. The possession of these organs of locomotion is a diagnostic feature of the class.

The insect's wings

The exoskeleton of an insect is tough, fibrous, flexible, light and impervious. The insect's wings are flattened extensions of the exoskeleton, stiffened by pleatings and by a network of thickened ridges, known as veins. Veins are concentrated at the leading edge of the wing, which can be seen to taper to the trailing edge when viewed in cross-section. In fact, the wing has the characteristics of an aerofoil. In flight the insect's wings produce lift by driving air downwards as they are moved forwards and backwards through the air in a path shaped like a figure of eight (Figure 22.8). The insect's body is moved forwards by the tilting of the wings as they beat.

The flight muscles of insects

An insect's flight muscles are striated muscles, resembling the skeletal muscles of vertebrates (p. 218). In a few insects (the butterflies and dragonflies, for example) the beating of the wings in flight is due to the contraction of muscles attached directly to the base of the wings. These are known as **direct flight muscles**.

In most flighted insects the wings are attached to the plates of the thorax wall, pivoted on the side plate (**pleuron**) of the thorax, with the base of the wing attached to the upper plate (**tergum**). Muscles attached to these plates, known as the **indirect flight muscles**, transmit their pull to the thoracic walls, distorting the flexible box structure of the thorax (Figure 22.8). When the longitudinal muscles contract the tergum is caused to bulge upwards, which creates the downstroke of the wings. When the dorso-ventral muscles contract the tergum is pulled down, which creates the upstroke of the wings. Thus the wings of most insects are operated by forces acting at the joint made with the body, but the twisting of the wings between upstroke and downstroke is caused by direct flight muscles attached at the wing bases.

Nervous control of flight

The frequency of wingbeats varies, from about 4 beats s^{-1} in some butterflies to about 1000 beats s^{-1} in certain midges. Insects with a slow wingbeat produce each muscular contraction by a nerve impulse. At the higher frequencies the nerves cannot work fast enough!

Indirect flight muscles have the special property of contracting if suddenly stretched, and relaxing again when released. This is a myogenic rhythm, because the contraction phase is triggered off by a state of tension in the muscle itself. Initial nervous stimulation of the dorso-ventral muscle (neurogenic contraction) distorts the thorax in such a way that the longitudinal muscles are stretched. Contraction of the longitudinal muscles causes stretching of the dorso-ventral muscles, which in turn causes their contraction, and so on until the stimulus fades away. The myogenic rhythm is initiated in this way, and is maintained as long as the insect flies, by periodic neurogenic contractions.

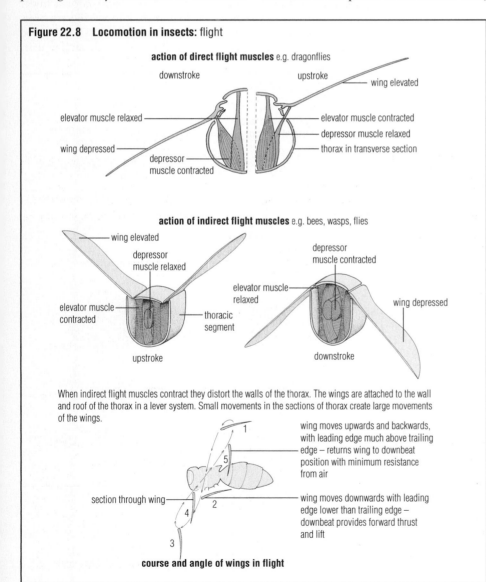

Figure 22.8 Locomotion in insects: flight

action of direct flight muscles e.g. dragonflies

downstroke

upstroke

wing elevated

elevator muscle relaxed

wing depressed

depressor muscle contracted

elevator muscle contracted

depressor muscle relaxed

thorax in transverse section

action of indirect flight muscles e.g. bees, wasps, flies

wing elevated

depressor muscle relaxed

elevator muscle contracted

thoracic segment

upstroke

depressor muscle contracted

elevator muscle relaxed

wing depressed

downstroke

When indirect flight muscles contract they distort the walls of the thorax. The wings are attached to the wall and roof of the thorax in a lever system. Small movements in the sections of thorax create large movements of the wings.

section through wing

wing moves upwards and backwards, with leading edge much above trailing edge – returns wing to downbeat position with minimum resistance from air

wing moves downwards with leading edge lower than trailing edge – downbeat provides forward thrust and lift

course and angle of wings in flight

Questions

1   If the nerve cord of an earthworm becomes severed, what type of movement is still possible? What type is prevented?
2   What are the advantages and disadvantages of an exoskeleton?
3   In insects, how is movement made possible despite the rigidity of the cuticle?
4   By labelled diagrams, show how muscles attached across a joint of an insect's leg can cause the leg to be alternately flexed and extended.

22.4 SUPPORT AND LOCOMOTION WITH AN ENDOSKELETON

■ The skeleton

We have already seen that an endoskeleton provides a rigid internal framework, maintaining the shape of the body. The bones of the skeleton protect many internal organs, and the whole skeleton provides numerous points of attachment for the voluntary, skeletal muscles of the body. Locomotion is made possible by the contraction of these muscles, acting across joints.

The skeleton of adult mammals (including humans) is made largely of bone, with cartilage tissue in certain parts. Bone is a rigid tissue. Cartilage is softer than bone, and less rigid. It is an important component of the joints between bones. The structures of cartilage and bone are described in Chapter 10, pp. 215–16.

There are physiological and biochemical roles for the bones of mammals, too.

For example, blood cells are produced in the marrow of many of the bones (p. 218). All bones act as reserve stores of calcium and phosphate ions (Ca^{2+} and PO_4^{3-}, p. 273).

The axial skeleton

The human axial skeleton consists of the skull and vertebral column (Figure 22.9).

The **skull** (Figure 22.11) is made of 22 bones, most of which are securely held together by immovable joints (**sutures**) with characteristic, crenate interlocking edges. The cranium encloses and protects the brain. It has a cubic capacity of about 1.5 litres. The bony wall of the cranium encloses the middle and inner ears (p. 453), and protects the olfactory organs in the roof of the nasal passage, and the eyes in sockets known as **orbits**.

In front and below the cranium are the facial bones, to which the upper jaw is fused. The lower jaw, the **mandible**, articulates with the cranium, and is held in position by muscles and ligaments. At the base of the cranium, on either side of the hole (foramen) through which the

spinal cord passes, are two small, smooth, rounded knobs (**condyles**). These articulate with the first bone of the vertebral column (the **atlas vertebra**), permitting the nodding movement of the skull.

The skull of the human fetus is initially formed from cartilage and fibrous tissues, which are slowly replaced by bone. At birth, this conversion is incomplete, and soft membranous spots, called fontanelles, remain. These fontanelles, including a noticeable one at the top of the head, allow slight compression of the cranium at birth, and thus facilitate parturition (p. 602). Fontanelles 'close' (become replaced by bone) within the first 12–24 months of life.

The **vertebral column** (backbone or spine, Figure 22.10) consists of a linear series of 33 bones, known as vertebrae. In total, they form a strong, flexible rod with many points for muscle attachment.

Figure 22.9   The human skeleton

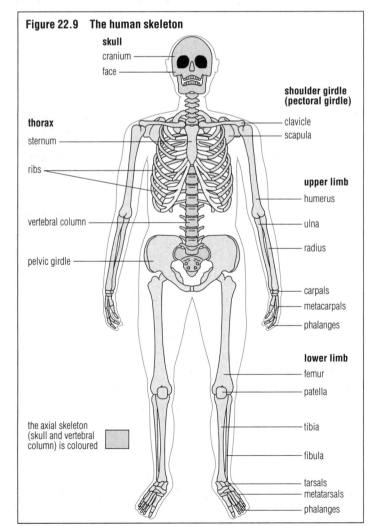

skull
— cranium
— face

thorax
— sternum
— ribs

vertebral column

pelvic girlde

shoulder girdle (pectoral girdle)
— clavicle
— scapula

upper limb
— humerus
— ulna
— radius

— carpals
— metacarpals
— phalanges

lower limb
— femur
— patella

— tibia
— fibula

— tarsals
— metatarsals
— phalanges

the axial skeleton (skull and vertebral column) is coloured

Figure 22.10   Vertebrae and the vertebral column

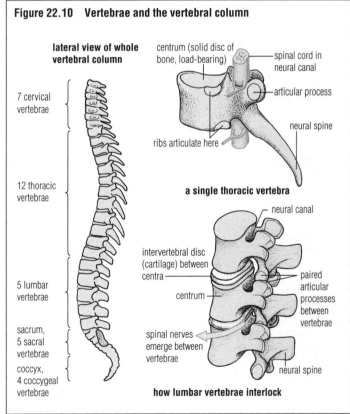

lateral view of whole vertebral column

7 cervical vertebrae

12 thoracic vertebrae

5 lumbar vertebrae

sacrum, 5 sacral vertebrae

coccyx, 4 coccygeal vertebrae

centrum (solid disc of bone, load-bearing)
spinal cord in neural canal
articular process
neural spine
ribs articulate here

a single thoracic vertebra

neural canal
intervertebral disc (cartilage) between centra
centrum
spinal nerves emerge between vertebrae
paired articular processes between vertebrae
neural spine

how lumbar vertebrae interlock

Questions

1 What functions does the skeleton of a mammal have?
2 How does the spinal column contribute to support and locomotion respectively?

Figure 22.11   The human skull: the skull consists of 22 bones

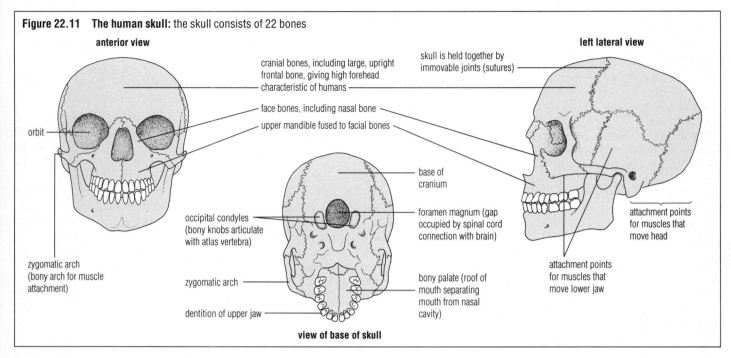

anterior view

left lateral view

cranial bones, including large, upright frontal bone, giving high forehead characteristic of humans

skull is held together by immovable joints (sutures)

face bones, including nasal bone

upper mandible fused to facial bones

orbit

base of cranium

foramen magnum (gap occupied by spinal cord connection with brain)

occipital condyles (bony knobs articulate with atlas vertebra)

zygomatic arch (bony arch for muscle attachment)

zygomatic arch

bony palate (roof of mouth separating mouth from nasal cavity)

dentition of upper jaw

view of base of skull

attachment points for muscles that move head

attachment points for muscles that move lower jaw

The whole vertebral column, with its attached muscles, supports the head and body. There are four distinct regions: cervical (neck), thoracic (chest), lumbar (small of back) and sacral (lower back), together with a vestigial tail or coccyx. There are important variations in the size and shape of the vertebrae in the different regions, but all have

▸ a body or **centrum** (a thick, solid, load-bearing part); above the centrum of each vertebra is a cartilaginous **intervertebral disc**, which functions as a shock-absorbing cushion, and allows movement between adjacent vertebrae

▸ a **neural canal** surrounded by the bony neural arch; this part surrounds and protects the spinal cord
▸ several bony **processes** to which muscles are attached, and with special surfaces that articulate with similar structures on the vertebrae in front and behind.

The rib-cage

In the thoracic region of the vertebral column is a bony cage formed by the ribs and sternum (breast bone). Twelve pairs of ribs support the wall of the thoracic cavity. Each rib articulates posteriorly with the corresponding thoracic vertebra. Anteriorly, cartilage connects the ribs to

the sternum. The marrow of the sternum is a site of red blood cell formation throughout life (p. 218). The rib-cage plays a key part in the ventilation of the lungs (p. 308).

Questions

1 What properties of the rib-cage are important in the ventilation mechanism of the thorax (p. 308)?
2 Carefully examine Figure 22.9, and then draw a simplified, labelled stick diagram of the human skeleton.

'SLIPPED DISCS'

The intervertebral discs between the fourth and fifth lumbar vertebrae are normally subject to the most severe compression forces in circumstances when the body is put under the greatest mechanical stress, perhaps due to the carrying of too heavy a load, or because of incorrect lifting and carrying posture. When a disc is 'slipped', its outer, fibrous coat becomes torn or weakened. The soft, elastic interior of the disc protrudes posteriorly, putting pressure upon the spinal cord or spinal nerve. This may cause acute pain.

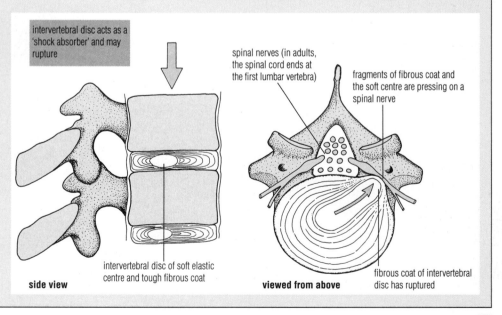

intervertebral disc acts as a 'shock absorber' and may rupture

spinal nerves (in adults, the spinal cord ends at the first lumbar vertebra)

fragments of fibrous coat and the soft centre are pressing on a spinal nerve

intervertebral disc of soft elastic centre and tough fibrous coat

fibrous coat of intervertebral disc has ruptured

side view

viewed from above

The appendicular skeleton

The appendicular skeleton consists of the limb girdles and limbs. The fore-limbs (arms) articulate with the pectoral girdle. The hind-limbs (legs) articulate with the pelvic girdle. The limb girdles are the connections between the axial skeleton and the limbs.

In mammals the pectoral girdle consists of two clavicles (collar bones) and two scapulae (shoulder blades). The clavicles articulate with the sternum ventrally, but the shoulder blades are held in position by a complex of muscles attached to the rib-cage.

The bones of the pelvic girdle are attached to the base of the vertebral column at the sacrum dorsally, and attached to each other ventrally at the pubic symphysis.

The limbs of vertebrates from the amphibians to the mammals all appear to be derived from a five-fingered limb, referred to as the **pentadactyl limb** (Figure 22.12). No living animal has exactly this type of limb skeleton. The ways in which limbs of vertebrates differ from the common plan reflect the mode of life adopted, and have important implications for evolution (p. 652).

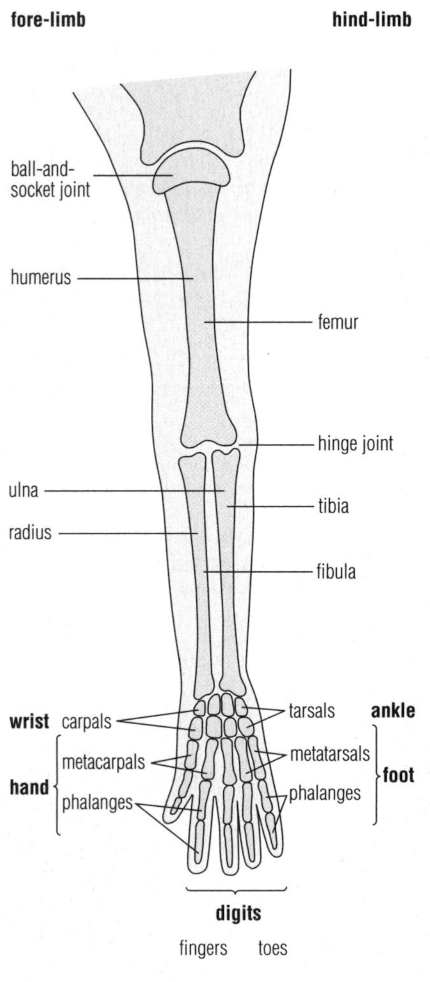

■ Joints

Joints occur in the skeleton where bones meet. Immovable joints, called **sutures**, are best seen between the bones of the cranium (Figure 22.11). Most joints are not rigid and immovable, but permit controlled movements of bones. In these cases, a smooth layer of cartilage covers the bone surfaces at the point at which they articulate, reducing the friction between their surfaces.

Joints are surrounded by a fibrous **capsule**, which is attached to the surrounding bone. The capsule consists of dense connective tissue, made of collagen fibres, and is flexible enough to permit movement. Special bundles of collagen fibres are also present; they function by holding bones together, limiting their movements, and resisting the dislocation of joints. These bundles of fibres are called **ligaments**.

At a **dislocated joint** a bone has become displaced from its socket. Dislocation causes damage to ligaments and sometimes to tendons also. Dislocations are most common between finger joints and at the shoulder.

Gliding joints that are movable through only a small range occur between the vertebrae, and between the bones of the wrist and ankle. These permit varying degrees of gliding movement. A **swivel joint** occurs between the first two cervical vertebrae, the atlas and the axis vertebrae, making possible the rotation of the head from side to side (Figure 22.14).

In freely movable joints, a fluid-filled space separates the articulating surfaces (Figure 22.13). The fluid therein, the **synovial fluid**, is secreted by the surrounding **synovial membrane**. Synovial fluid is formed from the blood plasma, and lubricates the joint as well as nourishing the articular cartilage. Synovial fluid also contains phagocytic cells that remove debris produced by wear of bone and cartilage at the joint. Examples of synovial joints are at the elbow and at the knee (hinge joints), and at the shoulder and hip (ball-and-socket joints).

Figure 22.12   The generalised pentadactyl limb

fore-limb   **hind-limb**

- ball-and-socket joint
- humerus
- femur
- hinge joint
- ulna
- tibia
- radius
- fibula
- **wrist** carpals
- tarsals   **ankle**
- metacarpals
- metatarsals   **foot**
- **hand**
- phalanges
- phalanges

digits

fingers   toes

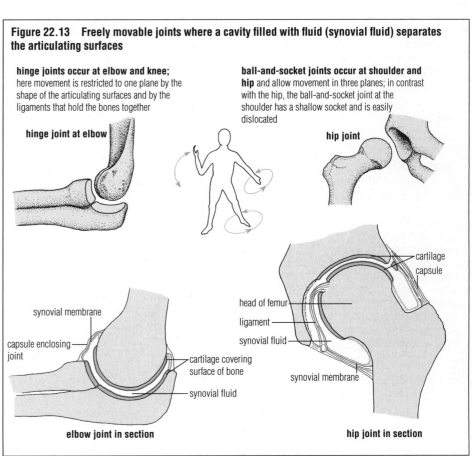

Figure 22.13   Freely movable joints where a cavity filled with fluid (synovial fluid) separates the articulating surfaces

hinge joints occur at elbow and knee; here movement is restricted to one plane by the shape of the articulating surfaces and by the ligaments that hold the bones together

hinge joint at elbow

ball-and-socket joints occur at shoulder and hip and allow movement in three planes; in contrast with the hip, the ball-and-socket joint at the shoulder has a shallow socket and is easily dislocated

hip joint

- synovial membrane
- capsule enclosing joint
- cartilage covering surface of bone
- synovial fluid

elbow joint in section

- cartilage
- capsule
- head of femur
- ligament
- synovial fluid
- synovial membrane

hip joint in section

A COMPARATIVE STUDY OF SKELETONS

All vertebrate skeletons possess the same basic structures of axial and appendicular skeletons. Comparison of the skeletons of different vertebrates discloses numerous examples of the way structures derived from common ancestors are modified in the process of meeting different functions (homologous structures, Figure 2.2, p. 23).

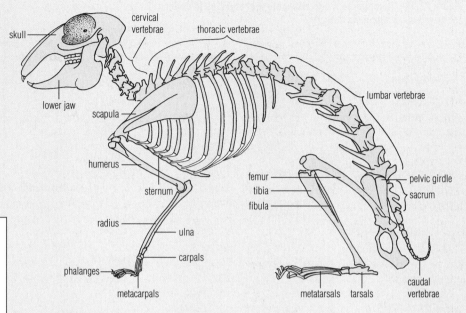

Figure 22.14   Joints where limited movement is possible

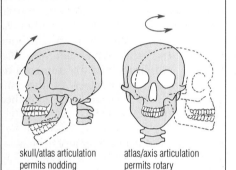

skull/atlas articulation permits nodding movement

atlas/axis articulation permits rotary movement

top view of atlas vertebra

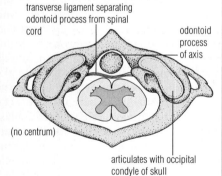

transverse ligament separating odontoid process from spinal cord

odontoid process of axis

(no centrum)

articulates with occipital condyle of skull

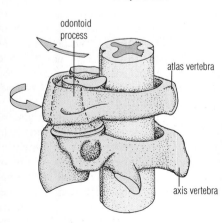

odontoid process

atlas vertebra

axis vertebra

Points of comparison	Rabbit	Human
Size	less than 0.3 m	1.5–1.75 m
Gait and limbs	quadruped with limbs adapted for running, leaping and digging	biped gait, with limbs very similar to the generalised pentadactyl limb; opposable fingers and thumb allow fine manipulation
Skull	relatively small cranium, lateral orbits for all-round vision, jaws and dentition for herbivorous ingestion	relatively large, prominent cranium with forward-directed orbits, permitting stereoscopic vision; dentition typical of omnivorous mammal
Vertebral column:		
cervical vertebrae	7	7
thoracic vertebrae	12–13	12
lumbar vertebrae	6–7	5
sacral vertebrae	4 (fused)	5 (fused)
caudal vertebrae	16	4 (fused)
	vertebrae have prominent extensions for muscle attachment; vertebral column allows for flexion/extension movements, aiding fast running	a supportive column along the line of the centre of gravity of the body

NEW HIP JOINTS

In Britain, surgeons replace about 30 000 hip joints damaged by arthritis every year. Current replacements are made of plastic and metal. Research into materials has led to the development of wrought titanium implants for the femur head (ball) and wrought titanium shell with polyethylene liner (socket). Titanium exhibits a degree of elasticity and load-bearing capability comparable with that of the natural bone into which it is implanted.

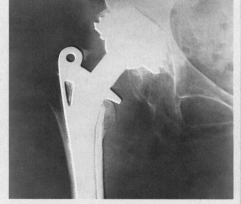

X-ray picture of a repaired hip joint

Figure 22.15 Arrangement and attachment of muscles to bone: most skeletal movement is controlled by antagonistic muscles

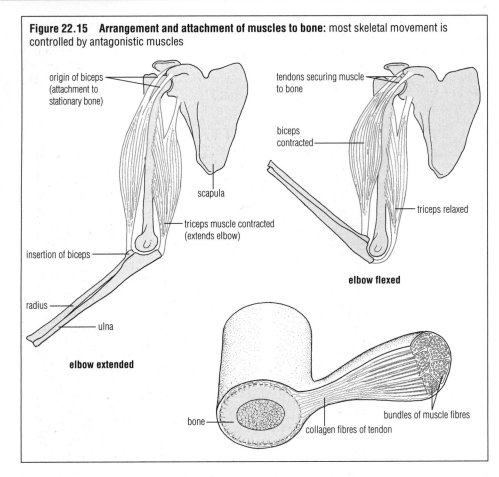

origin of biceps (attachment to stationary bone)

scapula

tendons securing muscle to bone

biceps contracted

triceps muscle contracted (extends elbow)

triceps relaxed

insertion of biceps

elbow flexed

radius

ulna

elbow extended

bone

collagen fibres of tendon

bundles of muscle fibres

LEVERS

We use machines to apply forces to objects in the world about us. A simple machine, commonly used, is a lever. The skeleton of an animal, together with the muscles attached across the joints, functions as a system of levers. Each joint acts as a pivot point, or fulcrum. The force applied (when the muscle contracts) is called the effort. The load or force to be overcome is known as the resistance. The further away from the fulcrum the effort is applied, the greater is the leverage: that is, the smaller the force that is required to raise the load.

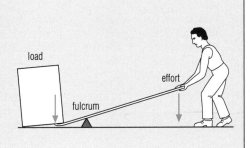

load

fulcrum

effort

This is the most common lever system in the body. The effort is greater than the load, but the distance moved by the effort (the muscle) is less than that moved by the load. Movement occurs with little shortening of muscle.

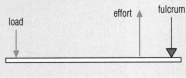

load

effort

fulcrum

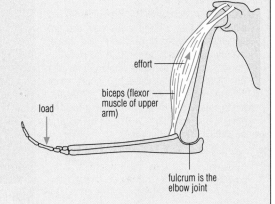

effort

biceps (flexor muscle of upper arm)

load

fulcrum is the elbow joint

■ Muscles and movement

You are already familiar with the three types of muscle tissue in mammals, namely **voluntary muscle** (striped or striated muscle), much of which occurs attached to bones, **involuntary muscle** (smooth or unstriped muscle), found in the walls of internal organs, and **cardiac muscle**, found only in the heart. The structure of muscle is described in Chapter 10 (p. 218).

Movements of the body are brought about by the contraction of voluntary muscles, attached to movable bones of the skeleton. Attachment of muscle to bone is often via tendons, although in certain cases muscle fibres are joined directly to bone. A **tendon** is a cord of dense connective tissue consisting of collagen fibres. In order for a muscle to act on a movable part of the skeleton to which it is attached, the other end of the muscle is fixed, across a joint, to a bone that, in this context, is non-movable. Skeletal muscles occur in antagonistic pairs: when one member of the pair contracts, the other is stretched (Figure 22.15).

■ The physiology of voluntary muscle

Voluntary (skeletal) muscle consists of large banded fibres, bound together in great numbers, and surrounded by special connective tissue (Figure 22.16). Individual **muscle fibres** vary in size and length (10–100 μm in diameter, and 2–3 cm in length), but all are long and thin. Very long fibres consist of shorter fibres connected in series. Each muscle fibre contains many nuclei, lying in the cytoplasm (**sarcoplasm**) just below the cell membrane, the **sarcolemma**. The cytoplasm contains extensive endoplasmic reticulum (p. 160), known as **sarcoplasmic reticulum**, and numerous mitochondria. The bulk of the muscle fibre consists of a mass of parallel, fibrous **myofibrils** running the length of the fibre. Myofibrils show a pattern of alternate light and dark bands (known as I and A bands respectively) when viewed under the light microscope – hence the alternative names of 'striped' or 'striated' muscle (Figure 22.16).

Under high-power electron microscopy the alternating dark and light bands are seen to be due to a system of interlocking filaments. Filaments are of two types, thick and thin, but both types are proteins. **Thick filaments** are somewhat shorter than thin filaments, and are confined to the dark bands. They are made of the protein **myosin**, and consist of a thick shaft with spirally arranged myosin molecules. Each myosin molecule has a bulbous head, forming a **cross-bridge** between the filaments, and carrying an actin-binding site and an ATP-binding site.

Thin filaments occur in the light bands, and also extend between the thick filaments. The thin filaments are composed of the protein **actin** and a second protein, the binding site blocking substance. Thin filaments are held together by transverse bands, the Z bands, and project in both directions from this point of anchorage. The length of the myofibril from one Z band to the next is described as a **sarcomere** (Figures 22.17 and 22.18, overleaf).

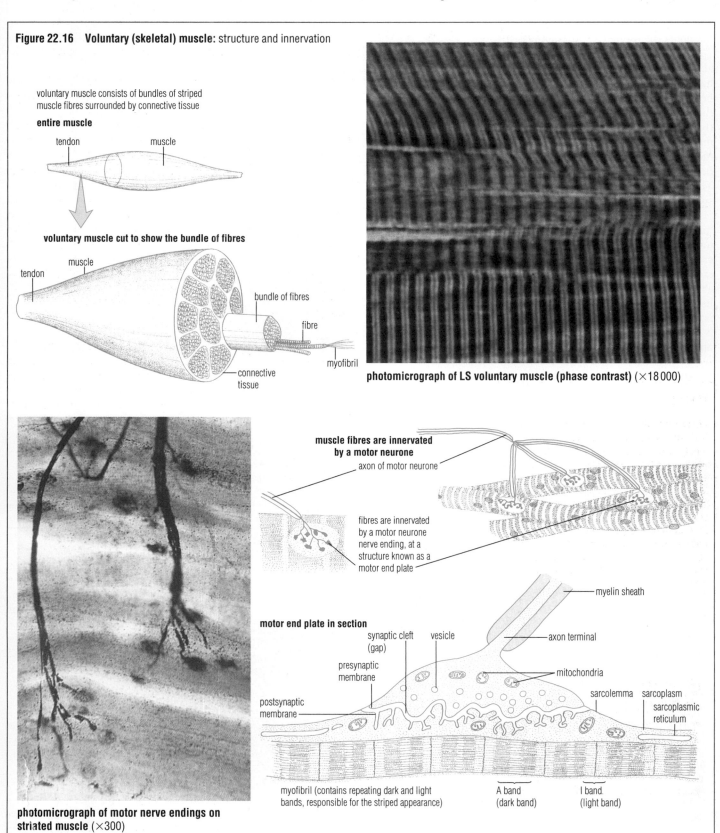

Figure 22.16   Voluntary (skeletal) muscle: structure and innervation

voluntary muscle consists of bundles of striped muscle fibres surrounded by connective tissue

entire muscle

tendon   muscle

voluntary muscle cut to show the bundle of fibres

tendon   muscle   bundle of fibres   fibre   myofibril   connective tissue

photomicrograph of LS voluntary muscle (phase contrast) (×18 000)

muscle fibres are innervated by a motor neurone

axon of motor neurone

fibres are innervated by a motor neurone nerve ending, at a structure known as a motor end plate

motor end plate in section

myelin sheath   axon terminal   synaptic cleft (gap)   vesicle   mitochondria   presynaptic membrane   postsynaptic membrane   sarcolemma   sarcoplasm   sarcoplasmic reticulum

myofibril (contains repeating dark and light bands, responsible for the striped appearance)   A band (dark band)   I band (light band)

photomicrograph of motor nerve endings on striated muscle (×300)

The sliding-filament hypothesis of muscular contraction

A hypothesis put forward concerning the mechanism of muscular contraction suggests that the shortening of the sarcomere is brought about by the thin protein filaments sliding inwards towards its centre. The lengths of the thin and thick protein filaments themselves do not change.

Contractions are produced by the action of the myosin cross-bridges, which have bulbous heads. Their bulbous heads become attached to the neighbouring thin (actin) filament and then swing through an angle, pulling the actin filament towards the centre of the sarcomere. Attachment of the bulbous heads occurs at specific sites, the binding sites. The heads of cross-bridges then detach, swing back, and re-attach further along.

This movement has been likened to a 'ratchet mechanism'. An alternative model visualises the activities of the cross-bridges as if they were the hands of crew members of a sailing ship hauling in a rope (the thin filament), hand over hand. However we visualise it, the result is that each sarcomere of the muscle fibre is shortened (Figures 22.17, 22.18 and 22.19).

Questions

1 A voluntary muscle fibre is described as a syncytium (p. 218). What does this mean?
2 Describe the relationship between a muscle, a muscle fibre, a myofibril and a myosin filament.
3 Explain why it is that, with a lever, the further away from the fulcrum the force is applied, the less force is required to lift the load.

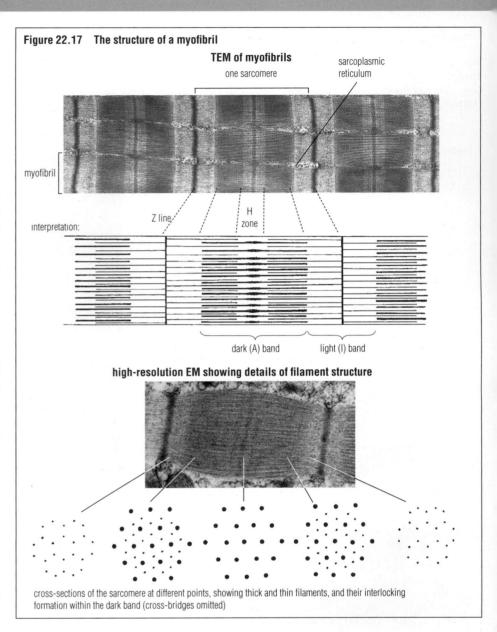

Figure 22.17   The structure of a myofibril

TEM of myofibrils

one sarcomere

sarcoplasmic reticulum

myofibril

interpretation:

Z line

H zone

dark (A) band   light (I) band

high-resolution EM showing details of filament structure

cross-sections of the sarcomere at different points, showing thick and thin filaments, and their interlocking formation within the dark band (cross-bridges omitted)

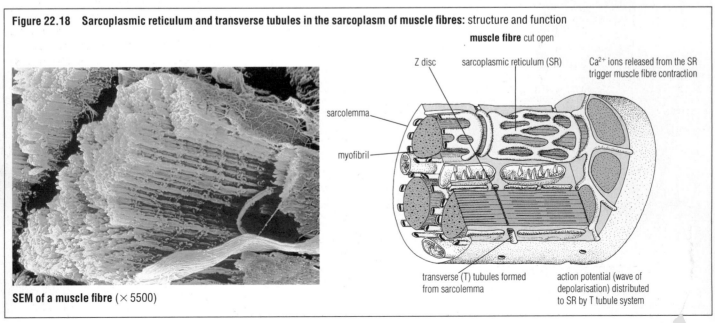

Figure 22.18   Sarcoplasmic reticulum and transverse tubules in the sarcoplasm of muscle fibres: structure and function

muscle fibre cut open

Z disc   sarcoplasmic reticulum (SR)

Ca^{2+} ions released from the SR trigger muscle fibre contraction

sarcolemma

myofibril

transverse (T) tubules formed from sarcolemma

action potential (wave of depolarisation) distributed to SR by T tubule system

SEM of a muscle fibre ($\times$ 5500)

Figure 22.19 The sliding-filament theory of muscle contraction

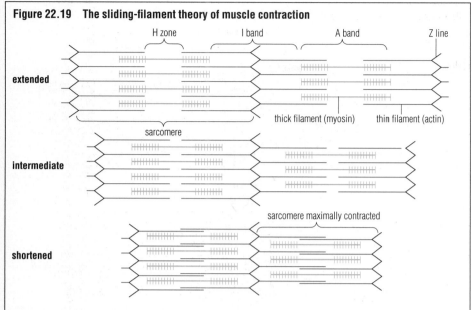

Figure 22.20 The response of a muscle fibre to a nerve impulse

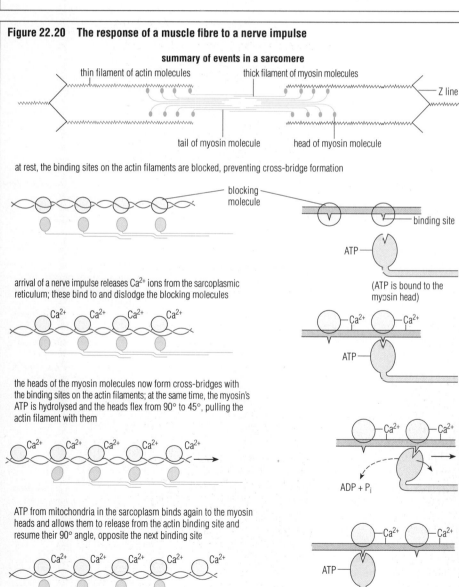

summary of events in a sarcomere

thin filament of actin molecules   thick filament of myosin molecules

Z line

tail of myosin molecule   head of myosin molecule

at rest, the binding sites on the actin filaments are blocked, preventing cross-bridge formation

blocking molecule

binding site

ATP

(ATP is bound to the myosin head)

arrival of a nerve impulse releases Ca^{2+} ions from the sarcoplasmic reticulum; these bind to and dislodge the blocking molecules

ATP

the heads of the myosin molecules now form cross-bridges with the binding sites on the actin filaments; at the same time, the myosin's ATP is hydrolysed and the heads flex from 90° to 45°, pulling the actin filament with them

$ADP + P_i$

ATP from mitochondria in the sarcoplasm binds again to the myosin heads and allows them to release from the actin binding site and resume their 90° angle, opposite the next binding site

ATP

Stages in voluntary muscle contraction

The stages of skeletal muscle contraction, from the point of nervous stimulation (Figure 22.20) to relaxation again, are listed below.

1 A nerve impulse (action potential or wave of depolarisation, p. 445) travelling along a motor neurone, reaches the motor end plate.

2 Acetylcholine (ACh) is released into the synaptic cleft at the end plate, diffuses to the sarcolemma, and combines with receptor sites.

3 When the threshold value of a generator potential is reached, an action potential is created in the muscle fibre. (ACh in the cleft region is then hydrolysed and the products re-absorbed into the motor end plate.)

4 The action potential is conducted to all microfibrils of the muscle fibre by the system of transverse tubules (T tubules, Figure 22.18), and spreads to the sarcoplasmic reticulum throughout the muscle fibre.

5 Calcium ions (Ca^{2+}) are released from the sarcoplasmic reticulum and bind to the blocking molecules of the actin filament, exposing the binding sites (ATP is involved in this movement).

6 The 'heads' of the cross-bridging myosin molecules attach to the newly exposed binding sites on the actin filaments. Release of energy by ATP hydrolysis accompanies cross-bridge formation.

7 As a result, the shape of the myosin bridge molecule changes. A 'bending' or 'rowing' action results. This is the power stroke. The thin filaments (actin molecules) are moved towards the centre of the sarcomeres, producing muscle cell contraction.

8 Fresh ATP attaches to the myosin head, releasing it from the binding site, and causing the cross-bridge myosin to straighten (it becomes 'cocked' for repeat movement).

9 Myosin heads become attached further along the actin chain, and repeat the movement sequence (referred to as a 'ratchet mechanism').

10 When nervous stimulation of the sarcolemma ceases calcium ions are rapidly pumped back into the sarcoplasmic reticulum, and the Ca^{2+} concentration falls below the threshold for contraction activity. The binding sites on the actin filaments are blocked again.

Energy supplies for muscle contraction

There is a small store of ATP in cells. In low level physical activity this is replaced by respiration as fast as it is used in muscle fibres. Energy for muscle contraction comes from the aerobic respiration of fatty acids and glucose present in the muscle tissue and supplied by the plasma solution. The ATP produced is used directly in muscle contraction. An energy store of glycogen is held in reserve in the muscle tissue (and elsewhere in the body; for example, in the liver). Glycogen is hydrolysed to glucose as needed.

At times of repeated muscle contraction there is the likelihood that voluntary muscle will be deprived of glycogen and oxygen. Muscle tissue holds reserves of energy in the form of **creatine phosphate**, and reserves of oxygen in the form of **oxymyoglobin**. Figure 22.21 illustrates the supply of resources for contraction.

If strong muscle contractions continue for a prolonged period the reserves of oxygen become exhausted. Eventually there may be insufficient oxygen to maintain **aerobic respiration**, and an 'oxygen debt' starts to build up (Figure 22.22). In this condition, musle switches to **anaerobic respiration** (lactic acid fermentation). How long the body's musculature can continue in this mode may depend upon the individual's tolerance to the resulting conditions, including the build up of lactic acid. (Tolerance can be increased by athletic 'training'.) The accumulation of lactic acid is often cited as a cause of muscle fatigue. Lactic acid dissociates into lactate and hydrogen ions. As H+ ions accumulate in the cytoplasm they may affect cell enzyme acitivty (p. 180–1).

Recovery – making good the oxygen debt

When a period of heavy exercise ends, oxygen uptake by the muscle tissue only slowly returns to resting levels (Figure 22.22). First the 'oxygen debt' of muscle tissue is repaid. Oxygen additional to resting-level demand is required to

▶ make good the normal levels of dissolved oxygen in tissues; in muscles, oxymyoglobin is recharged, too

▶ sustain the aerobic respiration required to reform creatine phosphate; the energy store of muscle is recharged

▶ sustain the additional metabolism as lactate is respired, or converted back to pyruvate and then to glucose and glycogen (p. 317). This occurs in the skeletal muscle tissue and other tissues of the body. it does not occur exclusively in the liver, as was once thought. During heavy exercise, heart muscle also converts lactate to pyruvate but then uses

it immediatley as a respiratory substrate. Carbon doixide and water are the waste products, and the energy transferred sustains muscle contraction

■ Muscles and posture

Muscles are involved in the maintenance of posture, as well as in movement. In fact there are two types of voluntary muscle fibre, with distinctly different physiological properties, known as the tonic (slow) fibres and the twitch (fast) fibres (Figure 22.23 and Table 22.1).

Tonic fibres are for sustained muscular contraction, and are essential in maintaining body posture. **Twitch fibres** are for fast contractions, and are used in locomotion and other purposeful body activity. In humans both types of muscle fibre occur in all muscles, in proportions that depend on their position in the body. The tonic fibres of the voluntary muscles are maintained in a state of sustained tension; otherwise we would collapse on the ground!

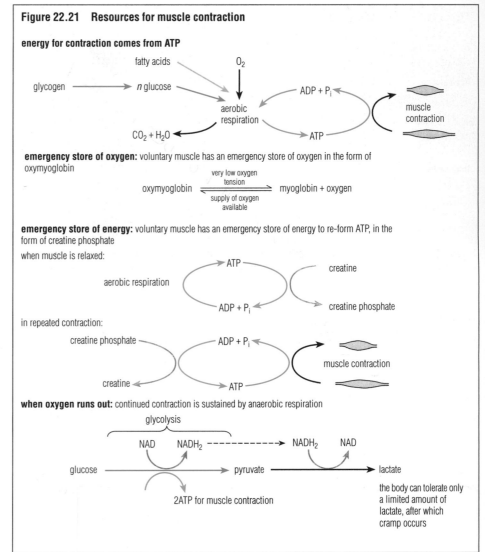

Figure 22.21   Resources for muscle contraction

energy for contraction comes from ATP

emergency store of oxygen: voluntary muscle has an emergency store of oxygen in the form of oxymyoglobin

emergency store of energy: voluntary muscle has an emergency store of energy to re-form ATP, in the form of creatine phosphate

when muscle is relaxed:

in repeated contraction:

when oxygen runs out: continued contraction is sustained by anaerobic respiration

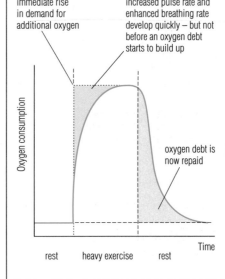

Figure 22.22   Oxygen uptake at rest, in heavy exercise and in recovery (making good the oxygen debt)

Figure 22.23   In an athlete, the proportions of tonic and twitch muscles depend on the type of his or her activity: sprinters (left) have more twitch fibres, while long-distance runners and weight-lifters (right) have more tonic fibres

Table 22.1   Properties of tonic and twitch muscles

	Tonic muscle fibres	Twitch muscle fibres
Function	sustained muscular contraction for maintenance of posture	fast muscular contraction, e.g. for locomotion
Structure	▶ many mitochondria ▶ red in colour, due to presence of myoglobin ▶ capillaries in close contact with fibres ▶ many motor end plates along fibres	▶ few mitochondria ▶ white in colour (little or no myoglobin) ▶ one or few motor end plates
Physiological activity	▶ obtain ATP from aerobic respiration ▶ require frequent stimulation to contract ▶ slow muscular contraction of long duration	▶ obtain ATP from anaerobic respiration once O_2 supply used up (oxygen debt builds up) ▶ membrane electrically excitable ▶ fast muscular contraction produced, but fatigues quickly

Maintaining posture

Muscles and joints contain tiny sense receptors, the proprioceptors (Table 21.1, p. 450). These internal sense organs keep the brain 'informed' of the body's position, and of the state of tension of the skeletal muscles; structures called **muscle spindles** (Figure 22.24) are examples. A steady discharge of impulses from muscle spindles all over the body maintains muscle tone and body posture.

If, for example, the body tends to sag under its own weight, the sensitive nerve endings in the muscle spindles are further stretched and trigger the stretch reflex, which adjusts muscle tone and body posture. At any time that a muscle contracts or relaxes and its length changes, the muscle spindles function as proprioceptors and 'inform' the central nervous system via sensory nerves. Of course, the brain is also 'informed' on body position and movement via sense organs such as eyes and touch receptors.

Antagonistic muscles of locomotion and the inhibitory reflex

Even the simplest of movements made by a part of the body involves many muscles. It is important to remember that most if not all of these muscles operate in antagonistic pairs (Figure 22.15). The nervous control of antagonistic muscle pairs is via reflex arcs, as shown in Figure 21.31, p. 462. The stimulus that initiates contraction of one muscle simultaneously inhibits the motor neurone that serves the opposing antagonistic muscle.

Figure 22.24   Muscle spindles and the maintenance of posture

Muscles contain stretch receptors known as muscle spindles. These are proprioceptors deep in the skeletal muscle. They are linked via nerves to the spinal cord.

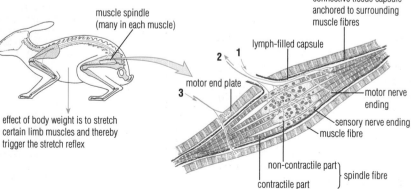

1 impulses from the CNS cause spindle fibres to contract or relax, thus maintaining the same tension on the spindle fibres
2 when muscle is stretched (or load on muscle is increased), additional impulses are sent to spinal cord via sensory neurones from spindle fibres
3 by reflex action, impulses are sent to tonic muscle fibres, increasing contraction and muscle tone

Questions

1 Trace the pathway of dietary carbohydrate, and the changes in it, from the point of digestion to its availability for muscle contraction.

2 What are the respective roles in muscle contraction of actin, myosin, ATP and calcium ions?

Muscles and locomotion

An animal moves forwards by pushing backwards against its surroundings. In terrestrial mammals the push is made downwards and slightly backwards against the ground. Pairs of antagonistic muscles are responsible for propulsion, many pairs being involved in the movement of each limb. The most important of the pairs of antagonistic muscles responsible for propulsion in the human leg are shown in Figure 22.25.

In the initiation of movement the centre of gravity is first swung away from the limb to be moved. As the body moves along its balance is aided by compensating arm movements.

In walking movement each leg in turn is moved forwards and backwards by alternate contraction of flexor and extensor muscles. Thus each leg acts as a lever, exerting a downwards and backwards force on the ground; this is resolved into lift and a slight forward component of thrust, which are transferred to the body. The sequence of events in walking is analysed in Figure 22.26.

In running, not only is the sequence of contractions speeded up, but for substantial periods neither foot is on the ground. The foot that drives the body forward when it meets the ground is already travelling backwards.

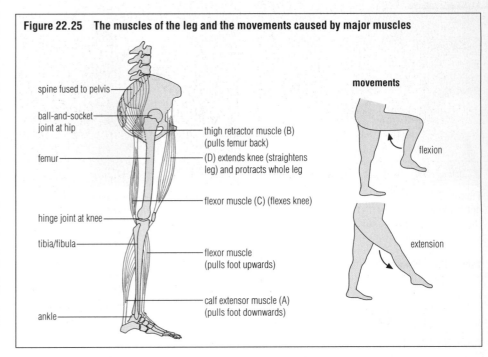

Figure 22.25   The muscles of the leg and the movements caused by major muscles

spine fused to pelvis

ball-and-socket joint at hip

femur

hinge joint at knee

tibia/fibula

ankle

thigh retractor muscle (B) (pulls femur back)

(D) extends knee (straightens leg) and protracts whole leg

flexor muscle (C) (flexes knee)

flexor muscle (pulls foot upwards)

calf extensor muscle (A) (pulls foot downwards)

movements

flexion

extension

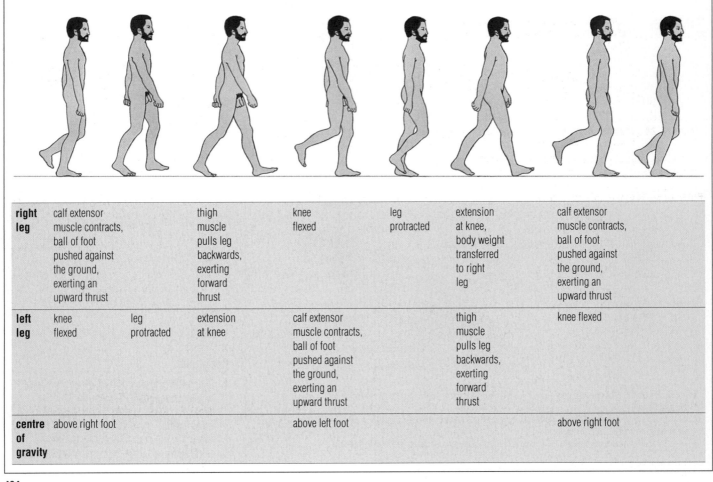

Figure 22.26   An analysis of human walking movements

right leg	calf extensor muscle contracts, ball of foot pushed against the ground, exerting an upward thrust		thigh muscle pulls leg backwards, exerting forward thrust	knee flexed		leg protracted	extension at knee, body weight transferred to right leg	calf extensor muscle contracts, ball of foot pushed against the ground, exerting an upward thrust
left leg	knee flexed	leg protracted	extension at knee		calf extensor muscle contracts, ball of foot pushed against the ground, exerting an upward thrust		thigh muscle pulls leg backwards, exerting forward thrust	knee flexed
centre of gravity	above right foot				above left foot			above right foot

Figure 22.27   Human running: when running, there are periods when neither foot is on the ground; the supporting leg is not straight (as it is in walking) but slightly bent

Figure 22.28   Changes in the roles of limbs and vertebral column in locomotion

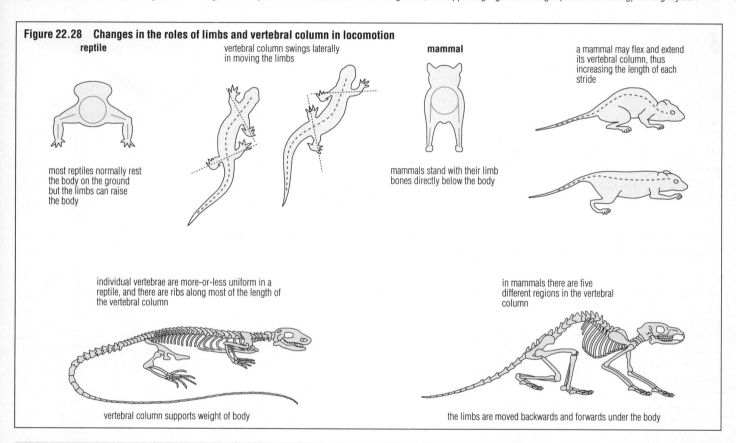

reptile

most reptiles normally rest the body on the ground but the limbs can raise the body

vertebral column swings laterally in moving the limbs

mammal

mammals stand with their limb bones directly below the body

a mammal may flex and extend its vertebral column, thus increasing the length of each stride

individual vertebrae are more-or-less uniform in a reptile, and there are ribs along most of the length of the vertebral column

vertebral column supports weight of body

in mammals there are five different regions in the vertebral column

the limbs are moved backwards and forwards under the body

Roles of limbs and backbone in terrestrial locomotion

Most terrestrial vertebrates move by using limbs as levers (**appendicular locomotion**). Likewise, the vertebral column of land animals functions as a support to the body. In the course of evolution, however, there has been a fundamental change in the roles of vertebral column and limbs in locomotion between reptiles and mammals (Figures 22.27 and 22.28).

We presume that when fish-like creatures first came on to land, fins evolved into limbs. These would have projected laterally, as the limbs of many amphibians still do. Subsequently, limbs evolved to a position partly below the body, enabling it to be raised above the ground surface, as we see in many reptiles. This process is completed in mammals, where the limbs are held parallel to the sides of the body directly below the body. Less muscular effort is expended holding the body off the ground, compared with a position in which the legs splay out at an angle.

Fundamental changes have also occurred in the vertebral column of terrestrial animals. In the first land vertebrates movements of the vertebral column were similar to the S-shaped movements made by fish as they swim (p. 496). In many modern amphibians and reptiles movement of the vertebral column is of this type. In mammals the role of the vertebral column in locomotion has changed. Mammals that run on four feet may flex and extend their vertebral column. By arching and extending the back the power and length of their strides are increased, as they move their limbs backwards and forwards under the body.

Questions

1   List the muscles identified in Figure 22.25 in the sequence in which they contract to move the leg forward in walking.

2   In a terrestrial animal that moves across the land surface, what advantage is conferred by having the limbs below and parallel to the sides of the body, rather than lateral to the body?

22.5 FISH: LOCOMOTION IN WATER

The largest animals are aquatic organisms. This is because water is a supportive medium. Water is also a relatively dense medium, and so fish require a streamlined shape in order to be able to move at all quickly in water. Whilst the density of water allows the body and median fins to push against the surrounding medium for locomotion, the drag that has to be overcome in movement also increases with density.

On average, the anterior of a fish is smooth and rounded, and the body long. The head merges into the body; there is no neck. The body reaches maximum girth about one-third of its length from the front and then tapers to the tail. The skin is covered by scales, over which is secreted a slimy coating. These features all tend to reduce drag by minimising the disturbance of water in the wake of the fish.

Fish swim by **wavelike movements** (undulations), which travel from anterior to posterior along the body, pushing the water backwards and the organism forwards (**axial locomotion**). There is a tendency for the fish to become unstable in water during movement, leading to **yawing**, **pitching** and **rolling** movements (Figure 22.29). These movements are effectively countered: yawing is opposed by the head and the dorsal and ventral (median) fins, pitching by horizontally held pectoral and pelvic fins (the paired fins) and rolling by the median and paired fins, which act as stabilisers.

Much of the trunk of the body and the whole of the tail consists of segmentally arranged muscle blocks (**myotomes**). Each myotome block is anchored to the vertebral column. The vertebral column is a chain of vertebrae, and functions as a flexible rod. The vertebral column resists compression when muscles contract but at the same time permits the bending movements of locomotion. Undulations are brought about by serial and alternating contraction of the right and left myotome blocks. Thus the muscles of the left and right sides of the body are antagonistic; when those on one side contract, those of the other side relax.

■ The herring

The herring (*Clupea* sp.) is an active swimmer. The posterior half of the body is thrown into alternating movements from side to side by waves of contraction

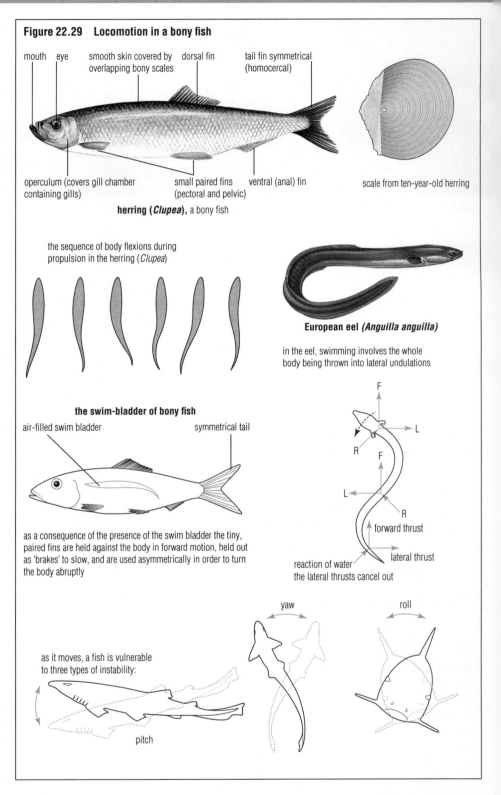

Figure 22.29 Locomotion in a bony fish

mouth eye smooth skin covered by overlapping bony scales dorsal fin tail fin symmetrical (homocercal)

operculum (covers gill chamber containing gills) small paired fins (pectoral and pelvic) ventral (anal) fin

scale from ten-year-old herring

herring (*Clupea*), a bony fish

the sequence of body flexions during propulsion in the herring (*Clupea*)

European eel (*Anguilla anguilla*)

in the eel, swimming involves the whole body being thrown into lateral undulations

the swim-bladder of bony fish

air-filled swim bladder symmetrical tail

as a consequence of the presence of the swim bladder the tiny, paired fins are held against the body in forward motion, held out as 'brakes' to slow, and are used asymmetrically in order to turn the body abruptly

forward thrust

lateral thrust

reaction of water the lateral thrusts cancel out

as it moves, a fish is vulnerable to three types of instability:

yaw

roll

pitch

and relaxation of myotome blocks. The greatest propulsion comes from the tail. In some bony fish, however, the entire body provides the thrust from S-shaped movements; the eel is an example (Figure 22.29). In other bony fish the body is not bent at all, and propulsion comes from movements of the tail only.

In many bony fish the role of the fins is influenced by the presence of the swim bladder. The swim bladder is a gas-filled outgrowth of the pharynx lying in the body cavity above the gut, and the fish is able to alter the volume of gas trapped within it. In some bony fish, such as the herring, the swim bladder remains 'open', and the air volume can be adjusted by swallowing air. In others, including the cod, the swim bladder is closed off from the pharynx and the gas volume can be adjusted only by means of blood capillaries in a gas gland projecting into the swim bladder. The ability to control the volume of gas in the swim bladder allows the fish to maintain its buoyancy when pressure of water changes with depth. The swim bladder helps to maintain the fish's position in the water without expenditure of energy.

Another consequence of a swim bladder is that the paired fins of bony fish may be small. In forward movement they are held pressed to the body, reducing drag. During manoeuvres the paired fins are used as stabilisers, and singly or together as 'brakes'. Used asymmetrically, the paired fins enable the fish to turn through 180° in its own length! As a consequence of the swim bladder, many bony fish swim quickly, and turn quickly and dart about with agility.

Questions

1 What features in the structure of a fish contribute to efficient locomotion?
2 How is a fish able to prevent sideways movements of the head during strong alternate contractions of the tail musculature?
3 How would the volume of the swim bladder change if a fish swam to a greater depth? What adjustments would be made by a cod in response to this change?
4 How can a sideways movement of a fish's tail generate forward propulsion?

22.6 BIRDS: LOCOMOTION IN THE AIR

Air has a low density, and provides little support or substance to push against. Despite this, the ability to fly (as opposed to mere gliding) **has evolved independently four times in the history of life**: in birds, insects (p. 483), certain mammals (the bats) and the extinct flying reptiles, the pterosaurs. Today, there are about 8500 living bird species. Birds are abundant in every climate and most habitats, from the polar ice caps to the tropics. Mastery of the air involves distinctive adaptations of structure and physiology.

■ Adaptions for flight in a bird

The feather-covered body of the bird is smooth and streamlined. The eyes are prominent (the sense of sight is extremely well developed), and there is no external ear to interrupt the smooth flow of air.

The skeleton is light, with much reduction and fusion of component parts (Figure 22.30). The bird's bones are hollow (tubular); in some species there is internal strutting where extra stress is experienced.

The central axis of the body is short, and modified to allow the weight of the body to be carried in two distinct ways, either on wings or on legs. To achieve this two plate-like girders are present, the sternum and the pelvis. Their large surfaces are for the attachment of the muscles that produce locomotion.

Feathers are horny outgrowths from the epidermis, composed of keratin (the protein that makes up mammalian hair and the scales of reptiles). Flight feathers occur on the wings and tail, contour feathers cover the body, providing a smooth surface more or less impervious to water, and down feathers next to the skin provide heat insulation.

The bird's fore-limbs (pentadactyl limb, p. 486) are modified as wings. Wings are essentially aerofoils, generating lift when moved through the air (Figure 22.31). Most of the surface is composed of flight feathers, which form a light, air-resistant but strong structure. Flight requires a large ratio of wing surface area to weight, so flying birds rarely achieve a large size. The strength and low density of the keratin of feathers is a key to the success of the class.

Flight demands a high metabolic rate. In birds this is sustained by a high and constant body temperature, 39–42 °C, which is higher than the average body temperature of mammals. As a consequence, flying birds consume a great deal of food. The brain of birds is larger, relative to the body, than in other vertebrate groups except the mammals. The parts of the brain that are well-developed are the optic lobes (sight), cerebellum (muscular coordination) and the underlying grey matter of the cerebral hemispheres (rather than the cerebral cortex, as in mammals). Bird behaviour is governed more by inherited and instinctive factors than is that of mammals.

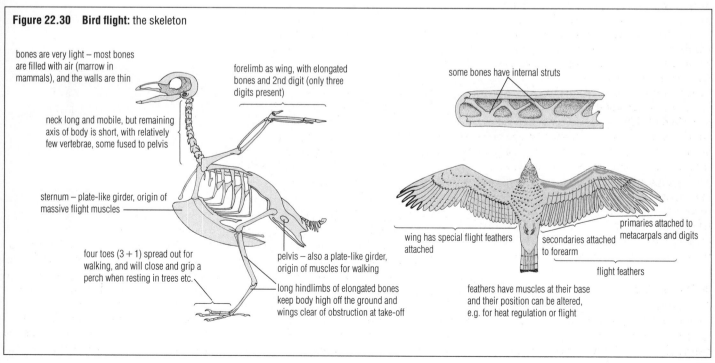

Figure 22.30   Bird flight: the skeleton

bones are very light – most bones are filled with air (marrow in mammals), and the walls are thin

forelimb as wing, with elongated bones and 2nd digit (only three digits present)

some bones have internal struts

neck long and mobile, but remaining axis of body is short, with relatively few vertebrae, some fused to pelvis

sternum – plate-like girder, origin of massive flight muscles

four toes (3 + 1) spread out for walking, and will close and grip a perch when resting in trees etc.

pelvis – also a plate-like girder, origin of muscles for walking

long hindlimbs of elongated bones keep body high off the ground and wings clear of obstruction at take-off

wing has special flight feathers attached

secondaries attached to forearm

primaries attached to metacarpals and digits

flight feathers

feathers have muscles at their base and their position can be altered, e.g. for heat regulation or flight

The essentials for forward flight

The wing, an aerofoil, is a slightly curved flat structure, convex above and concave below. In section, the front or leading edge is the widest part, tapering behind to the thin trailing edge. When a wing is moved forwards through the air the airflow is faster over the longer, upper surface than the shorter lower surface. This produces a fall in the pressure above the wing relative to the pressure below it, generating a lift force acting upwards (and a slight drag force acting backwards). For flight, the mean lift force must equal the weight of the bird.

The amount of lift generated by the wing depends upon the angle at which it is held in relation to the airstream. As the angle of attack is increased lift is increased, but eventually, at a critical angle, air ceases to flow smoothly over the upper surface of the wing. At and above an angle of 15° airflow separates from the wing and lift is reduced. This causes the bird to stall (drop out of the air).

Turbulent airflow is reduced if a second aerofoil (such as the bird's bastard wing) is held above and just ahead of the main aerofoil. The separating of the flight feathers at the end of the wing in some birds achieves a similar effect.

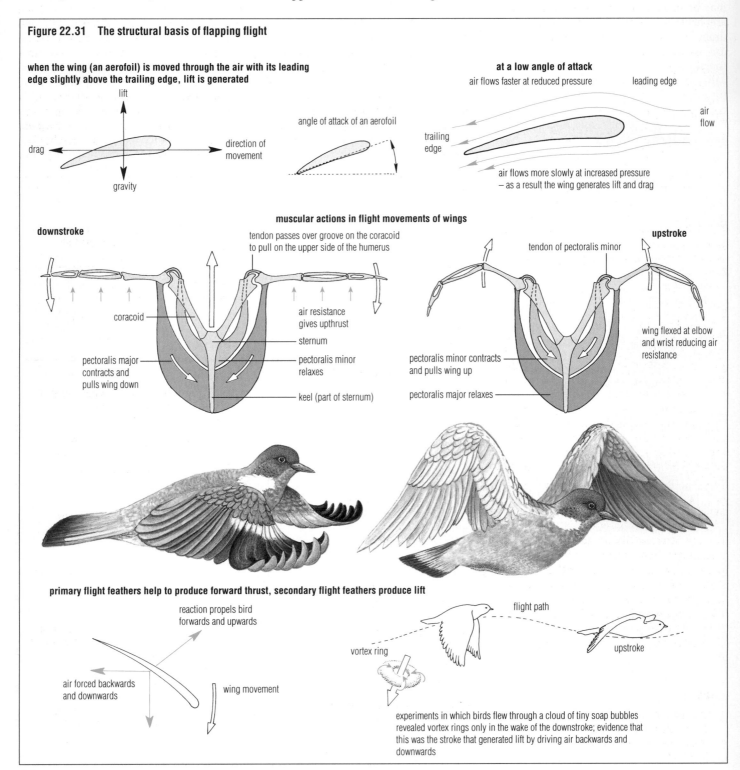

Figure 22.31   The structural basis of flapping flight

when the wing (an aerofoil) is moved through the air with its leading edge slightly above the trailing edge, lift is generated

lift

direction of movement

drag

gravity

angle of attack of an aerofoil

at a low angle of attack

air flows faster at reduced pressure   leading edge

trailing edge

air flow

air flows more slowly at increased pressure – as a result the wing generates lift and drag

muscular actions in flight movements of wings

downstroke

tendon passes over groove on the coracoid to pull on the upper side of the humerus

upstroke

tendon of pectoralis minor

coracoid

air resistance gives upthrust

sternum

pectoralis major contracts and pulls wing down

pectoralis minor relaxes

keel (part of sternum)

wing flexed at elbow and wrist reducing air resistance

pectoralis minor contracts and pulls wing up

pectoralis major relaxes

primary flight feathers help to produce forward thrust, secondary flight feathers produce lift

reaction propels bird forwards and upwards

air forced backwards and downwards

wing movement

vortex ring

flight path

upstroke

experiments in which birds flew through a cloud of tiny soap bubbles revealed vortex rings only in the wake of the downstroke; evidence that this was the stroke that generated lift by driving air backwards and downwards

Gliding and soaring flight

In gliding flight the bird holds its wing stationary, at a low angle of attack, and gradually loses height. Expert gliding birds such as the albatross lose very little height in still air because of the size and shape of their wings, and the continuous adjustments they make to the angle of attack of the wings. Gliding birds can often use upward air currents (either thermals, or winds deflected upwards by sloping ground or waves) to gain height. This type of flight is known as soaring flight. Albatrosses and vultures soar for hours at a time.

Flapping flight

In flapping flight the wings generate propulsion (forward thrust) as well as providing the lift. To do so, the wings are moved up and down rhythmically. The main power for flapping flight is provided by contractions of the large pectoral muscles. These muscles extend from the large sternum to the humerus, and make up 15–20% of the body mass of the bird (Figure 22.31). Pectoral muscles are red with myoglobin (p. 360) in birds with strong flight powers. (In a bird such as the chicken that flies only very briefly, they are white.)

The main power stroke is the downbeat of the wing, brought about by contraction of the pectoralis major. Forward thrust results from pushing the wing against the air. As the wing moves down the flight feathers of the wings are maintained in overlapping position, and the full surface of the wing is presented to the air.

The upstroke is powered by contraction of the pectoralis minor muscles. Resistance to the air is lowered by twisting the wing so that the leading edge is raised first, flexing the wrist so as partly to close the wing, and by the parting of the flight feathers, allowing air to pass between.

For take-off maximum lift is required. A bird may achieve additional lift by launching itself from a branch or by running into take-off. Failing this the first downstrokes are extra powerful. In landing the tail is used as a brake, and the wings are moved into a 'stall' position. The outstretched legs fold up below the bird, absorbing much of the landing impact.

Questions

1 Suggest a reason for the differences in size of the pectoralis muscles in Figure 22.31.
2 What is the role of myoglobin in muscle (p. 360)?

22.7 AMOEBOID AND CILIARY LOCOMOTION

A limited range of fundamental mechanisms of locomotion is found in the living world. There are only three basic mechanisms: muscular locomotion (discussed above), amoeboid locomotion involving pseudopodia, and ciliary locomotion using cilia or flagella.

■ Amoeboid locomotion

The protozoan *Amoeba* has given its name to a distinctive form of locomotion. Amoeboid movement is not only demonstrated by members of the phylum Rhizopoda (p. 28), to which *Amoeba* belongs; it is also the method of locomotion of the white blood cells of vertebrates (p. 364).

Amoeba moves by putting out pseudopodia (Figures 21.1 and 14.2). Locomotion is not maintained in any particular direction for long and *Amoeba* is constantly changing shape as it changes direction. Such movements often bring *Amoeba* into contact with food organisms (other small unicells) living on the mud of freshwater ponds.

Amoeboid movement is brought about by cytoplasmic streaming and by the interconversion of cytoplasm between a gel state (relatively stiff and jelly-like cytoplasm, as in the ectoplasm of *Amoeba*) and a sol state (fluid cytoplasm, as in the organism's endoplasm) (Figure 22.32). Energy from metabolism is involved in cytoplasmic streaming. It has been demonstrated that calcium ions are required for streaming of cytoplasm, and so too is ATP. The exact mechanism of movement is not understood.

Figure 22.32   Locomotion by amoeba

sol cytoplasm streams forward

gel to sol cytoplasm

sol to gel cytoplasm

gel cytoplasm

substratum

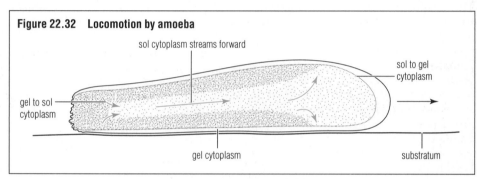

Figure 22.33   Locomotion by flagellum

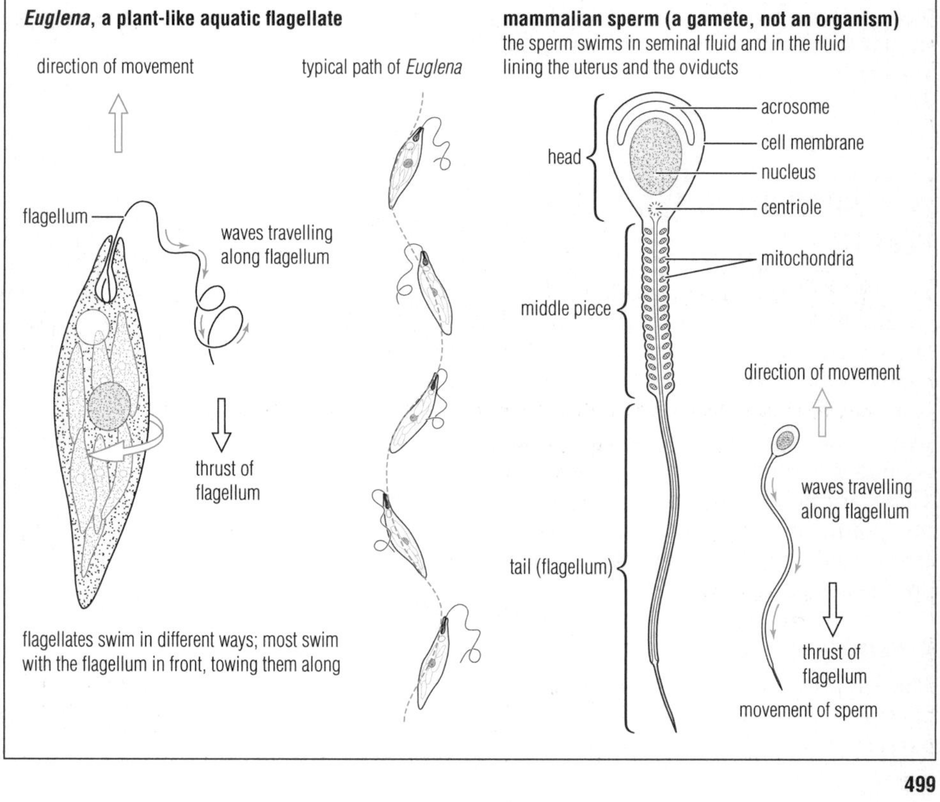

Euglena, a plant-like aquatic flagellate

direction of movement

typical path of *Euglena*

flagellum

waves travelling along flagellum

thrust of flagellum

flagellates swim in different ways; most swim with the flagellum in front, towing them along

mammalian sperm (a gamete, not an organism)
the sperm swims in seminal fluid and in the fluid lining the uterus and the oviducts

head
acrosome
cell membrane
nucleus
centriole

middle piece
mitochondria

direction of movement

tail (flagellum)
waves travelling along flagellum

thrust of flagellum

movement of sperm

■ Ciliary locomotion

Cilia and flagella are whip-like organelles (Figure 7.25, p. 165) which project from the surface of certain eukaryotic cells. Structurally cilia and flagella are different versions of the same organelle. If one or two of the organelles are present they are called flagella; if they are very many and short, they are called cilia. Cilia and flagella occur on certain single-celled organisms. For example, *Paramecium* is a ciliate, while *Euglena* moves by means of a flagellum. The locomotory organelle of the sperm of vertebrates is a flagellum, while cilia cover certain stationary cells of multicellular organisms – for example, the cells lining the vertebrate air passages are ciliated (p. 309). Cilia and flagella beat in different ways to effect locomotion (Figures 22.33 and 22.34).

Structure and functioning of cilia and flagella

The detailed structure of these organelles has become apparent by the use of electron microscopy. They consist of fine tubes composed of an extension of the cell plasma membrane, containing nine double microtubules around the circumference and two single microtubules in the centre. The tubules are made of a protein (tubulin). Each double microtubule is fringed by side projections. At the base, in the cytoplasm of the cell, is a basal body.

Movement of cilia and flagella involves no shortening of the microtubules; instead they are observed to slide past each other, in a movement very similar to that of the actin and myosin filaments of voluntary muscle (p. 490). ATP is involved in these actions.

22.8 SUPPORT IN FLOWERING PLANTS

The load on the stem is caused by its own weight, and by that of the leaves, flowers and fruits. The stem resists the resulting compression stress. In the wind the aerial system offers resistance; the windward side is under tension and the leeward side is under compression (Figure 22.1). There are possibilities of shear stress at the points of attachment of leaves to stems and in the main stem (wind, pp. 58–60).

■ Herbaceous plants

The support of the herbaceous stem is provided by the turgidity of the parenchyma cells (p. 223) contained

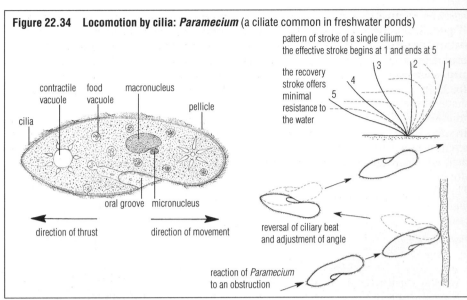

Figure 22.34   Locomotion by cilia: *Paramecium* (a ciliate common in freshwater ponds)

contractile vacuole
food vacuole
macronucleus
pellicle
cilia
oral groove
micronucleus

direction of thrust
direction of movement

pattern of stroke of a single cilium: the effective stroke begins at 1 and ends at 5

the recovery stroke offers minimal resistance to the water

reversal of ciliary beat and adjustment of angle

reaction of *Paramecium* to an obstruction

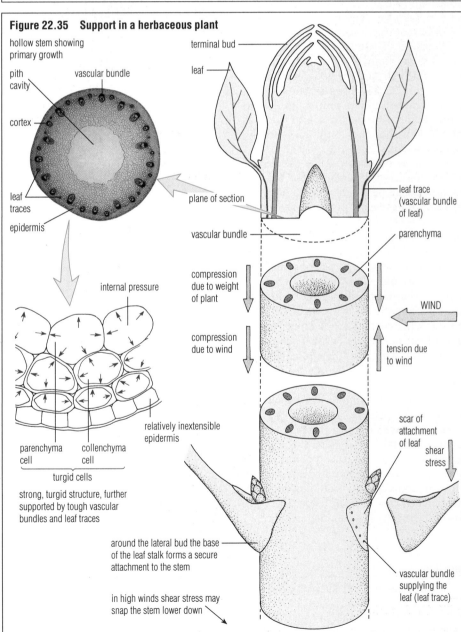

Figure 22.35   Support in a herbaceous plant

hollow stem showing primary growth
pith cavity
vascular bundle
cortex
leaf traces
epidermis

internal pressure

parenchyma cell
collenchyma cell
turgid cells

strong, turgid structure, further supported by tough vascular bundles and leaf traces

relatively inextensible epidermis

terminal bud
leaf
leaf trace (vascular bundle of leaf)
plane of section
vascular bundle
parenchyma

compression due to weight of plant
compression due to wind

WIND
tension due to wind

scar of attachment of leaf
shear stress

around the lateral bud the base of the leaf stalk forms a secure attachment to the stem

in high winds shear stress may snap the stem lower down

vascular bundle supplying the leaf (leaf trace)

within a relatively inextensible epidermis (Figure 22.35). The ability of the stem to resist compression and tension forces is enhanced by the vascular bundles and by any collenchyma and fibres present (p. 224).

■ Woody plants

In woody stems the support created by turgid parenchyma tissue contained within the epidermis is replaced by the rigidity of the secondary xylem which makes up the bulk of the stem (p. 226). The secondary xylem consists of massed xylem vessels and fibres. These occur as complete rings, visible in a cross-section, and known as annual rings (Figure 19.24, p. 412).

In branches and in leaning trunks the pith may be eccentric, with much wider growth rings on one side than on the other. A branch is like a cantilever beam in that it is fixed at one end only, and supports its own weight. The weight of the branch causes the lower side to be under compression and the upper side under tension. In transverse section, branches show reaction wood (Figure 22.36). In hardwood varieties (such as any dicotyledonous tree) this appears on the side under tension.

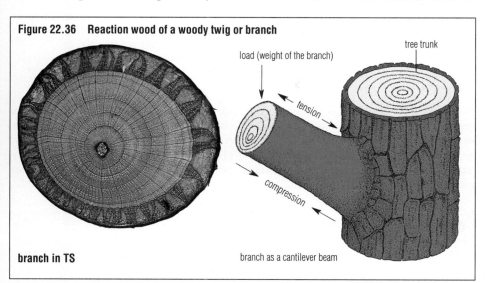

Figure 22.36   Reaction wood of a woody twig or branch

load (weight of the branch)

tree trunk

tension

compression

branch in TS

branch as a cantilever beam

22.9 LIFE STYLES AND SYMMETRY

The shape of an organism is related to its life style (Figure 22.37). For example, many animals that move actively have anterior, posterior, dorsal and ventral surfaces. One plane cuts their bodies into mirror-image halves (the sagittal plane). This type of symmetry (bilateral symmetry) is a characteristic of mobile animals. Often, their bodies are relatively compact and elongated in the direction of movement, a shape that offers least resistance to the surrounding medium. The anterior end of the organism experiences new environmental stimuli first, and the chief sense organs are located at this end; so too is the food-capture apparatus. The result is the evolution of a head (cephalisation) in most bilaterally symmetrical animals.

On the other hand, attached, non-motile animals (sessile life styles) experience the environment more or less equally on all sides. Some sessile animals are radially symmetrical. They evolve no distinct head. In the absence of locomotion, nervous systems and sense organs tend to be greatly reduced or underdeveloped. Many cnidarians have this passive, sedentary mode of existence; *Hydra* (p. 34) is an example. Most flowering plants are rooted and non-motile, and are radially symmetrical. Terrestrial plants in particular typically show an upright, radial structure.

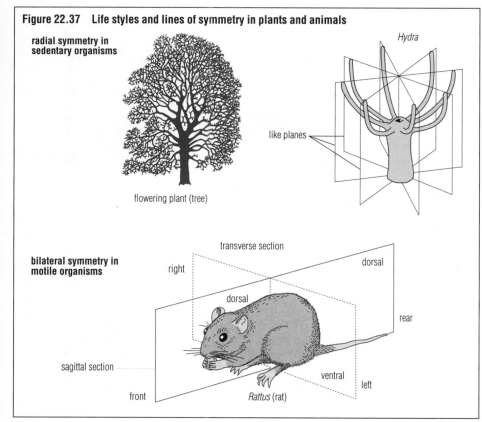

Figure 22.37   Life styles and lines of symmetry in plants and animals

radial symmetry in sedentary organisms

Hydra

flowering plant (tree)

like planes

bilateral symmetry in motile organisms

transverse section

right

dorsal

dorsal

rear

sagittal section

ventral

left

front

Rattus (rat)

Questions

1   In what ways may energy from respiration be involved in ciliary locomotion and amoeboid movements respectively?
2   What advantages are experienced by a sedentary animal from a body organisation that is radially symmetrical?
3   Explain why the forward and backward movements of a cilium in Figure 22.34 result in unidirectional movement.

FURTHER READING/WEB SITES

R McNeill Alexander (1992) *The Human Machine*. Natural History Museum Publication

23 Homeostasis and behaviour

- Homeostasis is the maintenance of a constant internal environment. When some internal condition deviates from a pre-set value, temporary corrective measures promptly return conditions to normal, by a process of negative feedback.
- In mammals, body core temperature and the composition of the blood (pH, blood sugar level, oxygen and carbon dioxide concentration, sodium ion concentration and water potential) are regulated precisely, in this way.
- Blood glucose level is regulated by hormones from the islets of Langerhans in the pancreas (insulin if blood glucose rises, glucagon if it falls). These hormones cause changes in carbohydrate metabolism in the cells of the liver and other organs, and thereby return the blood glucose level to normal.

- The liver is a key homeostatic organ, regulating the metabolism of carbohydrates, lipids and amino acids, the breakdown of haemoglobin from ageing red cells, storage of vitamins and iron, the formation of bile, and the destruction of toxins in the blood.
- Animals that regulate their own body temperature have a better chance of survival in fluctuating environments. Mammals and birds maintain a more-or-less constant body temperature by generating heat within the body and controlling heat loss.
- Behaviour in animals, the ways they respond to their environment, may be defined as instinctive (called innate behaviour) or learned, but in fact many animals display a range of these types of behaviour.

23.1 INTRODUCTION

The external environment of an organism may vary greatly, even over a brief period of time. We see examples of this when we study the effects of abiotic factors in the environments of living things. It is a characteristic of many animals that whatever the external environment, their internal condition changes remarkably little. **Homeostasis** is the name given to all the coordinated physiological processes by which an organism maintains itself in a **steady state**. The 'internal environment' is the fluid medium that bathes the cells and tissues, together with the transport medium (blood, for example) in the organism.

In fact, the word 'homeostasis' means 'staying similar'. The term is usually applied to vertebrates, and to birds and mammals in particular, for biologists know most about how the internal environment of these animals is regulated. In these 'higher' vertebrates, variables such as body temperature, blood composition (pH, blood sugar level, oxygen and carbon dioxide concentrations, water potential) and blood pressure are precisely **regulated**. That is, these variables are held nearly constant for the organism's survival. Others, such as sweating, urine formation, heart rate and breathing rate, vary quite widely but are **controlled** in order to maintain the regulated variables nearly constant.

The idea of homeostasis was first put forward by the French physiologist Claude Bernard, in 1859. He studied blood sugar concentration in various mammals that either had consumed a large meal or had been deprived of food. He found that blood sugar levels remained almost constant in his experimental animals, and established that

mammals are able to regulate the blood sugar level in the face of fluctuation in supply. Bernard surmised that, for an organism to function optimally, its cells and tissues must experience more-or-less constant physical conditions, be

surrounded by tissue fluid, and be supplied by blood of closely regulated composition. In 1929 the American physiologist Walter Cannon first used the term 'homeostasis' to describe this state of affairs (see panel).

CLAUDE BERNARD
Life and times
The parents of Claude Bernard (1813–78) were vineyard workers. From his early experiences Bernard retained an enthusiasm for the Beaujolais region of his native country and for the countryside in general, throughout a working life that kept him in a Paris laboratory.

He trained as an apothecary, and later as a doctor, but was no more than an average student. He never worked as a medical practitioner, but he did make his mark in experimental medicine.

Contributions to biology
Bernard's work centred on the physiology of mammals. Working with dogs and rabbits he established

- the presence of an enzyme in the gastric juice
- the nervous control of gastric juice secretion
- the digestion of dietary carbohydrates to sugar before absorption occurs
- that the digestion of fats involves bile and pancreatic juice
- that glycogen and sugar are interconverted in the liver
- that respiration produces heat in all body tissues.

Influences
Bernard was a 'modern' researcher, combining experimental skills with an appreciation of the theory behind his work. He was observant, noting any experimental results not in accord with existing ideas.

He is best remembered for his idea that animal life is dependent upon a constant internal environment – that cells function best in a narrow range of conditions of solute potential and temperature, and when bathed in a constant concentration of chemical constituents. His most celebrated dictum was the famous *'La fixité du milieu interior est la condition de la vie libre'*, meaning that for an organism to function optimally its component cells must be surrounded by a medium of closely regulated composition.

WALTER CANNON
Walter Cannon (1871–1945) was an American physiologist who worked at Harvard, and was professor of physiology there from 1906 until 1942.

Influences
Cannon introduced the first radio-opaque agent (he used bismuth, later barium salts) into the gut of a living animal to create a 'shadow' of the alimentary canal so that the mechanics of digestion could be observed (Figure 13.13, p. 282). This technique he devised first in 1896, the year after Röntgen discovered X-rays. The technique has been used in diagnostic radiography ever since.

Cannon developed Bernard's appreciation of the importance of the constancy of the internal environment and its regulation, to which he gave the name 'homeostasis', and he spent his life in the study of homeostatic mechanisms.

Self-regulation by servo mechanisms

In order to maintain an internal environment of constant conditions, an organism must be able to self-regulate. What does self-regulation entail?

We can start by looking at a familiar piece of laboratory equipment that does this, a constant-temperature water-bath (Figure 23.1). Human-manufactured (non-living) self-regulating systems, like the water-bath mechanism, are known as **servo mechanisms**.

The working water-bath illustrates the key points to an effective control system (Figure 23.2):

► a sensitive receptor or input device (a special thermometer) **detects** temperature change in the internal environment (heat loss or gain) and sends information to a control box

► in the control box this temperature is **compared** with the temperature set for the water-bath; if the temperature is below the **set point** the control box initiates an appropriate corrective **response**

► an effector mechanism or output device (a water heater) functions to counteract the change and return the system to normal conditions

► the thermometer detects the new temperature of the water-bath, and **feeds back** to the control box, influencing subsequent output.

As a general rule, a control system requires energy, and it involves repeated, step-by-step responses. This mode of function tends to work by 'trial and error', and it may involve some 'overshoot'. These may be features of biological systems too.

Positive and negative feedback

The analysis of how the water-bath works introduced the concept of 'feedback'. Information about the state of the system is fed back to the control centre. Feedback may be either negative or positive.

In **negative feedback** the effect of a deviation from the normal or set condition is to create a tendency to eliminate the deviation. Negative feedback is a part of almost all the control systems in living things. We shall look at several examples in the following pages. The effect of negative feedback is to reduce further corrective action of the control system once the set-point value is reached.

Figure 23.1 The laboratory water-bath, a self-regulating system

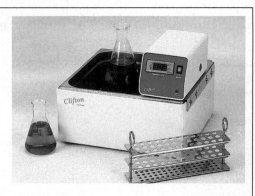

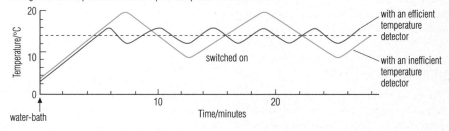

Change in water temperature when the required temperature is set to 14 °C

with an efficient temperature detector

switched on

with an inefficient temperature detector

water-bath

Figure 23.2 The principle of a control system

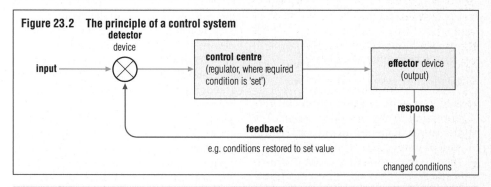

detector device

input

control centre (regulator, where required condition is 'set')

effector device (output)

response

feedback

e.g. conditions restored to set value

changed conditions

In **positive feedback** the effect of a deviation from the normal or set condition is to create a tendency to reinforce the deviation. Positive feedback intensifies the corrective action taken by a control system, so leading to a 'vicious circle' situation. Imagine a car in which the driver's seat was set on rollers (not secured to the floor!), being driven at speed. The slightest application of the foot brake causes the driver to slide and press harder on the brake as the car starts to slow, with an extreme outcome.

Biological examples of positive feedback are rare, but one can be identified at the synapse. When a wave of depolarisation (a nerve impulse) takes effect in the postsynaptic membrane (Figure 21.10, p. 448), the initial entry of sodium ions triggers the entry of further sodium ions at a greater rate. This is a case of positive feedback. The depolarised state is established, and the impulse moves along the postsynaptic neurone.

Homeostasis at population level

Homeostasis has been introduced here as a mechanism operating within the organism. But it can operate in living systems too. For example, competition for resources within an ecosystem is also a homeostatic mechanism, which regulates the size of populations (Figure 3.35, p. 65).

Gaia: homeostasis at planet level?

The scientist James Lovelock argues that the Earth with its atmosphere is 'alive', in the sense that the world shares with other living things the capacity to self-regulate (homeostasis). He calls his idea the Gaia hypothesis. You can read more about Gaia on p. 678.

Questions

1 In the body of a mammal, why is it essential that pH and blood pressure are regulated at nearly constant values (pp. 180 and 354)?

2 What are the essential differences between negative and positive feedback?

23.2 HOMEOSTASIS IN MAMMALS

Within the body of a mammal, many components of the internal environment are constantly and closely regulated. Regulation is by negative feedback control involving either the **endocrine glands** or the **nervous system**, or in some cases both. The outcome is a stable internal environment with minimum disturbance. Sudden or abrupt changes in the body's activity may disturb the concentration of metabolites, or physical conditions such as temperature or pH. Regulatory mechanisms activate the body machinery that maintains the *status quo*.

The nervous system maintains certain internal conditions within narrow limits. These conditions include

▶ the concentration (partial pressure) of the respiratory gases carbon dioxide and oxygen being transported in the blood (p. 358)
▶ the pressure of blood in the arteries (p. 352)
▶ the heart rate, although this is regulated within quite wide limits.

The endocrine system controls

▶ the osmotic concentration (solute potential) of the blood, via control of the water content and the concentration of ions (Chapter 18, p. 388)
▶ the concentrations of those essential metabolites that are present in the blood in substantial amounts, such as glucose (glucose levels are, like heart rate, regulated within quite wide limits; see below).

Body temperature is regulated by the combined actions of nervous and endocrine systems (p. 514).

■ Regulation of blood sugar levels

Glucose is a vital ingredient of the blood. The normal level of blood glucose is about $90\,mg\,100\,cm^{-3}$, varying between 70 mg (when the body has been without food for a prolonged period) and 150 mg (when a meal rich in carbohydrate has been digested and is being absorbed). Glucose diffuses into tissue fluid from the blood plasma (p. 353), and is actively (and quickly) absorbed into each cell across the plasma membrane. Mitochondria in every cell have the enzymes to respire glucose and produce ATP, the energy currency.

For most tissues glucose is the ideal food substrate for cellular energy. Glucose is the preferred fuel molecule for both cardiac and skeletal muscles. It is the only metabolic fuel molecule absorbed by the brain, for neurones are unable to store glucose as glycogen, or to respire lipid. In consequence, if the blood glucose level falls excessively the results can be disastrous: fainting, convulsions, coma and eventual death.

Responses to excess glucose

After a carbohydrate-rich meal has been digested, an excess of glucose is taken into the bloodstream at the small intestine (ileum). Blood leaving the ileum contains an excess of glucose above the set point (often referred to as the threshold level of glucose) of about $90\,mg\,100\,cm^{-3}$. From the ileum the blood goes first to the liver, via the hepatic portal vein.

In the liver the excess glucose is eventually absorbed and converted to glycogen. (The human liver can contain about 100 g of glycogen, which is sufficient to satisfy the body's needs for 24 hours.) If the rate of glucose intake from the gut exceeds the rate of oxidation or conversion to glycogen, then the glucose level in the blood may rise temporarily (**hyperglycaemia**). In the pancreas the raised glucose level triggers the secretion of the hormone insulin by the β-cells of the islets of Langerhans (Figure 23.5). Insulin circulates in the blood throughout the body and promotes the uptake of glucose by almost all its cells (although brain cells are not influenced by insulin), particularly by muscle cells, where it is stored as glycogen. The liver continues to take up glucose but converts it to fat or respires it.

When the blood glucose has returned to the threshold level the secretion of insulin is switched off (this is negative feedback). Circulating insulin is metabolised and then removed at the kidneys, and so rapidly disappears from the blood.

Responses to glucose deficiency

The glucose level in the blood falls (**hypoglycaemia**) during periods of prolonged abstinence from food and sweet drinks, particularly when associated with strenuous physical activity. Glycogen reserves of the muscles are used up in muscular contraction.

The fall in the concentration of blood glucose below the threshold value triggers the secretion of the hormone glucagon by the α-cells of the islets of Langerhans in the pancreas (Figure 23.5). As the glucose concentration falls, glucagon, together with several other hormones including adrenaline, stimulates the conversion of glycogen and amino acids to glucose not only in the liver but in cells all over the body. Glucose is thus added to the bloodstream. Severe hypoglycaemia, as we saw above, is a potentially life-threatening condition.

Muscles are virtually barred from contributing to the blood glucose pool. Once glycogen has been formed in the muscles it can only be used there, either as glycogen or as glucose, to facilitate contraction.

When the glucose concentration has returned to the threshold value the secretion of glucagon is inhibited (negative feedback).

Carbohydrate metabolism and glucose regulation are summarised in Figures 23.3 and 23.4.

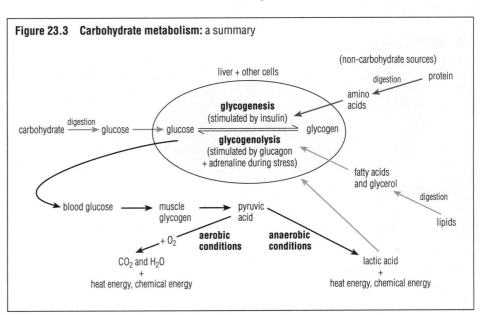

Figure 23.3 Carbohydrate metabolism: a summary

Figure 23.4   The sites and the processes of glucose regulation

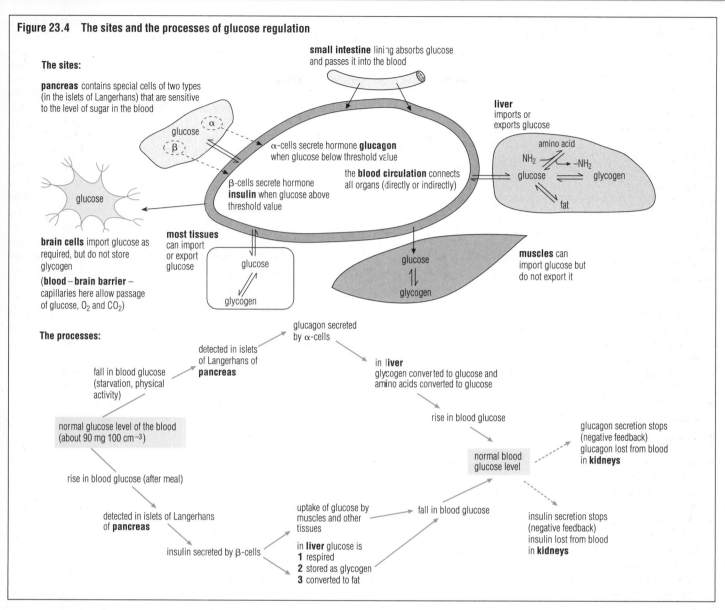

The sites:

pancreas contains special cells of two types (in the islets of Langerhans) that are sensitive to the level of sugar in the blood

small intestine lining absorbs glucose and passes it into the blood

liver imports or exports glucose

amino acid
$NH_2 \longrightarrow -NH_2$
glucose ⇌ glycogen
fat

α-cells secrete hormone **glucagon** when glucose below threshold value

β-cells secrete hormone **insulin** when glucose above threshold value

the **blood circulation** connects all organs (directly or indirectly)

glucose

α
β

brain cells import glucose as required, but do not store glycogen

(**blood – brain barrier** – capillaries here allow passage of glucose, O_2 and CO_2)

most tissues can import or export glucose

glucose ⇌ glycogen

glucose
glucose ⇌ glycogen

muscles can import glucose but do not export it

The processes:

glucagon secreted by α-cells

detected in islets of Langerhans of **pancreas**

in **liver** glycogen converted to glucose and amino acids converted to glucose

fall in blood glucose (starvation, physical activity)

rise in blood glucose

normal glucose level of the blood (about 90 mg 100 cm^{-3})

glucagon secretion stops (negative feedback) glucagon lost from blood in **kidneys**

normal blood glucose level

rise in blood glucose (after meal)

detected in islets of Langerhans of **pancreas**

uptake of glucose by muscles and other tissues

fall in blood glucose

insulin secretion stops (negative feedback) insulin lost from blood in **kidneys**

insulin secreted by β-cells

in **liver** glucose is
1 respired
2 stored as glycogen
3 converted to fat

Figure 23.5   The pancreas and hormone production

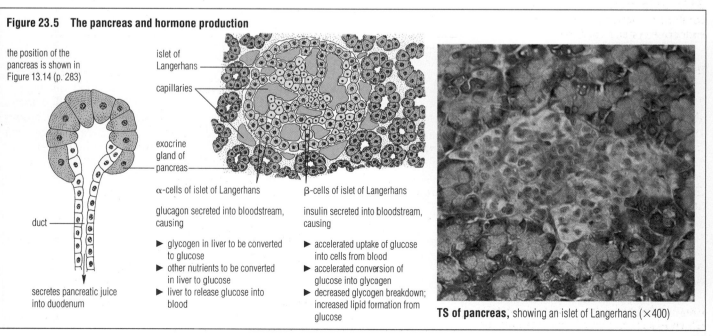

the position of the pancreas is shown in Figure 13.14 (p. 283)

islet of Langerhans

capillaries

exocrine gland of pancreas

duct

secretes pancreatic juice into duodenum

α-cells of islet of Langerhans

glucagon secreted into bloodstream, causing

► glycogen in liver to be converted to glucose
► other nutrients to be converted in liver to glucose
► liver to release glucose into blood

β-cells of islet of Langerhans

insulin secreted into bloodstream, causing

► accelerated uptake of glucose into cells from blood
► accelerated conversion of glucose into glycogen
► decreased glycogen breakdown; increased lipid formation from glucose

TS of pancreas, showing an islet of Langerhans (×400)

Hormones, the cell membrane, and glucose regulation

Insulin, glucagon and adrenaline are all small protein hormones that have a direct effect on glucose movement and metabolism. These hormones cannot cross the cell membrane (Figure 21.46, p. 473) and instead, have their effects by binding to specific receptors on the outside of cell membranes (liver cells, muscle fibres and other cells). We will look at their actions in turn (Figure 23.6).

Insulin, once it has bound to its receptor, makes cells more permeable to glucose by increasing the number of glucose-pump protein molecules active in the plasma membrane. It also enhances the rate at which glucose is respired in cells (in preference to other respiratory substrates, like fatty acids). In addition, it activates the main enzyme responsible for conversion of glucose to glycogen (glycogen synthetase). The outcome is that quite quickly after the intake of sugar in the ileum (Figure 13.18, p. 286) the blood sugar concentration falls to the threshold value again.

Glucagon, once it has bound to its receptor, stimulates the enzyme responsible for the breakdown of glycogen to glucose (glycogen phosphorylase). It also stimulates the conversion of lipids and amino acids to glucose in the liver. The liver quickly becomes a net exporter of glucose, and the blood glucose level starts to rise.

Adrenaline, on binding to its receptor, activates an enzyme immediately inside the cell called adenyl cyclase, which catalyses the production of cyclic AMP. This is a molecule of adenosine monophosphate (p. 176) where the phosphate is bonded to two points on the molecule of adenosine. Cyclic AMP, the 'second messenger' (the arrival of adrenaline on to the receptor is the first) activates another enzyme, which itself activates a third enzyme, and so on in a chain of reactions. Since each enzyme activates many others, this type of action is known as a 'cascade effect', which amplifies the original message. Adrenaline has many different effects in the body (p. 475), but its effect on liver cells is to stimulate them to break down glycogen to glucose and release it into the bloodstream.

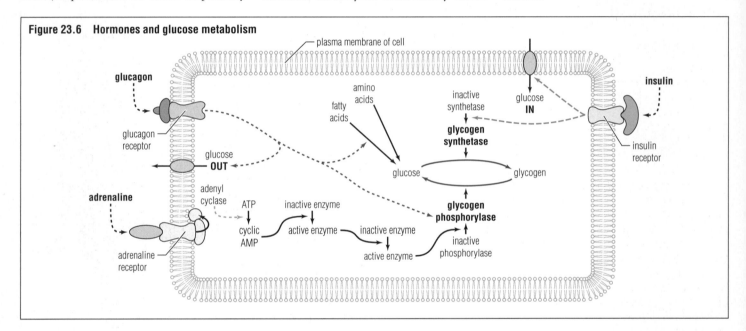

Figure 23.6   Hormones and glucose metabolism

Diabetes mellitus

Diabetes is not a single disease but a group of diseases, all manifesting as an elevation of glucose in the blood (hyperglycaemia) (Figure 23.7), excretion of glucose in the urine, increased urine production, excessive thirst and a tendency for excessive eating.

Type I diabetes ('early onset diabetes') develops abruptly, due to a sudden decline in the number of β-cells in the islets of Langerhans (caused by autoimmune destruction, p. 371). It appears most commonly in young people below the age of 20. People suffering from type I diabetes have the disease for the rest of their lives. Insulin has to be injected regularly, the amount matched to the amount of carbohydrate eaten and the amount of exercise taken. Type I diabetics often measure the concentration of glucose in their blood regularly throughout the day, using a miniaturised glucose meter, in order to match their insulin injections to body requirements.

Type II diabetes ('late onset diabetes') represents 90% of all cases of diabetes, and is common in people over 40, especially those who are overweight. Symptoms are mild, and the higher glucose levels in the blood can be corrected by diet alone. Type II diabetes occurs either because insulin secretions become deficient, or because the target cells fail to respond to the normal concentration of insulin.

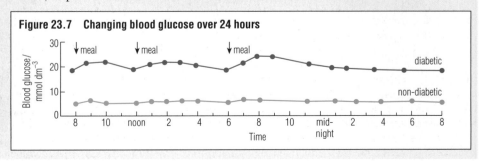

Figure 23.7   Changing blood glucose over 24 hours

23.3 THE LIVER, A HOMEOSTATIC ORGAN

The liver is a large organ located in the abdomen immediately below the diaphragm. It overlaps the stomach, and is attached to the diaphragm by a ligament. The gall bladder, which stores bile secreted by the liver, lies just below it, and is connected to the duodenum by the bile duct.

■ The anatomy of the liver

The liver consists of thousands of tiny polygonal blocks of cells, called lobules, each about 1 mm in diameter. Lobules are formed from radiating rows of vertical cords of liver cells, sometimes referred to as **acini** (singular, acinus). These cords are separated by blood spaces, the **sinusoids** (Figure 23.8). The sinusoids are open channels through which blood flows in direct contact with the liver cells, from the periphery of the lobule towards the centre. Situated around the periphery of the lobules are branches of the hepatic artery and the hepatic portal vein (pp. 287 and 351). Blood coming into the liver from these two vessels is mixed as it flows into the sinusoids. Thus liver cells receive a double supply of blood: oxygenated blood from the hepatic artery, and blood containing newly absorbed nutrients from the gut. At the centre of the lobule is a branch of the hepatic vein into which the blood from the liver drains.

Also present in the lobules are **bile canaliculi**. Each canaliculus forms as a blind tube between two rows of liver cells. Canaliculi merge to form bile ducts in which bile is drained from the liver.

The sinusoids are lined by numerous phagocytic cells known as **Kupffer cells** (these are also found in the spleen and bone marrow). These star-shaped (stellate) cells are part of the body's defence system (Figure 17.39, p. 364).

The liver cells (hepatocytes) are structurally undifferentiated, and are all identical. The cytoplasm contains mitochondria, and a prominent Golgi apparatus. Lysosomes, glycogen granules and fat droplets are common. The area of the surfaces of the liver cells facing the blood is increased by multiple infolding of the cell membrane (microvilli). As the blood flows past, liver cells extract oxygen and food substances, as well as various metabolites and any poisons present; they also secrete into the blood products manufactured by the liver cells for use elsewhere in the body.

Figure 23.8 The structure of the liver lobule

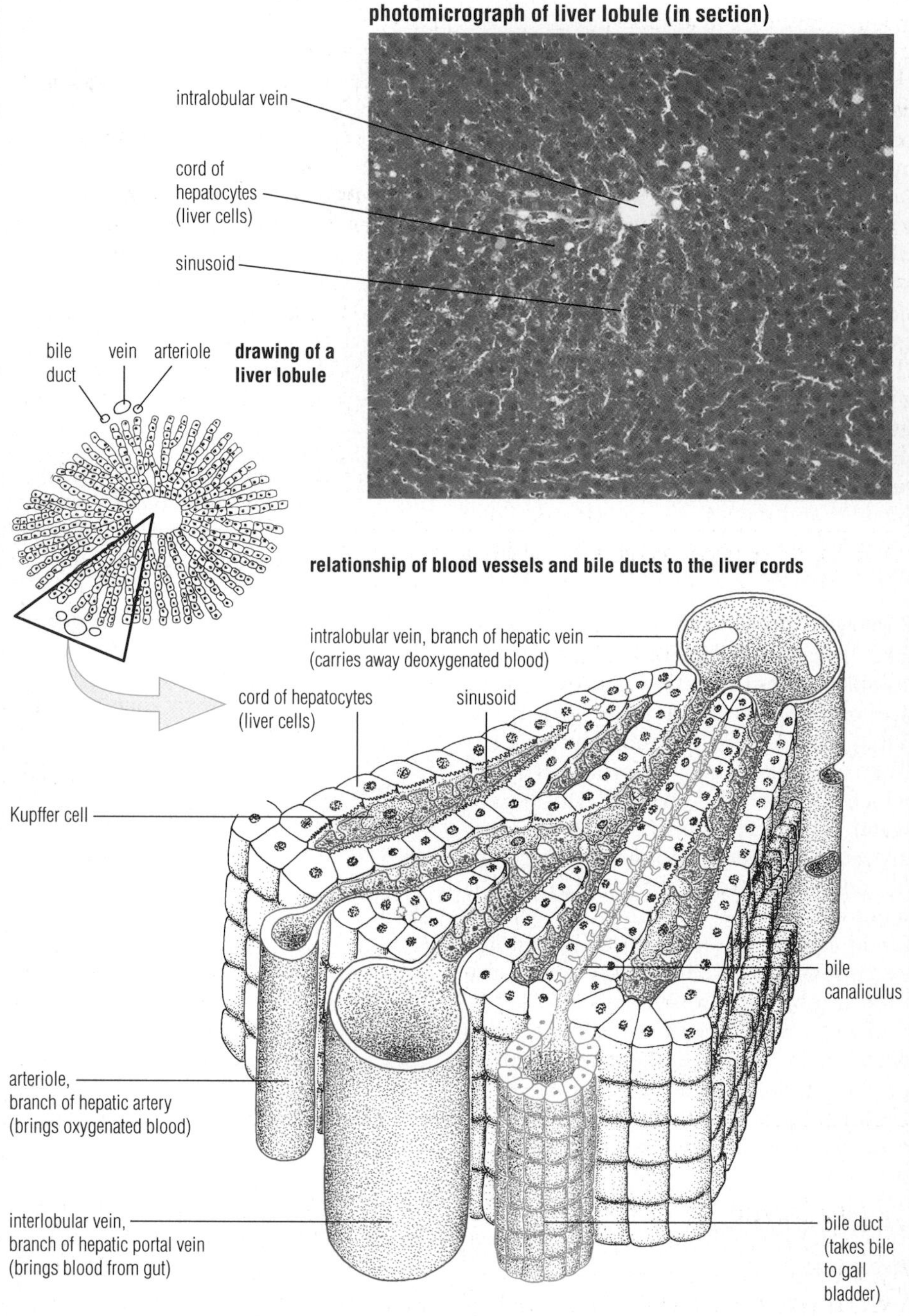

photomicrograph of liver lobule (in section)

intralobular vein

cord of hepatocytes (liver cells)

sinusoid

bile duct   vein   arteriole   **drawing of a liver lobule**

relationship of blood vessels and bile ducts to the liver cords

intralobular vein, branch of hepatic vein (carries away deoxygenated blood)

cord of hepatocytes (liver cells)   sinusoid

Kupffer cell

bile canaliculus

arteriole, branch of hepatic artery (brings oxygenated blood)

interlobular vein, branch of hepatic portal vein (brings blood from gut)

bile duct (takes bile to gall bladder)

■ The role of the liver

The many functions of the liver, all of which contribute directly to homeostasis, can be conveniently divided into three categories.

1 Production of bile

Bile is a yellow-green, alkaline, mucous fluid containing the **bile pigments**, which are excretory products formed from the breakdown of haemoglobin from worn-out red blood cells (these cells have a lifespan limited to about 120 days). Red cells that have reached the end of their useful lives are engulfed by the phagocytic cells (Kupffer cells). Their haemoglobin (Figure 17.28, p. 358) is broken down. The protein part, the globin, is converted to its constituent amino acids and added to the pool of amino acids in the bloodstream. They are then available for the synthesis of new protein. Iron is removed from the haem part of the molecule, and stored (see overleaf). The rest of the haem residue, the system of pyrrole rings, is converted to the bile pigments biliverdin (green) and bilirubin (red). (These are the same pigments that can be seen under the skin after a heavy blow has released blood internally, creating the green-yellow colours of a bruise.)

Bile also contains the **bile salts** (sodium glycocholate and sodium tauro-cholate), which are reclaimed by liver cells from the blood. Bile salts are used again and again in lipid absorption. They are secreted into the bile along with cholesterol and phospholipids. This cocktail of metabolites is essential for making soluble fat droplets in the duodenum and ileum, part-digesting them and facilitating their absorption.

Bile is secreted continuously, and up to a litre is formed each day. In the gall bladder, bile is concentrated (water is withdrawn). Bile is released intermittently by muscular contraction of the gall bladder and passes via the bile duct into the duodenum, under the control of hormones of digestion (p. 288).

2 Storage of vitamins and iron

The liver is the site of storage of fat-soluble vitamins (A, D, E and K), which are retained in quite large quantities. Vitamin A is also manufactured in the liver from the plant pigment carotene, which is absorbed in the diet. Most of the water-soluble vitamins of the B complex are retained in smaller quantities; vitamin B_{12} and folic acid (Table 13.2, p. 272), required for the manufacture of red blood cells in the marrow of bones, are exceptions, both being retained in the liver in substantial amounts.

The liver is the one site in the body where iron is stored, held attached to protein until required for manufacture of fresh haemoglobin. Potassium and many trace elements are also stored as ions in the liver.

3 Metabolite conversions

Carbohydrates

Glucose in excess of the body's immediate needs is converted to glycogen in the liver and stored. The liver contains the main store of glycogen, although muscles also store some. The conversion of glucose to glycogen (**glycogenesis**) is mediated by specific enzymes and takes place via intermediates (glucose phosphates), in the presence of insulin (Figure 23.6, p. 506).

Once the liver's capacity to store glycogen is reached any further glucose absorbed in the ileum is converted to lipid by the liver cells and despatched to other body cells, or stored under the skin.

When blood sugar levels start to fall the enzymic conversion of glycogen to

glucose (**glycogenolysis**) is triggered in the presence of the hormone glucagon. Under the 'flight or fight' response, the hormone adrenaline triggers the same conversion.

If the glycogen store in the liver is used up (in fasting or starvation, for instance) then the liver is able to convert non-carbohydrate sources to glucose to meet the body's needs. The chief source in this case is the amino acid pool of the blood. In the long term, the effect on the body is the depletion of proteins, causing the characteristic wasted appearance of a starving adult human.

Lipids

The products of lipid digestion are absorbed in the ileum (p. 286) and transported to the liver with the aid of bile salts. Liver cells are involved in the chemical changes to these lipids that are necessary for their utilisation, or their mobilisation and deposition around the body. For example, in the liver some fatty acids are converted into cholesterol (for membranes and the manufacture of steroid hormones) via the respiratory intermediate acetyl coenzyme A (p. 147) and others are converted into phospholipids (for membranes). Lipids are insoluble in water and have to be transported in association with proteins in components known as very low density lipoproteins (VLDL), low density lipoproteins (LDL) or high density lipoproteins (HDL), according to the proportions of proteins and lipids. In the body, some lipids are used as respiratory substrates (for example, fatty acids by the muscles), whilst others are stored as adipose tissues, laid down in connective tissue around the body organs.

Amino acids and proteins

The liver is the site of the **deamination** of excess amino acids (Figure 23.9). In

deamination the amino group (—NH$_2$) is removed and the remaining part of the molecule (a keto acid) is respired, or converted to glycogen. The amino group may be used either in the synthesis of organic bases (guanine, cytosine, thymine, uracil) used in nucleotides (p. 147), or added to a different keto acid in transamination (p. 209), or excreted in the form of urea. (Ammonia is poisonous, p. 377.) The cycle of changes by which urea is formed is called the ornithine cycle (Figure 18.6, p. 377).

In addition, the liver is the site of the **synthesis of proteins** of the blood plasma (blood-clotting agents prothrombin and fibrin, and plasma albumin and globulin, which includes the antibodies, p. 366).

Detoxification

Bacteria that enter the bloodstream via the gut are removed (engulfed) by the Kupffer cells. Drugs, toxins and poisons taken in with the diet, or produced by bacteria somewhere in the body and transported to the liver in the bloodstream, are biochemically broken down and rendered harmless. Some toxins and drugs can themselves harm the liver; for example, long-term excessive intake of alcohol causes cirrhosis of the liver in some people (see opposite).

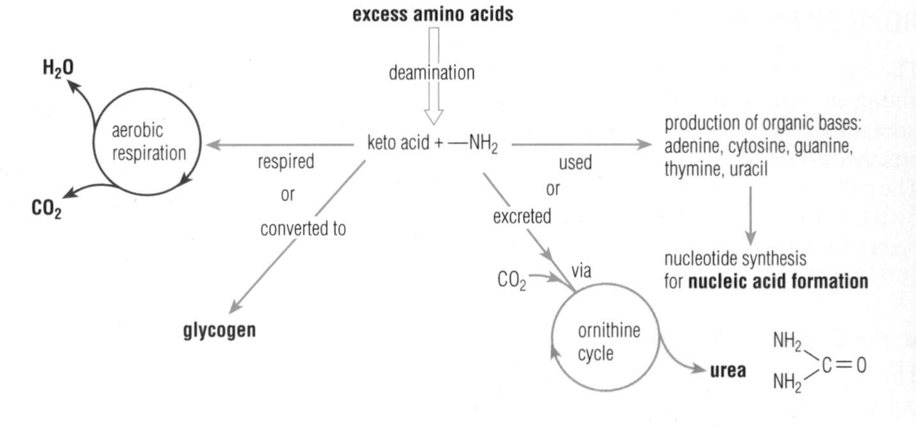

Questions

1 What do liver cells require from the blood in the hepatic artery that they do not get from blood from the portal vein?

2 Name the difference between the following three components of a liver lobule: a sinusoid, an acinus and a canaliculus.

3 **a** What do we mean by a 'phagocytic' cell?
  b What is the role of phagocytic cells in the liver?

Figure 23.9   The metabolism of excess amino acids: a summary

Liver and heat production

The liver is a very active organ metabolically, and most of the metabolic reactions occurring in the liver absorb heat (endothermic reactions). Claude Bernard (p. 502) published data showing the liver as a major exporter of heat in a mammal's body. But his values for the temperature of blood entering and leaving the liver were obtained at different times from different dogs.

Subsequent investigations, involving thermocouples placed in the hepatic portal vein and the hepatic vein, show that the liver is thermally neutral under normal climatic conditions. Nevertheless this misunderstanding about the liver as a heat-exporting organ has persisted.

Hepatitis

Hepatitis is a condition of inflammation of the liver, which may be caused by viruses; viral hepatitis A is a relatively mild disease but hepatitis B can cause chronic liver inflammation that may persist for years.

Cirrhosis

Cirrhosis is a condition in which the liver tissue is scarred. Normal liver cells are replaced by fibrous or adipose connective tissue. This leads to symptoms such as jaundice, and increased sensitivity to drugs. Cirrhosis may be caused by hepatitis, by liver parasites, and by certain chemicals and alcoholism.

Jaundice

If the liver is unable to remove bilirubin from the blood, large amounts of the bile pigment circulate in the blood and collect in the skin and in the sclera of the eye, producing a strong yellow colour.

Many newborn babies develop neonatal jaundice, since at this stage the liver functions rather ineffectively at bile production for some days.

Questions

1 Ammonia formed by deamination of amino acids is immediately combined with other substances, forming urea. Why is this conversion essential to the health of the liver cells?

2 What happens to blood insulin in the kidney tubule (p. 472)?

3 What does it mean to say that most reactions in the liver are endothermic?

23.4 TEMPERATURE REGULATION

Organisms require a certain temperature range for the optimum functioning of their enzymes. Consequently, temperature is an important environmental abiotic factor (p. 58). Note that the temperature fluctuations in water (aquatic environments) are much smaller than those that occur on land (Figure 23.10). Most living things adapted to life in terrestrial or in aquatic habitats tolerate a narrower temperature range than normally occurs in these habitats. However, some prokaryotes are adapted to the extreme cold experienced at the poles, and are able to complete their life cycle there. Others are able to live successfully in the hot springs found in certain parts of the world, both on land and in the ocean depths. Organisms able to thrive in extreme conditions are collectively called extremophiles.

■ How heat is gained and lost
Heat is transferred by

► **radiation** – the transfer of heat energy as (invisible) infra-red waves from a hotter to a cooler object
► **convection** – the transfer of heat energy by the upward movement of warm air or water (and the downward movement of cold air or water), transferring heat energy between warmer and cooler regions
► **conduction** – the transfer of heat from a hotter to a colder body in contact with it
► **evaporation** – the process by which liquid changes to vapour, accompanied by cooling (p. 126).

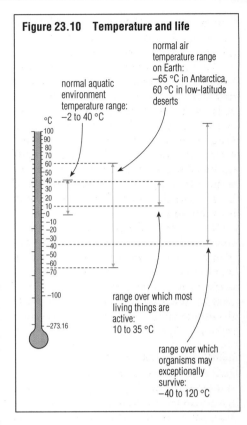

Figure 23.10 Temperature and life

normal air temperature range on Earth: −65 °C in Antarctica, 60 °C in low-latitude deserts

normal aquatic environment temperature range: −2 to 40 °C

°C

range over which most living things are active: 10 to 35 °C

range over which organisms may exceptionally survive: −40 to 120 °C

The **sources of heat** available to organisms are the Sun or the by-product of the reactions of metabolism such as cellular respiration and (in many animals) muscular contraction. Heat gain by an organism from the environment occurs by radiation, convection and conduction. Metabolic heat is distributed by the blood circulation.

Heat is lost by radiation, convection and conduction, and by evaporation of water vapour from the body surface (Figure 23.11).

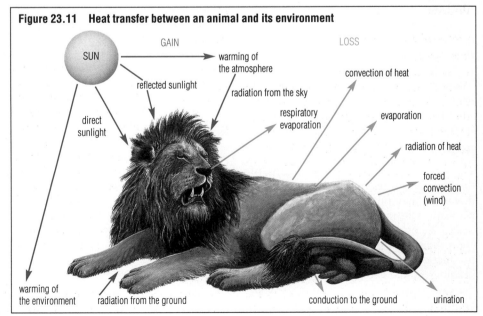

Figure 23.11 Heat transfer between an animal and its environment

GAIN LOSS

SUN

warming of the atmosphere

reflected sunlight

radiation from the sky

direct sunlight

respiratory evaporation

convection of heat

evaporation

radiation of heat

forced convection (wind)

warming of the environment radiation from the ground conduction to the ground urination

■ Temperature regulation in animals

Animals were once described as either 'cold-blooded' (**poikilothermic**), meaning that the body temperature fluctuates according to the temperature of the environment, or as 'warm-blooded' (**homoiothermic**), meaning that the body is maintained at a constant temperature. The evidence for this idea came from the work of Martin (1930), in which four contrasting vertebrates were exposed to ambient temperatures between 5 and 40 °C for a two-hour period, after which their body temperature (the 'core' temperature, taken in the rectum) was measured. The animals were contained in a restricted space, exposed to a lamp as heat source, and unable to move into shade. The results (Figure 23.12) suggested a fundamentally different response to external temperature by reptiles and mammals. The temperature of the reptile can be seen to be dictated by the environment, while that of the mammals is regulated within the animal's body.

The regulation of body temperature is more subtle than Martin's experiment suggests, however, and the terms 'cold-blooded' and 'warm-blooded' have been discarded.

'Ectotherms' and 'endotherms'

In nature, reptiles are capable of warming up to a good 'working' temperature by basking in the sun, and thereafter maintaining a constant high body temperature through the heat of the day by moving back and forth between sun and shade (behavioural thermoregulation, Figure 23.13). In the artificial experimental environment selected by Martin the possibility of behavioural thermoregulation was excluded.

Since the body temperatures of poikilotherms may, at times, be as high as those of homoiotherms, a more revealing distinction is to use the terms '**ectotherms**' for organisms obtaining heat from external sources, and '**endotherms**' for organisms that generate their body heat metabolically.

Metabolism: the heat source for endotherms

The basal metabolism of the organs of the body generates heat. Most birds and mammals maintain their core body temperature within the narrow range of 36–43 °C by using metabolic heat and controlling its loss, or by increasing heat production by muscles when the body becomes cold (Table 23.1). (Note that the organs making up less than 10% of the mass of the body produce more than

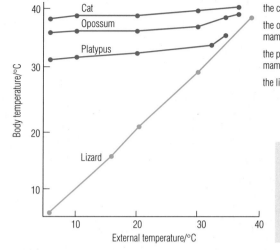

Figure 23.12   Variation of body temperature in a reptile and three mammals in response to environmental temperatures between 5 and 40 °C

the cat is a placental (eutherian) mammal

the opossum is a pouched (marsupial) mammal

the platypus is an egg-laying (monotreme) mammal

the lizard is a reptile

In fact, this experiment makes a false comparison since the conditions imposed did not interfere with the mechanism of temperature regulation adopted by mammals, but did so for the reptile (see below).

70% of the heat produced in the body at rest.) In the human, body temperature is normally held between 37 and 38 °C, but is not completely constant. Human body temperature does fluctuate, both on a regular basis through the 24-hour cycle, and erratically, as a result of exercise, at ovulation in the human female and following intake of food that is particularly hot or cold (Figure 23.14).

Table 23.1   Heat production in the human body at rest

Organ	Organ mass / % of body mass		Heat production at rest / % of total	
kidneys	0.45		7.7	
heart	0.45		10.7	
lungs	0.9		4.4	
brain	2.1	} 7.7	16.0	} 72.4
abdominal organs, not including kidneys	3.8		33.6	
skin	7.8		1.9	
muscle	41.5	} 92.3	15.7	} 27.6
other	43.0		10.0	
Total	100.0		100.0	

Figure 23.13   Behavioural thermoregulation in a reptile: the marine iguana of the Galapagos Islands hunts for food in cold sea water. First, however, it basks in hot sunshine on land, and warms its whole body. In short expeditions into sea water, heat is retained in inner organs and muscles of the iguana by vasoconstriction of the skin capillaries, and by lowering the heart rate

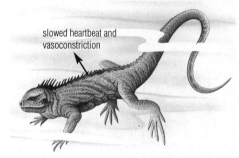

slowed heartbeat and vasoconstriction

Figure 23.14   Human body temperatures: taken in the mouth over a two day period

Figure 23.15 The structure of mammalian skin

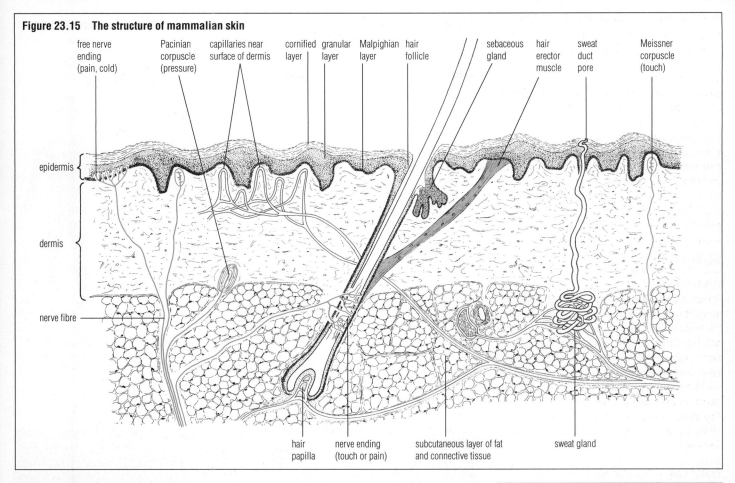

Temperature regulation and the skin

Skin structure

The body's regulation of heat loss is mediated through the skin. The structure of mammalian skin is shown in Figure 23.15.

A thin outer layer, the **epidermis**, consists of stratified epithelium (p. 213). The cells in the basal **Malpighian layer** of the epidermis constantly divide, pushing the cells above towards the skin surface. These cells progressively become flattened, and the contents are converted to the tough, fibrous protein keratin. The outermost cells are constantly rubbed off, by friction with the surroundings. Where friction is especially high, as it is on the ball and heel of the foot, the cornified layer of the epidermis is heavily thickened. In parts of the body, the keratinised layer of the skin has become modified as nails, claws, hooves, scales or horns, according to the species.

The **dermis**, below the epidermis, is a much thicker layer, consisting of elastic connective tissue containing blood capillaries, lymph vessels, hair follicles, muscle fibres and sweat glands.

Skin coloration is due to melanin, a dark brown pigment produced by specialised cells of the epidermis, called **melanocytes**, which are usually found between the cells of the basal layer and the upper layer of the epidermis.

Hairs are formed in invaginations of the dermis, the **hair follicles**. Near the base of each follicle an **erector pili** muscle is inserted, having its origin on the base (basement membrane) of the epidermis. Hairs are covered by a film of oil from the **sebaceous glands**, which keeps them supple and waterproof.

The **sweat glands** are coiled tubular exocrine glands connected to the surface of the skin by ducts. In human skin, sweat glands are common; in hairy mammals they are rare. The sweat produced is a weak solution of sodium chloride. When sweat passes to the skin epidermis, its evaporation cools the skin. Thus, blood in the skin is cooled.

A second type of sweat gland, found in the armpit and groin, secretes a different liquid, which contains pheromones (p. 477), and which smells unpleasant when decomposed by certain skin bacteria.

Blood capillaries occur in abundance in the dermis. Some vessels, called **shunts**, allow blood to fill or by-pass certain capillary beds and so help to regulate skin temperature.

THE MAIN FUNCTIONS OF THE SKIN

Skin functions may be summarised as follows:

▶ a physical barrier, preventing excessive water loss, physical abrasion of underlying tissues, damage by ultra-violet light, and the entry of foreign matter, including bacteria and viruses

▶ perception of external stimuli, since the skin is sensitive to temperature, touch and pressure (p. 450)

▶ synthesis of vitamin D by the action of sunlight on a precursor molecule (p. 272)

▶ temperature regulation, as detailed on these pages.

Questions

1 In a hot climate, reptiles can maintain a constant body temperature in the day by behavioural thermoregulation. What are the advantages and disadvantages of this mechanism compared with thermoregulation in mammals?

2 Above and below what limits would you judge your body temperature to be abnormal?

3 What examples of behavioural thermoregulation are commonly shown by mammals with which you are familiar?

4 In what ways is a stratified epithelium different from the majority of epithelia found in the body (p. 213)?

Figure 23.16   Responses of mammals to changes in ambient temperature

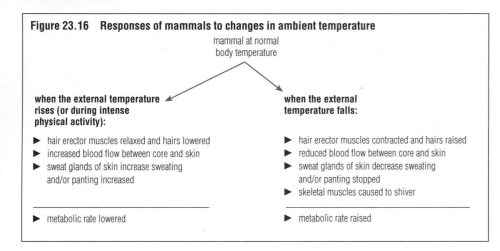

mammal at normal body temperature

when the external temperature rises (or during intense physical activity):

▶ hair erector muscles relaxed and hairs lowered
▶ increased blood flow between core and skin
▶ sweat glands of skin increase sweating and/or panting increased

▶ metabolic rate lowered

when the external temperature falls:

▶ hair erector muscles contracted and hairs raised
▶ reduced blood flow between core and skin
▶ sweat glands of skin decrease sweating and/or panting stopped
▶ skeletal muscles caused to shiver

▶ metabolic rate raised

Figure 23.17   Temperature distribution in the human body in cold and in heat

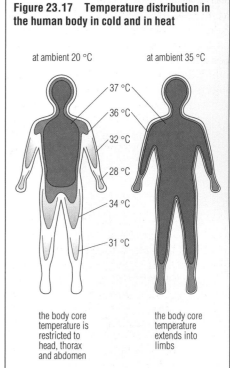

at ambient 20 °C   at ambient 35 °C

37 °C
36 °C
32 °C
28 °C
34 °C
31 °C

the body core temperature is restricted to head, thorax and abdomen

the body core temperature extends into limbs

Mechanisms to combat cold

Hairs are raised

Contraction of the erector pili muscles (under control of the autonomic nervous system, p. 462) raises the hairs to an almost vertical position (Figure 23.16). In a furry animal, the hairs in this position can trap a thick layer of stationary air next to the skin. Air is a relatively poor conductor of heat, so heat loss from the skin is reduced.

In humans, hair is much reduced, but in cold weather you can observe your hairs functioning in just the same way as those of a furry animal. The most obvious effect, however, is the 'goose pimples' caused by contraction of the erector pili muscles. (In birds the same insulating function is performed by fluffing up the feathers, the positions of which are also controlled by erector pili muscles (Figure 23.18).)

Reducing blood flow between core and skin

The blood transports heat around the body in the circulatory system by the process of being warmed in organs at higher temperatures, and warming the tissues that are at lower temperatures.

Narrowing the arterioles supplying the skin capillaries (**vasoconstriction**) can reduce heat flow to the skin and heat loss from the body (Figure 23.19). This cooling is more marked in the limbs than in the rest of the body (Figure 23.17).

Vasoconstriction is controlled by sympathetic nerves from the vasomotor centre in the brain under commands from the thermoregulatory centre in the hypothalamus.

Figure 23.18   Air trapped in feathers forms an insulating layer: robin in summer (left); robin in winter (right)

Figure 23.19   The role of skin capillaries in temperature regulation

arterio-venous by-pass closed

arteriole dilated

blood flow in capillaries

much heat lost

arterial blood gives up most of its heat to the vein before entering fingers

arterio-venous by-pass open

arteriole constricted

negligible heat loss

little or no blood flow in capillaries

Subcutaneous fat accumulation

Mammals store fat in adipose tissue below the skin surface. This tissue, with its limited blood supply, is not a good conductor of heat. Aquatic mammals inhabiting cold waters (such as seals, Figure 17.38, p. 363) have a thick layer of fat (blubber) under the skin. Terrestrial animals that remain active through the cooler season accumulate fat as part of their survival mechanisms, principally as a food reserve.

Brown fat

In mammals there are two types of adipose tissue (p. 138), white and brown. 'Brown fat' cells contain a great many mitochondria, and brown adipose tissue is well supplied by blood capillaries. It occurs under the skin of the upper back of many mammals (especially in young mammals), between the shoulder blades. Brown adipose tissue produces considerable amounts of heat, because there lipids are respired with little or no ATP formation; most energy is released as heat. This heat is transported in the blood to the body's organs. Adult humans have little or no brown fat, but it is present in newborn babies. Heat generation (**thermogenesis**) in brown fat is triggered by the activity of the sympathetic nervous system.

Other mechanisms

In chilly conditions, muscle tone tends to rise (initially this is under nervous control). If the cold threat continues, rhythmic involuntary muscle contractions (or shivering) begin, generating heat in the muscles. Under persistent cooler conditions the basal metabolic rate is raised (in the short term by the secretion of the hormone adrenaline and in the longer term by the secretion of thyroxine, p. 474), thus further increasing the amount of heat generated by the body.

Behavioural mechanisms to combat cold include moving into a warmer place (into the sun, for instance) and huddling close together with other individuals; humans also put on clothes, and engage in vigorous activity.

Hypothermia

In **hypothermia,** the body temperature of an endotherm falls well below normal. It may occur if heat is lost from the body faster than it is produced. In humans,

Figure 23.20 Panting accelerates heat loss

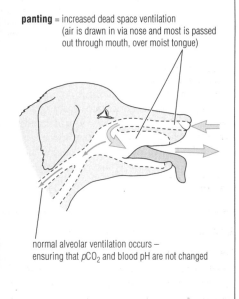

panting = increased dead space ventilation (air is drawn in via nose and most is passed out through mouth, over moist tongue)

normal alveolar ventilation occurs – ensuring that pCO_2 and blood pH are not changed

those most vulnerable are babies (large surface area:volume ratio) and the very old (rare failure of normal thermoregulatory mechanisms with age). The brain is rapidly affected (reducing the ability to take remedial actions), and eventually the lowered body temperature lowers metabolic rate (positive feedback). Death is likely if the human body temperature falls below 25 °C.

Mechanisms to combat overheating

Hairs are lowered

In warm conditions, the erector pili muscles relax and the hairs lie more or less smoothly against the skin (Figure 23.16), so reducing the amount of trapped air and thus the insulating effect of the hair or fur.

Increasing blood flow between core and skin

Dilation of the arterioles supplying blood to the skin capillaries (**vasodilation**) allows greater heat flow to the skin and thus increased heat loss from the body surface (Figure 23.19). Vasodilation, like vasoconstriction, is controlled by the sympathetic nervous system.

In addition, some large animals use their external ears as 'radiators' of heat. African elephants, for example, have much larger ears than elephant species from cooler climates. They flap their ears frequently in hot weather, and the blood flow to the ears is increased.

Smaller animals may also use the skin of the outer ear for heat loss. Evidence for this is seen in North American hares, in which there is good correlation between external ear size and the average temperature experienced by different species (Figure 23.22, overleaf).

Subcutaneous fat reduction

The proportion of subcutaneous fat in the body may be reduced over a period of time if hot conditions persist.

Other mechanisms

Behavioural mechanisms are employed, including moving into the shade, decreasing physical activity (thus lowering metabolic rate), keeping a good distance from other warm animals, and the licking of fur or immersion in water to increase evaporation. The practice of panting – an efficient method of losing heat from the mouth, tongue and breathing passages – is common in animals with restricted numbers of sweat glands (Figure 23.20). Humans wear fewer clothes and, having plentiful sweat glands, can also lose heat through the evaporation of sweat from the skin. The effectiveness of sweating in cooling the body depends on the temperature, air movement (wind) and the relative humidity of the air. Higher temperatures and moving air enhance evaporation, but if the relative humidity is 100%, then no evaporation occurs. This is why extremely humid climates are less tolerable than high temperature conditions with dry air.

Questions

1 What roles may hairs perform, other than the trapping of still air?

2 What evidence do we have that structures such as outer ears function as simple 'radiators' in some mammals (p. 514)?

The role of the hypothalamus

The thermoregulation centre in the hypothalamus in the forebrain acts as a thermostat, switching on and off heat loss mechanisms or heat conservation mechanisms, and stimulating additional heat gain if appropriate (Figure 23.21). The thermoregulation centre consists of a 'cold' and a 'hot' centre containing heat-sensitive neurones. The hypothalamus receives impulses from thermoreceptors of both the skin and the deep body tissues, and it senses the temperature of blood flowing through the brain. When tissue temperatures are lower than normal, the 'cold' centre triggers responses that increase heat production, decrease heat loss and inhibit activity by the 'hot' centre. When tissue temperatures are higher than normal the 'hot' centre triggers responses that decrease heat production, increase heat loss, and inhibit the activity of the 'cold' centre.

Structural adaptations in temperature regulation

We have seen that the heat that endotherms produce comes from various body organs (Table 23.1, p. 510). The total heat produced from this source depends upon the volume of the body. By contrast, the rate of heat loss depends upon the surface area. Now, as the size of animals increases the volume increases more rapidly than the surface area (Figure 15.1, p. 302). This is the explanation of the observation that the animals living in the cold regions of the world tend to be large animals, such as the polar bear (Figure 23.23). Smaller animals in cold regions need a high metabolic rate, and consequently require a regular and substantial food supply to survive.

Animals of hot regions typically have external ears richly supplied with blood capillaries, and carrying little fur (hair). The external ear functions as a radiator. A comparison of ear size in hares and rabbits with natural habitats at various latitudes on the North American continent (Figure 23.22) appears to support this. Very large animals in hot regions, such as elephants and hippopotamus, fan their ears in the heat, speeding up heat loss by radiation and convection.

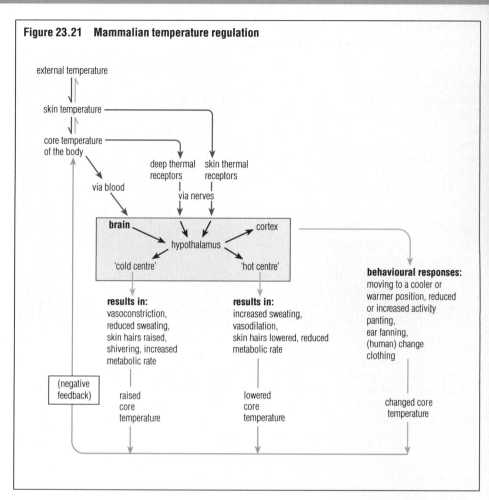

Figure 23.21   Mammalian temperature regulation

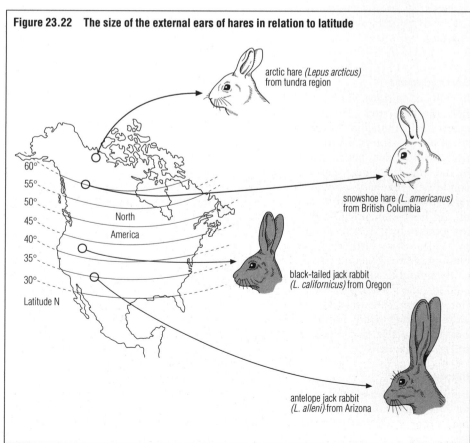

Figure 23.22   The size of the external ears of hares in relation to latitude

Figure 23.23   The polar bear in its natural habitat, avoiding frost bite by means of a countercurrent heat exchange mechanism

In the legs, the main artery and vein run parallel and very close together. All along the length of the leg heat is exchanged between the artery and vein.

The foot tissues are supplied with nutrients, and the tissue is held at a temperature that
1 prevents freezing, but
2 minimises the loss of heat by conduction.

The blood leaving the legs is at almost the same temperature as the arterial blood.

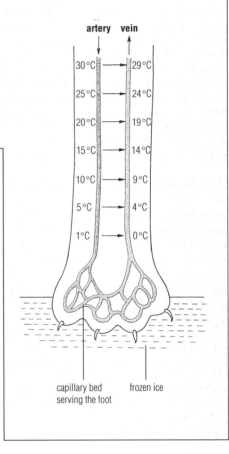

capillary bed serving the foot

frozen ice

Figure 23.24   The European hedgehog (*Erinaeus europaeus*): a study of blood pressure (BP), body temperature (T$_B$) and heart rate (HR) in hibernation and when naturally aroused (in springtime)

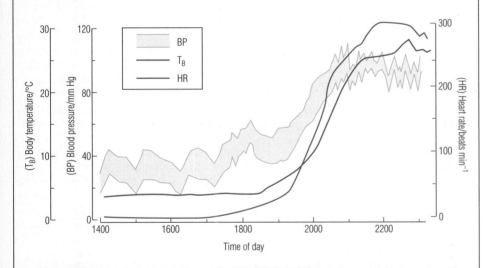

Behavioural adaptations in temperature regulation

When temperatures are low and food supplies are unavailable, many mammals adjust their behaviour (in effect they 'turn down' their pre-set temperature in the hypothalamus), and become torpid or dormant. In this state, they may be able to survive the unfavourable conditions. Their metabolic rate is lowered to the minimum needed to survive, so their respiration rate makes limited demands on body fat reserves. European bat species do this in both the summer in unexpected spells of low temperature, and also through the winter. The European hedgehog also hibernates regularly, from September until March (Figure 23.24).

23.5 BEHAVIOUR

Behaviour is broadly defined as the way organisms respond to their environment, and to other members of the same species. The study of behaviour within a zoological context is known as **ethology**. Modern studies in ethology are closely related to the disciplines of ecology, genetics and physiology. Psychology has close connections with animal behaviour, too.

The activities of animals enable them to survive, and to seek out a favourable environment. As an integral part of these activities, an organism receives information (sense data), allowing continuous adjustments of its responses. Behaviour is based upon 'feedback', and upon the machinery for response and coordination – that is, the nervous system and the effector organs. Thus it is appropriate to discuss animal behaviour at this point, following consideration of response and coordination in animals, and the phenomenon of homeostasis (control via 'feedback').

■ Interpreting animal behaviour

It is difficult to be totally objective when interpreting animal behaviour. Two errors commonly made in discussing behaviour can be recognised at the outset, however (forewarned is forearmed!).

First, it is all too easy to attribute human feelings to animals, and human intentions to their actions. This tendency, known as **anthropomorphism**, inevitably leads to misinterpretation of their behaviour. We should not expect an animal to have motivation or emotions similar to those of humans.

Secondly, it must be remembered that to claim that animals behave in a particular way in order to meet a particular need presupposes the concepts of purpose and planning in the 'mind' of the animal. Thus the cause of animal behaviour cannot be explained in terms of its outcomes. This would be **teleology**. Instead, we can explain actions in terms of conferring an advantage.

■ Nature of modern ethology

Modern studies of behaviour, based upon the observation of animals in their natural environment, were established by the work of Karl von Fritsch, Konrad Lorenz and Niko Tinbergen. They were collectively awarded the Nobel prize in physiology and medicine in 1973, for effectively establishing the new discipline of ethology within modern science.

Von Fritsch studied the behaviour of honey bees for over forty years. He established that bees use polarised light and magnetic fields to navigate by, and that they can see colours. He showed that bees communicate to fellow-workers the positions of food sources found at distances from the hive by means of a 'waggle dance' (p. 524).

Lorenz discovered the phenomenon of 'imprinting' in which very young animals fixate on another individual during a brief period in their early development (p. 521). By 'fixate' we mean to establish a new, stable relationship, and to do so quickly, after only a brief encounter. It became Lorenz's view that much animal behaviour is genetically determined and thus 'innate', not learned and flexible.

Tinbergen established that key stimuli function as 'releasers' for complex behaviour patterns. For example, a gull chick pecks at the red spot on the side of its parent's beak, and thereby elicits the regurgitation of food for itself (p. 519).

Today observers of animals in their natural habitats are aided by a range of resources and techniques that were mostly unavailable to pioneers of modern ethology. These include

- ▶ audio-visual equipment (colour photography, infra-red photography, tape-recorders)
- ▶ advanced techniques (cinematography, time-lapse photography, high-speed photography, multiple-flash photography)
- ▶ mechanical and electronic aids (automatic movement recorders, implanted signal generators used with remote-controlled tracking gear, radar tracking).

Using these, animals may be studied under natural conditions with minimal interference to their lives, often at great distances or during the absence of the observer. The result may be more impartial data, rather than subjective opinions on which to base interpretations.

An entirely different approach to the study of behaviour is the use of laboratory experiments conducted under controlled conditions. This is often the approach adopted by psychologists and neurobiologists. Typically, they study the role of the nervous system in controlling behaviour, learning behaviour (using mazes for example, p. 522) and the phenomenon of conditioned reflexes. For instance, pioneering work in this last-named field was carried out by Ivan Pavlov, studying the control of digestion in dogs (p. 521).

■ Types of animal behaviour

At one time it was common to identify animal behaviour as either **innate** (or **instinctive**) or **learned** behaviour. Use of the term 'instinctive' implies that the animal is genetically programmed with the response, which is automatically triggered in certain environmental circumstances. In fact, the division between 'innate' and 'learned' behaviour is not clear-cut. Instead, animals display a range of different types of activity. To begin, let us examine responses that are at the innate or 'instinctive' end of the spectrum.

23.6 INNATE BEHAVIOUR

■ Reflex actions

Reflex actions, such as the simple knee-jerk response (p. 461) and the response to treading upon a sharp tack, are stereotyped responses, the simplest form of behaviour of vertebrates such as mammals. Clearly, this type of response is largely determined by the pattern of neurones between a receptor (sense organ) and effectors (such as muscles). Although the central nervous system is not the instigator of a reflex action there is often a possibility of modification of a response by the brain.

Responses that are rapid yet short-lived, and are mediated by nervous inputs from sense organs, are classified as reflex behaviour. Reflexes are an essential part of many movements, particularly those involved in locomotion. They are frequently protective responses, such as responses to danger or pain, and important to survival of the animal. A good example is the escape response of the earthworm (p. 480).

■ Orientation behaviour

Orientation behaviour involves movement of motile organisms (and motile gametes) in response to external stimuli, and is critical in maintaining the individual in a favourable environment. You will have observed orientation behaviour in nature, when you have found nocturnal animals such as woodlice (*Oniscus*) sheltering under stones in damp places. Exposed to the light, they scatter and disappear under available cover. There they reposition themselves in a damp, dark microhabitat.

Orientation behaviour may be investigated in laboratory experiments, where single stimuli are applied under controlled conditions (Figure 23.25). Kinetic

Figure 23.25 Investigations of the orientation behaviour of woodlice

Studying orientation behaviour, using the choice chamber

1 'wet' and 'dry' pre-treatment

50 woodlice (*Oniscus asellus*), collected from typical wayside habitats, were adapted to 'wet' or 'dry' conditions for 3 hours, in darkness. To do this 25 woodlice were held in a Petri dish containing a thin film of water in the bottom (wet adapted), and 25 woodlice were similarly held above dry silica gel (dry adapted).

2 choice chamber observations

10 choice chambers (20 cm diameter) were set up with dry silica gel on one side and a dish of water on the other. Plastic gauze formed a platform above, and the sealed chambers were left for 30 minutes to establish a humidity gradient before 5 woodlice were added to each chamber and allowed to settle for 5 minutes. The dishes were held in dim light throughout the experiment.

results

'Dry' woodlice showed a marked preference for the wet half of the chamber. All were found in the wet half after 20 minutes.

'Wet' woodlice showed an initial preference for the dry half of the chamber, but this declined with time. After 75 minutes, 75% of these woodlice were in the 'wet' half of the chamber.

conclusions

The behaviour of the woodlice in the choice chamber depended on the humidity they experienced before the investigation.

It appears that woodlice move about to maintain a particular water balance within. As water becomes scarce the individual becomes inactive in areas of moisture.

Studying nocturnal activity

Woodlice emerge from hiding at dusk/after dark provided there is sufficient moisture available. During most of the hours of darkness they move about in search of food. Most return to their safe hiding place (normally the same place each day) well before sunrise.

By means of torch light, the number of woodlice moving on the surface of a marked length of a garden wall was noted on nine occasions, spaced out between dusk and dawn. The study was undertaken on a warm but humid summer night.

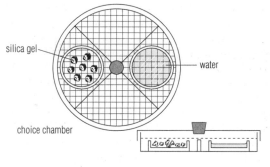

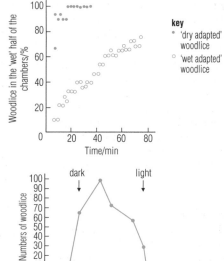

movements or **kineses** are random movements in which the rate of movement is related to the intensity of the stimulus but *not* to its direction. For example, woodlice move about quickly in dry conditions, but slow down and stop if their random movements bring them into an area of greater humidity.

Tactic movements or **taxes** (p. 422) occur when the direction of the stimulus determines the direction of the response. The flagellate *Euglena* (Figure 22.33, p. 499) swims towards light of moderate intensity, which is an example of positive phototaxis. Certain aerobic bacteria move towards sources of dissolved oxygen (Figure 12.11, p. 251), an example of positive chemotaxis. Woodlice (Figures 23.25 and 23.26) move directly away from bright light, an example of negative phototaxis.

■ Instinct

Charles Darwin defined instinctive behaviour as a series of complex reflex actions that are inherited (we would say they are genetically determined), and

therefore subject to natural selection. Instinctive behaviour, he believed, was adaptive and had evolved.

Ready-made behaviour patterns are essential in the lives of animals that have a short lifespan (they have no time for 'trial-and-error' learning), and for whom complex sequences of activity (such as nest-building, courtship and food selection) must be successfully completed in the right order if the individual and its offspring are to survive. Some animals lead solitary lives, with negligible opportunity to learn from others of the same species. In many life cycles there may be no period of parental care and 'training'. In all these situations appropriate 'instinctive' behaviour patterns are invaluable to survival of the individual.

Figure 23.26 The common shiny woodlouse (*Oniscus asellus*) is found in many habitats, but is common in compost heaps

Questions

1 How has modern audio-visual surveillance equipment changed the practice of ethology? Mention some examples from television programmes you have seen.

2 Describe how you would test the following hypotheses:
   a that a woodlouse's response to light is a taxis
   b that its response to humidity is a kinesis.

An example of instinctive behaviour

A good example of instinctive behaviour is given by the sand or digger wasps, which include species of the genus *Ammophila* (Figure 23.27). The female wasp digs a shaft in the sand. The burrow has a narrow neck but is widened below to form a flask-shaped cavity. She constructs several of these 'nests' at about the same time, in the same locality. Each is provisionally sealed, perhaps with a small pebble. The wasp then flies above the spot, apparently learning the position of the site for her to return to the spot. She then hunts for caterpillar prey, which, when found, she stings with a toxin that immobilises body wall muscles, but leaves the heart beating. She places the prey in the nest and deposits an egg beside it, and then seals the nest again. The larval stage develops from the egg, and feeds and grows at the expense of the immobilised prey. In some species the mother returns periodically, bringing fresh prey. The young adults that emerge from the nests repeat these processes without ever having met their parents.

If, on sealing the nest, the wasp is presented with another paralysed caterpillar she will re-open the nest and drag the prey inside. After laying a further egg, she reseals the nest. She will repeat this routine as long as additional caterpillars are provided, irrespective of how full the nest becomes. Here we assume that each step in the routine is a series of reflex actions, and that completion of one stage is the stimulus for commencement of the next – that is, the sight of a paralysed caterpillar is the 'trigger' stimulus for the next step of opening up the nest.

Nevertheless, there is evidence of a different type of behaviour operating, too. The sand wasp keeps several nests on the go at the same time, 'remembers' the location of each, and 'remembers' the stage of development of the larva in each nest. This information is not inherited; it must be learnt, remembered and used. The species that replenishes its nests also keeps some tally on the number of restocking visits paid to each nest, for there is equality of treatment for all its nests.

Instinctive behaviour and learning

Bird song is another species-specific behaviour pattern. In yellowhammers and corn buntings the song of isolated birds is exactly like that of birds raised normally, in the wild, so their song is innate. In chaffinches and blackbirds, on the other hand, the song has both an innate and a learned component.

Juvenile canaries, when they start to sing, produce a basic song known as subsong. Subsong is similar to the song of the parents but is less elaborate. As these birds mature they learn to perform the full song of their species. But a canary reared without any contact with other birds can perform subsong, establishing that this song is innate in origin. In nest parasites such as the cuckoo, a young bird is reared without experiencing the song of its parents, and yet the adult male cuckoo is fluent in its characteristic song. The sound of the cuckoo we hear in spring is an entirely instinctive behaviour pattern.

'Releasers' activate instinctive behaviour

Environmental clues that trigger a particular behaviour pattern are known as **sign stimuli** or **releasers**. Tinbergen's study of courtship in the three-spined stickleback (*Gasterosteus aculeatus*) established that the nuptial colours of the male stickleback are such a stimulus. In the spring, the male stickleback develops a red throat and belly and blue eyes, and establishes a territory. Intruding males that also have nuptial coloration are warned off. In Tinbergen's experiments, even crude models of the male were found to trigger aggression, provided they had a red belly on the lower surface (Figure 23.28). A fighting response was triggered at any time of the day, but only within the stickleback's territory. At the border of a territory the 'owner' was hesitant, and outside the territory, submissive.

Figure 23.27   A digger wasp at work

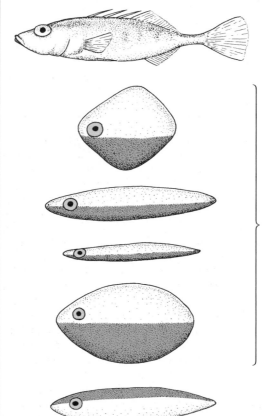

Figure 23.28   'Releasers' of stickleback aggression: models used to investigate aggression in males during courtship and breeding

model of male stickleback, lacking nuptial colours – elicited no fighting response in a rival male

models of a male stickleback that elicited the attack response in a rival male in nuptial colours; i.e. the 'releaser' does not need to be a whole fish shape, but must show the right colours in approximately the right orientation

model of a male stickleback with nuptial colours on its dorsal surface – this elicited no attack response in a rival male

Pecking response of young gulls

Tinbergen also interpreted bill pecking in newly hatched herring gull chicks (*Larus argentatus*) as instinctive. He observed that the parent flies in to the nest and 'mews'. The chick moves towards the parent and appears to 'beg' for food by pecking at the tip of its parent's beak. The parent then regurgitates a recently caught fish, which the chick feeds from. The adult bill is yellow, with a prominent red spot at the tip. Tinbergen tested the hypothesis that it is this spot that stimulates the chick to peck. In his study he presented to very young chicks in the nest two models of a gull's head, one in natural colours and the other identical save that the red spot was absent (Figure 23.29). The model with the red spot received all the pecks. Other experiments demonstrated that a contrast in colour between spot and beak (rather than the specific red colour) was the 'releaser'.

Subsequent work by Hailman indicated that gull chicks instinctively peck at conspicuous objects, but quickly learn to associate pecking accurately at their parents' beaks with the delivery of food. The chick is initially clumsily coordinated, but motivated by hunger. As it grows in strength and dexterity it learns to manipulate its parents for the reward of food. Thus the 'instinctive' behaviour of the gull chick is modified by a large measure of learning. Behaviour in gull chicks is therefore a product of innate behaviour patterns, powerful internal motivation forces such as hunger, maturation of coordination machinery (the nervous system) and environmental signals or releasers.

Innate behaviour patterns vary in their rigidity. Repetition of instinctive behaviour can lead to modifications derived from experience, causing instinctive acts to become more efficient. We now believe that an instinctive (possibly reflexive) behaviour, motivation derived from fundamental needs, and a capacity to learn, all work together to varying degrees in many of the so-called innate behaviour patterns observed in nature.

■ **Control in complex behaviour patterns**

The reproduction sequence of the stickleback falls into the stages of territory establishment, nest-building, courtship, fertilisation and parental care. Initiation of the courtship sequence arises from a changing environmental condition (day-length), which triggers increased levels of hormones (pituitary, thyroid and sex hormones) in both male and female fish. These hormone levels cause the male to develop a red throat and belly and blue eyes, and the female's abdomen to become swollen with eggs. At this stage the fish move into shallow water, where courtship commences.

In the sequence of steps of stickleback courtship, Tinbergen showed that most of the 'triggers' that initiate the successive steps come from the opposite partner (Figure 23.30, overleaf). Thus each response of the male is the stimulus for the next response of the female, and vice versa. The response chain is a framework, however, rather than a rigid schedule. Internal factors such as motivation may divert the sequence to an earlier step, or simply block the sequence for a while. The female may not immediately respond to the zigzag dance, perhaps because she is physiologically unready. The male may be diverted by a need to repair the nest, or to drive off an intruder. Once again, the components of 'instinctive' behaviour are more diverse and less regimented than they at first appear.

Figure 23.29   Studies of bill-pecking in gull chicks

normal feeding behaviour of gull chick

hand-held cardboard models of a gull's head, as used in Tinbergen's experiments

the response of gull chicks to beak models with differently coloured spots

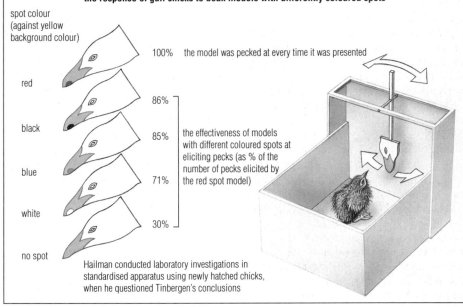

spot colour
(against yellow background colour)

red — 100%   the model was pecked at every time it was presented

black — 86%

blue — 85%   the effectiveness of models with different coloured spots at eliciting pecks (as % of the number of pecks elicited by the red spot model)

white — 71%

no spot — 30%

Hailman conducted laboratory investigations in standardised apparatus using newly hatched chicks, when he questioned Tinbergen's conclusions

Questions

1 Make a list of the characteristics of 'instinctive behaviour' as you now understand it.
2 What role can 'motivation' play in the behaviour of a gull chick?
3 What is the value to the male stickleback of adopting a non-aggressive posture on encountering another male when outside its territory?

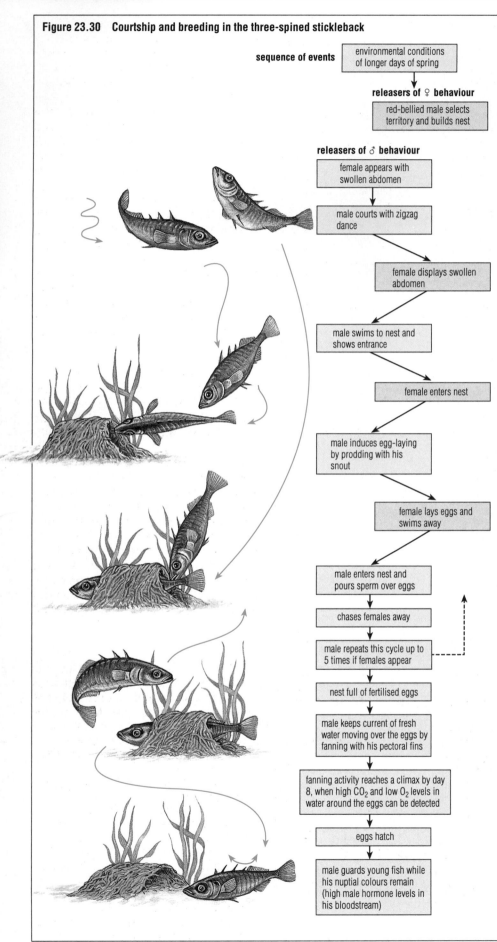

Figure 23.30 Courtship and breeding in the three-spined stickleback

sequence of events

environmental conditions
of longer days of spring

↓

releasers of ♀ behaviour

red-bellied male selects
territory and builds nest

↓

releasers of ♂ behaviour

female appears with
swollen abdomen

↓

male courts with zigzag
dance

↓

female displays swollen
abdomen

↓

male swims to nest and
shows entrance

↓

female enters nest

↓

male induces egg-laying
by prodding with his
snout

↓

female lays eggs and
swims away

↓

male enters nest and
pours sperm over eggs

↓

chases females away

↓

male repeats this cycle up to
5 times if females appear

↓

nest full of fertilised eggs

↓

male keeps current of fresh
water moving over the eggs by
fanning with his pectoral fins

↓

fanning activity reaches a climax by day
8, when high CO_2 and low O_2 levels in
water around the eggs can be detected

↓

eggs hatch

↓

male guards young fish while
his nuptial colours remain
(high male hormone levels in
his bloodstream)

23.7 LEARNING

When an animal changes its behaviour in response to some change in its environment the change may be attributable to learning. An impression of a certain experience is retained, and is used to vary behaviour on a future occasion. Of course, learned behaviour is fundamentally different from innate behaviour. Learning permits an animal to adapt quickly to changing circumstances. Learned behaviour is acquired by experience, and modified in the light of further experience. As a consequence, learned behaviour tends to reflect individual experiences (individual-specific behaviour), rather than being a characteristic of the whole species.

In primates, the basis of learning is 'memory', the ability to retain a mental image of experiences in useful form. We assume that memory consists of an altered pattern of neural connections as a conditioned reflex (see opposite). The facility of memory seems to have two parts: short-term memory, which for humans normally lasts for only a few seconds, and long-term memory, which may last very much longer, perhaps for a lifetime. Another important feature of learned behaviour is that it is not inherited; the facility to learn is inherited, however.

Various forms of learning behaviour are shown by animals. We will examine these in turn.

■ Habituation

Habituation is a form of learning in which repeated application of a stimulus results in decreased responsiveness. In habituation, animals learn not to respond to repeated stimuli that prove to be harmless or to which responses are unrewarding. Habituation is the simplest form of learned behaviour. The advantage of habituation is that whilst a novel stimulus may signify danger and so must be responded to as a precaution, a repeated stimulus that carries with it no further consequence is safe to ignore.

Habituation learning is observed throughout the animal kingdom. For example, touching a crawling snail with a leaf causes it to withdraw into its shell. Soon it re-emerges. Every time the leaf touches it, the snail withdraws for a shorter period, until the time comes when it does not respond protectively at all. A flock of pigeons may be driven away from crops when a 'gunshot' bird-scaring device is first installed, but later the birds feed peacefully despite the noises!

■ Imprinting

Imprinting is a type of learning that occurs during a very early, especially receptive stage in the life of birds and mammals. In this period, the young animal forms a more or less permanent bond with a larger, moving object that it first observes. The attachment quickly grows, especially when reinforced by rewards such as food, and perhaps warmth from body contact. Lorenz produced a spectacular demonstration of imprinting in the greylag goose (*Anser anser*) (Figure 23.31).

The obvious advantage of imprinting is the establishment of a more-or-less instantaneous bond (working relationship) with parents who will impart the essential skills for survival, such as feeding, communication and movement skills.

■ Conditioning (associative learning)

An example of a conditioned response is a behaviour pattern associated with a reward or punishment. Attention was drawn to the phenomenon of conditioning by work carried out on dogs in St Petersburg by the physiologist Ivan Pavlov, as part of an investigation of digestion, beginning in 1902. Subsequently, research into conditioning was pursued in order better to understand the learning process in animals.

Pavlovian conditioning

It is a common observation that a hungry dog salivates on sight or smell of food. Pavlov studied this response in laboratory dogs under conditions that ensured that they received no unintended stimuli (Figure 23.32). In his experiments a second stimulus not directly related to

food was introduced, immediately prior to feeding. This was the ringing of a bell (later the tick of a metronome was used). The bell alone produced no response at this stage. The dog continued to salivate on the introduction of food, and after several experiences of hearing the bell rung at feeding, the dog became conditioned to salivate whenever the bell was rung, even without food being provided. Pavlov called the new behaviour a **conditioned reflex**.

Biologists observe conditioned reflexes in nature. For example, several insects with yellow and black warning coloration either sting or have an unpleasant taste when eaten. Young predatory birds may mistakenly take such prey once, but rarely make the error again. Consequently all insects with this warning coloration are likely to be spared attack; predators are conditioned to avoid them.

Operant conditioning

Operant conditioning was the term applied by Professor B F Skinner to the response of hungry animals (pigeons and rats were mostly used), enclosed in a special experimental chamber that became known as a Skinner box. Inside was a lever that operated a food supply. When the lever was pressed by the occupant (accidentally, initially) a sample of food was delivered. After repeated accidental encounters with the lever the experimental animals learnt to press the lever when they were hungry. Skinner's observation established that 'trial-and-error' learning (p. 522) will take place more quickly when reinforced by reward.

Figure 23.31   Lorenz's imprinting experiment with greylag geese

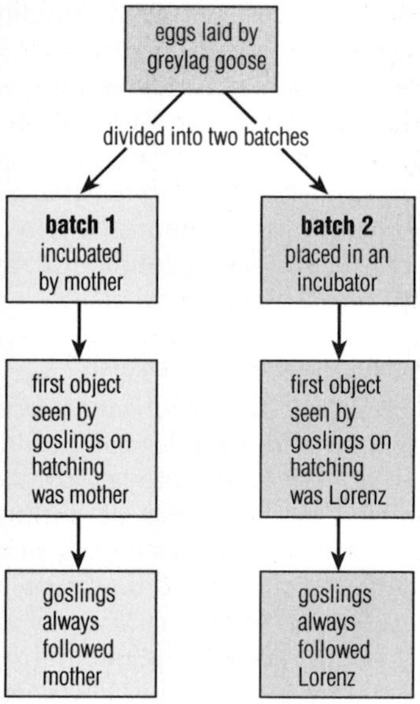

eggs laid by greylag goose

↓

divided into two batches

batch 1 incubated by mother → first object seen by goslings on hatching was mother → goslings always followed mother

batch 2 placed in an incubator → first object seen by goslings on hatching was Lorenz → goslings always followed Lorenz

EXPERIMENTS ON ANIMALS

The idea of carrying out an experiment that is expected to cause pain to an animal is unacceptable to most people. This revulsion has led to the introduction of laws that attempt to eliminate any 'unnecessary' suffering. Even so, some people believe that no experimentation on animals can ever be justified.

Pavlov's experimental procedure would not be acceptable to most biologists today, but it is described here in acknowledgement that much of our understanding of human physiology has been derived from experiments on animals.

Are there any circumstances in which you would accept a need to experiment with animals in a way that would cause them pain?

Figure 23.32   Pavlov's 1902 experiment on conditioned reflexes: this complex arrangement was devised so that the investigator was not present and that only two stimuli were presented in each experiment

---

Question

Tinbergen observed that a male stickleback fanned the developing eggs in the nest, starting at day 0. The periods of fanning built up steadily day by day, until a rate of 500 seconds spent fanning per 30 minute period was reached by day 8. Fanning then stopped, and the young fish hatched the next day.

In an experiment, the developing eggs were replaced by a batch of newly fertilised eggs on day 6. The fanning rate immediately became irregular and fell back to around 'day 3' levels, and then built up again in a similar pattern, this time reaching a maximum by day 13. On day 14 the young fish from the substituted batch of eggs hatched.

What evidence does the result provide that fanning by the male stickleback is controlled by both internal and external factors?

■ Trial-and-error learning

The ability of small animals to learn by trial and error is typically investigated in the laboratory using a maze. A maze is a series of pathways with one or more points where the animal has to choose which way to go. A wrong choice leads to a blind end and no reward. A series of correct choices leads to a reward (food for a hungry animal). The principle involved is that of **operant conditioning**. The simplest type of maze has a Y or T pattern. Animals with well-developed nervous systems can quickly learn a quite complex maze. (A maze is regarded as 'mastered' when the animal can consistently pass through without making a wrong turn.) For example, ants are found to learn a complex maze very quickly, although not as quickly as rats (Figure 23.33).

It is observed that the all-important exploratory behaviour of many animals, on which maze-learning is based, is at a high level in any new situation (a new maze, for example), even if the animal is not especially hungry. This level of interest is maintained in further explorations of the maze, provided 'success' is rewarded.

Figure 23.34   Insight learning

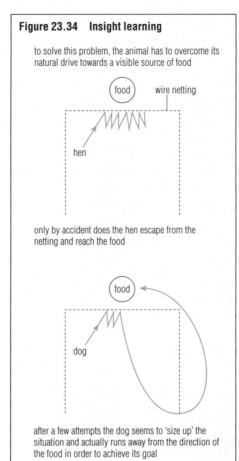

to solve this problem, the animal has to overcome its natural drive towards a visible source of food

only by accident does the hen escape from the netting and reach the food

after a few attempts the dog seems to 'size up' the situation and actually runs away from the direction of the food in order to achieve its goal

■ Insight learning or reasoning

When you solve a problem without recourse to a trial-and-error approach you are demonstrating **intelligent behaviour**, also referred to as **insight learning** or **reasoning**. This is the most sophisticated form of learning. In humans, this skill is assessed in terms of the ability to solve problems not previously experienced, sometimes by means of procedures known as 'intelligence tests'. People talk about intelligence readily enough, but professional psychologists find it hard to agree on a definition of intelligence.

Insight learning involves exploiting currently received sense data together with experiences held in memory (Figure 23.34). We abstract general principles (we call these concepts) from concrete experiences, and from other situations we have learnt about without necessarily having experienced them. We also use previous trial-and-error learning (some of this we may have called 'play'), and learn from operant conditioning experiences. We can analyse a new situation, and our perception of that situation is the basis of our response to it. All these mental faculties are based upon the efficient working of the brain.

What is memory?

Human memory is the ability to express or perform some previously learned piece of information or skill. Memory involves learning, storage, retention and retrieval processes. We have previously identified short-term memory and long-term memory. Short-term memory lasts seconds, whereas long-term memory lasts at least 24 hours and usually much longer.

It may be that short-term memory is held in interconnected nerve cells, and exists in the form of impulses passing around a circuit. If so, the loss of short-term memory could be by decay of the electrical impulse, or by interference from newer memory circuits acquired subsequently.

On the other hand, long-term memory is thought likely to take the form of a permanent change to brain cell chemistry. Neurones are especially rich in RNA, and there is some evidence that their RNA composition changes during the training of long-term memory.

As with the personal possessions we store about us, retrieval of information is as important a component of a storage system as acquisition and retention. An adequate theory of memory, when it comes, will have to account for the vagaries of our retrieval process.

Is reasoning practised by other animals?

Many animals can solve problems 'intelligently' by trial-and-error learning alone, provided they are sufficiently motivated. For example, the dropping of shellfish from a height to crack them open, practised by gulls and crows, and the smashing of snail shells against a rock anvil by thrushes, could have developed from experience, rather than by reasoning. (Given two alternative hypotheses, the one making the fewer assumptions is to be preferred – a principle attributed to William of Ockham, *c.*1300–*c.*1349.) Once discovered, these habits can be passed from parent to offspring during the period of parental feeding of fledglings.

Some non-human primates, such as chimpanzees, do show a facility to solve practical problems by insight learning, although a trial-and-error solution may be attempted first. Chimpanzees have demonstrated an ability to assemble boxes and stack them into a tower up which they will climb to reach food hung from the ceiling. In an alternative test, chimpanzees demonstrated the ability to reach food by inserting one stick into the end of another to make a long pole.

Figure 23.33   Trial-and-error learning investigated in a maze

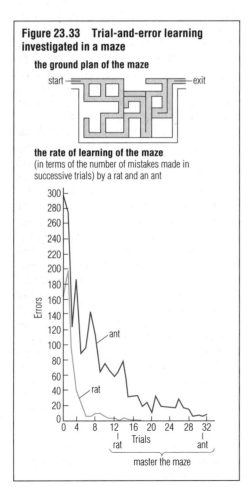

the ground plan of the maze

start — exit

the rate of learning of the maze
(in terms of the number of mistakes made in successive trials) by a rat and an ant

ant

rat

Errors

Trials

rat   ant

master the maze

23.8 SOLITARY *v.* SOCIAL ORGANISATION

Very few animals lead totally solitary lives; most gather together with others of the same species, if only during courtship and mating. Many species live in mutually tolerant and cooperative groups for all or part of their lives. Some species of fish form protective shoals (Figure 3.36, p. 65); starlings, chaffinches and many other species of bird form feeding flocks for a significant part of their lives.

Other animals live in groups on a more permanent and structured basis, and show varying degrees of differentiation in the roles of members. Primates such as the mountain gorilla (*Gorilla*) and the baboon (*Papio*, Figure 23.35) form permanent groupings in which members are differentiated by degrees of strength and varying duties. Here a social hierarchy or 'pecking order' is observed. Higher-ranking members control lower ones, their relative positions having been established in aggressive encounters, largely based upon ritualised threats to which the loser responds by appeasement behaviour.

Groups whose members are structurally differentiated include the remarkable insect communities of the ants, bees and termites, discussed below.

Issues that arise from these associations of individuals include courtship, territory and aggression, the complexity of insect communities, and the concept of altruism.

Figure 23.35 **Baboons: survival through social relationships.** Baboons are ground-dwelling primates living in troops of about 80 animals. They are largely vegetarian, feeding in a territory area of 10–15 km². When the troop moves away from trees the social structure becomes apparent, the females with infants protected by the most dominant male in the centre of the group, surrounded by less dominant adults and adolescents

■ Courtship

In nearly all bird species, male/female pairings are monogamous. Otherwise this condition is rare. For example, in mammals, whilst a female partner is soon detained in feeding the young of the union, the male has time to court and copulate with one or more other females, and frequently does. Courtships often involve elaborate routines. Why has courtship evolved as such a distinctive process, which is so expensive in terms of time and energy?

First, courtship has the initial function of attracting a partner who may be at a distance. Animals use sound (bird song is an example), coloration (as in the stickleback), elaborate movements, pheromones and body displays or even light (glow-worm, p. 173).

Secondly, courtship routines are genetically determined and species-specific. They thus ensure that pairing occurs between partners of the same species, most likely to produce fertile offspring.

Finally, sexual activity must be synchronised. The partners (male and female) must synchronise sexual development to achieve sexual maturity and readiness to breed. Synchronisations are achieved by the secretions of the master endocrine gland, the pituitary gland, and by the sex hormones from the gonads. (Synchronisation with the favourable season of the year is also essential, with the result that parents and young have access to sufficient food as they grow.)

■ Territory and aggression

The claiming of a territory and its defence is a behaviour pattern displayed by many animals, both vertebrates and non-vertebrates. A **territory** (a feeding zone or mating space, or both) is an area selected, demarcated and defended by an individual or by a group of individuals of the same species. Territories are defended from intruders of the same species by aggression, which is very often ritualised; a winner and loser emerge without damage. Territories may also be defended by scent marking (this is why dogs urinate on gate posts) and vocalisation (territorial defence is another function of bird song).

The advantage of territorial behaviour is that the population becomes spaced out in relation to the food supply – indeed, availability of food often determines territory size. The territory also gives enhanced safety from other males during courtship display and copulation.

■ Social insect communities

Social insects, such as honey bees and ants, form huge colonies, so large that it is hard to estimate them accurately. For example, a wood ant community (*Formica* spp.) may number 300 000.

Members of these insect communities show reproductive division of labour (Figure 23.36, overleaf). All members of a community are genetically related but reproduction is confined to a single queen and the male(s) that fertilise her. The queen lays thousands of eggs, and each egg is capable of developing into a worker or a fertile male or a queen. Diet may determine which is formed.

The workers are sterile individuals working on behalf both of themselves and of the fertile members of the community. The activities of the workers are controlled by communication between them. In ants the exchange of information is continual, achieved by the rubbing of heads and antennae. On expeditions, scent trails are laid and followed by the other workers. In bees, information is also exchanged between individuals, but foraging bees communicate to groups of workers, on their return, by means of the 'waggle dance' (Figure 23.37, overleaf). This tells fellow-workers the precise direction and approximate distance of a newly discovered food source.

Question

List the key differences between 'trial-and-error' and 'insight' learning.

Figure 23.36   The social life cycle of an ant community, a generalised account: ants are social insects, but only the females take part in community activities; males exist briefly, and take part in reproduction only

We are most aware of ants on sultry summer days ('flying ant days'). The communal nuptial flights are made by winged males and fully developed winged females. These ants come from many colonies over a wide area. In flight, each female (they become queens if they copulate) receives and stores sperms from one male sufficient to fertilise all her eggs, as these are laid during her working life in a colony. Ant colonies commonly have more than one queen.

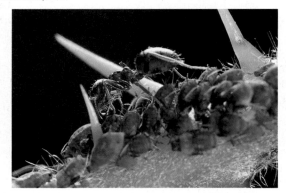

An ant colony is a labyrinth of communicating galleries in the soil. Eggs are laid there, and the larvae and pupae are reared. Eggs and larvae are moved about the galleries by worker ants who also clean the larvae and bring food for them. In return, larvae exude nutritive fluid eaten by the workers.

The majority of ants taking part in 'flying ant days' die at this stage, most being taken by birds.
Queens that survive return to earth, lose their wings, and go underground
► to an existing nest belonging to another colony (with an existing queen, perhaps); or
► to the colony she left on her nuptial flight (with an existing queen, perhaps); or
► to a new cavity, to start an entirely new nest.

The queen lays fertilised eggs that hatch into larvae, which are tended (initially by the queen, later by workers only). Most fertilised eggs form larvae that develop into wingless females, the worker ants. At a later stage in the life of a colony, the queen lays fertilised eggs that become winged females, and unfertilised eggs that become males. These ants eventually take off in their own nuptial flight. All species of ants found in the UK hibernate (larvae, workers and queens) over winter.

Figure 23.37   The honey bee community

Honey bees (*Apis mellifera*) live in communities, the majority being worker bees. They build wax combs of vertically hanging plates, 1.0–1.5 cm apart, made of thousands of tiny hexagonal 'cells'. Bee keepers arrange for bees to build their combs on wooden frames, but in the wild, bees construct nests in hollow trees or similar protected spaces.

The life history of a colony

When a new queen emerges in a colony, she kills other queen larvae. She makes her nuptial flight only after surveying the areas around the hive. A queen is fertilised by drones from her own or another queen's hive. Then she returns to the hive and starts laying fertilised eggs. These develop into new worker bees:

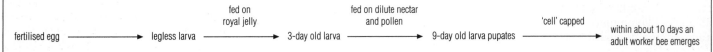

Worker bees serve the colony until they die.

The queen continues to lay fertilised eggs for most of her productive life. Pheromones from the queen maintain the identity and social structure of the colony. At some stage a queen also lays unfertilised eggs, from which the drones are formed. Also, fertilised eggs are placed in queen cells and fed royal jelly; new queens are formed. At this stage the old queen may leave the existing colony in a 'swarm' of workers and some drones. She then establishes a new colony.

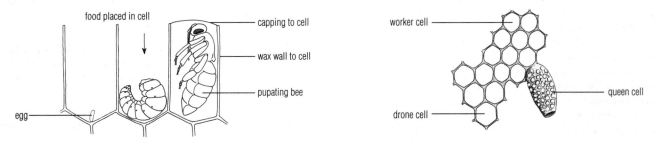

The waggle dance of a worker honey bee is the way it communicates the location of a new food source to other workers. The dance is performed on the vertical comb surface (or on the floor, at the hive entrance). It is a figure-of-eight dance, performed in darkness, surrounded by sister workers. The 'dancing' bee emits buzzing noises, vibrates its wings, and laterally vibrates its body.

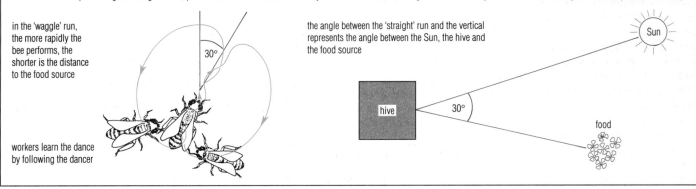

Life in ant communities

Ant communities as a whole are very stable and long-lasting. The worker ants are short-lived, but each queen lives for several years. By various mechanisms she ensures that only one queen, or at most a very few, are reared for continuance of the colony.

Ant food includes small insects, seeds and 'honey dew' discharged by aphids. Some ant species harbour aphids in their subterranean nests, and 'farm' the aphids feeding on plant roots. Thus aphids become 'ant cows', and the ant a 'farmer' that 'milks' the 'cows' by stroking their abdomens with antennae, stimulating the yield of honey dew. The 'parasol ants' of tropical South America are fungus-eaters that culture their own crop of fungal hyphae on beds of masticated leaf fragments in galleries below ground.

Some ant communities even exhibit 'social slavery'. One of the slave-making ant species in Britain is the red ant *Formica sanguinea*. Workers raid the nests of *F. fusa* and return with pupae, which are tended until the adults emerge. These then become slaves. Many other minibeasts are 'guests' in the ants' nest. They include scavenger and predatory beetles, which are often ignored by the ant community.

■ Altruism

In recognising that the behaviour of animals is directed towards their own survival, you may be surprised to learn that an occasional outcome is 'altruistic'. By **altruism** we mean behaviour that is beneficial to other animals of the same species. This may even involve self-destruction. Typically it is displayed in very many animal species by the parents of young offspring. In this case it increases the likelihood of the survival of the offspring, who themselves may inherit the tendency to care for their own young.

In social insects altruism is seen on the grand scale. Most members of the community are sterile workers who never themselves reproduce. They 'devote their lives' (often literally) to the feeding, caring and defence of the whole community. Although this behaviour does not benefit the 'altruistic' individual, it does help to ensure that 50% or more of their genes persist into the next generation. For example, in the honey bee hive the workers have the same genome as the queen, whose genes are passed on to the next generation. The altruism of the worker bee thus benefits her close genetic relatives (pp. 663–4).

23.9 RHYTHMICAL BEHAVIOUR PATTERNS

Some organisms carry out certain activities at regular intervals irrespective of the season or day-length. We describe this rhythmical behaviour as 'time biology', since the phenomenon appears to indicate the existence of a 'biological clock' within the organism. Rhythms that are controlled by biochemical and physiological changes within the organism are called **endogenous rhythms**. In animals, these involve the nervous and endocrine systems. The mechanism of the 'clock' is not understood.

We have seen that plant growth and development is closely coordinated with the seasons (p. 434). Many animals also show rhythmical behaviour patterns. Rhythms that are controlled by external changes such as the 24-hour cycle of light and dark (the photoperiod) are called **exogenous rhythms**. There are numerous examples of rhythmical behaviour patterns, and most are a blend of endogenous and exogeneous rhythms. A well-recorded example is the pattern of drinking and feeding in a group of laboratory rats exposed to a regime of 12 hours light followed by 12 hours of darkness, with food and water freely available throughout (Figure 23.38).

■ Annual rhythmical behaviour

Breeding seasons

Most animals, unlike humans, do not breed all the year round. They produce young in a season favourable for rearing and feeding. The hormonal control of sexual cycles is discussed on p. 594.

Biannual migration

The spectacular migratory journeys undertaken by the eel are illustrated on p. 391. A great many species of bird also have migratory life cycles. For example, the European swallow (*Hirundo rustica*) breeds in northern Europe but winters in central or southern Africa. It flies between 8000 and 11 000 km twice a year. Several insect species, including some butterflies, also migrate over great distances; the monarch butterfly of North America is one example.

Annual hibernation

The hedgehog is an example of an endothermic animal that survives very cold spells in winter by allowing its body temperature to fall to close to that of its environment. As a result, its heart rate slows, and its metabolism is maintained at a minimal level for the duration of the unfavourable parts of the season.

Daily (circadian) rhythms

Animals are active for only part of the 24-hour cycle. Some function at dusk or dawn (crepuscular), some in the night (nocturnal) and many in the day (diurnal).

When an organism with a marked daily rhythm (for example, the cockroach *Periplaneta*, which is nocturnal) is deprived of light and of the other clues of time of day, the animals are found to run on a clock that is close to the 24-hour cycle, but longer by about an hour. It seems that humans are not the only animals needing 25 hours in a day!

Question

How might a scientist, working with laboratory rats as described in Figure 23.38, seek to show there was an endogenous rhythm to the feeding pattern?

Figure 23.38   The pattern of feeding in rats exposed to cycles of 12 hours light/12 hours dark with unlimited food and water

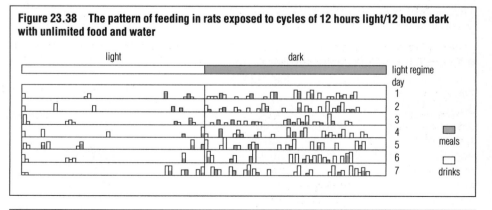

FURTHER READING/WEB SITES

Eric P Widmaier (1998) *Why Geese Don't Get Obese (And We Do)*. W H Freeman and Co

24 Health and disease

- Communicable diseases are due to agents such as viruses or particular bacteria, which are passed between organisms. Non-communicable diseases are self-inflicted or created by environmental conditions, such as dietary deficiency diseases or cancers.
- All viruses are potential pathogens because they cannot live outside a specific host cell. Human viral diseases include influenza and AIDS. The majority of bacteria are free-living saprotrophs, but pathogenic species cause disease, such as tuberculosis (*Mycobacterium tuberculosis*) or cholera (*Vibrio cholerae*).
- Certain protozoa are parasitic and cause disease, such as malaria in humans and other mammals (*Plasmodium*), and sleeping sickness (*Trypanosoma*). Pathogenic flatworms (*platyhelminths*) include the liver fluke (*Fasciola*) and the tapeworm (*Taenia*).

- Diseases caused by pathogens are not restricted to animals. Viruses, bacteria, fungi and other organisms may cause plant diseases, which are of ecological and economic importance because of the dependence of all living things upon plants.
- The non-communicable diseases of importance to human communities include malnutrition, cardiovascular and respiratory system diseases, cancers, genetic diseases, and diseases generated within the body by failure of the immune system.
- Degeneration of the body tissues with age may cause disease, such as short-sightedness or hardening of arteries. Damage due to accidents, such as a broken bone, for example, does not comprise a disease but contributes to an unhealthy condition of the body.

24.1 INTRODUCTION

This chapter is about human health and diseases. In addition, some important diseases of plants are discussed since these are of particular ecological and economic importance, given the dependence of all living things on green plants and on photosynthesis and plant metabolism.

'Health' and 'disease' are terms that are generally understood, but harder to define. Human health has been recognised as a state of physical, mental and social well being and not merely the absence of disease and infirmity (Figure 24.1). By 'infirmity' we refer to the degeneration of body tissues that occurs with age, and this may lead to ill health. So too may accidents. For example, a person with broken bones is obviously in a state of ill health, although being involved in an accident is not a condition described as diseased. Furthermore, a focus on disease is not necessarily a helpful way to approach health. Rather than merely warding off illness, we need to promote a life style that produces good health. Prevention is better than cure. The issue of preventive medicine is also introduced here.

■ Communicable (infectious) diseases

Communicable diseases are due to disease-causing agents. All the main groups of microorganisms – viruses, bacteria, fungi, algae and protozoa – include species that may cause disease in other organisms. A species that causes a disease by infecting a host is described as **pathogenic**. All viruses are potential pathogens, for a virus is not living outside its host cell (p. 548), although a virus can occur in host cells that show no symptoms. In contrast, very many bacterial species,

Figure 24.1 Issues in human health and disease

Health
'a state of complete physical, mental and social well-being, and not merely the absence of disease or infirmity'
World Health Organization (WHO)
(WHO is an agency of the United Nations with a global duty to promote health and to control communicable diseases.)

Non-communicable diseases
not 'caught' from other organisms

malnutrition, pp. 275–6

cardiovascular diseases, p. 354

respiratory diseases, p. 312

allergies and autoimmune diseases;
over-reaction of our immune system can cause unpleasant or dangerous illness e.g. asthma (p. 371), and Type I diabetes (p. 506)

genetic diseases, present at birth e.g. Down's syndrome, p. 626

cancer, e.g. breast cancer, lung cancer; due to uncontrolled division of cells, causing tumours, cancer is the most common cause of death in developed countries, after cardiovascular diseases

mental disease, e.g. wide range of disorders, many of which are now successfully treated, so that often 'care in the community' replaces periods of confinement in an asylum

Communicable diseases
due to a disease-causing agent (typically a microorganism or small organism) passed between organisms, directly or indirectly

viral disease,
e.g. influenza, rabies, poliomyelitis, AIDS due to HIV

bacterial disease,
e.g. tuberculosis, *Mycobacterium tuberculosis*; cholera, *Vibrio cholerae*; typhoid fever, *Salmonella typhi*

protoctistan disease,
e.g. malaria, due to *Plasmodium* sp.; sleeping sickness, due to *Trypanosoma* sp.

fungal disease,
e.g. athlete's foot, *Tinea pedis*; Thrush, *Candida albicans*

disease caused by platyhelminth worms,
e.g. *Schistosoma*, the blood fluke; *Taenia*, the tapeworm

Prions as disease agents,
e.g. Creutzfeldt-Jacob disease

Other issues with health implications

ageing disorders
e.g. hardening of the lenses, which makes focusing difficult

drug abuse

accidents

most algae, many fungi and most protozoa are not disease-causing parasites at all. Most are free-living and therefore non-pathogenic. Nevertheless, it is a common misconception that *all* microorganisms are harmful.

When we look at the causes of specific pathogenic diseases, we find that most **animal diseases** are caused by viruses or bacteria, whereas viruses and fungi cause most of the **diseases of plants**. In addition, several highly significant diseases are caused by species of protozoa, platyhelminths and insects. In this chapter we examine some examples of disease-causing organisms, and the ways by which they are transmitted to healthy individuals. With these examples we can illustrate some of the general principles of treatment and cure of diseases.

Major advances in the knowledge of diseases caused by microorganisms were due to the work of **Louis Pasteur** and **Robert Koch**. The contributions of these two European scientists are summarised in the panels shown here and on p. 530. The discovery of 'germs' led to the realisation that every infectious disease has an identifiable inductive agent with a life cycle that can be studied. The study of life cycles of causative organisms has enabled doctors and scientists to find treatments for many sufferers, and in many instances to prevent infections.

■ Non-communicable (non-infectious) diseases

In developed countries today infectious diseases have been very largely eliminated as causes of death. This contrasts sharply with the situation a few decades ago, and with the situation today in less-developed countries (Figure 24.2, overleaf).

Of the non-infectious diseases of importance today, we have already discussed coronary heart disease (p. 354) and the harm inflicted on the lungs and circulatory system by cigarette smoke (p. 312). Other groups of non-infectious diseases include those inherited from parents (congenital diseases, such as cystic fibrosis and haemophilia, p. 620), the diseases of malnutrition, and diseases such as diabetes that are due to malfunction of the endocrine system. Here we will also examine other important examples of non-infectious disease.

■ Cancer

Cancer is a disorder of the body's growth in which cells fail to respond to normal controls on their multiplication and enlargement. Cancer cells come to crowd out other, healthy tissues. The result is one or more **tumours**. A tumour, which may appear in any body tissue, is simply a mass of abnormally arranged cells. Cells of the tumour do not conform to

---

LOUIS PASTEUR, A PIONEER IN APPLIED BIOLOGY

Louis Pasteur, at the age of 18

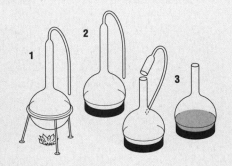

Pasteur's experiment in which broth (**1**) was sterilised, then either (**3**) exposed to air or (**2**) protected from air-borne spores in a swan-necked flask; only the broth in **3** became contaminated with bacteria

Life and times

Louis Pasteur (1822–95), was born in eastern France, the son of a tannery worker. He proved to be a most talented scientist, ambitious and industrious: 'In experimental science,' he said, 'chance favours the prepared mind'. He was also rather intolerant of those who disagreed with him. At school he was only modestly successful, but he did well enough to get to Paris, and eventually to the Ecole Normale Superior. Later he studied for his doctorate there.

Pasteur occupied successively the chairs of chemistry at Strasbourg and then at Lille, where he focused on local industrial interests. Then in 1857 he became Assistant Director at the Ecole Normale Superior in Paris where most of his work was undertaken. At the age of 46 he was partially paralysed by a stroke, yet he continued his studies until his death 27 years later by directing his research workers and assistants, the most notable being his wife.

Contributions to biology

Pasteur made too many discoveries to list in detail, but some are outlined below.

► In the field of crystallography and optical activity (rotation of polarised light, p. 133) Pasteur founded stereochemistry, and showed that microorganisms might metabolise only one of a pair of optical isomers.

► In studies of fermentation for the alcohol-producing industries he improved wine-making and brewing, and showed that fermentation was due to specific microorganisms and that putrefaction of the product was due to the entry of other microorganisms.

► He established that fermentation was a form of respiration, and that it could occur in the presence or absence of oxygen. Organisms obtain energy from food by two very different reactions, one requiring oxygen (aerobic respiration) and the other without oxygen (anaerobic respiration). Aerobic respiration makes more efficient use of food.

► He became convinced that spontaneous generation never took place. He recognised that putrefaction (and fermentation) were due to air-borne microorganisms. He introduced the practice of improving the keeping quality of certain drinks (milk and wine) by treating them to a brief period of moderate heating (now called pasteurisation).

► He developed the germ theory of disease, using as a model example the disease of anthrax, which afflicts both livestock and humans. He also established that many hospital practices led to contamination of wounds. Lister acknowledged his important contribution in the development of antiseptic surgery.

► Pasteur developed the technique of vaccination (originated by Jenner). Spectacular successes were obtained with vaccines against anthrax in livestock, and against rabies in many humans savaged by rabid dogs.

Influences

Pasteur rose to be one of the greatest figures in science. Although most of his work was of an applied nature, developed in response to industrial or health and disease problems, his studies led inescapably to the doctrine of the biochemical unity of life. As an applied biologist, Pasteur made a profound contribution to understanding biological principles.

---

the 'custom and practice' of normal cells of the tissue in which they appear. They continue to multiply in the absence of any need for new cells. Tumours are of two types.

A **benign tumour** consists of a slowly growing mass of cells. The adverse effects of this type of tumour are largely limited to those due to mechanical pressure on surrounding tissue.

A **malignant tumour** consists of fast-growing cells, which invade other tissues and may destroy them. Malignant tumour cells typically invade the lymph and blood systems, circulate therein and set up 'colonies' elsewhere in the body.

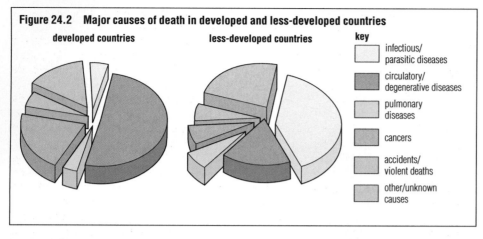

Figure 24.2 Major causes of death in developed and less-developed countries

developed countries   less-developed countries

key
- infectious/ parasitic diseases
- circulatory/ degenerative diseases
- pulmonary diseases
- cancers
- accidents/ violent deaths
- other/unknown causes

Types and incidences of cancers

The skin produces a greater variety and number of cancers than any other tissue, but most are curable. The commonest cause of death from cancer in males is lung cancer (40% of cancer deaths), whereas in females most cancer deaths are due to breast cancer (25%). In young people, the only group of malignant cancers commonly found are leukaemias, cancers of the blood-forming system. Leukaemia is difficult to treat successfully once it has spread about the body, but the prognosis is good for leukaemias that are diagnosed early.

Causes and treatment of cancer

The causes of cancer are many. They include exposure over a prolonged period to chemical carcinogens such as tobacco smoke, the hydrocarbons in tars, and asbestos, or to physical agents such as ultra-violet light and ionising radiation. Ionising radiation may be received from naturally occurring radon gas (present in some igneous rocks) or from artificial radioelements such as plutonium. Infection with certain viruses may also induce cancer.

Treatments that are effective are those that are applied at an early stage in the development of the disease; the early detection of cancer is critical, since treatment becomes difficult once malignant cells begin to move around the body. Tumours may be treated surgically, by excision of the tumour or by controlled directed doses of ionising radiation or the injection of chemicals (chemotherapy), both of which work by destruction of the DNA of dividing cells or by arresting mitosis.

Cancer genes

Carcinogens cause changes, known as mutations, in genes (p. 627). As a result, genes that are normally harmless are activated. These activated genes, known as **oncogenes** (p. 641), are altered forms of genes that regulate the growth and division of the ordinary cells of the body (somatic cells). Carcinogens alter their normal functioning, unleashing uncontrolled growth.

■ Mental health

Many life events are stressful: bereavements, accidents and examinations are just three common examples. People who temporarily feel unable to cope with their situation may respond in different, perhaps irrational ways. They may become irritable, or very tired, or extremely anxious and depressed. Some of these symptoms are similar to those that arise from the secretion of the hormones adrenaline and noradrenaline (p. 475) into the blood. In fact, stress often triggers the secretion of these hormones, preparing the body for a 'flight or fight' response to whatever has induced anxiety.

The mental disorders that some people experience at certain stages in their lives may develop from persistent exposures to stresses they feel they cannot overcome. Such disorders are known as **neuroses**. Most neuroses are mild conditions that are based on fears or anxieties that most people have at some time or another, but which have become disproportionate.

On the other hand, the mental disorders generally seen as 'madness' or 'insanity' are referred to as **psychoses**. These include manic depression and schizophrenia. Whilst psychoses are a more severe form of mental disorder, proper treatment by a psychiatrist (a doctor who deals with diagnosis and treatment of mental disorders) can lead to a recovery. There is a growing body of evidence that at least some psychoses result from biochemical malfunctions,

and may be relieved by drugs having specific effects on them.

Mental and behavioural disorders may account for half the referrals of patients to hospitals in a developed country like the United Kingdom, so mental illness is as important as physical disease (to which we often pay greater attention). There is much ignorance about the real causes of many mental problems. But in general society has come to accept mental ill-health as another form of disease that requires treatment, rather than something to conceal by 'locking away' the patient from the community. This is clearly a product of a 'scientific' approach to disease.

24.2 VIRAL DISEASES OF ANIMALS

Several highly significant human diseases are caused by viruses, including measles, rubella, mumps, chicken pox and shingles, the common cold and poliomyelitis (Table 24.1). The deadly viral infection smallpox was eliminated from all countries in 1977, by international medical action. The **human immunodeficiency virus** (HIV) is discussed on p. 370.

■ Influenza

Influenza is a common illness of humans and certain other animals. The disease is caused by a tiny virus consisting of a central strand of RNA coated in protein, surrounded by an outer lipoprotein coat (p. 27; a model is illustrated on p. 19). Three different forms of the virus, known as A, B and C, have been identified. Epidemics due to these strains occur at different intervals. For example, virus A is usually the cause of the major epidemics that sweep across continents every two or three years. Influenzas due to strains B and C are less frequent.

Viruses cannot multiply except in living cells. Influenza is an example of a virus that is not resistant to drying out in the air, outside the host's body. Consequently influenza is effectively transmitted via droplets. These may be inhaled into the airways, and the virus gains entry into the epithelium of the upper respiratory tract. The mechanism of entry into host cells is not precisely known, but cell membranes take in certain substances, including proteins, by infolding (invagination) of the plasma membrane to form vesicles (phagocytosis, p. 240). It is possible that the protein in the virus' outer coat induces the engulfing action of the plasma membrane.

The incubation period (the period between infection and the first appearance of disease signs and symptoms) lasts about 48 hours. Once inside a host cell, the virus normally sheds its outer coat, and the viral RNA directs the machinery of the cell to produce the components of the virus (lipid, protein and RNA), and to assemble them into countless new virus particles.

The signs and symptoms of influenza are fever, headaches, chills, pain in the bronchi and generalised muscular aches. The body temperature may reach 40 °C. These symptoms are caused by the damage to respiratory tract cells, by the toxins produced by the damaged cells that are released as the cells die, and by interferon. **Interferon** is a small protein molecule secreted by virus-infected cells, which limits the ability of the virus to multiply.

Once influenza is diagnosed, treatment to minimise the patient's discomfort should be provided. **Antibiotics are ineffective against viruses**, but may be prescribed to limit or prevent secondary infections of the respiratory tract. People can become extremely ill from influenza, but usually recover within 3–7 days. The severe complications that sometimes occur are caused by the secondary bacterial infections of pneumonia and bronchitis.

Some protection against infection can be given by the use of a vaccine (p. 368) made from methanal-inactivated virus particles. Virus is produced in bulk by culture on the membranes of hens' eggs. To be effective, of course, the vaccine needs to be derived from the viral strain responsible for the current outbreak. Producing large batches of vaccine takes time, and epidemics spread quickly. Influenza vaccine is usually only made available to 'at risk' members of the population, including care workers as well as patients.

■ **Rabies**

Rabies is a fatal disease, caused by a tiny bullet-shaped virus, transferred by direct contact with an infected animal. Typically (although rarely in the UK), humans contract the disease after being bitten by a rabid dog. In fact the virus is capable of infecting many mammals, both wild and domesticated. The signs and symptoms of rabies are headaches, nervousness, fever and paralysis. Painful spasms of the swallowing muscles arise from the sight of water and food. Convulsions and coma follow.

The incubation period varies between 14 days and many months, depending on how close to the neck or head the patient was bitten. This is because the virus spreads slowly from the site of the infection, travelling along nerve fibres until it reaches the brain. There is no effective treatment, once symptoms have appeared. Thus prompt and thorough cleansing of the initial wound is essential. It is necessary to find out whether the dog (or other animal) had rabies. If so, immediate injection of anti-rabies globulin is followed by injections of rabies vaccine. New vaccines are being developed, including one produced by genetic engineering.

A programme to eradicate rabies from western Europe is under way. A vaccine of live but weakened virus is fed to wild animals (mainly foxes) in baited carcases. As the foxes become immune to rabies the spread of infection is being reduced.

Diseases that are transmitted to humans from other vertebrate animals are known as **zoonoses**. Rabies is one important, if fairly rare, example of a zoonosis.

Table 24.1 Common viral diseases of humans

Disease	Virus/spread	Disease caused	Type of vaccine
1 *RNA viruses:*			
common cold	different rhinovirus, spread by touch (e.g. hand to eyes)	attacks upper respiratory tract, causing sneezing and coughing	inactivated virus vaccine – ineffective
mumps	a paramyxovirus, spread by droplet infection	swelling of salivary glands (and possibly testes, ovaries and pancreas); occurs mainly in children	attenuated virus
measles	a paramyxovirus, spread by droplet infection	mild fever, followed by small white spots in mouth, followed by rash spreading over the body; occurs mainly in children	attenuated virus
Rubella (German measles)	Rubella virus, spread by droplet infection	attacks respiratory passage and lymph nodes; crosses the placenta early in pregnancy, with 20% risk of blindness/deafness of baby	attenuated virus (given with mumps and measles vaccines) in MMR vaccine at 18 months, with booster especially to girls at 10–13 years
poliomyelitis (polio)	poliovirus, spread by droplet infection or by human faeces	attacks nerve fibres, leading to paralysis and muscle wasting	attenuated virus given orally
AIDS	HIV virus, spread by contaminated hypodermic needles, and during sexual intercourse	attacks CD4 lymphocytes (helper cells), eventually destroying immune system	no vaccine yet available
2 *DNA viruses:*			
smallpox	variola virus	(disease extinct – virus held in laboratories only)	eradication was possible because **1** disease was easily identified **2** virus does not change its surface antigens **3** vaccine was robust and stable **4** vaccine was applied by simple skin scratch technique
hepatitis B	virus spread by blood contact	attacks liver, causing jaundice, etc.	genetically engineered vaccine

24.3 BACTERIAL DISEASES OF ANIMALS

The general structure of a bacterium is illustrated on p. 551. Some common and important bacterial infections of humans include tuberculosis, diphtheria, tetanus, whooping cough, pneumonia, meningitis (p. 465), certain skin infections (spots and boils) and forms of food poisoning.

■ How bacteria cause harm to the body

Body tissues and systems can be damaged by pathogenic bacteria in two ways.

True infections

Here the pathogenic organism gains access to the body via a wound in the skin, or across an intact surface such as that of the gut. The bacterium may protect itself from being engulfed by white cells of the blood (p. 364) by means of a capsule around the cell wall. The bacteria may then adhere to host cells by pili or fimbriae (p. 551) and cause harm to the cells. An example is the bacterium *Listeria monocytogenes* which, in pregnant women, can cause the death of a fetus or a miscarriage, and which can make a newborn baby seriously ill.

Effects of toxins

Certain bacteria produce poisonous substances, or toxins. Some of these, known as **exotoxins**, are secreted by living bacteria and kill or injure the host cells. Other toxins, known as **endotoxins**, are components of the bacterial wall liberated only on the death and breakdown of the bacterial cell within the host.

■ Bacterial infection and the role of antibiotics

Many bacterial infections can be treated with antibiotics; that is, substances that in low concentration inhibit the growth of microorganisms (the term is generally restricted to substances produced by living organisms). The first antibiotic to be discovered, isolated and developed was penicillin, through the work of Alexander Fleming, Harold Florey and Ernst Chain (see panel opposite). Modern antibiotic production is discussed on p. 115.

The uses of antibiotics

Over four thousand different antibiotics have been isolated, but only about fifty have achieved wide usage. Many antibiotics have a high mammalian toxicity,

---

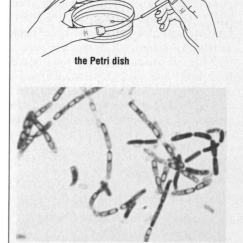
---

and thus are useless therapeutically. Antibiotics effective against a wide range of pathogenic bacteria are called 'broad-spectrum' antibiotics. This group includes chloramphenicol and the tetracyclines. Others, including penicillin and streptomycin, are effective against only a few pathogens. The use of antibiotics both with humans and in farmyard animals has brought very many bacterial diseases under control.

The prophylactic use (p. 107) of antibiotics in agriculture has also become widespread. Their use is particularly common in animals reared intensively. The practice has led to healthier animals that grow faster and reach marketable weights more quickly. To achieve this small quantities of antibiotic are often added to feed given to livestock and poultry. However, antibiotic residues may accumulate in the food chain.

With the increasing use of antibiotics, resistance has developed amongst many species of bacteria. **Resistance** to a chemical poison is the ability of an organism to survive exposure to a dose of the poison that would normally be lethal to it. Sometimes a few members of a population of microorganisms exposed to an antibiotic for the first time are able to resist its effects, a resistance acquired through mutation of an allele, p. 627. Excessive use of antibiotics accelerates the selection of resistant individuals, and their proportion in the population will increase. As a result, some pathogens are no longer suppressed by the antibiotics that once killed them. Antibiotics themselves must be adapted by genetic engineering of the microorganisms that produce them, and by enzyme technology, in order to maintain their therapeutic effects.

THE DISCOVERY OF ANTIBIOTICS

In 1929 **Alexander Fleming** (1881–1955), a Scottish bacteriologist, was studying the bacteria (*Staphylococci*) that cause boils and sore throats, at St Mary's Hospital, Paddington, London. Examining some older bacteriological plates he came across one in which a fungal colony had also become established. The bacteria were killed in areas surrounding the mould, which was identified as *Penicillium notatum*. He cultured the mould in broth, and he confirmed that a chemical from it – he named this substance penicillin – was bactericidal, and did not injure human body cells (he used white blood cells). He thought penicillin could be useful as a local antiseptic, but at that time no chemical methods were available for the preparation of concentrated penicillin.

With World War II there came a rapidly rising demand for an effective treatment for infected wounds. The leading figures in the development of penicillin for this purpose were the Australian pathologist **Harold Walter Florey** (1898–1968), the main figure in the introduction of penicillin as a useful antibiotic, and **Ernst Boris Chain** (1906–79), a German biochemist who emigrated to England in 1933 and joined the team that isolated penicillin in a stable form for therapeutic uses.

Penicillin was difficult to isolate because of its instability away from the fungus that secretes it. Teamwork at Oxford University, begun in 1938, studied the production, isolation and testing of penicillin. By 1941 the small quantity that had been extracted was sufficient to show

dramatic success with human patients. Large-scale production was transferred to the USA, well away from war-time hostilities. By 1944 penicillin was in use with the wounded of the Normandy battles, and on certain severely infected civilians.

Fleming, Florey and Chain jointly received the Nobel prize in 1945. Both Florey and Chain subsequently worked on variants of penicillin and on other antibiotics. Fleming had taken no further part in the penicillin story after his initial discovery, but antibiotics became so important that Fleming became regarded by the public as an almost legendary national hero.

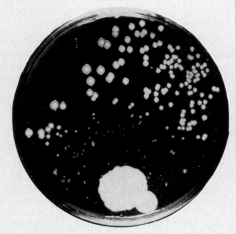

the Petri dish that Alexander Fleming noticed in his laboratory

■ Typhoid

Typhoid is a dangerous disease, caused by the bacterium *Salmonella typhi* taken into the body in food or drink. Typhoid is transmitted in domestic water supplies contaminated with faeces of people already suffering from the disease. Alternatively it can be transmitted on cooked food, especially meat, prepared under insanitary conditions – for example, by a person who is a carrier of the bacterium. The incubation period is between one and three weeks.

The typhoid patient develops a severe inflammation of the small intestine, causing diarrhoea, together with a high fever accompanied by a slow pulse rate. The bacterium reaches nearby lymph glands, and from there the bloodstream, via which it can infect many organs of the body. The spleen becomes enlarged.

Treatment is with powerful antibiotics, such as ampicillin or chloramphenicol. Protective measures are a clean water supply, the safe disposal of sewage and the hygienic preparation of food.

Questions

1 Tabulate the structural differences between prokaryote cells, eukaryote cells and viruses.
2 How do we define vaccination?
3 Why are Koch's postulates essential steps in establishing a causal relationship between the presence of a bacterium and the disease associated with it?

Table 24.2 Some bacterial diseases of humans

Disease	Bacterium/spread	Disease caused	Treatment e.g. vaccine or antibiotic
Botulism	*Clostridium botulinum* eaten in infected canned or smoked foods where the bacterium has grown and produced toxin. This requires anaerobic conditions, absence of high temperature or of salt, and a pH above 5.0. Modern food preservation avoids these dangers.	Within 24 hours, vomiting, constipation, thirst and paralysis of the muscles. High mortality rate (50% +).	Antitoxins to neutralise toxins, effective if the disease is diagnosed early enough.
Cholera	*Vibrio cholerae*, taken in water or food contaminated with the faeces of an infected person.	Toxins from the bacterium inflame the gut wall and cause severe diarrhoea. Fluid loss is so intense that dehydration results, with loss of ions, often causing death.	Immediate rehydration, e.g. water with salts and glucose, administered orally. Vaccine of killed bacterium may be effective. Antibiotics e.g. chloramphenicol, tetracyclines.
Tetanus	*Clostridium tetani*, from soil or dung into anaerobic wounds.	Violent muscular spasm over whole body, preventing breathing, drinking etc.	Vaccine of toxoid (inactivated toxin, still active as an antigen, causing antibody formation). Requires 'booster' injections periodically. Infection treated with serum containing antibodies.
Tuberculosis	*Mycobacterium tuberculosis*, a fungus-like bacterium, entering the body via infected milk or by droplet infection.	Many different organs vulnerable, but commonly the lungs. Causes body weight loss.	Vaccination (BCG) of attenuated bacterium, given only if not already immune or already infected. Antibiotics e.g. streptomycin. Current resurgence of the disease due to resistance to drugs by the bacterium.
Whooping cough	*Bordetella pertussis*, spread by droplet infection.	Severe coughing attacks in young children, with 'whoop' sound.	Vaccine of killed bacterium.

Checking the public water supply

In the water industry, pathogenic bacteria such as *Salmonella typhi* are rarely detected in water samples, even when the water source is known to have caused an outbreak of typhoid. This is because even contaminated samples usually contain pathogens of this type only intermittently and in very small numbers. Also, there is a substantial interval between the patient's initial infection and the subsequent appearance of the symptoms. This is why water is tested for harmless bacterial species common in human faeces (p. 99), rather than for specific pathogens.

■ *Salmonella* food poisoning

'Salmonella' food poisoning is caused by a motile, Gram-negative (p. 551) bacillus of the genus *Salmonella* (such as *S. enteritidis*, Figure 24.3), which can infect humans and many animals, including poultry, pigs, calves and horses. Some apparently healthy animals (and humans) are 'carriers'. *Salmonella* bacilli occur in their faeces.

Salmonella may be transmitted via fish and shellfish from sewage-polluted waters, fresh meats (including poultry) and contaminated cooked meats. Other sources of infection are duck eggs that have not been thoroughly cooked to destroy the bacteria, and damaged hens' eggs (where the shell has been cracked and the contents contaminated by hens' faeces). *Salmonella* has been found in intact hens' eggs where the birds have eaten feedstuff contaminated with the bacterium.

The signs and symptoms of *Salmonella* poisoning, which is rarely life-threatening, are due to an endotoxin released when a very large number of the bacteria has been ingested. Symptoms appear within 12–36 hours. Fever develops (temperatures of up to 39.5 °C), with diarrhoea and vomiting that may last for several days; normally, however, the patient recovers within three or four days.

How food poisoning occurs

All fresh foods contain microorganisms, and these enter the gut when they are eaten. Thorough cooking kills most microorganisms, but some may survive. The body normally copes with these levels of contamination via its range of natural defences. For example, the acidity of the stomach (p. 280) is sufficient to destroy most microorganisms. Of those that penetrate into the body, most are eliminated by the immune system (p. 366). The immune system works more quickly when it has encountered that particular invading pathogen before.

Occasionally, however, various factors may allow a pathogenic organism, taken in with food, to get the upper hand. This may occur under particular conditions.

Overwhelming numbers of pathogens

In ideal conditions, bacteria grow rapidly; a bacterium dividing every 20 minutes can produce hundreds of millions of cells in a few hours (p. 397). To avoid taking in a large dose of bacteria in food

► the food must be stored at a temperature too low for the growth of microorganisms, or be processed (pp. 109–11) to prevent the growth of any contaminating spores

► frozen meat joints that carry a risk of infection (typically chicken) must be totally thawed before cooking lest in the cooking time parts of the food, still frozen, are then only warmed, facilitating a huge increase in the contaminating bacteria they contain (most other frozen foods can be cooked direct from the freezer)

► food that is uncooked (or the juices from thawing food) must not come in contact with food that will be eaten without further cooking

► the kitchen utensils used with uncooked meat, and any working surfaces with which it comes into contact, must not also be used with cooked foods

► health regulations require caterers to wash thoroughly all clothes and utensils used and then to disinfect them with hypochlorite solution. Disposable towels are safer than dishcloths.

The presence of heat-resistant spores

Such spores are, of course, not killed by cooking, and if there is any possibility that they may be present, the food should be stored under conditions that prevent them from growing. For example, spores of *Clostridium botulinum* (the bacterium that produces botulin, one of the most potent poisons known) will not grow and manufacture toxin in the presence of oxygen, in acidic conditions, or in 8% salt solution.

A high vulnerability to infection

This may be the case with members of the following groups:

► the young whose immune system is not fully developed

► pregnant women, because of the extra demands the fetus makes

► people recuperating from other infections

► very elderly people whose defence systems may no longer be effective.

■ Sexually transmitted bacterial diseases

Most sexually transmitted (venereal) diseases (STDs) enter the body via the linings of the genital organs. The most important venereal diseases caused by bacteria are gonorrhoea and syphilis (Figure 24.5). They have been known for many centuries.

Gonorrhoea

The bacterium causing gonorrhoea, *Neisseria gonorrhoea*, is an obligate parasite of humans. The incubation period is short, only about five days. The signs and symptoms of gonorrhoea are distinctly different for the sexes. In males there is extreme discomfort in passing urine (micturition), and later a yellow discharge appears. Further complications in untreated cases include swelling of the prostate gland (p. 587), and possibly complete blockage of the urethra. In females the symptoms of infection are often minimal; if the disease is untreated, however, the bacterium may attack the oviducts, leading to sterility.

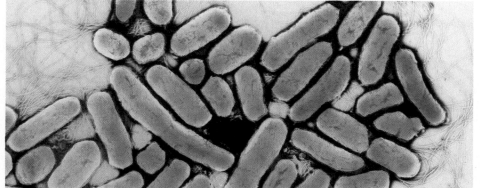

Figure 24.3 **TEM of *Salmonella enteritidis*** (×10 000), one of the bacteria that causes enteritis ('food poisoning')

Gonorrhoea is treated by chemotherapy, using drugs that either suppress the growth of the pathogenic organisms, or actually kill them. The first drugs to be used against gonorrhoea were chemicals known as sulphonamides (the formula of the parent substance is shown in Figure 24.4). These were developed over sixty years ago, long before antibiotics were available. Initially they proved extremely effective against many pathogenic bacteria, including those causing pneumonia and gonorrhoea, but strains of the pathogens resistant to these drugs then evolved. Today, successful chemotherapeutic treatment of gonorrhoea is by the antibiotic penicillin, but sulphanilamide may be employed where resistance to penicillin becomes apparent.

Figure 24.4   Sulphanilamide: this drug, and many related compounds, slows the growth and multiplication of certain bacteria, rather than killing them

Syphilis

Syphilis, like gonorrhoea, is caused by an obligate human parasite, in this case a spirochaete bacterium *Treponema pallidum* (Figure 24.6). The bacterium is acquired by direct contact through small abrasions of the skin of the genital organs of sexual partners; a fetus carried by an infected mother may acquire the disease by bacteria moving across the placenta. In this latter case the fetus may die, or may survive infected with latent syphilis, with the possibility of re-activation of the disease in later life.

Syphilis is a disease with three stages. In the primary stage the bacterium multiplies and spreads through the body during a period lasting between two and four weeks. The external signs of this infection are open, painless ulcers on the skin, known as chancres, typically but not necessarily exclusively found on the genital organs. The chancres may be unnoticed in the female partner.

The chancres slowly heal over in the following weeks, and the disease passes into a prolonged secondary phase lasting for about two years. This is a period of persistent slight fever, skin rash, a sore and ulcerated throat and enlarged lymph glands. In the third stage the patient is usually non-infectious. The bacterium lies dormant for between five and forty years before debilitating lesions (structural damage to an organ) may develop. Typically these may appear in the central nervous system, leading to insanity, or in the cardiovascular system, leading to an aortic aneurysm (ballooning of an aorta, p. 355).

Injections of penicillin normally eradicate the bacterium in the body, even if the treatment is delayed to the tertiary phase; by then, however, harmful lesions may have damaged body organs.

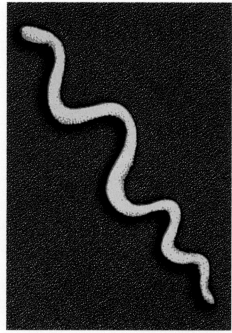

Figure 24.6   TEM of *Treponema pallidum*, the cause of syphilis in humans (×6000)

Questions

1 What steps are likely to be taken to find out whether a dog who has bitten someone was already infected with rabies?

2 A zoonosis is a disease transferred to humans from another vertebrate. What common sources of possible zoonoses are we often in contact with?

3 What are the differences between active and passive immunity (p. 368)?

4 How might food preserved in a can become a source of food poisoning? What steps in the canning process normally remove this danger (p. 109)?

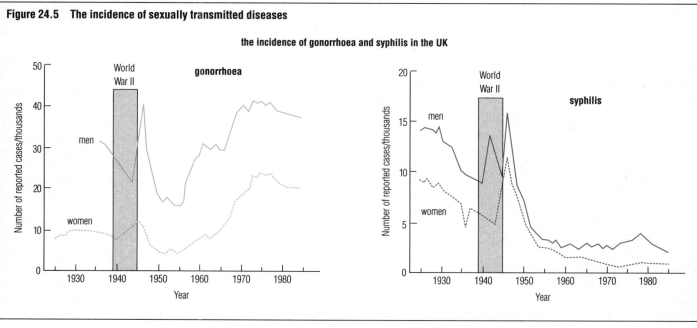

Figure 24.5   The incidence of sexually transmitted diseases

the incidence of gonorrhoea and syphilis in the UK

24.4 FUNGAL DISEASES OF ANIMALS

Relatively few fungi parasitise animals and very few cause disease in humans, compared, that is, with the numbers of viral and bacterial diseases animals may contract. Fungal infections are important in individuals whose immune system is damaged, however, particularly so in people suffering from AIDS.

Significant examples of skin infections caused by fungal pathogens are athlete's foot, due to 'ringworm' fungus (a common infection of those regularly using communal washing facilities), and ringworm of the skin, contracted by farm workers in contact with infected animals. Others are the infections caused by parasitic yeast.

■ Thrush

Candida albicans (Figure 24.7), a saprophytic fungus, is a normally harmless commensal, found in the mouth and also in faeces of most healthy individuals. Occasionally, however, this fungus can cause an irritating infection consisting of itchy or inflamed skin in the mouth (oral thrush) or on the body, or in the vagina (vaginal thrush). These infections mostly arise because of a reduction in resistance of the body, or because of locally changed circumstances. For example, the prolonged intake of antibiotics that can kill a wide range of microorganisms (broad-spectrum antibiotics) may destroy the delicate balance in the populations of commensal microorganisms present in the body, allowing parasitic yeasts (facultative parasites, p. 301) to flourish. The resulting disease can be cured by fungicides specific for the fungal parasite.

Vaginal thrush

The vagina has an acid environment (normally about pH 4.5), due to the presence of naturally occurring commensal Gram-positive bacilli. These feed on the glycogen-rich secretion of the uterus as it passes into the vagina, forming organic acids as a waste product. The uterus epithelium produces glycogen under the influence of the hormone oestrogen (p. 592). The resulting acidity keeps the vagina relatively free of pathogenic bacteria. *Candida albicans* survives in acid conditions, however, and it is favoured by a high concentration of carbohydrates such as glycogen and glucose. As a consequence, in conditions where excess glycogen reaches the vagina (this may sometimes happen in pregnant women, for example), a local infection may arise.

Questions

1. What do we mean by the term 'obligate' parasite? What name is given to organisms that may live parasitically in a host but are equally likely to occur outside a living organism (p. 301)?
2. How is it that the use of a powerful chemotherapeutic agent often speeds up the appearance of a resistant form of the disease?
3. How may other organisms of the mouth microflora inhibit the growth of the yeast that causes oral thrush?
4. How might syphilis be transmitted to a fetus?

24.5 PROTOZOAN DISEASES OF ANIMALS

Human diseases caused by protozoa are relatively few, but are individually of devastating consequence, since they include sleeping sickness (trypanosomiasis) and malaria – two diseases that have made swathes of the Earth's surface largely uninhabitable for productive, self-supporting human existence. On a less dramatic scale, an amoeba-like organism causes amoebic dysentery in humans.

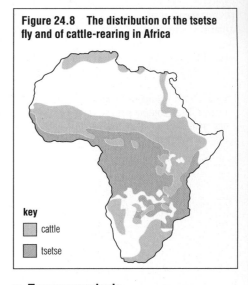

Figure 24.8 The distribution of the tsetse fly and of cattle-rearing in Africa

key
▨ cattle
▩ tsetse

■ Trypanosomiasis

Trypanosoma is a blood parasite that lives in the plasma, not in the blood cells. It causes sleeping sickness in humans and livestock, particularly cattle and horses. Since other vertebrate animals are a reservoir of infection for humans, sleeping sickness is a **zoonosis**.

There are three main species of *Trypanosoma*, causing different forms of sleeping sickness. Some are fatal within the first year after infection, others last for many years. In Africa the parasite is transmitted by the tsetse fly. Figure 24.8 shows the distributions of tsetse fly-ridden areas of Africa, and the areas where cattle-rearing is possible. Sleeping sickness prevents cattle-rearing in some of the most productive grassland (savannah) of the continent; in humans it causes periodic bouts of debilitating fever (Figure 24.9).

From the blood the parasite moves into the liver, lymphatic tissue and, in time, into the central nervous system. By the time this final stage is reached the patient has lapsed into mental apathy and persistent drowsiness, leading to eventual coma and death.

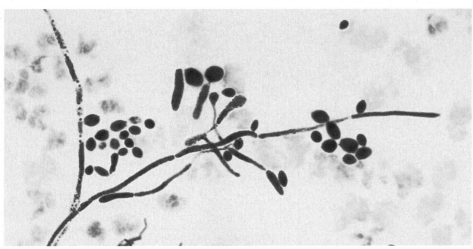

Figure 24.7 *Candida albicans*, a disease-causing yeast (×400): the budding yeast cells remain attached (a pseudomycelium) and numerous tiny, thick-walled spores are formed

Several drugs have been discovered to be effective in killing off this parasite during the early stages of an infection. Sustained efforts have also been made to reduce the tsetse fly population within restricted parts of Africa. The main thrust of research, however, has been towards the development of a vaccine that is effective against the parasite at the time when it enters the human body. A similar vaccine is required for cattle, if economically important areas of the African continent are to be opened up for this type of farming. An alternative approach is to 'farm' indigenous herbivore herds that are resistant to the diseases of the area where they live.

Figure 24.9   A young patient being treated for sleeping sickness

■ Malaria

Malaria is a major human disease, one of the most destructive in terms of the toll of human health and mortality. The life cycle of the parasite, *Plasmodium*, in the human body is illustrated on p. 29, and the female mosquito (secondary host and vector) is shown feeding on p. 293. (A **vector** is an organism, often an arthropod, which transports a pathogen.)

In 1955 the World Health Organization launched a campaign to achieve the eradication of malaria within ten years. This was attempted because malaria was then infecting hundreds of millions of people annually, millions of whom died. Despite early successes in this campaign, by 1970 it was clear that malaria was making a strong comeback, owing to resistance to insecticides built up in mosquitoes, and resistance to drugs in *Plasmodium* (Figure 24.10).

How malaria arises

The infection is acquired by a human when an adult female mosquito, already infected with the malarial parasite, inserts its mouthparts into the individual's skin and discharges saliva. (The saliva prevents coagulation of the blood meal in the insect's mouthparts, as it feeds.) Thousands of tiny spindle-shaped **sporozoites** are injected with the saliva. Within a few hours these sporozoites have entered liver cells of the host; here they feed at the expense of the contents of the cell, grow, and then divide by multiple fission (p. 29), producing large numbers of merozoites. The **merozoites** immediately invade red blood cells. Here they consume the contents of the host cell as they grow, and then divide to produce many more merozoites. As the cycle of infection is repeated, toxins from the parasitised cells accumulate in the bloodstream, causing the symptoms of fever. Eventually some of the merozoites develop into male and female **gametocytes**, which remain dormant in the bloodstream, unless drawn off into a visiting mosquito as part of its meal. If this happens the life cycle of *Plasmodium* is completed by sexual reproduction in the body of the mosquito to produce vast numbers of sporozoites. These migrate to the salivary glands and may be injected into a new host.

The effects of *Plasmodium* on the human host depend partly on the species of *Plasmodium* involved, and partly on how many previous bouts of malaria the patient has had, and hence the degree of immunity that has developed. During the peak of the infection, body temperature rises to 40–40.5 °C, with intense fever, and the spleen becomes enlarged and painful. Afterwards, the body temperature falls below normal, and there is profuse sweating.

Why the complexity of the parasite life cycle?

We should not imagine that the blood parasites are 'safe' within the primary host. Here they are the target of specific immune attack, from which they take temporary refuge by invading host cells. The complexity of a successful parasite's life cycle arises from 'hide and seek' between host and parasite. Use of a secondary host or vector to reach other primary hosts brings much the same problems. Often there is abrupt change of biochemical environment in the vector's body, which has to be mastered.

Control and prevention of malaria

Control measures are directed against the vector. For example, the air-breathing, aquatic mosquito larvae have been targeted by drainage of open water and by the oiling of water surfaces to deprive the larvae of oxygen. Elsewhere, predatory fish have been introduced to feed on the larvae. Adult mosquitoes have been killed by persistent insecticides such as DDT.

Control measures are also directed against *Plasmodium* whilst in the human bloodstream. No single drug can be used to kill all the stages of the parasite. Quinine is the oldest drug used, obtained from the cinchona tree, growing in South America and the Far East. This has now been replaced by synthetic antimalarial drugs, which are more powerful. At the time of writing, trials are proceeding of a malarial vaccine that kills the merozoite stage in the human bloodstream.

Figure 24.10   The world distribution of malaria

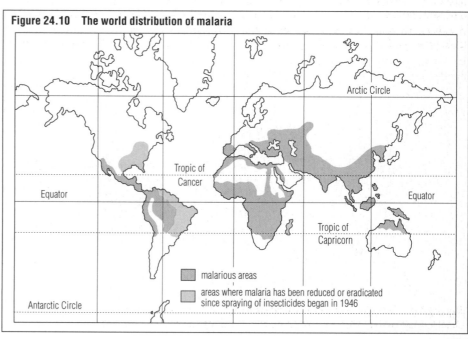

malarious areas

areas where malaria has been reduced or eradicated since spraying of insecticides began in 1946

24.6 ANIMAL DISEASES CAUSED BY PLATYHELMINTH WORMS

The flatworms or platyhelminths include two classes of animals whose members are all parasitic. These are the trematodes or flukes, and the cestodes or tapeworms. Several members of both these groups may infect humans.

■ The blood fluke, *Schistosoma*

Schistosoma is a trematode parasite that lives in the blood of its host. There are several species parasitic on humans, causing the disease Bilharzia or schistosomiasis, an extremely common illness of tropical and subtropical regions. Adult blood flukes live in blood vessels of the abdomen and pelvis. The male is smaller than the female, and at fertilisation the male is held in a groove in the female (Figure 24.11). The fertilised eggs are laid in the blood vessels, and they work their way out, some reaching the bladder or gut using the spines on their outer surface. Schistosomiasis is a very debilitating disease, due to the internal damage the eggs cause as they burrow through tissues, some lodging in the liver or spleen.

Eggs that reach the exterior (via urine or faeces) hatch into a motile form, which survives only if it reaches the secondary host: various species of aquatic snail. Within the snail tissues, multiplication of larval forms occurs, so that each infected snail eventually becomes the source of literally hundreds of thousands of motile larvae called cercariae, which have a distinctive forked tail. These swim strongly and those that reach humans may bore through the skin and reach the bloodstream, so completing the life cycle. Irrigation schemes to improve crops may incidentally proliferate habitats for aquatic snails and trigger infections of epidemic proportions. Irrigation engineers need to work with informed health workers, sewage engineers, and agriculturists to prevent schistosomiasis.

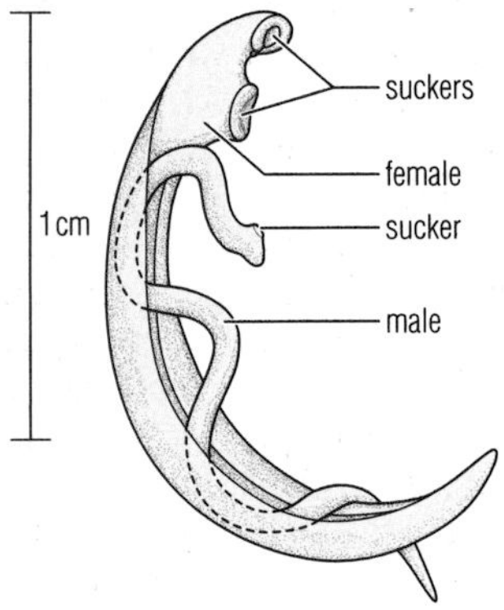

■ The liver fluke, another trematode

A simplified diagram of the liver fluke life cycle is shown on p. 35, and Figure 19.31 (p. 416) illustrates the various larval stages of the common liver fluke, *Fasciola hepatica*. *Fasciola* is a leaf-shaped animal about 1.5 cm in length (Figure 24.12). The fluke's primary hosts are vertebrates, including sheep, cattle and humans, but the secondary host is a non-vertebrate. The adult lives in the liver of the host, feeding on blood and liver cells. It does a good deal of harm; the infection is referred to as 'liver-rot'. The hermaphrodite adult produces eggs from which a motile larval stage emerges. These move to the secondary host, the small snail *Limnaea*. The miracidium larva enters the snail through the skin and undergoes multiplication and development into three further larval stages, finally emerging as a motile cercaria larva. For each miracidium about 10 000 cercariae are formed. These become attached to vegetation and form a protective membrane around themselves while they await ingestion by a primary host. Humans may be infected by eating wild watercress, growing where snails may infect it (commercial watercress is always grown in water that has no access to the infection). Sheep are infected by eating contaminated grass.

Once the fluke has reached the liver of the primary host, it is hard to eradicate. Infected humans are hospitalised during treatment because the drugs used carry some risks. The parasite is attacked at the secondary host stage, by draining pastures and by keeping sheep away from areas where ponds have dried up in high summer (just the time when cercaria escape from snails and encyst on vegetation). Ducks are useful predators of snails in damp meadows.

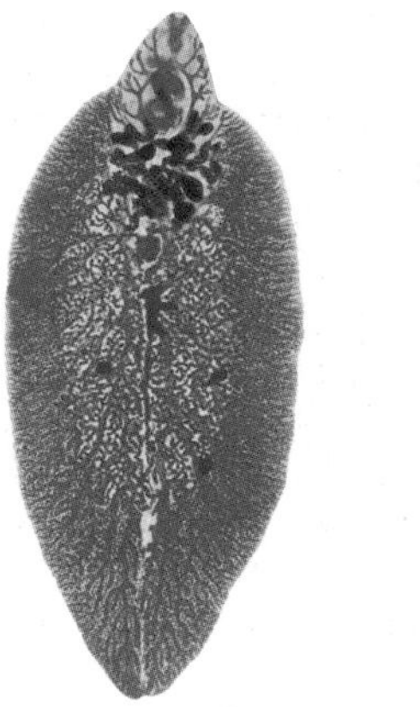

■ Pork and beef tapeworms

The flatworms of the class Cestoda are endoparasites of the gut of vertebrates. In Chapter 14 the parasitic mode of nutrition of the pork tapeworm (*Taenia solium*) is discussed, and its life cycle is illustrated in Figure 14.22 (p. 299). The beef tapeworm (*T. saginata*) has a structure and life cycle very similar to that of the pork tapeworm. Both are obligate human parasites.

People become aware of a tapeworm infection when distinctive creamy-white proglottids appear in the faeces. Outside the primary host the proglottids quickly die and disintegrate, but the embryos contained within protective shells survive for a prolonged period. The embryo of the beef tapeworm cannot develop further unless it is taken up into the secondary hosts, cattle. This may occur where grass is contaminated by untreated human sewage. The embryos of pork tapeworms will develop if taken up into the body of a pig, or they can infect humans directly; they can readily re-infect the human they came from, leading to the possibility of dangerous multiple infections of a single individual.

Both species of tapeworm can be eradicated from an infected human by the administration of drugs, once their presence is known. The spread of infection is prevented by the proper treatment of sewage (p. 101). Thorough cooking of meat will kill the form of the tapeworm (known as a bladder worm) found in infected meat products; in some parts of the world, however, people enjoy eating undercooked (rare) beef, and among these people infection by the beef tapeworm is common. Abattoir staff routinely examine pig and cattle carcases for the presence of the bladder worm, but they look only at 'prime cuts', whereas the bladder worm may lodge in more-or-less any tissues of its secondary hosts.

Figure 24.11 *Schistosoma*: male and female at the time of fertilisation, in the human host

Figure 24.12 **The adult liver fluke, *Fasciola* (×2)**

Question

From the examples you now know about, make a concise listing of the ways organisms causing diseases in animals may be transmitted. Include in your list appropriate references to bacterial contamination of water (p. 99), the feeding habits of the house fly (p. 293) and the transmission of the AIDS virus (p. 370).

24.7 PRIONS AS DISEASE AGENTS?

There is a group of fatal diseases, known as encephalopathies, that affect animals and humans. Afflicted organisms – humans (with **Creutzfeldt-Jacob disease**, or **CJD**), sheep (with scrapie), or cattle (with **bovine spongiform encephalopathy**, or **BSE**) – lose physical coordination. In humans, the patient loses memory as well as body control, prior to death. A post-mortem analysis of the brain in these cases shows tissue that has become spongy with holes where once groups of neurones occurred (Figure 24.13). Proteins called prions are now believed to be the agents that cause these diseases.

■ Discovered by cannibalism

The first 'prion' disease to be understood was called kuru or 'laughing disease' (because of the facial grimaces the patient developed). With time the sufferers lost coordination, and this was followed by dementia (a state of memory disorder, personality change and impaired reasoning). It occurred among people of a Papua New Guinea tribe. Their custom was to honour their dead by eating them; however, the men of the tribe ate only the muscle tissue, whilst the women and children received brain tissue. It was women and children that sometimes developed kuru, and it eventually became clear that this happened when the ancestor had died from kuru.

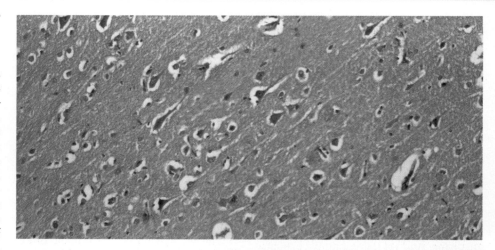

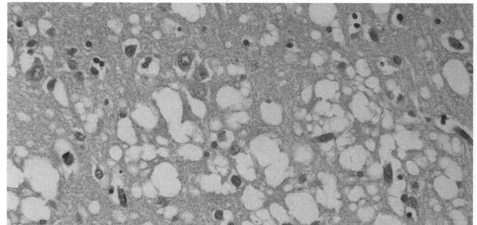

Figure 24.13 **Photomicrographs of thin sections of human cerebral cortex (post mortem):** normal tissue (×800)(above); patient who died with CJD – the tissue has a spongy appearance (×1200)(below)

Figure 24.14 **A cow in an advanced stage of BSE**

■ Scrapie and the 'species barrier'

Scrapie has been known for at least 250 years. Infected sheep are seen to rub off their wool against a fence post. The condition takes many years to develop, and it was once thought to be due to a slow-acting virus that might occasionally pass from infected to healthy sheep. Scrapie occurred in flocks on farms with other livestock, but the disease remained one of sheep alone. Then cows started to become infected with BSE (Figure 24.14) – the first reported case was in 1986. Recent evidence suggests that pituitary extracts prepared by vets were responsible. These were widely used in the UK in the late 1970s and early 1980s to boost milk yields and to stimulate multiple ovulation in prize cows. The infection was spread when cattle were fed infected meat and bone meal. The infection appears to have been able to cross the species barrier in certain circumstances.

■ Protein called prion is the infectious agent

In animals with an encephalopathy disease there is an accumulation in the brain of an abnormal form of prion protein, so these diseases are now known as prion diseases. Scrapie-infected material from a diseased sheep, which will cause scrapie in a healthy sheep, remains infectious even after treatment with ultra-violet light (which destroys DNA and RNA). The scrapie agent can be inactivated by treatments that denature protein. The name **prion** is derived from 'proteinaceous infectious particle'. Prion proteins are normal components of brain cells, but they can exist in two different shapes. In cases of infection, they show subtle shape changes. How they cause loss of brain tissue is not known. A major factor in deciding whether the disease can cross a particular species barrier is the degree of similarity of the amino acid sequences in the species' prions. The sequences of sheep and cattle prions differ by only seven amino acids. Between cattle and humans, prions differ by over 30 amino acids. Possibly, transmission of 'BSE' from cattle to humans will remain extremely unlikely, but the 'jury is still out' on this issue.

24.8 PREVENTIVE MEDICINE

By means of impressive detective work by many scientists, we have come to know about many of the inductive agents causing disease in humans and in economically important animals and plants. Many unpleasant diseases have been 'conquered', or at least reduced to manageable proportions, as a result.

Strangely, something of a problem has grown directly out of these successes. The building of scientific reputations of high standing has overshadowed the importance of rather mundane, preventive medicine. And yet there is ample evidence that preventive measures have made great inroads into the incidences and also the consequences of many diseases (Figure 24.15).

Preventive medicine seeks to establish environmental conditions and personal life styles that prevent ill health in the first place. These measures include

▶ the eating of good food in the correct proportions
▶ access to clean water
▶ effective measures for the safe disposal of sewage
▶ uncrowded living conditions
▶ a secure position in society and a personal sense of worth and achievement.

The achievement of these conditions (or at least most of them) for most people in Britain in the past hundred years has been sufficient largely to overcome many life-threatening diseases. For example, death rates due to tuberculosis, scarlet fever, diphtheria, whooping cough and measles in England and Wales since 1850 have decreased dramatically. These decreases can be directly attributed to the substantial improvements in public sanitation, housing and nutrition. Medical treatments involving drugs, including antibiotics, and compulsory immunisation have become available only in the past 50 years or so, and are responsible for relatively minor improvements by comparison.

The worldwide eradication of smallpox, achieved through concentrated action by the World Health Organization between 1967 and 1977, was on the other hand made possible only by the use of advanced medical techniques, such as vaccination.

Discussion point

Why do medical treatments with expensive drugs and by skilled surgery receive so much attention in our media, but virtually no consideration is given to the possibility of redirecting funds to more everyday measures to improve health and prevent diseases in the first place?

EPIDEMIOLOGY AND PREVENTIVE MEDICINE

The prevention of disease is a more effective use of resources than merely treating that disease once it has been diagnosed. Epidemiology, the study of epidemic disease, has a key part to play in prevention, for it is concerned with the twin questions of

▶ what is it that causes a community to have high or low rates of a particular disease?
▶ why do particular individuals develop certain diseases whilst others do not?

To answer these questions, the epidemiologist identifies communities at high and low risk. Statistically significant samples of individuals are quizzed about environmental factors that may predispose them to good health or to illness. These may include exposure to known carcinogens or to other industrial or workplace risks, their diets, the amount of exercise they take and their other leisure practices. Circumstantial evidence of a connection between such environmental factors and the incidence of a disease may suggest a hypothesis. This may be followed up, possibly by intervention trials in which the environmental danger is reduced or excluded for a sample of individuals who are thought to be at high risk.

THE EFFECTS OF CIGARETTE SMOKING ON HEALTH

The work of **Richard Doll**, in first showing the lethal link between cigarette smoking and lung cancer, established a pivotal role for epidemiology in modern medical science. Tobacco has been used in Europe for some 450 years, but only since the mass production of cigarettes began (over 100 years ago) did cigarette smoking become popular and widespread. Epidemics, among males, of the diseases now attributed to cigarette smoking (lung cancer, vascular disease, chronic bronchitis, emphysema), began in Europe in the 1920s and reached a peak about 25 years ago. Epidemics are now spreading to

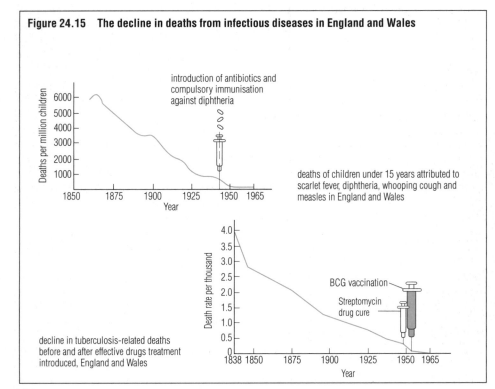

Figure 24.15   The decline in deaths from infectious diseases in England and Wales

introduction of antibiotics and compulsory immunisation against diphtheria

Deaths per million children

deaths of children under 15 years attributed to scarlet fever, diphtheria, whooping cough and measles in England and Wales

decline in tuberculosis-related deaths before and after effective drugs treatment introduced, England and Wales

Death rate per thousand

BCG vaccination

Streptomycin drug cure

Richard Doll at the age of 80

DIET AND CANCER

It is thought that a substantial proportion of cancers could be prevented by better diets. In 1991, a consortium of European epidemiologists from seven countries began monitoring some half a million citizens in a study that will last well into the twenty-first century. In the UK one of the coordinators is Professor Kay Tee Khaw, Professor of Gerontology at the University of Cambridge, working from Addenbrooke's Hospital. 75 000 British subjects, aged 40 years and over, have been selected. These recruits answer questions and keep 'food diaries' (over limited periods), so that researchers will be able to calculate how much animal and vegetable protein, saturated and unsaturated fats, sugar, starch and vegetable fibre have been consumed. In addition blood and urine samples have been taken. Tiny portions of these body fluids have been retained in liquid nitrogen for subsequent 'microanalyses' as new techniques are developed, and as the need for the data arises. Follow-up studies on the whole cohort will analyse causes of death attributed to cancer. The overall aim of this study is to be able to identify specific dietary patterns that may protect us from cancers, and to provide practical advice on how to prevent some widespread conditions associated with ageing.

Some dietary factors thought to be associated with common cancers

Cancer	Dietary factors: Adverse	Protective
breast	total fat	soya beans
	alcohol	β-carotene
large bowel/ rectum	total fat	calcium/vitamin D
	alchohol/beer	vitamin C/citrus
	meat	fruits
		green vegetables
		dietary fibre
stomach	nitrates/nitrites	vitamin C/
	alcohol	citrus fruits
	sodium/	green
	smoked food	vegetables
prostate	saturated fat	soya beans

Professor Kay Tee Khaw

women, and to peoples in other parts of the world, as cigarettes (most of which are manufactured by European and American companies) are marketed even more effectively, especially to consumers worldwide.

In 1947 the British Medical Research Council expressed doubt that the increasing mortality attributed to lung cancer was due solely to improved diagnosis and to the increasing proportion of the elderly in the population. They called for research, to which Austin Bradford Hill and Richard Doll responded.

Richard Doll had been trained as a doctor at St Thomas's Hospital, London, but he was equally interested in mathematics as a sixth former. Doll wrote an article in his medical school's *Gazette*, whilst still a student, recommending the use of statistical methods in medicine! This he did on reading claims for the efficacy of a treatment that were based on data from limited samples. He showed that the evidence, a relatively small difference in outcome between the treated patients and the controls, could be expected by chance alone.

Hill and Doll's initial study of smoking habits and health involved 3500 patients who had been admitted to hospital with a diagnosis of cancer or with some non-malignant disease. When, from their data, the lifelong non-smokers were taken as the baseline against which the risks of lung cancer at different levels of smoking could be compared, the positive correlation between the incidence of lung cancer and smokers became irrefutable.

To gain wider acceptance for their conclusion they studied smoking habits and health over a five-year period. Their 'guinea pigs', this time, were the doctors registered as general practitioners (GPs) with the BMA; 40 000 responded to the questionnaires. In only three years they had confirmed that mortality from lung cancer occurs roughly in proportion to the number of cigarettes smoked. This enquiry, which has been maintained over the years, established the risks to smokers of contracting vascular disease and bronchitis, too. Few GPs now smoke.

Proportion of men admitted to hospital with suspected lung cancer who had different smoking habits, as % of proportion among controls

| Discharge diagnosis | Most recent amount smoked/g tobacco per day | | | | |
	0	1–4	5–14	15–24	≥25
lung cancer confirmed	43	72	90	113	199
lung cancer excluded	94	127	99	103	75

Mortality from vascular diseases caused in part by smoking
(British doctors' data)

| Cause of death | Annual death rate per million men: | | Continuing cigarette-smokers | |
	Non-smokers	Continuing pipe-smokers	Any amount	≥25 a day
pulmonary heart disease	(61)*	0	6	9	20
aortic aneurysm	(320)	14	42	59	77
ischaemic heart disease	(6225)	546	601	862	988
myocardial degeneration	(829)	60	83	123	176
arteriosclerosis	(237)	23	23	40	71
cerebrovascular disease	(2594)	245	260	367	435

* Number of deaths observed in parentheses

24.9 AGEING AND HEALTH

Many animal and plant species show signs of ageing. In humans, ageing is a process of gradual decline after reaching maturity. It is a progressive failure of the body's ability to adapt homeostatically. The changes that occur in structure and function result in increasing vulnerability to environmental stress and disease. The study of ageing is called **gerontology** – a relatively new branch of medicine. Information comes from statistical studies of populations (as with preventive medicine, p. 538). These show, for example, that women live longer than men, and also that there is a major heritable component in ageing, since our life expectancy is largely linked to those of our parents.

Human life expectancy in the developed world, which in 1900 was about 50 years, has now gone up to 75 years. This trend is expected to continue (Figure 24.16). The genes that control factors influencing reproduction operate early in life and are subject to selection (p. 656). Meanwhile, genes that operate most in later life are almost certainly not subject to selection. In later life many more of our genes may operate in a harmful way, as for example cancer-causing genes do. It is a fact that the aged are mostly not important for the survival of the species.

■ Changes with ageing

Generally, our basal metabolic rate (p. 319) is high when we are young and it decreases with age. Several kinds of body cells, including heart cells, skeletal muscle cells and nerve cells are incapable of replacement when they wear out or are destroyed. DNA is a highly stable molecule, but lifelong exposure to background ionising radiation (X-rays, cosmic rays and radiation from radioactive sources, such as α, β and γ rays), and to non-ionising radiation (such as ultra-violet light) causes genes to mutate. Many of our genes control the day-to-day operation of the body and mutations of these genes (somatic mutations, p. 625) are mostly unfavourable. Genetic errors affecting cell and tissue function accumulate with time. Some important ageing changes in four of our body systems are detailed below.

1 Nervous and sensory system

▶ There is a slight but steady decline in the rate of nervous conduction, including transmissions across synapses.

▶ The ability to hear high-pitched sounds decreases with age.

▶ The lens hardens with age and the ability to accommodate for near vision (reading) is lost. This is corrected by biconvex lenses, for reading, for example.

▶ About 100 000 brain cells are lost per day and not regenerated, adversely influencing memory and intelligence.

▶ In extreme cases, mental deterioration is so pronounced that senile dementia sets in. One case of this is referred to as Alzheimer's disease in which the brain shrinks from loss of nerve cells, and changes to brain protein within and around the neurones (Figure 24.17).

2 Locomotory system

▶ Muscles weaken with time as lost fibres are replaced by connective tissue.

▶ Bones become thinner and less flexible, as calcium is lost faster than it is replaced. In extreme cases, **osteoporosis** (brittle bone disease) occurs. This is most common in females after menopause due to the fall in oestrogen in the blood circulation. Oestrogen maintains bone calcium, and the onset of osteoporosis is avoided by hormone replacement therapy (HRT) in which oestrogen is administered in pill or skin patch forms.

▶ Joints become stiff as the cartilage covering the articulated surfaces breaks down. This joint degeneration is due to the immune system when it attacks the joint tissue as 'foreign' matter, causing the disease known as rheumatoid arthritis.

3 Cardiovascular and respiratory systems

▶ The amount of blood pumped by the heart per minute decreases.

▶ Atherosclerosis (p. 355) increases in frequency, reducing the lumen of arteries, and leading to high blood pressure. General hardening of the artery walls (arteriosclerosis) leads to high blood pressure, too, causing hypertension (p. 354).

▶ There is a steady decline in lung elasticity, leading to a reduction in vital capacity and pulmonary ventilation.

▶ Efficiency of the immune system decreases, and autoimmune disease increases (p. 371).

4 Reproductive system

▶ Ovulation and the secretion of oestrogen ceases from menopause.

▶ The prostate gland enlarges, and secretion of testosterone decreases with age.

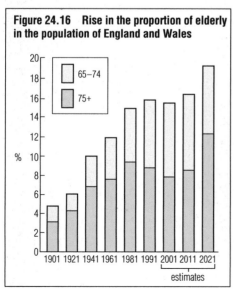

Figure 24.16   Rise in the proportion of elderly in the population of England and Wales

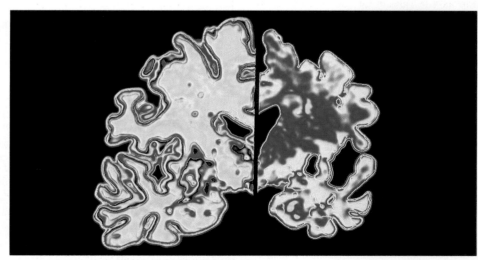

Figure 24.17   Computer-generated graphic of brain slices, from an Alzheimer's patient (right) and from a normal brain (left)

24.10 DISEASES OF PLANTS

Diseases of crop plants are important for obvious economic reasons. Most crop plants are grown in monocultures, in which diseases can be transferred from plant to plant quite easily. Because of the threat of infections from various plant pests the commercial grower often uses expensive pesticides (p. 90). These may be liberally applied at key stages in the plants' development, to reduce loss of productive capacity of the crop. This sometimes has dire consequences for that part of the living world not directly involved in food production. For example, wild plants and animals of the hedgerows are often lost too, or seriously reduced in numbers.

Figure 24.18   Tobacco mosaic virus (TMV) infection of *Nicotiana tabacum*: healthy leaves (left) and infected leaves (right)

■ Viral diseases in plants

The general structure of a virus is described on p. 27. Typical symptoms of viral diseases of plants are a general stunting of the aerial parts and the mottling of the foliage, often creating an impression of a mosaic, as with tobacco mosaic virus (TMV, Figure 24.18), or forming a more irregular collection of spots. Another symptom is the puckering or crinkling of leaf blades or the rolling up of stunted leaves, as in leaf roll of potato (Figure 24.20), a virus infection that drastically reduces the productivity of the potato plant. Similarly, the effects of TMV on tobacco leaves create havoc in the tobacco crop.

Virus transmission by aphids

The viruses mentioned above, as well as a great many other viruses of plants, are transmitted by aphids. Aphids are familiar as the 'green fly' of roses and the 'black fly' of bean plants, but in fact various aphid species parasitise the youngest shoots of a wide diversity of herbaceous and woody species. These insects may do more harm as vectors of viruses than because of the volume of sap they remove. They transmit viruses in two ways (Figure 24.19).

Direct infection occurs from the surface of the feeding stylet. As this stylet probes through infected plant tissues it picks up a coating of virus particles. These viruses may then be introduced into an uninfected plant, provided the cells of the plant are inoculated quite quickly, before the virus particles exposed on the stylet become inactivated (or before the aphid moults and sheds its cuticle).

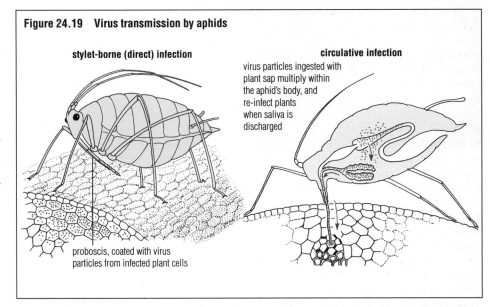

Figure 24.19   Virus transmission by aphids

stylet-borne (direct) infection

circulative infection

virus particles ingested with plant sap multiply within the aphid's body, and re-infect plants when saliva is discharged

proboscis, coated with virus particles from infected plant cells

Circulative infection involves virus particles that are taken up from an infected plant into the aphid with the phloem sap on which it feeds. The viruses multiply in the aphid's body and collect in the salivary glands. They are injected into other plants along with the saliva. Circulative viruses remain in the aphid throughout its life, and may be transmitted to the succession of plant primary hosts on which it feeds.

Figure 24.20   Leaf roll of potato. The presence of the virus in the leaves has caused the plant to be stunted and the leaves curled. (The plants just visible at the top are healthy.) Very few tubers are produced by a potato infected with leaf roll virus

■ Bacterial diseases of plants

Relatively few bacteria cause plant diseases. Many of these bacteria are rod-shaped bacilli (p. 550). Some of these are flagellated cells and consequently are motile organisms. Bacterial pathogens tend to enter plant hosts via existing gaps, holes and wounds. An example is the bacterium *Streptomyces scabies*, which in potatoes causes the tuber skin disease known as common scab (Figure 24.21). This rather unsightly surface infection reduces the commercial value of the crop, although it does not significantly reduce the food value of a tuber. (Potatoes may also become infected by a fungus that causes the disease known as black scab.)

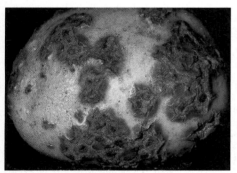

Figure 24.21   Common scab of potato: the cells of the scab are infected by a parasitic bacterium, *Streptomyces scabies*

■ Fungal diseases of plants

Fungal parasites of plants are very numerous indeed, and are classified into two groups according to the feeding relationship between host and fungus.

Necrotrophic fungi kill the host tissues and feed saprotrophically on the dead cells. These fungi normally have a wide host range, and can be cultured in laboratory conditions relatively easily. The fungus *Pythium debaryanum*, which causes 'damping-off' of seedlings, is an example of a necrotroph.

Biotrophic fungi derive their nutrients from individual live cells, and often cause only limited damage to the host tissue. The range of host species that is attacked is normally rather narrow, and most biotrophs are difficult to rear in laboratory conditions (on appropriately enriched agar plates). The rust fungi are examples of this group. Some rust species are plant parasites of great economic importance because they attack cereals (cultivated grasses), severely reducing the yield of grain. *Puccinia graminis*, for example, is the cause of black rust of wheat. Typically, *Puccinia graminis* has a complicated life cycle, in which it parasitises two hosts. It produces five different types of spore, two on the secondary host (the barberry plant), two on the wheat plant, and one in the soil and stubble in spring.

■ 'Damping-off' of seedlings

Pythium debaryanum is one of many soil-inhabiting species parasitic upon plants. The mycelium may gain entry to the seedling by forcing its way in between cells, and also by digesting the host cell walls by the secretion of enzymes. The attack may be so effective as to kill the seedlings before their emergence above ground. In other cases, the fungus enters and digests the stem tissue at ground level so the seedling first wilts, then collapses and falls over (Figure 24.22). The mycelium continues to feed on the dead plant tissues, and produces sporangia. The spores released may infect other host plants immediately. At the same time resistant oospores are also formed, and these may lie dormant in the soil until conditions favour development.

Other necrotrophs are the honey agaric or bootlace fungus, *Armillaria mellea* (p. 301), and the fungus that causes Dutch elm disease, *Ceratocystis ulmi*.

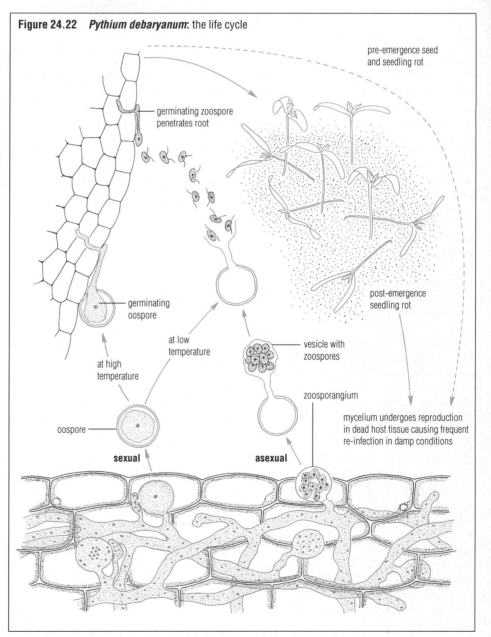

Figure 24.22   *Pythium debaryanum*: the life cycle

pre-emergence seed and seedling rot

germinating zoospore penetrates root

germinating oospore

at low temperature

at high temperature

vesicle with zoospores

post-emergence seedling rot

oospore

zoosporangium

mycelium undergoes reproduction in dead host tissue causing frequent re-infection in damp conditions

sexual   **asexual**

Questions

1   How does a pathogenic virus affect a plant cell once it has gained entry (pp. 370 and 549)?

2   For what reasons are surface virus particles short-lived on the proboscis of an aphid?

■ Dutch elm disease

Dutch elm disease is caused by the fungus *Ceratocystis ulmi*, which is introduced into healthy elm wood by the bark beetle *Scolytus* (p. 66). These insects are attracted (by the odour of decaying elm wood) to lay their eggs in elm trees already infected with the fungus. The young beetles that eventually emerge from the fertilised eggs carry spores (Figure 3.40, p. 66) of the fungus *Ceratocystis* on their bodies to other elm trees. The disease is quickly spread.

The pathogen *Ceratocystis ulmi* is a vascular wilt fungus. The symptoms of Dutch elm disease begin with wilting of the leaves, which then turn yellow and dry, and fall off. This occurs because the fungal hyphae grow through the pit membranes into the xylem vessels and partially block the vessels. The hyphae also secrete a toxin, which changes the pit walls so that balloon-like outgrowths of the surrounding xylem parenchyma, called **tyloses**, grow into the lumen of the xylem (Figure 24.23). They may eventually completely block the xylem vessels.

Dutch elm disease was first reported in the United Kingdom in 1927. By 1930 it was present all over Europe and the USA. Whilst many of the outbreaks were severe, killing numerous trees, some survived. Slowly the incidence of the disease decreased. During the late 1960s, however, it was re-introduced into Britain from Canada, in elm logs that also contained the *Scolytus* beetle. Subsequently, severe outbreaks were again reported, and since then there has been no evidence of recovery by the elms. It is assumed that a new, virulent strain of the pathogen has caused the present epidemic. Suckers may grow from the remains of the elm tree root stocks, but when the main trunks of these reach a diameter of about 10–15 cm the bark beetles again invade and lay eggs. Dutch elm disease soon follows.

■ Potato blight and the role of fungicides

Phytophthora infestans is an obligate parasite of the potato; its nutrition is discussed on p. 300 (Figure 14.23).

Despite the growing of resistant varieties of potato, there is still a danger of outbreaks of potato blight in seasons that are particularly wet and warm. The agricultural practice of varying the sequence of crops grown (crop rotation, p. 332) is largely ineffective at preventing the incidence of blight. This is because crop plants like potatoes are grown over many hectares, often with the plants at considerable density. As a consequence the crop is vulnerable to wind- and insect-borne fungal diseases. Once established, a pathogen may be transmitted between plants as their leaves rub together in the wind, but air- and splash-borne sporangia are more important in spreading the parasite. Fungicides (p. 90) are often needed to suppress epidemic outbreaks. Two types of fungicide are used against potato blight.

A **systemic fungicide** enters the plant and acts against an invading fungus from the inside. The chemical has to be specific for the pathogen and harmless to the host (and to humans that will feed on the tubers!). An example is benomyl (an organic systemic based on a substituted benzimidazole).

Bordeaux mixture is a surface-acting fungicide designed to kill spores as they 'germinate' on the plant epidermis. This fungicide is an aqueous mixture of copper sulphate and lime (calcium hydroxide) which forms a suspension of insoluble copper hydroxide on the plant cuticle. Fungal hyphae give out tiny quantities of organic acid as they grow. Such acids react with the copper hydroxide, releasing poisonous copper ions that quickly kill the fungus. Since the surface coating of Bordeaux mixture is easily dislodged by rain, spraying must be repeated in the damp conditions favourable for the growth of *Phytophthora* populations.

Figure 24.23 Destruction of an elm tree by Dutch elm disease

The hyphae of vascular wilt fungi may grow into the vascular tissue of the host flowering plant. These hyphae may block the xylem vessels and cause wilting, or their toxins may induce the formation of tyloses.

a young elm tree beginning to die due to infection by the hyphae of *Ceratocystis ulmi*

Blockage of xylem vessels by tyloses

parenchyma cell   vessel   pit

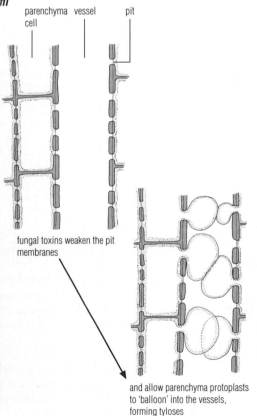

fungal toxins weaken the pit membranes

and allow parenchyma protoplasts to 'balloon' into the vessels, forming tyloses

FURTHER READING/WEB SITES

Scientific American articles and features: **www.scientificamerican.com**
New Scientist articles and features: **www.newscientist.com**

Questions

From the examples of plant pathogens you have read about and studied, construct a summary (in the form of a list) of the ways plant pathogens

a are transferred from host to host

b gain entry into a host plant.

EXAMINATION QUESTIONS SECTION 4

Answer the following questions using this section (chapters 19 to 24) and your further reading.

1 The graph shows the response of oat seedlings to various concentrations of externally applied auxin.

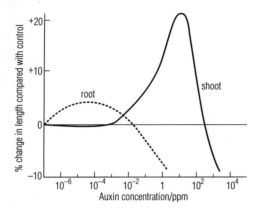

a Give **two** differences between the responses of shoots and roots shown by the graph. (2)

b Suggest a suitable control for this investigation. (1)

c Describe how the action of auxin on cells of the shoot promotes growth. (2)

d Explain why auxins may be used as selective weedkillers in cereal crops. (2)

NEAB AS/A Biology Module Test: Biology of Food Production (BY07) March 1998 Section A Q4

2 a State **two** functions of synapses. (2)

Calcium ions enter neurones through presynaptic membranes when action potentials arrive.

b i Explain how these ions enter the neurones. (2)

ii Describe the events which follow the entry of these ions until the depolarisation of the postsynaptic membrane. (5)

iii Explain the importance of acetylcholinesterase at synapses. (2)

The functioning of the synapse can be affected by the presence of drugs, as shown in the table.

Drug	Effect at synapse
curare	blocks the receptors on the postsynaptic membrane of a motor neurone–muscle junction
morphine	activates inhibitory receptors in the presynaptic membrane of sensory neurones.

c Suggest the consequences, on the action of the synapse, of

i curare; (2)

ii morphine. (2)

UoCLES AS/A Modular: Biology Foundation June 1998 Section A Q4

3 The diagram shows actin and myosin in a muscle.

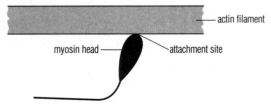

a During muscle contraction a change in position of the myosin head causes the actin filaments to move.

Describe the part played by each of the following in muscle contraction:

i calcium ions; (1)

ii ATP. (2)

b An investigation was carried out to find the effect of temperature on muscle contraction. The results are shown in the graph.

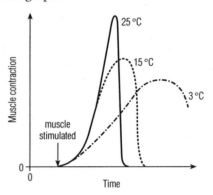

Describe and suggest an explanation for **one** effect of temperature on muscle contraction. (2)

AEB A Biology Module Paper 4 June 1998 Q1

4 The diagram shows the blood vessels which supply blood to and remove blood from the liver.

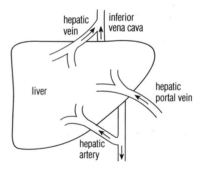

a Explain why the blood glucose concentration fluctuates more in the hepatic portal vein than in the hepatic vein. (2)

b i Which blood vessel has the lowest concentration of hydrogencarbonate ions? (1)

ii Give a reason for your answer. (1)

c Give the name of **one** excretory product found in bile. (1)

AEB A Biology Paper 1 June 1998 Q13

5 The illness diabetes mellitus occurs in two forms, early onset (Type I) and mature onset (Type II).

a Describe the differences between Type I and Type II diabetes mellitus. (3)

b Explain why it is important for people with Type I diabetes to eat regularly and to avoid large intakes of carbohydrate. (3)

c The graph that follows shows the relative risk of developing Type II diabetes in relation to the body mass index.

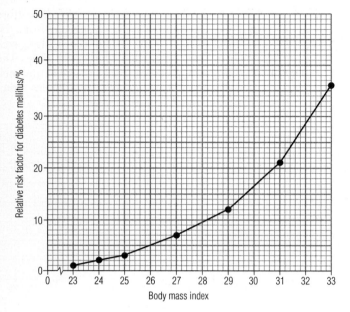

i Describe how the risk of developing diabetes changes in relation to the body mass index. (3)

ii Suggest why obesity may influence the development of Type II diabetes. (2)

*London A Biology and Human Biology Module Test B/HB4D Food and Health
June 1998 Q7*

6 The table below refers to microorganisms and the diseases they cause. Copy and complete the table by writing the appropriate word or words in the empty boxes.

Type of causative organism	Name of causative organism	Disease
bacterium	*Salmonella*	
	Staphylococcus	skin infection
		oral thrush
fungus	*Pythium*	

(5)

*London A Biology and Human Biology Module Test B/HB4A
Microorganisms and Biotechnology June 1998 Q1*

7 The drawing shows a section through part of a filamentous fungus.

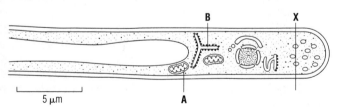

a i Name the features labelled **A** and **B** on the drawing. (2)

ii Give **two** features other than those labelled **A** and **B** on the drawing which show that this fungus is a eukaryote. (2)

b Calculate the diameter (in µm) of the part of the fungus shown at position **X** in the drawing. Show your working. (2)

*NEAB AS/A Biology Module Test: Microorganisms and Biotechnology (BY06) March 1998
Section A Q3*

8 a Describe the role of T cells in the immune response. (3)

b The graph below shows the development of an infection with human immunodeficiency virus (HIV) over a period of 10 years. Changes in the number of a type of T lymphocyte, T4 cells, are also shown.

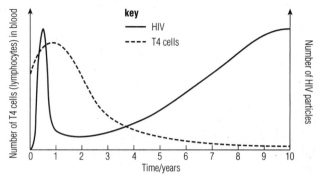

i Describe the changes in the number of HIV particles over the 10 year period. (3)

ii Describe **two** ways in which the curve for the T4 cells differs from that for the HIV particles. (2)

iii Suggest an explanation for the differences between the number of T4 cells in the blood and the number of HIV particles over the period. (3)

*London A Biology and Human Biology Module Test B/HB4A Microorganisms and
Biotechnology January 1998 Q6*

9 a i Explain what is meant by a *short-day plant*. (1)

ii Explain how flowering is controlled in short-day plants. (4)

Cocklebur (*Xanthium strumarium*) is a short-day plant. In an investigation using three groups of cocklebur plants, each with two branches, one branch (branch **A**) of each plant was exposed to continuous illumination. The other branch (branch **B**) was exposed to short-day illumination and had varying numbers of leaves removed.

The table shows the treatments used and the results obtained.

			Treatment	Number of leaves removed	Flowering response
Group 1		branch A	continuous illumination	none	flowers
		branch B	short-day illumination	none	flowers
Group 2		branch A	continuous illumination	none	flowers
		branch B	short-day illumination	all but $\frac{1}{8}$ of a single leaf	flowers
Group 3		branch A	continuous illumination	none	no flowers
		branch B	short-day illumination	all removed	no flowers

b With reference to the table, explain why
   i exposure of branch **B** to short-day length can produce flowers in branch **A**; (2)
   ii plants in Group 2 showed the same flowering response as those in Group 1; (2)
   iii there is no flowering response in Group 3. (2)
c For each of the following, state **two** advantages of the control of flowering by day length:
   i plants in their natural habitat; (2)
   ii commercial flower growers. (2)

UoCLES A Modular: Growth Development and Reproduction March 1997 Section A Q3

10 a Draw a labelled diagram to show the structure of a cell surface membrane. (5)

The diagram below shows the changes in membrane potential which occur during the transmission of a nerve impulse.

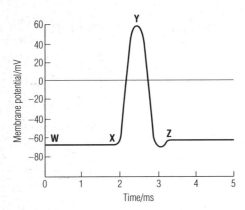

b **i** State which letters on the diagram correspond with the process of depolarisation of the axon membrane. (1)
   ii State the direction in which sodium ions will move across the membrane during depolarisation. (1)
   iii Explain how the impermeability of the axon membrane to sodium ions helps to maintain the resting potential at **W**. (2)

Mammals have myelinated axons whereas invertebrates, such as squids, have non-myelinated axons.
c Explain the advantage of having myelinated axons. (2)
The table shows the relationship between axon diameter and speed of conduction in an axon of a squid and that of a cat.

Axon	Diameter/μm	Conduction velocity/m sec^{-1}
squid	650	24
cat	4	26

d Suggest why it is possible for both animals to conduct impulses with similar velocity. (3)

UoCLES A Modular: Biology Foundation June 1997 Section A Q2

11 The yeast, *Candida utilis*, was grown in a liquid culture medium. Every hour, a 1 cm^3 sample was taken from the culture and diluted 100 times. One drop of the resulting suspension was then placed on a haemocytometer slide. The graph shows the mean number of yeast cells in each 0.004 mm^2 of the haemocytometer grid at different times after the yeast was added to the medium.

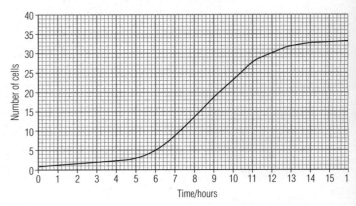

a Explain **one** reason for each of the following:
   i the slow initial rise in the number of yeast cells;
   ii the rapid rise in the numbers that followed. (2)
b Use the graph to calculate the number of yeast cells per cm^3 in the yeast culture at 13 hours. Show your working. (2)
c Describe how you could find the number of live yeast cells in 1 cm^3 of medium at 13 hours. (3)

AQA (NEAB) AS/A Biology: Microorganisms and Biotechnology (BY06) Module Test March 1999 Q4

12 Some wild varieties of potato have a natural resistance to potato blight, a disease caused by the parasitic fungus, *Phytophthora infestans*.
a Explain how this resistance may have evolved. (4)
The potato plant stores food materials in swollen regions of its underground stem, known as tubers. Spores of the fungus land on the leaves and germinate to produce threadlike structures known as hyphae. These penetrate the tissues of the leaves and, eventually, the tubers, releasing enzymes which allow them to penetrate the cells. Once established inside a cell, the hyphae release further enzymes which kill the cell and the fungus can then absorb the digested contents.
b Suggest **two** ways in which the hyphae might penetrate healthy leaves. (2)
c State the type of enzyme produced by the hyphae to penetrate
   i the cell wall
   ii the cell membrane. (2)
d Explain how infection by the fungus will affect the crop yield of the plant. (3)
Researchers have been able to use a bacterium to insert a resistance gene, known as a 'suicide' gene, into the potato cells.
e Describe how the 'suicide' gene is introduced initially into the bacterium. (4)

OCR (Cambridge Modular) AS/A Sciences: Central concepts in Biology June 1999 Q2

5 The Continuity of Life

Common blue butterflies (*Polyommatus icanus*) mating

25 Viruses, prokaryotes, protoctista and fungi

- Reproduction involves the copying of the instructions (as nucleic acid) that code for an organism's structure and behaviour. These are then shared between two or more individuals that originate from the parent organism and eventually replace it.
- Viruses are different from the organisms reviewed here in that they are not 'alive' outside a host. Once inside a host cell, they may take over the enzymic machinery and switch it to the production of new viruses.
- Bacteria are prokaryotes that grow rapidly under suitable conditions and reproduce frequently by splitting (binary fission), following replication of the circular chromosome present in the cytoplasm.

- The protoctista are single-celled eukaryotes and multicellular organisms closely related to them. Algae and protozoa are typical of this kingdom in showing diversity of structure, life cycle and reproduction.
- The fungi are eukaryotes consisting of a mycelium of thread-like hyphae with a wall made of chitin. Many feed saprotrophically, but others are parasites. Fungi typically reproduce sexually and asexually.
- Most of the structurally simple organisms in the diverse grouping discussed in this chapter are of great economic or ecological importance, and some are used by biotechnological industries and in the procedures of genetic engineering.

25.1 INTRODUCTION

In this chapter we review the reproduction processes within the life cycles of an artificial collection made up of viruses together with certain organisms that are structurally relatively simple, and which are classified as members of the prokaryotes, protoctista or fungi.

Continuation of life in all organisms is dependent upon **reproduction**. Reproduction involves the copying of the instructions that 'code' for the structure and way of life of the organism. As we have already seen, these instructions take the form of **nucleic acid** (almost always DNA, but RNA in certain viruses). The instructions are then shared between two or more individuals that originate from the parent organism and eventually replace it.

Of the organisms reviewed here, reproduction in the **viruses** is quite different from reproduction in living things. It is achieved by the virus nucleic acid taking over the enzymic machinery of the host cell (chiefly the protein synthesis machinery arranged around the ribosomes) and switching it into the exclusive production of virus particles. **All other organisms** replicate their DNA prior to mitotic or meiotic cell division as part of asexual or sexual reproduction, or both.

Asexual reproduction involves splitting (by fission or fragmentation), spore formation or budding, to form new individuals. It does not involve the union of gametes (sex) or of two separate cells to form a new individual. Asexual reproduction produces progeny that are genetically identical to the parent organism.

Sexual reproduction is a process that involves the production and fusion of special sex cells (gametes) to form a new individual. The progeny formed by sexual reproduction are similar to the parent organism(s), but not genetically identical.

25.2 VIRUSES

Viruses were discovered by the Russian biologist Iwanowski as an outcome of his studies of the plant disease we now refer to as **tobacco mosaic virus** (TMV, p. 541). He found that the disease could be transferred to a healthy plant by rubbing the leaves with juice from the tissue of a diseased plant, even after the extract had been filtered through unglazed porcelain. This type of filter is capable of holding back the smallest bacteria, and yet the agent causing TMV passed through! Later it was realised that tobacco mosaic virus was not an extremely small cell but rather a subcellular form of 'life' that can reproduce only inside a living cell. These agents of disease became known as viruses.

Essential features of viruses:

▶ they are extremely small when compared with bacteria; viruses are in the size range 20–400 nm (0.02–0.4 μm)
▶ they are not of cellular construction; viruses consist of nucleic acid (DNA or RNA) surrounded by a protein coat called a capsid (Figure 25.1). Some viruses also have an outer coating of lipoprotein
▶ they can reproduce inside specific host cells only, and therefore all viruses are endoparasites (p. 66)
▶ they are highly specific as to which type of cell they can invade; some viruses are parasitic to specific animal cells, some to specific plant cells and some to specific bacteria
▶ they can be classified by the type of nucleic acid they contain; some viruses contain DNA and others RNA. Nucleic acid may be present as a double strand or as a single strand, depending on the virus.

Viruses are not living organisms made of cells, yet they often cause the most serious of diseases for specific living organisms. This raises the question of exactly what viruses are. How did they evolve in the living world? This issue is discussed in Chapter 30, p. 666.

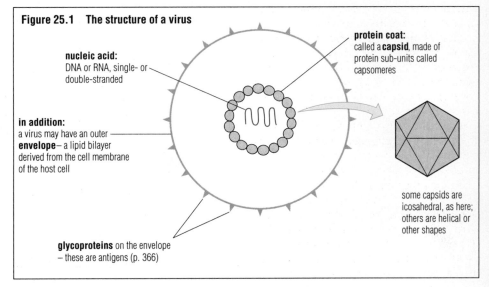

Figure 25.1   The structure of a virus

nucleic acid:
DNA or RNA, single- or double-stranded

protein coat:
called a **capsid**, made of protein sub-units called capsomeres

in addition:
a virus may have an outer **envelope** – a lipid bilayer derived from the cell membrane of the host cell

glycoproteins on the envelope – these are antigens (p. 366)

some capsids are icosahedral, as here; others are helical or other shapes

■ Bacteriophages

The viruses that parasitise bacteria, known as **bacteriophages** (or simply phages), are a category of viruses of more complex structure. They also contain lipoproteins. A phage particle has a spherical, ellipsoidal or polyhedral head containing nucleic acid, and a tail that is narrower than the head. The tail may consist of an inner tube and a contractile sheath. It ends in a baseplate, from which radiate fibres that attach the phage to a host cell. The structure of the bacteriophage that attacks the bacterium *Escherichia coli* (commonly referred to as *E. coli*, for short), known as T4 phage, is shown in Figure 25.2.

Entry into host and mode of replication

A bacteriophage first attaches to its host bacterium by means of its tail fibres. Then its viral DNA is injected into the host cell. Phage DNA takes over the ribosomal machinery of the host, causing the production of viral proteins that are enzymes. Viral enzymes cause the replication of viral DNA and the formation of viral protein and lipid. At the same time, the host's DNA is broken down. New viral particles are assembled and the remains of the host cell break down, releasing viruses, which may go on to repeat this life cycle in other host cells.

■ Other virus–host interactions

Since viruses are not 'alive' outside the host, entry to the host cell can be a major problem. For example, **plant viruses** have to cross the host cell wall. They are often accidentally introduced into the host by the action of aphids (p. 541), or enter via previously damaged cells at a wound site. Examples of plant viral diseases are discussed on p. 541. On the other hand, **animal viruses** tend to be taken in by the process of micropinocytosis. This follows upon an attraction between specific receptor sites on the host cell plasma membrane and antigens on the virus envelope. Examples of animal viral diseases are discussed on p. 528.

Once the virus is inside the host cell, the ensuing relationship between virus and host cell is either a transforming interaction or a lytic interaction.

In a **transforming** relationship, the viral DNA upon entry is incorporated into the host cell's DNA, attached to one of the chromosomes. It is passively replicated at the time the host chromosomes replicate in the cell cycle (p. 193) and is passed to all daughter cells. The viral DNA may

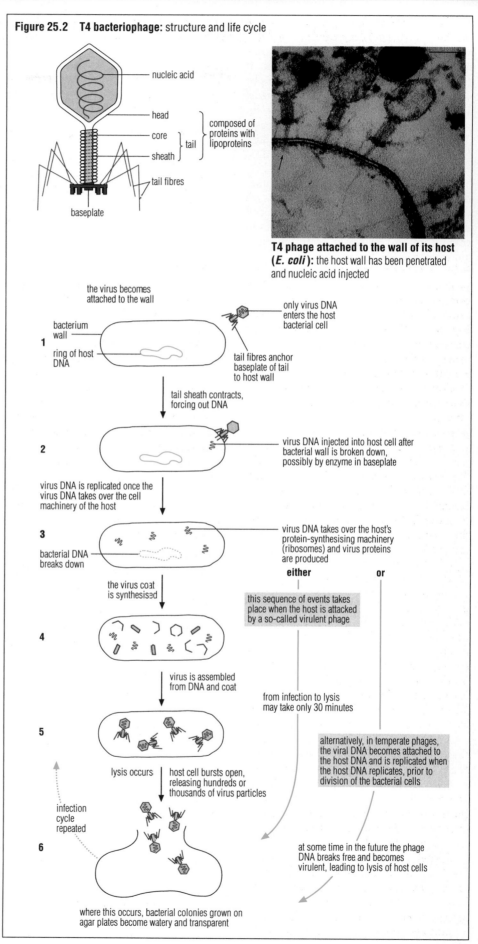

Figure 25.2   T4 bacteriophage: structure and life cycle

- nucleic acid
- head
- core
- tail
- sheath
- tail fibres
- baseplate

composed of proteins with lipoproteins

T4 phage attached to the wall of its host (*E. coli*): the host wall has been penetrated and nucleic acid injected

the virus becomes attached to the wall

only virus DNA enters the host bacterial cell

1 bacterium wall
ring of host DNA

tail fibres anchor baseplate of tail to host wall

tail sheath contracts, forcing out DNA

2 virus DNA injected into host cell after bacterial wall is broken down, possibly by enzyme in baseplate

virus DNA is replicated once the virus DNA takes over the cell machinery of the host

3 bacterial DNA breaks down

virus DNA takes over the host's protein-synthesising machinery (ribosomes) and virus proteins are produced

the virus coat is synthesised

either   **or**

this sequence of events takes place when the host is attacked by a so-called virulent phage

4

virus is assembled from DNA and coat

from infection to lysis may take only 30 minutes

5

alternatively, in temperate phages, the viral DNA becomes attached to the host DNA and is replicated when the host DNA replicates, prior to division of the bacterial cells

lysis occurs   host cell bursts open, releasing hundreds or thousands of virus particles

infection cycle repeated

6

at some time in the future the phage DNA breaks free and becomes virulent, leading to lysis of host cells

where this occurs, bacterial colonies grown on agar plates become watery and transparent

remain dormant (and possibly) undetected for months or years in this position. Herpes simplex (causing cold sores and genital herpes) and hepatitis B (causing serum hepatitis) are examples of viruses of this type. Eventually the virus DNA may become active and cause an unpleasant or life-threatening disease. This form of virus 'life cycle' is illustrated by HIV (which eventually causes AIDS) in Figure 17.46, p. 370.

In a **lytic** relationship, the virus immediately causes its own replication on entry to the host cell. The host cell machinery (for example, the ribosomes in the cytoplasm) is taken over to make enzymes needed for replication of viral nucleic acid and production of the capsid proteins. Once enzymes have assembled the virus particles, the remains of the host cell break down. This process, known as lysis, is often triggered by enzymes coded for by the viral nucleic acid. The new virus particles are released and may infect other host cells. Replication of rhinovirus, the cause of the common cold, is an example of a lytic interaction (Figure 25.3).

■ Retroviruses

Retroviruses are particular RNA viruses that, when they infect a host cell, bring with their proteins an enzyme called **reverse transcriptase**. HIV is one example (p. 370).

Reverse transcriptase catalyses the making of a double-stranded copy of DNA from the viral RNA. This DNA then becomes incorporated in the host DNA and is replicated from generation to generation (as in the transforming host–virus interaction).

Retroviruses are of special interest in genetic engineering because the reverse transcriptase enzyme allows a gene (a length of DNA coding for a particular enzyme) to be produced in the laboratory from the messenger RNA for that protein (p. 632).

Questions

1 How would you define the new term 'subcellular life form'?

2 For the replication of a virus in its host, it is not necessarily essential for the protein coat of the virus to enter the host cell. Why is this so?

3 What is the common microhabitat of the bacterium *E. coli* (p. 288)?

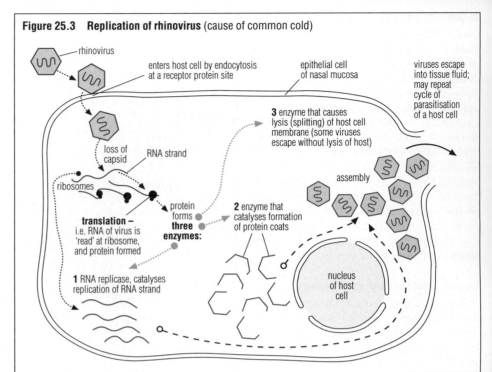

Figure 25.3   Replication of rhinovirus (cause of common cold)

rhinovirus

enters host cell by endocytosis at a receptor protein site

epithelial cell of nasal mucosa

viruses escape into tissue fluid; may repeat cycle of parasitisation of a host cell

loss of capsid

RNA strand

ribosomes

3 enzyme that causes lysis (splitting) of host cell membrane (some viruses escape without lysis of host)

assembly

translation – i.e. RNA of virus is 'read' at ribosome, and protein formed

protein forms **three enzymes:**

2 enzyme that catalyses formation of protein coats

nucleus of host cell

1 RNA replicase, catalyses replication of RNA strand

25.3 BACTERIA

The numbers of different bacteria are so great and their presence and metabolism of such ecological and economic importance that their study constitutes a separate branch of biology, namely **bacteriology.**

Bacteria are prokaryotes (p. 168); that is, they are organisms with very small cells and without an organised nucleus. Bacteria are unicells, but occur together in vast numbers, either as entirely separate cells or clumped together in groups. They are found virtually everywhere, in air, soil and water, on every surface around us, on our bodies, and even within them. Bacteria are biochemically very active and fast-growing (p. 397).

We have already seen that whilst some bacteria are parasitic and cause disease (p. 526), most are saprotrophic (p. 296) or autotrophic (p. 268), and are directly or indirectly extremely important in the maintenance of life on Earth.

■ Size, shape and structure of bacteria

Since bacterial cells are extremely small their characteristic shapes are often the most important visible clue to their identification. There are four major shapes: spherical bacteria (**coccus** forms), rods (**bacillus** forms), various forms of **spiral or curved rods**, and **filamentous bacteria**

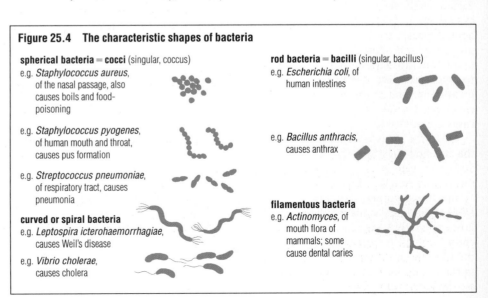

Figure 25.4   The characteristic shapes of bacteria

spherical bacteria = cocci (singular, coccus)
e.g. *Staphylococcus aureus*, of the nasal passage, also causes boils and food-poisoning

e.g. *Staphylococcus pyogenes*, of human mouth and throat, causes pus formation

e.g. *Streptococcus pneumoniae*, of respiratory tract, causes pneumonia

curved or spiral bacteria
e.g. *Leptospira icterohaemorrhagiae*, causes Weil's disease

e.g. *Vibrio cholerae*, causes cholera

rod bacteria = bacilli (singular, bacillus)
e.g. *Escherichia coli*, of human intestines

e.g. *Bacillus anthracis*, causes anthrax

filamentous bacteria
e.g. *Actinomyces*, of mouth flora of mammals; some cause dental caries

(Figure 25.4). The average diameter of a coccus may be $1 \mu m$ or less, whereas a bacillus or spirillum is typically $2-5 \mu m$ long (exceptionally rods may be $10 \mu m$ long), and $0.5-1.0 \mu m$ in diameter.

The internal structure of a generalised rod-shaped bacterium is shown in Figure 25.5. Certain of the features shown are common to all bacteria; others, such as the external capsule or slime layer, the flagella and pili, and the invagination of the plasma membrane to form meso-somes, are often seen but are not ubiquitous structures.

Conditions for bacterial growth

Bacterial growth is dependent upon the availability of adequate water and an appropriate food supply. In addition, the following external conditions are influential.

► **pH:** Most bacteria are favoured by slightly alkaline conditions (pH 7.4), but a few tolerate extremes of acidity or alkalinity.
► **Temperature:** The range of $25-45\,°C$ is favourable for most bacteria, although some are able to grow at temperatures as low as $0\,°C$ (awareness of these bacteria is important for food preservation) and others at above $80\,°C$ (in hot springs, for instance).
► **Air with oxygen:** Most bacteria flourish in air (they are aerobes), but many can survive in the absence of oxygen (facultative anaerobes). Some (obligate anaerobes) require the absence of oxygen.

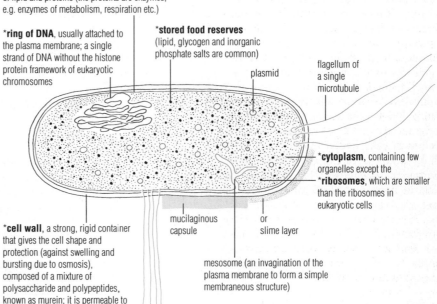

Figure 25.5 **The generalised structure of a bacterium** (structures asterisked* occur in all bacteria)

*plasma membrane, a fluid-mosaic membrane of lipid and proteins (the proteins are enzymes, e.g. enzymes of metabolism, respiration etc.)

*ring of DNA, usually attached to the plasma membrane; a single strand of DNA without the histone protein framework of eukaryotic chromosomes

*stored food reserves (lipid, glycogen and inorganic phosphate salts are common)

plasmid

flagellum of a single microtubule

*cytoplasm, containing few organelles except the
*ribosomes, which are smaller than the ribosomes in eukaryotic cells

*cell wall, a strong, rigid container that gives the cell shape and protection (against swelling and bursting due to osmosis), composed of a mixture of polysaccharide and polypeptides, known as murein; it is permeable to water and other small molecules and to ions, but proteins and nucleic acids cannot pass

mucilaginous capsule

or slime layer

mesosome (an invagination of the plasma membrane to form a simple membraneous structure)

pili (also known as fimbriae)

■ The Gram stain: Gram-positive and Gram-negative bacteria

Usually bacteria have to be stained to be identified, using an air-dried smear on a microscope slide. A staining technique, first used by Christian Gram in 1884, distinguishes two types of bacteria, Gram-positive and Gram-negative. The different staining properties of these two groups of bacteria are due to differences in the chemical composition of their walls (Figure 25.6).

Gram-positive bacteria have hardly any lipid along with the murein of the cell wall. These are bacteria that may form endospores (see overleaf). The Gram-positive bacteria include *Bacillus*, *Clostridium*, *Staphylococcus* and *Streptococcus* species.

Gram-negative bacteria, on the other hand, have substantial amounts of lipid in their wall. They are not adversely affected by the antibacterial enzyme lysozyme, and are resistant to the antibiotic penicillin (p. 530). Gram-negative bacteria include those of the genera *Salmonella*, *Escherichia* and *Azotobacter*.

Questions

1 What advantages may a capsule or slime layer confer on a bacterium (p. 201)?
2 Why is it unlikely that a colony of bacteria would maintain a growth rate producing 10^7 cells every 15 hours?
3 In what ways are bacteria ecologically important (pp. 61 and 72)?
4 In what ways are bacteria economically important (e.g. pp. 101–15)?

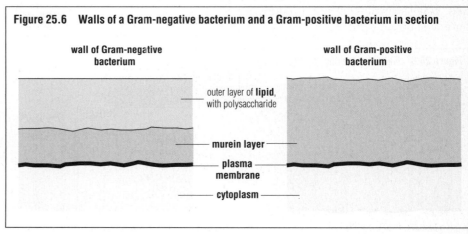

Figure 25.6 **Walls of a Gram-negative bacterium and a Gram-positive bacterium in section**

wall of Gram-negative bacterium

wall of Gram-positive bacterium

outer layer of **lipid**, with polysaccharide

murein layer

plasma membrane

cytoplasm

■ Growth and asexual reproduction of bacteria

Bacteria reproduce asexually by dividing into two. Cell division is preceded by replication of the DNA (Figure 25.7).

Bacteria grow and multiply extremely rapidly. In ideal conditions cell division may happen as often as once every 20 minutes, which would enable a single cell to give rise to 10^7 daughter cells every 15 hours, provided conditions remained favourable (exponential growth, Figure 19.1, p. 396). If the bacteria were growing on the surface of a nutrient agar plate, this number of cells would appear as a large colony spot. This rate of growth cannot be maintained indefinitely, but bacteria do occur in huge numbers; for example, a gram of garden soil may contain 10^9 bacteria. Because of the very large numbers of bacteria, populations have to be estimated from known dilutions from a measured sample in a method called the dilution plate technique (p. 99).

■ Endospores

Conditions for growth never remain ideal for long. When unfavourable conditions arise certain Gram-positive bacteria, including *Bacillus* and *Clostridium*, produce spores called **endospores**, within the existing cell wall (Figure 25.8). Endospores have great resistance to low temperature, high temperature, pH change, desiccation and the effects of chemicals.

■ Sexual reproduction in bacteria

A special and unusual form of sexual reproduction is found in certain bacteria. This involves exchange of genetic material, normally via bridges formed by fimbriae between two cells (**conjugation**, Figure 25.9).

In pneumococcal bacteria, DNA has been found to pass from one bacterium to another without conjugation, enabling cells that normally do not produce a capsule to do so (p. 201).

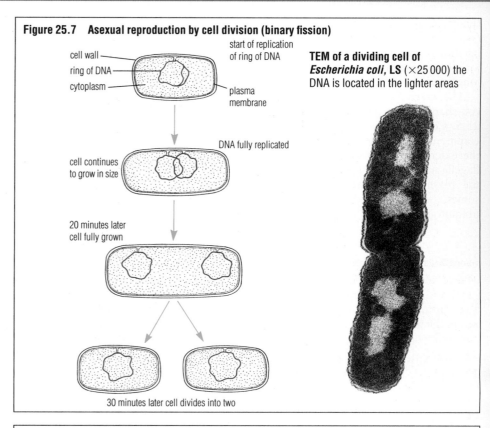

Figure 25.7 Asexual reproduction by cell division (binary fission)

cell wall
ring of DNA
cytoplasm
start of replication of ring of DNA
plasma membrane

cell continues to grow in size — DNA fully replicated

20 minutes later cell fully grown

30 minutes later cell divides into two

TEM of a dividing cell of *Escherichia coli*, LS ($\times 25\,000$) the DNA is located in the lighter areas

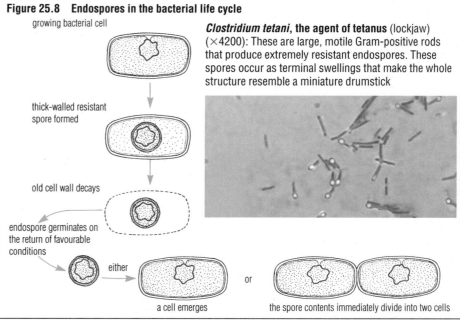

Figure 25.8 Endospores in the bacterial life cycle

growing bacterial cell

thick-walled resistant spore formed

old cell wall decays

endospore germinates on the return of favourable conditions

either a cell emerges or the spore contents immediately divide into two cells

***Clostridium tetani*, the agent of tetanus** (lockjaw) ($\times 4200$): These are large, motile Gram-positive rods that produce extremely resistant endospores. These spores occur as terminal swellings that make the whole structure resemble a miniature drumstick

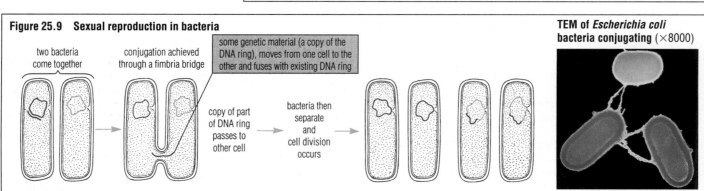

Figure 25.9 Sexual reproduction in bacteria

two bacteria come together

conjugation achieved through a fimbria bridge

some genetic material (a copy of the DNA ring), moves from one cell to the other and fuses with existing DNA ring

copy of part of DNA ring passes to other cell

bacteria then separate and cell division occurs

TEM of *Escherichia coli* bacteria conjugating ($\times 8000$)

25.4 PROTOCTISTA: ALGAE

Algae are eukaryotic cellular plants with a cell structure very similar to that of higher plants. The subkingdom consists of an array of photosynthetic organisms, divided into many phyla. The algae show great variety in plant form, ranging from unicellular organisms and simple filaments to giant seaweeds. Many unicellular forms are motile, possessing one or more flagella per cell. Algae usually live in water, although they occasionally grow on damp soil or moist surfaces. Classification of the algae is based on biochemical evidence (photosynthetic pigments, the nature of food reserves), and on structural features (cell wall structure, and the form of the flagella).

■ Chlamydomonas

Chlamydomonas is a genus of tiny, motile, freshwater algae belonging to the phylum Chlorophyta, the green algae (p. 30). The green algae have the same photosynthetic pigments as most terrestrial green plants. In structure they are mostly either unicellular or filamentous forms.

Chlamydomonas is common in water that is rich in ammonium ions (for example, ditch water contaminated by farmyard effluents). It moves through the water by the rapid beating of the two flagella that protrude through the cellulose wall at the anterior end (Figure 25.10). There is a coloured eyespot, believed to be sensitive to light intensity and direction, and a starch storage point or **pyrenoid**. Despite the presence of the cellulose cell wall, there are two contractile vacuoles at work in the cytoplasm, pumping out excess water (p. 378).

■ Chlorella

The unicellular alga *Chlorella* occurs in freshwater ponds, often colouring the water bright green. It grows best when the pH is slightly alkaline, and where the water is rich in nutrients from silt and decaying organic matter. The cells are minute spheres surrounded by a cell wall containing cellulose. Each has a large, cup-shaped chloroplast with a starch storage point, the pyrenoid, and colourless, vacuolated cytoplasm (Figure 25.11).

Asexual reproduction in Chlorella

During active growth the cell periodically divides into four protoplasts, which function as spores. Spores of this type, formed within the parent cell, are known as **autospores**. These then develop walls, enlarge, and break out from the mother

cell's wall. This is the only form of reproduction that *Chlorella* undergoes.

Chlorella as an experimental organism

Chlorella is a popular research organism with biochemists studying aspects of green plant metabolism. It is easily maintained in large and uniform populations, held under rigidly controlled environmental conditions. It has been extensively used in the investigations of the 'light' and 'dark' reactions of photosynthesis (p. 262), and in biotechnological research, where it has potential as an organism to use to harness the Sun's energy.

■ Pleurococcus

Pleurococcus, the most common green alga, occurs in small aggregations of cells on damp surfaces such as walls. *Pleurococcus* can be used as an indicator of damp conditions: it causes surfaces to appear bright green in wet weather, and look grey-green when dry. *Pleurococcus* reproduces by cell division (Figure 25.12).

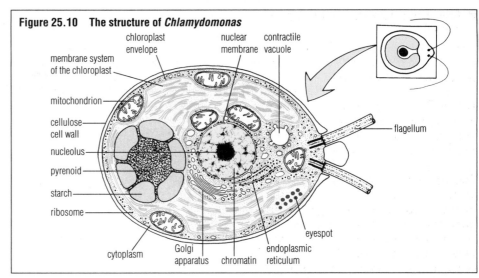

Figure 25.10 The structure of *Chlamydomonas*

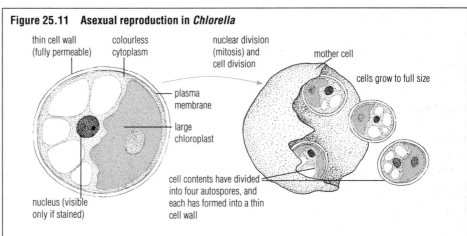

Figure 25.11 Asexual reproduction in *Chlorella*

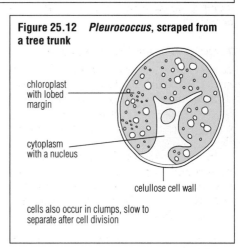

Figure 25.12 *Pleurococcus*, scraped from a tree trunk

cells also occur in clumps, slow to separate after cell division

Questions

1 Why are the algae and fungi now recognised as entirely separate Kingdoms (p. 24)?
2 What does *Chlamydomonas* require from its environment to sustain its nutrition?
3 What are the advantages of using a culture of *Chlorella* cells to investigate metabolism in a green plant cell?
4 What environmental hazards is the alga *Pleurococcus* likely to experience when it is growing on trees near an industrial area?

■ *Spirogyra*

Spirogyra is an unbranched, filamentous alga, found in ponds, lakes and slow-flowing streams. It grows in spring, attached to stones and other surfaces by means of the first cell in the filament, known as the holdfast. The filament is slimy to the touch, due to a layer of mucilage that coats the surface and holds the filaments together in a matted blanket. Bubbles of gas (mostly oxygen produced by photosynthesis) can become trapped, supporting the mass of filaments near the surface of the water, to obvious advantage photosynthetically. The chloroplasts are ribbon-shaped, with wavy edges. Each cell contains from one to seven chloroplasts, lying helically coiled in the layer of cytoplasm lining the long cell wall (Figure 25.13).

Life cycle of *Spirogyra*

A *Spirogyra* filament grows in length by cell division involving mitosis. Divisions occur anywhere along the length of a filament, which may become very long indeed. Break-up of the filament brings about asexual reproduction (Figure 25.15).

Sexual reproduction occurs at the end of the growing season, late in spring, when two filaments come together and undergo conjugation. Normally, one filament consists of 'male' cells; the contents of each cell migrate through a conjugation tube into a 'female' cell in the adjacent filament. The cell contents fuse, forming a zygote (Figure 25.14). Here the cell contents function as gametes. Since the cells are identical in size this is another example of isogamy (p. 582).

The zygote develops a thick, pigmented, protective wall to form a **zygospore**. This structure survives the unfavourable season at the bottom of the pond, where the parent cell walls rot away. When temperatures again rise the zygospore nucleus undergoes meiosis, but of the four nuclei produced, three abort. One haploid nucleus divides mitotically to form a fresh filament, which may be either 'male' or 'female' in its subsequent conjugation behaviour.

Figure 25.13   The vegetative structure of a *Spirogyra* filament

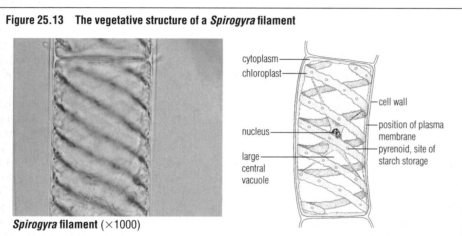

Spirogyra filament (×1000)

cytoplasm
chloroplast
cell wall
nucleus
position of plasma membrane
large central vacuole
pyrenoid, site of starch storage

Figure 25.14   Sexual reproduction in *Spirogyra*

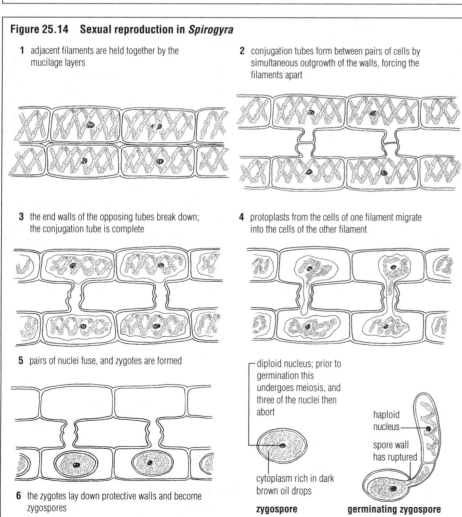

1 adjacent filaments are held together by the mucilage layers

2 conjugation tubes form between pairs of cells by simultaneous outgrowth of the walls, forcing the filaments apart

3 the end walls of the opposing tubes break down; the conjugation tube is complete

4 protoplasts from the cells of one filament migrate into the cells of the other filament

5 pairs of nuclei fuse, and zygotes are formed

6 the zygotes lay down protective walls and become zygospores

diploid nucleus; prior to germination this undergoes meiosis, and three of the nuclei then abort

haploid nucleus

spore wall has ruptured

cytoplasm rich in dark brown oil drops

zygospore

germinating zygospore

Figure 25.15   Life cycle of *Spirogyra*

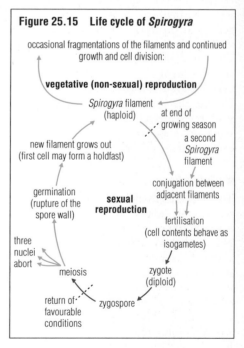

occasional fragmentations of the filaments and continued growth and cell division:

vegetative (non-sexual) reproduction

Spirogyra filament (haploid)

at end of growing season

a second *Spirogyra* filament

new filament grows out (first cell may form a holdfast)

conjugation between adjacent filaments

germination (rupture of the spore wall)

sexual reproduction

fertilisation (cell contents behave as isogametes)

three nuclei abort

meiosis

zygote (diploid)

return of favourable conditions

zygospore

Questions

1   How do you think the mucilage reaches the outer surface of *Spirogyra* filaments?

2   What is the significance of meiosis in the life cycle of *Spirogyra*?

3   What features of the structure and life cycle of *Fucus* (see opposite) can be described as being adaptations to its habitat?

Figure 25.16 *Fucus*: structures and sexual reproduction

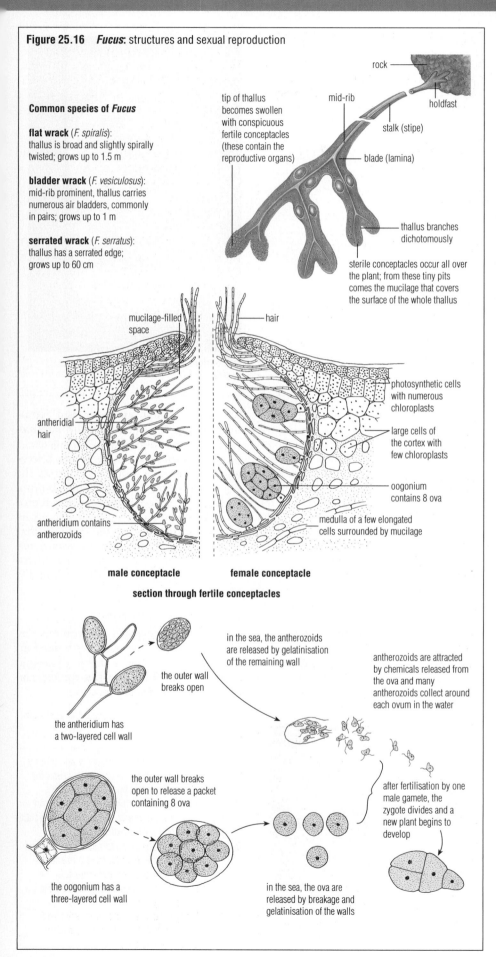

Common species of *Fucus*

flat wrack (*F. spiralis*):
thallus is broad and slightly spirally
twisted; grows up to 1.5 m

bladder wrack (*F. vesiculosus*):
mid-rib prominent, thallus carries
numerous air bladders, commonly
in pairs; grows up to 1 m

serrated wrack (*F. serratus*):
thallus has a serrated edge;
grows up to 60 cm

tip of thallus
becomes swollen
with conspicuous
fertile conceptacles
(these contain the
reproductive organs)

rock
mid-rib
holdfast
stalk (stipe)
blade (lamina)
thallus branches
dichotomously

sterile conceptacles occur all over
the plant; from these tiny pits
comes the mucilage that covers
the surface of the whole thallus

mucilage-filled
space
hair
photosynthetic cells
with numerous
chloroplasts
large cells of
the cortex with
few chloroplasts
oogonium
contains 8 ova
medulla of a few elongated
cells surrounded by mucilage
antheridial
hair
antheridium contains
antherozoids

male conceptacle   **female conceptacle**

section through fertile conceptacles

the antheridium has
a two-layered cell wall

the outer wall
breaks open

in the sea, the antherozoids
are released by gelatinisation
of the remaining wall

antherozoids are attracted
by chemicals released from
the ova and many
antherozoids collect around
each ovum in the water

the outer wall breaks
open to release a packet
containing 8 ova

after fertilisation by one
male gamete, the
zygote divides and a
new plant begins to
develop

the oogonium has a
three-layered cell wall

in the sea, the ova are
released by breakage and
gelatinisation of the walls

■ *Fucus*

Fucus ('wracks') is a genus of marine, intertidal algae common on the coasts of Europe, North America and parts of Asia. It belongs to the phylum Phaeophyta or brown algae. Several species occur on the rocks, breakwater posts and piers around the shores of Britain, including the three common wrack species described in Figure 25.16. The plants are adapted to withstand the limited exposure and mild desiccation that occurs between high tides (p. 48).

The plant body of a wrack is known as a **thallus**, and consists of a **holdfast**, a **stipe** (a 'stalk') and a **lamina** or blade (Figure 25.16). The holdfast grips on to an irregular surface, and is very resistant to being dislodged. The stipe extends from the holdfast into the lamina as the mid-rib, where it supports the lamina, the photosynthetic tissue. The thallus divides into equal branches (dichotomous branching) at regular intervals. The whole structure is tough and leathery, and because it is also covered by a film of mucilage, it is slippery. As a consequence it is resistant to damage by wave action.

Sexual reproduction in *Fucus* spp.

Gametes are formed in fertile **conceptacles**, which occur in conspicuous swollen tips of the thallus. Meiosis occurs during gamete formation. The female gametes (ova) are relatively large and non-motile. The male gametes (antherozoids) are small and motile. When gametes differ in this way the organism is described as **oogamous**, and this type of sexual reproduction is **oogamy**.

Gametes are released into the sea, where fertilisation occurs (Figure 25.17). Development follows immediately, and when a simple holdfast forms the tiny plant becomes attached as it grows.

Figure 25.17   The life cycle of *Fucus*

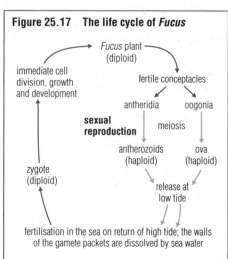

Fucus plant
(diploid)

immediate cell
division, growth
and development

fertile conceptacles

antheridia   oogonia

**sexual
reproduction**   meiosis

antherozoids   ova
(haploid)   (haploid)

zygote
(diploid)

release at
low tide

fertilisation in the sea on return of high tide; the walls
of the gamete packets are dissolved by sea water

25.5 PROTOCTISTA: PROTOZOA

Protozoans are a diverse group of eukaryotic, unicellular organisms. Most have heterotrophic nutrition, and their reproduction frequently involves fission of the cell after nuclear division.

■ Amoeba

The amoebae belong to the phylum Rhizopoda (p. 28). There are many different species of amoeba, some of which belong to the genus *Amoeba*, including one of the largest amoebae, the freshwater species *Amoeba proteus*. Most amoebae live in fresh or salt water, usually on a muddy substratum. A few occur in the soil. The parasite *Entamoeba histolytica* causes amoebic dysentery in humans.

The structure of *Amoeba proteus* is illustrated on p. 291. The movement of this organism involves putting out pseudopodia and flowing into them – for example, in the capture of food.

Amoeba reproduces asexually by binary fission (Figure 25.18), normally after a period of growth sustained by the assimilation of nutrients. It feeds by engulfing and digesting other, smaller microorganisms found in the mud around it.

No sexual reproduction has been observed in amoebae, and *Amoeba proteus* is not known to form any protective wall around itself when conditions in the environment become unfavourable.

■ Paramecium

Paramecium caudatum (p. 29) lives in stagnant or slow-flowing water rich in organic matter. It is an elaborately structured protozoan with a fixed body shape, roughly the shape of the sole of a shoe. *Paramecium*, the 'slipper organism', is a member of the phylum Ciliophora

Figure 25.18   Binary fission in *Amoeba*

Amoeba rounds off

nucleus divides by mitosis

cytoplasm divides into equal portions, each with a nucleus

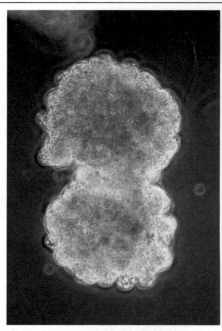

photomicrograph of *Amoeba* in binary fission (×25)

(ciliates), the body being covered by rows of cilia. The surface of the organism, the **pellicle**, is a stiff but flexible protein covering that looks as if it consists of tiny hexagonal plates. A single cilium is anchored at the centre of each plate. *Paramecium* is propelled through the water, rotating as it goes, by the coordinated, rhythmic beating of its cilia (p. 500). The bases of the cilia, in the outer plasmagel layer of the cytoplasm (ectoplasm), are connected by a system of fibres. Also in the ectoplasm, connected to the pellicle at the margin of the plates, is a system of sac-like **trichocysts**. Although these may discharge threads when stimulated, their function is unknown. Most of the cytoplasm is granular endoplasm; it contains large numbers of tiny green unicellular algae,

which live, photosynthesise and reproduce within the shelter of the *Paramecium*. *Paramecium* has two nuclei: a larger macronucleus and a tiny micronucleus. The macronucleus controls the organism's structure and metabolism, whereas the micronucleus controls reproductive activity. The oral groove is a permanent intucking of the pellicle that terminates as the **cytostome**. *Paramecium* feeds on bacteria and other tiny microorganisms living in the water around it (p. 291).

Reproduction in *Paramecium*

Paramecium normally reproduces by binary fission (asexual reproduction) (Figure 25.19), but a form of sexual reproduction, involving the conjugation of two individuals, also occurs.

Figure 25.19   Binary fission in *Paramecium*

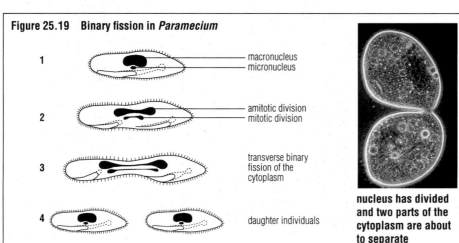

1 — macronucleus, micronucleus

2 — amitotic division, mitotic division

3 — transverse binary fission of the cytoplasm

4 — daughter individuals

nucleus has divided and two parts of the cytoplasm are about to separate

Questions

1 List all the materials that may be taken in by *Amoeba* and state the ways these cross the plasma membrane (Chapter 11, p. 228).

2 Why is water that is rich in suspended and dissolved organic matter likely to be especially favourable to *Paramecium*?

3 The macronucleus of *Paramecium* is derived from the micronucleus formed by sexual reproduction. Why is it important that the macronucleus has the same genes as the micronucleus?

4 How does the microstructure of cilia and flagella facilitate their role in locomotion (p. 165)?

25.6 FUNGI

Fungi comprise the moulds, yeast, mildews, mushrooms, puffballs and rusts. The study of fungi is called **mycology**.

Fungi have heterotrophic nutrition, many as saprotrophs, feeding on dead organic matter. Others are parasites, particularly of plants (p. 547). The fungal body, the **mycelium**, consists of fine, colourless branching threads called **hyphae**. Fungi are eukaryotes. A hypha has an outer wall, usually composed predominantly of chitin (Figure 25.20). Often the internal organisation is **coenocytic** – that is, the hyphae are not divided into uninuclear cells; instead, many nuclei occur together. In fact cytoplasm and nuclei tend to collect at the growing tips of the hyphae. Older hyphae contain a large vacuole and little cytoplasm.

At some stage in the life cycle spores are produced on a specialised part of the mycelium. Typically, vast numbers of spores are produced and dispersed. If these chance upon a favourable environment, a germ tube grows out and forms a branching hypha.

■ *Mucor*

Mucor is a pin mould fungus, so called because it produces spores in a **sporangium** at the tip of a fine aerial hypha called a **sporangiophore**, held high above the feeding mycelium. The whole structure looks not unlike a pin. Pin moulds are classified in the phylum Zygomycota (p. 31), named after the resistant spore formed by sexual reproduction.

Mucor commonly grows in the soil and on the dung of herbivorous animals. Other pin mould genera, such as *Rhizopus*, are common on mouldy bread. The hyphae secrete digestive enzymes on to the substrate on which they grow, digesting any proteins, lipids and polysaccharides present. The hyphal threads, particularly the tips of hyphae that are actively growing, absorb the amino acids, fatty acids and sugars formed.

Spore production and dispersal

Provided the environment remains humid and moist, a colony produces spores within a few days of becoming established on a food source. From within a sporangium, spores are produced and dispersed by air currents, insects and water droplets. When fungal spores in the air are sampled and grown, only about 0.2% are found to be pin mould spores, and most of these are *Rhizopus* spores. This finding can be correlated with the dispersal mechanisms adopted (Figure 25.21, overleaf).

Sexual reproduction

Sexual reproduction in *Mucor* can be observed in laboratory cultures, but is rare in nature. This is because there are two strains of many species of *Mucor* (**heterothallic** strains, which look very similar, designated + and –), which have to grow together before sexual reproduction can occur. The product of sexual reproduction is a zygospore, a resistant structure with a thick protective wall. Under favourable conditions the zygospore germinates and a sporangiophore with sporangium grows (Figures 25.22 and 25.23, overleaf).

Questions

1 What type of nutrition does *Mucor* exhibit?
2 What is the role of asexual spores in the life cycle of *Mucor*? What features of asexual reproduction support your answer?

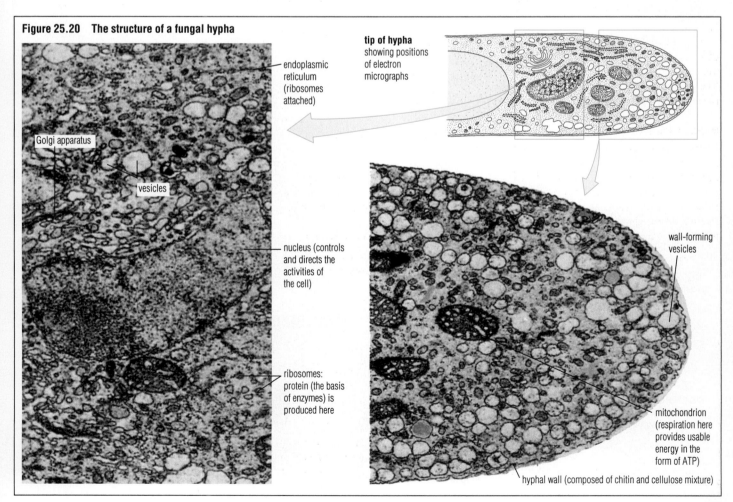

Figure 25.20   The structure of a fungal hypha

Golgi apparatus

vesicles

endoplasmic reticulum (ribosomes attached)

tip of hypha showing positions of electron micrographs

nucleus (controls and directs the activities of the cell)

ribosomes: protein (the basis of enzymes) is produced here

wall-forming vesicles

mitochondrion (respiration here provides usable energy in the form of ATP)

hyphal wall (composed of chitin and cellulose mixture)

Figure 25.21 **Asexual reproduction in *Mucor*, compared to that in *Rhizopus*:** in both fungi, the branching feeding mycelium starts asexual reproduction and spore formation soon after it becomes established in a new substrate

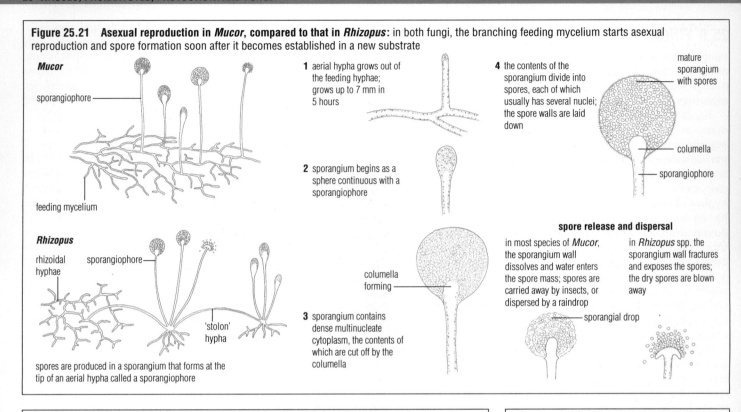

Mucor

sporangiophore

feeding mycelium

Rhizopus

rhizoidal hyphae

sporangiophore

'stolon' hypha

spores are produced in a sporangium that forms at the tip of an aerial hypha called a sporangiophore

1 aerial hypha grows out of the feeding hyphae; grows up to 7 mm in 5 hours

2 sporangium begins as a sphere continuous with a sporangiophore

columella forming

3 sporangium contains dense multinucleate cytoplasm, the contents of which are cut off by the columella

4 the contents of the sporangium divide into spores, each of which usually has several nuclei; the spore walls are laid down

mature sporangium with spores

columella

sporangiophore

spore release and dispersal

in most species of *Mucor*, the sporangium wall dissolves and water enters the spore mass; spores are carried away by insects, or dispersed by a raindrop

in *Rhizopus* spp. the sporangium wall fractures and exposes the spores; the dry spores are blown away

sporangial drop

Figure 25.22 **Sexual reproduction in *Mucor*, compared to that in *Rhizopus*:** most pin moulds require two strains to grow together before sexual reproduction can occur, but *Rhizopus sexualis* is an exception

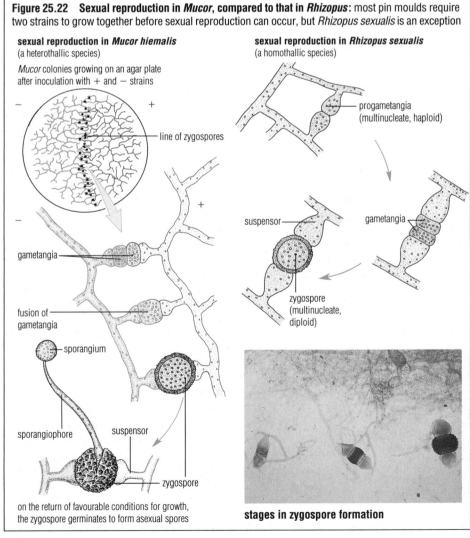

sexual reproduction in *Mucor hiemalis*
(a heterothallic species)

Mucor colonies growing on an agar plate after inoculation with + and − strains

line of zygospores

gametangia

fusion of gametangia

sporangium

sporangiophore

suspensor

zygospore

on the return of favourable conditions for growth, the zygospore germinates to form asexual spores

sexual reproduction in *Rhizopus sexualis*
(a homothallic species)

progametangia (multinucleate, haploid)

suspensor

gametangia

zygospore (multinucleate, diploid)

stages in zygospore formation

Figure 25.23 **The life cycle of *Mucor***

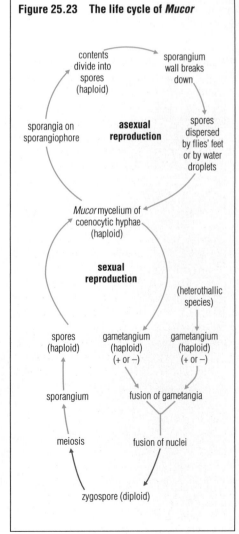

contents divide into spores (haploid)

sporangium wall breaks down

asexual reproduction

sporangia on sporangiophore

spores dispersed by flies' feet or by water droplets

Mucor mycelium of coenocytic hyphae (haploid)

sexual reproduction

(heterothallic species)

spores (haploid)

gametangium (haploid) (+ or −)

gametangium (haploid) (+ or −)

sporangium

fusion of gametangia

meiosis

fusion of nuclei

zygospore (diploid)

Neurospora

Neurospora is classified as a member of the phylum Ascomycota (the sac fungi, p. 31) because it produces special spores, **ascospores**, in a sac or **ascus**. The Ascomycota are the largest phylum within the fungal kingdom. In most members the asci are formed in a special fruiting body, which is a cup- or flask-shaped structure. *Neurospora* is a good example (Figure 25.24). (The yeasts, are also members of this phylum, however, and yet they form no fruiting body.)

Neurospora is an organism much favoured by experimental geneticists! The reasons for this are listed on p. 210. Most important among these is the fact that the ascospores (produced by meiosis and mitosis from a zygote nucleus) form a linear series in the narrow, tubular ascus, and can be separated and grown individually. Changes induced in their chromosomes can often be followed through to the organisms produced.

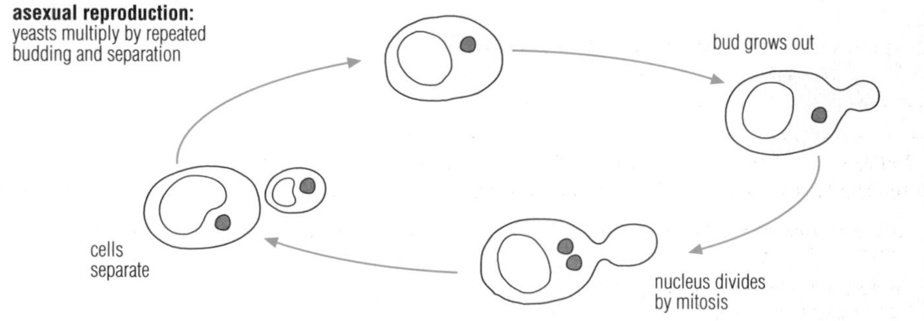

Saccharomyces

The name 'Saccharomyces' means 'sugar fungi'. *Saccharomyces* spp. are the yeasts. They are unicellular fungi commonly found in the liquids exuded on to the surface of fruits, flowers and leaves of plants. They also occur in soils, on animal faeces and in milk. The yeasts feed saprotrophically, mostly upon the sugars in the medium around them. Most yeasts respire anaerobically (p. 316), producing ethanol, carbon dioxide and a little energy. Yeasts are of great economic importance in biotechnological industries, both ancient and modern (p. 112). The structure of a yeast cell is shown in Figure 14.20 (p. 298).

Yeasts are classified as Ascomycota. This is because, under rather specialised conditions, many species of yeast can be induced to form ascospores. Most of the time, however, yeast cells bud repeatedly, separating into two cells, provided that conditions are favourable for growth. The new cells rapidly grow to full size and divide again. This cycle of growth and vegetative cell division represents asexual reproduction (Figure 25.25).

Agaricus

The common field mushroom, *Agaricus campestris* (Figure 25.26, overleaf), is a member of the Basidiomycota. The Basidiomycota produce special spores, **basidiospores**, on the surface of a club-shaped sporangium called a **basidium**. The spores form at the tips of tiny

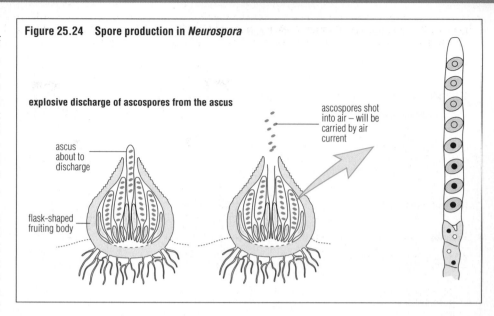

Figure 25.24 Spore production in *Neurospora*

explosive discharge of ascospores from the ascus

ascus about to discharge

flask-shaped fruiting body

ascospores shot into air – will be carried by air current

Figure 25.25 Asexual reproduction in baker's yeast

asexual reproduction: yeasts multiply by repeated budding and separation

bud grows out

cells separate

nucleus divides by mitosis

projections on the swollen end of the basidium, and each contains a single haploid nucleus. The spores are shot into the air, and are then carried away in air currents. The phylum Basidiomycota contains some of the largest fungi, including the horse mushroom (*A. arvensis*), large bracket fungi, and the puffballs and stinkhorns. The names 'mushroom' and 'toadstool' are often used for the umbrella-shaped fruiting bodies of many members of the phylum, but these terms have no precise meaning; they are alternative common names for a whole range of fascinating structures that generate much interest, in season.

The fungal feeding mycelium grows throughout the year, typically in the soil or in rotting wood. The mycelia of the Basidiomycota consist of colourless hyphal threads. As these grow they usually fuse, forming a three-dimensional network. The hyphal threads are often divided by septa into cells that each contain two nuclei within the cytoplasm. The mycelium absorbs nutrients and mobilises food reserves, concentrating these at points where a fruiting body forms. The fruiting body is built up from branching and interwoven hyphae, tightly packed together, and it is on the fruiting body that spores are produced and released into the air. Many animals nibble the fruiting bodies and inadvertently help in spore dispersal.

Spores germinate in suitable conditions. Initially there is a single nucleus per cell, but early on in growth, fusion of hyphae growing from another spore forms cells with two nuclei per cell. Later on, during reproduction, these nuclei fuse together to form a diploid nucleus, in each basidium in the fruiting body. The diploid nucleus undergoes meiosis, in the process of basidiospore formation (Figure 25.27, overleaf).

Questions

1 Why are microorganisms like *Neurospora* especially useful in experimental genetics?
2 Honey bees are sometimes inebriated when they return to the hive after foraging expeditions to flowers on warm sunny days. Why may this be so?
3 Why can cell division in yeasts be described as asexual reproduction, while cell division in *Spirogyra* (p. 554) is not?

Figure 25.26 Basidiospore formation and dispersal in *Agaricus*

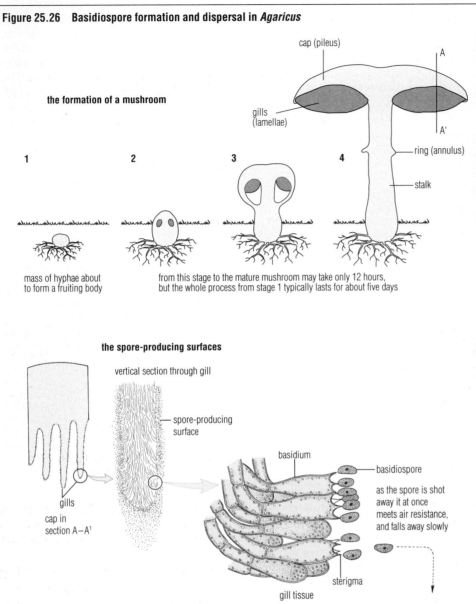

the formation of a mushroom

cap (pileus)

gills (lamellae)

ring (annulus)

stalk

1
2
3
4

mass of hyphae about to form a fruiting body

from this stage to the mature mushroom may take only 12 hours, but the whole process from stage 1 typically lasts for about five days

the spore-producing surfaces

vertical section through gill

spore-producing surface

gills

cap in section A–A¹

basidium

basidiospore

as the spore is shot away it at once meets air resistance, and falls away slowly

sterigma

gill tissue

the habitat of the common field mushroom, *Agaricus campestris*

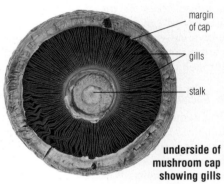

margin of cap

gills

stalk

underside of mushroom cap showing gills

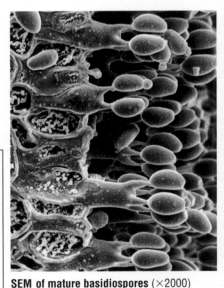

SEM of mature basidiospores (×2000)

Figure 25.27 The life cycle of *Agaricus*

Agaricus mycelium in the soil: network of cellular hyphae with two nuclei per cell (haploid)

hyphal fusion with compatible mycelium

mushroom fruiting body formed

mycelium of cellular hyphae with one nucleus per cell

gills with developing spore-producing surface (hymenium)

spore germination

binucleate basidia

basidiospore dispersal

nuclei fuse (diploid basidia)

four basidiospores per basidium (uninucleate, haploid)

meiosis

Questions

1   What type of nutrition is shown by the common field mushroom?
2   What are the essential differences between ascospores and basidiospores?
3   How are basidiospores discharged? How are they normally dispersed to new habitats?

Edible and poisonous Basidiomycota

Several fungi are very good to eat, although some need to be cooked first. A list of the best edible fungi is given in Roger Phillips's well-illustrated book *Mushrooms and Other Fungi of Great Britain and Europe*. Most 'mushrooms and toadstools' are harmless but worthless as food. But if you are considering the eating of wild mushrooms you *must* first learn to recognise and identify all you find. This is because several either are deadly poisonous, or have unpleasant effects if eaten. For example, *Amanita phalloides*, the death cap, is the most common of the lethally poisonous ones. If it is eaten, there is normally a lapse of 6–24 hours before the onset of symptoms. No antidote exists against its effects; the toxins cause failure of liver and kidneys, and frequently death.

A poisonous fungus much easier to identify is the fly agaric, *Amanita muscaria*. This fungus contains substances attacking the central nervous system, acting as a strong hallucinogen and intoxicant. The effects of consuming the fungus are unpredictable. Although not as poisonous as *A. phalloides*, it has caused the death of humans, and certainly should never be eaten. This colourful fungus is so named because, since medieval times, the cap has been sliced into a saucer of milk to create a mixture that killed flies!

Penny bun (*Boletus edulis*), edible

Chanterelle (*Cantharellus cibarius*), edible

Fly agaric (*Amanita muscaria*), poisonous

Death cap (*Amanita phalloides*), deadly poisonous!

25.7 WHY IS THIS AN IMPORTANT CHAPTER?

The organisms discussed in this chapter are of natural, intrinsic interest to biologists who seek to understand how all forms of life operate. In addition, they are of enormous importance in ecological processes, in health and disease, and in biotechnologies old and new. Many play important roles in modern genetics. It is therefore essential to know about their structures and life cycles, and about the conditions that favour their growth. Once the biology of an organism is understood it is often possible to control or prevent effects that are unhelpful, or to exploit the beneficial activities of the organism more effectively.

Some of the issues in which these organisms are involved or implicated are identified here, and a suggestion for further reading is given below.

■ Biodegradation and biodeterioration

Many of the bacteria and fungi have saprotrophic nutrition (p. 296), and consequently play a major part in recycling by digesting organic matter. The recycling of waste materials and dead organisms is the basis of the carbon cycle (p. 247) and several other biogeochemical cycles (p. 332). This is known as **biodegradation**, and it brings back into circulation (recycles) the elements needed by all organisms.

Sometimes these organisms attack materials that are not waste. When saprophytes cause decay of economically useful materials the process is known as **biodeterioration**. This may include the rotting of damp timbers of buildings, and damage to the metal of aircraft frames as bacteria cause decay of the fuel stored in the aircraft wings. These are all forms of saprotrophic nutrition.

■ Disease

The germ theory of disease can be traced to the work of Louis Pasteur and Robert Koch (p. 530). Many diseases of animals (including humans) and plants are caused by viruses, bacteria and fungi (Chapter 24).

■ Organisms exploited in biotechnology

Biotechnology is the industrial application of biological processes. Consult the index to locate descriptions of how viruses, prokaryotes, protoctists or fungi are implicated in the production of alcohol, the manufacture of foods such as cheese and butter, the baking of bread, the production of antibiotics and of single-cell protein, the microbial mining of metal ores, the extraction and immobilisation of enzymes for various purposes, sewage disposal, water purification, genetic engineering to bring about gene transfers between organisms and the exploitation of mycorrhizal fungi to sustain tree growth in otherwise unfavourable locations.

FURTHER READING/WEB SITES

J Postgate (1992) *Microbes and Man.* Cambridge University Press

26 Life cycles of green plants

- Reproduction in green plants involves the production of both spores and gametes. Spores are formed in a diploid phase of the life cycle, on the sporophyte plant. Gametes are formed in a haploid phase of the life cycle, on the gametophyte plant.
- Sporophyte and gametophyte phases alternate, each giving rise to the other. Consequently the life cycles of green plants show alternation of generations.
- In mosses (bryophytes), the gametophyte – the tiny moss plant – is the dominant phase. The sporophyte plant grows on the gametophyte and is dependent upon it.
- In ferns, the sporophyte – the fern plant – is the dominant phase. The gametophyte is a small, independent plant. The fern sporophyte has

stem, leaves and roots with vascular tissues (xylem and phloem), and is fully adapted to terrestrial conditions.
- Conifer trees and flowering plants (woody and herbaceous) are independent, sporophyte plants. The gametophyte phase of the life cycle is greatly reduced, and is retained within the spores of the sporophyte plant. Male spores (microspores) are known as pollen. The female spore (megaspore or embryo sac) remains on the sporophyte in an ovule.
- In flowering plants the reproductive organs are formed within flowers. Pollen is produced within anthers; ovules within ovaries. Many flowers attract insects, which transfer pollen from the male parts to the female parts of flowers.

26.1 INTRODUCTION

This chapter is about the life cycles and the reproductive processes in the organisms that make up the Kingdom Planta (the green plants). Most of these plants are terrestrial organisms, although certain species occur in aquatic habitats. The range of phyla in this kingdom is reviewed in the check-list of organisms on pp. 32–4.

The production of new individuals (reproduction) involves the transfer of genetic material (nucleic acid) from parent organism to offspring. This is true of the subcellular viruses and the unicellular organisms discussed in Chapter 25. Here we saw that reproduction may involve a **sexual process** (the production and fusion of gametes) or an **asexual process** (by spores, or when some part of the parent organism becomes cut off and grows independently of the parent). In fact, all organisms reproduce either sexually or asexually, or by both processes (Table 26.1). Asexual reproduction is not restricted to microorganisms. Most eukaryotes reproduce sexually, however, and sometimes some form of asexual reproduction is observed as well.

Sexual reproduction and meiosis

Organisms with sexual reproduction in their life cycle undergo meiosis at some stage. We can be certain of this because, in fertilisation, two cells (and their nuclei) fuse. At fusion, the number of chromosomes is doubled. Yet a species has a fixed number of chromosomes (p. 192) in the cells that make up the body structure (known as **vegetative** or **somatic cells**). As a consequence, a reductive division (meiosis) must take place at some point in the life cycle, either during the formation of the gametes or at some time after the gametes have fused.

Table 26.1 Sexual and asexual reproduction compared

	Sexual reproduction	Asexual reproduction
parents	normally two parents; parents unisexual (male or female) or hermaphrodite	single parent
gametes	special sex cells, called gametes, are formed; gametes fuse in pairs to form a zygote (fertilisation)	no special sex cells formed
meiosis	at fertilisation, the zygote nucleus (diploid) has double the number of chromosomes of the gametes (haploid); therefore meiosis must occur at some stage in the life cycle	no fusion of cells involved, therefore meiosis is not a necessary part of the process
offspring	offspring are never identical to the parents; variation arises in gamete formation (gametes do not contain identical chromosome sets) and also because fusion of gametes from different parents generates more variety in the genes of the offspring	offspring are genetically identical to the parent (apart from mutations that may arise in somatic cells, the 'sports' of horticulturalists)
occurrence	occurs in most plants and animals	common in plants, protoctists, prokaryotes and fungi
advantages	produces offspring similar but not identical to parents	may produce offspring quickly without a need for two parents

Types of life cycle

Meiotic cell division (p. 196) is a process that halves the chromosome complement of the cells formed to the basic or haploid number. The **haploid** set of chromosomes is a single set. It is present in gametes.

Fertilisation, the process of fusion of two gametes, restores the diploid number of chromosomes. The **diploid** set of chromosomes is a double set. The zygote contains a diploid set of chromosomes.

In a life cycle involving sexual reproduction there must be a haploid and a diploid phase. Furthermore, the timing of meiosis varies in the life cycles of different organisms; there are therefore different types of life cycle. That is, life cycles differ according to duration of the haploid and diploid phases (Figure 26.1).

▶ In a **haploid life cycle** meiosis occurs immediately after fertilisation, and the **haplontic condition** persists for most of the life cycle. The life cycle of *Spirogyra* (p. 554) is an example.

▶ In a **diploid life cycle**, meiosis occurs in the formation of the gametes. After fertilisation the **diplontic condition** persists for most of the life cycle. The life cycle of *Fucus* (p. 555) is an example.

▶ There is also a third type of life cycle, involving significant periods of both haploid and diploid phases, alternating with each other. This **haplodiplontic** type of life cycle is said to show an **alternation of generations**. The simplest green plants show pronounced alternation of generations in their life

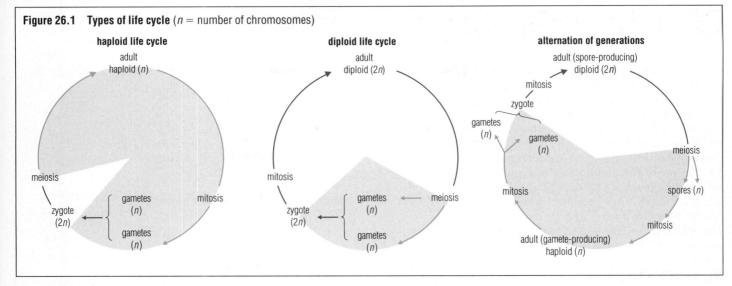

Figure 26.1 **Types of life cycle** (*n* = number of chromosomes)

cycles, but we shall see that *all* green plants show alternation of generations in their life cycles, in various forms. For example, in conifers and flowering plants (angiosperms) alternation of generations exists in a vestigial form (a feature from an earlier stage of evolutionary development).

26.2 PHYLUM BRYOPHYTA (BRYOPHYTES)

Bryophytes, the mosses and liverworts, are small, green terrestrial plants that tend to grow best in permanently moist conditions. Typically, the plant body consists of either a slender, relatively weak stem bearing whorls of leaves, as in the moss *Funaria* (Figure 26.3), or as a green, leaf-like thallus growing over the ground, as in *Pellia* (Figure 26.4) (see also Table 26.2, p. 565). These plants are anchored to the ground by delicate rhizoids (root-hair-like outgrowths of individual cells). Water is absorbed over the whole body surface, for there is no waxy cuticle present.

Bryophytes reproduce both sexually and asexually, and show alternation of generations. In fact the gametophyte and sporophyte generations have quite different structures, and the bryophyte type of life cycle is described as showing heteromorphic alternation of generations (Figure 26.2).

The moss or liverwort plant body just described is the haploid, gamete-producing generation. This plant, called the **gametophyte**, produces both male and female gametes.

The male gametes, called **antherozoids** (spermatozoids), are formed in large numbers in sac-like **antheridia**.

Antherozoids are tiny, biflagellate gametes that swim within films of water. Water is essential for fertilisation.

The female gametes (egg cells) are formed in flask-shaped **archegonia**. Each archegonium has a spherical base containing one egg cell, and a narrow, tubular neck. The cells occupying the neck canal break down before fertilisation, releasing chemicals that attract the antherozoids. These swim down the neck canal and one fuses with the egg cell, forming a zygote.

The zygote develops into a diploid, spore-producing plant, called the **sporophyte**. In bryophytes the sporophyte grows on the gametophyte plant and is dependent upon it (the sporophyte is sometimes described as being 'parasitic' upon the gametophyte). Ultimately the sporophyte produces spores within a capsule. Meiosis occurs during spore formation, and the haploid spores, after being dispersed, grow into new gametophyte plants.

■ The moss *Funaria hygrometrica*

Funaria hygrometrica is a moss that is often one of the first colonisers of soil made bare by the effects of fire. It may remain abundant at these sites in subsequent years. Mosses often grow in dense clumps or tufts, creating a humid microclimate for themselves and for numerous smaller organisms as well. The closely packed cushion habit of mosses allows them to survive and grow in places that would otherwise be too dry for them, since they have little or no protection against desiccation. Possibly this is why mosses are more numerous and more widely distributed than liverworts.

The structure of *Funaria* (Figure 26.3) is quite typical of mosses. The short, upright stem is clad in spirally arranged leaves. The leaves are one cell thick, except at the simple mid-rib. The central cells of the stem are more elongated than the outer cells, and have a conducting and supporting role. Mosses are anchored to the soil by multicellular rhizoids.

Figure 26.2 **Life cycle of *Pellia*:** the life cycle shows heteromorphic alternation of generations

spores germinate

Pellia, the liverwort, is the gametophyte plant (haploid, *n*)

during late spring/summer

capsule splits open: spores dispersed by the wind, with help from the elaters

archegonia (*n*)

antheridia (*n*)

spores (*n*)

antherozoids (*n*)

egg cells (*n*)

meiosis

zygote (2*n*)

capsule (2*n*)

elongates in the following spring

seta

foot

sporophyte plant (diploid, 2*n*) (parasitic on the gametophyte)

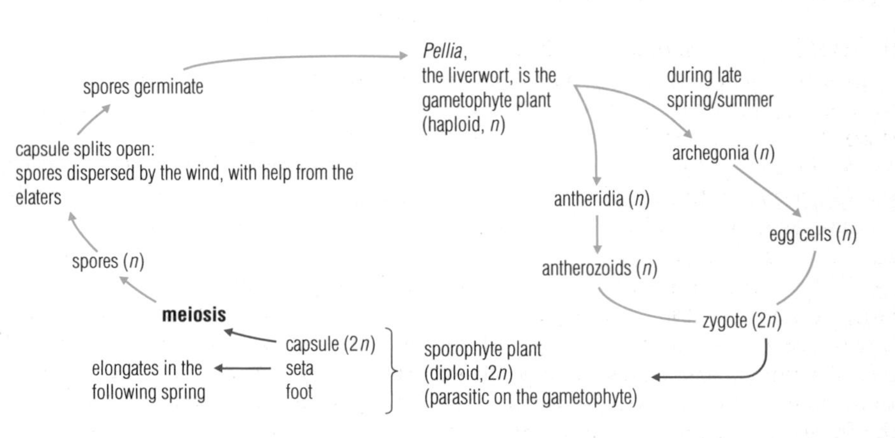

The moss plant produces antherozoids in antheridia, and egg cells in archegonia (Figure 26.3). These sex organs are produced either at the tops of the stems or at the tips of short side-branches. The sex organs are surrounded by a rosette of leaves, and are frequently interspersed with numerous hairs (**paraphyses**). Fertilisation occurs when antherozoids, released from antheridia, swim in the surface film of moisture or are splashed by rain on to the archegonia. Although all the archegonia of one stem tip may be fertilised, only one zygote develops further, forming a mass of cells that differentiate into an absorptive **foot**, which then becomes lodged in the gametophyte tissue. Above the foot the other cells differentiate into a **stalk** (or **seta**) and a **capsule**. The *Funaria* capsule has some photosynthetic tissue, and therefore the sporophyte generation is not totally dependent on the gametophyte generation for carbohydrates.

In the capsule the haploid spores are formed. Spores are dispersed by the wind in dry weather, the sporangium functioning rather like a pepper-pot (Figure 26.5). The spores germinate in moist conditions on the soil surface. From the spore grows a simple plant called a **protonema**, similar to a filamentous alga. On each protonema several 'buds' grow and each may develop into an independent moss plant.

■ The liverwort *Pellia*

Pellia grows in moist conditions on alkaline soils. The thallus is only about one centimetre wide, and dark green in colour (Figure 26.4).

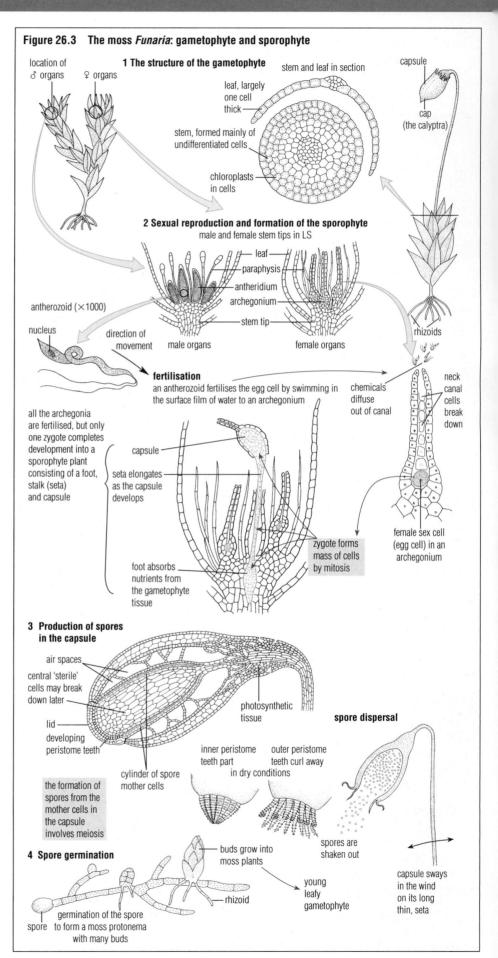

Figure 26.3 The moss *Funaria*: gametophyte and sporophyte

1 The structure of the gametophyte

location of ♂ organs
♀ organs
leaf, largely one cell thick
stem, formed mainly of undifferentiated cells
stem and leaf in section
chloroplasts in cells
capsule
cap (the calyptra)

2 Sexual reproduction and formation of the sporophyte
male and female stem tips in LS

antherozoid (×1000)
nucleus
direction of movement
leaf
paraphysis
antheridium
archegonium
stem tip
male organs
female organs
rhizoids

fertilisation
an antherozoid fertilises the egg cell by swimming in the surface film of water to an archegonium

chemicals diffuse out of canal
neck canal cells break down

all the archegonia are fertilised, but only one zygote completes development into a sporophyte plant consisting of a foot, stalk (seta) and capsule

capsule
seta elongates as the capsule develops
foot absorbs nutrients from the gametophyte tissue
zygote forms mass of cells by mitosis
female sex cell (egg cell) in an archegonium

3 Production of spores in the capsule

air spaces
central 'sterile' cells may break down later
lid
developing peristome teeth
the formation of spores from the mother cells in the capsule involves meiosis
cylinder of spore mother cells
photosynthetic tissue
inner peristome teeth part
outer peristome teeth curl away in dry conditions
spore dispersal
spores are shaken out
capsule sways in the wind on its long thin, seta

4 Spore germination

buds grow into moss plants
young leafy gametophyte
rhizoid
spore
germination of the spore to form a moss protonema with many buds

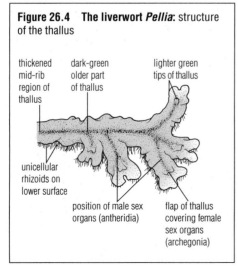

Figure 26.4 The liverwort *Pellia*: structure of the thallus

thickened mid-rib region of thallus
dark-green older part of thallus
lighter green tips of thallus
unicellular rhizoids on lower surface
position of male sex organs (antheridia)
flap of thallus covering female sex organs (archegonia)

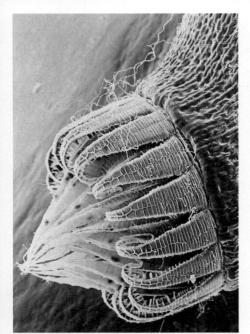

Figure 26.5 SEM of the peristome teeth of a moss capsule (×120)

Table 26.2 Liverworts and mosses compared

Liverworts, e.g. *Pellia*	Mosses, e.g. *Funaria*
rhizoids are unicellular	rhizoids are multicellular
the plant body is usually dorsiventrally flattened into a horizontally growing, branching, and more-or-less ribbon-shaped thallus (some liverworts are 'leafy', with deeply lobed or segmented leaves attached in rows to a stem)	plant body is usually radially symmetrical with a central stem supporting small, spirally arranged leaves
the seta lengthens rapidly after the capsule has reached full size	the seta elongates slowly as capsule develops
the seta is a colourless, semi-transparent structure	the seta at maturity is a slender, opaque, coloured and comparatively tough stalk
the capsule contains elongated, spirally thickened elaters amongst the spores (used in spore dispersal)	the capsule contains no elaters; the capsule is covered by a calyptra

■ The ecological significance of bryophytes

Mosses and liverworts are among the early colonisers of new habitats, sometimes with the lichens (p. 31). These habitats include rocks, walls, fence posts and bare soil, as well as the trunks and branches of trees (an **epiphytic habitat**, Figure 26.6), provided there is sufficient moisture available. In these situations they contribute to the early formation of soil, by aiding weathering of rocks, and to the build-up of organic matter there. They also provide microhabitats for many smaller organisms, and are the dominant primary producers in the food webs of these habitats.

Sphagnum (Figure 26.7) is a moss confined to acid waterlogged habitats.

Figure 26.6 Epiphytes on a woodland tree

Peat, the partially decomposed vegetation found in such habitats, very largely consists of a range of *Sphagnum* species. The leaves of *Sphagnum* include large, empty cells with thickened walls with pores. These cells act as reservoirs for rainwater, allowing the build-up of a raised bog. The cell walls contain organic compounds that exchange ions with the bog solutes, retaining metal cations and releasing hydrogen ions, and thus contributing to the acid environment.

Mosses and liverworts have no waxy cuticle, and they absorb water and ions over the whole plant surface. For this reason, they are vulnerable to air-borne pollution. For example, some bryophyte species are susceptible to the harmful effects of sulphur dioxide (as are most lichens). The presence or absence of these plants is an effective **biological indication** of the extent of atmospheric pollution.

Another consequence of the absence of a waxy cuticle is that bryophytes tend to accumulate pollutants in their cytoplasm. Samples of these plants may be collected periodically and analysed to determine the extent of contamination by toxic metal ions such as chromium, copper, lead and nickel from industrial sources. Bryophytes also accumulate radioactive ions from the air – for example, after accidental releases from nuclear power plants.

■ The evolutionary significance of bryophytes

Bryophytes may possibly have evolved from algal groups. The early form of a moss plant as it grows from a spore (the protonema) is evidence for this. The similarities between the chemistry of photosynthetic pigments and that of cell wall components of the two groups provide further evidence in support of this suggestion. There is, however, virtually no fossil record of the origin of the bryophytes to indicate the evolutionary steps involved.

Further, the bryophytes are thought to share a common ancestor with the ferns. We believe this may be so because the characteristic female sex organ of bryophytes, the archegonium, is also characteristic of the ferns (and of some other plant groups, too).

Although some structural features of bryophytes may have been established when plant life first moved on to dry land, further refinement to bryophyte form may have come much later, at the same time as the rise of the angiosperm forests. Only at that point in the history of the development of plant life did the necessary habitats for an epiphytic way of life become abundant. Evolution of modern bryophytes is almost certainly linked to the evolution of the angiosperm tree habit.

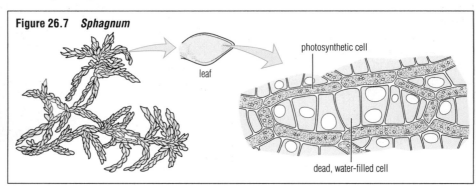

Figure 26.7 *Sphagnum*

leaf

photosynthetic cell

dead, water-filled cell

26.3 PHYLUM FILICINOPHYTA (THE FERNS)

Ferns grow in many habitats, ranging from mountain sides to woodlands, and from wall and rock faces to swamps, ditches and hedge banks. They are terrestrial plants consisting of stem, leaf (also known as a **frond**) and root. They contain vascular tissue in vascular bundles (p. 407) that serve all parts of the plant. The vascular bundles contain food-transporting cells (phloem) and water-transporting cells (tracheids). The aerial parts are covered by a waxy cuticle, enabling the plant to resist desiccation in dry conditions. The epidermis of the leaves and stem has stomata (p. 322) through which gaseous diffusion occurs. Thus, ferns show some of the structural complexity of the flowering plants.

The life cycle of ferns has two independent phases, with an alternation of generations between them. The dominant generation is the spore-bearing or sporophyte plant. Spores are the means by which the fern plant achieves both asexual reproduction and dispersal. A spore that lands in favourable conditions germinates and grows into a tiny, independent green plant, the gametophyte. Although fern gametophytes are quite small plants they are remarkably resistant to unfavourable environmental conditions, sometimes much more so than the corresponding sporophytes.

The gametophyte is the phase of sexual reproduction, for it bears the male and female sex organs (antheridia and flask-shaped archegonia respectively). Antheridia and archegonia are rarely produced simultaneously. An initial male phase is followed by a female phase, although a few antheridia may continue to be produced during the latter. The gametophyte is not protected against desiccation, and the male gamete is motile, requiring a surface film of water in which to swim to the archegonium. Thus ferns, despite their structural complexity, are not entirely independent of extremely moist conditions for the successful completion of their life cycle.

■ *Dryopteris filix-mas*, the male fern

The male fern has a short, semi-upright rhizome with numerous roots below ground, supporting a crown of leaves at the surface. It often grows to be a large plant, with leaves up to 200 cm long (Figure 26.10). It is found in shade or direct light, in banks, woodland and among rocks. Another fern, widespread throughout the world, is *Pteridium aquilinum* (bracken). *Pteridium* has a horizontal, branching rhizome whose growth enables it to invade new ground relatively quickly. This fern is an indicator of acidic soil conditions in Britain.

In mature ferns spores are produced on the underside of the leaves, in structures called **sori** (Figure 26.9). These consist of clusters of stalked sporangia, protected by a flap of tissue, the **indusium**. In *Pteridium* the sori occur around the margins of the leaves and the indusium is brown, whereas in *Dryopteris* the indusia are orange discs, situated over veins in the leaf tissue. Meiosis occurs during spore production. The haploid spores are ejected from the sporangia when these flick open and snap shut as they dry out (Figure 26.10), and are dispersed by the wind. Spore output by ferns is phenomenal; it has been estimated that 300 million spores are released by a typical bracken frond. Ferns have just one type of spore, and the leaves bearing the sporangia are often otherwise indistinguishable from the vegetative leaves.

The spores germinate on soil in moist conditions, and a tiny, heart-shaped plate of cells, the fern **prothallus**, is formed. The prothallus bears the sex organs, which, like the gametes they produce, are very similar in structure and behaviour to those of the bryophytes. Fertilisation is brought about in a similar manner. The zygote divides and grows, and an independent plant is formed (Figure 26.10). The life cycle of *Dryopteris* is summarised in Figure 26.8.

Figure 26.9 *Pteridium* (bracken) leaf, showing sori on the underside

Questions

1 How would you classify the response of antherozoids as the neck canal cells of an archegonium break down (p. 422)?

2 How does the 'parasitic' dependence of the bryophyte sporophyte differ from the usual definition of parasitism?

3 State the ways in which the 'cushion' habit of many moss species is advantageous to
   a the moss,
   b the microfauna of the same habitat.

4 What is the economic significance of the genus *Sphagnum*?

5 a How does a wind of dry air favour spore dispersal in *Funaria*?
   b How does this compare with the conditions necessary for male gamete mobility?

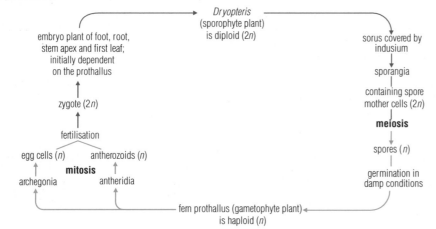

Figure 26.8 **Life cycle of *Dryopteris*:** the life cycle shows heteromorphic alternation of generations

Figure 26.10 The life cycle of *Dryopteris* (male fern)

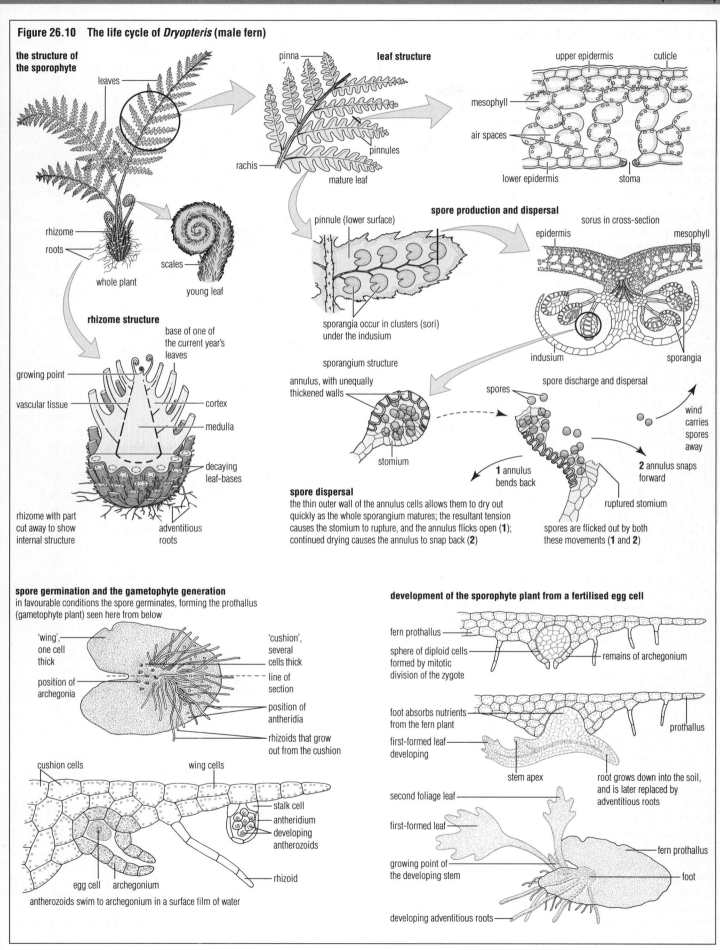

the structure of the sporophyte

leaves

rhizome

roots

whole plant

scales

young leaf

leaf structure

pinna

pinnules

rachis

mature leaf

upper epidermis

cuticle

mesophyll

air spaces

lower epidermis

stoma

rhizome structure

growing point

vascular tissue

base of one of the current year's leaves

cortex

medulla

decaying leaf-bases

adventitious roots

rhizome with part cut away to show internal structure

pinnule (lower surface)

sporangia occur in clusters (sori) under the indusium

spore production and dispersal

sorus in cross-section

epidermis

mesophyll

indusium

sporangia

sporangium structure

annulus, with unequally thickened walls

stomium

spore discharge and dispersal

spores

wind carries spores away

1 annulus bends back

2 annulus snaps forward

ruptured stomium

spores are flicked out by both these movements (**1** and **2**)

spore dispersal

the thin outer wall of the annulus cells allows them to dry out quickly as the whole sporangium matures; the resultant tension causes the stomium to rupture, and the annulus flicks open (**1**); continued drying causes the annulus to snap back (**2**)

spore germination and the gametophyte generation

in favourable conditions the spore germinates, forming the prothallus (gametophyte plant) seen here from below

'wing', one cell thick

position of archegonia

'cushion', several cells thick

line of section

position of antheridia

rhizoids that grow out from the cushion

cushion cells

wing cells

stalk cell

antheridium

developing antherozoids

egg cell   archegonium

rhizoid

antherozoids swim to archegonium in a surface film of water

development of the sporophyte plant from a fertilised egg cell

fern prothallus

sphere of diploid cells formed by mitotic division of the zygote

remains of archegonium

foot absorbs nutrients from the fern plant

first-formed leaf developing

prothallus

stem apex

root grows down into the soil, and is later replaced by adventitious roots

second foliage leaf

first-formed leaf

growing point of the developing stem

fern prothallus

foot

developing adventitious roots

26.4 PHYLUM CONIFEROPHYTA (CONIFERS)

The conifers are cone-bearing plants producing seeds. Most are large trees, many of them with needle-like leaves, and they constitute the dominant vegetation of the boreal forests (p. 45). Conifers survive on poor soils largely through their association with soil fungi in the formation of mutualistic mycorrhizal roots (p. 67).

■ *Pinus sylvestris* (the Scots pine)

The Scots pine is native to Britain, a dominant tree in parts of the Scottish Highlands, and common elsewhere, on sandy soils.

The tree grows to a height of about 35 metres and, at maturity, has a characteristic shape and outline with about three-quarters of the main trunk bare of branches (p. 33). As a young tree it has a very regular, symmetrical shape. The main trunk grows straight and its terminal growing point (protected by the terminal bud) persists from year to year. A whorl of lateral buds forms around the terminal bud, and these grow into whorls of lateral branches the following year. The main shoot and all lateral shoots show unlimited growth, year by year. All over these shoots and branches are scale leaves, spirally arranged. In the axils of these scales are tiny short shoots (called dwarf shoots) that terminate in pine needles. In *Pinus sylvestris* the needles occur in pairs.

The pine needles are 'evergreen'; that is, they remain on the tree for two or three years. The needles show many xeromorphic features (p. 375), which enable conifers to withstand drought and periods of limited water supply (for example, when the ground is frozen). Needles are compact structures with a small surface area:volume ratio. The stomata are sunken, and a very waxy cuticle covers the epidermis. Below the epidermis is a layer of fibres.

All conifers show secondary thickening (p. 410). The process is the same as in angiosperms, except that the xylem consists of tracheids (with bordered pits, p. 226) and fibres; no xylem vessels are produced. A feature of the tissues of conifers is the presence of a system of resin canals. Conifer tissue, when damaged or invaded, tends to ooze a resinous secretion, which seals off wounds and may overwhelm a bark invader. Pine wood is an important commercial timber (p. 412).

Life cycle of *Pinus*

The conifer plant is the sporophyte generation, producing male and female cones on the same tree in May each year (Figure 26.11). Male cones are small, and consist of numerous scale leaves (microsporophylls) arranged around a central axis, each bearing two pollen sacs (microsporangia). The pollen grains (microspores) develop large air-filled sacs, which aid in wind dispersal.

Male cones drop from the tree after the pollen has dispersed.

Female cones are larger than the male cones, and they remain attached for three years. On the upper surface of each megasporophyll are two naked ovules (not enclosed in an ovary). The ovules consist of a mass of cells arranged around a single megaspore mother cell. During the first year, the mother cell undergoes meiosis to form

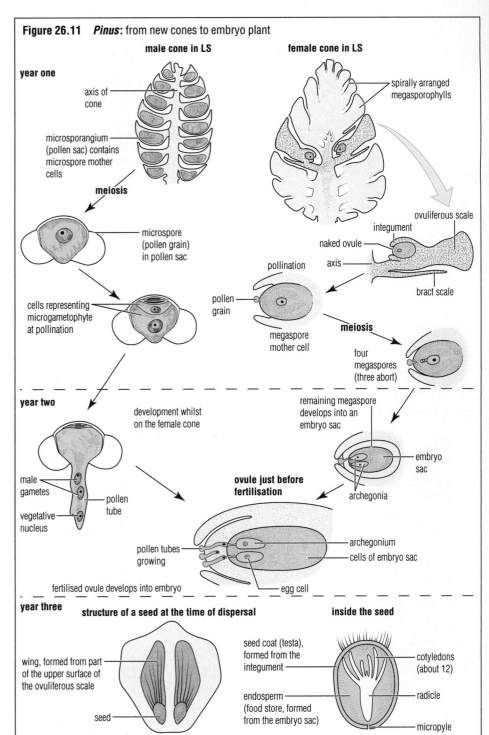

Figure 26.11 *Pinus*: from new cones to embryo plant

male cone in LS **female cone in LS**

year one

axis of cone

microsporangium (pollen sac) contains microspore mother cells

spirally arranged megasporophylls

meiosis

microspore (pollen grain) in pollen sac

ovuliferous scale

integument

naked ovule

axis

bract scale

pollination

pollen grain

cells representing microgametophyte at pollination

meiosis

megaspore mother cell

four megaspores (three abort)

year two

development whilst on the female cone

remaining megaspore develops into an embryo sac

embryo sac

male gametes

pollen tube

vegetative nucleus

ovule just before fertilisation

archegonia

archegonium

cells of embryo sac

pollen tubes growing

fertilised ovule develops into embryo

egg cell

year three   **structure of a seed at the time of dispersal**   **inside the seed**

wing, formed from part of the upper surface of the ovuliferous scale

seed coat (testa), formed from the integument

cotyledons (about 12)

endosperm (food store, formed from the embryo sac)

radicle

seed

micropyle

four megaspores, only one of which survives. In this condition, pollination occurs. The surviving megaspore develops into an embryo sac (mega-gametophyte) containing two archegonia. During the second year the pollen grains send out pollen tubes through the ovule tissue, and male nuclei fuse with the egg cells in the archegonia. As a result two or more embryos start to form in each embryo sac, but all but one abort. The embry-onic plant becomes surrounded by a food store. In the third year the forma-tion of a winged seed is completed, and the seeds are dispersed before the brown, scaly cones drop from the tree (Figure 26.12).

In comparing reproduction in conifers with that in flowering plants, it is possible to see a cone as the fore-runner of the inflorescence of flower-ing plants. Some angiosperms, such as alder, *Alnus glutinosa*, do have remarkably cone-like inflorescences. There is no fossil record of the origins of flowers, however, so this remains a speculative hypothesis.

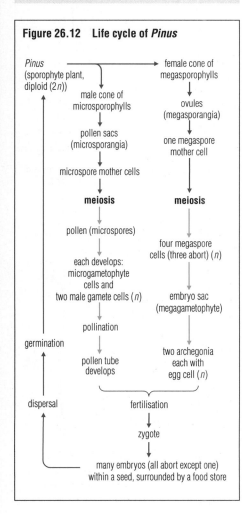

Figure 26.12 Life cycle of *Pinus*

26.5 PHYLUM ANGIOSPERMOPHYTA (FLOWERING PLANTS)

The angiosperms are the most successful of all terrestrial plants, in that they domi-nate almost all biomes (p. 45). There are more than 300 000 species, and they occur in all life forms (habits), such as trees, climbing plants, epiphytes, shrubs and herbaceous plants, in a variety of forms that far exceeds that of all other groups of terrestrial plants. The species of a single family of flowering plants, the Papilionaceae (the peas, beans, vetches and clovers, p. 575), far outnumber the surviving species of ferns, for example. Representatives of the angiosperms occur in every type of habitat, but angiosperms are particularly successful in the warmer parts of the Earth. A key feature of angiosperm success is their relationship with animal life. Animals only 'help them-selves' to other plants, but angiosperms have evolved the strategy of attracting animals to their flowers and feeding the visitors, and exploiting their mobility for pollination and seed dispersal.

We often refer to the flower as the organ of sexual reproduction, but we must return to this definition later (p. 575). Flowers are usually hermaphrodite structures (that is, they are both male and female), as in *Ranunculus* (buttercup) and the other flowers illustrated on the following pages. However some flowers are unisexual, either with male and female flowers carried on the same plant, as in *Quercus* (oak trees), or with separate male and female plants, as in *Mercurialis perennis* (dog's mercury). In angiosperms the female part, the ovule, is never exposed. It occurs boxed in the carpel, a structure formed from a modified, folded leaf. The full significance of this will be clear when we have examined how the flower works.

■ The structure of the flower and inflorescence

The flower develops from the tip of the shoot, and the parts of the flower may be regarded as modified leaves. In a few plants flowers are solitary; the tulip flower (*Tulipa*) is an example. More commonly they occur in groups on the same stem, an arrangement referred to as an **inflorescence**, which is probably more attractive to visiting insects than a soli-tary flower. An inflorescence in which the oldest flower is at the top of the stem is known as a **cyme**. This term strictly denotes a system of branching (vegetative or reproductive) in which the main axis ceases to be dominant, growth being con-tinued by the laterals. These normally repeat the pattern. Vegetative cymose branching is shown by chickweed (*Cerastium*). The *Ranunculus* inflores-cence is cymose.

An inflorescence in which the youngest flower is at the top of the stem is known as a **raceme**. The *Lathyrus* inflorescence is racemose. The specialised case of a raceme compressed into a compact platform is known as a **capitulum** (Figure 20.8, p. 421). This arrangement, typical of the Compositae, is especially conspicuous, and many flowers may be accessible to an insect in a single visit.

The parts of the flower are borne in whorls on the expanded tip (**receptacle**) of a flower stalk (the **pedicel**). At the base of the pedicel is a simple leaf, known as a **bract**.

The outermost parts of the flower, the **sepals**, form the **calyx**. The sepals are usually small and green. They enclose and protect the flower at the bud stage (flowers commence life as a flower bud).

The most conspicuous part of a flower is normally the **corolla**, which consists of the **petals**. Petals are often brightly coloured, and this may attract visitors to the flower.

Calyx and corolla are together known as the **perianth**; this is a useful term for flowers where sepals and petals look alike. In many flowers a part of the recep-tacle or the perianth forms a **nectary** pro-ducing a sugary solution (nectar).

The male parts of the flower, collec-tively called the **androecium**, are the **stamens**. Each stamen consists of an **anther** at the end of a stalk-like **filament**. The anther contains pollen grains in pollen sacs.

The female parts of the flower, collec-tively the **gynoecium**, are the **carpels**. Each carpel consists of an **ovary**, which contains one or more **ovules**, a **stigma**, which is the receptive surface for pollen, and the **style**, which connects the stigma with the ovary. A flower may contain a single carpel or many carpels, and carpels may be free or fused together.

The structure of the flower and inflo-rescence of a buttercup (*Ranunculus* sp.) is illustrated in Figure 26.13, overleaf. There is a similar basic structure to most flowers, but there is enormous variation too. For example, whilst flower parts often occur in fives (five sepals, five petals, and so on), they may occur in threes (as they do in the grasses, Figure

26.15) or in fours (in the Cruciferae, p. 576, for instance). Flower parts may be symmetrically arranged around the receptacle, as in *Ranunculus* (a regular or **actinomorphic** flower). An alternative arrangement has a single plane of symmetry (*Lathyrus* is an example), and the flower is described as irregular or **zygomorphic**. Another variation is that flower parts may fuse together to form tubes (as in white dead nettle, *Lamium album*; see *Study Guide*, Chapter 26). Again, the receptacle may be cone-shaped (as in *Ranunculus*), it may take the form of a disc or flask (as in *Prunus*), or it may completely surround the ovary (as in *Malus*, p. 576).

■ Adaptations for pollination

Pollen grains have no power of independent movement, and are usually carried from flower to flower either by means of insects (**entomophily**) or by means of wind (**anemophily**).

The structure of the sweet pea flower is shown in Figure 26.14. This flower is visited by long-tongued bees in search of nectar. As they alight the wing and keel petals of the flower are depressed, exposing the stigma to the underside of the bee's abdomen. The bee probes for nectar by passing its proboscis into the fused filament trough, below the one free anther.

The structure of the perennial ryegrass flower is shown in Figure 26.15. This flower is pollinated by the action of wind. Many plants with wind-pollinated flowers have distinct features, which are compared with those of insect-pollinated flowers in Table 26.3 (see *Study Guide*, Chapter 26).

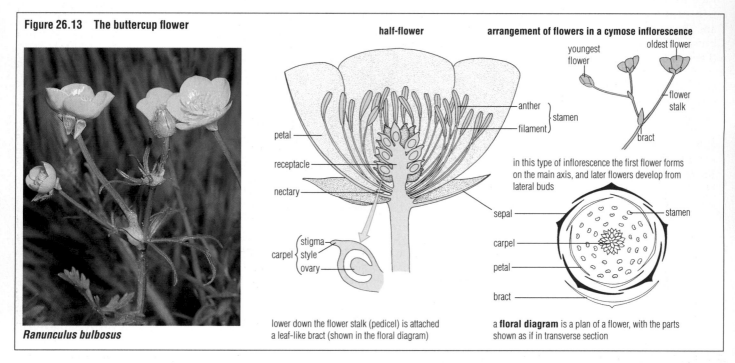

Figure 26.13   The buttercup flower

Ranunculus bulbosus

half-flower

petal
receptacle
nectary

carpel { stigma / style / ovary }

lower down the flower stalk (pedicel) is attached a leaf-like bract (shown in the floral diagram)

anther
filament } stamen

arrangement of flowers in a cymose inflorescence

youngest flower
oldest flower
flower stalk
bract

in this type of inflorescence the first flower forms on the main axis, and later flowers develop from lateral buds

sepal
carpel
petal
bract
stamen

a **floral diagram** is a plan of a flower, with the parts shown as if in transverse section

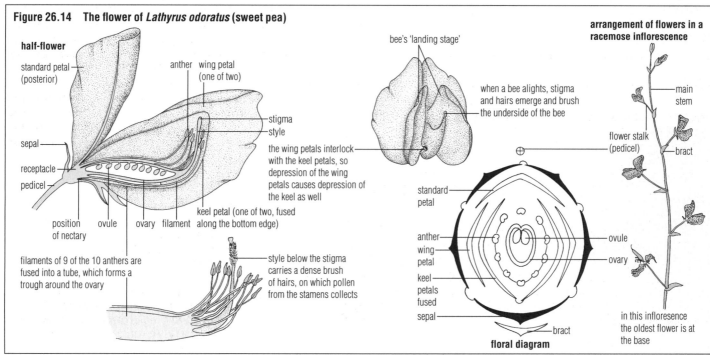

Figure 26.14   The flower of *Lathyrus odoratus* (sweet pea)

half-flower

standard petal (posterior)
sepal
receptacle
pedicel
position of nectary
ovule
ovary
filament
keel petal (one of two, fused along the bottom edge)

anther
wing petal (one of two)
stigma
style

filaments of 9 of the 10 anthers are fused into a tube, which forms a trough around the ovary

the wing petals interlock with the keel petals, so depression of the wing petals causes depression of the keel as well

style below the stigma carries a dense brush of hairs, on which pollen from the stamens collects

bee's 'landing stage'

when a bee alights, stigma and hairs emerge and brush the underside of the bee

standard petal
anther
wing petal
keel petals fused
sepal
ovule
ovary
bract

arrangement of flowers in a racemose inflorescence

main stem
flower stalk (pedicel)
bract

in this inflorescence the oldest flower is at the base

floral diagram

Figure 26.15 The flower of *Lolium perenne* (perennial rye-grass)

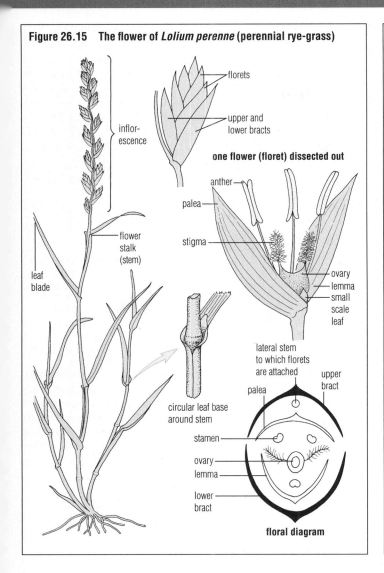

Figure 26.16 The stamen, and pollen grain formation

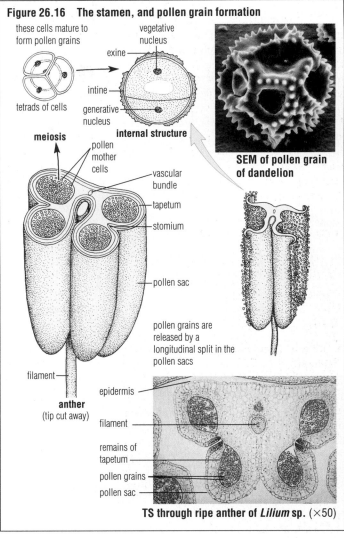

SEM of pollen grain of dandelion

TS through ripe anther of *Lilium* sp. (×50)

■ Stamens: development of the male gamete

The stamen (Figure 26.16) consists of an anther, containing four pollen sacs, supported by a filament. The inner layer of cells in the wall of the pollen sac, the tapetum, provides nutrients for the developing pollen grains. Meiotic division of microspore mother cells forms haploid cells in tetrads. These cells become the pollen grains. Later the pollen grain cell divides to form a generative cell and a vegetative nucleus.

Pollen grains are released by the rupture of the **stomium**, a group of fibrous cells at the junction of the pollen sacs. The grain is surrounded by a thin cellulose cell wall, the **intine**, and an outer wall, the **exine**, with a sculptured pattern characteristic of the species. The exine is composed of an extremely durable substance related to cutin and suberin. The pollen grain wall has thin areas through which a pollen tube may grow at a later stage.

Table 26.3 Pollination mechanisms and flower characteristics

Entomophilous flowers	Anemophilous flowers
insects are attracted by scent or colour, or both, and transport pollen from flower to flower on their bodies	winds (air currents) carry pollen, some of which may reach a flower of the same species
nectaries may be present (any part of a flower may form a nectary)	nectaries are not present
pollen grains variable in size; may be sticky, and readily become attached to an insect's body	pollen grains tend to be small and light
stamens may be protected from rain and, with other parts of the flower, also from visitors other than insects able to carry out pollination (insects for which flowers may show adaptations include flies, butterflies, moths, sawflies, wasps, ants and bees)	stamens (and stigmas) more likely to project from the flower when ripe/mature; anthers are versatile (attached near the middle, allowing movement)
flowers held in conspicuous positions on the plant	flowers often open before the foliage leaves do

Questions

1 What features of
   a the sweet pea flower
   b the buttercup flower
   make them efficient structures for achieving pollination?

2 What are the respective likely nutritional values of nectar and pollen to insects?

3 How does a bee harvest nectar (p. 296), and pollen respectively from a flower like the sweet pea?

Figure 26.17   The carpel, and formation of the embryo sac

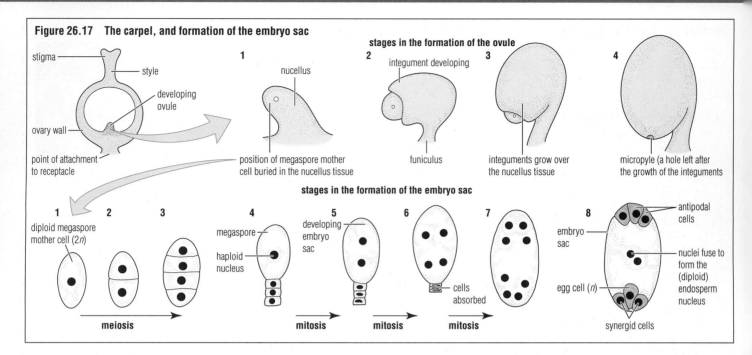

stigma
style
developing ovule
ovary wall
point of attachment to receptacle

1
nucellus
position of megaspore mother cell buried in the nucellus tissue

stages in the formation of the ovule

2 integument developing
3 integuments grow over the nucellus tissue
funiculus
4 micropyle (a hole left after the growth of the integuments

stages in the formation of the embryo sac

1 diploid megaspore mother cell (2n)
2
3
meiosis
4 megaspore / haploid nucleus
5 developing embryo sac
mitosis
6 cells absorbed
mitosis
7
mitosis
8 antipodal cells / embryo sac / nuclei fuse to form the (diploid) endosperm nucleus / egg cell (n) / synergid cells

■ Carpels: development of the female gamete

The carpel (Figure 26.17) consists of a stigma, style and ovary. Carpels may occur singly or in groups (free-standing or fused together) on the receptacle at the centre of the flower, or actually buried in the receptacle tissue. The ovary contains one or more ovules.

An ovule consists of parenchymatous cells called the **nucellus**, and is surrounded by one, two or three sheaths called the **integuments**, which grow over the nucellus but leave a hole, the **micropyle**, through which a pollen tube may later enter. Ovules are attached to the ovary wall at a point called the **placenta** by a short stalk, the **funiculus**.

Within the nucellus a single embryo sac mother cell develops. The mother cell undergoes meiotic division, and one of the four haploid cells produced becomes the embryo sac. Within the embryo sac eight nuclei are formed mitotically. These nuclei then migrate, and two fuse to form the endosperm nucleus. Now the mature embryo sac contains six haploid cells and the diploid endosperm nucleus. One of the haploid cells is the egg cell (female gamete). In this condition the ovule awaits fertilisation, which must be preceded by pollination.

■ Pollination

Pollination is the transfer of pollen from the stamens to the stigma. The germination of pollen from the stamen of a flower on its own stigma is **self-pollination**. If the pollen from the stamen of one flower reaches the stigma of another flower on the same plant and germinates there, this is also self-pollination. In these cases the progeny will have recombinations of the chromosomes of only one parent.

If pollen from the stamen of one flower reaches the stigma of another flower on a different plant of the same species and germinates there, this is **cross-pollination**. In cross-pollination the chromosomes are a new mixture from the chromosomes of two parents. As we have seen, pollination is commonly brought about by insects or the wind, although occasionally other agencies, such as birds or water, may be responsible.

Figure 26.18   After pollination . . .

changes in the pollen grain as the pollen tube begins to grow

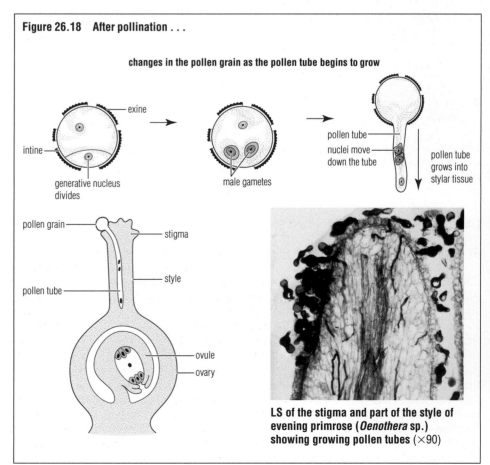

exine
intine
generative nucleus divides

male gametes

pollen tube
nuclei move down the tube
pollen tube grows into stylar tissue

pollen grain
stigma
style
pollen tube
ovule
ovary

LS of the stigma and part of the style of evening primrose (*Oenothera* sp.) showing growing pollen tubes (×90)

■ Barriers to self-fertilisation

Many plants have mechanisms that discourage or prevent self-pollination (see *Study Guide*, Chapter 26). For example:

▶ separate male and female flowers on different plants, as in *Ilex aquifolium* (holly)

▶ stamens and stigma mature at different times in hermaphrodite flowers; either the stamens ripen before the stigma (**protandry**), as in *Lamium album* (white dead nettle), or the stigma ripens before the stamens (**protogyny**) as in *Endymion non-scriptus* (bluebell)

▶ the structure of the flower may prevent self-pollination; for example, in *Iris pseudacorus* (iris) the stigma is a flap held above the stamens, and has a receptive surface that is exposed as the bee (or other insect) enters the flower to forage among the stamens, but not as the bee leaves

▶ self-incompatible mechanisms; if pollen of the same plant lands on its stigma, the pollen tube fails to grow, or fails to grow fast enough to reach the egg cell before a pollen tube from another individual does (Figure 30.22, p. 661). Incompatibility between style tissue and pollen tube may be caused by physiological mechanisms (such as differences in solute potential) or by biochemical mechanisms (such as the unavailability of essential nutrients), among other strategies. Self-incompatibility is used extensively in breeding hybrid varieties of crop plants.

Figure 26.19   Fertilisation and seed formation

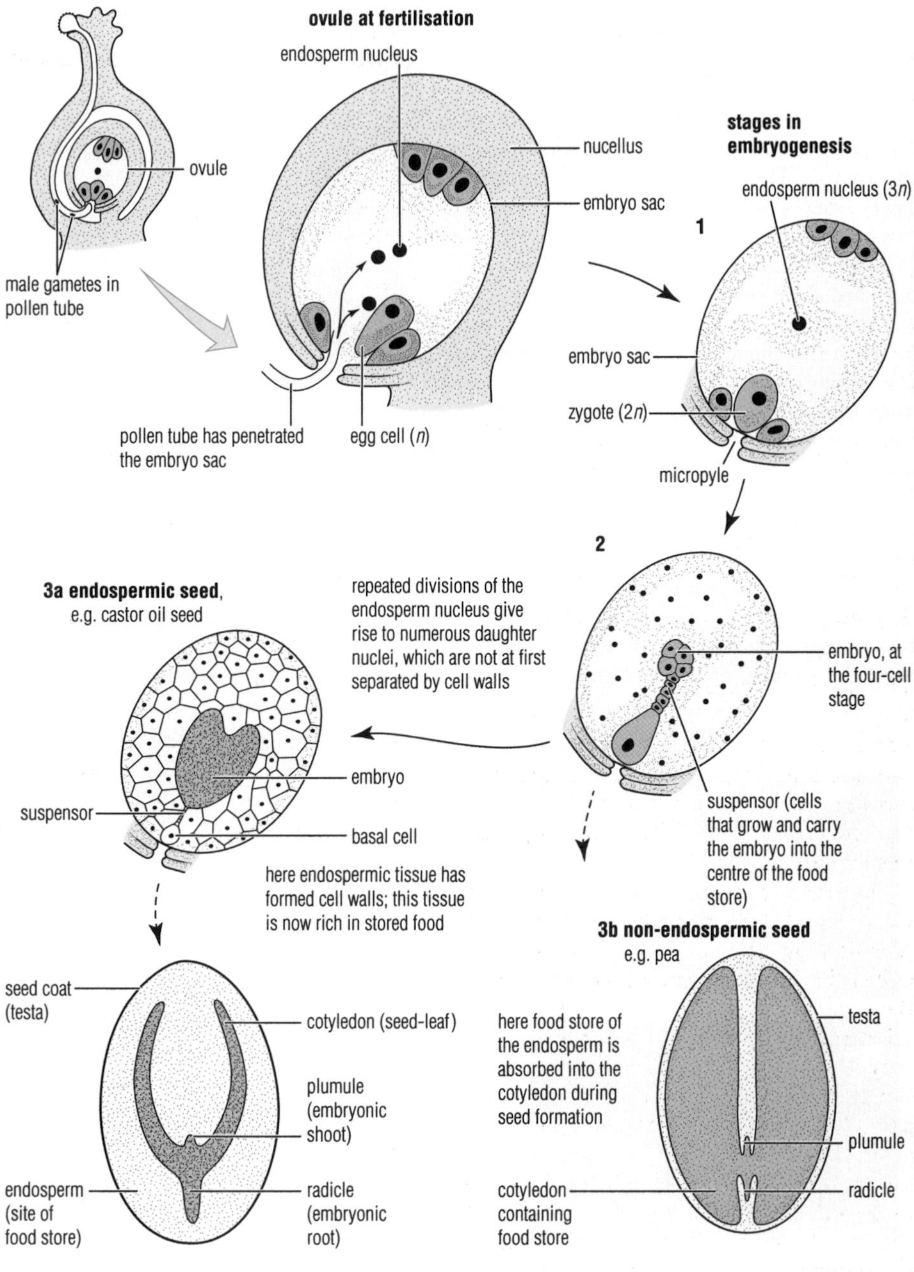

■ Fertilisation and seed formation

Once the pollen grain reaches the stigma, the generative nucleus divides to form two male gametes (if it has not already done so). Then a pollen tube grows out from one of the pores of the pollen grain wall (Figure 26.18). Growth of the tube is nourished by nutrients such as glucose, supplied by the stigma and style tissues.

Fertilisation is the fusion of the nucleus of a male gamete with the egg nucleus in the embryo sac to form a zygote. This becomes possible because the pollen tube grows through the style and penetrates the embryo sac. At the same time, the second nucleus enters the embryo sac and fuses with the endosperm nucleus. (This is referred to as **double fertilisation**; it is unique to the flowering plants.) Then the zygote divides to form a row of cells. The innermost of these cells becomes the embryo by rapid and repeated mitotic cell divisions.

The formation of the triploid endosperm nucleus triggers cell division, which leads to the formation of the endosperm, ensuring that a food store is produced for the embryo plant in the seed (Figure 26.19).

Questions

1 Distinguish between pollination and fertilisation.

2 What special features of pollen grains have enabled biologists to glean information of the flora of bygone times (p. 650)?

3 What mechanism of pollination ensures that endosperm (food store) is developed only in ovules that are fertilised?

4 Name a 'fruit' or 'vegetable' we eat derived from
   a one carpel containing a single seed
   b one carpel containing many seeds
   c several fused carpels each containing many seeds.

■ Fruits and seeds

We have seen that the seed develops from the fertilised ovule and contains an embryonic plant and a food store. The seed coat or **testa** develops from the integuments of the ovule. The testa carries a single scar, marking the point of attachment of the ovule to the placenta in the ovary. The fruit develops from the ovary and contains the seed or seeds (Figure 26.20). The fruit wall develops from the ovary wall, and is known as the **pericarp**. The outer layer of the pericarp has two scars, being the points of attachment to the placenta and to the style. A **true fruit** consists of the ovary alone; pea pods and rye-grass fruit are examples of true fruits. In a **false fruit** the receptacle tissue is fused to the ovary wall(s), as in the apple.

The seed is a form in which a plant survives the unfavourable season (**perennation**). It is also a form in which plants may be dispersed. There are biological advantages in having seeds carried away from the vicinity of the parent plant, and various mechanisms are used (Figure 26.21). Many seeds have no particular dispersal mechanism, however, apart from accidental dispersal by the careless action of animals as they feed on the fruits.

In the soil, seeds absorb water and, when conditions are suitable, many will germinate (p. 404), although others have a dormancy period that has to be overcome (p. 438).

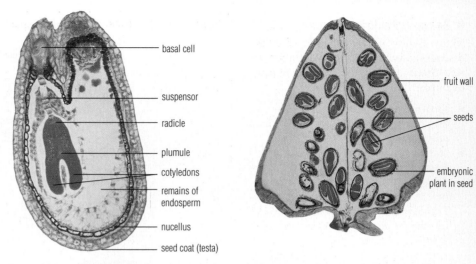

labels (left diagram): basal cell, suspensor, radicle, plumule, cotyledons, remains of endosperm, nucellus, seed coat (testa)

labels (right diagram): fruit wall, seeds, embryonic plant in seed

Figure 26.20 *Capsella bursa-pastoris* **(shepherd's purse):** the developing seed in LS (left, ×160); the fruit in LS (right, ×7)

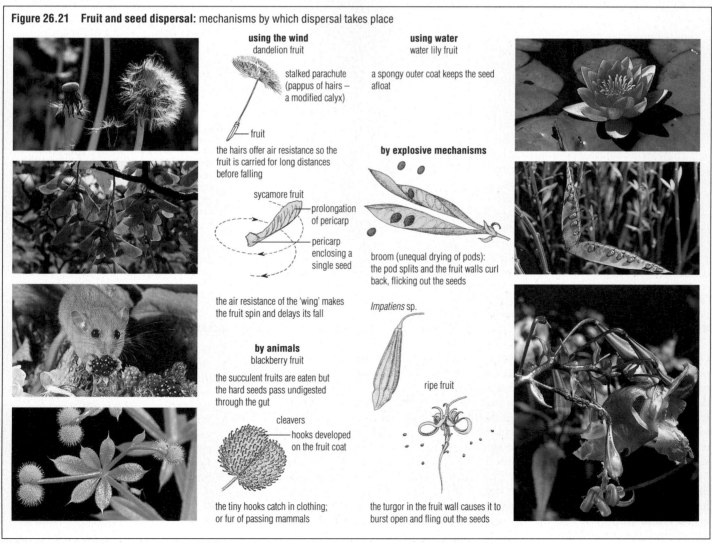

Figure 26.21 **Fruit and seed dispersal:** mechanisms by which dispersal takes place

using the wind
dandelion fruit

stalked parachute (pappus of hairs – a modified calyx)

fruit

the hairs offer air resistance so the fruit is carried for long distances before falling

sycamore fruit

prolongation of pericarp

pericarp enclosing a single seed

the air resistance of the 'wing' makes the fruit spin and delays its fall

by animals
blackberry fruit

the succulent fruits are eaten but the hard seeds pass undigested through the gut

cleavers

hooks developed on the fruit coat

the tiny hooks catch in clothing; or fur of passing mammals

using water
water lily fruit

a spongy outer coat keeps the seed afloat

by explosive mechanisms

broom (unequal drying of pods): the pod splits and the fruit walls curl back, flicking out the seeds

Impatiens sp.

ripe fruit

the turgor in the fruit wall causes it to burst open and fling out the seeds

■ The life cycle of the flowering plant

The life cycle of the flowering plant is summarised in Figure 26.22. Here the structures involved in sexual reproduction in angiosperms are placed in chronological sequence (bold type).

In this diagram the life cycle is also interpreted in terms of heteromorphic alternation of generations. Microspores are produced by microsporophylls (stamens). The pollen sacs of the anthers are microsporangia, producing microspores (the pollen grains). The microgametophyte in the angiosperms is reduced to the generative nucleus, and the male gametes are the two male nuclei.

The megaspore is formed within the megasporophyll (carpel). The embryo sac is a megaspore, and the megagametophyte is reduced to antipodal and synergid cells (Figure 26.17, p. 572). The female gamete is the egg cell.

On this analysis the flower is a spore-forming structure (asexual reproduction) of a plant producing two types of spore. Sexual reproduction is conducted within and between the micro- and megaspores (pollen grain and embryo sac). Similarities in life cycles among the groups of living green plants, together with the limited fossil evidence available, suggest that the evolutionary origin of angiosperms is among the 'seed ferns' of the late Palaeozoic (p. 649).

■ Diversity of floral morphology

Many of the events of reproduction, such as the formation of pollen, the existence of two male gametes that travel down the growing pollen tube, the structure of the embryo sac, and the steps in double fertilisation, are in principle more or less identical in all the flowering plants. At the same time there is an enormous range of colour, size and form of the flowers themselves. We can see something of this diversity if we look at members of three families of flowering plants.

The Papilionaceae

Many species of this family are a source of seeds for food, including pulses such as peas, beans and lentils. The flowers of the Papilionaceae (Figure 26.23) have five characteristic petals, consisting of the standard, two wings and two keel petals in a zygomorphic flower (bilaterally symmetrical). There are ten stamens, with the filaments joined into a tube. The fruit is a pod.

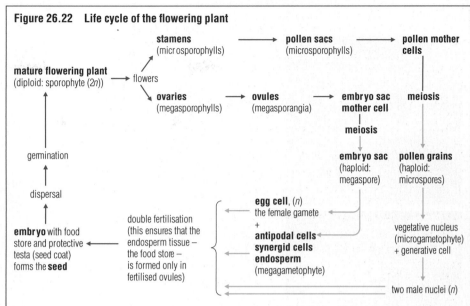

Figure 26.22 Life cycle of the flowering plant

Figure 26.23 Flowers of the Papilionaceae

Trifolium pratense (clover)
- insect (bee) lands on flower
- stamens and stigma are exposed and contact the underside of the bee's abdomen (reverts when bee departs)

Lotus corniculatus (bird's foot trefoil)
- insect (bee) lands
- keel joined along most of upper edge and along lower edge
- weight of bee forces the stigma and a plug of pollen through the hole at the tip of the keel

Sarothamnus scoparius (broom)
- bee alights
- pollen (only) is shed into the keel before the keel is 'exploded'
- weight of bee 'explodes' the keel; a shower of pollen dusts the bee's abdomen; the stigma collects pollen from the bee
- flower 'explodes' once only

Figure 26.24   Flowers of the Rosaceae

Rubus sp. (raspberry)

Prunus sp. (cherry)

Malus sp. (apple)

The Rosaceae

The rose family contains many fruit plants, such as apple, pear, peach, apricot, plum, cherry and strawberry. The flowers (Figure 26.24) are regular (actinomorphic), with four or five petals. In the flowers the receptacle varies widely in shape; it may be saucer-shaped or a deep cup, or completely fused to the carpels.

The Cruciferae

The family Cruciferae includes many economically important plants such as mustard, oilseed rape, cabbage, broccoli and cauliflower. The flowers are unusual in that they consist of four sepals and four petals, but normally have six stamens. The family is named after the cross-shaped arrangement of the petals (Figure 26.25). The ovary is divided into two chambers by an inner wall.

Some Cruciferae, such as the wall-flower (*Cheiranthus cheiri, Study Guide,* Chapter 26), have showy flowers in which nectar collects at the base of the sepals. These flowers are pollinated by bees. Others, such as shepherd's purse (*Capsella bursa-pastoris*), a common weed of open land, have smaller flowers and are self-pollinated.

Parthenocarpy

Parthenocarpy is the development of fruits without fertilisation, resulting in the production of fruits that lack seeds. It occurs naturally, as in some horticultural varieties of bananas, pineapples, cucumbers, tomatoes and figs. In these cases it seems that the placenta tissue in the unfertilised ovary produces auxin (IAA, p. 424), which is responsible for triggering fruit formation. Parthenocarpy is deliberately induced in certain species by application of plant growth substances (p. 440).

Questions

1   State which parts of a seed are
   a tissues of the parent plant
   b derived from the zygote
   c represent the megaspore
   d are the remains of the megagametophyte.
2   Construct a table to compare and contrast the following features of the life cycles of a bryophyte, a fern, *Pinus* and an angiosperm:
   a form of sporophyte plant
   b presence (or absence) of vascular tissue
   c typical habitat of sporophyte
   d whether homosporous or heterosporous
   e form of gametophyte plant
   f type of male gamete(s).

Figure 26.25   Flower of oilseed rape (*Brassica napus*) (Cruciferae)

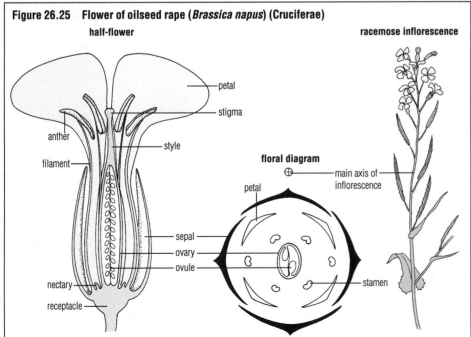

half-flower

racemose inflorescence

floral diagram

26.6 THE SUCCESS OF THE ANGIOSPERMS

Fossil angiosperms first appear in some numbers in the rocks laid down in the Lower Cretaceous (150 million years ago, p. 649). By the Upper Cretaceous (70 million years ago) the angiosperms were the dominant plant group, and many of the present-day flowering plant families were already established. The exact origin of the angiosperms remains obscure, but the features of sexual reproduction shown by all flowering plants (p. 667) suggest that the angiosperms had a single ancestor. Be that as it may, the speed of establishment of the angiosperms, the parallel eclipsing of other terrestrial plants (such as ferns and conifers), and the rate at which they achieved diversity may be attributed to several characteristic features as described below.

► Their sexual reproduction is rapid; the interval between flower production and setting of seed is usually a matter of weeks; in conifers the equivalent process may take at least one year.

► The closed ovary and the development of a style through which the pollen tube must grow not only protect the ovule to some extent, but also make possible an efficient incompatibility system, a device to exclude self-pollination (p. 661). Incompatibility mechanisms increase genetic diversity of the progeny, an essential factor for evolution by natural selection (p. 657).

► Their unique double fertilisation ensures that the parent plant invests a seed with a food store only if the ovum is fertilised.

► Their efficient vegetative metabolism and associated biochemistry has led to structures, storage products and chemical substances that can confer selective advantage to a plant adapting to new habitats and micro-climates.

► Their leaves are relatively succulent and decay rapidly on falling to the ground, producing humus from which ions are comparatively quickly released for re-use. Herbaceous epiphytes and plants growing in soil benefit from this. By contrast, conifer needles produce a slow-decaying leaf litter, unfavourable to the development of a diverse flora.

► The origin of flowers is bound up with the evolution of insects. Bees appear in the fossil record in the Oligocene, and butterflies and moths in the Eocene. The appearance of these insects can be correlated with the evolution of the angiosperm families having flowers with petals fused into a long tube. The evolution of the whole range of angiosperms has been a cooperative venture between insects and flowering plants to their mutual advantage.

26.7 CLASSIFYING ANGIOSPERMS ON STRUCTURE AND LIFE CYCLES

We have already seen that the angiosperms fall naturally into two classes, the dicotyledons and the monocotyledons ('dicots' and 'monocots', pp. 33–4). The chief differences between the members of these classes are summarised in Table 26.4. Dicotyledons are either herbaceous (non-woody) or woody plants, whereas monocotyledons are herbaceous (palm trees represent the single exception).

Another useful natural division of the angiosperms is based on the longevity of their life cycles, and on the way they survive an unfavourable season (known as **perennation**).

Ephemerals germinate and flower quickly, whenever conditions are appropriate. Normally ephemerals produce several generations in one year. Many temperate weed species are ephemerals; groundsel (*Senecio vulgaris*) and shepherd's purse (*Capsella bursa-pastoris*) are examples. Ephemerals also occur in desert and semi-desert habitats, and respond quickly to the limited rainfall. They survive unfavourable periods as seeds.

Annuals complete their life cycles from germination to seed production in one year. The seeds overwinter, and the cycle recommences in the spring.

Biennials, on the other hand, complete their life cycle in two years. In the first year the plants grow from seed, accumulate food reserves, and survive the first winter as a below-ground food store (typically in a swollen tap root, as in carrot or parsnip), with buds and a rosette of leaves. In the second year the plants flower and produce seeds, and then die. They survive the second winter as seeds.

Perennials survive for many years, and are of two types. Trees and shrubs are the **woody perennials**, surviving the unfavourable season with an aerial system that may shed its leaves (**deciduous** woody perennials) or as an 'evergreen' tree or shrub. Evergreens lose some leaves every year, but not all the leaves at any one time. **Herbaceous perennials** survive the unfavourable season as an underground stem (a rhizome, for example, as in *Iris*) or an underground elaborated bud (a bulb, as in the onion).

Table 26.4 Dicotyledons and monocotyledons: the angiosperms are divided into dicotyledonous and monocotyledonous plants, on the basis of the number of seed-leaves (cotyledons, which are structurally simpler than the normal leaves that develop later) in their embryos

	Dicotyledons, e.g. buttercup, oak	Monocotyledons, e.g. grasses, lilies
flowers	flower parts occur in pairs (or in multiples of 2) or in 5s	flower parts occur in 3s, or in multiples of 3
	outer parts of flower are (protective) sepals and (attractive) petals	outer parts of flower are bracts (as in grasses, lemma and palea) or perianth parts (as in lilies)
	flowers often insect-pollinated	flowers often wind-pollinated
seeds	embryo bears two seed-leaves	embryo bears one seed-leaf
leaves	leaves have broad blade (lamina)	leaves elongated, narrow and pointed
	lamina a dorsi-ventral leaf with veins arranged in a network	leaf isobilateral, with veins arranged in parallel rows
stem	vascular bundles in a ring	vascular bundles scattered
	cambium present in bundles, and secondary thickening may occur	cambium absent from bundles; virtually no secondary thickening
roots	tap root system with lateral roots	fibrous root system of numerous identical adventitious roots
	have a small number of protoxylem groups (typically 2–5) in the central stele	have a large number of protoxylem groups in the stele

Figure 26.26 Examples of vegetative reproduction

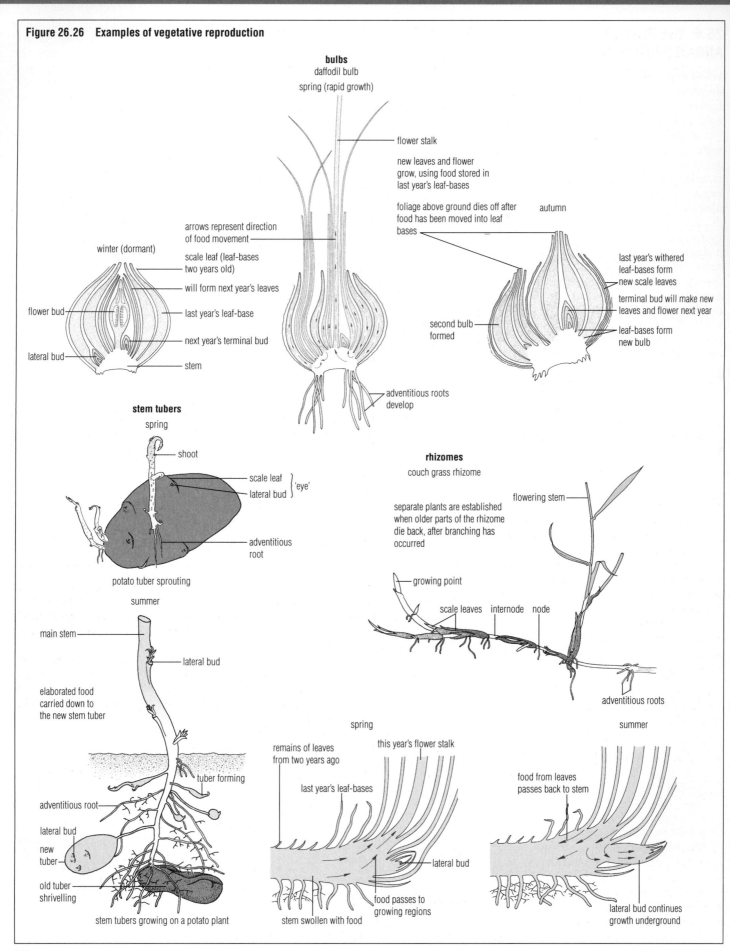

bulbs
daffodil bulb
spring (rapid growth)

flower stalk

new leaves and flower grow, using food stored in last year's leaf-bases

foliage above ground dies off after food has been moved into leaf bases

autumn

arrows represent direction of food movement

winter (dormant)

scale leaf (leaf-bases two years old)

will form next year's leaves

last year's leaf-base

next year's terminal bud

stem

flower bud

lateral bud

last year's withered leaf-bases form new scale leaves

terminal bud will make new leaves and flower next year

leaf-bases form new bulb

second bulb formed

adventitious roots develop

stem tubers
spring

shoot

scale leaf
lateral bud } 'eye'

adventitious root

potato tuber sprouting

rhizomes
couch grass rhizome

separate plants are established when older parts of the rhizome die back, after branching has occurred

flowering stem

growing point

scale leaves internode node

adventitious roots

summer

main stem

lateral bud

elaborated food carried down to the new stem tuber

tuber forming

adventitious root

lateral bud

new tuber

old tuber shrivelling

stem tubers growing on a potato plant

spring

remains of leaves from two years ago

this year's flower stalk

last year's leaf-bases

lateral bud

food passes to growing regions

stem swollen with food

summer

food from leaves passes back to stem

lateral bud continues growth underground

26.8 VEGETATIVE REPRODUCTION IN ANGIOSPERMS

Flowering plants reproduce vegetatively when a part (or parts) of the parent plant becomes separated and grows independently. Vegetative reproduction is an asexual process, and so the new plants formed are normally genetically identical to the parent plant. Often the plant organ involved is an organ of perennation (p. 574), and the parts separated are substantial structures. Plants that reproduce vegetatively also reproduce sexually, forming seeds. Some examples of vegetative reproduction are illustrated in Figure 26.26, and discussed below.

■ Bulbs

The bulb is an underground bud with swollen, fleshy leaf-bases attached to a short conical upright stem, with occasional axillary buds between the leaf-bases. The bulb is first and foremost an organ of perennation. Vegetative reproduction occurs when additional axillary buds form additional bulbs besides the existing bulb. The onion (*Allium*) and daffodil (*Narcissus*) are familiar examples.

■ Stem tubers

A stem tuber, such as a potato (*Solanum tuberosum*), is the swollen tip of an underground lateral stem, formed towards the end of the growing season at the expense of food manufactured in the aerial system. The stem tuber is a perennating organ, which may also sprout new shoots from its axillary buds in the subsequent growing season, forming several new plants.

■ Rhizomes

Rhizomes such as those of iris or couch grass (*Agropyron repens*) are horizontal stems that grow from buds at the tips of the rhizomes and branch by growth of axillary buds. Eventually the rhizome behind the new shoot dies back, leading to the formation of independent plants. The rhizome is also an organ of perennation

■ Micropropagation

Today, much of the propagation or cloning (p. 195) of higher plants is at least begun in laboratories by a process known as tissue culture or micropropagation. This is referred to as a 'micro' process because a few isolated cells or a tiny sample of tissue is all that is

Figure 26.27 Artificial propagation: budding and grafting

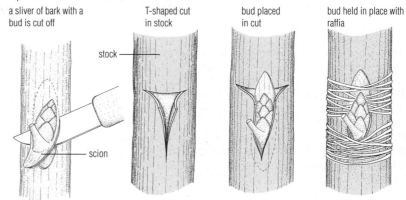

budding
a vegetative bud (the scion) is cut from the plant to be propagated; the bark of the stockplant is cut to expose the cambium and the bud is held in contact with it

a sliver of bark with a bud is cut off | T-shaped cut in stock | bud placed in cut | bud held in place with raffia

stock
scion

callus tissue forms at the junction, followed by formation of xylem and phloem tissues connecting the two parts; the stock is cut away above the bud before it sprouts

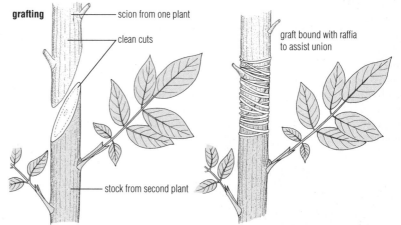

grafting
scion from one plant
clean cuts
graft bound with raffia to assist union
stock from second plant

the cambium tissue of two stems grows, forming a union between the stem of the scion and the stem of the stock

normally required. Cell or tissue samples (known as explants) are isolated (excised) and grown on sterile agar with essential nutrients and plant growth regulators, including cytokinins (p. 429).

The success of micropropagation lies in the phenomenon of **totipotency**. By this term we mean that all mature cells that retain a nucleus also retain the ability to grow into new organisms (or parts of organisms), provided suitable conditions are made available. A single mature cell has been isolated from a fully grown plant (initially this was shown using a carrot root cell) and grown into an entirely new plant, by being cultivated *in vitro*.

The advantages of micropropagation in horticulture include the following:
▶ speed; many new individuals can be

formed much more quickly than by conventional cultivation techniques
▶ disease-free conditions, particularly the absence of viruses, are much more easily achieved
▶ cells that have been genetically modified by various techniques are normally in a condition to be grown into entire plants by this method (p. 636).

Questions

1 In what ways is the biochemical diversity characteristic of flowering plants likely to be advantageous to them?
2 Why are methods of artificial propagation important in horticulture (Figure 26.27)?
3 What is the significant difference between life cycles that are described as haploid, diploid, and showing alternation of generations?

FURTHER READING/WEB SITES

C J Clegg (1984) *Lower Plants*. John Murray
C J Clegg and G Cox (1978) *Anatomy and Activities of Plants*. John Murray

27 Life cycles of animals

- Typically, animals have diploid life cycles. Many animals reproduce sexually only, e.g. mammals, but some also show asexual reproduction, e.g. by budding, as in *Hydra*, or by development of unfertilised diploid egg cells (parthenogenesis), as in aphids.
- Mammals are unisexual. The male produces sperms in two testes, and has ducts to transport the sperm cells to the outside of the body via the penis. The female has two ovaries in which egg cells are produced. An oviduct runs from a funnel-shaped opening held around each ovary, to join up with the central uterus. The uterus connects to the outside by the vagina.
- The testes and ovaries in mammals are also endocrine glands, producing hormones that regulate sexual development and gamete formation. In the female, a cycle of changes in ovaries (ovarian cycle) and uterus (uterine cycle) make up the oestrous cycle.

- Fertilisation, if it occurs, is internal in mammals, taking place in an oviduct. The fertilised egg becomes embedded in the lining of the uterus where the embryo develops into a fetus, inside the mother's body.
- The developing fetus obtains oxygen and nutrients from the mother's blood and disposes of waste substances via a placenta connected to the fetus via the umbilical cord. Mammalian young are eventually born in a relatively advanced state of development.
- Mammals (like birds) show pronounced parental care of their young. In mammals the early nutrition of the young is by milk produced in the mother's mammary gland.
- Human intervention in the reproductive process can prevent unwanted pregnancies, e.g. by 'barrier' or hormonal methods, and there are various ways by which childless couples may be helped to have a family, if they wish.

27.1 INTRODUCTION

We define the life cycle of an organism as the sequence of changes through which it passes during its life history, from its origin in reproduction until its death. Biologists have identified three basic types of life cycle, known respectively as haploid, diploid and haplodiplontic cycles (Figure 26.1, p. 563). The life cycles of green plants show various forms of alternation of generations (haplodiplontic cycles). In contrast, animals typically have **diploid life cycles**.

During its lifespan an organism may produce one or more new individuals that grow and repeat the life cycle process. Most offspring of many species that live in the wild, however,

die before reproducing. Reproduction, the ability to produce other individuals of the same species, is a fundamental characteristic of living things (p. 5), yet only limited numbers of each species actually succeed in reproducing (p. 657).

New individuals may be produced by either asexual or sexual reproduction, or in some species by both. In animals, asexual reproduction is far less common than it is in plants, protoctists and prokaryotes; in fact, many animal species reproduce only sexually. We will examine characteristic features of sexual and asexual reproduction in animals, before looking at mammalian reproduction in detail, using human reproduction as the example.

27.2 ASEXUAL REPRODUCTION IN ANIMALS

■ Budding

Budding is a form of asexual reproduction that is performed by certain species of animal. In budding, an offspring develops on the body of the parent organism and becomes cut off as a new individual. If these individuals remain attached to the parent organism for any reason, a **colony** of organisms results.

Budding as a form of sexual reproduction is comparable in outcome with the fragmentation of, say, *Spirogyra* (p. 554) and the binary fission of organisms like *Amoeba* (p. 556). The products of asexual reproduction are genetically identical to the parent (unless a mutation occurs in a body cell, p. 625). Asexual reproduction results in an increase in numbers of individuals, so it is a means by which large numbers of a successful organism can be built up, most often in an environment favourable for their further growth. Asexual reproduction may lead to a relatively rapid increase in numbers.

Budding in *Hydra*

An asexually produced hydra first appears as a bulge or 'bud' in the body wall of the parent (Figure 27.1). When food is plentiful, more than one may develop at the same time. The enteron of the 'bud' is continuous with that of the parent until the new individual has developed tentacles and a mouth. At this point, the bud becomes pinched off at its base, and forms an independent organism.

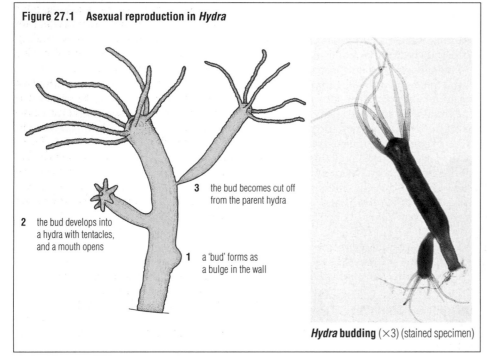

Figure 27.1   Asexual reproduction in *Hydra*

3 the bud becomes cut off from the parent hydra

2 the bud develops into a hydra with tentacles, and a mouth opens

1 a 'bud' forms as a bulge in the wall

***Hydra* budding** (×3) (stained specimen)

Taenia: a colony formed by budding?

The tapeworm *Taenia*, an endoparasite of the human gut (Figure 14.22, p. 299), consists of a long ribbon of proglottids with a head (scolex) at one end. The tapeworm can be considered as a colonial animal consisting of two kinds of individual. One, the scolex, reproduces asexually by transverse constrictions of the 'neck' region at its base. This process, a form of budding, cuts off the second type of individual, the proglottid. Thus the youngest proglottids are those that are closest to the scolex.

The proglottids reproduce only sexually, and each contains a complete set of male and female reproductive organs. By the time a ripe proglottid breaks off and passes out of the host with the faeces it is largely filled by the uterus. At this point the uterus is packed with the tiny embryos that have developed from fertilised eggs.

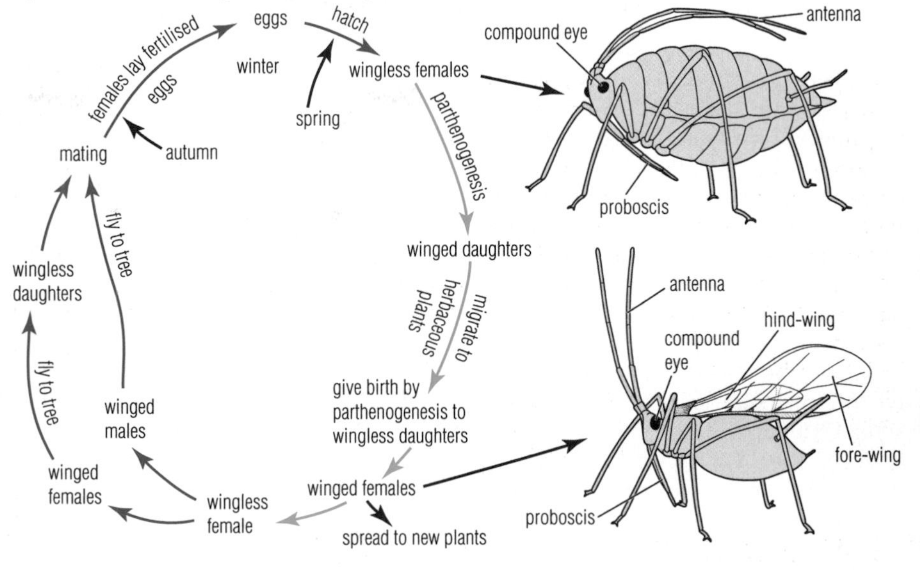

■ Parthenogenesis

Parthenogenesis is another form of asexual reproduction found in certain animals. It is in fact a modified form of sexual reproduction in which an egg cell (ovum) develops without fertilisation (and it is therefore an asexual process). Rotifers and aphids both reproduce

Figure 27.2   The life cycle of a bean aphid

naturally by parthenogenesis, and here we will examine the process in aphids because of the economic importance of these animals (p. 541).

There are very many species of aphid, all minute insects about 2–3 mm long when fully grown. Most aphids are green or brown in colour, with a narrow head and a bulbous abdomen (Figure 27.2). Aphids feed on juices in leaves and young shoots, obtaining their food from sieve tubes in the phloem tissue. Their reproductive capabilities are phenomenal, but at every stage in the life cycle vast numbers are eaten by predators, and many are killed by pesticides.

Aphid life cycles differ in detail from species to species, but typically an infestation starts with the development of fertilised eggs that have overwintered in cracks in the bark of a tree. These hatch into wingless females. Within these aphids the egg-producing apparatus forms eggs by mitosis, rather than by meiosis (**diploid parthenogenesis**). These immediately begin development inside the mother, so that live aphids (nymph stage) are 'laid' (Figure 27.3). The offspring feed, grow and reproduce parthenogenetically in their turn. Many generations of females are formed in this way, the later generations being winged females that fly to neighbouring plants, thus spreading the species. These winged females also reproduce parthenogenetically.

Aphid species show **facultative parthenogenesis**; that is, they can reproduce either by parthenogenesis or sexually. In late summer both male and female winged aphids are formed. These mate, and the females subsequently lay eggs on the twigs of trees. Thus the cycle of parthenogenesis followed by sexual reproduction is repeated the following season.

Figure 27.3   An aphid giving birth to live young (viviparity) (×8)

Questions

1 What evidence is there that only some individuals of any species succeed in reproducing themselves (p. 657)?

2 In what ways are parthenogenesis and sexual reproduction respectively advantageous in the aphid's life cycle?

■ Cloning in animals

The progeny produced by asexual reproduction by a single individual are described as a **clone**. The members of a clone are genetically identical. Cloning is common among certain species of non-vertebrates; for example, the individuals produced by budding from *Hydra* are a naturally occurring clone. Among vertebrates asexual reproduction does not occur naturally, but a body cell (somatic cell) may be taken from a frog, mouse or human, for example, and grown *in vitro* by tissue culture techniques similar to those used to grow plant cells (p. 636). Cells in culture can be induced to divide, and several generations of genetically identical cells can be grown in a few hours. Cloned cells are now used experimentally in medicine, agriculture, and the pharmaceutical industries, for example, to investigate the effects on cell metabolism of drugs, antibiotics, hormones, cosmetics and possible food additives. In some cases, tissue cultures of cloned cells are now used in place of laboratory animals in product development.

Another form of cloning involves extracting the nucleus from a skin or gut cell of a vertebrate, using microdissection equipment, and placing it in an egg cell from the same species in which the gamete nucleus has been destroyed by ultra-violet light. This engineered cell can be induced to grow and divide, and a complete new individual may be formed, genetically identical to that from which the mature nucleus was taken. This was first attempted with frogs. More recently it has been applied to mammals to produce Dolly the sheep (p. 642). All these experiments establish that a nucleus taken from a mature cell has all the genetic material needed to form a new individual. This phenomenon, **totipotency**, was first established with plants (pp. 429–31).

Natural cloning of humans occurs when identical twins are formed. About 1 in every 270 pregnancies in the UK produces identical twins. It occurs when the first-formed cells of the blastula (p. 413) divide into separate groups prior to implantation. Artificial cloning of humans was carried out in the USA in 1993, although the cells were cultured only briefly, to show that it can be done. Cloning of humans is banned in the UK, currently.

27.3 SEXUAL REPRODUCTION IN ANIMALS

Sexual reproduction is the production and fusion (**syngamy**) of male and female gametes to form a zygote, which grows and develops to form a new individual, very similar to the parents but not genetically identical to them. Gametes are produced in sex organs. In some animals the sex organs are quite elaborate structures, and reproduction involves complex behaviour patterns. Here we will discuss some general issues of importance in the evolution of sexual reproduction in animals. Later we will consider human reproduction in more detail.

■ The relative sizes of the gametes

In some organisms all the gametes are identical, a condition called **isogamy**; for example, isogamous gametes are found in *Spirogyra* (Figure 25.13, p. 554). In many organisms, however, gametes differ in size and structure and in their roles in reproduction. This condition is known as **heterogamy**. In a few organisms the gametes are of only slightly different sizes (**anisogamy**, seen in certain algae). In most cases of heterogamy there is a small and extremely motile 'male' gamete, and a 'female' gamete that is large and sedentary, normally due to the presence of stored food. This condition, known as **oogamy**, is characteristic of the green

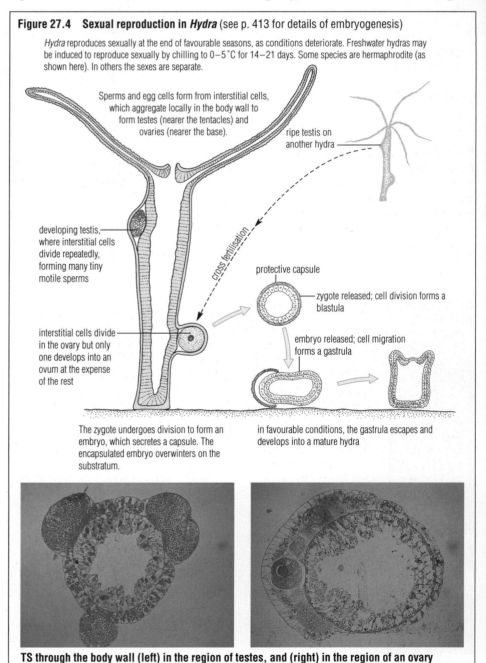

Figure 27.4   Sexual reproduction in *Hydra* (see p. 413 for details of embryogenesis)

Hydra reproduces sexually at the end of favourable seasons, as conditions deteriorate. Freshwater hydras may be induced to reproduce sexually by chilling to 0–5 °C for 14–21 days. Some species are hermaphrodite (as shown here). In others the sexes are separate.

Sperms and egg cells form from interstitial cells, which aggregate locally in the body wall to form testes (nearer the tentacles) and ovaries (nearer the base).

ripe testis on another hydra

cross fertilisation

developing testis, where interstitial cells divide repeatedly, forming many tiny motile sperms

interstitial cells divide in the ovary but only one develops into an ovum at the expense of the rest

protective capsule

zygote released; cell division forms a blastula

embryo released; cell migration forms a gastrula

The zygote undergoes division to form an embryo, which secretes a capsule. The encapsulated embryo overwinters on the substratum.

in favourable conditions, the gastrula escapes and develops into a mature hydra

TS through the body wall (left) in the region of testes, and (right) in the region of an ovary

plants, from bryophytes (Figure 26.3, p. 564) to angiosperms (Figure 26.19, p. 573).

In animals, too, sexual reproduction is oogamous. Sexual reproduction in *Hydra* (Figure 27.4) and in the frog (Figure 27.6) illustrates this. In both examples the female gamete carries a substantial store of food, used by the embryo in the early stages of its development. The male gametes are tiny and motile. They travel to the stationary egg cell, and there one fuses with it.

The evolutionary trend here is from isogamous gametes, regarded as the primitive condition, to oogamous reproduction as the most advanced. Mammals are oogamous, although mammalian eggs contain very little stored food; instead, the materials for development are obtained from the maternal blood supply through a special organ, the placenta (p. 599).

Cytoplasmic inheritance

In oogamous organisms, the male gamete contains very little cytoplasm. Moreover, the cytoplasm of the egg – a much larger quantity – remains with the zygote after fertilisation. As a result, the egg provides all or most of the cytoplasm of the zygote. One result is that organelles found in the cytoplasm of the zygote – mitochondria for example – all come from the egg cell, and none come from the sperm. All mitochondria contain small circular molecules of DNA (**mtDNA**), and on these occur mitochondrial genes. So, whilst 99% of our genetic inheritance (our DNA) occurs in our chromosomes (and comes 50% from the female parent and 50% from the male), 1% of our DNA comes from our female parent alone.

Mutations (abrupt changes) occur at a slow, steady rate in DNA, but alongside the chromosomal DNA in the nucleus enzymes occur that may repair these changes. However, these 'repair' enzymes are absent from mitochondria. As a consequence, mtDNA changes five to ten times faster than chromosomal DNA. Since there is also no mixing of mtDNA genes at fertilisation, the evidence about relationships from studying differences between samples of mtDNA is easier to interpret than that from chromosomal DNA. This has important repercussions in tracing the origins of species, and the evolutionary connections between them.

Figure 27.5   Sperm exchange in the earthworm

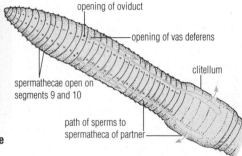

opening of oviduct
opening of vas deferens
clitellum
spermathecae open on segments 9 and 10
path of sperms to spermatheca of partner

two earthworms undergoing sperm exchange

1 For the exchange of seminal fluid, two worms lie side by side.

2 Seminal fluid passes from the seminal vesicles of each worm to the partner's spermathecae, along seminal grooves in the sides of the earthworms under the mucus.

3 The worms separate. Later, a mucus cylinder is secreted by the clitellum of each worm and is worked forwards by 'peristaltic' movements. As it passes segment 14, eggs from the ovary are laid in it. As it passes segments 9 and 10, sperms are expelled from the spermathecae and fertilise the eggs. When the worm has wriggled clear, the ends of the mucus tube close to form a cocoon enclosing the eggs.

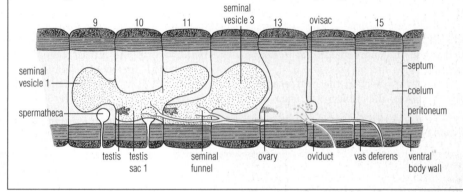

seminal vesicle 3
ovisac
seminal vesicle 1
spermatheca
9   10   11   13   15
septum
coelum
peritoneum
testis   testis sac 1   seminal funnel   ovary   oviduct   vas deferens   ventral body wall

---

■ Hermaphrodite or unisexual organisms?

In most animals the male and female sex organs are housed in different individuals. That is, the individuals are **unisexual**. In most flowering plants, however, and in some animals too (such as *Hydra*, Figure 27.4), male and female gametes are produced in each individual. The individuals are **hermaphrodite** or **bisexual**. The biological advantage of the hermaphrodite condition is that every individual is capable of forming fertilised eggs. In other words, hermaphrodite organisms have greater reproductive potential (fecundity) than unisexual organisms. A disadvantage is the possibility of self-fertilisation, leading to the production of offspring with reduced genetic variability.

Certain plants have mechanisms that reduce the possibility of self-fertilisation. This is often the case in hermaphrodite animals too. Just as hermaphrodite flowers may be protogynous or protandrous (p. 573), so in many hermaphrodite animals there are separate male and female phases. For example, the garden snail, *Helix*, starts the breeding season as a male organism. 'Male' snails pair up and exchange sperms, and only subsequently switch to egg production. The eggs formed are fertilised with sperms stored with the previous 'male' partner.

The earthworm is also a hermaphrodite organism, but here sperms and eggs are produced simultaneously. Sperms are exchanged above ground, at night. Two worms come together, facing in opposite directions. Segments 9 to 11 of one worm align with the clitellum of the other and vice versa, and the animals are held together by a tube of mucus secreted by the clitella. Seminal fluid is then exchanged, and stored. Later the acquired sperms are used to fertilise the worm's eggs, when these are being deposited in a cocoon of mucus, after the worms have separated (Figure 27.5).

Questions

1 List the sources of variation that arise between an offspring and its parents.

2 Why may oogamous reproduction be thought of as more advanced than isogamy?

3 How is self-fertilisation avoided in the earthworm?

■ Fertilisation, internal or external?

Fertilisation may take place outside the animal's body; many aquatic organisms, for instance, discharge their gametes directly into sea or fresh water. For example, cod (a demersal species, p. 102) releases huge quantities of sperms and eggs (seven million eggs during a female's lifetime!). In external fertilisation like this, where the gametes are quickly dispersed by the water, there is a strong possibility that many eggs will not encounter sperms. The chances that eggs and sperms will meet may be improved by chemotaxis (p. 517) and, in some species, by the simultaneous release of the male and female gametes.

In other examples of external fertilisation the sperms are discharged only in the vicinity of eggs. The male stickleback, for instance, first builds a nest as a 'nursery' (Figure 23.30, p. 520). The female is induced to lay her eggs within this confined space, and immediately afterwards the male sheds sperms over them.

When frogs meet in the ponds where they 'spawn', the union of eggs and

Figure 27.7 Copulating grasshoppers (p. 547 shows another insect species, the common blue butterfly, mating)

sperms is facilitated by a sexual embrace (amplexus, Figure 27.6). Before pairing begins, horny pads develop on the thumbs of the males. These pads enable the male to grip the female behind the forelimbs, and for several days he is carried about on the female's body, mostly in the water. Eventually eggs are

laid by the female (some 2000 each year), and the male immediately releases seminal fluid over them.

Internal fertilisation, on the other hand, ensures that all the sperms are deposited in the female's reproductive tract. Once sperms have entered the female reproductive system there is a

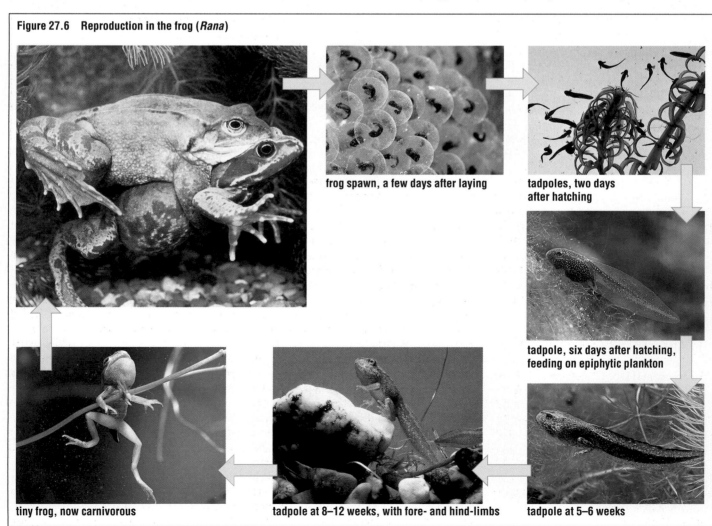

Figure 27.6 Reproduction in the frog (*Rana*)

frog spawn, a few days after laying

tadpoles, two days after hatching

tadpole, six days after hatching, feeding on epiphytic plankton

tadpole at 5–6 weeks

tiny frog, now carnivorous

tadpole at 8–12 weeks, with fore- and hind-limbs

good chance of fertilisation occurring, even though very nearly all sperms fail to reach an egg cell.

Internal fertilisation has evolved widely in the animal kingdom, and not only in the vertebrates. The process may involve an organ for introducing sperms into the female (an **intromittent** organ), such as a penis. For example, in insects, where the sexes are separate, the first step to fertilisation involves copulation (Figure 27.7). The tips of the abdomens are held close together. In many species the male has a pair of 'claspers', which hold the female's abdomen at the tip. Then a package (**spermatophore**) of sperms wrapped in a protein capsule is inserted into the female genital opening by means of the male insect's extensible penis. As the animals separate the sper-matophore may be ruptured and the fla-gellate sperms swim within the female reproductive system to a special chamber, off the vagina. Here they are stored, and later fertilise the eggs as these pass down the oviduct.

Insects, of course, are a large, highly successful group of non-vertebrate, mainly terrestrial animals (although the larvae of several species are aquatic). The vertebrates on the other hand, show an evolutionary transition from an aquatic to a terrestrial existence. Fish and most amphibia shed their gametes into the water; while amphibians can in general move efficiently on land, many must return to water to breed, and fertilisation is external. Reptiles, on the other hand, were the first group of vertebrates to overcome the problems of fertilisation (and the early development of the young) on land. Whilst in the reptiles and most birds the sperms are transferred to the female reproductive tract when the male and female genital openings are brought into close proximity, in mammals the male has a penis, and sperms are nor-mally deposited in the female's vagina.

■ Development of the zygote, internal or external?

The zygote develops by mitotic divisions to form a mass of cells. This first stage of development is called **cleavage**. As these cells continue to divide, they become arranged in discrete layers or blocks during **gastrulation**. Later the tissues and organs of the new organism are formed in a stage known as **organogeny**. The steps to growth and development of multi-cellular organisms are discussed in Chapter 19 (p. 396).

Figure 27.8 The egg-producing organs of a bird: only the left ovary and oviduct are developed in the mature female

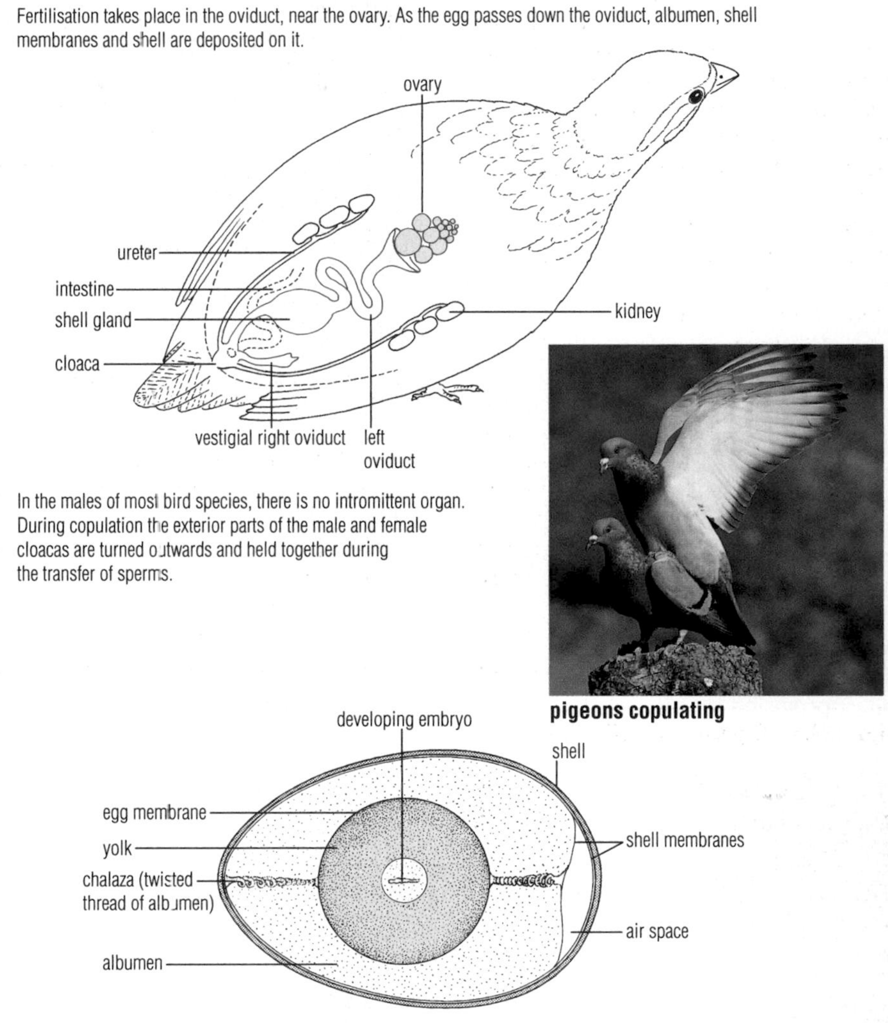

Fertilisation takes place in the oviduct, near the ovary. As the egg passes down the oviduct, albumen, shell membranes and shell are deposited on it.

In the males of most bird species, there is no intromittent organ. During copulation the exterior parts of the male and female cloacas are turned outwards and held together during the transfer of sperms.

pigeons copulating

bird's egg, as laid by the mother (part of the shell has been removed)

In many animals, including the frog (Figure 27.6), the early development of new individuals occurs outside the body of the parent. In insects, although fertili-sation is internal, the fertilised eggs are normally laid on a suitable food source, and the embryos also develop outside the body. Aphids are a notable exception; eggs that are produced by mitosis are retained in the body, and live young are born (Figure 27.3).

Among the vertebrates the gradual adaptation to life on land includes the evolution of an **amniote egg** in the rep-tiles and birds (**oviparity**). This has a fluid-filled cavity surrounded by a special membrane, the **amnion**. Outside the amnion is a protective shell, which encloses the embryo with the yolk sac and other membranes. Birds lay shelled eggs of this type (Figure 27.8), and then

incubate them. The embryo of the bird completes development outside the mother's body, and eventually the young bird hatches out (Figure 27.9).

Birds' eggs are laid soon after they are formed. But some animals, including many species of shark, produce shelled eggs that are retained within the body of the mother for some time, whilst early development of the embryo occurs. The young receive no additional nutrients from the mother, other than the substan-tial store of yolk laid down during forma-tion of the egg. The developing egg is held at body temperature, and provided with considerable protection. This type of reproduction is known as **ovoviviparity**.

By contrast, in the true mammals (Eutheria) the young are retained for a considerable time in the mother's uterus. The membranes of the amniote egg are present around the embryo in the uterus, but there is no shell. The embryo is nourished there from the mother's blood circulation, via the placenta, which is formed from an amalgam of embryonic membranes and the wall of the uterus. The young are born in a relatively advanced state of development. Reproduction of this kind is called **viviparity**; human reproduction is an example.

■ Parental care

Many animal species, both vertebrates and non-vertebrates, reproduce by forming and laying fertilised eggs, which are left to develop unattended. There is little or no parental care, though a parent may leave a food source beside the eggs for the young that will eventually hatch out (the digger wasp, p. 518, is an example). Often eggs are laid in a sheltered or protected position. Sticklebacks (and other species too) guard and protect their young while they develop (p. 520).

Pronounced parental care is typical of most species of birds (Figure 27.9) and mammals (p. 603). Parental care includes the provision of shelter from unfavourable environmental conditions, feeding, protection from predators, and in some species the 'training' of their offspring as they prepare for adult life.

27.4 REPRODUCTION IN AN ECOLOGICAL CONTEXT

Some species reproduce rapidly, produce vast numbers of offspring (high fertility) and have a short generation time. Ecologists define these organisms as having an 'opportunistic' way of life. Typically such species are pioneer organisms that move into new habitats, breed early and often achieve rapid colonisation. Migration and dispersal are important to their survival. Their mortality rates are high and their lifespans are short. They survive in environments that are variable and often unpredictable. Species adopting this reproductive strategy and life style are known to ecologists as *r*-species; many weed plants are examples.

On the other hand, there are very many species that reproduce slowly and have a long lifespan. The individuals move into habitats that are settled and stable, and survive because they compete successfully for the resources. Populations of these organisms are more or less constant in numbers for they occur at the **carrying capacity** for the habitat (where carrying capacity = maximum number of individuals of a particular species that can be supported indefinitely by a given habitat). Species with this type of life cycle are known to ecologists as *K*-species. Trees such as the oak are examples.

In reality, *r*- and *K*-strategies are two extremes. Most organisms fit on to a scale somewhere in between.

Questions

1 Development of an embryo is divided into the stages of cleavage, gastrulation and organogeny (p. 413). What are the key features of these stages?
2 To what extent does the stickleback show parental care (p. 520)? In what ways is this different from the care provided by typical garden birds for their young?
3 What constitutes an amniote egg, and in which groups of animals is it found?
4 In terms of *r*- and *K*-species, state how you would classify the following animals, and why:
  a wild rabbit   **c** badger
  b urban fox   **d** collared dove.
5 Very large numbers of eggs are normally laid by animals with external fertilisation, whereas many species with internal fertilisation produce relatively few eggs. Speculate on this difference.
6 What features of the life cycle of the frog are associated with an aquatic environment?

Figure 27.9 Parental care in birds: a summary (the events shown are typical of many species of bird, but not all; pheasant chicks, for example, leave the nest almost immediately)

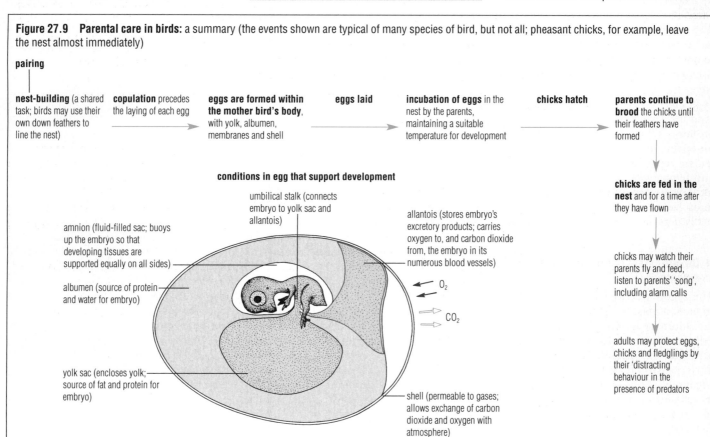

pairing

nest-building (a shared task; birds may use their own down feathers to line the nest) → **copulation** precedes the laying of each egg → **eggs are formed within the mother bird's body**, with yolk, albumen, membranes and shell → **eggs laid** → **incubation of eggs** in the nest by the parents, maintaining a suitable temperature for development → **chicks hatch** → **parents continue to brood** the chicks until their feathers have formed

chicks are fed in the nest and for a time after they have flown

chicks may watch their parents fly and feed, listen to parents' 'song', including alarm calls

adults may protect eggs, chicks and fledglings by their 'distracting' behaviour in the presence of predators

conditions in egg that support development

umbilical stalk (connects embryo to yolk sac and allantois)

amnion (fluid-filled sac; buoys up the embryo so that developing tissues are supported equally on all sides)

albumen (source of protein and water for embryo)

yolk sac (encloses yolk; source of fat and protein for embryo)

allantois (stores embryo's excretory products; carries oxygen to, and carbon dioxide from, the embryo in its numerous blood vessels)

O_2

CO_2

shell (permeable to gases; allows exchange of carbon dioxide and oxygen with atmosphere)

27.5 SEXUAL REPRODUCTION IN MAMMALS

All mammals (and most other vertebrates too) are unisexual, with separate male and female animals. In the reproductive system, the gametes are produced in special paired glands called **gonads**; the male gametes or sperms are produced in the testes, the female gametes or egg cells in the ovaries. The gonads are known as the **primary sexual organs**. Associated with the gonads is a system of tubes and accessory glands that help in the transport, reception and storage of gametes: the **secondary sexual organs**. The gonads are also endocrine glands, and the sex hormones they form regulate both the onset of sexual maturity and sexual activity.

27.6 MALE REPRODUCTIVE SYSTEM

The roles of the male reproductive system are

► to produce sperms in the testes
► to deliver sperms in a liquid medium (called semen) to the vagina during copulation
► to produce the chief male sex hormone, testosterone.

The male reproductive system (Figure 27.10) consists of two **testes**, with ducts that store and transport sperms to the outside of the body, via the **penis**, the intromittent organ. The penis consists of special connective tissue with numerous small blood vessels in it, and is covered by elastic skin. The tip of the penis is an expanded, sensitive region, the **glans**, covered by retractable skin, called the foreskin. Circumcision, the surgical removal of the foreskin, may be carried out for medical reasons (for example, if the foreskin is non-retractable), as a religious ritual (as in Judaism and Islam), and in some cultures as a puberty ritual. Associated with the reproductive ducts are certain accessory glands that add the secretions to sperms to make up semen (p. 594).

In the human male, and in most other male mammals (marine mammals and elephants are exceptions), the male gonads or testes are enclosed in an external sac, the scrotum, which is an extension of the abdomen wall. The testes actually start out in the abdominal cavity, but they descend into the scrotum before the birth. In this position the testes are maintained at a temperature about 2 °C lower than the body temperature, a condition essential for sperm production.

The testes are paired oval glands about 3 cm in length. Each testis consists of about a thousand **seminiferous tubules**. Seminiferous tubules are very long and thin, and are surrounded by connective tissue and held together by fibrous capsules. The connective tissue contains capillary networks, together with interstitial cells (**Leydig cells**), the endocrine cells of the testes (p. 589). All the seminiferous tubules in a testis connect with a single duct, the **epididymis**, via a system of tiny tubules. The epididymis is itself a much-coiled tube about 6 m long, which is connected to the **sperm duct** (vas deferens). The two sperm ducts connect with the urethra just below the bladder. Surrounding the urethra at this point is the **prostate gland**, and further down are the **Cowper's glands**. Above the junction point are paired **seminal vesicles**. All three types of gland contribute to the fluid part of semen.

In the male the urethra is the route by which both urine and semen pass to the exterior. During ejaculation the sphincter muscle at the base of the bladder is closed by sympathetic reflex action. Consequently, urine cannot be expelled during ejaculation (and semen cannot enter the bladder).

■ Vasectomy, male sterilisation

One method of birth control is achieved by the surgical severing of both sperm ducts. These tubes are reached via small incisions in the scrotum (under local anaesthetic). Each duct is tightly tied in two places and a short length removed. Sperm production continues, but the sperms cannot reach the urethra, and so they break down and are re-absorbed. The operation, known as vasectomy, has no effect on sexual performance, but achieves effective, permanent and normally irreversible sterilisation.

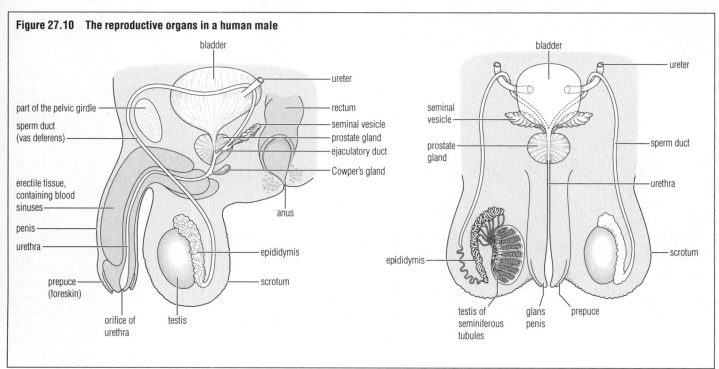

Figure 27.10 The reproductive organs in a human male

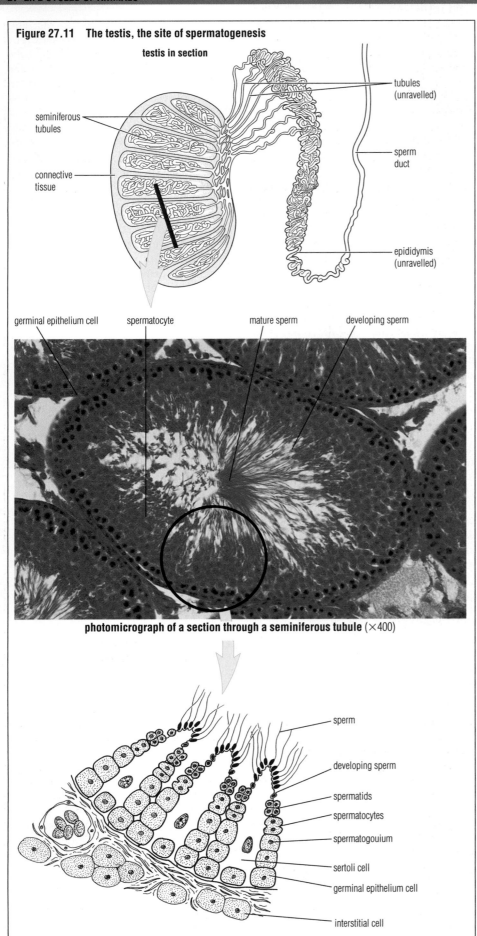

Figure 27.11 The testis, the site of spermatogenesis

testis in section

- seminiferous tubules
- connective tissue
- tubules (unravelled)
- sperm duct
- epididymis (unravelled)

germinal epithelium cell · spermatocyte · mature sperm · developing sperm

photomicrograph of a section through a seminiferous tubule (×400)

- sperm
- developing sperm
- spermatids
- spermatocytes
- spermatogouium
- sertoli cell
- germinal epithelium cell
- interstitial cell

■ Sperm production

The seminiferous tubules, the sites of sperm production, are lined by **germinal epithelial cells** (Figure 27.11), which divide repeatedly, forming cells called **spermatogonia**. These cells divide and grow to form the **primary spermatocytes**. These undergo the first meiotic division, to form **secondary spermatocytes**, and then the second meiotic division to produce **spermatids**, which thus are haploid (Figure 27.13). Also present amongst the cells lining the tubules are **nutritive cells** (known as Sertoli cells). Whilst positioned with the head inside these nutritive cells the spermatids develop tails and mature into sperms (Figure 27.12).

■ Sperm storage and the formation of semen

The sperms are immotile when first formed. As they are moved from the seminiferous tubules where they develop, through the fine tubules that lead to the epididymis, the nutritive fluid surrounding the sperms is re-absorbed. In the epididymis the sperms are retained for between two and ten days, progressively developing motility. Sperms are then moved into the sperm ducts by peristaltic contraction of the smooth muscle of the duct. Sperms are stored in the epididymis and, briefly before ejaculation, in the sperm duct. Those that have not been ejaculated within about four weeks may be re-absorbed into the body.

■ Disorders of the prostate gland

The prostate gland can become enlarged, and this is a common condition in men over 60. An enlarged prostate gland may obstruct the flow of urine down the ureter, and if it makes urination painful and difficult the part of the gland causing blockage is removed surgically.

Enlargement is occasionally due to a bacterial infection (which may be treated with antibiotics); but there are other causes, less well understood. In some cases, the obstruction may be due to a tumour, either benign or malignant. Currently, cancer of the prostate claims more lives of men (in their 50s and 60s) than does cervical cancer in women. A screening programme for this disease is possible, based upon rectal examination of the prostate gland, and a blood test for a protein released only by tumour cells.

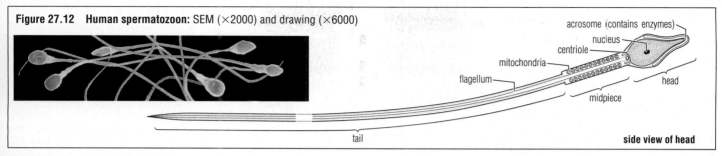

Figure 27.12 Human spermatozoon: SEM (×2000) and drawing (×6000)

acrosome (contains enzymes)
nucleus
centriole
mitochondria
flagellum
head
midpiece
tail
side view of head

■ The testis as an endocrine gland

The onset of sexual maturity (puberty) is triggered by **releasing hormone** from the hypothalamus of the brain (Figure 21.47, p. 474). This hormone, which travels by a small vein to the anterior pituitary immediately below it, stimulates the secretion of two **gonadotrophic hormones**. These hormones (named after the discovery of their roles in the female reproductive system) are **follicle-stimulating hormone (FSH)**, which initiates spermatogenesis, and **luteinising hormone (LH)**, which stimulates the production of the male hormone testosterone by the interstitial cells of the testes.

Both FSH and LH work on the cell membranes of cells of the testes by causing the release of **cyclic AMP** (adenosine monophosphate, the second messenger – see Figure 21.46, p. 473). The result, known as a **cascade effect**, is the amplification of the influence of these hormones on cell metabolism.

Testosterone (Figure 27.20, p. 592) is synthesised from cholesterol, drawn from the blood supplying the testes. Testosterone is largely responsible for initiating and maintaining the **secondary sexual characteristics** of the male. These include the growth of the sex organs and, in human males, the growth of body hair (facial and pubic, for example), enlargment of the larynx leading to a deeper voice, and general muscular development that results from the enhanced incorporation of protein into the muscles.

In addition, testosterone and FSH together control the continuous production of sperms.

An excess of testosterone in the blood circulation has an inhibitory effect on the secretion of LH by the anterior pituitary (and probably on the hypothalamus and the secretion of releasing hormone). Similarly, excess activity of the nutritive cells (excess production of sperms) results in the release of a hormone, **inhibin**, which inhibits secretion of FSH by the anterior pituitary (Figure 27.14).

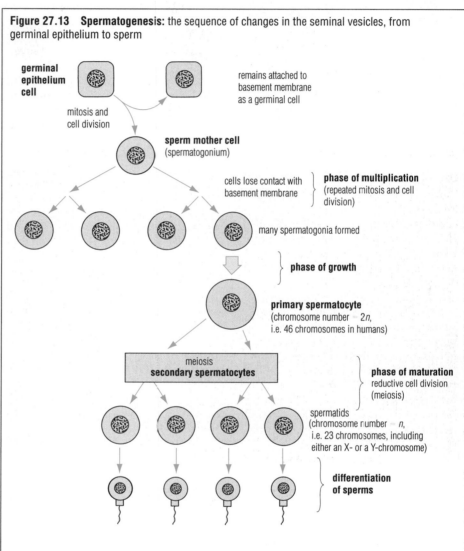

Figure 27.13 Spermatogenesis: the sequence of changes in the seminal vesicles, from germinal epithelium to sperm

germinal epithelium cell

mitosis and cell division

remains attached to basement membrane as a germinal cell

sperm mother cell (spermatogonium)

cells lose contact with basement membrane

phase of multiplication (repeated mitosis and cell division)

many spermatogonia formed

phase of growth

primary spermatocyte (chromosome number = 2n, i.e. 46 chromosomes in humans)

meiosis
secondary spermatocytes

phase of maturation reductive cell division (meiosis)

spermatids (chromosome number = n, i.e. 23 chromosomes, including either an X- or a Y-chromosome)

differentiation of sperms

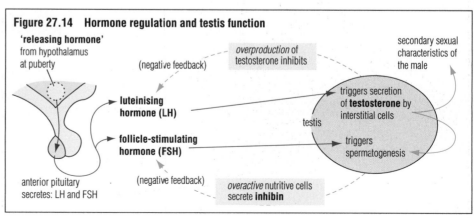

Figure 27.14 Hormone regulation and testis function

'releasing hormone' from hypothalamus at puberty

(negative feedback)

overproduction of testosterone inhibits

secondary sexual characteristics of the male

luteinising hormone (LH)

follicle-stimulating hormone (FSH)

triggers secretion of **testosterone** by interstitial cells

testis

triggers spermatogenesis

anterior pituitary secretes: LH and FSH

(negative feedback)

overactive nutritive cells secrete **inhibin**

589

27.7 FEMALE REPRODUCTIVE SYSTEM

The roles of the female reproductive system are

▶ to produce the female sex cells (egg cells) in the ovaries
▶ to deliver egg cells to the oviduct where fertilisation may occur
▶ to get the wall of the uterus into a condition favourable for implantation and development of an embryo
▶ to produce the chief female sex hormones, oestrogens.

The female reproductive system (Figure 27.15) consists of the primary sexual organs (the ovaries) and the secondary sexual organs (oviducts, uterus and vagina situated within the body, together with the external genitalia of the vulva).

The **ovaries** are small structures, about 3 cm long and 1–1.5 cm thick, held in position near the base of the abdominal cavity by ligaments. As well as producing egg cells, the ovaries are also endocrine glands. The **oviducts** are thin-walled, muscular tubes about 10 cm long, lined by a ciliated epithelium with gland cells that secrete mucus. At one end the oviducts open out as funnel-shaped structures close to the ovaries; at the opposite end, they lead into the **uterus**. The uterus is a thick-walled, muscular organ the shape and size of an inverted pear (about 7.5 × 5.0 cm), with a minute, fluid-filled cavity. The bulk of the uterus wall is of smooth muscle, but the inner lining, the **endometrium**, is made up of mucous membrane, richly supplied with arterioles. The **vagina** is a muscular tube connected to the uterus at the cervix, and opening to the exterior at the **vulva**. The vulva consists of folds of skin (the **labia**) and a small erectile structure, the **clitoris**, homologous with the penis.

■ Cervical cancer

The cervix is the base of the uterus, at the junction with the vagina. The cells formed at the surface here can become abnormal, appearing irregular and misshapen. This condition is detected by taking a smear from the tissue at the top of the vagina and examining it cytologically (mounting the smear on a microscope slide, and staining it for examination by light microscopy). Abnormality in these cells is not evidence of cancer, but irregular cells are more likely to become cancerous. Early diagnosis and treatment prevents development of this most common cancer of women. Smear tests should be taken every three years by all women between the ages of 25 and 64.

■ Egg cell production (oogenesis)

The production of egg cells begins in the ovaries of the fetus before birth, but the final development of the individual egg is completed only in adult life, anything from 11 to 55 years later – and then only after fertilisation (if and when this occurs).

During fetal development cells of the germinal epithelium (which covers the outer surface of the ovary) undergo repeated mitotic division, and numerous oogonia are formed. Oogonia migrate into the connective tissues of the ovary, and there grow and enlarge to form primary oocytes. These become surrounded by a layer of follicle cells, forming a structure known as a primary follicle (Figure 27.16). At birth the two ovaries each contain up to 200 000 primary follicles, which remain dormant until puberty. Less than 1% of follicles will complete their development; the rest will degenerate, and never produce eggs at all.

Between puberty (from the age of about 11 upwards) and the cessation of monthly ovulation (menopause, at age 45–55 years (p. 601)), primary follicles begin to develop further (Figure 27.17). In fact, many oocytes start to grow each month, but usually only one matures. The rest degenerate. Development involves the progressive enlargement of the ovarian follicle, which moves to the outer part of the ovary. The primary oocyte there undergoes the first meiotic division (meiosis I, p. 196), to form a secondary oocyte and a tiny polar body. (In other words, division of the cytoplasm is markedly unequal.) Then the second meiotic division (meiosis II) begins, and proceeds as far as metaphase II. In this condition the egg cell (secondary oocyte, Figure 27.18) is released from the ovary (**ovulation**), by rupture of the ovarian follicle and of the ovary wall. In the human female, ovulation occurs from one of the two ovaries about once every 28 days.

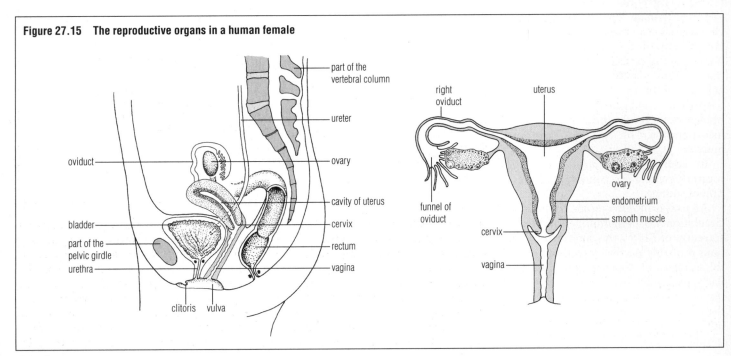

Figure 27.15   The reproductive organs in a human female

Figure 27.16   The ovary, showing stages in the development of an ovarian follicle and corpus luteum

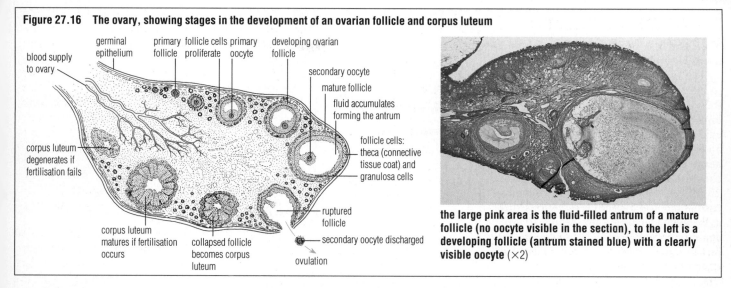

the large pink area is the fluid-filled antrum of a mature follicle (no oocyte visible in the section), to the left is a developing follicle (antrum stained blue) with a clearly visible oocyte (×2)

The egg, on release, is drawn into the oviduct funnel by ciliary action, and it moves down the oviduct. Development of the secondary oocyte does not proceed further unless a sperm fuses with it (fertilisation, see section 27.9). If the secondary oocyte is not fertilised it is lost from the body via the vagina about three days later. After the discharge of the egg the ruptured ovarian follicle develops into a yellowish mass of cells known as the **corpus luteum** (literally, 'yellow body'), which becomes an endocrine gland of pregnancy.

Figure 27.18   Secondary oocyte at the time of fertilisation

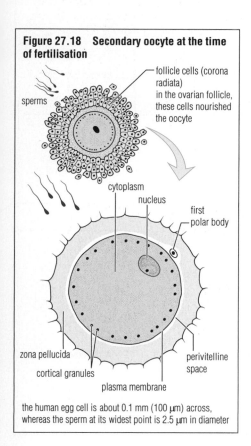

the human egg cell is about 0.1 mm (100 μm) across, whereas the sperm at its widest point is 2.5 μm in diameter

Figure 27.17   Oogenesis in humans: the sequence of changes in the ovary and oviduct, from germinal epithelium cell to ovum

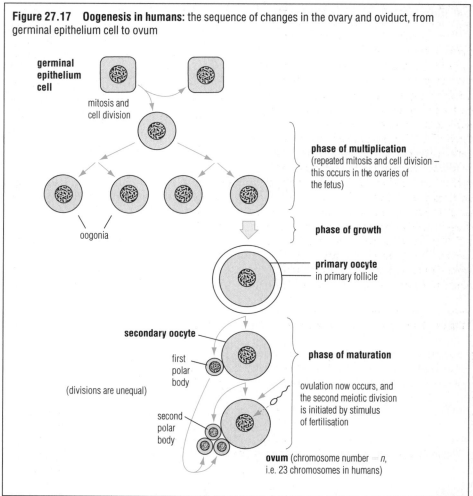

■ Oestrus and the menstrual cycle

In the human female a single egg (secondary oocyte) is normally produced and discharged (ovulation) from one of the two ovaries each month, from the time of puberty until the menopause. Ovulation is just one part of a cycle of changes, called the **menstrual cycle**. The menstrual cycle is often described as lasting 28 days, but normally it varies in length between 25 and 35 days. During the cycle, the lining of the uterus is made ready to accept the embryo that will grow in the uterus if the egg is fertilised. Thus the menstrual cycle involves a synchronised, recurring sequence of changes in the lining of the uterus (endometrium) of the non-pregnant female (the **uterine cycle**) linked to a sequence of changes in the ovaries (the **ovarian cycle**) (Figure 27.19). Note that the time when the egg(s) are released is referred to as **oestrus**.

The function of the menstrual cycle is to provide a favourable environment for the development of a fetus. But when implantation does not occur, the expanded endometrium is broken down and shed. This event, known as **menstruation** or a 'period', results in the discharge of a small amount (less than 20 cm³ per day) of blood, cell remains, tissue fluid and glandular secretions such as mucus. The cycle then begins all over again (Figure 27.19). By convention, the onset of menstruation is taken as **day 1** of the menstrual cycle.

■ Menstruation as a regular event

In most of human history, and for many women today, the experience of regular periods has been relatively rare. This is because, after reaching sexual maturity, they follow a rhythm of pregnancy–lactation–pregnancy–lactation for the greater part of their fertile lives. In Britain up to late Victorian times, this was the expectation of women in all social classes. Charles Darwin's wife, for example, gave birth to ten children between 1840 and 1856.

In the United Kingdom since about 1860 the birth rate has been decreasing. Here girls now reach puberty at an earlier age, which prolongs the time span of their sexual maturity. In addition, efficient means of contraception that are widely practised (the 'pill', the use of condoms, and male sterilisation) mean that in affluent societies many fewer children are born (more survive), and that most women menstruate regularly most of the time.

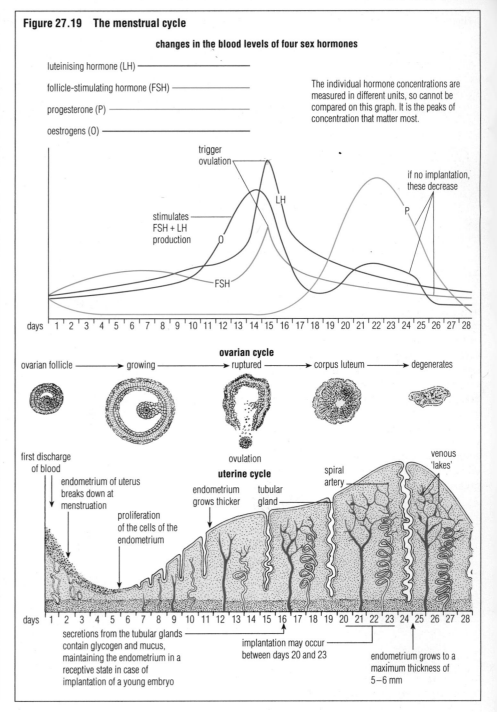

Figure 27.19   The menstrual cycle

changes in the blood levels of four sex hormones

luteinising hormone (LH) ———————

follicle-stimulating hormone (FSH) ———————

progesterone (P) ———————

oestrogens (O) ———————

The individual hormone concentrations are measured in different units, so cannot be compared on this graph. It is the peaks of concentration that matter most.

trigger ovulation

stimulates FSH + LH production

if no implantation, these decrease

ovarian cycle

ovarian follicle → growing → ruptured → corpus luteum → degenerates

first discharge of blood

ovulation

uterine cycle

endometrium of uterus breaks down at menstruation

proliferation of the cells of the endometrium

endometrium grows thicker

tubular gland

spiral artery

venous 'lakes'

endometrium grows to a maximum thickness of 5–6 mm

secretions from the tubular glands contain glycogen and mucus, maintaining the endometrium in a receptive state in case of implantation of a young embryo

implantation may occur between days 20 and 23

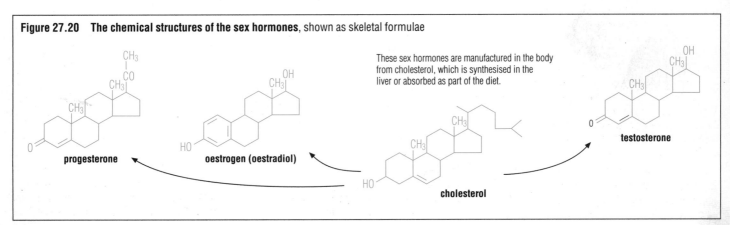

Figure 27.20   The chemical structures of the sex hormones, shown as skeletal formulae

These sex hormones are manufactured in the body from cholesterol, which is synthesised in the liver or absorbed as part of the diet.

progesterone

oestrogen (oestradiol)

cholesterol

testosterone

■ Endocrine control of the menstrual cycle

On day 1 of the menstrual cycle the ovaries are activated by the rising concentration of follicle-stimulating hormone (FSH), secreted by the anterior pituitary gland. This effect is augmented by luteinising hormone (LH), which is also present in the bloodstream, and which has a synergistic effect (p. 432). These hormones activate the growth and development of several ovarian follicles (more in mammals with numerous progeny in the litter). One of these follicles soon takes over and the remainder become dormant.

The wall (**theca**) of this ovarian follicle becomes an additional endocrine gland, and secretes an oestrogen hormone (oestradiol, Figure 27.20). The concentration of oestrogens in the blood rises quickly. The effects of this are, first, to inhibit the production and release of FSH by the anterior pituitary gland by a negative feedback mechanism (the level of LH remains the same), and then to initiate the rebuilding of the endometrium (Figure 27.21).

The concentration of oestrogens reaches a peak by day 11 of the cycle, and triggers a sudden surge in the production of both FSH and LH. Ovulation is triggered on day 14, and the empty ovarian follicle heals over and becomes an endocrine gland, this time producing the hormone progesterone (Figure 27.20). Meanwhile, the production of oestrogens falls off, only recommencing after the next ovulation. Progesterone and oestrogens together maintain the development of the endometrium, which contains stored glycogen and a little mucus. In this condition it awaits the implantation of an egg.

The higher concentrations of progesterone and oestrogens also inhibit the production of FSH and LH. The falling levels of these gonadotrophic hormones allows the corpus luteum to degenerate, and now the concentrations of progesterone and oestrogens also fall. As a result, the FSH and LH production by the anterior pituitary is no longer inhibited, and the events of the next cycle begin.

If an implantation occurs, the developing embryo and its placenta become an endocrine gland in their own right, maintaining the concentration of the female sex hormones, and the complex structure of the endometrium. The (unusual) case of fertilisation and implantation of an egg is considered in section 27.9.

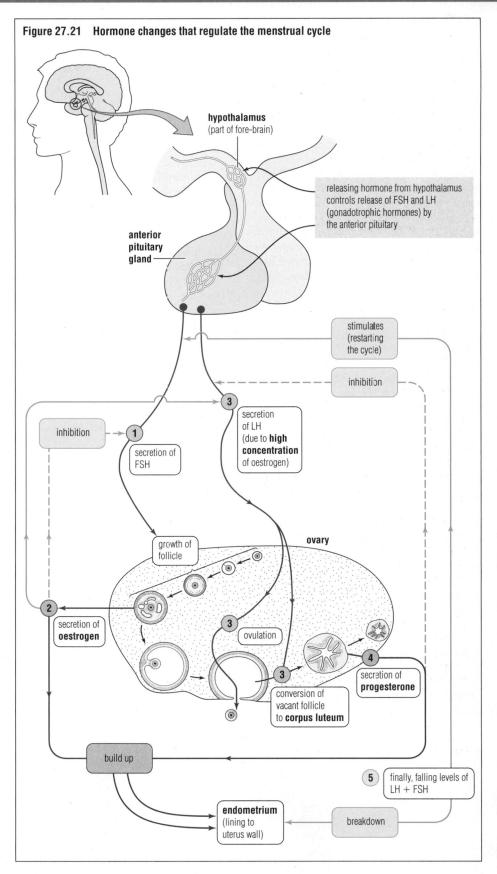

Figure 27.21 Hormone changes that regulate the menstrual cycle

hypothalamus
(part of fore-brain)

releasing hormone from hypothalamus controls release of FSH and LH (gonadotrophic hormones) by the anterior pituitary

anterior pituitary gland

stimulates (restarting the cycle)

inhibition

inhibition

3 secretion of LH (due to **high concentration** of oestrogen)

1 secretion of FSH

growth of follicle

ovary

2 secretion of **oestrogen**

3 ovulation

3 conversion of vacant follicle to **corpus luteum**

4 secretion of **progesterone**

build up

5 finally, falling levels of LH + FSH

endometrium (lining to uterus wall)

breakdown

Questions

1 Draw up a table to present concisely the contrast between the steps in gametogenesis in the human male and female.

2 What are the critical hormone changes in women that respectively trigger ovulation, and cause degeneration of the corpus luteum?

■ Patterns of sexual activity in mammals

A breeding season

Animals reproduce most effectively when there is a plentiful supply of food. Just as the timing of plant reproduction coincides with favourable environmental conditions (p. 434), most animals have a breeding season that normally enables their progeny to feed or be fed successfully. This timing of sexual activity is controlled by environmental changes that are detected in the sensory and nervous systems, and which alter the level of sex hormones. One environmental factor known to be responsible in many species is the seasonal variation in daylength. That is, a certain number of days with daylight lasting for more than a minimum period of time provides the environmental 'trigger' for breeding.

In mammals the duration of daily illumination (photoperiod) falling on the eyes controls the release of hormones by the hypothalamus, and thereby the onset of sexual function (Figure 27.14, p. 589 and Figure 27.21, p. 593). In both male and female mammals, FSH and LH have a key role in preparing the reproductive organs for sexual activity.

Many animals have a breeding season. For example in Britain, red deer (*Cervus elaphus*) come into 'season' (males into 'rut' and females into oestrus) as a result of the shortening of days in autumn (September/October), after the summer flush of vegetation. Day length is known to be the stimulus from observations on red deer that were transported to New Zealand by colonists from the UK, with the intention of breeding them for food. In New Zealand, short days occurred in June/July and long days in December/January. It took only 18 months for the imported deer to completely switch to being 'in season' in the autumn in New Zealand (February/March).

Being in season or 'on heat' (at the time of ovulation) typically involves the following characteristics.

▶ **In males:** an awareness of the presence of females, individually or as a herd with potential as a 'harem' (in the case of most herd animals, like deer). Males also compete (normally in ritualised conflict, p. 523) for the right to mate, and so become especially aware of competitor males.

▶ **In females:** typically restless, the animals may bellow or cry out, thereby drawing attention. Herd animals often mount other animals or allow themselves to be mounted. The external genitalia (the vulva) become swollen and red. Secretions from the vagina act as pheromones that may be detected by males of the same species, and stimulate moves by the male to copulate. The female 'on heat' will allow mating, and fertilisation is likely to follow.

During the breeding season, the male is fertile throughout and generally eager. Meanwhile, the female may come into oestrus once only (**monoestrous**) or many times (**polyoestrous**).

Continuous breeders

Among the mammals, quite a few species have become able to breed at more or less any time of the year (Table 27.1). However, even these mammals are most likely to breed when environmental conditions are most favourable.

Table 27.1 Breeding patterns in mammals

With a breeding season		Breeding at any time	
sheep (*Ovis aries*)	breeding season is late summer/early autumn; ewes are polyoestrous every 16–17 days, for 24 hours	**rat** (*Rattus rattus*)	oestrus occurs every 4–5 days, lasting for 8–20 hours
dog (*Canis familiaris*)	breeding seasons in early spring and autumn; bitch is monoestrous	**rabbit** (*Oryctolagus cuniculus*)	mating occurs throughout the year, depending only on availability of mates peak time for successful pregnancies is April/June; doe is 'induced ovulator': ovulation occurs 13 hours after mating, but fetus is reabsorbed if conditions become adverse
cat (*Felis catus*)	polyoestrous, with induced ovulation	**cow** (*Bos taurus*)	oestrus occurs every 20–22 days and lasts for about 24 hours
horse (*Equus caballus*)	polyoestrous	**pig** (*Sus scrofa*)	oestrus occurs every 20–21 days and lasts for 2–4 days

27.8 COURTSHIP AND COPULATION

In very many species, copulation is preceded and accompanied by **courtship behaviour**. The purposes of courtship are

▶ to attract a mate, possibly from a considerable distance, often by conspicuous or noisy behaviour

▶ to drive away other males competing for the female(s) at oestrus

▶ to ensure mating occurs between members of the same species, and between fit and healthy individuals

▶ to induce a comparable level of sexual arousal in both partners, making both equally ready and willing to copulate

▶ to synchronise the activities of both partners for successful copulation, leading to conception (fertilisation).

Copulation in humans is often referred to as sexual intercourse or **coitus**, and involves insertion of the erect penis into the vagina, and ejaculation of semen.

Erection of the penis is brought about by blood pressure, and occurs when the veins draining the penis are constricted and the arterioles delivering blood become dilated, under the control of the autonomic nervous system. Secretion of lubricating fluid by the vulva and vaginal lining facilitates the entry of the penis. Rhythmic movements generate friction between the penis and the walls of the vagina. Touch receptors in the swollen head of the penis and in the erect clitoris are stimulated and enhance the manifestations of sexual arousal, introduced by courtship activities. Sexual arousal involves certain reflexes of the autonomic nervous system, including increased heart rate, raised blood pressure, rapid breathing, dilation of capillaries in the skin of the face and chest, and pronounced sweating.

Continuing tactile stimulation of the penis induces contractions of the smooth muscle of the epididymis, sperm duct, seminal vesicles and the prostate and Cowper's glands. Seminal fluids accumulate in the ejaculatory duct and in the urethra at the base of the penis. Sphincter muscles at the base of the bladder close. These developments are controlled by reflex actions of the autonomic nervous system. Ejaculation is then brought about by rhythmic, wave-like contractions of the voluntary muscles of the ejaculatory duct and urethra, by a spinal reflex involving the voluntary nervous system. This occurs at a peak of sexual excitement, referred to as **orgasm**. In the female, orgasm is not a prerequisite condition for fertilisation, as it is in the male, but it is desirable. Coitus is associated with emotional and psychological phenomena, controlled by the higher centres of the brain.

27.9 FERTILISATION

About 3–5 cm³ of semen are deposited in the vagina in humans. Semen normally contains 40–150 × 10⁶ sperms cm⁻³. It is an alkaline fluid (pH 7.2–7.6) and it neutralises the natural acidity of the vagina, producing an environment of pH 6.0–6.5, which is apparently ideal for sperm mobility. Semen contains an enzyme that activates sperm after ejaculation, as well as an antibiotic that destroys bacteria. On ejaculation into the vagina semen coagulates into a clot, but within 5–20 minutes this clot has liquefied again. Coagulation and liquefaction are caused by enzymes secreted into semen from the prostate.

Fertilisation, the union of egg with sperm, occurs in the upper part of the oviduct. Sperms arrive there within a very brief period of time after the semen has liquefied. Only a few hundred of the sperms ejaculated reach this region of the oviduct, but it is most unlikely that they do so by their own unaided mobility. For example, the dense mucus normally present at the cervix thins out, leaving channels through which a few sperms can enter the uterus. These changes to the mucus are induced by the hormones that also induce ovulation. At the same time, waves of muscular contraction in the walls of the uterus and oviducts draw semen to the site of fertilisation.

Before any possible successful encounter with the secondary oocyte the sperms undergo 'capacitation', a process involving little-understood changes to the exterior surface that are a prelude to the 'acrosome reaction'. The acrosome at the tip of the sperm (Figure 27.12) is filled with hydrolytic enzymes (p. 187). Released at this stage, the enzymes of the acrosome digest a pathway for the sperm through the zona pellucida surrounding the secondary oocyte. Then the plasma membrane of the head of the sperm fuses with that of the secondary oocyte, and the sperm nucleus is engulfed (Figure 27.22). This fusion with a sperm triggers completion of meiosis II in the oocyte, leading to the formation of a tiny second polar body and the haploid egg cell. Polar bodies break down. The paternal and

maternal chromosomes come together at the equator of the spindle at the first mitotic division of the zygote. All this happens high up in the oviduct.

■ Implantation

The zygote is transported along the oviduct by ciliary action. Nuclear and cell divisions commence after a delay of 30 hours or so. The first division forms a two-celled embryo. Subsequent divisions follow immediately, and by the time the embryo reaches the uterus three or four days later it is a solid mass

of cells, called a **morula** (Figure 27.23, overleaf). These early divisions make up the **cleavage stage** of embryology, in which the number of cells increases, but there is no growth in size. The morula is still contained within the original zona pellucida.

Divisions continue, and the cells arrange themselves into a fluid-filled ball known as a **blastocyst**. The blastocyst consists of over 100 cells, and these cells are known as **blastomeres**. Implantation now commences (by day 7) and is completed by day 14.

Figure 27.22 The stages of fertilisation

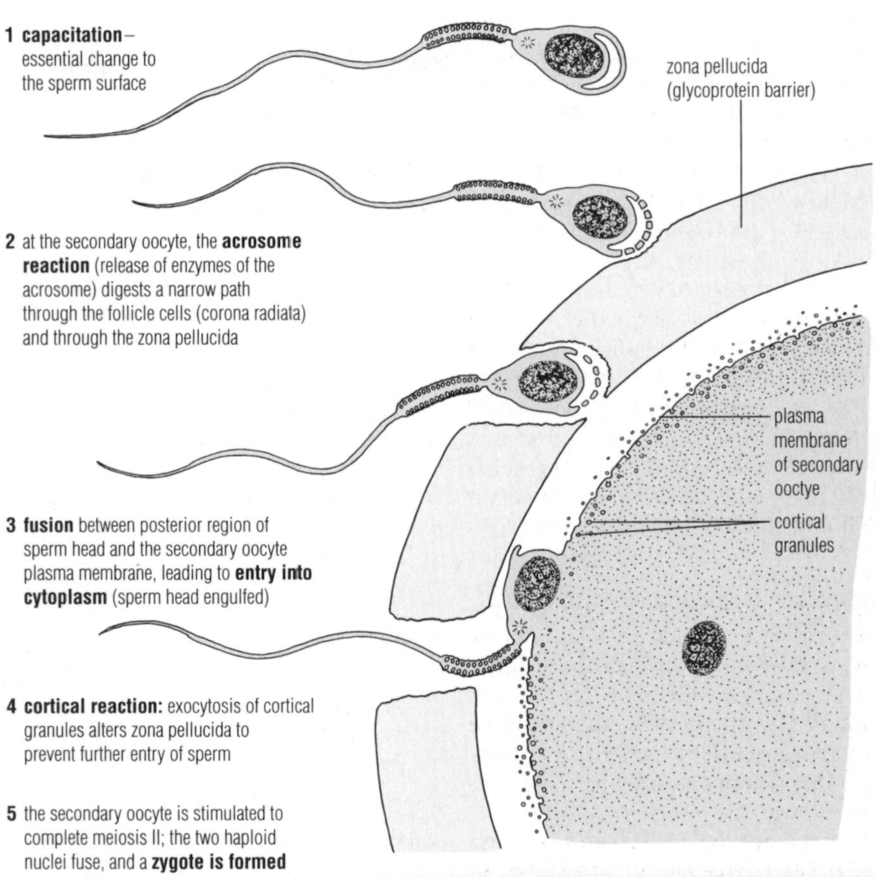

1 **capacitation** – essential change to the sperm surface

zona pellucida (glycoprotein barrier)

2 at the secondary oocyte, the **acrosome reaction** (release of enzymes of the acrosome) digests a narrow path through the follicle cells (corona radiata) and through the zona pellucida

plasma membrane of secondary oocyte

cortical granules

3 **fusion** between posterior region of sperm head and the secondary oocyte plasma membrane, leading to **entry into cytoplasm** (sperm head engulfed)

4 **cortical reaction:** exocytosis of cortical granules alters zona pellucida to prevent further entry of sperm

5 the secondary oocyte is stimulated to complete meiosis II; the two haploid nuclei fuse, and a **zygote is formed**

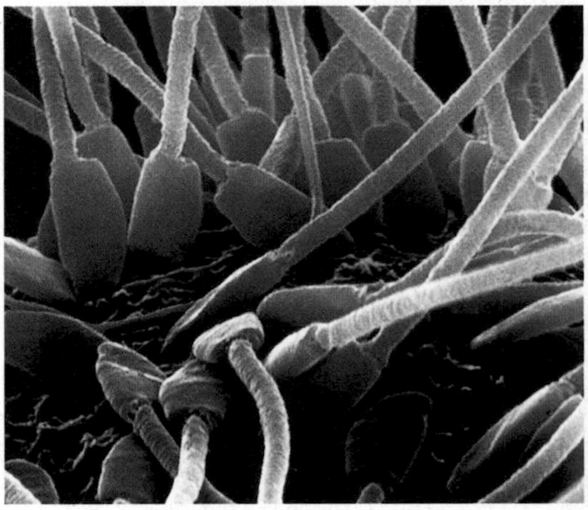

SEM (false colour) of sperms (pink) around a human egg (×4500)

27.10 PREGNANCY

Gestation is perhaps the most extraordinary period of the human life cycle. During the 40 weeks of pregnancy, the rates of growth and development of the new individual from a zygote to a baby far exceed those at any other stage of its life. Meanwhile, the mother's body also undergoes profound physiological changes; in particular, the uterus has to expand enormously to accommodate the growing fetus. Yet after the birth it returns to its normal condition and size relatively quickly, and the woman may be able to begin another pregnancy within only a few weeks. The processes involved in these complex changes are controlled by the endocrine systems of mother and child.

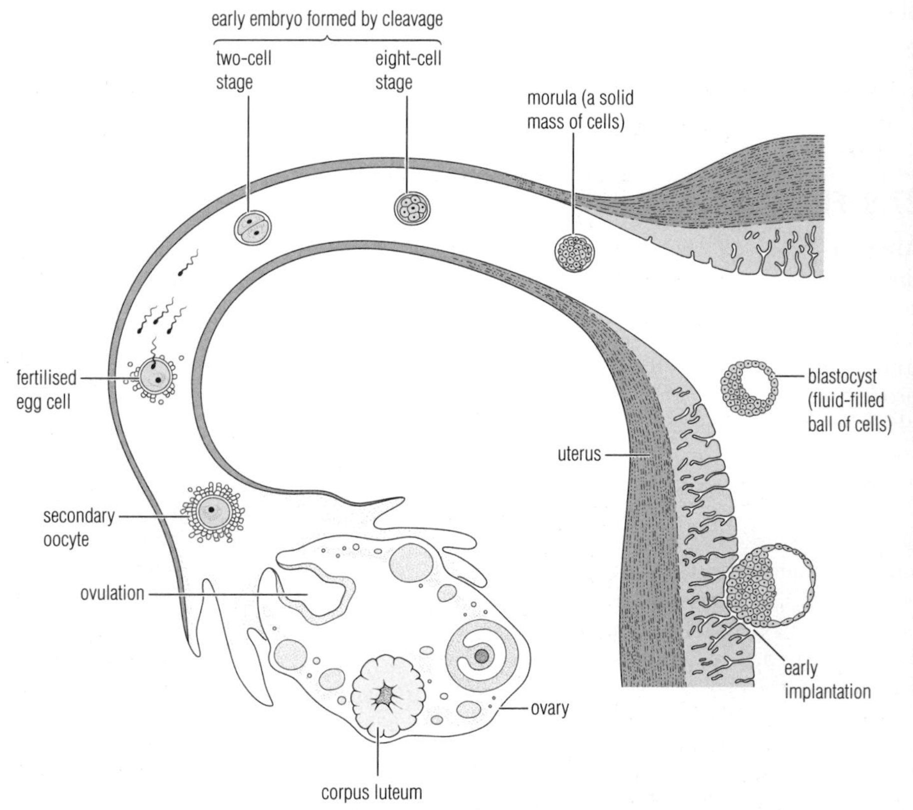

■ Embryonic development: formation of the fetus

At the time of implantation, about seven days after the fertilisation of the egg (Figure 27.23), the blastocyst is a ball of cells with a fluid-filled cavity, the **blastocoel**, and an **inner cell mass** (ICM). This tiny structure has been present in the cavity of the uterus (itself fluid-filled, and of only 7–10 µl capacity) for some 48 hours, whilst the zona pellucida breaks down and disappears. Then the blastocyst is free to invade the endometrium, and it becomes buried there. The outer cells of the blastocyst (the **trophoblast**) develop into the first of four membrane systems of the growing embryo (the **chorion**). The outer surface forms finger-like trophoblastic villi, which develop an intimate connection between the embryonic tissue and the endometrium, breaking down the maternal capillaries around them. As a result, the villi become surrounded by maternal blood in a system of sinuses called **lacunae** (singular, lacuna). Nutrients are made available to the developing embryo through these structures. Later this duty is taken over by a new structure, the placenta (p. 598).

Meanwhile, the cells of the ICM start to form the fetus proper. These cells divide, re-arrange themselves, and become differentiated into the three germ layers of the vertebrate embryology (ectoderm, mesoderm and endoderm, p. 414). This process, referred to as **gastrulation**, is an orderly one, influenced by hormone-like 'organisers' produced within the embryonic tissue. Unlike the cleavage stage of development, gastrulation involves cell interactions and growth

as well as cell division. From now on the embryo grows continuously and is constantly enlarging.

The outer cells and tissues of the embryo also grow and enlarge. They give rise to structures that support and protect the embryo proper, namely the placenta and membranes (Figure 27.24). The innermost membrane, the **amnion**, surrounds a fluid-filled cavity, the **amniotic cavity**, in which the fetus becomes suspended, cushioned against mechanical damage. Another membrane is the **yolk sac**, which has no obvious function in humans, and later becomes buried in the placenta. In reptiles and birds, however, the yolk sac is the structure by which the developing embryo obtains food from the yolk.

In the early stages of its development the implanted mammal embryo exchanges nutrients and waste products via the trophoblastic villi, but soon a fourth membrane develops, from the hind gut of the fetus. In humans this membrane, the **allantois**, develops at the same time as the yolk sac. The allantois grows in close contact with the chorion, and together they play an important part in the formation of the placenta.

■ The hormone control of pregnancy

As we have seen, when the egg cell is not fertilised (this is the outcome for most eggs) the corpus luteum persists as an endocrine gland only up to day 26 of the menstrual cycle. Then the secretion of progesterone and oestrogen ends, and the corpus luteum degenerates (p. 592). Without these hormones in the bloodstream, the endometrium breaks down.

If the egg is fertilised and an embryo starts to form, then as early as the blastocyst stage, the outer layer of cells of the embryo becomes an endocrine gland (even before it has implanted). The hormone human chorionic gonadotrophin (HCG) is secreted, and circulates in the bloodstream. The chief effect of HCG is to maintain the corpus luteum as an endocrine gland, at least for the first 16 weeks of the pregnancy. The maintenance of the corpus luteum means that the endometrium persists, to the benefit of the implanted embryo, and menstruation is prevented. (The premature demise of the corpus luteum, leading to degeneration of the endometrium during early pregnancy, is a possible cause of a miscarriage.)

Later on in the development process, the placenta develops.

Figure 27.23 Cleavage and implantation

early embryo formed by cleavage

two-cell stage

eight-cell stage

morula (a solid mass of cells)

fertilised egg cell

secondary oocyte

ovulation

corpus luteum

ovary

uterus

blastocyst (fluid-filled ball of cells)

early implantation

Figure 27.24   From blastocyst to fetus

1 blastocyst in the uterus, prior to implantation (7 days after fertilisation)

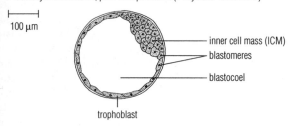

100 µm

- inner cell mass (ICM)
- blastomeres
- blastocoel
- trophoblast

2 developing embryo implanted in the endometrium (14 days after fertilisation)

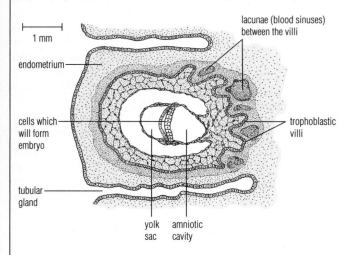

1 mm

- lacunae (blood sinuses) between the villi
- endometrium
- cells which will form embryo
- trophoblastic villi
- tubular gland
- yolk sac
- amniotic cavity

3 growing embryo has developed a placenta (5 weeks after fertilisation)

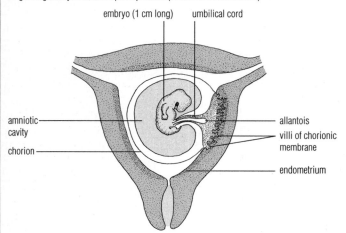

- embryo (1 cm long)
- umbilical cord
- amniotic cavity
- chorion
- allantois
- villi of chorionic membrane
- endometrium

4 developing human recognised as a fetus (10 weeks after fertilisation)

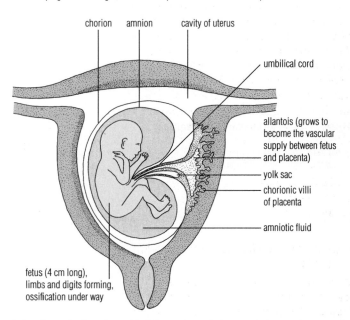

- chorion
- amnion
- cavity of uterus
- umbilical cord
- allantois (grows to become the vascular supply between fetus and placenta)
- yolk sac
- chorionic villi of placenta
- amniotic fluid
- fetus (4 cm long), limbs and digits forming, ossification under way

5 fetus (45 cm long) so advanced that if born prematurely, it has a good chance of survival (8 months after fertilisation)

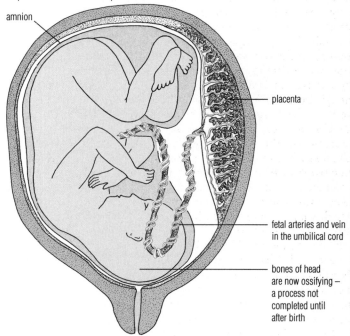

- amnion
- placenta
- fetal arteries and vein in the umbilical cord
- bones of head are now ossifying – a process not completed until after birth

Multiple births
Multiple births are extremely rare events (in the human) that may arise when several eggs are released at ovulation, and all are fertilised. This situation has sometimes arisen as a product of treatment for infertility which has inadvertently induced multiple ovulations. An alternative cause is the multiple division of the ICM, causing the formation of several embryos.

Identical and non-identical twins
Non-identical twins arise when two eggs are released simultaneously from the ovaries at ovulation, and both are fertilised and become implanted in the uterus.

    Identical twins normally arise when the ICM divides soon after it has formed, and develops into two embryos.

Siamese twins
These are identical twins that have incompletely separated during development and, as a consequence, share organs. Depending on the degree of separation, they may be difficult to divide surgically after birth, at least in such a way that both twins survive.

One role of the placenta is as an endocrine gland. When the corpus luteum does eventually break down, the placenta takes over the production of progesterone and oestrogens. These hormones prepare the mammary glands for lactation, as well as maintaining the endometrium. A third hormone released by the placenta is relaxin, which relaxes the elastin fibres that join the bones of the pelvic girdle, especially at the front (the symphysis pubis). It also helps to enlarge the cervix as the gestation period comes to a climax. The endocrine control of pregnancy is summarised in Figure 27.25.

■ The placenta

As the fetus grows and develops, it requires oxygen and nutrients, and waste products have to be removed. By the end of the fourth week, the embryo has developed a heart and a blood circulation.

As it continues to grow the villi of the trophoblast/chorion membrane are invaded by blood vessels from the embryonic blood circulation. By the end of the twelfth week, the placenta is a large, disc-shaped structure, and is the site of exchange between maternal and fetal blood circulations of nutrients and waste products. Structurally, the placenta is composed of the tissues of two different organisms: it arises from the fetal membranes (the chorion and the allantois) and also from part of the uterus wall. In the placenta the maternal blood is brought close to that of the fetus, but they normally do not mix. Finger-like projections of the allanto-chorion grow into the endometrium, and become bathed by maternal blood of the sinuses (lacunae) there (Figure 27.26). Maternal and fetal blood are brought into close proximity over a huge surface area.

The exchange of some substances is essential to the fetus:

▶ **Water** is exchanged across the placenta by osmosis.
▶ Respiratory gases, **carbon dioxide** and **oxygen**, in solution diffuse from regions of higher concentration to regions of lower concentration. Fetal haemoglobin has a higher affinity for oxygen than does the maternal haemoglobin (Figure 17.36, p. 363), a feature that accelerates exchange of oxygen.
▶ Nutrients, mainly **glucose, amino acids, fatty acids, essential ions**, and **vitamins**: these substances are mostly transported across the membranes of the placenta by active transport, involving protein pumps in the plasma membranes, driven by reaction with ATP. However, glucose is transported mostly by facilitated diffusion involving a carrier protein that opens a diffusion channel in the presence of the glucose molecule.
▶ Nitrogenous waste in the form of **urea** diffuses from fetal to maternal blood at the placenta.
▶ Antibodies produced by the mother are transported across the placenta and circulate in the fetal blood, conferring short-term passive immunity on the fetus (p. 368).

> Remember, the fetus is foreign tissue to that of the mother, and so the fetus carries antigens foreign to the mother. In response the mother produces antibodies against those proteins (if any) that cross the placenta.

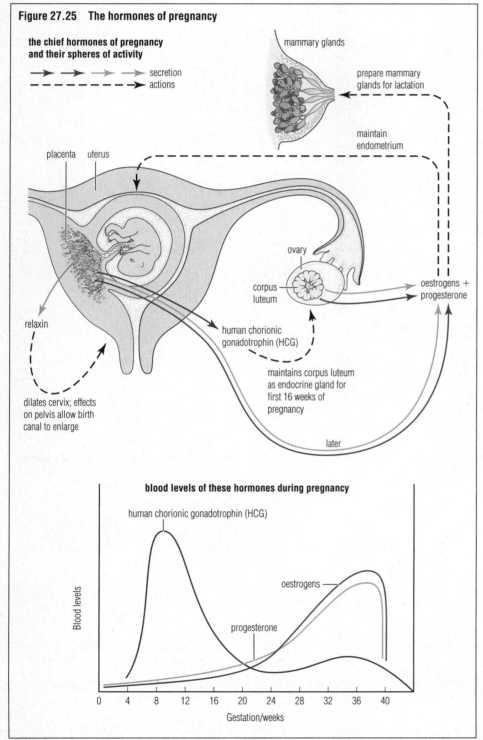

Figure 27.25 The hormones of pregnancy

the chief hormones of pregnancy and their spheres of activity

→ secretion
⇢ actions

mammary glands

prepare mammary glands for lactation

maintain endometrium

placenta   uterus

ovary

corpus luteum

oestrogens + progesterone

relaxin

human chorionic gonadotrophin (HCG)

dilates cervix; effects on pelvis allow birth canal to enlarge

maintains corpus luteum as endocrine gland for first 16 weeks of pregnancy

later

blood levels of these hormones during pregnancy

human chorionic gonadotrophin (HCG)

oestrogens

progesterone

Blood levels

0 4 8 12 16 20 24 28 32 36 40

Gestation/weeks

We have already noted (p. 369) that towards the end of pregnancy or during birth rhesus positive antigens are likely to pass from fetal to maternal blood. Consequently, in a rhesus negative mother, anti-rhesus antibodies are likely to be produced and will come to harm a subsequent fetus that is rhesus positive, unless the mother is treated immediately after the birth of the first offspring (Figure 17.45, p. 369).

Other substances may also pass across to the fetus:

▶ Harmful components of **cigarette smoke** may cross the placenta, mainly carbon monoxide (reducing the delivery of oxygen to the developing fetal tissues) and nicotine (restricting blood flow in the placenta and along the umbilicus). Statistical evidence suggests that offspring of a mother that smoked in pregnancy may carry a greater risk of death just before or just after birth. They are also more likely to have a lower birth weight, an increased risk of heart abnormalities as children, and a lowered reading age during early schooling. Statistical evidence is circumstantial, not conclusive. It suggests associations and links that need further, physiological and biochemical investigation. In the meantime, it is best not to take the risk.

▶ Drugs, including **alcohol, illegal drugs** and those taken as **medicines**, are all potentially harmful to the fetus. For example, a high level of alcohol intake by a pregnant woman may result in reduced brain function in the fetus, and reduced body growth rate. Once again, the evidence is circumstantial.

Heroin and cocaine, drugs that lead to addiction in adults, may also lead to addiction in the fetus and then in the baby after birth.

Medicines may contain drugs that cross the placenta and harm the fetus. For example, a drug called thalidomide was prescribed in the 1960s, to prevent the nausea of early pregnancy known as morning sickness, which some women experience intensely. The drug proved to cause limb deformities and sometimes defects in body organ systems, in some fetuses.

▶ **Viruses** (but not bacteria) are able to pass across the placenta. For example, Rubella (Table 24.1, p. 529) contracted early in pregnancy is highly likely to lead to congenital malformations in the fetus. Consequently, females aged 12–14 are normally given a booster vaccination to reduce the risk of contracting the disease. Other viruses that are able to cross the placenta, including hepatitis B, can lead to dangerous diseases in the newborn child.

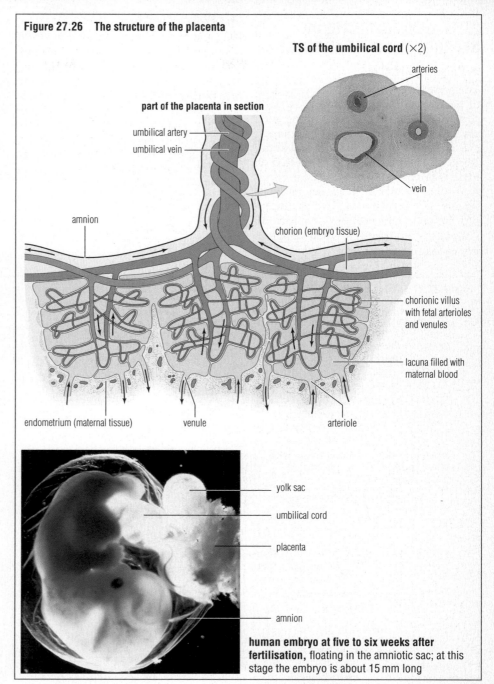

Figure 27.26 The structure of the placenta

TS of the umbilical cord (×2)

arteries

vein

part of the placenta in section

umbilical artery
umbilical vein

amnion

chorion (embryo tissue)

chorionic villus with fetal arterioles and venules

lacuna filled with maternal blood

endometrium (maternal tissue)   venule   arteriole

yolk sac

umbilical cord

placenta

amnion

human embryo at five to six weeks after fertilisation, floating in the amniotic sac; at this stage the embryo is about 15 mm long

Questions

1 Ethologists argue that courtship routines are of extra importance in animals that are aggressive carnivores, such as many species of spider. Speculate on why this may be so.

2 Continuity of nervous connections with the brain via the spine are not necessary for successful erection of the penis and for ejaculation, but the local connections with the nerves of the autonomic system are. Why?

3 Why, if the corpus luteum breaks down prematurely in pregnancy, is a miscarriage the likely result?

4 How and from where does the blastocyst obtain its nutrients before it has implanted?

5 Why is it imperative that the blood circulations of mother and fetus do not mix?

27.11 DEVELOPMENTS IN REPRODUCTION TECHNOLOGY

■ Pregnancy testing

The presence of HCG in urine is the basis of tests to detect pregnancy at an early stage. HCG appears in the urine of a woman after the first week of pregnancy, and continues to be excreted until about week 8. The concentration of HCG in urine is normally high enough for effective testing by 'home pregnancy test' kits (Figure 27.27) within 14 days of conception (about the time of a 'missed' period). Laboratory analysis of a blood sample is more sensitive, of course.

■ Family planning

The size of the human population worldwide and the rate at which it currently grows makes the issue of family planning of great importance. People can regulate the numbers and timing of a family by various methods of contraception (Figure 27.28). Many national governments work towards restriction of family size. The issues of regional differences in the birth rate and population growth around the world are discussed on p. 96.

Figure 27.28 Common methods of contraception

Barrier

Condom – thin, strong rubber sheath fitted to erect penis. Cheap, low failure rate, offers protection against sexually transmitted disease. Fitting may be disruptive in love-making.

Femidom• – female equivalent of condom, fitting inside vagina. Relatively new device.
Diaphragm cap• – fitted over cervix, and used together with spermicidal cream. Can be inserted well before intercourse. Must remain in place for at least 6 hours afterwards.

Hormonal

Pill• – contains oestrogen + progesterone. Taken daily for 3 weeks of cycle. Prevents ovulation by inhibiting secretion of FSH. Very reliable, and does not interfere with love-making.

Preventing implantation

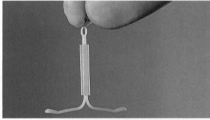

Coil• – plastic or metal device fitted in uterus by doctor, and left in place. Can be left in place for months/years, and remains effective.

'Natural'

Coitus interruptus – penis withdrawn prior to ejaculation. Requires self discipline by male partner. Has high failure rate.
Rhythm method – detection of ovulation event by rise in body temperature of female at this time. Intercourse avoided then.

'Sterilisation'

Vasectomy – vas deferens cut, so no sperms in semen. Difficult to reverse.
Tying of oviducts• – prevents oocytes from passing to the uterus. Not reversible.

(•Can be used without involvement of male so giving women control of contraception.)

Figure 27.27 The principle of the 'home pregnancy test' kit

A pregnant woman has a significant level of HCG hormone in her urine: a non-pregnant woman has a minute amount. The 'kit' contains a test strip with tiny coloured granules attached to HCG monoclonal antibodies in the lowest sector. Further up the strip are immobilised HCG antibodies. In the final sector there are immobilised antibodies to the HCG antibody–granule complexes.

thumb grip

up to two coloured bands appear in the result window

this end held in the stream of urine

overcap

immobile antibodies to coloured granule–HCG antibody complex

immobile HCG antibodies

mobile HCG antibodies complexed with coloured granules

urine sample containing HCG molecules

1

urine sample drawn up test strip

HCG attaches to granule–antibody complex

2

mobile complexes move up test strip

granule–HCG antibody complexes trapped by anti-antibodies

immobilised HCG antibodies trap the coloured complexes

3

one coloured band always appears in the result window showing the test has run correctly

a second coloured band appears in the result window and **confirms pregnancy**

urine sample without HCG

or: **3**

no second line indicates **non-pregnancy**

■ Menopause and HRT

Menopause is the time of cessation of the menstrual cycle, and normally occurs in women aged between 45 and 50. It is the result of a decreasing ability of the ovaries to respond to FSH and LH. As a consequence, the production of oestrogen by the ovaries decreases, and no more ovarian follicles are formed.

Oestrogen indirectly inhibits loss of calcium from skeletal bones, so after menopause, osteoporosis (p. 540) may become a problem in women. Hormone replacement therapy (HRT) is the technique of replacing the natural hormone lost in menopause, so slowing up loss in bone density (and reversing the general symptoms of menopause). Treatment may be continued for many years, and may involve tablets or the use of skin patches that permit absorption of this steroid hormone through the skin.

■ Learning about the fetus *in utero*

There are several techniques by which the developing fetus can be examined in the uterus. In ultrasonography, high-frequency, inaudible sound waves are directed into the abdomen and the waves reflected back are collected and interpreted as an image of the moving fetus, observed on a monitor. By such ultrasound scans the development and position of the fetus can be checked.

When samples of fetal tissue are required to test for conditions that are genetic in origin, amniocentesis or chorionic villus sampling may be undertaken (Figure 27.29).

■ Abortion

Natural (spontaneous) abortions occur from time to time. Induced abortions (known as terminations of pregnancy) are legal in the UK provided the fetus is less than 24 weeks old, but only if there is a risk to the physical or mental health of the woman or to her existing children. (A baby born prematurely after 24 weeks gestation has a reasonable chance of survival with modern medical care.) Alternatively, if there is risk to the life of the woman, or if the fetus has been found to be seriously abnormal, then an abortion is permitted at any stage. Two medical practitioners have to agree to the need, but the father has no right to prevent the abortion.

The ethical issue concerning abortion is 'When does human life begin?' If the answer is 'At conception', then abortion may be seen as unacceptable. However, in the absence of legal abortions, many illegal operations are carried out, often under unhygienic conditions, with dangerous outcomes. The debate continues.

■ Male infertility

Male infertility may arise when

► there are fewer than 20 million sperms cm⁻³ of semen
► there are 50% or more immobile sperms present in the semen
► more than 35% of sperms have abnormal morphology
► the chemical components (such as enzymes, and metabolites such as fructose) are present in the wrong proportions, or the pH of the semen is too high.

These problems may be due to treatable causes, such as certain diseases, obesity or drug use. Alternatively, fertilisation may be achieved by artificial insemination with donor sperm.

■ Female infertility

An intractable form of infertility arises in women whose oviducts are permanently blocked, preventing secondary oocytes from being fertilised, or indeed ever reaching the uterus. To overcome this, eggs have to be removed surgically from the ovary and fertilised externally before being returned to the uterus.

The patient is first treated with hormones (FSH and LH, p. 592) early in the menstrual cycle. This causes the production of several secondary oocytes (not just one). The ovary and ovarian follicles are detected by ultrasonography. Eggs are withdrawn immediately prior to their release, via a hollow needle inserted into individual follicles, normally nowadays via the wall of the vagina. These oocytes may be grown in fluid of similar composition to that in the oviducts and uterus (this is called 'in vitro culture'). Healthy sperms selected from those produced by the male partner are then added. The culture is microscopically observed for indications that fertilisation and cleavage have taken place.

If about three embryos are returned to the uterus at the 4–8 cell stage, there is a chance that one may become implanted. This embryo can continue to develop normally in the uterus, and in due time a healthy baby is born. The first such 'test tube baby' was born in Britain in 1978. Today, many more than 10 000 'IVF' babies have been born to otherwise infertile couples.

Unused embryos can be retained at low temperature (−172 °C) in case of subsequent need. 'Spare' human embryos may be used for research into infertility and into the detection of genetic disorders in the 'pre-implantation' embryo. The ethics of human embryo research are hotly debated, however. Currently the law in Britain permits studies on embryos up to 14 days post-fertilisation.

This time limit is set because this is when the 'primitive streak' first appears. The primitive streak marks out the cells that will form the embryo from those of the supporting membranes.

The ethics of *in vitro* fertilisation and human embryo research are discussed on pp. 643–4.

Figure 27.29 Learning about the fetus *in utero*

Amniocentesis
This is the technique of withdrawing some of the amniotic fluid in the period 16–20 weeks of gestation. This fluid contains cells that have floated away from the surface of the embryo, and which can be examined cytologically. Amniocentesis can be used to diagnose inherited biochemical defects and chromosomal disorders, such as those causing haemophilia or Down's syndrome.

Chorionic villus sampling
From early on during pregnancy (within 8–10 weeks, for example) tiny samples of the chorion (fetal tissue) can be withdrawn from the uterus via the vagina, without the need for surgery. This sample can be analysed quickly. (See Genetic counselling, p. 640.)

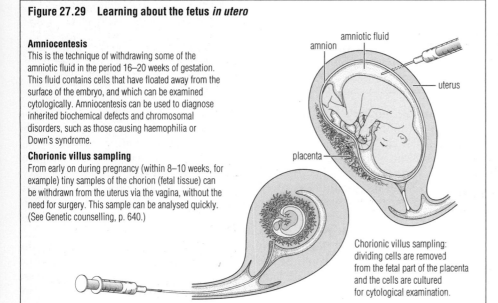

amniotic fluid
amnion
uterus
placenta

Chorionic villus sampling: dividing cells are removed from the fetal part of the placenta and the cells are cultured for cytological examination.

27.12 BIRTH

The length of time for which the off-spring of a mammal develops in the uterus is referred to as the **gestation period**. This differs among mammals; in humans, it lasts for 40 weeks.

At birth (parturition) the young are expelled from the uterus by intermittent waves of powerful muscular contractions that run through the muscles of the uterus wall, starting from the top of the uterus and ending at the cervix. In principle, these are not unlike waves of peristalsis. This process is known as 'labour'.

There are three main phases to birth. The **first stage** involves the dilation of the cervix, which has to be sufficiently wide for the infant's head (the widest part of its body) to pass through. The **second stage** is the expulsion of the baby from the uterus, involving the most powerful, frequent and often painful of the sequence of contractions. During the **final stage**, lesser uterine contractions separate the placenta from the endometrium, and cause the discharge of the placenta and the remains of the umbilicus (Figure 27.30).

■ Fetal circulation and changes at birth

The circulatory system of the fetus is similar to that of the adult, but there are some important differences (see Figure 17.35, p. 362). At birth, the roles of the placenta in gaseous exchange, nutrition and excretion are abruptly terminated, and are replaced by activities of the lungs, gut and kidneys. Also the liver, another important organ with a large blood supply, starts to function in metabolism. The by-pass blood vessels that had diverted the bulk of the circulation of fetal blood away from the liver (ductus venosus) and the lungs (ductus arteriosus) close as a result of pressure changes in the circulation, and become sealed off. The foramen ovale, the hole between the atria that allows the bulk of blood to avoid the pulmonary circulation, normally closes shortly after birth. Rarely, it does not; in these cases the hole in the baby's heart persists, and has to be repaired by surgery.

■ Hormonal control of birth

The changes in hormone concentration associated with birth, and the chief roles or effects of the many hormones, are well known. Speculation remains as to whether the trigger for initiation of birth

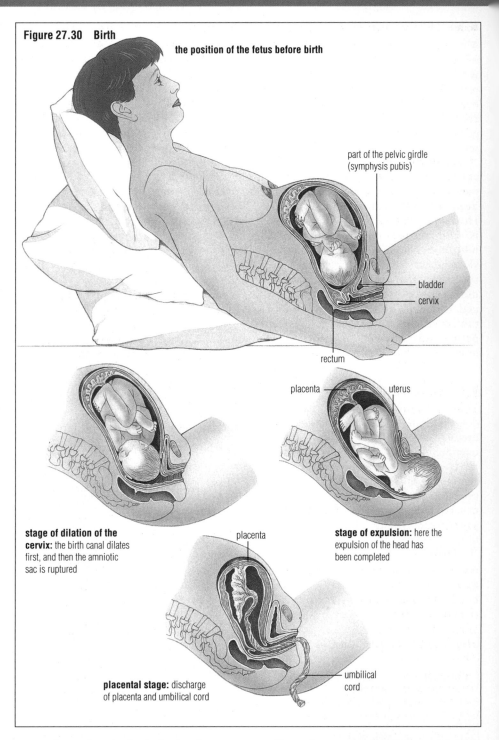

Figure 27.30 Birth

the position of the fetus before birth

part of the pelvic girdle (symphysis pubis)

bladder

cervix

rectum

placenta

uterus

stage of dilation of the cervix: the birth canal dilates first, and then the amniotic sac is ruptured

stage of expulsion: here the expulsion of the head has been completed

placenta

placental stage: discharge of placenta and umbilical cord

umbilical cord

comes from the mother's body alone, or whether a hormone released in the fetal pituitary gland causes a hormonal signal to be sent to the mother, across the placenta, that parturition should begin.

From week 16 of gestation oestrogen and progesterone are discharged into the bloodstream from the placenta in increasing amounts. Then, towards the end of gestation, more oestrogen than progesterone is released, and immediately before birth, the level of progesterone declines sharply. These developments cause other important changes.

First, the progesterone-derived inhibition of contractions by the muscles of the uterus wall is removed.

Secondly, the maternal posterior pituitary gland releases the hormone oxytocin. Oxytocin, one of the most important hormones regulating birth, stimulates the rhythmic contractions of the muscles of the uterus wall.

Thirdly, the release of a prostaglandin by the placenta is increased. Prostaglandins were first discovered in semen, and thought (erroneously) to be produced by the prostate gland, hence their name. They are formed from unsaturated fatty acids in various organs and tissues of the body, and they are extremely potent substances. Their principal effects are on involuntary (unstriped) muscle. The effect of this particular prostaglandin is to increase the force and frequency of uterine contractions.

■ The trauma of birth

At birth, the operation of the infant's body systems suddenly changes, as it switches from placenta and maternal blood supply as the source of nutrients, to establishing its own supplies of oxygen and nutrients. For example, the infant's lungs are at least partially filled with amniotic fluid when in the uterus. After delivery, gaseous exchange through the placenta ceases abruptly. The liquid in the lungs is absorbed. As the dissolved carbon dioxide level in the infant's blood rises, the respiratory centre in the medulla (p. 241) is stimulated, and triggers the drawing of the first breath by the baby. This is a reassuring moment for the parents, even though the first steps of inspiration frequently take the form of hearty cries.

27.13 PARENTAL CARE IN THE HUMAN FAMILY

■ Lactation

At birth, the mammary glands in both male and female humans are undeveloped. During puberty in the female oestrogen, released from the ovaries, stimulates the development of the lobes and lobules of the mammary glands. These glands are modified sebaceous glands, and will become the site of milk production (p. 241).

Throughout pregnancy the mammary glands are prepared for milk production by the action of the hormones oestrogen and progesterone. Milk production itself is actually inhibited by progesterone, however. Just before birth, when the progesterone level falls, lactation gets under way. **Lactation** is the production, secretion and ejection of milk. Milk constitutes the entire diet of a mammal during the earliest post-parturition stages of the offspring's existence. The first-formed milk, the **colostrum**, contains important antibodies that help the infant survive the first dangers from micro-organisms in its independent existence. Apart from these antibodies, colostrum

has no fat and little milk sugar (lactose). These substances appear in the milk, at concentrations the baby needs, three or four days after the birth.

The lactation process is controlled by the hormone prolactin, secreted by the anterior pituitary gland. Until the concentrations of oestrogen and progesterone fall at the end of gestation, the actions of prolactin are inhibited.

■ Development

Soon after birth the baby starts to suck from the breast. As well as providing the baby with the required diet, the process of breastfeeding is usually a happy and satisfying experience for the baby and the mother. The baby at birth may weigh roughly 3.5 kg, and be about 50 cm long. The body mass trebles in the first 12 months. The changing mass of the child as it grows to adulthood is shown in the graph in Figure 19.5 (p. 400).

The baby depends totally upon its parents for food, warmth and protection for a long time. The temperature-regulation system and most other systems of the baby's body are immature. Most body systems show important developments in the first few months, not least the nervous system. At birth the baby has many reflexes. Some of these, such as sucking and grasping, are lost by the end of the first year. Other reflexes persist, such as the turning of the head to sounds, and the stepping reflex.

The baby matures quickly, and can soon manipulate objects that it can grasp. By the end of the first 12–14 months it will have acquired head and hand control and learnt to sit, and can probably pull itself up, stand, and then begin to walk unaided. Whilst vocalisation starts within two months, speaking is established during the first year, and by the age of two the baby will have a vocabulary of around 200 words.

The total period of dependence of a child is very prolonged compared with other young mammals. Parental care involves processes we call education and training, and they are carefully planned. They amount in the human animal to an additional form of inheritance, deliberately communicated and consciously absorbed. It is the human's inheritance of a culture.

27.14 HUMAN SEXUALITY

In most animals living in the wild, sexual maturity is achieved by only a minority of juveniles. Many offspring fail to survive to adulthood. Those that survive will attempt to reproduce at the first opportunity. In humans, sex is more than a bodily function of this type. Nevertheless, our culture recognises the sex drive as one of the powerful motivating forces in our lives.

In human communities in the poorest developing countries, despite (or probably because of) hunger and deprivation, people have many children. Many children fail to survive to adulthood, but those that do are seen as a support and insurance for their parents in old age (p. 78). There is a strong cultural tradition in most human communities that obliges the young to care for their elders.

In human societies in developed countries, most people are relatively well fed. The effects of this privileged nutritional status is (probably) that fewer children are born, but that those children achieve puberty at an earlier age. Meanwhile, cultural traditions and economic realities may require young adults to delay permanent pair-bonding. The frustration this may generate in young adults may cause problems.

The psychological theories of Sigmund Freud proposed that the sexual urge provides the motivational energy for nearly all human activities. Whilst this is clearly an extreme position, we must acknowledge that there are many aspects of sex that are important in human life, over and above the basic structures and hormone control of human sexuality discussed in this chapter.

Questions

1 Explain the roles of the blood supply to the liver and the gut respectively
  a in the fetus before birth
  b in the baby after birth.
2 Why is colostrum valuable to the baby immediately after birth, although it lacks fat and lactose?

FURTHER READING/WEB SITES

The WWW Virtual Library – Developmental Biology: **sdb.bio.purdue.edu/Other/VL_D.B.html**

28 Variation and inheritance; genetics

- Genetics, the study of heredity, is concerned with genes and chromosomes.
- In diploid organisms, chromosomes occur in pairs, called homologous chromosomes, one coming originally from each parent. A gene occupies a position on a chromosome, its locus. Genes at a locus exist in two or more forms, called alleles.
- Modern genetics began with the rediscovery of the work of a monk, Gregor Mendel.
- Characteristics within a species vary between individuals. Discrete or discontinuous variables (like tall or short garden pea plants) are inherited. Continuous variables (like height in humans) may be genetically controlled or influenced by environment, or both.
- Mendel's monohybrid cross established that characteristics are controlled by a pair of alleles that segregate (separate) in equal numbers into different gametes. Segregation occurs in meiosis.

- Mendel's dihybrid cross established that two pairs of alleles assort independently in the formation of gametes. Independent assortment occurs in meiosis.
- Human inheritance is investigated by researching a family tree or 'pedigree', if appropriate records exist.
- Since Mendel, it has been appreciated that many genes are linked on the same chromosome and are inherited together. Crossing over of segments of chromosomes occurs, forming new combinations of alleles. Alleles may occur in more than two alternative forms.
- Most characteristics are controlled by many genes, known as polygenes. Both environmental factors and genes affect the characteristics of organisms.
- Mutations, or abrupt changes, occasionally occur to genes or chromosomes.

28.1 INTRODUCTION

Genetics is the study of heredity; that is, the transmission of characteristics or traits from one generation to another. We are well aware that organisms show all sorts of characteristics, and that their offspring show many of these, too. 'Like begets like,' says the proverb.

Modern genetics is concerned with the study of **genes**. Genes are the units of heredity that control the characteristics of organisms. We have already seen that a gene is made of DNA with a particular nucleotide sequence (p. 210). Geneticists investigate how these genes are transferred from one generation to the next, what their structure is and how they function to determine the characteristics of a living organism.

■ Genetics has ancient origins

Twenty thousand years ago humans were foragers and hunters, but by the end of the last ice age – ten thousand years ago – an agricultural revolution had begun, probably in several centres of the world, but certainly in that part of the Middle East known as the Fertile Crescent. This revolution, the Neolithic Revolution, produced most of the domesticated plant and animal species of modern agriculture. All these new species were derived from wild animals and wild plants by **selective breeding**. Today we describe this process as **artificial selection** (p. 644).

Cereals, the cultivated grasses, were the most important domesticated plants derived at that time because their fruits (grains) are a compact, concentrated form of starch, with protein and lipid, and can

be successfully stored. The most important of the domesticated animals were sheep, goats and cattle, because they feed by digesting cellulose (ruminant herbivores, p. 289) and do not compete for the grains that humans themselves use for food. These were herd animals (naturally sociable), and were valued for their meat and hides. As they became more docile they could be milked. Some became pack animals.

Domestication depended on the skills of killing the wildest animals in the herd for immediate consumption, breeding from the most docile, and deliberate **conservation of breeding stocks** (plants and animals) with favourable characteristics. All this was achieved without any understanding of the underlying processes of heredity.

Indeed, there was virtually no comprehension of the mechanism of heredity until the remarkable experimental work of Gregor Mendel was 'rediscovered' in 1900. Successful selective breeding of plants, livestock and pets has continued throughout human history, and Mendel was not the first to try to elucidate the underlying mechanism of inheritance. Other workers, however, looked at many characteristics in considering the overall similarities and differences between parent organisms and their progeny. The results were always confusing because progeny may resemble one or other parent in certain of their traits but neither parent in others. No underlying pattern emerged. The accepted (but largely erroneous) concept was that traits became 'blended' in offspring; Charles Darwin, for example, thought in terms of blending inheritance.

28.2 GREGOR MENDEL: THE FOUNDING OF MODERN GENETICS

It is remarkable, and quite sad, that whilst modern genetics really began with the exceptionally thorough and painstaking work of a single scientist, Gregor Mendel (see panel), no one realised the full significance of his discoveries in his lifetime. Mendel seems to have been unperturbed by this non-recognition; indeed, he experienced many disappointments about which he seems to have always remained philosophical.

■ Why was Mendel so successful?
Experimental technique and choice of organisms

Mendel was an imaginative and careful worker who planned his experiments on a large scale. He appreciated that by taking a large number of separate measurements he would eliminate mere chance effects. He was a pioneer in the practical application of statistics in biological work.

His choice of organism was critically important too. The garden pea plant (*Pisum sativum*) on which he carried out most of his work was ideal for the purpose, for it showed several sharply contrasting and easily recognised characteristics that were without intermediate form, and were relatively unaffected by environmental factors. The characters that caught Mendel's attention (Figure 28.1) were

- ▶ length of stem (tall or short)
- ▶ position of flowers (axial or terminal)
- ▶ colour of unripe pods (green or yellow)

- form of ripe pods (inflated or constricted)
- form of ripe seeds (round or wrinkled)
- colour of seed-leaves (cotyledons) (yellow or green)
- colour of seed coat (testa) (greybrown or white).

Discontinuous and continuous variables

A characteristic or trait found in individuals of a species can be described as a **variable**. For example, the characteristics that Mendel selected in the garden pea are variables. Variables like those listed above are **discontinuous** (or **discrete**) **variables**. Individual pea plants with these variables can be assigned to clearly defined categories or distinct groups, because there are no intermediate groupings; for example, Mendel's plants were all either tall or short, and there were none of medium height.

In complete contrast, the variable of height in humans is a **continuous variable**. A hundred men might all be of different heights; they would certainly not fall into two groups, but rather would form a **normal distribution** (Figure 28.2, overleaf). We will refer back to this difference later.

Handling data on variation

The 'middleness' or the central tendency of data can be expressed in three ways.

- The **mode** is the most frequent value in a set of values. For example, in the graph/histogram of human height in Figure 28.2 the mode value is 172–174 cm.
- The **median** is the middle value of the set of values when these are arranged in ascending order.
- The **average** or arithmetic **mean** is normally calculated by dividing the sum of the individual values by the number of values obtained. An alternative method is shown in Figure 28.3 (plotting a graph).

One good way of expressing the 'spread-out-ness' or measure of **dispersion** of the variables is as the **standard deviation** (SD) of the mean. The SD has the same units as the original observations. It is a useful measure of dispersion; a low SD indicates that the observations differ very little from the mean value. In fact, 68% of a similar sample can be expected to be within ±1 SD of the mean, and 95% of a sample to be within ±2 SDs of the mean.

GREGOR MENDEL

Gregor Mendel

Gregor Johann Mendel (1822–84) was born in a small village in Moravia, the son of a peasant farmer. The poverty his family experienced meant that Mendel had to work to support himself in his studies from an early age. It placed a strain on his health throughout his schooling. From the age of 16 he had to provide for himself entirely. At one point his younger sister renounced part of her dowry to finance his further study. At the age of 21 he was offered a place in the monastery at Brno. He accepted this offer, feeling 'compelled to enter a station in life which would free him from the bitter struggle for existence'. The monastery contained thirteen priests, chosen for their intellectual ability. They were free to travel, and to receive guests; the monastery was an important intellectual centre with an excellent cuisine.

From this base, Mendel taught biology for 14 years at a secular high school in Brno that had an able team of staff, and over 1000 pupils. He was a good teacher, friendly and humorous. He twice failed the examination to obtain a regular teaching licence but he kept his position, and the salary enabled him to repay his sister by supporting his nephews through medical school. It was during this period that Mendel conducted his studies on the mechanism of inheritance. He worked with a range of animals, but concentrated on hybridisation in several species of plants. He is best remembered for his work on the edible pea, *Pisum*. In 1868 his research was interrupted by his election to the position of abbot of the monastery. He continued to correspond with outside academics, but it is unlikely that his research papers were read with understanding in his lifetime. Mendel was not bitter about this; a friend heard him reflect *'Meine Zeit wird schon Kommen'* ('My time will soon come')!

Contributions to biology

Mendel studied the inheritance of seven contrasting characteristics of the garden pea. In experiments involving thousands of plants, these characters were shown not to blend on crossing but to retain their identity, and to be inherited in fixed mathematical ratios. He concluded that hereditary factors (we now call them genes) determined these characteristics, that they occurred in duplicate in the parents, but that the two segregated from each other in the formation of gametes.

Mendel was fortunate to choose a plant where these visible characteristics were each controlled by a single factor (gene), rather than by many. When the inheritance of two pairs of contrasting characteristics was studied he found that each of a pair of contrasting characteristics may be combined with either of another; that is, the factors show independent assortment.

Mendel never presented the results in the form of the two laws so often quoted as his, which may help to explain the difficulty others had in seeing the significance of his work. He published his results through his local scientific societies. He also wrote to an eminent German botanist, but was advised to obtain more data. Mendel's results were from 21 000 plants!

Influences

Mendel carried out his research in his spare time, motivated only by his devotion to science. He had no collaborators, and no students to carry it on. Mendel was convinced the larger the number of plants used, the more likely it was that chance effects would be excluded. He was a pioneer in applying statistical methods to biological research, and he overthrew the ancient and prevailing notion of blending inheritance by investigating the inheritance of discrete characteristics of parents and offspring.

He worked in his monastery's garden at the same time that Charles Darwin was working in his, near Bromley in Kent. He read Darwin's *Origin of Species* in a German translation, and knew of his ideas. Mendel's work was finally rediscovered 16 years after he died, by geneticists who were searching the literature for evidence of a mechanism of inheritance that would account for the origin of species.

Figure 28.1 Characters of the garden pea (*Pisum sativum*) that were studied by Mendel; the character of height (tall and dwarf plants) is illustrated in Figure 28.5

flower position   fruit shape   seed shape

round   wrinkled

inflated   constricted

fruit (pod) colour

cotyledon colour

green   yellow

seed coat (testa) colour

terminal   axial   green   yellow   grey   white

Figure 28.2 Continuous and discontinuous variation

continuous variation
variation in the height of adult male humans; the results cluster quite closely around one value, and show a normal distribution (in this type of distribution the mean, median and mode values coincide); for convenience in plotting, the heights are grouped into 13 arbitrary groups, each covering a height range of 2 cm

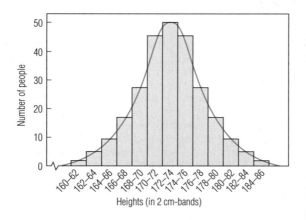

discontinuous (discrete) variation
variation in height above ground of pea plants, some of a dwarf variety and others of a tall variety; the heights of these plants fall into two discrete groups, in both of which there is a normal distribution of variation but between which there is no overlap

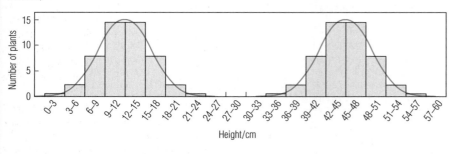

Figure 28.3 Working out standard deviation

Data on the height of two species of plant:

Height/cm	2	3	4	5	6	7	8	9	10	11	12	13	14	15	16
Number of plants of species A	2	8	17	27	30	27	17	8	2	–	–	–	–	–	–
Number of plants of species B	–	–	–	–	–	2	5	9	13	18	25	29	28	12	–

Mean and standard deviation of the sample of species A:
(Σ means 'the sum of')

Height of plants (x)	Frequency (f)	fx	Deviation of x from mean (d)	d^2	fd^2
2	2	4	−4	16	32
3	8	24	−3	9	72
4	17	68	−2	4	68
5	27	135	−1	1	27
6	30	180	0	0	0
7	27	189	+1	1	27
8	17	136	+2	4	68
9	8	72	+3	9	72
10	2	20	+4	16	32

$$\Sigma f = 138 \quad \Sigma fx = 828 \qquad \Sigma fd^2 = 398$$

$$\therefore \text{ mean } \bar{x} = \frac{828}{138} = 6$$

$$\text{standard deviation (SD)} = \sqrt{\frac{\Sigma fd^2}{\Sigma f}} = \sqrt{\frac{398}{138}} = \sqrt{2.88}$$

$$= 1.7 \text{ correct to 1 decimal place}$$

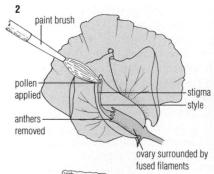

The garden pea plant is self-fertile

The flowers of *Pisum sativum* are self-pollinating (and self-fertilising) in nature, and so experimental geneticists can arrange either cross- or self-fertilisations, according to the breeding programme planned. Mendel selected the garden pea for this reason, after preliminary experiments with several members of the pea family (Papilionaceae, p. 575).

Figure 28.4 Cross-pollination of a *Pisum sativum* flower

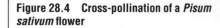

flowers of this family (Papilionaceae) have five petals:

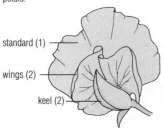

standard (1)

wings (2)

keel (2)

1 keel petals cut open to expose the stamens while these are still immature

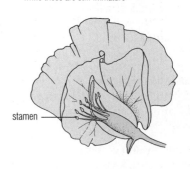

stamen

2

paint brush

pollen applied

anthers removed

stigma

style

ovary surrounded by fused filaments

3

muslin bag

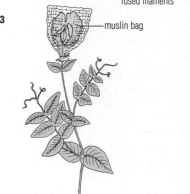

In order to make a cross between pea plants with contrasting characters (between tall and dwarf plants, say) self-fertilisation must be prevented. The anthers of flowers on plants of one group (perhaps the short plants) are dissected out whilst they are still immature, that is, before they have shed pollen. Then the stigmas of these flowers are dusted with pollen from mature anthers in the flowers of plants of the other group (tall plants), using a tiny paintbrush. These flowers are covered with a light muslin bag until the fruit has formed, so that no stray pollen reaches the stigma whilst it is still receptive (Figure 28.4). Subsequently the progeny self-fertilise naturally, which requires no tampering with the flowers.

The verification of pure lines and the analysis of progeny

The pea plants that Mendel used in his experiments were grown from seed that was obtained from plants that had 'bred true' when self-fertilised for at least two generations before the experiment. In this way Mendel could be confident that his plants were all **pure-breeding** (when **selfed** – that is, self-pollinated and self-fertilised – the offspring would always resemble the parents for the given trait).

In collecting the results, Mendel recorded the *numbers* of individuals in each class in the progeny, and established the *ratios* of the contrasting characters of many subsequent generations. This involved counting many thousands of seeds. Then, having formulated a hypothesis to explain the ratios obtained, he set out to confirm it by devising appropriate test crosses.

Figure 28.5   Mendel's experiment with tall and dwarf peas

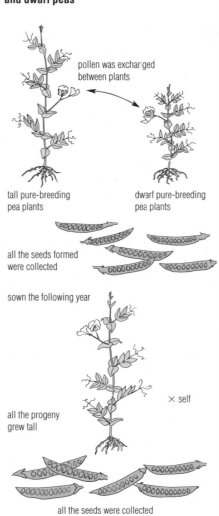

pollen was exchanged between plants

tall pure-breeding pea plants

dwarf pure-breeding pea plants

all the seeds formed were collected

sown the following year

× self

all the progeny grew tall

all the seeds were collected

the following season all the seeds were grown; the plants were in the ratio 3 tall to 1 short

Mendel's first series: the monohybrid crosses

Mendel restricted his first series of investigations to the inheritance of a single pair of contrasting characters, although the pea plants he was using would have shown all seven of the variables that originally attracted his attention. The term **monohybrid cross** refers to one that involves a single characteristic. Here we will consider the results of one such cross (Figure 28.5).

Tall × dwarf peas

Mendel took pure-breeding tall plants and crossed them with pure-breeding dwarf plants (the parents or **P generation**). He carried out **reciprocal crosses**; that is, in some experiments the pollen from tall plants was transferred to the stigmas of dwarf plants, while in others pollen from dwarf plants was transferred to the stigmas of tall plants. The progeny (the 'first filial' or F_1 generation, plants grown from the hundreds of seeds collected) were all tall, regardless of whether the parent plant was tall or dwarf. He called these **hybrids**.

Next season these tall plants were selfed, with the result that by the end of the season Mendel had collected no less than 1064 seeds (the 'second filial' or F_2 generation). When these seeds were germinated the following season, the F_2 generation included both tall and dwarf plants in the ratio of approximately 3 tall to 1 dwarf. Mendel continued with the selfing of the F_2 progeny for four more generations, to confirm his interpretation of the results.

Mendel's interpretation of the monohybrid cross

In the assumptions that Mendel spelt out to interpret his results, he made a conceptual breakthrough that now underpins the whole of modern genetics. He realised that, for example, because the 'dwarf' characteristic that disappeared in the F_1 had reappeared in the F_2, the controlling factor for 'dwarf' had remained intact and undiluted from generation to generation. It is never expressed, however, in the presence of a factor for 'tall'.

Table 28.1   Mendel's seven monohybrid crosses: many plants were grown in each cross, and therefore large numbers of F_2 progeny were obtained; Mendel recorded the numbers of plants of each type, and expressed the outcome as a ratio

Character	Progeny of F_1	Progeny of F_2: numbers	ratio
height: tall × dwarf	all tall	787 tall, 277 dwarf	2.84:1
flower position: axial × terminal	all axial	651 axial, 207 terminal	3.14:1
fruit colour: green × yellow	all green	428 green, 152 yellow	2.82:1
fruit shape: inflated × constricted	all inflated	882 inflated, 299 constricted	2.95:1
seed shape: round × wrinkled	all round	5474 round, 1850 wrinkled	2.96:1
cotyledon colour: yellow × green	all yellow	6022 yellow, 2001 green	3.01:1
testa: grey × white	all grey	705 grey, 224 white	3.15:1

Question

Using the data given in Figure 28.3:
a   plot a graph/histogram of the data on height of species A and B
b   state the mode, median and mean value of height of the sample of plants of species B
c   calculate the SD of the height of the sample of plants of species B.

Figure 28.6 The product of the binomial expression

the outcome of randomly combining two pairs of unlike factors (e.g. A, a) is the product of the binomial expression:

$$(A + a)(A + a) = 1AA + 2Aa + 1aa$$

this can be shown using a grid (known as a Punnett square, after its originator):

	factors from one parent	
	A	a
A	AA	Aa
a	Aa	aa

factors from other parent

possible combinations of offspring

the Punnett square notation shows all the possible combinations of random events

Mendel saw that there must be two independent factors for 'dwarf' and 'tall', otherwise the 'dwarf' characteristics would never reappear after the F_1 cross. The logical interpretation is that one factor comes from one parent and the other factor from the other parent. A gamete can contain only one of the factors; otherwise the constant 3:1 ratio would be impossible (the ratio could have any value at all).

Mendel's conclusions can be summarised as follows:

► within each organism are breeding 'factors' controlling characters such as 'tall' and 'dwarf'
► there are two of these factors in each cell
► one factor comes from each parent
► the factors in a parent separate in reproduction, and either can enter an offspring
► the factor for 'tall' is an alternative form of the factor for 'dwarf'
► the factor for 'tall' is dominant over the factor for 'dwarf'.

In interpreting his results Mendel made use of the mathematics (probability theory) he had learnt at university. The 3:1 ratio could be the product of the binomial expression derived from randomly combining two pairs of unlike elements (Figure 28.6). He rejected the concept of blending of the characters of the parents in the offspring.

Mendel subsequently repeated his monohybrid cross experiments, following the inheritance of each of the seven characteristics through the same sequence of crosses. His results (Table 28.1) confirmed his initial conclusions.

28.3 INTERPRETING MENDEL TODAY

■ The chromosome theory

Now that we know the structure of chromosomes, and about the way they behave during nuclear divisions, the interpretation of Mendel's experimental results presents no difficulties.

Mendel's 'factors' are interpreted as the **genes** found on the chromosomes in the nucleus of each cell (Figure 28.8). There are at least two forms of a gene, and these are called **alleles**. In Mendel's experiment, the allele in pea plants for 'tallness' was **dominant** over the allele for 'dwarfness', which was **recessive**. A pure-breeding tall pea plant is **homozygous** for the 'tall' allele, and a pure-breeding dwarf pea plant is homozygous for the 'dwarf' allele. The F_1 progeny are **heterozygous** tall, with one allele for 'tall' and one for 'dwarf'. In the heterozygous tall individual, we say that the recessive gene is not **expressed**. Conventionally, alleles are represented by the same letter to show that they are variants of a single gene. The dominant allele is given the capital letter.

The terms 'homozygous' and 'heterozygous' refer to the genetic constitution or **genotype** of the individual. The actual appearance of an organism is called its **phenotype**. The distinction between genotype and phenotype is important, since the environment may sometimes profoundly affect the appearance of an organism; for example, a 'tall' plant may be actually quite short if unfavourable conditions have stunted its growth. The environment does not alter the genotype of the organism, however.

The behaviour of the alleles for 'tallness' in Mendel's breeding experiment are shown in Figure 28.7.

■ The Law of Segregation

Mendel might have expressed his interpretations as a 'law': 'each characteristic of an organism is determined by a pair of factors of which only one can be present in each gamete.'

Today we are more precise in our statement of this 'Mendelian' First Law, the **Law of Segregation**: 'The characters of an organism are controlled by pairs of alleles, which separate in equal numbers into different gametes as a result of meiosis'.

■ The monohybrid test cross

In hybrids involving a single gene (heterozygotes), where one dominant allele masks the effect of a recessive allele, the appearance (phenotype) of the heterozygote will be identical with that of the homozygous dominant. How can they be distinguished?

Figure 28.7 The gene/allele interpretation of the monohybrid cross: the gene for 'height' in *Pisum sativum* has two alleles (T,t)

parental phenotype (P):	heterozygous tall		homozygous dwarf
parental genotype:	Tt	×	tt
	meiosis		meiosis
gametes:	(T)		(t)

gametes contain only one of the two alleles

| offspring (1) | genotype: | Tt |
| (F_1) | phenotype: | heterozygous tall |

(selfing)

gametes:	Tt	×	Tt
	(T) (t)		(T) (t)

each heterozygous parent produces two kinds of gamete with regard to the T allele

	$\frac{1}{2}T$	$\frac{1}{2}t$
$\frac{1}{2}T$	$\frac{1}{4}TT$	$\frac{1}{4}Tt$
$\frac{1}{2}t$	$\frac{1}{4}Tt$	$\frac{1}{4}tt$

possible combinations of genotypes

| offspring (2) genotypes (F_2): | TT | Tt | tt |
| genotype ratio: | 1 | 2 | 1 |

tall   dwarf

phenotype ratio: 3 : 1

these ratios are likely to be achieved only where a large number of offspring is produced

Figure 28.8   Genes are located on chromosomes: a homologous pair of chromosomes are shown here; in diploid organisms, chromosomes exist in pairs (homologous pairs), one of each pair having come from each parent

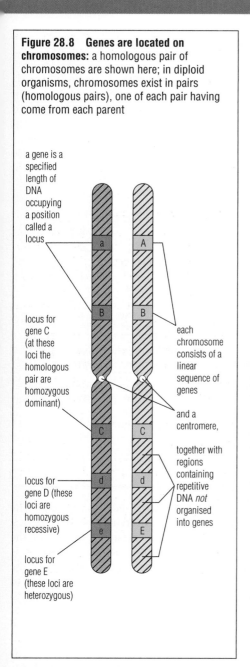

a gene is a specified length of DNA occupying a position called a locus

locus for gene C (at these loci the homologous pair are homozygous dominant)

locus for gene D (these loci are homozygous recessive)

locus for gene E (these loci are heterozygous)

each chromosome consists of a linear sequence of genes

and a centromere,

together with regions containing repetitive DNA *not* organised into genes

Figure 28.10   Two examples of monohybrid inheritance: many have been discovered since Mendel's time

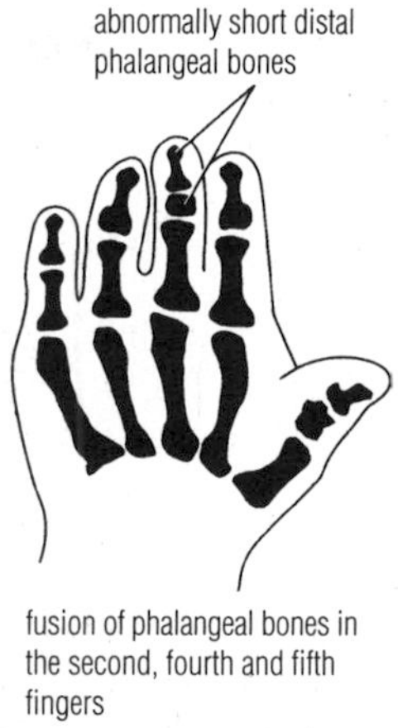

abnormally short distal phalangeal bones

fusion of phalangeal bones in the second, fourth and fifth fingers

Brachydactyly is due to a mutation in the gene for finger length. It is a rare dominant condition.

When the condition appears in the offspring of a family, about half the children are brachydactylous and half have normal hands. Thus:

parental phenotype:	brachydactylous parent	normal-fingered parent
parental genotype:	Dd ×	dd
gametes:	D d	d

	½d	½d
½D	¼Dd	¼Dd
½d	¼dd	¼dd

offspring genotypes:	Dd	dd
genotype ratio:	1 :	1
offspring phenotypes:	brachydactylous	normal
phenotype ratio:	1 :	1

This cob is the product of crossing two plants grown from purple fruit.

parental phenotype:	purple fruits	purple fruits
parental genotype:	Pp ×	Pp
gametes:	P p	P p

	½P	½p
½P	¼PP	¼Pp
½p	¼Pp	¼pp

offspring genotypes:	PP	Pp	pp
genotype ratio:	1 :	2 :	1
offspring phenotypes:	purple fruits		yellow fruits
phenotype ratio:	3 :		1

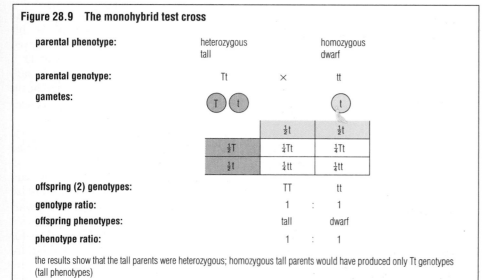

maize cob carrying yellow and purple fruit

Figure 28.9   The monohybrid test cross

parental phenotype:	heterozygous tall	homozygous dwarf
parental genotype:	Tt ×	tt
gametes:	T t	t

	½t	½t
½T	¼Tt	¼Tt
½t	¼tt	¼tt

offspring (2) genotypes:	TT	tt
genotype ratio:	1 :	1
offspring phenotypes:	tall	dwarf
phenotype ratio:	1 :	1

the results show that the tall parents were heterozygous; homozygous tall parents would have produced only Tt genotypes (tall phenotypes)

The difference between heterozygous and homozygous tall pea plants can be detected only by the difference in progeny produced from them in certain crosses. Mendel carried out a test that involved crossing with a homozygous dwarf pea plant (Figure 28.9). In this cross the heterozygous tall individual gives, on average, 50% dwarf progeny (homozygous recessives), whereas all the progeny of the homozygous tall plant are tall.

In crossing with the homozygous recessive, Mendel was using one parent plant of the heterozygous progeny, and so this cross was also known as a back-cross. This term is no longer in use but you meet it in older textbooks.

■ Other examples of single-gene inheritance

Many characters in many different species are controlled by single genes. For example, Beadle and Tatum identified genes that each control the production of a single enzyme in their 'one gene, one enzyme' experiments with *Neurospora* (p. 210). Two other examples of single-gene inheritance are shown in Figure 28.10.

■ Probability and chance in Mendelian inheritance

The presentation of a breeding experiment, such as that shown in Figure 28.7, is a prediction of the likely outcome. It represents the probable result, provided that

► fertilisation is random
► there is an equal opportunity of survival among the offspring
► large numbers of offspring are produced.

What is actually observed in a breeding experiment may not necessarily agree precisely with the prediction. For example, there is a chance that more pollen grains of one genetic constitution may fuse with egg cells than another. If the cross produces very few progeny, the progeny may not appear in the expected proportions.

In Mendel's 'tall' and 'dwarf' experiment the probability is that 1 in every 4 of the F_2 generation will be dwarf, and that 3 in every 4 of the F_2 generation will be tall. For every individual offspring in

the F_2 generation, the chance that it will be tall or dwarf is the same. The chance of a 3:1 ratio increases, however, as more progeny are produced.

If the actual result is not the same as the prediction, is the difference due to

chance, or is the prediction wrong? The larger the difference between the observed and the predicted outcomes, the more likely it is that the prediction is wrong. The smaller the difference, the more likely it is that the result is due to chance. The **chi-squared test** is used to estimate the probability that differences between the observed and expected results are due to chance.

The chi-squared test (χ^2) is a test of the significance of data that consists of discontinuous (discrete) variables. The χ^2 value is a measure of the size of the difference (deviation) between the observed result (O) and the expected result (E). The hypothesis to be tested (the **null hypothesis**) is that there is no significant difference between the observed and expected results. The formula for χ^2 is given in Figure 28.11. In this instance only two categories are being analysed (tallness and dwarfness) and so a correction factor (asterisked) is applied. We can see that $\chi^2 = 0.552$ for the results of Mendel's cross between tall and dwarf peas.

We now consult probability tables for χ^2 to find out what is the probability of obtaining this value. Probabilities are tabulated, as in Figure 28.11, with

Figure 28.12   The gene/allele interpretation of the dihybrid cross

parental phenotype:	homozygous round and yellow	homozygous wrinkled and green
parental genotype:	RRYY $\times$	rryy

offspring (1) phenotype: heterozygous round and yellow
genotype: RrYy
$\times$ self

gametes: RY Ry rY ry   RY Ry rY ry

offspring (2) genotypes:	$\frac{1}{4}$RY	$\frac{1}{4}$Ry	$\frac{1}{4}$rY	$\frac{1}{4}$ry
$\frac{1}{4}$RY	$\frac{1}{16}$RRYY round, yellow	$\frac{1}{16}$RRYy round, yellow	$\frac{1}{16}$RrYY round, yellow	$\frac{1}{16}$RrYy round, yellow
$\frac{1}{4}$Ry	$\frac{1}{16}$RRYy round, yellow	$\frac{1}{16}$RRyy round, green	$\frac{1}{16}$RrYy round, yellow	$\frac{1}{16}$Rryy round, green
$\frac{1}{4}$rY	$\frac{1}{16}$RrYY round, yellow	$\frac{1}{16}$RrYy round, yellow	$\frac{1}{16}$rrYY wrinkled, yellow	$\frac{1}{16}$rrYy wrinkled, yellow
$\frac{1}{4}$ry	$\frac{1}{16}$RrYy round, yellow	$\frac{1}{16}$Rryy round, green	$\frac{1}{16}$rrYy wrinkled, yellow	$\frac{1}{16}$rryy wrinkled, green

offspring phenotypes:	round, yellow	round, green	wrinkled, yellow	wrinkled, green
phenotype ratio:	9 :	3 :	3 :	1
offspring genotypes:	RRYY RRYy RrYY RrYy	RRyy Rryy	rrYY rrYy	rryy

Figure 28.11   The chi-squared test (for further explanation refer to the text)

$$\chi^2 = \sum \frac{(O-E)^2}{E}$$ where O = observed result
E = expected result and
Σ = 'the sum of'

*correction factor in cases where there are only two categories (subtract 0.5 from the value of $(O-E)$)

Category	Hypothesis	O	E	$O-E$	$(O-E)-0.5^*$ (ignore sign)	$(O-E^*)^2$	$\frac{(O-E^*)^2}{E}$
tall	3	787	798	+11	10.5	110.25	0.138
dwarf	1	277	266	−11	10.5	110.25	0.414
		total = 1064					$\Sigma = 0.552$

Thus $\chi^2 = 0.552$
There were two categories, and therefore only 1 degree of freedom (df).
Values of χ^2 for 1 degree of freedom (taken from statistical tables):

	Probability						cut-off point ↓		
	0.99	0.95	0.9	0.5	0.3	0.1	0.05	0.01	0.001
df:1	0.00016	0.004	0.016	0.455	1.074	2.71	3.841	6.635	10.83

↑
observed result

reference to numbers of **degrees of freedom** (df), which in this aspect of statistics are always one less than the number of categories in a sample. As there are only two categories in this case, there is only one degree of freedom. We find that $\chi^2 = 0.552$ falls between $P = 0.3$ and 0.5. In other words, it is likely that this deviation will apply in more than 30% and less than 50% of the times the experiment is repeated. The observed deviation from the expected is insignificant and the null hypothesis stands. The 'cut-off' point is $P = 0.05$ – that is, if the probability is less than 5% that the deviation is due to chance alone, then the prediction (and the underlying assumptions that produce it) is wrong.

■ Some exceptions that proved the rule

For a limited number of specific genes, the alternative alleles do not fit into the dominant/recessive categories observed so widely in nature.

Codominance

In the case of certain genes both allele effects are found to contribute to the phenotype. An example is the MN blood group stystem, in which three genotypes are possible:

▶ MM, which produces M antigen on the red blood cells (phenotype M)
▶ NN, which produces N antigen on the red blood cells (phenotype N)
▶ MN, which produces both M and N antigen (phenotype MN).

Thus in such **codominance** the heterozygote has the character of both parents.

Incomplete dominance

Incomplete dominance occurs where a single dominant allele does not make enough gene product to control the characteristic completely, as in the flower colour in the garden plant snapdragon (*Antirrhinum*). When homozygous red flowers and homozygous white flowers are crossed, the F_1 progeny have pink flowers. When plants grown from seeds of the pink flowers are 'selfed', the F_2 offspring segregate into red-, pink- and white-flowered plants in the ratio 1:2:1. Thus the heterozygote has a phenotype intermediate between the contrasting homozygous parent phenotypes. Pink flowers arise because the pigment produced by a single dominant allele is more dilute than that produced by the dominant homozygote.

■ The dihybrid cross

Mendel also investigated the simultaneous inheritance of two pairs of contrasting characters in the garden pea. These investigations he called **dihybrid crosses**. For example, he crossed pure-breeding pea plants grown from round seeds with yellow cotyledons with pure-breeding plants grown from wrinkled seeds with green cotyledons. All the progeny were round, yellow peas. When plants from these peas were grown and permitted to 'self' the following season, the resulting seeds (of which there were 556 to be

classified and counted!) were of four phenotypes, present as follows:

round seed, with yellow cotyledons	315
round, with green cotyledons	108
wrinkled, with yellow cotyledons	101
wrinkled, with green cotyledons	32

The ratio between the numbers of the difference types of seed is thus 315:108:101:32; this reduces to 9.8:3.38:3.15:1.0, or about 9:3:3:1 – another ratio of simple numbers.

Mendel also noticed that two new combinations, not represented in the parents, appeared in the progeny: either round or wrinkled seeds can turn up with either green or yellow cotyledons. Thus the two pairs of factors were inherited independently. He saw that either one of the pair of contrasting characters could be passed to the next generation. This meant that a heterozygote plant (round and wrinkled) must produce four types of gamete in equal numbers; that is, RrYy must yield RY, Ry, rY and ry equally often (Figure 28.12).

■ The dihybrid test cross

As with monohybrids, the appearance (phenotype) of the heterozygote will be identical to that of the homozygous dominant organism. (In heterozygotes the dominant alleles mask the effect of the recessive alleles.) How could they be distinguished?

Mendel predicted that if he were to cross the heterozygous dominant offspring of the F_1 generation with the dominant homozygous individual all the plants would be dominant, but that when the F_1 hybrid was crossed with the homozygous recessive the resulting progeny would show all the possible combinations of characters in the ratio 1:1:1:1 (Figure 28.13). This latter cross he described as his **test cross** by which the genotype of the F_1 offspring could be determined.

Questions

1 Many dwarf varieties of plants, including the dwarf pea, grow to be tall plants (phenotypically tall) if they are treated with the growth substance gibberellic acid (GA, p. 429) in low concentration. If phenotypically tall homozygous short pea plants are allowed to self-fertilise, what will be the genotype of the progeny?

2 Apply the chi-squared test to Mendel's results for the monohybrid cross of green and yellow pea fruits (Table 28.1, p. 607), and state whether χ^2 is significant at the 5% level.

Figure 28.13   The dihybrid test cross

	hybrid round, yellow peas	×	pure-breeding wrinkled, green peas	
	round, yellow peas	round, green peas	wrinkled, yellow peas	wrinkled, green peas
phenotype ratio:	1 :	1 :	1 :	1

The gene/allele interpretation

parental phenotypes:	heterozygous round, yellow	homozygous wrinkled, yellow	
parental genotypes:	RrYy ×	rryy	
gametes:	(RY) (Ry) (rY) (ry)	(ry)	

	¼RY	¼Ry	¼rY	¼ry
1ry	¼RrYy	¼Rryy	¼rrYy	¼rryy

offspring phenotypes: the offspring genotypes are shown in the Punnett diagram	round, yellow peas	round, green peas	wrinkled, yellow peas	wrinkled, green peas

Figure 28.14 **The physical basis of independent assortment:** a nucleus with four chromosomes (two pairs) is shown in selected stages of meiosis

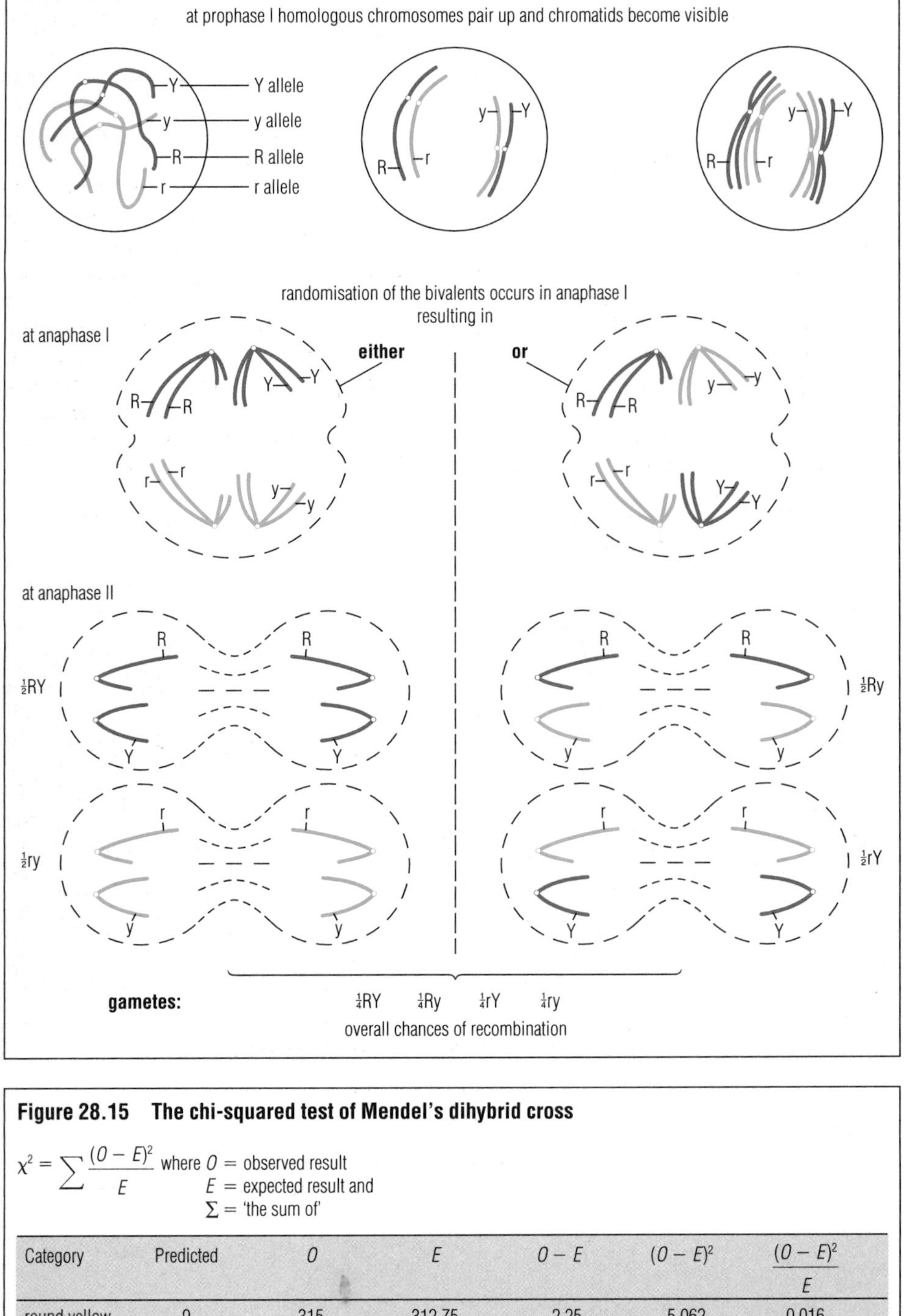

at prophase I homologous chromosomes pair up and chromatids become visible

Y — Y allele
y — y allele
R — R allele
r — r allele

randomisation of the bivalents occurs in anaphase I
resulting in

at anaphase I

either **or**

at anaphase II

$\frac{1}{4}$RY $\frac{1}{4}$Ry

$\frac{1}{4}$ry $\frac{1}{4}$rY

gametes: $\frac{1}{4}$RY $\frac{1}{4}$Ry $\frac{1}{4}$rY $\frac{1}{4}$ry
overall chances of recombination

Figure 28.15 **The chi-squared test of Mendel's dihybrid cross**

$$\chi^2 = \sum \frac{(O - E)^2}{E}$$ where O = observed result
E = expected result and
$\sum$ = 'the sum of'

Category	Predicted	O	E	$O - E$	$(O - E)^2$	$\dfrac{(O - E)^2}{E}$
round yellow	9	315	312.75	2.25	5.062	0.016
round green	3	108	104.25	3.75	14.062	0.135
wrinkled yellow	3	101	104.25	−3.25	10.562	0.101
wrinkled green	1	32	34.75	−2.75	7.562	0.218
total = 556						$\sum$ = 0.470

Thus $\chi^2 = 0.47$
There were four categories, and therefore only 3 degrees of freedom.
Values of χ^2 for 3 degrees of freedom (taken from statistical tables):

				Probability						
	0.99	0.95	0.9	0.7	0.5	0.3	0.1	0.05	0.01	0.001
df: 3	0.115	0.35	0.58	0.71	1.39	3.66	6.25	7.82	11.34	16.27

↑
observed result

The Law of Independent Assortment

Mendel had established that dissimilar pairs of factors that combined in a hybrid could separate from one another and come together in all possible combinations in subsequent generations – expressed colloquially, that either one of a pair of factors might combine with either one of another pair. Mendel was astonishingly thorough, and persisted with his crosses beyond any possible point of doubt about the outcomes. He also carried out similar studies of the inheritance of three other pairs of contrasting characters, and these studies further confirmed his conclusion.

Mendel did not express his discovery as a law, as such. But now that we know that Mendel's factors are alleles of genes on chromosomes, his discovery can be stated as his Second Law, the Law of Independent Assortment (Figure 28.14): 'Two or more pairs of alleles segregate independently of each other as a result of meiosis, provided the genes concerned are not linked by being on the same chromosome' (p. 616).

The chi-squared test again

The number of individuals of each type observed among the progeny of any cross normally deviates to some extent from the numbers expected. The chi-squared test can be applied to the numbers of progeny in the F_2 generation from Mendel's dihybrid cross of round/yellow with wrinkled/green peas (given above), in order to assess whether his results deviate significantly from the ratio of $9:3:3:1$ that we now expect from dihybrid crosses of this type.

The calculation of χ^2 is given in Figure 28.15. χ^2 is found to be 0.47. Thus P lies between a probability of 0.95 and 0.90. This means that a deviation of this size can be expected 90–95% of the times this experiment is conducted. Thus there is no significant deviation between the observed and expected results.

Question

In some cattle the alleles for red and white coat-colour are codominant. A cross between red bulls and white cows produces calves with red and white hairs in their coats. These are called 'roan'.

a If roan bulls and cows are mated together, what genotypes and phenotypes would you expect in the calves?

b What ratios would you expect from a large number of matings?

28.4 ILLUSTRATING MENDELIAN RATIOS IN *DROSOPHILA*

Drosophila melanogaster (the vinegar fly, Figure 28.16) is perhaps the best known of the organisms used by geneticists. *Drosophila* belongs to the order of insects known as Diptera (true flies). In these insects the hindwings are reduced to a pair of club-shaped **halteres** or 'balancers', leaving only one pair of membranous wings. Their mouthparts are suctorial (Figure 14.8, p. 293). *Drosophila* are attracted by fermenting vegetable material, and they are generally known as 'fruit-flies' because they are seen in numbers around rotting fruit. Fruit-flies are very widely distributed throughout temperate and tropical areas of the world.

The value of *Drosophila* in experimental genetics lies in the following features:

► Each pair of *Drosophila* produces hundreds of offspring, which quickly grow to adulthood. The life cycle is completed in only ten days at 25°C. Females mature sexually within 12 hours, and they lay eggs within two days of pupation.

► After mating, female insects can store sperms (p. 585) so it is essential to conduct crosses with virgin females. Virgin female *Drosophila* are easily recognised by their pale grey bodies and pupal faeces (a black spot) visible through the abdomen.

► It is relatively easy to handle *Drosophila*. They can be anaesthetised with ether and remain immobile for ten minutes or so during which period they can be sexed, the phenotypes counted or virgin females isolated, and new crosses set up (Figure 28.19).

► The chromosome complement of the body cells of *Drosophila* consists of only four pairs of chromosomes.

► The cells of the salivary glands, which can easily be dissected from the thorax, contain 'giant' chromosomes (Figure 28.17). Giant chromosomes are enlarged by repeated replications. The chromosomes are traversed by numerous dark bands, which incidentally help identification of particular parts of chromosomes. This banding is useful in the mapping of genes on chromosomes (p. 618), although the bands are not themselves the genes.

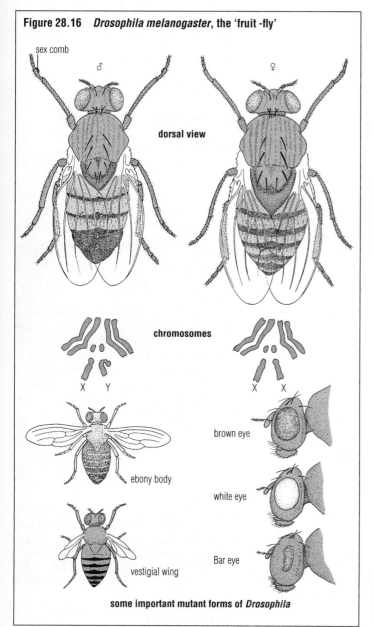

Figure 28.16 *Drosophila melanogaster*, the 'fruit-fly'

sex comb

♂   ♀

dorsal view

chromosomes

X   Y   X   X

brown eye

ebony body

white eye

vestigial wing

Bar eye

some important mutant forms of *Drosophila*

Figure 28.17 'Giant' chromosomes from a cell in the salivary gland of *Drosophila*, showing banding

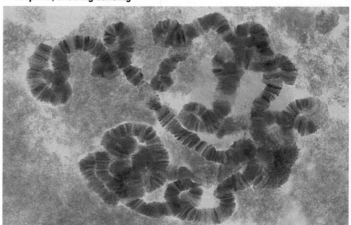

Figure 28.18 A monohybrid cross using *Drosophila*

a *Drosophila* with the mutant character curled wings

parental phenotypes (P):	homozygous straight wing	homozygous curled wing
genotypes:	SS ×	ss
gametes:	S	S
offspring (1) genotype:	Ss	
phenotype:	heterozygous straight wing	
sibling cross:	Ss ×	Ss
gametes:	S s	S s

	½S	½s
½S	¼SS	¼Ss
½s	¼Ss	¼ss

offspring (2) genotypes:	¼SS ½Ss	¼ss
phenotypes:	straight wing	curled wing
phenotype ratio:	3 :	1

The use of *Drosophila* in genetics research

Drosophila was introduced into genetics research by an American biologist, Thomas Hunt Morgan (1866–1945), whilst working at the California Institute of Technology. Using *Drosophila*, Morgan established the **chromosome theory of heredity** – the theory that Mendel's factors are in fact the linear series of genes on a chromosome. Morgan's work established the truth of Mendel's interpretation of his experiments at a time when these were still doubted by some.

Using *Drosophila* 'wild type' characters and various mutants, Morgan went on to establish

- ▶ sex linkage, involving the characters that are controlled by genes on the sex-determining chromosomes (p. 619)
- ▶ crossing over, the exchange of genes between chromosomes as a result of chiasmata formed during meiosis (p. 617)
- ▶ chromosome maps, the relative positions of many genes on the four chromosomes of *Drosophila* (p. 618).

Thomas Morgan was awarded a Nobel prize in 1933.

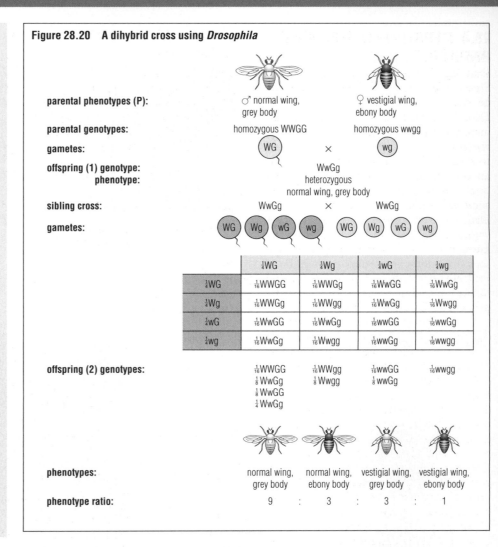

Figure 28.20   A dihybrid cross using *Drosophila*

parental phenotypes (P):	♂ normal wing, grey body	♀ vestigial wing, ebony body
parental genotypes:	homozygous WWGG	homozygous wwgg
gametes:	(WG) ×	(wg)

offspring (1) genotype: WwGg
phenotype: heterozygous normal wing, grey body

sibling cross: WwGg × WwGg

gametes: (WG) (Wg) (wG) (wg)   (WG) (Wg) (wG) (wg)

	¼WG	¼Wg	¼wG	¼wg
¼WG	1/16 WWGG	1/16 WWGg	1/16 WwGG	1/16 WwGg
¼Wg	1/16 WWGg	1/16 WWgg	1/16 WwGg	1/16 Wwgg
¼wG	1/16 WwGG	1/16 WwGg	1/16 wwGG	1/16 wwGg
¼wg	1/16 WwGg	1/16 Wwgg	1/16 wwGg	1/16 wwgg

offspring (2) genotypes:

1/16 WWGG	1/16 WWgg	1/16 wwGG	1/16 wwgg
⅛ WwGG	⅛ Wwgg	⅛ wwGg	
⅛ WwGG			
¼ WwGg			

phenotypes:	normal wing, grey body	normal wing, ebony body	vestigial wing, grey body	vestigial wing, ebony body
phenotype ratio:	9 :	3 :	3 :	1

Genetically controlled characters of *Drosophila*

A normal population of *Drosophila* consists typically of flies with grey bodies, long wings and red eyes. This form of *Drosophila* is the most common, and is often referred to as the **wild type** by geneticists (although this term is now avoided at introductory level).

Many mutant forms are also found in nature, as well. A mutant fly has an inherited departure from at least one of the characteristics of the normal form of *Drosophila*. The mutant characteristics are known by a name. For example:

mutation	*name*	
white eye	white	
ebony body	ebony	recessive
vestigial wing	vestigial	
brown eye	brown	
bar eye	bar	dominant

Recessive mutants are represented by lower-case letters, and dominant mutants by capitals.

In genetic cross experiments (for example, Figures 28.18 and 28.20) the normal allele at a particular locus has been represented by geneticists by a superscript '+', in the past. This symbol has been dropped at an introductory level, and the usual genetic notation is used for *Drosophila*. However the advanced reference books refer to 'wild type' features and use the geneticists' symbols.

Figure 28.19   A breeding experiment using *Drosophila*: the photograph shows sterilised milk bottles containing nutritious growth medium for the culturing of *Drosophila*. Larvae feed in the medium but pupate on the filter paper; adult flies are seen above the medium. Adult flies are temporarily anaesthetised and examined (sorted for new crosses), using a stereo light microscope

28.5 MENDELIAN INHERITANCE IN HUMANS

Studying Mendelian inheritance in humans by the traditional laboratory methods of geneticists (involving experimental crosses and requiring huge numbers of offspring) is out of the question.

Instead, the approach is to investigate the inheritance of a particular characteristic in humans by researching the family pedigree for the trait concerned. The difficulty for the geneticist here is that records of the characters and traits of the members of past generations (family histories) are often incomplete.

Where this approach is successful, the results are laid out in the form of a **pedigree chart**. This is a diagram of a family tree over several generations showing the descendants from particular ancestors, their relationships, and the presence or absence of the trait in all the members. Family histories are researched for as many previous generations as possible. In the chart, the males are represented by squares, the females as circles. Shading indicates the incidence of the particular phenotype under investigation. Each generation is set out along one line of the page, the birth sequence in the family running from left to right. Succeeding generations are shown on following lines.

Analysis of the pedigree chart enables us to detect the difference between a trait that is dominant from one that is recessive. A dominant trait (its phenotype) tends to occur in members of every generation. A recessive trait is seen infrequently, often skipping one or more generations. First-cousin marriages are more likely to reveal the presence of a recessive trait that has not appeared in members of the family for several generations.

For example, albinism is due to a recessive mutation in a single gene, causing a block in the body's biochemical pathways that produce the pigment melanin. An albino has white hair, very light-coloured skin and pink eyes. Most of the population are either homozygous or heterozygous for normal pigmentation. An albino is homozygous recessive for the allele that causes a defective pigment-producing mechanism. A pedigree chart for the inheritance of albinism is shown in Figure 28.21.

In Figure 28.10 the human condition brachydactyly, or very short fingers, was mentioned. This condition is also due to a single mutant gene, but this time a dominant one. Here the family tree shows an altogether different pattern of inheritance (Figure 28.22).

There are several genetics investigations that can be carried out using plants, animals and (sometimes) humans (see *Study Guide*, Chapter 28).

Figure 28.21 Albinism and its inheritance in humans, a recessive pattern of inheritance

Albino children may be the offspring of a cross between a carrier (a heterozygote in whom the recessive gene is not expressed) and an albino individual.

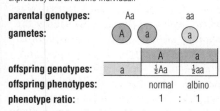

		A	a
parental genotypes:	Aa		aa
gametes:	A a		a
offspring genotypes:	a	$\frac{1}{2}$Aa	$\frac{1}{2}$aa
offspring phenotypes:		normal	albino
phenotype ratio:		1 : 1	

Alternatively, albino offspring may be the progeny of a cross between two carriers.

a typical pedigree diagram for recessive inheritance
(note that a first-cousin 'marriage', P + Q, is involved)

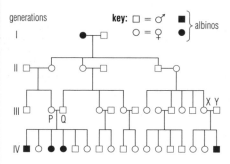

All offspring of generation I must be carriers, irrespective of whether the male parent was AA or Aa; they will all have the 'a' gene from their mother.

In generation III, neither of the parents in the PQ family or in the XY family was affected. In both these families, if one of the parents had been AA, there would have been no affected children. Since affected children were born into both families, P, Q, X and Y must all have been Aa; therefore they are all carriers of the albino allele.

Questions

1 Recombination of genes can occur during gamete formation in two ways. What ways are these?

2 Why is it necessary to select virgin female *Drosophila* flies in breeding crosses, but it is only necessary to use males that are fertile?

3 In *Drosophila*, mutants with scarlet eyes and vestigial wings are recessive to flies with red eyes and normal wings. Explain the phenotypic ratio to be expected in the F_2 generation when normal flies are crossed with scarlet eyes/vestigial wing mutants and sibling crosses of the F_1 offspring are then conducted.

4 Albinism is due to a recessive mutation. Brachydactyly is due to a dominant mutation.
   a What is meant by 'a dominant mutation'?
   b How may the pattern of inheritance of these conditions, shown in a pedigree chart, enable us to detect the difference between recessive and dominant mutations?

Figure 28.22 The inheritance of brachydactyly

from the pedigree diagram of an affected family, we can see that brachydactyly is controlled by a single dominant allele

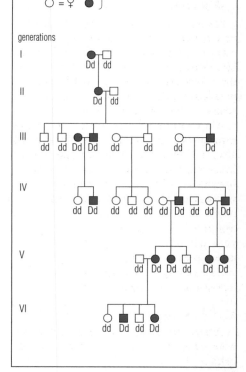

28.6 DEVELOPMENTS SINCE MENDEL

Mendel created modern genetics, virtually single-handedly. His success was based on his initial decision to follow the inheritance of only one or two pairs of contrasting characters in his experimental organisms, rather than the multitude of features that most organisms exhibit. But he was lucky, too. The contrasting characters he chose were each controlled by a single gene and not by several genes (as many human characteristics are, for example). Also, the pairs of contrasting characters he selected in order to investigate the dihybrid cross just happened to be controlled by single genes on different chromosomes (that is, they were not 'linked' on a single chromosome).

Mendelian genetics have been complemented and extended by the establishment of seven additional principles, which are discussed in sections 28.7 to 28.13. Taken together with Mendel's laws, these are the foundation of this all-important branch of modern biology.

28.7 LINKAGE

There are many thousands of genes per cell in an organism, whereas the number of chromosomes is often less than 50, and rarely exceeds 100 (an exception is the crayfish, *Astacus pallipes*, with 200 chromosomes per cell). Each chromosome, then, consists of many genes; it may be thought of as a linear sequence of genes that are all linked together. Genes on the same chromosome will tend to be inherited together. As a consequence, the Mendelian ratios we would expect for the inheritance of two pairs of contrasting characters will not be obtained if the genes are on the same chromosome.

An example of this was reported by experimental geneticists in 1906, soon after Mendel's work was rediscovered, and before linkage was properly understood. In studies of inheritance in the sweet pea plant (*Lathyrus odoratus*), plants homozygous for purple flower and elongated pollen grains were crossed with plants homozygous for red flower and rounded pollen. The progeny of the cross had purple flowers and long pollen. It was correctly concluded that the alleles for purple flower and elongated pollen were dominant.

When the progeny (the F₁ generation) were selfed it was anticipated that the two pairs of alleles would segregate

independently of each other, producing the four possible combinations of a dihybrid cross in the expected ratio of $9:3:3:1$.

This did not happen! Instead the progeny were largely of the 'parental types', that is, plants with purple flowers and elongated pollen grains and plants with red flowers and rounded pollen grains in the ratio of (approximately) $3:1$. Independent assortment had not occurred. It is now known that the reason for this is that the genes (alleles) concerned are linked on the same chromosome.

Figure 28.23   The formation of recombinants in linked genes

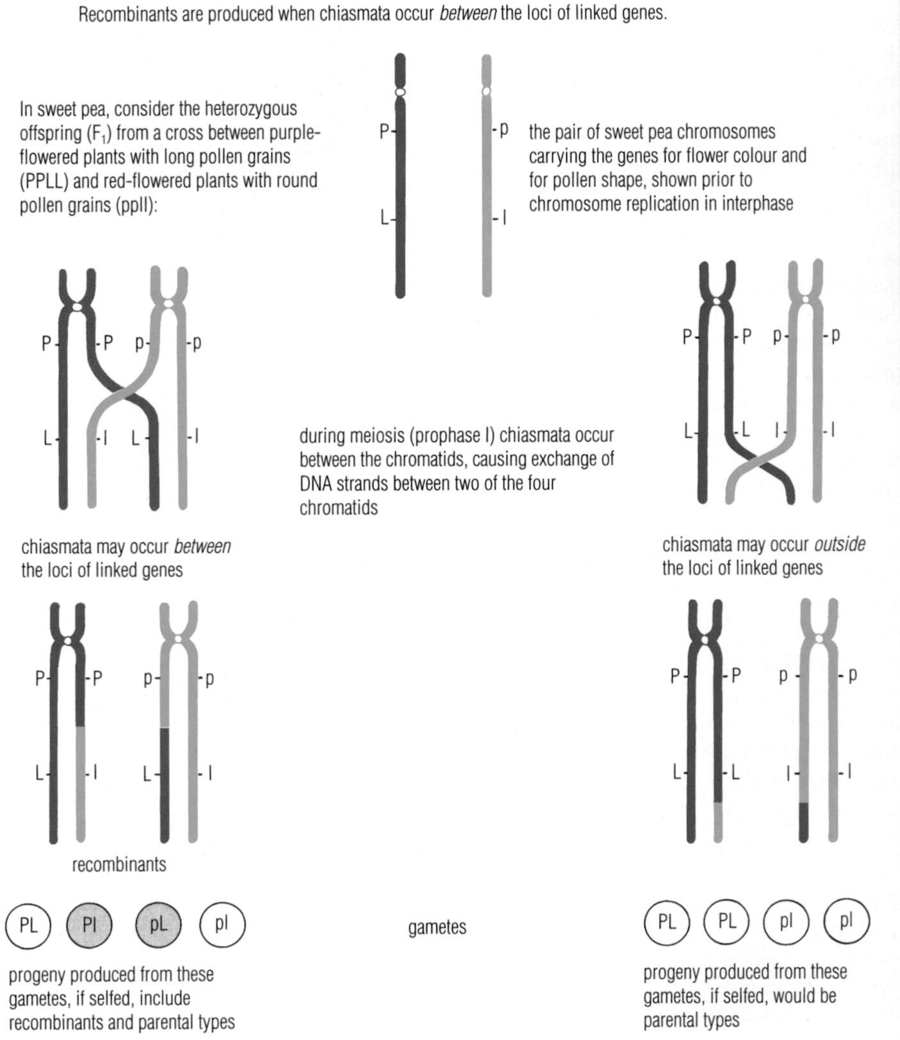

Recombinants are produced when chiasmata occur *between* the loci of linked genes.

In sweet pea, consider the heterozygous offspring (F₁) from a cross between purple-flowered plants with long pollen grains (PPLL) and red-flowered plants with round pollen grains (ppll):

the pair of sweet pea chromosomes carrying the genes for flower colour and for pollen shape, shown prior to chromosome replication in interphase

during meiosis (prophase I) chiasmata occur between the chromatids, causing exchange of DNA strands between two of the four chromatids

chiasmata may occur *between* the loci of linked genes

chiasmata may occur *outside* the loci of linked genes

recombinants

gametes

progeny produced from these gametes, if selfed, include recombinants and parental types

progeny produced from these gametes, if selfed, would be parental types

for example, in the F₂ offspring:

		phenotype	genotype
parental types:	{	purple flowers, long pollen grains	PPLL, PpLl
		red flowers, round pollen grains	ppll
recombinant types:	{	purple flowers, round pollen grains	PPll, Ppll
		red flowers, long pollen grains	ppLL, ppLl

Question

In *Drosophila* the gene loci for the recessive mutant characters scarlet eye and ebony body have been shown to be located on a single chromosome (chromosome 3). For an experimental cross between flies with normal (red) eyes and grey body, and mutant flies with scarlet eyes and ebony body, show the ratio of expected genotypes and phenotypes when sibling matings of the F₁ offspring are carried out (assuming 100% linkage).

28.8 CROSSING OVER AND LINKAGE

In the 1906 experiment just described, using sweet pea to demonstrate the linkage of genes on a chromosome, the progeny were in fact not quite all 'parental types'. There were a few purple-flowered plants with round pollen and some with red flowers and elongated pollen. These are known as the 'recombinant types', because they are new combinations of the parental characters. How had they arisen, considering that the genes for flower colour and pollen shape are located on the same chromosome? The answer is through the 'crossing over' between adjacent chromatids that occurs during meiosis in gamete formation (Figure 28.23).

■ The discovery of crossing over in *Drosophila*

Crossing over and the formation of chiasmata during meiosis (Figure 9.9, p. 199) were first reported by a European cytologist in 1909. T H Morgan, who was aware of this discovery, was able to explain the genetic significance of crossing over. Almost every pair of homologous chromosomes shows crossing over during prophase I (exceptions are described below). The significance was apparent from the results of crosses of normal ('wild type') *Drosophila* with flies

showing various mutant characteristics controlled by linked genes.

For example, when flies with ebony bodies and curled wings were crossed with normal flies (grey bodies and straight wings) the F₁ offspring were all heterozygous for grey body and straight wing (normal in appearance). When females of these heterozygotes were crossed (test cross) with male homozygous recessive parents, most of the offspring were of the parental types but about 20% were 'recombinant types' showing ebony bodies/straight wings or

grey bodies/curled wings. The results indicated, first, that the genes for body colour and wing shape were linked on the same chromosome; had they been on separate chromosomes Morgan would have obtained a phenotype ratio of 1:1:1:1 (of the dihybrid test cross, p. 611). Secondly, the presence of some recombinants among the offspring indicated that in a proportion of the homologous chromosomes concerned there was crossing over between the locus for body colour and that for wing shape. The outcome of this cross is explained in Figure 28.24.

■ Meiosis without chiasmata

Meiosis occurs without any crossing over in a limited number of species. In these organisms, in metaphase I the homologous chromosomes lie alongside each other at the equator of the spindle. They separate in the usual way in anaphase I, and meiosis proceeds normally thereafter. This type of meiosis is seen in a few insects, including the males of *Drosophila*, and in certain flowering plants. In all other species at least one cross-over is formed in each bivalent in prophase I.

The effect of the absence of cross-over in meiosis in the male of *Drosophila* can be quite dramatic. For example, if the cross shown in Figure 28.24 is repeated, this time using a male heterozygote for the test cross rather than the female, then complete linkage of the genes is shown, and no recombinant types appear in the progeny (Figure 28.25).

Figure 28.25 **An unusual meiosis in male *Drosophila*:** what happens when the test cross shown in Figure 28.24 is repeated with male heterozygotes

parental phenotypes:	♀ ebony body curled wings	♂ grey body, straight wings
parental genotypes:	ggss homozygous ×	GgSs heterozygous
		(no chiasmata formed)
gametes:	gs	GS gs

	GS	gs
gs	GgSs	ggss

F₁ offspring:	phenotypes	genotypes	offspring produced
parental types:	grey body, straight wings	GgSs	in the ratio 1:1
	ebony body, curled wings	ggss	

Figure 28.24 **Linkage and crossing over in *Drosophila***

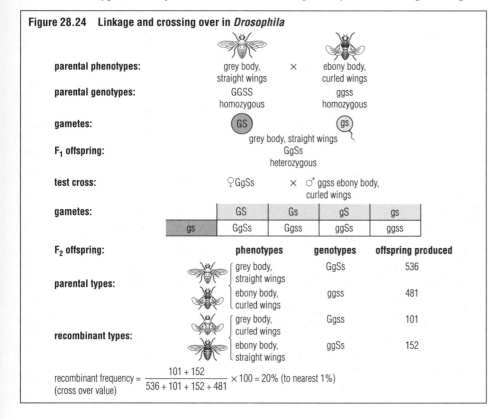

parental phenotypes:	grey body, straight wings	×	ebony body, curled wings	
parental genotypes:	GGSS homozygous		ggss homozygous	
gametes:	GS		gs	
	grey body, straight wings			
F₁ offspring:	GgSs heterozygous			
test cross:	♀ GgSs	×	♂ ggss ebony body, curled wings	

	GS	Gs	gS	gs
gs	GgSs	Ggss	ggSs	ggss

F₂ offspring:		phenotypes	genotypes	offspring produced
parental types:		grey body, straight wings	GgSs	536
		ebony body, curled wings	ggss	481
recombinant types:		grey body, curled wings	Ggss	101
		ebony body, straight wings	ggSs	152

$$\text{recombinant frequency} = \frac{101 + 152}{536 + 101 + 152 + 481} \times 100 = 20\% \text{ (to nearest 1\%)}$$
(cross over value)

■ Cross over value

Crossing over, then, is a process of exchange between homologous chromosomes during meiosis (normally associated with gamete formation) which gives rise to new combinations of characters. The frequency of recombination is known as the **cross over value**, and is calculated as a percentage from the expression:

$$\frac{\text{number of recombinant individuals produced}}{\text{total number of offspring}} \times 100$$

In organisms in which the incidence of crossing over has been extensively studied, such as maize and *Drosophila*, it has been shown that the proportions of recombinants produced in a particular cross are fairly constant. The proportions are different for other pairs of genes on the same chromosome, however. This is explained on the basis that every gene has a fixed location (its locus) on a particular chromosome, and that crossing over is more likely to occur between genes that are more widely spaced. Thus, in our first example of linkage, the genes for flower colour and pollen shape in the sweet pea, we can conclude that the loci are relatively close together, because very few recombinants were formed. In other cases of linked genes the recombinant frequency can have values up to 50%.

Figure 28.26   Gene mapping: an example of the process by which the recombinant frequencies (cross over values) of different alleles of pairs of genes located on the same chromosome are determined using breeding experiments (test crosses)

1 Genes A and B are found to have a cross over value of 8%, so we say that genes A and B are 8 chromosome map units apart:

2 Genes A and C are then found to have a cross over value of 20%, so A and C are 20 map units apart:

3 We deduce that the spacing of genes A, B and C is either

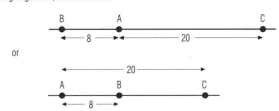

or

4 A test cross between B and C is carried out to find whether the cross over value is 12% or 28%.

Thus by means of three test crosses the relative positions of three genes are determined.

5 A complication may arise with genes lying great distances apart because there is a good chance that there may be two (or more) chiasmata between them; for example:

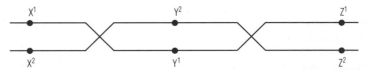

The presence of two chiasmata may apparently cancel out the effects of recombination. This problem is best overcome by the mapping of a large number of genes at short distances from each other (since double cross overs are unlikely between closely linked genes), thus allowing the composite picture to be constructed.

■ Gene mapping of chromosomes

Gene maps of chromosomes are based on the recombinant frequencies obtained from breeding experiments. To produce a gene map it is first necessary to establish which genes are linked together on the one chromosome. This is determined by means of test crosses to investigate whether the four types of progeny are produced in roughly equal numbers (that is, the recombinants are 50% of the progeny). If they are, this normally indicates that the genes are located on separate chromosomes, and the alleles demonstrate independent assortment. If the recombinant frequencies are less than 50%, then linkage of genes with some crossing over between them is indicated.

The second step is to establish the identity of as many genes as possible on any one chromosome. Once the different alleles at several gene loci of a particular chromosome are known it is possible to carry out test crosses between many or all of the different pairs of genes. The results of test crosses give recombinant frequencies (cross over values) for the genes concerned.

The cross over values obtained do not tell us the real distance between genes, but it is an agreed convention that 1% recombination is taken as 1 map unit of distance between the genes concerned. On this basis, chromosome 'maps' are produced. Two genes whose alleles give a cross over value of 10% are said to be 10 units apart. Once we have this information we can begin to construct the map. The steps in map production are illustrated in Figure 28.26.

Genetic maps that show the positions of a large number (not all) of the genes of the organism have been produced for many species, including plants such as maize, animals such as *Drosophila*, and many fungi, bacteria and viruses.

Questions

1 Why are recombinant frequencies never greater than 50%?

2 In the sweet pea, if an F_1 progeny with purple flowers and long pollen grains (PpLl) is test crossed with the homozygous recessive parental type (ppll), what possible F_2 progeny would you expect? Show why.

3 When *Drosophila* flies with normal wings and grey bodies (WwGg) were crossed with flies with vestigial wings and ebony bodies (wwgg), the progeny were as follows:

normal wing, grey body	586
vestigial wing, grey body	106
normal wing, ebony body	111
vestigial wing, ebony body	465

What is the cross over value? Show your working.

28.9 SEX DETERMINATION AND SEX LINKAGE

■ Sex determination

In human cells there are 23 pairs of chromosomes (p. 196); 22 pairs are described as **autosomal chromosomes**, and one pair as **sex chromosomes**. Females have two identical sex chromosomes designated as X; males have one X sex chromosome and also one Y sex chromosome, which is not found in females. In fact differentiation of male or female characteristics is controlled by several genes, many located on autosomal chromosomes rather than on the sex chromosome. The specific role of the sex chromosomes appears to be to switch on changes that direct the effects of other genes.

In the production of gametes the sex chromosomes segregate. All the egg cells have an X chromosome, whereas 50% of the sperms carry an X chromosome and 50% a Y chromosome. At fertilisation the egg cell may be fertilised by a sperm carrying an X chromosome, producing a zygote that has two X chromosomes and which will therefore develop into a female. It is equally likely to be fertilised by a sperm carrying a Y chromosome, to give a zygote with one X and one Y chromosome, which will develop into a male. So the sex of the offspring in humans (and in all mammals) is deter-mined by which sperm it is that fertilises the egg (Figure 28.27). We would expect equal numbers of male and female offspring to be produced (see overleaf).

How does the Y chromosome 'work'?

The role of the Y chromosome in mammals is to control the differentiation of testes from the undifferentiated gonads of the developing embryo. The Y chromosome carries a gene coding for a **testis determining factor** (TDF), which brings about this developmental change. In the absence of TDF the embryonic gonads develop into ovaries.

The functioning of the X chromosome in males and females is part of the issue of 'sex-linked' characters, discussed on p. 620.

Figure 28.27   The segregation of X and Y chromosomes: the determination of sex

parental phenotypes:	female ♀	male ♂
parental genotypes:	XX	XY
gametes:	Ⓧ Ⓧ	Ⓧ Ⓨ

	$\frac{1}{2}$X	$\frac{1}{2}$Y
$\frac{1}{2}$X	$\frac{1}{4}$XX	$\frac{1}{4}$XY
$\frac{1}{2}$X	$\frac{1}{4}$XX	$\frac{1}{4}$XY

offspring genotypes:	$\frac{1}{2}$XX	$\frac{1}{2}$XY
offspring phenotypes:	♀	♂

Barr bodies: X chromosome inactivation

The human Y chromosome is short (Figure 9.5), and the X chromosome carries many more genes. The female has two X chromosomes, so there appears to be a major imbalance in genes between males and females. In fact, in all mammals one of the two X chromosomes of the female is effectively 'shut down' very early in embryonic development. This chromosome then appears as a darkly staining body attached to the inside of the nuclear membrane. It is called the **Barr body**, after its discoverer. The detection of the presence of a Barr body in the nucleus of embryonic cells withdrawn from the uterus (see p. 601) is the way a fetus is sexed at an early stage of development.

As a result, the cells of the female have a single X chromosome operational throughout their lives. The X chromosomes are rendered inactive at random, and consequently the tissues are a mosaic of cells with either maternal or paternal X chromosomes. This can have strange consequences, as in the case of 'tortoise-shell' coat colour in female cats (Figure 28.28).

Figure 28.28   The inactivation of an X chromosome in mammals

One of the two X chromosomes in the cells of females is condensed throughout most of the life of a cell. Thus, in normal females (XX), one of the X chromosomes is reduced to a prominent Barr body, lying just below the nuclear membrane.

Condensation (formation of Barr bodies) occurs at about the 2000-cell stage of the embryo. This inactivation process is random, and so the cells of the female are a mosaic of her two X chromosomes.

The consequences of this random inactivation are visible in the coat colour of the domestic cat (Felis catus).

On the X chromosome of the cat there is a gene for coat colour.

In male cats (XY):
    genotype $X^BY \rightarrow$ phenotype 'black'
    genotype $X^bY \rightarrow$ phenotype 'orange'

In female cats (XX):
    genotype $X^BX^B \rightarrow$ phenotype 'black' (effectively X^B)
    genotype $X^bX^b \rightarrow$ phenotype 'orange' (effectively X^b)
but genotype $X^BX^b \rightarrow$ phenotype 'tortoise-shell'
        produced by a mosaic of cells with either the X^B or the X^b genes active in skin cells but randomly distributed

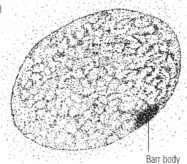

nucleus of somatic cell of XX female

Barr body

'tortoise-shell' cats are always female

black and orange cats may be either male or female

Sex determination in other organisms

The XY system is characteristic of most vertebrates, many insects (including *Drosophila*) and other non-vertebrate groups. It is also characteristic of those flowering plants where the sexes are separate (p. 569). In birds and in the butterflies and moths the males are XX and the females XY (or X0, meaning the second sex chromosome may be absent in the female). There is a different method of sex determination in the social insects.

The mechanism by which the sex chromosomes have their effects varies with the group of organisms. The picture is complicated, and is not fully understood. For example, in *Drosophila*, where the male is XY and the female XX, the Y chromosome is required for the sperms to function properly, but is not essential for the initial induction of 'maleness' (such as development of testes). Here sex is determined by the balance between the number of X chromosomes and the number of sets of autosomal chromosomes (*Drosophila* has three pairs of autosomal chromosomes) normally present (Figure 28.16, p. 613). Here XX (and also XXY) with three pairs of autosomal chromosomes produces a female, whereas X0 (and XYY) and three pairs of autosomal chromosomes produces a male. There is probably still much to be learnt about the control of sexuality in the living world!

The ratio of the sexes in humans

Equal numbers of X and Y sperms are produced, yet in practice slightly more males than females are conceived (or implanted). Perhaps the Y-bearing sperms are more efficient at fertilisation, or perhaps the male blastocyst has greater success at implantation in the endometrium of the uterus wall. Whatever the reason, the outcome is that about 106 boys are born for every 100 girls. Slightly more boys die in childhood than girls, however, so the ratio of males to females in the population reaches unity by the time that puberty is reached. Then, towards the end of their lives, females exceed males by 2:1, since the life expectancy of the male is about five years less than that of the female. In the human struggle for survival there are different winners at different stages!

■ Sex linkage

Sex-linked inheritance – the inheritance of genes carried on the sex chromosomes – was discovered in *Drosophila*, when the pattern of inheritance of certain traits was found to vary with the sex of parent and offspring. The discovery was made in an investigation of the inheritance of a mutant condition known as 'white eye'. White eyes are rare, and were shown to

be due to a recessive allele for eye colour (the normal condition, red eye, being dominant).

Reciprocal crosses of white- and red-eyed *Drosophila*, followed by sibling mating of the offspring, produced males and females in roughly equal numbers in all crosses, but different ratios of red and white eyes in the progeny. The pattern of inheritance of eye colour among the F_2 offspring could be explained if males and females had different chromosomes (X and Y in Figure 28.29) and the gene locus for eye colour was on the X chromosome.

Sex-linked conditions in humans

Haemophilia

The ability of the blood to clot is one of the defence mechanisms of human bodies; the steps in formation of a blood clot are shown in Figure 17.41 (p. 365). Haemophilia is a disease in which the blood of the affected person has a markedly reduced ability to clot, due to a deficiency of one of the blood clotting factors. If untreated, it is accompanied by internal bleeding into joints and muscles, as well as a risk of uncontrollable haemorrhages in the case of injury. Treatment is by administration of the clotting factor that the patient lacks.

Haemophilia is caused by a recessive gene carried on the X chromosome. The presence of a recessive allele for haemophilia can be masked by an allele for normal blood clotting in a female (this type of individual is described as a 'carrier'). In the male, the presence of an allele for haemophilia on the single X chromosome is sufficient to produce the disease, because the Y chromosome is much shorter and cannot carry an allele for normal blood clotting to mask its recessive counterpart.

Queen Victoria (1819–1901) must have received a mutant allele for haemophilia from one of her parents. Because of the social and political position of her family, personal details of individual members are well preserved in public records. The pedigree chart of the family (Figure 28.30) shows the features typical of a sex-linked recessive gene. These are that the disease affects mainly the males, and is normally transmitted by a female who is a carrier but who does not herself show symptoms of the disease. With respect to haemophilia, females may have the following genotypes:

$X^H X^H$ = normal
$X^H X^h$ = carrier
$X^h X^h$ = haemophiliac.

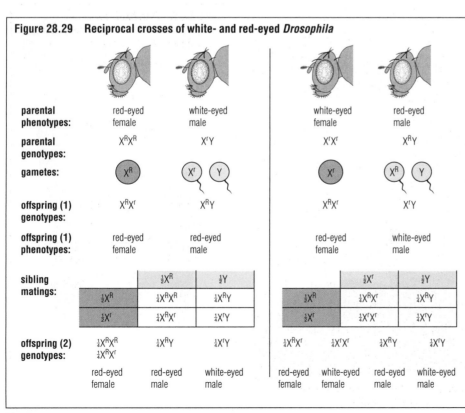

Figure 28.29 Reciprocal crosses of white- and red-eyed *Drosophila*

parental phenotypes:	red-eyed female	white-eyed male	white-eyed female	red-eyed male	
parental genotypes:	$X^R X^R$	$X^r Y$	$X^r X^r$	$X^R Y$	
gametes:	X^R	X^r Y	X^r	X^R Y	
offspring (1) genotypes:	$X^R X^r$	$X^R Y$	$X^R X^r$	$X^r Y$	
offspring (1) phenotypes:	red-eyed female	red-eyed male	red-eyed female	white-eyed male	

sibling matings:

	$\frac{1}{2}X^R$	$\frac{1}{2}Y$
$\frac{1}{2}X^R$	$\frac{1}{4}X^R X^R$	$\frac{1}{4}X^R Y$
$\frac{1}{2}X^r$	$\frac{1}{4}X^R X^r$	$\frac{1}{4}X^r Y$

	$\frac{1}{2}X^r$	$\frac{1}{2}Y$
$\frac{1}{2}X^R$	$\frac{1}{4}X^R X^r$	$\frac{1}{4}X^R Y$
$\frac{1}{2}X^r$	$\frac{1}{4}X^r X^r$	$\frac{1}{4}X^r Y$

offspring (2) genotypes:	$\frac{1}{4}X^R X^R$ $\frac{1}{4}X^R X^r$	$\frac{1}{4}X^R Y$	$\frac{1}{4}X^r Y$	$\frac{1}{4}X^R X^r$	$\frac{1}{4}X^r X^r$	$\frac{1}{4}X^R Y$	$\frac{1}{4}X^r Y$
	red-eyed female	red-eyed male	white-eyed male	red-eyed female	white-eyed female	red-eyed male	white-eyed male

Males may have the following genotypes:

X^HY = normal
X^hY = haemophiliac.

A haemophiliac daughter, X^hX^h, could be born to a haemophiliac father and a carrier (or haemophiliac) mother. Haemophilia affects only 0.004% (one in 25 000) of males in the population, however. Put in another way, the frequency of X chromosomes carrying the allele for haemophilia is 0.000 04 among human males. Among females, therefore, the occurrence of haemophilia cannot exceed $(0.000\ 04)^2$. Female haemophiliacs are much less common than that, because male haemophiliacs seldom breed and transmit the allele to their daughters (there is an exception to this among Queen Victoria's progeny, however!).

Muscular dystrophy

Duchenne muscular dystrophy (DMD, Figure 28.31) is a tragic, wasting disease of children (almost always of boys). To the affected child, physical activity presents problems from infancy. Progressive weakness develops as the muscles of the body are replaced by fibrous tissue. A child becomes restricted to a wheelchair (generally by the age of 10), and as muscular wasting continues, breathing itself is threatened. A person with this disease, at present, rarely survives beyond the age of 20.

The gene for DMD is sex-linked, and found on the X chromosome (Figure 28.32). It is believed that the allele for DMD codes for an enzyme that induces the replacement of muscle by fibre. DMD affects 1 in every 4000 male infants, in about one-third of these as a consequence of a spontaneous mutation in the gene (p. 625). In these cases the affected individual is born to a mother who is not herself even a carrier of the allele. Research in DMD is focused on the detection of carriers and on pre-natal diagnosis of the condition; ultimately, a solution may lie in some form of gene therapy (p. 639).

Question

Haemophilia results from a sex-linked gene. The disease is most common in males, but the gene is on the X chromosome. Explain this anomaly.

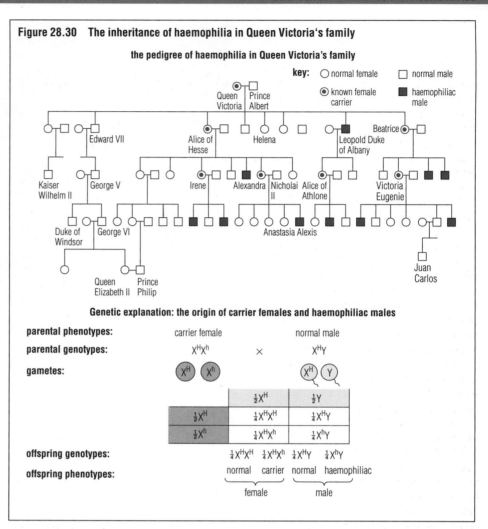

Figure 28.30   The inheritance of haemophilia in Queen Victoria's family

Genetic explanation: the origin of carrier females and haemophiliac males

parental phenotypes:	carrier female		normal male	
parental genotypes:	X^HX^h	×	X^HY	

	$\frac{1}{2}X^H$	$\frac{1}{2}Y$
$\frac{1}{2}X^H$	$\frac{1}{4}X^HX^H$	$\frac{1}{4}X^HY$
$\frac{1}{2}X^h$	$\frac{1}{4}X^HX^h$	$\frac{1}{4}X^hY$

offspring genotypes: $\frac{1}{4}X^HX^H$ $\frac{1}{4}X^HX^h$ $\frac{1}{4}X^HY$ $\frac{1}{4}X^hY$

offspring phenotypes: normal   carrier   normal   haemophiliac

female / male

Figure 28.31   Children with Duchenne muscular dystrophy (DMD)

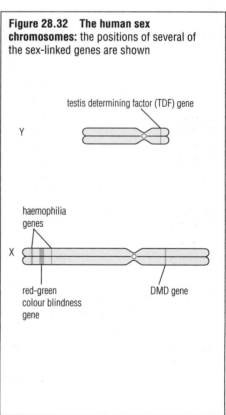

Figure 28.32   The human sex chromosomes: the positions of several of the sex-linked genes are shown

testis determining factor (TDF) gene

Y

haemophilia genes

X

red-green colour blindness gene

DMD gene

Red–green colour blindness

The red–green colour-blind person cannot distinguish the colours that others see as green, yellow, orange and red (they are seen as the same colour). This form of colour-blindness is detected using coloured diagrams such as the Ishihara test card (Figure 28.33), in which the detail can only be seen by a person with normal colour vision. People with normal vision see a clear line, where as those with red–green colour-blindness see a random pattern of dots. This condition, which cannot be treated, is due to a defect in one of the three groups of colour sensitive cone cells in the retina.

The genes for red–green vision are on the X chromosome, so all males with the allele for red–green colour-blindness are affected. A female can be a carrier of the colour-blindness allele but have normal vision; to be colour-blind, she must be homozygous for the recessive colour-blindness allele. Red–green colour-blindness affects about 8% of males but only about 0.4% of females.

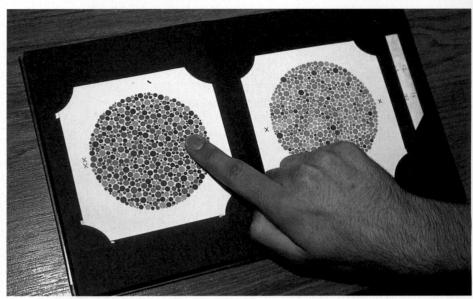

Figure 28.33   An Ishihara test card for red-green colour blindness: people with normal vision see a clear line; those with red–green colour blindness see a random pattern of dots

28.10 MULTIPLE ALLELES

The human blood group systems (part of the immune system, p. 366) are all genetically determined. For example, the Rhesus system is controlled by a single gene with two alleles (rhesus-positive and rhesus-negative), just like other characteristics we have met already. Inheritance of the Rhesus factor is a classical example of monohybrid inheritance.

The inheritance of the ABO system is different, and illustrates a new principle in genetic control of phenotypes. Here the gene for the ABO blood group exists as more than two alleles, although only two can occur in any one (diploid) organism. An individual's blood group is determined by the types of mucopolysaccharide that function as antigens (agglutinogens) on the plasma membrane of blood cells. Group A has only antigen A, group B has only antigen B, group AB has both antigens A and B, and group O has neither antigen.

The gene for the ABO system is conventionally represented by the symbol I. Alleles I^A and I^B code for the enzymes involved in the formation of the antigens A and B respectively, and the allele I^O for no known protein. A person has either two identical alleles (homozygous) or two different alleles (heterozygous), which results in six possible genotypes but four possible phenotypes (the familiar ABO blood groupings). Thus:

▶ genotypes I^AI^A and I^AI^O produce group A phenotype
▶ genotypes I^BI^B and I^BI^O produce group B phenotype
▶ genotype I^AI^B produces group AB phenotype
▶ genotype I^OI^O produces group O phenotype.

The alleles I^A and I^B jointly express themselves in the individual: they are codominant (p. 611). Both I^A and I^B are dominant to the recessive allele I^O.

28.11 POLYGENES

Many of the characteristics of an organism are not controlled by a single gene comprising two alleles, like the characteristics discussed earlier in this chapter. Rather, characteristics are controlled by two, three or more genes. These genes are often (but not necessarily) located on different chromosomes. In fact, the control of most characteristics is mulifactorial; single-gene control is relatively unusual. For example, in humans several different gene loci are involved in determining characteristics such as body height and body mass. The more loci and the more alleles that are concerned, the greater the number of phenotypic classes: each class becomes less distinct from the next. The number of gene loci controlling a characteristic does not have to be very large before the variation in phenotype becomes more or less continuous. The consequence is that no clear-cut

Mendelian ratios are found in tracing the inheritance of many characteristics. Nevertheless, the individual genes concerned are inherited in accordance with Mendelian principles.

The picture of hereditary control is further complicated by situations in which several of the gene loci interact with each other in determining a characteristic. Most traits are affected by gene interactions of some kind.

■ Complementary genes

Complementary genes, a very common form of gene interaction, were discovered in the sweet pea (*Lathyrus odoratus*) when two white-flowered varieties were crossed and all the F_1 offspring were found to bear purple flowers. The selfing of these progeny produced purple and white-flowered plants in the ratio 9:7. This result (Figure 28.34) arises because

▶ this characteristic is controlled by two different gene loci on different chromosomes
▶ one gene (we call this P) controls the production of a colourless pigment precursor
▶ the other gene (we call this C) controls the conversion of the precursor into a purple pigment
▶ unless the dominant alleles P and C are both present the flowers will be white, either because no precursor

was formed, or because it was formed but not converted to pigment

▶ in the original cross the purple flowers were the product of white-flowered strains, one of which was homozygous dominant for P and the other homozygous dominant for C.

■ Epistasis

Epistasis is a form of gene interaction in which a dominant allele at one locus totally inhibits the expression of an allele at a second locus. An example is the pronounced banding in the shell of the snail *Cepaea nemoralis* (Figure 28.35). The colour variations in the shell are also genetically controlled, and certain aspects of coloration are controlled by a single gene. But the banding, which exists in three common variants, is controlled by different genes on separate chromosomes. One of the genes determines whether or not bands are present on the shell. At this locus the allele for the absence of bands (A) is dominant to the allele for the presence of bands (a). A second gene determines the number and position of bands, and here a dominant allele codes for the 'mid-banded' condition, and a recessive allele codes for the 'five-banded' condition. This latter characteristic is not expressed unless the aa genotype occurs at the A locus.

The control of fur colour in mice is another example of epistasis (*Study Guide*, Chapter 28).

Questions

1   A red–green colour-blind boy discovered that one of his grandfathers was the last member of the family to have the defect. Was it his maternal or paternal grandfather?

2   State the possible blood groups of the offspring of
   a   a mother of group AB and a father of group O
   b   both parents of group AB.
       In what proportions might these blood groups be expected to appear among the offspring?

3   A woman whose blood group is O,Rh− marries a group AB,Rh+ man. What are the possible blood groups of their children? Explain your reasoning.

4   Describe the appearance (in terms of banding) of *Cepaea* snails having the respective genotypes: AaBB, aaBB and aabb.

Figure 28.34   Complementary gene action in the sweet pea

Two different white-flowered varieties of sweet pea (*Lathyrus odoratus*) were crossed. The F₁ offspring had purple flowers. Selfing of the progeny disclosed that this was a case of complementary gene action. The genes R and C are on different chromosomes.

parental phenotypes:	white flower	white flower
parental genotypes:	CCpp	ccPP
gametes:	(Cp) ×	(cP)
offspring (1) genotype:	CcPp	
offspring (2) phenotype:	purple flower ●	
gametes:	(CP) (Cp) (cP) (cp)	(CP) (Cp) (cP) (cp)

	¼CP	¼Cp	¼cP	¼cp
¼CP	1/16 CCPP ●	1/16 CCPp ●	1/16 CcPP ●	1/16 CcPp ●
¼Cp	1/16 CCPp ●	1/16 CCpp ○	1/16 CcPp ●	1/16 Ccpp ○
¼cP	1/16 CcPP ●	1/16 CcPp ●	1/16 ccPP ○	1/16 ccPp ○
¼cp	1/16 CcPp ●	1/16 Ccpp ○	1/16 ccPp ○	1/16 ccpp ○

offspring (2) phenotypes:	purple flowers	white flowers
phenotype ratio:	9	: 7
offspring (2) genotypes:	CCPP	CCpp
	CcPP	Ccpp
	CCPp	ccPP
	CcPp	ccPp
		ccpp

Cepaea nemoralis shells may be brown, pink or yellow, and may possess up to five dark bands.

All possible combinations of banding (and coloration) occur, although unbanded, mid-banded and five-banded are most commonly found.

The thrush (*Turdus philomelos*) selectively predates local populations, apparently according to how conspicuous are the shells among the background vegetation.

Figure 28.35   Banding in the shell of *Cepaea nemoralis*, a common snail of woods hedges and grassland: the species is one of the most variable in colour and banding, and these characteristics are genetically controlled

■ Simple gene interaction

The shape of the comb of the farmyard fowl is an example of a single character determined by two different genes, showing a rather unusual gene inter-action. The products of these interactions are four possible different phenotypes, known as pea, rose, walnut and single (Figure 28.36).

The genes involved, represented by the letters P and R, interact to form phenotypes as follows:

▶ pea   PPrr or Pprr
▶ rose   ppRR or ppRr
▶ walnut PPRR, PpRR, PPRr or PpRr
▶ single   pprr.

In other words, whilst pea and rose combs are the product of a dominant allele for the character in question combined with a recessive allele for the other, walnut comb is the product of interaction of these dominant alleles, and single comb is produced in the absence of any dominant alleles.

■ Pleiotropy

Pleiotropy is the situation in which one gene affects two or more unrelated characteristics. It arises, for example, when an enzyme for which the gene codes affects more than one phenotypic characteristic. For example, a mutation (facing page) in a certain human gene for chloride ion secretion in epithelial cells gives rise to the phenotype known as 'cystic fibrosis'. Cystic fibrosis patients have problems with breathing and digestion (among others). They produce rather viscous mucus, particularly in the gut, pancreas and lungs (Figure 28.37). This mucus is likely to dry up and block ducts, so the glands the mucus comes from (such as the exocrine glands of the pancreas, p. 284) may be destroyed. Typically, digestive enzymes have to taken with meals. Males are infertile.

Figure 28.37 Physiotherapy for a young cystic fibrosis patient: regular treatment is required to move copious viscous mucus from the lungs, to help breathing

Question

Using the standard notation, show the genotypes of the P, F₁ and F₂ generations responsible for the inheritance of the fowl comb, as illustrated in Figure 28.36.

28.12 ENVIRONMENTAL EFFECTS ON THE PHENOTYPE

Variations in the phenotype of offspring may also arise through the interaction of genotype and the environment. Here the influence of particular genes is determined by factors external to the organism. (In effect, 'nurture' has taken effect over 'nature', p. 397.) An extreme example of environmental influence is that of an organism with genes for tallness that, because of deprivation of sufficient resources, fails to grow to full size.

A different example of an environmental effect on gene expression occurs in the life cycle of the honey bee. Three phenotypes are present in the honey bee colony (Figure 23.37, p. 524): the workers, the drones and a queen (Figure 28.38). Males (the drones) arise from unfertilised eggs, but the workers and queens (both female) from fertilised eggs. The differences between the two last-named phenotypes arise from the diet the larvae receive between moults. For the first three days after hatching all female larvae receive 'royal jelly', a secretion from glands on the workers who rear the young. A queen is kept on this diet until she pupates, but the workers (who make up the majority of the colony) are switched to a diet of pollen and nectar after the first three days. So this dietary difference is responsible for these two sharply contrasting phenotypes.

Figure 28.36 The fowl comb as an inherited character

when fowl with pea comb and rose comb are crossed,

pea comb   and   rose comb
×

the progeny (F₁) have walnut combs   walnut comb

when sibling matings of fowl with walnut comb are made, the progeny (F₁) are of four types:

walnut comb   pea comb   rose comb   single comb
in the ratio   9   :   3   :   3   :   1

Figure 28.38 The honey bee community: a colony of three phenotypes but only two genotypes

The honey bee (*Apis mellifera*) colony is an organised community of 20 to 80 thousand individuals. All members of the colony are offspring of the queen.

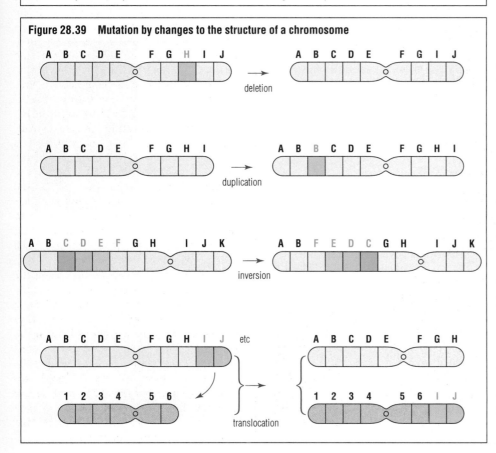

the queen lays eggs

most eggs laid are fertilised   occasionally, unfertilised eggs are laid

fed only on 'royal jelly' (protein-rich food)

briefly fed on 'royal jelly' followed by pollen + nectar

briefly fed on 'royal jelly' followed by pollen + nectar

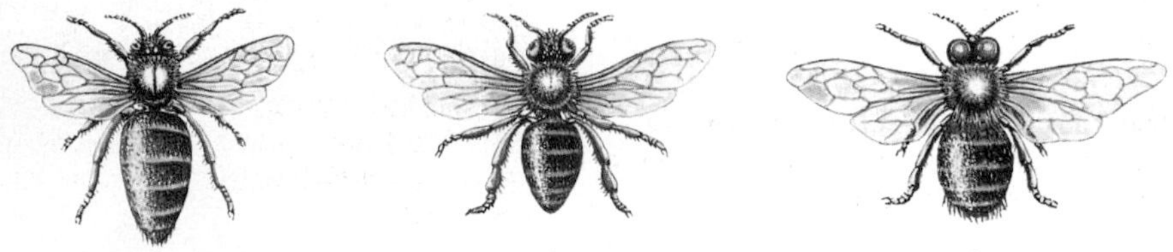

The queen lives for 2–5 years. She lays about 1500 eggs per day, each in its own wax cell. Subsequently the eggs are tended by worker bees.
Normally the queen mates only once. Sperms are sorted in her abdomen, and used to fertilise eggs.
Some eggs are laid unfertilised.

Workers are female bees whose reproductive organs do not function. Workers make up 99.9% of the colony. They live for about 6 weeks.
As they mature after emerging from the pupa they go through a sequence of duties often (but not necessarily) in this order:

► **nurse duties:** tending larvae with food, cleaning 'cells', secreting royal jelly, comb building
► **exploring and foraging:** surveying around the hive, guard duty at entrance, foraging for water and food.

Drones are fertile males that live for about 5 weeks.
They do not work in the hive.
Their sole function is to fertilise the queen.
They feed on reserves of pollen and nectar.

The life history of the colony and bee communication are discussed in Figure 23.32, p. 524.

Figure 28.39 Mutation by changes to the structure of a chromosome

A B C D E F G H I J → A B C D E F G I J deletion

A B C D E F G H I → A B B C D E F G H I duplication

A B C D E F G H I J K → A B F E D C G H I J K inversion

A B C D E F G H I J etc → A B C D E F G H and 1 2 3 4 5 6 I J translocation

28.13 MUTATIONS

The variations between parents and offspring discussed so far have resulted from independent assortment and crossing over in chromosomes, producing new combinations of alleles at fertilisation. **Mutations**, on the other hand, cause variation in an entirely different manner. A mutation is a change in structure (Figure 28.39), arrangement or quantity of the DNA of the chromosomes.

Mutations occur randomly and spontaneously. A mutation is detected as a marked difference in characteritics of an organism that arises from normal parents. This is caused by a change in the structure or number of chromosomes (a **chromosome mutation**) or in the structure of a gene (**gene mutation**), and certain of these mutations can be transmitted to offspring. An organism with characteristics changed by a mutation is called a **mutant**.

Somatic mutations occur in cells in the body (soma) other than the germ cells. Mutations of this kind are not transmitted to the next generation, but they may be significant in the life of an organism if they contribute to the malfunctioning of the body. Somatic mutations tend to accumulate with age.

Germ cell mutations, on the other hand, occur in the reproductive, germinal tissue (such as ovaries, testis, anthers or embryo sac) and result in changes to the genome of the gamete. Germ cell mutations may be passed on to the offspring.

■ Chromosome mutations

Chromosome mutations are of two types: euploidy and aneuploidy.

Euploidy

Euploidy involves an alteration in the number of whole sets of chromosomes. Remember, the number of chromosomes in the gametes is the haploid number, and at fertilisation the diploid number is established. Increases in numbers of sets of chromosomes above the diploid number are common in plants, but very rare in animals. When an organism has three or more sets of chromosomes it is known as a **polyploid**. More than 30% of flowering plants, including many cultivated crop plants and garden flowers, are polyploids or have polyploid forms. Polyploids have larger nuclei than their diploid equivalents. The cells are larger, and so too is the whole organism.

If the polyploid is based upon sets of chromosomes from the same species, it is an **autopolyploid**. Autopolyploids are most likely to have arisen by spontaneous doubling following the failure of a spindle to form or to function correctly at meiosis.

In **allopolyploids**, on the other hand, the additional set or sets of chromosomes have come from more than one species. The sets of chromosomes are then not completely homologous, and the plant is usually sterile. But if mitosis occurs in the polyploid cells, both sets of chromosomes double. Subsequently, pairing of chromosomes in meiosis is possible. A fascinating example of natural allopolyploidy occurred in the origins of modern bread wheat, during the Neolithic Revolution (Figure 28.40).

Aneuploidy

Changes to the chromosomes may affect only part of the set. **Aneuploidy** may arise when part of a chromosome is deleted, or duplicated, or broken and inverted in rejoining, or broken off and added to a different chromosome (Figure 28.39). Occasionally a chromosome may be duplicated, changing the total complement of chromosomes. Aneuploidy in humans may have serious consequences.

Down's syndrome

People with Down's syndrome (Figure 28.41) have 47 chromosomes in the nucleus of each cell: an extra chromosome 21 is present. They have some degree of learning difficulty, retarded physical development and a distinctive body and facial structure. The exact cause of this duplication of a single chromosome is not known, but the incidence of Down's syndrome is distinctly higher in children whose mothers are over 40 years old.

Aneuploids involving the sex chromosomes

There are three important (but fortunately very rare) aneuploid conditions of this type:

▶ XO (one X, but no Y chromosome), known as Turner's syndrome; an affected individual is female, but sexually underdeveloped and sterile
▶ XXY and XXXY, known as Klinefelter's syndrome; an affected individual is male, but is sterile, with some marked feminine features
▶ XYY, giving rise to an apparently normal fertile male, who is apt to be socially deviant, and may have a propensity for violence.

Figure 28.41 A young person with Down's syndrome: improved provision and education opportunities have enhanced the prognosis for people with Down's syndrome

---

Figure 28.40 The origin of bread wheat

Three wild grasses (forms of wild wheat), all diploid species, have each contributed a third of the genome of our modern wheat. Two natural hybridisations were involved. These occurred accidentally, about 9000 to 7000 years ago, at times when human civilisation was able to exploit the large inflorescences. These did not shatter when harvested, and required threshing; moreover, the heavy fruits did not blow away in the wind.

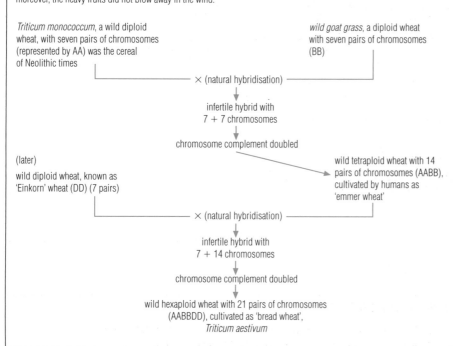

Triticum monococcum, a wild diploid wheat, with seven pairs of chromosomes (represented by AA) was the cereal of Neolithic times

wild goat grass, a diploid wheat with seven pairs of chromosomes (BB)

× (natural hybridisation)

infertile hybrid with 7 + 7 chromosomes

chromosome complement doubled

(later)
wild diploid wheat, known as 'Einkorn' wheat (DD) (7 pairs)

wild tetraploid wheat with 14 pairs of chromosomes (AABB), cultivated by humans as 'emmer wheat'

× (natural hybridisation)

infertile hybrid with 7 + 14 chromosomes

chromosome complement doubled

wild hexaploid wheat with 21 pairs of chromosomes (AABBDD), cultivated as 'bread wheat', *Triticum aestivum*

***T. monococcum* (left) and wild goat grass (right)**

T. aestivum

■ Gene mutations

A gene is a sequence of nucleotide pairs that codes for a sequence of amino acids (p. 206). Changes can occur to any base pair of any gene, anywhere along a chromosome. The coding length of a gene typically consists of between about 1200 and 1500 base pairs.

A change in one or more base pairs may occur spontaneously, as when a replicational error arises. It seems that the enzyme machinery occasionally inserts a base other than that coded for by the template DNA, for example. Alternatively, gene mutations may be induced by particular conditions affecting cells. The environmental agents (**mutagens**) include ionising radiation in the form of X-rays, cosmic rays, and α-, β- or γ-radiation from radioactive isotopes, any of which can break up the DNA. The rate of mutation induced by these mutagens is directly proportional to the dose received. Their effects are cumulative, so small doses over a prolonged period may be as harmful as a single, larger dose. Only replicational errors occurring at meiosis can affect gametes and offspring.

Non-ionising mutagens include ultraviolet light and numerous chemicals. These act by modifying the chemistry of the base pairs. The ways in which base composition of DNA may be altered and the consequences for coding of amino acids residues within the coded protein are shown in Figure 28.42.

The cause of sickle cell anaemia

One example of a gene mutation is that causing sickle cell anaemia. Haemoglobin (Figure 17.28, p. 358), the protein of oxygen transport, consists of four polypeptide chains, two α-chains (of 141 amino acid residues each), and two β-chains (of 146 amino acid residues). These are coded for by two genes on separate chromosomes. The difference between normal haemoglobin and sickle-cell haemoglobin lies in a single base pair in the gene controlling the β-chain (Figure 28.42). Sickle-cell haemoglobin (Hb^S) crystallises at low oxygen concentration, and it carries less oxygen.

In people heterozygous for Hb^S, the resulting Hb^S haemoglobin carried in red cells is less than 50% of the total. The patient is said to have the sickle cell **trait**, and is only mildly anaemic. In parts of the world where malaria (p. 535) is endemic, this condition is actually advantageous since *Plasmodium* does not live in red cells containing any Hb^S. Being homozygous for Hb^S is far more of a problem; heart and kidney failure are quite common in people so affected.

■ The genetic consequences of mutations

The effects of mutations are only occasionally beneficial. (After all, a mutation is a random change in the structure of the protein that a gene codes for.) In fact, many mutations are lethal, or at least disadvantageous. For example, most chromosome abnormalities are of this catagory. chromosome abnormality is the single most common cause of natural abortion of a human fetus in early pregnancy.

From mutations that are beneficial and may increase variation there are potential advantages. Most mutations are recessive to the normal allele, however (since in a heterozygote a normal allele of the gene is also present). A recessive mutant allele awaits replication in the gene pool over many generations before chance brings recessive alleles together, so they are expressed.

Questions

1 Suggest reasons why a hybrid may be sterile in the diploid condition but fertile when tetraploid (i.e. with 4 sets of chromosomes).
2 Why are microorganisms such as bacteria frequently used in the study of mutation?

Figure 28.42   Gene mutation

Outcomes of gene mutations that may lead to genetic change

1 **Mis-sense mutations** are caused by a base-pair substitution in a gene which results in a single amino acid residue being changed for another in a polypeptide chain; for example, haemoglobin (part of the β-chain) has the following amino acid residues:

Val—His—Leu—Thr—Pro—Glu—Glu— . . .

A mis-sense mutation substitutes valine for glutamine and causes haemoglobin of sickle-cell anaemia:

Val—His—Leu—Thr—Pro—Val—Glu— . . .

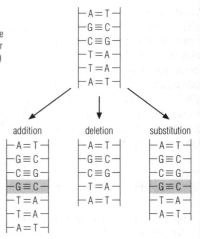

types of gene mutation

2 **Non-sense mutations** are caused by a base-pair substitution, addition or deletion that changes an amino-acid-specifying codon into a chain-terminating codon. A polypeptide chain is prematurely terminated.

3 **Frameshift mutations** are caused by a base-pair addition or deletion (other than multiples of three) that causes the whole gene to be misread. The change(s) produce an entirely new sequence of codons; for example, part of the normal code in mRNA reads:

| AAU | AUG | AUC | UUU | GUG | GGU | ARG . . . |
| Asn— | Met— | Ile— | Phe— | Val— | Gly— | Arg . . . |

Insertion of a base (U) causes frameshift:

| AAU | AUG | UAU | CUU | UGU | GGG | UAR . . . |
| Asn— | Met— | Tyr— | Leu— | Cys— | Gly— | stop. |

The change in one base has altered the sequence of amino acid residues, and terminated the peptide prematurely.

FURTHER READING/WEB SITE

C J Clegg (1999) *Illustrated Advanced Biology: Genetics and Evolution*. John Murray
Browse: **www.oml.gov.TechResource/Human_Genome/genetics.html**

29 Applications of genetics

- Genetic modification of useful plants and animals has occurred by selective breeding since early in human history, and continues. Today, genetic modification is also carried out by recombinant DNA technology (genetic engineering), and many microorganisms are used as well.
- DNA of chromosomes may be 'cut' to isolate individual genes (using restriction enzymes), which may then be attached to the DNA of another organism, using another enzyme (ligase). Gene vectors used to transfer DNA and introduce genes into other organisms include plasmids found in bacteria (and yeast), and certain viruses.
- Prokaryotes have been engineered to carry genes from human or from other sources, and then induced to secrete a required gene product, e.g. human insulin (for diabetics), human growth hormone, and interferon (for use against viruses).

- Genes have been transferred into animal and plant cells by direct injection, and into plant cells using the gall-inducing bacterium *Agrobacterium*, a natural gene-transferring organism, or by fusion of plant protoplasts. Genetically modified plants and animals are produced for agricultural and medical purposes, for example. Many future developments are planned.
- Gene therapy is the use of genetic engineering to overcome diseases due to defective genes, e.g. cystic fibrosis. Medicine also looks to the techniques of genetic engineering to provide a possible source of purpose-grown organs for transplants, in the future.
- Genetic engineering achievements and possible future applications raise enormously important economic, environmental and ethical issues, which require informed decisions by electorates and politicians.

29.1 INTRODUCTION

Early in human history, people unwittingly began the process of manipulating the genetic constitutions of organisms by their selection of plants and animals in the new activity of agriculture (Neolithic Revolution, p. 77). The breeding of domesticated species of plants and animals involves artificial selection and (more rarely) natural hybridisation between related species and the doubling of whole sets of chromosomes to give polyploids (p. 625).

In artificial selection, genetic modification is brought about when organisms with favourable characteristics that are genetically controlled are permitted to breed, and the progeny they produce are similarly sorted. Those with favourable characteristics are used as further breeding stock, and so on. Meanwhile, progeny without favoured characteristics are excluded from breeding. The result is a shift in the genotypes of new individuals away from those of the original population, and consequently a change in the gene pool. A sample of the gene pool forms the **genomes** (gene sets of individuals) of the next generation. The plants and animals of our farms and gardens and homes are all genetically modified by these processes. In section 29.6 (p. 644), current practice in artificial selection of plants and animals for agriculture and horticulture is reviewed.

Today a new type of genetic manipulation is in use, known as **genetic engineering** or **recombinant DNA technology**. The extremely long double strands of DNA found in the nucleus (p. 204) are functionally divided into shorter lengths, the **genes**. In recombinant DNA technology, genes from one organism are introduced into the genome of an unrelated organism. One outcome has been the 'engineering' of new varieties of organism, mostly microorganisms, which are used in the manufacture of useful chemicals by biotechnological processes. The work also extends our knowledge of the functioning gene, another aspect of the pure research we call molecular biology.

The economic, environmental and ethical issues of genetic engineering are reviewed on p. 634.

REMINDER

Gene: the basic unit of inheritance; a length of DNA on a chromosome.

Genome: the complement of genes of an organism (or of a cell).

Genotype: the genetic constitution of an organism; genes acting together with environmental factors produce the organism's appearance (phenotype).

29.2 RECOMBINANT DNA TECHNOLOGY

■ Role of restriction enzymes

Recombinant DNA technology became possible through the use of bacterial enzymes known as **restriction enzymes** (or, strictly, as **restriction endonucleases**, Table 29.1), which cleave double-stranded DNA molecules. It seems that virtually all species of bacteria synthesise restriction enzymes, retaining them in the cytoplasm. Their presence enables the destruction of alien DNA (such as that entering the bacterium as part of a virus parasite) before the protein synthesis machinery of the cell is taken over. Restriction enzymes were first discovered in the bacterium *Escherichia coli* (p. 552) by an American, W Arber, and subsequently confirmed by Hamilton Smith,

Table 29.1 The genetic engineer's tool kit of enzymes

Genetic engineering makes use of purified enzymes that enable DNA and RNA to be manipulated in precisely known ways.

Enzyme	Natural sources	Application in recombinant DNA technology
restriction enzymes (restriction endonucleases)	cytoplasm of bacteria – combats viral infection by breaking up viral DNA	breaks DNA molecules into shorter segments, at specific nucleotide sequences
DNA ligase	with nucleic acid in nuclei of cells of all organisms	joins together DNA molecules during replication of DNA
polymerase	with nucleic acid in nuclei of cells of all organisms	synthesises nucleic acid strands from nucleotides, guided by a template strand of nucleic acid
reverse transcriptase	found in particular types of RNA viruses called retroviruses	synthesises a DNA strand complementary to an existing RNA strand

who worked with *Haemophilus influenzae* at Johns Hopkins University in the early 1970s.

These enzymes were so named because they effectively restrict the range of host cells that a virus can parasitise successfully. Many hundreds of different restriction enzymes have been isolated. By convention they are named after the species and strain of bacterium in which they naturally occur. For example, restriction enzyme *Eco*RI is obtained from *Escherichia coli* (known as *E. coli*, for short), strain RY13.

■ 'Blunt' and 'sticky' ends

Restriction enzymes cleave DNA at specific sites. The reason is that each restriction enzyme 'recognises' a particular base sequence in a DNA double strand (a property of the active site of the enzyme), and cuts the strand at this sequence and nowhere else. Most of the sequences recognised by these enzymes are four, six or eight bases long.

Some restriction enzymes make a simple cut across both strands at a single point (as scissors do!), leaving what are called 'blunt ends'. With others the cut in the two strands is staggered, leaving short, single-stranded DNA overhanging at the cut ends. These ends are called 'sticky ends' because subsequent base pairing can stick cleaved ends together. Restriction enzymes are extracted from cultures of bacteria and purified before being used in genetic engineering experiments (Figure 29.1).

■ Ligase: joining complementary sticky ends

Any two strands of DNA that have complementary ends (that is, ends that show pairing of complementary bases) can be joined together. The joining process, called **annealing**, uses the enzyme ligase, which occurs in cells where it is part of the enzymic machinery available to repair the nucleic acid molecules damaged during, for example, transcription. Ligase re-forms covalent phosphodiester bonds between the sugar and phosphate groups of closely adjacent DNA backbones, under appropriate conditions of temperature and pH.

Complementary sticky ends arise in genetic experiments in two ways. Firstly, they are produced when two strands of DNA are cut by the same restriction enzyme. Secondly, certain restriction enzymes with different recognition sequences produce identical sticky ends. Examples include the three restriction enzymes

► *Bam*HI, which recognises GGATCC
► *Bgl*II, which recognises AGATCT
► *San*3A, which recognises GATC

all of which produce sticky ends –GATC. So DNA strand fragments produced by these three enzymes can be joined to each other, with the aid of ligase (Figure 29.2).

Figure 29.2   The joining of complementary sticky ends by ligase

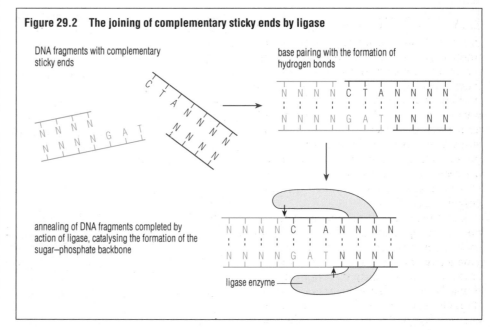

DNA fragments with complementary sticky ends

base pairing with the formation of hydrogen bonds

annealing of DNA fragments completed by action of ligase, catalysing the formation of the sugar–phosphate backbone

ligase enzyme

Figure 29.1   DNA extraction, and cutting of DNA with restriction enzyme

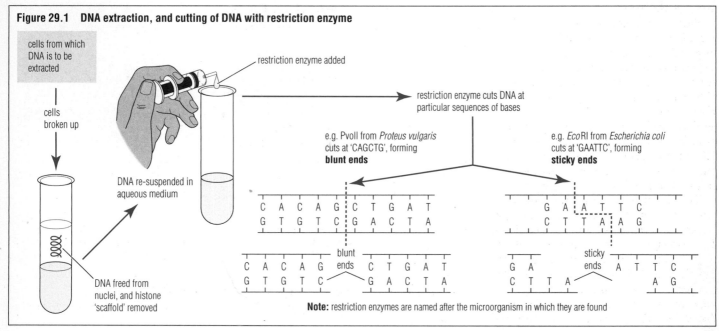

cells from which DNA is to be extracted

cells broken up

DNA re-suspended in aqueous medium

DNA freed from nuclei, and histone 'scaffold' removed

restriction enzyme added

restriction enzyme cuts DNA at particular sequences of bases

e.g. Pvoll from *Proteus vulgaris* cuts at 'CAGCTG', forming **blunt ends**

e.g. *Eco*RI from *Escherichia coli* cuts at 'GAATTC', forming **sticky ends**

Note: restriction enzymes are named after the microorganism in which they are found

■ Plasmids as vectors

Genes are most commonly cloned in bacteria. The structure of a typical bacterium is shown in Figure 25.5 (p. 551). The cytoplasm of bacteria contains numerous **plasmids**; that is, small rings of double-stranded DNA. Plasmids are additional to the main circular chromosome of the bacterium, and they replicate independently of the chromosome. Bacterial plasmids are popular **vectors** by which additional DNA is introduced into host cells (Figure 29.3).

The importance to humans of plasmids as vectors is that they can be extracted from bacterial cells, modified by recombinant DNA techniques, and returned to the bacterium. Back in the host the introduced gene may be multiplied by replication or expressed, and the gene product harvested. The rapidity of asexual reproduction in bacteria makes these organisms ideal for the quick production of large quantities of genes or their products.

The **plasmids are extracted** from bacteria by breaking up the cells and separating the contents by centrifugation. Once isolated, the plasmid rings are 'cut open' enzymically, using a particular restriction enzyme. The same restriction enzyme is used to cut the short length of DNA (a gene) to be inserted and replicated. As a consequence, the opened-up plasmids and the gene for cloning will have complementary base pairs exposed (sticky ends). Ligase is used to anneal the additional gene into the repaired plasmids.

These plasmids are returned to a culture of the bacteria, and many are **taken up into the bacterial cells**. The take-up process is effected by 'zapping' the cells with a brief electrical shock, which causes short-lived holes in the cell membrane that may let the plasmids in (a technique known more formally as electroporation). Alternatively, the bacteria may be treated with calcium chloride solution. The basic steps to gene cloning are summarised in Figure 29.3.

Questions

1 **a** What do we mean by the 'genome'?
  b Why do we say that DNA is 'double-stranded'?

2 What property of DNA is exploited in joining sticky ends?

3 What features of the inheritance/cell control machinery of prokaryotes (such as bacteria) make these organisms more appropriate for recombinant DNA technology than eukaryotes?

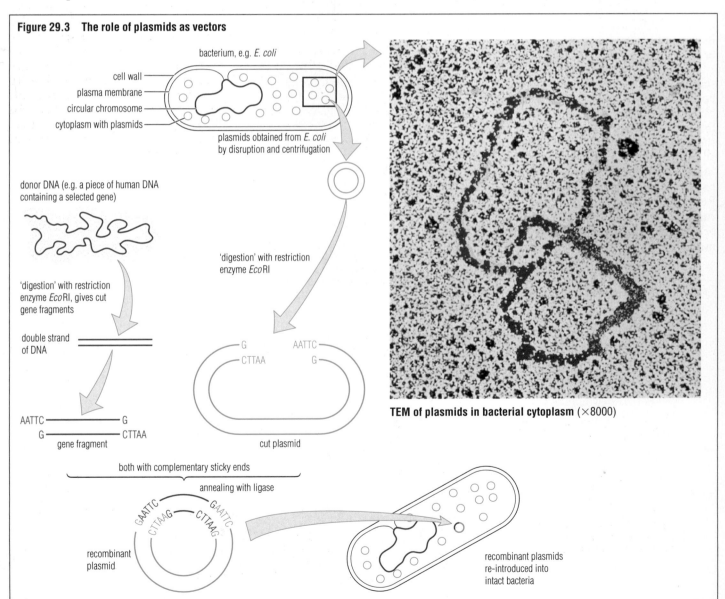

Figure 29.3   The role of plasmids as vectors

bacterium, e.g. *E. coli*

cell wall
plasma membrane
circular chromosome
cytoplasm with plasmids

plasmids obtained from *E. coli* by disruption and centrifugation

donor DNA (e.g. a piece of human DNA containing a selected gene)

'digestion' with restriction enzyme *Eco*RI, gives cut gene fragments

double strand of DNA

'digestion' with restriction enzyme *Eco*RI

G   AATTC
CTTAA   G

cut plasmid

AATTC ———— G
G ———— CTTAA
gene fragment

both with complementary sticky ends

annealing with ligase

GAATTC / CTTAAG   GAATTC / CTTAAG

recombinant plasmid

recombinant plasmids re-introduced into intact bacteria

TEM of plasmids in bacterial cytoplasm (×8000)

■ Selection and amplification via R-plasmids

The technique of using plasmids as vectors for gene transfer appears to be straightforward. However, there are three difficulties to be overcome.

▶ A restriction enzyme produces thousands of DNA fragments, all of different sizes, from the genome under investigation. Very few of these will contain the required gene. As a first step, the fragments have to be separated on size. This is done by 'sieving out' the fragments of required length, using electrophoresis (Figure 8.8, p. 175). (See also, gene probes, overleaf.)

▶ Only some of the plasmids will pick up recombinant DNA fragments when these are made available. Having been cut open by the restriction enzyme, a proportion of the plasmids simply close up again, unaltered, in the presence of ligase.

▶ Only some of the bacteria will take up the altered plasmids (containing the gene) at the end of the procedure.

Consequently there must be a selection step to isolate the relatively few bacteria that have acquired recombinant DNA. This is done by using plasmids that carry genes for resistance to certain antibiotics, called **R-plasmids**. The experiment is conducted so that only bacteria containing these plasmids, with recombinant DNA, are able to grow, due to the presence of antibiotics in the growth media. The procedure is carried out as described below (Figure 29.4).

1 R-plasmids are prepared containing a gene for ampicillin resistance and a gene for tetracycline resistance. Certain bacteria (for example, strains of *E. coli*) containing these plasmids can grow on a nutrient medium containing the antibiotics.

2 Next, the restriction enzyme *Bam*HI is used, which opens the plasmids. The 'cut' occurs in the middle of the tetracycline gene, thereby inactivating it. The same restriction enzyme is used to isolate the gene fragment, so both plasmids and fragments have complementary sticky ends and may join up and be annealed by ligase. Bacteria are then induced to take up these plasmids.

3 These modified bacteria are grown at a dilution that produces separate colonies for each bacterium, on nutrient agar containing the antibiotic ampicillin (plate A). Only bacteria that have taken up R-plasmids (with the gene for resistance to ampicillin) can survive on this plate.

4 Imprints (living cells) of these colonies are transferred to a second plate of nutrient agar, this time containing the antibiotic tetracycline (plate B). This process is called replica plating. On this second plate the colonies grown from a bacterium containing recombinant DNA inserted into the tetracycline gene of the plasmids cannot grow, of course, because the tetracycline-resistance gene has been inactivated by the installation of the recombinant DNA. So, by comparing the distribution of colonies on plates A and B, the few colonies containing recombinant DNA are located on plate A. Thus the bacteria with recombinant DNA have been selected.

Bacteria from these colonies are then cultured. All other bacteria from the experiment are disposed of.

Commonly, the selected bacteria then undergo a plasmid amplification phase (in which they are cultured with an inhibitor of protein synthesis called chloramphenicol). The effect is to greatly increase (by replication) the numbers of plasmids present in each bacterium (typically from 15 or so to about 3000). As a consequence, the amount of gene product the bacterial culture will yield is vastly increased. This is the amplification phase that follows 'selection'.

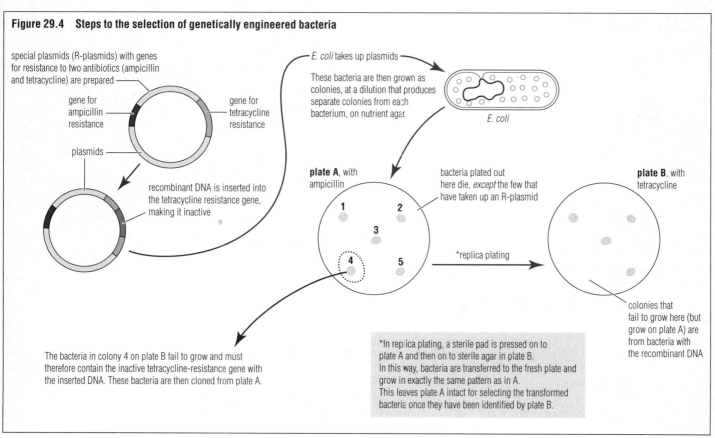

Figure 29.4   Steps to the selection of genetically engineered bacteria

special plasmids (R-plasmids) with genes for resistance to two antibiotics (ampicillin and tetracycline) are prepared

gene for ampicillin resistance

gene for tetracycline resistance

plasmids

recombinant DNA is inserted into the tetracycline resistance gene, making it inactive

E. coli takes up plasmids

These bacteria are then grown as colonies, at a dilution that produces separate colonies from each bacterium, on nutrient agar.

E. coli

plate A, with ampicillin

bacteria plated out here die, *except* the few that have taken up an R-plasmid

1   2
3
4   5

*replica plating

plate B, with tetracycline

colonies that fail to grow here (but grow on plate A) are from bacteria with the recombinant DNA

The bacteria in colony 4 on plate B fail to grow and must therefore contain the inactive tetracycline-resistance gene with the inserted DNA. These bacteria are then cloned from plate A.

*In replica plating, a sterile pad is pressed on to plate A and then on to sterile agar in plate B. In this way, bacteria are transferred to the fresh plate and grow in exactly the same pattern as in A. This leaves plate A intact for selecting the transformed bacteria once they have been identified by plate B.

■ Using a gene probe

A gene probe is a short length of DNA in which

▶ the sequence of nucleotides (and therefore the base sequence) is known

▶ the sugar–phosphate 'backbone' is radioactively labelled to aid detection of the probe; for example, by autoradiography (p. 339).

Gene probes are prepared in laboratories and used widely; for example, in DNA fingerprinting (p. 639), and in the human genome project (p. 639).

If the sequence of bases in the recombinant gene is known (or if the amino acid sequence of the protein for which that gene codes is known), then it is possible to make a **gene probe** ('marker') from the complementary bases, and use it to **locate colonies of bacteria containing the recombinant gene** (Figure 29.5). The process involves taking a replica of colonies growing on agar, breaking up the cells (lysis) and disposing of all contents except the DNA, which is treated to break the hydrogen bonds and expose its two strands. A radioactively labelled gene probe complementary to the base sequence of the recombinant gene is added and hybridises, and the recombinant DNA can be detected by autoradiography. When the autoradiograph is lined up with the original plate, the presence of the colonies that contain recombinant DNA is detected. These colonies can be subcultured, and all other colonies from the experiment discarded.

■ Bacteriophage as vector

The structure and life cycle of a bacteriophage (a virus that parasitises a bacterium) is shown in Figure 25.2, p. 549. The nucleic acid from the phage may be added to the circular chromosome and replicated with the bacterial DNA, when the bacterium divides. Genetic engineers have exploited the phage life cycle to transfer genes to bacteria (Figure 29.6). This mechanism is most useful in attempts to clone larger genes (longer lengths of double-stranded DNA).

■ Gene expression in genetic engineering

Genes are 'expressed' when their base sequence is being transcribed into mRNA for protein synthesis. Many genes do so only when activated ('switched on'). In prokaryotes such genes are arranged with a regulatory gene and a length of DNA

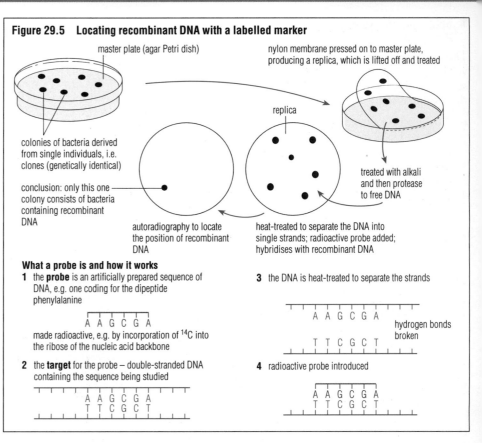

Figure 29.5 Locating recombinant DNA with a labelled marker

master plate (agar Petri dish)

colonies of bacteria derived from single individuals, i.e. clones (genetically identical)

conclusion: only this one colony consists of bacteria containing recombinant DNA

nylon membrane pressed on to master plate, producing a replica, which is lifted off and treated

replica

treated with alkali and then protease to free DNA

autoradiography to locate the position of recombinant DNA

heat-treated to separate the DNA into single strands; radioactive probe added; hybridises with recombinant DNA

What a probe is and how it works

1 the **probe** is an artificially prepared sequence of DNA, e.g. one coding for the dipeptide phenylalanine

A A G C G A

made radioactive, e.g. by incorporation of ^{14}C into the ribose of the nucleic acid backbone

2 the **target** for the probe – double-stranded DNA containing the sequence being studied

A A G C G A
T T C G C T

3 the DNA is heat-treated to separate the strands

A A G C G A   hydrogen bonds
                     broken
T T C G C T

4 radioactive probe introduced

A A G C G A
T T C G C T

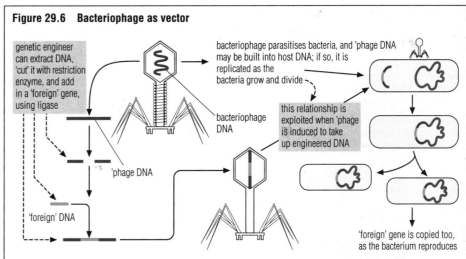

Figure 29.6 Bacteriophage as vector

genetic engineer can extract DNA, 'cut' it with restriction enzyme, and add in a 'foreign' gene, using ligase

bacteriophage parasitises bacteria, and 'phage DNA may be built into host DNA; if so, it is replicated as the bacteria grow and divide

bacteriophage DNA

this relationship is exploited when 'phage is induced to take up engineered DNA

'phage DNA

'foreign' DNA

'foreign' gene is copied too, as the bacterium reproduces

called an operator site. One such 'operon' is associated with the gene for lactose absorption and metabolism (Figure 9.25, p. 211), and this is exploited in the manufacture of insulin by recombinant DNA technology.

When the gene for insulin production is introduced into a prokaryote such as *E. coli*, it is attached to the lactose 'operon' mechanism, so that insulin production can be triggered from outside the cell by the presence of lactose in the medium.

Control of gene expression in eukaryotes is more complex, and it is less well understood.

■ Using reverse transcriptase

Genes are translated into messenger RNA in the process of protein synthesis (p. 208), so messenger RNA can itself be the template for manufacture of copies of the gene from which it was derived. In some cases, mRNA can be obtained from tissue much more easily than copies of the original gene. Now it is possible to use mRNA to obtain the gene because of the discovery of the enzyme 'reverse transcriptase', which is taken from certain viruses called retroviruses (p. 550). This technique is used commercially in insulin manufacture, for example (Figure 29.7).

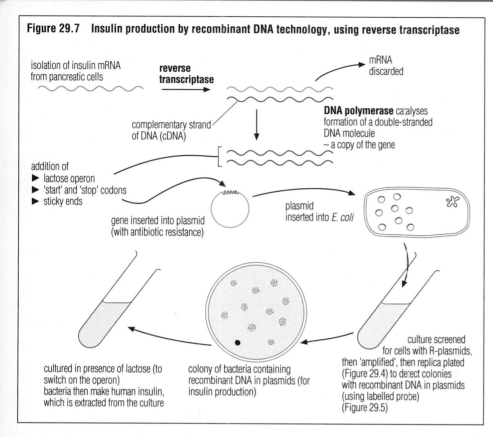

Figure 29.7 Insulin production by recombinant DNA technology, using reverse transcriptase

isolation of insulin mRNA from pancreatic cells

reverse transcriptase

mRNA discarded

complementary strand of DNA (cDNA)

DNA polymerase catalyses formation of a double-stranded DNA molecule – a copy of the gene

addition of
► lactose operon
► 'start' and 'stop' codons
► sticky ends

gene inserted into plasmid (with antibiotic resistance)

plasmid inserted into E. coli

culture screened for cells with R-plasmids, then 'amplified', then replica plated (Figure 29.4) to detect colonies with recombinant DNA in plasmids (using labelled probe) (Figure 29.5)

cultured in presence of lactose (to switch on the operon) bacteria then make human insulin, which is extracted from the culture

colony of bacteria containing recombinant DNA in plasmids (for insulin production)

The mRNA for insulin is obtained from pancreatic cells.

Using mRNA as the template, reverse transcriptase catalyses the synthesis of a single strand of DNA, complementary to the base sequence of the mRNA (this strand is called cDNA). Then the mRNA template is discarded. The enzyme DNA polymerase is introduced to catalyse the formation of a double-stranded DNA molecule, and so a copy of the gene is formed. The gene is then added to a vector plasmid and cloned for the manufacture of the gene product.

■ Polymerase chain reaction (PCR)

When a very small sample of DNA is all that is available for analysis or experiment, the polymerase chain reaction can be used to produce a large number of exact copies very quickly. So the DNA from a tiny sample of blood (left at the scene of a crime, for example) can be copied to produce a sample large enough to analyse (Figure 29.8).

PCR is an automated process, carried out by a machine, and having three steps:

► **strand separation** (breaking of hydrogen bonds) by heating to 93 °C
► **binding of primer** to one end of each strand (for which the base sequence here must be known, so that the primer can be synthesised to order), carried out at 55 °C to permit base-pairing and the formation of hydrogen bonds
► **synthesis of new strands**, starting from the primer molecules, carried out at 72 °C with a heat-stable polymerase, with excess nucleotides present. The polymerase is obtained from a bacterium found in hot springs.

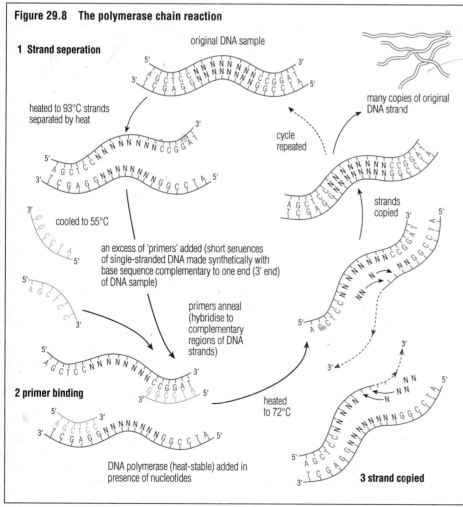

Figure 29.8 The polymerase chain reaction

1 Strand seperation

original DNA sample

heated to 93°C strands separated by heat

cooled to 55°C

an excess of 'primers' added (short seruences of single-stranded DNA made synthetically with base sequence complementary to one end (3' end) of DNA sample)

primers anneal (hybridise to complementary regions of DNA strands)

2 primer binding

DNA polymerase (heat-stable) added in presence of nucleotides

heated to 72°C

3 strand copied

cycle repeated

many copies of original DNA strand

strands copied

Questions

1 Automated machines have been devised that can be programmed to assemble DNA nucleotides into short lengths of DNA. Why is the sequence of amino acid residues of a particular protein equally useful to the machine programmer as knowledge of the base sequence of the gene coding for the protein?

2 What is the normal role of the enzyme reverse transcriptase in the replication of a retrovirus (p. 550)?

3 When locating a DNA sequence with a radioactive probe (Figure 29.5), why is it necessary to use a replica of the plate rather than the plate itself?

29.3 APPLICATIONS OF GENETIC ENGINEERING

The discovery of the necessary enzymes (Table 29.1, p. 628) and the development of the techniques of genetic engineering (which have all taken place since 1970) involved bacteria and viruses, rather than eukaryotes. Bacteria were the first genetically modified organisms (GMOs) produced. Bacteria are especially amenable to recombinant DNA technology for several reasons, as listed below.

▶ Their life cycles are short and they grow quickly under quite normal conditions.

▶ The process of transcription of DNA to mRNA is less complex in prokaryotes than in eukaryotes (Figure 29.9).

▶ The mechanism of gene expression in prokaryotes is largely understood, but that in eukaryotes is still almost unknown.

▶ Plasmids, the most useful vehicle for introducing recombinant DNA into cells, occur in bacteria but not in most eukaryote cells (with the exception of yeast cells, see opposite page).

▶ It has proved relatively simple to make bacteria 'competent' to receive engineered plasmids back into their cytoplasm (by 'zapping', or by treatment with calcium chloride), whereas the cell walls of plants present a barrier to entry.

Nowadays, the routine applications of recombinant DNA technology in horticulture, agriculture and medicine require molecular geneticists to 'engineer' the DNA of eukaryotes as well. The problems are receiving concerted attention. The current procedures for producing eukaryotic GMOs, together with some of their applications, are reviewed opposite, after successes with prokaryotes have been considered.

■ Prokaryotes as GMOs

Insulin manufacture

The daily need for insulin to treat insulin-dependent diabetes is enormous, for more than two million people across the world are afflicted with this disease. Insulin was for many years obtained from pancreases taken from cattle and pigs in slaughterhouses. But insulins from these species are subtly different from human insulin. As a consequence, some patients produce antibodies against the insulin they receive, resulting in the destruction of the hormone in the blood-

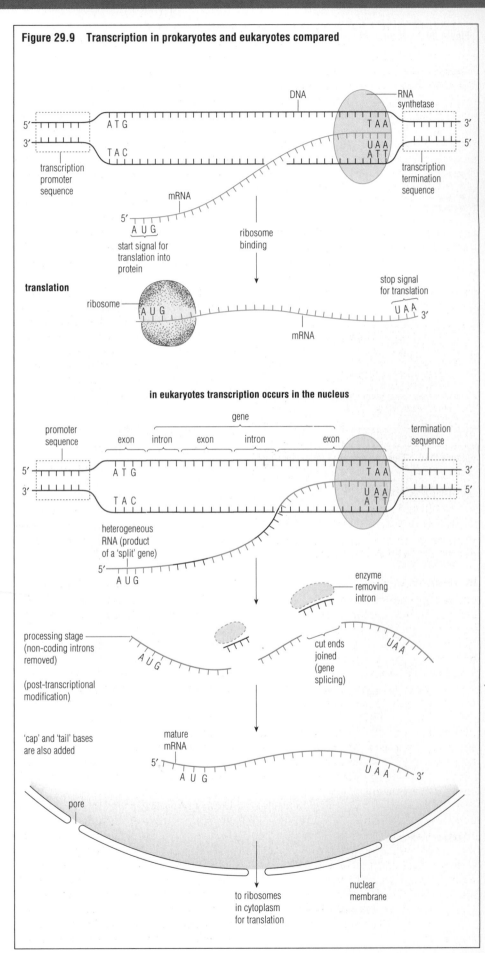

Figure 29.9 Transcription in prokaryotes and eukaryotes compared

stream before the body can use it. Two alternative sources of acceptable insulin exist at the moment.

Firstly, a biotechnological firm in Denmark has devised an economical process for the conversion of pig insulin into human insulin, using conventional chemical techniques. The available supply of pig insulin is insufficient to meet all demands, however.

Secondly, an American drug firm has devised a method of producing insulin by recombinant DNA technology (Figure 29.10). Insulin consists of two polypeptide chains (A and B), linked by disulphide bridges. The polypeptides are 'engineered' separately, and later brought together to make active insulin. The bacterium used is a strain of *E. coli* (Figure 29.7).

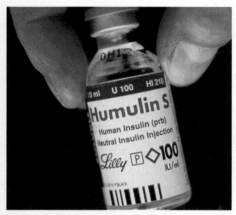

Figure 29.10 Human insulin produced through recombinant DNA technology, using *E. coli*

Human growth hormone

Human growth hormone, a protein, is secreted in the pituitary gland and regulates normal growth of our bodies. Abnormally low production of this hormone rarely occurs, but when it does, it leads to dwarfism.

This physical growth defect can be corrected with injections of human growth hormone (a similar molecule obtained from other mammals is not effective). Limited amounts of the hormone have been extracted from the pituitary glands of dead humans, but such extracts are not now used for fear of possible contamination with prions associated with CJD (p. 537). Today, this hormone (known as hGH) is produced by *E. coli* bacterial cultures, engineered to carry the human gene and secrete the human growth hormone into their growth medium. The technique of engineering *E. coli* to produce hGH is very similar to that for insulin production (Figure 29.7).

Bovine somatotrophin

Bovine somatotrophin (BST) is a growth hormone very similar in chemical structure to hGH. It is obtained from cows. This hormone is now available from genetically engineered bacteria, by a similar method to that shown in Figure 29.7. BST has been found to increase milk production in cows when injected in small doses at about two-week intervals throughout the lactation. It also increases muscle content (and carcass weight) of beef cows. Trials of this hormone have produced acceptable results in the USA, but use of BST in Europe has been banned. This is because there is a surplus of milk and beef production here already. There is concern for the health of cows treated with BST, in that they may be more susceptible to diseases such as mastitis (a disease of the udder). Europeans are also concerned that traces of this growth hormone, which may be taken in with food derived from BST-treated cattle, might be an unhealthy addition to their involuntary intake of additives.

■ Eukaryotes as GMOs

Genetic engineering with yeast

Yeast, unlike *E. coli*, has a discrete nucleus, which is bounded by a nuclear membrane and contains chromosomes (Figure 14.20, p. 298). Chromosomes undergo mitosis, prior to cell division. The bulk of the DNA of yeast is located in 17 chromosomes, rather than in the double-stranded ring of DNA found in a bacterial cell. Like bacteria, however, the cytoplasm of yeasts also contains DNA in the form of tiny plasmids. Yeasts can be grown in liquid culture, or on the surface of agar plates, in the same way as most bacteria are cultured. Consequently, yeasts are the most frequently used eukaryotes in gene cloning studies.

Recombinant DNA has been introduced into yeasts, using either bacterial plasmids or yeast's own plasmids as the vector. The yeast cell wall is not a barrier to entry of plasmids; yeast cells treated with lithium salts take plasmids into the cytoplasm. The human proteins that have been synthesised from yeast treated with recombinant DNA include an enzyme, **superoxide dismutase**, used to prevent tissue damage from free radicals in patients with osteoarthritis and in the post-operative care of heart patients. The gene for **a vaccine against hepatitis** has been engineered into yeast for commercial production. So, too, has the gene for 'hirudin', an anticoagulant produced by

leeches and used to aid their feeding mechanism, now used as a drug in the treatment of blood clotting disorders.

Genetic engineering with flowering plants

Members of the Papilionaceae (leguminous plants) form root nodules in which are housed bacteria of the genus *Rhizobium* (p. 334). Using a metalloprotein enzyme produced by the bacterium, these mutualistic organisms fix atmospheric nitrogen $(N_2 \rightarrow -NH_2)$, from which amino acids are then manufactured. These amino acids support growth of the host as well as of the bacterium.

Cereal crops are unable to form root nodules, and they do not develop a mutualistic relationship with *Rhizobium* (a bacterium that is common in soil). This facility in leguminous plants is genetically controlled.

The most important feat of genetic engineering in higher plants – one that is yet to be achieved – would be the introduction into the cultivated grasses of the ability to form root nodules. This would make the plants that yield the bulk of human nutrients self-sufficient for nitrogen fertiliser! The outcome would revolutionise the food supplies of the world.

The challenges facing the geneticists undertaking this task are enormous. About 20 genes are responsible for the enzyme machinery of nitrogen fixation. If these genes are successfully installed on chromosomes of cultivated grasses, it is found that they can be expressed only when all other fixed nitrogen products and molecular oxygen are excluded from the cells. None of the teams of molecular geneticists working on this challenge has yet had any significant success.

In the meantime, crop plants that are now GMOs (but mostly grown only in North America) include

▶ varieties of **wheat** that contain more proteins (glutinins) favourable for bread making
▶ varieties of **tomatoes** with resistance to virus attack, or resistance to insect attack.

Research also centres on the production of medically active compounds produced in plants, such as antibodies and vaccines.

Genetic engineering in flowering plants is currently achieved by one of three methods, as described on the following pages.

Using Agrobacterium tumifaciens

Agrobacterium tumifaciens is a soil-inhabiting bacterium that may invade growing plants at the junction of root and stem, where it can cause a cancerous growth known as a crown gall. The bacterium, which infects dicotyledonous plants only, contains a plasmid (known as a **Ti plasmid**) that carries the genes for tumour formation. When the bacterium invades the host cells the Ti plasmid enters the host nuclei. Part of the Ti plasmid becomes inserted into a chromosome, introducing genes for the synthesis of a plant hormone (causing proliferation of host cells) and secretion of special amino acids called **opines**, on which the bacteria in the host and in the surrounding soil feed.

In effect, the bacterium carries out genetic engineering on its own behalf. Molecular geneticists have exploited the Ti plasmid by inserting a short length of DNA (a gene) in place of an existing gene in the plasmid. This recombinant DNA becomes integrated with the DNA of a chromosome of the host plant (Figure 29.11), and the engineered gene is subsequently expressed. The method has been used to engineer a gene for herbicide resistance into crop plants. Then, when the field was sprayed with herbicide the weeds were killed but the crop survived.

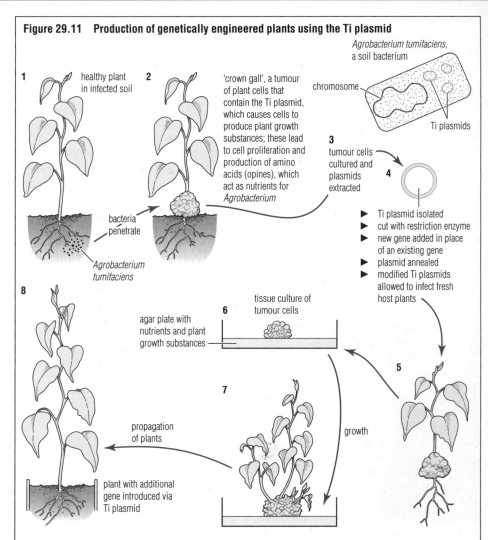

Figure 29.11   Production of genetically engineered plants using the Ti plasmid

1 healthy plant in infected soil

Agrobacterium tumifaciens

bacteria penetrate

2 'crown gall', a tumour of plant cells that contain the Ti plasmid, which causes cells to produce plant growth substances; these lead to cell proliferation and production of amino acids (opines), which act as nutrients for *Agrobacterium*

Agrobacterium tumifaciens, a soil bacterium

chromosome

Ti plasmids

3 tumour cells cultured and plasmids extracted

4
- ▶ Ti plasmid isolated
- ▶ cut with restriction enzyme
- ▶ new gene added in place of an existing gene
- ▶ plasmid annealed
- ▶ modified Ti plasmids allowed to infect fresh host plants

6 tissue culture of tumour cells

agar plate with nutrients and plant growth substances

5

7 growth

8

propagation of plants

plant with additional gene introduced via Ti plasmid

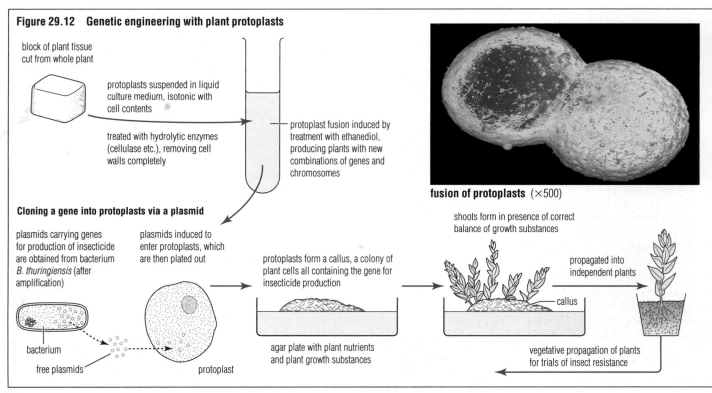

Figure 29.12   Genetic engineering with plant protoplasts

block of plant tissue cut from whole plant

protoplasts suspended in liquid culture medium, isotonic with cell contents

treated with hydrolytic enzymes (cellulase etc.), removing cell walls completely

protoplast fusion induced by treatment with ethanediol, producing plants with new combinations of genes and chromosomes

fusion of protoplasts (×500)

Cloning a gene into protoplasts via a plasmid

plasmids carrying genes for production of insecticide are obtained from bacterium *B. thuringiensis* (after amplification)

bacterium

free plasmids

plasmids induced to enter protoplasts, which are then plated out

protoplast

protoplasts form a callus, a colony of plant cells all containing the gene for insecticide production

agar plate with plant nutrients and plant growth substances

shoots form in presence of correct balance of growth substances

callus

propagated into independent plants

vegetative propagation of plants for trials of insect resistance

By protoplast production

The plant cell wall is a rigid structure and potentially a barrier to the introduction of new genes. But plant cell walls can be removed by the gentle action of enzymes, including cellulase, hemicellulase and pectinase (Figure 7.27, p. 166). Plant cells from which the wall has been removed are called **protoplasts**. They can be cultured in liquid medium provided the water potential of the medium is appropriately regulated (p. 235). Under these conditions the protoplast quickly regenerates a new cell wall, but before it can do so there is an opportunity for the molecular geneticist to adjust the genome of the denuded cell (Figure 29.12).

For example, plasmids of bacterial origin can be induced to enter by using the 'zapping' (electroporation) technique to open pores in the plasma membrane. The bacterium *Bacillus thuringiensis* has a plasmid-based gene for the production of a protein that is toxic to many insect larvae (a naturally occurring insecticide). Copies of these plasmids have been introduced into protoplasts of potato, cotton and tobacco species. From these genetically engineered protoplasts, plants have been produced with natural defences against a range of insect pests.

Another use of protoplasts is in cell fusion experiments. By appropriate treatment (using ethanediol, for example) protoplasts from the same species or from differing species may be induced to fuse. In the latter case, the product may be a case of induced allopolyploidy (rather than a naturally occurring polyploid from different species, p. 625).

Using 'biolistics'

A biolistics machine (Figure 29.13) literally fires genes into tissues. To operate it in the UK you must have a firearms licence! The genes are coated on to tiny gold or tungsten 'bullets' and fired into a tissue sample. Plant or animal tissues are routinely successfully engineered to carry additional genes by this method.

Genetic engineering in mammals

The molecular geneticist seeking a specific gene product often needs to be able to engineer eukaryotic genes into eukaryotic cells, rather than into bacteria. For example, when a eukaryotic gene is to be expressed, eukaryotic cells have the enzymes needed to convert heterogeneous mRNA (consisting of introns and exons, Figure 29.9) prior to translation. Prokaryotic cells do not contain these enzymes (their genes do not contain introns). In addition, a newly synthesised protein often has to be modified by the action of products of other genes, before it becomes effective: again, a prokaryote into which a gene has been engineered is unlikely to have the additional genes for modification.

How can genes be added into the genome of a mammalian cell, apart from by firing them in (using biolistics)?

By direct uptake of DNA into mammalian cells?

The DNA of a gene to be introduced is extracted from donor cells and replicated many times over, using the polymerase chain reaction (PCR, p. 633), for example. The technique of PCR was perfected in 1989 by Dr Kary Mullis, working in California. It requires knowledge of the DNA sequence on either side of the gene (the promoter sequence and the termination sequence). Then a copy of the gene is made, using a DNA polymerase that specifically recognises the promoter and terminator sequences of the gene to be replicated, and no other. The hydrogen bonds between the DNA template and the complementary strand are broken by the action of heat, and both are then used as templates by the action of fresh DNA polymerase. In this way the original gene is duplicated many times.

The purified DNA may be made available to cells under conditions that induce its uptake. This is relatively inefficient, for the uptake rate is low. Alternatively, the gene may be injected directly into the nucleus using microdissection

Figure 29.13 A 'biolistics' machine

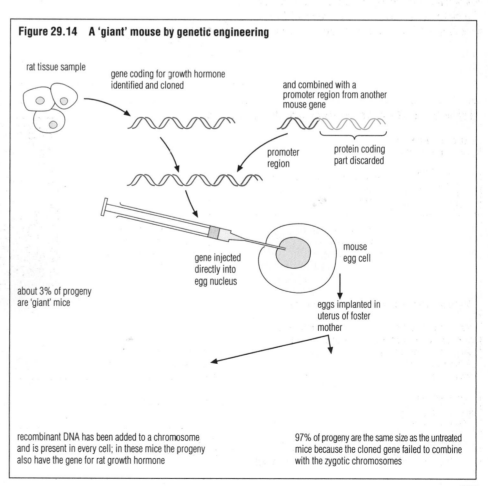

Figure 29.14 A 'giant' mouse by genetic engineering

rat tissue sample

gene coding for growth hormone identified and cloned

and combined with a promoter region from another mouse gene

promoter region

protein coding part discarded

gene injected directly into egg nucleus

mouse egg cell

eggs implanted in uterus of foster mother

about 3% of progeny are 'giant' mice

recombinant DNA has been added to a chromosome and is present in every cell; in these mice the progeny also have the gene for rat growth hormone

97% of progeny are the same size as the untreated mice because the cloned gene failed to combine with the zygotic chromosomes

equipment. For example, a gene coding for a growth-promoting hormone was injected into fertilised egg cells of a mouse. The eggs were subsequently implanted in the uterus of a foster mother. (The growth hormone gene had to be attached to a promoter region of another gene first, so that it could be switched on by means of a diet change.) The final product was a 'giant' mouse (Figure 29.14).

By replacement of treated cells?

Bone marrow cells can be withdrawn from a mammal's body, cultured *in vitro*, and subsequently returned to the body where they will continue to grow. Whilst in tissue culture, recombinant DNA may be added into the DNA of the nucleus. Dr M J Cline of the University of California has genetically engineered a gene for resistance to the harmful effects of an antitumour drug and introduced it into bone marrow cells of the mouse. When these cells are returned to the mouse they survive, and actually thrive if the animal is given large doses of the antitumour drug (Figure 29.15).

This experiment may have significance for humans. In order to cure bone marrow cancer, the patient requires heavy doses of antitumour drug; in some cases it is necessary to use dosages that can threaten the survival of healthy marrow cells. If a sample of unaffected bone marrow cells is first withdrawn from the patient, cultured and 'engineered' to acquire resistance to the antitumour drug, and then returned to the patient's body before drug therapy, the chances of survival of healthy bone marrow and of eventual full recovery are greatly increased.

By the use of plasmids?

Bacterial plasmids have been introduced into mammalian cells, although the plasmids are not normally replicated here. In some cases, however, the genes introduced via plasmids may be taken up into the DNA of individual cells. This is the case in the production of GM sheep, able to secrete the human protein AAT (α-1-antitrypsin) (Figure 29.16). This protein occurs in our blood (some white cells secrete it) and has a role in inhibiting the enzyme elastase, present in the lungs. If uninhibited, elastase destroys the elastic fibres there, during the normal turnover of tissues by which the lungs are maintained. People with emphysema (p. 312) are deficient in AAT and benefit from treatment with this protein.

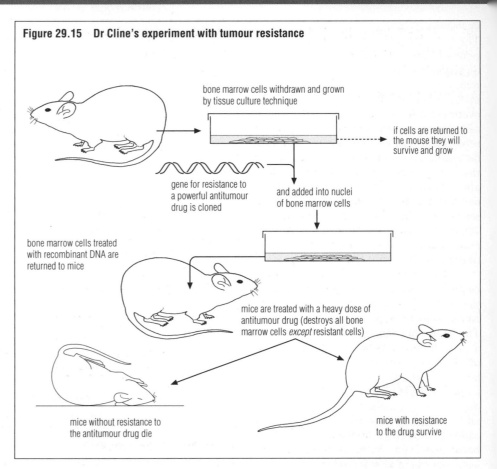

Figure 29.15   Dr Cline's experiment with tumour resistance

bone marrow cells withdrawn and grown by tissue culture technique

if cells are returned to the mouse they will survive and grow

gene for resistance to a powerful antitumour drug is cloned

and added into nuclei of bone marrow cells

bone marrow cells treated with recombinant DNA are returned to mice

mice are treated with a heavy dose of antitumour drug (destroys all bone marrow cells *except* resistant cells)

mice without resistance to the antitumour drug die

mice with resistance to the drug survive

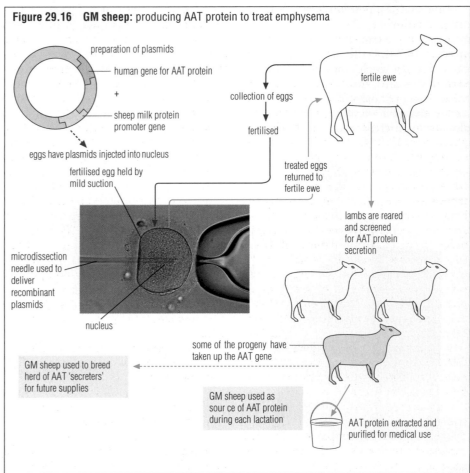

Figure 29.16   GM sheep: producing AAT protein to treat emphysema

preparation of plasmids

human gene for AAT protein

+

sheep milk protein promoter gene

collection of eggs

fertile ewe

fertilised

eggs have plasmids injected into nucleus

treated eggs returned to fertile ewe

fertilised egg held by mild suction

lambs are reared and screened for AAT protein secretion

microdissection needle used to deliver recombinant plasmids

nucleus

some of the progeny have taken up the AAT gene

GM sheep used to breed herd of AAT 'secreters' for future supplies

GM sheep used as source of AAT protein during each lactation

AAT protein extracted and purified for medical use

29.4 OTHER APPLICATIONS OF RECOMBINANT DNA TECHNOLOGY

■ Genetic fingerprints

About 90% of the DNA of human chromosomes does not code for protein and has no known function. These non-functional sequences are not subject to evolutionary pressure to change; random changes in non-functional sequences are not selected against because they have no effects on the phenotype. (Changes to functional genes, on the other hand, are often harmful and tend to be selected against when they come to be expressed as a mutant character.) Thus individual organisms accumulate sequences of non-functional DNA. These vary in length, but consist of sequences of bases, many 20–40 bases long, often repeated very many times. A typical sequence might read

C G A T T C A G C G A T T C A G

These highly variable and virtually unique lengths of non-functional DNA are passed on to offspring along with the complement of genes (they are inherited). It is this part of DNA that is used in DNA 'fingerprinting'.

To produce 'genetic fingerprints' from a sample of blood cells (or any other type of cell), the DNA present must be extracted and enzymically 'cut' into fragments by a restriction enzyme. The fragments are then separated into bands on the basis of their size, using electrophoresis (p. 175). The invisible pattern of the fragment bands so produced is copied on to a nylon membrane and radioactive DNA probes (p. 632) are applied to this replica surface. Complementary probes bind by hydrogen bonding, and the unbound probes are washed away. In order to visualise the pattern of DNA fragments, an X-ray film is laid on the nylon membrane. This is fogged by the radioactive components of the pattern, producing the DNA 'fingerprint'. This is then compared with fingerprints from other individuals to identify and analyse any common features (Figure 29.17). DNA fingerprinting is used to settle disputes over parentage and other relationships, and to identify criminals from blood and other cells left at the scene of a crime. Where a tiny sample of DNA is all that is available, the sample is first copied repeatedly, by means of the polymerase chain reaction (PCR, p. 633).

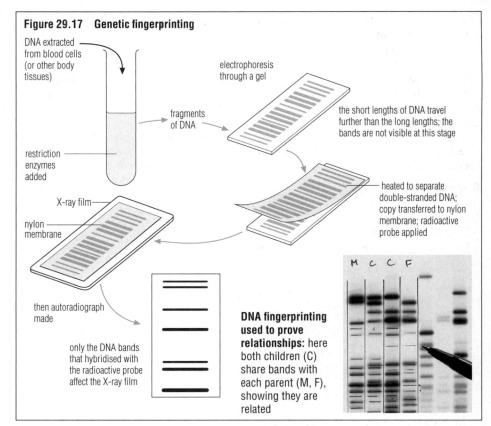

Figure 29.17 Genetic fingerprinting

DNA extracted from blood cells (or other body tissues)

restriction enzymes added

fragments of DNA

electrophoresis through a gel

the short lengths of DNA travel further than the long lengths; the bands are not visible at this stage

heated to separate double-stranded DNA; copy transferred to nylon membrane; radioactive probe applied

X-ray film

nylon membrane

then autoradiograph made

only the DNA bands that hybridised with the radioactive probe affect the X-ray film

DNA fingerprinting used to prove relationships: here both children (C) share bands with each parent (M, F), showing they are related

■ Human genome project

A human's genome consists of about three billion base pairs. There is a coordinated, international project with the aim of mapping this entire human genome on the 46 chromosomes, begun in 1990. The outcome, which is apparently now feasible, will involve the eventual recording of the entire sequence of base pairs that are the instructions to make a human being. The work has been shared out between 250 laboratories around the world, to avoid unnecessary duplication of effort. It is the largest collaborative project attempted in biology.

The investigation involves establishing the positions of genes relative to others on the same chromosome. Then the base sequence of the genes has to be determined. The DNA of chromosomes is sliced into pieces, the positions of the fragments in the chromosome as a whole are worked out, and the base sequence of the fragments analysed. The task is a huge one.

Many currently incurable diseases arise from defects in genes; possible advantages of the human genome project include the finding of cures. This depends on identifying the relevant genes and understanding how they are regulated.

This work also provides one more example of our successful quest for knowledge about ourselves as humans, our world, and how life has come about. The evolution of humans from other hominids may become clearer.

An argument against the human genome project is that it is expensive and time consuming, and progress towards the prevention of genetic diseases might be much quicker if efforts were focused exclusively on relatively few 'problem-causing' genes. In addition, the outcome of the project raises problematical issues regarding ownership of information about an individual's genome, and how that information may be used by the state, by businesses (like insurance companies), and by doctors.

■ Genetic diseases and gene therapy
Genetic diseases in humans

About 4000 disorders of humans have a recognised genetic basis. These diseases affect between 1 and 2% of the human population (that is, more than one in every 100 children born carries a serious genetic defect). More than half of the known genetic diseases are due to a mutation involving a single gene. The mutant allele may be recessive or (more rarely) dominant, and may be found on the autosomal chromosomes or on the sex chromosomes (p. 619). Some examples of genetic diseases are discussed overleaf.

Huntington's disease is due to an autosomal dominant allele on chromosome 4. Affected individuals are almost certainly heterozygous for the defective gene. The disease is extremely rare (1 case per 20 000 live births). Appearance of the symptoms is usually delayed until the age of 40–50 years, by which stage the affected person – unaware of the presence of the disease – may have passed the dominant allele to one or more of his or her children.

The disease takes the form of progressive mental deterioration, which is accompanied by involuntary muscle movements (twisting, grimacing and staring in 'fear'). The disease is named after the American doctor who first investigated the condition. There is no known treatment.

Cystic fibrosis is due to an autosomal recessive allele. People suffering from cystic fibrosis are homozygous for the gene causing the disease, so that both the parents must be carriers. This disease is quite common; it occurs among Caucasians at the rate of about 1 per 2000 live births.

The product of the normal gene is a component of the mucus secreted by epithelia (p. 312). An affected person produces mucus that is thick and unable to flow, causing problems in the lungs (they become congested with mucus), the pancreas (the pancreatic duct becomes blocked and food digestion is incomplete) and the intestine (the protective lining of the gut is defective). Most patients can be helped by daily physiotherapy for their lung problems, and periodic treatment with pancreatic extract; even so, most become severely handicapped and may die before they are 30.

Another example of genetic disease due to an autosomal recessive allele is **sickle-cell anaemia** (p. 210).

Haemophilia (p. 620) and **Duchenne muscular dystrophy** (p. 621) are both due to a sex-linked (X chromosome) recessive allele. They normally affect only males (XY), because the incidence of both diseases is so low that the chance of a female getting two X chromosomes that carry the recessive allele is extremely slight.

Gene therapy

The aim of **gene therapy** is to treat genetic disease by replacing defective genes in the patient's body with copies of the healthy gene. As a branch of applied molecular biology it is a very recent development. Whilst the ultimate goal of gene therapy is to cure all genetic disease, this eagerly awaited outcome is still a long way off. Remember, too, that the vast majority of human diseases are 'environmental' (due to a parasite, a toxin or malnutrition, perhaps), and cannot be prevented or cured by a genetic engineer.

Theoretically at least, genes may be added into germ cells (eggs or sperms) or into body cells (somatic cells). Adding genes to germ cells would mean that the genome of future individuals was being changed. Tampering with the genes of human sex cells is outlawed, in fact.

The available technologies for getting genes into cells (and in the chromosomes of eukaryotes) have already been introduced in this chapter. These are the approaches the genetic engineer seeks to adapt for human gene therapy. The therapist concentrates on disorders caused by a single gene, since these present the best chances of success. The healthy gene has to be located, isolated and cloned to make it available for 'transplantation'. Target cells in the body need to be robust enough to survive while a sample is withdrawn, cultured, modified genetically, and then returned to the body. They must then survive for a substantial period (whilst the healthy gene is expressed). Molecular biologists have concentrated on bone marrow cells, liver cells and skin cells, so far. Ideally, the transplanted cells ought to divide *in situ*, replicating the implanted, healthy gene in the patient's body.

Gene therapy for cystic fibrosis

Here the approach has been to 'wrap' cloned healthy genes in microscopic lipid envelopes (forming **liposomes**) and periodically to spray these on to the surface of the lungs of a diseased animal where mucus production is failing due to a defective gene. Application is by means of an aerosol spray. The expectation is that many of the tiny liposome packages will be taken into the cells of the lung surface (by membrane merger), and the genes may enter the chromosomes and be expressed. In trials using laboratory mice with induced cystic fibrosis, this technique has appeared to be largely successful. It is now to be tried out with human sufferers.

Gene therapy for familial hypercholesterolaemia (FH)

The surface membranes of liver cells carry receptors for low-density lipoproteins (LDL). LDLs circulate in the blood and are normally taken into cells and metabolised. The membrane receptors in FH sufferers are defective, and consequently LDLs accumulate in the blood circulation. In FH, the lipids are laid down in arteries as cholesterol (leading to atherosclerosis and coronary heart disease), and under the skin.

Samples of liver tissue may be removed and the cells genetically modified and returned (Figure 29.18). (The human body quickly regenerates missing liver tissue.) Liver cells are relatively easy to handle and many genetic diseases may eventually be corrected by modifying liver cell function.

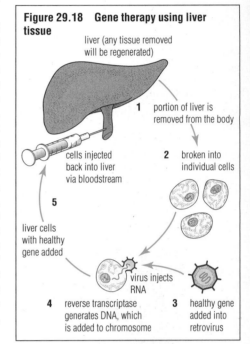

Figure 29.18 Gene therapy using liver tissue

liver (any tissue removed will be regenerated)

1 portion of liver is removed from the body

2 broken into individual cells

3 healthy gene added into retrovirus

virus injects RNA

4 reverse transcriptase generates DNA, which is added to chromosome

liver cells with healthy gene added

5 cells injected back into liver via bloodstream

Genetic counselling

Unaffected members of a family with a history of a genetic defect can consult a **genetic counsellor** for advice on the risk of having an affected child. Genetic counselling may also be sought by parents with an affected child who wish to know the likelihood that any future child they conceive may be affected. Advice is based on knowledge of the family's pedigree and the frequency of the faulty gene in the national population, and on whether the parents are closely related. For some conditions it is possible to examine blood samples of the parents to check whether they carry a faulty gene (for example, for cystic fibrosis).

Genetic counsellors may also advise on the actual genetic status of the fetus, knowledge of which is obtained using techniques such as chorionic villus sampling (p. 601). Typically this advice is sought by or offered to parents at particular risk.

Will genetic diseases die out?

Genetic diseases are so clearly harmful that you may wonder why the genes responsible for them are not 'lost' by natural selection.

One point to remember here is that most harmful genes are recessive and only have any effect in a person who is homozygous for the gene. Since selection acts only on the phenotype, most defective genes are passed from generation to generation without any selection 'pressure' on heterozygotes.

Remember too that most harmful genes are the product of a gene mutation, possibly due to a substitution of one or more bases in the DNA. Mutations arise spontaneously, at a low but persistent rate. Many mutations appear at about the same rate as they are also being excluded by selection.

Finally, even though a faulty allele may cause a genetic disease when present as the double recessive (homozygous), it may actually confer an advantage when present as a heterozygous allele. An example is the allele for sickle-cell anaemia, which in the homozygous condition causes this dangerous and painful condition. However, in the heterozygous condition this allele confers resistance to malaria (since the sickle cells do not provide satisfactory nutrition for the malaria parasite). The advantage occurs in areas where malaria is endemic (Figure 29.19).

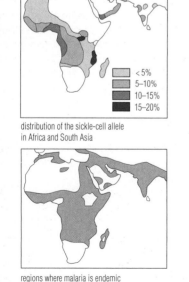

Figure 29.19   Geographical distributions of sickle-cell trait (top) and of malaria (bottom)

distribution of the sickle-cell allele in Africa and South Asia

< 5%
5–10%
10–15%
15–20%

regions where malaria is endemic

Cancer

'Cancer' is a diverse collection of diseases, rather than being a single disease, but all cancers are diseases of cells where specific damage has occurred to the DNA of the nucleus. In cancer, the rate at which cells multiply exceeds the rate at which they die. This uncontrolled proliferation of cells produces a 'lump', but most cancers are more than just lumps of cancer cells. They are made up of a mixture of abnormal cells and normal cells. The abnormal cells are under the influence of specific genes (see below). The normal cells are a mixture of those attempting to destroy mutant cells (for example, lymphocytes and macrophages) and those recruited by cancer cells to aid growth (for example, connective tissue cells, cells that make new blood vessels). The result is that a cancer has some of the properties of a wound that will not heal.

To understand the significance of DNA changes in cancer cells, we need to look at life and death of normal cells. Multicellular animals grow by cell division and cell proliferation. These events are triggered by instructions from neighbouring cells. The trigger is usually a small protein, which binds to a receptor on the cell membrane (Figure 29.20). The result is that specific genes in the targeted cell are activated. Many of the genes in cells are switched on or off by this type of mechanism. As a result, the cell divides, or the cell differentiates to fulfil a particular role, or the cell dies in a controlled and coordinated way. This means there are specific genes that bring about these changes, if and when activated. In controlled cell death, the cell shrinks, and then the DNA is packaged and digested. Finally, the membranes and organelles are digested by surrounding macrophages. Cell death may be switched on when cells are no longer useful, or are in the wrong place, or there are too many of one type. (Note that this controlled cell death is quite different from the destruction of a cell when exposed to a poison or to mechanical damage. In this latter case, the cell enlarges and disintegrates, discharging its contents. Neighbouring cells are usually damaged by some of the contents, such as hydrolytic enzymes, from the broken cell.)

Two types of genes are involved in cancer: tumour suppressor genes, and oncogenes.

1 We have noted that cancer is caused by damage to DNA. If the damage is to a gene controlling cell growth, then the cell may grow out of control. However, when any of the DNA of a cell is damaged the cell attempts to repair it. For example, a protein known as p53 or 'the guardian of the genome' stops the copying of damaged DNA so that other enzymes that repair DNA can work. In cancer cells, the genes that code for the repair processes (for example, the gene coding for p53) are also damaged and non-functioning. The gene for p53 has been found to be damaged in about 40% of cancers. This p53 gene is one of several such genes – all examples of **tumour suppressor genes** – that when damaged, allow a cancer to grow out of control.

2 The proteins coded by **oncogenes** are mostly internal signalling substances that control cell behaviour. If damage (mutation) in an oncogene occurs, abnormal cell behaviour, including abnormal growth, may result.

The development of a cancer is likely to start with a single cell in which DNA damage as described above has occurred and leads to multiplication, forming a small, visible clump of cells. These are essentially pre-cancerous cells. If the ball of cells grows and forms a tumour, and cells from it invade surrounding tissues, a state of cancer is established. The abnormal cells spread via blood and lymph, and may form satellite tumours in other tissues of the body.

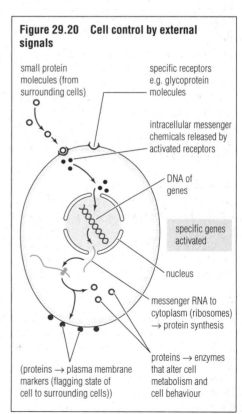

Figure 29.20   Cell control by external signals

small protein molecules (from surrounding cells)

specific receptors e.g. glycoprotein molecules

intracellular messenger chemicals released by activated receptors

DNA of genes

specific genes activated

nucleus

messenger RNA to cytoplasm (ribosomes) → protein synthesis

(proteins → plasma membrane markers (flagging state of cell to surrounding cells))

proteins → enzymes that alter cell metabolism and cell behaviour

Treatment of cancer may involve destruction of the tumour by surgery, and this can be effective if the cancer is detected early enough (before excessive enlargement and any spread of the cells). Once cancers have spread, the cells are harder to kill. The use of drugs that kill cancer cells (chemotherapy) has unpleasant side effects because so many normal cells of the body are also actively dividing and are therefore 'hit' as well. (Can you list all the places in the body where normal, healthy active cell division continues throughout life?)

Cancer cell membranes carry proteins that differ from those of normal cells. In future, can the genetic engineer deliver to these cell membranes alone, chemicals that will trigger the natural cell death procedures? Alternatively, can the immune system be caused to recognise cancer cells as 'foreign' and so mount an immune response against them?

■ Issues in transplant surgery

Replacement of damaged or diseased organs by grafting a healthy replacement organ is known as transplant surgery. Initially, the major issue here was the rejection of the transplanted organ, due to the immune response (p. 366). This has largely been overcome as a problem.

Today, the major issue in transplant surgery in developed countries is finding sufficient healthy organs. The success of this technology has produced long waiting lists for transplants of kidneys, hearts or livers, for example.

The resulting organ shortage has led to research into the use of animals (**xeno-transplantation**) specially bred as a supply of compatible organs. Incompatible organs from a potential donor organism such as a pig are identified in the recipient's body by the proteins or glycoproteins on the cell membranes of the donated organ. There is a particular

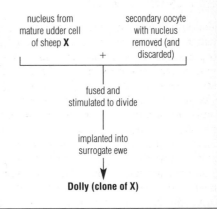

Figure 29.21   Dolly the sheep, cloned by Ian Wilmut and colleagues at Roslin Institute, Edinburgh

nucleus from mature udder cell of sheep **X**

secondary oocyte with nucleus removed (and discarded)

+

fused and stimulated to divide

implanted into surrogate ewe

Dolly (clone of X)

'signal' on the cells of a human organ that prevent an outright rejection reaction when transplanted into another human. This regulator protein is known as RCA, and it is coded for by a gene called HDAF. This gene has been isolated and cloned. It has now been added to a pig, and a herd of piglets has been produced, all carrying this gene. These genetically modified animals, all HDAF-carriers, are available for organ transplantation in clinical trials. The issue of xenotransplantation is under consideration by regulatory bodies.

Pig transplants may never be safe (even if they become generally acceptable to the public), for it may be impossible to breed pigs totally free of potentially harmful retroviruses (that is, viral disease) that might adapt to parasitise humans.

Meanwhile, another avenue has opened up in the search for sources of donor organs. Dr Ian Wilmut and colleagues at the Roslin Institute, Edinburgh, bred a mammal (Dolly the sheep, Figure 29.21) that had been grown **from the genome of a mature** (udder) **cell**. This experiment confirms **totipotency** (that every nucleus contains information for the development of *all* body tissues,

p. 211). It also creates a new possibility, known as **therapeutic cloning**. This would use the technique that created Dolly to grow cells, tissues and (eventually) organs for transplants, using the proposed recipient's own cells to provide the genome (Figure 29.22). There would be no rejection. This idea, that people requiring transplants could have organs grown to order from their own cloned cells, is currently put on hold by the Government, perhaps through uncertainty over public reaction. There are enormous technical problems remaining, too, even if the research were to be encouraged.

Questions

1   Explain the differences between sickle-cell trait and sickle-cell anaemia (p. 210).
2   X-carried sex-linked diseases due to a recessive gene are rare in human females. Why?
3   What are the principles of gene therapy, as proposed for humans?
4   In the treatment of bone marrow cancer, why is it desirable to 'engineer' the patient's own cells rather than inject treated cells from another source?

Figure 29.22   The principle of a 'body repair kit'

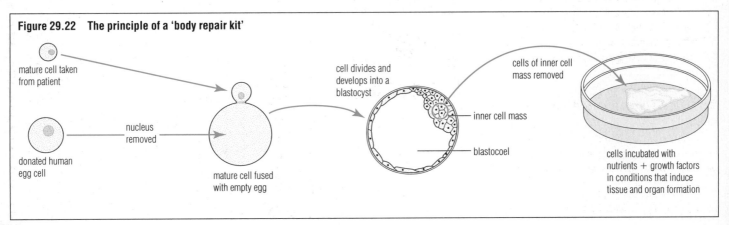

mature cell taken from patient

donated human egg cell

nucleus removed

mature cell fused with empty egg

cell divides and develops into a blastocyst

inner cell mass

blastocoel

cells of inner cell mass removed

cells incubated with nutrients + growth factors in conditions that induce tissue and organ formation

29.5 GENETIC ENGINEERING – ISSUES OF CONCERN

Geneticists often produce new organisms when genes are transferred between organisms. Reproduction biologists work with human embryos *in vitro*, apparently far removed from natural conditions. Consequently, these activities are potentially sources of hazards for humans and certainly generate concerns. You will have noticed very contrasting headlines in the media regarding the work of genetic engineers, depending on perceptions about the intended outcomes. Work towards 'replacement body parts', for example, is more likely to be hailed as an 'innovation' than the production of 'frankenstein foods'. If the genes for root nodule formation and the ability to fix nitrogen are successfully engineered into the grass family (p. 635) then headlines about genetic engineering may be very favourable indeed.

■ Economic issues to consider

▶ Genetic engineering is a costly technology, of most benefit to people of developed nations. If funds made available for genetic engineering were diverted to solving the more basic problems of housing, health, employment and nutrition of the poor worldwide, would the money not benefit vastly more humans, immediately?

■ Environmental issues to consider

▶ Might a gene, added to a genome, function in an unforeseen manner, perhaps even triggering cancer in the recipient, for example?

▶ Might an introduced gene for resistance to adverse conditions get transferred from a crop plant or farm animal into a weed species or to some predator?

▶ Could a harmless organism, such as the human gut bacterium *E. coli*, with recombinant DNA technology, be transformed into a harmful pathogen that escapes into the population at large?

■ Ethical issues to consider

▶ Is there an important overriding principle to be held to, that humans should not tamper with 'nature'?

▶ Are those who object to 'unnatural' intervention in human fertilisation right, especially when this is arranged outside the mother's body?

THE SUNDAY TIMES
INNOVATION

22 AUGUST

Laboratories unveil plans to grow hearts, livers and kidneys

A £3 billion worldwide project to create artificial organs is about to begin constructing its first man-made heart, writes *Roger Dobson*.

In the most ambitious tissue-engineering programme the world has seen, three teams of specialists are planning to grow a heart, liver and kidney.

The technology they hope to use has already been successfully tested on human cartilage but the funding has not been available to complete an entire organ.

Team leaders have been appointed for the 10-year project and a business plan is being finalised for presentation early next year to governments in Europe, America, Canada and Japan next February.

Professor Michael Sefton, director of the Institute of Biomaterials and Biomedical Engineering at the University of Toronto, and his colleagues are now finalising their plan to grow the organs.

"We are focusing on vital organs – the heart, liver and kidney – which represent the next generation of tissue development. The project will also address the problems of a shortage of donors," he says. Techniques will involve polymer scaffolding made to the shape of the organ. The heart will probably be grown in four parts – valve, muscle, blood vessels and conduits – with the components then being grown together. The advantage of growing the parts separately would be that as each one became viable it could be used on patients. The heart problems of some patients, for example, could be solved by having a new heart valve fitted.

MEDICINE

"We could make the heart in one or in four with cells. With the help of a biochemical soup, the cells would start to grow and spread along the scaffolding until they reached the end when they would stop growing. Once the tissue was in place, the scaffolding would degrade, leaving the finished organ.

It has also been suggested by a team of scientists and doctors at Harvard that whole limbs eventually may be grown in the laboratory, ready for fitting to patients who have suffered amputations.

Skin and cartilage have been successfully grown already; bone is expected to be not too far behind.

Patients have had cartilage cells taken from damaged knee joints and grown in the laboratory before being transplanted back into the knee. Cartilage cells have also been implanted to boost muscles and combat incontinence.

Researchers have been drawing up plans for growing breast tissue for transplanting into mastectomy patients to achieve better cosmetic results.

With the scaffolding technique, a porous structure is made to the shape of the organ or part of the organ and is then seeded with cells. With the help of components by using scaffolding and that is the simplest approach," says Sefton. "Growing in bits makes it easier for thinking about the spin-off parts of the project. But we may have four parallel approaches for the first few years until we make a decision on which way to go."

The main challenge with growing whole organs is that unlike skin and cartilage they contain more than one type of tissue and the scaffolding has to be seeded with different cells. The key to success is to get cells growing together with different biochemical solutions and growth factors to mimic the shape of the real organ.

The project is being sold on the basis that industrialised nations spend millions of pounds every year on the consequences of diseases of vital organs and that off-the-shelf replacements will save health services huge amounts of money as well as tens of thousands of lives.

The heart, kidney and liver have always been major targets of the tissue growers because of the numbers of people who die from diseased or failed organs.

"The business plan we are finalising takes account of various possibilities and the next step is to get seed funding. We will be approaching governments, probably in February. I think it will be a 10-year project," says Sefton.

643

The United Kingdom Genetic Manipulation Advisory Group publishes guidelines and 'polices' experimental work in the interests of the public. Gene therapy experiments are restricted to specialist laboratories with advanced facilities for containing mutant microorganisms. Experiments are restricted to 'disabled' species of bacteria that can grow only under specialised laboratory conditions, not available outside the laboratory. All research programmes have to be approved. Genetically engineered crops are subject to independent trials concerning their safety and usefulness. All these mechanisms of control need to be maintained and improved, as experience shows necessary.

In *Advanced Biology Study Guide*, Chapter 29, is a role-play activity on 'Genetic engineering raises ethical issues', in which groups can explore their ideas about the acceptability of genetic engineering, and review interventions in genetic disorders from the viewpoints of family members who may be involved.

29.6 ARTIFICIAL SELECTION

It is clear that the skills of selective breeding were exercised by humans well before written history began. Animals and plants were probably first domesticated in those eastern Mediterranean lands known as the Fertile Crescent (Figure 29.23); the earliest animal domestication was apparently that of the dog (*Canis familiaris*), from a wild wolf. This happened in many human communities, but in the Fertile Crescent it seems to have been under way about 13 000 years ago.

■ The principles of selective breeding

Plants are more important as food sources than animals, because of their mode of nutrition (autotrophic, p. 246), and their consequent location in the food chain. Agriculture is basically a series of food chains that commence with green plants and end with humans!

The object of **plant breeding** is to improve crop yields, by increasing the size of the whole plant, or by selectively increasing the edible part of the plant, or by the selection and development of disease-resistant varieties. The principle of plant breeding is the combining together of superior genes. The end-product must be an organism that is not only superior in identified qualities, but for which there is a recognised market. Human choice and preference are important factors in the economics of both plant and animal breeding.

Plant breeding is carried out very largely by crossing different varieties, and then selecting progeny with particular features from which to breed further. Plant breeders are looking for qualities in the progeny such as:

► fast spring germination and/or growth
► good response to applied fertilisers
► dwarfed stem growth (in cereals, for example) and enhanced leaf growth
► efficient capture of solar energy
► efficient inflorescence and flower formation (such as long 'ear' and many grains in cereals)
► tolerance of occasional unfavourable conditions, such as drought
► effective resistance to diseases and common pests
► superior nutritive value of the grain/fruit/leaf product yielded.

The combining together of whole sets of chromosomes in the formation of polyploids is another branch of plant breeding (pp. 625–6).

The object of **animal breeding** is to improve the yield or quality of wool or hide, or to increase milk production, or to increase meat quality (protein at the expense of fat, perhaps) or litter size, according to the animal species concerned and according to local needs. Animal breeding makes use of artificial insemination techniques (p. 601), as well as the newest developments of genetic engineering (p. 634).

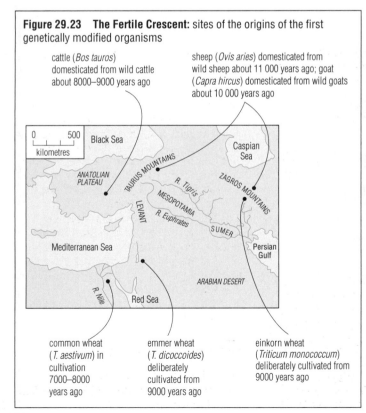

Figure 29.23 The Fertile Crescent: sites of the origins of the first genetically modified organisms

cattle (*Bos taurus*) domesticated from wild cattle about 8000–9000 years ago

sheep (*Ovis aries*) domesticated from wild sheep about 11 000 years ago; goat (*Capra hircus*) domesticated from wild goats about 10 000 years ago

0 — 500 kilometres

Black Sea

Caspian Sea

ANATOLIAN PLATEAU

TAURUS MOUNTAINS

ZAGROS MOUNTAINS

R. Tigris

MESOPOTAMIA

LEVANT

R. Euphrates

SUMER

Mediterranean Sea

Persian Gulf

ARABIAN DESERT

R. Nile

Red Sea

common wheat (*T. aestivum*) in cultivation 7000–8000 years ago

emmer wheat (*T. dicoccoides*) deliberately cultivated from 9000 years ago

einkorn wheat (*Triticum monococcum*) deliberately cultivated from 9000 years ago

Figure 29.24 Field trials for new varieties of wheat

Semen is collected in an artificial vagina, and then appropriately diluted and stored in quantities suitable for individual inseminations. Storage of sperms at low temperature ($-172\,^{\circ}$C) provides viable sperms over a prolonged period. The diluted semen sample is injected into the uterus of a cow at oestrus, by means of a long, fine tube.

The animal breeder has a clear idea of the kind of animals wanted. For example, the manager of a milking herd of cows keeps records of both quality and quantity of the milk produced by the heifers and cows of the herd. Those with the better milk yields are used as mothers for the next generation, inseminated by semen from a bull known to sire good milkers.

■ Selective breeding, the process

Plant and animal breeders may work in laboratories but they are just as likely to be found on experimental farms or working on field trials (Figure 29.24). Conventional (and highly successful) methods of plant breeding are not unlike the early experiments of Mendel (p. 605). Today, of course, these methods are augmented by work with isolated plant protoplasts (p. 636), in which protoplasts from different sources are fused, or in which new genes are introduced directly into a cell. From this successful implant the new crop organism is grown.

29.7 GENETIC CONSERVATION

When a particular species becomes extinct, its genes are permanently lost, and the total pool of genes on which life operates is diminished. One consequence is that plant and animal breeders and genetic engineers are deprived of potential sources of useful genes.

The loss of major habitats around the world, such as destruction of forests or permanent water meadows, effectively decreases the genetic heritage (Figure 29.25). So too, does the careless or excessive use of pesticides, so that much adjacent wildlife is obliterated along with the 'pests' within the crop.

Genetic conservation seeks to prevent such loss by regulating pesticide use effectively, acting to preserve vulnerable habitats, and by establishing 'gene banks' of various types; for example, seed banks and sperm banks. The ways in which we care for the world's biodiversity are important to genetic engineering too.

Figure 29.25 A lakeland valley once covered by trees, now cleared as human housing, farming and transport have been developed

---

PLANT AND ANIMAL BREEDING: SOME COMMON TERMS

Variety: a genetically distinct population within a plant species. When a plant variety is under intense cultivation it is also known as a **cultivar**.

Pure line: a variety produced by intensive selfing so that the progeny are homozygous at many more gene loci than is normal. In effect, pure lines are inbred lines (see below).

Breed: the animal equivalent of a plant variety.

Inbreeding occurs when the gametes of close relatives (animal or plant) fuse. Typically inbred plants are from species that are naturally self-fertile. In animals inbreeding is achieved by sibling and test-cross matings.

Inbreeding promotes homozygosity. Many cultivated species that have been inbred over many generations show a degree of loss of vigour, size and fertility. It thus appears that some diversity of alleles is essential to maintaining vigour in populations.

Outbreeding occurs via the crossing (mating) of unrelated varieties. In plant species that are naturally outbreeding there is present some incompatibility mechanism (p. 573). Outbreeding promotes heterozygosity.

Hybridisation is the crossing of carefully selected varieties in order to bring together desirable qualities from both parents in the offspring. The offspring sometimes (not always) grow more vigorously, a characteristic known as **hybrid vigour**. Hybrid vigour arises when the new sets of chromosomes not only produce differences but are complementary in their effects.

Interspecific hybridisation: crosses between plants of different species. This has occurred naturally in the evolution and selection processes that have produced modern wheats (p. 626).

Conservation of genetic resources: all modern food organisms of agriculture and horticulture have come from wild plants or animals. One effect of plant and animal breeding has been increasing genetic uniformity, and the loss of rarer alleles. Important qualities previously neglected by breeders (such as disease resistance, flavour and resistance to cold, perhaps) may be added back into highly cultivated varieties, using the wild plants and animals of our environments as a gene bank. Human degradation of the global environments and habitats threatens our ability to do this.

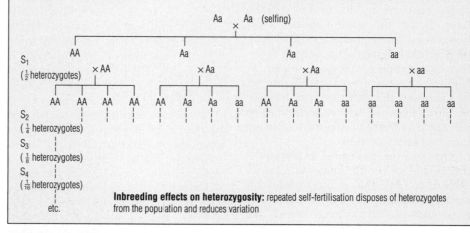

Inbreeding effects on heterozygosity: repeated self-fertilisation disposes of heterozygotes from the population and reduces variation

---

Question

The serious metabolic disorder known as phenylketonuria (PKU) is due to the absence of an enzyme to convert excess of the amino acid phenylalanine to tyrosine. Instead, it is converted into a toxic compound, which damages the developing brain. (Brain damage can be avoided by keeping to a phenylalanine-free diet from early childhood.) PKU is inherited as a recessive character controlled by a single gene. The frequency of the homozygous recessive is 1 in 10 000 of the population in Europe.

a   What is the frequency of the dominant allele in the population?

b   What is the frequency of 'carriers' in the population?

FURTHER READING/WEB SITES

GM foods debate, reviewed from all sides:
www.ncbe.rdg.ac.uk/NCBE/GMFOOD/menu.html
gmworld.newscientist.com

30 Evolution

- In biology, 'evolution' refers to the development of life in geological time. Life has been transformed from its earliest beginnings to the great diversity of forms we know about today, living and fossilised.
- The evidence for evolution comes from studies of the geographical distribution and comparative anatomy of living things, of fossil forms, of comparative biochemistry and of systematics (classification), and also from experiments in artificial selection.
- Charles Darwin and Alfred Wallace independently suggested that evolution occurs by natural selection of chance variations found among progeny that compete to survive and reproduce. Darwin presented evidence for evolution clearly, and proposed a mechanism for the origin of species that could be tested.

- Neo-Darwinism is a restatement of the concept of evolution by natural selection in terms of Mendelian and post-Mendelian genetics. This is one of the organising principles of modern biology.
- Evidence for speciation comes from studies of change in the gene frequencies in populations, which may be due to various 'disturbing influences' including emigration or immigration, mutation, selective predation and random genetic drift.
- If evolution has occurred, it may also account for the origin of vertebrates from non-vertebrates, the sudden rise of flowering plants, the origin of primates and humans, the origin of multicellular forms from unicells, and the origin of life and the first cells.

30.1 INTRODUCTION

By **evolution** we mean the development of life in geological time. The term comes from the Latin word *evolvere*, meaning 'to unroll', originally in the context of opening records by unrolling a scroll. This early usage became redundant when books replaced scrolls in the Middle Ages. Today the word implies 'origin from earlier forms' and is used widely in the English language, not only in reference to a complex biological process. In biology the term is used specifically for the processes that have transformed life on Earth from its earliest beginnings to the vast diversity of fossilised and living forms we know of today.

The idea of biological evolution is closely linked to the name of Charles Darwin (p. 11), but in fact it was discussed and championed by several biologists and geologists long before the publication of Charles Darwin's controversial theory (and Thomas Huxley's supportive battles) shocked Victorian England, in 1858.

30.2 THE HISTORY OF EVOLUTIONARY IDEAS

Some of the early Greek thinkers talked about the gradual evolution of life. But Aristotle, that meticulously careful observer of living things, did not support this concept. His 'ladder of nature' (p. 10), on which living things could be arranged on a scale of increasing complexity, held no spaces to be filled by new forms of life appearing in future, and he implied no movement or mobility on it. To Aristotle, species were fixed and did not change.

From early biblical times the account of the Creation given in the first four chapters of the First Book of Moses ('Genesis') was generally accepted as an authoritative description of the origin of species. Associated with this was the idea that, since species had been individually created by God, they were unchanging (immutable). Also, from the implied chronology of the biblical account, life on Earth was quite recent – a mere few thousand years old – and (by implication) the Earth's history was also relatively short. For example, in 1654 Archbishop Ussher calculated that the Creation had taken place in 4004 BC. This timescale was very widely accepted, in Europe at least, until well into the nineteenth century.

For many, these convictions were a product of faith alone, but for others they were underpinned by observations of the diversity of life. Thus William Paley (1802), in his book *Natural Theology*, argued that if one stumbled across a watch, from its intricate structure one would deduce there was a watchmaker! For him, the great complexity of living things was sufficient evidence of the work of a Creator. This hypothesis was known as 'special creation'.

Figure 30.1 Erasmus Darwin (1731–1802)

Figure 30.2 Sedimentary rocks with fossil remains of extinct species

The early geologists and a new view of time

James Hutton (1726–97), in a varied career as doctor, farmer and experimental scientist, also became the founder of geology. He reflected on the on-going but extremely slow erosion of rocks by water. Hutton realised that the sedimentary rocks of many existing mountain ranges had once been the beds of lakes and seas and, before that, had been the rocks of even older mountains. He made no estimates of the age of the Earth from these observations, but he appreciated that, in contrast to estimates based on biblical ideas, the Earth timescale virtually had 'no beginning and no end'. His contribution became known as **gradualism** in the development of the Earth. Later the idea was developed into a theory called **uniformitarianism** by the geologist Charles Lyell (1797–1875).

Subsequent geologists have been able to estimate the age of the Earth, and today we think of the Earth as being 4500 million years old, and life as having originated 3500 million years ago. Now we can see there has been ample time for gradual change to have significant effects.

The fossil puzzle

Hutton observed the fossils found in many sedimentary rocks, as did others before and after him (Figure 30.2). **Georges Cuvier** (1769–1832, p. 22) attempted to explain fossils in creationist terms. He was able to show, sometimes from the merest fragments of bones, the relationship between living organisms and a fossil of an extinct form of life – part of the basis of modern palaeontology.

Cuvier studied the succession of fossil species found in the sedimentary rocks of the Paris basin. Each stratum was characterised by a distinct set of fossils. The deepest strata had fossils with least in common with living flora and fauna (as we would expect, today). From his studies, Cuvier realised that there were vast numbers of extinct species, and that many new forms of life had appeared. He explained this in terms of repeated mass extinctions in the history of life. That is, periodic catastrophes had happened. In his version of **catastrophism**, the devastated areas were 'localised' and eventually repopulated by species from elsewhere on the Earth. Other catastrophists believed that God created life anew after each global catastrophe.

Species do change, but they degenerate

Georges-Louis Buffon (1707–88), a doctor, mathematician and scientist, wrote a beautiful and popular *Natural History* in 44 volumes. In it he provided a survey of the living world, paying attention to those major features of organisms that could be traced through series of living things. This was in contrast to the minute differences between species on which Linnaeus was focusing in devising his binomial system of classification (p. 22). Buffon gradually realised that the idea that species were fixed and unchanging was wrong. Organisms may change if structures fall into disuse and degenerate. Thus, on the pig's foot, lateral toes that are never used are degenerate or **vestigial forms** of toes that were once used (Figure 30.3). The alteration of species with time can sometimes be seen in their degenerate structures. Buffon, who claimed the ape as a degenerate human, also wrote of common ancestors for species that resemble each other.

An outright evolutionist

Erasmus Darwin (1731–1802), grandfather of Charles Darwin, was a famous physician, poet and inventor (Figure 30.1). Erasmus believed in the evolution of the living world ('organic evolution'), and wrote convincingly on the subject. In fact, Archdeacon Paley wrote his powerful book *Natural Theology* in order to counteract the influence of Erasmus, who was seen by many as a 'dangerous atheist'. Meanwhile, Erasmus, the first of five generations of Darwins to become a Fellow of the Royal Society (between 1761 and 1962), robustly ignored opposition to his views.

Erasmus trained at the Edinburgh Medical School and set up a practice in the Midlands. He was an energetic but portly man with the motto 'Eat or be eaten!' His instantaneous professional success was based on his sympathetic and cheerful approach as a conscientious, thorough and kindly doctor. He had many wealthy patients, and often treated poor people for free. He joined with many influential friends to set up a scientific discussion group that met during periods of full moons (for safety of travel), and hence was called the Lunar Society.

Erasmus Darwin had read Buffon and came to believe in evolution by variation and 'improvement' of species by their own inherent activities. He believed, too, that characteristics acquired by the parent may be transferred to the offspring. He reported these ideas in a book *Zoonomia: or the Laws of Organic Life*. He also wrote about the evolution of life from 'tiny specks in the primeval sea' to form fish, amphibians, reptiles and human beings. His books went out of fashion as 'evolutionary' ideas became dangerous – but his grandson Charles read them as a young man.

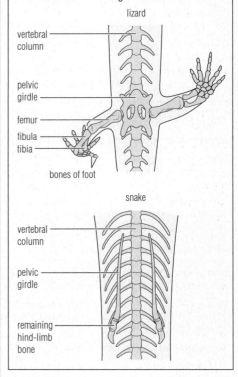

Figure 30.3 Vestigial forms: the pelvic limb girdle and limb bones of the snake, compared to those of other living reptiles, are so reduced as to have lost their original function

Questions

1 Of what significance for evolution theory was the realisation by geologists that the Earth was more than a few thousand years old?

2 Why can we expect that, of all the fossils found in a sedimentary rock, those of the lowest strata may bear the least resemblance to present-day forms?

Figure 30.4 The evidence for evolution: a summary

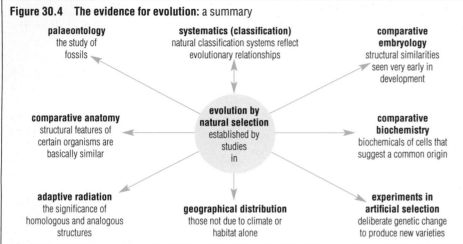

palaeontology
the study of fossils

systematics (classification)
natural classification systems reflect evolutionary relationships

comparative embryology
structural similarities seen very early in development

comparative anatomy
structural features of certain organisms are basically similar

evolution by natural selection
established by studies in

comparative biochemistry
biochemicals of cells that suggest a common origin

adaptive radiation
the significance of homologous and analogous structures

geographical distribution
those not due to climate or habitat alone

experiments in artificial selection
deliberate genetic change to produce new varieties

Figure 30.5 Steps to fossil formation by petrification

1 dead remains of organisms may fall into lake or sea, and become buried in silt/sand, in anaerobic, low temperature conditions

2 hard parts of skeleton/lignified plant tissues may persist and become impregnated by silica/carbonate ions, hardening them

3 remains hardened in this way become compressed, in layers of sedimentary rock

4 after millions of years, upthrust may bring rocks to the surface, and erosion of these rocks commences

5 land movements may expose some fossils, and a few are discovered by chance, but of the few organisms fossilised, very few will ever be found by humans

Figure 30.6 Radiometric dating

Dating rocks using radioactive potassium (^{40}K)

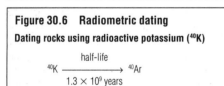

$$^{40}\text{K} \xrightarrow[\text{1.3} \times \text{10}^9 \text{ years}]{\text{half-life}} {}^{40}\text{Ar}$$

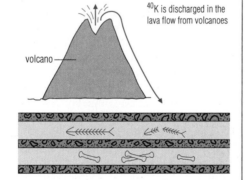

^{40}K is discharged in the lava flow from volcanoes

volcano

^{40}K is discharged in the lava flow from volcanoes (at this temperature argon gas 'boils' away and is lost). Only when lava cools is argon (^{40}Ar), released by radioactive decay, trapped in the rock. Geochemists measure the ratio of ^{40}Ar to ^{40}K in lava deposits, and so estimate the age of the lava layers that were laid down. Potassium/argon dating is accurate with materials more than half a million years old, with a reliability to the nearest 50 000 years.

In the East African Rift Valley, where volcanoes have been active for millions of years, layers of sedimentary rock (containing fossils, including human fossils) alternate with layers of solidified lava. Thus by dating the lava deposits, the age of the fossils can be estimated.

Dating fossils using radioactive carbon (^{14}C)

The bulk of carbon on Earth is ^{12}C (a stable isotope), but a tiny part is ^{14}C, formed constantly, at a constant rate, by the action of background cosmic radiation. The ^{14}C formed is radioactive:

$$^{14}\text{C} \xrightarrow[\text{5.6} \times \text{10}^3 \text{ years}]{\text{half-life}} {}^{14}\text{N}$$

^{14}C is used for dating fossils up to about 50 000 years old (5.0×10^4).

Living organisms take up carbon and maintain the same ratio of ^{14}C:^{12}C that is in the environment.

After death, no more carbon is taken in, and the ^{14}C present decays in the remains of the organism. That is, the ratio of ^{14}C:^{12}C is less than that in the equivalent living tissues.

For example, a fossilised skull with only 50% of the ^{14}C of living skull bones is 5.6×10^3 years old.

30.3 EVIDENCE FOR EVOLUTION

Evidence for evolution comes from many sources, summarised in Figure 30.4.

■ Evidence from palaeontology

Palaeontology is the study of plants and animals of the geological past, as represented by their fossil remains. Fossils are the dead remains of plants and animals, preserved in sedimentary rocks (mostly in limestones, chalks and clays, rather than in sandstones or gravels, which are too porous), in waterlogged peat or in the sticky gum that exudes from certain trees. Fossils are virtually the only source of information about forms of life that are now extinct. But only a tiny proportion of the organisms living at any time will become fossilised; of those fossils that are formed, very few will ever be discovered by a human and their secrets recorded for posterity.

Fossil formation

Fossils are formed by chance, depending on where the organism dies and on the extent to which its remains escape being eaten by scavengers, rapid decay by microorganisms and dispersal by wind or rain. Fossils are normally formed from the hard parts of organisms, which typically make up 5–20% of most organisms. So fossilised parts are the exoskeletons of arthropods, molluscs or echinoderms, or the endoskeletons and teeth of vertebrates. In plants it is hardened parts or organs, such as the woody tissues (xylem and fibres) or entire seeds, spores or pollen grains, with their protective outer surfaces, which are preserved. Occasionally 'fossils' take the form of moulds of footprints, of animal burrows or of the outline of the body or soft organs. Exceptionally, whole organisms have been frozen in the ice of previous ice ages, and are later discovered in the permafrost. In successful fossilisation the organic matter may be rapidly impregnated with mineral salts such as silica or carbonates so that it hardens into 'stone' or crystallises, as it becomes incorporated into the rock (Figure 30.5).

Dating rocks accurately

The early geologists had to guess at the ages of rocks. Now, however, rocks can often be dated precisely by **radiometric dating** techniques. Radioactivity results from the spontaneous disintegration of atomic nuclei. Radioactive isotopes decay at a constant rate irrespective of environmental conditions, the parent

Table 30.1 The geological timescale and some biological events (the climate refers to Britain; 'mya' stands for 'million years ago')

Era	Period	Epoch	mya	Climate	Animal life	Plant life
Cainozoic	Quaternary	Holocene	0.01	Postglacial	Historic time, dominance of humans	Flora of modern Britain
		Pleistocene	2	Ice Ages	Origin of humans	
	Tertiary	Pliocene Miocene Oligocene Eocene Palaeocene	65	Part of south east England submerged; climate warm to subtropical	Development of most mammal groups, and of pollinating insects	Development of angiosperms
Mesozoic	Cretaceous		135	Climate cool; sea covers much of England (chalk deposited); fresh water covers south east England	Worldwide extinction of many large reptiles; extinction of ammonites; beaked birds appear; mammals all small	Flowering plants appear
	Jurassic		200	Climate warm and humid	Large dinosaurs dominate; mammals all small	Floating phytoplankton abundant
	Triassic		250	Climate hot, with alternating wet and dry periods	Adaptive radiation of reptiles; first mammals appear	Development of conifers and related plants
Palaeozoic	Permian		290	Desert conditions	Insects diversify on land and in fresh water	First conifers
	Carboniferous		355	Equatorial climate; coal measures laid down	Reptiles and insects appear; development of amphibians	Widespread 'coal forests' and swamps, i.e. tree-like ferns
	Devonian		405	Climate warm to moderate; sandstones laid down	Amphibians appear; development of bony fish	Development of fern-like plants
	Silurian		440	Climate warm to moderately warm	Invasion of land by arthropods	First vascular land plants
	Ordovician		500	Climate warm to moderately warm	First vertebrates (jawless fish)	Marine algae abundant
	Cambrian		580	Climate uncertain	Origin of many non-vertebrate phyla	Many algae
	Precambrian {75% of Earth's history}		700 1500 3000 5000	? ? Origin of Earth?	Origin of first animals Origin of eukaryotes Oldest fossils (prokarotes)	First photosynthesis

isotope forming stable 'daughter' isotopes. The rate of decay is expressed in terms of the **half-life**, which is the time taken for the amount of that radioactive isotope to fall by half. The shorter the half-life, the faster is the rate of decay.

Many isotopes in the Earth's crust have been found to have half-lives of the order of 10^9 to 10^{10} years. By comparing the abundance of these isotopes to that of their decay products, it is possible to estimate the age of very ancient rocks (the Earth is believed to be 4.5×10^9 years old). Typically, rocks may be dated by the potassium/argon method. Fossils themselves are more commonly dated by the radiocarbon method (Figure 30.6).

The geochronological record

The sequence of geological periods (grouped into eras) shown in Table 30.1 is the product of many years of detective work by geologists and palaeontologists. Superimposed layers (strata) of rock and the fossils sandwiched within them in one part of the world have been correlated with strata at different locations by the presence of similar fossils. The fossils in a stratum probably represent a local community of organisms, living at a particular time (although it is likely to be a very incomplete picture). The ages of fossils and rocks have been determined relatively accurately by radiometric dating studies. As a result, the fossil record also tells us something about the sequence in which groups of species have evolved, and the sequence and timing of the appearance of the major phyla. The fact that fossilised remains form a succession in the sedimentary strata, with those of the earliest/lowest strata showing least in common with most present day forms, is evidence for organic evolution. For example, should a discovery of a fossilised mammal (or flowering plant) in carboniferous rocks occur, this would be contrary evidence (and none have been found).

Intermediate forms

Intermediate forms of animal and plant organisation may be interpreted as the links between related groups of species. Although the fossil record might be expected to provide evidence of the evolutionary origin of new forms, it is in fact frustratingly incomplete. Very few intermediate forms have been found.

Of interest is the fossil record of the evolution of the modern horse (*Equus*, Figure 30.7, overleaf). Between *Eohippus* of the Eocene period and modern *Equus* there lived innumerable intermediate species, all now extinct. By selective arrangement of fossilised bones and teeth, a hypothetical picture of horse evolution can be constructed showing a sequence of increasing size of the animal and the length of its limbs, reduction in lateral toes and strengthening of the third (middle) digit, and the development of molar teeth of the upper and lower jaw as an interlocking 'grinding mill'.

Another possible example of an intermediate form, discovered in the fossil record, is *Archaeopteryx* (Figure 30.8, overleaf), a 'missing link' between reptiles and birds.

Questions

1 The ages of fossils sandwiched between layers of solidified volcanic lava can be accurately dated. Why is this so? How is it done?

2 How may the spore/pollen record give clues to climate in the history of life on Earth (p. 650)?

Figure 30.7 The evolution of *Equus*

The wide range of fossil remains of extinct 'horse-type' mammals has enabled palaeontologists to piece together an indirect pathway (with many intermediate species) by which the modern horse may have evolved over a period of 54 million years.

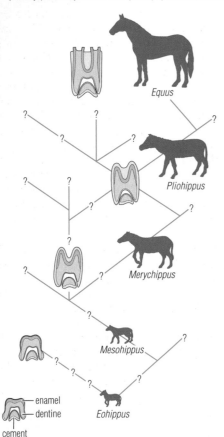

modern horse (1.6 m high), existing from Pleistocene; lived on dry grassland, feeding by chewing; molar teeth with large surface area, enamel exposed to make an efficient grinding surface; fast-running animal, with metacarpals and metatarsals lengthened, one-toed foot with hoof

Equus

Pliocene horse with three digits (second and fourth much reduced) and hoof; thickened metacarpal and metatarsal of main digit

Pliohippus

Miocene 'horse' of open, dry prairie grassland; fast-running, about 1 m high; premolar and molar teeth enlarged, with cement (strengthening) over the surface

Merychippus

Oligocene 'horse' of forest and prairie; height about 0.6 m; walked on three digits, with central digit enlarged

Mesohippus

dog-sized mammal (0.4 m high) of Eocene, living around streams; broad 'paw' for support on soft ground; four digits to fore-limb, three digits to hind-limb; teeth for crushing soft vegetation

Eohippus

— enamel
— dentine
cement

? = other forms, now extinct, which may or may not be related, as shown

Figure 30.8 *Archaeopteryx*, a missing link

reconstruction of *Archaeopteryx* skeleton

bird skeleton

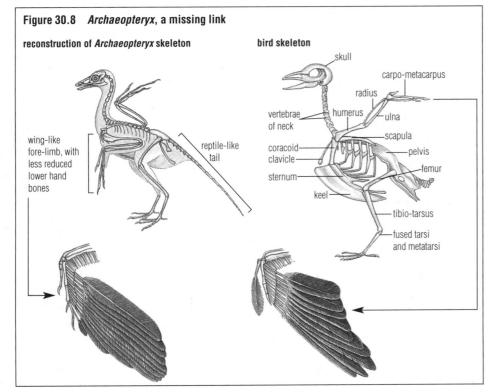

wing-like fore-limb, with less reduced lower hand bones

reptile-like tail

skull
carpo-metacarpus
radius
vertebrae of neck
humerus
ulna
scapula
coracoid
clavicle
pelvis
sternum
femur
keel
tibio-tarsus
fused tarsi and metatarsi

■ Studying the past by pollen analysis

Identifiable microfossils of pollen grains and spores are found preserved in peat and in lake sediments. They have an extremely long geological life, especially if they become rapidly submerged below the water table. The outer part of the wall of a pollen grain (the exine, p. 571) effectively resists decay. The general shapes of grains and spores, and the sculpturing of their outer surfaces, enable palaeobotanists to identify them (mostly by using reference collections of living and fossil forms), sometimes to species or genus and sometimes to family, according to how distinctive the surface characteristics are. Studies of pollen found in sediments can thus shed light on the vegetation that prevailed in the area in the past (Figure 30.9). Spores are known that date back to the Silurian period, more than 450 mya. The oldest pollen was formed in the Carboniferous period, about 330 mya.

Figure 30.9 Pollen analysis

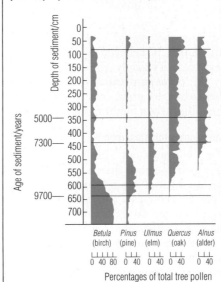

pollen prepared from ancient peat

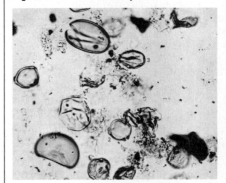

Percentages of total tree pollen

Betula (birch) *Pinus* (pine) *Ulmus* (elm) *Quercus* (oak) *Alnus* (alder)

■ Evidence from comparative biochemistry

All living things have DNA as their genetic material, with a genetic code (p. 206) which is almost universal. The processes of 'reading' the code and protein synthesis, using RNA and ribosomes, are very similar in prokaryotes and eukaryotes, too (Figure 29.9, p. 634).

Physiological processes such as respiration involve the same types of step and similar or identical biochemical reactions in all species. ATP is the universal energy currency. Among autotrophic organisms, the biochemistry of photosynthesis is virtually identical, as well.

Thus biochemical analysis suggests a common origin for living things. The biochemical differences between the living things of today are limited. This makes it possible to measure the relatedness of different groups of organisms by the amount of difference between specific molecules such as DNA, proteins and enzyme systems.

'Genetic distance': differences between DNAs

Comparing DNA from a variety of living species can indicate quantitatively the degree of relatedness of their genes. To quantify the relatedness of two species (for example, humans and chimpanzees), DNA strands from both species are extracted and separated (hydrogen bonds between complementary base pairs are broken by heat treatment). Then restriction enzymes (p. 646) are used to cut single-stranded DNA into fragments about 500 nucleotides long. If the fragments from the two species are mixed, double-stranded DNA is formed by hydrogen-bond formation between complementary base pairs. Some of the double strands are of DNA from the same species but others are from different species (**hybrid DNA**). On heating, double-stranded DNA will separate; the hybrid DNA, however, separates at a lower temperature. The closer the temperature at which hybrid DNA separates to the temperature at which single-species DNA separates, the more closely related are the two species. The degree of relatedness of various primates to humans estimated in this way gives results as follows:

chimpanzee (an ape) 97.6%
rhesus monkey 91.1%
vervet monkey 84.2%
galago (a prosimian) 58.0%

These results tie in with other evidence about the evolution of the primates (p. 668).

The composition of key proteins

Another way of quantifying the degree to which species are related is to find the number of amino acid sequences they have in common in an essential protein such as cytochrome c (Figure 15.23, p. 316) or the protein of haemoglobin (Figure 17.28, p. 358) which occurs widely in vertebrates. Variations in amino acid sequences of proteins arise from rare, spontaneous mutations in the genes that code for them. The more distantly related two organisms are the more differences they will show in protein composition, because more mutations will have taken place over the geological time that separates them.

The composition of large proteins is quickly and accurately determined by semi-automated laboratory apparatus. It turns out that the sequences of amino acid residues of such proteins are similar in all living things in which they occur. Between the human and the chimpanzee, our nearest primate relative, we have no differences in the amino acid sequence of our cytochromes and haemoglobin!

Immunological evidence

Immunology provides a method of detecting differences in proteins, and so quantifying relatedness of species.

Human serum is used to induce the production of antibodies in a rabbit, and the product, human antigens (anti-human antibodies or rabbit antibodies to human proteins), are obtained as a suspension. Serum from other animals may be tested against a suspension of human antigens. The more closely related the animal is to humans, the greater is the precipitation reaction that results (Figure 30.10). Taking the precipitate produced by human serum reacting with human antigens as 100%, the following percentage values indicate the degree of evolutionary relationship with humans:

chimpanzee	African	97
gorilla	apes	97
gibbon	Asian ape	92
baboon	Old World monkey	75
spider monkey	New World monkey	58
lemur	prosimian	37
hedgehog	insectivorous mammal	17
pig	even-toed hoofed mammal	8

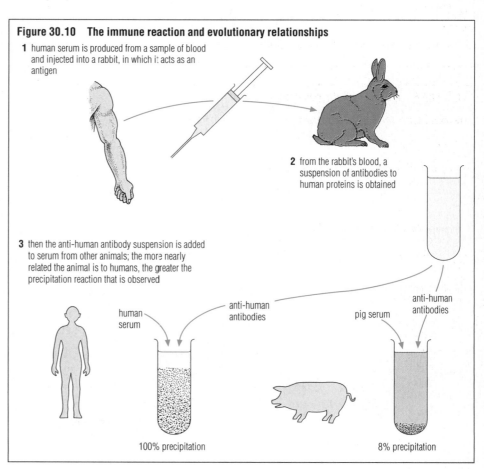

Figure 30.10   The immune reaction and evolutionary relationships

1 human serum is produced from a sample of blood and injected into a rabbit, in which it acts as an antigen

2 from the rabbit's blood, a suspension of antibodies to human proteins is obtained

3 then the anti-human antibody suspension is added to serum from other animals; the more nearly related the animal is to humans, the greater the precipitation reaction that is observed

human serum

anti-human antibodies

pig serum

anti-human antibodies

100% precipitation

8% precipitation

Figure 30.11   Continental drift (plate tectonics): the continents of the Earth are parts of the crust that 'float' on the underlying mantle; their movements may be the result of convection currents in the mantle (Figure 3.1 shows the Earth in section, and Figure 3.2 shows the present-day distribution of the continents, p. 45)

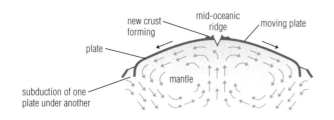

1 **about 200 million years ago (mya)**, it is believed that a single giant continent, known as Pangaea, comprised all the land masses of the Earth

2 **about 180 mya**

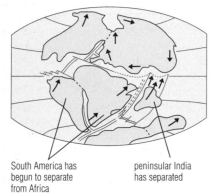

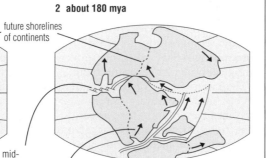

3 **about 155 mya**

4 **about 65 mya**

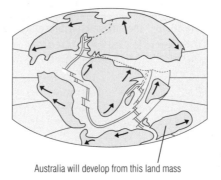

South America has begun to separate from Africa

peninsular India has separated

Australia will develop from this land mass

5 **present day: the distribution of the continents today is shown in Figure 3.2**

6 **in 50 million years from now**, continental drift will have continued, with mid-oceanic ridges still causing plate movement

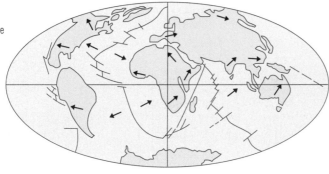

■ Evidence from geographical distribution

Countries with similar climates and habitats do not necessarily have the same flora and fauna. Why?

To answer this we will take, as examples, South America and Africa, since both have a very similar range of latitudes, and their habitats include tropical rain forests, savanna, and mountain ranges. Both these land masses have certain common fossil remains. These include a fossil dinosaur, known as Mesosaurus, living on both continents in the Jurassic period (from about 200 mya). However, the faunas (and floras) of the land masses now differ. For example, South America now supports New World monkeys (they have tails), llamas, tapirs, pumas (a mountain lion) and jaguars, but not the fauna of Africa. The present day fauna of Africa includes Old World monkeys, apes, African elephants, dromedarys, antelopes, giraffes, and lions, but not the fauna of South America.

The explanation is that these land masses were once joined, and they then shared a common flora and fauna. However, about 150 mya they literally drifted apart (plate tectonics, Figure 30.11). For the past 100 million years, the organisms of South America and Africa have evolved, isolated from each other.

We believe species originate in a given area and disperse out from that point, coming to occupy favourable habitats wherever they find them. But the huge ocean that opened up between South America and Africa formed an impossible barrier, and organic evolution took separate paths here, producing quite different modern faunas and floras.

■ Evidence from comparative anatomy

Studies of comparative anatomy show that many structural features of different organisms are basically similar. For example, the bones of all mammalian fore-limbs are similar (Figure 2.2, p. 23) in spite of differences in shape (which are largely due to specialisation for a particular function); in fact, the limbs of all vertebrate groups, from amphibia to mammals, are clearly of the pentadactyl plan (p. 486). This may be regarded as evidence of a common evolutionary history; the alternative is to assume that the limbs of the various species were created independently and their resemblance is a mere coincidence.

Another example is the flask-shaped archegonium that contains the female gamete (egg cell) in the mosses (Figure 26.3, p. 564). The presence of an archegonium on the gametophytes of ferns (Figure 26.10, p. 567) indicates to us that ferns and mosses share a common ancestry. Other plant groups, including plants known as the club mosses, all well established in the Carboniferous period, show this type of reproductive organ, and must be related.

■ Evidence from adaptive radiation

The idea of homologous and analogous structures in organisms was introduced in Figure 2.2, p. 23. The similarities of homologous structures are due to their common ancestry, whereas those of analogous structures are due to a common function only. Homology is illustrated by the mouthparts of insects. Their modifications to cope with different diets illustrate the process of adaptive radiation.

It can be argued that the basic plan (that is, the mouthparts of an ancestral insect) consisted of a **labrum** (upper lip), a pair of **mandibles, hypopharynx** (the floor of the mouth), a pair of maxillae and the **labium** (a fused, second pair of mandibles). These structures are thought to have originated from **paired limbs**, one pair on each of the segments that developed into the head. Within the Class Insecta the characteristic mouthparts have been modified in a variety of ways. For example, liquid food can be efficiently sucked up (as in the house fly, Figure 14.10, p. 293, and also in the butterfly, Figure 14.7, p. 292), skin can be pierced and blood removed (female mosquito, Figure 14.8, p. 293) or vegetation may be chewed (grasshopper, Figure 14.13, p. 295).

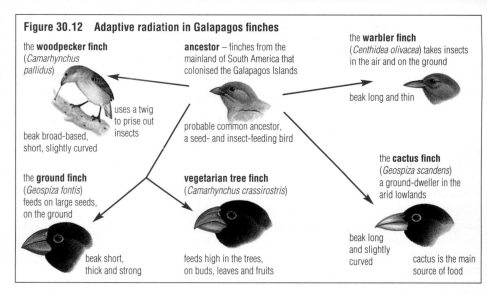

Figure 30.12   Adaptive radiation in Galapagos finches

the **woodpecker finch** (*Camarhynchus pallidus*) uses a twig to prise out insects — beak broad-based, short, slightly curved

ancestor – finches from the mainland of South America that colonised the Galapagos Islands

probable common ancestor, a seed- and insect-feeding bird

the **warbler finch** (*Centhidea olivacea*) takes insects in the air and on the ground — beak long and thin

the **cactus finch** (*Geospiza scandens*) a ground-dweller in the arid lowlands — beak long and slightly curved — cactus is the main source of food

the **ground finch** (*Geospiza fontis*) feeds on large seeds, on the ground — beak short, thick and strong

vegetarian tree finch (*Camarhynchus crassirostris*) feeds high in the trees, on buds, leaves and fruits

The **range of form in finches** that Darwin observed in the Galapagos Islands during the voyage of the *Beagle* was mentioned on p. 11. There are 14 species, differing greatly in size and other features, including beak morphology. The variations in beak morphology reflect **differences in feeding habits**. Darwin suggested that these Galapagos finches were all closely related, in spite of their differences, and the ornithologist David Lack followed up Darwin's observations during an extended study of his own, published in 1947. Lack suggested that soon after the islands were formed they were populated by a flock of finches from the mainland. These then evolved to fill the available niches (Figure 30.12).

Different lines of evolution appear among faunas (and floras) that become separated when land masses part and continents form. A good example comes from comparisons of the natural history of Australia with that of North America and Europe (Figure 30.13). Australia has

many species of pouched mammals (marsupials), but virtually no true (placental or eutherian) mammals. Yet Australia has a climate and habitats highly favourable to placental mammals, for when these are introduced there they have prospered; the sheep and rabbit, for example, have been conspicuously successful.

The explanation lies in the geographical isolation of Australia, the result of continental drift (plate tectonics, Figure 30.11). Australia became isolated at about the time the marsupial and eutherian mammals diverged from a common ancestor (in the Cretaceous, about 120 million years ago (mya). In Australia, marsupials have developed most successfully without competition from eutherian mammals. And comparison of Australian marsupials with the eutherian mammals of North America, for example, reveals dramatic examples of parallel adaptive radiation. For instance, the thylacine (the Tasmanian

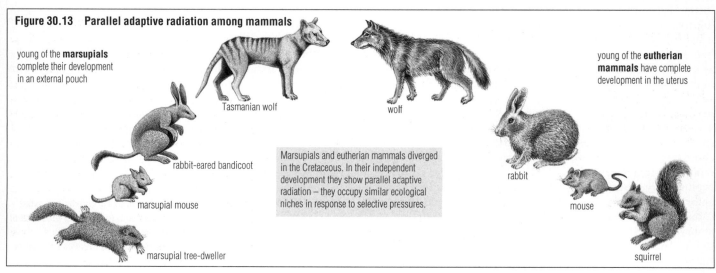

Figure 30.13   Parallel adaptive radiation among mammals

young of the **marsupials** complete their development in an external pouch

Tasmanian wolf

wolf

young of the **eutherian mammals** have complete development in the uterus

rabbit-eared bandicoot

marsupial mouse

Marsupials and eutherian mammals diverged in the Cretaceous. In their independent development they show parallel adaptive radiation – they occupy similar ecological niches in response to selective pressures.

rabbit

mouse

marsupial tree-dweller

squirrel

wolf, a marsupial dog) and the wolf or feral dog (a eutherian mammal) are extremely similar in gross appearance and in skeletal structure, including their skulls and teeth). Their reproductive systems, however, are quite dissimilar, and indeed they are fundamentally different mammals. Here similar 'designs' have evolved from differing starting points, in response to selective pressure (Figure 30.13).

Analagous structures and convergent evolution

Throughout the world, there are also many organisms, living in the same or very similar environmental conditions (opportunities and constraints), which have acquired similar adaptive structures and lifestyles. The tendency of unrelated organisms to acquire similar structures is known as evolutionary **convergence** (Figure 30.14).

The principle of convergence is illustrated by the different vertebrates that have taken to life in the sea, sometimes as a secondary adaptation. Take, for example, a cartilaginous fish, an aquatic mammal such as the dolphin, and a bird such as the penguin. These are only distantly related animals (though they are all vertebrates) which have mastered movement in a marine environment from different starting points. As a result their bodies, with a streamlined shape with fins or paddles, resemble each other in their outline and function, but are derived from fundamentally different structures. Structures that fulfil the same roles but have differing origins are called **analogous structures**.

■ Evidence from comparative embryology

A study of the development of an organism from egg to adult (its **ontogeny**) may throw light on its evolution (its **phylogeny**). Embryological stages may give insight into what the young stages were like. For example, all vertebrates start life as a single cell and their development into multicellular, complex organisms has similarities. In certain cases, similarities may represent stages in evolution. All vertebrate embryos have a notochord (p. 414) yet it is only the primitive protochordates (Figure 30.30, p. 666) that retain this structure in the adult. But the evolutionary significance of similarities in embryology must be treated with caution. Some may be adaptations to embryonic environment. The notochord, though not retained, could be a vital support structure for vertebrate embryos.

Figure 30.14 Convergent evolution, illustrated by comparison of vertebrates that move in water by reference to the structure of their fore-limbs

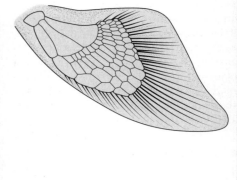

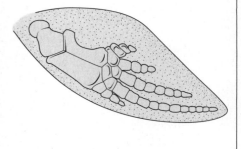

shark, a cartilaginous fish

dolphin, an aquatic mammal

penguin, an aquatic bird no longer able to fly

Nevertheless, the arrangement of the arteries and the structure of the developing heart in early vertebrate embryos follow similar patterns (p. 345). The 30-day human embryo has pharyngeal gill pouches reminiscent of those of a developing fish. In the chick embryo these pouches open, only to close again at a later stage. Clearly, there are plausible implications here for the pathway of evolution.

Ernst Haeckel (1834–1919), a contemporary of Charles Darwin, took this idea to a misleading extreme in his generalisation: 'ontogeny repeats phylogeny'. The evidence that he published in support was fraudulent, too. Haeckel admitted his offence to a university court in Jena, Prussia. Unfortunately, cases of fraudulent evidence occur from time to time, often perpetrated by scientists (like Haeckel) who otherwise have done excellent work.

Questions

1 Apart from environmental constraints, what other factors might determine the presence of a species in a favourable niche?

2 What are the essential differences between marsupials and eutherian mammals (p. 41)?

■ Evidence from artificial selection

The breeding of domesticated plants and animals has created varieties with little external resemblance to their wild ancestors. From among a family of organisms the breeder selects those that show a particular characteristic – for example, pigeons that have slightly fanned tail feathers. The breeder mates these together and selects again from the offspring for more widely fanned tail feathers. This **artificial selection** results, after only a few generations, in a true-breeding variety of pigeons with fan tails. Artificial selection produces marked changes in a short time.

Charles Darwin started breeding pigeons (Figure 30.15) as a direct result of his interest in variation in organisms. He noted in *The Origin of Species* that there were more than a dozen varieties of pigeon which, had they been presented to an ornithologist as wild birds, would have been immediately recognised as separate species. All of these pigeons were descendants from the rock dove, a common wild bird.

Darwin argued that if so much change can be induced in so few generations, then species must be able to evolve into other species by the gradual accumulation of minute changes, as environmental conditions alter and select some progeny and not others.

■ Evidence from systematics

The scheme of taxonomy established by Linnaeus (p. 22) gave each species a two-part (binomial) Latin name. Linnaeus, who originated the concepts of species and genus, also adopted a hierarchical filing system for species, placing them in increasingly general categories. He was convinced that all species had been created and none had become extinct ('God created, Linnaeus classified', he claimed). Nevertheless his system of classification paved the way for the development of one based on evolutionary relatedness.

Evolution implies that species have common ancestors and, taken to its logical conclusions, that living things originated from non-living matter at some remote time in the past. In systematics, the study of the diversity of living things (p. 20), an attempt is made to base our taxonomy of the whole living world on evolutionary relationships. The evolutionary history of related groups of organisms is known as **phylogeny**.

In fact we are some way from having a completely **phylogenetic classification**. (A phylogenetic classification is alternatively called a **natural classification**.) Within our scheme of classification (see pp. 27–41) some of the groupings are of organisms that merely show similar levels of structural complexity. The members of such groupings are derived from more than one ancestral form and are said to have a **polyphyletic** origin.

It seems highly likely that the members of the subkingdom Protozoa are polyphyletic, their only common feature being that they consist of a single cell. Other groupings of organisms with highly significant structural features in common are also polyphyletic. For example, the huge phylum Arthropoda (animals with segmented bodies and relatively hard external skeletons, p. 37), includes very diverse subgroupings such as the insects, spiders, crustaceans, millipedes and centipedes. In fact, the phylum Arthropoda is really a 'level of organisation' that was reached by several separate, more or less parallel evolutionary lines.

On the other hand, all the organisms of the phylum Angiospermophyta (p. 33) produce seeds inside an ovary (which becomes a fruit). More importantly, they all show 'double fertilisation' in the embryo sac (p. 573). These features imply they have a common evolutionary history. Similarly, the birds certainly share a common evolutionary origin. The similarities and differences shown by the members of such groupings are best explained as adaptations by closely related organisms to particular environmental conditions. Groupings like these clearly have a **monophyletic** origin, and classification within these phyla is based on homologous structures.

The ability to classify a significant proportion of living things within a phylogenetic scheme is evidence for evolution.

■ Evidence from computer simulation

Richard Dawkins of Oxford University has demonstrated the practical effects of accumulating small changes in the development of entirely new forms and structures. He used a computer program with two components (procedures), called DEVELOPMENT and REPRODUCTION. DEVELOPMENT uses the nine 'genes' passed to it to draw 'progeny' on the monitor screen. These take the form of outlines resembling other organisms, and are called 'biomorphs'. The operator then selects one biomorph (because survival and breeding opportunities are not random) as the parent of the next generation. The REPRODUCTION component then takes the selected progeny, and passes its genes down to the next generation, but only after it has introduced a minor random error (mutation). In the program, a mutation consists in a +1 or a −1 being added to the value of any one gene. The resulting (slightly modified) genome is then handed back to DEVELOPMENT to draw the new generation of biomorphs on the screen. The very high 'mutation' rate (one every 'reproduction') that is built into the program is distinctly unbiological (the probability of a mutation is often about 1 in a million). Otherwise, the program simulates the way change may occur in nature. The rate of production of new biomorphs is extraordinary. Starting with an 'ancestral' dot, a host of biomorphs (not unlike many living species) have been produced at random (*Study Guide*, Chapter 30).

Figure 30.15 The rock dove (*Columbia livia*), the ancestral species from which all domesticated pigeons have been bred

jacobin

fantail

pouter

From his breeding of pigeons, Darwin noted that there were more than a dozen varieties that, had they been presented to an ornithologist as wild birds, would have been classified as separate species.

Questions

1 What is meant by 'genetic difference'?
2 What is the difference between a variety and a species?

30.4 THE MECHANISM OF EVOLUTION

Thus it is generally accepted that the present-day flora and fauna have arisen by change ('descent with modification'), perhaps mostly gradual change, from pre-existing forms of life. The shorthand term for this idea is 'organic evolution'. But what is the mechanism of organic evolution? In modern times, four mechanisms have been suggested. These we will consider next.

■ The inheritance of acquired characteristics

This idea was developed by Jean Baptiste de Monet de Lamarck (1744–1829). Lamarck was an able, but often (then and since) lightly esteemed, French biologist, although Charles Darwin described him as 'justly celebrated'. From about 1800 onwards, Lamarck began to discuss the idea that species were not 'fixed' (Darwin used the term 'immutable').

Lamarck explained the mechanism of change in his 'law of use and disuse'. Since changes in the environment made special demands on certain organs, these organs became especially well-developed. At the same time, disuse of other organs caused them to atrophy. These developments in organisms were transmitted (at least in part) to the offspring. He suggested that the giraffe evolved from

Figure 30.16 Giraffe feeding

ground-feeding herbivores that experienced a shortage of ground-level vegetation, and so took to feeding from trees (Figure 30.16). The constant straining of the neck over many generations resulted, he believed, in the accentuation of this feature and its inheritance by the progeny.

Table 30.2 Lamarck's theory of evolution

First law
Organisms tend to increase in size and complexity with time.
Second law
New organs develop in response to an organism's specific need for them.
Third law
Organs vary in size and efficiency in direct proportion to use.
Fourth law
All that is acquired in an organism's lifetime may be transmitted to the offspring in reproduction.

■ Continuity of germ plasm

The central idea of Lamarckian inheritance, on which his whole hypothesis hangs, is that acquired characteristics are inherited (Table 30.2). August Weismann (1834–1914) was the most influential scientist in establishing the modern rejection of the concept. He suggested that the substances (gametes) passed from parents to offspring (he called this 'germ plasm') were not changed by the surrounding body cells ('soma'), merely nourished by them and passed to the next generation (Figure 30.17).

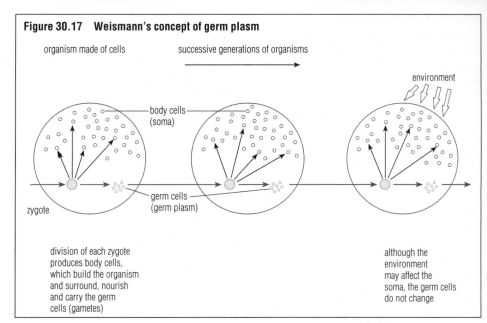

Figure 30.17 Weismann's concept of germ plasm

organism made of cells   successive generations of organisms

environment

body cells (soma)

zygote   germ cells (germ plasm)

division of each zygote produces body cells, which build the organism and surround, nourish and carry the germ cells (gametes)

although the environment may affect the soma, the germ cells do not change

■ Natural selection

The contributions of Charles Darwin to biology are introduced in the panel on p. 11. Darwin's unique contribution to the debate about evolution was, firstly, to state the evidence for evolution clearly, and to argue from this basis that species were not individually created in their present forms. Secondly, he reviewed the objections to the idea of evolution, and showed they were answerable. Thirdly, he proposed a mechanism that was testable, and which excluded supernatural forces. This mechanism he called natural selection. Ultimately, he convinced most of the scientific community (and many other people) that species have evolved from earlier forms of life.

Darwin began to compile notes on evolution in 1837 (he called it 'transmutation'), and made a first draft of his theory in 1842. The stimulus to publish came when a comparable hypothesis reached him in a private letter from Alfred Russel Wallace (1823–1913, Figure 30.18). A joint paper was read to a Linnean Society meeting in 1858, but it had very little impact. Then Darwin went on to develop his ideas in his book *On the Origin of Species by Natural Selection, or the Preservation of the Favoured Races in the Struggle for Existence*, published by John Murray of Albemarle Street, London, in 1859. The book, always known as *The Origin of Species*, sold out on its first day of publication, and has proved one of the most influential books of all time (despite Darwin's obsessive fears that it would be ignored or that he would be ostracised by society). The book has been in print ever since.

The theory of natural selection

The arguments of *The Origin of Species* can be summarised in four statements (S) and three deductions from those statements (D):

S1 Organisms produce a far greater number of progeny than ever give rise to mature individuals.

S2 The numbers of individuals in species remain more or less constant.

D1 Therefore there must be a high mortality rate.

S3 The individuals in a species are not all identical, but show variations in their characteristics.

D2 Therefore some variants will succeed better and others less well in the competition for survival, so the parents for the next generation will be selected from among those members of the species better adapted to the conditions of the environment.

S4 Hereditary resemblance between parents and offspring is a fact.

D3 Therefore subsequent generations will maintain and improve on the degree of adaptation of their parents, by gradual change.

The argument is persuasive. But although D1 is a valid deduction from S1 and S2, D2 is in fact a hypothesis which needs to be (and has been) checked by experimental means. D2 was restated by Herbert Spencer (a popular philosopher who extended the concept of evolution to many other fields) as 'nature guarantees the survival of the fittest'. This phrase Darwin took up, and it appears in later editions of The Origin. To avoid the criticisms that 'survival of the fittest' is a circular statement (how can fitness be judged except in terms of survival?), the term 'fitness' is applied in particular contexts. Thus the 'fittest' of the wildebeest of the African savanna (hunted herbivore) might be those with the acutest senses, the quickest reflexes and the strongest leg muscles (for efficient escape from the predatory lions).

Finally, the mechanism behind S4 was unknown in Darwin's time, for Mendel's work was not rediscovered and reassessed until 1900. In the meantime there was some confusion at the heart of Darwin's view of natural selection, for it was based on the idea that 'like begets like, but not always'! The mechanism of inheritance is central to natural selection.

Alfred Russel Wallace

Figure 30.18   Down House: Darwin lived here and wrestled in private with his Theory of Natural Selection for over 10 years before similar ideas, presented in a letter by Wallace, forced him to publish

'Blending inheritance', in vogue in Darwin's time, could only lead to a dilution of variation.

■ Neo-Darwinism

Neo-Darwinism is the restatement of the concept of evolution by natural selection in terms of Mendelian (and post-Mendelian) genetics (Table 30.3).

The fundamental origins of genetic variation are mutation (chromosome mutation, p. 625, and gene mutation, p. 626) and random assortment of maternal and paternal homologous chromosomes (p. 199), and recombination of segments of individual maternal and paternal homologous chromosomes during crossing over (p. 617). Once genetic differences are established in an organism, they are likely to be expressed as phenotypic variations. Some phenotypes may be better adapted than others to survival and reproduction in a particular environment. Then, when natural selection operates, there is a change in the proportion of genetic variation in a population of the species. This may lead to the formation of new varieties and species. Neo-Darwinism differs from Lamarckism in its view that variations *arise spontaneously* and are selected by environmental pressure, rather than arising *in response to* environmental pressure.

An important development in the understanding of evolutionary theory was the birth of population genetics, and the realisation that continuous variation may be genetically controlled. Population genetics made it possible to analyse changes in the genetic composition of populations and to demonstrate the forces responsible for evolutionary change. These forces are discussed overleaf.

Table 30.3   The ideas of Neo-Darwinism

Genetic variations arise via:
▶ mutations (chromosome mutations and gene mutations), pp. 625–6
▶ random assortment of parental chromosomes in meiosis, p. 191
▶ recombination of segments of parental homologous chromosomes during crossing over, p. 616 (and via the random fusion of male and female gametes in sexual reproduction – understood in Darwin's time)

Then, when genetic variation has arisen in organisms:
▶ it is expressed in their phenotypes
▶ some phenotypes are better able to survive and reproduce in a particular environment
▶ natural selection operates, causing changes in the proportions of particular genes in an isolated population
▶ it may lead to new varieties and new species

The subsequent development of the analytical techniques of population genetics (p. 658) has allowed natural selection to be detected in populations.

Figure 30.19 The sequence of mass extinctions

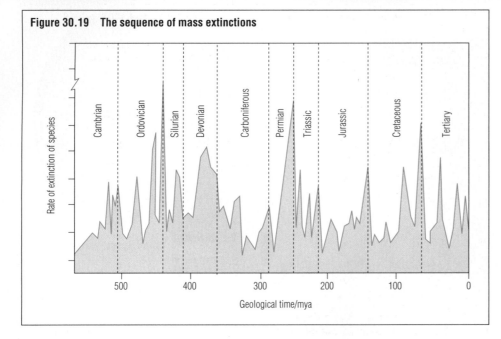

Punctuated equilibria

The death of the dinosaurs about 65 mya has been widely discussed. It is less well known that at that time almost half the genera of marine non-vertebrates also became extinct. Moreover, this was only one of many extinction events. The extinction at the end of the Permian period, 250 mya, elimated over 80% of marine non-vertebrate genera! The sequence of mass extinctions, obtained by plotting the number of genera that die out in a geological period against the duration of that period, is shown in Figure 30.19. The peaks stand out against the 'background level' of extinctions; the pattern does not look like one of 'gradual change', as predicted by the theory of evolution by natural selection.

Observations like these were addressed by Niles Eldridge and Stephen Gould (1972) in their book *Punctuated Equilibria: an Alternative to Phylogenetic Gradualism*. Their argument builds on two points.

Firstly, there are still relatively few clear examples among fossils of long-term gradual changes. Instead, new species appear relatively quickly (but they are talking about periods of 5000–50 000 years!). These species then remain virtually unchanged, often for millions of years, and then rather abruptly disappear, to be replaced by new species (according to the limited fossil record we have to go on). Eldridge and Gould refused to accept the explanation that the fossil record is incomplete. Instead, they said the fossil record accords with their theory of the origin of species. Secondly, they claimed that most anatomical changes that come about as new species are formed are compressed into a relatively short geological time-span. Brief periods of speciation seem to punctuate long periods of population stability.

What often happens when the environment changes for the worse is that species move to more congenial conditions. Organisms migrate, as well as evolving! So an organism's first response to an unfavourable habitat is to change it if possible. Those that cannot move may die out at this stage. When the changes are violent or profound, mass extinctions result. Nevertheless, populations at the fringe of the range, including those that have moved in response, often include small, isolated reproductive communities. Populations at the margins of species are already adapting to a different environment. If the isolated community survives it may establish a new species relatively quickly.

At this point we are left with the issue: gradual or puntuational evolution? There is evidence for both processes in the geological history of life. But the two alternatives are not necessarily exclusive. Both may operate simultaneously.

Questions

1 How did Charles Darwin's theory of evolution differ from that of his grandfather (p. 647)?
2 Summarise the importance of modern genetics to the theory of the origin of species by natural selection.

30.5 MICROEVOLUTION: THE FORMATION OF SPECIES

Population genetics

In nature, organisms are members of a local population. A **population** is a group of organisms that may freely interbreed. Most species are distributed as a large number of local populations. An isolated population of Inuit in Greenland and another isolated population of Amazonian Indians in the Brazilian jungle are two examples of local populations. The individual members belong to the same species but have no practical possibility of interbreeding. A small isolated population is known as a **deme**. The individuals of a deme are not exactly alike, but they resemble one another more closely than they resemble members of other demes. This similarity is to be expected, partly because the members are closely related genetically (similar genotypes), and partly because they experience the same environmental conditions (which affect the phenotype). Examples of isolated human populations have to be drawn from such groups, since so many human populations are not truly isolated, today. In many ways, many of us live in a 'global village'!

By contrast, the populations of most plants and slow-moving animals are much smaller groupings, even though the boundaries of any one population may be hard to define. The bluebells (*Endymion non-scriptus*) growing in a particular wood, or the garden snails (*Helix aspersa*) found living around an old compost heap, are likely to constitute breeding populations.

Some populations are 'open', with pronounced immigration of genes from overlapping populations; for example, pollinating insects linking plants with others from more distant communities, or snails moving in from a nearby hedge population. At the opposite extreme are 'closed' populations, cut off by effective barriers, such as an expanse of insecticide-treated cash crop with few pollinators surviving, or a wide dry roadway across which snails cannot travel.

Gene pools

Population genetics is the study of the genes in populations. In any of these populations, the total of the genes located in reproductive cells make up the **gene pool**. From these genes a sample forms the genomes of the next generation.

Allele frequency in gene pools may change

The composition of alleles making up a gene pool may change with time for one or more of the following five reasons:

1 Selective predation, where individuals of the population of a particular genotype are especially vulnerable or particularly advantaged. For example, the shell coloration of some snails (which is genetically controlled) may leave certain snails either particularly visible or effectively camouflaged in relation to predators.

2 Emigration or immigration, where successful migration depends on a genetically controlled characteristic, such as a superior locomotory musculature, or enhanced ability to withstand moisture loss during journeys. These characteristics might apply to members of a population of snails, making certain of them more likely to survive than the snails without the alleles involved.

3 Mutation, when a random, rare, spontaneous change in the genes occurs in gonads of a breeding member of the population, leading to the possibility of a new characteristic in the offspring of this organism. The changed characteristic might be the ability to inactivate a pesticide molecule, for example. This may greatly influence survival chances of the progeny, causing the mutant allele to spread through the population quickly.

4 Random genetic drift, where sudden hostile physical conditions, such as an intense cold spell, flooding, or a period of drought, may sharply reduce a natural population to very few survivors. On the return of a favourable environment, numbers of the affected species would quickly return to normal due to reduced competition for food sources. However, the new population would have been built up from a small sample of the original population, with many more 'first cousin' and backcross matings than normal. This causes fewer heterozygotes and more homozygotes (p. 645), and some alleles would be lost altogether.

5 Founder effect, when a barrier arises within a population, instantly isolating a small sample of the original population. This sample is likely to carry an unrepresentative selection of the gene pool, yet it would be the basis of a new population. This is another form of genetic drift. The barriers that may arise may be physical or behavioural (p. 661).

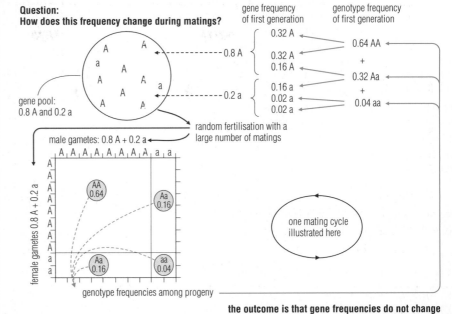

Figure 30.20 **The fate of genes in a gene pool**, where there are two alleles A and a, initially present in the frequencies 0.8 and 0.2 respectively

Question:
How does this frequency change during matings?

gene pool: 0.8 A and 0.2 a

male gametes: 0.8 A + 0.2 a

random fertilisation with a large number of matings

gene frequency of first generation

0.8 A { 0.32 A, 0.32 A, 0.16 A }

0.2 a { 0.16 a, 0.02 a, 0.02 a }

genotype frequency of first generation

0.64 AA + 0.32 Aa + 0.04 aa

one mating cycle illustrated here

genotype frequencies among progeny

the outcome is that gene frequencies do not change

We identify all these 'forces for change' in the allele frequency by the collective name of 'disturbing factors', and will return to them later.

Allele frequencies may remain constant

In fact, provided that all disturbing factors are absent, the breeding population is a large one, and all genotypes (individuals) have an equal opportunity of contributing gametes, random matings will perpetuate the original proportions of alleles in the population. In these circumstances, the allele frequency does not change. We can demonstrate this by setting two alleles of a gene at some arbitrary frequency in an infinitely large gene pool, and then following the frequency of possible genotypes these will produce by random matings over several generations.

The example of two alleles, A and a, present in the population at frequencies of 0.8 A and 0.2 a, is illustrated in Figure 30.20. When $0.8 + 0.2$ male gametes randomly fertilise $0.8 + 0.2$ female gametes, the progeny (AA + Aa + aa) will occur in the following proportions:

AA 0.64   Aa 0.32   aa 0.04

These genotypes will become the parents of the next generation and, at this stage, contribute their genes to the gene pool. At this point the frequencies are now:

$A = AA + \frac{1}{2}Aa = 0.64 + 0.16 = 0.8$

and

$a = \frac{1}{2}Aa + aa = 0.16 + 0.04 = 0.2$

Thus the frequency of the genes has not changed after one random mating. In fact, this frequency will never change, however many matings there may be, unless there are what a geneticist calls 'disturbing factors' at work.

■ The Hardy–Weinberg principle

The Hardy–Weinberg principle states that, in a large, randomly mating population, gene and genotype frequencies remain constant (in the absence of migration, mutation and selection). We have derived this principle from our understanding of the behaviour of alleles during meiosis and fertilisation (in Figure 30.20). The mathematicians G H Hardy and W Weinberg independently discovered this 'law', which is fundamental to modern genetics, in 1908. To them it was a particular consequence of the binomial expression. (You may remember that it was Mendel's familiarity with the product of the binomial expression from randomly combining two pairs of unlike 'factors' that enabled him to interpret his 1 : 2 : 1 ratio as due to 'particulate factors' in organisms, controlling characters, pp. 607–8.)

It happened this way. Dr R C Punnett (he of the Punnett square) was lecturing to doctors about the new science of genetics when he was asked why, if brown eyes were dominant to blue eyes, the population did not become progressively brown-eyed. Not confident of the answer, Punnett sought out his colleague

G H Hardy (a Cambridge mathematician whom he knew well because they played cricket together!). Presented with this problem as a mathematical one, Hardy came up with the explanation. In the same year the German doctor Weinberg published the derivation of the law, which has been named after them both. This is what they showed.

Let p and q represent the respective frequencies of the dominant and the recessive alleles of any gene.

The frequency of all alleles must add up to unity (1), and therefore

$$p + q = 1 \text{ (and } q = 1 - p)$$

The random combination of the alleles p and q in any mating is given by

$$(p + q) \times (p + q) = (p + q)^2$$

Now $(p + q)^2$ is the binomial expression:

$$(p + q)^2 = p^2 + 2pq + q^2$$

where p^2 = the frequency of one homozygote

$2pq$ = the frequency of the heterozygote

q^2 = the frequency of the other homozygote

So in the next generation, what will be the frequencies of the two alleles?

The frequency of p

$$= p^2 + \frac{2pq}{2}$$
$$= p^2 + p(1 - p) \text{ (since } q = 1 - p)$$
$$= p^2 + p - p^2$$
$$= p$$

The frequency of q

$$= q^2 + \frac{2pq}{2}$$
$$= q^2 + q(1 - q)$$
$$= q^2 + q - q^2$$
$$= q$$

Thus the gene frequencies of p and q do not change, in the absence of **disturbing factors** (see below).

Applications of the Hardy–Weinberg Law

There are several applications of the Hardy–Weinberg Law that are of use to the population geneticist.

Firstly, if the frequency of one of the alleles in a gene pool is known, the Hardy–Weinberg equation can be used to calculate the expected proportions of the genotypes in the population.

For example, consider the case of the genetically controlled failure to produce the pigment melanin (**albinism**, p. 615).

In a large population, only one person in 10 000 is an albino. From this we understand that the frequency of the homozygous recessive (or q^2) = 1 in 10 000.

Thus $q^2 = 0.0001$
$q = \sqrt{0.0001} = 0.01$

So the frequency of the non-melanin-secreting allele in the population = 0.01 (or 1%).

Since $p + q = 1$
$p = 1 - q$
$= 1 - 0.01$
$= 0.99$

The frequency of the melanin-secreting allele in the population is thus 0.99 (or 99%).

Since $p = 0.99$ and $q = 0.01$
$2pq = 2 \times 0.99 \times 0.01$
$= 0.02 \text{ (to 2 places of decimals)}$

Thus carriers of albinism (who possess the recessive allele, but show normal pigmentation) are about 1 in 50 of the population.

The Hardy–Weinberg Law demonstrates that a large proportion of recessive alleles exists in the heterozygotes. Heterozygotes are a reservoir of genetic variability, for good or ill (the latter in the case of albinism).

Secondly, if the frequency of the genes or genotypes in a population is known we can calculate whether this proportion is to be expected from the Hardy–Weinberg principle (or whether there is evidence of a disturbing factor, such as natural selection, at work).

For example, in an isolated human population the numbers of people of the MN blood groupings was:

M 82, MN 47, N 9: total = 138

Do these frequencies accord with the Hardy–Weinberg equilibrium? Consider Table 30.4.

In the case of a population of 138 we would expect:

frequency of MM genotype = p^2
$= 0.76^2 = 0.58$

actual numbers = $0.58 \times 138 = 80$

frequency of MN genotype = $2pq$
$= 2p(1 - p) = 1.52 \times 0.24 = 0.36$

actual numbers = $0.36 \times 138 = 50$

frequency of NN genotype = q^2
$= (1 - p)^2 = 0.24^2 = 0.06$

actual numbers = $0.06 \times 138 = 8$

For the genotypes	MM	MN	NN
observed numbers	82	47	9
expected numbers	80	50	8

Thus there is such close correspondence between the observed and expected numbers of each genotype that it may be concluded that there are no disturbances to the MN blood grouping gene pool in this community.

The assumptions of the Hardy–Weinberg Law

The assumptions of the Hardy–Weinberg Law are that

▶ reproduction in the population is sexual and the organism is diploid (if haploid, then genotype frequencies = allele frequencies)

▶ the population is a large one, and matings are random

▶ the allele frequency does not change over time.

'Disturbing factors'

We have already seen that several factors can act to change allele frequency. These factors, which include random genetic drift, mutations, migrations and also natural selection, may lead to genetic change and the eventual formation of new species. We will look at speciation next.

Table 30.4

Pheno-types	Numbers	Geno-type	Frequency of M allele	Frequency of N allele
M	82	MM	164	0
MN	47	MN	47	47
N	9	NN	0	18
frequency =			211	65
frequency as a percentage =			0.76	0.24
Total no. of alleles =	211 + 65 = 276			

Question

In a small, isolated population the incidence of the following genotypes was investigated and found to be:

AA 109
Aa 252
aa 39

What was the frequency of the alleles A and a in this local population?

■ Steps to speciation

Speciation, the process by which new species arise, may occur when a local population becomes isolated from the main bulk of the population or from other demes, and the gene pool remains isolated for hundreds or thousands of years. Speciation by this process normally takes a very long time to become apparent. In plants, however, there is an alternative pathway to speciation, and the mechanism involved, **polyploidy**, occurs almost instantly (although very rarely). (A polyploid is an individual or species whose chromosome number is a multiple, other than a doubling, of the haploid number, p. 625.)

Finally, natural selection is a powerful force for change in allele frequency and speciation in *certain circumstances*.

Geographical isolation between demes

Isolation of a deme may result where natural geographical barriers restrict the movements of organisms, as when a few members of a population move over a mountain range and become permanently isolated (Figure 30.21). The outcome is the same when continents drift apart (plate tectonics, p. 652), or when a mountain range rises between two lowland populations, or when desertification spreads across a previously productive region. Barriers such as these separate whole communities (and thus all the populations they contain) over a long period of time. Lesser obstacles arise more often, and separate specific populations, according to the barrier concerned, as when a stream bursts its banks and cuts a new pathway, fragmenting an existing terrestrial population. Similarly, aquatic organisms can become separated by land barriers, and freshwater species by brackish or saline waters.

The effective distance needed to separate two populations depends upon the size of the organisms and their dispersal abilities. Snails and other relatively small, slow-moving animals are more easily isolated than flying vertebrates. Entirely sessile organisms such as rooted plants or surface-bound fungi may nevertheless be part of a huge population (gene pool) through the efficient dispersal of their seeds or spores.

Thus some barriers effectively prevent gene exchange between demes. Such a barrier is known as an **isolating mechanism**. Isolated demes diverge (evolve) as they accumulate genetic differences (see below). Ultimately, these differences may prevent successful mating if members of the two demes come together again. By this process an ancestral species gives rise to two or more descendent species, which grow increasingly unlike as they evolve. Geographical isolation is the most common way in which populations become isolated. The formation of new species that follows the spatial separation of populations is known as **allopatric** (= 'different countries') **speciation**.

Reproductive isolation of demes

Isolating mechanisms may develop that are strong enough to prevent interbreeding between two demes, living within the same geographic area. Such mechanisms of **reproductive isolation**, by preventing the production of fertile, hybrid offspring, may also lead to the formation of new species over a prolonged period. Species formation occurring in demes in the same geographical area is known as **sympatric** (= 'same country') **speciation**.

The barriers to breeding that may be found within a population in the same geographical area are discussed below. Compared to geographic isolation, there is less convincing evidence that the individual mechanisms of reproductive isolation are consistently effective as a force for speciation. It is likely, however, that several of these mechanisms, working together, may stop gene flow and facilitate speciation.

Ecological or habitat isolation

This mechanism will operate between subspecies that live within different habitats in the same area and do not normally meet. For example, ecological isolation will certainly affect many parasite populations.

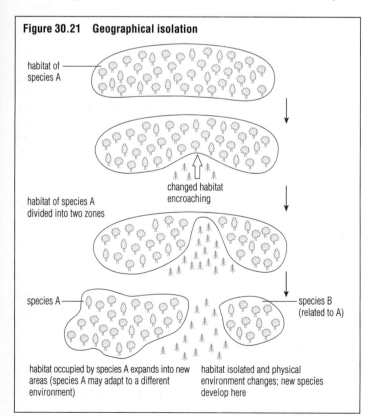

Figure 30.21 Geographical isolation

habitat of species A

changed habitat encroaching

habitat of species A divided into two zones

species A

species B (related to A)

habitat occupied by species A expands into new areas (species A may adapt to a different environment)

habitat isolated and physical environment changes; new species develop here

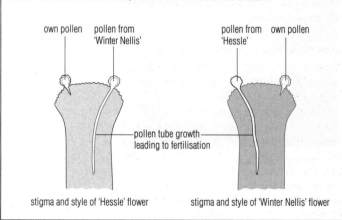

Figure 30.22 Incompatibility

'Incompatibility' in flowering plants is the name given to physiological mechanisms that may make fertilisation impossible by preventing the growth of pollen tubes on the stigma or in the style. It is one of the strategies by which self-fertilisation is avoided. Sexual incompatibility mechanisms are found in a wide range of plant species. In these species, pollination is effective only if there is a difference in alleles at the gene loci that control incompatibility.

Some varieties of pear are incompatible, and this has to be considered when a commercial pear orchard is planted. Trees of a particular variety are produced by vegetative propagation from a parent plant. Thus they all have the same genetic constitution. Some pear varieties, such as 'Conference' are self-fertile. Other varieties, including some valued especially for characteristics such as flavour or keeping qualities, are not self-fertile and must be planted with compatible varieties to ensure fertilisation and fruit production. For example, the variety 'Hessle' will fertilise 'Winter Nellis', and 'Onward' will fertilise 'Durandeau'.

own pollen   pollen from 'Winter Nellis'

pollen from 'Hessle'   own pollen

pollen tube growth leading to fertilisation

stigma and style of 'Hessle' flower

stigma and style of 'Winter Nellis' flower

661

Temporal or seasonal isolation

This mechanism will operate when two subspecies can be crossed artificially to produce hybrids, but in the wild produce gametes at different times of year. Examples are subspecies of plants that flower at different times, and animals with oestrus in different seasons; for instance, rainbow trout breed in the spring, whereas brown trout breed in the autumn.

Behavioural isolation

This operates where the steps of an elaborate courtship or mating routine of one subspecies fail to attract or to elicit the necessary 'next step' response in a potential partner of another subspecies.

Mechanical isolation

This mechanism involves structural differences between the genitalia of animals, which prevent copulation between members of different subspecies, or differences between flower structures involved in pollen transfer.

Gametic isolation

This mechanism isolates organisms that are reproductively mature at the same time, but in which physiological incompatibility of the gametes prevents the production of hybrids (Figure 30.22, p. 661). For example, sperms may fail to survive in the oviduct of the partner, and pollen may fail to grow on the stigma of a particular partner.

Hybrid inviability or hybrid sterility

This may arise if a hybrid zygote is formed, but fails to develop or at least to achieve sexual maturity, or if the hybrid individual formed is sterile.

Speciation by polyploidy

Cord grass (*Spartina anglica*) is an important plant of estuaries and salt-marshes. It was first recorded in Southampton Water in the 1870s, and it is still spreading around the coasts of Britain at a remarkable rate.

The origins of this plant have been worked out by cytological examination (Figure 30.23) although, of course, the exact time it occurred cannot be known. It is an allopolyploid (p. 626) that arose by hybridisation between a foreign species (*S. alterniflora*), probably introduced from America, and the native cord grass (*S. maritima*). The product, a sterile hybrid, then underwent immediate, spontaneous chromosome doubling, forming the fertile allotetraploid (*S. anglica*). All polyploids are instantly isolated from the parent plants by a barrier of sterility.

Polyploid formation is quite common in many plant species, yet the process in animals is rare. One reason seems to be that the sex chromosome system of animals normally breaks down if a polyploid forms, producing completely sterile progeny that immediately die out (although this is only a partial

explanation since many animals, including fish and reptiles, do not have sex chromosomes).

Natural selection, genetic change and speciation

Natural selection operates in populations in nature. It is a process that brings about adaptation and evolution. Through uneven rates of reproduction (that is, reproductive success or failure) the gene frequencies of populations are changed. If an animal or plant does not operate as effectively as others of the same species in the environment, then it will have less chance to perpetuate its genes through reproduction. Thus natural selection brings about adaptation to environment.

The three recognised types of natural selection are illustrated in Figure 30.24.

The first type, described as **directional selection**, is seen as operating on the peppered moth in a polluted environment (see Industrial melanism, opposite). Here selection favours one extreme of the distribution of inherited variability. This type of selection arises from a changing environment and results in reduced variance, and also in a shift in the population mean for the selected character. This process results in the organism becoming adapted to a changed environment.

The second type we describe as **stabilising selection**. Here the selective pressure favours the existing mean and leads to a reduction in variance about that mean –

Figure 30.23   Speciation by polyploidy and hybridisation

cord grass (*Spartina anglica*) is a halophyte; it is a C$_4$ plant, which spreads by means of sexual reproduction (seed production) and also by vegetative reproduction (growth of the rhizome)

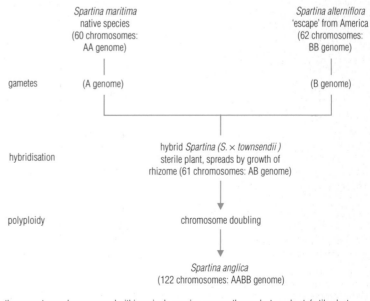

Spartina maritima
native species
(60 chromosomes:
AA genome)

Spartina alterniflora
'escape' from America
(62 chromosomes:
BB genome)

gametes (A genome) (B genome)

hybridisation hybrid *Spartina* (*S.* × *townsendii*)
sterile plant, spreads by growth of
rhizome (61 chromosomes: AB genome)

polyploidy chromosome doubling

Spartina anglica
(122 chromosomes: AABB genome)

these events may have occurred within a single growing season; the product, a robust, fertile plant, rapidly established itself and is progressively replacing the native species

that is, variants are selected out. This type of selection is typical of an unchanging environment. Figure 30.24 also shows the birth weights of babies born, and of those who died, in London between 1935 and 1946, illustrating graphically the higher death rates of the lightest and the heaviest. It is evident that stabilising selection favours a birth weight of 3.55–3.60 kg.

A third type of selection favours the extremes of variance, and is called **disruptive selection** because it leads to divergence of the phenotype – that is, to the emergence of two distinct phenotypes. This form of selection is rare but it is important, for the divergent process underpins the origin of new species. Disruptive selection might arise in a species adapting to two different habitats, for example. It is the result of conditions that favour more than one type of phenotype within a population. The effect is to split the population into two subpopulations, leading to polymorphism. If gene flow between the subpopulations is prevented then ultimately each population may give rise to a new species.

Industrial melanism

During the Industrial Revolution in the early part of the nineteenth century, air pollution by gases (such as sulphur dioxide, p. 86) and solid matter (mainly soot) was distributed over the industrial towns and cities and the surrounding countryside. The lichens and mosses on brickwork and tree trunks were killed off and the surfaces were blackened. The numbers of dark varieties of some 80 species of moth increased in these habitats during this period. This rise in the proportion of darkened forms is known as **industrial melanism**; it provides a good example of natural selection.

Organisms that exist in two forms in significant numbers are examples of **polymorphism**. In **transient polymorphism**, one form is gradually being replaced by another. (In **balanced polymorphism**, both forms coexist within a population in equilibrium; male and female forms are balanced polymorphs.) The peppered moth, *Biston betularia*, is an example of polymorphism for it exists both in a speckled white form and in a darkened or **melanic** form. The peppered moth is common in Britain. The moth normally rests on branches, where it depends upon cryptic coloration to blend with its background. The moths are preyed upon by birds, which peck them off the trees.

In the peppered moth, coloration is controlled by a single gene with two alleles. The moths of genotype CC and Cc are melanic, and moths of genotype cc are pale. Melanic forms have been found to predominate in polluted areas, and the relative incidence of melanic and pale forms is due to natural selection (Figure 30.25).

Kin selection

Darwin argued that natural selection acted only on the individual (*not* at species level; that is, not 'for the good of the species'). But he made an exception in the case of social insects such as ants and bees (p. 524). These organisms exist in huge communities consisting of many sterile individuals, genetically closely

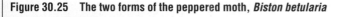

Figure 30.24   Types of selection: the three modes of selection operating on continuous phenotypic variation

directional selection is illustrated by the response of peppered moth populations to polluted environments

stabilising selection is illustrated by human birth weight and the percentage mortality at different weights

disruptive selection is illustrated by populations adapting to contrasting habitats

directional selection and disruptive selection can lead to speciation, whereas stabilising selection does not

Figure 30.25   The two forms of the peppered moth, *Biston betularia*

melanic form (above) and the normal, pale form of *Biston betularia* **(below)**

The pale form is protected from predation by insectivorous birds while it rests with wings outstretched on lichen-covered surfaces (and the melanic form whilst resting on soot-polluted surfaces). The evidence for this comes from two kinds of study, of which typical results are shown here.

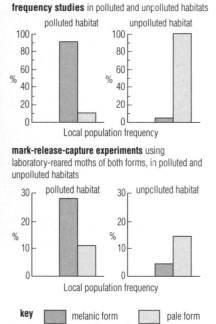

frequency studies in polluted and unpolluted habitats

mark-release-capture experiments using laboratory-reared moths of both forms, in polluted and unpolluted habitats

key   melanic form   pale form

related to the fertile members on whose behalf they toil. Self-destructive altruism is also a feature of these communities (p. 525).

Acts of altruism may be explained in terms of kin selection. Organisms that share many of the same genes help to perpetuate these when they risk self-destruction, provided they do so for the benefit of their close relatives. Thus altruism confers a selective advantage, provided the loss of an individual is offset by the benefits received by a relative. Brothers and sisters have about half their genes in common; when the distinguished and eccentric geneticist JBS Haldane (1892–1946) was asked if he would sacrifice his own life for his brother, he answered that he would do so for two brothers or four cousins!

■ Sexual selection

Sexual selection is the name Darwin gave to the struggle between individuals of one sex (normally the males) for the possession of individuals of the opposite sex. The outcome of the struggle, as far as the loser is concerned, is few or no offspring. Victory in the struggle depends on the use made of special features, such as the colourful plumage of some male birds, the antlers of male deer or the huge mandibles of the male stag beetles. In fact, when the males and females of an animal species occupy identical niches but differ in size, structures or colours, it appears their differences often arose due to sexual selection. Most of the differences are not otherwise beneficial, and they may attract predators. They enhance the chance of reproductive success, however, and so they have been selected for.

Questions

1 Explain the key difference between natural and artificial selection.
2 What is meant by saying that the environment can 'direct' natural selection?
3 When members of isolated demes are brought together again by chance, breed and produce fertile offspring, what is the likely consequence in terms of evolution?
4 What is meant by the statement 'all polyploids are instantly isolated from the parent plant by a barrier of sterility'?
5 Why must mutations be regarded as the most important forces for speciation?
6 How might events in a population of rats on a farm lead to changes in gene frequency?

30.6 ISSUES IN MACROEVOLUTION

If evolution is a fact of life, then Neo-Darwinism should also account for the major steps in evolution (often called macroevolution). These include the origin of life and of the first cells, the origin of the eukaryotic cell, the origin of multicellular organisation, the rise of flowering plants, the origin of vertebrates from non-vertebrates and the origin of viruses.

■ The origin of life

No one can be certain how the Earth began, but it is known that the Earth is one of the smallest planets grouped in the Solar System around a central star, the Sun. It seems to be unique within the Solar System in supporting a variety of living organisms. The fact that the planets all revolve in the same plane around the Sun supports the theory that the Sun and planets were all formed from the condensation of a single revolving disc of gaseous matter.

It is likely that the Earth originated as a mass of molten rock with no atmosphere. As it cooled a crust formed into a rugged, restless surface disturbed by volcanic activity (as hot gases escaped from the molten interior), and by folding and fractures as it contracted. As the surface cooled to 100 °C an atmosphere developed, consisting of compounds such as water, ammonia, carbon dioxide and methane, formed from the lighter elements, and also hydrogen. It was virtually without oxygen; this is indicated by the presence in the oldest rocks, already formed at this time, of elements such as iron in reduced form (iron(II)) only.

As the Earth continued to cool the water vapour in the atmosphere condensed and returned to the surface as rain. Rivers, lakes and seas formed. Now the processes of erosion began to mould the landscape, and the eroded debris became the first sedimentary rocks.

The appearance of life

In the 1920s both A J Oparin, a Russian plant physiologist, and JBS Haldane (a physiologist as well as a geneticist) independently argued from theoretical considerations that life originated in the sea from a 'primeval soup', itself the product of substances present in Earth's atmosphere. This, they believed, occurred in three stages:

▶ formation of 'small molecules' (p. 135) such as amino acids, organic bases and monosaccharides
▶ formation of larger polymers, such as proteins, lipids, nucleotides and polysaccharides
▶ formation of organisms (simple prokaryote cells) from polymers.

In 1935 Stanley Miller, working in California, devised an apparatus to simulate supposed early Earth conditions (Figure 30.26), and thus test the first two steps in the hypothesis. Using an electric

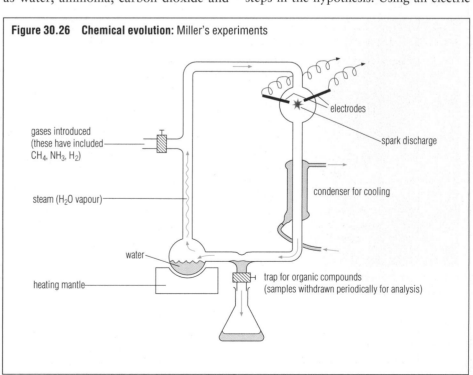

Figure 30.26 Chemical evolution: Miller's experiments

gases introduced (these have included CH₄, NH₃, H₂)

electrodes

spark discharge

steam (H₂O vapour)

condenser for cooling

water

heating mantle

trap for organic compounds (samples withdrawn periodically for analysis)

spark (an analogue for lightning) in the high-energy chamber he found he could synthesise traces of amino acids, the organic base adenine and the monosaccharide ribose. Experiments of this type have been repeated with slightly differing ingredients, according to prevailing speculations on Earth conditions at the time, and a wider range of biochemicals have been formed, including all 20 amino acids, many sugars, lipids, short lengths of polynucleotides and even ATP. Thus before life appeared its chemical building blocks may have accumulated naturally and, washed from the atmosphere by rain, collected (perhaps concentrated by evaporation) in surface pools and lakes.

Miller's critics claim that the 'reducing atmosphere' of the young Earth contained carbon dioxide, rather than methane. Carbon atoms in this compound will not 'break out' to make larger, organic molecules as Miller suggested.

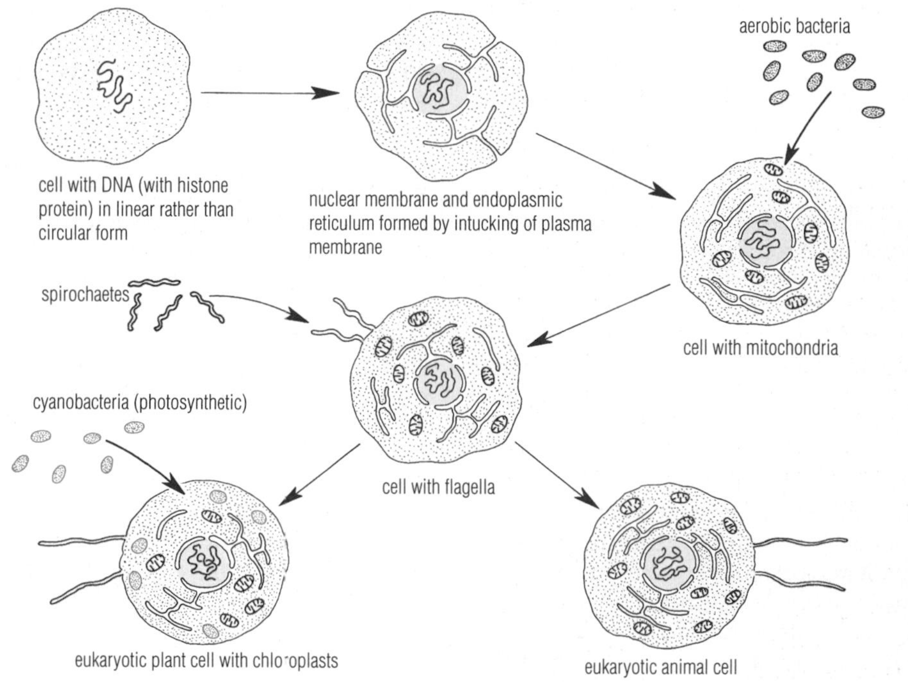

The origin of protobionts

Nothing is known for certain about the origin of functioning cells from the 'primeval soup'. In a living cell the synthesis of macromolecules by reactions between smaller molecules involves enzymes. The polymers of membranes may have been synthesised abiotically when dilute solutions of amino acids and other small molecules washed over hot clay particles and the water evaporated, concentrating the reactants. Clay particles may have functioned as

inorganic catalysts (p. 178). There are charged sites on clay particles, and metal ions are held there. Other theorists have suggested that life originated around undersea volcanoes, where extremely hot gases, including carbon dioxide and hydrogen sulphide, escape from the ocean floor. The iron of the Earth's crust may have reacted with the hydrogen sulphide, forming crystalline pyrites (iron sulphide) and hydrogen ions. Perhaps the

pyrites provided a catalytic surface on which the hydrogen began the reduction of carbon.

Among the early polymers formed may have been molecules that catalysed the synthesis of other polymers. The formation of an 'RNA-world' has been proposed as the beginning of life – that is, the first genes to be formed may have been short lengths of RNA. RNA is certainly a more robust molecule than an equivalent length of DNA, and molecules of RNA are now known to function both as enzymes and as self-replicating introns (p. 634). Relatively short pieces of RNA (50–60 nucleotides long) can catalyse the synthesis of other (complementary) RNA molecules.

It can be imagined that small droplets of primeval soup contained in polymers became encased by a layer of lipid to form 'protobionts', the forerunners of cells. Protobionts could exploit the steady supply of primeval soup in an early form of heterotrophic nutrition. Early 'reproduction' might have taken the form of mechanical fragmentation as protobionts broke apart under pressure from the environment. Ultimately an elaborate system of nucleic acid, coding for specific enzymes, would have given the 'progeny' a unique advantage.

Clearly, we can only speculate about these essential changes. The origin of the first living, replicating cells will remain a

Figure 30.28 The origin of the eukaryotic cell: a hypothesis

cell with DNA (with histone protein) in linear rather than circular form

nuclear membrane and endoplasmic reticulum formed by intucking of plasma membrane

aerobic bacteria

cell with mitochondria

spirochaetes

cyanobacteria (photosynthetic)

cell with flagella

eukaryotic plant cell with chloroplasts

eukaryotic animal cell

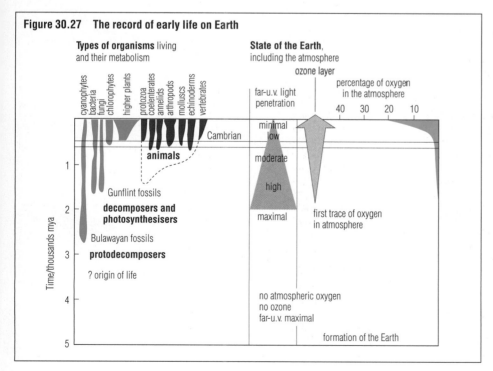

Figure 30.27 The record of early life on Earth

Types of organisms living and their metabolism

cyanophytes
bacteria
fungi
chlorophytes
higher plants
protozoa
coelenterates
annelids
arthropods
molluscs
echinoderms
vertebrates

Cambrian

animals

Gunflint fossils

decomposers and photosynthesisers

Bulawayan fossils

protodecomposers

? origin of life

Time/thousands mya

State of the Earth, including the atmosphere

ozone layer

far-u.v. light penetration

percentage of oxygen in the atmosphere

40 30 20 10

minimal
low

moderate

high

maximal

first trace of oxygen in atmosphere

no atmospheric oxygen
no ozone
far-u.v. maximal

formation of the Earth

mystery. But unless we are to assume the intervention of supernatural forces (divine intervention), living organisms must have arisen from non-living matter. Despite all the experiments and accumulated circumstantial evidence, however, the mechanism is unclear and the subject is one of much controversy.

Moving from speculation to the geological record, the oldest fossils are of aquatic prokaryotes, and date to 3500 mya (Figure 30.27, p. 665). The release of oxygen gas into the environment from photosynthetic organisms began around 2300 mya. From this time, too, iron oxide deposits are common in rocks. The presence of free oxygen gas indicates that photosynthetic bacteria and cyanobacteria were photosynthesising and releasing oxygen as a waste product (p. 246). By 2000 mya, photosynthetic organisms were so numerous that free oxygen gas began to accumulate in the atmosphere. At this stage life was still restricted to the seas, but as oxygen built up and ozone accumulated in the upper atmosphere, life could survive on land with reduced danger from ultraviolet radiation (p. 250). As the concentration of atmospheric oxygen reached 1%, aerobic respiration operated more efficiently than anaerobic respiration, and aerobic organisms evolved. Immediately after this point the fossil record becomes abundant and diverse.

■ Origin of the eukaryotic cell

The fossil evidence shows that not only were the first cells prokaryotic, but that it was thousands of millions of years before eukaryotic cells joined them – in fact, not until photosynthesis had evolved and oxygen had begun to accumulate in the atmosphere. The difference in the levels of organisation of these two types of cell is greater than that between plants and animals, so this was a major milestone in the development of the complexity of living things. Once again, there is no fossil record of the event. All we can do is speculate on the origin, and seek corroborative evidence for the more plausible hypotheses.

The eukaryotic cell is characterised by a range of organelles not present in prokaryotes, most being membranous structures. The nuclear membrane and the membranes of the endoplasmic reticulum and Golgi apparatus are believed to have been derived from the plasma membrane by intucking and invagination. But chloroplasts and mitochondria almost certainly have an entirely different origin. In modern eukaryotic cells these complex organelles are believed to be remnants of prokaryotes that have been 'captured' and retained in the cytoplasm as mutualistic organisms. Thus the chloroplast is seen as a descendant of a photosynthetic prokaryote, and the mitochondrion as a descendant of an aerobic bacterium.

Perhaps these organisms were initially captured as prey, but then some of the prey successfully resisted digestion and, instead, established a relationship of mutual benefit (Figure 30.28, p. 665). The relative sizes of these organelles are consistent with this theory, as is the presence within them of a ring of DNA, typical of the prokaryotic cell.

■ The origins of viruses

Viruses are complex biochemical molecules, rather than a form of life. An isolated virus is totally inactive, yet within a host cell it becomes a highly active genetic programme that takes over the nuclear machinery of the host. Viruses are specific to particular host species, and perhaps this is a clue to their origins in the history of life. It is a fact that, from time to time, fragments of chromosomes break off during nuclear division, and sometimes the fragments have been left behind in cytokinesis following the division of the nucleus. Isolated fragments of an organism's genome may have become a parasite specific to the species from which they originated. They would be dependent upon being introduced into living cells in order to parasitise the host.

■ Origin of multicellular organisation

Single-celled organisms constitute roughly half the total biomass of living

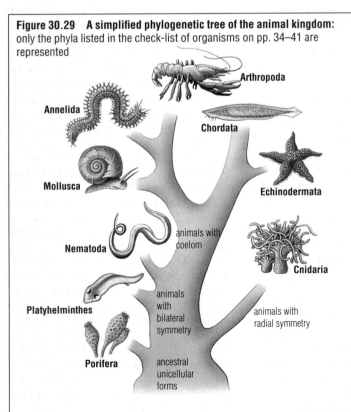

Figure 30.29   A simplified phylogenetic tree of the animal kingdom: only the phyla listed in the check-list of organisms on pp. 34–41 are represented

Arthropoda
Annelida
Chordata
Mollusca
Echinodermata
Nematoda
animals with coelom
Cnidaria
Platyhelminthes
animals with bilateral symmetry
animals with radial symmetry
Porifera
ancestral unicellular forms

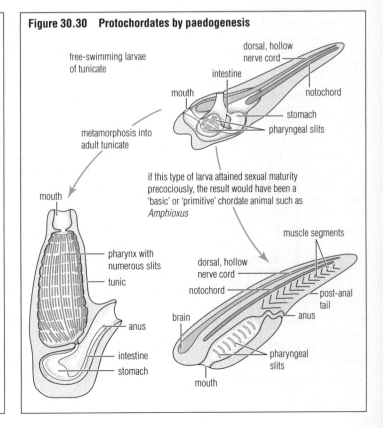

Figure 30.30   Protochordates by paedogenesis

free-swimming larvae of tunicate

dorsal, hollow nerve cord
intestine
mouth
notochord
stomach
pharyngeal slits

metamorphosis into adult tunicate

if this type of larva attained sexual maturity precociously, the result would have been a 'basic' or 'primitive' chordate animal such as *Amphioxus*

mouth
pharynx with numerous slits
tunic
anus
intestine
stomach

muscle segments
dorsal, hollow nerve cord
notochord
brain
post-anal tail
anus
pharyngeal slits
mouth

Figure 30.31   A simplified phylogenetic tree of land plants: the cross-section of the evolutionary lineages is roughly proportional to the number of living species

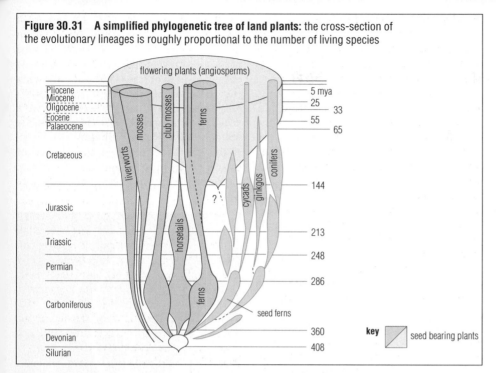

things, and no other form of life has anything like as long an evolutionary history. Multicellular organisation has evolved, however, and it is clear there are benefits to the organism of being multicellular. These include sophisticated mobility (speed and distance), and the ability to maintain a favourable internal environment under more extreme, adverse external conditions. The advantages of multicellularity are based on division of labour between the component cells.

There is no known fossil evidence of the way the multicellular condition arose. Perhaps the nucleus of a unicellular organism divided repeatedly, forming a multinuclear organism. (Tiny, multinuclear protoctists still exist today.) At some later stage such organisms may have developed a multicellular structure by formation of cell membranes.

Another possible route to multicellular organisation might have been through the failure of daughter cells of unicellular organisms (such as bacteria, yeasts or algae, pp. 550, 559 and 553) to separate after nuclear division and cytokinesis. Initially, this would generate a mere mass of undifferentiated cells. At a later stage such aggregates might have been transformed into multicellular organisms by specialisation of individual cells for particular tasks. This division of labour then requires that the activities of the whole organism are integrated by a control and command system. Specialisation leads to complexity of organisation.

■ The origin of vertebrates from non-vertebrates

The incompleteness of the fossil record obscures the path of the origin of the earliest vertebrates from a non-vertebrate group. We are left to speculate on their origin from studies of the comparative anatomy and embryology of existing vertebrates and of non-vertebrates that appear most closely related to them.

The major branches of the animal kingdom are introduced in the check-list of organisms on pp. 34–41, and are shown in Figure 30.29. You will notice that early in the phylogenetic tree the bulk of the animal phyla are grouped according to body symmetry (radial or bilateral, Figure 22.37, p. 501), and that the bulk of the phyla of bilaterally symmetrical animals have a body cavity known as the coelom (p. 345).

The coelomate phyla divided into two major groupings some time during the Cambrian (Table 30.1, p. 649). Members of these groups are distinguished by several fundamental differences of anatomy and embryology. One of these groups includes the annelids, molluscs and arthropods, and thus constitutes the bulk of non-vertebrate animals. The other group contains two major phyla, the echinoderms (starfish, brittle stars and sea urchins) and the chordates, a group which includes the vertebrates.

Zoologists believe that the first chordates (**protochordates**) arose from free-swimming larvae, like that of living tunicates, by attaining sexual maturity

precociously rather than proceeding to metamorphosis and the formation of a sessile adult form (Figure 30.30). The attainment of sexual maturity in a larval form is known as **paedogenesis**.

■ The rise of the flowering plants

The kingdom of green plants is reviewed in the check-list of organisms (pp. 32–4). The life cycles of these plants show a sporophyte (spore-bearing) phase alternating with a gametophyte (gamete-producing) phase, known as alternation of generations. Among the structurally more complex green plants (the tracheophytes, p. 32), the sporophyte is the dominant phase; the gametophyte generation has decreased in size and in importance in the life cycle.

Angiosperms appear in the fossil record quite suddenly, with little evidence of transitional forms that might establish links to common ancestors with other green plants, to which they are obviously related. The oldest fossil angiosperms are found in rocks that were laid down in the early Cretaceous, about 120 mya. By the end of the Cretaceous, 65 mya, the angiosperms were the dominant terrestrial plants, with most modern families represented in the fossil record.

The angiosperms of the Cretaceous grew among ferns, gymnosperms and a group of the now extinct plants known as 'seed ferns', from which the angiosperms seem to have evolved (Figure 30.31). In the seed ferns the sporophyte produced microspores and megaspores (heterosporous condition, as seen today in the club moss *Selaginella*, p. 32). Some seed ferns came to retain the megaspore on the sporophyte, and the microspore was carried (by wind) to 'pollinate' and allow fertilisation. Seed ferns with these features are believed to be the common ancestors of angiosperms and of gymnosperms such as the Scots pine. In plants of both groups, pollen grains have replaced motile gametes as the mechanism for delivery of a male nucleus to the egg cell. Fertilisation is possible in dry conditions. In the angiosperms the pollen tube has to grow through the style tissue to reach the ovule; in gymnosperms, however, the pollen grain germinates in a drop of liquid on the ovule surface, and grows directly into the ovule. In both, the zygote develops into an embryo, which is retained with a food store within a protective structure, the seed. The seed replaces the spore of other tracheophytes as the dispersal phase in the life cycle.

30.7 HUMAN EVOLUTION

The primates are a diverse order of mammals, varying from mouse-sized lemurs to huge gorillas. The group includes humans as a biological species, together with many extinct species. Virtually all the primates are restricted to the tropics, with many living in tropical rain forests, but with others in tropical savanna or scrubland habitats. Only *Homo sapiens* ranges over the entire surface of the globe.

■ The characteristics of the primates

The characteristic features of the primates include

▶ adaptation for a tree-living existence, with prehensile (grasping) hands and feet, five fingers and five toes, often with opposability of the thumb and big toe to the other digits

▶ nails instead of claws, and sensitive touch pads at the ends of the digits

▶ a progressive shortening of the snout, with reduction in the apparatus and function of smell

▶ an increasingly forward-facing position of the eyes with the development of binocular vision and a tendency towards the bony enclosure of the eyes' orbits

▶ an expansion and elaboration of parts of the brain, especially the parts associated with muscular coordination, vision, tactile senses, memory, thought and learning

▶ a tendency to hold the head erect and to sit upright (truncal uprightness)

▶ an omnivorous diet, often of fruits, leaves, insects and other small prey, and associated with this a reduction in the number of teeth and the retention of a simple molar cusp pattern

▶ a reproductive strategy that usually involves a single infant each pregnancy, and extended parental care

▶ prolonged prenatal existence, with progressive elaboration of the placenta, and increasing intimacy of the blood circulation between fetus and the mother

▶ a prolonged postnatal period of dependency during which the young are carried about holding on to the mother, and are fed from thoracically placed mammary glands

▶ existence in persisting social groupings that develop a complex hierarchical social structure.

■ Origin and diversity

The primates appeared 60–65 mya, at the time when the angiosperms (including trees) were becoming the dominant vegetation. Their common ancestor is likely to have been a member of the mammalian order of insectivores (Figure 30.33), something like a present-day tree-shrew.

The primitive primate stock almost immediately diverged into the anthropoids (apes and monkeys) and the prosimians (meaning 'before the monkeys') (Figure 30.32). Later came the separation between New World and Old World monkeys, which shared a common evolutionary history until the Oligocene (35–40 mya), when the process of continental drift (plate tectonics) divided them permanently. It is exclusively from the Old World stock that apes, then 'human–apes' and, finally, hominids developed.

■ From apes to humans in 35 million years

The earliest fossils that can be confidently identified as apes have been found at many sites in Africa and date from about 35 mya (Figure 30.34). Apes are distinguished from monkeys by several important features, which are reflected in their fossil remains. One of the most important is brachiation (progression between trees by arm-swinging). To be able to do this the shoulder joint is extended beyond the body wall, allowing the joint to be rotated through 360°. Also, the elbow can be extended in a straight line, and the lower arm can be rotated through 180° (Figure 30.35, p. 670). We have acquired this degree of dexterity from our ape-like ancestors; movements in monkeys are far more limited.

Figure 30.33 **Senegal Bush Baby** (*Galago senegalensis*), an insect and fruit eater, living in trees; these animals have features similar to those of primate ancestors

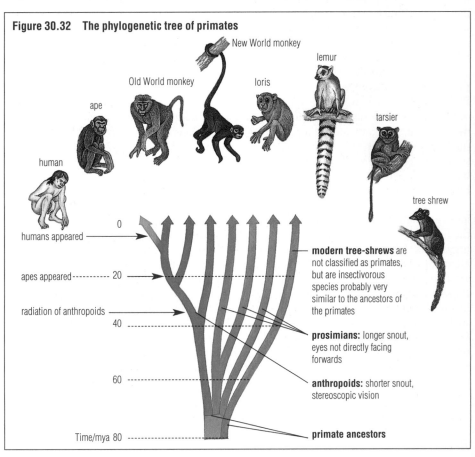

Figure 30.32 **The phylogenetic tree of primates**

New World monkey

Old World monkey

lemur

loris

ape

tarsier

human

tree shrew

humans appeared → 0

apes appeared ------- 20

radiation of anthropoids → 40

60

Time/mya 80 -----------

modern tree-shrews are not classified as primates, but are insectivorous species probably very similar to the ancestors of the primates

prosimians: longer snout, eyes not directly facing forwards

anthropoids: shorter snout, stereoscopic vision

primate ancestors

Figure 30.34   The human story, over 35 million years, as an illustrated time-line

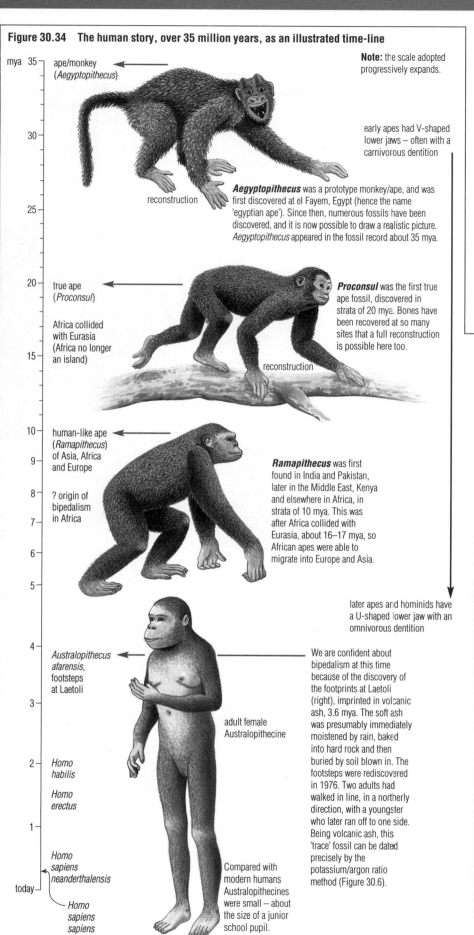

Note: the scale adopted progressively expands.

early apes had V-shaped lower jaws – often with a carnivorous dentition

reconstruction

Aegyptopithecus was a prototype monkey/ape, and was first discovered at el Fayem, Egypt (hence the name 'egyptian ape'). Since then, numerous fossils have been discovered, and it is now possible to draw a realistic picture. *Aegyptopithecus* appeared in the fossil record about 35 mya.

mya 35 — ape/monkey (*Aegyptopithecus*)

30

25

20 — true ape (*Proconsul*)

Africa collided with Eurasia (Africa no longer an island)

15

Proconsul was the first true ape fossil, discovered in strata of 20 mya. Bones have been recovered at so many sites that a full reconstruction is possible here too.

reconstruction

10 — human-like ape (*Ramapithecus*) of Asia, Africa and Europe

9

8 — ? origin of bipedalism in Africa

7

6

5

Ramapithecus was first found in India and Pakistan, later in the Middle East, Kenya and elsewhere in Africa, in strata of 10 mya. This was after Africa collided with Eurasia, about 16–17 mya, so African apes were able to migrate into Europe and Asia.

later apes and hominids have a U-shaped lower jaw with an omnivorous dentition

4 — *Australopithecus afarensis*, footsteps at Laetoli

3

adult female Australopithecine

2 — *Homo habilis*

Homo erectus

1

Homo sapiens neanderthalensis

today —

Homo sapiens sapiens

We are confident about bipedalism at this time because of the discovery of the footprints at Laetoli (right), imprinted in volcanic ash, 3.6 mya. The soft ash was presumably immediately moistened by rain, baked into hard rock and then buried by soil blown in. The footsteps were rediscovered in 1976. Two adults had walked in line, in a northerly direction, with a youngster who later ran off to one side. Being volcanic ash, this 'trace' fossil can be dated precisely by the potassium/argon ratio method (Figure 30.6).

Compared with modern humans Australopithecines were small – about the size of a junior school pupil.

At the time that apes and human–apes were evolving, the climate was changing dramatically. One reason for this was the drifting movements of the continents (Figure 30.11, p. 652). For example, it was only about 16–17 mya that Africa collided with Eurasia and took up a position close to its current one. A direct consequence was that, over this time, the extent of the African rain forests changed: at some times they covered the continent, but at others they were reduced to a few isolated pockets, about the size and in the position of modern Zaïre. Environmental pressures of these proportions would have driven those apes that could adapt to move on to the savanna, a factor that must have contributed to the development of bipedalism (Figure 30.36, overleaf).

the Laetoli footprints

Once Africa touched the continents of Eurasia, the migrations of human–ape forms and early humans could begin. The fossil record of the later stages of human development has come from a much wider area of the Earth than the home of the original African fossils. This distribution of fossils has contributed to the debate about whether the evolution of modern humans occurred exclusively in Africa – a debate that continues; you may have read of it in journals like the *New Scientist*.

The sequence of hominids that appeared in the fossil record since the human–ape *Australopithecus*, some 3.5 mya, is reviewed in Figure 30.37. The story of these changes is fascinating and well worth further reading. There are good reference books and other sources available (p. 671). New finds and fresh evidence are coming to hand all the time. This is yet one more branch of science where the 'frontier' is entirely open. You can follow the debate from now on. Perhaps you may contribute to it too.

Figure 30.35   Brachiation and the structure of the skeleton

ape

Swinging from arm to arm (brachiation) is made possible when:

1 pectoral girdle (shoulder blade) and clavicle (collar bone) protrude beyond rib cage, so ball-and-socket joint of arm allows 360° movement

2 elbow joint allows arm to straighten completely

3 lower arm is able to twist by maximum rotation of radius and ulna

4 wrist allows extensive movement of hand.

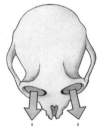

monkey

Compared with an ape the monkey has limited movements of the arm. The monkey walks among the canopy, reaching down from branches strong enough to take its weight, and sometimes jumping, but not brachiating.

Rapid forward movement, high in trees also requires binocular vision (for distance judging), and the reduction in the size of the muzzle for near vision. These are features of many primates.

eyes moved to front of head, reduction in size of snout/muzzle

Figure 30.36   The development of bipedalism

Apes run on four legs, or 'knuckle walk', but when they stand on hind-legs they can only 'waddle'.

Bipedalism frees the fore-limbs for carrying and for use of the hands. In overhead sunlight it allows the body to be slightly cooler, particularly the head and brain.

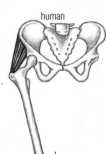

ape attempting to stride along on hind-legs

Bipedalism requires changes to the bones and musculature of the back, hip (pelvic girdle) and legs and leg joints.

A biped loses heat faster because more of its surface area is higher above the ground (air temperature falls the further it is away from a sunlit surface).

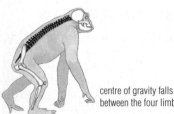

ape knuckle-walking

A quadruped gains more heat in direct sunlight because it presents a greater surface area to radiated energy.

centre of gravity falls between the four limbs

ape

this angle causes waddle

human

centre of gravity falls between feet

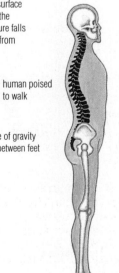

human poised to walk

this angle permits walking

pelvic girdle and upper leg of ape compared with that of human

Figure 30.37   Steps from human–ape to *Homo sapiens*

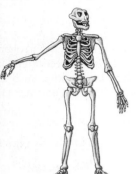

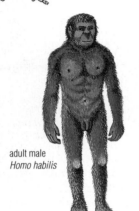

adult male
Homo habilis

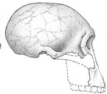
Australopithecus

▶ found between 3.5 and 2.5 mya
▶ earliest known hominid species
▶ living in woodland or savanna
▶ height 1–1.5 m
▶ walked bipedally with rolling gait
▶ long arms and curled fingers suggest that
  Australopithecus also brachiated
▶ cranial volume 380–450 cm³

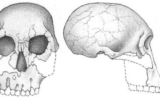
Homo habilis (handy human)

▶ found between 2.5 and 1.5 mya
▶ the first makers of stone tools
▶ cranial volume 750–800 cm³
▶ manual dexterity typified by power grip and
  precision grip

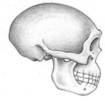 
Homo erectus (upright human)

▶ found between 1.6 and 0.5 mya
▶ height 1.5–1.8 m (i.e., up to 6 feet high)
▶ used fire
▶ cranial volume 850 cm³
▶ lived in bands of 30 or more
▶ team hunting imposed selection pressure of
  bipedalism and on intelligence/ingenuity

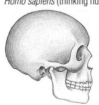 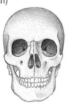
Homo sapiens (thinking human)

▶ found since 0.3 mya
▶ cranial volume 1350 cm³
▶ *Homo sapiens neanderthalensis* was present
  until the last ice age
▶ practised ritualised burial of the dead
▶ hunter-gatherers (until the Neolithic
  Revolution)
▶ *Homo sapiens sapiens* has dominated since
  the last ice age

Questions

1 What properties of enzymes make them essential for the functioning of a self-replicating cell?

2 What are the specific differences between prokaryotic and eukaryotic cells that are more profound than the differences between animal and plant cells (p. 169)?

3 In view of the evolutionary longevity of unicellular organisms, discuss the specific advantages that eventually result from multicellularity.

4 Why would the presence of oxygen gas in the atmosphere have led to the loss of the 'primeval soup'?

5 What features of chordates are *not* found in non-vertebrate organisation?

6 Why can we say the angiosperms are 'obviously related' to other groups of green plants?

7 What features of human structure and behaviour have enabled them to escape geographical restriction to the tropics?

8 Which of the listed features of primates in general are no longer typical of humans?

9 What are the most likely effects on ape/human evolution of the long-term suppression of tropical rain forest in the history of Africa?

FURTHER READING/WEB SITES

C J Clegg (1999) *Illustrated Advanced Biology: Genetics and Evolution.* John Murray
S Tomkins (1998) *The Origins of Humankind.* CUP
How relevant is *The Origin of Species* today: **www.bbc.co.uk/education/darwin**

EXAMINATION QUESTIONS SECTION 5

Answer the following questions using this section (chapters 25 to 30) and your further reading.

1 The diagram shows an ovule from a dicotyledonous plant, immediately *after* self-fertilisation has occurred.

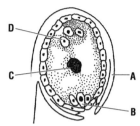

a i The structures within the ovule are haploid (n), diploid (2n) or triploid (3n). Copy and complete the table, naming the structures **A** to **D**, indicating the number of chromosome sets present and the % contribution to each structure made by the male and/or female parts of the parent plant.

Name of structure	Number of chromosome sets	% Contribution made to each structure by:	
		male part of parent plant	female part of parent plant
A			
B			
C			
D			

(6)

ii Describe the *structural* changes which occur after fertilisation, leading to the development of the seed and fruit. (5)

b Describe an experimental procedure for investigating the development of the embryo in a flowering plant species. (3)

c State **three** reasons why not all seeds germinate immediately after dispersal even though conditions are favourable for germination. (3)

UoCLES A Modular: Growth, Development and Reproduction November 1998
Section A Q4

2 The diagram shows the sequence of events in the development of a mature ovarian (Graafian) follicle and corpus luteum.

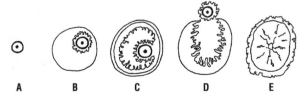

a What is the main hormone produced by the ovary when stage **B** is present? (1)

b i Which **two** of stages **A** to **E** would you expect to find in the ovary of a woman during the early stages of pregnancy? (1)

ii Give reasons for your answer. (2)

c i Some oral contraceptives contain only oestrogen. Which of the stages **A** to **E** would you expect to find in the ovary of a woman who had been taking such an oral contraceptive for a prolonged period of time? (1)

ii Give reasons for your answer. (2)

NEAB AS/A Biology Module Test: Social Biology (BY09) March 1998 Section A Q3

3 Below is a diagram of the human female reproductive tract, indicating the position of the endometrium.

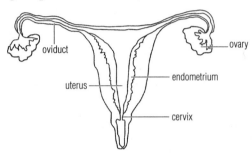

a i Outline the changes which take place in the endometrium during the menstrual cycle. (3)

ii List **three** ways in which the normal sequence of events of the menstrual cycle, in sexually mature females, may be altered. (3)

iii State what happens to the muscle tissues of the uterus wall and the cervix during the birth of a baby. (3)

Endometriosis is a condition which arises when cells from the endometrium pass into the oviducts and implant around the ovaries. The condition can result in damage to ovarian tissue, causing reduced fertility.

The graph shows the results of a three-year study to compare the levels of successful conception in couples in which the female was either normal or affected by endometriosis.

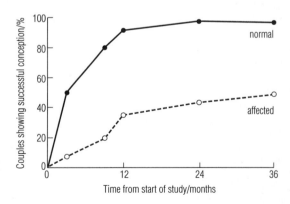

b With reference to the graph and the information given, state the effect that the condition has on the chances of successful conception. (2)

Women with a severe form of endometriosis are unlikely to conceive naturally and may be offered *in vitro* fertilisation (IVF).

c State **three** ethical objections to IVF. (3)

UoCLES A Modular: Growth, Development and Reproduction November 1998
Section A Q3

4 The stems of tomato plants can be different colours (green or purple) and may be hairy or without hairs (hairless). Each character is controlled by a different gene. Each gene has two alleles.

The following symbols are used to represent the alleles involved.

A dominant allele for colour
a recessive allele for colour
B dominant allele for hairiness of stem
b recessive allele for hairiness of stem

In an experiment, a homozygous tomato plant with a green, hairless stem was crossed with a homozygous plant with a purple, hairy stem. The F$_1$ seeds were collected and grown. All the resulting F$_1$ seedlings had purple, hairy stems.

a State the genotypes of each of the parent plants and of the F$_1$ seedlings in this cross. (3)

b The F$_1$ plants were self-pollinated (interbred). The phenotypes and numbers of offspring are given in the table below.

Phenotypes	Number of offspring
purple, hairy stem	293
purple, hairless stem	15
green, hairy stem	12
green, hairless stem	98

   i In this dihybrid cross, what would be the expected ratio of the phenotypes in the offspring? (2)

   ii Explain the differences between the expected ratio and the numbers shown in the table. (3)

c A tomato grower wants to find out which of the purple, hairy plants are homozygous for both characters.

   i State the genotype of the plant which should be crossed with the purple, hairy plants in this test cross. (1)

   ii Explain why this genotype should be used. (2)

London AS/A Biology and Human Biology Module Test B/HB1 January 1998 Q6

5 In tomatoes, the allele for red fruit, **R**, is dominant to that for yellow fruit, **r**. The allele for tall plant, **T**, is dominant to that for short plant, **t**. The two genes concerned are on different chromosomes.

a A tomato plant is homozygous for allele **R**. Giving a reason for your answer in each case, how many copies of this allele would be found in

   i a male gamete produced by this plant, (1)

   ii a leaf cell from this plant? (1)

b A cross was made between two tomato plants.

   i The possible genotypes of the gametes of the plant chosen as the male parent were **RT**, **Rt**, **rT** and **rt**. What was the genotype of this plant? (1)

   ii The possible genotypes of the gametes of the plant chosen as the female parent were **rt** and **rT**. What was the phenotype of this plant? (1)

   iii What proportion of the offspring of this cross would you expect to have red fruit? Use a genetic diagram to explain your answer. (3)

NEAB AS/A Biology Module Test: Continuity of Life (BY02) March 1998 Section A Q5

6 Night blindness is a condition in which affected people have difficulty seeing in dim light. The allele for night blindness, **N**, is dominant to the allele for normal vision, **n**. (These alleles are *not* on the sex chromosomes.)

The diagram shows part of a family tree showing the inheritance of night blindness.

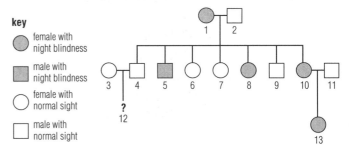

a Individual 12 is a boy. What is his phenotype? (1)

b What is the genotype of individual 1? Explain the evidence for your answer. (2)

c What is the probability that the next child born to individuals 10 and 11 will be a girl with night blindness? Show your working. (2)

AEB A Biology Paper 1 June 1998 Q6

7 The polymerase chain reaction is a process which can be carried out in a laboratory to make large quantities of identical DNA from very small samples. The process is summarised in the flowchart.

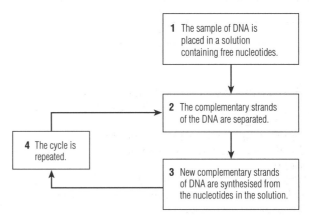

a   i At the end of one cycle, two molecules of DNA have been produced from each original molecule. How many DNA molecules will have been produced from one molecule of DNA after 5 complete cycles? (1)

   ii Suggest **one** practical use to which this technique might be put. (1)

b Give **two** ways in which the polymerase chain reaction differs from the process of transcription. (2)

c The polymerase chain reaction involves semi-conservative replication. Explain what is meant by *semi-conservative* replication. (2)

NEAB AS/A Biology Module Test: Continuity of Life (BY02) March 1998 Section A Q4

8 The diagram overleaf shows a map of pBr322, a small piece of double-stranded, circular DNA found in a bacterium in addition to the bacterial chromosome.

The genes for ampicillin resistance (Amp^{res}) and tetracycline resistance (Tet^{res}) are indicated.

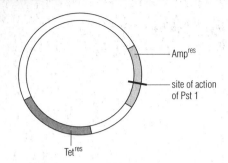

Pst 1 is a restriction endonuclease (enzyme) that has its effect at the site shown.

Pst 1 recognises the base sequence 5′ C–T– G–C–A–G
                                         G–A– C–G–T–C 5′
and acts on the DNA between guanine and adenine bases.

a i State the name given to such a piece of circular DNA. (1)

ii Explain the use of such DNA in genetic engineering. (3)

b Using the information given,
  i explain what is meant by the term *restriction endonuclease*; (2)
  ii explain what is meant by the term *sticky ends*; (4)
  iii describe how foreign DNA could be inserted into pBr322 using Pst 1; (3)
  iv describe how it could be shown that bacteria in liquid culture had taken up the modified pBr322. (3)

UoCLES A Modular: Microbiology and Biotechnology June 1997 Section A Q3

9 Animals in which genes have been altered by the technique of recombinant DNA technology are referred to as *transgenic animals*.

This technique may be used, for example, to enable a sheep to produce milk which contains proteins normally produced by humans. A human gene for the required protein can be substituted for a similar gene in a sheep chromosome.

The diagram below shows the sequence of events in this process.

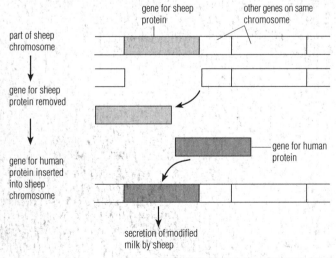

adapted from Freeland, Micoorganisms in Action

a i Name the enzyme which would be used to remove the gene for sheep protein from the sheep chromosome. (1)

ii Name the enzyme which would be used to attach the gene for the human protein to the sheep chromosome. (1)

b Describe how the modified DNA might be reintroduced into a sheep. (2)

c Suggest **two** advantages of using transgenic animals rather than microorganisms to produce human proteins. (2)

London AS/A Biology and Human Biology Module Test B/HB4D Food and Health January 1998 Q3

10 In Africa, south of the Sahara and north of the Zambezi, the sickle cell allele Hb^s is very common. In some ethnic groups the proportion of newborn babies that are homozygous recessive can be as high as 0.053 (5.3%). These babies suffer from sickle cell anaemia.

a Calculate the **frequency** of the sickle cell allele in these ethnic groups. (2)

b Calculate the **percentage** of the population that are carriers of the sickle cell allele. (1)

c Outline the reasons for the high frequency of the sickle cell allele in these ethnic groups, despite the serious consequences of sickle cell anaemia. (2)

International Baccalaureate Biology Higher Level Paper 3 November 1998 D2

11 *Bacillus thuringiensis* produces a protein that is toxic to leaf-eating caterpillars. This protein has been used by farmers as a natural insecticide. Recently, the gene that codes for the toxin has been genetically engineered into several crop plants.

a Outline how the gene that codes for the toxin could have been isolated. (5)

The gene, once isolated, is inserted into a host plant cell either by using the bacterium *Agrobacterium tumifaciens* as a vector to infect a plant cell or using a particle gun which shoots DNA-coated pellets into a plant cell.

b i Suggest why a plasmid vector cannot be used to insert the gene into plant cells. (1)

ii Explain why it is important to insert the gene into a single isolated plant cell rather than into a cell within a whole plant. (2)

c State **two** environmental implications of genetically engineered pest resistance in plants. (2)

It has been suggested that the integration of the gene for toxin production from *B. thuringiensis* into a wide range of crop plants could result in a loss of the effectiveness of the toxin.

d Outline how this loss of effectiveness of the toxin might occur. (2)

OCR (Cambridge Modular) A Sciences: Microbiology and Biotechnology March 1999 Q3

12 a The diagram shows the structure of a human sperm.

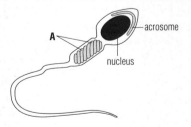

i Explain the part played by the organelles labelled **A** in the processes leading to fertilisation. (2)

ii The acrosome contains an enzyme that breaks down proteins. Describe the function of this enzyme in the processes leading to fertilisation. (1)

b Give **two** ways in which cell division in sperm formation in humans differs from cell division in egg formation. (2)

AQA (AEB) A Human Biology: Paper 1 June 1999 Q12

6 The Nature of Science

Head of a bison, Niaux Cave,
Tarascon-sur-Ariège, France

31 What is this thing called Biology?

The achievements of modern science in offering explanations about the way the world is, and in providing opportunities to improve our physical environment, are very impressive. It is therefore important to keep in mind that

▶ humans have demonstrated scientific potential from a very early time in their evolutionary history

▶ science has origins and a practical history in all human cultures

▶ whilst the subject matter of science is distinct from other disciplines of study, the methods of science are difficult to define simply

▶ the common scientific approach of reducing complex problems to narrowly defined aspects amenable to controlled investigation is criticised by those who believe that 'the whole is greater than the sum of the parts'.

This short chapter examines these issues, and offers further reading about them.

31.1 THE ANCIENT ORIGINS OF BIOLOGY

Throughout this book references have been made to the history of biological ideas. Occasionally we have traced the development of ideas back to the Greeks, but more often we have recalled recent figures, typically scientists of Europe or America, who have carried out pioneering developments in particular subject areas. As a consequence, you might have come to feel that a science such as biology (or chemistry or physics) is a comparatively recent product of 'western' civilisation. In fact, scientific activity has been a feature of every culture, growing naturally out of the innate curiosity of humans, as they gather knowledge from day-to-day activities, working with plants, animals and other humans and their physical environment. Humans have demonstrated outstanding practical talents from time to time throughout their history, starting in earliest times. A consequence has been the steady evolution of a tradition of empirical (experimental) analytical science, with roots in the experiences of all world cultures.

We have already noted the cave art of some of the early *Homo sapiens* (pp. 9 and 675). Palaeolithic cave paintings, works of great force and splendour,

represented many of the herbivorous animals used for food, together with a few of their predators. Animals were drawn on a large scale, on cave walls, and coloured with pigments of extraordinary permanence. The painters' only lighting would probably have been lamps made from small, hollowed stones, and fuelled by animal fat. Seen today under full, artificial illumination, these paintings are often chaotically superimposed on the limited surfaces suitable for drawings.

Great care was taken over details, and so it has been possible to classify to species level many of the animals shown. The accuracy of the representation of some of the extinct species has been established from frozen carcasses discovered in the ice in recent times. The pictures demonstrate perspective representation. (At one time this was regarded as a technique invented first in Renaissance Florence.)

These pictures were created by small groups (bands or tribes) of hunter–gatherer peoples, about the time of the last ice age. For over a million years *Homo sapiens* lived by hunting and collecting plants, before the Neolithic Revolution (p. 77) led on to the first agricultural societies. Subsequently, the first human civilisations appeared. Almost all of the great civilisations of the ancient world originated in river valleys, were sustained by trade with neighbouring peoples, and led to the formation of cities. For example, some 5000 years ago, the uniting of the peoples of the Upper and Lower Nile created the ancient Egyptian empire, centred around cities beside the Nile, such as Memphis and Thebes.

31.2 THE MULTICULTURAL ORIGINS OF BIOLOGY

Early civilisations were the achievements of city-dwellers, but their success sprang from rural technologies. Domestication of animals and plants (p. 604) required a substantial population engaged in agriculture, and led to a food surplus. This surplus freed a privileged few within hierarchical societies to develop other activities. For example, in ancient Egypt mathematics was developed to a level that sustained the construction of the great pyramids. Extensive irrigation systems were designed and maintained to

sustain productivity. Efficient administration systems achieved standardised weights and measures. Hieroglyphics were converted into alphabet writings on clay tablets and papyrus. Considerable knowledge of human anatomy was gleaned from the practice of embalming, a preparation for an imagined 'afterlife'. Medicine consisted largely of home remedies, but led to shrewd observations of the functioning body; the concept of a blood circulation was not achieved at this stage, but the record that 'the heart speaks in several parts of the body' is an advanced observation. Yeast was used in brewing (Figure 31.1) as well as for baking, so biotechnology was initiated, too.

If space allowed, we could go on to explore the achievements of the other great ancient civilisations to illustrate this important point that the roots of modern science are ancient and diverse. You may care to read more about their unique achievements. In fact it is fascinating to discover how many of the concepts that scientists use every day originated at these times. For example, from ancient India we get the decimal point and decimal fractions (amongst many other important skills). The idea of the doctor and the hospital as honourable institutions originated in ancient Persia. Dr Joseph Needham's pioneering study of Chinese science and technology has shown just how advanced that culture has been throughout most of its history. Traders moving along the 'silk road' to the west disseminated many of the skills and technologies that we now take for granted, and which were first developed in China. All of these developments were under way long before the Greeks of Athens and Alexandria made their own distinctive contributions to the philosophy and mathematics in which western science traditionally finds its roots.

31.3 THE DEVELOPMENT OF SCIENTIFIC METHOD

Whilst modern science was heir to all these movements, further developments in Europe were few during the Middle Ages (about AD 400–1500). Not until the scientific revolution that began in the sixteenth and seventeenth centuries did the idea of science as an identifiable process emerge.

■ Baconian 'induction'

Francis Bacon (1561–1626), a founder of the Royal Society of London – still the premier scientific society of Britain – was more of a lawyer and politician (under James I) than a scientist. He effectively advocated the 'experimental method', however, saying that if we wanted to understand the natural world, that was what we should study, not the works of Aristotle and other ancient Greeks, as was the practice. Bacon was one of the first people to advocate experimental science for the betterment of humans, believing nature could be controlled. He suggested there was a method in science which consisted of

► open-minded accumulation of data (with careful exclusion of extraneous observations)
► the development of a hypothesis to explain the data
► hypothesis testing by experiments
► confirmed hypotheses becoming scientific laws and contributing to the body of scientific knowledge.

The distinctive step in this sequence is the formulation of a hypothesis by generalisation from individual observations, and is known as the **induction process**. Bacon's popular view of the scientific method as 'induction' has been widely accepted. It is the foundation of the persisting view of science as 'authoritative', meaning that scientific knowledge is proven because it is derived in a rigorous way from the facts of our experiences.

■ Popper and 'falsification'

The soundness of this unified view of the nature of science was shattered by the Scottish philosopher, **David Hume** (1711–76), who showed there was no logical basis for the truth of statements that were derived from a large number of observations of individual cases. We cannot be certain that all swans are white, however many white swans are observed. **Karl Popper** (1902–94), physicist, mathematician and philosopher, sought to answer this 'problem of induction' for science in his book *The Logic of Scientific Discovery*, published in 1934. Popper's arguments were fourfold.

Firstly, scientific work very rarely, if ever, starts with the open-minded accumulation of data. Data are usually collected with a view to establishing a particular theory or explanatory notion, of which the observer is already largely convinced. The results are nearly always anticipated.

Figure 31.1 Beer drinking in Egypt, between 4000 and 5000 years ago: scenes from a tomb wall

Secondly, theories do not originate mechanically by induction, but rather by a creative process. The work of many scientists is as imaginative and creative as that of artists, musicians or novelists. Very often, too, the sequence of events by which a theory or hypothesis is formulated may be a mystery even to the person who originated the explanation.

Thirdly, scientific theories are tested to find out if they can be shown to be false. We can never prove the hypothesis that all swans are white, but we can disprove it if we find just one black swan. Thus, scientific laws can only be tested to prove them wrong. In fact, theories can only be scientific if we know what it would need to prove them false.

Finally, any theory or law that has not yet been falsified is accepted as a working hypothesis for the time being. Scientific knowledge consists of credible working hypotheses that are currently accepted.

■ Kuhn and the nature of scientific method

One problem that arises with 'falsification' is that people cannot always accept that a hypothesis must be rejected when evidence arises against it. The history of science abounds with examples of rejections of observation statements and retention of theories with which they clash. John Steinbeck the novelist and Ed Rickets the marine biologist, in the book *The Sea of Cortez*, put it this way:

There is one great difficulty with a good hypothesis. When it is completed and rounded, the corners smooth and the content cohesive, it is likely to become a thing in itself, a work of art . . . One hates to disturb it. Even if subsequent information should shoot a hole in it, one hates to tear it down because it once was beautiful and whole.

Thomas Kuhn (1922–96), an American physicist, philosopher and historian of science, was struck by the marked discrepancy between the theories of the methods of practical science, such as those of Bacon and Popper, and the way scientists are seen to proceed.

Kuhn recognised that in any established subject area or field of science there tended to be a generally accepted model, a theory, that underlies the work at any time. This accepted model he called a **paradigm**. For example, in molecular biology the underlying paradigm is the DNA double helix and the mechanism by which it dictates the protein components of the cell. A paradigm is accepted by everyone (or almost everyone) working in the field. The paradigm determines the issues that are investigated, and the questions that are asked, and suggests the techniques by which they are answered.

Kuhn also noted that issues in any branch of science moved from what he called an **immature** or **revolutionary phase**, characterised by controversy, in which two or more groups of workers (we might call them schools) compete among themselves as to whose suggested explanation is correct. Eventually, the ideas of one of the schools are more generally accepted and the immature or revolutionary phase comes to an end. The subject now moves into a phase of normal science. The **normal phase**, during which a single paradigm holds sway, very often persists for a long time, at least whilst the leading team remains active. Their work, and that of supporting groups, checks, extends and investigates the paradigm. A paradigm is resistant to change; only when persistently inconsistent results accumulate will a paradigm start to break down. When it does so, the subject moves back to a revolutionary phase. Usually it is young workers who lead in the search for a new explanation, the new paradigm.

An excellent example of a long-surviving paradigm is Darwin's theory of the origin of species by natural selection. When he originally introduced it he was far from confident that it would be widely accepted. In the final chapter of *The Origin of Species* he wrote:

Although I am fully convinced of the truth of the views given in this volume . . . , I by no means expect to convince experienced naturalists whose minds are stocked with a multitude of facts all viewed, during a long course of years, from a point of view directly opposite to mine . . . A few naturalists endowed with much flexibility of mind, and who have already begun to doubt the immutability of species, may be influenced by this volume; but I look with confidence to the future, to young and rising naturalists, who will be able to view both sides of the question with impartiality.

■ A reductionist *v.* a holistic approach to science?

Science proceeds by the subdivision of complex situations into component issues simple enough to be investigated. This approach is described as reductionist, and there is a school of thought that argues that scientists should have reservations about the reductionist approach. They should seek, instead, a holistic approach to investigations where this is possible. The Gaia hypothesis is an example.

Gaia

A scientist, James Lovelock, argues that the world with its atmosphere can be thought of as 'alive', much as a tree is alive and responding to the environment, even though 99% of the tree is made of dead wood. He uses the word 'alive' in the sense that the world shares with other living things the capacity to self-regulate (homeostasis, p. 502). At the suggestion of his friend William Golding, the novelist, he called his idea the **Gaia hypothesis**.

The Gaia hypothesis postulates that the physical and chemical conditions of the Earth, the atmosphere and the oceans have been and are actively made fit and comfortable by the processes of life itself. This idea contrasts with the idea that life has adapted to the planetary conditions. The Earth has remained a comfortable place for the 3–5 billion years since life began, despite a 25% increase in the output of heat from the Sun, so Lovelock argues.

The atmosphere is an unstable mixture of reactive gases, yet its composition remains constant and breathable for long periods and for whoever are the inhabitants. Organisms have always kept their planet fit for their own life. The theory sees the evolution of species closely coupled with the evolution of their physical and chemical environment.

Together, organisms and Earth constitute a single evolutionary entity.

Most scientists have been sceptical about the theory. Dr Lovelock views modern science as having developed so rapidly that it has fragmented into a collection of independent disciplines, with little interdisciplinary thinking. He claims that the idea of the Earth as alive is as old as the human race. The eighteenth-century scientist James Hutton, founder of geology, claimed the Earth as a superorganism. Hutton compared the circulation of the blood (discovered by William Harvey, p. 346) with the circulation of the nutrient elements of the Earth, as illustrated by the water cycle (p. 75), for example. Today there is so much information to be collected and sorted in our various investigations, that we may have lost sight of a whole picture of our world. It is hard to think of Earth as a single entity.

'Gaia' was the name the Greeks gave to their 'Earth goddess'. The modern Gaia hypothesis may be useful because it has consequences for our view of the ways the planet will respond to the current ecological crisis. Dr Lovelock suggests that our current environmental problems arise because we are city-dwellers, obsessed with human problems. We are overconcerned about the risk of shortening our own life expectancy through, for example, cancers caused by industrial hazards, rather than about the vastly greater dangers from degradation of the natural world by deforestation and climate change.

Lovelock claims that 'the three deadly Cs' are cars, cattle and chainsaws. Can you see why?

And, if Lovelock is right in this, what is to be done?

FURTHER READING/WEB SITES

The ancient history of science

C A Roan (1984) *The Illustrated History of the World's Science.* Cambridge University Press
J Bronowski (1974) *The Ascent of Man.* BBC Publications

The nature of science

A F Chalmers (1982) *What Is This Thing Called Science?* Oxford University Press
B Magee (1987) *The Great Philosophers* (especially Chapter 7, on David Hume). BBC Publications
B Magee (1982) *Popper.* Fontana
J E Lovelock (1987) *Gaia; a New Look at Life on Earth.* Oxford University Press

ACKNOWLEDGEMENTS

The following have provided photos or given permission for copyright photos to be reproduced:

FRONT COVER AND TITLE PAGE: The Stock Market

CHAPTER 1
p. 1 Stephen Dalton/NHPA; **p. 4** *Fig 1.4* Stephen Dalton/NHPA; **p. 5** *Fig 1.5* D Phillips/Science photo Library, *Fig 1.6 left* Biophoto Associates, *right* Heather Angel; **p. 9** R Sheridan/Ancient Art and Architecture; **p. 10** Ancient Art and Architecture; **p. 11** MEPL; **p. 13** Andrew Purcell/Bruce Coleman Ltd; **p. 17** *top* Dr Persephone Borrow, *bottom* Steve Day.

CHAPTER 2
p. 19 *top* M Wurtz/Science Photo Library, *middle and bottom* Biophoto Associates; **p. 21** *left* M J Thomas/FLPA, *middle* David Scharf/Science Photo Library, *right* Michael Leach/Oxford Scientific Films; **p. 22** MEPL; **p. 26** *Fig 2.5 lizard* Rod Planck/NHPA, *tit* Stephen Dalton/NHPA, *crocodile* Heather Angel; **p. 31** *left* Heather Angel, *right* Adrian Davies/Bruce Coleman Ltd; **p. 33** *top* Heather Angel, *bottom* Bob Gibbons/Holt Studios; **p. 37** Jane Burton/Bruce Coleman Ltd; **p. 38** *top* G I Bernard/NHPA, *bottom* Kim Taylor/Bruce Coleman Ltd; **p. 42** C J Clegg.

CHAPTER 3
p. 49 *Fig 3.9 both* David Woodfall/NHPA, *bottom left* Stephen Kraseman/NHPA, *bottom right* Jules Cowan/Bruce Coleman Ltd; **p. 52** Heather Angel; **p. 53** *Fig 3.12 top left* Laurie Campbell/NHPA, *top right* Mark Mattock/Planet Earth Pictures, *bottom left* Roy Waller/NHPA, *bottom right*, Planet Earth Pictures; **p. 54** *Fig 3.4 and 3.6* Heather Angel; **p. 56** *Fig 3.18 top* Simon Fraser/Science Photo Library, *bottom* Mike Lane/Science Photo Library; **p. 58** *Fig 3.21 both* Biophoto Associates, *Fig 3.21* C J Clegg; **p. 60** *Fig 3.25 left* Dr Jeremy Burgess/Science Photo Library, *right* Hans Rheinhard/Bruce Coleman Ltd; **p. 61** *Fig 3.27* C J Clegg; **p. 65** *top left* E A Janes/NHPA, *top right* Kim Taylor/Bruce Coleman Ltd, *bottom left* Gerard S Cubitt/Bruce Coleman Ltd, *bottom right* Peter Scoones/Planet Earth Pictures; **p. 66** *bottom left* John Burbridge/Science Photo Library, *top right* Jane Burton/Bruce Coleman Ltd, *bottom right* Gordon Langsbury/Bruce Coleman Ltd; **p. 67** Heather Angel; **p. 68** *Fig 3.42* Roger Tidman/NHPA; **p. 69** *Fig 3.44* David Dennis/Oxford Scientific Films; **p. 70** *Fig 3.46* Chris Newton/FLPA.

CHAPTER 4
p. 81 *Fig 4.7* R A Janes/NHPA, *Fig 4.8* Martin Harvey/NHPA; **p. 82** *Fig 4.9* Richard M Matthews/Planet Earth Pictures; **p. 83** *Fig 4.11* Sinclair Stammers/Science Photo Library; **p. 85** *Fig 4.14* Gene Feldman/Science Photo Library; **p. 87** *Fig 4.18* R Bender/FLPA; **p. 89** *Fig 4.20* Science Photo Library; **p. 90** *Fig 4.21 left* Laurie Campbell/NHPA, *right* Allan Potts/Bruce Coleman Ltd; **p. 91** Roger Tidman/NHPA; **p. 93** *Fig 4.25* Simon Fraser/Science Photo Library; **p. 95** Stephen Dalton/NHPA; **p. 96** *Fig 4.27 left* Andy Rouse/NHPA, *right* Hanumantha Rao/NHPA.

CHAPTER 5
p. 100 *Fig 5.3 top* David Woodfall/NHPA, *bottom* Martin Bond/Science Photo Library; **p. 104** *Fig 5.8* Laurie Campbell/NHPA, *Fig 5.9* Ivor Edmonds/Planet Earth Pictures; **p. 106** *top* D G Mackean, *Fig 5.11* Jurgen Dielenschneider/Holt Studios; **p. 107** *left* Silvestris/FLPA, *top right* Patrick Fagot/NHPA, *bottom right* Joe Blossom/NHPA; **p. 108** *left* Martin Chillmaid/Oxford Scientific Films, *right* Peter Menzel/Science Photo Library; **p. 109** *Fig 5.12* John Townson/Creation; **p. 109** *Fig 5.13* The Stockmarket; **p. 110** *Fig 5.14* Britstock-IFA; **p. 111** Damien Lovegrove/Science Photo Library; **p. 112** *Fig 5.16* Rob Cousins/Oxford Scientific Films; **p. 113** Scimat/Science Photo Library; **p. 114** *Fig 5.17* Fisher Scientific; **p. 116** *Fig 5.22 and Fig 5.23* Science Photo Library; **p. 117** *Fig 5.24* John Townson/Creation; **p. 118** Sean Avery/Planet Earth Pictures; **p. 119** *Fig 5.25* Sean Sprague/Panos Pictures, *Fig 5.26* David Hall/Science Photo Library.

CHAPTER 6
p. 123 David Parker/Science Photo Library; **p. 128** *Fig 6.4* Don Mackean, *Fig 6.5* Fisher Scientific; **p. 138** *Fig 6.17 left* Professor John Dodge/Royal Holloway and Bedford Hospital, *right* Oxford Scientific Films.

CHAPTER 7
p. 148 *both* Biophoto Associates; **p. 150** *Fig 7.2 left* Biophoto Associates, *right* Gene Cox; **p. 152** *both* Gene Cox; **p. 153** *Fig 7.5 all* Biophoto Associates; **p. 154** *Fig 7.7 left* Ron Boardman/Life Science Images, *right* Biophoto Associates; **p. 155** *Fig 7.9 and 7.10 all* Biophoto Associates; **p. 158** *Fig 7.13* Science Photo Library, *Fig 7.14 bottom* Biophoto Associates; **p. 161** *Fig 7.18 left* Ron Boardman/Life Science Images, *right* Biophoto Associates; **p. 162** *Fig 7.19* Biophoto Associates; **p. 163** *Fig 7.21 left* Dawn Fawcett/Science Photo Library, *right* Ron Boardman/Life Science Images, *Fig 7.22* Science Photo Library; **p. 164** *Fig 7.23 top* Biophoto Associates, *bottom* Science Photo Library, *Fig 7.24* Science Photo Library; **p. 165** *Fig 7.25 top and bottom* Science Photo Library; **p. 166** *Fig 7.27 both* Science Photo Library; **p. 167** *Fig 7.29 top* Professor Preston University of Leeds, *bottom* Biophoto Associates; **p. 169** *Fig 7.31 both* Gene Cox.

CHAPTER 8
p. 171 MEPL; **p. 173** *Fig 8.4* Heather Angel; **p. 174** *Fig 8.5* Peter Gould, *Fig 8.7* Sinclair Stammers/Science Photo Library; **p. 175** *Fig 8.8* C J Clegg; **p. 189** *Fig 8.29* Unilever; **p. 190** *Fig 8.30 both* Unilever Research Lab, Port Sunlight; **p. 191** *Fig 8.33* Faye Norman/Science Photo Library

CHAPTER 9
p. 194 *Fig 9.3 all* Gene Cox; **p. 196** *Fig 9.5 both* Biophoto Associates; **p. 203** *top* MEPL, *Fig 9.13* Biomedical Sciences Dept./Kings College London; **p. 204** *Fig 9.14* Biophoto Associates; **p. 210** *Fig 9.24* Gene Cox.

CHAPTER 10
p. 214 *Fig 10.3* Gene Cox, *Fig 10.4 top* Gene Cox, *Fig 10.4 and 10.5* Biophoto Associates; **p. 215** *Fig 10.7 and 10.8,* Gene Cox; **p. 217** *Fig 10.11* Gene Cox; **p. 223** *Fig 10.19* Gene Cox, **p. 224** *Fig 10.20 and 10.21* Gene Cox; **p. 225** *Fig 10.23* Gene Cox; **p. 226** *Fig 10.24* Gene Cox, *Fig 10.25 right* Biophoto Associates, *left* Ron Boardman/Life Science Images, *Fig 10.26* Gene Cox.

CHAPTER 12
p. 245 John Shaw/NHPA; **p. 249** *Fig 12.7 both* Biophoto Associates; **p. 255** *Fig 12.18* Dr Jeremy Burgess/Science Photo Library, *Fig 12.19 top* Laurie Campbell/NHPA, *bottom* Biophoto Associates; **p. 257** *Fig 12.22* Nigel Cattlin/Holt Studios; **p. 258** *Fig 12.25* Biophoto Associates; **p. 262** *Fig 12.32* Corbis; **p. 269** *Fig 12.41 both* Science Photo Library.

CHAPTER 13
p. 279 C J Clegg; **p. 280** *Fig 13.11*; **p. 281** *Fig 13.12* Gene Cox; **p. 282** *Fig 13.13* Science Photo Library; **p. 285** *Fig 13.17* David Scharf/Science Photo Library.

CHAPTER 14
p. 291 *Fig 14.1* Trevor McDonald/NHPA; **p. 292** *Fig.14.6* David Maitland/Planet Earth, *Fig 14.7* Anthony Bannister/NHPA; **p. 293** *Fig 14.9* Heather Angel, *Fig 14.10* Biophoto Associates; **p. 294** *Fig 14.12* Holt Studios; **p. 295** *Fig 14.13* FLPA; **p. 298** *Fig 14.20* A Dowsett/Science Photo Library, *Fig 14.21* Wadworth Brewery; **p. 301** *Fig 14.24* Cameron Read/Planet Earth Pictures.

CHAPTER 15
p. 303 *Fig 15.2 both* Gene Cox; **p. 305** *Fig 15.5* Biophoto Associates; **p. 310** *Fig 15.11 and Fig 15.12* Gene Cox; **p. 312** *Fig 15.15* Gene Cox; **p. 313** *Fig 15.17 left* Gene Cox, *right* Damien Lovegrove/Science Photo Library.

CHAPTER 16
p. 322 *Fig 16.4 both* Gene Cox; **p. 326** *Fig 16.12 and 16.13* Gene Cox; **p. 327** *Fig 16.14* Gene Cox; **p. 329** *both* Nigel Cattlin/Holt Studios; **p. 331** *Fig 16.21* CJ Clegg; **p. 333** Dr Tony Brain/Science Photo Library; **p. 334** Dr Jeremy Burgess/Science Photo Library; **p. 335** *Fig 16.29 both* Claude Nuridsany and Marie Perennou/Science Photo Library; **p. 337** *Fig 16.31 top* Dr Richard Johnson and Guy/Aberdeen University; **p. 338** *Fig 16.34 left* Science Photo Library, *right* from M Richardson's *Translocation in Plants*

CHAPTER 17
p. 342 *Fig 17.2* Ken Lucas/Planet Earth Pictures; **p. 347** *Fig 17.8 left* Science Photo Library, *right* Biophoto Associates; **p. 351** *Fig 17.16* Gene Cox; **p356** *Fig 17.24* Science Photo Library; **p. 357** *Fig 17.27 both* Dr Zivin/American Scientific Journal; **p. 363** *Fig 17.38* Laurie Campbell/NHPA; **p. 365** *Fig 17.41* Science Photo Library.

CHAPTER 18
p. 373 Don Mackean; **p. 374** *Fig 18.1* Dr Jeremy Burgess/Science Photo Library, *Fig 18.2* Dr Mike Jackson/IACR Long Ashton; **p. 375** *Fig. 18.3 left* Jim Bain/NHPA, *right* John Buckingham/NHPA; **p. 376** *Fig 18.4* Gene Cox, *Fig 18.5* Nigel Cattlin/Holt Studios; **p. 382** *Fig 18.12* Biophoto Associates, *Fig 18.13 both* Gene Cox; **p. 383** *Fig 18.14* Biophoto Associates; **p. 386** *Fig 18.18 both* Gene Cox; **p. 390** *Fig 18.25 both* C J Clegg.

CHAPTER 19
p. 395 Stephen Dalton/NHPA; **p. 401** *Fig 19.6* Nigel Cattlin/Holt Studios; **p. 406** *Fig 19.15* Biophoto Associates; **p. 409** *top left, top right and bottom left* Gene Cox, *bottom right* Ron Boardman/Life Science Images; **p. 410** *Fig 19.20* Gene Cox; **p. 412** *Fig 19.24* Gene Cox; **p. 415** *Fig 19.29* D P Wilson/FLPA; **p. 416** *Fig 19.30 both* Heather Angel; **p. 417** *Fig 19.32* Claude Nuridsany and Marie Perennou/Science Photo Library.

CHAPTER 20
p. 418 *Fig 20.1* Nigel Cattlin/Holt Studios; **p. 419** *Fig 20.3 both* Dr Eckart Pott/Bruce Coleman Ltd; **p. 420** *Fig 20.4* Don Mackean, *Fig 20.6* Nigel Cattlin/Holt Studios; **p. 421** *Fig 20.8 left* Hans Rheinhard/Bruce Coleman, *right* Bob Gibbons/Ardea; **p. 428** *Fig 20.21 both* Professor Wilkins/University of Glasgow; **p. 431** *both* Nigel Cattlin/Holt Studios; **p. 432** *Fig 20.25* Nigel Cattlin/Holt Studios, *Fig 20.26* Don Mackean; **p. 436** *Fig 20.30 and 20.31* Nigel Cattlin/Holt Studios; **p. 438** *Fig 20.33 left* Biophoto Associates; **p. 441** *Fig 20.38 top* Gene Cox.

CHAPTER 21
p. 447 *Fig 21.9* Manfred Kage/Science Photo Library; **p. 448** *Fig 21.10* Professor Cinti/Science Photo Library; **p. 465** *Fig 21.34 top* Larry Muvehill/Science Photo Library, *bottom* James Homes/Science Photo Library, *Fig 21.35* Mehau Kulyk/Science Photo Library; **p. 468** *Fig 21.39 both* Biophoto Associates; **p. 469** *Fig 21.40* Stephen Hopkin/Planet Earth Pictures; **p. 471** *Fig 21.44* David Scharf/Science Photo Library; **p. 477** *Fig 21.52 top* Hans Christian Heap/Planet Earth Pictures, *bottom* Biophoto Associates.

CHAPTER 22
p. 487 Science Photo Library, *Fig 22.16 top and bottom* Gene Cox; **p. 490** *Fig 22.17 top* Pradeep K Luther/Imperial College of Science, *bottom* Ron Boardman/Life Science Images, *Fig 22.18* Science Photo Library; **p. 493** *Fig 22.23 left* Gary

ACKNOWLEDGEMENTS

Mortimor/Allsport, *right* Chris Cole/Allsport; **p. 494** *Fig 22.27* The Stockmarket; **p. 500** *Fig 22.35* Gene Cox; **p. 508** *Fig 22.36* Gene Cox.

CHAPTER 23

p. 502 Science Photo Library; **p. 503** *Fig 23.1* Fisher Scientific; **p. 505** *Fig 23.5* Gene Cox; **p. 507** *Fig 23.8* Science Photo Library; **p. 510** *Fig 23.13* Georgette Douwma/Planet Earth Pictures; **p. 512** *Fig 23.18 left* Robert Maier/Bruce Coleman *and right* David Hosking/FLPA; **p. 513** *Fig 23.20* C J Clegg; **p. 515** *Fig 22.23* Johnny Johnson/Bruce Coleman Ltd, *Fig 23.24* Manfred Danegger/NHPA; **p. 517** *Fig 23.26* Stephen Hopkin/Planet Earth Pictures; **p. 518** *Fig 23.27* Geoff de Feu/Planet Earth Pictures; **p. 519** *Fig 23.29* R Wilmshurst/FLPA; **p. 523** *Fig 23.31* Sybille Kalas; **p. 524** *Fig 23.36* Kim Taylor/Bruce Coleman Ltd.

CHAPTER 24

p. 527 Science Photo Library; **p. 530** Biophoto Associates; **p. 532** *Fig 24.3* Alfred Pasieka/Science Photo Library; **p. 534** *Fig 24.7* Bruce Iverson/Science Photo Library; **p. 535** *Fig 24.9* Stills Picture Library; **p. 536** *Fig 24.12* Science Photo Library; **p. 537** *Fig 24.13 both* Dr Mann/University of Manchester; **p. 537** *Fig 24.14* Farmers Weekly; **p. 538** Professor Richard Doll; **p. 539** Matt Cook; **p. 540** *Fig 24.17* Alfred Pasieka/Science Photo Library; **p. 541** *Fig 24.18 left* Nigel Cattlin/Holt studios, *right* Science Photo Library, *Fig 24.20* The Institute of Scottish Research, *Fig 24.21* Biophoto Associates; **p. 543** *Fig 14.23* C J Clegg.

CHAPTER 25

p. 547 C J Cambridge/NHPA; **p. 549** *Fig 25.2* Dr Lee D Simon/Science Photo Library; **p. 552** *Fig 25.7* Biophoto Associates, *Fig 25.8* Gene Cox, *Fig 25.9* Dr L Caro/Science Photo Library; **p. 554** *Fig 25.13* Biophoto Associates; **p. 556** *Fig 25.18 and 25.19* Biophoto Associates; **p. 557** *Fig 25.20 both* C E Backer; **p. 558** *Fig 25.22* Biophoto Associates; **p. 560** *Fig 25.26 all* Biophoto Associates; **p. 561** *far right and middle left* C J Cambridge/NHPA, *far left* Laurie Campbell/NHPA, *middle right* William S Paton/Bruce Coleman Ltd.

CHAPTER 26

p. 565 *Fig 26.5* Gene Cox, *Fig 26.6* Heather Angel; **p. 566** *Fig 26.9* Ron Boardman/Life Science Images; **p. 570** *Fig 16.13* GE Hyde/FLPA; **p. 571** *Fig 26.16 both* Gene Cox; **p. 572** *Fig 26.18* Gene Cox; **p. 574** *Fig 26.20 and 26.21* Gene Cox, *Fig 26.21 top left* Tony Bennett/Planet Earth Pictures, *middle top left* Tony Wharton/FLPA, *middle bottom left* Heather Angel, *bottom left* Jane Burton/Bruce Coleman Ltd, *top right* Peter Steyn/Ardea, *middle right* John Bracegirdle/Planet Earth Pictures, *bottom right* Heather Angel; **p. 575** *top* N A Callow/NHPA, *middle* Ardea, *bottom* Jim Bain/NHPA; **p. 576** *Fig 26.24 top* Photos Horticultural, *middle* Bob Gibbons/Ardea, *bottom* Silvestris/FLPA.

CHAPTER 27

p. 580 *Fig 27.1* Gene Cox; **p. 581** *Fig 27.3* Stephen Dalton/NHPA; **p. 582** *Fig 27.4 both* Biophoto Associates; **p. 583** *Fig 27.5* David Thompson/Biophoto Associates; **p. 584** *Fig 27.6 top left* Biophoto Associates, *top middle* Stephen Dalton/NHPA, *top right* John Mason/Ardea, *middle right* W Meinderts/FLPA, *bottom right* G I Bernard/NHPA *bottom middle* Planet Earth Pictures, *bottom left* Stephen Dalton/NHPA; **p. 584** *Fig 27.7* Claude Nuridsany and Marie Perennou/Science Photo Library; **p. 585** *Fig 27.8* Manfred Danegger/NHPA; **p. 588** *Fig 27.11* Andrew Syred/Science Photo Library; **p. 589** *Fig 27.12* Science Photo Library; **p. 591** Science Photo Library; **p. 595** Science Photo Library; **p. 599** *Fig 26.26 top* Gene Cox, *bottom* Science Photo Library; **p. 600** *Fig 27.28 top* Gary Parker/Science Photo Library, *middle and bottom* Science Photo Library.

CHAPTER 28

p. 605 Science Photo Library; **p. 609** *Fig 28.10* Chris Howes/Planet Earth Pictures; **p. 613** *Fig 28.17* Andrew Syred/Science Photo Library; **p. 614** *Fig 28.19* Dr Jeremy Burgess/Science Photo Library; **p. 615** *Fig 28.21* Nancy Hamilton/SPL; **p. 619** *Fig 28.28 all* Marc Henrie; **p. 621** *Fig 28.31* P Plailly/ Science Photo Library; **p. 622** *Fig 28.33* Andrew McClenaghan/Science Photo Library; **p. 623** *Fig 28.33* Jane Burton/Bruce Coleman Ltd; **p. 624** *Fig 28.36 top left* The Roslin Institute, *top right* Jean-Paul Ferrero/Ardea, *middle* Jane Burton/Bruce Coleman Ltd, *bottom far right* Reinhard/Bruce Coleman Ltd, *Fig 28.37* Simon Fraser/Science Photo Library; **p. 626** *Fig 28.40 top left and top right* C J Clegg, *bottom* Nigel Cattlin/Holt Studios, *Fig 28.41* Hattie Young/Science Photo Library *top right* Simon Fraser/Science Photo Library.

CHAPTER 29

p. 630 *Fig 29.3* Science Photo Library; **p. 635** *Fig 29.10* Julia Kamlish/Science Photo Library; **p. 635** *Fig 29.12* Dr Jeremy Burgess/Science Photo Library; **p. 637** *Fig 29.13* Bio-Rad; **p. 638** *Fig 29.16* Science Photo Library; **p. 639** *Fig 29.17* David Parker/Science Photo Library; **p. 642** *Fig 29.21* Rex Features Ltd; **p. 644** *Fig 29.2* Ardea; **p. 645** *Fig 29.25* C J Clegg.

CHAPTER 30

p. 646 *Fig 30.1* Science Photo Library, *Fig 30.2* Planet Earth Pictures; **p. 650** *Fig 30.9* R G West; **p. 654** *Fig 30.14 top* NHPA, *middle* FLPA, *bottom* Hans Rheinhard/Bruce Coleman Ltd; **p. 656** *Fig 30.16* Ardea; **p. 657** *Fig 30.18 left* C J Clegg, *right* Science Photo Library; **p. 663** *Fig 30.25* Stephen Dalton/NHPA; **p. 668** *Fig 30.33* Anthony Bannister/NHPA; **p. 669** *Fig 30.34* John Reader/Science Photo Library.

CHAPTER 31

p. 675 Sheridan Photo Library/Ancient Art and Architecture Collection; **p. 677** *Fig 31.1* Ancient Art and Architecture.

The following are sources from which artwork and data have been adapted and redrawn.

p. 10 *Urinogenital system of mammal*, from Fig. 23, C Singer, *A Short History of Scientific Ideas*, Oxford University Press, 1959. By permission of Oxford University Press

p. 10 *Ladder of Nature*, from Fig. 22, C Singer, *A Short History of Scientific Ideas*, Oxford University Press, 1959. By permission of Oxford University Press

p. 18 *Fig. 2.1*, from Figs. 206 and 207, G Monger and M E Tilstone (eds), *Introducing Living Things: Revised Nuffield Biology Text 1*, Longman Group for the Nuffield Foundation, 1976

p. 52 *Panel on ants*, from M V Brian, *Ants*, 1997, by permission of HarperCollins Publishers

p. 43 *Bar graphs data*, from C A Edwards, 'Soil Pollutants and Soil Animals', *Scientific American*, April 1969

p. 45 *Fig. 3.2*, from *Life: An Introduction To Biology* by Colin S. Pittendrigh, George G. Simpson, and Lewis H. Tiffany, copyright © 1957 by George Gaylord Simpson, Colin S. Pittendrigh, and Lewis H. Tiffany and renewed 1985 by Anne R. Simpson, Joan Simpson Burns, Ralph Tiffany, Helen Vishniac, and Elizabeth Leonie S. Wurr, reproduced by permission of Harcourt Inc.

p. 54 *Fig. 3.13*, from Figs. 2 and 3, S M Evans and J M Hardy, *Investigations in Biology: Seashore and Sand Dunes*, Heinemann, 1970; reprinted by permission of Butterworth Publishers, a division of Reed Educational and Professional Publishing Ltd

p. 58 *Table 3.3*, from Fig. 3.6, reproduced from G Williams, *Techniques and Fieldwork in Ecology*, 1987, by permission of HarperCollins Publishers

p. 59 *Fig. 3.23*, from A J Willis, *Introduction to Plant Ecology*, George Allen and Unwin, 1973

p. 59 *Fig. 3.23*, from T Lewis and L R Taylor, *Introduction to Experimental Ecology*, Academic Press, 1972

p. 60 *Fig. 3.25*, from M Cannell and M Coults, 'Growing in the Wind', *New Scientist*, January 1988

p. 63 *Earthworm data*, from C A Edwards and J R Lofty, 'Effects of earthworm inoculation upon root growth of direct drilled wheat', *Journal of Applied Ecology*, 17, 533, Blackwell Science Ltd, 1980

p. 64 *Fig. 3.33*, from *Life: An Introduction To Biology* by Colin S. Pittendrigh, George G. Simpson, and Lewis H. Tiffany, copyright © 1957 by George Gaylord Simpson, Colin S. Pittendrigh, and Lewis H. Tiffany and renewed 1985 by Anne R. Simpson, Joan Simpson Burns, Ralph Tiffany, Helen Vishniac, and Elizabeth Leonie S. Wurr, reproduced by permission of Harcourt Inc., and from *Fundamentals of Ecology, Third Edition* by Eugene P. Odum, copyright © 1971 by Saunders College Publishing, reproduced by permission of the publisher

p. 65 *Fig. 3.37*, from *Fundamentals of Ecology, Third Edition* by Eugene P. Odum, copyright © 1971 by Saunders College Publishing, reproduced by permission of the publisher

p. 66 *Fig. 3.42*, from Fig. 31, T J King, *Selected Topics in Biology*, Thomas Nelson, 1980

p. 72 *Fig. 3.49*, from 'Ecologists Build Pyramids Again', *New Scientist*, July 1985. *Other Pyramids*: D Hulyer, Education Office, The Wildfowl and Wetlands Trust

p. 73 *Fig. 3.50*, from M A Tribe et al, *Ecological Principles: Basic Biology Course*, Cambridge University Press, 1974 and J Phillipson, *Ecological Energetics: Arnold Studies in Biology No. 1*, Cambridge University Press, 1966

p. 76 *Fig. 4.1*, from J Janek et al. *Plant Science: An Introduction to World Crops*, © 1974 W H Freeman and Company

p. 77 *Fig. 4.2*, from *World Population Reference Sheet*, Population Reference Bureau, 1994

p. 78 *Fig. 4.3*, from N Keyfitz, 'The Growing Human Population', *Scientific American*, 261 (3), 1989, pp. 70–7

p. 79 *Fig. 4.4*, from *Agricultural Production Indices* and *Population*, UN Food and Agriculture Organisation Statistical Database, 1997

p. 79 *Fig. 4.5*, from W D Ruckalshaw, 'Towards a Sustainable World', *Scientific American* 261 (3), 1989, pp. 114–20

p. 80 *Fig. 4.6*, from W O Pruitt, *Boreal Ecology: Arnold Studies in Biology No 91*, Cambridge University Press, 1978

p. 81 *Fig. 4.8*, from Fig. 1, D O Hall, 'Are there Commercial Prospects for Solar Energy', *New Journal of Chemistry*, 11 (2)

p. 83 *Table 4.1*, from Table 5, Bennett, *Elements of Soil Erosion*, McGraw-Hill, 1947. Reproduced with the kind permission of McGraw-Hill Publishing Company

p. 84 *Fig. 4.13*, adapted from S M Woodwell, 'The CO_2 Question' *Scientific American* 238 (1), 1978, p.37; with new data from The Scripps Institute

p. 85 *Fig. 4.15*, adapted from data from P D James, 'The Climate of the Past 1000 Years', *New Scientist*, 14 (1), 1990 pp. 129–36; with new data from Climate Research Unit of the University of East Anglia, *Scientific American*, 261 (3), pp. 38–47

p. 86 *Fig. 4.16*, data from UK Review Group of the Institute of Terrestrial Ecology

p. 86 *Fig. 4.17*, from Fig. 1, J Lee, 'Acid Rain', *Biological Sciences Review*, 1.1, Philip Allan Publishers, 1988, pp. 15–18

p. 91 *Egg shell graph*, from I Manton, Institute of Terrestrial Ecology

680

p. 91 *Fig. 4.22*, from G R Conway, MAFF. Crown copyright material is reproduced with the permission of the Controller of Her Majesty's Stationery Office

p. 93 *Table 4.2*, from *fig. 4.24*, from R Armson, 'Turning Rubbish into Heat and Electricity', *New Scientist*, November 1985

p. 95 *Barn owl data*, from Fig. 1, J Fisher, 'The Barn Owl', *Biological Sciences Review*, 2.4, Philip Allan Publishers, 1990, pp. 6–8

p. 102 *Fig. 5.6*, data from Dr R Cook, FRS Marine Laboratory (personal communication)

p. 103 *Fig. 5.7*, from A Milne, 'Pollution and Politics in the North Sea', *New Scientist*, November 1987

p. 106 *Fig. 5.10*, data from MAFF Statistics (Census and Surveys) Division and J Bingham (1983) Plant Breeding Institute, Cambridge

p. 118 *Sugar cane data*, from J Coombs, 'Sugar Cane', *Biologist*, 30 (1), 1983, pp. 15–22, Institute of Biology

p. 183 *Fig. 8.19*, from Professor Tom Steitz, Yale University

p. 201 *Table 9.4*, from E J Wood and W R Pickering, *Introducing Biochemistry*, John Murray, 1982

p. 236 *Fig. 11.15*, from Fig. 4.1, J L Hall and D A Baker, *Cell Membranes and Ion Transport*, © Longman Group Limited 1977. Reprinted by permission of Pearson Education Limited

p. 250 *Fig. 12.8*, from Figs 3.6 and 3.8, D O Hall and K K Rao, *Photosynthesis* 4th edn., Cambridge University Press, 1992

p. 253 *Fig. 12.15*, from J L Monteith, *Neth. J. Agric. Sci.*, 10(5), 1962, pp. 334–6

p. 266 *Fig. 12.37*, from P Moore, 'The Varied Ways Plants Tap the Sun', *New Scientist*, February 1981

p. 268 *Fig. 12.40*, from D O Hall, 'Solar energy conversion through biology: Part 1', *Biologist*, 26 (1), 1979, pp. 16–22, Institute of Biology

p. 271 *Table 13.1*, data from *Manual of Nutrition* 10th edn., MAFF. Crown copyright material is reproduced with the permission of the Controller of Her Majesty's Stationery Office

p. 273 *Graph*, from Fig. 97, J Ricketts and C Wood-Robinson (eds), *Living Things in Action: Revised Nuffield Biology Text 2*, Longman Group for the Nuffield Foundation, 1975

p. 275 *Fig. 13.4*, from *Food Production Yearbook*, Food and Agricultural Organisation of the United Nations, 1983

p. 275 *Fig. 13.4*, adapted from A F Walker *Cambridge Social Biology Topics: Human Nutrition*, Cambridge University Press, 1990, p.74

p. 288 *Fig. 13.20*, from Fig. 125, J H Green, *An Introduction to Human Physiology* 4th edn., Oxford University Press, 1984. By permission of Oxford University Press.

p. 322 *Fig. 16.6*, from Fig. 4, W O James, *An Introduction to Plant Physiology*, Oxford University Press, 1931. By permission of Oxford University Press

p. 325 *Fig. 16.11*, from Figs 9.19 and 9.25, V A Grenlach, *Plant Function and Structure*, 1973

p. 327 *Fig. 16.15*, from Fig. 6.1, J Sutcliffe, *Plants and Water: Arnold Studies in Biology No. 40*, 1979

p. 330 *Fig. 16.19*, from Fig. 7.10, J Bonner and A W Galston, *Principles of Plant Physiology*, © 1952 W H Freeman and Company

p. 331 *Fig. 16.20*, adapted from J E Harper et al, *Exploitation of Physiological and Genetic Variability to Enhance Crop Productivity*, American Society of Plant Physiologists, 1985

p. 338 *Fig. 16.33*, from Fig. 7.1, J Bonner and A W Galston, *Principles of Plant Physiology*, © 1952 W H Freeman and Company

p. 338 *Fig. 16.35*, from Fig. 1.7, A J Peel, *Transport of Nutrients in Plants*, Butterworths, 1974; reprinted by permission of Butterworth Publishers, a division of Reed Educational and Professional Publishing Ltd

p. 342 *Fig. 17.1*, from Fig. 3.1, R H Peters, *The Ecological Implications of Body Size*, Cambridge University Press, 1983

p. 360 *Fig. 17.32*, from Fig. 4.5, G Chapman, *The Body Fluids and their Functions: Arnold Studeis in Biology No. 8*, Cambridge University Press, 1978

p. 378 *Fig. 18.7*, from Fig. 10.1, K Schmidt-Nielson, *Animal Physiology: Adaptation and Environment*, Cambridge University Press, 1983

p. 378 *Fig. 18.8*, from Fig. 12.41, R Eckert with D Randall, *Animal Physiology Mechanism and Adaptation*, © 1983 W H Freeman and Company

p. 380 *Table 18.3*, reprinted from O C J Lippoid and F R Winston, *Human Physiology*, 1979, by permission of the publisher Churchill Livingstone

p. 390 *Fig. 18.24*, from Fig. 9.13, K Schmidt-Nielson, *Animal Physiology: Adaptation and Environment*, Cambridge University Press, 1983

p. 397 *Fig. 19.2*, from Table 7.1, reproduced from J I Williams and M Shaw, *Micro-organisms 2nd edn*, 1982, by permission of HarperCollins Publishers

p. 400 *Fig. 19.5*, from Table 12, R Gliddon (ed), *The Perpetuation of Life: Revised Nuffield Biology Text 4*, Longman Group for the Nuffield Foundation, 1975

p. 403 *Fig. 19.9*, from J M Tanner, *Growth at Adolescence*, Blackwells, 1962

p. 404 *Table 19.1*, from Table 1.1, J A Bryant, *Seed Physiology: Arnold Studies in Biology No. 165*, Cambridge University Press, 1985

p. 425 *Fig. 20.16*, from Figs 7.5 and 7.6 in P F Waring and I D J Philips, *Growth and Differentiation in Plants 3rd edn*, Pergamon, 1982. Reprinted by permission of Butterworth Heinemann Publishers, a division of Reed Educational and Professional Publishing Ltd

p. 428 *Fig. 20.20*, from R D Firn, 'Phototropism', *Journal of Biological Education*, 24 (3), 1990, pp. 153–7 and A B Hall et al, 'Auxin and shoot tropisms: tenuous connection?', *Journal of Biological Education*, 14 (3), 1980, pp. 195–9, Institute of Biology

p. 428 *Fig. 20.21*, from Fig. 3, M Wilkins, 'How Does Your Garden Grow?', *Biological Sciences Review*, 33, Philip Allan Publishers, 1991, pp. 22–6

p. 439 *Fig. 20.34*, from Table 2.1, J A Bryant, *Seed Physiology: Arnold Studies in Biology No. 165*, Cambridge University Press, 1985

p. 441 *Fig. 20.39*, from Fig. 17.22, J Bonner and A W Galston, *Principles of Plant Physiology*, © 1952 W H Freeman and Company

p. 467 *Fig. 21.38*, from *Life: An Introduction to Biology* by Colin S. Pittendrigh, George G. Simpson, and Lewis H. Tiffany, copyright © 1957 by George Gaylord Simpson, Colin S. Pittendrigh, and Lewis H. Tiffany and renewed 1985 by Anne R. Simpson, Joan Simpson Burns, Ralph Tiffany, Helen Vishniac, and Elizabeth Leonie S. Wurr, reproduced by permission of Harcourt Inc.

p. 514 *Fig. 23.22*, from J Z Young, *Life of Mammals*, Oxford University Press, 1990. By permission of Oxford University Press.

p. 515 *Fig. 23.24*, from Fig. 23.23, N Reeve, *Hedgehogs*, T and A D Poyser Ltd, Natural History Series, Academic Press Ltd, 1994

p. 518 *Fig. 23.28*, from Fig. 20, N Tinbergen, *Study of Instinct*, Oxford University Press, 1951. By permission of Oxford Universtiy Press

p. 519 *Fig. 23.29*, from J P Holiman, 'How an Instinct is Learned', in T Einsner and E O Wilson, *Animal Behaviour Readings from Scientific American*, © 1975 W H Freeman and Company

p. 520 *Fig. 23.30*, from Fig. 47, N Tinbergen, *Study of Instinct*, Oxford University Press, 1951. By permission of Oxford University Press.

p. 522 *Fig. 23.33*, from *Life: An Introduction to Biology* by Colin S. Pittendrigh, George G. Simpson, and Lewis H. Tiffany, copyright © 1957 by George Gaylord Simpson, Colin S. Pittendrigh, and Lewis H. Tiffany and renewed 1985 by Anne R. Simpson, Joan Simpson Burns, Ralph Tiffany, Helen Vishniac, and Elizabeth Leonie S. Wurr, reproduced by permission of Harcourt Inc.

p. 525 *Fig. 23.38*, from F Toates, *Biology*, from *Brain and Behaviour Book 5: Control of Behaviour*, The Open University, 1992

p. 528 *Fig. 24.2*, data adapted from A D Lopez, 'Causes of death in the industrialized and the developing countries: estimates for 1985' in D T Jamison and H Mosley (eds) *Disease Control Priorities in Developing Countries*, OUP, 1993

p. 533 *Fig. 24.5*, from D Llewellyn-Jones, *Sexually Tranmitted Diseases*, Faber and Faber, 1990

p. 538 *Fig. 24.15*, from D Saunders, *The Struggle for Health: Medicine and the Politics of Underdevelopment*, Macmillan 1985

p. 539 *Smoking deaths data*, from R Doll and A B Hill, *Brit. Med. J.*, 2, 739, 1950 and 1952, pp. 1271–86

p. 540 *Fig. 24.16*, from Fig. 2, G J Lithgow, 'The Biology of Ageing', *Biological Sciences Review 10.3*, Philip Allan Publishers, 1998, pp. 18–21

p. 592 *Fig. 27.19*, from Fig. 7.2, J W Buckle, *Animal Hormones: Arnold Studies in Biology No. 158*, Cambridge Universtiy Press, 1983

p. 595 *Fig. 27.22*, from Fig. 7.1, J Cohen, *Reproduction*, Butterworths, 1977; reprinted by permission of Butterworth Publishers, a division of Reed Educational and Professional Publishing Ltd

p. 600 *Fig. 27.27*, from Unipath Ltd., Bedford

p. 633 *Fig. 29.8*, from Fig. 1, B Brown, 'PCR (polymerase chain reaction)', *Biological Sciences Review*, 11.1, 1998, pp. 18–19

p. 643 *Innovation panel*, from R. Dobson, 'Laboratories unveil plans to grow hearts, livers and kidneys', *The Sunday Times Innovations Section*, 22 August 1999. Roger Dobson/Sunday Times Newspapers Ltd, 2nd May 1988

p. 650 *Fig. 30.9*, from Fig. 30.11, R West, *Studying the Past by Pollen Analysis*, Oxford Biology Reader, 1971. By permission of Oxford University Press

p. 663 *Fig. 30.24*, data from H B D Kettlewell, 'The Phenomenon of Industrial Melanism in the Lepidoptera', *Ann. Rev. Entomol.*, 6, 1961, pp. 245–62

p. 665 *Fig. 30.28*, from M A Tribe, A J Morgan and P A Whittaker, *The Evolution of the Eukayotic Cell: Arnold Studies in Biolgy No. 131*, Cambridge University Press, 1981

p. 667 *Fig. 30.31*, M Ingrouille, *Diversity and evolution of Land Plants*, Chapman and Hall, 1992, with kind permission from Kluwer Academic Publishers

Exam questions have been reproduced with kind permission from the following examination boards:

- AQA: The Associated Examining Board (AEB), and Northern Examinations and Assessment Board (NEAB)
- International Baccalaureate Organisation (IB)
- London Examinations, a division of Edexcel Foundation
- OCR

Answers

The following are only *guides to the answers*; explanations in the book are referred to when appropriate. (Figures, tables and headings are referred to in **bold type**.)

CHAPTER 1

Page 3

1 Green plants (**Interdependence within the ecosystem**, p. 46), apart from contribution from chemosynthetic bacteria (p. 269).

2 **Figure 1.2** (p. 3).

3 Both involve oxidation, and in both, chemical energy becomes heat energy; respiration involves many intermediates and many enzymes, and occurs at much lower temperatures.

4 Yeasts (p. 559); voluntary muscle under oxygen debt (p. 360); plant roots in waterlogged soil.

Page 5

1 Sensitivity, movement and locomotion.

2 Asexual reproduction.

3 **a** $y - x\,g$ = fresh mass increase due to water uptake.

 b Growth (and development) due to an increase in number of cells.

4 Catabolism: oxidation of carbohydrate, digestion of protein, deamination of protein, anaerobic respiration. Anabolism: production of starch from glucose, protein formation, photosynthesis.

Page 7

1 Plant cells with chloroplasts carry out photosynthesis, animal cells cannot make their own food; plant cells are surrounded by a cellulose cell wall and become rigid with turgor (p. 234), animal cells are less rigid.

2 Examples:
 a epithelium cell of villus, **Figure 13.17** (p. 285).
 b muscle fibres (p. 218).
 c collenchyma cell, **Figure 10.20** (p. 224).
 d neurone, **Figure 10.15** (p. 220).

3 Unable to divide; soon dies, unable to regulate its metabolism.

4 **Figure 11.1** (p. 228).

Page 9

1 Accurate observation, careful recording, production of tools and materials to facilitate the recording.

2 Contrast roles of cells shown in **Figure 1.9** (p. 7), **Figure 1.10** (p. 8) and **Figure 10.11** (p. 217).

3 **Division of labour** (p. 8).

4 Literature search and review, formulation of a hypothesis, design of experiments (or observation sequence), recording of results, analysis of data (including statistical analysis), interpretation of data (to decide whether the hypothesis is supported or rejected), reporting the study to them in sufficient detail for others to repeat the work, and the negotiation of further financial support from sponsors, industry or other agency. (All or many of these.)

5 **Tissues and organs** (p. 8).

Page 11

1 Detailed study of individual organisms as representative of the species; recording much data; not involved in numerical records, but rather in morphology, anatomy and 'ecology'.

2 **a** Living things originate from inorganic matter, e.g. pond organisms from mud; Aristotle (p. 10), Leeuwenhoek (p. 148), Pasteur (p. 527).
 b Efficient microscopy.

3 Because it asserts that all living things are related, directly or indirectly.

Page 15

1 Independent variable = hydrogen peroxide concentration;
 dependent variable = volume of oxygen produced;
 controlled variables = temperature, volume of the reaction mixtures, enzyme concentration.

2 Possible examples:
 (i) zinc ions (dissolved from the wire by rainwater) are poisonous to grass plants;
 (ii) the drops of rainwater, falling from the wire above, mechanically damage grass and eventually kill it.
 Test the effects of dilute solutions of zinc ions on the growth of grass seedlings.

3 Light is required for starch formation; the stencil prevents CO_2 from reaching part of the leaf; light is required for the deposition of starch in leaf cells.

4 Remove a nucleus from another egg cell and then return it to the cell. Is it the damage done when a nucleus is removed that prevents further development of an egg?

CHAPTER 2

Page 19

1 **Viruses** (p. 27), **How many kingdoms of living things?** (p. 24).

2 Continual discoveries of previously unknown living things (and fossils) that are then fitted into schemes of classification.

3 Presence (or absence) of true nucleus, by electron microscopy (p. 154).

Page 21

1 Species, genus, family, class, phylum, kingdom.

2 Precisely defined and internationally agreed, they facilitate cooperation between observers.

3 Examples: vegetables (of root, leaf, fruit or seed), flowers (spring-, summer- or autumn-flowering).

Page 23

1 Established a flexible scheme of classification that facilitated comparisons (p. 22).

2 Based on detailed, accurate observations of a very large number of species; after the demise of ancient Greek culture the practice of observing nature became uncommon (until the scientific revolution, p. 676).

3 Homologous: pentadactyl limbs of other vertebrates. Analogous: wings of insects.

4 **a** Members of a species can normally interbreed to produce fertile offspring.

b Ability to interbreed is not the sole or essential diagnostic characteristic of a species.

Page 25

1 Examples:
 Artificial: animal or height (for example) at or below 35 cm – cat; animal of height greater than 35 cm – sheep.
 Natural: eutherian mammal (p. 41) well equipped for meat-eating (e.g. teeth adapted for catching and killing prey, including curved and pointed canine teeth), with 4–5 clawed digits on each limb – cat; eutherian mammal well equipped for grass-eating (p. 289), with 2 hoofed digits on limbs – sheep.

2 **How many kingdoms of living things?** (p. 24), **Kingdom Fungi** (p. 30).

Page 26

1 The definition of a species is now also based on cytological and biochemical differences (p. 23).

2 Structural (morphological and anatomical), cytological, biochemical and perhaps behavioural/ecological differences.

3 By comparison of anatomical, physiological and biochemical features (p. 651).

Page 42

1 **a** You may learn little about the unknown organism or the reasons for its classification.
 b **We can identify by comparison** (p. 41).

2 Examples: providing evidence of levels of pollution in the past compared with those in present-day species from similar habitats (e.g. **eggshell thickness change due to organochlorine insecticides**, p. 91), or changes in anatomy with time (e.g. numbers of stomata per unit area of leaf of plants grown in pre-Industrial Revolution atmospheres, *Study Guide*, Chapter 18).

CHAPTER 3

Page 47

1 Spores of bacteria and fungi.

2

Tropical rain forest	Temperate deciduous woodland
experiences a hot, wet climate	experiences cold winters and warm summers
soil nutrient content low, e.g. dead organisms decay rapidly, ions quickly removed by plant roots and carried back to canopy (or leached away)	soil rich in humus and nutrients (brown-earth soil type)
huge diversity of species present, flora and fauna largely occur in the tree canopy	a range of species present, including field and shrub species, particularly around young trees

3 Plants largely determine climate locally (and globally); plants provide 'cover' and habitats; plants are the food source, directly or indirectly.

4 Examples: playing fields (or park), walls, hedgerows; light, humidity, topography, soil (or substratum).

5 Population = all frogs; ecosystem = lake; habitat = mud of lake; abiotic factors = water flow, temperature variation; community = plants and animals of lake; biomass = mass of vegetation.

Page 48

1 Food chains and food webs (p. 47).

2 Niche = position occupied *and* life style/function; habitat = position occupied.

Page 51

1 Analysing the shore community by transects (p. 50).

2 Whether the substratum is being worn away or added to by the action of the sea.

3 Conditions (e.g. salinity) on shore are similar to those of the sea for part of every 24 hours, but always very different from conditions on dry land.

4 As shown in **Figure 3.10**.

5 Figure 3.12 (p. 53).

6 Figure 3.11 (p. 51).

7 Table 6.2 (p. 126).

8 Hot day: e.g. a raised temperature, dehydration, salinity increase.
Rain storms: e.g. decreased salinity.

Page 55

1 Ability to absorb light of particular wavelengths (**Figure 3.12**), and to resist desiccation (**Figure 3.13**).

2 Site where a secure attachment can best be achieved.

3 6%.

4 a Move to more favourable positions; close shells.

b Protective outer structures (**Figure 3.15**).

5 The conditions that immediately surround an organism; e.g. rock pool, the holdfast of a large seaweed, an empty shell.

6
```
plankton ──────► barnacle

seaweed ──────► limpet ──────► dog whelk
                  │
                  ▼
winkle  ──────► shore crab ──────► sea bird
```

7 Rock pool has very variable salinity, and is normally periodically submerged (part of sea bed) in every 24 hours.

8 The shorter, less-branched structure with fewer air bladders offers less resistance.

Page 61

1 Mean temperature + rainfall (**Figure 3.22**, p. 58), light (**Table 3.3**, p. 58), soil (**Figure 3.31**, p. 63).

2 Figure 3.20 (p. 57) and **Figure 4.6** (p. 80).

3 Soil, mean temperature, rainfall, availability of seeds, presence of herbivorous animals.

4 Presence of animals dependent on plants of field layer and shrub layer for food and for protection and cover (habitat).

5 Factors affecting transpiration rate (p. 324).

6 An indirect effect on most factors (**Figure 3.24**, p. 59), e.g. weathering, water content, organic matter, soil air, pH and soil nutrients, soil organisms; **Figure 3.30** (p. 63).

7 Good aeration, satisfactorily drained, sufficient water content, favourable temperature, availability of essential nutrients.

8 Soil components; mineral skeleton (p. 60).

Page 63

1 Set up 'wormery' apparatuses (as shown below, and see *Study Guide*, Chapter 3), containing layers of loam, sandy and clay soils, and with/without dead leaves on the surface. Observe the effect on the soil of the addition of deep-burrowing and shallow-burrowing species of earthworm (e.g. *Lumbricus terrestris* and *Allolobophora chlorotica* respectively) compared with that in a wormery without additional earthworms.

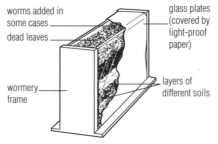

worms added in some cases
dead leaves
wormery frame
glass plates (covered by light-proof paper)
layers of different soils

2 pH 6.5–7.

Panel questions

1 Increased root growth (the effect of added deep-burrowing worms was greater than that of the shallow-burrowing worms and the normal soil worm population); increased yield of grain.

2 Examples: increased incorporation of organic matter (e.g. straw), which may (indirectly) increase humus content, drainage, water retention, ion availability, soil temperature.

3 Most other soil organisms, including the saprotrophic bacteria and fungi, many competitors and predators of earthworms.

Page 65

1 Selection (survival and reproduction) of the more vigorous (healthier?) members of the species (intraspecific competition), and selection of the species best equipped to exploit the local resources whilst these are available (interspecific competition, p. 64).

2 Almost 48 hours.

3 Competition and niches (p. 65).

4 Figure 3.35 (p. 65); **Positive and negative feedback** (p. 503).

Page 67

1 Symbiosis (p. 66).

2 Examples: *Trypanosoma* (protozoan), causes 'sleeping sickness' (p. 534); *Tinea pedis* (fungus), causes ringworm of the feet, known as 'athlete's foot' (p. 534); *Mycobacterium tuberculosis* (bacterium), causes tuberculosis (p. 531); *Herpes zoster* (virus), causes chicken pox and shingles.

3 Contribute to the early formation of soil; contribute to the chemical weathering of rocks; provide organic matter in the first-formed soil; food for herbivorous mammals during winter, including deer.

4 Appearance and survival of elm trees with resistance to the fungus *Ceratocystis*.

5 a Figure 3.4 (p. 46).

b Interdependence within the ecosystem (p. 46).

c Primary producer = green plant matter; omnivore = caddis fly larvae; primary consumer = e.g. water flea; carnivore = e.g. water boatman; secondary consumer = small fish; tertiary consumer = perch.

Page 69

1 a Favourable abiotic factors such as mild winter, good supplies of food; favourable biotic factors such as absence of competitor species, parasites, etc.

b Intraspecific competition, as population grows.

2 From less than 10 to 1000.

3 Marking of samples may enhance losses due to predation; disturbances of habitats may drive members of the population to migrate.

4 Factors in population change (p. 68).

Page 71

1 Their seeds are efficiently dispersed to diverse habitats by wind action.

2 Destructive grazing of shrub and tree species as seedlings.

3 Primary succession (p. 70), **secondary succession** (p. 70).

4 Constant high water table. Water flowing at speed through the lake.

Page 73

1 a Carnivorous snake, e.g. python.

b Filter-feeding whale, e.g. blue whale feeding on plankton.

2 $(44\,090/4.6 \times 10^6) \times 100\%$ = less than 10%.

3 Level 2 when eating food of plant origin (e.g. bread, fruit, vegetables); level 3 when eating products of animals fed on plant products (grain, hay, etc.).

4 Nitrates, phosphates, calcium ions (all needed by the plankton); nutrients are taken up by primary producers in growth, but released after death and decay.
Figure 3.50 May.

Page 75

1 Secondary productivity (p. 74).

2 Sun; heat passed out into space.

3 Involving organisms: exchange pool, in organisms, etc.
Independent of organisms: reservoir pool, in oceans, etc.

4 Energy flow is unidirectional; the flow of nutrients is a cycle.

CHAPTER 4

page 77

1 Implies increasing diversity of life. In fact, the numbers of living species have varied erratically at certain times in Earth history (**Figure 30.19**, p. 658).

2 Origin of life (p. 664).

3 Food sources are shared with many other living things; productivity of natural (non-cultivated) 'crops' is low.

Page 79

1 1985 (the most recent period for which data are given), because the difference between death rate and birth rate is greatest at this time.

2 5 × 450 millions in the less-developed countries, and 5 × 70 millions in the developed countries.

3 The distribution of food does not match population distribution. Many people are too poor to buy food. Free distribution of surplus may undermine existing production mechanisms. Because of cultural differences, the food of countries in surplus may be of limited use to those at starvation level.

Page 83

1 Minimal ground, field and shrub layers supported, resulting in little or no biodiversity within a coniferous plantation, compared with that in a mixed woodland.

2 Examples: aesthetic and environmental considerations, if these are addressed by the community, possibly following representations by environmental pressure groups; awareness of a market for hardwood products within the timespan of the growth of deciduous broad-leaved trees.

3 Water vapour loss to the atmosphere, causing clouds and precipitation within the area; retention of rainwater in stabilised soil of the forest, providing a more-or-less constant supply of moisture.

4 a Nuclear energy, hydroelectric energy (with solar and wind power sources, perhaps).
b Biomass (p. 119), together with increasing contributions from those in **a** above.

5 The concentration of atmospheric carbon dioxide (p. 84).

6 Principal causes: expansion of land for industry (including agriculture), transport, homes; exploitation of forests for timber resources; climate change leading to desertification. Harmful effects: loss of habitats of species, loss of gene pool, climate change, loss of resource bank for sustained forestry in the future.

7 Flooding as a consequence of deforestation (p. 83); **Soil erosion and desertification** (p. 83).

8 The absence of sound conservation practice in forestry elsewhere in the world, by those offering advice; the need to pay for imports with currency earned from the sale of a limited range of primary produce on world markets (**World trade and poverty**, p. 97).

Page 85

1 The 'greenhouse effect' (p. 84); **Other 'greenhouse' gases** (p. 84).

2 Yes, but much less so than that due to current consumption of fossil fuel. The destruction of trees for firewood decreases the number of photosynthesising plants (although the scale of industrial and domestic consumption was slight by current standards). Note **Carbon cycle**, **Figure 12.2** (p. 247).

Page 87

1 Acid rain (p. 86).

2 Figure 3.29 (p. 62).

3 Oxides of nitrogen, carbon and sulphur, and hydrocarbon vapour (unburnt fuel); **Low-level ozone** (p. 87).

4 Acid-tolerant species increased whilst acid-susceptible species decreased.

Page 89

1 Low-level ozone and **The ozone layer** (pp. 87–9).

2 The maintenance of the ozone layer (pp. 88–9).

3 As above.

Page 92

1 Pesticides; **Recognition of the problems** (p. 90).

2 To overcome the possibility of harm to other organisms in the food chain, other than to the targeted pest.

3 Biological control; Pesticides today (p. 92).

4 Pesticides (p. 90); The leaves, seeds, fruits, stems and roots of many plants are significantly predated. Many plants, however, have substances or structures that deter predators. Some leaves have spines, prickles or 'stings' that may deter. The common fern bracken (*Pteridium aquilinum*) is, perhaps, an extreme example. Bracken contains a cocktail of four chemicals that harm animals that consume the leaves.

Page 96

1 Diversion of the water supply of the reservoir; contamination of the water table with pesticides, and with fertilisers if these are applied to the trees.

2 Authoritative data on protected species that may be potentially threatened by the development.

3 Paper and other organic matter. Reduction in packaging; the practice of composting vegetable waste to produce 'compost' to use in place of manufactured fertilisers.

4 Landfill: methane generated, and escaping into atmosphere (**Methane**, p. 84), harmful solutes leached into water supplies.
Incineration: **Toxin production by the incinerator** (p. 93).

5 An international challenge (p. 94).

6 Open oceans: e.g. **Human activities and the sea** (p. 56).
Dry deserts: e.g. **Soil erosion and desertification** (p. 83).

Page 97

1 Example: **Maintenance of representative ecosystems** (p. 94).

2 Appropriate methods of development (p. 97).

CHAPTER 5

Page 101

1 Figure 5.2 (p. 99).

2 5×10^3 cm^{-3}.

3 Spores of pathogenic bacteria, present in the soil at very low concentration, may have increased in number to a dangerous level.

4 Nitrate and phosphate ions are absorbed by aquatic green plants. This results in enhanced growth of aquatic plants, especially phytoplankton (quick growing). This 'algal' bloom is short-lived. The quantity of suspended (non-living) organic matter in the water increases. 'Population explosion' of saprotrophic (aerobic) bacteria and fungi follows. Dissolved oxygen concentration in water falls to zero. Aerobic organisms (including fish) are asphyxiated. Anaerobic decay of dead organic matter now ensues. Hydrogen sulphide (poisonous) released from anaerobic decay of sulphur-containing proteins, threatens all living things.

5 Waste water and industrial effluent treatment (p. 101).

Page 103

1 Ions in water (fresh or sea) are available for absorption by aquatic plants. They subsequently pass through food webs, and accumulate in the 'top' omnivores and carnivores.

2 a The need for conservation (p. 103).
b The need for marine biological research (p. 103).

3 The world's fishing grounds (p. 102).

Page 106

1 Contribution: **Table 5.2** (p. 105).
Identification: **Table 26.4** (p. 577).

2 Fertiliser: **The application of fertilisers** (p. 105).
Pesticide: **The application of pesticides** (p. 105).
Selective breeding: **The principles of selective breeding** (p. 644).
Irrigation: **Water stress** (p. 325).

3 Example: **Economic aspects of photosynthesis** (p. 268).

Page 111

1 Animals kept outdoors are under conditions similar to those of their natural environment; **Developments in animal husbandry** (p. 107).

2 Recycling in garden compost (p. 297).

3 Life cycle of the cow in **Milk production and the milking parlour** (p. 108).

4 Example: difficulty of protection of grain from biodeterioration (p. 561) and pests.

5 Preserving and storing foods (p. 109).

6 Examples:
Antioxidants – biscuits, margarine
Emulsifiers – salad-dressing, processed cheese
Salt – meat products, crisps.

Page 119

1 Cheese manufacture (p. 112).

2 Examples: monitoring the process of product synthesis; checking that the culture is not contaminated by other microorganisms, causing alternative reactions to occur.

3 ATP synthesis (**Microbial mining**, p. 113).

4 Examples: reduced 'side effects'; effective against a wider range of microorganisms; increased 'shelf life'.

5 Biological fuel generation (p. 118).

6 See the discussion on p. 628; **Insulin manufacture** (p. 634).

SECTION TWO

CHAPTER 6

Page 125

1 **The chemical make-up of living things** and **Figure 6.1** (p. 124), **The ionisation of water** (p. 126), **Ionic and covalent compounds** (p. 128).

2 a **Table 6.1** (p. 124).

  b Examples: carbohydrates, lipids and proteins.

3 **Water** (p. 125) and **Figure 6.1** (p. 124).

Page 127

1 **The properties of water** (very high specific heat capacity) (p. 126).

2 **Table 6.2** (p. 126).

3 Hydrochloric acid + sodium hydroxide → sodium chloride + water

  $HCl(aq) + NaOH(aq) → NaCl(aq) + H_2O(l)$

4 $C_6H_{12}O_6$:

  $(6 × 12) + (12 × 1) + (6 × 16)$ g = 180 g

  KNO_3: $39 + 14 + (3 × 16)$ g = 101 g

Page 129

1 Because pH is expressed as the *reciprocal* of the H^+ concentration; the greater the concentration of H^+, the lower the pH. Thus pure water at 25 °C has $[H^+] = 10^{-7}$ mol dm^{-3} and pH = 7.0, whereas a hydrochloric acid solution might have $[H^+] = 10^{-1}$ mol dm^{-3}, and pH = 1.0.

2 **The ionisation of water** (p. 126), **Ionic and covalent compounds** (p. 128).

3 The pH scale is logarithmic, i.e. pH 6 is ten times more acid than pH 7, pH 5 is a hundred times more acid than pH 7 (**pH**, p. 128).

Page 131

1 **The chemistry of carbon** (p. 129), **The important chemical properties of carbon** (p. 130).

2 a The part of the organic molecule other than the functional group (**Figure 6.9**, p. 131).

  b R—OH; R—COOH; R—CHO; R—CO—CH$_2$OH.

Page 133

1 **Hexose sugars**, and **Figure 6.10** (p. 132).

2 Structural and optical isomerism (p. 132).

3 **Pentose and triose sugars** (p. 133).

4 **Figure 6.12** (p. 133); **Figure 6.10** (p. 132).

Page 136

1 **Condensation and hydrolysis reactions** (p. 131).

2 'Malt extract' is obtained from partially germinated barley, and is principally used in brewing and food manufacture. **Figure 14.21, The role of yeast in brewing** (malting, cracking and mashing steps) (p.298). **Sugar cane as an energy crop** (p. 118).

3 'Small' molecules *v.* the macromolecules (p. 135).

4 Contribute to the strength of the cellulose microfibrils by binding parallel fibres together.

Page 139

1 a **The fats and oils** (p. 137).

  b **Phospholipids** (p. 139).

2 a **Roles of triacylglycerols in living things** (p. 138), **Carbohydrates and fats, and the supply of energy** (p. 271), **Figure 18.24** (p. 390).

  b Cholesterol is produced in the body from unsaturated fats. **Saturated and unsaturated fats** (p. 137); **Saturated fats and cholesterol** (p. 138). **Factors that predispose to coronary heart disease** (p. 356), **Figure 27.20** (p. 592).

3 a **Figure 6.15 The formation of triacylglycerols** (p. 137).

  b In the fatty acids (R groups) from which they are formed.

4 **Lipids** (p. 137).

Page 141

1 a **Figure 6.9** (p. 131) and **Figure 6.20** (p. 139).

  b **Amino acids** (p. 139).

2 **Peptides** (p. 140), **Figure 6.22** (p. 141).

3 The substances organisms contain are those their cells have specific enzymes to metabolise. The enzymes that metabolise amino acids are specific to the L-forms.

4 a, b **Amino acids as polar molecules** (p. 140).

Page 145

1 **Table 6.3 The functions of proteins** (p. 144), **Transport across the cell membrane** (p. 159), **How active transport occurs** (p. 236).

2 a **Hydrolysis of proteins** (p. 144).

  b, c **Pancreatic juice** (p. 284), and **Table 13.4** (p. 287).

3 **Denaturing of proteins** (p. 144), **Investigating membrane permeability** (p. 229).

Page 147

1 **Figure 6.30** (p. 145) and **Figure 6.29** (p. 145).

2 **Nucleic acids** (p. 145).

3 Nucleotides.

CHAPTER 7

Page 149

1 He worked alone, avoiding the sharing of his techniques with others, even with qualified students who might have continued his approach either under him, or possibly elsewhere in Europe, for example. Perhaps he had a secretive personality? He made his results public, though.

2 **Cell theory** (p. 148).

3 Showed that some water-dwelling microorganisms can survive drying up, and 'reappear' when water returns, rather than being formed in some mysterious way.

Page 151

1 a Plasma membrane, nucleus and cytoplasm (with many of the organelles in common).

  b

Plant cells	Animal cells
cell wall (containing cellulose) present	no cell wall present
typically, large permanent vacuole(s) present	no permanent vacuole present
green cells (photosynthetic) contain chloroplasts	no chloroplasts present
no centrosome present	centrosome of two centrioles occurs just outside the nucleus

2 The advantage of division of labour between cells are discussed on p. 212.

3 Of the cells represented on pp. 150–51, which is typical of the others?

Page 153

1 **Using vital dyes** (p. 152), **Preserved and stained tissue; staining** (p. 152).

2 **Fixation** (p. 152).

3 a Example: to investigate animal and plant tissue structures by light microscopy.

  b **Phase-contrast microscopy** (p. 153).

  c **Oil-immersion microscopy** (p. 153).

Page 155

1 $(× 10) × (× 5) = ×50$.

2 **Magnification and resolution**, p. 154.

Page 156

1 By causing the cytoplasmic components to appear a different size or shape from that in the living cell.

2 **Artefacts in microscopy** (p. 156).

3 a, b **Electron microscopy** (p. 154).

4 a, b, c **Figure 12.26** (p. 259).

Page 159

1 Protein molecules tend to have no fixed position, but rather move about between the lipid molecules, which are themselves also in motion to some extent.

2 Lipid bilayer is lipid only (**Figure 7.13**). 'Double' membranes consist of a 'sandwich' of two fluid mosaic membranes of lipid and protein (**Figure 7.15**), side by side.

3 Hydrocarbon chain (**Lipids**, p. 137).

Page 161

1 Exit point for mRNA (**Figure 9.19**, p. 208).

2 **Figure 9.10** (p. 200) Chromosome and chromatin (p. 160).

3 a, b **Ribosomes** (p. 161).

Page 163

1 **Figure 7.19** (p. 162).

2 **Lysosomes** (p. 162).

3 a **Mitochondria** (p. 163).

  b Metabolism is 'funded' by energy (and resourced by metabolites) made available by pyruvic acid oxidation (part of aerobic respiration).

4 Enzyme(s) packaged within a membraneous sac. Peroxisomes contain catalase, rather than hydrolytic enzymes in an acidic medium.

Page 165

1 a Light into chemical energy.

  b Grana of the chloroplast: **a summary of photosynthesis** (p. 265).

2 **Microtubules** (p. 164).

3 **Flagella and cilia** (p. 165). Examples: lining of oviduct, movement of ovum (p. 595).

Page 167

1 a, b, c **Figure 7.28** (p. 166), **Cell walls** (p. 166).

2 **Figure 7.30** (p. 167).

3 **Cellulose** (p. 135), **Figure 7.29** (p. 167).

Page 168

1 **Table 7.2** (p. 169).

2 The cell wall protects the cell from mechanical damage situations where the water potential of the surrounding medium is less negative than that of the cell (**Figures 11.10** and **11.11**, p. 234).

3 **Origin of the eukaryotic cell** (p. 666).

4 **Figure 7.30** (p. 167), and **Desmosomes**, **Tight junctions** and **Gap junctions** (p. 168).

CHAPTER 8

Page 171
1 **Photosynthesis in outline** (p. 246), **Aerobic respiration** (p. 302).
2 **Figure 1.3** (p. 4).

Page 173
1 **Energy transformation** (p. 172).
2 **Figure 8.2** (p. 172).
3 **Figure 8.3** (p. 173).
   Endergonic reactions: synthesis of ATP from ADP and P$_i$ (**Figure 8.10**, p. 176), glucose + orthophosphate = glucose phosphate (p. 176).
   Exergonic reactions: hydrolysis of ATP, hydrolysis of GTP (p. 177).
4 They continue at identical rates.

Page 177
1 There are other common pathways in which it is an intermediate.
2 **Figure 7.11** (p. 156).
3 The energy of a molecule is that required for its synthesis. We therefore say that the energy occurs in the arrangement of atoms that make up that molecule.
4 **Tissue respiration** (p. 314).
5 The linking together of an energy-releasing (exergonic) reaction with an energy-requiring (endergonic) reaction so that one 'drives' the other (p. 176).

Page 179
1 **The properties of catalysts** (p. 178).
2 **How do enzymes work?** (p. 179) and **Figure 8.11** (p. 178).
3 As above.

Page 181
1 To ensure that, when the reactants are mixed, the reaction occurs at the (known) temperature.
2 **Figure 8.15** (p. 181).
3 Working with an excess of substrate molecules (a relatively high concentration of substrate), the effect of an increase in the concentration of enzyme is to increase the rate of reaction. This is because, at any moment, proportionately more substrate molecules are in contact with an enzyme molecule.

Page 183
1 *In vitro* the products of a reaction accumulate, the substrate concentration decreases, the enzyme (the quantity of which is fixed) is progressively denatured, and the conditions (such as pH, etc.) are progressively changed. (In cells these conditions are almost certainly held constant.)
2 Not at all (**Figure 8.17**, p. 182).
3 **Figures 8.18** and **8.19**, (p. 183).

Page 187
1 Each step of a pathway (e.g. respiration, **Figure 15.21**, p. 315) is catalysed by a different enzyme.
2 Hydrolase (**The naming and classification of enzymes**, p. 187).
3 **Control of secretion of gastric juice** (p. 288).
4 By lowered formation of the previous intermediate (precursor molecule) or of the enzymes that catalyse it, or by allosteric inhibition (**Figure 8.26**, p. 186).

5 The effects of the allosteric inhibitor are reversible, e.g. by raising the substrate concentration.

CHAPTER 9

Page 193
1 **Figure 7.16** (p. 160).
2 **Figure 9.1** (p. 192).
3 a All the reactions of metabolism, plus protein synthesis.
   b Chromosome replication, and mRNA synthesis.

Page 195
1 **Prophase, metaphase, anaphase** and **telophase** (p. 195) and **Table 9.2** (p. 199).
2 Examples: formation of **Erythrocytes** (**red blood cells**) (p. 217); epithelial cells of small intestine (**Villi**, p. 283); basal layer of Malpghian layer (**Skin structure**, p. 511).
   Examples: meristematic cells of stem and root apex, and of vascular cambium (**Meristems and plant growth**, p. 406).
3 **Microtubules** (p. 164).

Page 199
1 **Homologous chromosomes** (p. 196).
2 A diploid nucleus becomes haploid via meiosis. The members of each homologous pair are shared between daughter nuclei.
3 **Table 9.2** (p. 199).
4 Random assortment of maternal and paternal homologous chromosomes. Recombination of segments of maternal and paternal chromosomes during crossing over.
5 **Prophase** (p. 198).

Page 201
1 **Chromosome protein** (p. 200).
2 **a, b Figure 9.11** (p. 201).

Page 202
Hershey and Chase would have expected that only the bacteria infected with virus labelled with ^{35}S would have subsequently produced radioactive virus (they were not aware that the protein coat of the virus remains outside the host cell).

Page 205
1 **Nucleic acids** (p. 145).
2 **X-ray diffraction patterns** (p. 202).
3 **Watson and Crick's model** (p. 204).
4 **Base ratio analysis** (p. 203).

Page 207
1 **Step One: Transcription** (p. 208).
2 **Evidence [for evolution] from comparative biochemistry** (p. 651).
3 Ser—Tyr—His—Gln—Lys.
4

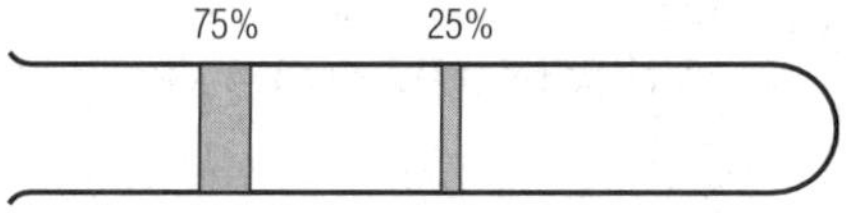

Page 209
Figure 9.15 (p. 205).

Page 211
1 **An experiment using *Neurospora*** (p. 210).
2 Some genes relate to particular stages of development, others to particular cellular tissues. An organism needs to activate certain genes in limited sites for restricted periods.

3 The proteins (globins, **Figure 17.28**, p. 358) in haemoglobin are composed of a linear sequence (a chain) of about 150 amino acid residues. Sickle-cell anaemia arises because one amino acid in the chain is replaced by another (**Sickle-cell anaemia**, p. 210).

CHAPTER 10

Page 214
1 a Examples: leaf, parenchyma, procambial strand cell (p. 222).
   b Examples: skin (p. 511), stratified epithelium, cell of generative layer (p. 214).
2 **Differentiation** (p. 212).
3 a Example: ciliated epithelium of bronchi (p. 309).
   b Example: stratified epithelium of skin (p. 511).
   c Example: squamous epithelium of lungs (p. 309).
4 **Implantation** (p. 595).

Page 216
1 Dense, semi-fluid ground substance with tough elastic fibres, makes supportive 'packaging' material. Nerves and blood vessels facilitate communication/supply to tissues. Macrophagous cells combat infection (engulf invading bacteria).
2 **Bone, Cartilage** (p. 215).
3 **Blood cells** (p. 217).
4 a, c **Cartilage** (p. 215).
   b Largely as in adult, apart from the ends of the diaphyses (**Figure 10.10**, p. 216).

Page 218
1 **Factors affecting enzyme action, The effect of pH** (p. 180).
2 Radioactive elements (e.g. strontium 90) are taken up into the mineral structure of a bone: they may irradiate (and therefore harm) dividing cells. **Gene mutations** (p. 627), Cancer of the blood: the disease leukaemia is a cancer of the leucocyte-forming cells in the bone marrow. In acute leukaemia there is uncontrolled production of immature leucocytes, most of which fail to reach maturity. As a result the normal bone marrow cells are crowded out, and production of erythrocytes and platelets is prevented. Death may result from internal haemorrhaging in vital brain centres, and infection, which cannot be resisted by the body because of the lack of normal leucocytes.

Page 219
1 **Muscle** (p. 218).
2 Site of ATP synthesis. Blood circulation delivers glucose and oxygen, and removes carbon dioxide and metabolic products.

Page 221
1 a **Types of neurone, afferent neurones** (p. 220), **Nerves** (p. 221).
   b By determining whether it connects a receptor or an effector with the CNS.
2 a **Nervous tissue** (p. 220).
   b Length and degree of branching of fibres (dendron, dendrites and axon), position of cell body.

Page 223

1 **Meristematic cells and meristems** (p. 222).
2 a Example: chlorenchyma (**Mesophyll tissue of leaves**, p. 223).
  b Example: starch storage.

Page 225

1 Via pits (p. 225).
2 Annular and spiral vessels can be stretched in extension growth (cf. reticulate and pitted vessels, which are too massively lignified).

Page 227

1 **Xylem role** (p. 225) and **Transpiration** (p. 321).
2 **Phloem role** (p. 227), **Xylem role** (p. 225).

CHAPTER 11

Page 229

1 **The cell surface membrane (plasma membrane)** (p. 158).
2 **How do substances move across membranes?** (p. 229).
3 a To remove the contents of damaged cells.
  b Approximately 50 °C.

Page 231

1 The higher the temperature the faster molecules of gases and liquids move, and so they diffuse faster at higher temperature (**Table 11.1**, p. 230).
2 Diffusion continues in both directions.
3 a Ammonium ion (alkaline pH) exists in an aqueous medium.
  b To avoid gas movements in the tube other than those due to diffusion.
  c All the ammonia molecules were initially concentrated at one end.

Page 233

1 a The dilute sucrose solution.
  b The concentrated sucrose solution.
  c The solution with the lowest water potential (concentrated sucrose solution).
2 a Solution B.
  b Yes.

Page 235

1 **Figure 11.12** (p. 234).
2 More water molecules are diffusing in than out (net inflow of water).
3 a Enhanced activity by the contractive vacuole.
  b Mechanical damage to the plasma membrane.
4 Plasmolysis of root cells is possible. (Enhanced uptake of ions is also likely.)

Page 237

1 a K^+.
  b Na^+.
2 **Adenosine triphosphate and metabolism** (p. 176).
3 Examples: in plants **Table 16.1** (p. 328); in animals **Table 13.3** (p. 273).
4 a Differences in the usefulness of these ions in metabolism. Entry of K^+ is linked to exit of Na^+.
  b Na^+.

Page 239

1 **Figure 11.16** (p. 237), **Figure 11.18** (p. 238).
2 **Facilitated diffusion** (p. 231).
3 All cells are relatively close to the bathing medium, experience comparable physiological conditions, and are likely to respond similarly (uniform, reproducible outcome).

Page 241

1 Diffusion, osmosis, active uptake, phagocytosis.
2 **Bulk transport across the cell membrane** (p. 240).
3 a RER is the site of milk protein. SER is site of lipid formation.
  b Site of 'translation' in protein synthesis (**Translation**, p. 208).

SECTION THREE

CHAPTER 12

Page 247

1 **Photosynthesis, a summary** (p. 246).
2 a, b **Figure 12.1** (p. 246).

Page 251

1 **Absorption spectra and action spectra** (p. 251).
2 The chlorophyll molecule, once extracted from the lipids and proteins of the thylakoid membranes, is no longer stable. The energy in light causes it to be chemically changed and the green colour disappears.

Page 253

1 The experiment in **Figure 12.13** can be carried out at different temperatures, but more appropriate would be the adaptation of the experiment shown in **Figure 12.16** (p. 254).
2 a In order to reproduce exactly all the conditions in (A) (e.g. temperature, humidity) so that only the CO_2 is varied.
  b Example: sugar (made from starch stored in root or stem) translocated into the leaf and deposited there as starch.

Page 254

Because temperature should be a controlled variable in such an experiment, and the light source (a lamp) is also a source of heat.

Page 256

1, 2 **Table 12.1** (p. 256).

Page 258

a Developing is a chemical reaction (temperature dependent), but exposure of the photographic emulsion is a photochemical reaction, which is largely temperature insensitive.
b **The effect of temperature on the rate of photosynthesis** (p. 258).

Page 259

a To minimise autolysis (p. 162); **The effect of temperature** (pp. 179–80).
b The organelles of the cytoplasm.
c The water potential of the medium is adjusted to that of the cytoplasm so that the organelles are not structurally damaged by osmotic effects, causing biochemical changes too.

Page 265

1 a, b **Mechanism of the light-dependent stage** (p. 260).
2 **Figure 12.29** (p. 260).
3 Because it shows that 6 molecules of oxygen (O_2) are produced yet only 6 molecules of water (H_2O) are used. Thus, according to this (summary) equation, at least 50% of the oxygen atoms in the water must come from the carbon dioxide. An alternative, less misleading way of summarising photosynthesis is:
$$6CO_2 + 12H_2O + \text{LIGHT ENERGY} \rightarrow C_6H_{12}O_6 + 6O_2 + 6H_2O.$$

4 Because it is dependent on the products of the light-dependent stage (**Figure 12.34**, p. 263).
5 **The path of carbon in photosynthesis** (p. 263).
6 a The size of the molecules, and the fact that carbon is in a lower oxidation state in the products of photosynthesis.
  b **A summary of photosynthesis** (p. 265).
7 **Figure 12.33** (p. 263) and **Figure 12.35** (p. 264).

Page 267

The C_4 mechanism maintains the concentration of CO_2 that is above normal atmospheric levels around the chloroplast of the inner sheath cells as it is; for C_3 plants, a raised concentration of CO_2 in the atmosphere around the plant will quickly reach the chloroplasts and enhance photosynthesis by suppressing photorespiration.

Page 269

1 **Figure 12.40** (p. 268).
2 Ions (e.g. NO_3^{3-}, PO_4^{3-} etc.).
3 Synthesise ATP from ADP and P_i, and build up carbohydrates from CO_2.

CHAPTER 13

Page 271

1 a They are heterotrophs (**Heterotrophic nutrition**) (p. 270).
  b **Holozoic nutrition, Saprotrophic nutrition, Parasitic nutrition** (p. 270).

2
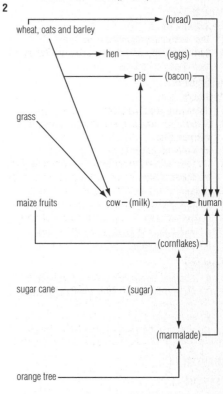

3 30 g of potato crisps contain:
$30 \times 60/100 = 1.8$ g protein
$30 \times 36/100 = 10.8$ g fat
$30 \times 50/100 = 15.0$ g carbohydrate.
The energy content of these components is
from protein: 1.8×17 kJ = 30.6 kJ
from fat: 10.8×37 kJ = 399.6 kJ
from carbohydrates: 15.0×16 kJ = 240.0 kJ
Total energy content of 30 g of
crisps = 670.2 kJ.

Page 273

1 The experiment might be repeated with two groups of young rats, one of which is fed a balanced diet, and the other a similar diet specifically lacking the vitamins present in milk. The mass of both groups should be monitored. When that of the rats without milk vitamins falls markedly, this experiment group should be provided with the essential vitamins to check that they regain the mass of the correctly fed rats.

2 To avoid the release of ammonium ions in cells, since solutions of ammonium ions are alkaline (and NH_3 itself is highly toxic). Remember, NH_3 is only excreted by organisms having access to an excess of water (p. 377).

Page 274

1 a Beer (alcoholic drink), ham, bread roll.
  b Bread roll and potatoes.
  c Milk, wholemeal bread.

2 None.

Page 276

1 Intake of too high a proportion of fats and carbohydrates in the diet. A lack of exercise.

2 **Steroids** (p. 139).

Page 278

1 **The mouth and buccal cavity** (p. 278).

2 **Mastication in the buccal cavity** (p. 278).

3 **Compact bone, Spongy bone** (p. 215).

Page 282

1 **Swallowing** (p. 280).

2 **The functioning of the stomach** (p. 280).

3 **Saliva** (p. 278); **The oesophagus** (p. 280); mucus-secreting cells (p. 282).

Page 284

1 **Bile** (p. 284).

2 **'Intestinal juice'** (p. 284).

3 Glucose; glucose and fructose; amino acids (20 different types); fatty acids and glycerol.

4 a From slightly alkaline in the mouth, to pH 1 in the stomach, to pH 8–9 in the duodenum.
  b Inactivation.

Page 285

1 To prevent autodigestion of the cells of the gastric glands in which they are formed.

2 a **Absorption of digested food** (p. 285).
  b Water, minerals, vitamins.

Page 288

1 a Glucose, amino acids, fatty acids and glycerol, vitamins, with some salts (as ions) and water.
  b Water and mineral salts.

2 Bile pigments (**Production of bile**, p. 507).

CHAPTER 14

Page 291

1 a, b **Lysosomes** (p. 162).

2 Similarities = food vacuoles formed, digestion, assimilation of products. Differences = *Amoeba*; **food vacuoles formed by pseudopodia and *Paramecium*; vacuoles formed at cytostome** (p. 290).

Page 293

1 **Use of the mouthparts of aphids as a micropipette** (p. 339).

2 a **Adaptations for pollination** (p. 570) and **Table 26.3** (p. 571).

b Butterflies can feed in flowers where the corolla tube (fused petals) is much longer than flowers from which bees can feed.

3 a Examples: mosquito, housefly.
  b Example: worker honeybees.
  c Example: garden spider.

4 Liquid feeders; aphids on plants, (female) mosquitoes on blood.

Page 295

1 **Active transport of metabolites** (p. 236).

2 Increases the surface area of food matter for enzyme action, and releases the nutritional contents of cells from within the cellulose wall, for digestion.

3 Sensing food matter, and possibly for communication between grasshoppers.

4 Soil organisms (p. 62).

5 Part of digestion in *Hydra* occurs in food vacuoles; all of digestion in *Amoeba* does so.

6 Action of the radula against the hard glass surface, coated with algae.

Page 297

Discussion point In silage, air is very largely excluded, so that aerobic decay is prevented. Instead, bacterial fermentation occurs under anaerobic conditions. As a consequence, some of the cellulose of grass cell walls is turned into nutritious sugars and other soluble metabolites, and plant cell contents are more accessible when the silage is later eaten by cows. Sometimes, industrially produced enzymes are added, to facilitate digestion of the grass in the clamp.

Page 300

1 **Figure 14.16** (p. 296).

2 **Recycling in garden compost** (p. 297).

3 **Yeast, an economically important saprotroph** (p. 298).

4 **Figure 14.20** (p. 298), **Figure 25.5** (p. 551), **Table 7.2** (p. 169).

Page 301

1 **Saprotrophic nutrition, Parasitic nutrition** (p. 272).

2 Adaptations for parasitism:
  ► attachment to or penetration of host (*Pediculus humanus*, p. 66; bacteriophage, p. 549)
  ► degeneration of redundant body structures (comparison of body of free-living flatworms, p. 35, with adult liver flukes, p. 536)
  ► mechanisms aiding dispersal to new hosts (*Phytophthora*, p. 300)
  ► mechanisms resisting attack by host defences (mucus is secreted by the outermost cells of body parasites, e.g. tapeworm; stages of malarial parasite in host bloodstream move into liver cells and red blood cells, avoiding immune reactions of host).

CHAPTER 15

Page 303

1 Examples: large surface area, thin epithelium, rich capillary supply, efficient ventilating mechanism (**Gaseous exchange**, p. 303, with **Table 11.1**, p. 230, and **Figure 11.6**, p. 231).

2 **The ADP–ADT cycle and metabolism** (p. 176).

3 **Carbon dioxide** (p. 253).

4 **Gaseous exchange in plants** (p. 303).

Page 305

1 **The earthworm (*Lumbricus terrestris*)** (p. 304).

2 Diffusion: air movement in the tracheal system; O_2 into the tissues from tracheae; CO_2 from the tissues (via haemocoel and cuticle).
  Ventilation: air sacs functioning as bellows (some insects).

Page 307

1 **Oxygen from water** (p. 307).

2 **Figure 15.6** (p. 306).

3 One-way flow of the respiratory medium over the respiratory surface, achieved by muscular contraction acting on skeletal components.

Page 309

1 **Alveolar structure and gaseous exchange** and **Figure 15.10** (p. 309).

2 **Ventilation of the lungs** and **Figure 15.9** (p. 308).

Page 316

1 **Figure 15.21** (p. 315); **Figure 15.22** (p. 315); **Figure 15.23** (p. 316).

2 NAD (point 5 of **Six points about redox reactions**, p. 314).

3 a The addition or removal of CO_2 from a compound.
  b The removal or addition of hydrogen to a compound.

Page 317

1 The reserves (the 'pool') of hydrogen acceptor (NAD) will run short as $NADH_2$ accumulates and cannot be re-oxidised (**Figure 15.23**, p. 316) in substantial quantities.

2 Pyruvic acid + reduced NAD → lactic acid + NAD
  $CH_3COCOOH + NADH_2 \rightarrow CH_3CHOHCOOH + NAD$

3 **Figure 15.26** (p. 318).

4 a **Table 19.1** (p. 404).
  b **Brown fat** (p. 513).

CHAPTER 16

Page 321

1 **The movement of water in plants** (p. 320).

2 **Components of water potential** (p. 232).

3 **The role of transpiration** (p. 321).

Page 323

1 Guard cells of the stomata only (**Stomata: the mechanism for opening and closing**, p. 324).

2 **Diffusion through pores** (p. 323).

3 Mount pieces of leaf epidermis containing stomata in:
  (i) water (or a solution of water potential less negative than that of the cell);
  (ii) a solution of water potential more negative than that of the cell.
  In (i), net diffusion of water (osmosis) occurs into the leaf cells, which become fully turgid. The stomata open (even under water, under the cover slip). In (ii), net osmosis occurs from the leaf cells, which become flaccid. The stomata close.

4 **Xerophytes** (pp. 375–6).

Page 325

1 a **The opening and closing of stomata** and **Figure 16.8** (pp. 322–3); **Figure 16.9** (p. 324).

b Loss of turgor in the guard cells due to external conditions favouring transpiration (**The rate of transpiration**, pp. 324–5).

2 a **Measuring transpiration rate** (p. 324).

b With difficulty. Dry cobalt chloride paper turns pink as it picks up water vapour from the air (*Study Guide*, Chapter 16). The time taken for the colour change in cobalt chloride paper supported against the leaf surface (upper and lower surfaces can be compared) is a measure of the *rate* of water vapour loss.

3 **Table 18.2** (p. 375).

4 a From before 8.00 until about 17.00, i.e. at least 9 hours.

b Sufficient wilting had occurred to cause guard cells to become flaccid.

Page 327

1 The outermost surface of a suberised root is impervious to liquids (and gases).

2 **Water uptake by roots** (p. 326).

3 Blocks the apoplast route at the endodermis (**Role of the Casparian strip**, p. 327).

Page 329

1 To avoid the unintentional delivery of significant quantities of useful micronutrients (or of poisonous ions) as impurities.

2 **Table 16.1** (p. 328).

3 The growing points and youngest leaves are generally supplied with ions from the older parts of the plant. Calcium ions, however, become immobilised, e.g. in cell walls, and are not free to be translocated to the growing points.

4 Nitrogen as nitrates, ammonium salts and amino acids; sulphur as sulphates and in certain amino acids.

Page 331

1 **Figure 11.15** (p. 236).

2 **The process of ion uptake** (p. 330).

3 a **The path of carbon in photosynthesis** (p. 263).

b The proteins of enzymes that are no longer in use are broken down to amino acids, which are then used to make other proteins, those stored in fruit (in this case) (**Figure 6.27**, p. 144).

Page 335

1 a *Rhizobium* **in the nodules of leguminous plants** (p. 334).

b Clovers (*Trifolium* spp.).

2 **Chemosynthetic bacteria** (p. 269).

3 Green plants (often those having root nodules) grown to plough back into the ground in order to increase the combined nitrogen content in the soil, when released by aerobic decay.

4 Atmospheric nitrogen gas; respiration (via $NADH_2$); ATP directly, also indirectly from sunlight.

Page 336

1 **Special methods of obtaining nutrients** (p. 335).

2 a, b **Ectotrophic mycorrhiza** (pp. 335–6).

Page 339

1 **Movement of elaborated food** (p. 337); **Transpiration** (p. 321).

2 **Formation of vascular tissue** (p. 407).

3 **Figure 16.35** (p. 338).

4 Thin sections of the stem tissue are dried flat and subsequently held against X-ray film in the dark for a period of time. Radiation 'fogs' the emulsion in a manner comparable to light. When the film is removed, developed and fixed, the presence of radioactive substance in the plant tissue section is disclosed by the 'exposed' parts of the film.

Page 341

1 High temperature causes the denaturation of proteins (including enzymes); trichloromethane affects the lipids in membranes; cyanide combines irreversibly with coenzymes containing metal ions such as iron (e.g. cytochromes).

2 **The pressure flow hypothesis** (p. 340).

3 By the selective removal of ions from one side of a sieve plate, and their addition to the other side.

CHAPTER 17

Page 343

1 Examples: glucose, amino acids, ions.

2 a Larger organisms have more body tissues to maintain.

b Maintenance of a (high) constant body temperature is achieved by a higher rate of metabolism than that of ectotherms.

Page 345

1 a Example: blood cannot be directed to a respiratory surface immediately before (or after) servicing tissues that are metabolically active (with a high respiration rate).

b Less efficient transport, since the delivery of metabolites and waste products is not directed.

2 Capillary beds (**Figure 17.18**, p. 353), and the formation of tissue fluid (p. 353).

3 Circulated around the body at a higher pressure than in a single circulation system.

4 Examples: air-breathing lungs; limbs for crawling/walking; resistance to desiccation.

Page 346

1 **Voluntary muscle** (p. 218); **Cardiac or heart muscle** (p. 219).

2 **The human heart** (p. 346).

3 Blood that should be pumped to the lungs (pulmonary circulation) may be diverted to the other side of the heart and pass to the body without being fully oxygenated (systemic circulation).

4 Chordae tendinae keep the heart valve flaps pointing in the direction of blood flow. (They stop the valves turning inside out when the pressure rises in the ventricles.)

Page 349

1 **The heart as a pump** (p. 348).

2 Detection of heart disease or malfunction due to damaged or abnormal valve structures.

3 a, b **The control of heartbeat** (p. 349).

Page 352

1 **Regulation of heartbeat** (p. 350).

2 **The adrenal medulla** (p. 475).

3 **Figure 17.14** (p. 351); **Assimilation of absorbed food** (p. 286).

4 **Figure 17.13** (p. 350).

Page 353

1 **The role of the hypothalamus** (p. 514) and **Figure 23.21** (p. 514).

2 a, b **Tissue fluid; exchange between blood and cells** (p. 353).

3 **Tissue fluid: exchange between blood and cells** (p. 353).

Page 356

1 **Atherosclerosis** (p. 355); **Embolisms and thromboses** (p. 355); **Clotting of blood** (p. 364).

2 a To drive the circulation of blood through arteries, narrow capillaries and veins, back to the heart.

b **Hypertension** (p. 354).

3 **Factors that predispose to coronary heart disease** (p. 356).

Page 359

1 a, b A fall of about 20% in each case.

2 **Haemoglobin and oxygen transport** (p. 358); **Oxygen tension and oxyhaemoglobin formation** (p. 359).

3 **Respiratory disease** (p. 312); **Factors that predispose to coronary heart disease** (p. 356).

Page 361

1 **Transport of oxygen** (p. 358); **Transport of carbon dioxide** (p. 361).

2 Carbonic anhydrase, within the red cells.

3 Very low, below the unloading tension of its haemoglobin.

4 **The properties of catalysts** (p. 178).

Page 363

1 **Circulation in the human fetus** and **Changes at parturition** (p. 362).

2 Because the atmospheric pressure is lower; **Oxygen tension and oxyhaemoglobin formation** (p. 359).

3 **Mammals at high altitude** (p. 362).

Page 365

1 **Atherosclerosis** (p. 355).

2 a, b, c **Protection of the lungs** (p. 309), and **Figure 15.11** (p. 310).

Page 367

1 a **Immunity and the immune system** (p. 366).

b **Defence against disease** (p. 364).

2 a, b See **a** above.

3 a **Figure 17.39** (p. 364).

b **Figure 17.43** (p. 367).

Page 369

1 **Blood transfusions** (p. 369).

2 **The rhesus factor and pregnancy** (p. 369).

Page 370

1 **AIDS, the Acquired Immune Deficiency Syndrome** (p. 370).

2 **Organ transplantation and the immune response** (p. 371).

3 Skin from another part of the patient's body does not carry any 'foreign' antigens.

CHAPTER 18

Page 373

1 **Introduction** (p. 372).
2 In **a**, **c** and **d**.
3 Because they do not respire an excess of protein, and therefore do not produce ammonia (or have other forms of combined nitrogen to dispose of, **Excretion** p. 377).

Page 376

1 **a** Water loss by transpiration exceeds water uptake (**Figure 16.11**, p. 325).
   b Loss of water via the cuticle is minimal (**The role and importance of stomata**, p. 322).
2 **Table 18.2** (p. 375).
3 **a**, **b** **Xerophytes** (pp. 375–6).
4 **Hydrophytes** and **Figure 18.1** (p. 374).

Page 377

1 **Uric acid** (p. 377).
2 **Water relations of plant cells** and **Water relations of animal cells** (p. 235).
3 **Birds** and **Figure 18.27** (p. 391).
4 Carbon dioxide, water and ammonia are the chief excretory products of Cnidaria such as the freshwater *Hydra*. Almost all the cells are in contact with the surrounding water, and excretion occurs by diffusion. The cells of *Hydra* are hypertonic to fresh water, and yet the cells do not have osmoregulatory organelles such as a contractile vacuole. It is believed that ions are excreted into the enteron (p. 343) by the cells of the body layers. This lowers the water potential in the enteron; water therefore flows by osmosis from the cells into the enteron, and is periodically expelled from the mouth.

Page 381

1 **a** **Figure 18.11** (p. 381).
   b **Figure 18.10** (p. 380).
2 **a** **Cortical and juxtamedullary nephrons** (p. 381).
   b **Figure 18.15** (p. 384).

Page 383

1 **a, b, c** **Blood pressure in the glomerulus** and **Figure 18.14** (pp. 382–3).
2 **Ultrafiltration in the renal capsule; The sieve** (p. 382).
3 Brush border = microvilli. Microvilli provide a vastly increased surface area for the movement of substances (compare **Absorption of digested food**, p. 285).

Page 385

1 Because the volume of fluid leaving the loop of Henle is reduced. But whilst the filtrate entering the loop is isotonic with blood plasma, the liquid leaving the loop is now hypotonic because more salt than water has been withdrawn.
2 The principle of a countercurrent system is of two separate, adjacent flows (liquid or gas), moving in opposite directions. Exchange occurs between the moving substances, e.g. heat energy from the hotter to the cooler flow (**Figure 23.23**), or metabolites, e.g. O_2 from oxygenated water to deoxygenated blood. Because the directions of flow are opposite in the two tubes, a gradient is maintained along the entire exchange surface (Figure 15.6). The exchange is more efficient than in parallel flow.

3 **Quantifying water relations in the medulla** (p. 385).

Page 387

1 **The importance of pH in biology** (p. 128); **The effect of pH** (p. 180); **Homeostasis in mammals** (p. 504).
2 **a, b** **Water re-absorption in the collecting duct** (pp. 386–7).

Page 389

1 A countercurrent exchange system (see answer to **Q2**, p. 385).
2 **Osmosis**, (p. 232).
3 For example because some of the conditions that trigger blood clotting are generated at the site of a tapped fistula (**Figure 18.23**, p. 389, and **Figure 17.41**, p. 365). Blood clots may cause an embolism (**Embolisms and thromboses**, p. 355).
4 Because identical twins have the same genetic constitution (i.e. the same histo-compatibility factors in their cells), thus avoiding rejection.

Page 391

Self-regulating mechanisms (p. 503).
Homeostasis in mammals (p. 504).

SECTION FOUR

CHAPTER 19

Page 399

1 **a, b** **Introduction** (p. 396); **Growth of multicellular organisms** (p. 397).
2 **Analysing data on growth** (the lag phase) (p. 400).
3 The curve of exponential growth, when presented as a log plot, becomes a straight line. The straight line makes interpretation easier.
4 By the dilution plate method of counting unicells, since only cells that are alive and give rise to a fresh colony on a dilution plate are 'counted' (**Figure 5.2**, p. 99).
5 **Measuring the growth of a multicellular organism, Fresh mass *v.* dry mass** (p. 398).

Page 401

1 **Relative growth rate** (p. 400).
2 **Amount of protein** (p. 399).

Page 403

1 **'Intermittent' growth of insects** (p. 402).
2 **Limited *v.* unlimited growth** (p. 402).
3 Examples: ecdysis may temporarily interrupt feeding; there may be a slight loss in mass when the old cuticle is shed.

Page 405

1 **a** **Germination: the necessary conditions** (p. 404).
   b CO_2, lactic acid, and ethanol (**Fate of the products of anaerobic respiration**, p. 317).
2 **a** **Mobilisation of stored food** (p. 404).
   b **The seed** (p. 404).
3 **a** **Water uptake by the roots** (p. 326).
   b **Epigeal and hypogeal germination** (p. 405).

Page 407

1 **a, b** **Meristems and plant growth** (p. 406).
2 Region of elongation (**Figure 19.18**, p. 408).
3 **Xylem vessels, Structure and occurrence** (p. 225).

4 **Parenchyma, Role** (p. 223); **Collenchyma, Role** (p. 224).
5 **a, b** **Figure 10.26** (p. 227); **Figure 10.22** (p. 225).

Page 410

1 Leaf traces branch from vascular bundles and serve the leaves (**Figure 19.19**).
2 **The primary growth of the stem** (p. 406); **The primary growth of the root** (p. 408).
3 **a, b** **Support in flowering plants** (p. 500); **Sclerenchyma** (p. 224), **Xylem** (p. 225).
4 **Origin of leaves and lateral buds** (p. 408); **The origin of lateral roots** (p. 410).

Page 412

1 **Primary growth of the plant** (p. 406); **Secondary growth of plants** (p. 410).
2 **a, b** **Annual rings** (p. 412); **The cork cambium** (p. 410).
3 The base of the trunk has undergone more years of secondary thickening (i.e., increase in girth) than the more recently formed upper regions, so the base is always wider than the apex.

Page 415

1 **Fertilisation and cleavage** (p. 413); **Gastrulation** (p. 414).
2 It is a food store to be drawn upon for early growth.
3 **Development of the triploblastic gastrula** (p. 414).
4 **Metameric segmentation** (pp. 414–15). In annelids the segments of the body are largely identical, with limited specialisation or adaptation of particular regions of the body. In adult insects on the other hand, the basic segmentation is overlaid by divisions of the body into head, thorax and abdomen, and the development of specialised organs of locomotion attached to the thorax (p. 38).

Page 417

1 **a, b** **Ecosystem** (p. 44); **Competition between organisms** (p. 64). Example: the butterfly larva (caterpillar) and adult.
2 **Two types of insect life cycle** (p. 417).
3 **The roles of the larval stage in the life cycle** (pp. 415–16).
4 **a** **Two types of insect life cycle** (p. 417).
   b Metamorphosis involves a drastic change in form. The adjustments in size and structure that occur between tadpole and adult frog take place quite gradually. In many insects, metamorphosis involves the abrupt restructuring of the organisms during pupation.

CHAPTER 20

Page 419

1 **Plant and animal sensitivity compared** (p. 418).
2 Examples: the types of stimulus that are effective in inducing responses, and how strong/weak stimulation needs to be; the effects of external conditions that may influence physiological processes, such as light/darkness, and ambient temperature over a 'physiological' range; the changes in the chemistry or structure of the cells involved that can be detected following a response; any detectable change in properties of the membranes of the cells involved.
3 Example: reduced damage by leaf-eating animals.

Page 421

1 ▶ Light is necessary for photosynthesis, including photosynthetic photophosphorylation (light-induced ATP formation), chlorophyll formation (in angiosperms), straight growth of the stem tip
   ▶ Light enhances leaf expansion
   ▶ Light inhibits extension growth of stems (so that unilateral light on a stem results in a bending growth towards the light source).
2 Examples: growth of grass or weeds when covered by an upturned bucket for several days; growth of potted plants in poorly illuminated rooms compared with that of the same species grown in a conservatory.
3 Example: using mature dandelion plants (grown in plant pots) with open flowers, observe growth movement of the flowers when they are transferred from light to darkness at a constant ambient temperature of (for example) 25 °C. The experiment should then be repeated, but at a much lower ambient temperature of (for example) 10 °C.

Page 423

1 18°, equivalent to 0.23 mg dm^{-3}.
2 Examples: use larger samples of coleoptiles than Darwin's; seek to prevent bacteria from contaminating the cut surface of coleoptiles (using antibiotic); decapitate and replace some tips (to see whether damage prevents curvature).
3 **Went and the bioassay of auxin** (p. 422).

Page 425

1 An amino acid typically has two functional groups, an amino group (—NH$_2$) and a carboxylic acid group (—COOH), both of which are ionised in cells (**Figure 6.20**, p. 139); IAA does not have —NH$_2$ (**Figure 20.12**, p. 424).
2 In continuous light, the coleoptile splits open and the leaves emerge.
3 IAA, if allowed to accumulate in plant tissue, may cause abnormal growth.
4 Lowest: junction of plumule and radicle. Highest: apices of coleoptile and root.
5 By measuring the phototropic response of stem or coleoptile to light of different wavelengths.

Page 427

1 Examples: use of an appropriate antibiotic in agar blocks to prevent contamination by microorganisms; batches of many plants subjected to each treatment, so that the statistical significance of the results can be checked; all plants in the batches should be the same age and at the same stage in development; environmental conditions should be uniform; a 100% relative humidity environment should be maintained to prevent the drying up of the agar blocks.
2 **Geotropism** (p. 426).
3 In conifers such as Scots pine (at least, when a young tree), lodgepole pine and larch, the terminal buds continue each year, giving a characteristic 'Christmas tree' outline. In willow, beech and lime the terminal buds frequently wither, and are replaced by lateral buds. The result is the branched outline of most broad-leaved trees.

Page 431

1 IAA is synthesised from tryptophan, an amino acid that may be released in significant quantities when protein is hydrolysed.
2 **The role of IAA in tropism is controversial** (p. 428).
3 1.28 μg of GA in alcoholic solution, applied to a plant leaf.
4 **Mobilisation of stored food** (p. 404).
5 An organic base in nucleotides, and in ATP (pp. 145–7).
6 a **Plant and animal hormones compared** (p. 430).
   b Plant hormones, e.g. IAA (pp. 427–8); animal hormones, e.g. adrenaline (p. 475).

Page 433

1 **The role of ethene in development** (p. 432).
2 Synergists: IAA and GA in stem elongation. Antagonists: IAA and ABA in stem elongation.
3 IAA: inhibition of lateral bud break; cytokinins: release of lateral buds from inhibition; abscisic acid: nearby buds remain dormant.
4 Stocks of ripe and unripe fruit are divided into small batches (about five fruit) of very similar mass. Then several trials are set up, each consisting of a sample of ripe and a sample of unripe fruit contained within a transparent plastic box. The time taken for the unripe fruit to ripen is recorded and compared with the time to ripening of samples of unripe fruit on their own.

   The effects of light and temperature could be investigated by holding duplicate trials, firstly at different ambient temperatures and secondly at the same temperature but with some samples kept in continuous light and others in the dark.

Page 435

1 Examples: annual, shepherd's purse (flowers January to December); biennial, mullein (flowers June to August of second year); woody perennial, oak tree (flowers April to May); herbaceous perennial, nettle (flowers June to September).
2 **Figure 20.28** (p. 435).
3 **Figure 20.29** (p. 435).

Page 437

1 **The hormone control of flowering** (p. 435), and **Figure 20.30** (p. 436).
2 **Figure 20.28** (p. 435).
3 Examples: because florigen does not exist (flowering being triggered by the accumulation of a particular *balance* of existing metabolites); because florigen exists only in concentrations too small to detect; because although florigen exists, it is a substance too unstable and 'short-lived' to detect.

Page 439

1 Leaves, efficient organs of photosynthesis in spring and summer, become potential liabilities to plants in the autumn and winter (transpiration when liquid water is scarce, increased vulnerability to snow damage and wind throw). By withdrawing the valuable metabolites, cutting off the leaf and sealing the junction, the plant 'cuts its losses' (it survives the winter, it is less vulnerable to water loss or the entry of spores of disease microorganisms).

2 This facilitates mechanisation of cultivation, and achieves maximum productivity whilst growing conditions are ideal.
3 Dead empty cells with suberised (waxy), impervious walls.
4 Winter wheat requires cold treatment (vernalisation) in order to complete vegetative growth, and switch to 'readiness to flower' on return of long days. Protective buds of woody perennials require cold treatment before they will break open and develop either vegetatively or into flowers, according to the type of bud. Mechanisms that ensure development occurs in the appropriate seasons.
5 Freshly collected seeds fail to germinate if intact or if the fruit/seed coat is removed (cf. sycamore).

Page 441

1 Some are more effective than naturally occurring substances, are easily manufactured, are stable in environmental conditions favourable for growth, and can be more selective in the range of species affected.
2 Reduction in the numbers of fruits and seeds formed, for example, when environmental conditions are unfavourable.
3 Loss of soil nutrients to broad-leaved weeds and competitors is reduced.

CHAPTER 21

Page 443

1 a *Animals*
   ▶ rapid responses involving complex sense organs, elaborate nervous system, and hormones
   ▶ animals adapt by modifying their behaviour.
   Plants
   ▶ slow responses (plants are anchored) involving growth movements, and plant growth regulators
   ▶ plants adapt by modifying their growth.
   b Not necessarily. Sensitivity is a characteristic of all living things. Degrees of sensitivity depend upon the thresholds of stimulation in particular cells.
2 Examples: chemicals diffusing from the 'food'; movements of flagella; electrical stimuli, the product of membrane polarity.
3 Sense receptor in the retina → sensory neurone (retina to brain) → relay neurone in brain → motor neurone serving muscles of eyelid.

Page 447

1 **Myelin sheaths** (p. 446).
2 **The refractory period** (p. 446).
3 **Figure 21.7** (p. 446).
4 Dominoes set up = setting up of resting potential.
   Fall of first domino = initiation of an action potential.
   Falling dominoes = transmission of the action potential.
   Direction of domino fall = only one direction of transmission.
   Time needed to reset dominoes = refractory period.

Page 449
1 Facilitated diffusion.
2 Formation of ATP and the generation of vesicles of transmitter substance.
3 **Drugs and the synapse** (p. 450).

Page 451
1 Initiating the feedback of information on posture and movement.
2 **The role of the synapse** (p. 449).
3 **Figure 21.36** (p. 466).

Page 453
1 **Taste and smell** (p. 452).
2 a, b **Hearing and balance** (p. 452).
3 **The mechanism of hearing** (p. 454).
4 Movements (vibrations) of the round window; also as electrical impulses (transduced) in the auditory nerve.

Page 455
1 a, b **How does the ear discriminate between sounds?** (p. 454).
2, 3 **Balance** (p. 454).

Page 457
1, 2 **Accommodation; focusing of the image**, and **Figure 21.22** (pp. 456–7).
3 A reduction in the size of the pupil (equivalent to the iris diaphragm of a camera) increases the depth of focus (**Figure 21.23**).
4 **Sight**, and Figure 21.21 (p. 456).

Page 460
1 Use **Rod cells**, **Cone cells** and **Figure 21.24** (p. 458).
2 **Stereoscopic vision** (p. 459).

Page 462
1 The autonomic nervous system (p. 462) supplies nerve fibres to internal organs not under voluntary control. We say the autonomic nervous system mediates involuntary responses. (All reflex actions are 'involuntary', however, in the sense that they happen without prior thought, including reflexes of the somatic nervous system.)
2 a This is a voluntary response. Retinal receptors (rods, cones) → bipolar neurone → ganglion cell → optic nerve → relay neurone in brain → motor neurones in spinal cord → motor neurones in spinal nerves to leg muscles.
   b This is an involuntary (reflex) response. Light touch receptors of the throat/bronchus → sensory neurone to spinal cord → relay neurone of spinal cord → motor neurone in spinal nerves to abdominal wall muscles (and possibly to intercostal muscles).
3 Maintenance of muscle tone, and prevention of injury from overstretching of muscles.
4 Five.
5 **The autonomic nervous system** (p. 462).

Page 465
1 **Nerves** (p. 221).
2 **Nerve cord and simple 'brain' of Lumbricus** (p. 469); the development of a head in different groups of animals has always been associated with some type of 'brain' (a 'receiving' and 'processing' centre, integrating responses by the organism).
3 **The role of the human brain** (p. 464).

Page 467
1 **The functions of the parts of the human brain** (p. 465).
2 The medulla oblongata (in the hindbrain) contains control centres for vital body functions (p. 466).
3 a, b The parts of the sensory and motor cortical areas given over to processing impulses to and from the face and fingers are disproportionately larger than those concerned with the remainder of the body. Consequently the sensitivity of the face and the dexterity of the fingers are greater than those of many other parts of the body.
4 **The forebrain** (p. 466).
5 With four neurones arranged as shown below, (i) excitation of the postsynaptic fibre will occur when **a**, or **b**, or **a** + **b**, or **a** + **b** + **c** fire; (ii) no excitation of the postsynaptic fibre will occur when **a** + **c**, or **b** + **c** fire.

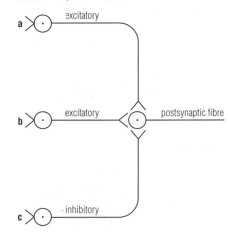

Page 469
1 **The nerve net of Hydra** (p. 468).
2, 3 **Nerve cord and simple 'brain' of Lumbricus** (p. 469); **Nervous control of locomotion in the earthworm** (p. 480); **The diameter of the axon** (p. 446).

Page 471
1 In insects, the thorax is typically associated with flight and walking organs. The nervous systems of insects have enlarged ganglia in the thorax as well as in the head. Locomotion in annelids is a function 'delegated' to most body segments, and so here the control is more diffuse.
2 More light may reach the retinal cells (in low light), but probably a more diffuse image is produced (p. 471).

Page 473
1 **The endocrine system, Mode of action of hormones** (p. 472).
2 The effect of arrival of a nerve impulse is depolarisation or enhanced polarisation of a cell membrane, leading to movement of ions. The effect of the arrival of a hormone at its receptors is to promote release of chemicals (for example, cyclic AMP or enzyme activators) inside the membrane (**Mode of action of hormones**, p. 472, and **Figure 21.46**, p. 473).
3 **Positive and negative feedback** (p. 503).
4 **The speed of transmission of impulses** (p. 446); **Mode of action of hormones** (p. 472).

Page 475
1 Portal veins, such as the hepatic portal vein (**Figure 17.14**, p. 351), begin and end in capillary networks (ordinary veins end by joining up and discharging blood into the heart).
2 **The hypothalamus and the pituitary gland** (p. 474).
3 a **Control of secretion of thyroxines** (p. 474) and **Figure 21.49** (p. 475).
   b **Endocrine control of the menstrual cycle** (p. 593).

Page 477
1 It may lead to the formation of a (very small) pupa at the next moult.
2 Setting up of pheremone-baited traps (small containers into which insects can enter), which also contain sticky surfaces to trap the insects, or lethal insecticide.

CHAPTER 22

Page 479
1 The shape of most organisms is vital to movement and to other bodily functions, such as breathing, feeding, reproduction, etc., as well as in recognition by other members of the species and the community at large.
2 Compression due to the mass of the organism, and the pressure from the surrounding environment; tension due to wind (**Figures 22.35** and **22.36**, pp. 500–501).
3 **Cartilage** (pp. 215–16); **Compact bone** (pp. 215–16).

Page 483
1 Sequential movement of segments can continue, yet the speedy transmission of impulses along the length of the body is impossible.
2 **Efficiency of an exoskeleton** (p. 481).
3 **Structure of the insect exoskeleton** (p. 482).
4

Page 484
1 **Support and locomotion with an endoskeleton, The skeleton** (pp. 484–7); **Blood cells** (p. 217).
2 **The axial skeleton** (p. 484) and **Figure 22.28** (p. 495).

Page 485
1 **The rib-cage** (p. 485) and gaseous exchange in mammals (p. 308).

2

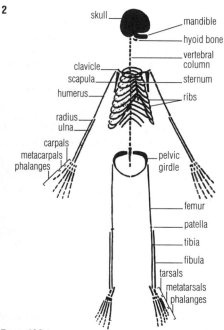

skull
mandible
hyoid bone
vertebral column
clavicle
scapula
sternum
humerus
ribs
radius
ulna
carpals
metacarpals
phalanges
pelvic girdle
femur
patella
tibia
fibula
tarsals
metatarsals
phalanges

Page 490

1 **Muscle** (pp. 218–19).

2 A voluntary muscle is composed of bundles of muscle fibres wrapped in connective tissue (**Figure 22.16**, p. 489), with each fibre made up of many tiny fibrous myofibrils (**Figure 22.17**, p. 490), consisting of overlapping contractile protein filaments arranged as shorter, thick filaments (made of myosin, **Figure 22.19**, p. 491) and longer, thin filaments.

3 In **Levers**, p. 488, the person is exerting an effort force against a load, using a lever. The effort force is smaller than the load force because the effort is applied at a greater distance from the pivot than the load. The lever is magnifying the force; mechanical advantage = (load ÷ effort).

Page 493

1 Digestion of dietary carbohydrate such as starch (p. 136) commences in the mouth (p. 278) and is completed in the small intestine where absorption into the blood occurs (pp. 285–6). The blood sugar level is regulated (pp. 504–6), so that muscles receive sugar on demand and hold a store of glycogen, too (p. 492).

2 **Stages in voluntary muscle contraction** (p. 491).

Page 495

1 Contraction of C to flex the knee slightly, plus contraction of D to swing the leg forward.

2 Bulk of the body is raised above the land surface, reducing friction in movement. Muscular effort in keeping the body off the ground is less than that expended when the limbs are splayed out (cantilever position; reptile; **Figure 22.28**).

Page 497

1 **Fish: locomotion in the water** (p. 496).

2 Effect of median fins (and anterior part of body) combating yaw (**Figure 22.29**, p. 496).

3 Volume would decrease as a result of the increased pressure. Cod's body will increase the volume of air retained in the bladder, thereby decreasing the density of the body, allowing movement at greater depth with less muscular effort.

4 **Figure 22.29** (p. 496).

Page 499

1 The pectoralis major muscle is responsible for moving the wing down during the 'power stroke', the pectoralis minor for the 'return stroke' (minimal resistance).

2 **Myoglobin and haemoglobin** (p. 360).

Page 501

1 **Structure and function of cilia and flagella** (p. 500); **Amoeboid locomotion** (p. 499).

2 Ability of response to resources reaching the organism from any direction, e.g. plant and light (**Life styles and symmetry**, p. 501).

3 **Figure 22.34** (p. 500).

CHAPTER 23

Page 503

1 **Homeostasis in mammals** (p. 504), **Enzymes – the effect of pH** (p. 180), **Hypertension** (p. 354).

2 **Positive and negative feedback** (p. 503).

Page 508

1 An adequate supply of dissolved oxygen for aerobic respiration.

2 **The anatomy of the liver** (p. 507).

3 a **Figure 11.23** (p. 240).

   b To engulf pathogenic bacteria and degenerating red cells (**Detoxification**, p. 508).

Page 509

1 **Excretion** (p. 377).

2 It is removed from the bloodstream by ultrafiltration, and appears in the urine.

3 An endothermic chemical reaction is one in which heat energy is absorbed (rather than given out, as in an exothermic reaction).

Page 511

1 Advantages: a low requirement for food as 'fuel' to warm the body and deliver a constant temperature; the likelihood of being in an active state when the majority of prey organisms are also active.

Disadvantages: vulnerability to predators such as mammals at times when the body is not 'warmed up' by the environment; inability to exploit food resources available at times when the ambient temperature is too low (after dark, for example).

2 Above 37.5 °C, below 35.5 °C.

3 Examples: cats and dogs move between full sun and shade in warm weather; dogs pant, exposing their long, damp tongues, when hot.

4 **Epithelia** (pp. 213–14).

Page 513

1 Examples: protection, camouflage, sensory.

2 A correlation between ear size and the normal range of ambient temperatures in the natural environment of mammals such as elephants, foxes and hares (**Figure 23.22**).

Page 517

1 **Nature of modern ethology** (p. 516). Examples: behaviour of urban foxes recorded by infra-red photography; activities of bats followed by ultrasonic equipment; movements of fish shoals detected by sonar equipment; behaviour of hedgehogs monitored by radio transmitters.

2 **Orientation behaviour** (pp. 516–17). Observation and recording of the movement of groups of woodlice whilst in a choice chamber (**Figure 23.25**),

   a experiencing full illumination/shade, or

   b humid/dry conditions.

Page 519

1 Instinctive behaviour: automatic, genetically programmed; not learnt behaviour, automatically triggered by environmental clues; reflex action, not automatically involving higher centres of the CNS.

2 **Pecking response of young gulls** (p. 519).

3 Avoidance of damaging conflict, determination of the bounds of mutual territories without excessive energy investment.

Page 521

See figure below.

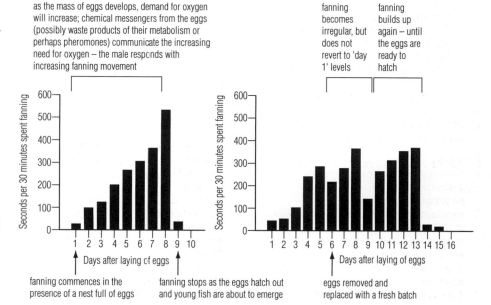

as the mass of eggs develops, demand for oxygen will increase; chemical messengers from the eggs (possibly waste products of their metabolism or perhaps pheromones) communicate the increasing need for oxygen – the male responds with increasing fanning movement

fanning becomes irregular, but does not revert to 'day 1' levels

fanning builds up again – until the eggs are ready to hatch

fanning commences in the presence of a nest full of eggs

fanning stops as the eggs hatch out and young fish are about to emerge

eggs removed and replaced with a fresh batch

Page 523

Trial and error: the principle is one of operant conditioning (p. 522); that is, a 'correct' choice leads to a reward, an 'incorrect' choice to none.
Insight: intelligent behaviour or reasoning shown in the solving of problems not yet experienced, using incoming sense data plus experience held in memory. Can be based, in part, on earlier 'play' (p. 522).

Page 525

By varying the environmental clues to the passing of time (for instance, the pattern of light = day, and night = dark), and observing no or little change in the feeding pattern displayed.

CHAPTER 24

Page 531

1 **Table 7.2** (p. 169), **Viruses** (pp. 548–9).
2 **Types of immunity** (p. 368).
3 Microorganisms, including many pathogenic bacteria, occur widely as spores (or possibly as active cells) as casual members of a microflora. The presence of a particular pathogen in an infected organism is not proof that it is the causative agent.

Page 533

1 Isolation of the dog; culturing and identification of the microorganism taken in swabs from the dog; investigation of the dog's natural environment to gauge the likelihood of prior infection with rabies.
2 Dogs and cats, carrying toxocariasis worms (nematode worms similar to *Ascaris*, p. 36); cattle carrying 'ringworm' (fungus infection).
3 **Types of immunity** (p. 368).
4 **Canning** (p. 109); **How food poisoning occurs** (p. 532).

Page 534

1 **Obligate *v.* facultative parasites** (p. 301).
2 A case of artificial selection (p. 644) in which unresistant individuals are killed off, reducing the competition for resources, thus facilitating the growth and reproduction of resistant strains.
3 Competition for resources (**Competition between organisms**, p. 64).
4 **Syphilis** (p. 533).

Page 536

Pathogens may be transmitted via
▶ drinking water (p. 99)
▶ undercooked foods (e.g. pp. 299 and 531–2)
▶ insect bites (p. 292)
▶ insect contamination of food (p. 293)
▶ airborne spores (p. 309)
▶ droplet infections entering via nasal passages
▶ wounds and breaks in the skin (pp. 364–5)
▶ close physical contact (pp. 532 and 370)
▶ contacts with pets/farm animals (p. 534).

Page 543

a Transmission of plant pathogens from host to host occurs by
▶ transport of spores in the air
▶ splashing by water droplets
▶ infected seeds
▶ during feeding by insects
▶ contacts between damaged cells of adjacent plants
▶ growth and invasion from a base of nearby infected material

b Plant pathogens gain entry to the host by
▶ direct growth/invasion of host tissues
▶ insect feeding
▶ contacts between damaged cells of adjacent plants.

SECTION FIVE

CHAPTER 25

Page 550

1 In the ways that you define viruses, i.e.
▶ consisting of nucleoproteins of high M_r
▶ extremely small (ultramicroscopic), passing through filters that retain even the smallest bacteria
▶ disease-producing agents, capable of multiplying in host cells only.
2 The nucleic acid of the viruses codes, inter alia, for the protein coat. Inside the host cell fresh 'coat' proteins are made at the host cell's ribosomes.
3 **The large intestine** (p. 287).

Page 551

1 Protection from the defensive mechanisms (such as antibodies or enzymes) of a host organism (**An experiment with pneumonia-causing bacteria**, p. 201).
2 **Measuring the growth of unicellular organisms** (p. 397).
3 **Organic matter**, (p. 61); **The cycling of nutrients** (p. 72).
4 **Waste water and industrial effluent treatment** (p. 101); **Preserving and storing foods** (p. 109); **Microbal mining** (p. 113); **Antibiotics** (p. 115); **Insulin manufacture** (p. 634).

Page 553

1 **How many kingdoms of living things?** (p. 24); **Kingdom Protocista** (p. 28); **Kingdom Fungi** (p. 30).
2 The requirements of all green plants: water, light, carbon dioxide and essential ions.
3 *Chlorella* cells are easily cultured, under identical controlled conditions, and representative samples can be easily withdrawn (cf. attempting to work with green leaf cells).
4 **Most atmospheric pollutants come from fossil fuels** (p. 86).

Page 554

1 Produced in Golgi apparatus vesicles, secreted across the *Spirogyra* cell plasma membrane (reverse pinocytosis) as part of the wall laid down by the cytoplasm.
2 **Meiosis** (p. 196), and **Figure 25.15** (p. 554).
3 **Adaptations to the conditions of the intertidal zone** (p. 53); holdfast, texture of thallus, mucilage (*Fucus*, p. 555; and **Figure 25.16**, p. 555); motile male gametes, and aquatic fertilisation (**Figure 25.17**).

Page 556

1 Entry into *Amoeba* of:
▶ water, by osmosis
▶ oxygen, by diffusion
▶ ions (e.g. K^+, HPO_4^{2-}) by active transport (ion pumps)
▶ prey (e.g. smaller protozoa, algae), by phagocytosis involving a pseudopodial cup.

2 Supports the bacterial community on which *Paramecium* chiefly feeds (p. 291).
3 Between them, the macro- and micro-nuclei control and direct the activities of the organism.
4 **Figure 7.25** (p. 165); **Structure and functioning of cilia and flagella** (p. 500).

Page 557

1 **Saprotrophic nutrition** (p. 296).
2 *Mucor*, **spore production and dispersal** (p. 557), **Figure 25.21** (p. 558); **Table 26.1** (p. 562).

Page 559

1 **An experiment using *Neurospora*** (p. 210).
2 Action of yeasts (present in flowers) on nectar (a dilute sugar solution) produces ethanol as a waste product.
3 In the latter case, cell division does not lead to an increase in the number of organisms (unless a filament of *Spirogyra* fragments accidentally).

Page 560

1 **Saprotrophic nutrition** (p. 296); *Agaricus* (p. 559).
2 **Phylum Ascomycota** and **Phylum Basidiomycota** (p. 31).
3 **Figure 25.26** (p. 560).

CHAPTER 26

Page 566

1 Chemotaxis (**Tropism, nasties and taxes: a summary**, p. 422).
2 Tissues of the sporophyte and gametophyte generation of bryophytes share the same genes, whereas a true parasite is an entirely different species from the host organism.
3 **a, b** The cushion maintains a microhabitat, which retains water, ions and mineral particles on surfaces often otherwise barren. These resources are advantageous to the moss and to other organisms that live in the cushion.
4 *Sphagnum* is used as a water-retaining packaging material around living plants in transit from nurseries, and in window-box/hanging basket 'gardens'; *Sphagnum* remains form the bulk of peat, used as a fuel and as a soil 'improver'.
5 **a, b Figure 26.3** (p. 564).

Page 571

1 **a, b** The features of *entomophilous flowers*, listed in **Table 26.3** (p. 571), which they exhibit (see **Figures 26.13** and **26.14**).
2 Sugar (carbohydrate), and some protein and lipid (pollen grains only).
3 **Worker honeybees; feeding on nectar and pollen** (p. 292).

Page 573

1 **Pollination** (p. 572), **Fertilisation and seed formation** (p. 573).
2 **Stamens; development of the male gamete** (p. 571); **Studying the past by pollen analysis** (p. 650).
3 **Fertilisation and seed formation** (p. 573).
4 Examples:
   a peach, plum
   b runner bean, mangetout pea
   c tomato, cucumber.

Table 1

Examples	Pellia, Funaria	Dryopteris	Pinus	angiosperm
a	foot (in gametophyte), seta and capsule	perennial rhizome with roots and leaves	woody perennial tree, with long shoots and short shoots with pine needles	annual, biennial or perennial (woody or herbaceous) plants of stem, leaves and roots
b	absent	present	present	present
c	parasitic on gametophyte	terrestrial (independent) plant	common on sandy soils	most are terrestrial plants, of a range of habitats
d	homosporous	typically homosporous	heterosporous	heterosporous
e	independent plant	small independent plant	reduced to a few cells in microspore and embryo sac	tube nucleus and antipodal and synergid cells
f	biflagellate antherozoids	biflagellate antherozoids	nuclei in pollen tube	nuclei in pollen tube

Page 576

1 a testa
   b embryo
   c embryo sac
   d antipodal cells, synergid cells, endosperm.
      (See **The life cycle of the flowering plant** and **Figure 26.22** (p. 575).

2 See Table 1, above.

Page 579

1 The ability to make structures or substances to take advantage of environmental opportunity or with which to protect themselves from predators (**The success of the angiosperms**, p. 577).

2 These produce plants genetically identical to the parent stock.

3 **Types of life cycle** (p. 563).

CHAPTER 27

Page 581

1 Many offspring are produced by most species, yet the sizes of their populations are usually more or less constant for long periods.

2 Parthenogenesis permits rapid increase in numbers when conditions are favourable. Sexual reproduction introduces genetic variability in the offspring.

Page 583

1 Gametes (haploid) have a random assortment of maternal and paternal homologous chromosomes. Further variation arises from recombinations of segments of individual maternal and paternal homologous chromosomes during crossing over. Fusion of gametes from two parents introduces more genetic variation.

2 The new individual may develop at the expense of the food stored in the female gamete. Male gametes are specialised for motility, and they compete to fertilise the egg.

3 Sperm exchange between two individuals is a separate step from the laying of the eggs and their fertilisation (with sperms from the partner, which have been stored in the spermatheca) (**Figure 27.5**).

Page 586

1 **Growth and development in multicellular animals**, p. 413.

2 **'Releasers' activate instinctive behaviour** (p. 518) and **Figure 23.30** (p. 520). Sticklebacks undertake the nest building, apparently guard the nest, and enhance aeration of the eggs, but the young fish fend for themselves when they hatch.
   Birds feed their young, and a degree of 'training' is built into this relationship.

3 An egg with an amnion membrane; that is, one that encloses the developing embryo in a fluid-filled sac during development.
   Embryos of the 'higher' vertebrate groups (reptiles, birds and mammals) are enclosed within an amnion during development (**Figure 27.9**, p. 586).

4 **Reproduction in an ecological context**
   a *r*
   b *K – r*
   c *K(– r?)*
   d *r – K*.

5 **Fertilisation, internal or external?** (p. 584).

6 External fertilisation in fresh water; development via stages as tadpoles (aquatic-dwelling juvenile forms), e.g. gill breathing.

Page 593

1 See Table 2, below.

2

Oestrus and the menstrual cycle (p. 591). **Endocrine control of the menstrual cycle** (p. 593).

Page 599

1 An innate behavioural tendency of carnivorous animals is to attack and eat any organism that comes close enough to catch. An innate courtship routine may be essential to overcome this dominant tendency, permitting fertilisation without predation.

2 Erection and ejaculation may be triggered either by the arousal of higher brain centres or by reflex action in response to local stimulation.

3 **The hormone control of pregnancy** (p. 596).

4 Uterine secretions replace the follicle cells as the source of nutrients, as the blastocyst grows (**Figure 27.18**, p. 591).

5 The fetus is foreign tissue to the mother (for example, they may not be of the same blood group), and carries antigens foreign to her (p. 598).

Page 603

1 a Delivery of oxygen and nutrients to aid growth and development of these organs. In the fetus, the liver is a site of blood cell production, so the blood vessels draining the liver carry new blood cells to the whole fetal circulation.

   b Normal roles of the liver in metabolism are established (**The liver, a homeostatic organ**, p. 507). The hepatic artery, hepatic portal vein and the hepatic vein facilitate these. Normal roles of the gut are established (movement of ingested substances, secretion of digestive juices, absorption). The blood supply facilitates these roles, and the maintenance of the gut tissues.

2 **Lactation** (p. 603).

CHAPTER 28

Page 607

1 a

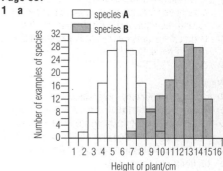

Table 2

	Male	**Female**
Germinal epithelium	of seminiferous tubules, active from puberty throughout life	of ovaries, active in fetus only
Phase of multiplication	few spermatocytes die	most oocytes die
Phase of maturation meiosis; (1st and 2nd division)	each primary spermatocyte divides to form two secondary spermatocytes (1st division), and then four spermatids; spermatids mature into sperms (**Figure 27.13**, p. 589)	each primary oocyte divides to form a secondary oocyte and a 1st polar body (1st division); further division is delayed until fertilisation by a single sperm (**Figure 27.17**, p. 591)

b Mode = 13; median = 11; mean = 12.14 (12, to the nearest whole number by which measurements were taken).

c See Table 3, below.

Table 3

Height of plants (x)	Frequency (f)	fx	Deviation d from mean ($x - \bar{x}$)	d^2	fd^2
7	2	14	−5	25	50
8	5	40	−4	16	80
9	9	81	−3	9	81
10	13	130	−2	4	52
11	18	198	−1	1	18
12	25	300	0	0	0
13	29	377	1	1	29
14	28	392	2	4	112
15	12	180	3	9	108
	$\sum f = 141$	$\sum fx = 1712$			$\sum fd^2 = 530$

Mean $(\bar{x}) = {}^{1712}/_{141} = 12.14$ (i.e. 12)

Standard deviation $= \sqrt{\left(\dfrac{\sum fd^2}{\sum f}\right)} = \sqrt{\dfrac{530}{141}} = 1.9$ to 1 decimal place

Page 611

1 Genotypically dwarf.

2

categories	hypothesis	O	E	O-E	O-E-0.5*	(O-E)²/E*
green	3	428	435	−7	6.5	0.097
yellow	1	152	145	+7	6.5	0.291
		580				0.388

$\chi^2 = 0.388$, i.e. it falls below $p = 0.5$.

Thus there is no significant difference between the observed and the expected result.

(* = correction factor in cases of only two categories.)

Page 612

a, b

Genotypes	Phenotypes	Ratio
RR	red	1
RW	roan	2
WW	white	1

Page 615

1 By independent assortment (**The law of Independent Assortment**, p. 612), and by crossing over between genes linked on the same chromosome (**Crossing over and linkage**, p. 617).

2 **Illustrating Mendelian ratios in *Drosophila*** (p. 613).

3 When homozygous normal flies (red eye/normal wing) are crossed with homozygous flies with scarlet eye and vestigial wing, and the progeny (heterozygous normal flies) are crossed, the F₂ generation will occur in the ratio:

9 normal flies

3 red-eyed, vestigial-winged flies

3 scarlet-eyed, normal-winged flies.

1 scarlet-eyed and vestigial-winged fly.

The reason are those shown by the cross in **Figure 28.12** (p. 610).

4 **a** The characteristic is due to an allele that is expressed even in the heterozygous condition.

b The characteristic due to a dominant mutation is likely to occur in individuals produced in every generation.

Page 616

Normal wild-type flies have grey bodies and red eyes.

parental phenotypes	grey body red eyes	ebony body scarlet eyes
parental genotypes	(GGRR) ×	(ggrr)
gametes	(GR)	(gr)
F₁ offspring		(GgRr)
sibling crosses	(GgRr) ×	(GgRr)
gametes	(GR)(gr) ×	(GR)(gr)

	½(GR)	½(gr)
½(GR)	¼(GGRR)	¼(GgRr)
½(gr)	¼(GgRr)	¼(ggrr)

F₂ offspring

genotypes	GGRR	GgRr	ggrr
genotype ratio	1 :	2 :	1
phenotypes	grey body /red eye		ebony body /scarlet eye
phenotype ratio	3 :		1

Page 618

1 If greater than 50% the genes are segregating independently (not linked).

2

parental phenotypes:	purple flower, × long pollen grain	white flower, short pollen grain
parental genotypes:	PpLl ×	ppll
gametes:	¼PL ¼pL ¼Pl ¼pl	pl
offspring:	PpLl ppLl Ppll ppll	

purple flower, long pollen grain	PpLl	25%
white flower, long pollen grain	ppLl	25%
purple flower, short pollen grain	Ppll	25%
white flower, short pollen grain	ppll	25%

3 Recombinant (cross over) frequency:

$$\frac{106 + 111}{586 + 106 + 465 + 111} \times 100\% = 17.1\%$$

Page 621

1 **Haemophilia** (pp. 620–21).

Page 623

1 Maternal grandfather, because the boy's colour blindness allele is on his X chromosome.

2 **a** A or B.

b A, AB or B

1 : 2 : 1.

3 The mother's genotype = I⁰I⁰ Rh−Rh−, and she produces gametes I⁰ Rh− only.

The father's genotype may be IᴬIᴮRh+Rh+ orIᴬIᴮRh+Rh−, and he produces gametes:

IᴬRh+, IᴬRh−, IᴮRh+ and IᴮRh−.

The possible blood groups of the children are: A, Rh+; A, Rh−; B, Rh+; B, Rh−.

4 AaBB = unbanded; aaBB = mid-banded; aabb = five-banded.

Page 624

Inheritance of comb form in the fowl:

parental phenotypes:	pea	×	rose
genotypes:	PPrr		ppRR
gametes:	Pr		rP
F₁ offspring genotype:		PpRr	
phenotype:		walnut	
		× siblings	
gametes:		PR Pr pR pr	PR Pr pR pr

	¼PR	¼Pr	¼pR	¼pr
¼PR	¹⁄₁₆PPRR	¹⁄₁₆PPRr	¹⁄₁₆PpRR	¹⁄₁₆PpRr
¼Pr	¹⁄₁₆PPRr	¹⁄₁₆PPrr	¹⁄₁₆PpRr	¹⁄₁₆Pprr
¼pR	¹⁄₁₆PpRR	¹⁄₁₆PpRr	¹⁄₁₆ppRR	¹⁄₁₆ppRr
¼pr	¹⁄₁₆PpRr	¹⁄₁₆Pprr	¹⁄₁₆ppRr	¹⁄₁₆pprr

offspring phenotypes:	walnut	pea	rose	single
ratio:	9	3	3	1
genotypes:	PPRR	PPrr	ppRR	pprr
	PPRr	Pprr	ppRr	
	PpRR			
	PpRr			

Page 627

1 Chromosomes from different organisms may be unable to pair (prophase 1 of meiosis) in the diploid state, but doubling of the chromosomes will certainly permit pairing between homologous chromosomes.

2 All mutations are expressed in prokaryotes (single strand of DNA). In diploid eukaryotes, only the dominant mutations will be immediately expressed (in a heterozygous state). They also have very rapid rates of reproduction.

CHAPTER 29

Page 630

1 **a** **Introduction** (p. 628).

b **Figure 9.14** (p. 204).

2 **Base ratio analysis** (p. 203), and **Ligase; joining complementary sticky ends** (p. 629).

3 **Eukaryotes and GMOs** (p. 635). Also because all genes of prokaryotes are capable of being expressed at any time since the DNA of prokaryotes is a single strand; that is, in prokaryotes genes do not exist as paired alleles, as they do in all eukaryotes.

Page 633

1 **The triplet code** (p. 206).

2 **AIDS, the Acquired Immune Deficiency Syndrome** (p. 370).

3 Because a sample of the DNA of the bacteria needs to be heat-treated to break it into single strands for treatment with the probe, but some of the original bacteria are needed subsequently for further culturing.

Page 642

1 **The cause of sickle-cell anaemia** (p. 627).

2 They require an individual to be homozygous for the recessive allele before the trait it controls is expressed (**Sex-linked conditions in humans; Haemophilia**, p. 620).

3 **Gene therapy** (p. 639).

4 To avoid any rejection of foreign cells (**Immunity and the immune system**, p. 366).

Page 645

a, **b** Let p and q represent the respective frequencies of the dominant and the recessive allele (Hardy–Weinberg Law); $p + q = 1$.
We know that the frequency of the homozygous recessive (PKU-sufferer) = 1 in 10 000; that is, $a^2 = 1/10\,000$ or $\underline{0.0001}$.
Therefore $q = \sqrt{0.0001} = 0.01$
Since $p + q = 1$
$p = 1 - 0.01$
$p = 0.99$.
So the frequency of the dominant allele is 0.99. According to the Hardy–Weinberg formula the frequency of the heterozygote = $2pq$
$= 2 \times 0.01 \times 0.99$
$= 0.0198$
Thus about 20 in every 1000 of the population can be expected to be heterozygous (carriers) for PKU, or 1 in 50 of the population.

CHAPTER 30

Page 647

1 **The early geologists and a new view of time** (p. 647).

2 Fossils laid down in the deeper strata of sedimentary rock are older than fossils of the higher (more recently laid down) strata.

Page 649

1 **Dating rocks accurately** and **Figure 30.6** (p. 648).

2 The dominant vegetation at any time is very often an excellent guide to the prevailing climate. For one thing, the dominant plants often play a part in determining the climate (**The biosphere and biomes**, p. 45), **Studying the past by pollen analysis** (p. 650).

Page 654

1 Proximity of parent stock or the absence of competition from other species.

2 **Class Mammalia** (p. 41).

Page 655

1 Genetic 'distance'; differences between DNAs (p. 651).

2 **Defining a species** (p. 23). A variety is a grouping within a species whose members differ in some significant respect from other members of the species.

Page 658

1 Charles had developed the concept of natural selection (see **An outright evolutionist**, p. 647).

2 **Neo-Darwinism** (p. 657).

Page 660

A = 0.5875; a = 0.4125 (**Gene frequencies in populations**, p. 659).

Page 664

1 **Evidence from artificial selection** (p. 655); **Natural selection** (p. 656).

2 Changes in the environment are a major selecting 'agent'.

3 Any tendency for speciation to occur will be reversed (**Reproductive isolation of demes**, p. 661).

4 Polyploids are unable to produce gametes that can form a zygote with gametes from their parents (**Speciation by polyploidy**, p. 662).

5 They may be a source of entirely new alleles (**Mutations**, p. 659).

6 Examples: selection, due to the effects of a new form of rat poison; migrations and 'founder effect' may operate, leading to an early stage of reproductive isolation.

Page 671

1 **The properties of catalysts** (p. 668).

2 **Table 7.2** (p. 169).

3 Multicellularity provides opportunities for increases in size and complexity (division of labour). Structural diversity may lead to
▶ the mastery of a wide range of habitats (for instance, escape from aquatic habitats)
▶ the regulation of the internal environment of an organism, sometimes in hostile external conditions.

4 The organic molecules of a 'primeval soup' would certainly be oxidised to simpler substances in the presence of oxygen.

5 **Phylum Chordata** (p. 39).

6 **Kingdom Plantae (green plants)** (pp. 32–4).

7 Manual dexterity (construction of clothing and shelter); large brain, permitting problem solving; memory and speech for passing on information; their inventiveness with fire; an increasing tendency to manipulate their environment.

8 Specific adaptations for a tree-living existence.

9 **From apes to humans in 35 million years** (p. 678).

Index